Bridging Concepts and Calculus

Brief Calculus Concepts, Second Edition

A groundbreaking text that employs modeling, real-life data and technology in an innovative approach to the mathematics of change.

Take a look at how this text helps students build the necessary bridges between calculus and its concepts.

- **Concept Objectives** at the beginning of each chapter identify the key concepts to be presented and skills to be mastered within the chapter.
- **Concept Inventories** at the end of each section summarize the key concepts and skills developed within that section.
- **Concept Checks** at the end of each chapter summarize the key concepts and skills presented within the chapter. In addition, sample activities that correspond to each of the concepts and skills allow students to assess their understanding of the chapter content.
- **Concept Review/Chapter Tests** at the end of every chapter provide practice problems that reinforce the concepts and skills presented within the chapter. Answers to the Concept Review/Chapter Tests can be found in the back of the text.

Technology Guides for Excel and Graphing Calculators show students how to solve certain examples in the text using their particular technology. Each guide includes instructions for Texas Instruments calculators as well as for Excel. Sections of the manuals are referenced in the text by a technology icon next to the particular example discussed in the technology guide.

Integrated Web Resources are integrated throughout the text with a special icon sending students to the text-specific website where they can find resources such as algebraic tutorials, additional practice problems, links to data updates, a glossary of terms, ACE self-tests, and more.

- **SMARTHINKING.com** live online tutoring comes FREE with every new text purchased from Houghton Mifflin, providing students with a valuable tutoring resource outside of class.

Building Bridges to Better Learning

- **Instructor's Annotated Edition** contains margin notes with explanations of content or approach, teaching ideas, indications of where a topic appears in later chapters, indications of topics that can be easily omitted or streamlined, suggestions for alternate paths through the book, warnings of areas of likely difficulty for students, and helpful references to topics in the *Instructor's Resource Guide.*
- **Video Series** contains video lectures that can be viewed by students who miss a class or who would benefit from seeing another teacher explain a particular topic. These videos could also be used as training tools for graduate teaching assistants.
- **Listserve for instructors** provides an online environment where instructors teaching applied calculus can share teaching ideas and build community. E-mail **college_math@hmco.com** or contact your local Houghton Mifflin sales representative for more information.

SECOND EDITION
INSTRUCTOR'S
ANNOTATED EDITION

Calculus Concepts

An Informal Approach to the Mathematics of Change

Donald R. LaTorre
Clemson University

John W. Kenelly
Clemson University

Iris B. Fetta
Clemson University

Cynthia R. Harris
Reno, Nevada

Laurel L. Carpenter
Lansing, Michigan

Houghton Mifflin Company
Boston New York

Editor-in-Chief: Jack Shira
Sponsoring Editor: Paul Murphy
Editorial Associate: Marika Hoe
Senior Production/Design Coordinator: Carol Merrigan
Senior Manufacturing Coordinator: Jane Spelman
Senior Marketing Manager: Michael Busnach
Marketing Assistant: Nicole Mollica

Cover photograph by Gary Land

Printed in the U.S.A.

Library of Congress Control Numbers

Student Text: 2001131524
Brief Student Text: 2001088406

International Standard Book Numbers

Student Text: 0-618-12175-7
Instructor's Annotated Edition: 0-618-12177-3
Brief Student Text: 0-618-12170-6
Brief Instructor's Annotated Edition: 0-618-12176-5

2 3 4 5 6 7 8 9-DOW-05 04 03 02

Contents

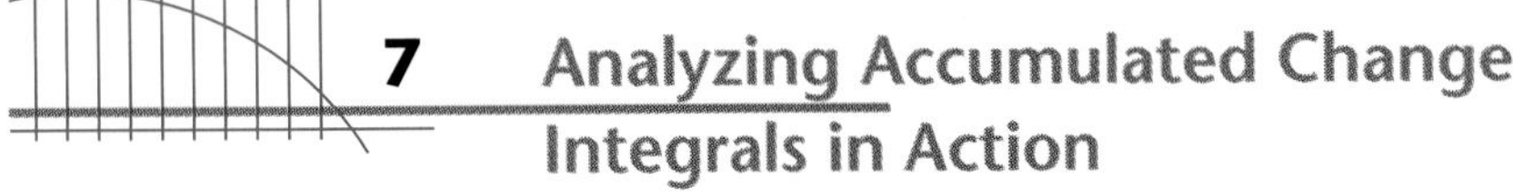

Chapter 11 and the Appendix are available via the Internet.
To view and download this material, please visit http://college.hmco.com.

11 Dynamics of Change: Differential Equations and Proportionality

Preface

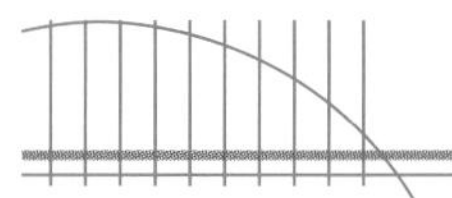

Bridging Concepts

PHILOSOPHY

This book presents a fresh, new approach to the concepts of calculus for students in fields such as business, economics, liberal arts, management, and the social and life sciences. It is appropriate for courses generally known as "brief calculus" or "applied calculus."

Our overall goal is to improve learning of basic calculus concepts by involving students with new material in a way that is different from traditional practice. The development of conceptual understanding coupled with a commitment to make calculus meaningful to the student are guiding forces. The material in this book involves many applications of real situations through its data-driven, technology-based modeling approach. It considers the ability to correctly interpret the mathematics of real-life situations of equal importance to the understanding of the concepts of calculus in the context of change.

Data-Driven

Many everyday, real-life situations involving change are discrete in nature and manifest themselves through data. Such situations often can be represented by continuous or piecewise continuous mathematical models so that the concepts, methods, and techniques of calculus can be utilized to solve problems. Thus we seek, when appropriate, to make real-life data a starting point for our investigations.

The use of real data and the search for appropriate models also expose students to the reality of uncertainty. We emphasize that sometimes there can be more than one appropriate model and that answers derived from models are only approximations. We believe that exposure to the possibility of more than one correct approach or answer is valuable.

Modeling Approach

We consider modeling to be an important tool and introduce it at the outset. Both linear and nonlinear models of discrete data are used to obtain functional relationships between variables of interest. The functions given by the models are the ones used by students to conduct their investigations of calculus concepts. It is the connection to real-life data that most students feel shows the relevance of the mathematics in this course to their lives and adds reality to the topics studied.

Interpretation Emphasis

This book differs from traditional texts not only in its philosophy but also in its overall focus, level of activities, development of topics, and attention to detail. Interpretation of results is a key feature of this text that allows students to make sense of the mathematical concepts and appreciate the usefulness of those concepts in their future careers and in their lives.

Informal Style

Although we appreciate the formality and precision of mathematics, we also recognize that this alone can deter some students from access to mathematics. Thus we have sought to make our presentations as informal as possible by using nontechnical terminology where appropriate and a conversational style of presentation.

PEDAGOGICAL FEATURES

- *Chapter Opener* Each chapter opens with a real-life situation and several questions about the situation that relate to the key concepts in the chapter. Many of these applications have been rewritten so that they correspond to and reference an activity in the chapter.
- *Concept Objectives* The Concept Objectives appear at the beginning of each chapter and list the goals of the chapter. The objectives are divided into two categories: concepts to be understood and skills to be learned.
- *Concept Inventory* A Concept Inventory listed at the end of each section gives students a brief summary of the *major ideas developed in that section.*
- *Section Activities* The Activities at the end of each section cement concepts and allow students to explore topics using, for the most part, actual data in a variety of real-world settings. Questions and interpretations pertinent to the data and the concepts are always included in these activities. The activities do not mimic the examples in the chapter discussion and thus require more independent thinking on the part of the students. Possible answers to odd activities are given at the end of the book.
- *Chapter Summary* A Chapter Summary connects the results of the chapter topics and further emphasizes the importance of knowing these results.
- *Concept Check* A check list is included at the end of each chapter summarizing the main concepts and skills taught in the chapter along with sample odd activities corresponding to each item in the list. The sample activities are to help students assess their understanding of the chapter content and identify on which areas to focus their study.
- *Chapter Review Test* A Chapter Review Test at the end of each chapter provides practice with techniques and concepts. Complete answers to the Chapter Review Tests are included in the answer key located at the back of the text.
- *Projects* Projects included after each chapter are intended to be group projects with oral or written presentations. We recognize the importance of helping students develop the ability to work in groups, as well as hone presentation skills. The projects also give students the opportunity to practice the kind of writing that they will likely have to do in their future careers.

CONTENT CHANGES IN THE SECOND EDITION

This new edition contains pedagogical changes intended to improve the presentation and flow of the concepts discussed. It contains many new examples and activities. In addition, many data sets have been updated to include more recent data.

- ***Algebraic Rigor*** The authors have sought to increase the algebraic rigor of some sections of *Calculus Concepts* as a means of making the book more usable in those courses demanding a certain level of algebraic difficulty. In doing so, however, the authors have not diminished the reform flavor of their approach, and all added algebraic rigor can be easily omitted without loss of continuity.
- ***New Topics*** Some new topics (such as limits, classical optimization, and related rates) have been added to increase the book's appropriateness for some courses. Most new topics can be easily omitted. The inclusion of algebraic rigor and new topics expand your choices in customizing a course that best suits the needs of your students.
- ***Chapter 11 Dynamics of Change: Differential Equations and Proportionality*** For those of you who may want to integrate coverage of these topics into your course, this chapter is available to you and your students in downloadable PDF format at the *Calculus Concepts* Web Site (accessible through **college.hmco.com**).

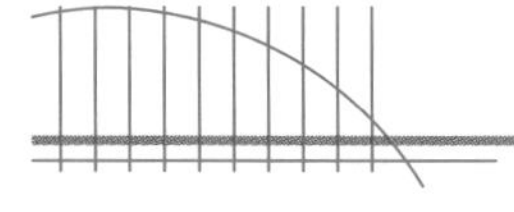

Bridging Technology

TECHNOLOGY FOCUS

Graphing Calculators and Spreadsheets

Calculus has traditionally relied upon a high level of algebraic manipulation. However, many nontechnical students are not strong in algebraic skills, and an algebra-based approach tends to overwhelm them and stifle their progress. Today's easy access to technology in the forms of graphing calculators and computers breaks down barriers to learning imposed by the traditional reliance on algebraic methods. It creates new opportunities for learning through graphical and numerical representations. We welcome these opportunities in this book by assuming continual and immediate access to technology.

This text requires that students use graphical representations (scatter plots of data and graphs of functions) freely, make numerical calculations routinely, and find functions to fit data. Thus continual and immediate access to technology is essential. Because of their low cost, portability, and ability to personalize the mathematics, the use of graphing calculators is appropriate. The concepts in the text have also been successfully taught using computer software (such as Maple and Excel).

Technology Guides

Because it is not the authors' intent that class time be used to teach technology, we provide two Technology Guides for students: a ***Graphing Calculator Instruction Guide*** containing keystroke information adapted to materials in the text for the TI-83, TI-86, and TI-89 models, and an ***Excel Manual*** providing basic spreadsheet instruction. In the student text, technology icons (formerly open book icons in the first edition) refer readers to applicable sections within the appropriate technology guide.

It is worth noting that different technologies may give different model coefficients than those given in this book. We used a TI-83 graphing calculator to generate the models in the text and the answer key. Other technologies may use different fit criteria for some models than that used by the TI-83.

Eye on the Internet

The ***Calculus Concepts* Web Site** (accessible through **college.hmco.com**) provides an exceptional variety of valuable resources for instructors and students alike. Password-protected resources for instructors include worksheets, PowerPoint slides, additional projects, data sets categorized by type for use on tests and quizzes, and other resource materials.

Instructors can also participate in the *Calculus Concepts* Listserve—a regularly scheduled e-mail service facilitated by the text authors that will provide instructors with links to teaching resources and suggestions for integrating these materials into classroom instruction. For more information about the *Calculus Concepts* Listserve, please e-mail **college_math@hmco.com** or contact your local Houghton Mifflin Sales Representative.

For students, the web site provides a glossary of terms, skill and drill quizzes, and graphing calculator keystroke information. In addition, the web icon found in the margins of the textbook directs students to the book-specific web site for further information on algebra review, additional practice problems, practice tests, and links to sites for obtaining data updates.

NEW! On-line tutoring provided by **SMARTHINKING** will be available free of charge for all students. This service provides one-on-one, real-time tutorials with qualified instructors, seven days a week.

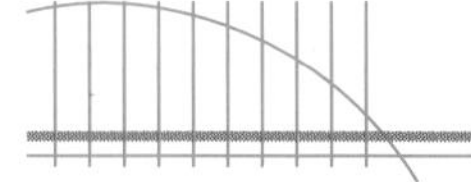

Building Bridges to Better Learning

RESOURCES FOR INSTRUCTORS

In addition to the resources found on the web site, the printed ***Instructor's Resource Guide with Complete Solutions*** gives practical suggestions for using the text in the manner intended by the authors. It gives suggestions for various ways to adapt the text to your particular class situation. It contains sample syllabi, sample tests, ideas for in-class group work, suggestions for implementing and grading projects, and complete activity solutions.

The *Instructor's Annotated Edition* is the text with margin notes from the authors to instructors. The notes contain explanations of content or approach, teaching ideas, indications of where a topic appears in later chapters, indications of topics that can be easily omitted or streamlined, suggestions for alternate paths through the book, warnings of areas of likely difficulty for students based on the authors' years of experience teaching with *Calculus Concepts*, and references to topics in the *Instructor's Resource Guide* that may be helpful.

LEARNING RESOURCES FOR STUDENTS

1. The ***Calculus Concepts Video Series*** contains chapter-by-chapter lectures by a master teacher. The video series can be used by students who miss a class or by students who think they would benefit from seeing another teacher explain a particular topic. These videos can also be used as training tools for graduate teaching assistants.

2. The ***Student Solutions Manual*** contains complete solutions to the odd activities.

3. The ***Graphing Calculator Instruction Guide*** contains keystroke information adapted to material in the text for the TI-83, TI-86, and TI-89 models.

4. An ***Excel Manual*** provides basic instruction on this spreadsheet program.

These two Technology Guides contain step-by-step solutions to examples in the text and are referenced in this book by a technology icon.

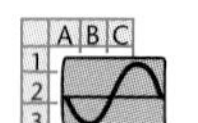

5. The ***Calculus Concepts*** Web Site (accessible through **college.hmco.com**) contains extra practice problems, help with algebra, links to updated data needed for certain activities, a glossary of terms, practice quizzes, and other assistance. The web icon in the text directs you to the book-specific web site when appropriate.

6. Students can also access free on-line tutoring available through **SMARTHINKING**. Through a special agreement with Houghton Mifflin, this on-line tutoring service provided by **SMARTHINKING** will be available free of charge. This service provides one-on-one, real-time tutorials with qualified instructors, seven days a week.

For more information about **SMARTHINKING**, please visit the *Calculus Concepts* Web Site at **college.hmco.com**.

ACKNOWLEDGMENTS

We gratefully acknowledge the many teachers and students who have used this book in its preliminary and first editions and who have given us feedback and suggestions for improvement. In particular, we thank the following reviewers whose many thoughtful comments and valuable suggestions guided the preparation of this second edition revision.

Patricia Beaulieu—*University of Louisiana at Lafayette*
Sherry Biggers—*Clemson University*
Joseph Cieply—*Elmhurst College*
Mark Clark—*Palomar College*
John Haverhals—*Bradley University*
Michael Helinger—*Clinton Community College*
Larry Johnson—*Metropolitan State College of Denver*
Christine Joy—*University of Oklahoma*
Greg Klein—*Texas A&M University*
Biyue Liu—*Monmouth University*
Diana McCoy—*Florida International University*
Lawrence Merbach—*North Dakota State College of Science*
Carl Mueller—*Georgia Southwestern State University*
Patrick O'Brien—*Mesa Community College*
David Platt—*Mesa Community College*
Richard Porter—*Northeastern University*
Carol Rardin—*Central Wyoming College*
Helen Read—*University of Vermont*
Bernd Rossa—*Xavier University*
Charlotte Simmons—*University of Central Oklahoma*
Donna Stein—*University of Tennessee, Knoxville*
Paul Vaz—*Arizona State University*

Special thanks to Frank Wilson for his careful work in checking the text and answer key for accuracy. The authors express their sincere appreciation to Charlie Hartford, who first believed in this book, and to Paul Murphy, Mary Beckwith, Marika Hoe, Kathy Deselle, and their associates at Houghton Mifflin Company for all their work in bringing this second edition into print.

To Students

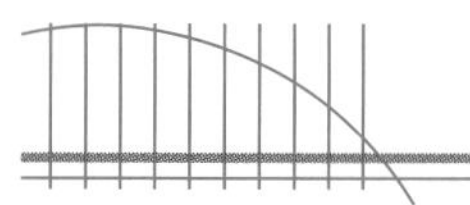

Bridging Concepts

WHAT THIS BOOK IS ABOUT

This book is written to help you understand the inner workings of how things change and to help you build systematic ways to use this understanding in everyday real-life situations that involve change. Indeed, a primary focus of the material is on change, since calculus is the mathematics of change.

Even if you have studied calculus before, this book is probably different from any other mathematics textbook that you have ever used. It is based on three premises:

1. Understanding is as important as the mastery of mathematical manipulations. Algebraic skill and the ability to manipulate expressions must be regularly practiced, or they will fade away. If you understand concepts, you will be able to explain some things in your life forever.

2. Mathematics is present in all sorts of real-life situations. It is not just an abstract subject in textbooks. In real life, mathematics is often messy and not at all like the tidy, neat equations that you were taught to factor and solve. Speaking of equations, where do they come from? Nature seldom whispers an equation into our ears.

3. The new graphics technology in today's calculators and computers is a powerful tool that can help you understand important mathematical connections. Like many tools in various fields, technology frees you from tedious, unproductive work; enables you to engage situations more realistically; and lets you focus on what you do best . . . think and reason.

HOW TO USE THIS BOOK

- Begin by throwing away any preconceived notions that you may have about what calculus is and any notion that you are "not good" in mathematics.
- Make a commitment to learn the material—not just a good intention, but a genuine commitment.
- Study this book. Notice that we said "study," not "read." Reading is a part of study, but study involves much more. You should not only read (and reread) the discussions but work through each example to understand its development.
- Use paper, pencil, and your graphing calculator or computer when you study. These are your basic tools, and you cannot study effectively without them.
- Find a study partner, if at all possible. Each of you will be able to help the other learn. Communicating within mathematics, and about mathematics, is important to your overall development toward understanding mathematics.

- Write your solutions clearly and legibly, being certain to interpret all of your answers with complete sentences using proper grammar. Careful writing will help you sort through your ideas and focus your learning.
- Make every effort not to fall behind. You know the dangers, of course, but we remind you nevertheless.
- Finally, remember that there is no substitute for effective study. You have your most valuable resource with you at all times—your mind. Use it.

HOW TO GET HELP

Despite your best efforts, you may find yourself needing additional help with the ideas and skills presented in this book. We have designed several aids to assist you.

The *Answer Key* following the last chapter contains the answers to all the odd activities.

In addition, the following supplements provide further explanation and practice on the concepts of calculus. Please refer to pages xvi–xvii of the Preface for an in-depth description of the supplements listed below. You may purchase the printed supplements by visiting Houghton Mifflin's on-line College Store at **college.hmco.com** or in your local bookstore.

- Student Solutions Manual (0-618-12180-3)
- Brief Student Solutions Manual to accompany the brief version of *Calculus Concepts* (0-618-12181-1)
- The *Calculus Concepts* Video Series (0-618-12467-5)
- The *Calculus Concepts* Web Site—visit our site at **college.hmco.com**
- SMARTHINKING Free On-Line Tutoring available through the *Calculus Concepts* Web Site
- Graphing Calculator Instruction Guide (0-618-12179-X)
- Excel Manual (0-618-12178-1)

Calculus Concepts

Instructor's Annotated Edition

1

Ingredients of Change: Functions and Linear Models

Concept Application

Robert Brenner/PhotoEdit

Chlorofluorocarbons (CFCs) released into the atmosphere are believed to thin the ozone layer. A worldwide protocol calling for the phasing out of all CFC production was ratified in 1987. Here are some questions about the release of CFCs into the atmosphere that can be answered mathematically by using functions and calculus:

- What was the amount of CFCs released into the atmosphere in 1975? in 1995?
- At what rate was the release of CFCs declining between 1974 and 1980?
- On the basis of data accumulated since 1987, in what year will there no longer be any CFCs released into the atmosphere?

This chapter gives you the tools that are necessary to answer such questions. The information needed to answer these questions is found in Activity 27 of Section 1.5.

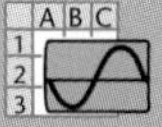

This symbol directs you to the supplemental technology guides. The different guides contain step-by-step instructions for using graphing calculators or spreadsheet software to work the particular example or discussion marked by the technology symbol.

This symbol directs you to the Calculus Concepts *web site, where you will find algebra review, links to data updates, additional practice problems, and other helpful resource material.*

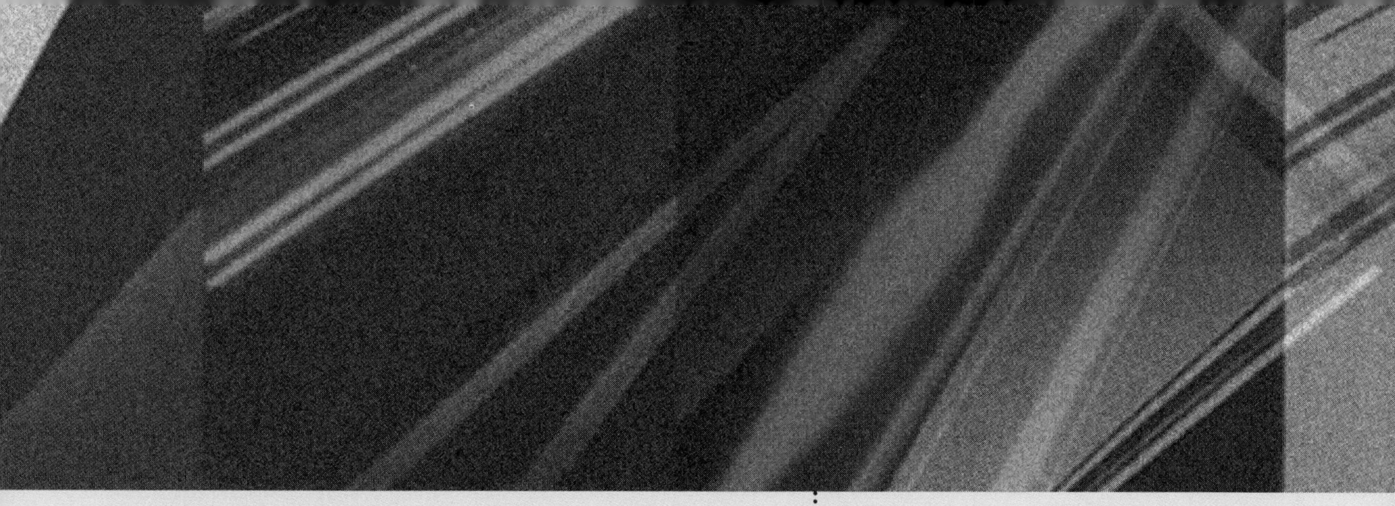

Concept Objectives

This chapter will help you understand the concepts of

- Mathematical model
- Function
- Discrete versus continuous
- Function composition
- Inverse function
- Limits and end behavior
- Continuity
- Linear function, slope, and rate of change

and you will learn to

- Set up and appropriately label a model
- Find inputs and outputs of functions
- Interpret graphs
- Solve simple business problems
- Find composite and inverse functions
- Work with piecewise continuous functions
- Estimate limits graphically and numerically
- Find limits algebraically
- Use technology to fit a linear model to data
- Determine the rate of change of a linear model

The primary goal of this book is to help you understand the two fundamental concepts of calculus—the derivative and the integral—in the context of the mathematics of change. This first chapter is therefore devoted to a study of the key ingredients of change: functions and mathematical models. Functions provide the basis for analyzing the mathematics of change because they enable us to describe relationships between variable quantities.

This chapter introduces you to the process of building mathematical models. In later chapters, we apply methods and techniques of calculus to the equations the models provide as we investigate rates of change and their implications in a variety of everyday situations.

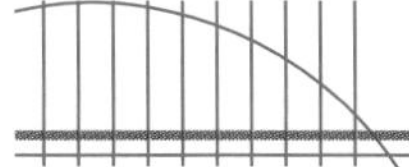

1.1 Models, Functions, and Graphs

Point out to students that calculus is applied primarily to equations (not to data) and that this fact motivates the modeling approach (converting data to equations) in this book.

Calculus is the study of change—how things change and how quickly they change. We begin our study of calculus by considering how we describe change. Let us start with something that many Americans desire to change: their weight. Suppose Joe weighs 165 pounds and begins to diet, losing 2 pounds a week. We can describe how Joe's weight changes in several ways.

With a table, such as Table 1.1:

TABLE 1.1

Weeks after diet starts	0	1	2	3	4	5
Weight (pounds)	165	163	161	159	157	155

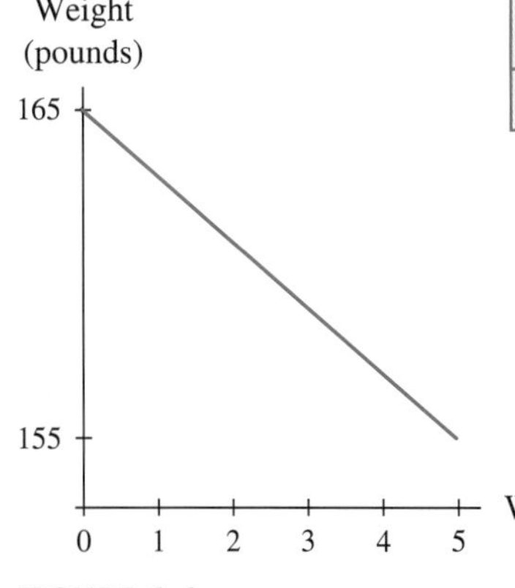

FIGURE 1.1

With a graph, such as Figure 1.1.

With words:

> $W(t)$ is Joe's weight (in pounds) t weeks after he begins to diet if he weighs 165 pounds at the start and loses 2 pounds each week.

With an equation:

> $W(t) = 165 - 2t$ pounds, where $W(t)$ is Joe's weight after t weeks of dieting

Mathematical Models

Note that we use the word *model* to refer to an equation with labeled input and output. We use the term *equation* and sometimes *function* when referring to only a mathematical formula apart from its interpretation in a particular context.

Although all four representations are helpful, only the equation and the graph enable us to apply calculus concepts to a particular situation in order to study change. The process of translating a real-world problem into a usable mathematical equation is called **mathematical modeling,** and the equation (with the variables described in the problem context) is referred to as a **model.** We use models to describe numerical data or verbal information. Then we use the model to answer a variety of questions about the situation, interpreting our answers and stating them in numerical and verbal form. This process of translating a real-life situation into a usable equation, applying mathematics to the equation to answer questions about the situation, and translating the answers back into the real-world situation is the focal point of this text.

EXAMPLE 1 *Building a Tunnel*

A civil engineer is planning to dig a tunnel through a mountain as shown in Figure 1.2. (Note that Figure 1.2 is not drawn to scale.) The tunnel will begin 575 feet above sea level and will be constructed with a constant upward slope of 5%; that is, the tunnel will rise vertically 5 feet for every 100 feet of horizontal distance. Table 1.2 shows the amount of vertical rise for several horizontal distances.

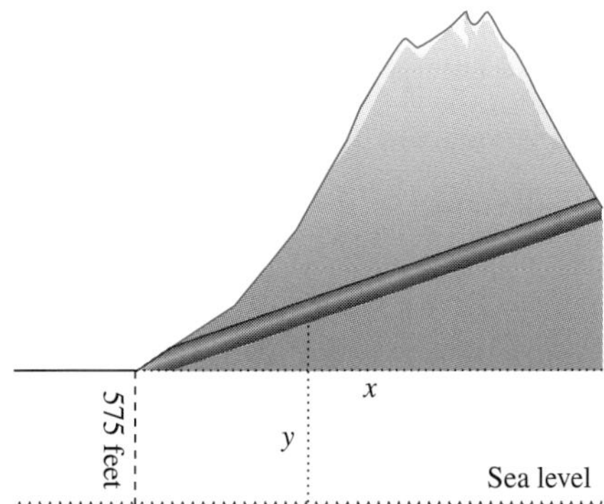

FIGURE 1.2

TABLE 1.2

Horizontal increase (feet)	Vertical increase (feet)
100	5
200	10
300	15
400	20
500	25

a. Use the verbal description and the table of data to write a model for the elevation above sea level of the tunnel in terms of the horizontal distance from where the tunnel begins at the base of the mountain.

b. Find the elevation of the tunnel at a horizontal distance of 2500 feet from the starting point.

c. If the tunnel exits the mountain at a horizontal distance of 7000 feet from where it began, what is the elevation of the tunnel when it emerges from the mountain?

d. If the tunnel will cost \$120 per foot to construct, what will be the cost (to the nearest thousand dollars) of building the tunnel?

Solution

a. The elevation of the tunnel begins at 575 feet and rises 5 feet for every 100 horizontal feet. We write this as

Elevation of tunnel = 575 + 5 feet for every 100 horizontal feet

Using y as the elevation and x as the horizontal feet in hundreds, we convert this statement to the equation

$$y = 575 + 5x \text{ feet above sea level}$$

where x is the horizontal distance from the starting point in hundreds of feet. This equation (with the variable descriptions) is a model for the elevation of the tunnel.

b. Any of the following three methods can be used to determine the elevation. *Algebraically:* Substitute $x = 25$ in the equation to obtain 700 feet above sea level. *Graphically:* Draw a graph of the equation $y = 575 + 5x$ with x-values between 0 and 50 and suitable y-values. Read the y-value from the graph that corresponds to $x = 25$. It is $y = 700$. *Numerically:* Use the fact that the tunnel rises 25 feet for each 500 feet of horizontal increase to extend Table 1.2 as shown in Table 1.3.

TABLE 1.3

Horizontal increase (feet)	500	1000	1500	2000	2500
Vertical increase (feet)	25	50	75	100	125

Because the elevation at the starting point is 575 feet, the height of the tunnel 2500 feet from the starting point is $125 + 575 = 700$ feet above sea level.

c. Using any of the methods described in part *b*, we find that the elevation of the tunnel when it emerges from the mountain is 925 feet.

d. Using the Pythagorean Theorem, we find that

This symbol indicates that additional help with the algebra required for this example is available on the *Calculus Concepts* web site.

algebra

$$\text{Length of tunnel} = d = \sqrt{7000^2 + 350^2} = \sqrt{49{,}122{,}500} \approx 7009 \text{ feet}$$

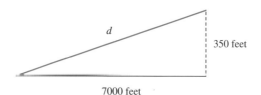

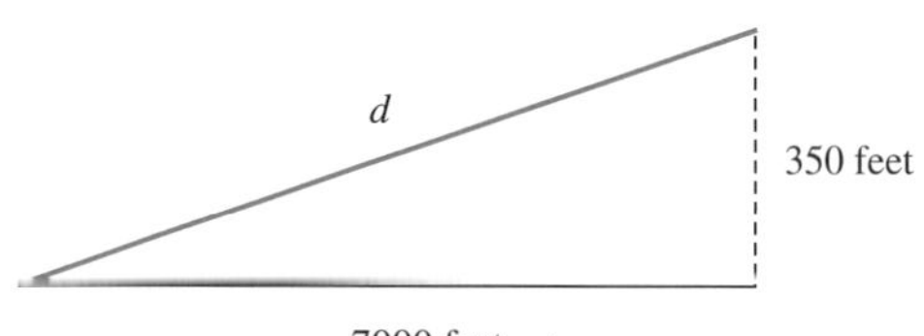

FIGURE 1.3

The cost of construction is (7009 feet) ($120 per foot), or approximately $841,000.

Example 1 serves as a preview to our use of mathematical models in calculus. It uses the available data to produce a mathematical equation (model) that describes the relationship between the variable quantities of interest (elevation of the tunnel above sea level and horizontal distance from where the tunnel enters the mountain). It then uses the model to answer associated questions.

Your students' ability to understand and identify input and output will be essential in their ability to understand and interpret derivatives (how quickly output changes as input changes).

Functions

How Joe's weight changes during the course of his diet and how the tunnel's height above sea level changes as the horizontal distance changes are examples of *functions.* The four representations (data, words, graph, and equation) all describe the same function. Informally, a function is a description of how one thing, called the **output**, changes (Joe's weight or the tunnel's elevation) as something else, called the **input**, changes (the number of weeks on the diet or the horizontal distance). To understand functions thoroughly, however, we need a more precise definition.

Think of Joe's weight function as a *rule* that tells you what he weighs when you know how long he has been dieting. We visualize the relationship between weight and time spent dieting with the **input/output diagram** shown in Figure 1.4. A rule is a function if each input produces exactly one output. If any particular input produces more than one output, then the rule is not a function.

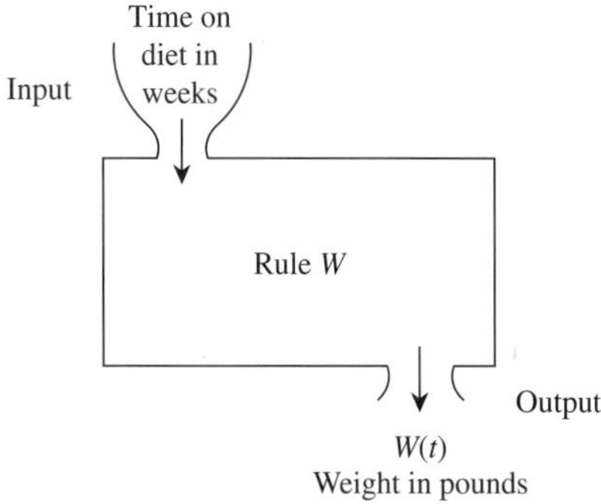

FIGURE 1.4

> A **function** is a rule that assigns exactly one output to each input.

To verify that the rule for Joe's weight is a function, we must ask, "Can Joe have more than one weight after a certain number of weeks on his diet?" Assuming that Joe's weight is measured at a fixed time on the same scale once a week, there will be only one weight that corresponds to each week. Thus Joe's weight is a function of the number of weeks he has been on his diet.

Many students mistakenly think that two different inputs having the same output prevents a relation from being a function. Point out that in Example 2, Q is a function although two students have the same grade. You may wish to introduce the term *one-to-one* to your students and indicate that not all functions are one-to-one.

You probably recall from previous math courses than the standard terms for the set of inputs and the set of outputs of a function are *domain* and *range*, respectively. Other terms for input and output include *independent variable* and *dependent variable* and *controlled variable* and *observed variable*. In this book, however, we use the terms *input* and *output*.

In the table representation of Joe's weight function, the set of inputs is $\{0, 1, 2, 3, 4, 5\}$ and the set of outputs is $\{165, 163, 161, 159, 157, 155\}$. In the graph representing Joe's weight, the set of inputs is all real numbers (not just integers) between 0 and 5 ($0 \le$ weeks ≤ 5) and the set of outputs is all real numbers between 155 and 165 ($155 \le$ weight ≤ 165). We call this graph continuous because it can be drawn without lifting the writing instrument from the page. We will give a more precise definition of *continuous* in Section 1.4.

EXAMPLE 2 *Student Grades*

Suppose $Q(s)$ is the grade (out of 30 points) that student s scored on the first quiz in a course attended by the five students listed in Table 1.4.

TABLE 1.4

Student, *s*	J. DeCarlo	S. Dyers	J. Lykin	E. Mills	G. Schmeltzer
Grade, *Q*(*s*)	28	12	25	21	25

a. Is Q a function of s?

b. Identify the set of inputs and the set of outputs.

Solution

a. Q is a function of s because each name (input) corresponds to only one quiz grade (output). The input/output diagram for Q is shown in Figure 1.5.

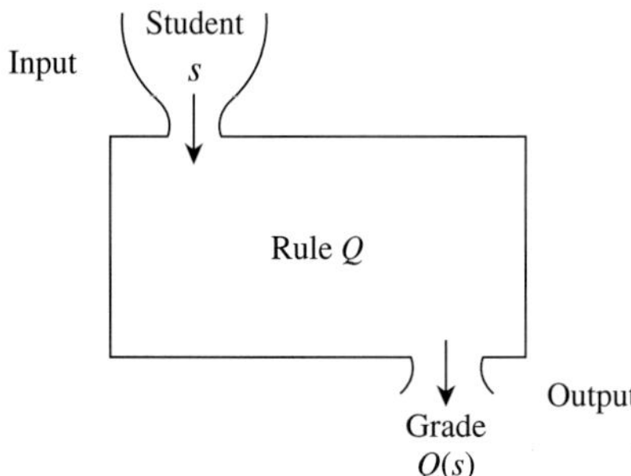

FIGURE 1.5

b. The set of inputs is simply the students' names, and the set of outputs is $\{12, 21, 25, 28\}$.

Determining Outputs

In the example of Joe's diet, t is the number of weeks on the diet (the input) and W is the name of the rule. The notation for the output is $W(t)$. The t is inside the parentheses to remind us that t is the input, and the W is outside to remind us that it is the rule that gives the output. Thus $W(4) = 157$ means that when Joe has been on his diet for 4 weeks (the input), he weighs 157 pounds (the output).

The way you find the output that corresponds to a known input depends on how the function is represented. In a table, you simply locate the desired input in the input row or column. The output is the corresponding entry in the adjacent row or column. For example, the output corresponding to an input of 1999 is in color in Tables 1.5 and 1.6.

TABLE 1.5

Year	1997	1998	1999	2000	2001	2002
Amount	7.3	5.9	5.8	6.4	3.2	3.2

TABLE 1.6

Year	Amount
1998	496,321
1999	527,379
2000	546,108
2001	599,630
2002	654,127
2003	683,044

In a function represented by a graph, the input is traditionally on the horizontal axis. Locate the desired value of the input on the horizontal axis, move directly up (or down) until you reach the graph, and then move left (or right) until you encounter the vertical axis. (You will find a see-through ruler helpful.) The value at that point on the vertical axis is the output. In Figure 1.6, the graph of average faculty salaries at a private liberal arts college, the output is approximately \$56,000 when the input is 1999.

Notice that we do not show a break in the axes when the point of intersection of the two axes is not the origin. This is true throughout the book. Students are probably accustomed to seeing graphs in this manner as a result of using graphing calculators or graphing software, but you may wish to warn them not to assume that the axes intersection shown corresponds to (0, 0).

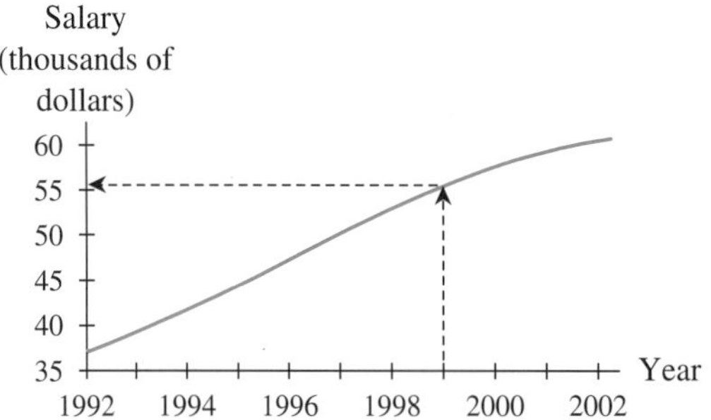

FIGURE 1.6

Finally, if a function is represented by a formula, you simply substitute the value of the input everywhere that the variable appears in the formula and calculate the result. To use the formula $W(t) = 165 - 2t$ to find Joe's weight after 3 weeks, substitute 3 for t in the formula: $W(3) = 165 - 2(3) = 165 - 6 = 159$ pounds.

Units of Measure

Sometimes the units of measure are the same as what is being measured. For instance, in Example 3 we are measuring the number of bacteria and the units of measure are *bacteria.* Students should avoid giving an answer such as "6000 number of bacteria."

In understanding how functions and graphs model the real world, it is important that you understand the **units of measure** of the input and output of functions. In the weight loss example, the unit of measure of the input is *weeks* and the unit of measure of the output is *pounds*. The units of measure in Figure 1.6 can be read from the graph. The unit of measure of the output is *thousands of dollars*, and the unit of measure of the input is *years*. Note that the unit of measure is a word or short phrase telling *how* the variable is measured, not an entire description telling what the variable represents. For example, it would be incorrect to say that the unit of measure of the output in Figure 1.6 is "average faculty salary in thousands of dollars." This is a description of the output variable, not the output units.

EXAMPLE 3 *Bacteria Growth*

Notice that the terms *increasing, decreasing,* and *constant* are defined in the solution of this example.

We all have bacteria on our skin, and they reproduce at a rapid rate. The graph in Figure 1.7 describes the growth of a colony of bacteria on the human body.

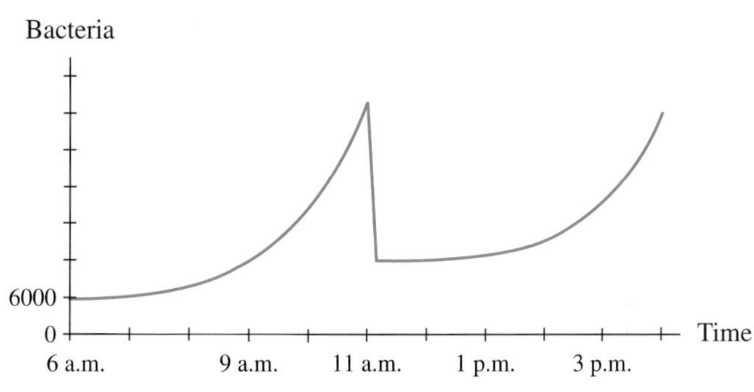

FIGURE 1.7

a. Describe a situation that might result in the graph in Figure 1.7 for the growth of bacteria on your skin.

b. Is the vertical axis the input or the output axis? What are the units on this axis? Fill in values for the unlabeled tick marks on the vertical axis.

c. Describe the horizontal axis variable. In what units is the distance between two consecutive tick marks measured?

d. Use the graph to estimate the number of bacteria on your skin at 9 a.m. and at 3 p.m.

e. At what times is the concentration of bacteria on your skin the highest? Give an estimate of the maximum number of bacteria on your skin at these times.

f. Estimate the intervals of time during which the number of bacteria is increasing.

g. Estimate the intervals of time during which the number of bacteria is decreasing.

Solution

a. There are many possible descriptions you could give. Consider this one. While you are sleeping in your air-conditioned apartment, the rate of growth of the bacteria is slow. After you arise at 8 a.m., the colony grows faster, and its growth really accelerates when you play tennis between 9 a.m. and 11 a.m. You then take a 15-minute shower and wash off most of the pesky critters! You relax, eat lunch in your apartment, and go jogging at about 2 p.m. After jogging, you remain outside to talk with a friend until approximately 3:30 p.m. In the humid, moist atmosphere created by jogging, the number of bacteria grows rapidly.

b. The vertical axis measures output, and the units are bacteria. The graph indicates that each tick mark represents 6000 bacteria. Thus, reading upward, the remaining tick marks are labeled 12,000, 18,000, 24,000, 30,000, 36,000, and 42,000.

c. The horizontal axis variable is time. The distance between displayed tick marks is measured in hours.

d. The number of bacteria on your skin at 9 a.m. is near 12,000. One possible visual estimate is 11,500. (See Figure 1.8.) The number at 3 p.m. is between 18,000 and 24,000. A possible answer is 21,000 bacteria.

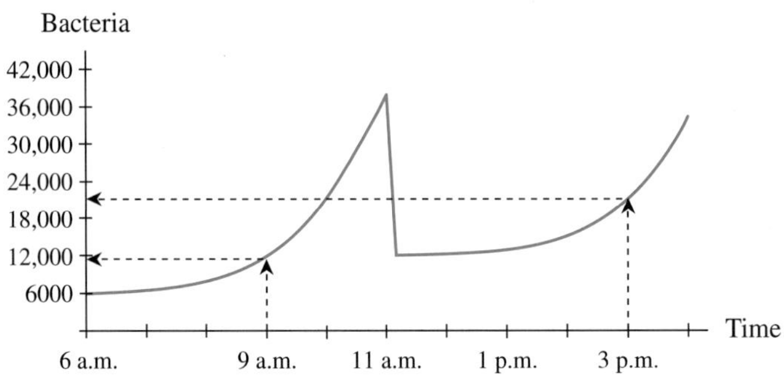

FIGURE 1.8

e. The times at which the concentration of bacteria on your skin appears to be the highest in relation to nearby times are 11 a.m. and around 4 p.m. The maximum number of bacteria is a value on the vertical axis. The values corresponding to 11 a.m. and 4 p.m. are approximately 38,000 bacteria and 36,000 bacteria, respectively. We estimate that the greatest number of bacteria was 38,000, occurring at 11 a.m.

f. If a graph is rising as you move from left to right along the horizontal axis, then it is said to be **increasing.** This graph is increasing between approximately 6 a.m. and 11 a.m. It is also increasing from about 11:15 a.m. until approximately 4 p.m.

g. A graph is said to be **decreasing** if it is falling as you move from left to right along the horizontal axis. This graph is decreasing between 11 a.m. and approximately 11:15 a.m. A graph that neither rises nor falls is called a **constant** graph. This graph is never constant.

Recognizing Functions

Consider the following input/output tables:

TABLE 1.7

t	1	2	0	3	1	4
$F(t)$	7	8	9	10	11	12

TABLE 1.8

q	1	2	3	4	5	6
$C(q)$	10.6	8.3	10.6	8.3	10.6	8.3

To determine whether these tables represent functions, we must decide whether each input produces exactly one output. Table 1.7 does not represent a function because when the input is $t = 1$, the output can be either 7 or 11. In Table 1.8, each input has only one associated output, so this table does represent a function. This is true even though different inputs ($q = 1, 3, 5$) correspond to the same output $[C(q) = 10.6]$.

It is easy to determine whether a graph represents a function. Examine the two graphs in Figure 1.9. The top graph does not describe y as a function of x because each positive x-value produces two different y-values. The bottom graph does show P as a function of t because every input produces only one output. The quickest way to find out whether a graph represents a function is to apply the **Vertical Line Test.**

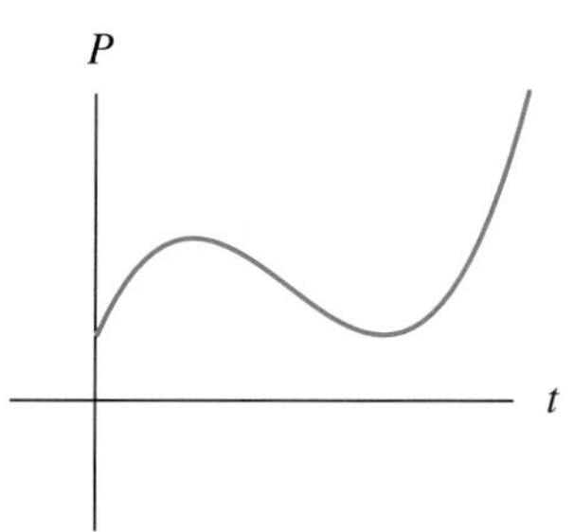

FIGURE 1.9

> **Vertical Line Test**
>
> Suppose that a graph has inputs located along the horizontal axis and outputs located along the vertical axis. If at any input you can draw a vertical line that crosses the graph in two or more places, then the graph does not represent a function.

Figure 1.10 shows a graph that fails the Vertical Line Test.

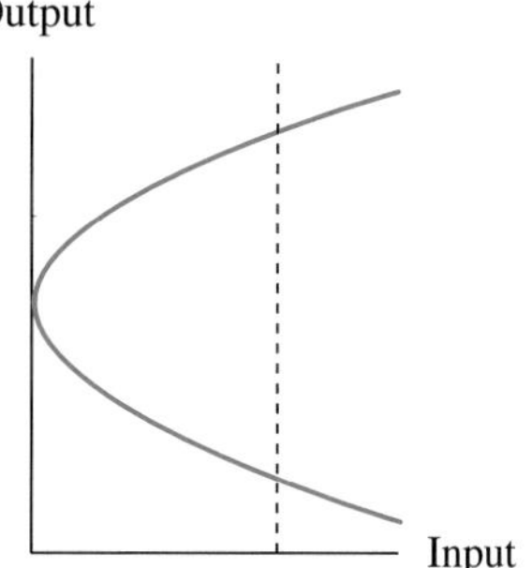

FIGURE 1.10

To determine whether a formula represents a function, first graph the equation and then apply the Vertical Line Test.

EXAMPLE 4 *Land Value*

This symbol indicates that instructions specific to this example for using your calculator or computer are given in a technology supplement.

The value of a piece of property between 1980 and 2000 is given by the equation

$v(t) = 3.622(1.093^t)$ thousand dollars

where t is the number of years since the end of 1980.

a. Describe the input variable. What are its units?

b. Describe the output variable. What are its units?

c. Graph the equation. Explain why the graph represents a function.

d. What was the land value in 1995?

e. When did the land value reach $20,000?

Solution

a. The variable t is the number of years since the end of 1980. Its units are years.

b. The output $v(t)$ is the value of a piece of property. Its units are thousand dollars.

Throughout this book, you may choose to emphasize algebraic equation-solving, or you may allow the students to rely on technology to solve equations. In Section 2.2 we discuss logarithmic solutions for exponential equations needed to algebraically solve the equation in solution *e* of this example.

See the *Instructor's Resource Guide* for a discussion of how to help your students properly interpret input values corresponding to years. This is a tricky skill for students, and if you help them with it here, where it is first an issue, then you will be doing them a favor in later chapters. (The page following the table of contents in the *Instructor's Resource Guide* contains a list of text pages with margin notes that reference the guide and the corresponding *Instructor's Guide* page numbers.)

c. In order to draw an appropriate graph of the value equation, we should first consider which values of t and v we want to see. Because the equation describes the value between the years 1980 and 2000, we want to view the graph with a minimum value of t as 0 and the maximum value of t as 20. In order to determine the output values corresponding to those input values, evaluate the equation at $t = 0$ and $t = 20$. The output values are $v(0) = 3.622$ and $v(20) \approx 21.446$. If you graph the equation with t varying from 0 to 20 and v varying from 3.5 to 21.5, then you should see the graph in Figure 1.11, showing the property value between 1980 and 2000.

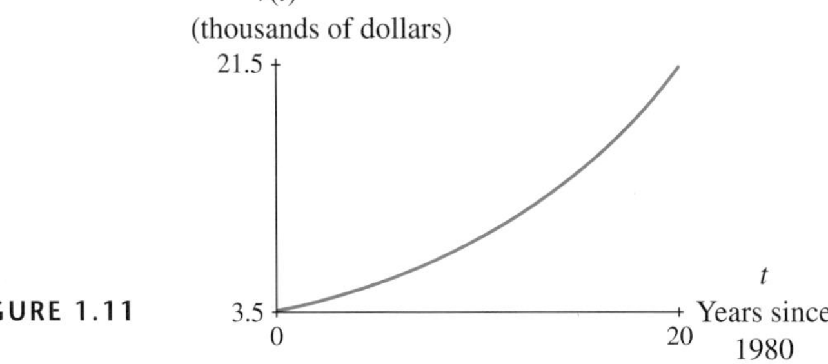

FIGURE 1.11

This graph represents a function because any vertical line drawn through the graph crosses it at only one point. In terms of land value, the function gives only one value for the land at any particular time.

d. The input $t = 15$ corresponds to 1995, so the value of the land in 1995 was

$v(15) = 3.622(1.093^{15}) \approx \$13,748.$

e. In this question we know the output, and we need to find the corresponding input. This requires us to solve for t in the equation $20 = 3.622(1.093^t)$. You can either solve the equation algebraically (using logarithms) or use technology to solve it. In either case, you should find that $t \approx 19.2$. Note that because t is defined as the number of years since the *end* of 1980, $t = 19$ corresponds to the end of 1999, so $t = 19.2$ corresponds to early in the year 2000. Thus the land reached a value of $20,000 in 2000.

Modeling in Business

Throughout this text, we assume that all students have a thorough understanding of these basic business terms. You should take the time to explain them here.

Modeling in the business world often involves terms that are easily understood. Examples of such business terms are *fixed costs* (also called *overhead*), *variable costs, total cost, revenue, profit*, and *break-even point*. To help you understand these terms, consider an enterprising college student who opens a dog-grooming business. An initial capital investment is needed to purchase equipment, buy a business license, set up a lease agreement for a shop, and so on.

Once the business officially opens, the owner sets aside money at the beginning of each week to pay rent, utilities, salaries, and the like. These costs remain the same regardless of the number of dogs that are groomed during the week and are called **fixed costs.** Other costs change with the number of customers. These **variable costs** include expenditures such as advertising, supplies, laundry and custodial fees, and overtime. The fixed costs plus the variable costs give the **total cost** of doing business.

$$\text{Total cost} = \text{fixed costs} + \text{variable costs}$$

When the commodity produced (dogs groomed, in this example) can be measured in units, then the total production cost divided by the number of units produced is known as the **average cost.**

$$\text{Average cost} = \frac{\text{total cost}}{\text{number of units produced}}$$

Businesses receive **revenue,** which is, in general, the quantity of a commodity sold times the price of that commodity. In the pet-grooming business, revenue is the number of dogs groomed times the grooming price. **Profit** is revenue minus total cost.

$$\text{Profit} = \text{revenue} - \text{total cost}$$

The graph in Figure 1.12 shows the weekly profit for this business as function of the number of dogs groomed.The places where a graph crosses (or touches) the horizontal and vertical axes are called **intercepts.** The intercepts often have interpretations in the context of the situation that the graph represents. The horizontal axis intercept in Figure 1.12 is approximately 31, and the vertical axis intercept is about -380. Therefore, from this graph we can conclude that the fixed costs each week are approximately $380 (the cost associated with grooming no dogs). Profit is negative until 31 dogs are groomed. This point on the graph is called the **break-even point,** the point at which revenue equals total cost so that profit is zero.

The **break-even point** occurs when revenue equals total cost and profit equals 0.

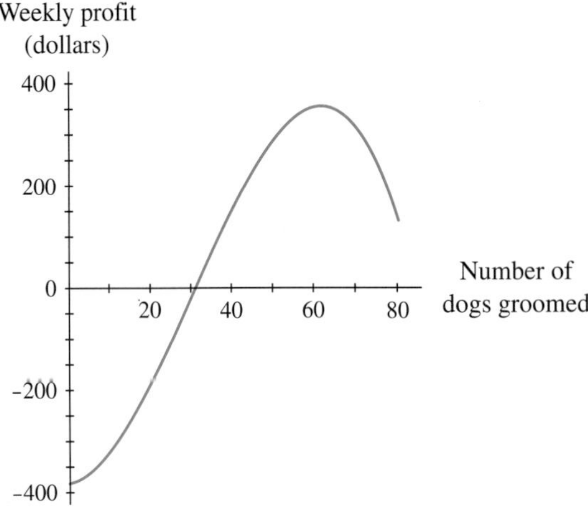

FIGURE 1.12

We recommend encouraging students to work on activities in pairs. You may even wish to assign this first activity set to be completed by pairs of students. While you may risk one student doing the work for another, you may also be helping several of your students to get into the habit of studying math with a partner. The authors believe that the benefits of encouraging students to discuss concepts and work together on homework outweigh the risks.

Profit appears to peak at approximately \$350. This occurs when about 62 dogs are groomed each week. Thus 62 is the optimal number of dogs for this business to groom each week. If more than 62 dogs were groomed each week, profit would actually decline, perhaps because of the need to pay overtime or hire another employee. Cost, profit, revenue, and break-even points are fundamental concepts that are important for business and non-business students to understand. We will refer to them regularly throughout this text.

In this section we have considered models, functions, and graphs. These three concepts form the foundation for the remainder of this text. Because calculus is the study of how functions (and the models and graphs representing functions) change, it is vital that you have a thorough understanding of functions. The remainder of this chapter and the next chapter are devoted to a further exploration of functions.

If you are a first-time user of this text, we strongly recommend that you work through (at least, read through) all of the activities in a section before making a homework assignment.

1.1 Concept Inventory

- *Mathematical modeling*
- *Function*
- *Inputs and outputs*
- *Units of measure*
- *Graph terms: continuous, increasing, decreasing, constant, axes intercepts*
- *Vertical Line Test*
- *Business terms: fixed costs, variable costs, average cost, total cost, revenue, profit, break-even point*

Even activities do not directly correspond to odd activities. The first activities are not necessarily the easiest.

The activities in this text do not parallel those in a more traditional mathematics book. In particular, there are activities for which there is no corresponding example in the text discussion.

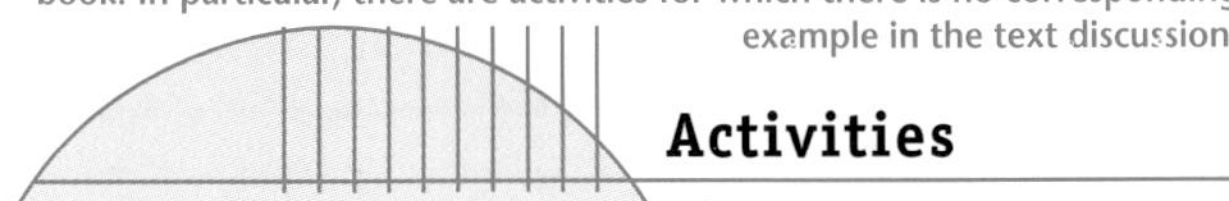

Activities

For each of the rules in Activities 1 through 10, specify the input and output, the input and output variables, and the input and output units. Determine whether each rule is a function.

1. $R(w)$ = the first-class domestic postal rate (in cents) of a letter weighing w ounces

2. $G(t)$ = your score out of 100 on the first test in this course when you studied t hours the week before the test

3. $B(x)$ = the number of students in this class whose birthday is on the xth day of the year (assuming it is not a leap year)

4. $D(p)$ = the distance from the origin to a point p on the circle $x^2 + y^2 = 1$

5. $A(m)$ = the amount (in dollars) you pay for lunch on the mth day of any week

6. $C(m)$ = the amount of credit (in dollars) that Citibank Visa will allow a 20-year-old with a yearly income of m dollars

7. a. $B(t)$ = the amount in an investment account (in dollars) after t years, assuming that no deposits or withdrawals are made during the t years

   b. $A(t)$ = the amount in an investment account (in dollars) after t years, assuming that deposits and withdrawals are permissible.

8. $D(r)$ = the number of years it takes for an investment to double if the annual percentage rate (APR) is r%

9. $P(x)$ = the hair color of movie starlet x during 2002

10. $H(a)$ = your height in inches when you were age a years old

Determine whether the tables in Activities 11 through 14 represent functions. Assume that the input is in the left column.

11. **TABLE 1.9**

Age[1]	Percent with flex schedules at work
16–19	20.7
20–24	22.7
25–34	28.4
35–44	29.1
45–54	27.3
55–64	27.4
over 64	31.0

12. **TABLE 1.10**

Military rank[2] (10 years service)	Basic monthly pay in 1997 (dollars)
Second Lieutenant	2171
First Lieutenant	2752
Captain	3449
Major	3620
Lt. Colonel	3764
Colonel	4260
Brigadier General	5796
Major General	6694
Lt. General	7010
General	7912

13. **TABLE 1.11**

Person's height	Person's weight (pounds)
5′ 3″	139
6′ 1″	196
5′ 4″	115
6′ 0″	203
5′ 10″	165
5′ 3″	127
5′ 8″	154
6′ 0″	189
5′ 6″	143

14.

TABLE 1.12

Year	Total wagers in Nevada on the Super Bowl[3] (millions)
1992	$50.3
1993	$56.8
1994	$54.5
1995	$69.6
1996	$70.9
1997	$70.9
1998	$77.3

1. *Statistical Abstract*, 1998.

2. *Information Please Almanac*, 1998.

3. As reported in the *Reno Gazette–Journal*, January 18, 1999, p. 1A.

15. $P(m)$ is the median sale price[4] (in thousands of dollars) of existing one-family homes in metropolitan area m in 1997. Write the following statements in function notation.
    a. The median sale price in Honolulu was \$307,000.
    b. In Springfield, MA, the median sale price was \$107,000.
    c. \$152,000 was the median sale price in Portland, OR.

16. $E(t)$ is the value of cotton exports[5] in millions of dollars in year t. Write sentences interpreting the following mathematical statements.
    a. $E(1988) = 1975$
    b. $E = 1999$ when $t = 1992$

17. Which of the following graphs represent functions? (The input axis is horizontal.)

a.

FIGURE 1.1.1

b.

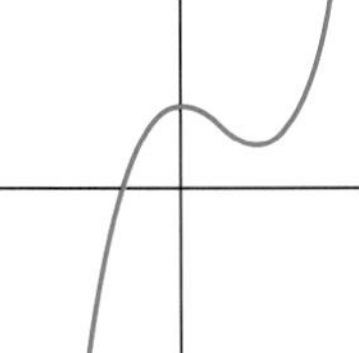

FIGURE 1.1.2

c.

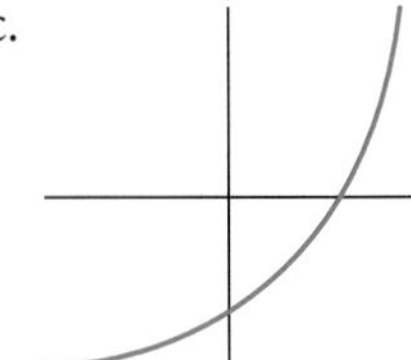

FIGURE 1.1.3

18. Which of the following graphs represent functions? (The input axis is horizontal.)

a.

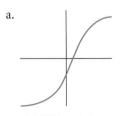

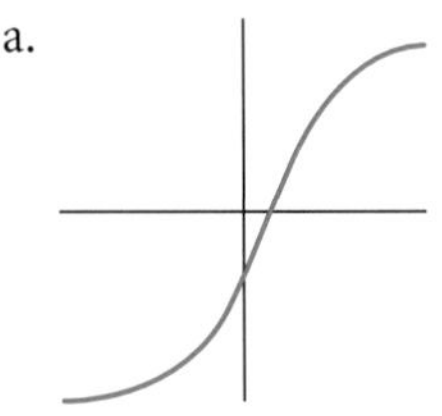

FIGURE 1.1.4

b.

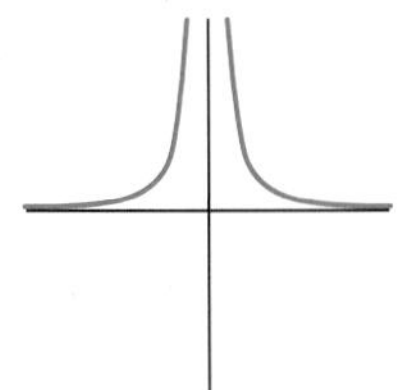

FIGURE 1.1.5

c.

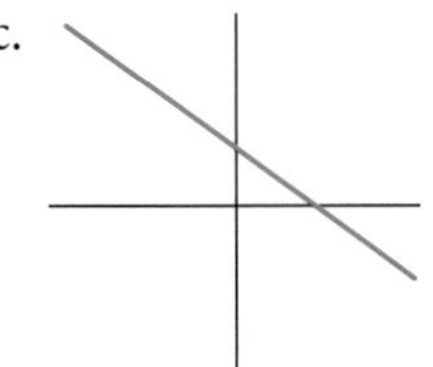

FIGURE 1.1.6

19. The graph in Figure 1.1.7 shows the cost of buying compact discs at a music store that offers one free CD with the purchase of four CDs priced at \$12 each (tax included).
    a. What is the cost of 6 CDs?
    b. How many CDs could you buy if you had \$36?
    c. How many CDs could you buy if you had \$80?
    d. What is the average price of each CD if you buy 3 CDs? 6 CDs?

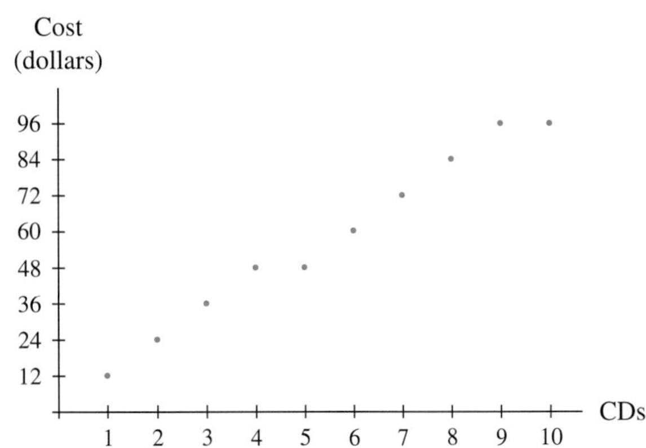

FIGURE 1.1.7

20. A fraternity is selling T-shirts on the day of a football game. The shirts sell for \$8 each.
    a. Complete Table 1.13.

TABLE 1.13

Number of shirts sold	Revenue (dollars)
1	
2	
3	
4	
5	
6	

    b. Construct a revenue graph by plotting the points in Table 1.13.
    c. How many T-shirts can be purchased with \$25?

4. *Statistical Abstract*, 1998.
5. *Statistical Abstract*, 1994.

d. If an 8% sales tax were added, how many T-shirts could be purchased with $25?

21. You are interested in buying a used car and in financing it for 60 months at 10% interest. As a special promotion, the dealer is offering to finance with no down payment. The graph in Figure 1.1.8 shows the value of the car as a function of the amount of the monthly payment.

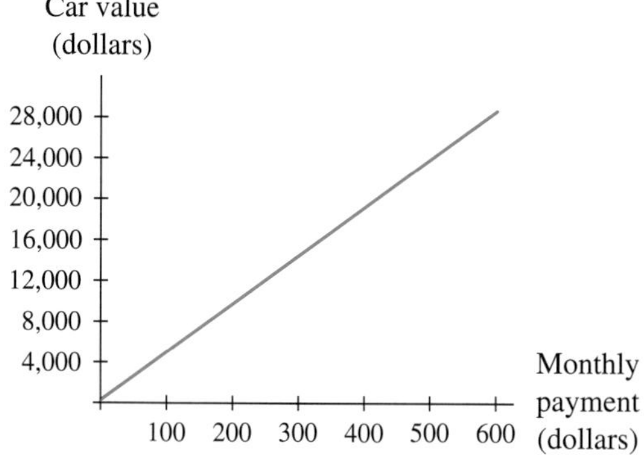

FIGURE 1.1.8

a. Estimate the value of the car you can buy if your monthly payment is $200.

b. Estimate the monthly payment for a car that costs $16,000.

c. Estimate the amount by which your monthly payment will increase if you buy a $20,000 car rather than a $15,000 car.

d. How would the graph change if the interest rate were 12.5% instead of 10%?

22. You have decided to purchase a car for $18,750. You have 20% of the purchase price to use as a down payment, and the purchase will be financed at 10% interest. The graph in Figure 1.1.9 shows the monthly loan payment in terms of the number of months over which the loan is financed.

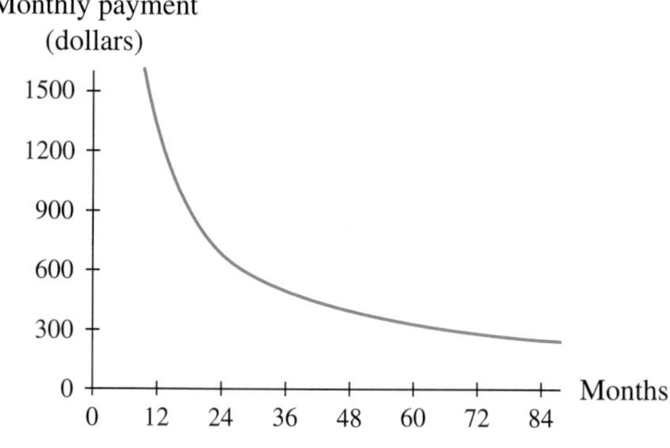

FIGURE 1.1.9

a. What is the actual amount being financed?

b. Estimate your monthly payment if you finance the purchase over 36 months.

c. Estimate the number of months you will have to pay on the loan if you can afford to pay only $300 a month.

d. If you decrease the time financed from 48 to 36 months, by how much will your payment increase?

e. How would the graph change if you were buying a $20,000 car?

23. The graph[6] in Figure 1.1.10 shows the percentage by which Social Security checks increased as a result of cost-of-living adjustments every other year from 1988 through 1998.

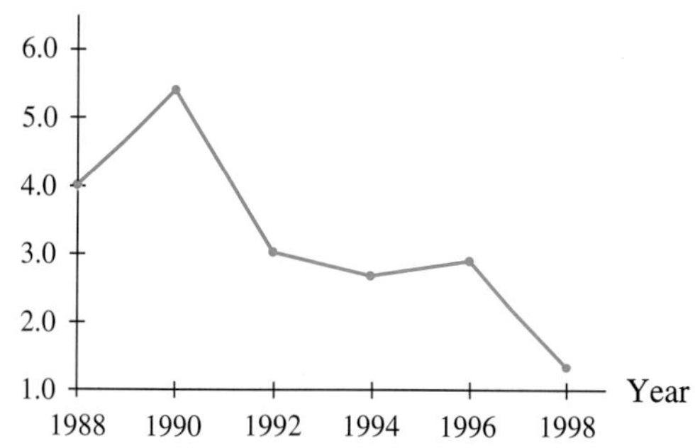

FIGURE 1.1.10

a. What was the cost-of-living increase in 1992?

b. When was the cost-of-living increase the greatest? Estimate the cost-of-living increase in that year.

c. When was the cost-of-living increase 2.8%?

d. Did Social Security benefits increase or decrease between 1996 and 1998? Explain.

24. The graph[7] in Figure 1.1.11 compares the minimum wage to its adjusted value in constant 1996 dollars.

6. Social Security Administration.
7. Bureau of Labor Statistics.

FIGURE 1.1.11

a. Estimate the minimum wage in 1980 and the adjusted minimum wage in that same year.
b. During the years when the minimum wage did not increase, what happened to the buying power of the minimum wage?
c. In what years was the adjusted minimum wage below $5.75?

25. A baby weighing 7 pounds at birth loses 7% of her weight in the 3 days after birth and then, over the next 4 days, returns to her birth weight. During the next month, she steadily gains 0.5 pound per week. Sketch a graph of the baby's weight from birth to 4 weeks. Accurately label both axes.

26. Harry Eagle is putting on the 18th hole of a golf course he has never played. His approach shot has landed on the green, and his putts are described by the graph in Figure 1.1.12. (The dot on the vertical axis represents the distance from the hole that Harry's approach shot landed on the green.)

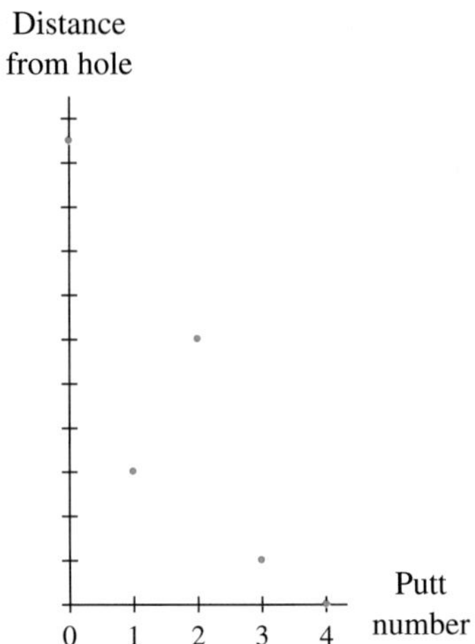

FIGURE 1.1.12

a. Assign appropriate units to the tick marks on the vertical axis.
b. What do you think happened on Harry's second putt?
c. If Harry's approach shot that landed on the green was his third shot, what was his score on the hole?

27. The plot in Figure 1.1.13 depicts a certain girl's height plotted according to her age.
a. What are the units along the vertical axis?
b. How long was the girl when she was born?
c. Approximately how much did she grow during the first year?
d. At approximately what age did she reach her full height? How tall did she become?
e. Did the girl grow faster during her first 3 years or during the last 3 years before she attained her full height?
f. Would you expect the graph to increase or decrease after age 20? Why?

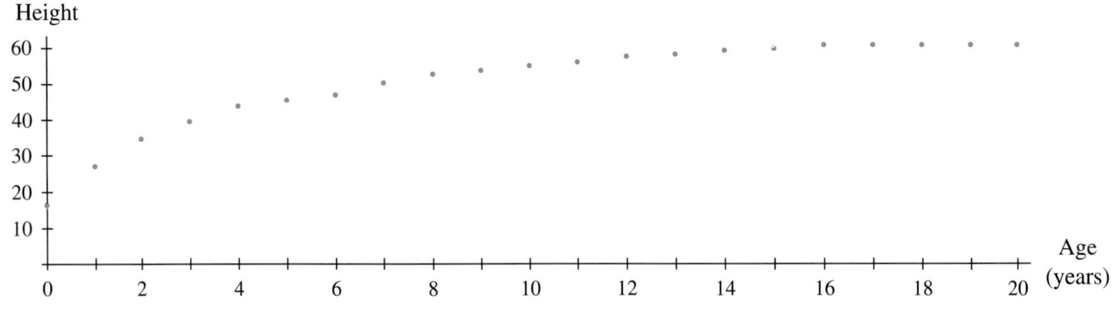

FIGURE 1.1.13

28. A graph depicting the average monthly profit for Slim's Used Car Sales for last year is shown in Figure 1.1.14. Slim needs your help in interpreting this graph.

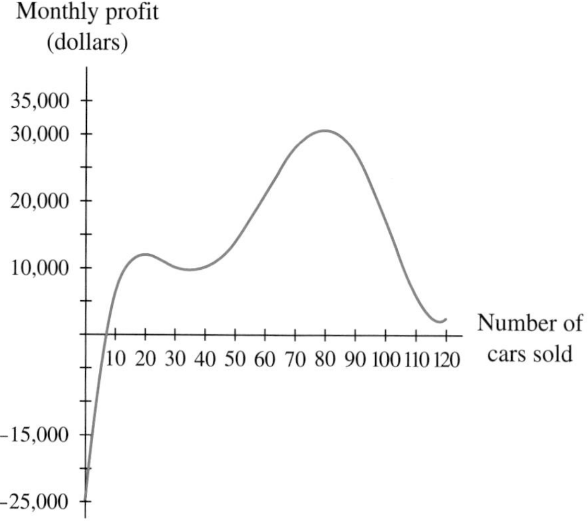

FIGURE 1.1.14

a. Slim pays rent on the lot where the cars to be sold are displayed, and he pays himself and his two salespeople a fixed salary at the beginning of each month. There is also the cost of electricity, water and sewer, health insurance, and so on. These costs remain the same regardless of the number of cars sold during the month. Estimate Slim's fixed costs at the beginning of each month.

b. Slim also has variable costs, such as the prices he pays for the cars. Other variable costs include paying someone to wash and maintain the cars on the lot, minor repairs to the cars, and so on. Slim receives revenue from the customer for each car he sells. Where is Slim's total cost more than his revenue?

c. What is Slim's monthly break-even point?

d. Between what numbers of cars sold is Slim's profit increasing?

e. The monthly profit peaks at two places. Estimate the number of cars sold and the monthly profit generated from the sale of those cars at these two points.

f. Give plausible reasons why Slim's profit sometimes decreases.

g. On the basis of the information presented in the graph, how many cars should Slim try to sell each month to maximize his monthly profit?

29. Donner Pass in the Sierra Nevada of northern California is the most important transmontane route between Reno and San Francisco. Donner Pass is named after George and Jacob Donner, who in 1846 attempted to lead a party of more than 80 immigrants through the pass to the Sacramento Valley. They were unable to make it through the pass before they became snowbound. A little over half the party survived the winter and finished their journey.

The graph[8] in Figure 1.1.15 represents the depth of the snow cover at Donner Memorial State Park from December 1993 through February 1994.

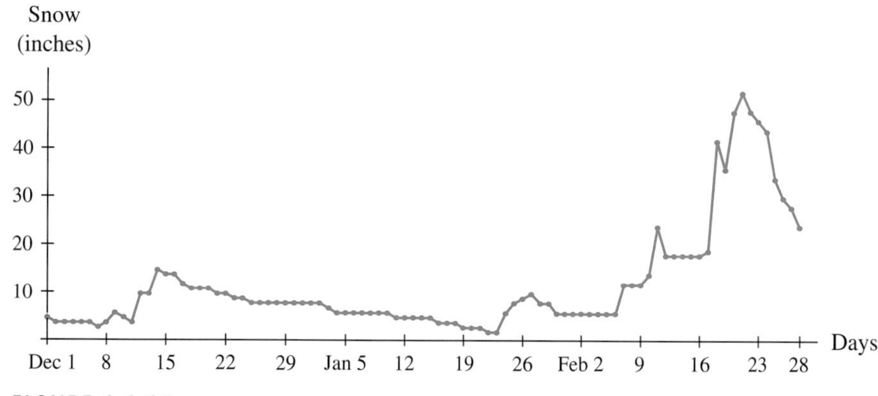

FIGURE 1.1.15

8. Compiled from data in *Monthly Climatological Data: California*, vol. 97, no. 12 (December 1993) and vol. 98, nos. 1 and 2 (January/February 1994).

a. How deep was the snow on December 1, 1993?
b. For approximately how many days after December 2 did the snow remain at the same depth?
c. What do you think happened on December 8–9, December 12–15, January 23–27, February 9–11, February 17, and February 19–21?
d. Why might there have been an immediate decrease in the depth of the snow cover after each one of the peaks?
e. On what day was the snow deepest? How deep (in feet and inches) was the snow at that time?
f. What event may have taken place to cause the rapid decrease of the graph after the last peak?
g. During what day(s) did the most snowfall probably occur?

30. A patient is instructed by her doctor to take one pill containing 500 milligrams of a drug. Assume that her body uses 20% of the original amount of the drug daily. Let y equal the number of milligrams of the drug remaining in the patient's body after x days.
a. Complete Table 1.14 to describe numerically the elimination of the drug from the patient's body.

TABLE 1.14

x	y
0	500
1	400
2	
3	
4	
5	

b. Find a model that describes the elimination of the drug from the patient's body.
c. What are the smallest and largest values of x that make sense in this problem? What are the smallest and largest values of y that make sense in this problem? Graph your model.
d. Calculate the x- and y-intercepts for the graph of your model. Interpret these values in the context of this problem.
e. For which values of x is y decreasing?
f. Use a graph of the model to estimate how much of the drug is left in the patient's body after 3.5 days. Use your model to find the exact value.
g. Use a graph of the model to estimate when the concentration of the drug will be 60 milligrams. Use your model to find the exact value.

31. Repeat Activity 30, assuming that each day the patient's body uses 20% of the amount of drug that was in her body *the previous day.*

practice

For Activities 32 through 35, find the output of the function corresponding to each input value given.

Clarify whether you expect your students to use technology in Activities 36–43 or to solve the equations algebraically. It is the authors' intention that these activities be done using technology. If you expect your students to solve these algebraically, note that the cubic equation in Activity 39 cannot be factored.

32. $y = 7x + 3$; $x = 20, x = -2$

33. $R(w) = 39.4\,(1.998^w)$; $w = 3, w = 0$

34. $S(t) = \dfrac{120}{1 + 3.5e^{-2t}}$; $t = 10, t = 2$

35. $Q(x) = 0.32x^3 - 7.9x^2 + 100x - 15$; $x = 12, x = 3$

For Activities 36 through 39, find the input of the function corresponding to each output value given.

36. $y = 7x + 3$; $y = 24, y = 13.5$

37. $R(w) = 39.4\,(1.998^w)$; $R(w) = 78.8, R(w) = 394$

38. $S(t) = \dfrac{120}{1 + 3.5e^{-2t}}$; $S(t) = 60, S(t) = 90$

39. $Q(x) = 0.32x^3 - 7.9x^2 + 100x - 15$;
$Q(x) = 515, Q(x) = 33.045$

For each of the functions in Activities 40 through 43, determine whether an input or an output value is given, and find its corresponding output or input.

40. $g(x) = 49x^2 + 32x - 134$; $g(x) = 5086$

41. $A = 3200e^{0.492t}$; $t = 15$

42. $p = \dfrac{100}{1 + 25e^{-0.37m}}$; $p = 95$

43. $y = (39.4)(\ln 1.998)(1.998^x)$; $y = 2.97$

practice

This symbol indicates that additional practice problems are available on the *Calculus Concepts* web site.

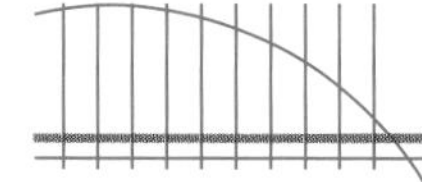

1.2 Functions: Discrete and Continuous

As we have seen, functions can be represented with tables, graphs, words, or equations. Although each is a valid way to describe functions, there are important differences between these representations. Table 1.1 in Section 1.1 gives Joe's weight at six different moments in time: when he begins dieting, 1 week later, 2 weeks later, and so on. In contrast, the graph in Figure 1.1 shows Joe's weight at every time during the 5 weeks of his diet. We use the terms **discrete** and **continuous** to distinguish between the type of information given in Table 1.1 and that given in Figure 1.1. All the information we encounter in tables is discrete, but it is possible to record continuous information. Continuous information is recorded in some type of graph, such as an EKG strip chart or output from a seismograph.

Table 1.1 contains discrete information: Joe's weight at six different moments in time, not at all values during the first 5 weeks. In Figure 1.13, we plot these six data points and get a discrete graph called a **scatter plot.**

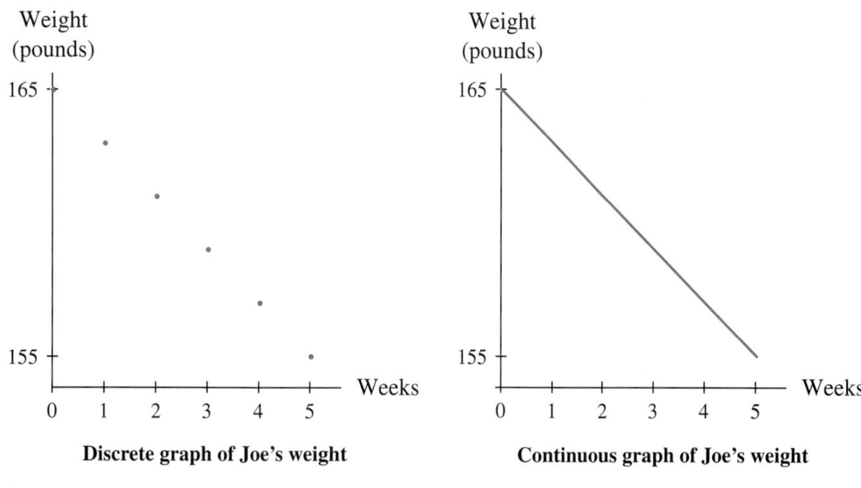

Discrete graph of Joe's weight

FIGURE 1.13

Continuous graph of Joe's weight

FIGURE 1.14

Once we draw a line through the points on the scatter plot, we change the discrete graph to a continuous one showing Joe's weight constantly changing as he diets. See Figure 1.14. The continuous graph has a weight associated with every possible time during the 5 weeks of Joe's diet and can be drawn without lifting the writing instrument from the page. The continuous graph is a better description of how Joe's weight changes, because he doesn't lose weight all at once at the end of each week. An equation [in this case, $W(t) = 165 - 2t$] that represents a continuous graph is a continuous equation.

Continuous Functions with Discrete Interpretation

There are times when we use a continuous function, represented by a continuous graph, in a situation that is not actually continuous. Perhaps the best example of this is compounded interest. Suppose you invest \$100 at a 7% annual percentage rate. We call the initial amount invested the *principal.* If interest is compounded annually, then the formula for the amount in your account t years after you made the investment is $A(t) = 100(1.07^t)$ dollars. Its graph is shown in Figure 1.15.

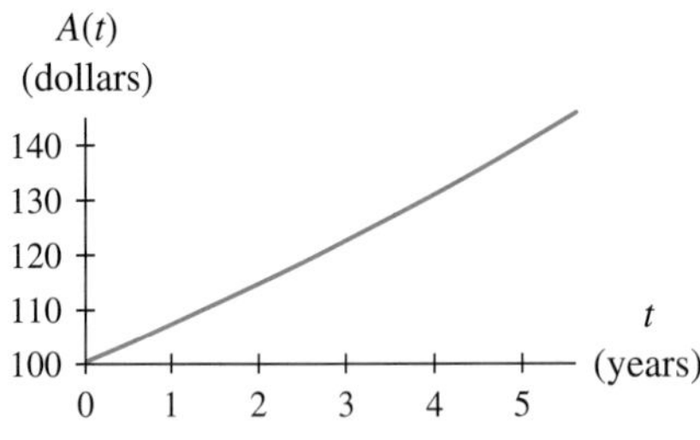

FIGURE 1.15

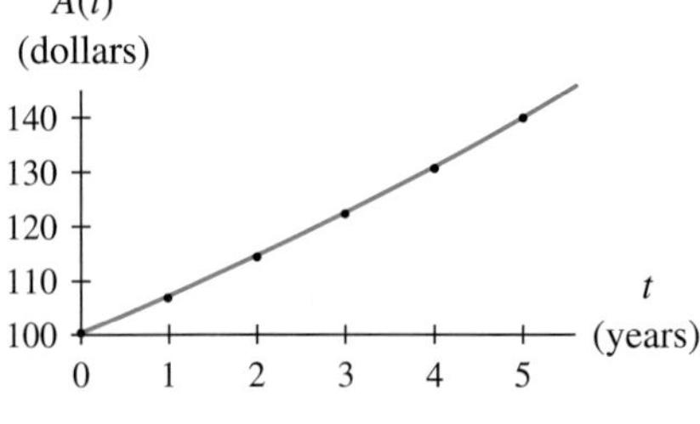

FIGURE 1.16

Although the equation and graph are continuous (they are defined at all values of t, and the graph can be drawn without lifting the writing instrument off the page), they make sense only for integer values of $t \geq 0$ in the context of annually compounded interest.

Suppose that 6 months after you invest the \$100, you have to withdraw the money. According to the equation, when $t = \frac{1}{2}$ year, the amount in the account should be $100(1.07^{\frac{1}{2}}) \approx \$103.44$. However, because interest is compounded once a year (at the end of the year), your money has earned no interest after 6 months, and the amount in the account is only \$100. The only points on the continuous graph that have meaning in this investment situation are those corresponding to positive integer inputs, some of which are shown in black in Figure 1.16. Because A, which is a continuous function, has meaning in this context only at distinct points, we say that A is a **continuous function with discrete interpretation.**

Continuous models are important because we need them in our study of calculus, but it is imperative that students have a clear understanding of whether the underlying situation is actually continuous. This issue of discrete interpretation comes up repeatedly throughout the text.

Continuous Functions with Discrete Interpretation

A continuous function whose interpretation makes sense only at certain distinct points is said to be continuous with discrete interpretation.

A more accurate graph of the amount, which shows the balance changing only at positive integer input values, is given in Figure 1.17. (The open circles at the end of each year indicate that the amount is the value of the solid dot at the beginning of the next year.)

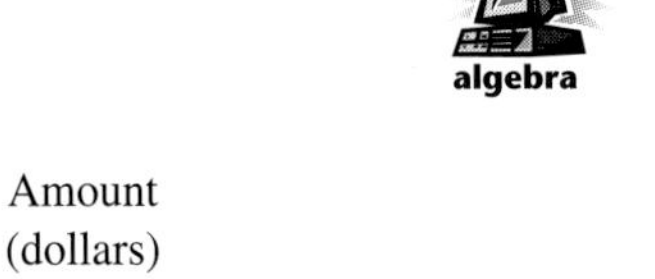

FIGURE 1.17

The equation for the graph in Figure 1.17 describes the balance after t years and can be written as

$$B(t) = \begin{cases} \$100 & \text{when } 0 \leq t < 1 \\ \$107 & \text{when } 1 \leq t < 2 \\ \$114.49 & \text{when } 2 \leq t < 3 \\ \$122.50 & \text{when } 3 \leq t < 4 \\ \$131.08 & \text{when } 4 \leq t < 5 \\ \$140.26 & \text{when } t = 5 \end{cases}$$

We call this representation of the amount a *piecewise continuous function.* We will discuss functions of this type in Section 1.3. This function is not a continuous function, and its graph is not a continuous graph; however, we do not use the term *discrete* to describe either the equation or the graph. We reserve the word *discrete* for scatter plots, tables of data, and the interpretation of continuous functions in some situations.

Inputs and Outputs of Continuous Functions with Discrete Interpretation

One of the most important ideas you can impart to your students is that of using their reasoning ability to work with functions rather than following a fixed set of rules. The methods of determining inputs illustrated in this discussion would not work if the input variable were defined differently. Only by carefully thinking about the variable definitions in the context can students be confident that they will always arrive at the correct solution. Resist the temptation to give your students hard and fast rules. Instead, encourage them to think and reason.

You may wish to show your students a graph of A and discuss how to use the graph to help answer the questions posed.

When we are working with a continuous model that has discrete interpretation, it is important that we understand how to interpret both input and output values correctly. Let us reconsider the example of an account in which you invest \$100 at a 7% annual percentage rate with the account balance given by the function $A(t) = 100(1.07^t)$ dollars after t years. As we have already discussed, $A(t)$ makes sense only for positive integer values of t. What do we do when we need to answer questions such as "If you close the account after 5 years and 3 months, how much money will be withdrawn?" or "When will your account balance first exceed \$125?"

In order to answer the first question, we must remember that 5 years and 3 months is represented as $t = 5.25$ years and that the output of this function does not have an interpretation at this input. However, we know that the balance will be the same amount 3 months into the sixth year as it is at the end of the fifth year. The appropriate input to use to answer this question is $t = 5$. The output at $t = 5$ is $A(5) = 100(1.07^5) \approx 140.26$ dollars. When the account is closed after 5 years and 3 months, the amount withdrawn is \$140.26.

Now consider the second question, "When will your account balance first exceed \$125?" Because \$125 is an output of the function A, we set $A(t)$ equal to \$125 and solve for the input t: $100(1.07^t) = 125$. In this case, $t \approx 3.298$. However, t makes sense in the context only if it is a positive integer. We cannot simply round the value of t. If we were to round to $t = 3$, then the amount in the account would be $A(3) = \$122.50$, which does not exceed \$125. We must instead consider what will happen the next time interest is compounded after $t \approx 3.298$. Interest is next compounded at $t = 4$, and at that time the balance is $A(4) = \$131.08$. Thus the first time the amount in the account exceeds \$125 is at the end of 4 years.

The following example further illustrates how to interpret the input and output of a continuous function with discrete interpretation.

EXAMPLE 1 *Quarterly Sales*

A model for the quarterly sales of a corporation over a 2-year period is

$$q(t) = 0.07t^2 - 0.65t + 2.38 \text{ million dollars sold}$$

during the tth quarter. Figure 1.18 shows a continuous graph of this function.

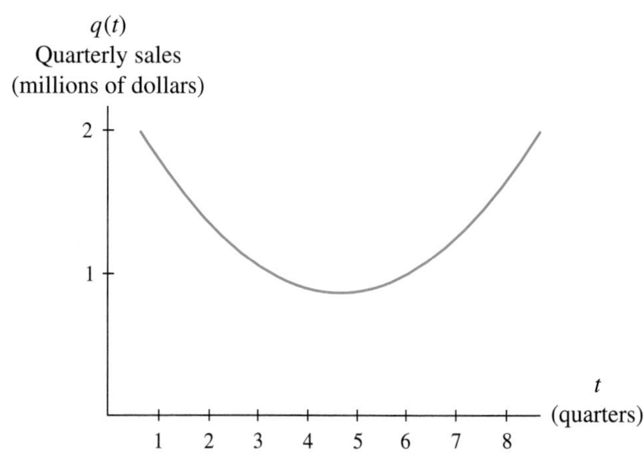

FIGURE 1.18

a. Why must this continuous function be interpreted discretely in the context of quarterly sales?

b. According to the model, when did quarterly sales dip below \$1,500,000?

c. When did quarterly sales rise above \$1 million?

d. In what quarter were sales at their worst, and what was the lowest quarterly sales figure?

Solution

One way to determine if a continuous function describing a quantity can be interpreted continuously is to ask whether the "counter" recording output is reset to zero at periodic intervals. If so, then the function must be interpreted discretely. See the *Instructor's Resource Guide* for more detail on distinguishing between continuous and discrete situations.

a. When we calculate quarterly sales, the sales are zero at the beginning of each quarter, rising to the quarterly total at the end of the quarter. Then sales begin at zero again, rising to the next quarter's total. The continuous function does not describe this behavior of sales. The function values for inputs other than positive integers have no meaning in relation to this corporation's sales. For this reason, the model must be interpreted discretely at positive integer values only.

b. We are asked to find the input (quarter) that corresponds to a certain output (\$1,500,000). Note that the output units of q are *million* dollars, so we must write \$1,500,000 as \$1.5 million. We set the function q equal to 1.5 and solve for t:

$$0.07t^2 - 0.65t + 2.38 = 1.5$$

There are two solutions: $t \approx 1.645$ and $t \approx 7.640$. We can ignore the second solution, because the graph indicates that sales are not falling around this point. Remember, when considering the first solution ($t \approx 1.645$), that t must be a positive integer in the context of the quarterly sales; thus, we consider what happens at the integer values on either side of $t \approx 1.645$. When $t = 1$ the quarterly sales are \$1.8 million, and at $t = 2$ the quarterly sales are \$1.36 million. We conclude that the second-quarter sales dropped below \$1.5 million.

c. We first need to find the input that corresponds to an output of \$1 million. Set the function q equal to 1 and solve for t:

$$0.07t^2 - 0.65t + 2.38 = 1$$

The two solutions are $t \approx 3.286$ and $t = 6$. We are looking for a time in which quarterly sales are rising. Quarterly sales are declining, rather than rising, around $t \approx 3.286$, so this cannot be the correct answer. However, quarterly sales are rising around $t = 6$. Do not assume, just because this value of t is a positive integer, that it is the answer we seek. Carefully consider the question: When did sales rise above \$1 million? At $t = 6$ sales were \$1 million. We can conclude that quarterly sales were *above* \$1 million at $t = 7$. Thus seventh-quarter sales (sales for the third quarter of the second year) were above \$1 million.

d. To answer the question "When were quarterly sales the worst?", we need to consider the lowest point on the graph of the function q. This lowest point appears to be between $t = 4$ and $t = 5$. We find the corresponding outputs by evaluating the function at these input values:

$$q(4) = \$0.9 \text{ million} \qquad \text{and} \qquad q(5) = \$0.88 \text{ million}$$

The lowest point on the function q that has a valid interpretation is at $t = 5$ when $q(t) = 0.88$. Thus the lowest quarterly sales amount was \$0.88 million. This was the sales amount for the first quarter of the second year.

As the previous discussion and example illustrate, when you are working with a continuous function with discrete interpretation, it is important to consider carefully what is happening to both the input and the output. You must take special care not simply to round input values in these situations without thinking about the context of the question you are answering and the nature of the input and output of the model.

Other Continuity Considerations

There are situations that are not technically continuous but for which we use a continuous model that is not discretely interpreted. For example, consider the cumulative sales of the corporation in Example 1. The quarterly sales and cumulative sales at the ends of the first four quarters are given in Table 1.15.

TABLE 1.15

Quarter	1	2	3	4
Quarterly sales (millions of dollars)	1.80	1.36	1.06	0.90
Total sales (millions of dollars)	1.80	3.16	4.22	5.12

It is common in business as well as many areas of scientific study to use continuous functions interpreted continuously to describe step-wise change. Using a continuous model in this way is most often done when the step size is extremely small in relation to the total magnitude of the quantity being measured. One of the best examples of relatively small step size is population growth.

Table 1.15 contains discrete information, and a scatter plot of the data is a discrete graph. We also know that if it were possible to record cumulative sales of the corporation at every moment in time, the graph would not be smooth. It would "jump" up as sales were made (or down in the case of refunds) because sales values are always numbers that have two decimal places. For this reason, a graph of sales at every moment in time is not continuous. However, because the "jumps" in the graph are insignificant in the context of total sales of the corporation over an extended period of time, we use a continuous function, interpreted as a continuous function, to model total sales. Situations similar to total sales include cumulative totals of anything and growth of populations.

The models used throughout this text are either continuous or piecewise continuous. The question to ask yourself in every modeling situation is "For this model, do inputs of any value make sense in context?" If the answer is yes, then the model is **continuous without restriction.** If the answer is no, then the model must have discrete interpretation. In the case in which a piecewise continuous function is used, we say *piecewise continuous without restriction* or *piecewise continuous with discrete interpretation.* To illustrate, consider the four function situations we have discussed:

1. *Joe's weight:* Because Joe has a weight at every possible moment, and because the model gives an approximation of his weight at all times, we use the continuous model without restriction.
2. *Annually compounded interest:* Although there is an account balance at every possible time, and although the output of the continuous function exists at every possible time, the output from the function does not describe the account balance for inputs other than positive integers. For this reason, if we use the continuous function, it must be discretely interpreted at positive integer values only.
3. *Quarterly sales:* Because quarterly sales exist only at the ends of quarters, a continuous function would not have meaning for any inputs other than those associated with ends of quarters. For this reason, a continuous function modeling quarterly sales would have discrete interpretation.

4. *Cumulative sales:* Because cumulative sales exist at all times and because a continuous function is an approximation of those sales, we use the continuous function without restriction.

Examples of situations that can be represented by continuous functions but have discrete interpretation include interest that is not compounded continuously and anything that is a yearly total, such as sales, profit, revenue, number of live births, number of overseas telephone calls, and yearly high price of gasoline.

EXAMPLE 2 *Cost of Public Defender Services*[9]

Determine whether each of the following function representations is discrete, continuous without restriction, or continuous with discrete interpretation.

a. **TABLE 1.16**

Year	1987	1989	1991	1993	1995	1997	1999
Percent of fees collected for public defender services	41	53	50	44	37	30	11

b.

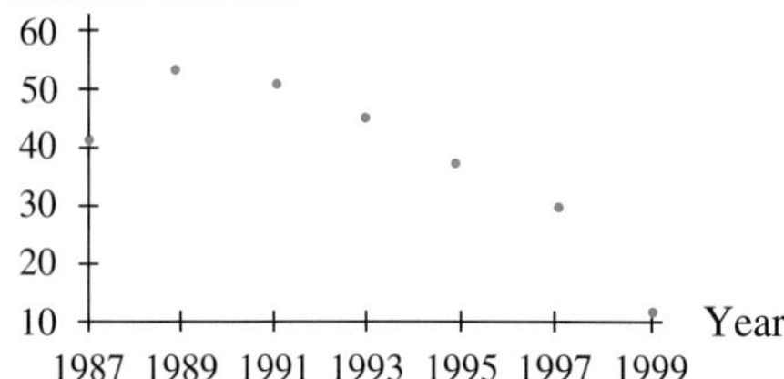

FIGURE 1.19

c.

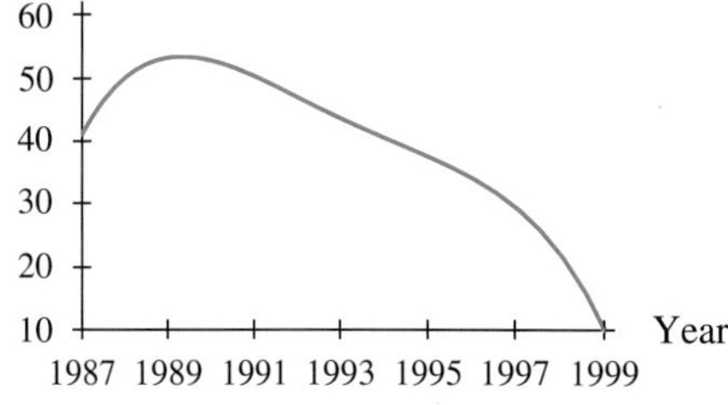

FIGURE 1.20

d. $C(t) = -0.0175t^4 + 0.441t^3 - 4.015t^2 + 12.415t + 40.987$ percent of fees collected by public defenders t years after 1987

9. For Washoe County, NV, Public Defender's Office.

e. Total fees collected in 1999 (thousands of dollars)

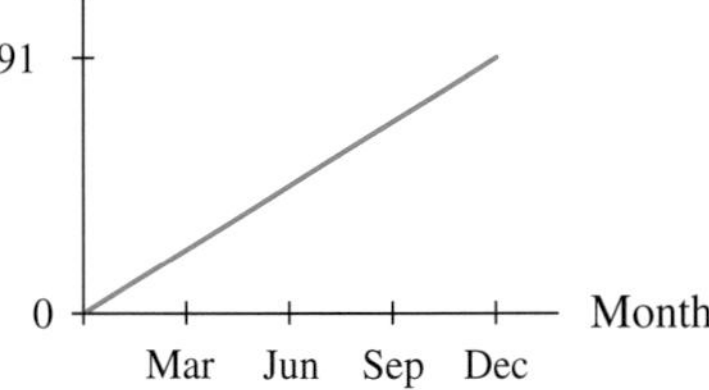

FIGURE 1.21

Solution The table of data in part *a* and the scatter plot in part *b* are discrete. The graph in part *c* and the equation in part *d* are both continuous functions with discrete interpretations, because they are defined only at values of t that correspond to the end of each year. The graph in part *e* is a continuous function that can be used without restriction. This function can give us the total fees collected at any time throughout 1999.

Because calculus is primarily the study of continuous change rather than discrete change, the ability to transform discrete data into a continuous function (a model) is a vital part of our study of calculus. It is important, however, to understand for what input values the model makes sense in context.

1.2 Concept Inventory

- *Discrete and continuous*
- *Scatter plot*
- *Continuous with discrete interpretation*
- *Continuous without restriction*

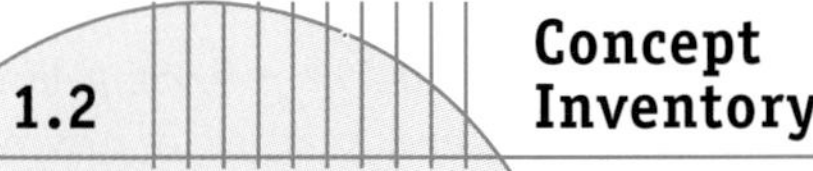

1.2 Activities

1. Which of the following statements describe continuous situations, and which describe discrete situations?
   a. $H(a)$ is your height at age a years.
   b. $W(t)$ is the percentage of the popular vote that the winning candidate for president received in year t.
   c. $A(t)$ is the amount in a savings account after the principal has been invested for t years at 7% compounded continuously.
   d. $P(x)$ is the number of passengers traveling on an airline during the xth month of 1997.

2. If a continuous function is used to model the following function descriptions, can it be used without restriction or must it be discretely interpreted?
   a. $P(m)$ is the amount, in millions of dollars, of property value loss resulting from an earthquake of magnitude m centered in a particular major metropolitan area.
   b. $H(x)$ is the number of hits at a certain web site on the xth day of 2001.
   c. $Q(t)$ is your body temperature in degrees Celsius t hours after you take 500 mg of Tylenol.
   d. $C(r)$ is the number of children who register for summer day camp when the registration fee is set at r dollars.

In Activities 3 through 10, categorize each function representation as discrete, continuous without restriction, or continuous with discrete interpretation.

3. **TABLE 1.17**

Year	Average basic cable rate[10] (dollars per month)
1970	5.50
1980	7.69
1985	9.73
1990	16.78
1995	23.07
1996	24.41
1997	26.48

4.

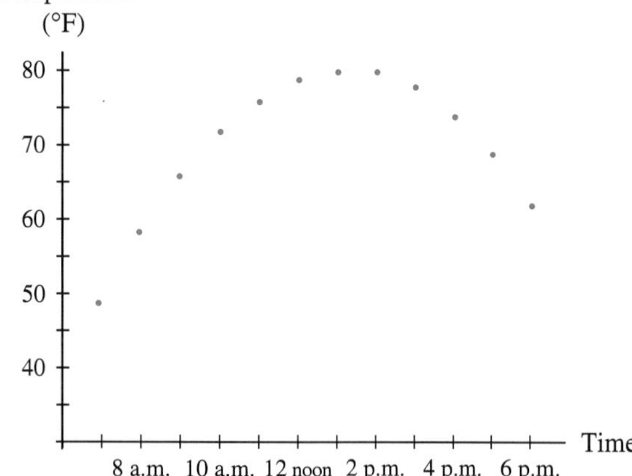

FIGURE 1.2.1

5. On the basis of data recorded in the spring of each year between 1988 and 1997, the number of students of osteopathic medicine in the United States can be modeled[11] as

$$O(t) = 0.027t^2 - 4.854t + 218.929$$

thousand students in year $1900 + t$.

6.

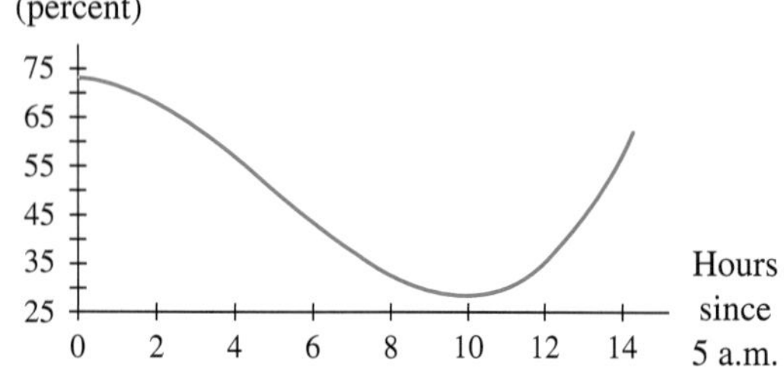

FIGURE 1.2.2

7.

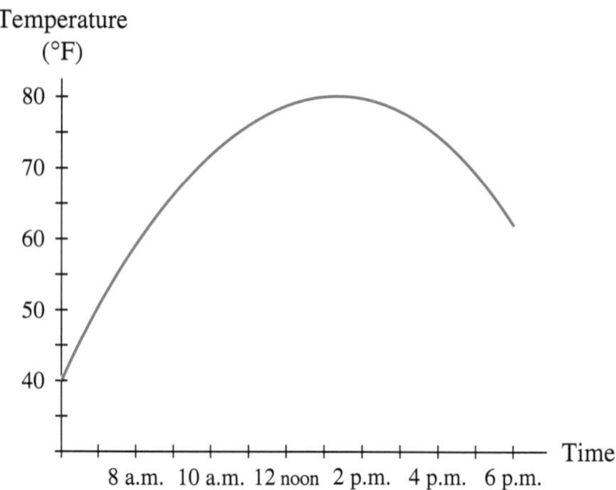

FIGURE 1.2.3

8.

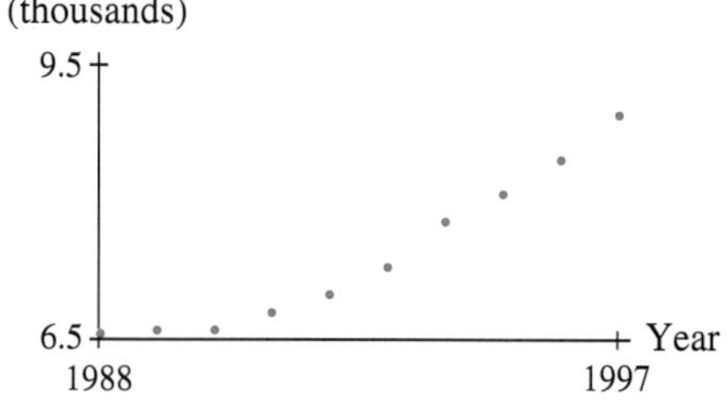

FIGURE 1.2.4[12]

9. $G(m) = (2.798 \cdot 10^{-4})m^2 - 0.041m - 9.956$ gallons of gas in a car's gas tank after m minutes of driving

10.

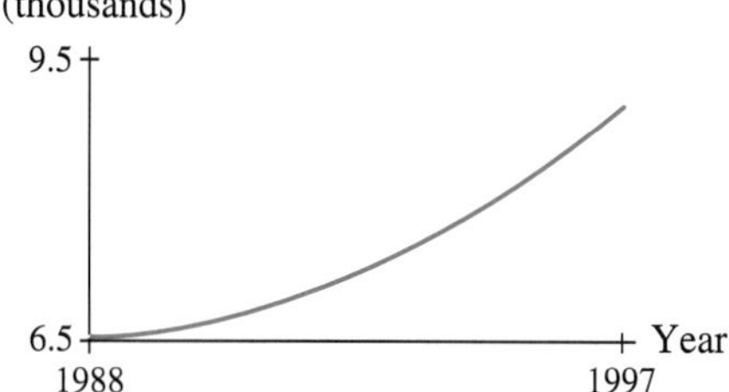

FIGURE 1.2.5[13]

11. Table 1.18 shows the predicted amount of rainfall as a function of the wind speed of a tropical storm.

10. *Statistical Abstract*, 1998.
11. Based on data from *Statistical Abstract*, 1998.
12. *Statistical Abstract*, 1998.
13. Based on data from *Statistical Abstract*, 1998.

Notice in part *d* of Activity 11 that the question refers to the *representation* of the function as a table of data, not to the underlying relationship between wind speed and rainfall. The table representation is discrete, but the underlying relationship is continuous. A continuous (without restriction) model for this data is found in Activity 12.

TABLE 1.18

Wind speed (mph)	10	8	5	3
Predicted rainfall (inches)	10	15	23	33

a. Draw a scatter plot of these data.

b. Use your scatter plot and the Vertical Line Test to verify that Table 1.18 represents a function.

c. Can you determine, using the given information, the rainfall predicted when the wind speed is 6 mph? If so, find the amount. If not, explain why not.

d. Is this function representation discrete or continuous? Give reasons for your answer.

12. The predicted amount of rainfall for a tropical storm can be modeled by the equation

$$R(v) = 54.1(0.847^v) \text{ inches}$$

where v is the wind speed (in miles per hour) of the storm.

a. Describe the input and output of the function R.

b. State the units of measure of the input and output.

c. Is R an increasing, decreasing, or constant function for wind speeds between 3 mph and 10 mph?

d. Can you determine, using the given information, the rainfall predicted when the wind speed is 6 mph? If so, find the amount. If not, explain why not.

e. Is this model discrete, continuous without restriction, or continuous with discrete interpretation? Give reasons for your answer.

f. According to this function, at what wind speed will there be 28 inches of rain from the storm?

13. The average daily profit resulting from telephone marketing calls by an employee of a certain company can be modeled by

$$P(c) = 0.043c^3 - 3.129c^2 + 71.133c - 315.524 \text{ dollars}$$

when c calls are made daily.

a. Find $P(25)$ and interpret the answer.

b. Find the value of c between 0 and 25 for which $P(c) = 180$, and interpret the answer.

c. Graph this function for 0 to 30 calls per day. Use the graph to estimate when the greatest daily profit occurs.

Activity 15 part *e* cannot be answered using the function *O*. The point of this activity is to help the students understand the limitations of a particular model.

14. Funds awarded[14] to arthritis researchers by the Arthritis Foundation each year between 1990 and 2000 can be modeled by the equation

$$R(t) = 0.95\,(1.408^t) + 9 \text{ million dollars}$$

t years after 1990.

a. Write "In 1999 the funds awarded to arthritis researchers were approximately \$29,660,000" using function notation.

b. Explain the meaning of the notation $R(11.5)$. Does $R(11.5)$ make sense in the context of this problem? Explain.

c. Find the value of t for which $R(t) = 18$. Interpret your answer.

15. On the basis of data for the years 1988 through 1997, the number of students of osteopathic medicine[15] in the United States, measured in the spring of the year $1900 + t$, can be modeled as

$$O(t) = 0.027686t^2 - 4.854t + 218.58318 \text{ thousand students}$$

a. Describe the input and output of the function O.

b. State the units of measure of the input and output.

c. Should this continuous model be discretely interpreted? Why or why not?

d. Use the model to estimate the number of osteopathy students in the spring of 1998.

e. How many osteopathy students attended summer school in 1998?

f. According to the model, when will the number of students studying osteopathic medicine exceed 10,000?

16. The number of yearly kidney transplants[16] performed in the United States between 1992 and 1996 can be modeled by the equation

$$K(t) = 9088.859 + 1697.717 \ln t \text{ transplants}$$

t years after 1990.

a. Describe the input and output of the function K.

b. State the units of measure of the input and output.

14. Arthritis Foundation.
15. *Statistical Abstract*, 1998.
16. Based on data from *Statistical Abstract*, 1998.

c. Should this continuous model be discretely interpreted? Why or why not?

d. Use the model to estimate the number of transplants in 1995.

e. Find $K(2.5)$ and interpret the answer.

Part e has no interpretation in this context.

f. According to the model, when did the number of yearly transplants first exceed 13,000?

17. The price of Amazon.com stock on a particular trading day can be modeled by

$$P(h) = -2.295h + 125.4375 \text{ dollars per share}$$

h hours after the market opened.

a. Describe the input and output of the function P.

b. State the units of measure of the input and output.

c. Should this continuous model be discretely interpreted? Why or why not?

d. What was the price of this stock after 3 hours of trading?

e. Did the price reach \$100 per share during this particular day? (The market opens at 9:30 a.m. and closes at 4:00 p.m.)

18. The customer base in the late 1990s for Amazon.com[17] can be modeled by

$$A(t) = 2.3(4.652^t) \text{ million customers}$$

t years after 1998.

a. Describe the input and output of the function A.

b. State the units of measure of the input and output.

c. Should this continuous model be discretely interpreted? Why or why not?

d. Estimate the size of the customer base at the end of 2000.

e. When was the customer base 9 million?

19. The yearly per capita consumption of milk[18] from 1980 to 1996 can be modeled by

$$C(x) = -0.2138x + 44.8434 \text{ gallons}$$

where x is the number of years after 1900.

a. Describe the input and output of the function C.

b. State the units of measure of the input and output.

c. Should this continuous model be discretely interpreted? Why or why not?

d. When was the yearly per capita consumption 25 gallons?

20. The total number of passengers of a certain airline who have flown into the Seattle–Tacoma International Airport during 2002 can be modeled by

$$P(d) = 86.86d \text{ passengers}$$

d days into 2002.

a. Describe the input and output of the function P.

b. State the units of measure of the input and output.

c. Should this continuous model be discretely interpreted? Why or why not?

d. Estimate the number of passengers during the first 6 months of 2002.

e. Estimate the total numbers of passengers in 2002 and in 2005.

f. When during 2002 did the number of passengers reach 20,000?

1.3 Constructed Functions

In Sections 1.1 and 1.2, we examined what a function is and how functions can be represented. Now we consider constructing more complicated functions.

Combining Functions

We can combine two or more functions to create a new function by adding or subtracting. Consider Tables 1.19 and 1.20, which show sales and costs for a corporation from 1996 through 2001.

17. Based on data from "Amazon.com takes loss for quarter," *Reno Gazette–Journal*, July 23, 1999, p. 4B.

18. Based on data from *Statistical Abstract*, 1998.

TABLE 1.19

Year	1996	1997	1998	1999	2000	2001
Sales (millions)	\$3.570	\$3.945	\$4.359	\$4.817	\$5.323	\$5.881

TABLE 1.20

Year	Costs (thousands)	Year	Costs (thousands)
1996	\$1061	1999	\$2329
1997	\$1562	2000	\$2594
1998	\$1984	2001	\$2782

Both tables represent functions, and by subtracting costs from sales, we can get a new function representing profit. Before doing so, however, we need to ask ourselves two questions: "Is the set of inputs the same for both functions?" and "Are the outputs given in the same units?" The answer to the first question is yes, but the answer to the second is no. However, we can easily change one of the tables to achieve the same output units. The resulting profit function is shown in Table 1.21.

TABLE 1.21

Year	1996	1997	1998	1999	2000	2001
Sales (millions)	\$3.570	\$3.945	\$4.359	\$4.817	\$5.323	\$5.881
Cost (millions)	\$1.061	\$1.562	\$1.984	\$2.329	\$2.594	\$2.782
Profit (millions)	\$2.509	\$2.383	\$2.375	\$2.488	\$2.729	\$3.099

The equation $S(t) = 3.570(1.105^t)$ gives the sales in millions of dollars t years after 1996, and the equation $C(t) = -39.2t^2 + 540.1t + 1061.0$ gives the costs in thousands of dollars t years after 1996. As we did with the tables, we can find the profit equation by subtracting costs from sales as long as the output units correspond and the set of inputs is the same for both functions.

To construct an equation P for profit, we must first multiply C by 0.001 to convert the output units from thousands of dollars to millions of dollars. Then profit can be written as

$$\begin{aligned} P(t) &= S(t) - 0.001C(t) \\ &= 3.570(1.105^t) - 0.001\,(-39.2t^2 + 540.1t + 1061.0) \text{ million dollars} \end{aligned}$$

where t represents the number of years since 1996.

Further, we can now use P to find the profit in any given year. For instance, the profit in 1998 was $P(2) \approx 2.375$ million dollars.

Product functions are created by multiplying two functions. Again, this makes sense only if both functions have the same set of inputs and if the output units are compatible so that, when multiplied, they give a meaningful result.

Suppose that $S(x)$ is the selling price (in dollars) of a gallon of milk on the xth day of last month, and $G(x)$ is the number of gallons of milk sold. The input/output diagrams are shown in Figure 1.22.

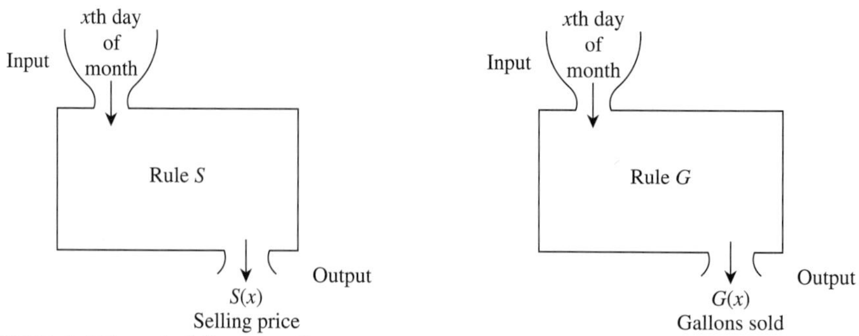

FIGURE 1.22

The inputs of both functions are the same, and when the outputs are multiplied, with units (\$ per gallon)(gallons), the result is the sales in dollars on the xth day of last month. The input/output diagram for the new function $T = S \cdot G$ is given in Figure 1.23.

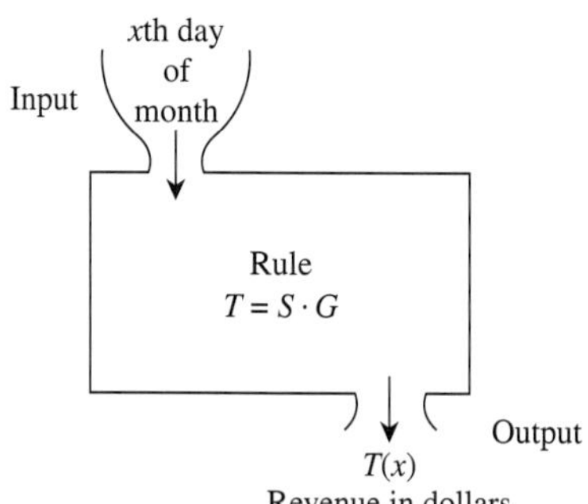

FIGURE 1.23

$S(x) = 0.007x + 1.492$ dollars per gallon and $G(x) = 31 - 6.332(0.921^x)$ gallons are functions that model the selling price and the number of gallons of milk sold on the xth day of last month. Thus milk sales can be modeled by the function

$$T(x) = [S(x)]\cdot[G(x)]$$
$$= [0.007x + 1.492]\cdot[31 - 6.332(0.921^x)] \text{ dollars}$$

on the xth day of last month.

An understanding of function composition is needed for students to understand the Chain Rule in Chapter 4.

Another way in which we combine functions is called **function composition.** Consider Tables 1.22 and 1.23, which give the altitude of an airplane as a function of time and give air temperature as a function of altitude, respectively.

TABLE 1.22

t = time into flight (minutes)	$F(t)$ = feet above sea level
0	4500
1	7500
2	13,000
3	19,000
4	26,000
5	28,000
6	30,000

TABLE 1.23

F = feet above sea level	$A(F)$ = air temperature (degrees Fahrenheit)
4500	72
7500	17
13,000	-34
19,000	-55
26,000	-62
28,000	-63
30,000	-64

It is possible to combine these two data tables into one table showing air temperature as a function of time. This process requires using the output from Table 1.22 as the input for Table 1.23. The only restriction is that both the numerical values and the units in the output of one function match those in the input of the other function. The resulting new function is shown in Table 1.24.

TABLE 1.24

t = time into flight (minutes)	0	1	2	3	4	5	6
$A(F)$ = air temperature (degrees Fahrenheit)	72	17	-34	-55	-62	-63	-64

Because we used the output of Table 1.22, $F(t)$, as the input for Table 1.23, the new function output is commonly written as $A(F(t))$. The A outside the parentheses is to remind us that the output is air temperature, and putting $F(t)$ inside the parentheses reminds us that altitude from Table 1.22 is the input. It is common to refer to A as the *outside function* and to F as the *inside function*. The mathematical symbol for the composition of an inside function F and an outside function A is $A \circ F$, so

$$(A \circ F)(t) = A(F(t))$$

The altitude data can be modeled as $F(t) = -222.22t^3 + 1755.95t^2 + 1680.56t + 4416.67$ feet above sea level, where t is the time into the flight in minutes. The air temperature can be modeled as $A(F) = 277.897(0.99984^F) - 66$ degrees Fahrenheit, where F is the number of feet above sea level. The composition of these two functions is

$$\begin{aligned}(A \circ F)(t) &= A(F(t)) \\ &= 277.897(0.99984^{(-222.22t^3 + 1755.95t^2 + 1680.56t + 4416.67)}) - 66\ ^\circ\text{F}\end{aligned}$$

where t is the time in minutes into the flight.

Let us consider another example to help make function composition clearer.

EXAMPLE 1 *Lake Contamination*

Consider the word descriptions of the following two functions and their input/output diagrams shown in Figure 1.24.

$C(p)$ = parts per million of contamination in a lake when the population of the surrounding community is p people

$p(t)$ = the population in thousands of people of the lakeside community in year t

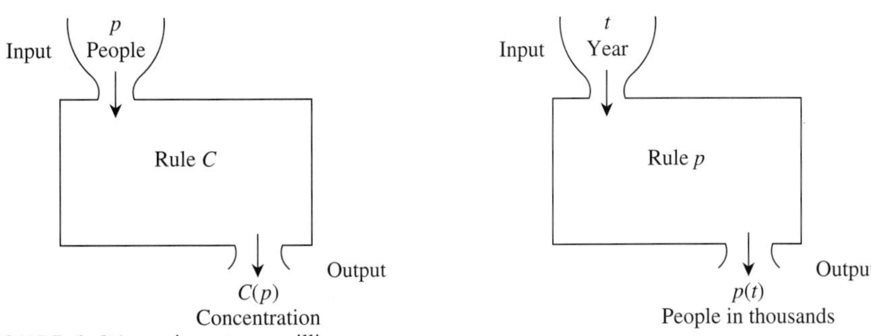

FIGURE 1.24

a. Draw an input/output diagram for the composition function that gives the contamination in a lake as a function of time.

b. The contamination in the lake can be modeled as $C(p) = \sqrt{p}$ parts per million when the community's population is p people. The population of the community can be modeled by $p(t) = 0.4t^2 + 2.5$ thousand people t years after 1980. Write the function that gives the lake contamination as a function of time.

Solution

a. Note that the output of the second function is almost the same as the input of the first function. If we multiply $p(t)$ by 1000 to convert the output from thousands of people to people, we have the function $P(t) = 1000p(t)$ people in year t. Now we can compose the two functions C and P to create the new function $C \circ P$ whose input/output diagram is shown in Figure 1.25.

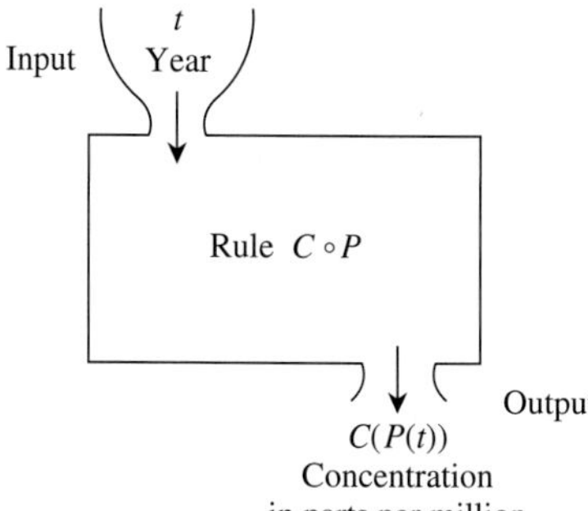

FIGURE 1.25

b. To convert the output of the function p to people, multiply by 1000 to obtain the function $P(t) = 1000(0.4t^2 + 2.5)$ people t years after 1980. Thus the contamination in the lake t years after 1980 can be modeled as

$$(C \circ P)(t) = C(P(t)) = \sqrt{1000(0.4t^2 + 2.5)} \text{ parts per million}$$

We formalize the process of function composition as follows: Given two functions f and g, we can form their composition if the outputs from one of them, say f, can be used as inputs to the other function, g. In terms of inputs and outputs, we have those shown in Figure 1.26. In this case, we can replace the portion of the diagram within the blue box by forming the composite functions $g \circ f$ whose input/output diagram is shown in Figure 1.27.

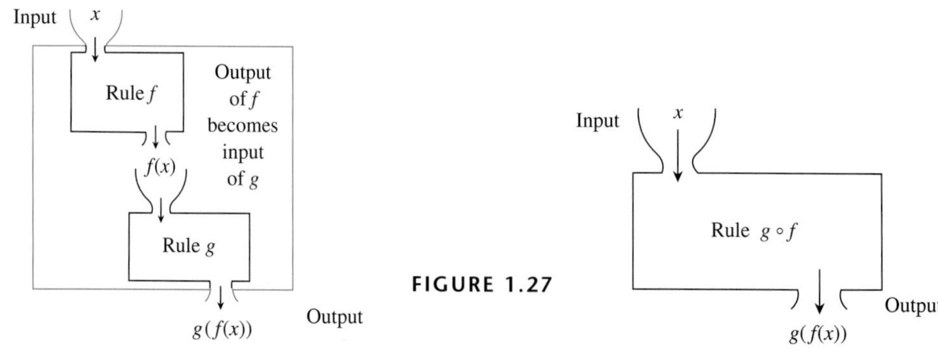

FIGURE 1.26

FIGURE 1.27

Note that the inputs to the composite function $g \circ f$ are the inputs to the inside function f and that the outputs are from the outside function g. We use composite functions more in later chapters.

Inverse Functions

The purpose of our inclusion of inverse functions is to solidify the concepts of *function, input,* and *output.*

Given a function, we can sometimes create a new function by reversing the input and output of the original function. We call this new function an **inverse function.** To illustrate, we begin by examining the quiz grade function from Table 1.4 in Section 1.1.

Student, s	J. DeCarlo	S. Dyers	J. Lykin	E. Mills	G. Schmeltzer
Grade, $Q(s)$	28	12	25	21	25

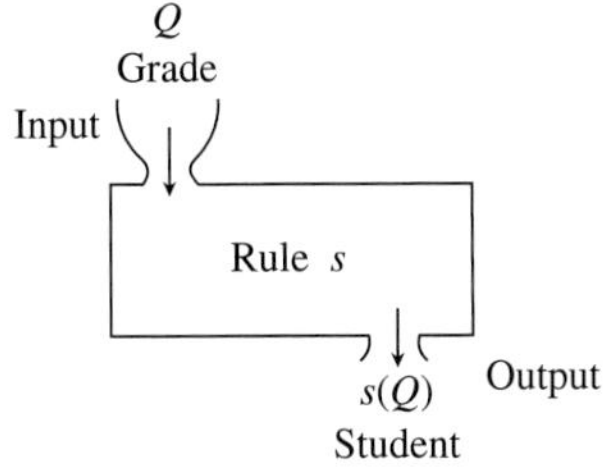

FIGURE 1.28

Let us turn the relationship around and let the input be the grade on the quiz and the output be the student who made that grade. The input/output diagram is shown in Figure 1.28.

The set of inputs is now {12, 21, 25, 28} and the set of outputs is {J. DeCarlo, S. Dyers, J. Lykin, E. Mills, G. Schmeltzer}. Is s a function of Q? This is the same as asking, "Does each grade correspond to only one student?". The answer is no. If the input grade is 25, the output is both J. Lykin and G. Schmeltzer. This relationship is not a function. If, however, the quiz grades had been as shown in Table 1.25,

TABLE 1.25

Student, s	J. DeCarlo	S. Dyers	J. Lykin	E. Mills	G. Schmeltzer
Grade, $Q(s)$	28	12	30	21	25

then swapping the inputs and outputs *would* have resulted in a function because no two students had the same grade. When the input and output of a function are reversed, we obtain a new rule. If this rule is also a function, we call it the **inverse** of the original function.

EXAMPLE 2 *New-Home Prices*[19]

a. Does Table 1.26 show new-home prices as a function of the year?

TABLE 1.26

Year	1970	1980	1985	1990	1995
Average sale price of a new home	\$23,400	\$64,600	\$84,300	\$122,900	\$133,900

b. If inputs and outputs are reversed, is the result an inverse function?

19. *Statistical Abstract*, 1998.

Solution

a. Because each year (input) corresponds to exactly one average price (output), Table 1.26 shows the average new-home price as a function of the year.

b. If we swap the inputs and outputs, we get

TABLE 1.27

Average sale price of a new home	\$23,400	\$64,600	\$84,300	\$122,900	\$133,900
Year	1970	1980	1985	1990	1995

Each average price (input) corresponds to only one year (output), so Table 1.27 also represents a function, the inverse of the original function.

EXAMPLE 3 *Office Space*

The scatter plot in Figure 1.29 shows the percentage of vacant office space[20] in Chicago from 1987 through 1993.

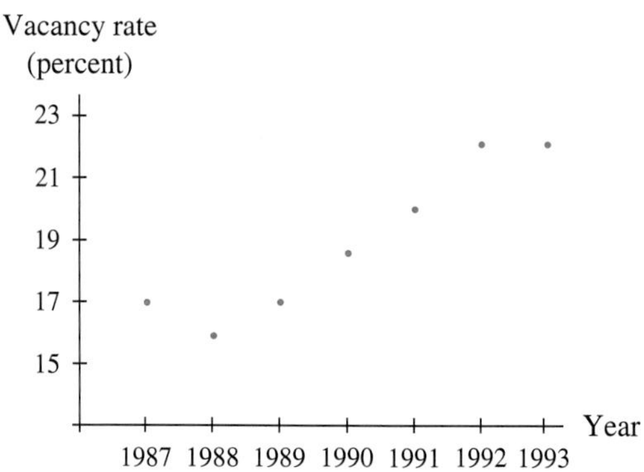

FIGURE 1.29

Does the discrete function represented by this scatter plot have an inverse function?

If you discussed one-to-one functions in Section 1.1, this is the place to make the connection between a function being one-to-one and its inverse being a function.

Solution Swapping the inputs and outputs for this scatter plot requires that we put years on the vertical axis and vacancy rates on the horizontal axis.

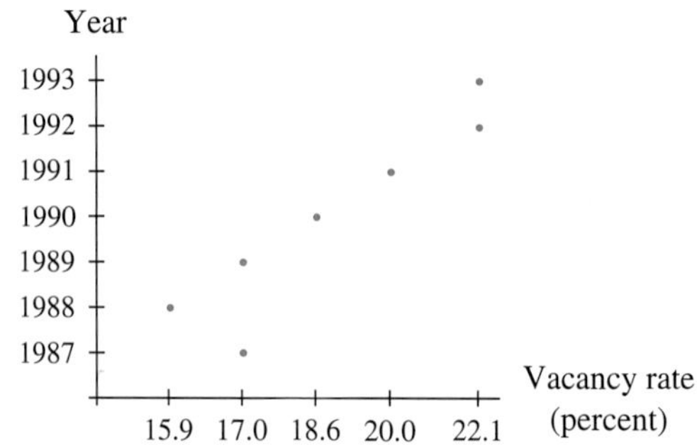

FIGURE 1.30

20. *Statistical Abstract*, 1994.

Note in Figure 1.30 that a rate of 17.0% corresponds to two different years (a vertical line through 17.0% passes through two points). This also occurs at 22.1%. The year is not a function of the vacancy rate, so the function represented by the scatter plot in Figure 1.29 does not have an inverse function.

Algebraically Finding Inverse Functions

If you are de-emphasizing algebraic skill, you may wish to omit this discussion of finding inverse functions algebraically.

Finding an algebraic expression for an inverse function requires that you solve for the input variable of a function in terms of the output variable. Sometimes this is simple, sometimes it is difficult, and sometimes it is impossible. Although determining inverse functions algebraically is not the focus of our discussion, looking at this technique will help reinforce your understanding of the concept of inverse functions. It will also enable us to discuss an important algebraic property of inverse functions. You may recall from previous math courses that when a function f is used as the input for its inverse function g, the result is simply the input variable. That is, when a function and its inverse are combined to create a composite function, the composition has the result of "undoing" the effects of the two functions.

Composition Property of Inverse Functions

If f and g are inverse functions, then

$$f(g(x)) = (f \circ g)(x) = x \quad \text{and} \quad g(f(x)) = (g \circ f)(x) = x$$

EXAMPLE 4 *Cab Fares*

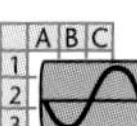

The fares for a cab company are determined by the function

$$F(d) = 1.8d + 2.5 \text{ dollars}$$

where d is the distance traveled in miles.

a. Find the inverse function of F.

b. Find the distance traveled that corresponds to a fare of \$33.10.

c. Find the fare for a 3-mile ride.

d. Use F and its inverse to illustrate the Composition Property of Inverse Functions.

Solution

a. Because the input and output of the original function are swapped in the inverse function, we need to find a function with the cab fare as input and the distance traveled as output. To begin, we replace $F(d)$ with the variable f and solve for d:

We choose to use lower case for the independent variable, but students should understand that this is not necessary.

$$f = 1.8d + 2.5$$

$$f - 2.5 = 1.8d$$

$$d = \frac{f - 2.5}{1.8}$$

We rewrite this equation using standard function notation as

$$D(f) = \frac{f - 2.5}{1.8} \text{ miles}$$

where f is the cab fare in dollars.

b. To answer the question of distance given the fare, we use the inverse function found in part *a*.

$$D(33.1) = \frac{33.1 - 2.5}{1.8} = 17 \text{ miles}$$

A cab ride that costs $33.10 corresponds to a distance of 17 miles.

c. To answer the question of fare given distance, we use the original function F.

$$F(3) = 1.8(3) + 2.5 \approx \$7.90$$

A 3-mile cab ride costs $7.90.

d. The Composition Property of Inverse Functions can be illustrated with the following equations:

$$(D \circ F)(d) = D(F(d)) = \frac{F(d) - 2.5}{1.8} = \frac{(1.8d + 2.5) - 2.5}{1.8} = d$$

$$(F \circ D)(f) = F(D(f)) = 1.8D(f) + 2.5 = 1.8\left(\frac{f - 2.5}{1.8}\right) + 2.5 = f$$

If you are allowing students to rely on technology, then you may wish to omit Example 5 and the preceding discussion. It is possible to teach the log model in Section 2.1 without this background. Another option is to cover this portion of this section in Section 2.1.

Two particular inverse functions that will be of interest to us later in this book are the exponential function $f(x) = e^x$ and the natural logarithmic function (or natural log function) $g(x) = \ln x$. These functions should be familiar to you from algebra, but they will be discussed in more detail in Chapter 2. The Composition Property of Inverse Functions applied to these two functions gives

$$e^{\ln x} = x \qquad \text{and} \qquad \ln(e^x) = x$$

You may recall from algebra that when e^x in the second equation is replaced with b^x, the result is not x but is a constant multiple of x. In particular,

$$\ln(b^x) = x \ln b$$

These important properties of exponential and logarithmic functions will be useful to us in Chapter 2.

Exponential and Logarithm Properties

1. $e^{\ln x} = x$
2. $\ln(e^x) = x$
3. $\ln(b^x) = x \ln b$

We illustrate the use of Property 3 in Example 5.

EXAMPLE 5 *Investment*

Suppose \$100 is invested in a savings account that pays 7% annual interest. An equation for the balance in the account after t years of investment is

$$A(t) = 100(1.07^t) \text{ dollars}$$

a. When will the balance in the account be doubled?

b. Find the inverse function of A.

c. Use the inverse function to determine when the balance will surpass \$400.

Solution

a. To determine when the balance will double, we replace $A(t)$ with \$200 and solve for t. The solution can be found using technology, but we show the algebraic solution.

$$200 = 100(1.07^t)$$
$$2 = 1.07^t$$

We now apply the natural log function to both sides of the equation.

$$\ln 2 = \ln(1.07^t)$$

Using Property 3, we have

$$\ln 2 = t \ln 1.07$$
$$\frac{\ln 2}{\ln 1.07} = t$$
$$t \approx 10.24 \text{ years}$$

Because A is a continuous function with discrete interpretation, our answer, a value of t, must be a positive integer. Recall from the discussion in Section 1.2 that we cannot simply round the answer. If the amount function is \$200 when $t \approx 10.24$ years and A is an increasing function, then the amount is less than \$200 after 10 years and more than \$200 after 11 years. See Figure 1.31.

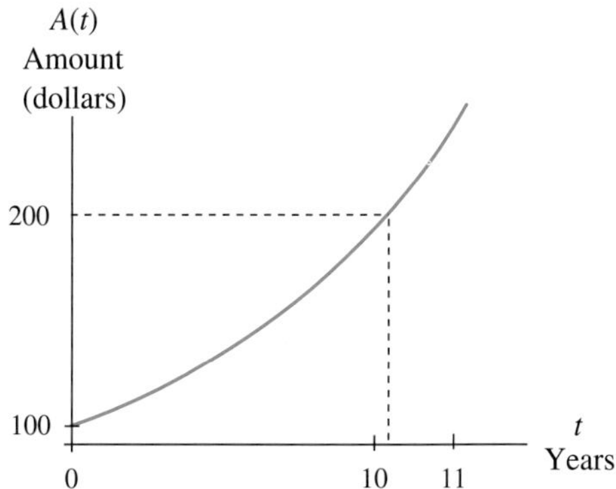

FIGURE 1.31

We can conclude that the balance in the account will never be exactly \$200 but that the balance will exceed \$200 at the end of 11 years.

b. To find the inverse function, we replace $A(t)$ with a and solve for t. Solving for t requires the same algebraic steps as in part *a*.

$$\begin{aligned} a &= 100(1.07^t) \\ \frac{a}{100} &= 1.07^t \\ \ln\left(\frac{a}{100}\right) &= \ln(1.07^t) \\ \ln\left(\frac{a}{100}\right) &= t \ln 1.07 \\ \frac{\ln\left(\frac{a}{100}\right)}{\ln 1.07} &= t \end{aligned}$$

We now rewrite this equation in standard function notation.

$$T(a) = \frac{\ln\left(\frac{a}{100}\right)}{\ln 1.07} \text{ years}$$

where a is the amount in dollars in the account.

c. To answer the question of when the balance will exceed \$400, we can simply substitute 400 for a in the inverse function formula.

$$T(400) = \frac{\ln\left(\frac{400}{100}\right)}{\ln 1.07} = \frac{\ln 4}{\ln 1.07} \approx 20.5 \text{ years}$$

Again, we must keep in mind that only positive integer year values make sense in this context. Using similar reasoning to that in part *a*, we conclude that the amount will exceed \$400 at the end of 21 years.

Piecewise Continuous Functions

One final way in which we will construct functions is putting two or more functions together to create a **piecewise continuous function**. In Section 1.2 we saw that the investment function in Example 5 over the first 5 years of investment could be better represented by the piecewise continuous function

$$B(t) = \begin{cases} \$100 & \text{when } 0 \le t < 1 \\ \$107 & \text{when } 1 \le t < 2 \\ \$114.49 & \text{when } 2 \le t < 3 \\ \$122.50 & \text{when } 3 \le t < 4 \\ \$131.08 & \text{when } 4 \le t < 5 \\ \$140.26 & \text{when } t = 5 \end{cases}$$

t years after the \$100 was invested. A graph of the function is shown in Figure 1.17 on page 22.

We call this equation a piecewise continuous function because it is made up of portions of six different continuous functions and is defined for all values of t between 0 and 5. The values of t where the formula changes (here $t = 1, 2, 3, 4, 5$) are called the **break points** of the function. Piecewise continuous functions will be especially helpful in modeling situations in which a quantity demonstrates a distinct and sudden change in behavior. Such a situation is illustrated in Example 6.

EXAMPLE 6 *Population of West Virginia*

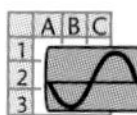

The population of West Virginia from 1985 through 1993 can be modeled[21] by

$$P(t) = \begin{cases} -23.514t + 3903.667 \text{ thousand people} & \text{when } 85 \leq t < 90 \\ 7.7t + 1098.7 \text{ thousand people} & \text{when } 90 \leq t \leq 93 \end{cases}$$

Ask your students why a strict inequality is necessary on only one portion of the function *P*. Connect this concept with the graph in Figure 1.32 and the result of the Vertical Line Test if the two dots are both closed.

where t is the number of years since 1900.

a. Graph P.

b. According to the model, what was the population of West Virginia in 1987? in 1992?

c. Find $P(85)$ and $P(90)$. Interpret your answers.

Solution

a. The graph of P is shown in Figure 1.32.

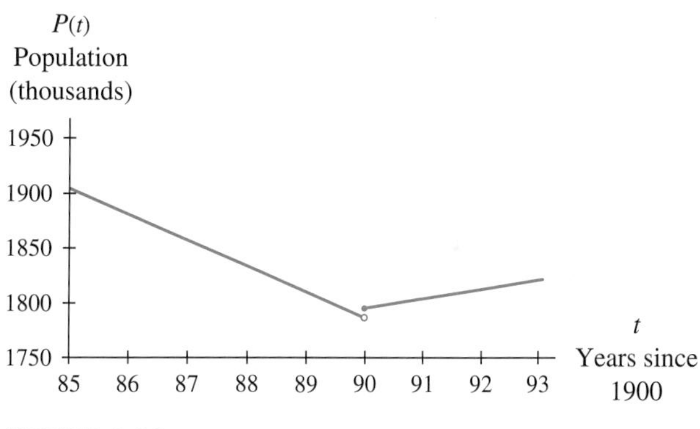

FIGURE 1.32

b. To determine the population in 1987, we substitute 87 for t in the top function.

$$P(87) = -23.514(87) + 3903.667 \approx 1858 \text{ thousand people}$$

To determine the population in 1992, we use the bottom function.

$$P(92) = 7.7(92) + 1098.7 \approx 1807 \text{ thousand people}$$

c. $P(85) = -23.514(85) + 3903.667 \approx 1905$ thousand people. In 1985 the population of West Virginia was approximately 1,905,000 people.

$P(90) = 7.7(90) + 1098.7 \approx 1792$ thousand people. In 1990 the population of West Virginia was approximately 1,792,000 people.

Calculus is the mathematics of change that occurs in continuous portions of functions. This is why we have reviewed the basic ideas of functions in the first three

21. Based on data from *Statistical Abstract*, 1998.

sections of this text. We have reviewed the definition of a mathematical function, and we have looked at a few techniques for constructing new functions from basic functions.

Throughout these sections, we have considered mathematical functions as they model the real world. As we embark on our study of calculus, we will consider modeling primarily as a tool for developing the functions we use to analyze the change that occurs in real-world problems. In order to apply calculus to situations that occur in business, health science, agriculture, and the like, we must first be able to describe the situations as continuous or piecewise continuous functions (some having discrete interpretation). Section 1.5 and Chapter 2 deal with using different functions to model real-world data. Once we can model data with a function, we can use calculus to analyze that model—especially to determine rates of change and identify maxima, minima, and inflection points.

If you are using the two-semester version of this text and integrating Chapter 8 (trigonometry) throughout Chapters 1–7, you should insert Section 8.1 either after Section 1.3 or Section 1.4. See the *Instructor's Resource Guide* for more information about this topic.

1.3 Concept Inventory

- *Adding, subtracting, and multiplying functions*
- *Function composition*
- *Inverse functions*
- *Composition Property of Inverse Functions*
- *Exponential and log properties*
- *Piecewise continuous functions*

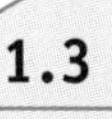

1.3 Activities

1. The number of kidney and liver transplants[22] in the United States between 1992 and 1996 can be modeled by

$$K(x) = 9088.859 + 1697.717 \ln x \quad \text{kidney transplants}$$

$$L(x) = 2424.764 + 915.025 \ln x \quad \text{liver transplants}$$

where x is the number of years since 1990.

   a. Construct a model for the number of kidney and liver transplants between 1992 and 1996.

   b. Use the model to estimate the number of kidney and liver transplants in 1995.

2. The salary of one of Compaq Computer Corporation's senior vice presidents[23] from 1996 through 1998 can be modeled by

$$S(t) = 69{,}375t + 380{,}208 \text{ dollars}$$

t years after 1996. His other, nonsalary compensations during the same period can be modeled by

$$C(t) = -31.67t^2 + 137.15t + 233.5 \text{ thousand dollars}$$

t years after 1996. In addition, each year he received an average bonus of \$650,000.

   a. Construct a model for this VP's total yearly salary package, including nonsalary compensation and bonuses.

   b. Estimate the VP's 1997 total salary package.

3. Per capita milk consumption[24] in the United States between 1980 and 1996 can be modeled by

$$M(t) = -0.2138t + 44.8434 \text{ gallons per person per year}$$

and consumption of whole milk for the same period can be modeled by

$$W(t) = -0.5472t + 60.4767 \text{ gallons per person per year}$$

In both models, t is the number of years since 1900. Use the models to construct a model giving the per capita consumption of milk other than whole milk.

4. State lotteries have gained in popularity over the past 20 years. The cost[25] of the lotteries to the states (prizes and expenses) between 1980 and 1997 can be modeled by

$$C(x) = 35.302x^2 + 740.223x + 1207.223 \text{ million dollars}$$

where x is the number of years after 1980. The profit realized by the states as a result of the lotteries can be modeled by

$$P(x) = -2.281x^3 + 57.900x^2 + 320.625x + 976.390 \text{ million dollars}$$

where x is the number of years after 1980. Find an equation for the revenue the states earned as a result

22. Based on data from *Statistical Abstract*, 1998.
23. Compaq's 1999 Notice of Annual Meeting.
24. Based on data from *Statistical Abstract*, 1998.
25. Ibid.

of lotteries, and estimate the total lottery revenue in the year 2000.

5. The percentages of Iowa corn farmers in two communities who had heard about, and who had planted, hybrid seed corn t years after 1924 can be modeled[26] as follows:

   Percentage hearing:
   $$h(t) = \frac{100}{1 + 128.0427e^{-0.7211264t}} \text{ percent}$$
   Percentage planting:
   $$p(t) = \frac{100}{1 + 913.7241e^{-0.607482t}} \text{ percent}$$
   Write an equation for the percentage of Iowa corn farmers who had heard about, but who had not yet planted, hybrid seed corn t years after 1924.

6. Table 1.28[27] shows two functions: $I(t)$ is the amount of natural gas imports in trillions of cubic feet in year t, and $E(t)$ is the amount of natural gas exports in billions of cubic feet in year t. Show how you could combine the two functions to create a third function giving net trade of natural gas in year t. (Net trade is negative when imports exceed exports.)

TABLE 1.28

t	1975	1980	1985	1990	1995
$I(t)$	0.953	0.985	0.950	1.532	2.841
$E(t)$	73	49	55	86	162

7. The number of births[28] during the 1980s to women who were 35 years of age or older can be modeled as

   $n(x) = -0.034x^3 + 1.331x^2 + 9.913x + 164.447$ thousand births x years after 1980

   The ratio of cesarean-section deliveries[29] performed on women in the same age bracket during the same time period can be modeled as

   $p(x) = -0.183x^2 + 2.891x + 20.215$ deliveries per 1000 live births x years after 1980

   Write an expression for the number of cesarean-section deliveries performed on women 35 years of age or older during the 1980s.

26. Based on data from Ryan and Gross, "The Diffusion of Hybrid Seed Corn in Two Iowa Communities," *Rural Sociology*, March 1943.
27. *Statistical Abstract*, 1998.
28. Based on data from *Statistical Abstract*, 1992.
29. Ibid.

8. Table 1.29[30] represents two functions: $R(t)$ is the average price in dollars of a gallon of regular unleaded gasoline in year t, and $P(t)$ is the purchasing power of the dollar as measured by consumer prices in year t, using 1983 as the base year. Indicate how to combine the two functions to create a third function showing the price of gasoline in constant 1983 dollars.

TABLE 1.29

Year	$R(t)$	$P(t)$
1990	1.22	0.77
1992	1.19	0.71
1994	1.17	0.68
1996	1.29	0.64
1997	1.29	0.62
1998	1.19	0.61
1999	1.19	0.60

Determine whether the pairs of functions in Activities 9 through 12 can be combined by function composition. If so, give function notation for the new function, and draw and label its input/output diagram.

9. $P(c)$ is the profit in dollars generated by the sale of c computer chips.

   $C(t)$ is the number of computer chips a manufacturer has produced after t hours of production.

10. $C(t)$ is the number of cats in the United States at the end of year t.

    $D(c)$ is the number of dogs in the United States at the end of year c.

11. $C(t)$ is the average number of customers in a restaurant on a Saturday night t hours after 4 p.m.

    $P(c)$ is the average amount in tips in dollars generated by c customers.

12. $R(x)$ is the revenue in deutsche marks from the sale of x soccer uniforms.

    $D(r)$ is the dollar value of r deutsche marks.

In Activities 13 through 18, rewrite each pair of functions as one composite function.

13. $f(t) = 3e^t \qquad t(p) = 4p^2$

30. Energy Information Administration; *Statistical Abstract*, 1998; and Economagic. com.

14. $h(p) = \frac{4}{p}$ $\quad p(t) = 1 + 3e^{-0.5t}$

15. $g(x) = \sqrt{7x^2 + 5x - 2}$ $\quad x(w) = 4e^w$

16. $s(q) = -14 + 7455 \ln q$ $\quad q(p) = 6.439(0.07^p)$

17. $g(t) = 3$ $\quad t(m) = 4m + 17$

18. $c(x) = 3x^2 - 2x + 5$ $\quad x(t) = 4 - 6t$

Swap the inputs and outputs for each of the functions in Activities 19 through 24. Draw an input/output diagram for each new rule. Then write out the new rule in words. Which of the new rules are inverse functions?

19. $R(w)$ = the first-class domestic postal rate (in cents) of a letter weighing w ounces

20. $G(t)$ = your score out of 100 on the first test in this course when you studied t hours the week before the test

21. $B(x)$ = the number of students in this class whose birthday is on the xth day of the year, assuming it is not a leap year

22. $A(m)$ = the amount (in dollars) you pay for lunch on the mth day of any week

23. $B(t)$ = the amount in an investment account (in dollars) after t years, assuming no withdrawals are made during the t years

24. $D(r)$ = the number of years it takes for an investment to double if the annual percentage rate (APR) is r%

Reverse the inputs and outputs in the function tables in Activities 25 through 28, and determine whether the results are inverse functions.

25. **TABLE 1.30**

Age[31]	Percent with flex schedules at work
16–19	20.7
20–24	22.7
25–34	28.4
35–44	29.1
45–54	27.3
55–64	27.4
over 64	31.0

26. **TABLE 1.31**

Military rank[32] (10 years service)	Basic monthly pay in 1997 (dollars)
Second Lieutenant	2171
First Lieutenant	2752
Captain	3449
Major	3620
Lt. Colonel	3764
Colonel	4260
Brigadier General	5796
Major General	6694
Lt. General	7010
General	7912

27. **TABLE 1.32**

Person's height	Person's weight (pounds)
5′ 3″	139
6′ 1″	196
5′ 4″	115
6′ 0″	203
5′ 10″	165
5′ 3″	127
5′ 8″	154
6′ 0″	189
5′ 6″	143

28. **TABLE 1.33**

Year	Total wagers in Nevada on the Super Bowl[33] (millions of dollars)
1992	50.3
1993	56.8
1994	54.5
1995	69.6
1996	70.9
1997	70.9
1998	77.3

29. Table 1.34 gives tuition rates at a certain university[34] in the southeast. Note that a full-time student is one taking 12 or more credit hours.

31. *Statistical Abstract*, 1998.

32. *Information Please Almanac*, 1998.

33. As reported in the *Reno Gazette–Journal*, January 18, 1999, p. 1A.

34. Clemson University 1999–2000 *Undergraduate Announcements*.

TABLE 1.34

Schedule of academic charges	Resident	Nonresident
Full-time academic fee	$1735	$4728
Part-time academic fee (per hour)	138	388
Auditing academic fee (per hour)	69	194
Graduate assistant fee	512	512

a. Complete Table 1.35 for a resident undergraduate.

TABLE 1.35

Credit hours	Tuition	Credit hours	Tuition
1		10	
2		11	
3		12	
4		13	
5		14	
6		15	
7		16	
8		17	
9		18	

Let $T(c)$ = tuition (in dollars) for a resident undergraduate taking c credit hours.

b. Find a formula for $T(c)$.

c. Graph T.

d. Is T a function?

e. Give the set of inputs and the set of outputs of T.

f. Draw an input/output diagram for the rule obtained by swapping inputs and outputs of T.

g. Is the rule in part f an inverse function? Why or why not?

practice

The functions given in Activities 30 through 36 all have inverses. Find an algebraic formula for each inverse function.

30. $g(r) = \ln r$

31. $s(v) = e^v$

32. $f(x) = 7.36x^3 - 12.9$

33. $g(t) = \sqrt{4.9t + 0.04}$

34. $h(x) = 0.0063 + \frac{4.7913}{0.019x}$

35. $c(s) = 1034.926(0.0062^s)$

36. $j(y) = 73{,}000(2.639^y)$

Skip Activities 30–42 if you omitted the discussion of finding inverse functions algebraically.

37. The number of health insurance complaints received in 2001 by a state's board of health can be modeled by

$$C(m) = 319.6m + 801 \text{ complaints}$$

in the mth month of 2001.

a. Can C be used without restriction in this context or must it be discretely interpreted?

b. Find the inverse of C.

c. In what month were 2399 complaints filed?

d. How many complaints were filed in the first quarter of the year 2001?

e. Use C and its inverse to illustrate the Composition Property of Inverse Functions.

38. The time it takes for an average 8- to 14-year-old athlete to swim the 100-meter freestyle[35] can be modeled by the equation

$$T(a) = -5.5a + 137 \text{ seconds}$$

where a is the age of the swimmer in years.

a. Can T be used without restriction in this context or must it be discretely interpreted?

b. Find the inverse of T.

c. How long does it take an average 9-year-old to swim the 100-meter freestyle?

d. If a young person can swim 100 meters freestyle in 1 minute 20 seconds, what is the person's probable age?

e. Use T and its inverse to illustrate the Composition Property of Inverse Functions.

39. The number of livestock that can be supported by a crop of rye grass[36] can be modeled by

$$L(r) = \frac{9.52(15{,}000)}{r} \text{ animals per hectare}$$

where r is the annual energy requirement of the animal in megajoules per animal.

35. *Swimming World*, August 1992.

36. R. S. Loomis and D. J. Connor, *Crop Ecology: Productivity and Management in Agricultural Systems* (Cambridge, England: Cambridge University Press, 1992).

a. Find the inverse of L.

b. If milking cows require 64,000 megajoules per cow, how many cows can this crop of rye grass support?

c. If the crop is supporting 7 animals per hectare, what is the energy requirement of the animals?

d. Use L and its inverse to illustrate the Composition Property of Inverse Functions.

40. The heat loss[37] experienced by the human body exposed to a 16-mile-an-hour wind can be modeled by

$$H(t) = 34.45(33 - t) \text{ kilogram-calories per square meter of body surface area}$$

when the temperature is t°C.

a. Find the inverse of H.

b. What is the temperature if the heat loss of the body is 1550 kilogram-calories per square meter?

41. The population of Mexico[38] between 1920 and 1990 can be modeled as

$$P(t) = 7.394(1.0269^t) \text{ million people}$$

t years after 1900.

a. Can P be used in this context without restriction or must it be discretely interpreted?

b. Find the inverse of P.

c. In what year was the population 40 million?

d. What was the population in the middle of 1930?

42. The yearly emissions of carbon monoxide[39] in the United States between 1980 and 1992 can be modeled by

$$E(t) = 115.0326(0.9687^t) \text{ million metric tons}$$

t years after 1980.

a. Can E be used in this context without restriction or must it be discretely interpreted?

b. Find the inverse of E.

c. What were emissions in 1987?

d. In what year were emissions 82 million metric tons?

43. Yearly sales of water skis in the United States (excluding Alaska and Hawaii) between 1985 and 1992 can be modeled[40] by

$$S(x) = \begin{cases} 12.1x - 905.4 \text{ million dollars} \\ \quad \text{when } 85 \le x \le 88 \\ -14.8x + 1414.9 \text{ million dollars} \\ \quad \text{when } 89 \le x \le 92 \end{cases}$$

where x is the number of years since 1900.

a. Find the values of $S(85)$, $S(88)$, $S(89)$, and $S(92)$.

b. Use the answers to part a to sketch a graph of S.

c. Is S a function of x? Explain.

44. The population of North Dakota between 1985 and 1996 can be modeled[41] by

$$P(t) = \begin{cases} -7.35t + 1301.05 \text{ thousand people} \\ \quad \text{when } 85 \le t \le 91 \\ 2.5t + 403.5 \text{ thousand people} \\ \quad \text{when } 91 < t \le 96 \end{cases}$$

where t is the number of years since 1900.

a. Find $P(85)$, $P(89)$, $P(94)$, and $P(96)$.

b. According to the model, what was the population in 1991?

c. Use the answers to parts a and b to sketch a graph of P.

d. Is P a function of t? Explain.

45. A mail-order company charges a percentage of the amount of each order for shipping. For orders of up to \$20, the charge is 20% of the order amount. For orders of greater than \$20 up to \$40, the charge is 18%. For orders of greater than \$40 up to \$75, the charge is 15%. For orders of greater than \$75, the charge is 12%.

a. Why would it benefit a company to assess the shipping charges described here?

b. What will shipping charges be on orders of \$17.50? \$37.95? \$75.00? \$75.01?

c. Write a formula for the shipping charge for an order of x dollars.

d. Sketch a graph of the formula in part c.

37. W. Bosch and L. G. Cobb, "Wind Chill," UMAP Module 658, *The UMAP Journal*, vol. 5, no. 4 (Winter 1984), pp. 477–492.
38. SPP and INEGI, Mexican Census of Population 1921 through 1990 as reported by Pick and Butler, *The Mexico Handbook*: Westview Press, 1994).
39. Based on data from *Statistical Abstract*, 1994.
40. Based on data from *Statistical Abstract*, 1994.
41. Based on data from *Statistical Abstract*, 1998.

e. Which of the following representations of the shipping charge function do you believe it would be best for the company to put in its catalog? (i) a word description, (ii) a formula as in part *c*, (iii) a graph, or (iv) some other representation (specify)? Explain the reasons for your choice.

46. A music club sells CDs for \$14.95. When you buy five CDs, the sixth one is free.

    a. Sketch a graph of the cost to buy x CDs for values of x between 1 and 12.

    b. Write a formula for the graph you sketched in part *a*.

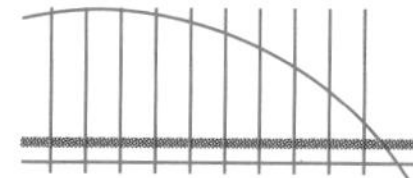

1.4 Limits: Describing Function Behavior

You may wish to shorten this section by focusing on those topics that will be of most importance later in the text. There are many ways to do so. Please see the *Instructor's Resource Guide* for specific suggestions for streamlining this section.

We have already stated that calculus is the study of change. There are algebraic descriptions of how quantities change, but what sets calculus apart from algebra is the use of *limits* to describe change. Limits enable us to describe change in ways that would otherwise be impossible. Limits set the stage for understanding the calculus concepts of *derivative* and *integral* that are later presented in this text.

In addition to their usefulness in calculus, limits give us a mathematical way to describe the behavior of a function's output. It is this use of limits as descriptions of function behavior that is our focus in this section.

An Introduction to Limits

Before we define and use limits, we want to give you an opportunity to explore limits as they are applied to a variety of functions and to become intuitively comfortable with this important concept. We consider numerical and graphical estimation of limits, as well as algebraic analysis of limits.

Consider the function $r(t) = \frac{t^2 - 16}{t - 4}$. This function is not defined at $t = 4$ (that is, $r(4)$ does not exist), because the denominator of r is 0 when $t = 4$. However, rather than discuss the function value *at* $t = 4$, we wish to explore what happens to the output of the function at input values *near*, and on either side of, 4. To do so, we evaluate the function at values increasingly close to $t = 4$. Table 1.36 shows $r(t)$ values as t approaches 4 from the left (this is denoted $t \to 4^-$), and Table 1.37 shows $r(t)$ values as t approaches 4 from the right (this is denoted $t \to 4^+$).

TABLE 1.36

$t \to 4^-$	$r(t)$
3.8	7.8
3.9	7.9
3.99	7.99
3.999	7.999

TABLE 1.37

$t \to 4^+$	$r(t)$
4.2	8.2
4.1	8.1
4.01	8.01
4.001	8.001

In Table 1.36 it appears that as t approaches 4 from the left, the output is becoming closer and closer to 8. We symbolize this idea by writing

$$\lim_{t \to 4^-} r(t) = 8$$

which is read "the limit of $r(t)$ as t approaches 4 from the left is 8." On the basis of Table 1.37, we make a similar statement: "The limit of $r(t)$ as t approaches 4 from the right is 8," which we write as

$$\lim_{t \to 4^+} r(t) = 8$$

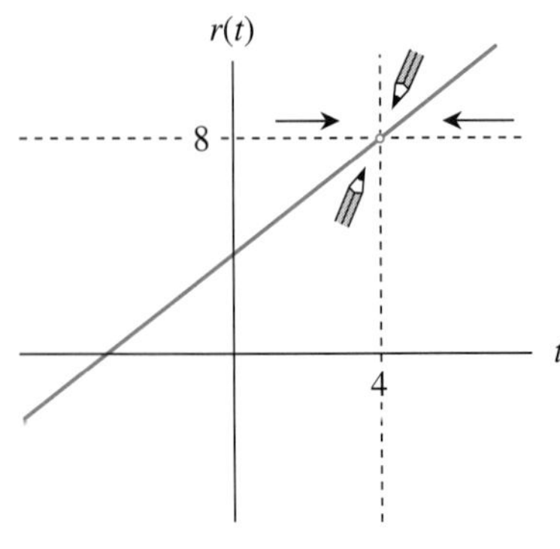

FIGURE 1.33

Note that the input values in the tables were arbitrarily chosen. You should arrive at the same conclusion if you choose other values increasingly close to $t = 4$. Because the limits from the left and right of 4 are the same, we conclude that $\lim_{t \to 4} r = 8$; that is, the limit of $r(t)$ as t approaches 4 is 8.

To support our conclusion that 8 is the limit, we consider the graph of r shown in Figure 1.33. The graph of this function is simply a line with a hole in it at $t = 4$. To estimate $\lim_{t \to 4} r(t)$ using the graph, place the tips of two pencils on the graph on either side of $t = 4$ and, keeping the pencils on the graph, move them toward each other. The pencil tips should move toward the hole in the function. Indeed, as the input becomes closer and closer to 4 from either direction, the graph becomes closer and closer to an output of 8. The fact that $r(t)$ is never actually 8 is of no consequence. This fact does not affect the limit, because when determining limits, we are interested in where the function is headed, not in whether it ever actually arrives there.

> When answering limit questions, we are concerned only with where a function is headed, not with whether the output ever arrives at that value.

Later in this section, we will see that when a function is defined at a point and there are no jumps or breaks in the function at that point, the limit can be found simply by evaluating the function at the point. For example, consider again the function $r(t) = \frac{t^2 - 16}{t - 4}$ that is defined and has no jumps or breaks at all input values for which $t \neq 4$ (see Figure 1.33). We calculate the limit of $r(t)$ as t approaches -2 as follows:

$$\lim_{t \to -2} \frac{t^2 - 16}{t - 4} = \lim_{t \to -2} \frac{(-2)^2 - 16}{-2 - 4} = \frac{4 - 16}{-6} = \frac{-12}{-6} = 2$$

Before proceeding with other examples, we define the limit of a function at a point.

Definition of *Limit*

A **limit** is a number to which the output of a function becomes closer and closer as the input becomes closer and closer to a stated value. We use the notation $\lim_{x \to c} f(x)$ to denote the limit of $f(x)$ as x approaches c, and we make the following informal definition.

> **The Limit at a Point**
>
> The limit of $f(x)$ as x approaches a constant c is the number L; that is,
>
> $$\lim_{x \to c} f(x) = L$$
>
> if $f(x)$ becomes arbitrarily close to L as x becomes arbitrarily close to c.

As we saw for the function r, limits can be estimated numerically or graphically, and some limits can be calculated algebraically. When estimating the limit of $r(t)$ as t approaches 4, we considered both left and right limits of the function because r was not defined at $t = 4$. It may be the case that a function is defined at a point but the limit does not exist there. If the left and right limits are different at a point, then we say that the limit does not exist at that point. This is often the case at the break point(s) for portions of piecewise continuous functions. It is important to keep in mind the necessity of checking limits from both directions for such functions, and it is also helpful to view a graph.

Left and Right Limits

If the left and right limits are not the same, then the limit does not exist. That is,

For a constant c, if $\lim_{x \to c^-} f(x)$ does not equal $\lim_{x \to c^+} f(x)$, then $\lim_{x \to c} f(x)$ does not exist.

If the left and right limits are the same number, then the limit does exist and is that number. That is,

If L is a number and $\lim_{x \to c^-} f(x) = \lim_{x \to c^+} f(x) = L$, then $\lim_{x \to c} f(x) = L$.

Before we further discuss methods for finding limits, we consider two situations where limits do not exist.

Limits That Do Not Exist

If you are caught speeding in a certain state, the fine is \$35 for the first 5 mph (or a portion thereof) over the limit and an additional \$50 for the next 5 mph (or a portion thereof) over the limit. For each additional 2 mph above 10 mph over the speed limit, the fine increases by \$2. Thus the function that represents the fine for speeding in this state is

$$s(x) = \begin{cases} 35 \text{ dollars} & \text{when } 0 < x \le 5 \\ 85 \text{ dollars} & \text{when } 5 < x \le 10 \\ 85 + 2(x - 10) \text{ dollars} & \text{when } x > 10 \end{cases}$$

where x is the number of miles per hour the vehicle is traveling over the speed limit. Does the limit of $s(x)$ as x approaches 5 exist? Note that $s(5) = 35$, but because s is a piecewise continuous function, we need to consider the left and right limits as x approaches 5 in order to answer this question. To both the left and the right of $x = 5$, s is constant. If you choose values increasingly close to $x = 5$ from the left, s remains at 35, so $\lim_{x \to 5^-} s(x) = 35$. For values increasingly close to $x = 5$ from the right, s remains at 85, so $\lim_{x \to 5^+} s(x) = 85$. Because the limits from the left and right of 5 are not the same, we conclude that the limit of $s(x)$ as x approaches 5 does not exist.

Other types of functions for which limits do not exist are functions whose outputs become larger and larger as the input approaches a certain value. We say that the output of a function f is **increasing without bound** as x approaches c if the output continues to increase in height infinitely. If the output of a function f continues to decrease infinitely as x approaches c, we say that the output is **decreasing without**

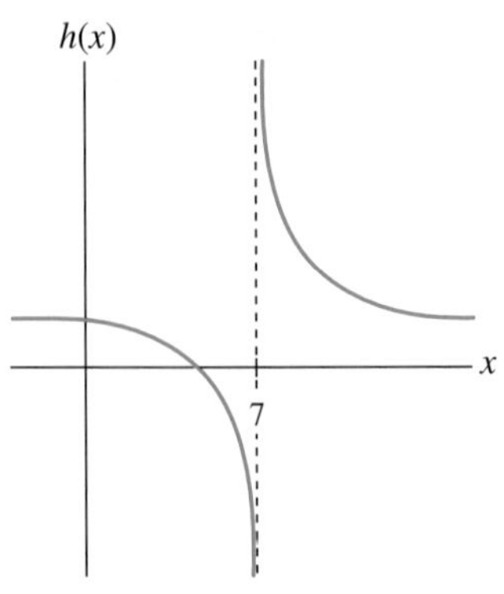

FIGURE 1.34

bound as x approaches c. These descriptions can apply to the limits of the function from the left, from the right, or from both directions. Consider, for example, the function shown in Figure 1.34. Although the function does not have a limiting value as x approaches 7, we use the notation $\lim_{x \to 7^-} h(x) \to -\infty$ to indicate that the output $h(x)$ is decreasing without bound as x approaches 7 from the left. As x approaches 7 from the right, we use the notation $\lim_{x \to 7^+} h(x) \to \infty$ to indicate that the output $h(x)$ is increasing without bound. Neither of these limits exists.

Graphically Estimating Limits

Suppose that a small nation is ruled by a ruthless dictator who wishes to rid a part of his country of all people of a certain ethnic background. For 10 years he practices subtle ethnic cleansing, virtually unnoticed by the world. After 10 years he suddenly begins such violent purging that many of the people flee the country or are killed in a matter of a few weeks. The extreme violence ends, but over the next year some of the people continue to seek refuge in neighboring countries. At the end of the 11th year of his reign, the dictator is assassinated and a democratic government is established. Immediately the refugees flood back into the country. The graph in Figure 1.35 shows the approximate size of the population of the affected region of the country for the 20 years after the dictator's rule began.

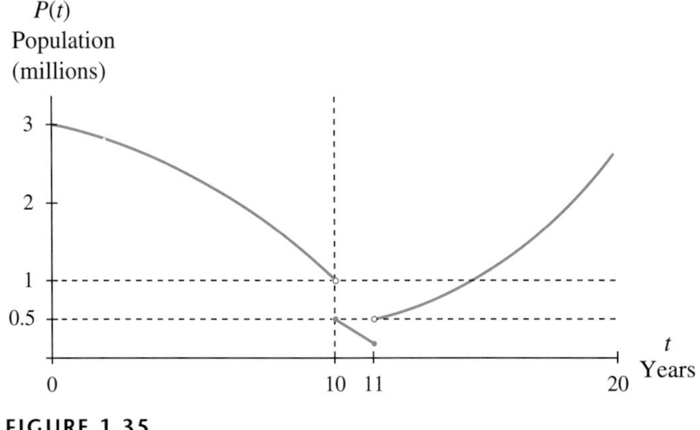

FIGURE 1.35

When considering the behavior of the population as t approaches 10 years, we must first consider two pieces of the function: the one to the left of $t = 10$ and the one immediately to the right of $t = 10$. Between 0 and 10 years, the population decreases and becomes closer and closer to 1 million people. That is, as t approaches 10 from the left, $P(t)$ approaches 1 million people: $\lim_{x \to 10^-} P(t) = 1$ million people. Recall that the fact that the function does not actually have the value of 1 million when $t = 10$ does not affect our answer.

Similarly, we consider the portion of the function immediately to the right of $t = 10$ to determine the limit of $P(t)$ as t approaches 10 from the right. To do so, place the tip of a pencil on the middle portion of the function graph near $t = 11$. Move the pencil on the graph from *right to left* toward $t = 10$ to see that the graph rises and becomes closer and closer to $\frac{1}{2}$ million. We say that the limit of $P(t)$ as t approaches 10 from the right is $\frac{1}{2}$ million people: $\lim_{x \to 10^+} P(t) = \frac{1}{2}$ million people. Thinking backwards in time from 11 years to 10 years, the population rose to $\frac{1}{2}$ million people.

We have found that $\lim_{x \to 10^-} P(t) = 1$ million people and $\lim_{x \to 10^+} P(t) = \frac{1}{2}$ million people. Because the behavior of the output of the function is different as t becomes closer to 10 from the left than it is as t becomes closer to 10 from the right, we say that $\lim_{t \to 10} P(t)$ does not exist. Similar reasoning reveals that $\lim_{t \to 11} P(t)$ does not exist.

EXAMPLE 1 *Finding Limits Graphically*

Consider the graph of g shown in Figure 1.36.

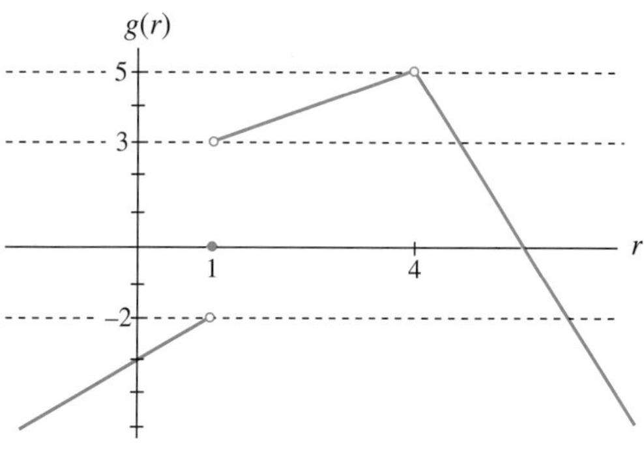

FIGURE 1.36

a. Find $g(1)$, $\lim_{r \to 1^-} g(r)$, $\lim_{r \to 1^+} g(r)$, and $\lim_{r \to 1} g(r)$.

b. Find $g(4)$, $\lim_{r \to 4^-} g(r)$, $\lim_{r \to 4^+} g(r)$, and $\lim_{r \to 4} g(r)$.

Solution

a. From the graph it appears that $g(1) = 0$, $\lim_{r \to 1^-} g(r) = -2$, and $\lim_{r \to 1^+} g(r) = 3$. Thus $\lim_{r \to 1} g(r)$ does not exist.

b. It appears from the graph that $g(4)$ does not exist because there is not a point on the graph where $r = 4$. An examination of the graph also indicates that $\lim_{r \to 4^-} g(r) = 5$ and $\lim_{r \to 4^+} g(r) = 5$. Thus $\lim_{r \to 4} g(r) = 5$.

Numerically Estimating Limits

This discussion provides an introduction to numerical estimation, which is used extensively in later discussions of derivatives and integrals.

We illustrated how to estimate limits numerically in our opening example, where we considered $\lim_{t \to 4} \frac{t^2 - 16}{t - 4}$. In that example, we chose values on either side of 4 and increasingly close to 4. We looked for a trend in the output values—that is, a value to which the output was becoming closer and closer. We again illustrate this process of numerical investigation in Example 2.

EXAMPLE 2 *Estimating Limits Numerically*

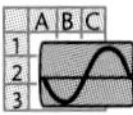

For $u(x) = \dfrac{3x^2 + 3x}{9x^2 + 11x + 2}$, numerically estimate

a. $\lim\limits_{x \to -1} u(x)$ b. $\lim\limits_{x \to -2/9} u(x)$

Does a graph of u suggest that your answers are correct?

Solution

a. Note that $u($ $1)$ is not defined because the denominator of u is zero at $x = -1$. To estimate this limit numerically, we need to consider values of x increasingly close to -1 from both the left and the right. The calculations are shown in Table 1.38.

TABLE 1.38

$x \to -1^-$	$u(x)$
-1.1	0.41772
-1.01	0.42736
-1.001	0.42845
-1.0001	0.42856
-1.00001	0.42857

$x \to -1^+$	$u(x)$
-0.9	0.44262
-0.99	0.42981
-0.999	0.42869
-0.9999	0.42858
-0.99999	0.42857

From the table, we estimate that $\lim\limits_{x \to -1^-} u(x) \approx 0.429$ and $\lim\limits_{x \to -1^+} u(x) \approx 0.429$. Because the left and right limits appear to be the same, we estimate that $\lim\limits_{x \to -1} u(x) = 0.429$.

b. Again, $u(x)$ is not defined at $x = \frac{-2}{9}$ because the denominator is zero when $x = \frac{-2}{9}$. The procedure we follow is similar to that shown in part *a*. Consider values of x increasingly close to $\frac{-2}{9} \approx -0.222222$ from both the left and the right as shown in Table 1.39.

TABLE 1.39

$x \to \frac{-2}{9}^-$	$u(x)$
-0.23	9.9
-0.223	95.6
-0.2223	952.7
-0.22223	9524.1

$x \to \frac{-2}{9}^+$	$u(x)$
-0.21	-5.7
-0.219	-22.7
-0.2219	-229.6
-0.22219	-2298.5

It appears that the output values of u are becoming increasingly large as x moves closer and closer to $\frac{-2}{9}$ from the left. Also, the output values of u are negative, and their absolute values seem to become increasingly large as x moves closer and closer to $\frac{-2}{9}$ from the right. On the basis of this numerical investigation, we conclude that the limits as x approaches $\frac{-2}{9}$ from the left and right do not exist. To indicate that $u(x)$ increases without bound as x approaches $\frac{-2}{9}$ from the left, we use the notation $\lim\limits_{x \to -2/9^-} u(x) \to \infty$. We also write $\lim\limits_{x \to -2/9^+} u(x) \to -\infty$ to indicate that $u(x)$ decreases without bound as x approaches $\frac{-2}{9}$ from the right.

Keep in mind that this numerical method gives only an estimate. Figure 1.37 shows a graph of the function u.

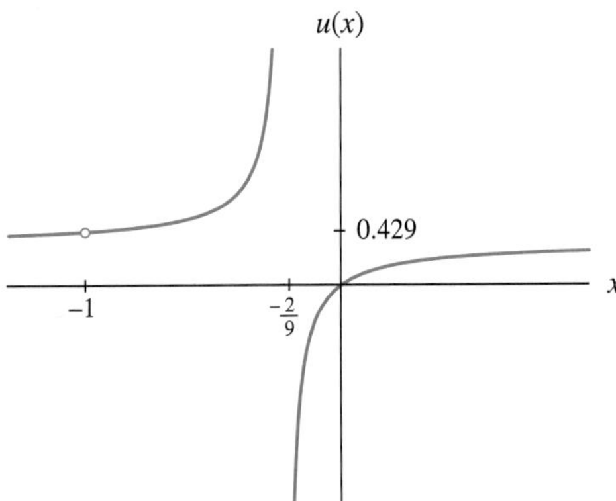

FIGURE 1.37

The graph seems to support our estimates that $\lim_{x \to -1^-} u(x) \approx 0.429$, $\lim_{x \to -1^+} u(x) \approx 0.429$, $\lim_{x \to -2/9^-} u(x) \to \infty$, and $\lim_{x \to -2/9^+} u(x) \to -\infty$.

Algebraically Determining Limits

The speeding fine illustration presented earlier in this section indicates that the limit of a constant function is that constant. There are many other rules and properties that apply to limits of functions. We state three here.

- **The Constant Rule:** The limit of a constant is a constant.

  $\lim_{x \to c} k = k$, where c and k are constants

- **The Sum Rule:** The limit of a sum is the sum of the limits.

  $\lim_{x \to c}[f(x) + g(x)] = \lim_{x \to c} f(x) + \lim_{x \to c} g(x)$, where c is a constant

- **The Constant Multiplier Rule:** The limit of a constant times a function is the constant times the limit of the function.

  $\lim_{x \to c} kf(x) = k \lim_{x \to c} f(x)$, where c and k are constants

Based on these limit rules we can state the following rules, which give us our first method of algebraically determining limits. The Replacement Rules enable us to find limits of *polynomial functions* and *rational functions*, which are quotients of polynomials.

Replacement Rules

1. If f is a polynomial function and a is a real number, then

$$\lim_{x \to a} f(x) = f(a)$$

2. If f is a rational function and a is a valid input of f, then

$$\lim_{x \to a} f(x) = f(a)$$

The Replacement Rules enable us to determine certain limits by evaluating a polynomial or rational function at a constant. For example,

$$\lim_{x\to 0.46}(x^2 + 2x - 5.2) = 0.46^2 + 2(0.46) - 5.2 = -4.0684$$

$$\lim_{p\to -1}\frac{3}{p} = \frac{3}{-1} = -3$$

$$\lim_{t\to 2}\frac{t^2 - 16}{t - 4} = \frac{4 - 16}{2 - 4} = \frac{-12}{-2} = 6$$

If you plan to cover Section 4.2 on algebraically finding derivatives, you should thoroughly discuss the Cancellation Rule presented here.

In some instances, we can algebraically determine the limit of a rational function at an input value for which it is not defined. In this section, we have seen two rational functions with a single point removed. The function $r(t) = \frac{t^2 - 16}{t - 4}$ is a line with the point (4, 8) removed, and $u(x) = \frac{3x^2 + 3x}{9x^2 + 11x + 2}$ is missing the point at which $x = -1$ and $u(-1) \approx 0.439$. Note that in both cases, when we factor the numerator and denominator, we see a common factor:

$$r(t) = \frac{t^2 - 16}{t - 4} = \frac{(t + 4)(t - 4)}{t - 4}$$

and

$$u(x) = \frac{3x^2 + 3x}{9x^2 + 11x + 2} = \frac{3x(x + 1)}{(9x + 2)(x + 1)}$$

The presence of a common linear factor in the numerator and denominator of a function is often an indication that a single point is missing from the graph of the function. The algebraic act of reducing the fraction by canceling the common factor creates a new function identical to the original function except that the single point is filled in. For example, if we reduce $r(t) = \frac{(t + 4)(t - 4)}{t - 4}$ to $s(t) = t + 4$, we obtain the line shown in Figure 1.33 with the hole filled in. Similarly, if we reduce $u(x) = \frac{3x(x + 1)}{(9x + 2)(x + 1)}$ to $v(x) = \frac{3x}{9x + 2}$, we have a new function identical to the one shown in Figure 1.37 except that the point corresponding to $x = -1$ is no longer missing. It is important to understand that r and s are not the same function and that u and v are not the same function. However, because the graphs of each pair of functions differ by only a single point, their limits at all points are the same. Therefore, by reducing and applying the Replacement Rules, we can say

$$\lim_{t\to 4}\frac{t^2 - 16}{t - 4} = \lim_{t\to 4}(t + 4) = 4 + 4 = 8$$

and

$$\lim_{x\to -1}\frac{3x(x + 1)}{(9x + 2)(x + 1)} = \lim_{x\to -1}\frac{3x}{9x + 2} = \frac{-3}{-7} = \frac{3}{7} \approx 0.429$$

These limits found algebraically confirm our previous estimates. Note that in the second case, the algebraic method gives us the exact answer, $\frac{3}{7}$, which we could not determine from the numerical or graphical investigations. These are two examples of a general rule that will be important in our consideration of limits in calculus in Chapter 4.

Cancellation Rule

If the numerator and denominator of a rational function share a common factor, then the new function obtained by algebraically canceling the common factor has all limits identical to the original function.

The principle that filling in a point does not affect limits can also be applied to break points of piecewise continuous functions. Consider the function

$$f(x) = \begin{cases} 3x + 2 & \text{when } x > 0 \\ 5x - 3 & \text{when } x \le 0 \end{cases}$$

Replacement Rule 1 enables us to find the limit as x approaches zero from the right by substituting 0 for x in $5x - 3$. If we remove the point (0, -3) and add the point (0, 2), we change the function to

$$g(x) = \begin{cases} 3x + 2 & \text{when } x \ge 0 \\ 5x - 3 & \text{when } x < 0 \end{cases}$$

Modifying the function by a single point does not affect the limits. We determine the limit of g as x approaches zero from the left by substituting 0 for x in $3x + 2$. Thus we find left and right limits at break points by substituting the input value at the break point into the portions of the function on either side of the break point as long as each portion can be evaluated at that input value.

Piecewise Continuous Rule

If f is a piecewise continuous function with $x = a$ as a break point, then the existence of $\lim_{x \to a} f(x)$ can be determined by using the Left and Right Limits Rule. If they exist, the left and right limits can be found by substituting a for x in the portion of f to the left and right of $x = a$.

Definition of *Continuous*

It is possible to omit this formal definition of continuity and rely on the graphical understanding of continuity as a graph that can be drawn without lifting the writing instrument from the page. An informal understanding of continuity is all that is required for the remainder of the topics in the text.

Graphically, we understand that a function is continuous if its graph can be drawn without lifting the writing instrument from the page. The functions r and u whose limits we determined using the Cancellation Rule are not continuous functions. Because calculus is the study of continuous portions of functions, it is important to have a complete understanding of what makes a function continuous or discontinuous. Limits enable us to define *continuous* more precisely.

Consider the four functions shown in Figure 1.38.

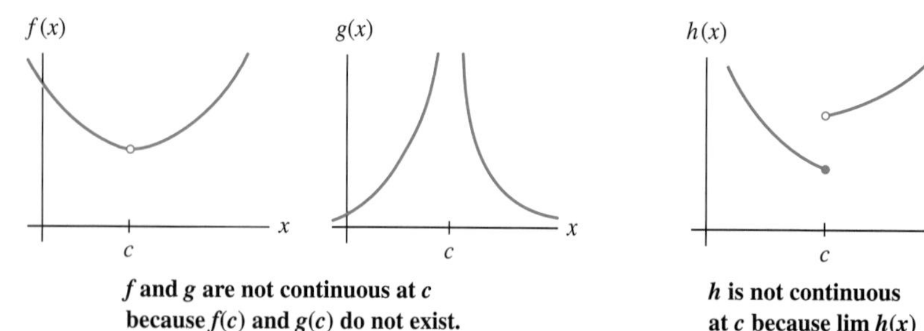

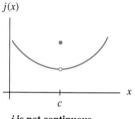

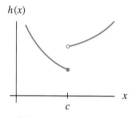

f and *g* are not continuous at *c* because *f*(*c*) and *g*(*c*) do not exist.

h is not continuous at *c* because $\lim_{x \to c} h(x)$ does not exist.

$j(x)$ x c

j is not continuous at *c* because $\lim_{x \to c} j(x) \neq j(x)$.

FIGURE 1.38

These graphs illustrate three criteria for determining whether a function is continuous at an input value *c*: The function must be defined at *c*, its limit must be defined as the input approaches *c*, and the limit must be the same as the output value at *c*. Intuitively, continuity at a point occurs when the function arrives at the point to which it is headed. These observations lead us to the definition of *continuous.*

Continuous Function

A function $y = f(x)$ is continuous at $x = c$ if and only if

1. $f(c)$ exists
2. $\lim_{x \to c} f(x)$ exists
3. $\lim_{x \to c} f(x) = f(c)$

A function is continuous on an open interval if these three conditions are met for every input value in the interval.

A function is continuous everywhere if it meets all three conditions for every possible input value. We refer to such functions as **continuous functions.**

Note that the Replacement Rules are simply restatements of the definition of continuous. We find the limit at an input value for which a function is continuous by finding the output of the function at that input value. Example 3 illustrates how to apply the definition to determine whether a function is continuous.

EXAMPLE 3 *Determining Continuity*

Consider the function

$$Q(p) = \begin{cases} \dfrac{p-2}{p^2-2p} & \text{when } p < 4 \\ 0 & \text{when } p = 4 \\ \dfrac{1}{8-p} & \text{when } 4 < p \le 7 \\ p-5 & \text{when } p > 7 \end{cases}$$

Determine whether Q is continuous at $p = 0, 2, 4,$ and 7.

Solution Q is not continuous at $p = 0$ or $p = 2$ because those input values make the denominator of $\frac{p-2}{p^2-2p}$ zero. Thus $Q(0)$ and $Q(2)$ do not exist, which violates the first condition for continuity.

The function does exist at $p = 4$ and $p = 7$: $Q(4) = 0$ and $Q(7) = \frac{1}{8-7} = 1$. In order to determine whether the function is continuous at these input values, we must determine the corresponding limits. To find the limit as p approaches 4, we use the Piecewise Continuous Rule and Replacement Rule 2 to find the limits from the left and right:

$$\lim_{p\to 4^-} Q(p) = \lim_{p\to 4^-} \frac{p-2}{p^2-2p} = \frac{4-2}{4^2-2(4)} = \frac{2}{8} = \frac{1}{4}$$

$$\lim_{p\to 4^+} Q(p) = \lim_{p\to 4^+} \frac{1}{8-p} = \frac{1}{8-4} = \frac{1}{4}$$

Because the left and right limits are the same, we conclude that $\lim_{p\to 4} Q(p) = \frac{1}{4}$. Thus we have

1. $Q(4) = 0$
2. $\lim_{p\to 4} Q(p) = \frac{1}{4}$
3. $\lim_{p\to 4} Q(p) \neq Q(4)$

and we conclude that Q is not continuous at $p = 4$ because the limit is not the same as the function value.

The limits from the left and right of 7 also can be found using the Piecewise Continuous Rule and the Replacement Rules:

$$\lim_{p\to 7^-} Q(p) = \lim_{p\to 7^-} \frac{1}{8-p} = \frac{1}{8-7} = 1$$

$$\lim_{p\to 7^+} Q(p) = \lim_{p\to 7^+} p - 5 = 7 - 5 = 2$$

Because the left and right limits are not the same, we conclude that $\lim_{p\to 7} Q(p)$ does not exist and that the function is not continuous at $p = 7$.

The graph in Figure 1.39 seems to support that Q is not continuous at $p = 0, 2, 4,$ or 7.

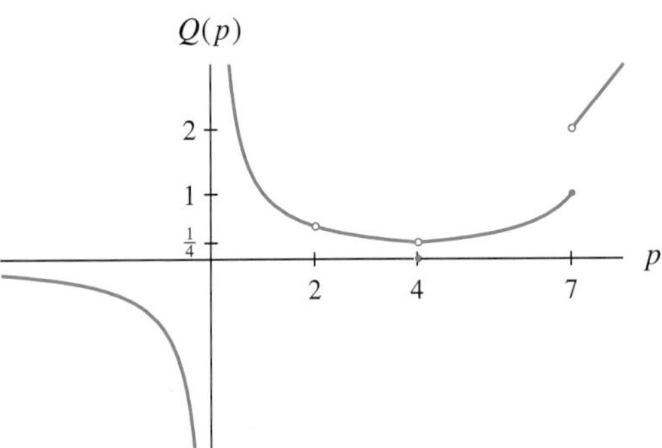

FIGURE 1.39

Limits Describing End Behavior

We use limits to describe the end behavior of the model types presented in Chapter 2. Even if you do not cover the specifics of finding end-behavior limits, you should at least introduce the concepts of end behavior and horizontal asymptotes.

How the output of a function behaves as the input becomes more and more positive (increases without bound) or more and more negative (decreases without bound) is called the **end behavior** of the function. For example, consider the graph of the function shown in Figure 1.40.

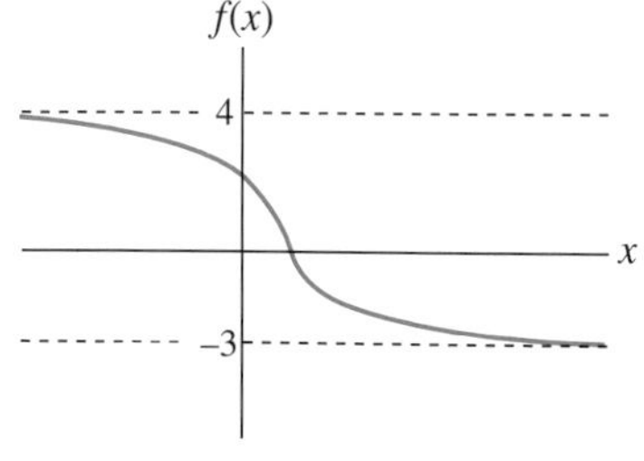

FIGURE 1.40

As x becomes arbitrarily large (that is, as we move from left to right without bound along the graph of the function), we say that x is approaching positive infinity, and we write $x \to \infty$. As x approaches infinity, the function outputs decline while becoming closer and closer to -3. As x becomes more and more negative (moving from right to left without bound on the function graph), we say that x approaches negative infinity, and we write $x \to -\infty$. As x approaches negative infinity, the function outputs become closer and closer to 4. We express these two facts mathematically by writing

$$\lim_{x\to\infty} f(x) = -3 \qquad \text{and} \qquad \lim_{x\to-\infty} f(x) = 4$$

When a function $y = f(x)$ approaches a number k as the input increases or decreases without bound, we call the horizontal line that has equation $y = k$ a **horizontal asymptote.** The function depicted in Figure 1.40 has two horizontal asymptotes, $y = -3$ and $y = 4$.

In addition to estimating end behavior from a graph, it is also possible to estimate the end behavior of a function numerically by evaluating the function at increasingly large values of the input variable. This process is illustrated in Example 4.

EXAMPLE 4 *Numerically Estimating End Behavior*

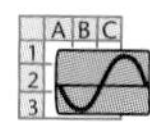

For $u(x) = \dfrac{3x^2 + x}{3x + 11x^2 + 2}$, numerically estimate

a. $\lim_{x\to\infty} u(x)$ b. $\lim_{x\to-\infty} u(x)$

Solution

a. To estimate the positive end behavior of u numerically, choose increasingly large values of x, as shown in Table 1.40.

TABLE 1.40

$x \to \infty$	100	1000	10,000	100,000
$u(x)$	0.272008	0.272653	0.272712	0.272727

It appears that $\lim_{x\to\infty} u(x) \approx 0.273$.

b. Next, consider the negative end behavior of u by choosing increasingly negative values of x, as shown in Table 1.41.

TABLE 1.41

$x \to -\infty$	-100	-1000	-10,000	-100,000
$u(x)$	0.273495	0.272802	0.272735	0.272728

It also appears that $\lim_{x \to -\infty} u(x) \approx 0.273$. Figure 1.41 shows the graph of u.

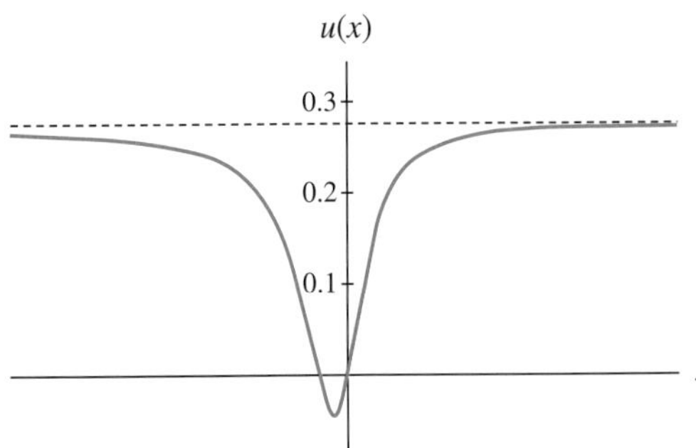

FIGURE 1.41

Note that the end behavior of the function cannot be estimated as accurately from the graph as from the numerical investigation.

If a function approaches the same value when x becomes increasingly positive as it does when x becomes increasingly negative (that is, if $\lim_{x \to \infty} f(x) = L$ and $\lim_{x \to -\infty} f(x) = L$) we describe this end behavior in one statement by writing $\lim_{x \to \pm\infty} f(x) = L$. In Example 4, we say $\lim_{x \to \pm\infty} u(x) \approx 0.273$.

> If a function f has the same limit L when x approaches positive infinity as it does when x approaches negative infinity, then we write $\lim_{x \to \pm\infty} f(x) = L$.

Caution: Do not confuse the statement "$x \to \pm\infty$" with the symbols used to indicate approaching an input value from the right or from the left ($x \to c^+$ or $x \to c^-$). A plus or minus sign placed *in front of* a number or the infinity symbol indicates whether the number or infinity is positive or negative, whereas a plus or minus sign written *after* a number indicates a specific direction from which that number is to be approached.

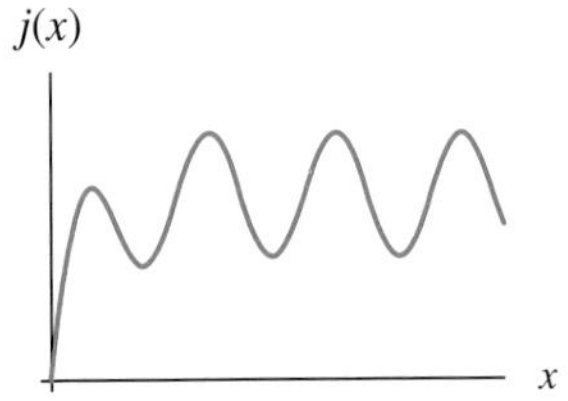

$j(x)$

x

$\lim_{x \to \infty} j(x)$ does not exist.

FIGURE 1.42

It is possible for a function to oscillate and not approach any specific output value as x approaches infinity. Such is the case with the function j in Figure 1.42. In this case, we say that the limit of j as x approaches infinity *does not exist.*

Algebraically Determining End Behavior

It is easy to determine end behavior for quotients of polynomials. Most of the functions in this section are in this category. As the input increases or decreases without bound, the terms in the numerator and denominator with the greatest power dominate the behavior of the function. We call a term **dominant** if it is the term in the expression with the highest power of the variable. For example, in the function $Q(p) = \frac{p-2}{p^2-2p}$, the dominant terms on the top and bottom are p and p^2, respectively. Thus the end behavior of Q is the same as the end behavior of the quotient of the dominant terms: $\frac{p}{p^2}$. That is,

$$\lim_{p\to\infty} \frac{p-2}{p^2-2p} = \lim_{p\to\infty} \frac{p}{p^2} = \lim_{p\to\infty} \frac{1}{p} = 0$$

and

$$\lim_{p\to-\infty} \frac{p-2}{p^2-2p} = \lim_{p\to-\infty} \frac{p}{p^2} = \lim_{p\to-\infty} \frac{1}{p} = 0$$

In the activities at the end of this section, you will be asked to support graphically and numerically the statements that $\lim_{p\to\pm\infty} \frac{1}{p} = 0$. These statements are special cases of the following rule concerning limits at infinity.

Limits at Infinity

If c is any constant and n is any positive number, then

1. $\lim_{c\to\pm\infty} c = c$
2. $\lim_{x\to\infty} \frac{c}{x^n} = 0$
3. $\lim_{x\to-\infty} \frac{c}{x^n} = 0$, provided that x^n is defined for $x < 0$

EXAMPLE 5 *Algebraically Determining End Behavior*

Determine the end behavior of $u(x) = \frac{3x^2 + x}{3x + 11x^2 + 2}$.

Solution The terms with the largest powers in the numerator and denominator of u are $3x^2$ and $11x^2$, respectively. Thus the end behavior of u is determined by considering the end behavior of $\frac{3x^2}{11x^2} = \frac{3}{11}$. Because this quotient is a constant, u approaches $\frac{3}{11}$ as x increases or decreases without bound; that is, $\lim_{x\to\pm\infty} \frac{3x^2 + x}{3x + 11x^2 + 2} = \frac{3}{11}$.

Thus the function u has one horizontal asymptote with equation $y = \frac{3}{11} \approx 0.27273$. Note that, unlike the numerical method used in Example 4, this algebraic method gives the exact value.

Summary of Limit Methods

One of the easiest ways to estimate limits is to examine the graph of a function. Graphs, however, cannot always provide information that is accurate enough to identify a limit. The method of creating a table of values to estimate a limit is also helpful. This numerical estimation process often can provide a more accurate answer than an estimate obtained from a graph.

Although graphical and numerical methods are important in understanding the concept of a limit, they do have limitations. Algebraic methods give accurate limit answers, whereas graphical and numerical methods yield only estimates. Algebraic methods include evaluating a function at a given input value when the function is continuous there, factoring the numerator and denominator of a function to look for common factors, and considering the dominant terms in a quotient of polynomials. Our final example combines all three methods.

EXAMPLE 6 *Using Multiple Methods for Limits*

Consider the function represented by the graph shown in Figure 1.43.

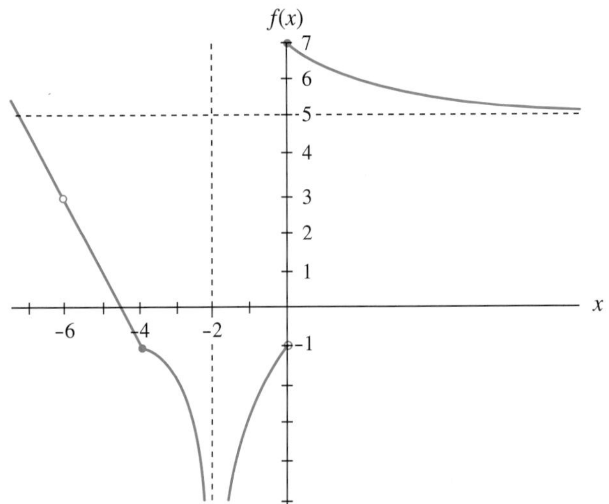

FIGURE 1.43

The equation for the function is

$$f(x) = \begin{cases} \dfrac{-2x^2 - 21x - 54}{x + 6} & \text{when } x \leq -4 \\[2ex] \dfrac{-4}{(x+2)^2} & \text{when } -4 < x < 0 \\[2ex] \dfrac{x+4}{2+7x} + 5 & \text{when } x \geq 0 \end{cases}$$

a. Use the graph to estimate

   i. $\lim\limits_{x \to -6} f(x)$ ii. $\lim\limits_{x \to -2} f(x)$ iii. $\lim\limits_{x \to -4} f(x)$

   iv. $\lim\limits_{x \to 0} f(x)$ v. $\lim\limits_{x \to \infty} f(x)$ vi. $\lim\limits_{x \to -\infty} f(x)$

b. Numerically estimate the limits in parts *ii* and *v* of part *a*.

c. Algebraically determine the limits in parts *i*, *iii*, *iv*, and *v* of part *a*.

Solution

a. From the graph we make the following estimates:

i. $\lim_{x\to -6} f(x) = 3$

ii. $\lim_{x\to -2} f(x) \to -\infty$

iii. $\lim_{x\to -4} f(x) = -1$

iv. $\lim_{x\to 0^+} f(x) = 7$ and $\lim_{x\to 0^-} f(x) = -1$, so $\lim_{x\to 0} f(x)$ does not exist.

v. $\lim_{x\to \infty} f(x) = 5$

vi. $\lim_{x\to -\infty} f(x) \to \infty$

b. ii. Numerically estimate $\lim_{x\to -2} f(x)$ using $f(x) = \frac{-4}{(x + 2)^2}$ (because x is approaching a number between -4 and 0) and the values shown in Table 1.42.

TABLE 1.42

$x \to -2^-$	f(x)	$x \to -2^+$	f(x)
-2.1	-400	-1.9	-400
-2.01	-40,000	-1.99	-40,000
-2.001	-4,000,000	-1.999	-4,000,000
-2.0001	-400,000,000	-1.9999	-400,000,000

We conclude that $\lim_{x\to -2} f(x) \to -\infty$.

v. Table 1.43 shows the function values for increasingly large input values. The formula $f(x) = \frac{x + 4}{2 + 7x} + 5$ was used for these calculations. Table 1.43 enables us to estimate numerically the right end behavior of the function as $x \to \infty$.

TABLE 1.43

$x \to \infty$	1000	10,000	100,000	1,000,000
f(x)	5.143388	5.142910	5.142862	5.142857

From this table we conclude that $\lim_{x\to\infty} f(x) \approx 5.143$. This estimate is more accurate than our graphical estimate of 5 in part *a*.

c. i. $\lim_{x\to -6} f(x) = \lim_{x\to -6} \frac{-2x^2 - 21x - 54}{x + 6}$

$$= \lim_{x\to -6} \frac{(-2x - 9)(x + 6)}{x + 6}$$

$$= \lim_{x\to -6} (-2x - 9) = -2(-6) - 9 = 12 - 9 = 3$$

Therefore, $\lim_{x\to -6} f(x) = 3$.

iii. Because f is a piecewise continuous function with a break point at $x = -4$, we must consider the limits of f as x approaches -4 from the left and right. Replacement Rule 2 indicates that we do so by evaluating the portions of the function to the left and right of $x = -4$ at -4:

$$\lim_{x\to-4^-} f(x) = \lim_{x\to-4^-} \frac{-2x^2 - 21x - 54}{x + 6} = \frac{-2(-4)^2 - 21(-4) - 54}{-4 + 6} = \frac{-2}{2} = -1$$

$$\lim_{x\to-4^+} f(x) = \lim_{x\to-4^+} \frac{-4}{(x + 2)^2} = \frac{-4}{(-4 + 2)^2} = \frac{-4}{(-2)^2} = \frac{-4}{4} = -1$$

Because the limit from the right is the same as the limit from the left, we conclude that $\lim_{x\to-4} f(x) = -1$.

iv. Again, we must consider the two portions of the function on either side of $x = 0$.

$$\lim_{x\to0^-} f(x) = \lim_{x\to0^-} \frac{-4}{(x + 2)^2} = \frac{-4}{4} = -1$$

$$\lim_{x\to0^+} f(x) = \lim_{x\to0^+} \left(\frac{x + 4}{2 + 7x} + 5\right) = \frac{4}{2} + 5 = 7$$

Because these limits are different, we conclude that $\lim_{x\to0} f(x)$ does not exist.

v. To determine the end behavior as x increases without bound, we first identify the dominant terms in the expression $\frac{x + 4}{2 + 7x}$. The end behavior of $\frac{x + 4}{2 + 7x}$ is the same as the end behavior of $\frac{x}{7x} = \frac{1}{7}$. On the basis of the first statement in the Limits at Infinity box, we can say that the end behavior of $\frac{x + 4}{2 + 7x} + 5$ is $\frac{1}{7} + 5$. That is,

$$\lim_{x\to\infty}\left(\frac{x + 4}{2 + 7x} + 5\right) = \lim_{x\to\infty}\left(\frac{x}{7x} + 5\right) = \lim_{x\to\infty}\left(\frac{1}{7} + 5\right) = 5\tfrac{1}{7}$$

Compare this answer with the numerical estimate of 5.143 from part *b* and the graphical estimate of 5 from part *a*. This algebraic limit is exact.

As we will see in subsequent chapters, limits are a foundational idea in all of calculus. Your ability to understand limits conceptually will greatly enhance your ability to understand the calculus concepts of derivative and integral.

1.4 Concept Inventory

- *Limits at a point*
- *Left and right limits*
- *Increasing and decreasing without bound*
- *Graphically and numerically estimating limits*
- *Algebraically determining limits*
- *Replacement Rules*
- *Cancellation Rule*
- *Piecewise Continuous Rule*
- *Continuous function*
- *End behavior*
- *Horizontal asymptote*

1.4 Activities

Use the graphs shown in Activities 1 through 6 to estimate the limits given.

1. a. $\lim_{x\to\infty} h(x)$ b. $\lim_{x\to-\infty} h(x)$ c. $\lim_{x\to2} h(x)$

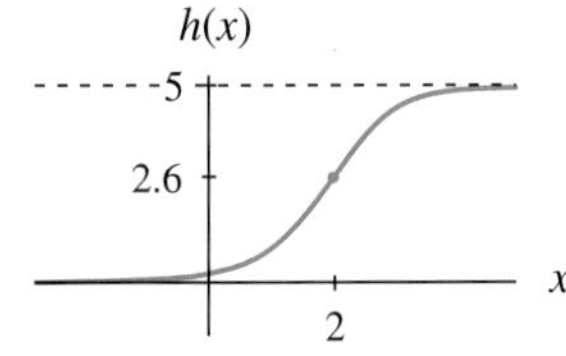

2. a. $\lim_{x\to\infty} s(x)$ b. $\lim_{x\to-\infty} s(x)$ c. $\lim_{x\to 0} s(x)$

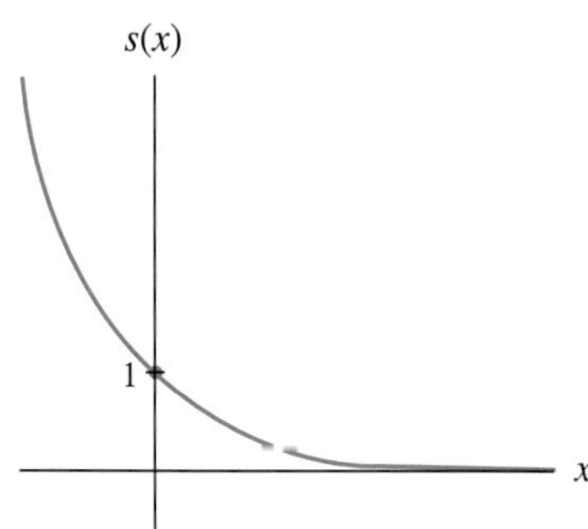

3. a. $\lim_{t\to 1^-} g(t)$ b. $\lim_{t\to 1^+} g(t)$ c. $\lim_{t\to 1} g(t)$
d. $\lim_{t\to\infty} g(t)$

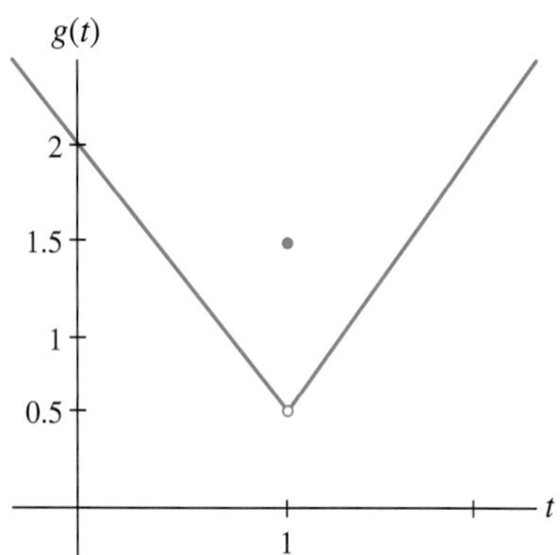

4. a. $\lim_{t\to 1^-} m(t)$ b. $\lim_{t\to 1^+} m(t)$ c. $\lim_{t\to 1} m(t)$
d. $\lim_{t\to\infty} m(t)$ e. $\lim_{t\to-\infty} m(t)$

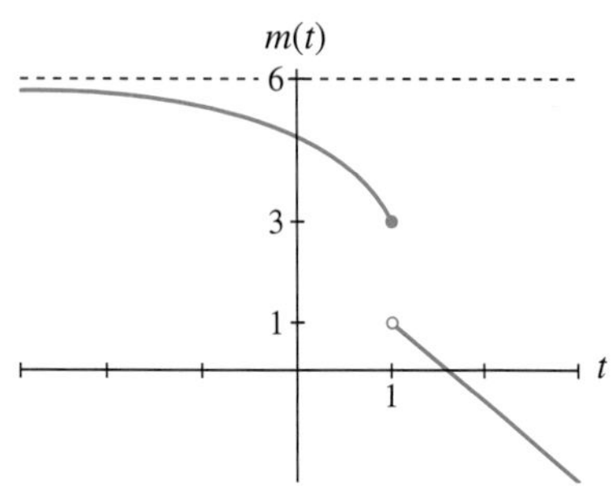

5. a. $\lim_{x\to 2} f(x)$ b. $\lim_{x\to -1} f(x)$ c. $\lim_{x\to\infty} f(x)$
d. $\lim_{x\to-\infty} f(x)$

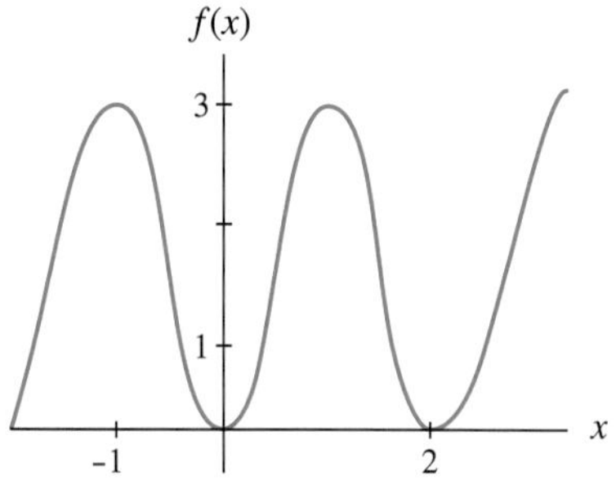

6. a. $\lim_{p\to -2} t(p)$ b. $\lim_{p\to\pm\infty} t(p)$

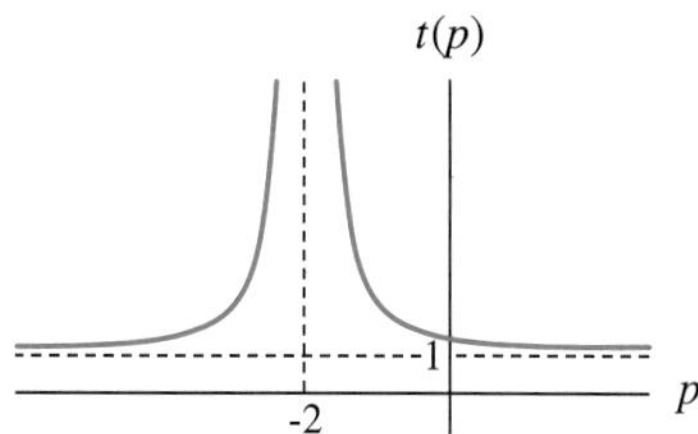

7. A function g is such that $\lim_{x\to 3^-} g(x) = 20$, $\lim_{x\to 3^+} g(x) = 20$, and $g(3) = 14$.
   a. Does $\lim_{x\to 3} g(x)$ exist? Explain your reasoning.
   b. Is g continuous at $x = 3$? Explain.
   c. Sketch a possible graph of g near $x = 3$.

8. A function f is such that $f(-6)$ does not exist, $\lim_{t\to -6^-} f(t) = 9$, and $\lim_{t\to -6^+} f(t) = 9$.
   a. Does $\lim_{t\to -6} f(t)$ exist? Explain your reasoning.
   b. Is f continuous at $t = -6$? Explain.
   c. Sketch a possible graph of f near $t = -6$.

9. A function h is such that $\lim_{x\to\infty} h(x) = 2$, $\lim_{x\to-\infty} h(x) = -2$, $\lim_{x\to 2} h(x) \to -\infty$, and $\lim_{x\to -2} h(x) \to \infty$.
   a. Is h continuous at $x = 2$ and $x = -2$? Explain.
   b. Sketch a possible graph of h.
   c. Discuss the end behavior of this function.

10. If f is a continuous function, is it possible for $\lim_{x\to c^-} f(x) = L_1$ and $\lim_{x\to c^+} f(x) = L_2$, where L_1 and L_2 are not the same number? Explain.

11. If f is a continuous function, is it possible for there to be an input value of f for which the limit does not exist? Explain.

Algebraically find the limits given in Activities 12 through 17.

12. $\lim_{x\to 4} 7$

13. $\lim_{w\to 3} (3.78w - 4.3)$

14. $\lim_{h\to 4} \dfrac{h}{h^2 + 4h}$

15. $\lim_{t\to -3} \dfrac{t^2 - 4t - 21}{t + 3}$

16. $\lim_{t\to 4} \dfrac{t^2 - 4t}{t - 4}$

17. $\lim_{x\to 2} \dfrac{2x^3 - 2x^2 + x}{4x^2 - x^3}$

18. Consider the function

$$f(x) = \begin{cases} \dfrac{6x}{x^2 - 6x} & \text{when } x < 6 \\ x - 7 & \text{when } x \geq 6 \end{cases}$$

Find the following limits:

a. $\lim_{x\to -\infty} f(x)$ b. $\lim_{x\to \infty} f(x)$ c. $\lim_{x\to 0} f(x)$

d. $\lim_{x\to 6^-} f(x)$ e. $\lim_{x\to 6^+} f(x)$ f. $\lim_{x\to 6} f(x)$

19. Consider the function

$$g(t) = \begin{cases} 2^t - 4 & \text{when } t < 2 \\ 0 & \text{when } 2 \leq t < 4 \\ \dfrac{t^2 - 16}{t - 4} & \text{when } t > 4 \end{cases}$$

Find the following limits:

a. $\lim_{t\to -\infty} g(t)$ b. $\lim_{t\to 2^-} g(t)$ c. $\lim_{t\to 2^+} g(t)$

d. $\lim_{t\to 2} g(t)$ e. $\lim_{t\to 4^-} g(t)$ f. $\lim_{t\to 4^+} g(t)$

g. $\lim_{t\to 4} g(t)$ h. $\lim_{t\to \infty} g(t)$

20. Refer to the function f in Activity 18.

a. Are there any horizontal asymptotes for this function? If so, give the equation(s) of the asymptote(s). If not, explain why not.

b. Discuss whether the function is continuous for the following inputs:

i. $x = 2$ ii. $x = 6$ iii. $x = 10$

21. Refer to the function g in Activity 19.

a. Are there any horizontal asymptotes for this function? If so, give the equation(s) of the asymptote(s). If not, explain why not.

b. Discuss whether the function is continuous for the following inputs:

i. $x = -5$ ii. $x = 2$ iii. $x = 3$ iv. $x = 4$

22. Refer to the function s on page 49 that gives the fine for speeding in a certain state.

a. Draw a graph of this function, and use the graph to estimate these limits:

i. $\lim_{x\to 5^-} s(x)$ ii. $\lim_{x\to 5^+} s(x)$

b. Estimate numerically these limits:

i. $\lim_{x\to 10^-} s(x)$ ii. $\lim_{x\to 10^+} s(x)$

c. Discuss the continuity of s at these inputs:

i. $x = 1$ ii. $x = 5$ iii. $x = 10$ iv. $x = 20$

Numerically estimate the limits given in Activities 23 through 26.

23. $\lim_{x\to 3} \dfrac{1}{x - 3}$

24. $\lim_{x\to 3} \dfrac{1}{(x - 3)^2}$

25. $\lim_{t\to \infty} \dfrac{14}{1 + 7e^{0.5t}}$

26. $\lim_{q\to -\infty} (-2^q + 6)$

Numerically estimate the limits in Activities 27 through 30.

27. $\lim_{x\to -1/3} \dfrac{x^3 + 6x}{3x + 1}$

28. $\lim_{x\to 3/7} \dfrac{2.8x^3 - 4x}{7x - 1}$

29. $\lim_{x\to 2/3} \dfrac{3x^2 + 1.6x - 2.4}{11x - 6 - 3x^2}$

30. $\lim_{x\to 0.65} \dfrac{8x^3 + 5.2x^2 - 3.38x - 2.197}{4x^2 - 1.69}$

31. The U.S. Postal Service priority mail may be used for packages weighing up to 70 pounds and not exceeding 108 inches in length and girth combined. Fractions of a pound are rounded up to the rate for the next full pound. Table 1.44 gives priority mail rates for parcels up to and including 5 pounds.[42]

TABLE 1.44

Weight, w, not exceeding	Mailing fee, $f(w)$
1 lb	\$3.50
2 lb	\$3.95
3 lb	\$5.15
4 lb	\$6.35
5 lb	\$7.55

42. United States Postal Service.

a. Find the priority mail fee for packages weighing 1.5 pounds and packages weighing 3.28 pounds.

b. Sketch a graph of the mailing fee as a function of the weight for $0 \le w \le 5$.

c. Find, if possible, each of the following limits. If any limit does not exist, explain why not.

i. $\lim_{w\to 1.3^-} f(w)$ ii. $\lim_{w\to 1.3^+} f(w)$ iii. $\lim_{w\to 1.3} f(w)$

iv. $\lim_{w\to 2^-} f(w)$ v. $\lim_{w\to 2^+} f(w)$ vi. $\lim_{w\to 2} f(w)$

vii. $\lim_{w\to 0^+} f(w)$ viii. $\lim_{w\to 3} f(w)$

32. The accumulated amount in an account that is opened with an initial deposit of \$P and earns interest at an annual rate of $100r\%$ for t years can be expressed as

$$A(x) = P(1 + rx)^{t/x} \text{ dollars}$$

where $x = \frac{1}{n}$ for n compounding periods each year.

a. Discuss the continuity of A.

b. For $r = 0.1$ and $t = 10$, numerically estimate $\lim_{x\to 0^+} A(x)$. Interpret your answer.

c. For $r = 0.05$ and $t = 3$, numerically estimate $\lim_{x\to 0^+} A(x)$. Interpret your answer.

33. The cost of cleaning up $100x\%$ of a small chemical spill from a mountain stream is given by

$$C(x) = \frac{5}{1 + 5e^{-0.4x}} \text{ thousand dollars}$$

a. What is the cost of removing 50% of the chemical spill?

b. Find $\lim_{x\to\infty} C(x)$. Does this limit have meaning in the context of this problem? Explain.

c. What is the cost of cleaning up all of the spill?

34. An urban hospital parking garage charges \$3 for parking 1 hour or less plus \$0.50 for each additional half-hour or fraction thereof.

a. How much would you pay for parking 45 minutes in this garage?

b. How much would you pay for parking 2 hours and 50 minutes?

c. Draw a graph of f whose output is the fee for parking t hours, where t is between 0 and 5.

d. Is the graph of f continuous?

e. Find $\lim_{t\to 3.5} f(t)$

f. Find $\lim_{t\to 4^-} f(t)$.

g. Find $\lim_{t\to 4^+} f(t)$.

h. What can you say about $\lim_{t\to 4} f(t)$?

35. A solicitation for donations to a scholarship fund is mailed to graduates of a small college. A model describing the total response is $R(t) = 38(1 - e^{-0.7472t})$ percent of graduates responding, where t is the number of weeks after the solicitations were mailed.

a. Find and interpret $R(6)$.

b. What percent of the graduates will eventually respond?

36. Ultram is a non-narcotic pain reliever. Suppose a person takes two 50-milligram tablets of Ultram. A model for the amount in her blood t hours after the initial dose is

$$U(t) = 100(e^{-1.3t} - e^{-9.5t}) \text{ milligrams}$$

a. Draw a graph of U, and describe the concentration of Ultram in this person's blood.

b. Find $\lim_{t\to\infty} U(t)$.

c. When do you feel it is safe for this person to take another dose of Ultram?

37. The price of Microsoft Corporation stock between 1989 and 1998 can be modeled by

$$P(x) = 1.0324(1.5184^x) \text{ dollars per share}$$

where x is the number of years after 1989. Find, if possible, each of the following limits. If it is not possible to find the limit, explain why not.

a. $\lim_{x\to 0^+} P(x)$ b. $\lim_{x\to 8} P(x)$ c. $\lim_{x\to\infty} P(x)$

d. $\lim_{x\to 0^-} P(x)$ (Note that 1989 was the first year that Microsoft Corporation stock was offered.)

38. A grocery store puts its brand of cola on sale for one day according to the model

$$C(x) = \begin{cases} \frac{1.29}{x} \text{ dollars} & \text{when } 1 \le x \le 3 \\ 1.19 \text{ dollars} & \text{when } x > 3 \end{cases}$$

where x is the number of bottles of cola purchased.

a. How much do 3 bottles of this cola cost on sale?

b. How much do 6 bottles of this cola cost on sale?

c. Find, if it exists, $\lim_{x\to\infty} C(x)$. If the limit does not exist, explain why not.

39. U.S. factory sales of electronics[43] between 1986 and 1990 can be modeled by

$$S(t) = 220 - 44.43e^{-0.3912t} \text{ billion dollars}$$

t years after 1986.

a. Find and explain the difference between $S(4)$ and $\lim_{t\to 4} S(t)$.

b. Find $\lim_{t\to\infty} S(t)$, and interpret it in this context.

c. Discuss how your answer to part b is related to the term *horizontal asymptote.*

40. Suppose that $\frac{ax^m}{bx^n}$ (with $a, b \neq 0$) is the reduced quotient of the dominant terms of a rational function. There are three situations that can result for this reduced quotient:

i. $m = n$ ii. $m < n$ iii. $m > n$

Discuss $\lim_{x\to\infty} \frac{ax^m}{bx^n}$ in each of these three cases.

Determine the end behavior of the functions in Activities 41 through 46 as the input increases and decreases without bound.

41. $t(m) = \dfrac{3m^3 + m}{m^5 - 3}$

42. $s(t) = \dfrac{3t^2 - 12t}{t - 4t^2}$

43. $f(x) = \dfrac{2x^3 - 2.8x^2 + x}{4x^2 - 3.5x^3}$

44. $g(w) = \dfrac{3.4w - 12}{w^3 - 4.6w}$

45. $h(t) = \dfrac{t^2 + 7.3}{1.6 - 2t}$

46. $f(x) = \dfrac{-15.6x^5 + 3x^2 - 1}{4x^4 + 3.6x^5 - 108}$

1.5 Linear Functions and Models

Having explored the concept of a function earlier in this chapter, we now turn our attention to several specific types of functions that will be helpful as we seek to describe real-life situations with mathematical models. Our goal here and in the next chapter is to give you an understanding of the behavior underlying certain functions in order to help you determine what function is appropriate in a particular modeling situation.

Representations of a Linear Model

We begin with the simplest of all functions: the linear function. A **linear function** is one that repeatedly and at even intervals adds the same value to the output. The output values form a line when graphed; thus we use the term *linear* to describe the data. For example, consider a newspaper delivery team that makes weekly deliveries of newspapers and devotes their Saturday mornings to selling new subscriptions. Suppose that the team starts with 80 customers and that 5 new customers are added each week. The data and scatter plot showing the number of customers over a 7-week period give us a picture of the increase in the number of customers. (See Table 1.45 and Figure 1.44a.)

TABLE 1.45

Weeks	0	1	2	3	4	5	6	7
Number of customers	80	85	90	95	100	105	110	115

43. Based on data from *Statistical Abstract*, 1992.

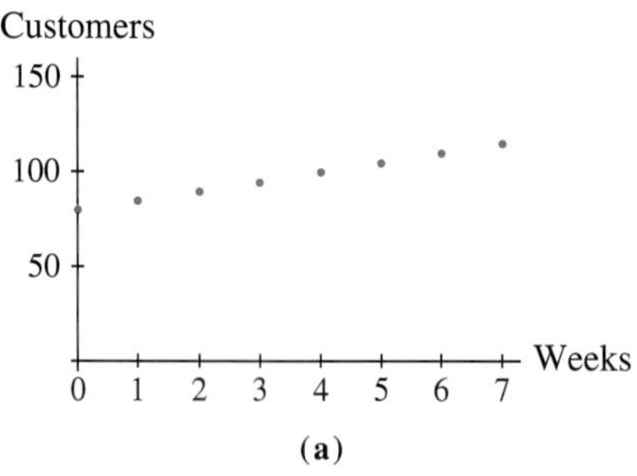

(a)

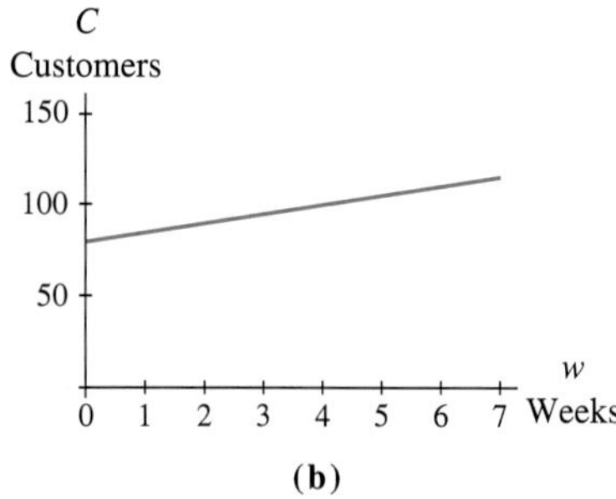

(b)

FIGURE 1.44

Let us also consider a continuous model of the situation given by the equation $C = 5w + 80$ customers, where w stands for the number of weeks since the team began the subscription drive. The model gives us a comprehensive description of the situation and is available to help answer questions about the number of customers. A graph of the model is the line that results when the points of the scatter plot are connected. (See Figure 1.44b.)

Note how each of the four representations (data, scatter plot, graph, and equation) provides its own particular insight into the nature of the customer base. The data in Table 1.45 show the regular pattern of the customer counts and its steady increase. The scatter plot and graph in Figure 1.44 show the steady climb in the number of subscribers and the constant nature of the growth. Finally, the algebraic expression given in the mathematical model captures the mathematical relationship between the two variable quantities (weeks and number of customers).

The model can be used in many ways. At the start of the campaign to recruit new customers, team members could take the model as their planned objective and then make plots of each week's result to check their progress. The model could also be used to answer a variety of "what if" questions. For example, if the newspaper had a strike and the team could receive at most 130 papers to deliver, how many weeks could the strike continue before the team ran into supply problems?

Where should you look for linear models? In general, linear models are appropriate when things are left undisturbed and are allowed to undergo regular, incremental change. In these situations, momentum usually takes hold, and things change by a regular, constant amount; that is, the data are linear.

The Parameters of a Linear Model

A linear equation is determined by two parameters: a starting value and the amount of the incremental change. All linear functions graph as lines and algebraically appear as

$$f(x) = ax + b$$

where a is the incremental change and b is the starting value.[44]

The magnitude of the incremental change is the steepness of the line, and the starting value indicates the placement of the line on a coordinate axis. The *slope* of a line is a measure of its steepness or how rapidly the line is rising or falling. The starting value is the point at which the line crosses the vertical axis. This value on the vertical axis is known as the vertical axis *intercept*. The concept of the steepness of a

44. $y = a + bx$ is an alternative form of the linear equation. This form is commonly used by statisticians.

graph is of primary importance in our study of calculus. As we shall see in Chapter 3, the slope of most functions is determined using calculus; however, the slope of a line can be calculated more simply.

The directed horizontal distance from one point on a graph to another is called the *run*, and the corresponding directed vertical distance is called the *rise*. The quotient of the rise divided by the run is the **slope** of the line connecting the two points. Consider again the newspaper subscription graph. We have chosen the points on the graph that correspond to 0 and 7 weeks. The rise and run corresponding to those two points are shown in Figure 1.45.

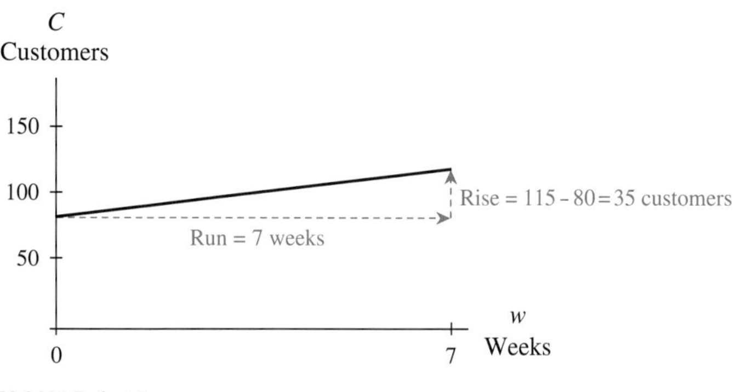

FIGURE 1.45

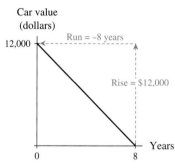

One of your goals in teaching this section needs to be helping your students understand that the rate of change of a line is defined as the line's slope. Linking rate of change with slope now will benefit the students in Chapters 3 and 4.

The slope is calculated as $\frac{\text{rise}}{\text{run}} = \frac{35 \text{ customers}}{7 \text{ weeks}} = 5$ customers per week. This slope value will be the same regardless of which two points are chosen for the calculation. The steeper a line, the greater the magnitude of the value of the slope. Lines that fall rather than rise have negative slope. Unlike the slope, the steepness of a line does not depend on whether it rises or falls.

The slope of a graph of a function at a particular point is a measure of how quickly the output is changing as the input changes at that point. We call this measure the **rate of change** of the function at that point and discuss it more thoroughly in Chapters 3 and 4. Because linear functions are characterized by constant incremental change, their slope (rate of change) is constant at all points. Other types of functions have a different slope at every point; that is, their rate of change is not constant.

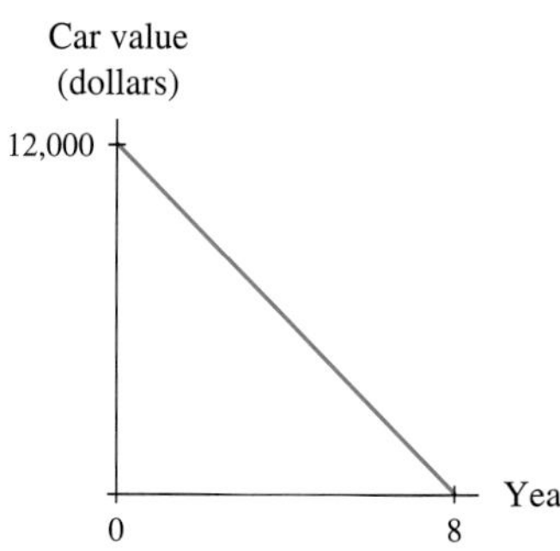

FIGURE 1.46

As another example, consider the resale value of a used car represented graphically as shown in Figure 1.46. Recall that the places where the line crosses the horizontal and vertical axes are called **intercepts**. The vertical axis intercept is 12,000 and corresponds to the value of the car, in dollars, when it was purchased. The horizontal axis intercept is 8 and indicates the first year in which the car had essentially no value.

To travel from the years intercept 8 to the value intercept 12,000 on the graph requires that you move 12,000 units up (+\$12,000 = rise) and 8 units to the left (-8 years = run). See Figure 1.47. The quotient of the rise divided by the run is $\frac{\$12{,}000}{-8 \text{ years}} = -\$1500$ per year. If you traveled from the value intercept 12,000 to the years intercept 8, the run would be positive and the rise negative. The quotient of the rise divided by the run would be $\frac{-\$12{,}000}{8 \text{ years}} = -\$1500$ per year. Note that the quotient $\frac{\text{rise}}{\text{run}}$ is the same either way you compute it. The quotient $\frac{\text{rise}}{\text{run}}$ is equal to the slope of the linear model and is the rate of change of the value of the car. We say that the car depreciated at a rate of \$1500 per year.

FIGURE 1.47

The equation of the line that represents the value of the car has slope -1500 and starting value 12,000. If we let the variable t represent the number of years since the car was purchased, the equation for the value of the car is

$$\text{Value} = -1500t + 12{,}000 \text{ dollars}$$

Although the slope of a particular linear model never changes, the graph of the model may look different when the horizontal or vertical scale is changed. You should therefore always use the same horizontal and vertical views when comparing two different graphs. The importance of this is shown in Example 1.

EXAMPLE 1 Energy Production and Consumption

The two graphs in Figure 1.48 show energy production and energy consumption[45] in the United States from 1975 through 1980.

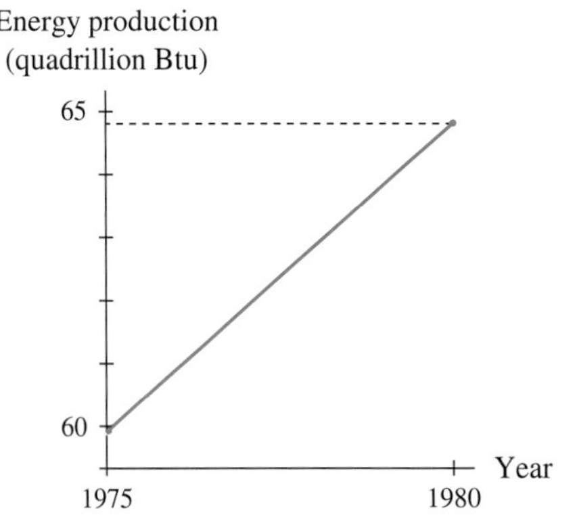

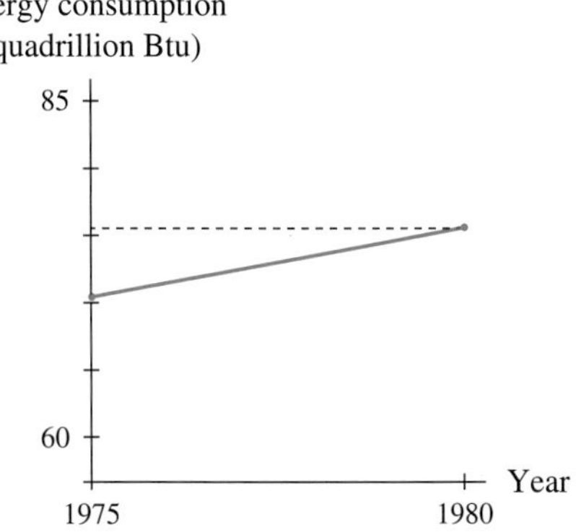

FIGURE 1.48

a. Which of the two graphs appears to be steeper?

b. Calculate the slope of each graph.

c. Which of the two graphs is steeper?

d. Write a linear equation for each graph.

e. What were the rates of change of energy production and energy consumption between 1975 and 1978?

f. What was the rate of change of energy production in 1976? in 1978?

Solution

a. The energy production graph appears to be steeper.

b. Slope of energy production

$$\frac{\text{rise}}{\text{run}} \approx \frac{64.8 - 59.9 \text{ quadrillion Btu}}{5 \text{ years}}$$

Slope of energy consumption

$$\frac{\text{rise}}{\text{run}} \approx \frac{76.0 - 70.6 \text{ quadrillion Btu}}{5 \text{ years}}$$

45. Based on data from *Statistical Abstract*, 1994.

c. The energy consumption graph is the steeper graph because it has a greater positive slope than the energy production graph.

d. The starting value b for energy production is about 59.9. The slope a is approximately 0.98, so the equation is

$$\text{Energy production} = 0.98x + 59.9 \text{ quadrillion Btu}$$

where x is the number of years since 1975. Similarly,

$$\text{Energy consumption} = 1.08x + 70.6 \text{ quadrillion Btu}$$

where x is the number of years since 1975.

e. The rate of change of a linear function is simply the slope of the line. The rate of change of energy production between 1975 and 1980 was approximately 0.98 quadrillion Btu per year, and the rate of change of energy consumption was about 1.08 quadrillion Btu per year.

f. The rate of change of a linear model is constant at all input values for which the model applies. Therefore, the rate of change of energy production in 1976 and 1978 was 0.98 quadrillion Btu per year.

We summarize our discussion of linear equations as follows:

Linear Model

A linear model is one that has a constant rate of change and has an equation of the form

$$f(x) = ax + b$$

where a and b are constants. A linear function forms a line when graphed. The parameter a in the equation is the constant rate of change of the output and is the slope of the line. The parameter b is the vertical axis intercept.

Finding a Linear Model

Suppose you are sole proprietor of a small business that has seen no growth in sales for the last several years. You have noticed, however, that your federal taxes have increased as shown in Table 1.46. A scatter plot of these data is shown in Figure 1.49.

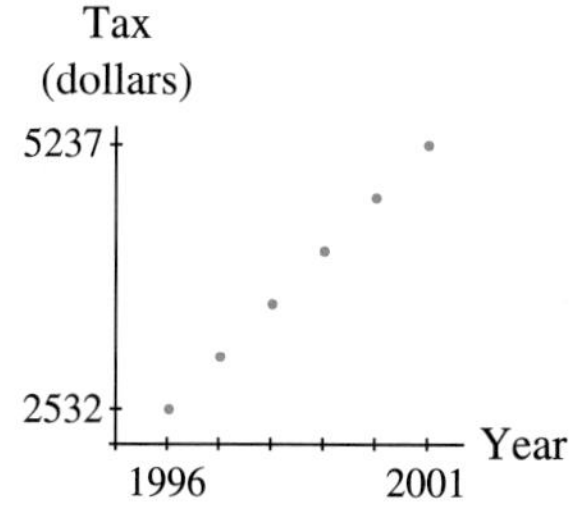

FIGURE 1.49

TABLE 1.46

Year	1996	1997	1998	1999	2000	2001
Tax	\$2532	\$3073	\$3614	\$4155	\$4696	\$5237

The scatter plot appears to show a linear pattern. Upon close examination of the data, you note that taxes increased by the same amount every year.

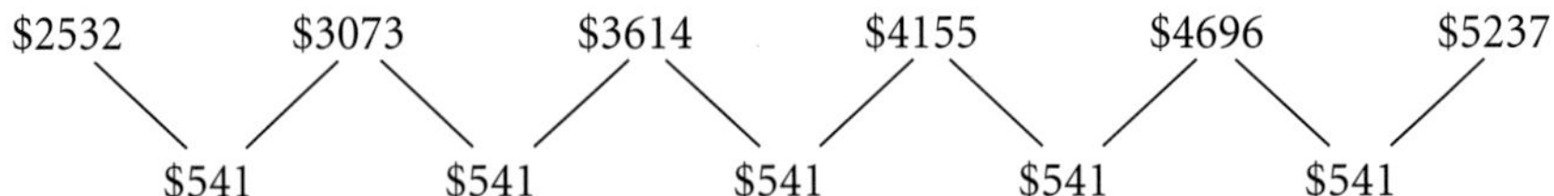

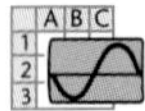

The changes in successive output, which are called **first differences,** are constant. This constant increase is the incremental change where the increment is one year. Because the incremental change is constant, we know that the data represent a linear increase in taxes. Because the data show the tax amount each year, this incremental change is the rate of change of the tax amount and the slope of the underlying linear model. If the data had shown taxes every other year, the first differences would be constant, indicating a linear pattern, but the value of the first differences would be twice the slope. Slope values are always expressed per unit increase in input.

On the basis of the calculation of first differences, we make the following equivalent statements:

The equivalence of these statements cannot be overemphasized.

Taxes increased by $541 per year.
The slope of the linear function described by the data is $541 per year.
The rate of change of the tax amount is $541 per year.

Some students want to state that the slope is "$1082 per two years." While this is not technically incorrect, emphasize that slope values are commonly expressed per single input unit. Sometimes students will say "$541 per years." In this case, emphasize that the word following the "per" should be singular.

If we consider 1996 to be our starting point, then the starting tax value is $2532. That is, the tax amount began in 1996 at $2532 and increased each year by $541. Thus the parameters in the linear function for the tax amount are $a = \$541$ and $b = \$2532$, and the linear model for the tax amount is

$$\text{Tax} = 541t + 2532 \text{ dollars}$$

where t represents the number of years since 1996. Such models are important because they often enable us to analyze the results of change. With certain assumptions, they may even allow us to make cautious predictions about the short-term future. For example, to predict the tax amount owed in 2004, we substitute $t = 8$ into the tax model.

$$\text{Tax} = 541(8) + 2532 = \$6860$$

Admittedly, in this instance, an equation is not necessary to make such a prediction, but there are many situations in which it is difficult to proceed without a model.

It is important to understand that when we use mathematical models to make predictions about the short-term future, *we are assuming that future events will follow the same pattern as past events.* This assumption may or may not be true. That does not mean that such predictions are useless, *only that they must be viewed with extreme caution.*

A Word of Caution

When you predict output values for input values that are within the interval of input data in your scatter plot, you are using a process called **interpolation.** Predicting output values for input values that are outside the interval of the input data is called **extrapolation.** Because you do not know what happens outside the range of given data, **estimates obtained by extrapolation must be viewed with caution and may result in misleading predictions.**

We have already noted that the rate of change of the tax amount is \$541 per year, which is how much taxes increase during a 1-year period. Thinking about rate of change in this way helps us answer questions such as

- If you pay taxes twice a year, how much will taxes increase each time?

$$(\$541 \text{ per year})\left(\tfrac{1}{2}\text{ year}\right) = \$270.50$$

Emphasizing these calculations now will help your students understand tangent line approximations (differentials) and marginal analysis when it is presented in Chapter 5.

- How much will taxes increase each time if you pay taxes quarterly?

$$(\$541 \text{ per year})\left(\tfrac{1}{4}\text{ year}\right) = \$135.25$$

- How much will taxes increase during the next 3 years?

$$(\$541 \text{ per year})\ (3 \text{ years}) = \$1623$$

EXAMPLE 2 *Business Survival*

TABLE 1.47

t (years)	*P* (percent)
5	50
6	47
7	44
8	41
9	38
10	35

Table 1.47 shows the percent P of companies that are still in business[46] after t years in operation.

a. Find the constant rate of change in the percent of businesses surviving.

b. Find a model for the percentage of companies still in business as a function of years of operation.

c. Is the function in part *b* discrete, continuous without restriction, or continuous with discrete interpretation? Explain.

d. If the rate of change remains constant, in what year of operation will the percent drop below 25%?

Solution

a. Calculating the first differences of the data in Table 1.47 indicates that the percent of businesses surviving decreased by 3 percentage points per year. We could also say that the rate of change of the survival percentage is -3 percentage points per year.

Caution: Note that this is not the same as a rate of change of -3% per year. When the output of a function is a percent, units on the rate of change must be expressed in terms of percentage points per input unit. More will be said about this in Chapter 3.

b. The data reveal that the percent begins with 50% and that each year after the first table value, the percent declines by 3 percentage points. We write this mathematically as

$$P(t) = -3t + 50 \text{ percent of businesses}$$

t years after the fifth year in operation.

Be sure to point out to your students that slopes involving outputs that are percentages must have units stated as "percentage points per year." This situation occurs repeatedly throughout the text. See the *Instructor's Resource Guide* for a good method of illustrating the difference between "percent per year" and "percentage points per year."

46. Cognetics, Cambridge, MA, 1998.

c. The function for the percent of businesses in operation is continuous, and it can be interpreted continuously without restriction. For example, $P(6.5) \approx 31$ can be interpreted as "About 31% of all businesses are still in operation after eleven and a half years" or as "After eleven and a half years, about 69% of businesses will no longer be in operation." Similarly, $P(8.083) \approx 26$ can be interpreted as "About 26% of businesses last 13 years and 1 month."

d. Solving the equation $25 = -3t + 50$ gives $t \approx 8.3$ years after the 5th year. The end of the 13th year corresponds to $t = 8$, so the survival percentage will drop below 25% in the 14th year of operation (in fact, during April of the 14th year). That is, less than 25% of companies will still be in business after 14 years of operation.

Point out that first differences (and second differences and percentage differences as seen in Chapter 2) are meaningless unless the input data are evenly spaced. Also, while constant first differences indicate that a linear model is appropriate, non-constant first differences do not indicate that a linear model is not appropriate. Make sure your students understand that it is rare for real-life data to perfectly conform to a certain model.

Examples 1 and 2 illustrate methods of finding a linear model of the form $f(x) = ax + b$ for data points that fall on a line. However, real-life data values are seldom perfectly linear. For instance, the tax data we considered earlier are not likely to occur in real-life situations because tax rates and revenues of most businesses change from year to year. Consider the following modification to the tax data:

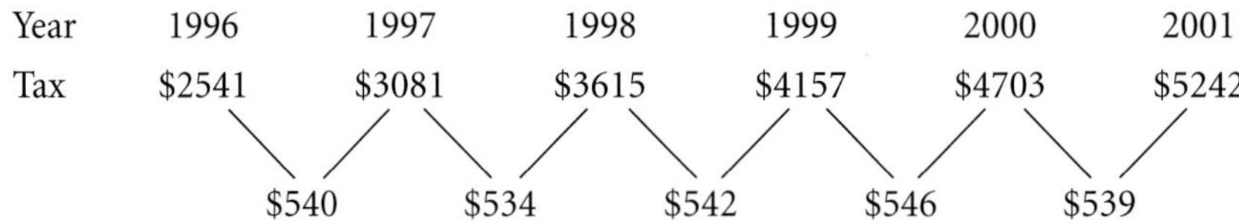

Year	1996	1997	1998	1999	2000	2001
Tax	\$2541	\$3081	\$3615	\$4157	\$4703	\$5242

The first differences are not constant but are "nearly constant." A linear model may be used if first differences are close to constant. Be sure to calculate first differences only for data that are evenly spaced. How do we get a linear model in this situation? Your calculator or computer will find an equation for the linear model that best fits the data.

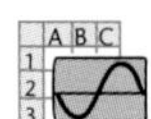

Use your calculator or computer to construct a scatter plot of these data. An examination of the scatter plot reinforces our earlier observation, from first differences, that the data are close to being linear. Next, use the linear regression routine that is built into your calculator or computer to find a linear equation that fits the data. You should find the equation to be

$$\text{Tax} = 540.371t - 1{,}076{,}042.467 \text{ dollars}$$

where t is the year.

We strongly recommend that you do not establish a "rule for alignment" for your class. While such a rule may make life easier for you, it also removes the opportunity for students to make their own aligning choices. Some situations may suggest a particular alignment. As we will see in Chapter 2, some alignments may produce computational error in the calculation of regression equations. (This second edition includes logarithmic models for which we cannot use an input of zero.) We deliberately have not always aligned in the same manner throughout the text and answer key.

If we are willing to assume that the rate of change of the tax remains constant at about \$540.37 per year given by this model, then we can use the linear model to predict future tax amounts. For instance, the model predicts that the 2002 tax will be $540.371(2002) - 1{,}076{,}042.467 \approx \$5781$ and that the 2003 tax will be $540.371(2003) - 1{,}076{,}042.467 \approx \$6322$. (These values were calculated using the unrounded model coefficients.)

Aligning Data

When the input data are years, it is often desirable to modify how the years are numbered to reduce the number of digits you have to enter, as well as to reduce the magnitude of some of the coefficients in the model equation. We refer to the process of renumbering data as **aligning** the data. For example, if we renumber the years in the tax data so that 1996 is year 0, 1997 is year 1, and so on, we obtain the data shown in Table 1.48.

TABLE 1.48

Aligned year	0	1	2	3	4	5
Tax	\$2541	\$3081	\$3615	\$4157	\$4703	\$5242

The aligned years are described as "the number of years since 1996." Using a calculator or computer to obtain a linear model for this aligned data, we have

$$\text{Tax2} = 540.371t + 2538.905 \text{ dollars}$$

where t is the number of years since 1996. Note that this equation and the equation for the unaligned data have the same slope (rate of change) but differ in the vertical axis intercept (b value). Aligning the data has the effect of shifting the data (usually to the left). This does not change the slope of the line, but it does change where the line crosses the vertical axis.

Emphasize that aligning does not change the slope of a line. Make the connection between aligning data and horizontally shifting a graph.

It is also possible to align the data by using the number of years since 1900. See Table 1.49. For some data, it may make more sense to align by using the number of years since 2000.

TABLE 1.49

Aligned year	96	97	98	99	100	101
Tax	\$2541	\$3081	\$3615	\$4157	\$4703	\$5242

Using this alignment shown in Table 1.49, we obtain the model

$$\text{Tax3} = 540.371t - 49{,}336.752 \text{ dollars}$$

where t is the number of years since 1900.

The three models we have obtained by aligning the data differently are

$$\text{Tax1} = 540.371t - 1{,}076{,}042.467 \text{ dollars, where } t \text{ is the year}$$

$$\text{Tax2} = 540.371t + 2538.905 \text{ dollars, where } t \text{ is the number of years since 1996}$$

$$\text{Tax3} = 540.371t - 49{,}336.752 \text{ dollars, where } t \text{ is the number of years since 1900}$$

These models may look different, but they are equivalent. They all share the same slope (rate of change) but have different vertical axis intercepts and different input variable descriptions.

When using a model obtained from aligned data, it is important that you keep in mind at all times the description of the input variable. To estimate the tax in 2003, we use a value of $t = 2003$ in the first model, a value of $t = 7$ in the second model (the number of years that 2003 is from 1996), and a value of $t = 103$ in the third model:

$$\text{Tax1} = 540.371(2003) - 1{,}076{,}042.467 \text{ dollars} \approx \$6322$$

$$\text{Tax2} = 540.371(7) + 2538.905 \text{ dollars} \approx \$6322$$

$$\text{Tax3} = 540.371(103) - 49{,}336.752 \text{ dollars} \approx \$6322$$

All these models give the same output values. They are equivalent. Note that of the three models, the one with the lowest-numbered input (Tax2) has the smallest intercept value. In general, the smaller the magnitude of the input data, the smaller the magnitude of the coefficients.

To summarize, when the input is years, it is desirable to align data to make the data entry faster and to reduce the magnitude of some of the coefficients. The Tax2 model also has the advantage that the constant term ($2538.905) is the approximate beginning amount shown in the table. Two models obtained from different data alignments are equivalent models. Take care to define the input variable accurately.

EXAMPLE 3 *Hepatitis A*

Table 1.50 shows the incidents of hepatitis A per 100,000 people in Clark County, Nevada, in the mid-1990s.[47]

TABLE 1.50

Year	1993	1994	1995	1996	1997
Incidents per 100,000	7.1	12.8	20.2	24.4	32.3

a. Use a calculator or computer to find three linear models for these data by aligning the years in three different ways.

b. Use each model in part *a* to estimate the incidents of hepatitis A in 1999.

c. Is the calculation in part *b* interpolation or extrapolation?

Solution

a. Three possible models are

$H1(t) = 6.2t - 12{,}349.64$ incidents per 100,000 in year t

$H2(t) = 6.2t - 569.64$ incidents per 100,000 t years after 1900

$H3(t) = 6.2t + 6.96$ incidents per 100,000 t years after 1993

Note that in each model, the slope value is the same and the intercept value is different. Also note that the smaller the input values, the smaller the magnitude of the intercept value. The third model has the advantage that the vertical intercept corresponds to the first value shown in the table.

Some students may think that it is improper to report the number of hepatitis A cases to the tenths place, rather than to the nearest integer. Explain to them that this value is calculated as the total number of cases divided by the population size in hundred thousands. Although both of these values must be integers, the resulting quotient does not have to be.

b. To estimate the incidents of hepatitis A in 1999, we substitute 1999 for t in $H1(t)$, 99 for t in $H2(t)$, and 6 for t in $H3(t)$.

$H1(1999) \approx 44.2$ cases per 100,000 people

$H2(99) \approx 44.2$ cases per 100,000 people

$H3(6) \approx 44.2$ cases per 100,000 people

We estimate that in 1999, there were approximately 44.2 cases of hepatitis A per 100,000 people in Clark County, NV.

c. The estimate in part *b* is an extrapolation because 1999 is outside the year values given in the data. This estimate may or may not be valid. If health officials wanted to predict accurately the number of hepatitis A cases in a particular year, they would take many factors other than past incidence into account.

47. Nevada Health Division.

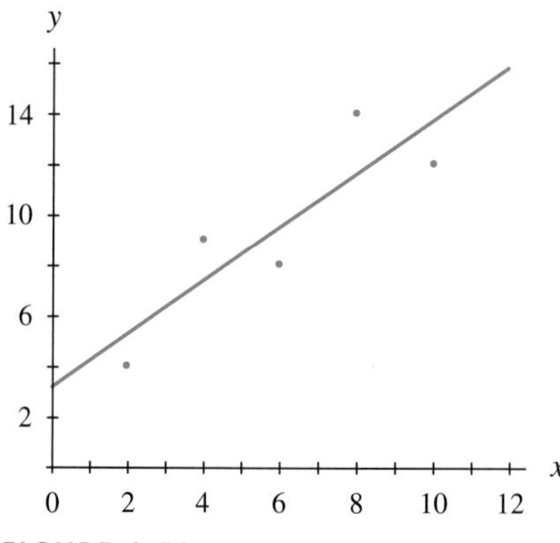

FIGURE 1.50

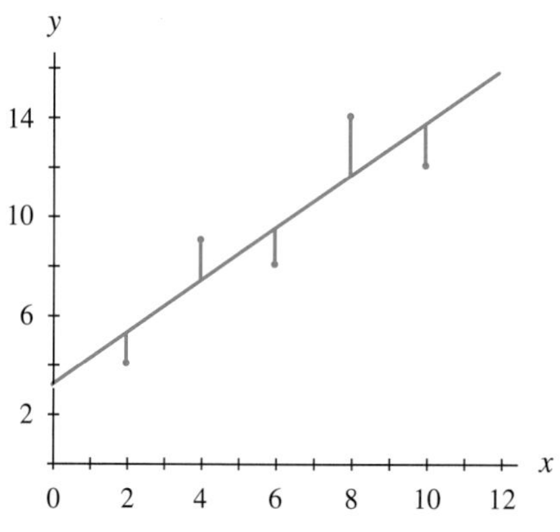

FIGURE 1.51

What Is "Best Fit"?

Whenever you use a calculator or computer to fit a linear equation to data, you should always use the routine that produces the line of best fit. How do we decide that a particular line best fits the data? After all, there are many lines that can be drawn through the data. Figure 1.50 shows a line fit to some data points.

A visual indication of how well the line fits the data is the extent to which the data points deviate from the line. We determine a numerical measure of this visual observation in the following way. Calculate the amount by which the line misses the data; that is, calculate the vertical distance (the *deviation*) of each data point from the line. These deviations are simply the lengths of the vertical segments shown in Figure 1.51. Each deviation measures the *error* between the associated data point and the line.

$$\text{Deviation} = \text{error} = y_{\text{data}} - y_{\text{line}}$$

To obtain a single numerical measure we square the errors and sum:

$$\begin{aligned}\text{SSE} &= \text{the sum of squared errors} \\ &= (\text{first error})^2 + (\text{second error})^2 + \cdots + (\text{last error})^2\end{aligned}$$

This number, denoted by **SSE (sum of squared errors)**, is an overall measure of how well the line fits the data. The following example demonstrates how to calculate SSE.

EXAMPLE 4 *Consumer Price Index*

Table 1.51 shows the consumer price index[48] (CPI) from 1987 through 1997. The CPI is 100 times the ratio obtained by comparing the current cost of a specified group of goods and services to the cost of comparable items determined at an earlier date. An index is a ratio of quantities with the same units; thus CPI is a unitless quantity.

TABLE 1.51

Year	CPI (1982–1984 = 100)
1987	113.6
1989	124.0
1991	136.2
1993	144.5
1995	152.4
1997	160.5

a. Explain the meaning of the CPI of 152.4 in 1995.

b. Find the deviation of each data value from the line $y = 4.7x + 82.3$, where y is the CPI and x is the number of years since 1980.

c. Calculate SSE for $y = 4.7x + 82.3$.

Solution

a. Goods and services that cost \$100 from 1982 to 1984 would have cost \$152.40 in 1995.

48. *Statistical Abstract*, 1998.

b. **TABLE 1.52**

Year	CPI	y-value from model	Deviation
1987	113.6	115.2	1.6
1989	124.0	124.6	0.6
1991	136.2	134.0	-2.2
1993	144.5	143.4	-1.1
1995	152.4	152.8	0.4
1997	160.5	162.2	1.7

c. Obtain SSE by squaring the deviations and summing.

$$\text{SSE} = 2.56 + 0.36 + 4.84 + 1.21 + 0.16 + 2.89 = 12.02$$

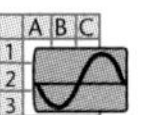

See the *Graphing Calculator Instruction Guide* or the *Calculus Concepts* web site for a program that is helpful in illustrating the method of least squares.

The key idea to the correct use of SSE is this: Smaller values of SSE arise from lines where the errors are small, and larger values of SSE arise from lines where the errors are large. Thus a common strategy for choosing the best-fitting line is to choose the line for which the sum of squared errors (SSE) is as small as possible. Such a line is designated as the line of "best fit," and the procedure for choosing it is called the **method of least squares**. The line obtained by using the method of least squares for the data in Example 4 is $y = 4.686x + 115.105$. The SSE associated with this line is 11.84. This is the smallest SSE possible; therefore, $y = 4.686x + 115.105$ is the line that best fits the CPI data.

Although it is possible to use calculus to find a line of best fit according to the method of least squares, we will use the routine in a calculator or computer whenever we wish to fit a line to data. In the next chapter, we will look at other functions that can be used to fit data. The method of least squares is the accepted method of finding a best-fitting line, but there are other methods of fitting different kinds of functions to data in which SSE does not play a role. Even though it is possible to compute SSE for each of these functions, SSE should *not* be used to compare models of different types. Such comparison is not a valid statistical procedure.

Numerical Considerations

When dealing with numerical results, it is important to understand their accuracy and how precise the results need to be. For instance, imagine that you have a bank account with a balance of \$18,532.71 paying interest at a rate of $17\frac{1}{4}\%$ compounded annually. At the end of the year, the bank calculates your interest to be $(18{,}532.71)(0.1725) = \$3196.892475$. Would you expect your next bank statement to record the new balance as \$21,729.60248? Obviously, when reporting numerical results dealing with monetary amounts, we do not consider partial pennies. The bank reports the balance as \$21,729.60. The required precision is only two decimal places.

Students often will give as many decimal places as shown by their calculator or computer. Emphasize that this is poor mathematics because it ignores the interpretation of the answer in context. We place as much importance on correct interpretation as on a correct numerical answer.

> You should always round numerical results in a way that *makes sense* in the context of the problem.

TABLE 1.53

Year	Net sales (millions of dollars)
1998	115.6
1999	80.6
2000	45.7

If a situation allows us to report answers to the nearest cent, the value \$170.533333333 million is not appropriate because this answer indicates \$170,533,333.333 with the final digit representing a fraction of a cent.

Generally, results that represent people or objects should be rounded to the nearest whole number. Results that represent money usually should be rounded to the nearest cent or, in some cases, to the nearest dollar. Consider, however, a company that reports net sales as shown in Table 1.53

A linear model for these data is $y = -34.95x + 69{,}945.683$ million dollars, where x is the year. If we wished to estimate net sales in 2001, we would substitute $x = 2001$ in the model to obtain $y =$ \$10.73333333 million. Should we report the answer as \$10,733,333.33 or \$10,733,333? Neither! When we are dealing with numerical results, our answer can be only as accurate as the least accurate output data. In this case, the answer would have to be reported as \$10.7 million or \$10,700,000.

> You should round numerical results to the same accuracy as the given output data.

You cannot overemphasize the importance of never rounding intermediate calculations or model coefficients. You should repeatedly stress that, when you round model coefficients in class (on a board or overhead), you are still using all the coefficients when calculating.

Although the correct rounding and reporting of numerical results is important, it is even more important to calculate the results correctly. Because you must sometimes round your answers, it is tempting to think that you can round during the calculation process. Don't! Never round a number unless it is the final answer that you are reporting. Rounding during the calculation process may lead to serious errors. Your calculator or computer is capable of working with many digits. Keep them all while you are still working toward a final result.

When you use your calculator or computer to fit a function to data, it finds the parameters in the equation to many digits. Although it may be acceptable to your instructor for you to round the coefficients when reporting a model, make sure that you use all of the digits while working with the model. This helps reduce the possibility of round-off error.

For example, suppose that your calculator or computer generates the following model equation for a set of data showing weekly profit for an airline for a certain route as a function of the ticket price.

$$\text{Weekly profit} = -0.00374285714285x^2 + 2.5528571428572x - 52.71428571429 \text{ thousand dollars}$$

where x is the ticket price in dollars. In the answer key, you would see the model reported to three decimal places:

$$\text{Weekly profit} = -0.004x^2 + 2.553x - 52.714 \text{ thousand dollars}$$

However, if you used the rounded model to calculate weekly profit, your answers would be incorrect because of round-off error. Table 1.54 shows the inconsistencies between the rounded and unrounded models.

TABLE 1.54

Ticket Price	Profit from rounded model	Profit from unrounded model
\$200	\$298 thousand	\$308 thousand
\$400	\$328 thousand	\$370 thousand
\$600	\$39 thousand	\$132 thousand

> Never use a rounded model to calculate, and never round intermediate answers during the calculation process.

It would be extremely tedious for everyone if we reported all the digits found by all possible models of graphing calculators or computer software or even listed all digits for one particular type of technology! We therefore adopt the following convention for reporting models and statements of models in the text:

> We feel it is important to print the fewest digits possible under the condition that (a) the rounded model visua. fits the data, and (b) the rounded model gives values fairly close to answers obtained with the full model. In particular, if there is a difference between the results from the full model and those from the rounded model, that difference should appear only in the last digit for which we can claim accuracy.

Thus the number of decimal places shown in models in the text will vary. However, whenever we calculate with a model, we use all the digits available at that point in the text. If the data are given, the unrounded model will be used in calculations. When a model found earlier is used in a later section and the data are not repeated, all calculations will be done with the rounded model given in that later section. For convenience and consistency, the answer key will report all models with three decimal places in coefficients and six decimal places in exponents.

Also keep in mind that when you are writing models, it is important to write down what the input and output variables represent. Look back to the models we obtained for the small business tax. The model written as Tax2 $= 540.37143t + 2538.905$ is incomplete because no label is given to indicate the units on the tax. Is the tax measured in dollars, cents, or thousands of dollars? Tax is *what* is measured, and dollars is a label telling *how* it is measured. Always label a model equation with output units to tell how the output is measured.

It is equally important to describe the input variable. If all that had been written was Tax2 $= 540.37143t + 2538.905$ dollars, you might have erroneously predicted that your federal taxes in 2003 would be $540.37143(103) + 2538.905 = \$58{,}197$ or even $540.37143(2003) + 2538.905 = \$1{,}084{,}903$! However, there is a statement with the equation that reads "where t is the number of years since 1996." Because 2003 is the 7th year since 1996, you should correctly predict your federal taxes in 2003 to be $540.37143(7) + 2538.905 = \$6322$.

The tax model was obtained from data between 1996 and 2001. It is important, when giving a model that will be used without the data from which it was obtained, that you indicate the interval of input values over which the model is valid. This is the only way that other people who use the model will know whether they are interpolating or extrapolating. Here is the convention we will adopt: When a model is presented without the data from which it was obtained, we will indicate the range of input data used. For example, we would report the tax model by saying, "The tax between 1996 and 2001 can be modeled as Tax $= 540.37143t + 2538.905$ dollars t years after 1996." This statement incorporates the four elements of a model.

Be merciless in requiring that your students incorporate all these elements when writing a model. Accurate and effective communication of results is an extremely important job skill. A good quiz to give the day after you cover this section is one that asks the student a question such as, "The current temperature is 72°F and is increasing at a constant rate of 1.5°F per hour over the next 4 hours. Write a model for the temperature h hours from now." When grading the quiz, give no credit for an answer of $T(h) = 72 + 1.5h$. This answer is an equation, not a model. Grading in this manner should convince your students that you (and we, the authors) take the matter of properly labeling a model very seriously. Of course, the correct answer should be of the general form "$T(h) = 72 + 1.5h$ °F h hours from now. This model applies for values of h from 0 to 4."

There are four important elements of every model:

1. An equation
2. A label denoting the units on the output
3. A description (including units) of what the input variable represents
4. An indication of the interval of input values over which the model is valid. This information should be given whenever the model is presented or would be used apart from the data from which it was obtained.

Finally, keep in mind that a number by itself is likely to be absolutely worthless as an answer to a question. Consider the reaction of a manager to a memo that reads

> Sykes:
>
> 30
>
> Paul

Sykes would probably have no idea what Paul was attempting to communicate. Suppose the next memo from Paul read

> Sykes:
>
> Regarding your question about how much overtime my division puts in per week, we typically put in 30.
>
> Paul

Again, Sykes would probably not understand. 30 what? 30 minutes? 30 hours? If Paul had 60 people in his division, it might even mean 30 days. *A number is useless without a label that clearly indicates the units involved!*

In summary, keep in mind the following general guidelines for numerical results.

Guidelines for Numerical Results

1. When *using* a calculator- or computer-generated model equation to obtain answers, do not round the parameters in the model during the calculation process. When you are *reporting* the model, it may be acceptable to your instructor if you round the parameters to three or four decimal places.
2. When reporting a model, always label units on the output variable and state what the input variable in the model represents. A model is meaningless without a clear description of the variables.
3. If a model is presented without the data from which it was obtained, indicate the interval of input values of the data. This allows someone using the model to know whether they are interpolating or extrapolating.
4. Numerical results are worthless if you do not understand what they mean. The first step in understanding is to state the units clearly on all numerical answers. Numerical answers are incomplete without appropriate units.
5. Round final numerical results in a way that *makes sense* in the context of the problem.
6. When you are dealing with numerical results, an answer can be only as accurate as the least accurate piece of output data.

1.5 Concept Inventory

- *Algebraic form of a linear model: $f(x) = ax + b$*
- *Rate of change (slope) of a linear model is constant*
- *Calculating slope as $\frac{rise}{run}$*
- *First differences*
- *Aligning data*
- *Interpolating and extrapolating*
- *The method of least squares*
- *Rounding rules*
- *Labeling units on answers*
- *Four elements of a model*

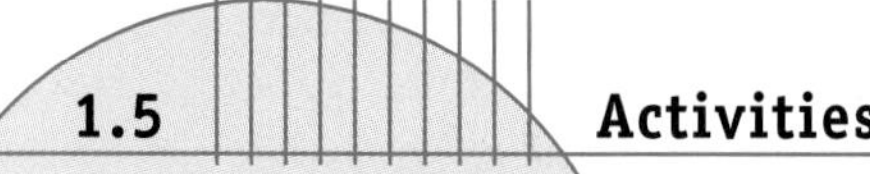

1.5 Activities

1. The graph in Figure 1.5.1 shows a corporation's profit in millions of dollars over a period of time.
   a. Estimate the slope of the graph, and write a sentence explaining the meaning of the slope in this context.

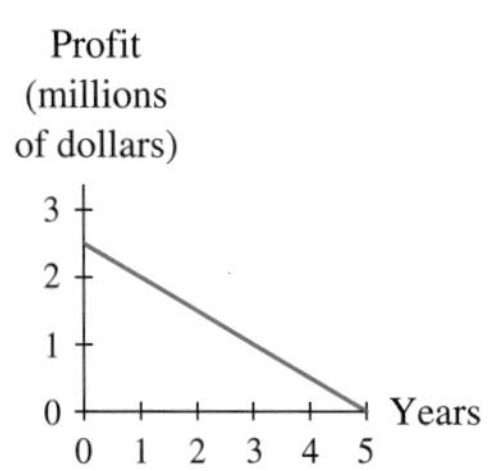

FIGURE 1.5.1

   b. What is the rate of change of the corporation's profit during this time period?
   c. Identify the horizontal and vertical axis intercepts, and explain their significance to this corporation.

2. The air temperature in a certain location from 8 a.m. to 3 p.m. is shown in the graph in Figure 1.5.2.
   a. Estimate the slope of the graph, and explain its meaning in the context of temperature.
   b. How fast is the temperature rising between 8 a.m. and 3 p.m.?

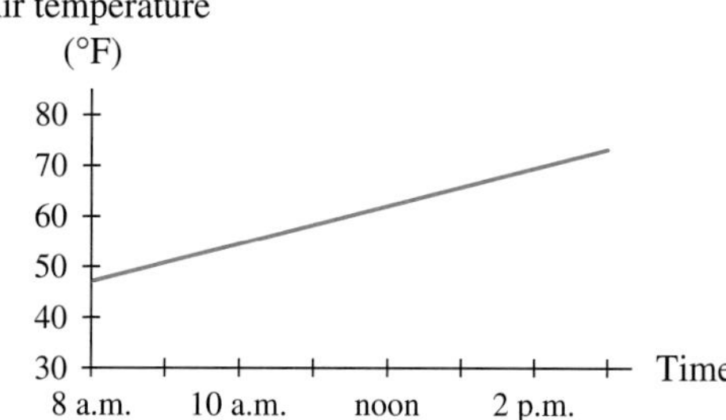

FIGURE 1.5.2

c. What would be the meaning of a horizontal axis intercept for this temperature graph?

3. The number of organ donors[49] in the United States between 1988 and 1996 can be modeled by

$$D(t) = 382.5t + 5909 \text{ donors}$$

t years after 1988.

a. According to the model, what is the rate of change of the number of organ donors?

b. Sketch a graph of the model. What is the slope of the graph?

c. Find the vertical axis intercept, and explain its significance in context.

4. Total Chapter 13 bankruptcy filings[50] between 1994 and 1997 can be modeled by

$$B(t) = 48.6t + 240 \text{ thousand filings}$$

t years after 1994.

a. What is the rate of change of the number of Chapter 13 bankruptcy filings?

b. Sketch a graph of the model. What is the slope of the graph?

c. Find the vertical axis intercept, and explain its significance in context.

5. The population of a certain country can be modeled by $P(t) = 6.98t + 12.39$ million. Estimate the population of that country in 2007. Activities 5 and 6 cannot be answered using the information given.

6. The absorption of a certain drug into the body can be modeled by $A(x) = -0.076x + 0.3$ grams. Find the amount of this drug that has been absorbed 3 hours after the drug was taken.

7. The revenue for International Game Technology[51] was \$744.0 million in 1997 and \$824.1 million in 1998. Assume that revenue was increasing at a constant rate.

a. Find the rate of change of revenue.

b. By how much did revenue increase each quarter of 1998?

c. Assuming that the rate of increase remains constant, complete Table 1.55.

TABLE 1.55

Year	Revenue (millions of dollars)
1997	
1998	
1999	
2000	

d. Find an equation for revenue in terms of the year.

8. Suppose you bought a Mazda Miata in 1999 for \$24,000. In 2001 it was worth \$18,200. Assume that the rate at which the car depreciates is constant.

a. Find the rate of change of the value of the car.

b. Complete Table 1.56.

TABLE 1.56

Year	Value
1999	
2001	
2002	
2003	
2004	

c. Find an equation for the value in terms of the year.

d. How much will the value of the car change during a 1-month period? Round your answer to the nearest dollar.

9. Table 1.57 gives the number of gallons of oil in a tank used for heating an apartment complex t days after January 1 when the tank was filled.

TABLE 1.57

t	Oil (gallons)
0	30,000
1	29,600
2	29,200
3	28,800
4	28,400

49. Based on data from United Network for Organ Sharing.
50. Based on data from Administrative Offices of U.S. Courts.
51. International Game Technology Annual Report, 1998.

a. What is the rate of change of the amount of oil?

b. How much oil can be expected to be used during any particular week in January?

c. Predict the amount of oil in the tank on January 30. What assumptions are you making when you make predictions about the amount of oil? Are the assumptions valid? Discuss.

d. Find and graph an equation for the amount of oil in the tank.

e. When should the tank be refilled?

10. A jewelry supplier charges a \$55 ordering fee and \$2.50 per pair of earrings regardless of the order size.

a. If a retailer orders 100 pairs of earrings, what is the cost?

b. What is the average cost per pair of earrings when 100 pairs are ordered?

c. If the retailer marks up the earrings 300% of the average cost, at what price will the retailer sell the 100 pairs of earrings? (Sale price = cost + 300% of cost)

d. Find an equation for cost as a function of the number bought.

e. Find an equation for the average cost per pair as a function of the number bought.

f. Find an equation for the sale price based on a 300% markup.

g. If 100 pairs of earrings are purchased, how many must the retailer sell at a 300% markup to recover the cost of the earrings (break-even point)?

11. Table 1.58 shows the amount of motor vehicle taxes[52] collected by a certain county between 1994 and 1998.

TABLE 1.58

Year	1994	1995	1996	1997	1998
Amount collected (millions of dollars)	108	122	137	155	169

a. Calculate the first differences of the collected amounts. Would you classify the differences as constant, nearly constant, or neither?

b. Find three models for the data using three different alignments.

c. Estimate the amount collected in 2000.

d. According to your models, when will the amount collected be \$300,000,000? Under what conditions will your estimate be accurate?

12. In 1996 the average ATM transaction fee for U.S. banks[53] was \$0.97. In 1999 the average fee was \$1.37.

a. Assuming that ATM fees increased linearly between 1996 and 1999, calculate the rate of change in the average fee.

b. By aligning the years differently, find three linear models for the amount of the average ATM fee between 1996 and 1999.

c. Use your models to estimate the average fee in 1998 and 2003.

d. Are the estimates in part *c* interpolations or extrapolations?

13. Table 1.59 shows the enrollment for a university in the southeast from 1965 through 1969.

TABLE 1.59

Year	1965	1966	1967	1968	1969
Students	5024	5540	6057	6525	7028

a. Find three linear models for these data by aligning the years differently.

b. Use all three models to estimate the enrollment in 1970.

c. The actual enrollment in 1970 was 8038 students. How far off was your estimate? Do you consider the error to be significant or insignificant? Explain your reasoning.

d. Would it be wise to use these models to predict the enrollment in the year 2000? Explain.

14. You and several of your friends decide to mass-produce "I love calculus and you should too!" T-shirts. Each shirt will cost you \$2.50 to produce. Additional expenses include the rental of a downtown building for a flat fee of \$675 per month, utilities estimated at \$100 each month, and leased equipment costing \$150 per month. You will be able to sell the T-shirts at the premium price of \$14.50 because they will be in such great demand.

a. Give the equations for monthly revenue and monthly cost as functions of the number of T-shirts sold.

52. Nevada Legislative Counsel Bureau.

53. U.S. Public Interest Research Group.

b. How many shirts do you have to sell each month to break even? Explain how you obtained your answer.

15. The price[54] of tickets to a college's home football games is given in Table 1.60.

TABLE 1.60

Year	Price (dollars)	Year	Price (dollars)
1981	10	1989	16
1983	12	1991	18
1985	13	1994	20
1987	15		

a. Find a linear model for the price of tickets.

b. Use the model to find the ticket price in 1984, 1992, and 1999. (Round to the nearest dollar.) Were you interpolating or extrapolating to find these prices?

c. In 1999, tickets cost $25. Was your prediction accurate?

d. Suppose you graduate, get married, and have a child who attends that college. Predict the price you will have to pay to attend a football game with your child during his or her freshman year. Describe in detail how you arrived at your answer.

e. Repeat part *d* assuming that you have a grandchild who attends the same college.

f. What assumptions are you making when you use the linear model to make predictions? Are these assumptions reasonable over long periods of time? Explain.

g. What conclusions can you draw about the validity of extrapolating from a model?

16. You are an employee in the summer at a souvenir shop. The souvenir shop owner wants to purchase 650 printed sweatshirts from a company. The catalog contains a table and directions to call the company for costs for orders more than 350. The catalog table is shown in Table 1.61.

The shop owner, who has tried unsuccessfully for a week to contact the company, asks you to estimate the cost for 650 shirts.

TABLE 1.61

Number purchased	Total cost (dollars)	Number purchased	Total cost (dollars)
50	250	250	700
100	375	300	825
150	500	350	950
200	600		

a. Find a linear model to fit the data.

b. Use the model to predict the cost for 650 shirts. Note that all costs in the table are integer multiples of $25.

c. Determine the average cost per shirt for 650 shirts.

d. The shop owner is preparing a newspaper advertisement to be published in a week. If the standard markup is 700%, what should the advertised price be?

e. How many of the 650 shirts will need to be sold at the price determined in part *d* in order to pay for the cost of all 650 shirts (break-even point)?

17. The percentage[55] of funding for public elementary and secondary education provided by the federal government during the 1980s is given in Table 1.62.

TABLE 1.62

School year	% Funding	School year	% Funding
1981–82	7.4	1986–87	6.4
1982–83	7.1	1987–88	6.3
1983–84	6.8	1988–89	6.2
1984–85	6.6	1989–90	6.1
1985–86	6.7		

a. Find a linear model for the percentage of funding.

b. What is the rate of change of the model you found?

c. On the basis of the model, estimate the percentage of funding provided by the federal government in the 1993–94 school year.

d. Judging by the model, determine in what school year federal government funding will first be less than 5% of educational funding.

18. Consumer credit[56] in the United States was $838.6 billion in 1993 and $1235.8 billion in 1997.

54. Clemson University Athletic Department.

55. Based on data from *Statistical Abstract*, 1992 and 1994.

56. *Statistical Abstract*, 1998.

Assume that consumer credit increases at a constant rate.

a. Find the rate of increase.

b. On the basis of the rate of increase, estimate consumer credit in 2000.

c. Is the assumption that consumer credit increases at a constant rate valid? Explain.

d. Assuming a constant rate of change, when will consumer credit reach $2 trillion?

e. Do you expect consumer credit to reach $2 trillion sooner or later than your answer in part *d* indicates? Why?

19. A house sells for $73,000 at the end of 1990 and for $97,500 at the end of 2002.

a. If the market value increased linearly from 1990 through 2002, what was the rate of change of the market value?

b. If the linear increase continues, what will the market value be in 2005?

c. In what year might you expect the market value to be $75,000? $100,000?

d. Find a model for the market value. What does your model estimate for the market value in 1999? Do you believe this estimation? What assumption was made when you created the model? Is this assumption necessarily true?

20. Twenty-eight percent of the births[57] in 1990 were to unmarried women. The percent in 1995 was 32.

a. Find the rate of change of the percentage of births, assuming that it is constant.

b. Estimate the percentage of births to unmarried women in 1996.

21. Table 1.63 shows year 2001 first-class domestic postage[58] for mail up to 9 ounces.

TABLE 1.63

Weight not exceeding	Postage
1 oz	$0.34
2 oz	0.55
3 oz	0.76
4 oz	0.97
5 oz	1.18
6 oz	1.39
7 oz	1.60
8 oz	1.81
9 oz	2.02

a. Draw a scatter plot of these data to determine visually whether a linear model is appropriate.

b. Verify your observations in part *a* by calculating first differences in the postal rates.

c. Find a formula for the postage in terms of weight. Be specific about what the variables represent.

22. The 1999–2000 tuition changes[59] for students attending a certain university in the southeast are given in Table 1.64. A full-time student is a student who enrolls in 12 or more credit hours.

TABLE 1.64

Schedule of academic charges	Resident	Nonresident
Full-time academic fee	$1735	$4728
Part-time academic fee (per credit hour)	138	388
Auditing academic fee (per credit hour)	69	194
Graduate assistant fee	512	512

Find formulas for the amount of tuition that each of the following students pays when she or he registers for x credit hours.

a. Resident student

b. Nonresident student

c. Graduate assistant

23. Explain the difference between using a model for interpolation and using it for extrapolation. Does interpolation always give an accurate picture of what is happening in the real world? Does extrapolation? Why or why not?

24. The population of West Virginia[60] from 1985 through 1993 is shown in Table 1.65.

TABLE 1.65

Year	Population (thousands)	Year	Population (thousands)
1985	1907	1990	1793
1986	1882	1991	1798
1987	1858	1992	1806
1988	1830	1993	1816
1989	1807		

57. *Statistical Abstract*, 1998.

58. United States Postal Service.

59. Clemson University *Undergraduate Announcements.*

60. *Statistical Abstract*, 1998.

a. Observe a scatter plot of the data. What year is the dividing point that should be used to create a piecewise continuous function?

b. Divide the data in the year you determined in part *a*. Include the dividing point in both data sets. Find linear models for each set of data, and write the function in correct piecewise continuous function notation. Compare your model with the one given in Example 6 in Section 1.3.

c. Use your model to estimate the population in 1997. Compare your estimate with the actual population of 1,816,000. What does this comparison tell you about extrapolation?

25. The population of North Dakota[61] between 1985 and 1996 is shown in Table 1.66.

TABLE 1.66

Year	Population (thousands)	Year	Population (thousands)
1985	677	1993	635
1987	661	1994	640
1989	646	1995	641
1991	633	1996	643

a. Observe a scatter plot of the data. What year is the dividing point that should be used to create a piecewise continuous function?

b. Divide the data in the year you determined in part *a*. Include the dividing point in both data sets. Find linear models to fit each set of data, and write the function in correct piecewise continuous function notation.

c. Use your model to estimate the population in 1997. Compare your estimate to the actual population of 641,000. Why is this estimate significantly better than the one in part *c* of Activity 24?

26. Membership in the Girl Scouts[62] of the U.S.A. in selected years from 1970 through 1992 is shown in Table 1.67.

a. Look at a scatter plot of the data. Describe the behavior of the data from 1970 through 1980 and from 1985 through 1992.

b. Find linear models for the 1970–1980 data and for the 1985–1992 data. Write the two models as one piecewise model.

TABLE 1.67

Year	Membership (thousands)	Year	Membership (thousands)
1970	3922	1988	3052
1975	3234	1989	3166
1980	2784	1990	3269
1985	2802	1991	3383
1986	2917	1992	3510
1987	2947		

c. Use the models to estimate the membership in 1973 and in 1993. Categorize your estimates as interpolation or extrapolation.

27. Chlorofluorocarbons (CFCs) released into the atmosphere are believed to thin the ozone layer. In the 1970s, the United States banned the use of CFCs in aerosols, and the atmospheric release of CFCs declined. The decline was temporary, however, because of increased CFC use in other countries. In 1987, the Montreal Protocol, calling for a phasing out of all CFC production, was ratified. Table 1.68 shows atmospheric release[63] of CFC-12, one of the two most prominent CFCs, from 1974 through 1992.

TABLE 1.68

Year	CFC-12 Atmospheric release (millions of kilograms)
1974	418.6
1976	390.4
1978	341.3
1980	332.5
1982	337.4
1984	359.4
1986	376.5
1988	392.8
1990	310.5
1992	255.3

a. Look at a scatter plot of the data. Does the behavior of the data seem to support the foregoing description of the atmospheric release of CFCs since the 1970s? Explain.

61. *Statistical Abstract*, 1998.
62. *Statistical Abstract*, 1994.

63. *The True State of the Planet*, ed. Ronald Bailey (New York: The Free Press, for the Competitive Enterprise Institute, 1995).

b. Use three linear models to create a piecewise model for CFC-12 atmospheric release.

c. Sketch a graph of the data and the three-piece model.

d. Use your model to answer the questions in the chapter opener on page 2.

28. To promote Be Kind To Animals Week®, the American Humane Association ran a notice in the *Anderson Independent-Mail* in the first week of May 1993. The notice read, "Every hour in the United States, more than 2000 dogs and 3500 cats are born. . . . Add these animals to an existing pet population of 54 million dogs and 56 million cats and the total exceeds one billion!" For the questions that follow, consider only dog and cat births, not deaths.

a. What is the rate of change of the number of dogs in the United States?

b. What is the rate of change of the number of cats in the United States?

c. Find an equation for the number of dogs t days after the beginning of 1993.

d. Find an equation for the number of cats t days after the beginning of 1993.

e. How many total animals (cats and dogs) do you estimate there were at the end of 1993? Does this agree with the prediction in the article?

29. In 1999, the U.S. Bureau of the Census estimated that the world's population had reached 6 billion people. A newspaper article[64] about the population stated that "Despite a gradual slowing of the overall rate of growth, the world population is still increasing by 78 million people a year. . . . [T]he number of humans on the planet could double again to 12 billion by 2050 if the current growth rate continues."

a. According to the article, what is the rate of change of the world's population?

b. Find a linear model for the population of the world, assuming that the population was 6 billion at the beginning of 2000.

c. According to the model in part *b*, when will the world's population be 12 billion? Does this agree with the estimate given in the article?

d. What assumption did you make when you made the prediction in part *c*? What can you conclude about the assumptions made by the Census Bureau in making their estimate of when the population would reach 12 billion?

Summary

Mathematical Modeling

Mathematical modeling is the process by which we construct a mathematical framework to represent a real-life situation. We analyze this framework with mathematical methods and techniques and then translate and interpret the results back into the context of the real-life situation.

As we apply it in this book, mathematical modeling usually means fitting a line or curve to data. The resulting equation, including output label and input description, which we often refer to as the mathematical model, provides a representation of the underlying relationship between the variable quantities of interest.

You should keep in mind that the data from which the models are constructed are sometimes incomplete, often of a size that is smaller than we would prefer, and perhaps of questionable accuracy. Thus the models themselves necessarily reflect these shortcomings. Nevertheless, the models still convey important and valuable information.

Functions and Graphs

The concept of a function is a fundamental mathematical idea that arises in almost every area of mathematics. Informally, a function is a description of how one thing changes (output) as something else changes (input).

64. "Population of world ready to hit 6 billion," *Chicago Tribune*, October 21, 1999, p. A1.

More precisely, a function is simply a rule (a formula, prescription, or procedure) that assigns to each input exactly one output. If two different outputs are assigned to a single input, then the assignment rule is not a function. Examples of functions are everywhere, and we encounter functions in four ways: through tables of data, graphs, simple word descriptions, and mathematical equations.

Discrete and Continuous

The word discrete is used to describe data and graphs of data, which are called scatter plots. It also refers to input/output relationships in which the input consists of certain distinct values rather than an interval of values. The word continuous refers to a graph that is an unbroken curve whose set of inputs includes all the values in an interval. We also call the functions that describe such graphs continuous.

We often begin with data and the associated scatter plot and build a continuous model to describe the underlying situation. Although we can construct continuous models from discrete data, a model may have meaningful interpretation for only certain input values. We call such models continuous with discrete interpretation. Great care must be taken when working with continuous models with discrete interpretation.

At other times a situation is actually continuous—such as the temperature at any given time throughout the day. We use a continuous model in such cases and use it without restriction on the input.

Constructed Functions

The easiest way to combine two functions is to add or subtract them. In order for this to make sense, we must first be certain that both functions have the same set of inputs and that their outputs are given in the same units. Product functions are obtained by multiplying outputs. Again, this gives a meaningful result only when both functions have the same set of inputs and the output units are compatible.

Function composition is a more involved way of combining functions. In order to form the composite $g \circ f$ of two functions f and g (note that f is the inside function and g is the outside function), we require that the outputs of f be used as inputs to g. Thus, the inputs to $g \circ f$ are the inputs to the inside function f. Outputs from $g \circ f$ are outputs from the outside function g.

Inverse functions are created by reversing the input and output of a function. Sometimes we can algebraically solve for the input of a function in terms of the output in order to find an algebraic expression for an inverse function. At other times, a new function is defined as the inverse function. For example, the inverse of the exponential function $y = e^x$ is called the natural log function and is denoted $y = \ln x$. When a function and its inverse are combined to form a composite function, the result is simply the input variable.

Portions of two or more continuous functions can be combined to form piecewise continuous functions. In piecewise continuous functions, the outputs over different intervals of inputs are determined by different functions. When dealing with piecewise continuous functions, be sure that you use the correct portion of the function when determining the output associated with a given input.

Limits

Limits are a tool used to describe the behavior of functions. They also form a basis for our discussion of calculus concepts in subsequent chapters. A limit is a value to which the output of a function becomes closer and closer as the input becomes closer and closer to a given number. Limits also can describe what happens to the output of a function as the input becomes increasingly large in the positive or negative direction. We call this the end behavior of a function.

We can estimate limits graphically or numerically and can, in some cases, calculate limits algebraically. If a function is discontinuous, care must be taken at the point of discontinuity to consider the limit from the right and the limit from the left. The limit at the point of discontinuity exists only if the limits from the right and the left are equal at that point.

Linear Functions and Models

A linear function models a constant rate of change. Its underlying equation is that of a line:

$$y = ax + b$$

In terms of the line, the parameter a is called the slope of the line and is calculated as

$$\text{Slope} = \frac{\text{rise}}{\text{run}}$$

Because the slope of a line is a measure of its rate of increase or decrease, the slope is also known as the rate of change for the linear model. The parameter b appearing in the linear model $y = ax + b$ is simply the output of the model when the input is zero.

When inputs to a linear model are evenly spaced, the corresponding outputs have constant first differences. In fact, this feature characterizes linear models. Although we rarely encounter real-life situations where the first differences are perfectly constant, it is not uncommon to find nearly constant first differences in processes that are devoid of external influences. This accounts for the widespread occurrence of linear models.

Numerical Considerations

In dealing with numerical results, it is always important to consider the accuracy and meaning of these results. Round the numerical results to the same accuracy as the given data and only in a way that makes sense in the context of the particular problem. Serious errors can influence results when rounding is done during the calculation process or when a rounded model is used to calculate results. Every model has four key elements: an equation, output units, a description of what the input variable represents, and a statement indicating over what interval of input values the model is valid. Without these elements, a model is incomplete.

The Role of Technology

In order to construct mathematical models from data, we must use appropriate tools. Normally, these tools are graphing calculators or personal computers. To attempt to build the models without these tools—by paper-and-pencil methods only—is often impractical.

You should clearly understand that our use of technology will simply be a tool in the service of mathematics and that no tool is a substitute for clear, effective thinking. Technology carries only the graphical and numerical computational burden. You will have to perform the mathematical analyses, interpret the results, make the appropriate decisions, and then communicate your conclusions in a clear and understandable manner.

The *Concept Check* provides students a quick way of checking on which topics they need to spend more time. The activities listed here are generally simple ones, and the student should be warned that being able to do these activities does not guarantee that they fully understand all the concepts—especially not if they only read through the activity and look at the answer in the back of the book without actually working through the solution themselves.

Concept Check

Can you	To practice, try	
• Identify functions?	Section 1.1	Activities 1, 5, 11, 13, 17
• Correctly interpret graphs?	Section 1.1	Activity 27
• Understand business terms?	Section 1.1	Activity 28
• Accurately work with functions?	Section 1.1	Activities 33, 37
• Distinguish between discrete and continuous functions?	Section 1.2	Activities 1–11 odd
• Correctly work with continuous functions with discrete interpretation?	Section 1.2	Activity 15
• Combine two functions?	Section 1.3	Activities 3, 7
• Find and use inverse functions?	Section 1.3	Activity 37
• Estimate limits graphically?	Section 1.4	Activity 5
• Estimate limits numerically?	Section 1.4	Activities 25, 29
• Find limits algebraically?	Section 1.4	Activities 15, 43
• Determine continuity?	Section 1.4	Activity 9
• Interpret the parameters of a linear model?	Section 1.5	Activity 3
• Construct and work with a linear model?	Section 1.5	Activity 11
• Construct and work with a piecewise-continuous model?	Section 1.5	Activity 25

Review Test

You may wish to discuss with your students how one of your tests may differ from this practice test in terms of length, problem selection, and problem difficulty. In some classes, this review test will be excellent preparation for the actual test, but in other classes, it will not be good preparation.

1. The graph in Figure 1.52 shows the yearly gain of wetlands[65] in the United States between 1987 and 1994.

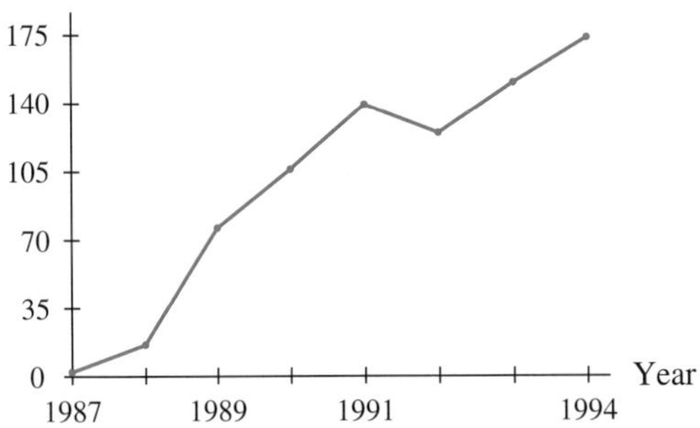

FIGURE 1.52

   a. Did the number of acres of wetlands in the United States increase or decrease between 1991 and 1992? Explain.

   b. Estimate the slope of the portion of the graph between 1989 and 1991. Interpret your answer.

2. Let $T(d)$ be the number of tickets sold by the Majestic Theater box office for the Broadway musical *Phantom of the Opera* on the dth day of 2000.

   a. Draw and label an input/output diagram for T.

   b. Is T a function of d? Why or why not?

   c. When the inputs and outputs of T are reversed, is the result an inverse function? Why or why not?

3. Of the three top causes of death (heart disease, cancer, and stroke), cancer is the only one for which the death rate is increasing. According to the National Cancer Institute, the percent of the U.S. population dying from cancer in 1973 was 18% and in 1995 was 23%. Assume that the percent of the population dying from cancer between 1973 and 1995 increased at a constant rate.

   a. Find the rate of increase of the percent of deaths due to cancer between 1973 and 1995.

   b. On the basis of the increase found in part *a*, estimate the percent of cancer deaths in 2002. Under what conditions is this estimate valid?

4. Consider the function

$$f(x) = \begin{cases} 2x + 5 & \text{when } x < 1 \\ -2x + 9 & \text{when } 1 < x < 2 \\ \dfrac{2(x^2 - x - 2)}{x^2 + x - 6} & \text{when } x \geq 2 \end{cases}$$

   a. Determine the following limits:

   i. $\lim_{x \to 1} f(x)$ ii. $\lim_{x \to 2} f(x)$ iii. $\lim_{x \to \infty} f(x)$

   b. Is f continuous at $x = 1$ and $x = 2$? Explain why or why not.

5. In *The True State of the Planet*, the statement is made that "the amount of arable and permanent cropland worldwide has been increasing at a slow but relatively steady rate over the past two decades." Selected data[66] between 1970 and 1990 reported in *The True State of the Planet* are shown in Table 1.69.

TABLE 1.69

Year	Cropland (millions of square kilometers)
1970	13.77
1975	13.94
1980	14.17
1985	14.31
1990	14.44

   a. Find a linear model for the data.

   b. Is the model in part *a* discrete, continuous with discrete interpretation, or continuous without restriction?

   c. What is the rate of change of your model in part *a*? Write a sentence interpreting the rate of change.

   d. Do the data and your model support the statement quoted above? Explain.

   e. According to your model, what was the amount of arable and permanent cropland in 1995?

65. *The True State of the Planet*, ed. Ronald Bailey (New York: The Free Press, for the Competitive Enterprise Institute, 1995).
66. Ibid.

Project 1.1 Tuition Fees

Specific information regarding the implementation of projects can be found in the *Instructor's Resource Guide* and at the *Calculus Concepts* web site.

Setting

Nearly all students pursuing a college degree are classified as either part-time or full-time. Many colleges and universities charge tuition for part-time students according to the number of credit hours they are taking. They charge full-time students a set tuition regardless of the number of credits they take, except possibly in an overload situation.

Tasks

1. Find the tuition charges at your college or university. Some schools have different tuition rates for different classifications of students (for instance, residents may be charged a different tuition than nonresidents). Pick one classification to model. Write a piecewise linear model for tuition as a function of the number of credit hours a student takes. Consider as part of the college tuition charges such as matriculation and activity fees that all students must pay.
2. Use your model to generate a table of tuition charges for your school. Compare your table with the published tuition charges. Are there any discrepancies? If so, explain why they might have occurred.
3. Even though you would not pay tuition if you registered for 0 credit hours, your model may have an interpretation at this point. What is that interpretation?

Reporting

Write a letter addressed to the Committee on Tuition. You should explicitly define to whom the model applies and what the variables in the model represent. Include an explanation of how to use the model, a statement about which input values are valid and which are invalid for use in the model, and a discussion of the preceding tasks.

Project 1.2 Finding Data

Setting

Newspapers, journals, the Internet, and government documents are good sources of data.

Tasks

1. Find a set of data with more than 7 data points that would be reasonably modeled by a linear function.
2. Find a linear model to fit the data.
3. What do your function's slope and intercept mean in the context of the data?

Reporting

1. Write a report in which you describe the meaning of the data, discuss the linear model you used to fit the data, and answer the question posed in Task 3. In your report, you should explicitly state the model, define the variables in the function, and state what input values are valid for this function. You should also include a scatter plot and graph of your function. Properly cite the source of your data using correct bibliographic form. Attach a photocopy or a printout of the data and the page on which the title of the article or document appears.
2. Prepare a 3- to 5-minute oral presentation of your report. You should incorporate the use of visual aids to enhance your oral presentation.

2

Ingredients of Change: Nonlinear Models

Concept Application

Eric Fowke/PhotoEdit

The national gender ratio is the number of males per 100 females in the United States. Factors such as immigration, war, and advances in health care affect the gender ratio. The Census Bureau keeps track of gender ratio data. By examining the behavior of the data, we can choose an appropriate function to model the gender ratio. The model then can be used to answer questions about the gender ratio such as

- For what age is the gender ratio 100?
- For what age are there twice as many women as men?
- What is the expected future trend in the gender ratio?

Answers to questions such as these are important to government agencies, health care providers, retailers, and others. The information needed to answer these questions is found in Activities 27 and 29 of Section 2.4.

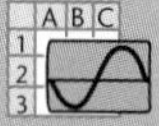
This symbol directs you to the supplemental technology guides. The different guides contain step-by-step instructions for using graphing calculators or spreadsheet software to work the particular example or discussion marked by the technology symbol.

This symbol directs you to the Calculus Concepts *web site, where you will find algebra review, links to data updates, additional practice problems, and other helpful resource material.*

Concept Objectives

This chapter will help you understand the concepts of

- Exponential and logarithmic models
- Compound interest, present value, and future value
- Exponential growth and decay
- Logistic models
- Quadratic and cubic models
- Concavity and inflection points

and you will learn to

- Find one of six equations to fit data
- Combine two equations to form a piecewise continuous model
- Use models to answer many questions
- Answer financial investment questions
- Work with exponential growth and decay models
- Choose an appropriate model on the basis of a scatter plot and how the model will be used

Although linear functions and models are among the most frequently occurring ones in nonscience settings, nonlinear models apply in a variety of situations.

This chapter begins with a study of exponential and logarithmic functions and includes a section on their applications. We also consider situations in which various influences act to restrict exponential growth in some way. Such situations often can be modeled by an S-shaped curve known as a logistic function. The final functions we consider are quadratic and cubic, along with their applications to real-world situations.

We conclude the chapter with a summary of the steps taken in examining a set of data and determining which model is best suited to describe the data and answer questions about changes in the underlying context.

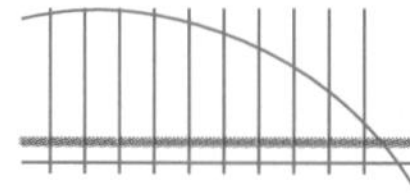

2.1 Exponential and Logarithmic Functions and Models

In Section 1.5 we studied linear functions whose output resulted from the repeated *addition* of a constant at regular intervals. We now turn to a function whose output is the result of repeated *multiplication* by a constant at regular intervals.

You may not realize it, but the balance of a savings account is determined by repeated multiplication. For example, suppose that you make an initial deposit of \$100 into an account that pays 5% per year with interest compounded annually, and you make no other deposits and no withdrawals. You will have yearly balances at the anniversary dates of your account as shown in Table 2.1.

TABLE 2.1

Year	0	1	2	3	4
Balance	\$100.00	\$105.00	\$110.25	\$115.76	\$121.55

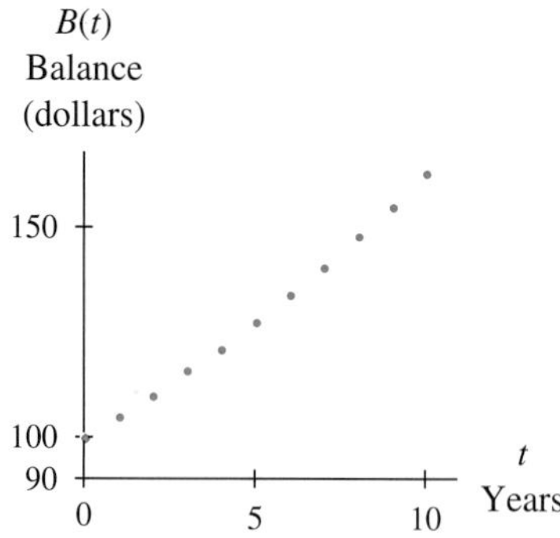

FIGURE 2.1

A little mathematical investigation reveals that each balance is 1.05 times the prior balance. This suggests that the mathematical model that connects the balance B and t, the number of years since the start of the account, is

$$B(t) = 100(1.05^t) \text{ dollars}$$

When we examine a scatter plot of the account balance over the first 10 years that the account is in existence, we see that the points certainly do not fall on a line. (See Figure 2.1.) The account balance grows more rapidly in the later years. Why? It is because the multiplier operates on an increasingly larger balance as the years advance. The repeated multiplication by 1.05 makes the year variable t appear as an exponent on the 1.05 term in the model; that is, the year value counts the number of times that 1.05 has been used as a multiplier. Because in the equation $B(t) = 100(1.05^t)$ the variable t appears in the exponent, we call the equation an **exponential equation.**

Any change in a quantity that results from repeated multiplication generates an exponential function. That is, if we start with an amount a and multiply by a constant positive factor b each year, then the quantity that we have at the end of t years is given by the exponential equation

$$Q(t) = ab^t$$

defining Q as a function of t.

Linear functions exhibit a constant rate of change, but exponential functions exhibit a constant *percentage change* (such as the 5% annual growth in an investment). Because calculus is so closely associated with the mathematics of change, exponential models and linear models are important in calculus.

Where do we find constant percentage change—that is, exponential change? We find it in situations where things feed on themselves—when change is determined by the current size. For instance, when a new product is introduced into the economic marketplace, word-of-mouth advertising often takes place. Each satisfied customer immediately tells another potential customer about the exciting new product. For that reason, exponential models are often used to analyze a product's sales growth.

Exponential models are also useful in population studies such as those where we are interested in how often a bacterial culture doubles. For instance, in the case of a bacterial culture that starts with 10,000 cells and doubles every hour, we have a population at the end of t hours that is given by the model

$$P(t) = 10{,}000(2^t) \text{ cells}$$

Here we start with the initial 10,000 cells and multiply the number of cells by 2 every hour to account for the exploding population. The resulting population growth is called **exponential growth.**

Exponential equations also arise in the study of radioactive material when we see a *decreasing* amount of the original substance. In this case, the multiplier b in the exponential equation $Q(t) = ab^t$ is between 0 and 1. Such a situation, wherein an amount diminishes by a constant multiplier, is referred to as **exponential decay**. We investigate exponential decay in more detail in Section 2.2.

In the case of exponential growth, the multiplier b in the equation $Q(t) = ab^t$ is greater than 1. A graph of the equation curves upward, increases without bound as the input approaches ∞, and approaches 0 as the input approaches $-\infty$. See Figure 2.2a. In the case of exponential decay, the multiplier b is less than 1. A graph of the equation curves upward, approaches 0 as the input approaches ∞, and increases without bound as the input approaches $-\infty$. See Figure 2.2b.

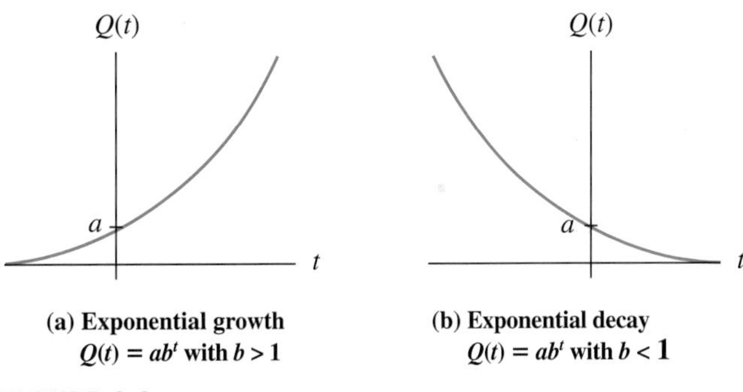

(a) Exponential growth $Q(t) = ab^t$ **with** $b > 1$

(b) Exponential decay $Q(t) = ab^t$ **with** $b < 1$

FIGURE 2.2

Finding Exponential Models

We use the terms *percentage difference* and *percentage change* interchangeably.

Emphasize to your students that input data must be evenly spaced in order to calculate percentage differences to help determine whether the data are exponential.

Often we do not know what the proper repeated multiplier should be in order to write an exponential model. However, if we are given data, sometimes we can look at percentage differences to determine the repeated multiplier. **Percentage change** is calculated from data with increasing input values by dividing each first difference by the output value of the lesser input value and multiplying by 100.

We look at how to calculate percentage differences and how to construct an exponential model by considering a small town's dwindling population. According to the town's records, the population data from 1994 through 2003 are as shown in Table 2.2.

TABLE 2.2

Year	Population	Year	Population
1994	7290	1999	4805
1995	6707	2000	4420
1996	6170	2001	4067
1997	5677	2002	3741
1998	5223	2003	3442

A scatter plot of the data is shown in Figure 2.3.

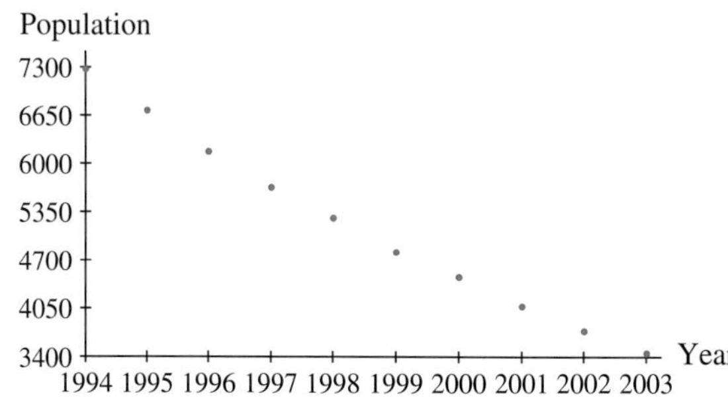

FIGURE 2.3

Let us now develop an algebraic description of the declining population. Because the inputs are evenly spaced, examine the output data in greater detail by calculating first differences:

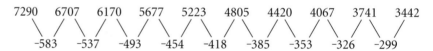

7290		6707		6170		5677		5223		4805		4420		4067		3741		3442
	-583		-537		-493		-454		-418		-385		-353		-326		-299	

The first differences show us that the change in population from year to year is not constant. However, if we calculate the percentages that these yearly changes represent, we notice a pattern. From 1994 to 1995, the population decreased by 583 people. This represents an 8% (approximate) decrease from the 1994 population of 7290 people.

$$\frac{-583 \text{ people}}{7290 \text{ people}} \approx -0.07997 \approx -8\%$$

From 1995 to 1996, the population decreased by 537 people. This also represents about an 8% decrease from the previous year's (1995) population.

$$\frac{-537 \text{ people}}{6707 \text{ people}} \approx -0.08007 \approx -8\%$$

In fact, every year the population decreased by approximately 8%.

$$\text{1996 to 1997:} \quad \frac{-493}{6170} \approx -0.07990 \approx -8\%$$

$$\text{1997 to 1998:} \quad \frac{-454}{5677} \approx -0.07997 \approx -8\%$$

1998 to 1999: $\dfrac{-418}{5223} \approx -0.08003 \approx -8\%$

1999 to 2000: $\dfrac{-385}{4805} \approx -0.08012 \approx -8\%$

2000 to 2001: $\dfrac{-353}{4420} \approx -0.07986 \approx -8\%$

2001 to 2002: $\dfrac{-326}{4067} \approx -0.08016 \approx -8\%$

2002 to 2003: $\dfrac{-299}{3741} \approx -0.07993 \approx -8\%$

Each year the population is 92%, or $\frac{92}{100}$, of what it was the previous year. In other words, each year the population is 0.92 times the previous year's population. The equation for the population decay must show the 1994 population of 7290 people repeatedly multiplied by 0.92. Thus an exponential equation that models the population data is

$$P(x) = 7290(0.92^x) \text{ people}$$

where x is the number of years since 1994.

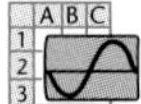

We can also use technology to determine an equation for the data. Enter the data points (aligning the years with $x = 0$ in 1994, $x = 1$ in 1995, $x = 2$ in 1996, etc.), and have your calculator or computer fit an exponential equation. You should obtain the equation

$$T(x) = 7290.25032(0.91999^x) \text{ people}$$

where x is the number of years since 1994.

Why is this equation not identical to the one we developed previously? Remember that as we computed percentage change for each year, we rounded, so the percentage decline was not exactly 8% each year. This rounding contributed to the difference between the models. Note that the technology-generated equation indicates an 8.001% decline.

If we assume that the population continues to decline by the same percentage (approximately 8%), we can predict that during 2004 (that is, from the end of 2003 through the end of 2004), the town's population will be 92% of the population in 2003. Thus the town's population in 2004 is approximately

$$3442(0.92) \approx 3167 \text{ people}$$

This prediction can also be computed using the unrounded equation for population with input $x = 2004 - 1994 = 10$:

$$T(10) = 7290.25032(0.91999^{10}) \approx 3167 \text{ people}$$

This estimate from the model agrees with the one calculated using only the data.

Percentage Change in Exponential Functions

Note that the percentage differences calculated from a set of exponential data are equal to the percentage change in the underlying exponential model only if the step size of the input data is 1 unit and the output data are not rounded.

For the exponential equation $f(x) = ab^x$ we determine the constant percentage change by calculating $(b - 1)100\%$. For example, for the bank account whose balance was modeled by $B = 100(1.05^t)$ dollars t years after the initial investment was made, the constant percentage change of the balance is

$$(1.05 - 1)100\% = (0.05)100\% = 5\%$$

That is, each year the balance of the account increases by 5%. Likewise, for the bacterial growth model $P(t) = 10{,}000(2^t)$ cells, the constant percentage change is $(2-1)100\% = 100\%$; that is, the bacterial population increases by 100% each hour. Finally, for the small town with population given by $P(t) = 7290(0.92^t)$ people t years after 1994, the constant percentage change is $(0.92-1)100\% = (-0.08)100\% = -8\%$. Thus the population of the town decreases by 8% each year.

Exponential Model

An exponential model has a constant percentage change and has an equation of the form

$$f(x) = ab^x$$

where $a \neq 0$ and $b > 0$. The percentage change is $(b-1)100\%$, and the parameter a is the output corresponding to an input of zero.

Note that the equation in the box defines a function f with input x. From now on, when we refer to an exponential equation or function, we will consider the equation to be of the form $f(x) = ab^x$.

When considering a set of data, percentage changes in the output data vary from point to point unless every data point falls on the exponential equation that is fit to the data. In that case and when the input values are 1 unit apart, the percentage changes will all be equal to $(b-1)100\%$. A similar situation arises with linear models. It is only when each data point falls on the linear equation and the input values are 1 unit apart that the first differences are all equal to the slope of the linear equation.

EXAMPLE 1 *Dr. Laura*

In the late 1990s, Dr. Laura Schlessinger hosted the fastest-growing national radio program[1] in history. The program debuted in July of 1994 with 20 affiliate stations, and the number of stations grew by 86.4% per year.

a. Why is an exponential model appropriate for the number of affiliate stations?

b. Find a model for the number of affiliate stations.

c. How many affiliate stations carried the "Dr. Laura" program in July of 1999?

Solution

a. An exponential model is appropriate because the percentage change is constant.

b. The number of affiliate stations can be modeled by the equation

$$L(x) = 20(1.864^x) \text{ stations}$$

x years after July of 1994.

1. "Dr. Laura celebrates five years," *Reno Gazette–Journal*, July 27, 1999, p. 2D.

c. According to the model, the number of affiliate stations in July of 1999 was

$$L(5) = 20(1.864^5) \approx 450 \text{ stations}$$

EXAMPLE 2 *Automotive Tire Sales*

Table 2.3 shows sales data for a tire manufacturer.

TABLE 2.3

Year	Tire sales (millions of dollars)
1974	23
1978	38.4
1982	64
1986	107
1990	179
1994	299
1998	499
2002	833

a. What is the constant percentage change in tire sales?

b. Use the percentage change to predict tire sales in 2006.

c. Use technology to find an exponential equation for tire sales. Write the model for tire sales. What is the percentage change in this equation?

d. Graph the equation in part *c* together with a scatter plot of the data. In terms of limits, describe the end behavior of the function as time increases.

e. Use the model in part *c* to predict tire sales in 2006.

Solution

a. Calculating percentage changes, we see that every 4 years, sales have increased by approximately 67%.

1974 to 1978: $\frac{15.4}{23} \approx 0.66957 \approx 67\%$

1978 to 1982: $\frac{25.6}{38.4} \approx 0.66667 \approx 67\%$

1982 to 1986: $\frac{43}{64} \approx 0.67188 \approx 67\%$

1986 to 1990: $\frac{72}{107} \approx 0.67290 \approx 67\%$

1990 to 1994: $\frac{120}{179} \approx 0.67039 \approx 67\%$

1994 to 1998: $\frac{200}{299} \approx 0.66890 \approx 67\%$

1998 to 2002: $\frac{334}{499} \approx 0.66934 \approx 67\%$

Thus 67% is the approximate percentage change of the tire sales every 4 years.

Recall that our convention requires that output answers be rounded to the accuracy of the least accurate output data value. In this example, although the table shows a value of 38.4, the least accurate values are integers. Thus we report this answer as an integer multiple of one million.

b. If the percentage growth remains constant, then sales in 2006 will increase 67% over the 2002 sales.

Amount of increase (rounded to the nearest million) = 67% of 2002 sales = (0.67)(\$833) ≈ \$558 million

Prediction for 2006:

2002 sales + amount of increase = 833 + 558 = \$1391 million

c. Entering the tire sales data points into a calculator or computer (aligning the years with $t = 0$ in 1974, $t = 4$ in 1978, $t = 8$ in 1982, etc.) and having the calculator or computer fit an exponential equation produces the model

$$S(t) = 22.98235(1.13683^t) \text{ million dollars}$$

t years after 1974.

Note that the percentage change in this exponential function is calculated as $(1.13683 - 1)100\% \approx 14\%$. Were you expecting to see 67%? What happened? Recall that in part *a* we calculated the percentage growth to be 67% *every 4 years*. The percentage growth in the model is the *annual* percentage growth; that is, it is a growth of 14% *per year*. If we multiply 1.13683 (the repeated multiplier in the function S) by itself 4 times, we obtain 1.67: $1.13683^4 \approx 1.67$ representing a 4-year growth of 67%.

d. A graph of the equation and the tire sales data is shown in Figure 2.4.

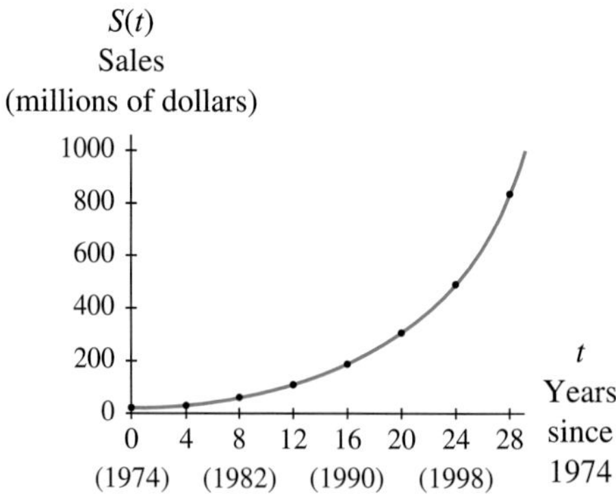

FIGURE 2.4

As the years increase, the output $S(t)$ increases without bound. We say $\lim_{t\to\infty} S(t) \to \infty$.

e. Using the exponential model S, we calculate the sales for 2006 as $S(32) \approx \$1392$ million.

Note the slight difference between the answers to part *b* and part *e* in Example 2. Two factors are involved in this difference. First, when calculating \$1391 million, we used a rounded percentage change: 67% every four years instead of the percentage change in the technology-generated model. Second, the result \$1391 million was calculated from a data point instead of from the sales in 2002 generated by the model.

Although both of these predictions are valid, for consistency we adopt the following rule of thumb:

> Once an equation has been fitted to data, we will use the equation to answer questions rather than using data points or rounded estimates.

Numerical Considerations

Whenever you use a calculator or computer to fit an exponential equation to data, it is important to align the input values so that they are small in magnitude. If you fail to align the input values appropriately, the technological numerical computation routine may return an invalid result because of improper scaling, numerical overflow, or round-off errors. For instance, if you use $x = 1994, 1995, 1996$ instead of $x = 0, 1, 2$ in the small town population example, the model returned is likely to be $y = ab^x$, where $a \approx 1.187 \cdot 10^{76}$. This huge value for a could cause problems computationally.

Warn your students that exponential models of the form $y = ab^x$, where $a \neq 0$, cannot have an output of zero. Also, most exponential regression routines built into technology do not accept negative output. Should a data set that exhibits exponential behavior have negative or zero output values, the data set could be vertically shifted in order to find a exponential model for the data. Section 2.5 discusses vertical shifting.

It is also important that the input values be aligned in the tire sales example to small numbers such as $t = 0, 4, 8, \ldots$ instead of as $t = 1974, 1978, 1982, \ldots$. If you use the actual year as input, then the equation that is returned is likely to be $y = ab^t$, where $a \approx 0$. It is also possible that an overflow error will result and no equation will be returned.

As you work with exponential models, keep in mind the following principle:

> **Aligning Exponential Data**
>
> When using technology to find the equation for an exponential model, align the input data using reasonably small values to avoid numerical computation errors.

Logarithmic Models

A mathematical idea created to solve one problem often can be applied to solve many different types of problems. Such is the case with logarithms that were developed in the early 1600s to perform numerical calculations involving large numbers. With the development and easy availability of calculators and computers, this use of logarithms has declined. As the following application illustrates, **logarithmic functions** (or **log functions**) currently play an important role in modeling certain physical and biological phenomena.

Altitude and air pressure are intrinsically related. Consider the data shown in Table 2.4.

TABLE 2.4

Altitude (thousands of feet above sea level)	Air pressure (inches of mercury)
20	13.76
40	5.56
60	2.14
80	0.82
100	0.33

A scatter plot suggests an exponential function, and a calculation of percentage differences (an approximate 60% decrease in air pressure per 20,000 feet of altitude) confirms that an exponential model describes the relationship between altitude and air pressure. An exponential model for the data is

$$P(a) = 35.54(0.9542^a) \text{ inches of mercury}$$

where a is the altitude in thousands of feet. A graph of the function and a scatter plot are shown in Figure 2.5.

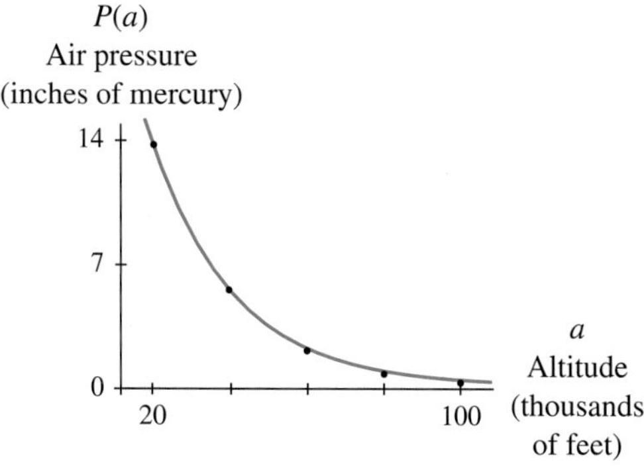

FIGURE 2.5

Altimeters are instruments that determine altitude by measuring air pressure. In the function P, the input is altitude, and the output is air pressure. For an altimeter, the input is air pressure, and the output is altitude. Recall from Section 1.3 that reversing the input and output of a function produces an inverse and that a logarithmic (or log) function is the inverse of an exponential function. Thus a model for altitude as a function of air pressure must be a log model. Table 2.5 shows the reversed data, and Figure 2.6 shows a scatter plot of the reversed data.

TABLE 2.5

Air pressure (inches of mercury)	Altitude (thousands of feet)
13.76	20
5.56	40
2.14	60
0.82	80
0.33	100

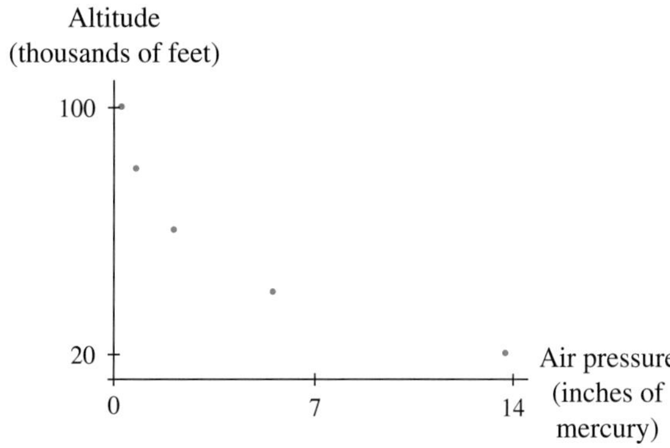

FIGURE 2.6

Although the scatter plot has the basic declining, concave-up appearance of an exponential function, our knowledge of inverse functions assures us that a log function is, in fact, the underlying function.

The log equation used by most technologies is

$$f(x) = a + b \ln x$$

This equation defines a function f with input x. The b-term in this equation determines whether the function increases or decreases and how rapidly the increase or decrease occurs. The a-term determines the vertical shift of the function. From now on, when we refer to a log equation or function, we will consider the equation to be of the form $f(x) = a + b \ln x$.

If you are emphasizing algebraic solutions in your course, you may wish to assign Activity 38, which asks the student to find the inverse of $f(x) = a + b \ln x$ algebraically.

Logarithmic Model

A logarithmic (log) model has an equation of the form

$$f(x) = a + b \ln x$$

This log function is the inverse of the exponential function $y = AB^x$, where $A = e^{-a/b}$ and $B = e^{1/b}$.

Figures 2.7a and 2.7b show graphs of typical log functions.

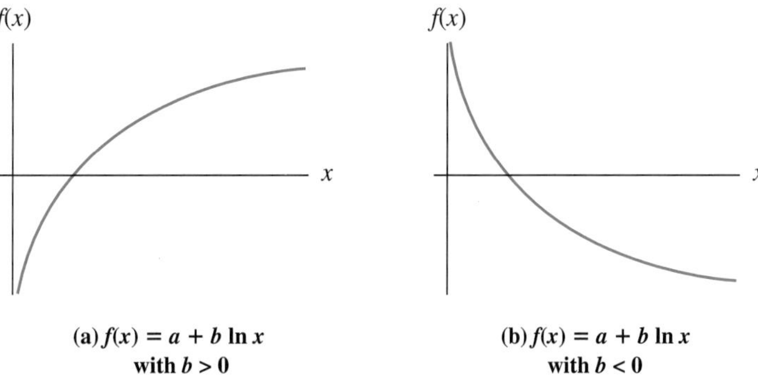

(a) $f(x) = a + b \ln x$ **with** $b > 0$

(b) $f(x) = a + b \ln x$ **with** $b < 0$

FIGURE 2.7

The end behavior of log functions is particularly important in determining when they are appropriate to use to fit data. As shown in Figure 2.7, an increasing log function increases without bound as the input increases, and a decreasing log function decreases without bound as the input increases. Also, log functions are not defined for negative or zero input, and as the input approaches zero from the right, the function either increases or decreases without bound. We summarize the end behavior of log functions with the following limit statements:

For $f(x) = a + b \ln x$, with $b > 0$	For $f(x) = a + b \ln x$, with $b < 0$
$\lim_{x \to 0^+} f(x) \to -\infty$	$\lim_{x \to 0^+} f(x) \to \infty$
$\lim_{x \to \infty} f(x) \to \infty$	$\lim_{x \to \infty} f(x) \to -\infty$

Consider the graphs in Figure 2.7. Note that log functions exhibit increasingly smaller changes in output for constant changes in input. This slow growth or decline characterizes log functions. The characteristic that makes log functions different from any other functions we will study is that their growth or decline becomes increasingly slower but their output never approaches a horizontal limiting value, as do the declining exponential function and, as we will see later, the logistic function.

Returning to the altitude example, we use technology to find the log equation that fits the data in Table 2.5:

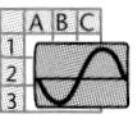

$A(p) = 76.174 - 21.331 \ln p$ thousands of feet above sea level

where p is the air pressure in inches of mercury. Figure 2.8 shows a graph of the function A and a scatter plot of the data in Table 2.5.

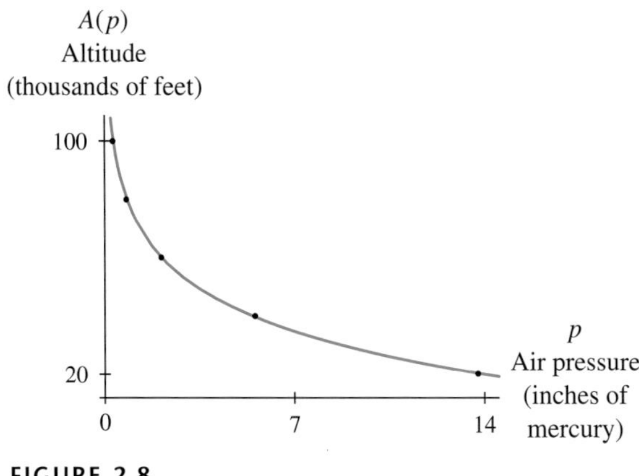

FIGURE 2.8

EXAMPLE 3 *International Investment*

An international investment fund manager models bond rates of countries as a tool when making investment decisions. He uses the data[2] in Table 2.6 to create a yield curve for Germany, where long-term bond rates are higher than short-term rates.

2. *Investment Digest*, VALIC, vol. 12, no. 1, 1998.

TABLE 2.6

Time to maturity (years)	German bond rate (percent)	Time to maturity (years)	German bond rate (percent)
1	3.60	6	4.65
2	4.10	7	4.75
3	4.25	8	4.80
4	4.40	9	4.90
5	4.50	10	4.95

a. Sketch a scatter plot of the data.

b. Find a log model for the data.

c. This investment manager estimates a 5.25% rate for 20-year bonds and a 5.50% rate for 30-year bonds. How closely does the log model match these estimates?

Solution

a.

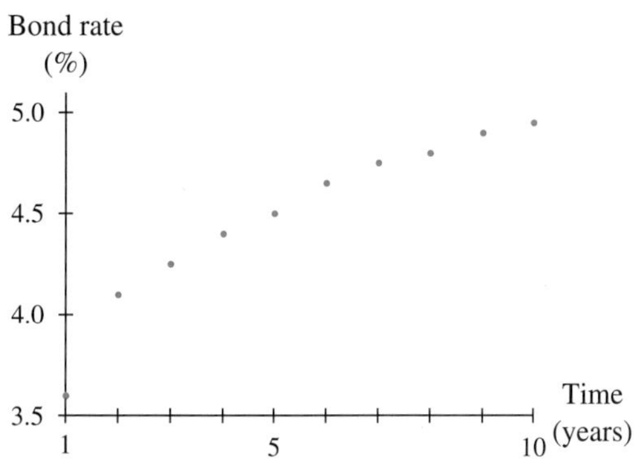

FIGURE 2.9

The scatter plot in Figure 2.9 suggests the slow growth modeled by a log function.

b. A log model for the data is

$$R(t) = 3.6296 + 0.5696 \ln t \text{ percent}$$

for a maturity time of t years.

c. The model in part b predicts the following rates for 20- and 30-year bonds:

$$R(20) \approx 5.34\% \quad \text{and} \quad R(30) \approx 5.57\%$$

These predictions are slightly higher than the estimates made by the fund manager.

Numerical Considerations

Recall that it is often convenient to align yearly data to small input values when finding a linear model so that the coefficients in the equation will not be unnecessarily large values. Although it is not necessary to align large input values when fitting log equations to data, you may want to do so for convenience, as with linear equations. However, if you align the data so that the first input value is 0, your calculator or computer will return an error message when you attempt to find a log equation. The reason is mathematical, not technological. Recall that exponential functions of the form $y = ab^x$, with positive a and b, lie above the horizontal axis; that is, their output is positive. Log functions are the inverses of exponential functions, and inverse functions reverse output and input. Thus the input of log functions must be positive. Zero is not a permissible input for a log function. If you choose to align the data, then all aligned inputs must be positive values.

Aligning Logarithmic Data

Whenever you align input data before fitting an equation for a log model, all aligned input values must be numbers greater than zero.

Exponential functions with their constant percentage change are common in finance, health and life sciences, social sciences, and many other areas. Although log functions are less common, their importance as the inverses of exponential functions leads us to include them in our discussion. As we continue our exploration of calculus concepts, you will become increasingly familiar with these two functions.

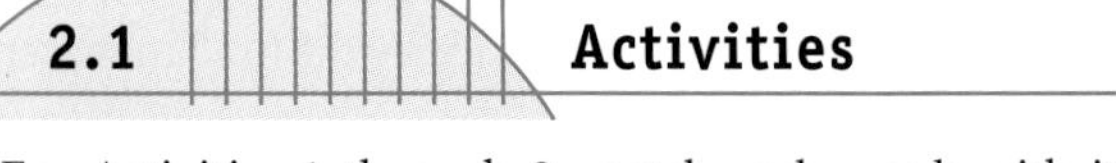

2.1 Concept Inventory

- *Exponential function: $f(x) = ab^x$ (characterized by constant percentage change)*
- *Exponential growth and decay*
- *Finding percentage change*
- *Importance of aligning data*
- *Log function: $f(x) = a + b \ln x$ (inverse of exponential function)*
- *End behavior of exponential and log functions*

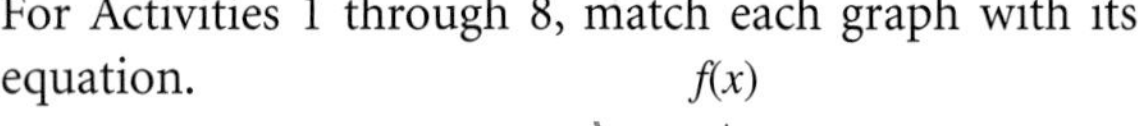

2.1 Activities

For Activities 1 through 8, match each graph with its equation.

1. $f(x) = 2(1.3^x)$

   $f(x) = 2(0.7^x)$

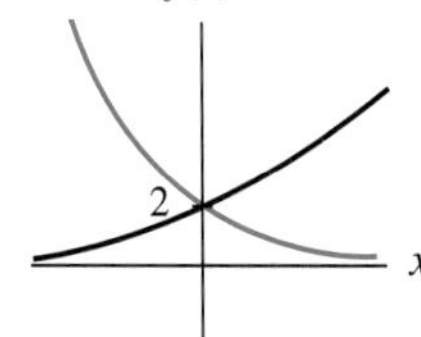

2. $f(x) = 2(1.3^x)$

   $f(x) = -2(1.3^x)$

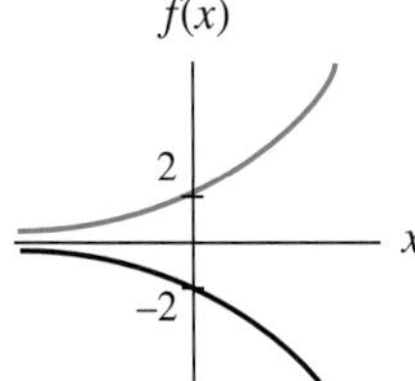

3. $f(x) = 3(1.2^x)$

   $f(x) = 3(1.4^x)$

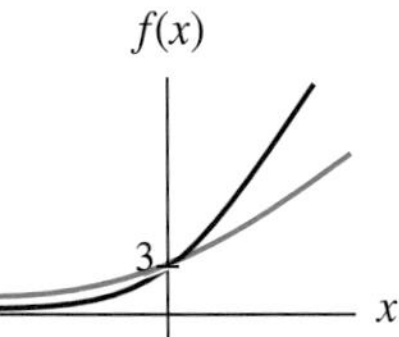

4. $f(x) = 2(0.8^x)$

   $f(x) = 2(0.6^x)$

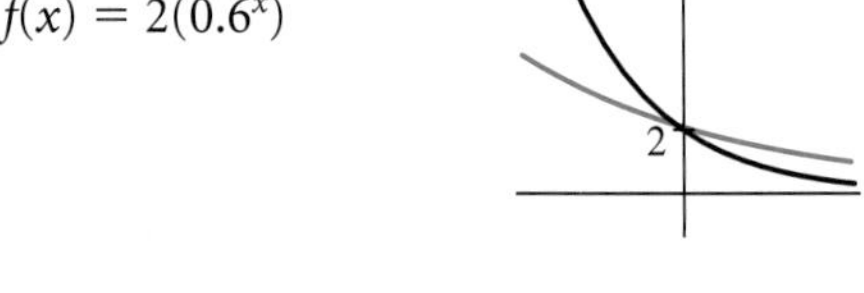

5. $f(x) = 2 \ln x$

   $f(x) = -2 \ln x$

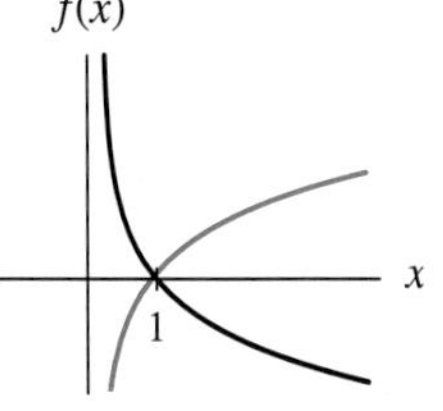

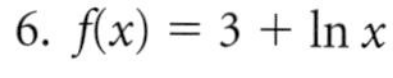

6. $f(x) = 3 + \ln x$

   $f(x) = \ln x$

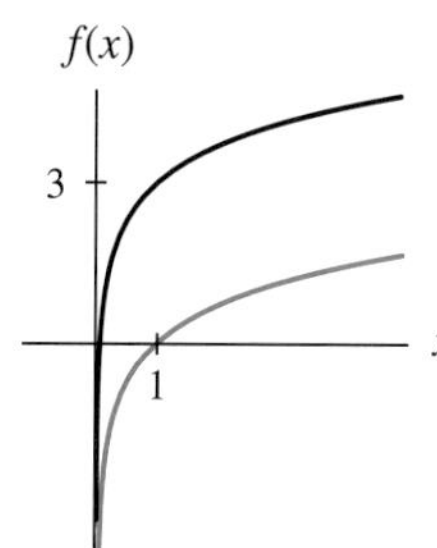

7. $f(x) = 2 \ln x$

   $f(x) = 4 \ln x$

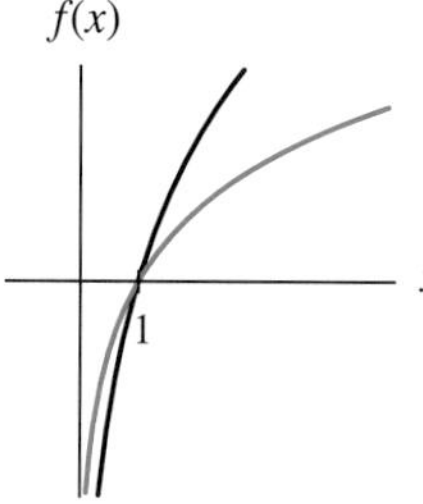

8. $f(x) = -4 \ln x$

   $f(x) = -2 \ln x$

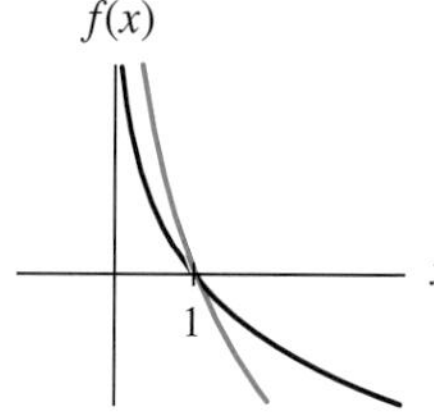

For Activities 9 through 12, tell whether the function describes exponential growth or decay, and give the constant percentage change.

9. $f(x) = 72(1.05^x)$
10. $K(r) = 33(0.92^r)$
11. $y(x) = 16.2(0.87^x)$
12. $A(t) = 128.57(1.035^t)$

For Activities 13 and 14, find the constant percentage change, and interpret it in context.

13. After h hours, the number of bacteria in a Petri dish during a certain experiment can be modeled as $B(h) = 100(0.61^h)$ thousand bacteria.

14. The membership of a popular club can be modeled by $M(x) = 12(2.5^x)$ members by the end of the xth year after its organization.

15. The Consumer Federation of America reported that credit card spending for Visa, MasterCard, American Express, and Discover was $296.3 billion in 1989 and grew by 15.1% per year between 1989 and 1999.
    a. Find a model for credit card spending between 1989 and 1999.
    b. According to the model, when was spending $1 trillion?

You may wish to remind your students that in Activities 15–19, they should round output answers to the same accuracy as the output value (not the percentage change) given in the problem statement.

16. In 1994, Charles Schwab & Co. had approximately $135 billion of assets in customer accounts, and throughout the late 1990s that value grew by approximately 39% per year.
    a. Find a model for Schwab's customer account assets in the late 1990s.
    b. Use the model to estimate Schwab's customer account assets in 2000.

17. A New York City subway token cost $1.50 in 1997. It has been estimated that the cost will rise by 7.46% per year through 2010.
    a. Find a model for New York City subway token prices from 1997 through 2010.
    b. What does the model predict as the 2010 price?

18. In 1975 the EPA emissions standard for cars was 3.1 grams of nitrogen oxide per mile of driving. Assume that the EPA standard decreased by 9.3% per year between 1975 and 2000.
    a. Find a model for the emission standard between 1975 and 2000.
    b. Estimate the EPA standard in 2000.

19. According to the *1999 World Almanac*, the number of workers per beneficiary of the Social Security program was 8.432 in 1955 and has been declining at a rate of 1.94% per year. The *Almanac* estimates that the percentage decline will remain constant through 2030.
    a. Find a model for the number of workers per beneficiary from 1955 through 2030.
    b. What does the model predict the number of workers per beneficiary will be in 2030? How will this number affect your life?

20. According to the back of a milk carton sold by Model Dairy, the number of days that milk will keep when stored at various temperatures is as shown in Table 2.7.

TABLE 2.7

Temperature (degrees Fahrenheit)	Days
30	24
38	10
45	5
50	2
60	1
70	0.5

In Activity 19, many students will assume that the output units are *workers* and that they should therefore round their answer to the nearest integer. However, the output is measured in *workers per beneficiary* and can be a decimal value. In fact, because the one output value given in the problem statement is reported to three decimal places, the answer to part *b* should be reported to three decimal places.

a. Find an exponential model for the data.

b. If a refrigerator is adjusted to 40°F from 37°F, how much sooner will milk spoil when stored in this refrigerator?

21. Carefully read the accompanying newspaper article (from the *Chicago Tribune*, October 21, 1999), and answer the questions that follow. You may wish to assign this activity as a team project.

6,000,000,000 people today

World population, which did not reach 1 billion until 1804, will pass 6 billion at 8:24 p.m. EDT Sunday. Years it took to add each billion since 1804:

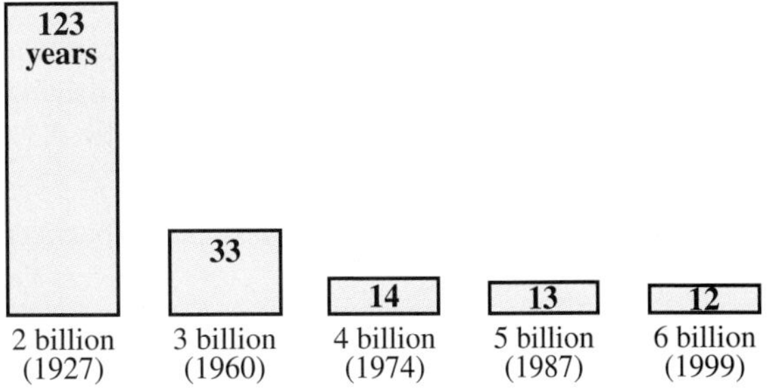

Population of world ready to hit 6 billion

LOS ANGELES TIMES

WASHINGTON–Call this Y6B: The year of 6 billion, a milestone the world's population is expected to reach this weekend.

The birth of the planet's 6 billionth inhabitant, projected by the U.S. Census Bureau, also will mark another historic first: The world's population has doubled in less than 40 years.

Despite a gradual slowing of the overall rate of growth, the world population is still increasing by 78 million people a year. That's the equivalent of adding a city nearly the size of San Francisco every three days, or the combined populations of France, Greece and Sweden every year, according to a coalition of environmental and population groups.

"It took all of human history for the world's population to reach 1 billion in 1804, but little more than 150 years to reach 3 billion in 1960. Now, not quite 40 years later, we are twice that number," said Amy Coen, president of Population Action International.

Even with a decelerating growth rate, the number of humans on the planet could double again to 12 billion by 2050 if the current growth rate continues, the coalition projects.

Source: Data from U.S. Bureau of the Census, compiled by *USA Today.*

Your local newspaper is a valuable source of data pertinent to your students' lives. Your class will be greatly enriched by the inclusion of problem scenarios of local interest.

a. Use the data shown in the figure accompanying the article to find an exponential model for world population from 1927 through 1999. Examine a scatter plot of the data with a graph of the exponential function. Discuss how well the function fits the data.

b. The article gives a current rate of change of population. On the basis of that rate of change and the 1999 population, find a linear model for world population.

c. Rewrite the equation in part *b* so that its input matches that of the function in part *a*. For example, if the input in part *a* is years since 1900 and the input in part *b* is years since 1999, then use the relationship

Linear equation input = exponential equation input − 99

to transform the linear equation to one whose input matches that in the exponential equation.

d. Graph the linear function together with the scatter plot and exponential function from part *a*. Discuss what you observe about the two graphs.

e. Look at the graph in part *d* from 1900 through the year 2050. (Be sure to adjust the vertical axis appropriately.) Comment on the two models and their use in estimating the population from 1900 through 2000 and predicting the population beyond 2000.

f. Use the two models to estimate the population in 2000 and 2050. What assumptions do you make about population growth when you predict the population using these two models?

g. Compare the predictions from part *f* for 2050 with the prediction given in the article. What does this tell you about the rate-of-change assumptions the U.S. Census Bureau used in making their 2050 prediction?

22. When a company ceases to advertise and promote one of its products, sales often decrease exponentially, provided that other market conditions remain constant. At the time that publicity was discontinued for a newly released popular animated film, sales were 520,000 videotapes per month. One month later, videotape sales had fallen to 210,000 tapes per month.

a. What was the monthly percentage decline in sales?

b. Assuming that sales decreased exponentially, give the equation for sales as a function of the number of months since promotion ended.

c. What will sales be 3 months after the promotion ends? 12 months after?

d. Walt Disney has been known to produce videotapes of some of its films for a limited time before completely stopping production. Do you believe that this is a wise business decision?

e. What would be necessary to ensure that the exponential decrease in sales of a product did not occur?

23. In 1992, CDs passed cassettes as the most popular format for pre-recorded music. Table 2.8 gives the numbers of CD singles[3] (in millions) sold during the mid 1990s.

TABLE 2.8

Year	Number of CD singles sold (millions)
1993	7.8
1994	9.3
1995	21.5
1996	43.2
1997	66.7

a. Find an exponential model for the data.

b. According to the model in part *a*, what was the yearly percentage growth in sales of CD singles from 1993 through 1997?

24. An appliance manufacturer that prides itself on the reliability of its washing machines conducts an extensive survey of 3200 of its customers. The survey asked for the number of years that the washing machine remained in working condition before requiring a major service call (not including routine maintenance). Partial results are given in Table 2.9.

a. Discuss the appropriateness of linear and exponential models for these data.

b. Find the better model for the data.

c. Use the model to estimate the number of customers surveyed who had washing machines that lasted between 30 and 35 years before requiring a major service call.

TABLE 2.9

Years before servicing	Number of customers
5	787
10	619
15	487
20	383
25	301
30	237

25. The population of Mexico,[4] in millions of people, in selected years from 1921 through 1990 is given in Table 2.10.

TABLE 2.10

Year	Population (millions)
1921	14.335
1930	16.553
1940	19.654
1950	25.791
1960	34.923
1970	48.225
1980	66.847
1990	81.250

(*Note:* Even though population censuses are normally taken every 10 years in Mexico, the census at the beginning of the 1920s was taken in 1921, not in 1920.)

a. Find an exponential model for the population.

b. What percentage growth rate does the exponential model represent?

c. Use the model to estimate the population of Mexico in 1997.

d. *The 1998 Information Please Almanac* reports the 1997 population of Mexico to be 97,563,364. How close is this to your estimate?

26. The per capita consumption of bottled water[5] in the United States has increased dramatically in the past 20 years. Table 2.11 shows selected years and the per capita bottled water consumption in those years.

3. *Statistical Abstract*, 1998.

4. SPP and INEGI, Mexican Censuses of Population 1921 through 1990 as reported in Pick and Butler, *The Mexico Handbook*, Westview Press, 1994.

5. *Statistical Abstract*, 1998.

TABLE 2.11

Year	Per capita bottled water consumption (gallons per person per year)
1980	2.4
1985	4.5
1990	8.0
1995	11.6
1996	12.4

a. Find linear and exponential models for the data. Graph the equations for these models on a scatter plot of the data. Which model do you think better describes the per capita bottled water consumption?

b. Give the rate of change of the linear model and the percentage change of the exponential model.

c. Use the two models to estimate bottled water consumption in 2000.

d. According to each model, when will per capita bottled water consumption exceed 20 gallons per person per year?

27. The number of U.S. farms with milk cows[6] has been declining. See Table 2.12.

TABLE 2.12

Year	Farms (thousands)	Year	Farms (thousands)
1980	334	1994	149
1985	269	1995	140
1990	193	1996	127
1992	171	1997	117
1993	159		

a. Find an exponential model for this data. What is the percentage change indicated by the model?

b. Express the end behavior of the model as time increases using a limit statement. Does this end behavior reflect what you believe will happen in the future to the number of farms with milk cows?

c. Give at least two reasons for the decline in the number of farms with milk cows.

6. Ibid.

28. The amounts of chlorofluorocarbons[7] (CFCs) released into the atmosphere during the first half of the 1990s are given in Table 2.13.

TABLE 2.13

Year	CFCs (1000 metric tons)	Year	CFCs (1000 metric tons)
1990	223	1994	113
1991	201	1995	105
1992	173	1996	73
1993	149		

a. Find an exponential model for the release of CFCs. What is the percentage change indicated by the model?

b. Use the model to estimate CFC release in 2000. What does this estimate assume about the decline in CFC release?

29. Table 2.14 gives the total number of personal computers in use in China.[8]

TABLE 2.14

Year	PCs (millions)	Year	PCs (millions)
1985	0.09	1993	1.26
1988	0.25	1994	1.90
1989	0.36	1995	2.80
1991	0.63	2000	12.70*
1992	0.87		

* Projection

a. Find an exponential model for the number of PCs in China between 1985 and 2000. Graph the equation for the model on a scatter plot of the data. How well does it appear to fit the data?

b. Use the model to estimate the number of PCs in China in 1990 and in 2005. In which of these estimates can you have the most confidence and why?

30. During the last three decades, computing power has grown enormously. Table 2.15 gives the number of transistors[9] (in millions) in Intel processor chips.

7. Ibid.
8. *The Time Almanac*, 1999.
9. **Intel.com** (Accessed 9/15/99).

TABLE 2.15

Processor	Year	Number of transistors (millions)
4004	1971	0.0023
80286	1982	0.134
386DX	1986	0.275
486DX	1989	1.2
Pentium®	1993	3.1
Pentium® Pro	1995	5.5
Pentium® II	1997	7.5
Celeron	1999	19.0

a. Find an exponential model for the data.

b. According to the model found in part *a*, what is the annual percentage increase in the number of transistors used in an Intel computer processor chip?

c. Is it reasonable to expect this exponential growth rate to continue beyond the year 2000? Explain your reasoning.

31. The pH of a solution, measured on a scale from 0 to 14, is a measure of how acidic or how alkaline that solution is. The pH is a function of the concentration of the hydronium ion, H_3O^+. Table 2.16 shows the H_3O^+ concentration and associated pH for several solutions.

TABLE 2.16

Solution	H_3O^+ concentration (moles per liter)	pH
Cow's milk	$3.981 \cdot 10^{-7}$	6.4
Distilled water	$1.000 \cdot 10^{-7}$	7.0
Human blood	$3.981 \cdot 10^{-8}$	7.4
Lake Ontario water	$1.259 \cdot 10^{-8}$	7.9
Seawater	$5.012 \cdot 10^{-9}$	8.3

a. Find a log model for pH as a function of the H_3O^+ concentration.

b. What is the pH of orange juice with H_3O^+ concentration $1.585 \cdot 10^{-3}$?

c. Black coffee has a pH of 5.0. What is its concentration of H_3O^+?

d. A pH of 7 is neutral, a pH less than 7 indicates an acidic solution, and a pH greater than 7 shows an alkaline solution. Beer has H_3O^+ concentration $3.162 \cdot 10^{-5}$. Is beer acidic or alkaline?

32. The body weight of mice used in a drug experiment is recorded by the researcher. The data[10] are given in Table 2.17.

TABLE 2.17

Age beyond 2 weeks (weeks)	Weight (grams)
1	11
3	20
5	23
7	26
9	27

a. Find a log model for the data.

b. Express the end behavior of the function in terms of limits.

c. Does the end behavior fit the context? What does this tell you about using your model to extrapolate?

d. Estimate the weight of the mice when they are 4 weeks old.

33. The American Association of Pediatrics has stated that lead poisoning is the greatest health risk to children in the United States. Because of past use of leaded gasoline, the concentration of lead in soil can be described in terms of how close the soil is to a heavily traveled road. Table 2.18 shows some distances and the corresponding lead concentrations[11] in parts per million.

TABLE 2.18

Distance from road (meters)	Lead concentration (ppm)
5	90
10	60
15	40
20	32

a. Find a log model for these data.

10. Estimated from information given in "Letters to Nature," *Nature*, vol. 381 (May 30, 1996), p. 417.

11. Estimated from information in "Lead in the Inner City," *American Scientist*, January-February 1999, pp. 62–73.

b. An apartment complex has a dirt play area located 12 meters from a road. Estimate the lead concentration in the soil of the play area.

c. Find an exponential model for the data. Compare this model to the one found in part *a*. Which of the two models better displays the end behavior suggested by the context?

34. The concentration of a drug in the blood stream increases the longer the drug is taken on a daily basis. Table 2.19 gives estimated concentrations (in micrograms per milliliter) of the drug piroxicam taken in 20-mg doses once a day.

TABLE 2.19

Days	Concentration (μg/mL)	Days	Concentration (μg/mL)
1	1.5	11	6.5
3	3.2	13	6.9
5	4.5	15	7.3
7	5.5	17	7.5
9	6.2		

a. Find a log model for the data.

b. Express the end behavior of the equation using limits.

c. Does the end behavior of the equation fit the end behavior suggested by the context?

d. Estimate the concentration of the drug after 2 days of piroxicam doses.

35. The fund manager in Example 3 on page 106 also models the yield curve for New Zealand, in which long-term bond rates are lower than short-term rates. He uses the data given in Table 2.20.

TABLE 2.20

Time to maturity (years)	New Zealand bond rate (%)
0.25	9.40
2	7.90
3	7.65
4	7.50
6	7.30
10	7.10

a. Find a log model for the bond rate data.

b. The fund manager estimates 15-year rates at 7.00%. How close does your model come to this estimate?

36. Table 2.21 shows the projected number of homes with a cable system offering Internet access.[12]

TABLE 2.21

Year	Homes (millions)	Year	Homes (millions)
1998	19.1	2002	51.0
1999	29.0	2003	57.3
2000	39.0	2004	62.8
2001	45.4	2005	67.4

a. Find a log model for the number of homes as a function of the number of years since 1990.

b. According to the model in part *a*, when will the number of homes reach 75 million?

c. What end behavior is suggested by the model equation as time increases? Write this end behavior using limit notation.

37. Table 2.22 shows the average salaries[13] of public elementary and secondary school teachers from 1980 through 1997.

TABLE 2.22

Year	Average salary	Year	Average salary
1980	$15,970	1995	$36,685
1985	$23,600	1996	$37,716
1990	$31,367	1997	$38,562
1994	$35,737		

a. Find a log model for the average salary as a function of the number of years since 1970.

b. Use the log model to estimate the average salary in 1993. What does the log model predict for the average salary in 2000? In which of these two estimates can you be more confident and why?

c. According to the model, when will the average salary reach $40,000?

38. Find the inverse function of $f(x) = a + b \ln x$. Compare your answer with the statement in the box on page 105.

12. Paul Kagan Associates, Inc., *Cable TV Technology*.
13. *Statistical Abstract*, 1998.

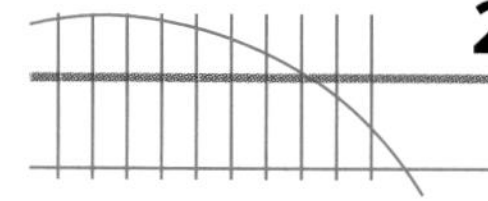

2.2 Exponential Models in Action

Compound Interest

If you are concerned about not having enough time to finish all the sections required for your specific course, you may wish either to skip portions or all of this section or to streamline the section by allowing your students to use technology rather than finding solutions algebraically. See the *Instructor's Resource Guide* for a list of the activities throughout the book that contain references to topics in this section. You will need to omit or modify these activities if you skip this section.

In everyday life, exponential models occur in investment situations. You probably are familiar with the *interest formula*

$$A = P(1 + r)^t$$

where A is the dollar amount accumulated after t years when P dollars are invested at an interest rate of $100r\%$ compounded annually. (Note that r is a decimal number.) If interest is compounded more often than once a year, then the general **compound interest formula** should be used:

Compound Interest Formula

The amount accumulated in an account after t years when P dollars are invested at an annual interest rate of $100r\%$ compounded n times a year is

$$A = P\left(1 + \frac{r}{n}\right)^{nt} \text{ dollars}$$

Here $\left(\frac{r}{n}\right)100\%$ is the interest rate that is applied to the balance in the account at the end of each compounding period, and nt is the total number of compounding periods in t years.

Consider \$1000 invested at a 5% annual interest rate compounded monthly. The equation for the accumulated amount $A(t)$ after t years is

$$A(t) = 1000\left(1 + \frac{0.05}{12}\right)^{12t} \text{ dollars}$$

This equation can be rewritten in the exponential form

$$A(t) = 1000b^t = 1000(1.004166667^{12})^t \approx 1000(1.05116^t) \text{ dollars}$$

What is the meaning of the value 1.05116 in this formula? Remember, for an exponential equation $f(x) = ab^x$, the constant percentage change is $(b - 1)100\%$. Thus in the amount formula $A(t) = 1000(1.05116^t)$ dollars, the constant percentage increase of the amount is $(1.05116 - 1)100\% = (0.05116)100\% = 5.116\%$. That is, the amount invested increases by 5.116% every year. In the investment context, this percentage is called the **effective rate** or **annual percentage yield** (APY). The advertised 5% **annual percentage rate** (APR) is called the **nominal rate.** When comparing compound interest investment opportunities, you should always consider annual percentage yields because the nominal rates do not reflect the effect of the compounding periods.

You should be aware that even though the compound interest formula is a continuous function, it has a discrete interpretation because the amount changes only at the actual times of compounding. For instance, we can use the monthly compounding function $A(t) = 1000(1.05116^t)$ dollars after t years to find the amount of the investment at the end of the 3rd month of the 6th year by calculating $A(6.25) \approx \$1365.95$. (This value was calculated using the unrounded value of b.) However, it would be incorrect to use $t = 6.2$ to calculate the amount in the account on the 14th day of the 3rd month of the 6th year, because interest is compounded monthly, not daily.

EXAMPLE 1 *Comparing Annual Percentage Yields*

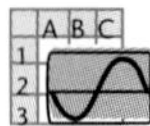

Consider two investment offers: an APR of 6.9% compounded quarterly or an APR of 6.7% compounded monthly.

a. Which offer is a better deal?

b. Compare the time it will take an investment to double for each offer.

Solution

a. For a given principal P, we need to compare the effective rates in the formulas

$$Q(t) = P\left(1 + \frac{0.069}{4}\right)^{4t} \text{ dollars and } M(t) = P\left(1 + \frac{0.067}{12}\right)^{12t} \text{ dollars}$$

after t years. Because $\left(1 + \frac{0.069}{4}\right)^4 \approx 1.07081$ and $\left(1 + \frac{0.067}{12}\right)^{12} \approx 1.06910$, we see that the annual percentage yield for the 6.9% quarterly deal is about 7.08%, compared to only 6.91% APY for the 6.7% monthly deal. You should choose the first of these two options.

b. To determine doubling time, we first set the amount functions in part *a* equal to $2P$ (double the amount invested).

$$2P = P\left(1 + \frac{0.069}{4}\right)^{4t} \qquad 2P = P\left(1 + \frac{0.067}{12}\right)^{12t}$$

Dividing each side of each equation by P, we see that the result does not depend on the amount invested.

$$2 = \left(1 + \frac{0.069}{4}\right)^{4t} \qquad 2 = \left(1 + \frac{0.067}{12}\right)^{12t}$$

You may use technology to solve for t or solve the equations algebraically by applying the natural log function to each side of each equation.

$$\ln 2 = \ln\left(1 + \frac{0.069}{4}\right)^{4t} \qquad \ln 2 = \ln\left(1 + \frac{0.067}{12}\right)^{12t}$$

$$\ln 2 = 4t \ln\left(1 + \frac{0.069}{4}\right) \qquad \ln 2 = 12t \ln\left(1 + \frac{0.067}{12}\right)$$

$$t = \frac{\ln 2}{4 \ln\left(1 + \frac{0.069}{4}\right)} \qquad t = \frac{\ln 2}{12 \ln\left(1 + \frac{0.067}{12}\right)}$$

$$t \approx 10.13 \text{ years} \qquad t \approx 10.37 \text{ years}$$

In the first case, recall that interest is compounded quarterly. The solution $t \approx 10.13$ years lies between the end of the last quarter of the 10th year (when the amount is not quite doubled) and the end of the first quarter of the 11th year (when the amount is more than doubled). Thus the amount will never be exactly doubled, but it will first be more than doubled at the end of the first quarter of the 11th year. In the second case, interest is compounded monthly. The solution $t \approx 10.37$ years is between 4 months ($t \approx 10.33$ years) and 5 months ($t \approx 10.42$ years). We conclude that the amount first exceeds twice the investment at the end of the 5th month of the 11th year after the investment is made.

Continuous Compounding and the Number *e*

For a given nominal rate, the only factor that influences the annual percentage yield is the number of times the interest is compounded each year. To see the effect of the number of compoundings, consider the following:

Suppose that you invest \$1 for 1 year at a nominal rate of 100% compounded n times during the year: Amount $= \$1\left(1 + \frac{1}{n}\right)^n$. As n increases, how does the original \$1 grow? Consider Table 2.23, which shows the relationship between n and the accumulated amount.

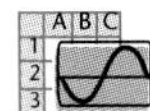

TABLE 2.23

Compounding occurs	n	Amount
yearly	1	\$2
semiannually	2	\$2.25
quarterly	4	\$2.44
monthly	12	\$2.61
weekly	52	\$2.69
daily	365	\$2.71
every hour	8760	\$2.72
every minute	525,600	\$2.72
every second	31,536,000	\$2.72

Does it surprise you that the amount does not grow infinitely large? In fact, your \$1 will not grow to more than \$2.72 during the year, regardless of how often interest is compounded. As n gets larger and larger, the unrounded amount is getting closer and closer to the number that we call e, and we say

$$\lim_{n\to\infty}\left(1 + \frac{1}{n}\right)^n = e$$

The number e is an *irrational* number, which means that its decimal expansion never repeats and never terminates (that is, never ends). A 12-digit decimal approximation to e is 2.71828182846.

The limit $\lim_{n\to\infty}\left(1 + \frac{1}{n}\right)^n = e$ represents the unrounded amount to which 1 dollar would grow in a year if the nominal rate were 100% and the number n of compounding periods per year increased without bound; that is, the interest is *compounded continuously*. What if we continuously compounded P dollars for t years at a nominal interest rate of $100r$%? In other words, what is $\lim_{n\to\infty} P\left(1 + \frac{r}{n}\right)^{nt}$? A little algebra goes a long way. To simplify the algebra, we let $m = \frac{n}{r}$ so that $mr = n$. Then

$$\left(1 + \frac{r}{n}\right)^{nt} = \left(1 + \frac{1}{m}\right)^{mrt}$$
$$= \left[\left(1 + \frac{1}{m}\right)^{m}\right]^{rt}$$

Also, for a fixed r, because $m = \frac{n}{r}$, $m \to \infty$ as $n \to \infty$. Thus

$$\begin{aligned}\lim_{n\to\infty} P\left(1+\frac{r}{n}\right)^{nt} &= \lim_{m\to\infty} P\left[\left(1+\frac{1}{m}\right)^{m}\right]^{rt} \\ &= P\left[\lim_{m\to\infty}\left(1+\frac{1}{m}\right)^{m}\right]^{rt} \\ &= Pe^{rt}\end{aligned}$$

This result is known as the **continuously compounded interest** formula.

Continuously Compounded Interest Formula

The amount accumulated in an account after t years when P dollars are invested at a nominal interest rate of $100r\%$ compounded continuously is

$$A = Pe^{rt} \text{ dollars}$$

Note that A is a continuous function with continuous interpretation. Unlike the general compound interest formula (which is a continuous function with discrete interpretation), it is meaningful here to ask questions about the amount in the account at any instant in time.

What is the APY for this type of account? With continuous compounding, the annual percentage yield (APY) is $(e^r - 1)100\%$. For example, if \$1000 is invested at a 5% nominal interest rate compounded continuously, then the accumulated amount, $A(t)$, after t years is

$$\begin{aligned}A(t) &= 1000e^{0.05t} \\ &= 1000(e^{0.05})^t \\ &\approx 1000(1.05127^t) \text{ dollars}\end{aligned}$$

The annual percentage yield is $(1.05127 - 1)100\% \approx 5.13\%$.

It is easy to convert from an exponential function f of the form $f(t) = ae^{kt}$ to an exponential function f of the form $f(t) = ab^t$. We just saw an example of this conversion:

$$1000e^{0.05t} \approx 1000(1.05127^t)$$

Note that the value of a remains the same in either form and that $b = e^k$.

This conversion from ae^{kt} to ab^t gives us a simple way to find the constant percentage change of the function f where $f(t) = ae^{kt}$. (Recall that the constant percentage change of $f(t) = ab^t$ is $(b - 1)100\%$.) We illustrate this concept in Example 2.

EXAMPLE 2 *APY of Continuous Compoundings*

What is the annual percentage yield (APY) on an investment that has an advertised nominal rate of 12.6% compounded continuously?

Solution For a given principal investment P, we use the continuously compounded interest formula $A = Pe^{rt}$ to determine the amount at any time t (in years). In this case,

$$A(t) = Pe^{0.126t} \text{ dollars after } t \text{ years}$$

Converting from the form ae^{kt} to the form ab^t, we obtain the following equivalent equation:

$$A(t) = Pe^{0.126t} = P(e^{0.126})^t \approx P(1.13428^t) \text{ dollars after } t \text{ years}$$

Thus the APY is approximately 13.4%.

It is also possible to convert from an exponential function f of the form $f(t) = ab^t$ to one of the form $f(t) = ae^{kt}$. In order to convert a function, say $f(t) = 120(1.05^t)$, to the form $f(t) = ae^{kt}$, we apply Property [1] from page 38 and rewrite 1.05 as $e^{\ln 1.05}$. Thus

$$f(t) = 120(1.05^t) = 120(e^{\ln 1.05})^t \approx 120e^{0.04879t}$$

EXAMPLE 3 *Determining APR*

a. An investment that has interest compounded continuously has an APY of 9.2%. What is its APR?

b. Suppose you are 25 years old and have $10,000 to invest for retirement. What APR compounded monthly is needed for your money to grow to one million dollars in 40 years?

Solution

a. For an investment of P dollars, we set up the amount function A as

$$A(t) = P(1.092^t) \text{ dollars after } t \text{ years}$$

Converting this to the form $A(t) = ae^{kt}$, we get

$$A(t) = P(1.092^t) = P(e^{\ln 1.092})^t \approx Pe^{0.08801t}$$

dollars after t years. Thus the APR (when interest is compounded continuously) is approximately 8.8%.

b. To answer the question posed, substitute 40 for t and 1,000,000 for $A(t)$ in the formula $A(t) = 10{,}000\left(1 + \frac{r}{12}\right)^{12t}$, and solve for r.

$$1{,}000{,}000 = 10{,}000\left(1 + \frac{r}{12}\right)^{12(40)}$$

$$100 = \left(1 + \frac{r}{12}\right)^{480}$$

You can use technology to solve this equation, or you can solve it algebraically. To begin the algebraic solution, raise each side to a power of $\frac{1}{480}$ and simplify.

$$100^{1/480} = \left[\left(1 + \frac{r}{12}\right)^{480}\right]^{1/480}$$

$$100^{1/480} = 1 + \frac{r}{12}$$

Now subtract 1 and multiply by 12 to solve for r.

$$12(100^{1/480} - 1) = r$$

$$r \approx 0.11568$$

An APR of approximately 11.57% compounded monthly is needed for $10,000 to grow to a million dollars in 40 years.

Present and Future Value

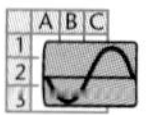

When money is invested in an account that earns interest, we are concerned with its **future value**—that is, its value at some time in the future. The compound interest formulas $A = P\left(1 + \frac{r}{n}\right)^{nt}$ and $A = Pe^{rt}$ are used to determine future values.

EXAMPLE 4 *Future Value*

Suppose that on July 1 of each year, you deposit $1500 into an account that earns interest at the nominal rate of 7.5% compounded monthly. What is the future value of these investments 1 year after the fifth and final deposit is made?

Solution

The initial $1500 will grow to $1500\left(1 + \frac{0.075}{12}\right)^{60} \approx \$2179.94$

The second $1500 will grow to $1500\left(1 + \frac{0.075}{12}\right)^{48} \approx \$2022.90$

The third $1500 will grow to $1500\left(1 + \frac{0.075}{12}\right)^{36} \approx \$1877.17$

The fourth $1500 will grow to $1500\left(1 + \frac{0.075}{12}\right)^{24} \approx \$1741.94$

The last $1500 will grow to $1500\left(1 + \frac{0.075}{12}\right)^{12} \approx \$1616.45$

The future value of these investments is the sum

$$\$2179.94 + \$2022.90 + \$1877.17 + \$1741.94 + \$1616.45 = \$9438.40$$

Often, however, we are interested in how much money should be invested now in order to achieve a desired future value. The amount to be invested now is called the **present value** of the investment. The compound interest formulas are also used to determine present values, as illustrated in Example 5.

EXAMPLE 5 *Present Value*

Find the present value of the $9438.40 in Example 4, assuming that the money is invested at a nominal rate of 7.5% with

a. monthly compounding for 5 years.

b. continuous compounding for 5 years.

Solution

a. Using an interest rate of 7.5% compounded monthly, we solve the equation

$$9438.40 = P\left(1 + \frac{0.075}{12}\right)^{60}$$

for the present value P. Because this equation is just

$$9438.40 = P(1.453294408)$$

we see that the present value is $P = \frac{9438.40}{1.453294408} \approx \$6494.49$. In other words, you would need to invest \$6494.49 now at 7.5% compounded monthly to have \$9438.40 in 5 years.

b. On the other hand, if we seek the present value of \$9438.40 assuming a rate of 7.5% compounded continuously over 5 years, then we must solve

$$9438.40 = Pe^{0.075(5)} = Pe^{0.375} = P(1.454991415)$$

for P to obtain $P \approx \$6486.91$. Again, this is the amount you would need to invest at 7.5% continuously compounded to have \$9438.40 in 5 years.

Exponential Decay

Certain chemical elements are known to decay at a constant percentage rate; that is, the amount of the substance decreases over time in an exponential manner. It is common to express exponential decay in terms of how long it takes for half of the substance to decay. This value is known as the **half-life** of the substance.

The half-life of polonium-12 is 140 days. If a storage container contains 70 grams of polonium-12, then it will contain only 35 grams after 140 days, 17.5 grams after 280 days, 8.75 grams after 420 days, and so on. To find an exponential equation modeling the decay of polonium-12, use two of the known data points and technology to fit an equation to those points. We choose the points $(0, 70)$ and $(140, 35)$ corresponding to the initial 70 grams and the 35 grams that exists after 140 days. The exponential model is

$$A(t) = 70(0.995061^t) \text{ grams after } t \text{ days}$$

Recall that when the parameter b is less than 1, the output of the equation is decreasing. In this case we see a percentage change of $(0.995061 - 1)100\% \approx -0.49\%$ per day.

The model can be used to answer questions such as "When will there be only 2 grams of the polonium remaining?" To answer this question, we set $A(t)$ equal to 2 and solve for t.

$$2 = 70(0.995061^t)$$

Again, you can use technology to solve this equation, or you can solve it algebraically as follows:

$$\frac{1}{35} = (0.995061^t)$$

$$\ln\left(\frac{1}{35}\right) = \ln(0.995061^t)$$

$$\ln\left(\frac{1}{35}\right) = t \ln 0.995061$$

$$t = \frac{\ln\left(\frac{1}{35}\right)}{\ln 0.995061} \approx 718 \text{ days}$$

Thus, there will be only 2 grams of the polonium-12 remaining after approximately 718 days.

Exponential decay also occurs when drugs are taken into the body. Once the drug is absorbed, its level in the blood stream is at a peak. Each time the blood passes through the kidneys, liver, spleen, and other organs, a percentage of the drug is removed. Pharmaceutical companies report half-life information in terms of either elimination half-life or plasma concentration half-life. *Elimination half-life* is the time it takes for half of the drug to be removed from the body. *Plasma half-life* is the amount of time it takes for the concentration of the drug in the plasma to reach half of its peak concentration. Plasma concentration is usually measured in micrograms per milliliter (μg/mL). The use of an exponential model in this type of situation is illustrated in Example 6.

EXAMPLE 6 *Dilantin Absorption*

Dilantin[14] is a drug used to control epileptic seizures. On Monday a patient takes a 300-mg Dilantin tablet at 4 p.m. Eight hours later the Dilantin reaches its peak plasma concentration of 15 μg/mL. The average plasma half-life of Dilantin is 22 hours.

a. Write a model for the concentration of Dilantin in the patient's plasma as a function of the time after peak concentration is reached.

b. What is the concentration of Dilantin in this person's plasma at 8 a.m. on Tuesday?

c. The minimum desired concentration of Dilantin is 10 μg/mL. Will the concentration dip below that level before the patient takes a second tablet on Tuesday at 4 p.m.?

Solution

a. The input of the model is the time after the peak concentration that occurs at midnight on Monday, and the initial amount we consider is the peak concentration of 15 μg/mL. After 22 hours, the concentration will be half its peak, or 7.5 μg/mL. Using the two data points (0, 15) and (22, 7.5), we obtain the model

$$C(h) = 15(0.96898^h) \ \mu\text{g/mL}$$

h hours after midnight Monday.

b. Eight hours after midnight, the plasma concentration is

$$C(8) = 15(0.96898^8) \approx 11.7 \ \mu\text{g/mL}$$

c. To determine when the plasma concentration reaches 10 μg/mL, solve the equation $10 = 15(0.96898^h)$ using technology or algebra. You should obtain $h \approx 12.9$ hours. It takes approximately 12.9 hours after the peak concentration at midnight for the concentration to reach 10 μg/mL. This corresponds to approximately 1 p.m. on Tuesday. Thus the concentration will be below 10 μg/mL when the patient takes another tablet at 4 p.m. on Tuesday.

14. Based on information obtained from **www.parke-davis.com** (Accessed 6/11/00).

2.2 Concept Inventory

- *Compound interest formula:* $A = P\left(1 + \frac{r}{n}\right)^{nt}$
- *Nominal rate or annual percentage rate (APR)*
- *Effective rate or annual percentage yield (APY)*
- *The number e*
- *Converting from e^{kt} to b^t*
- *Converting from b^t to e^{kt}*
- *Continuously compounded interest formula:* $A = Pe^{rt}$
- *Present value and future value*
- *Exponential decay*
- *Half-life*

2.2 Activities

For Activities 1 through 4, tell whether the exponential function is increasing or decreasing, convert the function from the form $f(x) = ae^{kx}$ to the form $f(x) = ab^x$, and find the constant percentage change of the function.

1. $f(x) = 100e^{0.98x}$
2. $f(x) = 1.92e^{1.02x}$
3. $A(t) = 1000e^{-0.21t}$
4. $P(t) = 294.3e^{-0.97t}$

For Activities 5 through 8, tell whether the exponential function is increasing or decreasing, and convert the function from the form $f(x) = ab^x$ to $f(x) = ae^{kx}$.

5. $f(x) = 39.2(1.29^x)$
6. $f(x) = 62.4(0.93^x)$
7. $h(t) = 1.02(0.62^t)$
8. $g(t) = 0.93(1.05^t)$

9. You invest \$2000 at 4.5% APR compounded quarterly.
   a. Give the formula of the form $A = P\left(1 + \frac{r}{n}\right)^{nt}$ to determine the balance in this account after t years. Also write the formula in the form $A = ab^t$.
   b. Calculate the amount in the account after 2 years.
   c. Find the amount of the investment after 3 years and 9 months.
   d. Calculate the amount in the account after 2 years and 2 months.

10. You invest \$10,000 at 8.125% APR compounded monthly.
    a. Give the formula of the form $A = P\left(1 + \frac{r}{n}\right)^{nt}$ to determine the balance in this account after t years. Also write the formula in the form $A = ab^t$.
    b. Calculate the amount in the account after 1 year.
    c. Find the amount of the investment after 5 years.

11. Your credit card statement indicates a finance charge of 1.5% per month on the outstanding balance.
    a. What is the nominal rate (APR), assuming that interest is compounded monthly?
    b. What is the effective rate of interest (APY)?

12. In order to offset college expenses, at the beginning of your freshman year you obtained a nonsubsidized student loan for \$15,000. Interest on this loan accrues at a rate of 0.739% each month. However, you do not have to make any payments against either the principal or the interest until after you graduate.
    a. Write a function giving the total amount you will owe after t years in college.
    b. What is the nominal rate?
    c. What is the effective rate?

13. Suppose that six years ago you invested \$1400 in an account with a fixed interest rate compounded continuously. You do not remember the interest rate, but your end-of-the-year statements for the first five years show the amounts given in Table 2.24.

TABLE 2.24

Year	Dollar amount
1	1489.55
2	1584.82
3	1686.19
4	1794.04
5	1908.80

   a. Using the data, estimate the constant percentage change. What is the financial term for this percentage? On the basis of this percentage change, determine the balance at the end of the sixth year.
   b. Find an equation of the form $A = Pe^{rt}$ that models the data. What is the financial term for $100r$%?
   c. How much money will be in the account at the end of the 6th year? Under what circumstances might this amount change?

14. How long would it take an investment to double under each of the following conditions?
   a. Interest is 6.3% compounded monthly.
   b. Interest is 8% compounded continuously.
   c. Interest is 6.85% compounded quarterly.

15. You have \$1000 to invest, and you have two options: 4.725% compounded semiannually or 4.675% compounded continuously.
   a. Determine the better option by calculating the annual percentage yield for each.
   b. Verify your choice of option by comparing the amount the first option would yield with the amount the second option would yield after 2, 5, 10, 15, 25, and 50 years. Does your choice of option depend on the number of years you leave the money invested?
   c. By how much would the two options differ after 10 years?

16. Suppose that to help pay for your college education, 15 years ago your parents invested a sum of money that has grown to \$25,000 today. How much did they originally invest if
   a. the investment earned 7.5% interest compounded monthly?
   b. the investment earned 7.5% interest compounded daily?
   c. the investment earned 7.5% interest compounded continuously?

17. The formula that is used to calculate the monthly payment of a loan is

$$\frac{r}{12}A = m\left[1 - \left(1 + \frac{r}{12}\right)^{-n}\right]$$

where: A = loan amount
r = interest rate expressed as a decimal
n = number of months of the loan
m = monthly payment

Suppose you are considering buying a \$13,500 car. You have \$1200 for a down payment and have two options for loans.

Option A: 10.9% compounded monthly for 3.5 years

Option B: 10.5% compounded monthly for 4 years

   a. Determine the monthly payment for each loan.
   b. Determine the total amount paid for each loan.
   c. Determine the total amount of interest paid for each loan.
   d. What are the advantages and disadvantages of each loan? Which loan would you choose and why?

18. Your parents are hoping to buy a house 5 years from now when you finish college. Because of rising inflation, prices on the housing market are expected to increase by 3% each year over the next 5 years. Estimate how much your parents will have to pay in 5 years for a house that is currently priced at \$85,000.

19. A certificate of deposit (C.D.) is bought for \$2500 and held for 3 years. What is the future value of the C.D. at the end of the 3 years if it earns interest compounded quarterly at a nominal rate of 6.6%?

20. You were able to save \$3000 from your summer job. If you invest it now in an account with 6.7% annual interest compounded monthly, how much will your account be worth when you graduate?

21. You are saving for a down payment on a car. You worked all summer and want to invest part of your earnings at 6.2% APR compounded monthly in order to make the down payment when you graduate. How much of your summer earnings should you invest now in order to have \$2000 in 2 years?

22. Next summer, you plan to put part of your summer earnings in the bank in order to meet your January tuition payment. How much should you invest in August at 5.5% APR compounded daily if you need \$3600 at the beginning of January?

23. a. How much money would you have to invest now at 10% APR compounded monthly in order to accumulate a nest egg of a quarter of a million dollars over the next 45 years?
   b. Using the answer from part *a* as the principal amount invested at 10% APR compounded monthly, determine the amount to which the principal will grow after 45 years. Why is the answer not \$250,000?

24. What is the present value of an investment that will be worth \$10,000 in 5 years, assuming that the effective rate is 7.1%?

25. Bill Gates, CEO of Microsoft, had a net worth of approximately \$85 billion on October 1, 1999.
   a. If Bill Gates's wife donated \$2 million to her alma mater at 8 a.m. that morning, how long did

it take for the amount to be recouped if their net worth was growing at a rate equivalent to 6.2% APY?

b. If you solved part *a* using technology, explain how it could be solved using logarithms.

26. The number of radio stations carrying the "Dr. Laura" program between 1994 and 1999 can be modeled by

$$L(x) = 20(1.864^x) \text{ stations } x \text{ years after July 1994}$$

a. Find a model for the year as a function of the number of stations.

b. Which model should be used to answer the following questions?

i. When were 100 stations carrying the program?

ii. How many stations carried the program in July of 2000?

27. A bicyclist weighing 120 pounds and riding a bike weighing 30 pounds begins coasting on flat ground at 20 miles per hour. The cyclist's speed is given by the equation

$$v(t) = v_0\, e^{-2800t/w}$$

where v_0 is the speed in mph when coasting began, w is the combined weight in pounds of the bike and rider, and t is the time in hours of coasting. Note that the units in which speed and time are expressed must be compatible.

a. Find the rider's speed after 2 minutes.

b. When is the rider's speed 5 mph?

c. Graph the function v. From the graph estimate $\lim\limits_{t\to\infty} v(t)$.

d. What does the limit in part *c* tell you about the rider's speed as described by the function v?

28. The distance coasted by the cyclist described in Activity 27 is given by the equation

$$s(t) = \frac{v_0\, w}{2800}\,(1 - e^{-2800t/w})$$

where s is the distance traveled in miles, v_0 is the speed in mph when the cyclist begins coasting, w is the combined weight in pounds of the bicycle and cyclist, and t is the coasting time in hours.

a. How far has the cyclist coasted in 1 minute?

b. When will the cyclist have coasted half a mile?

29. Carbon-14, ^{14}C, has a half-life of approximately 5580 years. If a sample of an artifact contains 0.027 gram of ^{14}C, how long ago did it contain 0.05 gram of ^{14}C?

30. An abandoned building is found to contain radioactive radon gas. Thirty hours later, 80% of the initial amount of gas is still present.

a. Find the half-life of this radon gas.

b. Give a model for the amount of radon gas present after t hours.

c. What is the limit of the function in part *b* as time approaches infinity? Interpret this answer in the context of the radon gas.

31. An equation known as Newton's Law of Cooling describes how hot objects cool off and cold objects warm up. Newton's Law of Cooling states that the temperature of an object after time x is

$$T(x) = R + (I - R)10^{-kx}$$

where I is the initial temperature of the object, R is the temperature of the surrounding substance (air, water, or whatever), and k is a constant that depends on the object and the surrounding substance.

a. When a pizza comes out of an oven in a restaurant kitchen, its temperature is 400°F. If the temperature of the restaurant remains at a constant 75°F and it takes 5 minutes for the pizza to cool to 150°F, what is the constant k?

b. It takes 90 seconds for a waiter to remove a pizza from the oven and carry it to a customer's table. The waiter tells the customer, "Be careful, it's hot." Just how hot is it?

c. A demanding customer wants her pizza served to her at no less than 300°F. How long does the waiter have to take this customer's pizza from the oven to her table?

32. Refer to Newton's Law of Cooling in Activity 31. An object at 100°F is immersed in a liquid that is assumed to remain at a constant temperature of 32°F.

a. Solve for k, and write the specific cooling equation for this object if, after 1 minute, the temperature of the object is 92°F.

b. What is the temperature of the object after 5 minutes?

33. The elimination half-life of a certain type of penicillin is 30 minutes.
    a. Write a model for the amount of this penicillin left in a person's body if the initial dose is 250 mg.
    b. If it is safe to take another dose of this penicillin once the amount in the body is less than 1 mg, when should another dose be taken?

34. If a person takes Digoxin, a heart stimulant, the concentration in the person's blood stream t hours after the Digoxin reaches its peak concentration is

$$D(t) = D_0 e^{-0.0198t} \ \mu\text{g/mL}$$

where D_0 is the peak concentration in micrograms per milliliter. Find the half-life of Digoxin.

Be aware that some technologies do not have a built-in logistic regression routine. Students who use such technologies should obtain a logistic curve-fitting program. See the technology supplements or the *Calculus Concepts* web site for more information.

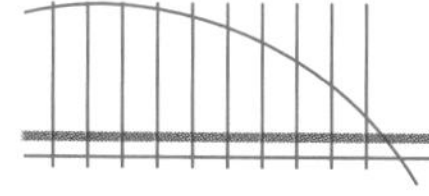

2.3 Logistic Functions and Models

Exponential Growth with Constraints

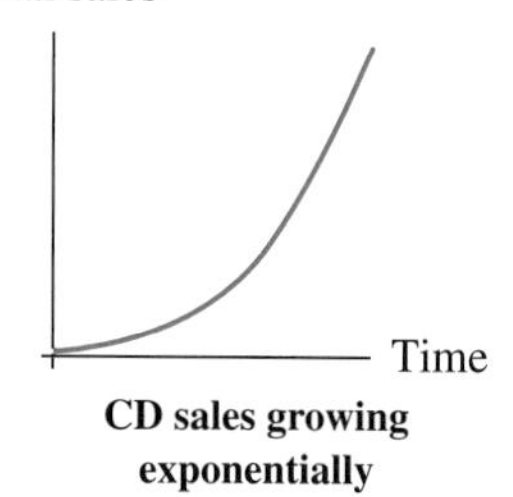

CD sales growing exponentially

FIGURE 2.10

Although exponential models are common and useful, it is sometimes unrealistic to believe that exponential growth can continue forever. In many situations, there are forces that ultimately limit the growth. Here is a situation that may seem familiar to you:

You and a friend are shopping in a music store and find a new compact disc that you are certain will become a hit. You each buy the CD and rush back to campus to begin spreading the news. As word spreads, the total number of CDs sold begins to grow exponentially as shown in Figure 2.10.

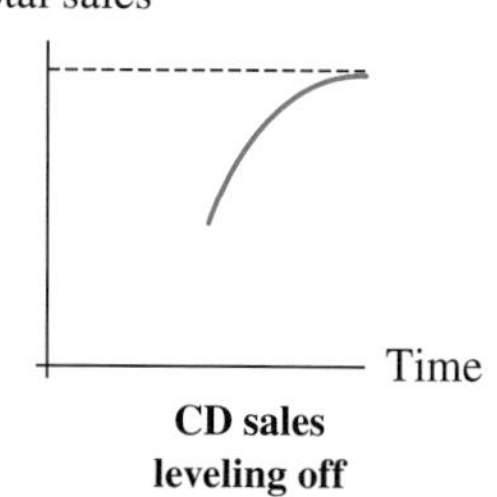

CD sales leveling off

FIGURE 2.11

However, this trend cannot continue indefinitely. Eventually, the word will spread to people who have already bought the CD or to people who have no interest in it, and the rate of increase in total sales begins to decline (Figure 2.11). In fact, because there is only a limited number of people who will ever be interested in buying the CD, total sales ultimately must level off. The graph representing the total sales of the CD as a function of time is a combination of rapid exponential growth followed by a slower increase and ultimate leveling off. See Figure 2.12.

S-shaped behavior such as this is common in marketing situations, the spread of disease, the spread of information, the adoption of new technology, and the growth of certain populations. A mathematical function with such an S-shaped curve is called a **logistic function**. Its equation is of the form

$$N(t) = \frac{L}{1 + Ae^{-Bt}}$$

Total sales of a CD

FIGURE 2.12

From now on, when referring to a logistic equation or function, we will consider an equation of this form.

The number L appearing in the numerator of a logistic equation determines a horizontal asymptote $y = L$ for a graph of the function N. In fact, a logistic function has two horizontal asymptotes. If we look at a graph of a positive, increasing logistic function (as in Figure 2.13a), we see that the graph appears to be trapped between the horizontal axis ($y = 0$) and the horizontal asymptote $y = L$. In fact, for an increasing logistic function I, where $I(x) = \frac{L}{1 + Ae^{-Bx}}$ with $B > 0$,

$$\lim_{x \to -\infty} I(x) = 0 \quad \text{and} \quad \lim_{x \to \infty} I(x) = L$$

Similarly, for a decreasing logistic function D, where $D(x) = \frac{L}{1 + Ae^{-Bx}}$ with $B < 0$, as shown in Figure 2.13b, the end behavior can be described by the limit statements

$$\lim_{x \to -\infty} D(x) = L \qquad \text{and} \qquad \lim_{x \to \infty} D(x) = 0$$

Thus any logistic function has two horizontal asymptotes. We will refer to $y = 0$ as the *lower asymptote* and to $y = L$ as the *upper asymptote*. We will also refer to L as the *limiting value* of the function.

In social science and life science applications, the limiting value is often called the *carrying capacity* or the *saturation level*. In other applications it is sometimes referred to as the *leveling-off value*.

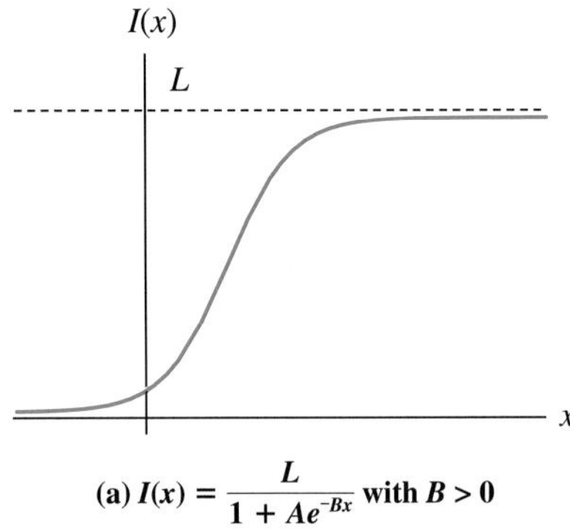

(a) $I(x) = \frac{L}{1 + Ae^{-Bx}}$ with $B > 0$

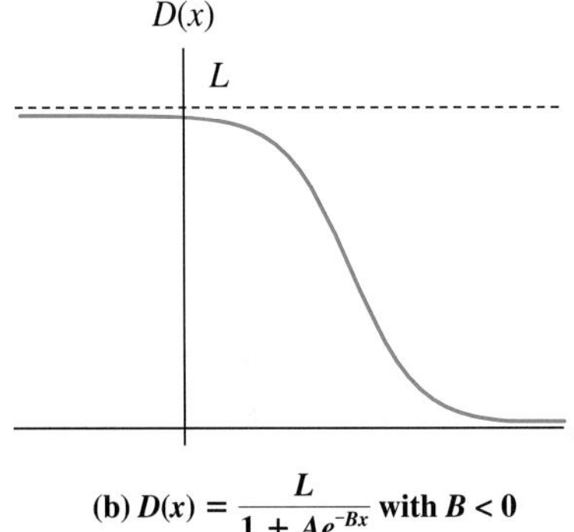

(b) $D(x) = \frac{L}{1 + Ae^{-Bx}}$ with $B < 0$

FIGURE 2.13

Logistic Model

A logistic model has an equation of the form

$$f(x) = \frac{L}{1 + Ae^{-Bx}}$$

The logistic function f increases if B is positive and decreases if B is negative. The graph of a logistic function is bounded by the horizontal axis and the line $f(x) = L$. We refer to L as the limiting value of the function.

Finding Logistic Models

The *Instructor's Resource Guide* contains a discussion of how to incorporate this example in an engaging way by using a random-number generator to help model the spread of a virus.

Suppose that 10 people work in an office. Someone comes to work with a highly contagious virus that causes the person to sneeze once every hour. If an infected person sneezes near an uninfected person, then the uninfected person becomes infected instantly and begins sneezing within the next hour.

Imagine that the spread of the virus is as summarized in Table 2.25 and graphed in Figure 2.14.

TABLE 2.25

Time	Total number of people infected
0 (before 8 a.m.)	1
1 (8 to 9 a.m.)	2
2 (9 to 10 a.m.)	4
3 (10 to 11 a.m.)	6
4 (11 a.m. to noon)	8
5 (noon to 1 p.m.)	9

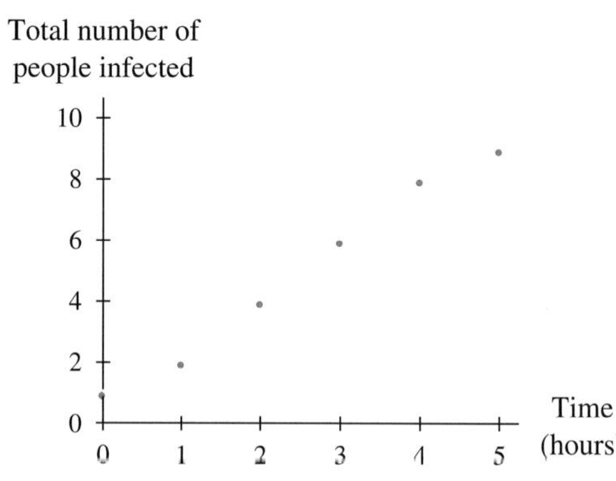

FIGURE 2.14

Assuming that no one else comes into the office when someone is sneezing and that no one leaves the office and sneezes, there are only 10 people who can get the virus. The maximum potential number of people infected is therefore 10. This means that as time increases, the number infected can approach, but never exceed, 10. Thus $L = 10$ would make sense as a leveling-off, or limiting, value for this situation. However, a logistics equation calculated by a calculator or computer using a best-fit technique[15] is

$$I(t) = \frac{9.98632}{1 + 9.33812e^{-0.89511t}} \text{ people infected}$$

t hours after 8 a.m. Note that the leveling-off value here is $L = 9.98632$, which is approximately, but not exactly, 10 people.

A graph of the function I is shown in Figure 2.15. The dotted, horizontal line portrays the upper asymptote. Note that the curvature on the left side of this graph is **concave up** ‿ whereas the curvature on the right side is **concave down** ⁀. The point on the graph at which the concavity changes is called the **inflection point**. The inflection point on the logistic graph modeling the spread of the virus (Figure 2.15) is marked with a black dot. In some situations, inflection points have very important interpretations. We later use calculus to find and help interpret these special points.

As with exponential models, the output of a logistic model of the form $y = \frac{L}{1 + Ae^{-Bx}}$ cannot be zero, and the regression routines in most technologies will not accept negative output values. It is possible to shift a set of data vertically in order to find a logistic model for data with nonpositive output values. See Section 2.5 for details.

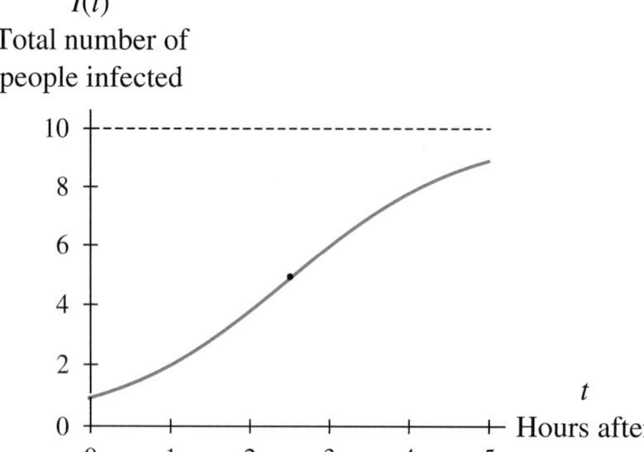

FIGURE 2.15

15. Different calculators and computer software may use slightly different methods for calculating a "best-fit" equation. This results in slightly different coefficients and exponents in the models. Recall that our goal in fitting curves to data is to develop models so that we can use calculus to study the general behavior of the data. Slightly different equations will not significantly affect our study of the general behavior of the data. We report logistic models with equations obtained from a TI-83 calculator.

When you use a calculator or a computer program to find an equation and construct the graph of a logistic model for a set of data, be sure to align the input data. The same numerical computation problems that we discussed for exponential models can occur with logistic models because of the exponential term in the denominator.

EXAMPLE 1 *Swimsuit Sales*

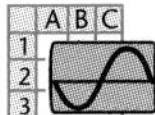

Table 2.26 shows the total number of swimsuits that a department store sold during the first three quarters of last year.

TABLE 2.26

End of month	Total number of swimsuits
January	4
February	12
March	25
April	58
May	230
June	439
July	648
August	748
September	769

Find a logistic model that fits the data. What is the end behavior of the model as time increases?

Solution A scatter plot of the data (Figure 2.16) suggests that a logistic model is appropriate.

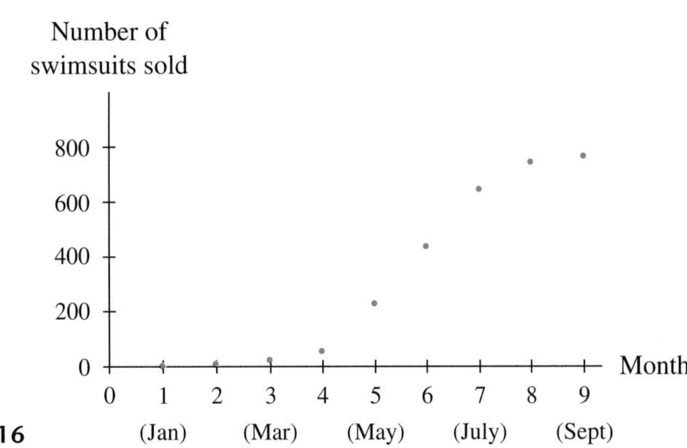

FIGURE 2.16

A possible logistic model is

$$S(x) = \frac{786.70445}{1 + 1464.70161e^{-1.26110x}} \text{ swimsuits}$$

by the end of month x, where $x = 1$ in January, $x = 2$ in February, and so on.

The limiting value of this model is $L = 786.70445$ or, in context, approximately 787 swimsuits. The graph of S on a scatter plot appears to fit fairly well (see Figure 2.17). The dotted line in Figure 2.17 denotes the upper asymptote. As time increases, the number of swimsuits sold approaches 787.

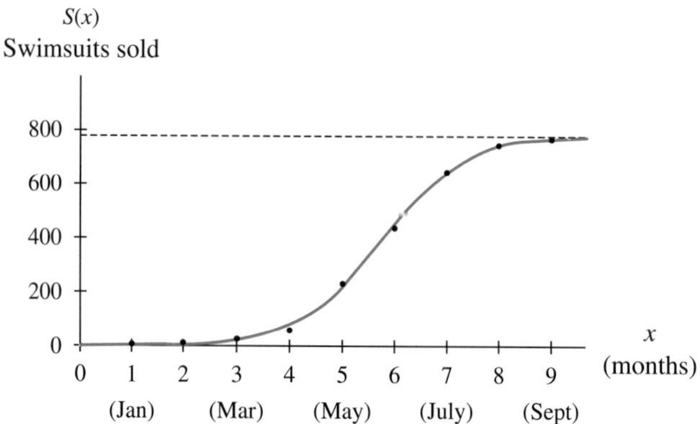

FIGURE 2.17

The preceding logistic curves begin near zero and then increase toward a limiting value L. As previously mentioned, there are also situations where the curve begins near its limiting value and then decreases toward zero.

EXAMPLE 2 *Heights of College Men*

Of a group of 200 college men surveyed, the number who were taller than a given number of inches is recorded in Table 2.27.

TABLE 2.27

Inches	Number of men
65	198
66	195
67	184
68	167
69	139
70	101
71	71
72	43
73	26
74	11
75	4
76	2

Find an appropriate model for the data. What is the end behavior of the model as height increases?

Solution A scatter plot of the data is shown in Figure 2.18.

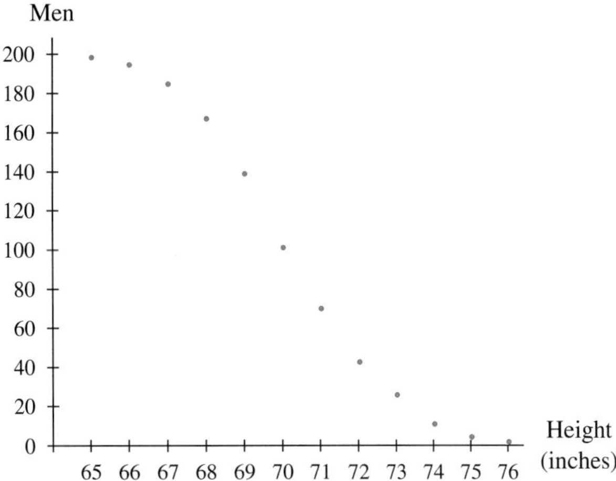

FIGURE 2.18

A logistic model for the data is

$$M(x) = \frac{205.66709}{1 + 0.02998e^{0.69528x}} \text{ men out of 200 men surveyed}$$

who are taller than $x + 65$ inches. The two horizontal asymptotes of a graph of this function are $y = 0$ and $y \approx 205.7$. Because the function decreases, the end behavior as height increases is given by the lower asymptote; that is, as the heights increase, the number of men taller than a particular height approaches zero.

Recall that a logistic function is of the form $f(x) = \frac{L}{1 + Ae^{-Bx}}$. What happened to the negative sign before the B in Example 2? In that case, $B = -0.69528$. When the formula was written, the negatives canceled.

$$\text{Number of men} = \frac{205.66709}{1 + 0.02998e^{-(-0.69528)x}}$$

$$= \frac{205.66709}{1 + 0.02998e^{0.69528x}}$$

This will always be the case for a logistic function that decreases from its upper asymptote.

It is likely that a logistic model will have a limiting value that is lower or higher than the one indicated by the context. For instance, in Example 2 only 200 men were surveyed, but the limiting value is $L \approx 206$ men. This does not mean that the model is invalid, but it does indicate that care should be taken when extrapolating from logistic models.

2.3 Concept Inventory

- *Logistic function:* $f(x) = \frac{L}{1 + Ae^{-Bx}}$ *(exponential in nature with two horizontal asymptotes)*
- *Concave up and concave down*
- *Inflection point*
- *Limiting value*
- *Importance of aligning data*

2.3 Activities

Identify the scatter plots in Activities 1 through 6 as linear, exponential, logarithmic, logistic, or none of these. If you identify the scatter plot as none of these, give reasons.

1. 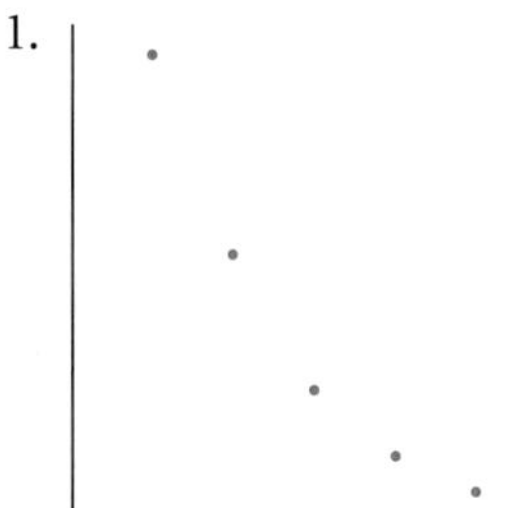

2. 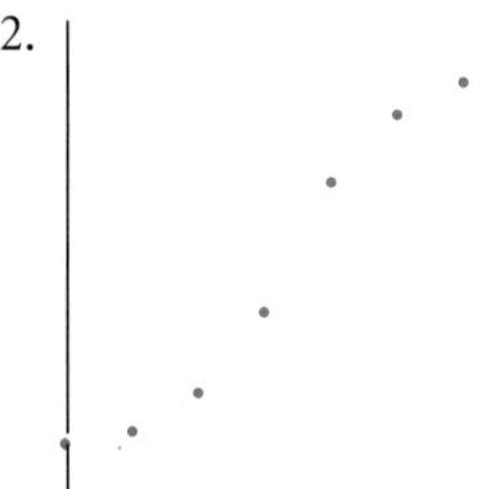

3. 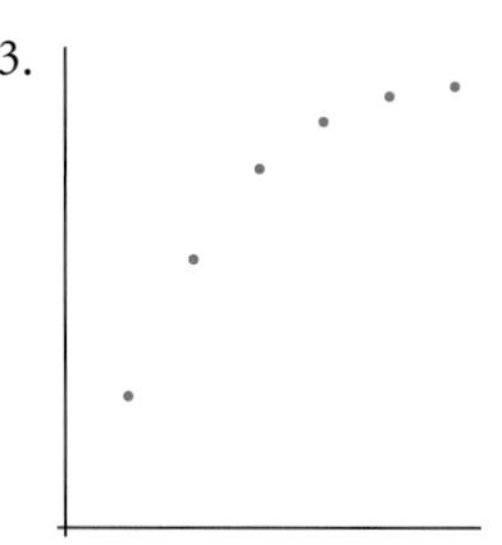

4. 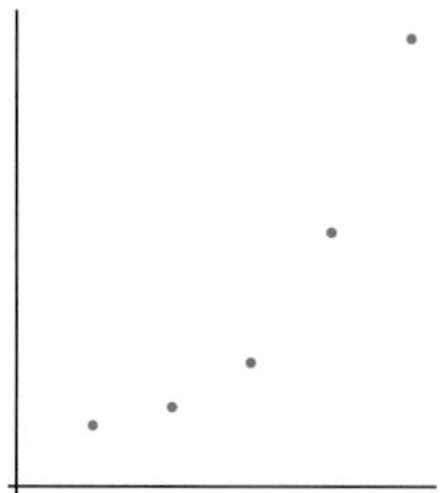

5. 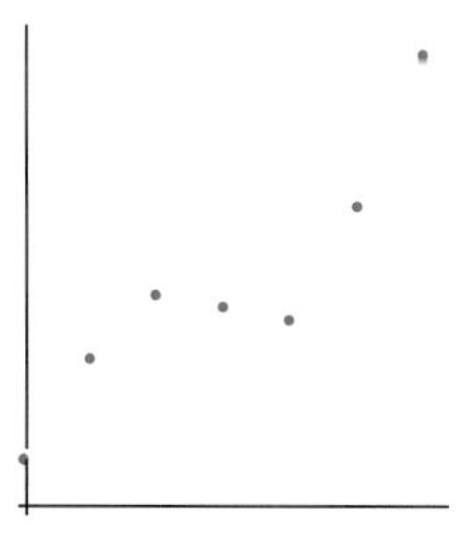

6. 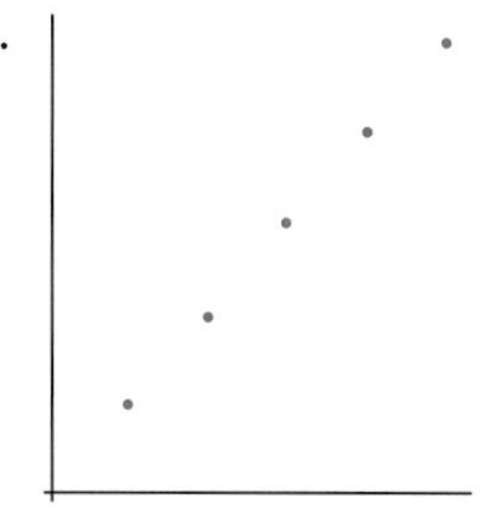

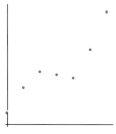

For Activities 7 through 12, tell whether the function is an increasing or a decreasing logistic function. Also identify the limiting value of the function and the two horizontal asymptotes.

7. $f(x) = \dfrac{100}{1 + 9.8e^{-0.98x}}$

8. $A(t) = \dfrac{1925}{1 + 321.1e^{1.86t}}$

9. $h(g) = \dfrac{39.2}{1 + 0.8e^{0.325g}}$

10. $k(x) = \dfrac{16.5}{1 + 1.86e^{-0.43x}}$

11. $g(x) = \dfrac{42}{1 + 5e^{-0.6x}}$

12. $m(p) = \dfrac{6.73}{1 + 39.67e^{0.0043p}}$

13. Table 2.28 gives the number of European, North American, and South American countries that issued postage stamps[16] from 1840 through 1880.

16. "The Curve of Cultural Diffusion," *American Sociological Review*, August 1936, pp. 547–556.

TABLE 2.28

Year	Total number of countries
1840	1
1845	3
1850	9
1855	16
1860	24
1865	30
1870	34
1875	36
1880	37

a. Find a logistic model for the data, and discuss how well the equation fits.

b. Sketch the graph of the equation in part *a*, and mark where the curve is concave up and where it is concave down. Label the approximate location of the inflection point.

14. Table 2.29 gives the total number of states associated with the national P.T.A. organization[17] from 1895 through 1931.

TABLE 2.29

Year	Total number of states	Year	Total number of states
1895	1	1915	30
1899	3	1919	38
1903	7	1923	43
1907	15	1927	47
1911	23	1931	48

a. Find a logistic model for the data.

b. What is the maximum number of states that could have joined the PTA by 1931? How does this number compare to the limiting value given by the model in part *a*?

15. The approximate numbers of patents for plow sulkies[18] beginning in 1865 are given in Table 2.30.

a. What is a *plow sulky*, and to what modern farm implement was it a precursor?

b. Find a logistic model for the data.

17. Hamblin, Jacobsen, and Miller, *A Mathematical Theory of Social Change* (New York: John Wiley & Sons, 1973).

18. Ibid.

TABLE 2.30

Year	Cumulative number of plow sulky patents
1871	200
1877	340
1883	980
1889	1800
1895	2200
1901	2400
1907	2500
1913	2550
1919	2620
1925	2700

c. Discuss why it is logical that the total number of patents for a new invention would increase according to a logistic equation.

16. The personnel manager for a contracting company keeps track of the total number of labor hours spent on a construction job each week during the construction. Partial data appear in Table 2.31.

TABLE 2.31

Weeks after start of the construction project	Cumulative labor hours
1	25
4	158
7	1254
10	5633
13	9280
16	10,010
19	10,100

a. Discuss why a logistic model is appropriate for the data.

b. Find a logistic model for the data.

c. In which week is the 8000-labor-hour point reached?

d. If the company charges $50 per labor hour, what would the total labor cost be for this construction job?

17. In the fall of 1918, an influenza epidemic hit the United States Navy. It spread to the Army, to American civilians, and ultimately to the world. It is estimated that 20 million people had died from the epidemic by 1920. Of these, 550,000 were Americans—over 10 times the number of World War I

battle deaths. Table 2.32 gives the total numbers of Navy, Army, and civilian deaths[19] that resulted from the epidemic in 1918.

TABLE 2.32

	Total deaths		Total civilian deaths in 45 major cities
Week ending	Navy	Army	
August 31	2		
September 7	13	40	
September 14	56	76	68
September 21	292	174	517
September 28	1172	1146	1970
October 5	1823	3590	6528
October 12	2338	9760	17,914
October 19	2670	15,319	37,853
October 26	2820	17,943	58,659
November 2	2919	19,126	73,477
November 9	2990	20,034	81,919
November 16	3047	20,553	86,957
November 23	3104	20,865	90,449
November 30	3137	21,184	93,641

a. Find logistic equations to fit each set of data. In each case, graph the equation on the data.

b. Write models for each of the data sets.

c. Compare the limiting values of the models in part *b* with the highest data values in the table. How many more people do the models indicate died from the epidemic after November 30? Do you believe that the limiting values indicated by the models accurately reflect the ultimate number of deaths? Explain.

18. Again consider the virus in the office example, except this time look at it from the viewpoint of how many people have *not* yet been infected by time *x*.

a. Using the data given in Table 2.25 on page 128, fill in the table at the top of the next column.

b. Examine a scatter plot of the "not infected" data, and describe its curvature.

c. Is the scatter plot increasing or decreasing? How does this compare with the scatter plot of the original "infected" data?

Time	Total number of people not infected
0 (before 8 a.m.)	
1 (8 to 9 a.m.)	
2 (9 to 10 a.m.)	
3 (10 to 11 a.m.)	
4 (11 a.m. to noon)	
5 (noon to 1 p.m.)	

d. Find a logistic model for the "not infected" data.

e. Are the horizontal asymptotes for the "not infected" data the same as those for the "infected" data? Why?

f. Is the value of B in the logistic model positive or negative? How is the sign of B related to the curvature of the logistic curve?

19. In 1949 the United States experienced the second worst polio epidemic in its history. (The worst was in 1952.) Table 2.33 gives the cumulative number of polio cases[20] diagnosed on a monthly basis.

TABLE 2.33

Month	Total number of polio cases
January	494
February	759
March	1016
April	1215
May	1619
June	2964
July	8489
August	22,377
September	32,618
October	38,153
November	41,462
December	42,375

a. Observe a scatter plot of the data from January through June. Describe the concavity and estimate the location of the inflection point indicated by the scatter plot. Does this portion of the data appear to be logistic?

19. A.W. Crosby, Jr., *Epidemic and Peace 1918* (Westport, CT: Greenwood Press, 1976).

20. *Twelfth Annual Report*, The National Foundation for Infantile Paralysis, 1949.

b. Observe a scatter plot of the entire data set given. Does the entire data set appear to be logistic?

c. Find a logistic model for the data.

d. Observe the model in part *c* graphed on a scatter plot of the data between January and June. Discuss how well the graph fits this portion of the data.

20. The total numbers of visitors to an amusement park that stays open all year are given in Table 2.34.

TABLE 2.34

Month	Cumulative number of visitors by the end of the month (thousands)
January	25
February	54
March	118
April	250
May	500
June	898
July	1440
August	1921
September	2169
October	2339
November	2395
December	2423

a. Find a logistic model for the data.

b. The park owners have been considering closing the park from October 15 through March 15 each year. How many visitors will they potentially miss by this closure?

21. A chemical reaction begins when a certain mixture of chemicals reaches 95°C. The reaction activity is measured in Units (U) per 100 microliters (100μL) of the mixture. Measurements[21] during the first 18 minutes after the mixture reaches 95°C are listed in Table 2.35.

a. Examine a scatter plot of the data. Estimate the limiting value. Estimate at what time the inflection point occurs.

b. Find a logistic model for the data. What is the limiting value for this logistic function?

21. David E. Birch et al., "Simplified Hot Start PCR," *Nature*, vol. 381 (May 30, 1996), p. 445.

TABLE 2.35

Time (minutes)	Activity (U/100μL)	Time (minutes)	Activity (U/100μL)
0	0.10	10	1.40
2	0.10	12	1.55
4	0.25	14	1.75
6	0.60	16	1.90
8	1.00	18	1.95

c. Use the model to estimate by how much the reaction activity increased between 7 and 11 minutes.

22. Table 2.36 records the volume of sales (in thousands) of a popular movie for the first 18 months after it was released on videocassette.

TABLE 2.36

Months after release	Number of cassettes sold each month (thousands)
1	580
2	565
3	527
4	467
5	321
6	291
7	204
8	131
9	79
10	61
11	31
12	17
13	9
14	4
15	3
16	3
17	2
18	2

a. Examine a scatter plot, and describe its curvature.

b. Explain why this type of data could reasonably be modeled by a logistic equation.

c. Would the value B of a logistic model be positive or negative in this case?

d. Find a logistic model for the data.

23. San Francisco Giants legend Willie Mays's cumulative number of stolen bases between 1951 and 1963 are as shown in Table 2.37.

TABLE 2.37

Year	Cumulative stolen bases
1951	7
1952	11
1953	11
1954	19
1955	43
1956	83
1957	121
1958	152
1959	179
1960	204
1961	222
1962	240
1963	248

a. Find a logistic model for the data. Comment on how well the logistic equation fits the data.

b. What is the interpretation of first differences in this context?

c. Use the model in part *a* to estimate the number of bases that Mays stole in 1964. Compare this estimate with 19, the actual number of stolen bases in 1964.

24. Total emissions of lead[22] into the atmosphere for selected years between 1970 and 1995 are as shown in Table 2.38.

TABLE 2.38

Year	Lead emissions (millions of tons)
1970	220.9
1975	159.7
1980	74.2
1985	22.9
1990	5.0
1995	3.9

a. Find a logistic model for these data.

b. Use limit notation to express the two horizontal asymptotes of the logistic function.

c. What is the end behavior of this function as time increases?

d. Why have lead emissions in the environment declined since 1970?

25. A 1998 United Nations population study reported the world population between 1804 and 1987 and projected the population through 2071. These populations are as shown in Table 2.39.

TABLE 2.39

Year	Population (billions)	Year	Population (billions)
1804	1	1999	6
1927	2	2011	7
1960	3	2025	8
1974	4	2041	9
1987	5	2071	10

a. Find a logistic model for world population. Discuss how well the equation fits the data.

b. Do you consider the model appropriate to use in predicting long-term world population behavior? According to the model, what will ultimately happen to world population?

c. Do you believe the model should be used to estimate the world population in 1850? in 1990? Explain.

26. The percentages of all banks that levied surcharges on transactions at automated teller machines[23] between 1996 and 1999 are as shown in Table 2.40.

TABLE 2.40

Year	1996	1997	1998	1999
ATM surcharges (percent)	15	45	71	93

a. Examine a scatter plot of the data. Discuss the curvature suggested by the scatter plot.

b. Would a logistic model be appropriate in this situation? Why or why not?

c. Find a logistic model for the data. Express the limiting value of the function using a limit statement. What does the limiting value tell you about using the model to extrapolate?

22. *Statistical Abstract*, 1998.

23. U.S. Public Interest Research Group National Survey.

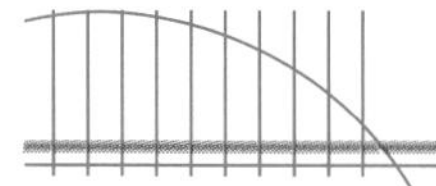

2.4 Polynomial Functions and Models

It is possible to streamline this section by drawing on your students' knowledge of quadratic and cubic functions from previous math courses and omitting much of the discussion in this section. The basics of this section are covered in the three examples. You may wish to mention the role of second differences in examining quadratic data because second differences are discussed in Section 2.6.

Polynomial functions and models have been used extensively throughout the history of mathematics. Their successful use stems from both their presence in certain natural phenomena and their relatively simple application. Prior to the availability of inexpensive computing technology, ease of use made polynomials the most widely applied models because they were often the only ones that could be calculated with pencil-and-paper techniques. However, we no longer have to deal with this restriction, and we call on polynomials only when they are appropriate.

Even though higher-degree polynomials are useful in some situations, we limit our discussion of polynomial functions and models to the linear, quadratic, and cubic cases.

Quadratic Modeling

A roofing company in Miami keeps track of the number of roofing jobs it completes each month. The data from January through June follow.

Month	January	February	March	April	May	June
Number of Jobs	12	14	22	37	58	84

The first differences and percentage differences (percentage changes) are not close to being constant, so we conclude that some model other than a linear model or an exponential model is appropriate in this case. Note, however, that the differences between the first differences are nearly constant.

Number of Jobs	12		14		22		37		58		84
First Differences		2		8		15		21		26	
			6		7		6		5		

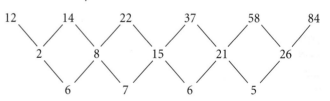

Remind your students that first and second differences should be calculated only if the input data are evenly spaced.

We call these **second differences.** When first differences are constant, the data can be modeled by a linear equation. When second differences are constant, the data can be modeled by the **quadratic function** $f(x) = ax^2 + bx + c$. A quadratic equation of this form defines a function f with input x. From now on, when we refer to a quadratic equation or function, we will consider an equation of the form $f(x) = ax^2 + bx + c$.

The graph of a quadratic function is a **parabola.** By examining a scatter plot of the roofing company data with January aligned as month 1, we observe a plot (see Figure 2.19) that suggests a portion of the familiar parabolic shape.

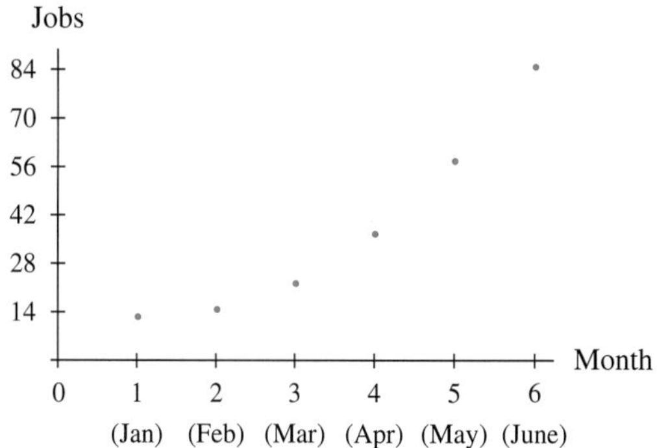

FIGURE 2.19

For this illustration, the rate of change in roofing jobs measures how rapidly the number of jobs is increasing (or possibly decreasing if the trend changes) each month. Unlike the case for a linear function, the rate of change for a quadratic function is *not* constant. We later discuss in detail the rate of change of a quadratic function.

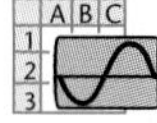

How do we obtain a quadratic model in this situation? After noting that the second differences are nearly, but not quite, constant and observing that the scatter plot suggests that the data are close to quadratic, we use technology to obtain a quadratic equation that fits the data. You should find the quadratic model to be

$$J(x) = 3.07143x^2 - 7.01429x + 15.8 \text{ jobs}$$

where $x = 1$ in January, $x = 2$ in February, and so on. Note in Figure 2.20 that the function J provides an excellent fit to the data.

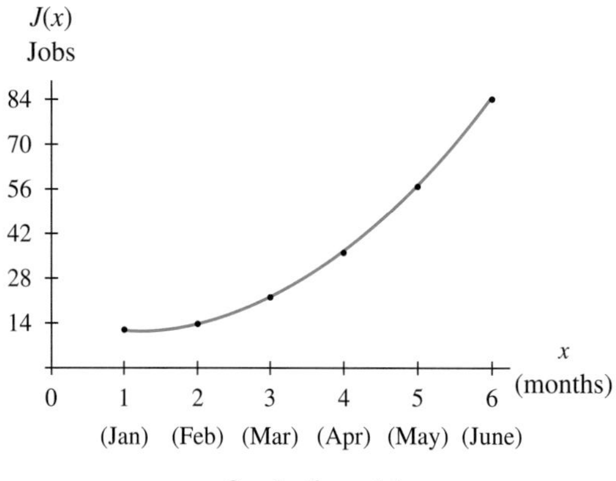

FIGURE 2.20

If we are willing to assume that this quadratic model continues to model the number of roofing jobs for the next 3 months, how many jobs would we predict the company will have in August? Substituting $x = 8$ into the quadratic equation yields

$$J(8) = 3.07143(8)^2 - 7.01429(8) + 15.8 = 156.2571 \text{ jobs}$$

Recall from the numerical guidelines established in Section 1.5 that output obtained from a model can be only as accurate as the output data from which the model was obtained. Because the roofing job data were reported as integers, we must report the August prediction as 156 jobs.

EXAMPLE 1 *Fuel Consumption*

The average fuel consumption[24] of passenger vehicles, buses, and trucks from 1970 through 1995 is given in Table 2.41.

24. *The Time Almanac*, 1999.

TABLE 2.41

Year	1970	1975	1980	1985	1990	1995
Average fuel consumption (gallons per vehicle per year)	830	790	712	685	677	700

a. Find a quadratic model for the data.

b. Compare the minimum of the parabola with the minimum of the data.

Solution

a. First, align the years so that 1970 is 0, 1975 is 5, and so on. A scatter plot of the aligned data suggests a concave-up shape with a minimum around 1990. See Figure 2.21.

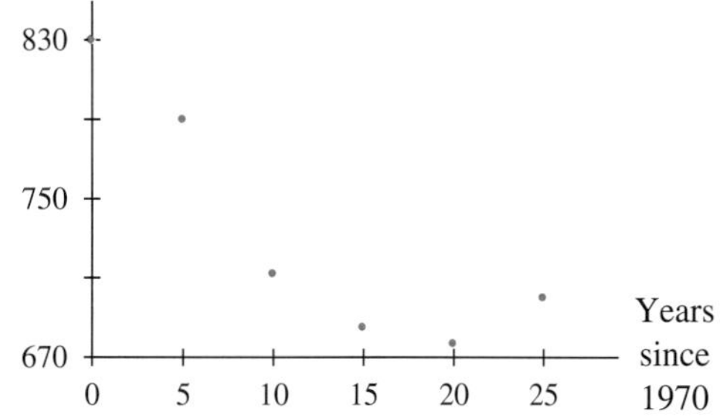

FIGURE 2.21

A quadratic model for the data is

$$F(t) = 0.425t^2 - 16.431t + 840.321 \text{ gallons per vehicle per year}$$

where t is the number of years since 1970.

b. Looking at the quadratic function graphed on the scatter plot (see Figure 2.22), we observe that the minimum of the parabola is slightly to the left of and higher than the minimum data point. This means that the model slightly overestimates minimum fuel consumption and estimates that the minimum consumption will occur slightly before it does in the data table.

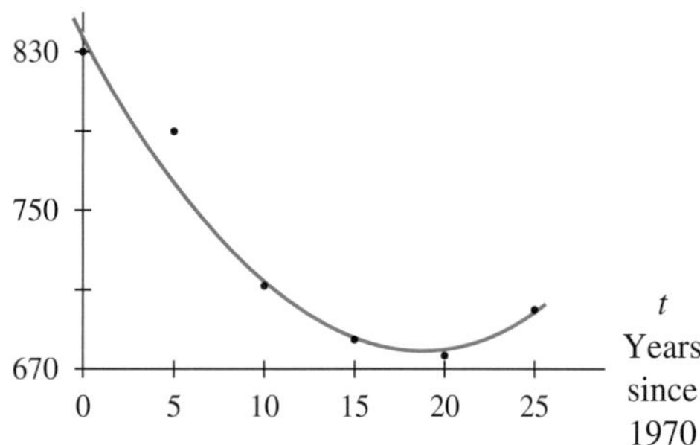

FIGURE 2.22

Quadratic or Exponential?

The fuel consumption data in Example 1 have an obvious minimum, so it is easy to identify the data as quadratic. However, not all quadratic data sets have an obvious minimum (or maximum). Sometimes all we see is the left side or right side of a parabola.

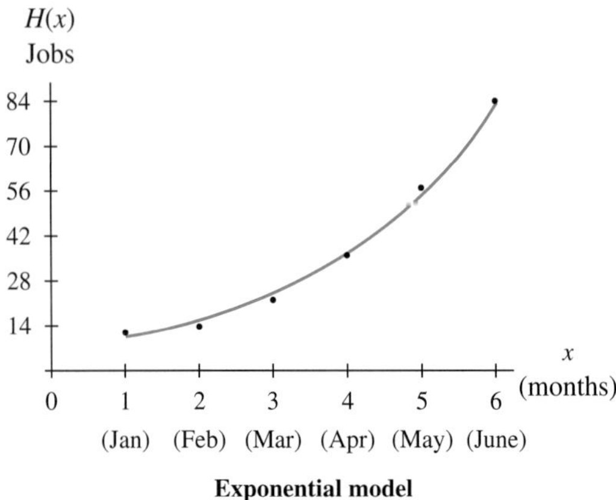

FIGURE 2.23

You may have noticed that the scatter plot of the roofing job data (Figure 2.19) looks as if an exponential curve might fit, even though the percentage differences are not perfectly constant. If you found an exponential model, it would be

$$H(x) = 6.92293(1.51388^x) \text{ jobs}$$

where $x = 1$ in January, $x = 2$ in February, and so on. A graph of the exponential function H on the scatter plot is shown in Figure 2.23. The exponential equation and the quadratic equation (Figure 2.20) both appear to fit the roofing job data well.

In cases in which two equations appear to fit a set of data equally well, you need to consider carefully which one you would prefer to use. (See the hints for model selection in Section 2.6.) Often, however, it is obvious that one model is more appropriate than the other. This is the case in Example 2.

EXAMPLE 2 *United States Population*

Table 2.42 shows the population[25] of the contiguous states of the United States.

TABLE 2.42

Year	Population (millions)	Year	Population (millions)
1790	3.929	1910	91.972
1810	7.240	1930	122.775
1830	12.866	1950	150.697
1850	23.192	1970	203.302
1870	39.818	1990	248.718
1890	62.948		

Find an appropriate model for the data.

Solution An examination of the scatter plot shows an increasing, concave-up shape (see Figure 2.24).

25. *Statistical Abstract*, 1998.

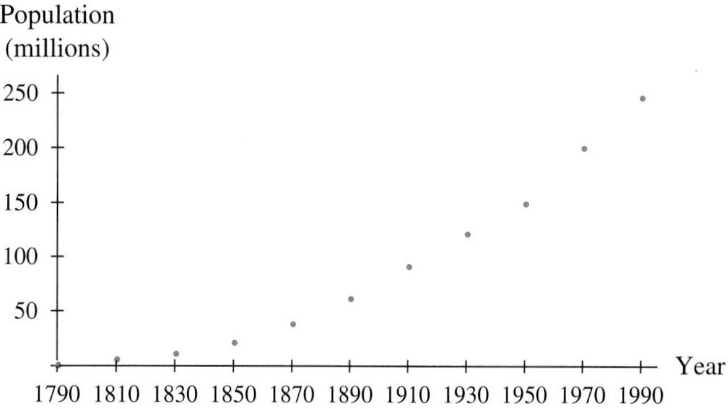

FIGURE 2.24

At first, it seems logical to try an exponential model for the data. An exponential model is

$$E(t) = 5.74677\ (1.02096^t) \text{ million people}$$

where t is the number of years since 1790. However, this equation does not seem to fit the data very well, as Figure 2.25 shows.

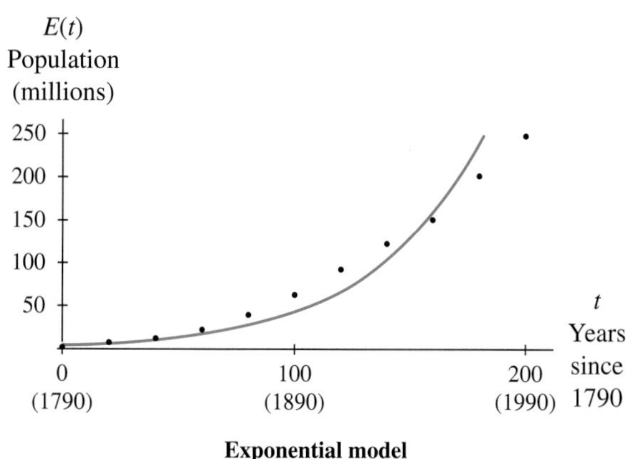

FIGURE 2.25

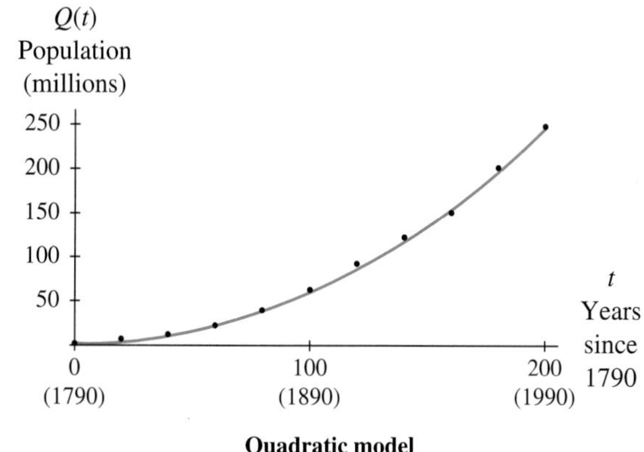

FIGURE 2.26

Another possibility is a quadratic model. The right half of a parabola that opens upward could fit the scatter plot. A quadratic model is

$$Q(t) = 0.00640t^2 - 0.06588t + 4.87697 \text{ million people}$$

where t is the number of years since 1790. When graphed on the scatter plot, this equation appears to be a very good fit (see Figure 2.26). It is the more appropriate model for the population of the contiguous United States using the given data.

In the foregoing examples, you have seen data sets that appear to be quadratic. That is, they may be reasonably modeled by a quadratic equation. In each case, the equation had a positive coefficient before the squared term. This value is called the **leading coefficient** because it is usually the first one that you write down in the equation $f(x) = ax^2 + bx + c$. Also, in each case, the graph of the equation appeared to be part of a parabola opening upward. Remember that we call such curvature *concave up*.

Table 2.43 gives the population of Cleveland, Ohio,[26] from 1900 through 1980.

TABLE 2.43

Year	Population
1900	381,768
1910	560,663
1920	796,841
1930	900,429
1940	878,336
1950	914,808
1960	876,050
1970	750,879
1980	573,822

A scatter plot of the data suggests a parabola opening downward (or *concave down*). The scatter plot and the graph of a quadratic equation fitted to the data are shown in Figure 2.27.

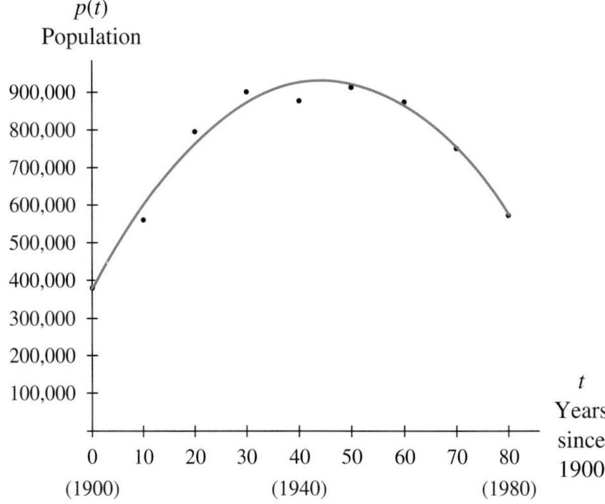

FIGURE 2.27

Would you expect the leading coefficient of the equation to be positive or negative? The population of Cleveland is given by

$$p(t) = -279.995t^2 + 24{,}919.057t + 374{,}959.903 \text{ people}$$

where t is the number of years since 1900.

When the graph of a quadratic function is concave up, its leading coefficient is positive; when the graph of a quadratic function is concave down, its leading coefficient is negative. Curvature will be important in later discussions that involve concepts of calculus.

We summarize quadratic models as follows:

26. *Statistical Abstract*, 1998.

> **Quadratic Model**
>
> A quadratic model has an equation of the form
>
> $$f(x) = ax^2 + bx + c$$
>
> where $a \neq 0$. The graph of a quadratic function is a concave-up parabola if $a > 0$ and is a concave-down parabola if $a < 0$. (See Figures 2.28a and 2.28b.)

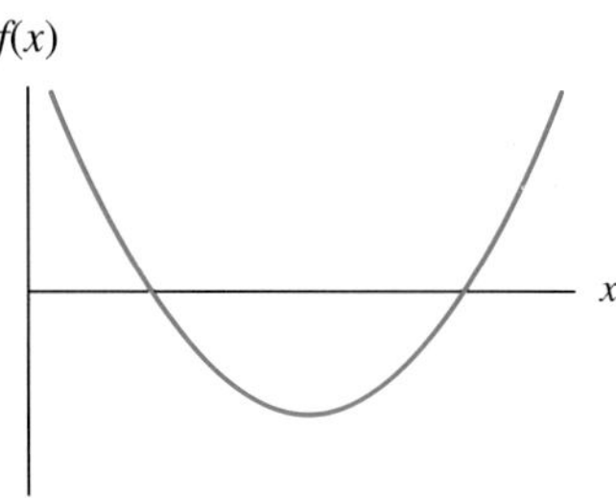

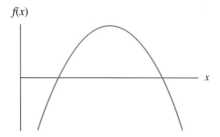

(a) $f(x) = ax^2 + bx + c$ with $a > 0$
$\lim_{x \to \pm\infty} f(x) \longrightarrow \infty$

f(x)

x

(b) $f(x) = ax^2 + bx + c$ with $a < 0$
$\lim_{x \to \pm\infty} f(x) \longrightarrow -\infty$

FIGURE 2.28

Cubic Modeling

We saw that when the first differences of a set of evenly spaced data are constant, the data can be modeled perfectly by the linear equation $y = ax + b$. Likewise, when the second differences of evenly spaced input data are constant, the data can be modeled perfectly by the quadratic equation $y = ax^2 + bx + c$. It is possible for the third differences to be constant. In this case, the data can be modeled perfectly by a cubic equation. A cubic equation of the form $f(x) = ax^3 + bx^2 + cx + d$ is a function with input x. From now on, when we refer to a cubic equation or function, we consider the equation to be of this form.

Because in the real world we are extremely unlikely to encounter data that are perfectly cubic, we will not look at third differences. Instead, we will examine a scatter plot of the data to see whether a cubic model may be appropriate. Figure 2.29 shows the graphs of some cubic equations.

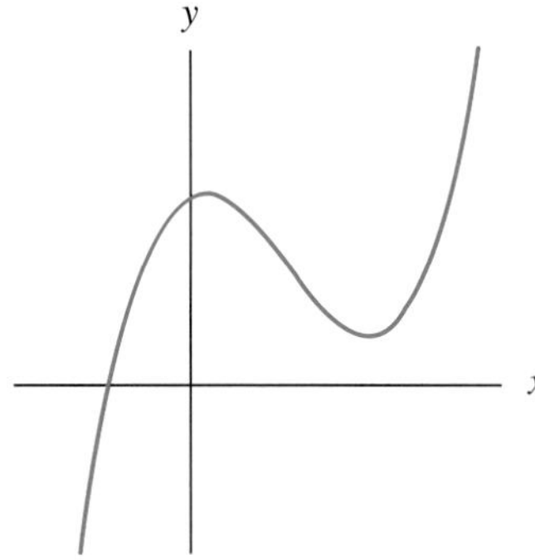

FIGURE 2.29a

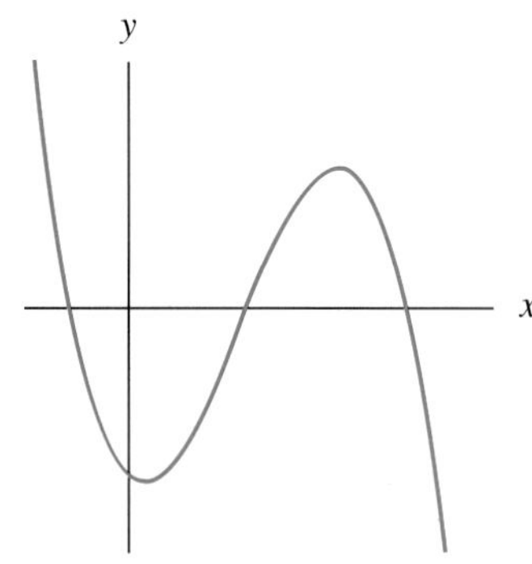

FIGURE 2.29b

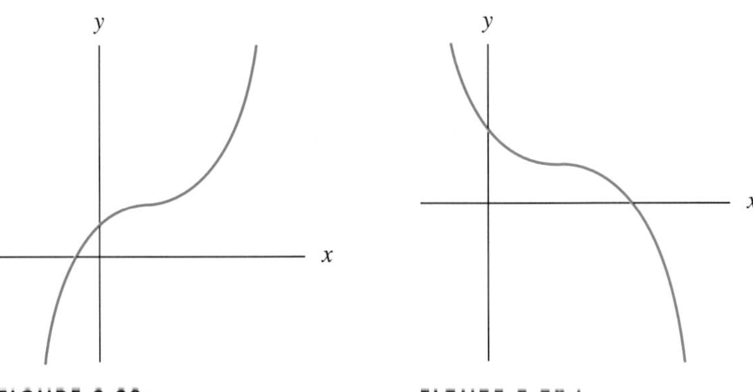

FIGURE 2.29c FIGURE 2.29d

The four graphs in Figure 2.29 are typical of all cubic equations. That is, every cubic function has a graph that resembles one of these four. Figures 2.29a and 2.29c correspond to equations in which $a > 0$, and Figures 2.29b and 2.29d are graphs of equations in which $a < 0$. For a cubic equation $f(x) = ax^3 + bx^2 + cx + d$ with $a > 0$, the end behavior is described by the limit statements

$$\lim_{x \to -\infty} f(x) \to -\infty \quad \text{and} \quad \lim_{x \to \infty} f(x) \to \infty$$

Similarly, if $a < 0$, then

$$\lim_{x \to -\infty} f(x) \to \infty \quad \text{and} \quad \lim_{x \to \infty} f(x) \to -\infty$$

Figure 2.30 shows scatter plots of data sets that could be modeled by cubic equations.

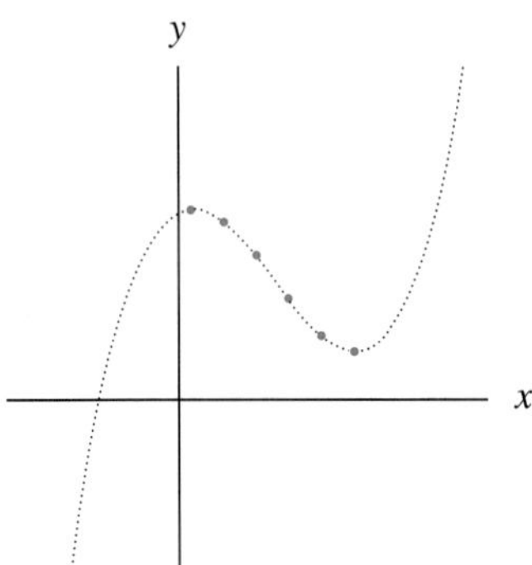

FIGURE 2.30a

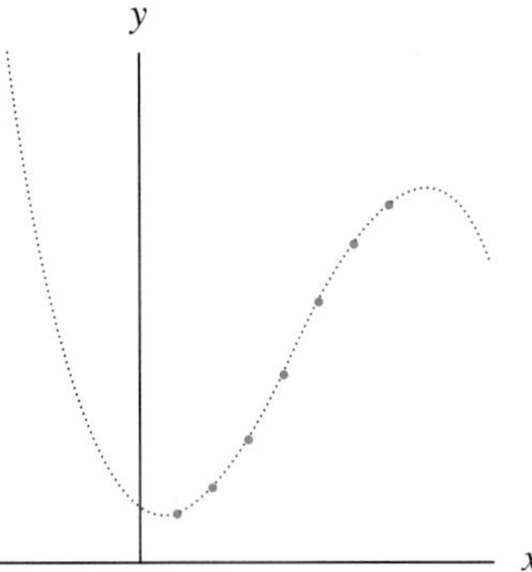

FIGURE 2.30b

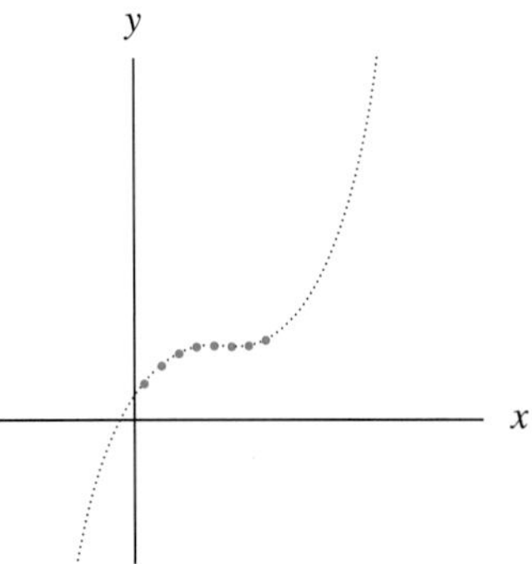

FIGURE 2.30c

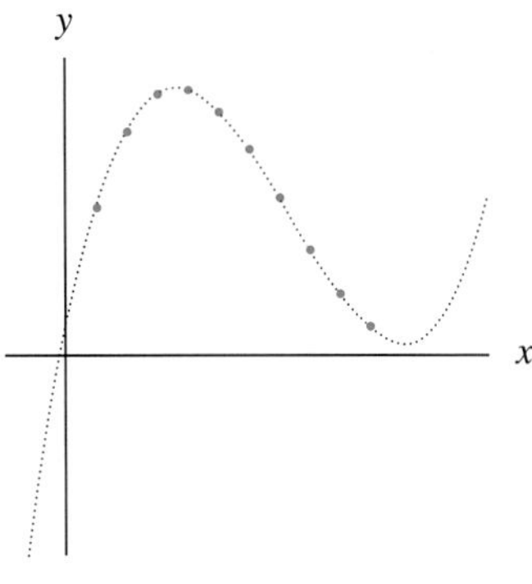

FIGURE 2.30d

You may have already noticed that in every cubic function, the curvature of the graph changes once from concave down to concave up, or vice versa. As we noted with the graph of a logistic curve, the point on the graph at which concavity changes is called the *inflection point.* All cubic functions have one inflection point. The approximate location of the inflection point in each of the graphs in Figure 2.31 is marked with a dot. In Chapter 5, we see how calculus can be used to determine the exact location of the inflection point of a cubic function.

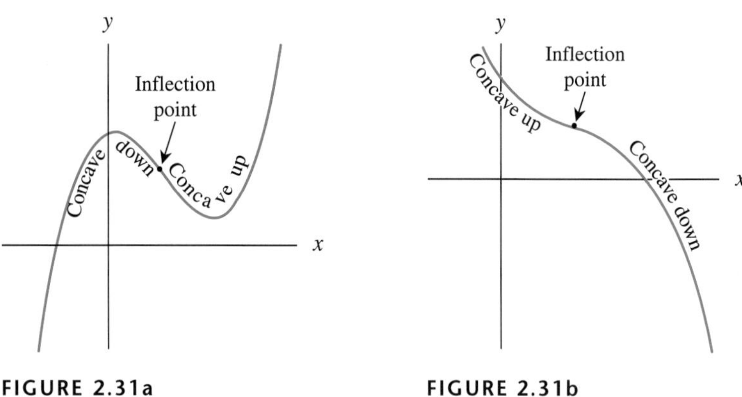

FIGURE 2.31a

FIGURE 2.31b

It is often the case that a portion of a cubic function appears to fit extremely well a set of data that can be adequately modeled with a quadratic function. In an effort to keep things as simple as possible, we adopt the following convention:

Even though it is true that a function of order $n + 1$ will always statistically fit a set of data better than a function of order n, we choose to apply the principle of "Keep it Simple, Students!" and use the model of lowest degree that adequately fits the data and describes the underlying situation.

> If the scatter plot of a set of data fails to exhibit an inflection point, then it is not appropriate to fit a cubic equation to the data.

We must be extremely cautious when using cubic models to extrapolate. For the data sets whose scatter plots are shown in Figures 2.30a and 2.30c, the functions indicated by the dotted curves appear to follow the trend of the data. However, in Figure 2.30b, it would be possible for additional data to level off (as in a logistic model), whereas the cubic function takes a downward turn. Also, additional data in Figure 2.30d might continue to get closer to the x-axis, whereas the cubic function that is fitted to the available data begins to rise.

Cubic Model

A cubic model has an equation of the form

$$f(x) = ax^3 + bx^2 + cx + d$$

where $a \neq 0$. The graph of a cubic function has one inflection point and no limiting values.

EXAMPLE 3 *Natural Gas Prices*

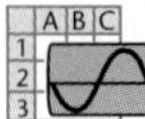

The average price in dollars per 1000 cubic feet of natural gas[27] for residential use in the United States from 1980 through 1997 is given in Table 2.44.

TABLE 2.44

Year	Price (dollars)	Year	Price (dollars)
1980	3.68	1995	6.06
1982	5.17	1996	6.34
1985	6.12	1997	6.93
1990	5.77		

a. Find an appropriate model for the data. Would it be wise to use this model to predict future natural gas prices?

b. Use the model in part *a* to estimate the price in 1994.

c. According to the model, when did the price of 1000 cubic feet of natural gas first exceed \$6.00?

Solution

a. An examination of a scatter plot shows that a cubic model is appropriate. Note that the scatter plot shown in Figure 2.32 appears to be mostly concave down between 1980 and 1990 but then is concave up between 1990 and 1997. That is, there appears to be an inflection point (a change of concavity) near 1990.

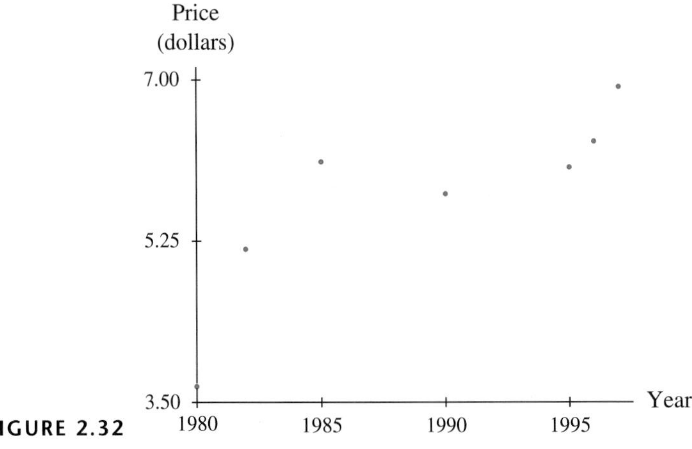

FIGURE 2.32

A cubic model for the price of natural gas is

$$P(x) = 0.004265x^3 - 0.11816x^2 + 0.9674x + 3.681 \text{ dollars}$$

where x is the number of years since 1980. A graph of the equation over the scatter plot is shown in Figure 2.33.

Note that the graph is increasing to the right of about 1993. Natural gas prices will probably not continue to rise indefinitely as the cubic function does, so it is unwise to use the model to predict future prices of natural gas. Additional data should be obtained to see the pattern past 1997.

27. *Statistical Abstract*, 1992 and 1998.

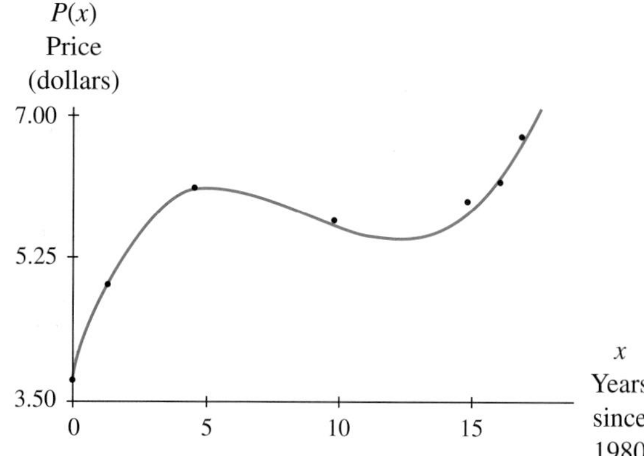

FIGURE 2.33

b. In 1994 the average price of 1000 cubic feet of natural gas was $P(14) \approx \$5.77$.

c. To determine when the average price first exceeded \$6.00, solve the equation $P(x) = 6$. In Figure 2.34, a dotted line is drawn at $P(x) = \$6.00$.

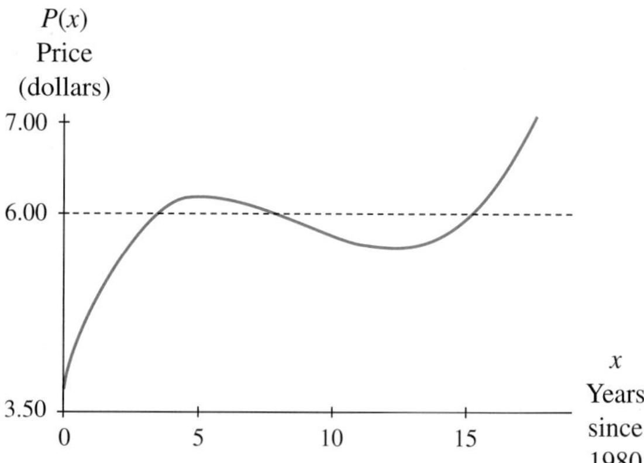

FIGURE 2.34

The line intersects the graph at the three places, so there are three solutions to the equation $P(x) = 6$. The solutions are $x \approx 4.33, 8.38$, and 15.00. We seek the smallest solution ($x \approx 4.33$), which corresponds to the first time the average price is \$6.00. However, because the data represent yearly averages, we must interpret the model discretely. In 1984 the average price was less than \$6.00, and in 1985 the average price was more than \$6.00. Therefore, the average price first exceeded \$6.00 in 1985.

We can also model the natural gas prices in Example 3 with a cubic equation by renumbering the years so that x is the number of years since 1900:

$$\text{Price} = 0.004265x^3 - 1.14182x^2 + 101.7663x - 3013.7594 \text{ dollars}$$

Although the two models have different inputs, they yield the same results. Notice that the coefficient on x^3 is the same in both models. This will be the case for any alignment.

If you are using the two-semester text and integrating Chapter 8 throughout Chapters 1–7, you should insert Section 8.2 between Sections 2.4 and 2.5. See the *Instructor's Resource Guide* for more information about this topic.

2.4 Concept Inventory

- *First and second differences*
- *Quadratic function: $f(x) = ax^2 + bx + c$ (constant second differences, no change in concavity)*
- *Parabola*
- *Cubic function: $f(x) = ax^3 + bx^2 + cx + d$ (one change in concavity)*
- *Inflection point*

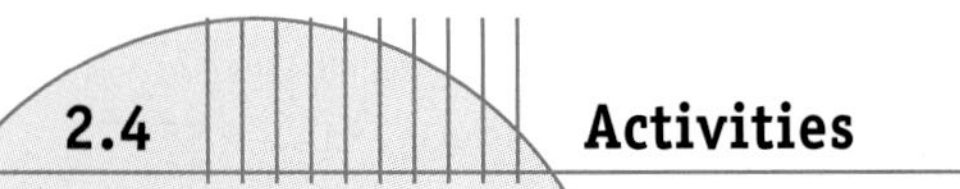

2.4 Activities

Identify the curves in Activities 1–6 as concave up or concave down. In each case, indicate the portion of the horizontal axis over which the part of the curve that is shown is increasing or decreasing.

1.

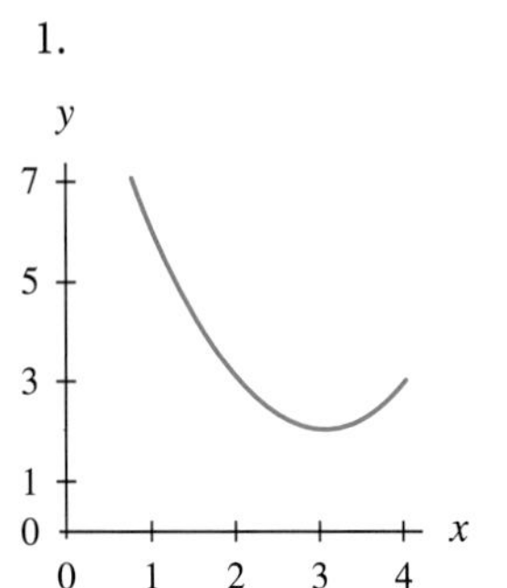

2.

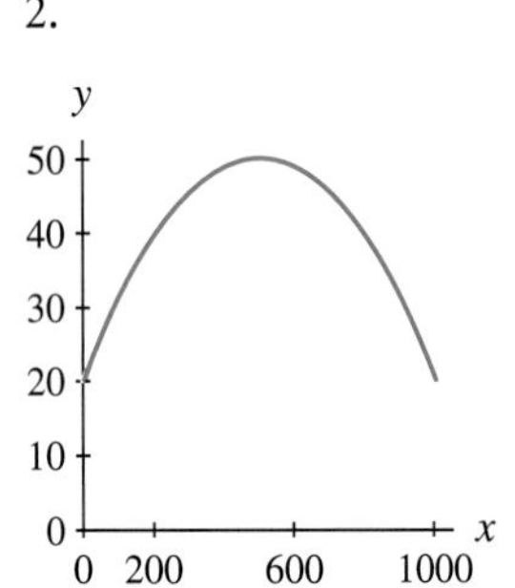

3.

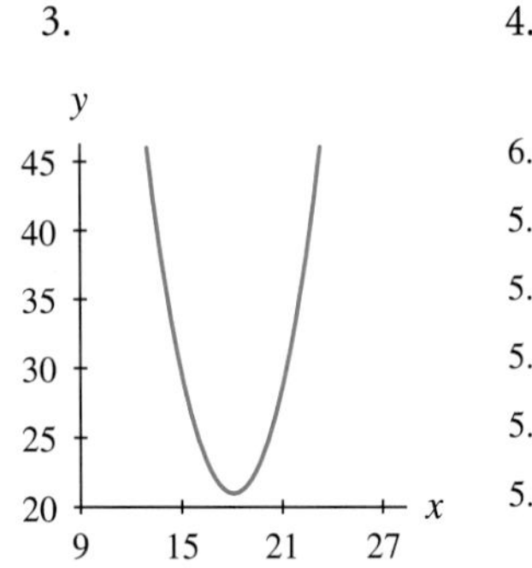

4.

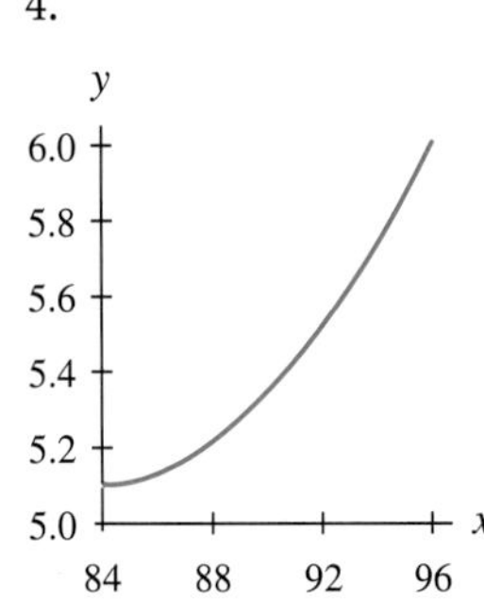

Do not allow your students to try every possible model and then choose the one with the best visual fit. This practice does not encourage them to examine the context of the situation and the overall pattern of the data.

5.

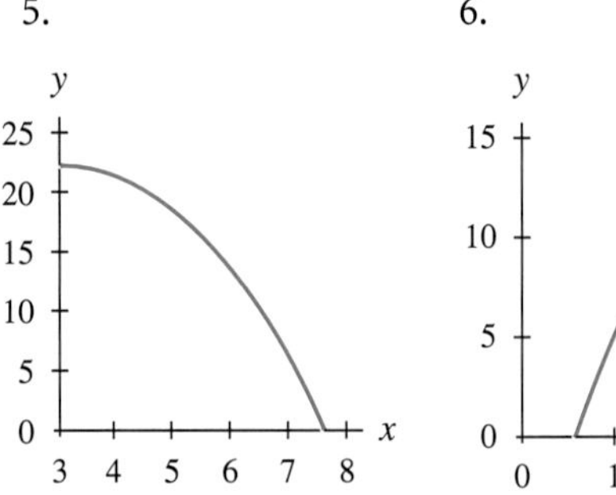

6.

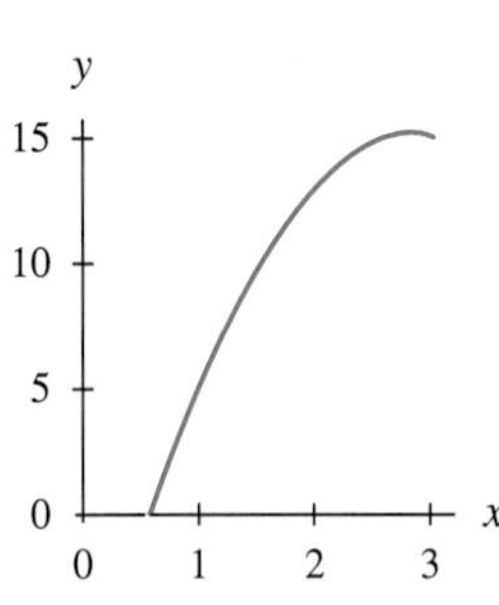

7. During a training mission in the South Pacific, a Tomahawk Cruise missile misfires. It goes over the side of the ship and hits the water. Suppose the data showing the height of the missile are collected via telemetry as shown in Table 2.45.

TABLE 2.45

Seconds from launch	Feet above water
0	128
0.5	140
1	144
1.5	140
2	128
2.5	108
3	80
3.5	
4	

a. Without graphing, show that the data in Table 2.45 are quadratic.

b. Without finding an equation, fill in the missing values.

c. Find a quadratic model for the data.

d. Use the model to determine when the missile hits the water.

8. A discount store has calculated from past sales data the weekly revenue resulting from the sale of blenders. These amounts are shown in Table 2.46.

a. Why does selling 19 blenders result in less revenue than selling 16 blenders?

b. Examine a scatter plot of the data in Table 2.46, and find a quadratic model for the data.

c. What is the revenue when 17 blenders are sold? 18 blenders?

TABLE 2.46

Number of blenders sold	Revenue (dollars)
4	114
7	247
10	356
13	418
16	451
19	446
22	412

d. How could the data in the table and the answers in part *c* be beneficial to the manager of the store?

9. Suppose that the roofing company discussed in this section operates in Michigan instead of Miami.

a. Using the quadratic model

$$J(x) = 3.07143x^2 - 7.01429x + 15.8 \text{ jobs}$$

where $x = 1$ in January, $x = 2$ in February, and so on, determine the number of jobs in December.

b. Do you feel this answer is reasonable? Why or why not?

c. Would your answer to part *b* be different if the roofing company operated in your home state?

10. Table 2.47 gives the price in dollars of a round-trip ticket from Denver to Chicago on a certain airline and the corresponding monthly profit, in millions of dollars, for that airline.

TABLE 2.47

Ticket price (dollars)	Profit (millions of dollars)
200	3.08
250	3.52
300	3.76
350	3.82
400	3.70
450	3.38

a. Is a quadratic model appropriate for the data? Explain.

b. Find a quadratic model for the data.

c. As the ticket price increases, the airline should collect more money. How can it be that when the ticket price reaches a certain amount, profit decreases?

d. At what ticket price will the airline begin to post a negative profit (a net loss)?

11. As listed in *The 1999 World Almanac*, the median age (in years) at first marriage of females in the United States is shown in Table 2.48.

TABLE 2.48

Year	1960	1970	1980	1990
Age	20.3	20.8	22.0	23.9

a. Refute or defend the following statement: "The data are perfectly quadratic."

b. Without finding an equation, estimate the age that corresponds to the year 2000.

c. Find a quadratic model for the data.

d. Use the model to estimate the age in the year 2000. Is it the same as your answer to part *b*?

e. Find the most current data available for the median age at first marriage of women. Compare the model value with the actual value.

f. When you add the new data to Table 2.48, is a quadratic model still appropriate? Explain.

12. The monthly profit P (in dollars) from the sale of x mobile homes at a dealership is given by Table 2.49.

TABLE 2.49

x	P (dollars)	x	P (dollars)
7	43,700	14	63,000
8	48,000	15	63,750
11	57,750	19	61,750
13	61,750	22	55,000

Examine a scatter plot of the data. Find a quadratic model to fit the data, and graph it on the scatter plot. Does it appear that the quadratic function is a good fit?

13. A factory makes 7-millimeter aluminum ball bearings. Company analysts have determined how much it costs to make certain numbers of ball bearings in a single run. These costs are shown in Table 2.50.

TABLE 2.50

Number of ball bearings	Cost (dollars)
500	3.10
1000	4.25
3000	8.95
4500	12.29
7000	18.45

a. Should the data be modeled by using a linear or a quadratic equation? What is the model?

b. How much is the overhead for a single run?

c. How much will it cost to make 5000 ball bearings?

d. Ball bearings are made in sets of 100. If the company is planning to make 5000 ball bearings in a run, how much more would it cost them to make one extra set of 100 in that same run? What term is used in economics to express this idea?

e. Ball bearings are sold in cases of 500. Rewrite the cost equation in terms of the number of cases of ball bearings.

14. The factory discussed in Activity 13 sells ball bearings in cases of 500. It charges the prices shown in Table 2.51.

TABLE 2.51

Number of cases	Price (dollars)
1	78.47
10	168.85
20	263.83
100	817.88
200	999.96

a. Find a quadratic model for the data.

b. According to the model, how much revenue is made on 10 cases of ball bearings?

c. Using the cost model found earlier, $C(u) = 1.17627u + 1.87986$ where u is the number of cases of ball bearings, and the revenue model you just found, write an expression to represent the profit made on u cases of ball bearings.

d. How much profit does the factory make on the sale of 500, 1000, and 10,000 ball bearings, respectively?

e. How many ball bearings must be sold to make a profit of $800?

f. How many ball bearings must the factory sell at one time to a single customer before it no longer makes a profit from that sale?

15. The numbers of AIDS cases diagnosed[28] in the United States from 1994 through 1997 were as shown in Table 2.52.

TABLE 2.52

Year	1994	1995	1996	1997
Number of cases	78,279	71,547	66,885	58,493

a. Find a quadratic model for the data, and examine the equation graphed on a scatter plot of the data.

b. Do you believe the equation in part *a* is a good fit? Explain.

c. The number of AIDS cases diagnosed in 1998 had not yet been tabulated when the data in Table 2.52 was published. What does the quadratic model give as the number of cases in 1998 Do you believe that this estimate is reliable?

d. According to your model, in what year were less than 40,000 AIDS cases diagnosed?

e. Do you believe this model gives a good or a poor representation of what will happen? Support your answer with more current data.

16. Table 2.53 lists the death rates[29] (number of deaths per thousand people whose age is x) for the United States.

TABLE 2.53

x	Death rate (deaths per thousand people)
40	2.2
45	3.0
50	5.0
55	8.0
60	12.6
65	18.7

28. *Statistical Abstract*, 1998.
29. *Information Please Almanac*, 1992.

We recommend that you assign Activity 18. (See Project 1.1 for more specific instructions.) The results of this assignment will give you a collection of data sets that you can use in test and quiz questions throughout the semester. This activity can be extended to a semester-long project. Details for such a project are available at the *Calculus Concepts* web site.

a. Find a quadratic model for the data in Table 2.53. Discuss how well the equation fits the data.

b. Use the model to complete Table 2.54:

TABLE 2.54

Age	Model prediction	Actual rate
51		5.5
52		6.0
53		6.6
57		9.6
59		11.5
63		16.2
66		20.1
68		23.7
70		28.2
75		42.4
80		65.5

c. What can you conclude about using a model to make predictions?

17. Lead was banned as an ingredient in most paints in 1978, although it is still used in some specialty paints. Lead usage in paints[30] from 1940 through 1980 is as shown in Table 2.55.

TABLE 2.55

Year	Lead usage (thousands of tons)
1940	70
1950	35
1960	10
1970	5
1980	0.01

a. Examine a scatter plot of the data. Find quadratic and exponential models for lead usage. Comment on how well each equation fits the data.

b. What is the end behavior suggested by the data? Which function in part *a* has the end behavior suggested by the context?

c. Which model would be more appropriate to use to estimate the lead usage in 1955? Use that model to estimate the usage in 1955.

30. Estimated from information in "Lead in the Inner Cities," *American Scientist*, January-February 1999, pp. 62–73.

18. From a magazine, newspaper, or some other source, collect data that you feel may be modeled by one of the models we have studied. Find a model, and discuss why you chose the model and how well it fits the data. Do you feel that predictions based on this model are likely to be realistic? Why or why not?

For the graphs in Activities 19 through 24, describe the curvature by indicating the portions of the displayed horizontal axis over which each curve is concave down or concave up. Mark the approximate location of the inflection point on each curve.

19.

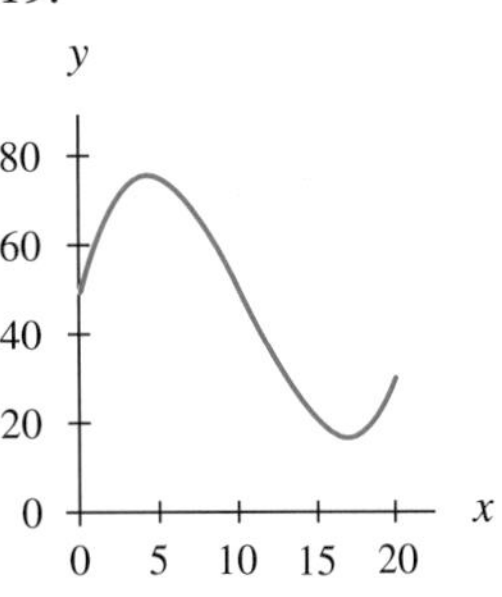

20.

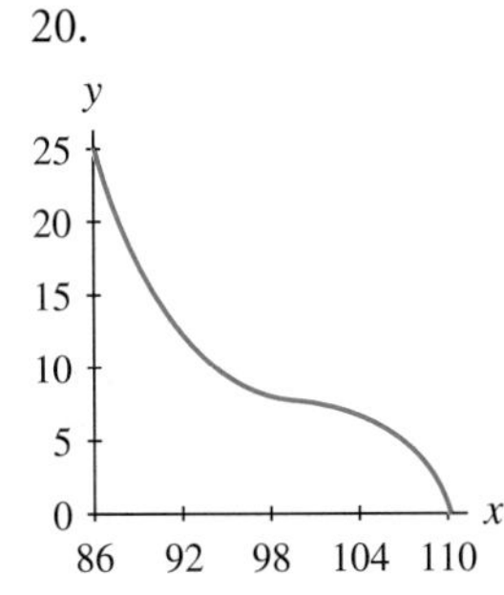

21.

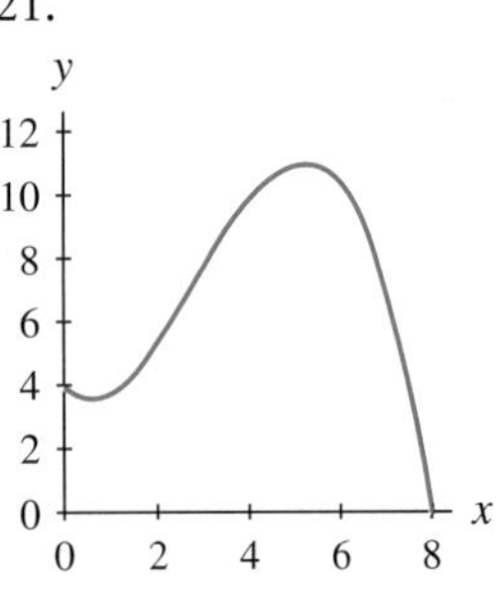

22.

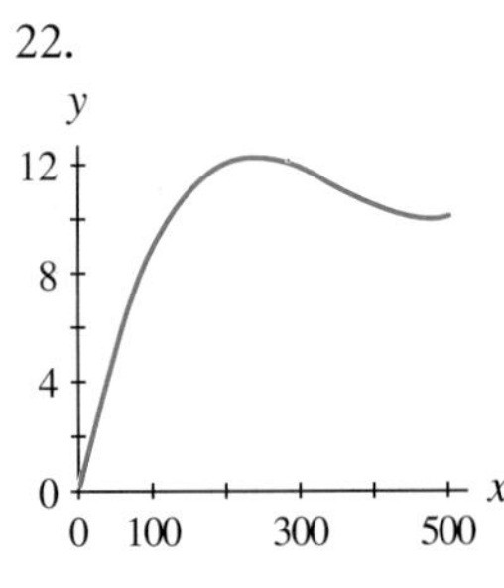

23.

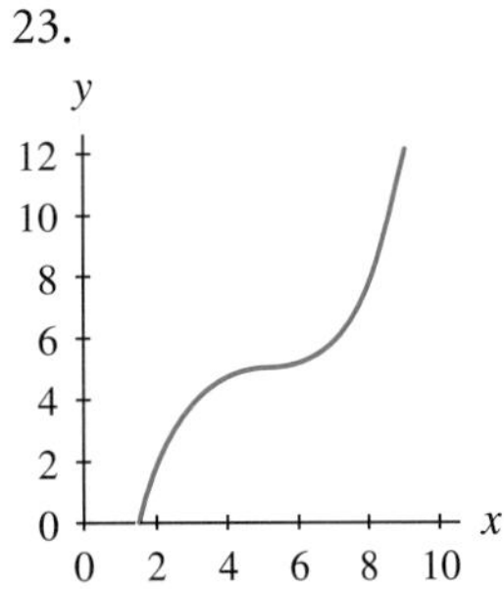

24.

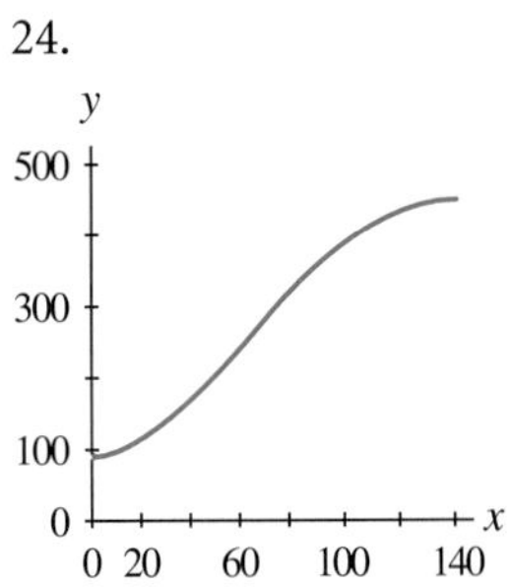

25. Table 2.56 shows the amounts spent on reducing sizes of first-grade through third-grade public school classes in a certain state.[31]

TABLE 2.56

Year	Amount (millions)
1990	$3
1992	$31
1994	$37
1996	$42
1998	$66

a. Examine a scatter plot of the data. How does a scatter plot indicate that a cubic model might be appropriate?

b. Find a cubic model for the amount spent on class size reduction.

c. Use the model to estimates the amounts in 1993 and 1999. In which of these estimates can you have more confidence?

d. Compare the estimates from part *c* with the actual amounts of $34 million spent in 1993 and $99 million spent in 1999. Does this comparison support the statements you made in part *c* concerning reliability of the estimates?

26. *The 1998 Information Please Almanac* lists the numbers of live births between the years 1950 and 1995 in the United States to women 45 years of age and older as given in Table 2.57.

TABLE 2.57

Year	Births	Year	Births
1950	5322	1975	1628
1955	5430	1980	1200
1960	5182	1985	1162
1965	4614	1990	1638
1970	3146	1995	2727

a. Examine a scatter plot of the data in Table 2.57 Find a cubic model for the data, and graph the equation on the scatter plot. Discuss how well the equation fits the data.

b. What trend does the model indicate beyond 1995? Do you believe the trend is valid? Explain.

c. Use the model to estimate the number of live births in 2000.

d. Do you believe your model would be an accurate predictor of the number of live births to women 45 years of age and older for the current year? Why or why not?

27. Table 2.58 shows the number of males per 100 females in the United States[32] calculated using census data. This number is referred to as the *gender ratio.*

TABLE 2.58

Year	Males per 100 females
1900	104.6
1910	106.2
1920	104.1
1930	102.6
1940	100.8
1950	98.7
1960	97.1
1970	94.8
1980	94.5
1990	95.1

a. Examine a scatter plot of the data. What behavior of the data suggests that a cubic model is appropriate?

b. Find a cubic model for the gender ratio. What trend does the model indicate for the years beyond 1990? Do you believe this is a good predictor of future gender ratios? Why or why not?

c. What factors might have contributed to changes in the gender ratio in this country?

28. Table 2.59 shows a manufacturer's total cost (in hundreds of dollars) to produce from 1 to 33 forklifts per week.

a. Examine a scatter plot of the data in Table 2.59. What characteristics of the scatter plot indicate that a cubic model would be appropriate?

b. Find a cubic model for total manufacturing cost.

c. What does the model predict as the cost to produce 23 forklifts per week? 35 forklifts per week?

d. Convert the cubic equation in part *b* for total cost to one for average cost.

31. Nevada Department of Education.

32. United States Bureau of the Census.

TABLE 2.59

Weekly production	Total cost (hundreds of dollars)
1	18.5
5	80
9	125
13	160
17	185
21	210
25	225
29	245
33	280

e. Graph the average cost function.

f. How could the average cost graph help a production manager make production decisions?

29. The gender ratio in the United States is different for different age groups. Table 2.60 shows gender ratios[33] corresponding to age.

TABLE 2.60

Age	Males per 100 females
under 1	104.6
10	105.3
20	104.0
30	100.3
40	98.2
50	95.6
60	88.8
70	76.5
80	56.6
90 and over	31.4

a. According to the data, for what age are the numbers of males and females equal?

b. Find quadratic and cubic models for the gender ratio as a function of age. Compare the fit of the two equations. Which equation do you believe is better for modeling these data?

c. Use the model you chose in part *b* to find the age for which there are twice as many women as men. What does this information tell you about death rates of men and women?

30. Table 2.61 gives the numbers of imported passenger cars[34], in thousands, sold in the United States from 1984 through 1994.

TABLE 2.61

Year	Imported passenger cars sold (thousands)
1984	2439
1985	2838
1986	3245
1987	3196
1988	3004
1989	2699
1990	2403
1991	2038
1992	1937
1993	1776
1994	1735

a. Examine a scatter plot of the data in Table 2.61, and discuss its curvature. Should the data be modeled by a quadratic or a cubic equation? What is the model?

b. What does the model estimate as the number of imported passenger cars sold in the United States in 1995 and 1996? Compare the model estimates to the actual sales of 1506 thousand cars in 1995 and 1273 thousand cars in 1996.

c. Discuss why you feel the model you chose would or would not be a good estimator of the number of imported passenger cars sold in the United States in the year 2000.

d. Find more recent data on imported passenger car sales, and add the values you found to the table. Is the model you chose appropriate for the expanded data set? Explain.

31. The per capita utilization of commercially produced fresh vegetables[35] from 1970 through 1992 is shown in Table 2.62.

33. Ibid.

34. *Statistical Abstract*, 1998.

35. *Statistical Abstract*, 1994.

TABLE 2.62

Year	Vegetable consumption (pounds per person)
1970	88.1
1975	88.6
1980	92.5
1985	103.0
1988	111.5
1989	115.5
1990	113.3
1991	110.4
1992	109.3

a. Observe a scatter plot of the data in Table 2.62, and explain why a piecewise model is needed to describe this data set.

b. Divide the data into two subsets, each sharing a common data point. Find an equation for each subset of data, and combine the equations to form a piecewise continuous model.

c. Use your model to estimate the per capita use of fresh vegetables in 1973, 1983, and 1993.

d. List as many reasons as you can for the change in the per capita use of fresh vegetables in the early 1990s.

e. What behavior does your model indicate for per capita use of fresh vegetables beyond 1992?

f. Find more recent data on per capita use of fresh vegetables, and add it to Table 2.62. Is the model you chose appropriate for the expanded data set? Explain.

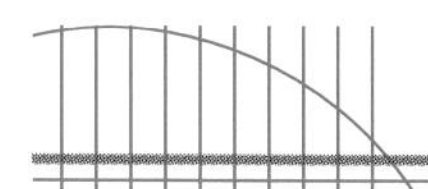

32. Consider the average hospital stays[36] for selected years between 1980 and 1996 as shown in Table 2.63.

TABLE 2.63

Year	Average stay (days)	Year	Average stay (days)
1980	7.3	1993	6.0
1985	6.5	1994	5.7
1990	6.4	1995	5.4
1991	6.4	1996	5.2
1992	6.2		

a. Observe a scatter plot of the data. Comment on the curvature suggested by the scatter plot.

b. Find a cubic model for average hospital stay. Does the curvature of a graph of the function match that of the scatter plot?

c. Divide the data into two sets: from 1980 through 1990 and from 1990 through 1996. Include the 1990 data point in both sets. Find models for each set of data, and use the models to write a piecewise continuous model for average hospital stay. How well does this piecewise function match the curvature of the data?

d. Use the model in part *c* to estimate the average hospital stay in 1987 and in 1998.

f. List as many factors as you can that contribute to changes in the average hospital stay.

g. What current issues in our health care system may affect average hospital stay in the future?

2.5 Shifting Data

Some data sets strongly suggest one of the models that we have studied, but the technology-generated function is a poor fit to the data. In many cases, shifting the data produces a better-fitting model. Our experience with this aspect of curve fitting led us to include this topic.

We have discussed aligning data input values in order to reduce the magnitude of coefficients in model equations and to prevent numerical computation error. This alignment is vital for exponential and logistic models. We now consider situations in which aligning the output values is necessary or desirable. Aligning input values shifts data horizontally; similarly, aligning output values shifts the data vertically.

Some of the functions that we have considered have both vertical and horizontal shifts built into them; that is, if they fit a certain set of data well, then they will fit those data equally well when the data are shifted vertically or horizontally. This is true of all the polynomial functions—linear, quadratic, and cubic. We list the functions we have studied, along with the types of shifts they account for.

36. *Statistical Abstract*, 1998.

If you are concerned about not having enough time to finish all the sections required for your specific course, you may wish to skip this section. See the *Instructor's Resource Guide* for a list of the activities throughout the book that contain references to topics in this section. You will need to omit or modify these activities if you skip this section.

Function		*Shift*
Polynomial functions:		
Linear	$f(x) = ax + b$	Horizontal and vertical
Quadratic	$f(x) = ax^2 + bx + c$	Horizontal and vertical
Cubic	$f(x) = ax^3 + bx^2 + cx + d$	Horizontal and vertical
Log function	$f(x) = a + b \ln x$	Vertical
Exponential function	$f(x) = ab^x$	Horizontal
Logistic function	$f(x) = \frac{L}{1 + Ae^{-Bx}}$	Horizontal

The addition of a constant term in a model indicates that the function accounts for vertical shift. Note that the polynomial and log functions all have a constant term and, therefore, have a built-in vertical shift. The exponential and logistic functions do not have vertical shift, but they do account for horizontal shifts. In this section we focus on those functions that do not account for both horizontal and vertical shifts.

Some technologies use functions in different forms from those we present. For example, some technologies use a logistic function of the form $f(x) = \frac{L}{1 + Ae^{-Bx}} + D$. The parameter D is the vertical shift parameter. If the technology you use produces a logistic function of this form, then there is no need for you to shift logistic data in order to obtain a better fit. You need to know your technology and the functions it uses well enough to know whether the following discussions of shifting data apply to you.

Shifting Exponential and Logistic Data

We have seen two functions whose end behavior is to approach a number or numbers. The output of an exponential function approaches zero as the input becomes increasingly positive or negative, and the output of a logistic function approaches zero in one direction and a nonzero number in the other direction. These functions are useful, but they have limitations. Consider, for example, the data shown in Table 2.64.

TABLE 2.64

x	0	2	4	6	8
f(x)	101	104	116	164	356

A scatter plot of the data (Figure 2.35) suggests an exponential function.

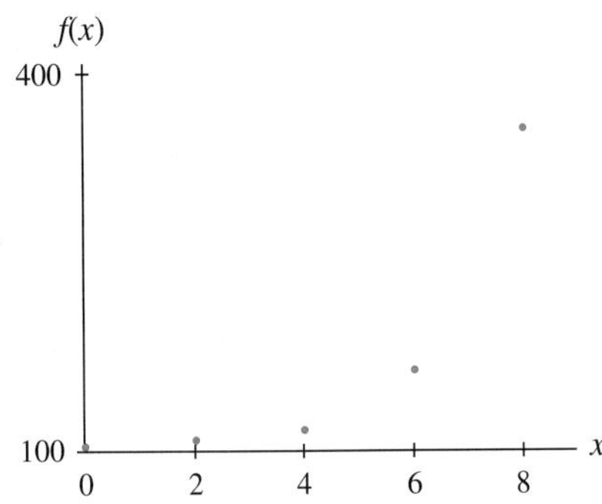

FIGURE 2.35

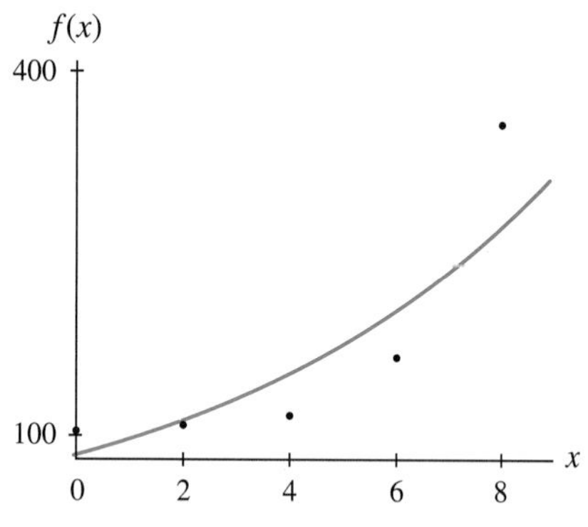

FIGURE 2.36

When we find an exponential equation to fit the data, we obtain the function $f(x) = 81.6597(1.1604^x)$. A graph of this function on a scatter plot of the data is shown in Figure 2.36. The equation is a poor fit for the data and is not what we expect to see.

The problem is that the output data are not approaching zero to the left. They are shifted up by 100, so they approach 100. However, the exponential regression routine built into most technologies does not allow for a vertical shift. We can compensate for this limitation by shifting the data ourselves, modeling them, and adjusting the model accordingly. We modify the data in Table 2.64 by subtracting 100 from all the output values, but we do not change the input values. See Table 2.65.

TABLE 2.65

x	0	2	4	6	8
f(x) – 100	1	4	16	64	256

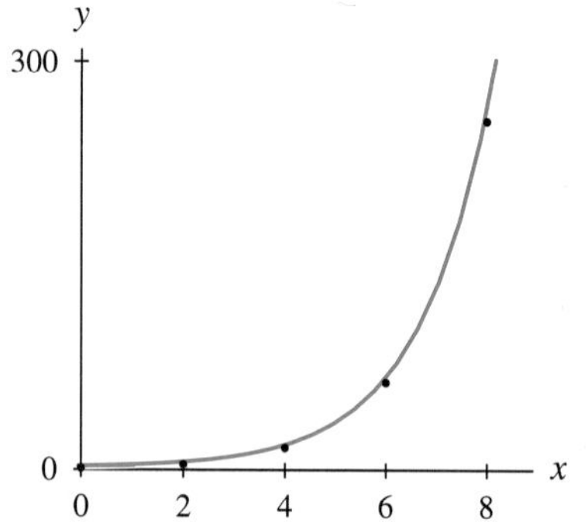

FIGURE 2.37

A scatter plot appears exactly like that in Figure 2.35 except that the vertical axis values are different. An exponential function that fits these adjusted data is $y = 2^x$. A graph of this function on a scatter plot of the shifted data is shown in Figure 2.37.

The fit is perfect, because the data are perfectly exponential. We now must compensate for the shifted data by reversing our shift. We originally subtracted 100, so we now reverse the process by adding 100 to the function for the shifted data. Our final equation for the data in Table 2.64 is

$$g(x) = 2^x + 100$$

Vertically shifting data may be appropriate when a scatter plot suggests a curve with a horizontal limiting value but the technology-generated equation for that curve is a poor fit to the data and does not reflect the observed behavior of the data. Example 1 gives another situation in which vertical shifting is necessary.

EXAMPLE 1 *Investment Clubs*

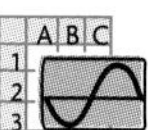

Table 2.66 shows the number of investment clubs[37] in existence during the 1990s. A scatter plot of the data is shown in Figure 2.38.

TABLE 2.66

Year	Number of clubs	Year	Number of clubs
1990	7085	1995	16,054
1991	7360	1996	25,409
1992	8267	1997	31,828
1993	10,033	1998	36,500
1994	12,429		

37. National Association of Investors Corp.

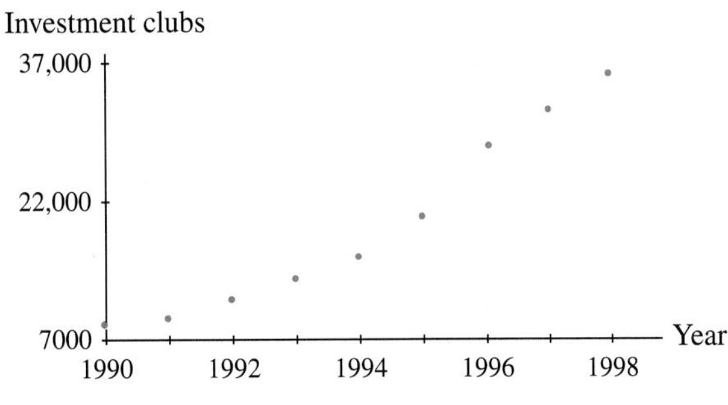

FIGURE 2.38

Find an appropriate model for the data, and use it to estimate the number of investment clubs in 1999.

Solution

The scatter plot strongly suggests a logistic curve because the data exhibit an obvious inflection point and suggest two horizontal asymptotes. We align the input values to years since 1990 in order to avoid computational error. When we fit a logistic equation to the aligned data, we obtain

$$N(t) = \frac{244{,}096.932}{1 + 48.227e^{-0.2721t}} \text{ investment clubs}$$

t years after 1990. This model should look suspicious to you, because the limiting value of 244,097 is much greater than the one suggested by the data. A graph of the function on the scatter plot is shown in Figure 2.39.

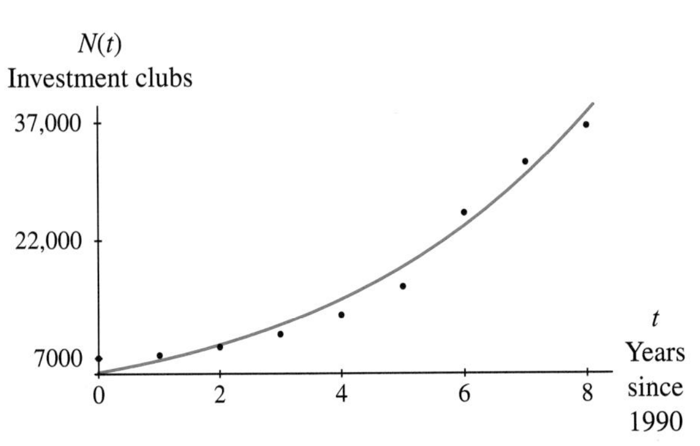

FIGURE 2.39

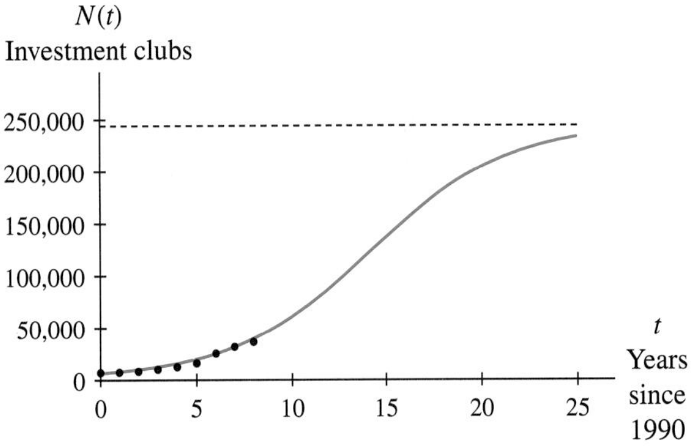

FIGURE 2.40

Although at first glance it might seem that this function is a reasonable fit, look at it more closely. The logistic graph does not exhibit an inflection point that coincides with the inflection point suggested by the data, and the graph does not level off where we expect it to. A view of the graph from a different perspective (see Figure 2.40) shows that the data are modeled by only the concave-up portion of this logistic function.

The underlying problem is that the logistic function N approaches zero on the left, but the data do not approach zero on the left. They appear to approach a value closer to 7000. To compensate for this difference, we shift the data down, closer to

Your students will probably struggle with knowing how far to shift data. There is no rule to follow. You should be willing to allow for many correct answers. Experience is the only effective teacher. Activity 14 gives the students the opportunity to explore shifting these investment club data by different amounts and to compare the results.

zero. We choose to do so by subtracting 7000 from the output data. This value is arbitrarily chosen, and other values will produce similar results. Table 2.67 shows the shifted data. The scatter plot will look exactly like the one in Figure 2.38 except that the vertical axis labels will be different.

TABLE 2.67

Aligned year	Number of clubs − 7000
0	85
1	360
2	1267
3	3033
4	5429
5	9054
6	18,409
7	24,828
8	29,500

When we fit a logistic equation to this shifted data, we obtain the model

$$f(t) = \frac{34{,}280.654}{1 + 207.434e^{-0.8991t}} \text{ investment clubs above 7000}$$

t years after 1990. A graph of this function is shown in Figure 2.41.

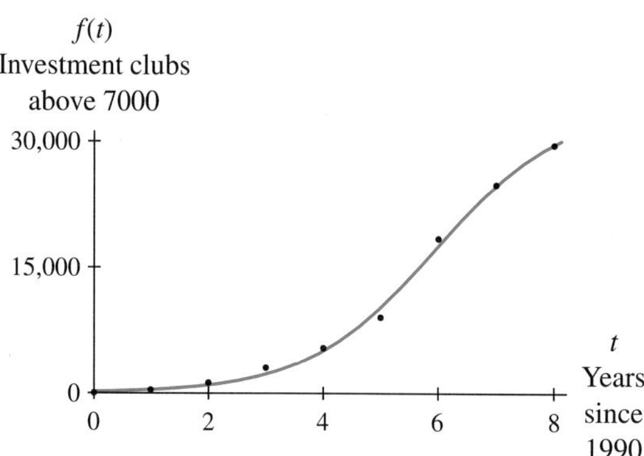

FIGURE 2.41

This model has horizontal asymptotes and an inflection point that better match those indicated by the shifted data.

To compensate for the fact that this model is for the number of clubs above 7000, we simply add 7000 to the output $f(t)$ to obtain a model for the original aligned data.

$$I(t) = \frac{34{,}280.654}{1 + 207.434e^{-0.8991t}} + 7000 \text{ investment clubs}$$

t years after 1990.

Using the two models to estimate the number of clubs in 1999, we see that the difference in the estimates is significant:

$$N(9) \approx 47{,}248 \text{ clubs} \quad \text{and} \quad I(9) \approx 39{,}235 \text{ clubs}$$

The estimate from the second model is probably more accurate, provided that the logistic pattern continued into 1999.

Note in Example 1 that the horizontal asymptotes of the function I are 7000 on the left and $34{,}280.654 + 7000 = 41{,}280.654$ on the right. The limiting value is 41,280.654. According to the model, the number of investment clubs will level off at approximately 41,281. It is the issue of end behavior that should determine when it is appropriate to shift data vertically before fitting an equation.

Shifting Log Data

Recall that log functions are inverses of exponential functions. It follows that if vertically shifting output data can result in a better-fitting exponential equation, then horizontally shifting input data can sometimes produce a better-fitting log equation. The log function is the only function we have studied for which aligning input data differently changes the fit of the equation. When shifting input data for a log function it is particularly important to keep in mind that log functions cannot have negative or zero input.

EXAMPLE 2 *Voting Turnout*

A community in central Indiana has been funding a local campaign to increase voter turnout in local elections. The campaign began in 1996. The percentages of all eligible voters voting in the yearly local elections for years between 1995 and 2000 are shown in Table 2.68.

TABLE 2.68

Year	1995	1996	1997	1998	1999	2000
Voter turnout (%)	2.0	4.1	5.3	6.2	6.8	7.4

a. Examine a scatter plot of the data. What characteristics of the data indicate that a log model may be appropriate?

b. Find a log model for the data with the number of years since 1900 as the input. Examine a graph of the model on a scatter plot of the data, and comment on the fit.

c. Find a log model for the data with the number of years since 1994 as the input. How does the fit of this model compare to that of the one in part *b*?

d. Why did we not use the number of years since 1995 as the input?

Solution

a.

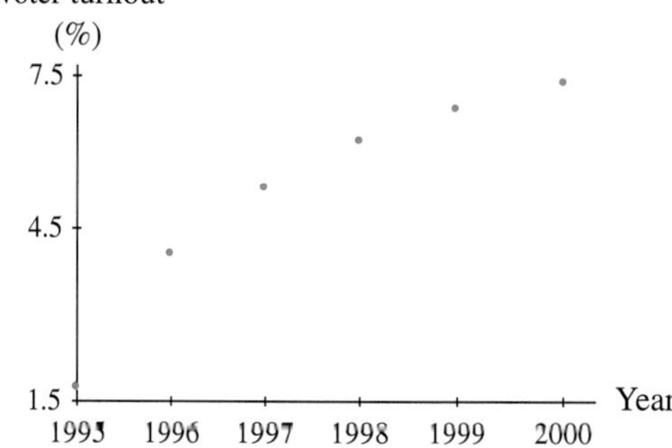

FIGURE 2.42

The scatter plot in Figure 2.42 shows an increasing, concave-down pattern. Because there is no indication that the data will decline, a log model is more appropriate than a quadratic model.

b. $V(t) = -454.737 + 100.451 \ln t$ percent turnout when t is the number of years since 1900.

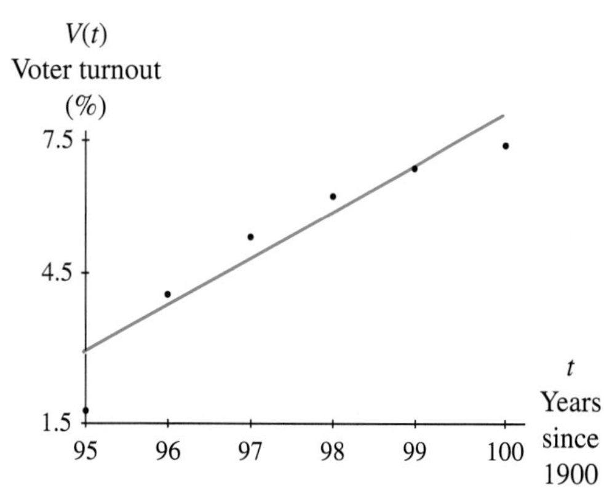

FIGURE 2.43

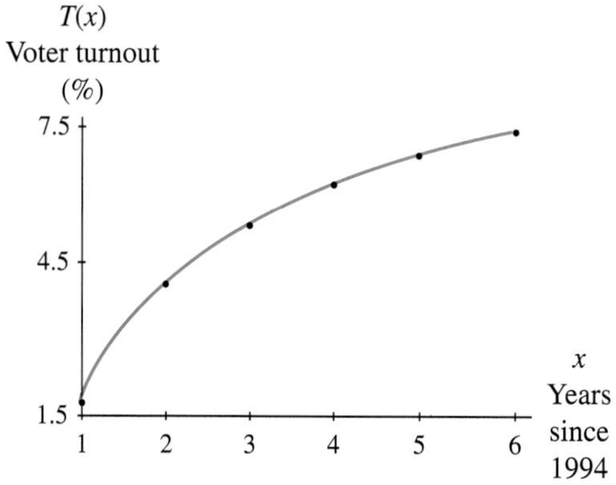

FIGURE 2.44

A graph of the function V (shown in Figure 2.43) appears to be linear compared with the curved data.

Note that we use a different input variable in T to emphasize that we are realigning the input data. You may wish to present the model in part c in the equivalent form

$P(t) = 2 + 3 \ln(t - 94)$

percent turnout when t is the number of years since 1900.

c. $T(x) = 2 + 3 \ln x$ percent turnout when x is the number of years since 1994. A graph of this function (shown in Figure 2.44) appears to be an excellent fit to the data.

d. If we had aligned the input as years since 1995, then the first input value would have been 0. Zero is not a valid input for a log function.

It is important to realize that there is no one correct way to shift data either horizontally or vertically. The more experience you have fitting equations to data, the better equipped you will be to make decisions about shifting.

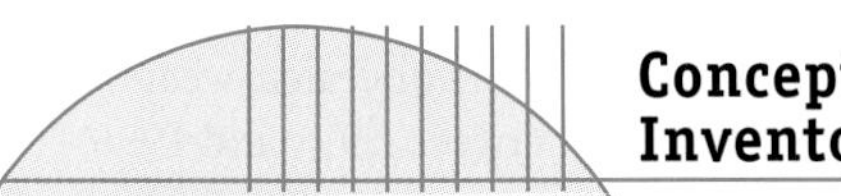

Concept Inventory

- *Shifting data vertically: exponential and logistic data*
- *Shifting data horizontally: log data*

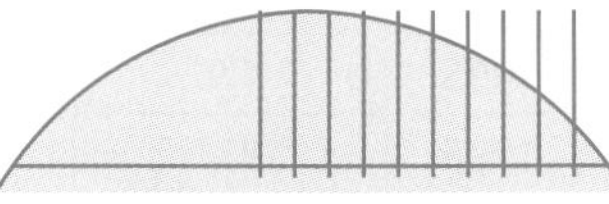

Activities

List all horizontal asymptotes for the functions in Activities 1 through 4.

1. $f(x) = 2.39(0.4332^x) + 1.69$
2. $g(t) = 98{,}765(1.5^t) - 11{,}000$
3. $j(r) = \dfrac{0.023}{1 + 0.15e^{0.967r}} - 0.003$
4. $h(y) = \dfrac{17.5}{1 + 3.3e^{-0.5y}} + 3.5$
5. Consumer credit card debt,[38] in billions of dollars, during the 1990s was as shown in Table 2.69.

TABLE 2.69

Year	Debt (billions of dollars)	Year	Debt (billions of dollars)
1990	172	1995	349
1991	183	1996	410
1992	200	1997	445
1993	229	1998	460
1994	280		

a. Observe a scatter plot of the data. Describe the concavity suggested by the data. In what year do you estimate the concavity changed? Which equations could be used to model these data?

b. Find a logistic model for the data. Comment on how well the equation fits the data.

c. Shift the data down by 165, and find a logistic model for the shifted data. Comment on how well the equation fits the shifted data. Compare the fit of this equation with that of the logistic equation in part *b*.

d. On the basis of the model in part *c*, write a model for credit card debt.

e. What are the two horizontal asymptotes for the function in part *d*? Compare these with the horizontal asymptotes for the original logistic function in part *b*.

6. A regional transportation commission estimates that the daily traffic counts for a certain intersection[39] will be as shown in Table 2.70.

TABLE 2.70

Year	Traffic count (vehicles)
1999	33,700
2010	47,600
2020	66,200
2030	74,000

a. Align the input data as the number of years since 2000. Observe a scatter plot of the data. Estimate the inflection point and horizontal asymptotes values suggested by the scatter plot.

b. Find a logistic model for the data. How well does it match the behavior indicated by the scatter plot? What is the limiting value of this logistic function?

c. Shift the output data down by 33,000. Find a logistic model for the shifted data. How well does this new equation fit the shifted data?

d. Use the model in part *c* to write a new model for traffic counts. What are the two horizontal asymptotes of this new function? What does this model assume about the population of the surrounding metropolitan area?

e. According to the model in part *d*, what will the daily traffic counts be in 2005? In what year will traffic counts reach 60,000 vehicles per day?

7. The population of Iowa[40] for selected years between 1987 and 1997 is given in Table 2.71.

TABLE 2.71

Year	Population (thousands)	Year	Population (thousands)
1987	2767	1993	2820
1989	2771	1995	2841
1991	2791	1997	2852

38. Consumer Federation of America.

39. Reno/Sparks, NV, Regional Transportation Commission for McCarren Blvd. East at Prater Way.

40. *Statistical Abstract*, 1998.

a. Align the input as the number of years since 1980, and shift the output data down by 2700. Observe a scatter plot of the aligned and shifted data. Why is a logistic model appropriate?

b. Find a logistic model for the aligned and shifted data. Discuss how well the equation fits these data.

c. Subtract 60 from the shifted output data in part *a*, and find a logistic model for these new data. How well does this equation fit these data?

d. Using the model in part *c*, write a model for the population of Iowa from 1987 through 1997. What are the two horizontal asymptotes of this logistic function?

8. The data in Table 2.72 show how postal rates have increased for first-class letters weighing up to 1 ounce.

TABLE 2.72

Year of rate increase	Rate (cents)
1958	4
1963	5
1968	6
1971	8
1974	10
1975	13
1978	15
1981	20
1985	22
1988	25
1991	29
1995	32
1999	33
2001	34

a. Find a logistic model for the postal rate data. How well does the equation fit the data? What is the limiting value of this function?

b. Shift the postal rates down by 3, and find a logistic model for the shifted data. Does this equation fit the shifted data better than the equation in part *a* fits the original data?

c. Use the model in part *b* to write a model for postal rates. What are the two horizontal asymptotes of this function? Do you believe the model accurately describes what will happen to postal rates in the future?

9. Table 2.73 shows the number of concealed-weapons permit renewals in Utah[41] between 1991 and 1996.

TABLE 2.73

Year	Permit renewals	Year	Permit renewals
1991	146	1994	328
1992	195	1995	476
1993	215	1996	831

a. Find an exponential model for the data. Graph the equation on a scatter plot.

b. Shift the number of renewals down by 135, and find an exponential model for the shifted data. Examine a graph of this model on a scatter plot of the shifted data. Does this equation appear to fit the shifted data better than the equation in part *a* fits the original data?

c. Using the model in part *b*, write a model for permit renewals. Use the model to predict when the number of permit renewals will exceed 2000. What are the limits of this function as the input increases and decreases without bound?

d. Use the model in part *c* to estimate the numbers of renewals in 1997 and 1998. Compare your estimates with the actual numbers of renewals: 4,968 in 1997 and 8,478 in 1998. What does this tell you about extrapolating with an equation that fits data well?

10. Table 2.74 gives, for selected years, the percentage of children living with their grandparents.[42]

TABLE 2.74

Year	1970	1980	1992	1997
Percent	3.2	3.6	4.9	5.5

a. Find an exponential model for the data. Examine the graph of the function on a scatter plot of the data. How well does the equation appear to fit the data?

b. Shift the percentages down by 3, and find an exponential model for the shifted data. How does the fit of this equation compare with the fit of the equation in part *a*? What are the limits of this shifted function as the input increases and decreases without bound?

41. Utah Department of Public Safety/Bureau of Criminal Investigation.

42. U.S. Bureau of the Census.

c. Using the model in part *b*, write a model for the percentage of children living with their grandparents. According to this model, when will the output reach 8%? Do you believe the model accurately reflects the future trend of this percentage?

11. The numbers of children living with their grandparents[43] during selected years between 1970 and 1997 are as shown in Table 2.75.

TABLE 2.75

Year	1970	1980	1992	1997
Number (millions)	2.2	2.3	3.3	3.9

a. Find an exponential model for these data. How well does the equation appear to fit the data?

b. By vertically shifting the data, find a better exponential model for the number of children living with their grandparents.

12. The consumer price index (CPI) values for refuse collection[44] between 1990 and 1997 are shown in Table 2.76 (1982–1984 = 100).

TABLE 2.76

Year	CPI	Year	CPI
1990	171.2	1994	231.4
1991	189.2	1995	241.2
1992	207.3	1996	246.0
1993	220.5	1997	250.5

a. Align the input data as the number of years since 1900, and find a log model for the data. How well does this equation fit the data?

b. Realign the input data as the number of years since 1988. How well does this equation fit the data?

c. What does this model predict will happen to CPI values beyond 1997? Do you believe that this model accurately describes future CPI for refuse collection?

13. The values of manufacturer shipments of compact disc albums[45] between 1993 and 1997 are as shown in Table 2.77.

a. Find a log model for the value of shipments as a function of the number of years since 1900. Comment on how well the equation fits the data.

TABLE 2.77

Year	CD Shipments (millions of dollars)
1993	6511
1994	8465
1995	9377
1996	9935
1997	9915

b. Realign the input as the number of years since 1992, and find a new log model for the data. How does the fit of this equation compare to that of the equation in part *a*?

c. Realign the input as the number of years since 1993 plus 0.5, and find another log model for the data. How well does this equation fit compared with those in parts *a* and *b*?

d. What is the behavior of the function in part *c* beyond 1997? Does the data suggest this same behavior?

e. Why can we not align the input data as the number of years since 1993 if we wish to use a log model?

Be warned that Activity 14 is time-consuming. You may want to assign it as a project.

14. Consider again the investment club data in Example 1.

a. From a scatter plot of the data, estimate the year corresponding to the inflection point. Also estimate the two horizontal asymptotes.

b. Shift the output data down by 1000, 3000, 5000, 6000, 6500, 7000, and 7084. For each shifted data set, do the following:

i. Find a logistic function. Add the appropriate constant to construct a model for the original data.

ii. State the two horizontal asymptotes.

iii. The input of the inflection point for a logistic function $f(x) = \frac{L}{1 + Ae^{-Bx}}$ is given by the formula $x = \frac{\ln A}{B}$. Using this formula, find the input of the inflection point.

iv. Find the output values of the function corresponding to the inputs in Table 2.66. Compare these function values with the actual numbers of investment clubs given in the table.

c. On the basis of your answers to parts *a* and *b*, which shift do you believe produces the best-fitting model?

43. Ibid.
44. *Statistical Abstract*, 1998.
45. Ibid.

2.6 Choosing a Model

General Guidelines

In the preceding material, we discussed several types of models that we use in this book to describe two-variable data: linear, quadratic, cubic, exponential, logarithmic, and logistic. We also have discussed how to combine two or more functions to form a piecewise continuous model. Of course, there are others that we have not mentioned and will not use. Although most of our work so far has been concerned with applying a particular type of equation to a given data set, it is important to understand that in many real-life situations, it is not always clear which model to apply. Sometimes, none of the models is appropriate. However, there are some general, common-sense guidelines that we should keep in mind.

1. First of all, *produce a scatter plot* using appropriate scaling for both horizontal and vertical axes. It is important that neither horizontal nor vertical view is too narrow or too wide. If either view is too narrow, then you may not be able to see all the data points. On the other hand, if either view is too wide, then you may be unable to identify accurately the underlying curvature of the scatter plot.

2. Perhaps the most obvious guideline is to *examine the scatter plot carefully* with an eye toward its general underlying shape relative to the shapes of the functions that you know. Try to imagine a smooth curve fitted to the scatter plot:

   a. Does this curve appear to be a straight line? If so, try a linear model.

   b. If the curve does not appear to be a line, then as you look from left to right, does the curve appear to be always concave up or concave down? If so, then a quadratic, exponential, or log model may be appropriate.

   c. What if the curve does not appear to be a line and is not always concave in one direction (up or down)? If there is a single change in concavity, then a cubic or logistic model may be suggested because the graphs of these functions have a concavity change (inflection point).

   d. In the case of exponential and logistic models, vertically shifting the data (by changing the output values) may produce a better-fitting equation. Horizontally shifting data (by changing the input values) may produce a better-fitting log equation.

   e. If the data demonstrate a dramatic change in behavior, then it may be appropriate to combine two or more functions to form a piecewise continuous model.

3. Suppose that we have narrowed our choices to two models. Which one should we choose? If one equation fits the data significantly better than the other one, it should be chosen. If the two equations appear to fit the data equally well, then the key is to examine the end behavior of the scatter plot to see whether limiting values are suggested in the context of the situation. Remember, exponential models have one limiting value, logistic models have two limiting values, and linear, log, cubic, and quadratic models have none.

Examining Scatter Plots

Look at the scatter plots in Figure 2.45. None of them appears to be linear, quadratic, log, or exponential. Can you determine which are cubic and which are logistic?

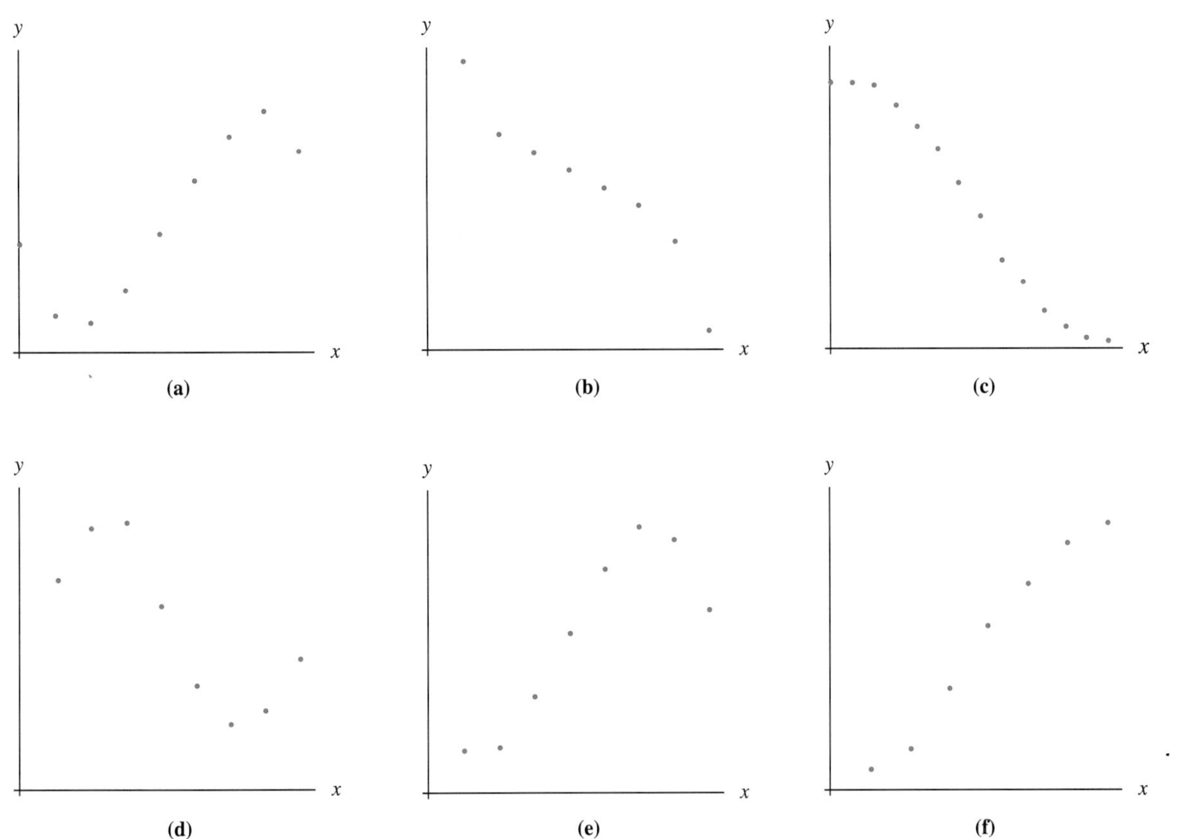

FIGURE 2.45

Plots a and b are cubic (no limiting values). Plots d and e are also cubic because they do not suggest limiting values. Plots c and f could be either logistic or cubic. Here you must ask yourself two questions:

1. Are you more concerned with end behavior or with how well the equation fits the data?
2. If you are concerned with end behavior, what would the end behavior be?

If you are mainly interested in interpolating, then it would be appropriate to choose the equation that better fits the data points. However, if you wish to extrapolate or examine end behavior, you should choose the equation that you think would better follow the end behavior pattern, provided that the fit is satisfactory. In general, if the context or scatter plot suggests that the data have two limiting values, then use a logistic model. If the data are likely to rise or fall again, then a cubic model would be appropriate.

Consider the scatter plots in Figure 2.46, and see whether you can determine which types of functions are appropriate.

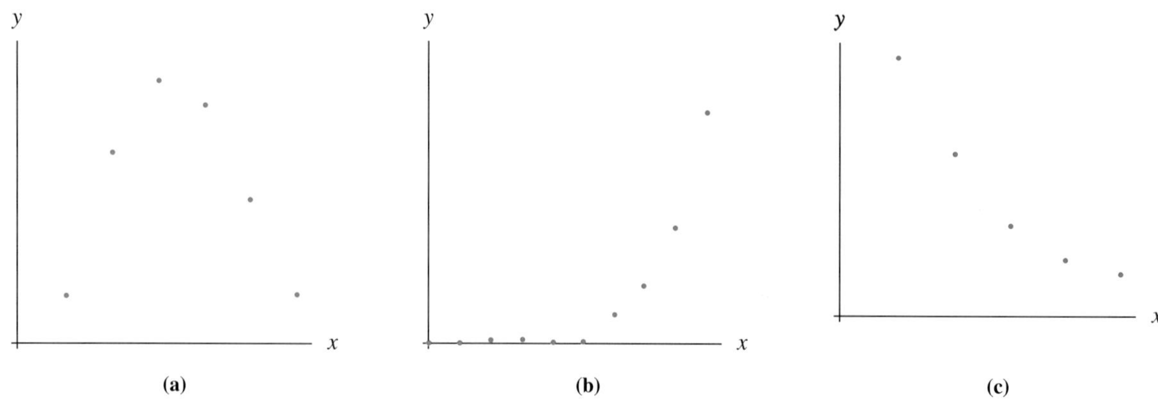

FIGURE 2.46

Plot a seems to be quadratic, whereas plot b appears to be exponential. Plot c could be exponential, quadratic, or logarithmic. Again ask yourself those two questions: (1) Are you more concerned with end behavior or with how well the equation fits? (2) If you are concerned with end behavior, what would the end behavior be? Recall that a quadratic curve quickly becomes infinitely large (either negatively or positively) at each end, whereas both exponential and logarithmic curves tend quickly to infinity (or negative infinity) at one end and exhibit slow growth or decline at the other end. The exponential curve has a limiting value, whereas the log curve has no limiting value even though it exhibits increasingly slow growth or decline.

Finally, consider the two scatter plots shown in Figure 2.47.

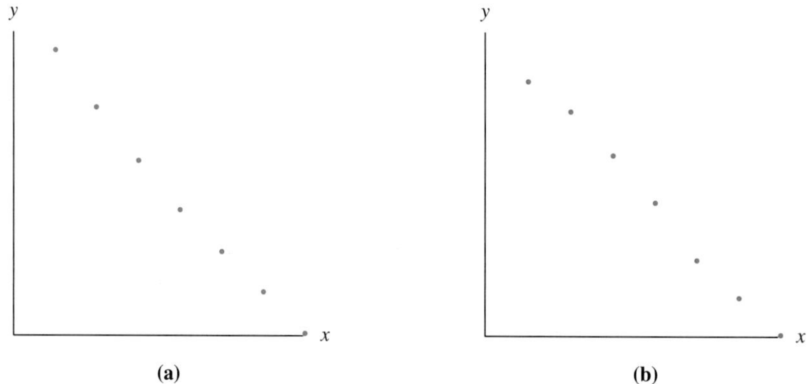

FIGURE 2.47

They both appear to be linear, so it makes sense to try linear functions. Examine lines graphed on the scatter plots, as shown in Figure 2.48, and notice that slight curvature can be seen in both cases. You may wish to try one of the other functions, but do so with caution. You may sacrifice an easy-to-use linear model without gaining much in the way of increased precision.

When you are considering higher-order polynomial models, our advice is *the simpler the better*. For instance, although scatter plots a and c in Figure 2.46 could be modeled by either a quadratic or a cubic, you should choose the quadratic. Do not consider a cubic model unless the data exhibit an inflection point.

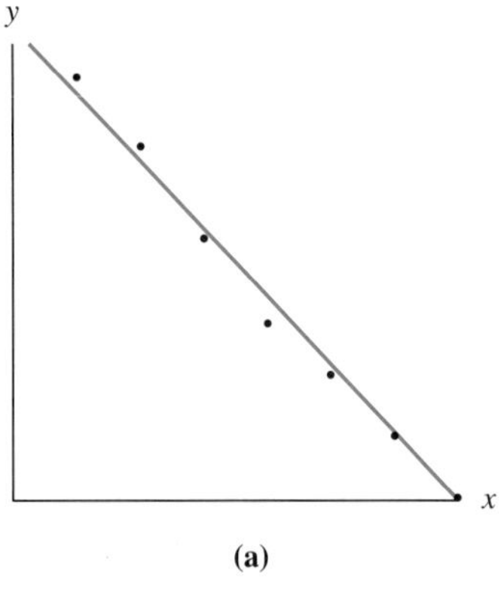

(a)

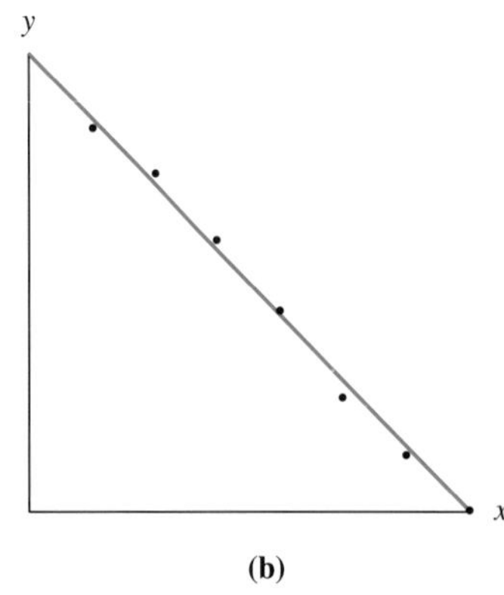

(b)

FIGURE 2.48

Four Steps in Choosing a Model

See the guidelines in the *Instructor's Resource Guide* for fitting piecewise models to data.

Steps in Choosing a Model

1. **Look at the curvature of a scatter plot of the data.**
   - If the points appear to lie in a straight line, try a linear model.
   - If the scatter plot is curved but has no inflection point, try a quadratic, an exponential, or a log model.
   - If the scatter plot appears to have an inflection point, try a cubic and/or a logistic model.
   - If the scatter plot indicates curvature not described by one of the functions we have studied, then consider combining two or more functions to form a piecewise continuous model.

   (*Note*: If the input values are equally spaced, it might be helpful to look at first differences versus percentage differences to decide whether a linear or an exponential model would be more appropriate, or to compare second differences with percentage differences to decide whether a quadratic or an exponential model would be better.)
2. **Look at the fit of the possible equations.** In Step 1, you should have narrowed the possible models to at most two choices. Compute these equations, and graph them on a scatter plot of the data. The one that comes closest to the most points (but does not necessarily go through the most points) is normally the better model to choose. Vertically aligning the data may produce a better-fitting exponential or logistic equation. Horizontally aligning the data may produce a better-fitting log equation.
3. **Look at the end behavior of the scatter plot.** If Step 2 does not reveal that one model is obviously better than another, consider the end behavior of the data, and choose the appropriate model.
4. **Consider that there may be two equally good models for a particular set of data.** If that is the case, then you may choose either.

Let us look at some examples where we must apply these steps in choosing a model.

EXAMPLE 1 *Age of Marriage*

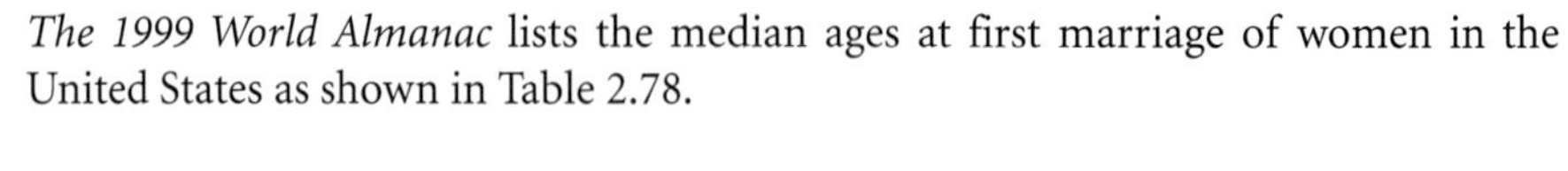

The 1999 World Almanac lists the median ages at first marriage of women in the United States as shown in Table 2.78.

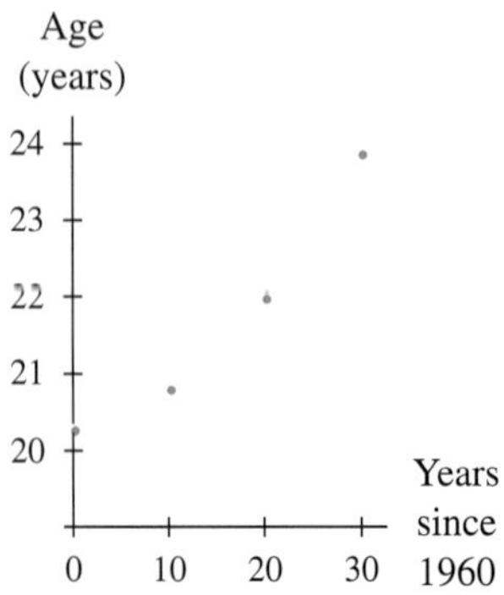

FIGURE 2.49

TABLE 2.78

Year	1960	1970	1980	1990
Age	20.3	20.8	22.0	23.9

a. Find the best model for the data.

b. Use the model in part *a* to estimate the median age in 1997. Compare this estimate with the reported age of 25.0 years.

c. Use the model in part *a* to predict the median age at first marriage of women in the year 2005.

d. Is the part *c* prediction valid?

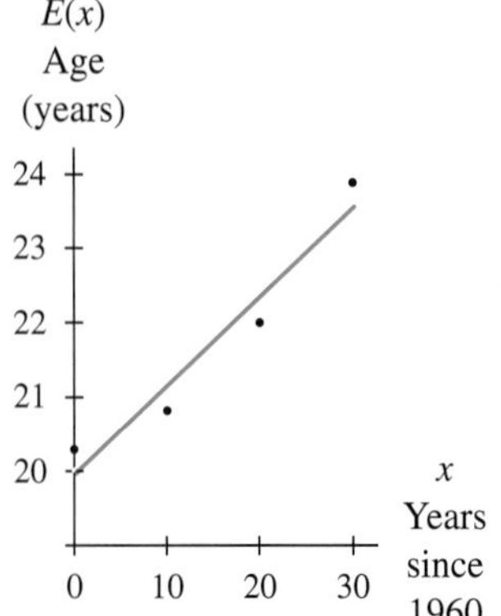

Exponential model

FIGURE 2.50

Solution

a. ***Step 1:*** Consider a scatter plot of the data.

The scatter plot shown in Figure 2.49 is increasing and concave up, so we should try either an exponential model or a quadratic model.

Step 2: Compare the fit of the equations.

An exponential model for the data is

$$E(x) = 20.00024(1.00547^x) \text{ years of age}$$

where x is the number of years after 1960. However, when graphed on the scatter plot, it is obviously a poor fit to the data (see Figure 2.50). If we shift the data down by 20, we obtain the model

$$S(x) = 0.32145(1.08993^x) + 20 \text{ years of age}$$

x years after 1960. This exponential equation is a much better fit to the data than the original one. See Figure 2.51. Other vertical shifts may result in even better-fitting equations.

A quadratic model for the data is

$$Q(x) = 0.0035x^2 + 0.015x + 20.3 \text{ years of age}$$

where x is the number of years since 1960. As Figure 2.52 shows, this model fits the data perfectly.

We could have avoided the process of comparing exponential and quadratic models if we had first calculated second differences and percentage differences.

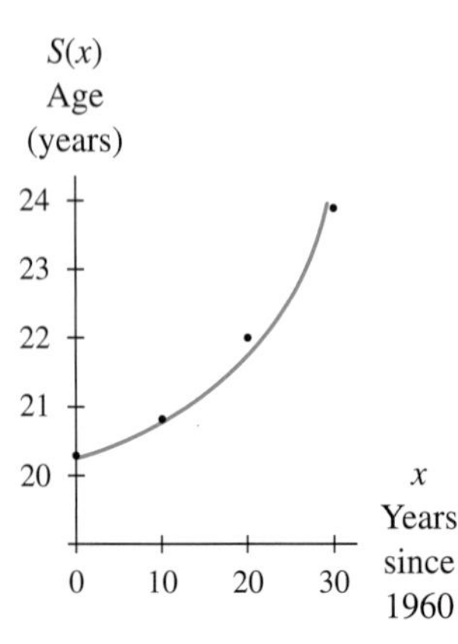

Shifted exponential model

FIGURE 2.51

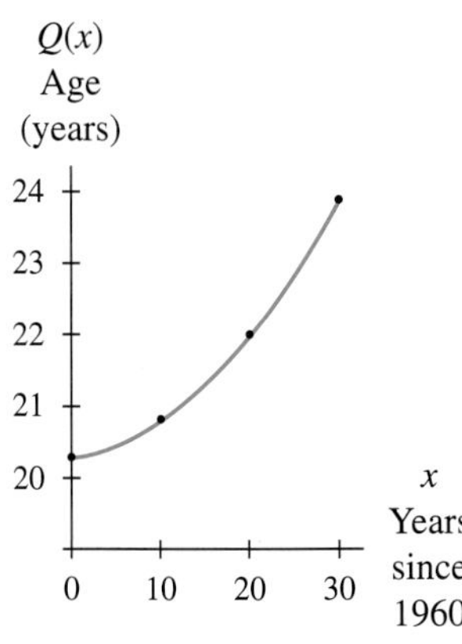

Quadratic model

FIGURE 2.52

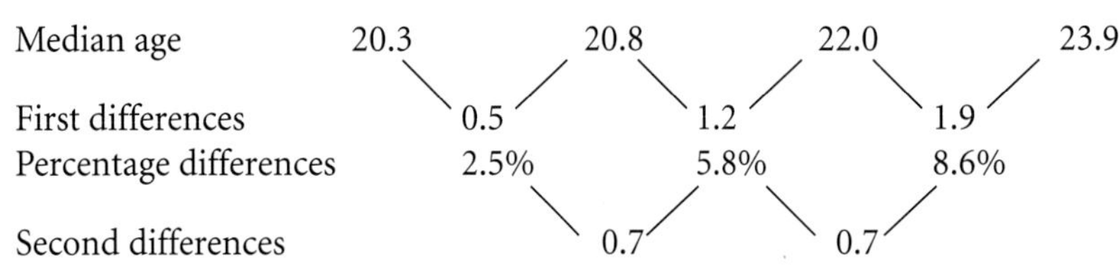

Median age 20.3 20.8 22.0 23.9
First differences 0.5 1.2 1.9
Percentage differences 2.5% 5.8% 8.6%
Second differences 0.7 0.7

The constant second differences indicate that the data are perfectly quadratic. Thus there is no need to consider an exponential model.

Steps 3 and 4 are not necessary here, because we found the best model in Step 2. However, we must keep in mind that this model may not be valid for the years before 1960 or after 1990 because its end behavior does not describe that of the context.

b. The median age of marriage in 1997 is estimated to have been

$$Q(37) = 0.0035(37)^2 + 0.015(37) + 20.3 \approx 25.6 \text{ years}$$

Note that we report the answer to one decimal place because that is the accuracy to which the data were reported. The estimate exceeds the known value by 0.6 year. We can speculate that the model will overestimate in any extrapolations. Note that the shifted exponential model gives the estimate as $S(37) \approx 27.8$ years. This model estimates even higher ages than the quadratic model beyond the range of the data. This is further confirmation that the quadratic model is the better choice.

c. According to the model Q, the median age of marriage in the year 2005 will be

$$\begin{aligned} Q(45) &= 0.0035(45)^2 + 0.015(45) + 20.3 \\ &\approx 28.1 \text{ years of age} \end{aligned}$$

d. This prediction is probably not valid. Even though the model fits the data perfectly, there is no indication that it is a good predictor of future events and trends. There are many factors other than time that affect the median age of first marriage.

Sometimes it is not quite so easy to choose which model is best, as the next example illustrates.

EXAMPLE 2 *Countries Issuing Postage Stamps*

Table 2.79 gives the numbers of European, North American, and South American countries that issued postage stamps[46] from 1840 through 1880.

46. "The Curve of Cultural Diffusion," *American Sociological Review*, August 1936, pp. 547–556.

TABLE 2.79

Years	Number of countries
1840	1
1845	3
1850	9
1855	16
1860	24
1865	30
1870	34
1875	36
1880	37

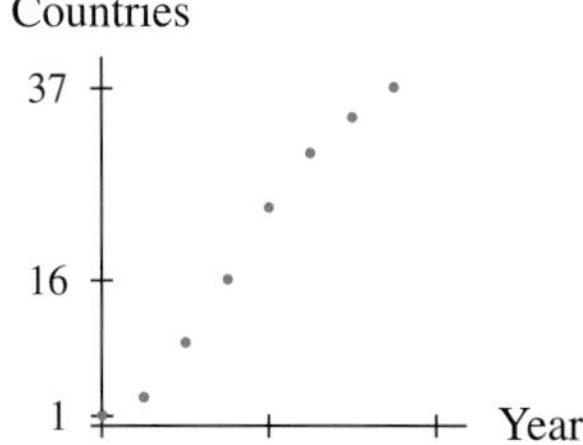

FIGURE 2.53

Find an appropriate model for the data.

Solution

Step 1: Consider the curvature of a scatter plot.

The scatter plot in Figure 2.53 appears to be concave up between 1840 and 1850 or 1855 and then is concave down between 1855 and 1880. Thus there is an apparent inflection point, so either a cubic or a logistic model could be appropriate.

$L(x)$
Countries
37
16
1
0
20
40
x Years since 1840

Logistic model

FIGURE 2.54

Step 2: Compare the fit of the two equations.

A logistic model is

$$L(x) = \frac{37.19488}{1 + 21.37367e^{-0.18297x}} \text{ countries}$$

where x is the number of years since 1840. It appears to fit the data reasonably well (see Figure 2.54).

A cubic model is

$$C(x) = -0.00120x^3 + 0.06027x^2 + 0.40435x + 0.44444 \text{ countries}$$

where x is the number of years since 1840. The cubic model also appears to fit the data well (see Figure 2.55).

If all you are concerned about is fit, you should choose the cubic model. However, if you would like to predict how many countries may have issued stamps in 1890, you should also consider end behavior.

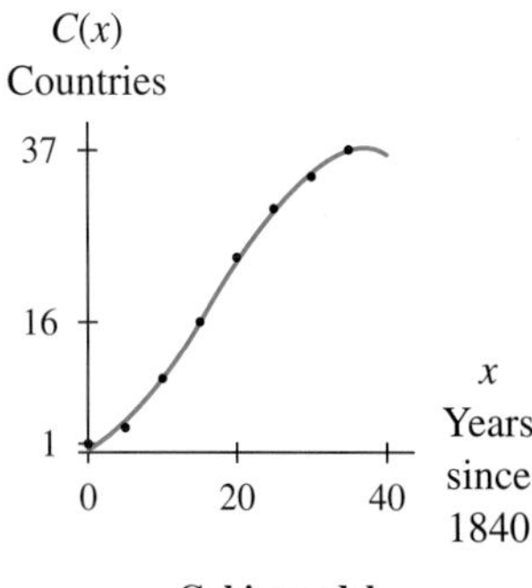

FIGURE 2.55

Step 3: Consider end behavior.

Because the only countries under consideration are those in Europe, North America, and South America, there must be an upper limit on the number of countries issuing stamps. Thus it would be logical to think that a model for these data should approach some limiting value. This end behavior would be better described by the logistic function.

In conclusion, if your only concern is the better-fitting equation, and you do not wish to use the model to extrapolate, then you should choose

$C(x) = -0.00120x^3 + 0.06027x^2 + 0.40435x + 0.44444$ countries

where x is the number of years since 1840.

If you are concerned about end behavior more than about a better fit, you should choose

$$L(x) = \frac{37.19488}{1 + 21.37367e^{-0.18297x}} \text{ countries}$$

where x is the number of years since 1840.

As we mentioned in Step 1 of "Steps in Choosing a Model" on page 167, it might be helpful to consider differences when input values are equally spaced. We consider such data in Example 3.

EXAMPLE 3 *Depreciation*

The value of a certain industrial machine depreciates according to the length of time it has been used. Table 2.80 shows the value of this machine after x years of use.

TABLE 2.80

Years of use	0	2	4	6	8	10
Value (dollars)	40,000	36,700	33,670	30,900	28,340	26,000

Find the most appropriate model for the data.

Solution Consider the curvature of a scatter plot shown in Figure 2.56. The scatter plot appears to be fairly straight, so we try a linear function. A linear model for this data is

$V(x) = -1397.85714x + 39{,}590.95238$ dollars

where x is the number of years that the machine has been in use. However, when V is graphed on the scatter plot (see Figure 2.57), it appears that the data may be slightly concave up.

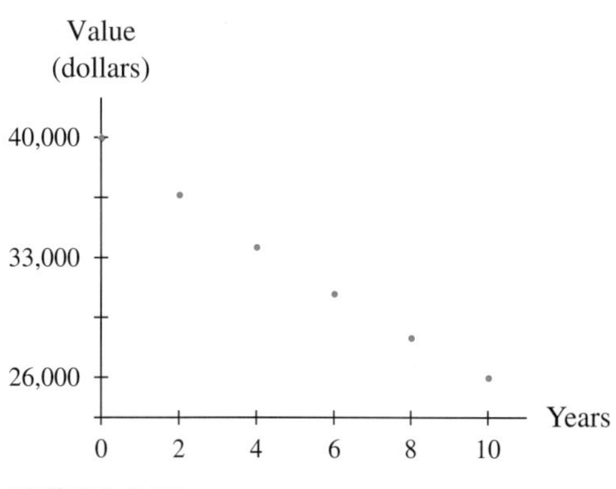

FIGURE 2.56

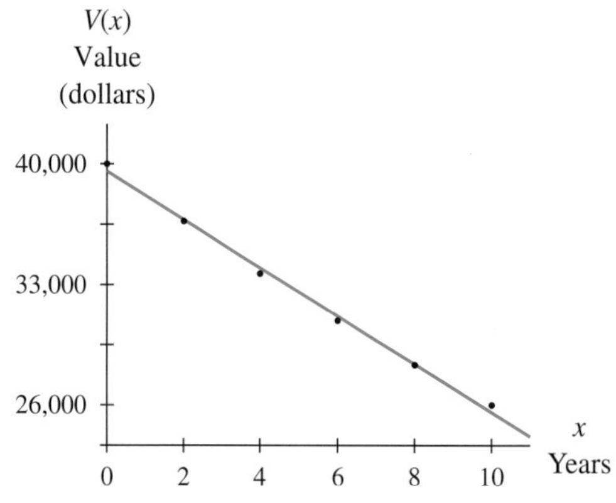

FIGURE 2.57 Linear model

Because the inputs of the data are equally spaced, we can look at first differences, second differences, and percentage differences to see whether there is sufficient reason to try a curved model.

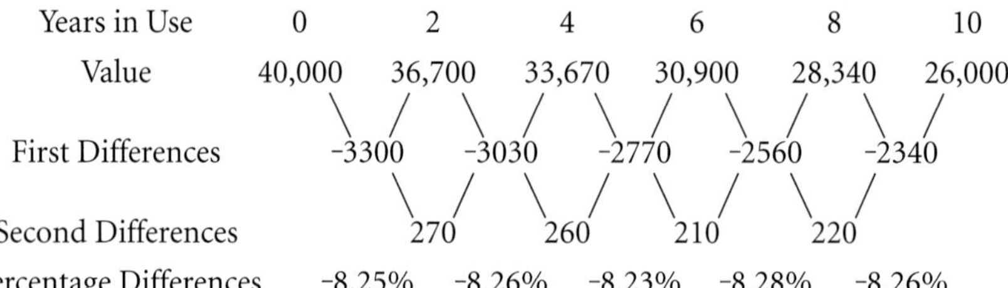

Years in Use	0		2		4		6		8		10
Value	40,000		36,700		33,670		30,900		28,340		26,000
First Differences		-3300		-3030		-2770		-2560		-2340	
Second Differences			270		260		210		220		
Percentage Differences		-8.25%		-8.26%		-8.23%		-8.28%		-8.26%	

Even though none of these are perfectly constant, the percentage differences do seem to be closer to remaining constant than either the first or the second differences. This indicates that an exponential model may be more appropriate than the linear model.

An exponential model is

$$E(x) = 40{,}002.49083(0.95784^x) \text{ dollars}$$

when the machine has been in use x years. This equation appears to be a good fit to the data (see Figure 2.58).

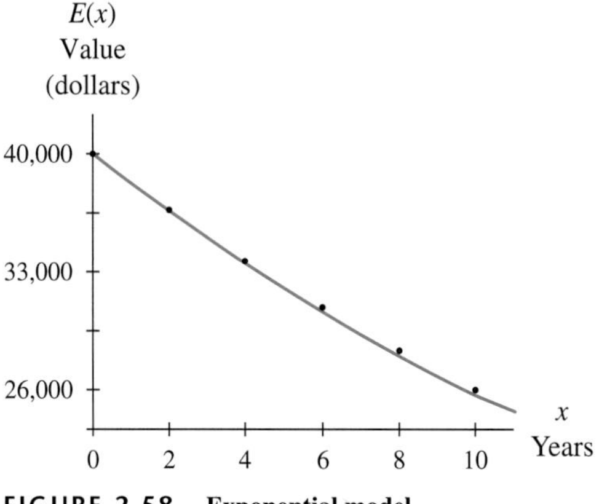

FIGURE 2.58 **Exponential model**

Therefore, the exponential model is the most appropriate one for the data.

Be open to the possibility that none of the models we have studied is appropriate for a set of data. Indeed, many sets of data in the real world do not fit neatly into one of our categories. Modeling, even in the simplest sense, is as much an art as a science. Sometimes a creative mind can be as helpful as a mathematical one. In any case, the most beneficial way to learn how to determine which model to use for a data set is by practicing.

Although the first two chapters have focused on the construction of elementary mathematical models, it is important that you understand that modeling is *not* our primary objective. In reality, mathematical modeling can be a complicated and highly sophisticated endeavor. We are using basic modeling with elementary curve fitting strictly as a way to obtain functional relationships between variables. Because most information in the real world is collected in discrete form (tables of data), it is important in the study of calculus to obtain continuous functional relationships between the variables in order to study their changing behavior. Our main objective is to apply concepts and methods of calculus to these functions.

Because experience is the best teacher, we recommend that you assign all even and odd activities in this section. If students work in pairs or groups on these activities, they are likely to benefit from the differing opinions of others.

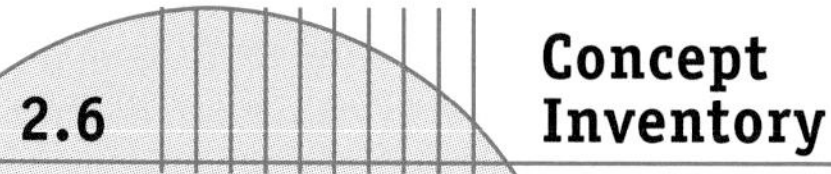

2.6 Concept Inventory

- *Choosing a model:*
  - Consider curvature and calculating differences
  - Compare the fit of possible equations
  - Consider shifting or realigning the data
  - Consider end behavior
  - Consider that there may be two valid choices
- *Sometimes no model we have studied is appropriate*

2.6 Activities

For each of the scatter plots in Activities 1 through 6, state which function or functions are candidates to fit the data. Explain why those functions are appropriate, and also explain why the other types of functions are not appropriate.

1. 2.

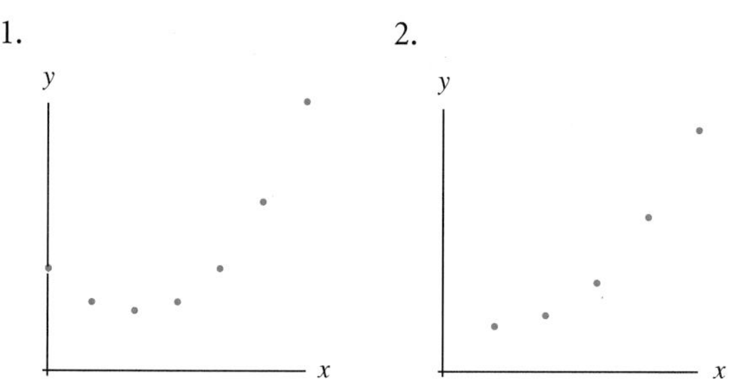

3. 4.

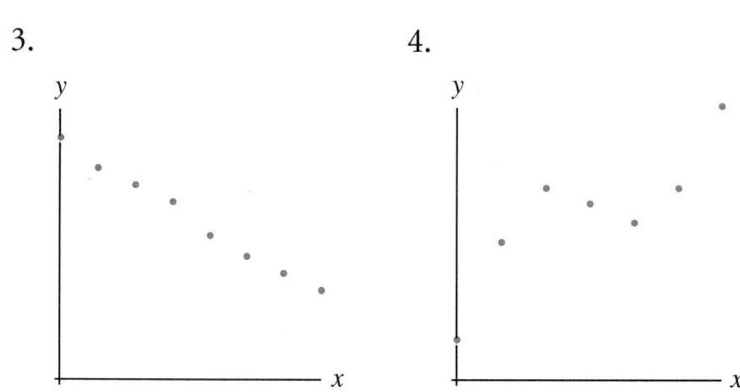

5. 6.

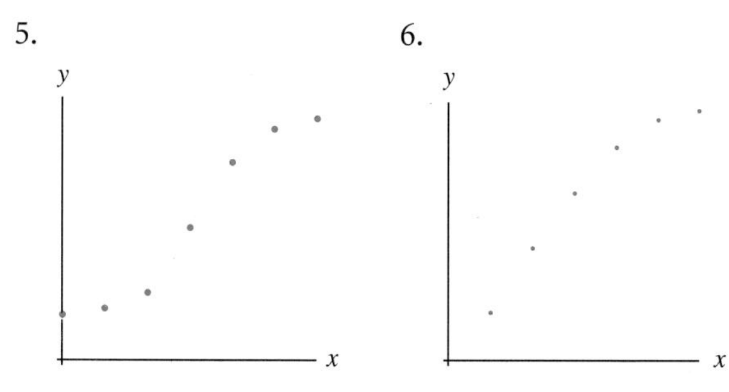

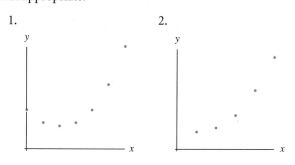

7. Given a data set in which the input values are evenly spaced, discuss how you could determine whether the data are
   a. linear.
   b. quadratic.
   c. exponential.

For each of the data sets in Activities 8 through 12, examine the data set, and determine the best model. Give the equation, define the variables you use, include units of measure, and explain why you chose that model.

8. The *Kelly Blue Book Official Guide* lists 1999 resale values for the Cadillac 4D Sedan DeVille in excellent condition. The data are shown in Table 2.81.

TABLE 2.81

Model year	Resale value (dollars)
1990	6435
1992	9310
1994	16,510
1996	23,575
1998	32,140
1999	37,225

9. *Swimming World* (August 1992) lists the time in seconds that an average athlete takes to swim 100 meters freestyle at age x. The data are shown in Table 2.82.

TABLE 2.82

Age, x (years)	Time (seconds)	Age, x (years)	Time (seconds)
8	92	22	50
10	84	24	49
12	70	26	51
14	60	28	53
16	58	30	57
18	54	32	60
20	51		

10. The 1998 *Statistical Abstract of the United States* gives the number of overseas telephone calls (in millions) billed in the United States. See Table 2.83.

TABLE 2.83

Year	Number of calls (millions)	Year	Number of calls (millions)
1985	411	1993	1926
1989	835	1994	2313
1990	984	1995	2821
1991	1371	1996	3485
1992	1643		

11. Table 2.84 shows yard waste[47] as a percentage of all waste generated in the United States.

TABLE 2.84

Year	Yard waste (percent)
1980	18.2
1985	18.2
1990	17.1
1995	14.1
1996	13.4

12. The birth rates[48] for 15- to 19-year-olds from 1991 through 1997 are shown in Table 2.85.

TABLE 2.85

Year	Birth rate (births per 1000 women)
1991	16.3
1992	15.9
1993	15.5
1994	15.2
1995	14.8
1996	14.7
1997	14.6

13. For each of the situations described in Activities 8 through 12, determine a question that could be asked about the data or a prediction that could be made on the basis of the model you found. Determine a reasonable answer to each of your questions.

14. Table 2.86 gives death rates due to lung cancer (deaths per 100,000 males in the United States) from 1930 through 1990.

TABLE 2.86

Year	Death rate (deaths per 100,000 men)
1930	5
1940	11
1950	21
1960	39
1970	59
1980	66
1990	67

a. Discuss briefly why each of the six models we have studied is or is not appropriate for this data set.

b. Find the model you believe to be the best for these data.

c. Use your equation in part *b* to estimate the death rate in 2000. Do you believe your model is appropriate for this or other extrapolations?

15. Table 2.87 gives the value[49] of manufacturers' shipments of recordings in millions of dollars (based on list price) between 1990 and 1997.

TABLE 2.87

Shipments of LP and EP albums		Shipments of CD albums	
Year	Value (millions of dollars)	Year	Value (millions of dollars)
1990	86.5	1990	3452
1991	29.4	1991	4338
1992	13.5	1992	5327
1993	10.6	1993	6511
1994	17.8	1994	8465
1995	25.1	1995	9377
1996	36.8	1996	9935
1997	33.3	1997	9915

a. Find models for each set of data.

b. Use the models to estimate values of shipments of LP/EP albums and CD albums in 1999 and 2000.

47. *Statistical Abstract*, 1998.
48. Ibid.
49. Ibid.

16. Table 2.88 gives the average weight of the brain in human males as a percentage of body weight.

TABLE 2.88

Age (years)	Percentage of body weight accounted for by brain
0	11.0
2	8.0
4	7.0
6	6.0
8	5.0
10	4.5
12	4.0
14	3.5
16	3.25

a. Describe in words what happens to the percentage of body weight accounted for by the brain as a male child grows.

b. Find quadratic, exponential, and log equations to fit the data. Note that you must shift the ages to the right in order to use a log equation. Shifting the percentages down will produce a better-fitting exponential equation. Write models using each equation.

c. Use each model in part *b* to determine the age (in months) at which brain weight is 10% of body weight.

d. Use each model in part *b* to predict the percentage brain weight for an 18-year-old man.

e. Discuss which equation best models the data and why.

17. A college student works for 8 hours without a break assembling mechanical components. Her supervisor records every hour how many total components, *N*, the student has completed. These figures, which are cumulative, are given in Table 2.89.

TABLE 2.89

Hour	*N*	Hour	*N*
1	9	5	44
2	15	6	51
3	23	7	56
4	33	8	58

a. How many components did the student assemble during the fourth hour?

b. If this student started assembling components at 9:00 a.m., how many components were assembled after 1 p.m.?

c. Find a cubic equation for the data. Does this equation provide a good fit to the data?

d. According to the cubic model, how many hours will it take the college student to assemble 40 components? (Report your answer in hours and minutes.)

e. What does the model indicate will happen to the number of components after 8 hours? Do you believe the model is a good indicator of future events? Explain.

f. Find a logistic equation for the data. Does this equation provide a good fit to the data?

g. What does the model indicate will happen to the number of components after 8 hours? Do you believe the model is a good indicator of future events? Explain.

h. According to the trend exhibited in the logistic model, would it be worthwhile for the student's supervisor to have her work more than 8 hours? Explain.

i. Compare the fit of the logistic equation to the fit of a cubic equation. Besides fit, what other factors should you consider when deciding which model would be better?

j. Which model would you choose? Why?

18. The numbers of AIDS cases[50] diagnosed in the United States from 1985 through 1997 were as shown in Table 2.90.

TABLE 2.90

Year	Cases (thousands)	Year	Cases (thousands)
1985	8.2	1994	78.3
1990	41.6	1995	71.5
1991	43.7	1996	66.9
1992	45.5	1997	58.5
1993	103.5		

50. Ibid.

a. Examine a scatter plot of the data, and discuss its curvature. What type of model would be most appropriate for the data?

b. Find what you believe is the best model for the data.

19. Table 2.91 gives the numbers of used cars,[51] in millions, sold in the United States by franchised dealers from 1998 through 2002.

TABLE 2.91

Year	Used car sales (millions)
1998	15.92
1999	16.38*
2000	16.47*
2001	16.40*
2002	16.35*

*projected

a. Examine a scatter plot of the data, and discuss its curvature. What type of equation should you try to fit to these data? What is the model?

b. Use your model from part *a* to estimate the number of used cars sold by franchised dealers in the United States in 2003.

20. Table 2.92 shows the population[52] of South Carolina between 1790 and 1990.

TABLE 2.92

Year	Population (thousands)
1790	249
1900	1340
1950	2117
1980	3122
1990	3487

a. Examine a scatter plot of the data, and discuss its curvature. What types of curves might be appropriate for such curvature?

b. Find the two models that are appropriate according to your analysis in part *a*.

c. Discuss the fit of each equation from part *b* to the data.

d. Use each model to estimate the 1997 population. Compare your estimates with the actual population of 3,760,000 people.

e. Which of the models from part *b* would be more appropriate to model the South Carolina population? Discuss any issues (besides curvature) that lead you to this decision.

21. Chlorofluorocarbons (CFCs) are believed to thin the ozone layer and were banned from use in aerosols in the United States in the 1970s. Table 2.93 shows the estimated atmospheric release[53] of CFC-12 (one of the two most prominent CFCs) between 1958 and 1978.

TABLE 2.93

Year	CFC-12 atmospheric release (millions of kilograms)
1958	66.9
1960	89.1
1962	114.5
1964	155.5
1966	195.0
1968	246.5
1970	299.9
1972	349.9
1974	418.6
1976	390.4
1978	341.3

a. Observe a scatter plot of the data. What year do you believe the ban on CFCs in the United States took effect?

b. Divide the data into two subsets that have one data point in common. Find an equation to fit each set of data, and combine the two to form a piecewise continuous model.

c. Do you believe that the trend beyond 1978 indicated by your model is valid?

d. If your school's library has *The True State of the Planet*, look on page 429 for a more extensive data set. Summarize what happened to CFC-12

51. "Used Car Transactions in the Dust," *The Salt Lake Tribune*, August 8, 1999, p.E3.

52. *Statistical Abstract*, 1998.

53. Ronald Bailey, ed. *The True State of the Planet* (New York: The Free Press for the Competitive Enterprise Institute, 1995).

release after 1978 and the reasons for the behavior that occurred.

22. The amounts of fish produced by fisheries[54] in the United States for food in selected years between 1970 and 1990 are given in Table 2.94.

TABLE 2.94

Year	Amount (millions of pounds)
1970	2537
1972	2435
1975	2465
1977	2952
1980	3654
1982	3285
1985	3294
1987	3946
1990	7041

a. Discuss the curvature of the data. Why is a cubic function not appropriate? Why is a single quadratic function not appropriate?

b. Divide the data into two subsets that share one data point. Find equations to fit each set of data, and combine them to form a piecewise continuous model.

c. Discuss factors that might have contributed to the behavior exhibited by the data.

d. Use your model to estimate fish production in 1973, 1983, and 1993.

23. Population data[55] for Iowa in the 1980s and 1990s are given in Table 2.95.

TABLE 2.95

Year	Population (thousands)
1980	2914
1985	2830
1987	2767
1989	2771
1991	2791
1993	2820
1995	2841
1997	2852

a. Examine a scatter plot of the data. Discuss the curvature.

b. Discuss the appropriateness or inappropriateness of using a cubic equation to fit this data set.

c. Find a cubic equation to fit the data, and graph it on the scatter plot. Does the graph support or contradict your conclusion in part *b*?

d. Divide the data into two subsets that share one point. Find a piecewise continuous model for the population of Iowa.

e. Update the data to the most recent year for which population data are available. Is the model you chose appropriate for the expanded data set?

data

24. Suppose that you are required to model a set of data that has evenly spaced input values and for which second differences and percentage differences in the outputs are constant. How might you determine which model is best?

25. For which models is reducing the magnitude of the input values important and why?

26. Discuss when it is necessary to shift data vertically.

54. *Statistical Abstract*, 1994.

55. *Statistical Abstract*, 1998.

Summary

Exponential and Logarithmic Functions and Models

Second in importance to linear functions and models are exponential functions and models. Based on the familiar idea of repeated multiplication by a fixed positive multiplier b (the base), the basic exponential function is of the form

$$f(x) = ab^x$$

The parameter a appearing in the equation is the output when the input is zero.

Whereas linear functions model situations that have constant rates of change, exponential functions model constant percentage change. In terms of the function $f(x) = ab^x$, exponential growth occurs when b is greater than 1, and exponential decline takes place when b is between 0 and 1. In fact, the constant percentage growth or decline is $(b - 1)100\%$.

The inverse of the exponential function is the log or logarithmic function. The basic form of the log function that we use is

$$f(x) = a + b \ln x$$

The input of this function must be a value greater than zero. The log function is useful for situations in which the output grows or declines at an increasingly slow rate.

Exponential Models in Finance

The general compound interest formula is

$$A = P\left(1 + \frac{r}{n}\right)^{nt} \text{ dollars}$$

where A is the amount accumulated in an account after t years when P dollars are invested at an annual interest rate of $100r\%$ compounded n times each year. The advertised annual interest rate $100r\%$ is called the annual percentage rate (APR) or the nominal rate. For a given nominal rate $100r\%$ and number n of compounding periods per year, we are also interested in the annual percentage yield (APY), or effective rate, $\left(1 + \frac{r}{n}\right)^n - 1$. If you are an investor, then the annual percentage yield gives a true picture of your actual interest earnings. If you are a borrower who is paying interest, then the effective rate gives a true picture of your actual cost in terms of interest.

If we imagine that the number n of compounding periods per year increases without bound, then we are led to consider the notion of continuous compounding, described by the limit

$$\lim_{n\to\infty} P\left[1 + \frac{r}{n}\right]^{nt} - Pe^{rt}$$

This gives the continuously compounding interest formula $A = Pe^{rt}$. With continuous compounding, the effective rate is $e^r - 1$.

The term future value describes the value of an investment at some future time, and the term present value describes how much must be invested now to achieve a desired future value. We return to future value and present value in a later chapter and consider them in a more sophisticated context using calculus.

Exponential Models in Science

Exponential functions with repeated multiplier less than 1 are useful in modeling exponential decay. Exponential decay is often expressed in terms of the half-life of a substance. As long as we know the quantity of a substance at some time and the half-life, we can use technology to obtain an exponential model for the quantity of the substance as a function of time. Exponential decay applications include radioactive decay, Newton's Law of Cooling, and drug absorption rates, as well as other applications in the life sciences and health sciences.

Logistic Functions and Models

A common situation is to have initial exponential growth followed by a leveling-off approach toward a limiting value L. This behavior is characteristic of logistic growth, modeled by the logistic equation

$$y = \frac{L}{1 + Ae^{-Bx}}$$

These S-shaped curves model such everyday occurrences as the accumulated sales of a new fad item and the spread of information. In situations such as these, the parameter B that appears in the exponential term of the denominator is positive, a reflection of initial exponential growth. Logistic curves also occur as inverted, or upside-down, S-shaped curves. In these cases, the parameter B is negative, a reflection of the decline in output toward the horizontal axis as the input values increase.

Graphs of increasing logistic functions show curvature that changes from concave up to concave down, whereas graphs of decreasing logistic functions show curvature that changes from concave down to concave up. In each case, the point where the concavity changes is known as an inflection point. We later develop calculus techniques that enable us to find such points.

Polynomial Functions and Models

Polynomial functions and models have a well-established role in calculus.

We consider three cases:

linear functions: $f(x) = ax + b$

quadratic functions: $f(x) = ax^2 + bx + c$

cubic functions: $f(x) = ax^3 + bx^2 + cx + d$

Quadratic equations have graphs known as parabolas. The parabola with equation $f(x) = ax^2 + bx + c$ opens upward (is concave up) if a is a positive and opens downward (is concave down) if a is negative.

Cubic equations have graphs that resemble one of the four types shown in Figure 2.29 on pages 143–144. Like logistic models, cubic models show a change of concavity at an inflection point. We must be especially careful in using cubic models when extrapolating beyond the range of data values from which the models are constructed.

Shifting Data

When fitting exponential and logistic equations to data, it is extremely important that you align the input data to values that are reasonably small. This will avoid numerical computation errors. For polynomial equations, aligning the data may produce smaller coefficients in the equation. When fitting a log equation to data, you must sometimes align the input data to ensure that the input values are greater than zero. Aligning input data has the effect of shifting the data horizontally.

When fitting exponential and logistic equations to data, it is sometimes helpful to shift the output data. This vertical shift is particularly useful when the data appear to approach a value other than zero. The goal in shifting is to move the data closer to the horizontal axis. Once a model is obtained for the shifted data, be sure to adjust the equation to compensate for the shift.

Choosing a Model

Although it is not always clear which (if any) of the functions we have discussed apply to a particular real-life situation, it helps to keep in mind a few general, common-sense guidelines. (1) Given a set of discrete data, begin with a scatter plot. The plot will often reveal general characteristics that point the way to an appropriate model. (2) If the scatter plot does not appear to be linear, consider the suggested concavity. One-way concavity (up or down) often indicates a quadratic model, but keep in mind that upward concavity may also suggest an exponential or log model. Downward concavity may also suggest a log model. If input values are evenly spaced, then second differences that are nearly constant indicate a quadratic model, whereas percentage differences that are nearly constant suggest an exponential model. (3) When a single change in concavity seems apparent, think in terms of cubic or logistic models. But remember, the graphs of logistic models tend to become flat on each end, whereas cubics do not. Also, logistic models are suggested in cases where the data are cumulative over time with bounds on the extent of the accumulation. Never consider a cubic or a logistic model if you cannot identify an inflection point.

Concept Check

Can you	To practice, try	
Find an exponential model and determine its percentage change?	Section 2.1	Activity 23
Find and use a log model?	Section 2.1	Activity 37
Calculate APR and APY?	Section 2.2	Activity 11
Calculate present and future value?	Section 2.2	Activities 19, 21
Solve exponential growth and decay problems?	Section 2.2	Activity 29
Find and use a logistic model?	Section 2.3	Activity 21
Find and use a quadratic model?	Section 2.4	Activity 15
Find and use a cubic model?	Section 2.4	Activity 25
Find and use a piecewise continuous model?	Section 2.4	Activity 31
Vertically shift data?	Section 2.5	Activity 11
Identify concavity and inflection points?	Section 2.6	Activity 5
Observe a scatter plot and choose a model?	Section 2.6	Activities 11, 19, 21

Review Test

1. According to the U.S. Census Bureau, the number of children living with their grandparents was 2.2 million in 1970 and 3.9 million in 1997. Assume that the number grew exponentially between 1970 and 1997.
   a. Find an exponential model for the number of children living with grandparents as a function of the number of years since 1970.
   b. What is the percentage change indicated by the model?
   c. According to the model, when will the number of children reach 5 million?
   d. Do you believe an exponential model accurately reflects future change in the number of children living with grandparents? Explain. If not, what do you think would be a more appropriate model?

2. a. What is the effective rate (APY) for 7.3% compounded monthly?
   b. What is the largest possible effective rate for a nominal rate of 7.3%?

3. The yearly high stock prices[56] for Microsoft Corporation for selected years between 1989 and 1998 are shown in Table 2.96.

TABLE 2.96

Year	Price (dollars)	Year	Price (dollars)
1989	0.74	1996	15.02
1990	2.11	1997	31.59
1992	4.38	1998	54.19
1994	6.44		

   a. Examine a scatter plot of the data. Discuss the curvature of the data and the possible models that could be used for this data set.
   b. Find the model that you believe is best for this data set.
   c. What issues in our current economy do you believe will affect the price of Microsoft stock in the future? Do you think the model you chose is a good description of the future price?

56. Hoover's Online Guide, November 1999.

4. On August 28, 1993, the *Philadelphia Inquirer* reported temperatures from the previous day. Table 2.97 lists these temperatures (in degrees Fahrenheit) from 5 a.m. to 5 p.m.

TABLE 2.97

Time	°F	Time	°F
5 a.m.	76	noon	90
6 a.m.	75	1 p.m.	91
7 a.m.	75	2 p.m.	93
8 a.m.	77	3 p.m.	94
9 a.m.	79	4 p.m.	95
10 a.m.	83	5 p.m.	93
11 a.m.	87		

a. Examine a scatter plot of the data. Explain why the curvature indicates that a cubic model is appropriate.

b. Find a cubic model for the data.

c. According to the model, what was the temperature at 5:30 p.m.?

d. According to the model, at approximately what times (on August 27) was the temperature 90 degrees Fahrenheit?

5. A 1998 United Nations study projected that the world population would stabilize at 10.73 billion just after the year 2200. The study's projections include those shown in Table 2.98.

TABLE 2.98

Year	Population (billions)
1999	6
2011	7
2025	8
2041	9
2071	10

a. Is the following statement true or false? Explain.

"Because the first differences of the data are constant, a linear model is appropriate."

b. Examine a scatter plot of the data. Discuss the curvature suggested by the scatter plot and the functions that could be used to model the data.

c. Align the input as the number of years since 1900, and find quadratic, logistic, and log models for the projected population. Discuss how well each equation fits the data.

d. Use limits to express the end behavior of the three equations in part *c*. Which equation best exhibits the end behavior suggested by the United Nations study?

Project 2.1 Compulsory School Laws

Setting

In 1852, Massachusetts became the first state to enact a compulsory school attendance law. Sixty-six years later, in 1918, Mississippi became the last state to enact a compulsory attendance law. Table 2.99 lists the first 48 states to enact such laws and the year each state enacted its first compulsory school law.

TABLE 2.99

State	Year	State	Year	State	Year
MA	1852	SD	1883	IA	1902
NY	1853	RI	1883	MD	1902
VT	1867	ND	1883	MO	1905
MI	1871	MT	1883	TN	1905
WA	1871	IL	1883	DE	1907
NH	1871	MN	1885	NC	1907
CT	1872	ID	1887	OK	1907
NM	1872	NE	1887	VA	1908
NV	1873	OR	1889	AR	1909
KS	1874	CO	1889	TX	1915
CA	1874	UT	1890	FL	1915
ME	1875	KY	1893	AL	1915
NJ	1875	PA	1895	SC	1915
WY	1876	IN	1897	LA	1916
OH	1877	WV	1897	GA	1916
WI	1879	AZ	1899	MS	1918

Source: Richardson, "Variation in Date of Enactment of Compulsory School Attendance Laws," *Sociology of Education* vol. 53 (July 1980), pp. 153–163.

Tasks

1. Tabulate the cumulative number of states with compulsory school laws for the following 5-year periods:

1852–1856	1887–1891
1857–1861	1892–1896
1862–1866	1897–1901
1867–1871	1902–1906
1872–1876	1907–1911
1877–1881	1912–1916
1882–1886	1917–1921

2. Examine a scatter plot of the data in Task 1. Do you believe a logistic model is appropriate? Explain.

3. Find a logistic model for the data in Task 1.

4. What do most states in the third column of the original data have in common? Why would these states be the last to enact compulsory education laws?

5. The seventeen states considered to be southern states (below the Mason-Dixon Line) are AL, AR, DE, FL, GA, KY, LA, MD, MO, MS, NC, OK, SC, TN, TX, VA, and WV. Tabulate cumulative totals for the southern states and the northern/western states for these dates:

Northern/Western States Years	Southern States Years
1852–1856	1891–1895
1857–1861	1896–1900
1862–1866	1901–1905
1867–1871	1906–1910
1872–1876	1911–1915
1877–1881	1916–1920
1882–1886	
1887–1891	
1892–1896	
1897–1901	
1902–1906	

6. Examine scatter plots for the two data sets in Task 5. Do you believe that logistic models are appropriate for these data sets? Explain.

7. Find logistic models for each set of data in Task 5. Comment on how well each equation fits the data.

8. It appears that the northern and western states were slow to follow the lead established by Massachusetts and New York. What historical event may have been responsible for the time lag?

9. One way to reduce the impact of unusual behavior in a data set (such as that discussed in Task 8) is to group the data in a different way. Tabulate the cumulative northern and western state totals for the following 10-year periods:

   1852–1861
   1862–1871
   1872–1881
   1882–1891
   1892–1901
   1902–1911

10. Find a logistic model for the data in Task 9. Comment on how well the function fits the data. Compare models for the data grouped in 10-year intervals and the data grouped in 5-year intervals (Task 6). Does grouping the data differently significantly affect how well the equation fits? Explain.

11. Find an equation that fits the data for the southern states better than the logistic equation. Write the model using the better-fitting equation. Explain your reasoning.

Reporting

1. Prepare a written report of your work. Include scatter plots, models, and graphs. Include discussions of each of the tasks in this project.

2. (Optional) Prepare a brief (15-minute) presentation on your work.

Project 2.2 Fund-Raising Campaign

Setting

In order to raise funds, the mathematics department in your college or university is planning to sell T-shirts before next year's football game against the school's biggest rival. Your team has volunteered to conduct the fundraiser. Because several other student groups have also volunteered to head this project, your team is to present its proposal for the fund drive, as well as predictions about its outcome, to a panel of mathematics faculty.

Note: This project is also used as a portion of Project 5.2 on page 377.

Tasks

1. Develop a slogan and a design for the T-shirt. Keep in mind that good taste is a concern. Decide on a target market, and determine a strategy to survey (at random) at least 100 students who represent a cross section of the target market to determine the demand for T-shirts (as a function of price) within that market. It is important that your sample survey group properly represent your target market. If, for example, you polled only near campus dining facilities at lunch time, your sample would be biased toward students who eat lunch at such facilities.

   The question you should ask is "How much would you be willing to spend on a T-shirt promoting the big football rivalry? $20, $18, $16, $14, $12, $10, $8, or not interested?" Keep an accurate tally of the number of students who answer in each category. In your report on the results of your poll, you should include information such as your target market; where, when, and how you polled within that market; and why you believe that your polled sample is likely to be a representative cross section of the market.

2. a. From the data you have gathered, determine how many students from your sample survey group would buy a T-shirt at $8, $10, $12, and so on.

   b. Devise a marketing strategy, and determine how many students within your target market you can reasonably expect to reach. Assuming that your poll is an accurate indicator of your target population, determine the number of students from your target market who will buy a T-shirt at each of the given prices.

   c. Taking into account the results of your poll and your projected target market, develop a model for demand as a function of price. Keep in mind that your model must make sense for all possible input values.

3. Estimate the cost that you will incur per T-shirt from the partial price listing in Table 2.100. Use the demand function from Task 2 to create equations for revenue, total cost, and profit as functions of price. (Revenue, total cost, and profit may not be one of the basic models that were discussed in class. They are sums and/or products of the demand function with other functions.)

Reporting

1. Prepare a written report summarizing your survey and modeling. The report should include your slogan and design, your target market, your marketing strategy, the results from your poll (as well as the specifics of how you conducted your poll), a discussion of how and why you chose the model of the demand function, a discussion of the accuracy of your demand model, and graphs and equations for all of your models. Attach your questionnaire and data to the report as an appendix.

2. Prepare a 15-minute oral presentation of your survey, modeling, and marketing strategy to be delivered before a panel of mathematics faculty. You will be expected to have overhead transparencies of all graphs and equations as well as any other information that you consider appropriate as a visual aid. Remember that you are trying to sell the mathematics department on your campaign idea.

TABLE 2.100 The T-Shirt Company: Partial Price Listing

Minimum order	Number of colors					
	1	2	3	4	5	6
25	\$4.35	\$5.10	\$5.70	\$6.30	\$7.00	\$7.50
50	\$4.25	\$5.05	\$5.65	\$6.25	\$6.95	\$7.45
100	\$4.15	\$4.95	\$5.55	\$6.15	\$6.85	\$7.35
250	\$4.00	\$4.80	\$5.40	\$6.00	\$6.65	\$7.20
500	\$3.75	\$4.50	\$5.10	\$5.70	\$6.30	\$7.00
750	\$3.50	\$4.25	\$4.90	\$5.50	\$6.00	\$6.50
1000	\$3.45	\$4.00	\$4.65	\$5.00	\$5.50	\$6.00
1500	\$3.25	\$3.80	\$4.45	\$4.85	\$5.30	\$5.80
2000	\$3.15	\$3.70	\$4.30	\$4.65	\$5.10	\$5.60
2500	\$3.10	\$3.60	\$4.20	\$4.55	\$5.00	\$5.45
5000	\$3.05	\$3.55	\$4.15	\$4.50	\$4.95	\$5.40
7500	\$3.05	\$3.50	\$4.10	\$4.45	\$4.90	\$5.35
10,000	\$3.00	\$3.45	\$4.05	\$4.40	\$4.85	\$5.30

Source: Based on data compiled from 1993 prices at Tigertown Graphics, Inc., Clemson, SC.

3

Describing Change: Rates

Concept Application

Roger Tully/Tony Stone Images

The Centers for Disease Control (CDC) in Atlanta, Georgia, is concerned with following the spread of diseases in the United States. CDC information is valuable to doctors, hospitals, pharmaceutical companies, and medical researchers. The CDC has closely followed the AIDS epidemic. Mathematics can help the CDC answer such questions as

- On average, how quickly did the number of AIDS patients increase between 1980 and 1990?
- How quickly was the number of AIDS patients growing in 1990?
- By what percentage was the number of AIDS patients growing in 1996?

This chapter will provide you with some of the tools that make it possible to answer questions like these. You will have the opportunity to estimate the answers to the above questions in Activity 22 of Section 3.3.

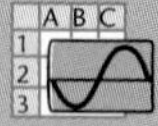

This symbol directs you to the supplemental technology guides. The different guides contain step-by-step instructions for using graphing calculators or spreadsheet software to work the particular example or discussion marked by the technology symbol.

This symbol directs you to the Calculus Concepts *web site, where you will find algebra review, links to data updates, additional practice problems, and other helpful resource material.*

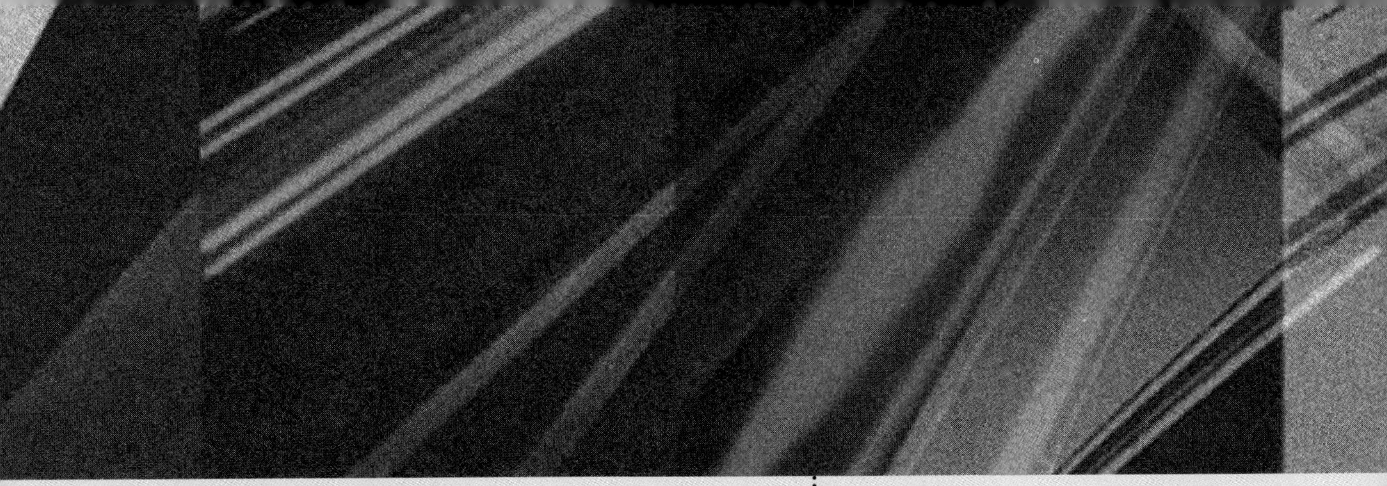

Concept Objectives

This chapter will help you understand the concepts of

- Change, percentage change
- Average rate of change
- Instantaneous rate of change and derivative
- Secant line and tangent line and their relationship
- Percentage rate of change
- Slope graph

and you will learn to

- Find descriptions of change using data, graphs, and equations
- Interpret descriptions of change
- Sketch and find slopes of secant and tangent lines
- Estimate rates of change using tangent lines
- Estimate rates of change using symmetric difference quotients
- Interpret derivatives
- Sketch rate-of-change graphs

Change is everywhere around us and affects our lives on a daily basis. From the explosive growth in the use of personal computers to the decline in the number of family-owned farms, from the shrinking value of a dollar to the growing national debt, change impacts us as individuals, in families, through businesses, and in our government.

In this chapter we consider several ways to describe change. Starting from the actual change in a quantity over an interval, it is a simple step to describe the change as a percentage change or as an average rate of change over that interval. The notion of average rate of change, when examined carefully in light of the underlying geometry of graphs, leads to the more subtle and challenging concept of instantaneous change. Indeed, the precise description of instantaneous change in terms of mathematics is one of the principal goals of calculus.

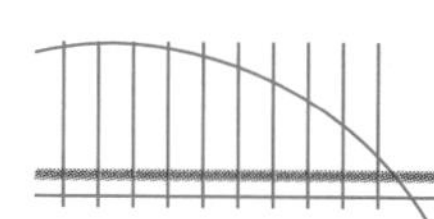

3.1 Change, Percentage Change, and Average Rates of Change

One of the primary goals of calculus is the description of change. In preparation for learning how calculus describes change, we introduce three numerical ways of reporting change. We begin with a simple business example.

The yearly revenue for a large department store declined from \$1.4 billion in 1998 to \$1.1 billion in 2001. There are three common ways to express this change in revenue. We could use a negative sign to indicate the decrease in revenue and say that the **change** in revenue was -\$0.3 billion over 3 years. We could also express this change by saying that the revenue declined by \$0.3 billion (or \$300 million) over 3 years.

Point out to your students that this method of calculating percentage change is the same as the way we calculated percentage change using data in Section 2.1.

It is often helpful to express the change as a percent of the 1998 revenue. This **percentage change** (or percent change) is found by dividing the change by the revenue in 1998 and multiplying by 100:

$$\text{Percentage change} = \frac{-\$0.3 \text{ billion}}{\$1.4 \text{ billion}} \cdot 100\% \approx -21.4\%$$

The company saw a 21.4% decline in revenue between 1998 and 2001. You should already be familiar with percentage change from the discussion of exponential models in Chapter 2.

The third way to express change involves spreading the change over the time interval to obtain the **average rate of change.** This is done by dividing the change by the length of the interval. In the case of the department store revenue, the length of the interval is 3 years:

$$\text{Average rate of change} = \frac{-\$0.3 \text{ billion}}{3 \text{ years}} = -\$0.1 \text{ billion per year}$$

On average, the revenue declined at a rate of \$0.1 billion per year (or \$100 million per year) between 1998 and 2001.

We summarize these three ways to describe change as follows:

Change, Percentage Change, and Average Rate of Change

If a quantity changes from a value of m to a value of n over a certain interval, then

- The change in the quantity is found by subtracting the first value from the second.

$$\text{Change} = n - m$$

- The percentage change is the change divided by the first value and then multiplied by 100.

$$\text{Percentage change} = \frac{\text{change}}{\text{first value}} \cdot 100\% = \frac{n-m}{m} \cdot 100\%$$

- The average rate of change is the change divided by the length of the interval.

$$\text{Average rate of change} = \frac{\text{change}}{\text{length of interval}} = \frac{n-m}{\text{length of interval}}$$

Interpreting Descriptions of Change

Correctly calculating these descriptions of change is important, but being able to state your result in a meaningful sentence in the context of the situation is equally important. When interpreting a description of change, you should answer the questions *when, what, how,* and *by how much.*

Interpreting Descriptions of Change

When describing change over an interval, be sure to answer the following questions:

- When? Specify the interval.
- What? Specify the quantity that is changing.
- How? Indicate whether the change is an increase or a decrease.
- By how much? Give the numerical answer labeled with proper units:

Description	*Units*
change	output units
percentage change	percent
average rate of change	output unit per input unit

When stating an average rate of change, use the word *average* or the phrase *on average.*

Encourage your students not to memorize this list of unit forms. Instead, encourage (and consistently *demonstrate* in all your calculations in class) keeping units attached to all values used during any calculation.

EXAMPLE 1 *Air Temperature*

Consider Table 3.1, which shows temperature values on a typical May day in a certain midwestern city. Find the following descriptions of change, and write a sentence giving the real-life meaning of (that is, interpret) each result.

TABLE 3.1

Time	Temperature (°F)
7 a.m.	49
8 a.m.	58
9 a.m.	66
10 a.m.	72
11 a.m.	76
noon	79
1 p.m.	80
2 p.m.	80
3 p.m.	78
4 p.m.	74
5 p.m.	69
6 p.m.	62

a. The change in temperature from 7 a.m. to 1 p.m.

b. The percentage change in temperature between 3 p.m. and 6 p.m.

c. The average rate of change in temperature between 8 a.m. and 5 p.m.

Solution

a. To find the change in temperature, subtract the temperature at 7 a.m. from the temperature at 1 p.m.

$$\text{Change} = 80°\text{F} - 49°\text{F} = 31°\text{F}$$

The interpretation statement answers the questions *when* (1 p.m. and 7 a.m.), *what* (the temperature), *how* (was greater than), and *by how much* (31°F): The temperature at 1 p.m. was 31°F greater than the temperature at 7 a.m.

b. To find percentage change, first find the change in temperature between 3 p.m. and 6 p.m.

$$\text{Change} = 62°\text{F} - 78°\text{F} = -16°\text{F}$$

Next, divide the change by the temperature at 3 p.m. (the beginning of the time interval under consideration), and multiply by 100.

$$\text{Percentage change} = \frac{-16°\text{F}}{78°\text{F}} \cdot 100\% \approx -20.5\%$$

The temperature declined by about 20.5% between 3 p.m. and 6 p.m. In this statement, "between 3 p.m. and 6 p.m." answers the question *when*, "the temperature" again identifies *what*, and "declined by about 20.5%" tells *how and by how much.*

c. Begin the calculation of the average rate of change by finding the change in the temperature from 8 a.m. to 5 p.m.

$$\text{Change} = 69°\text{F} - 58°\text{F} = 11°\text{F}$$

Next, divide the change by the length of the time interval (9 hours).

$$\text{Average rate of change} = \frac{11°\text{F}}{9 \text{ hours}} \approx 1.2°\text{F per hour}$$

In interpreting this average rate of change, we must use the word *average* in our sentence in addition to answering the four questions. We state the interpretation as follows: Between 8 a.m. and 5 p.m., the temperature rose at an average rate of 1.2°F per hour.

Although these descriptions of change are useful, they have limitations. It appears from the answer to part *c* of Example 1 that the temperature rose slowly throughout the day. However, the average rate of change does not describe the 22°F rise in temperature followed by the 11°F drop in temperature that occurred between 8 a.m. and 5 p.m.

Finding Percentage Change and Average Rate of Change Using Graphs

You may have noticed that calculating the average rate of change is the same as calculating slope. This observation allows for the easy calculation of average rates of change if you are given a graph. For instance, when plotted, the May daytime temperatures in Example 1 fall in the shape of a parabola (see Figure 3.1).

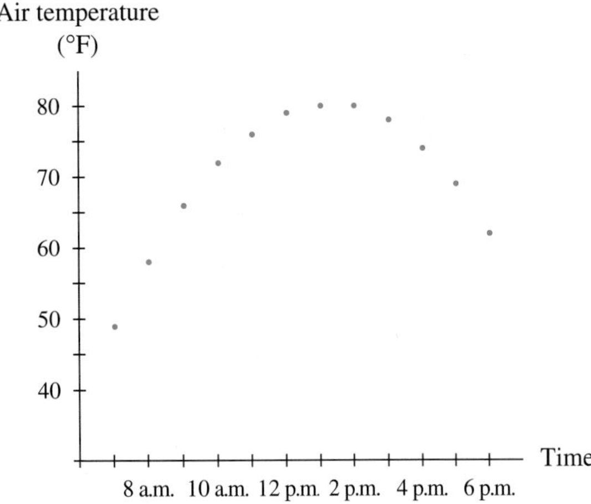

FIGURE 3.1

To find the average rate of change between 9 a.m. and 4 p.m., first use a straightedge to draw carefully a line connecting the points at 9 a.m. and 4 p.m. (see Figure 3.2). We call this line connecting two points on a scatter plot or a graph a **secant line** (from the Latin *secare*, "to cut"). Next, approximate the slope of this secant line by estimating the rise and the run for a portion of the line (see Figure 3.3).

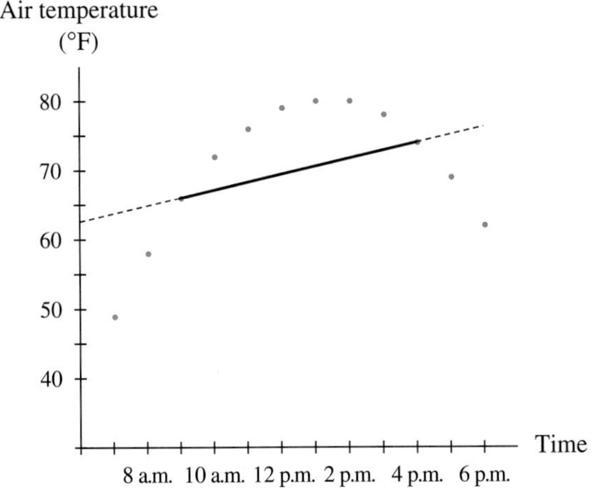

FIGURE 3.2

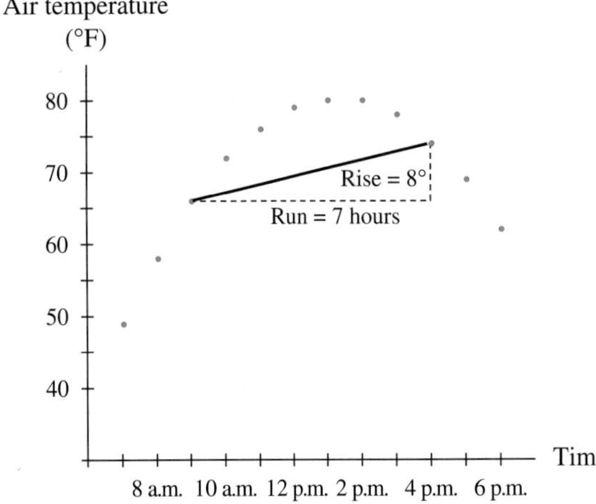

FIGURE 3.3

$$\text{Average rate of change} = \frac{\text{rise}}{\text{run}} = \frac{\text{change in temperature}}{\text{change in time}}$$

$$= \frac{8°\text{F}}{7 \text{ hours}} \approx 1.1°\text{F per hour}$$

Between 9 a.m. and 4 p.m., the temperature rose at an average rate of 1.1°F per hour. Note that the rise is the change in temperature. To calculate percentage change from the graph, divide the rise by the estimated output of the first point.

$$\text{Percentage change} = \frac{\text{change in temperature}}{\text{temperature at 9 a.m.}} \approx \frac{8°\text{F}}{66°\text{F}} \cdot 100\% \approx 12.1\%$$

Accepting variations in answers does not mean that you accept sloppy sketching. Accurate sketching will be especially important in Section 3.2 when students must sketch tangent lines.

It is important to note that this graphical method of calculating descriptions of change is imprecise if you are given only a scatter plot or a graph. It gives only approximations to change, percent change, and average rate of change. The method depends on drawing the secant line accurately and then correctly identifying two points on the line. Slight variations in sketching are likely to result in slightly different answers. This does not mean that the answers you obtain are incorrect. It simply means that descriptions of change obtained from graphs are approximations.

Example 1 and the subsequent discussion of air temperature use the term *between* two time values. There are other ways to describe intervals on the input axis, and we take a moment now to discuss one of them.

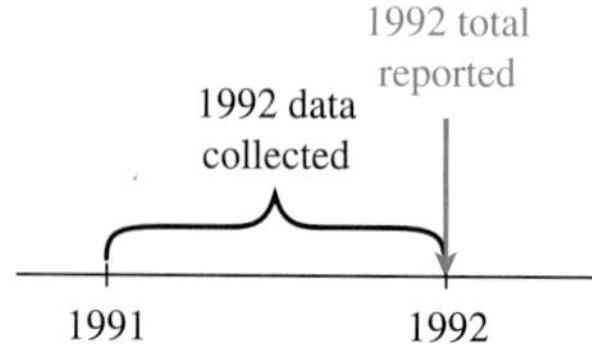

FIGURE 3.4

When we use data that someone else has collected, we often do not know when the data were reported or recorded. It seems logical to assume that yearly (or monthly or hourly and so forth) totals are reported at the *end* of the intervals representing those periods. For instance, a 1992 total covers the period of time from the end of 1991 through the end of 1992 (see Figure 3.4).

Therefore, we adopt the convention that "from a through b" on the input axis refers to the interval beginning at a and ending at b. If a and b are years and the output represents a quantity that can be considered to have been measured at the end of the year, then "from a to b" means the same thing as "from the end of year a through the end of year b." The phrases "between a and b" and "from a to b" have the same meaning as "from a through b." We use this terminology in the remainder of the text.

EXAMPLE 2 *Fuel Consumption*

The average annual fuel consumption[1] of personal passenger vehicles, buses, and trucks is shown in Figure 3.5. A smooth curve connecting the points is also shown.

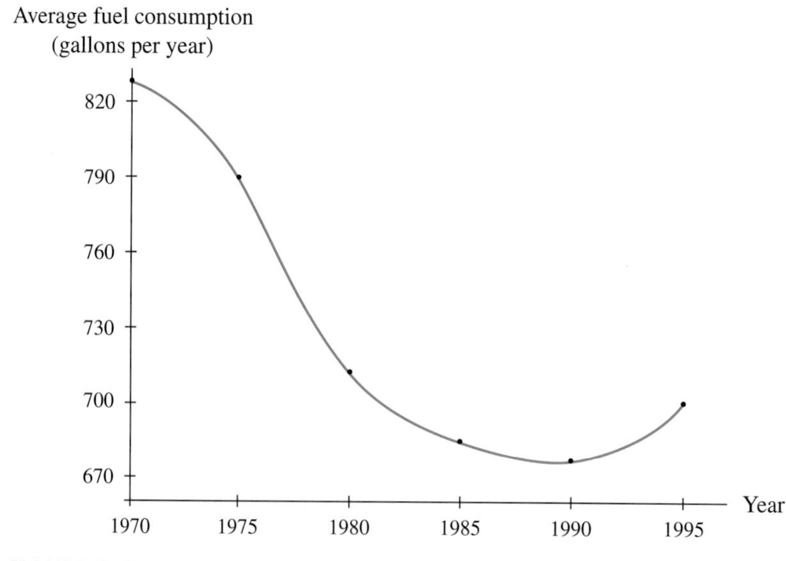

FIGURE 3.5

1. *Statistical Abstract*, 1998.

a. Estimate the change, percentage change, and average rate of change in average annual fuel consumption between 1975 and 1985. Write a sentence interpreting each answer in context.

b. Estimate the change, percentage change, and average rate of change in average annual fuel consumption between 1990 and 1995. Write a sentence interpreting each answer in context.

Solution

a. Begin by estimating from the graph the fuel consumption in 1975 and 1985. One possible estimate is 790 gallons per year in 1975 and 685 gallons per year in 1985. To calculate change, subtract the 1985 value from the 1975 value.

$$\text{Change} = 685 - 790 = -105 \text{ gallons per year}$$

Average fuel consumption fell by 105 gallons per year from 1975 to 1985. It is also correct to say that average *annual* fuel consumption fell by 105 gallons from 1975 to 1985. The use of the word *annual* implies that the value is *per year*. The word *yearly* could be used in a similar manner.

Convert this change to percentage change by dividing by the estimated fuel consumption in 1975.

$$\text{Percentage change} = \frac{-105 \text{ gallons per year}}{790 \text{ gallons per year}} \cdot 100\% \approx -13.3\%$$

The average annual fuel consumption decreased by 13.3% between 1975 and 1985.

Convert the change to an average rate of change by dividing the change by the length of the interval between 1975 and 1985.

$$\begin{aligned}\text{Average rate of change} &= \frac{-105 \text{ gallons per year}}{10 \text{ years}} \\ &= -10.5 \text{ gallons per year per year}\end{aligned}$$

The average fuel consumption decreased an average of 10.5 gallons per year per year between 1975 and 1985. It is also correct to say that the average *annual* fuel consumption decreased by 10.5 gallons per year between 1975 and 1985. The average rate of change is the slope of the secant line through the point corresponding to 1975 and the point corresponding to 1985 shown in Figure 3.6.

b. Begin by estimating the output values in 1990 and 1995. We estimate 680 gallons per year in 1990 and 700 gallons per year in 1995. Following the same procedure as in part *a*, we have

$$\begin{aligned}\text{Change} &\approx 700 - 680 = 20 \text{ gallons per year} \\ \text{Percentage change} &\approx \frac{20 \text{ gallons per year}}{680 \text{ gallons per year}} \cdot 100\% \approx 2.9\% \\ \text{Average rate of change} &\approx \frac{20 \text{ gallons per year}}{5 \text{ years}} = 4 \text{ gallons per year per year}\end{aligned}$$

Between 1990 and 1995, the average annual fuel consumption rose by about 20 gallons. This represents a percentage increase of about 2.9%. On average, the fuel consumption rose by approximately 4 gallons per year per year during this

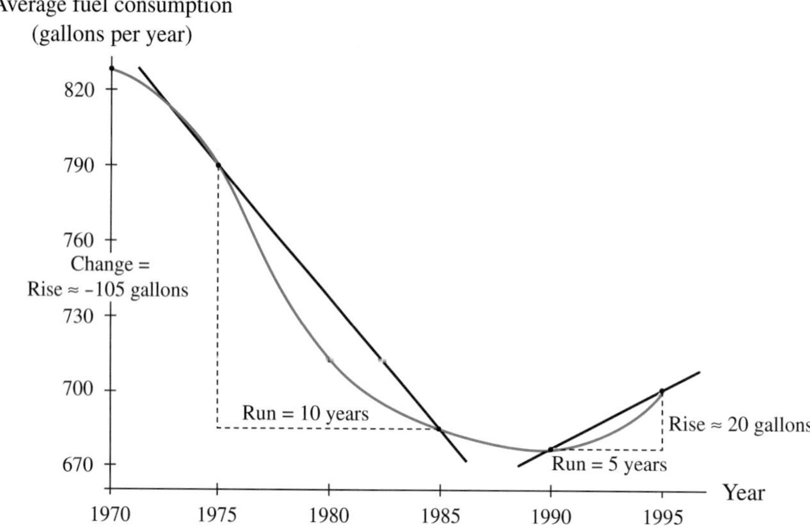

FIGURE 3.6

time. Again, the average rate of change is the slope of the secant line connecting the points on the graph that correspond to 1990 and 1995. (See Figure 3.6.)

Determining Percentage Change and Average Rate of Change Using an Equation

It is also possible to determine descriptions of change when we are given only an equation. A model for the temperature data on a typical May day in a certain midwestern city is

$$\text{Temperature} = -0.8t^2 + 2t + 79\ °\text{F}$$

where t is the number of hours after noon.

Calculate the percentage change and average rate of change between 11:30 a.m. and 6 p.m. as follows:

1. Note that at 11:30 a.m., $t = -0.5$ and that at 6 p.m., $t = 6$.
2. Substitute $t = -0.5$ and $t = 6$ into the equation to obtain the corresponding temperatures.

   At 11:30 a.m.: $\text{Temperature} = -0.8(-0.5)^2 + 2(-0.5) + 79 = 77.8°\text{F}$

   At 6 p.m.: $\text{Temperature} = -0.8(6)^2 + 2(6) + 79 = 62.2°\text{F}$

3. To calculate percentage change, divide the change in temperature by the temperature at 11:30 a.m. and multiply by 100.

   $$\frac{62.2°\text{F} - 77.8°\text{F}}{77.8°\text{F}} \cdot 100\% \approx -20\%$$

4. Divide the change in temperature by the change in time, subtracting the first temperature from the second temperature.

   $$\frac{62.2°\text{F} - 77.8°\text{F}}{6 - (-0.5)} = \frac{-15.6°\text{F}}{6.5 \text{ hours}} = -2.4°\text{F per hour}$$

Thus, between 11:30 a.m. and 6 p.m., the temperature fell by 15.6°F. This represents a 20% decline and an average rate of decline of 2.4°F per hour. Note that the temperature fell 15.6°F in 6.5 hours; however, when finding an average rate of change, we state the answer (in this case) as the number of degrees per *one* hour.

EXAMPLE 3 *Population Density*

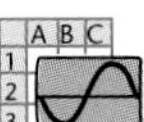

The population density of Nevada[2] from 1950 through 1997 can be approximated by the model

$$p(t) = 0.1617(1.04814^t) \text{ people per square mile}$$

where t is the number of years since 1900.

a. Find the average rate of change of the population density from 1950 through 1980. Interpret your answer.

b. Find the average rate of change of the population density between 1980 and 1997.

c. Use the fact that the land area of Nevada is 110,540 square miles to convert both average rates of change to people per year.

d. On the basis of your answers, what can you conclude about growth in Nevada?

e. Calculate the percentage change in the population density between 1950 and 1997. Interpret your answer.

Solution

a. The average rate of change from 1950 through 1980 is

$$\frac{p(80) - p(50)}{80 - 50} \approx \frac{6.95 - 1.70 \text{ people/mi}^2}{30 \text{ years}} \approx 0.18 \text{ person per square mile per year}$$

Between 1950 and 1980, the population density of Nevada increased at an average rate of 0.18 person per square mile per year.

b. The average rate of change between 1980 and 1997 is

$$\frac{p(97) - p(80)}{97 - 80} \approx \frac{15.47 - 6.95 \text{ people/mi}^2}{17 \text{ years}} \approx 0.50 \text{ person per square mile per year}$$

As always, in calculating these answers, we used unrounded values from the function *p*. Of necessity, we show rounded intermediate values.

c. From 1950 through 1980:

$$\frac{0.18 \text{ person/mi}^2}{\text{year}} \cdot 110{,}540 \text{ mi}^2 \approx 19{,}372 \text{ people per year}$$

From 1980 through 1997:

$$\frac{0.50 \text{ person/mi}^2}{\text{year}} \cdot 110{,}540 \text{ mi}^2 \approx 55{,}347 \text{ people per year}$$

d. In the 17-year period from 1980 through 1997, the population grew by almost three times as much as it had grown in the 30-year period from 1950 through 1980.

2. Based on data from *Statistical Abstract*, 1992, 1995, and 1998.

e. The percentage change between 1950 and 1997 is

$$\frac{15.47 - 1.70}{1.70} \cdot 100\% = 811.4\%$$

The population density of Nevada increased by about 811.4% between 1950 and 1997.

All three descriptions of change (change, percentage change, and average rate of change) are valuable, but the concept of average rate of change will be the bridge between the algebraic descriptions of change examined in this section and the calculus description of change that we begin exploring in the next section. The graphical interpretation of the average rate of change as a slope of a secant line will be particularly useful.

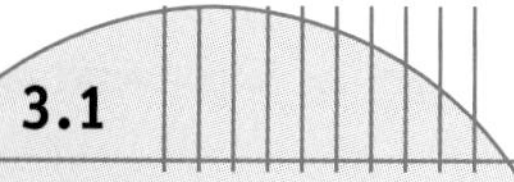

Concept Inventory

- *Change*
- *Percentage change*
- *Average rate of change*
- *Secant line*
- *Slope of a secant line = average rate of change*
- *Interpreting descriptions of change*

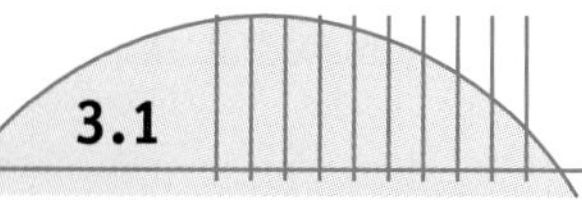

Activities

Rewrite the sentences in Activities 1 through 5 to express how rapidly, on average, the quantity changed over the given interval.

1. In five trading days, the stock price rose $2.30.
2. The nurse counted 32 heart beats in 15 seconds.
3. The company lost $25,000 during the past 3 months.
4. In 6 weeks, she lost 17 pounds.
5. The unemployment rate has risen 4 percentage points in the past 3 years.

For Activities 6 through 9, calculate the change, percentage change, and average rate of change over the interval specified. Write a sentence interpreting each description of change.

6. The average cost for a 20-year, level-premium, term life insurance policy[3] for a 35-year-old male non-smoker was $448 in 1991 and $180 in 1999.
7. The national ACT college test composite average[4] for females was 20.3 in 1990 and 20.9 in 1999.
8. The American Indian, Eskimo, and Aleut population[5] in the United States was 362 thousand in 1930 and 2387 thousand in 1999.
9. The number of Internet users[6] in China grew from 0.9 million in 1997 to 12.0 million in 2000.
10. a. On October 1, 1987, 193.2 million shares were traded on the stock market.[7] On October 30, 1987, 303.4 million shares were traded. Find the percent change and average rate of change in the number of shares traded per trading day between October 1 and October 30.
    b. The scatter plot in Figure 3.1.1 shows the number of shares traded each day during October of 1987. On the scatter plot, sketch a line whose slope is the average rate of change between October 1 and October 30.
    c. The behavior of the graph on October 19th and 20th has been referred to as "October Madness." Write a sentence describing how the number of

3. *Insurance Information*, based on analysis of more than 600 companies.
4. ACT, Inc.
5. U.S. Bureau of the Census.
6. BDA (China), The Strategis Group.
7. *The Dow Jones Averages 1885–1990*, ed. Phyllis S. Pierce (Homewood, IL: Business One Irwin, 1991).

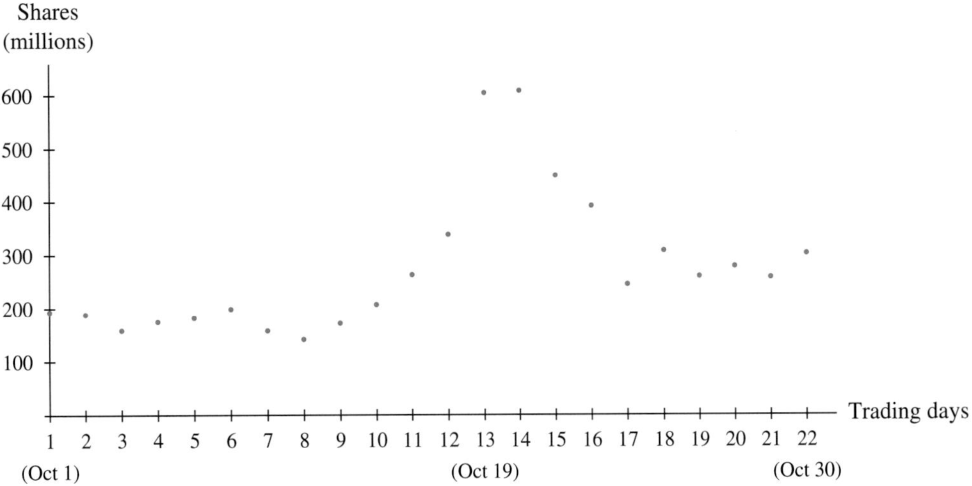

FIGURE 3.1.1

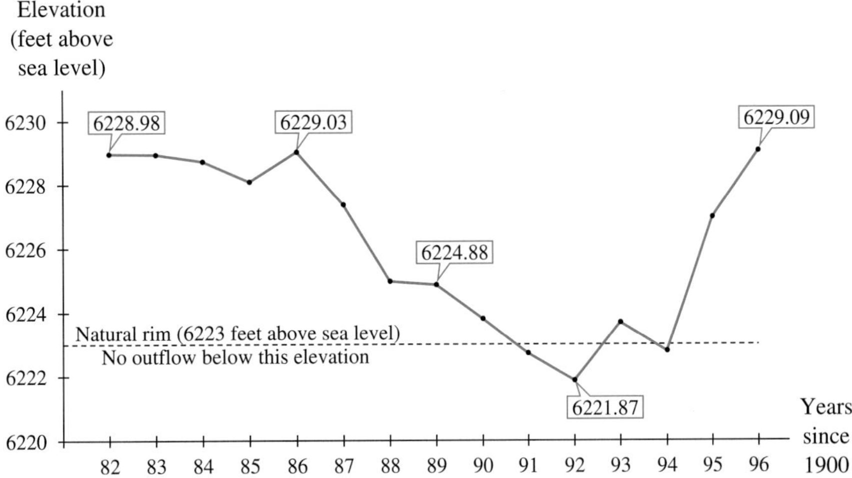

FIGURE 3.1.2

shares traded changed throughout the month. How well does the average rate of change you found in part *a* reflect what occurred throughout the month?

11. The graph in Figure 3.1.2 shows the highest elevations above sea level attained by Lake Tahoe[8] (located on the California-Nevada border) from 1982 through 1996.

    a. Sketch a secant line connecting the beginning and ending points of the graph. Find the slope of this line.

    b. Write a sentence interpreting the slope in the context of Lake Tahoe levels.

    c. Write a sentence summarizing how the level of the lake changed from 1982 through 1996. How well does your answer to part *b* describe the change in the lake level as shown in the graph?

12. Refer once again to the roofing jobs example in Section 2.4. The data are shown in Table 3.2.

TABLE 3.2

Month	Number of jobs
January	12
February	14
March	22
April	35
May	58
June	84

8. Data from Federal Watermaster, United States Department of the Interior.

a. Use the data to find the change and percent change in the number of roofing jobs from January through June.

b. Use the data to find the average rate of change in the number of roofing jobs from January through June.

c. Find a model for the data.

d. Use the equation to find the change and the average rate of change in the number of roofing jobs from January through June.

e. Are the answers obtained from the data more accurate than those obtained from the equation?

13. Imagine that 6 years ago you invested \$1400 in an account with a fixed interest rate and with interest compounded continuously. You do not remember the interest rate, but your end-of-the-year statements for the first 5 years yield the data shown in Table 3.3.

TABLE 3.3

End of year	Amount at end of year
1	\$1489.55
2	\$1584.82
3	\$1686.19
4	\$1794.04
5	\$1908.80

a. Use the data to find the change and percent change in the balance from the end of year 1 through the end of year 5.

b. Use the data to find the average rate of change of the balance from the end of year 1 through the end of year 5. Interpret your answer.

c. Using the data, is it possible to find the average rate of change in the balance from the middle of the fourth year through the end of the fourth year? Explain how it could be done or why it cannot be done.

d. Find a model for the data, and use the equation to find the average rate of change over the last half of the fourth year.

14. Table 3.4 gives the price in dollars of a round trip flight from Denver to Chicago on a certain airline and the corresponding monthly profit (in millions of dollars) for that airline on that route.

TABLE 3.4

Ticket price (dollars)	Profit (millions of dollars)
200	3.08
250	3.52
300	3.76
350	3.82
400	3.70
450	3.38

a. Find a model for the data.

b. Estimate the average rate of change of profit when the ticket price rises from \$200 to \$325.

c. Estimate the average rate of change of profit when the ticket price rises from \$325 to \$450.

15. Refer to the influenza epidemic data in Table 2.31, Activity 17 of Section 2.3.

a. How rapidly (on average) did the number of deaths from influenza increase in the Navy between August 31, 1918, and October 12, 1918?

b. How rapidly did civilian deaths increase from October 5, 1918, through November 2, 1918?

c. Compare the average rates of increase in the Navy and Army between October 12, 1918, and October 19, 1918.

16. A travel agent vigorously promotes cruises to Alaska for several months. The number of cruise tickets sold during the first week and the total (cumulative) sales every 3 weeks thereafter are given in Table 3.5.

TABLE 3.5

Week	Total tickets sold
1	71
4	197
7	524
10	1253
13	2443
16	3660
19	4432
22	4785
25	4923

a. Find the first differences in the numbers of tickets sold, and convert them to average rates of change.

b. When were ticket sales growing most rapidly? How rapidly (on average) were they growing at that time?

c. If the travel agent made a \$25 commission on every ticket sold, how rapidly did the agent's commission revenue increase between weeks 7 and 10?

17. The population of Mexico[9] between 1921 and 1997 is given by the model

Population $= 7.391(1.02695)^t$ million persons

where t is the number of years since 1900.

a. How much did the population change from 1940 through 1955? Convert the change to percentage change.

b. How rapidly was the population changing on average from 1983 through 1985?

18. The number of AIDS cases diagnosed[10] from 1994 through 1997 can be modeled by

Cases diagnosed $=$
$-x^3 + 16.1x^2 - 90.7x + 247.5$ thousand cases

where x is the number of years since 1990. Find the percent change and the average rate of change in cases diagnosed between 1995 and 1997.

19. The graph in Figure 3.1.3 shows the path of the misfired missile from Activity 7 of Section 2.4.

a. Use a secant line to estimate the average rate of change in position between 0 seconds and 2 seconds.

b. Use a secant line to estimate the average rate of change in position between 2 seconds and 3 seconds.

c. Convert your answer in part b to miles per hour.

20. The graph in Figure 3.1.4 shows the median age at first marriage[11] for men in the United States. Estimate by how much and how rapidly the median marriage age grew from 1988 through 1997. What is true about the average rate of change between any two points on a linear graph?

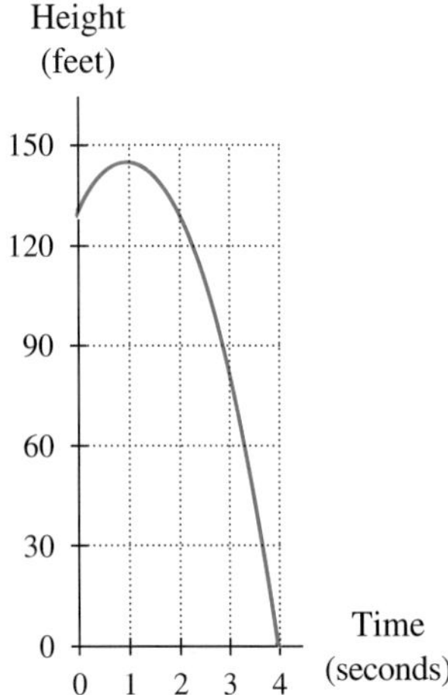

FIGURE 3.1.3

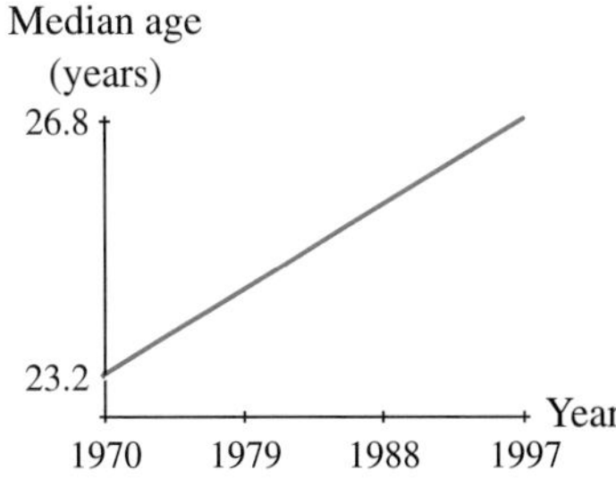

FIGURE 3.1.4

21. A graph of the equation for a model for the yearly interest income[12] earned between 1990 and 1998 by Kelly Services, Inc., a leading global provider of staffing services, is shown in Figure 3.1.5.

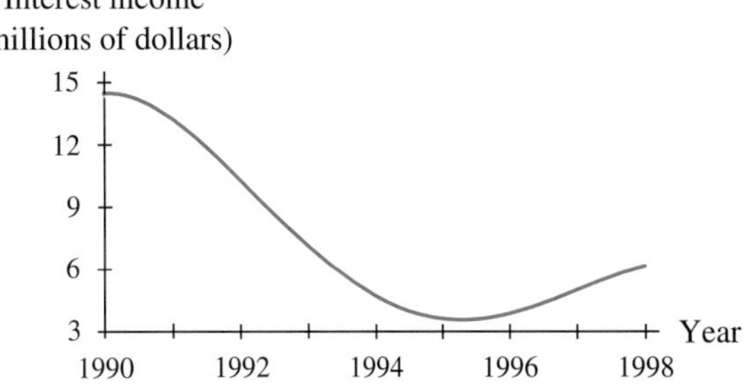

FIGURE 3.1.5

9. Based on data from SPP and INEGI, Mexican Censuses of Population 1921 through 1990 as reported by Pick and Butler, *The Mexico Handbook*, Westview Press, 1994; and on data from *Statistical Abstract*, 1998.
10. Based on data from *Statistical Abstract*, 1998.
11. Based on data from *Statistical Abstract*, 1998.
12. Kelly Services, Inc. 1994 and 1998 *Annual Report*.

a. Use the graph to estimate the average rate of change in Kelly's interest income between 1990 and 1994. Interpret your answer.

b. Estimate the percent change in the interest income between 1996 and 1998.

c. According to the graph, was Kelly losing money between 1990 and 1995?

22. The graph in Figure 3.1.6 models the number of states associated with the national P.T.A. organization[13] from 1895 through 1931.

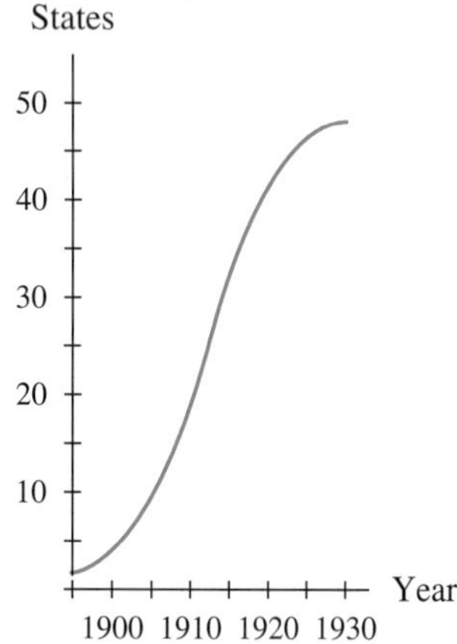

FIGURE 3.1.6

a. Approximately how rapidly was the membership growing from 1905 through 1915?

b. Approximately how rapidly was the membership growing from 1920 through 1925?

23. The percentage of all banks between 1996 and 1999 that levied surcharges on the use of automated teller machines[14] can be modeled by the equation

$$s(t) = \frac{102.665}{1 + 9729.989e^{-1.2613t}} \text{ percent}$$

t years after 1990. Find the percentage change in the percent of banks assessing surcharges on ATMs between 1996 and 1999.

24. The consumer price index values (1982–1984 = 100) for refuse collection[15] between 1990 and 1997 can be modeled by the equation

$$R(t) = 131.496 + 55.096 \ln t$$

t years after 1988. Find the average rate of change in the CPI values for refuse collection between 1992 and 1999.

25. Consider the linear function $y = 3x + 4$.

a. Find the average rate of change of y over each of the following intervals:

i. From $x = 1$ to $x = 3$

ii. From $x = 3$ to $x = 6$

iii. From $x = 6$ to $x = 10$

b. Find the percentage change of y for each of the following intervals:

i. From $x = 1$ to $x = 3$

ii. From $x = 3$ to $x = 5$

iii. From $x = 5$ to $x = 7$

c. On the basis of the results in part *a* and your knowledge of linear functions from Chapter 1, what generalizations can you make about percentage change and average rate of change for a linear function?

26. Consider the exponential function $y = 3(0.4^x)$.

a. Find the percentage change and average rate of change of y for each of the following intervals:

i. From $x = 1$ to $x = 3$

ii. From $x = 3$ to $x = 5$

iii. From $x = 5$ to $x = 7$

b. On the basis of the results in part *a* and your knowledge of exponential functions from Chapter 2, what generalizations can you make about percentage change and average rate of change for an exponential function?

27. The annual costs[16] of President Clinton's 1999 proposed prescription drug plan for Medicare patients are given in Table 3.6.

TABLE 3.6

Year	2000	2002	2004	2006	2008	2009
Cost (billions of dollars)	0.57	8.57	12.86	15.00	16.43	17.14

a. Draw a scatter plot of the data, and discuss the curvature.

b. Find a model for the data.

13. Based on data from Hamblin, Jacobson, and Miller, *A Mathematical Theory of Social Change* (New York: Wiley, 1973).
14. Based on data from U.S. Public Interest Research Group National Survey.
15. Based on data from *Statistical Abstract*, 1998.
16. Based on data from the Congressional Budget Office.

c. Find the change in the proposed annual costs between 2000 and 2005.

d. How quickly are the proposed costs rising on average between 2000 and 2005?

e. What is the percentage change in the proposed annual costs between 2000 and 2005?

28. Funds awarded[17] to arthritis researchers by the Arthritis Foundation for selected years are given in Table 3.7.

TABLE 3.7

Year	1990	1992	1994	1996	1997	2000
Research funds (millions of dollars)	9.8	10.9	14.6	15.6	18.4	37.5*

*Proposed

a. Use the data to find the average rate of change of research funds awarded between 1994 and 1997. Interpret your answer.

b. Find a model for the data.

c. Use the equation to estimate the percentage change in research funds awarded between 1998 and 2000.

d. If the Arthritis Foundation's goal for research dollars is determined by the model in part *b*, in what year did the actual amount of funds awarded exceed the projected goal by the greatest amount? By how much did it exceed the goal in that year?

29. The multiple-birth rate[18] for births involving more than twins jumped 19.7% between 1995 and 1996 and 344% between 1980 and 1996.

a. If the birth rate for births involving three or more babies was 152.6 per 100,000 births in 1996, find the multiple-birth rates in 1995 and 1980.

b. Use the information presented in Table 3.8 to find a model for the multiple-birth rate between 1971 and 1996.

TABLE 3.8

Year	1971	1976	1981	1986	1991	1996
Multiple-birth rate (births per 100,000)	29.1	31.7	39.4	49.9	80.2	152.6

c. Use the equation to estimate the multiple-birth rates in 1995 and 1980. How close are those values to the results of part *a*? Are these estimates found with interpolation or extrapolation?

d. Suggest reasons why the multiple-birth rate has been rising rapidly.

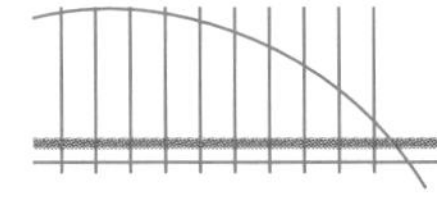

3.2 Instantaneous Rates of Change

In Section 3.1 we considered the average rate of change over an interval. Now we consider the concept of the rate of change at an instant in time. The most common example of an instantaneous rate of change is as close as the nearest steering wheel. Suppose that you begin driving north on highway I-81 at the Pennsylvania/New York border at 1:00 p.m. As you drive, you note the time at which you pass each of the indicated mile markers (see Table 3.9).

These data can be used to determine average rates of change. For example, between mile 0 and mile 19, the average rate of change of distance is 67.1 mph. In this context, the average rate of change is simply the average speed of the car. Average speed between any of the mile markers in the table can be determined in a similar manner. Average speed will not, however, answer the following question:

TABLE 3.9

Time	Mile marker
1:00 p.m.	0
1:17 p.m.	19
1:39 p.m.	42
1:54 p.m.	56
2:03 p.m.	66
2:25 p.m.	80
2:45 p.m.	105

17. Arthritis Foundation.
18. *The Greenville News*, July 1, 1998, p. A1.

> If the speed limit is 65 mph and a highway patrol officer with a radar gun clocks your speed at mile post 17, were you exceeding the speed limit by more than 10 mph?

The only way to answer this question is to know your speed at the instant that the radar locked onto your car. This speed is the **instantaneous rate of change** of distance with respect to time, and your car's speedometer measures that speed in miles per hour.

Just as an average rate of change measures the slope between two points, an instantaneous rate of change measures the slope at a single point. How we measure instantaneous rate of change and why it is useful are important aspects of calculus.

Figure 3.7 shows a continuous graph of air temperature as a function of time.

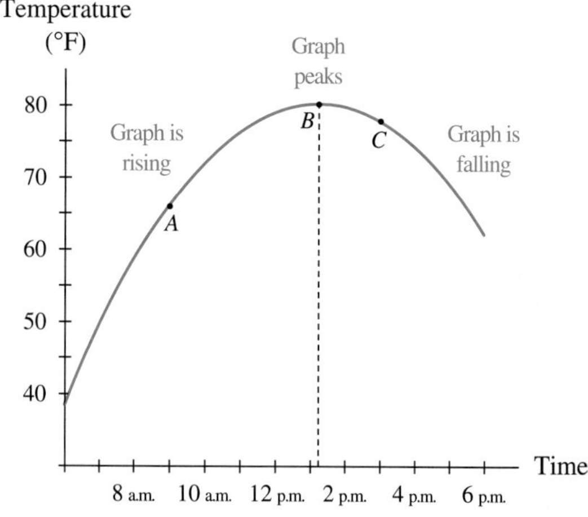

FIGURE 3.7

The graph reaches its peak value at approximately 1:15 p.m. Reading from left to right, the graph is rising until it reaches its peak at 1:15 p.m. and then is falling after 1:15 p.m. The slope of the graph is positive at each point on the left side of the peak and is negative at each point on the right side. The slope of the graph is zero at the top of the parabola. Note that the graph levels off as you move from 7 a.m. to 1:15 p.m. It is not as steep at 1 p.m. as it is at 7 a.m., so the slope at 1 p.m. is smaller than the slope at 7 a.m. In fact, at each point on the graph there is a different slope, and we need to be able to measure that slope in order to find the instantaneous rate of change at each point.

Instantaneous Rate of Change

The **instantaneous rate of change** at a point on a curve is the slope of the curve at that point.

In precalculus mathematics, the concept of slope is intrinsically linked with lines. In terms of lines, slope is a measure of the tilt of a line. Now we wish to measure how tilted a graph is at a point. It may come as no surprise to you to learn that we still must rely on lines to measure the slopes at points on a graph.

Local Linearity and Tangent Lines

With current technology, we can easily obtain close-up views of a portion of a graph. For any smooth, continuous graph, we will eventually see a line as we look closer and closer. For example, consider the temperature graph in Figure 3.7. Close-ups of the graph 1/100 unit away from each labeled point in both horizontal and vertical directions are shown in Figure 3.8.

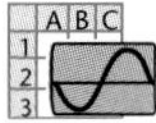

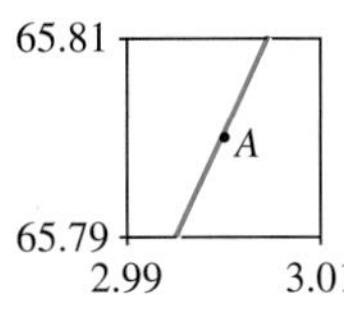

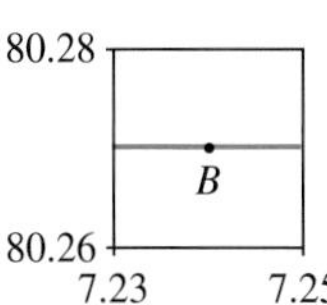

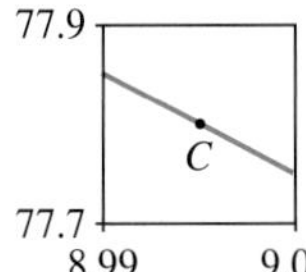

FIGURE 3.8

The graphs in Figure 3.8 illustrate the principle of **local linearity**. As we zoom in on a point P on a smooth (no sharp points), continuous curve, the curve will look more and more like a line. We call this line the **tangent line**, and its slope is the instantaneous rate of change of the curve at the point P, the **point of tangency.** A tangent line (from the Latin word *tangere*, "to touch") at a point on a graph touches that point and is tilted exactly the way that the graph is tilted at the point of tangency. The slope of the tangent line at a point is a measure of the slope of the graph at that point.

A wonderful example of local linearity is the short-term growth of the U.S. population. Long-term population growth is exponential, but the short-term growth usually appears to be linear. The U.S. Census Bureau web site (**www.census.gov**) gives the population of the United States and of the world, updated every minute. Using data taken from these population "clocks," you can demonstrate to your students that short-term growth appears linear. Make the connection between a model based on these data and local linearity.

Local Linearity

If we look closely enough near any point on a smooth curve, the curve will look like a line, which is called the tangent line at that point.

The tangent lines at points A, B, and C are shown in Figure 3.9. Do you see that these are the same as the lines in Figure 3.8?

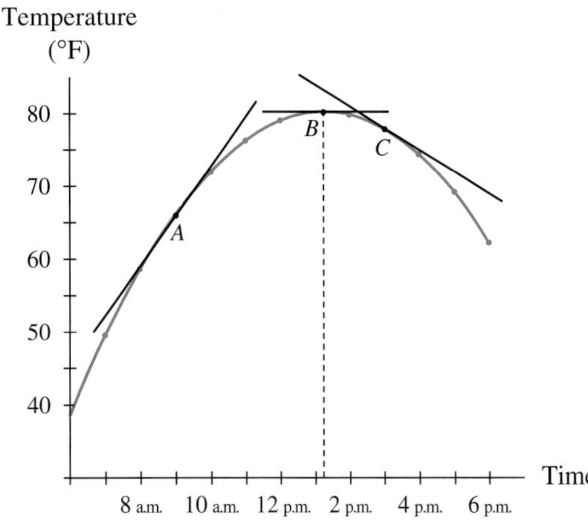

FIGURE 3.9

The slope of a graph at a point is the slope of the tangent line at that point.

In Figure 3.10, tangent lines are drawn on the temperature graph at 7 a.m., noon, and 4 p.m. Can you see that the tangent lines are tilted to match the tilt of the graph at each point? Can you also see that the tangent lines are the lines we would see if we zoomed in at each point? The tangent lines at points *D* and *E* are tilted up, so the slope at these points is positive. The slope at point *F* is negative, because the tangent line at point *F* is tilted down. Even though the tangent line at point *F* has the least slope of these three tangents, it is *steeper* than the tangent line at point *E* because the magnitude (absolute value) of its slope is larger than that of the line tangent to the curve at point *E*. That is, the temperature is falling faster at 4 p.m. than it is rising at noon.

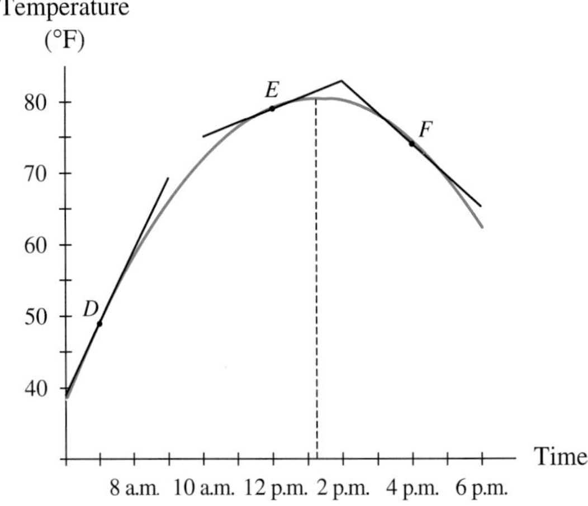

FIGURE 3.10

Examine Figure 3.10 carefully. The slope of the tangent line at point *D* is 10°F per hour. (A method for calculating this slope will be discussed later.) Therefore, the slope of the graph at 7 a.m. is also 10°F per hour. This is the same as saying that the instantaneous rate of change of the temperature at 7 a.m. is 10°F per hour. In other words, at 7 a.m., the temperature is rising 10°F per hour.

Similarly, the following statements can be made.

- The slope of the tangent line at point *E* is 2°F per hour.
- The slope of the graph at noon is 2°F per hour.
- The instantaneous rate of change of the temperature at noon is 2°F per hour.
- At noon, the temperature is rising 2°F per hour.

And,

- The slope of the tangent line at point *F* is −4.4°F per hour.
- The slope of the graph at 4 p.m. is −4.4°F per hour.
- The instantaneous rate of change of the temperature at 4 p.m. is −4.4°F per hour.
- At 4 p.m., the temperature is falling 4.4°F per hour.

We summarize the results of this discussion in the following way:

> Given a function f and a point P on the graph of f, the instantaneous rate of change at point P is the slope of the graph at P and is the slope of the line tangent to the graph at P (provided the slope exists).

EXAMPLE 1 *NRA Membership*

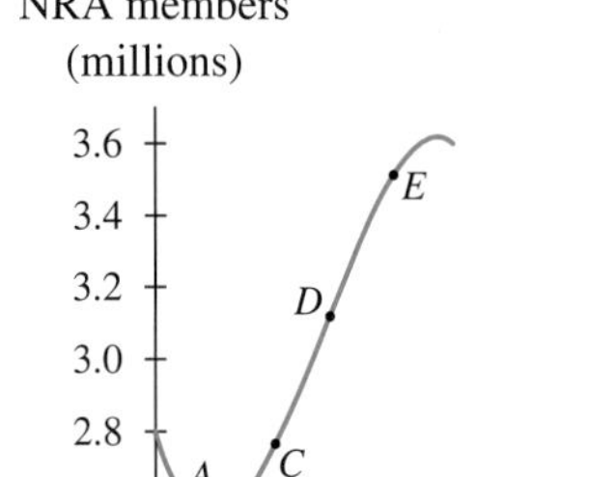

FIGURE 3.11

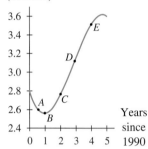

Figure 3.11 shows the National Rifle Association membership between 1990 and 1995. Consider the following statements:

- The slope of the line tangent to the graph at A is -0.24 million members per year.
- The slope of the graph is zero at point B.
- The instantaneous rate of change of the NRA membership at point C is 340,000 members per year.
- NRA membership is increasing the fastest at point D. The rate of fastest increase is 0.42 million members per year.
- The slope of the line tangent to the graph at point E is 260,000 members per year.

Using this information, answer the following questions.

a. At which of the indicated points is the slope of the graph (i) the greatest? (ii) the least?

b. At which of the indicated points is the steepness of the graph (i) the greatest? (ii) the least?

c. What is the instantaneous rate of change of the NRA membership at point A? at point E?

d. What is the slope of the tangent line at point C? at point D?

Solution

a. The numerical values of the slopes, in million members per year, at the indicated points are

A: -0.24 B: 0 C: 0.34 D: 0.42 E: 0.26

The greatest value occurs at point D (the inflection point) and the least where the only negative slope occurs, at point A.

b. The steepness of the graph is a measure of how much the graph is tilted at a particular point. The direction of tilt is considered in the slope but not when describing steepness. Thus the steepness at each of the indicated points is

A: 0.24 B: 0 C: 0.34 D: 0.42 E: 0.26

The graph is steepest at point D. The degree of steepness is least at point B.

c. The instantaneous rate of change at A is -0.24 million, or $-240{,}000$, members per year. At E the rate of change is 0.26 million, or 260,000, members per year.

d. The slope of the tangent line at C is 0.34 million members per year. The slope at D is 0.42 million members per year.

Secant and Tangent Lines

In addition to understanding tangent lines in terms of local linearity, it is helpful to understand the relationship between secant lines and tangent lines.

Recall that a tangent line at a point on a graph touches that point and is tilted exactly the way the graph is tilted at that point. A secant line, on the other hand, is a line that passes through two points on a graph. We illustrate the relationship between secant lines and tangent lines with an example on a simple curve.

EXAMPLE 2 *Secant and Tangent Lines*

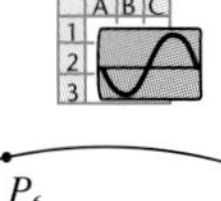

a. On the curve shown in Figure 3.12, draw secant lines through P_1 and Q, P_2 and Q, P_3 and Q, and P_4 and Q.

b. Which of the four secant lines drawn in part *a* appears to be tilted most like the curve at Q?

c. Again, on the curve shown in Figure 3.12, draw secant lines through P_6 and Q, P_7 and Q, P_8 and Q, and P_9 and Q. Which of these four secant lines appears to most closely match the tilt of the curve at Q?

d. Where could you place a point P_5 so that the secant line through P_5 and Q would be even closer than the secant lines in parts *a* and *c* to tilting the same way the curve does at Q?

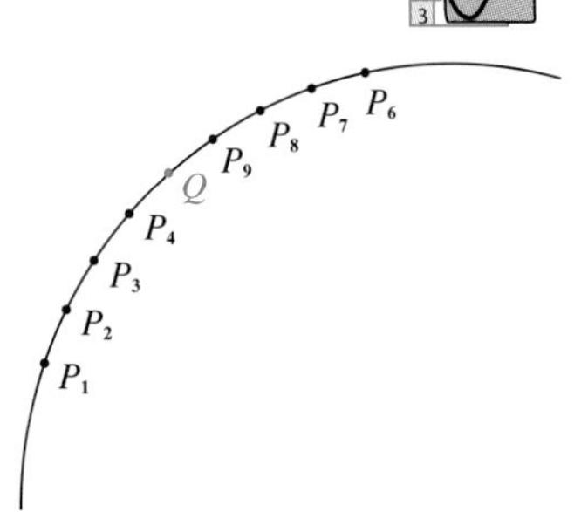

FIGURE 3.12

Solution

a. The four secant lines are shown in Figure 3.13.

b. The secant line through P_4 and Q appears almost to hide the curve near Q. That is, it appears to match the tilt of the curve at Q better than any of the other three secant lines.

c. The four secant lines are shown in Figure 3.14. Again, we see that the secant line through Q and the point closest to Q (P_9 in this case) appears to match the tilt of the curve at Q most closely.

d. The point P_5 should be placed even closer to Q than P_4 and P_9 in order for the tilt of the secant line to match the tilt of the curve at Q more closely.

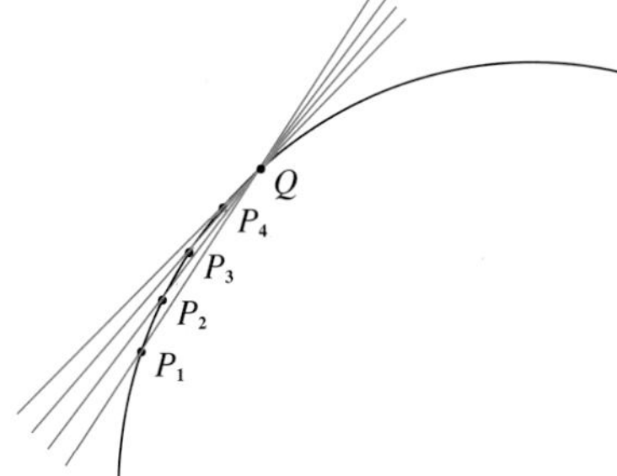

FIGURE 3.13

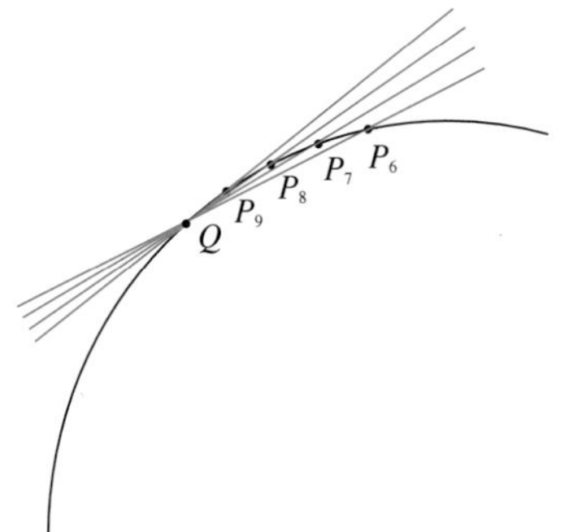

FIGURE 3.14

We generalize the results of Example 2 by saying that the tilt of the secant line through P and Q becomes closer and closer to the tilt of the curve at Q as P becomes closer and closer to Q. Indeed, if you draw a secant line through the point Q and a point P on the curve near Q, then the closer P is to Q, the more closely the secant line approximates the tangent line at Q. You can think of the tangent line at Q as the limiting form of the secant lines between P and Q as P gets closer and closer to Q.

> **Line Tangent to a Curve**
>
> The tangent line at a point Q on a smooth, continuous graph is the limiting position of the secant lines between point Q and a point P as P approaches Q along the graph (if the limiting position exists).

Sketching Tangent Lines

Sketching tangent lines may seem tedious, but this skill is required in later chapters.

Although thinking of a tangent line as a limiting position of secant lines is vital to your understanding of calculus (and is a subject to which we will return in Chapter 4), it is important for you to have an intuitive feel for tangent lines and to be able to sketch them without first drawing secant lines.

Consider the curve in Figure 3.15 and the lines through points A, B, and C.

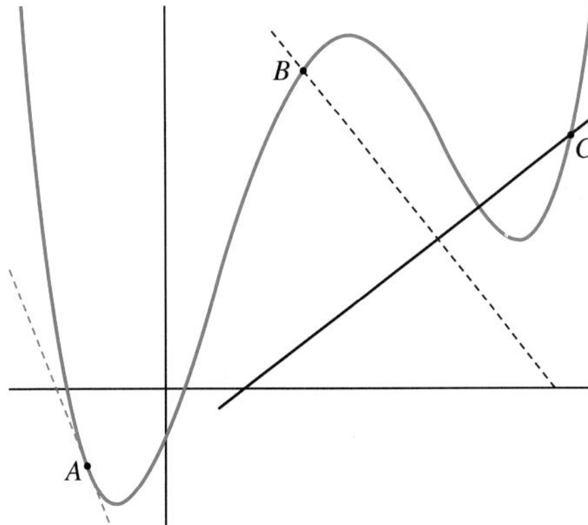

FIGURE 3.15

If you understand a tangent line as the line you would see if you zoomed in close to the point of tangency, then it is easy to see that the lines through points B and C are not tangent lines. The slopes of those two lines do not match the slopes of the graph at points B and C.

> Recall that tangent lines must be tilted in the same way that the graph is tilted at the point of tangency.

The lines through B and C also violate a general rule about tangent lines:

> **General Rule for Tangent Lines**
>
> Lines tangent to a smooth curve do not "cut through" the graph of the curve at the point of tangency and lie completely on *one side* of the graph near the point of tangency except in the following two cases:
>
> 1. At an inflection point on a smooth curve
> 2. At any point on a line

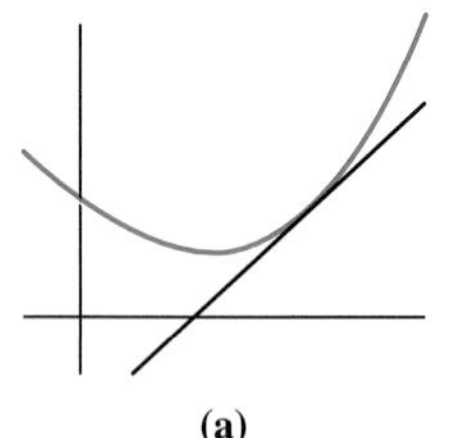

(a)
If the curve is *concave up* at the point of tangency, then the tangent line will lie *below* the curve near the point of tangency.

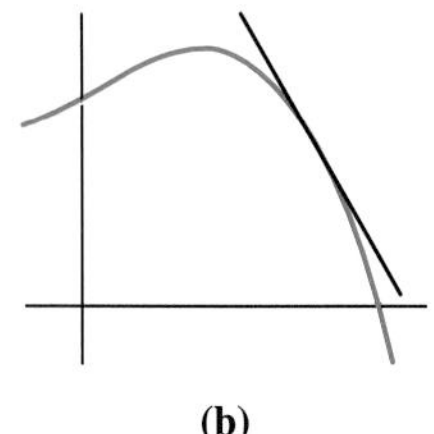

(b)
If the curve is *concave down* at the point of tangency, then the tangent line will lie *above* the curve near the point of tangency.

FIGURE 3.16

These two exceptions are dealt with after Example 3. For cases in which the exceptions do not apply, we can determine on which side of the curve the tangent line should lie by noting the concavity. If the curve is concave up at the point of tangency, then the tangent line will lie below the curve near the point of tangency. If the curve is concave down at the point of tangency, then the tangent line will lie above the curve near the point of tangency. See Figure 3.16.

Refer again to Figure 3.15. At point C, the curve is concave up. The tangent line should lie below the curve. However, to the left of C, the tangent line lies above the curve, so the line through C is not a tangent line.

The line at A is the only tangent line of the three. It touches the curve only at A; it is tilted in the same way that the graph is tilted at A; and it lies below the curve which is concave up at A. It is the line we would see if we zoomed in on the curve at point A.

The correct tangent lines at A, B, and C are shown in Figure 3.17.

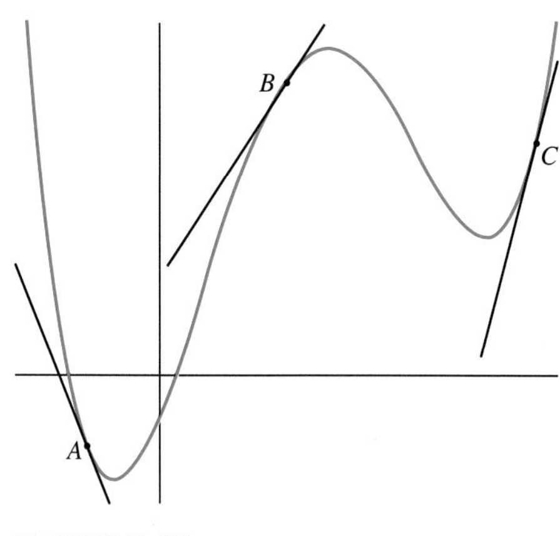

FIGURE 3.17

EXAMPLE 3 *Weight Loss*

A woman joins a national weight-loss program and begins to chart her weight on a weekly basis. Figure 3.18 shows the graph of a continuous function of her weight from when she began the program through 7 weeks into the program.

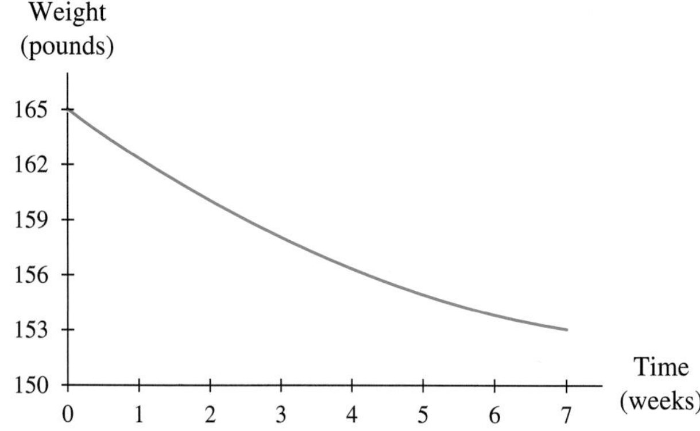

FIGURE 3.18

Your students should expect variations in answers when calculating slopes of sketched tangent lines. You should accept variations in answers, but do not accept sloppy sketching. It is always important to stress precision in mathematics.

a. Carefully sketch a line tangent to the curve at 5 weeks.

b. Estimate the slope of the tangent line at 5 weeks.

c. How quickly was the woman's weight declining 5 weeks after the beginning of the program?

d. What is the slope of the curve at 5 weeks?

Solution

Warn your students not to sketch the tangent line or the triangle used to determine the slope of the tangent and not to read values from the coordinate axes without using a straightedge. Not using a straightedge almost always leads to inaccuracies.

a.

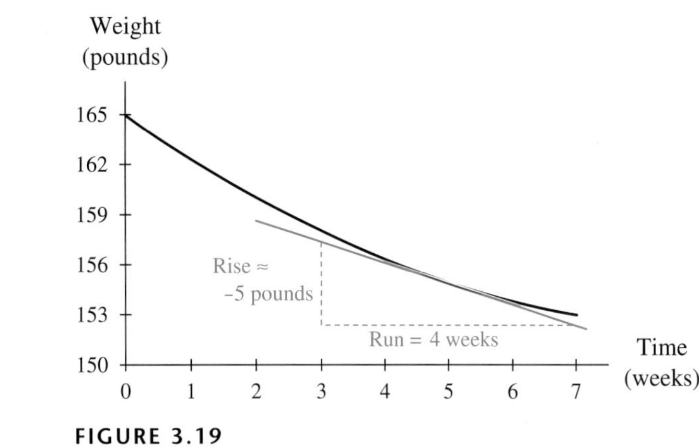

FIGURE 3.19

b. Slope $= \frac{\text{rise}}{\text{run}} \approx \frac{-5 \text{ pounds}}{4 \text{ weeks}} = -1.25$ pounds per week

c. The woman's weight was declining by approximately 1.25 pounds per week after 5 weeks in the program.

d. The slope of the curve at 5 weeks is approximately -1.25 pounds per week.

As we mentioned in the general rule for tangent lines, there are exceptions to the principle that tangent lines do not cut through the graph and lie on only one side of the graph. At a point of inflection, the graph is concave up on one side and concave down on the other. As you might expect, the tangent line lies above the concave-down portion of the graph and below the concave-up part. To do this, the tangent line must cut through the graph. It does so at the point of inflection. When drawing tangent lines at inflection points, be careful to make sure that the tangent line is tilted to match the tilt of the graph at the point of tangency. See Figure 3.20.

Let us also examine the case where the graph itself is a line. Consider the graph of a line with point P on that line. Recall that the tangent line at P touches the graph at P and is tilted in the same direction as the tilt of the graph. The only way to draw such a tangent line is to draw the line itself. Because every line has the same slope at every point, every point has the same tangent line, and the slope of the tangent line is exactly the slope of the original line.

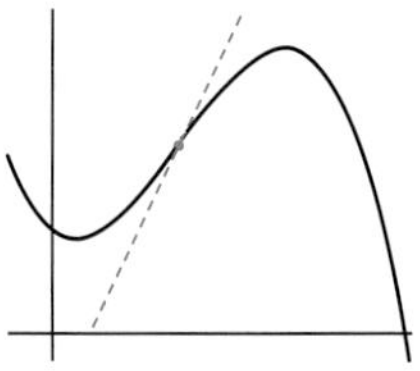

(a)
Although this line lies above the concave-down portion and below the conve-up part of the graph, it is not a tangent line because it is not tilted in the same way that the graph is tilted at the point.

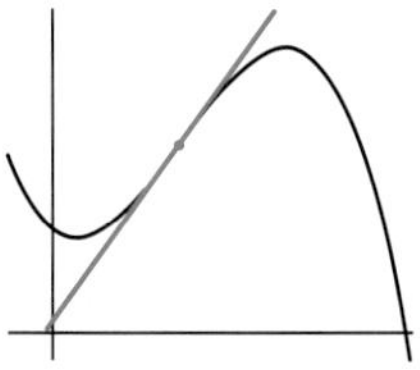

(b)
This tangent line is correctly drawn at an inflection point.

FIGURE 3.20

Where Does the Instantaneous Rate of Change Exist?

Our discussion of tangent lines would not be complete without a mention of piecewise functions. Consider the graph of a model for the population of Indiana[19] from 1980 through 1997 (see Figure 3.21).

19. Based on data from *Statistical Abstract*, 1998.

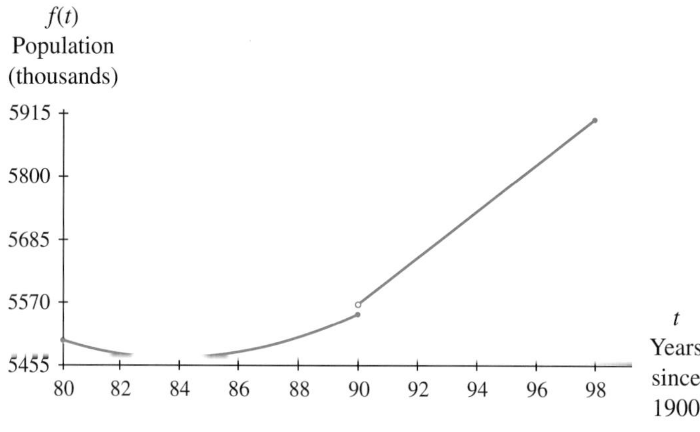

FIGURE 3.21

We make a distinction between the instantaneous rate of change of the function and the instantaneous rate of change of the underlying output quantity. It is naïve to conclude that because the instantaneous rate of change of this specific function used to model the population does not exist, the instantaneous rate of change in the context of this real-life situation does not exist. Although it may be mathematically comfortable to state, "We cannot calculate this instantaneous rate of change," it is more beneficial for your students to discuss ways to estimate the instantaneous rate of change of the underlying situation. One method of estimating the instantaneous rate of change is discussed at the end of this section.

The equation of the graph in Figure 3.21 is

$$f(t) = \begin{cases} 2.32t^2 - 389t + 21{,}762 \text{ thousand people} & \text{when } 80 \le t \le 90 \\ 44.515t + 1549.156 \text{ thousand people} & \text{when } 90 < t \le 98 \end{cases}$$

where t is the number of years after 1900.

Consider the tangent line at the point where the function f is not continuous (1990). If we use the idea of a limiting position of secant lines, then we would have to conclude that we cannot draw a line tangent to the graph at $t = 90$, because secant lines drawn using points to the left of $t = 90$ are not approaching the same position as those drawn using points to the right of $t = 90$.

If we use the principle of local linearity and zoom in close to the point at $t = 90$, then we see something similar to Figure 3.22. Recall that if we zoom close enough to a point on a smooth curve, we see a line that is, indeed, the tangent line. In this case, we do not have a smooth curve, and we see two lines. Again we conclude that there is no tangent line at $t = 90$. Does this mean that there is no instantaneous rate of change in the population of Indiana in 1990? No, it simply means that we cannot use our piecewise continuous model to calculate the rate of change in 1990.

It is possible for a piecewise function to have a tangent line at a point where the pieces join. This occurs only if a limiting position of secant lines exists; in other words, zooming in close to the point reveals a single line. This is a rare occurrence, and the rarity of it is one of the limitations of piecewise models. It is usually not possible to construct a tangent line at the point(s) where the function is divided into pieces.

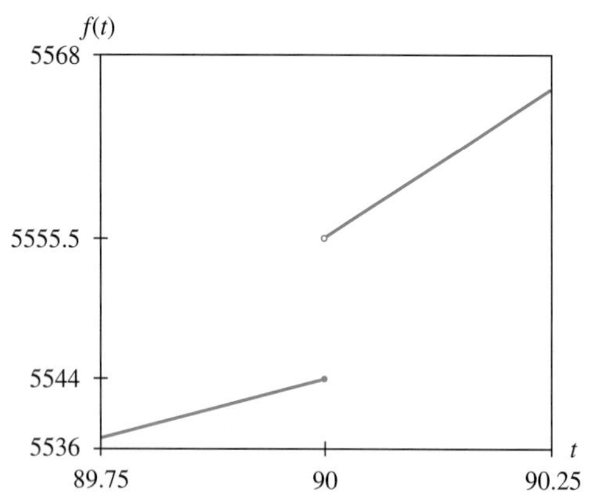

FIGURE 3.22

As we saw in the model for Indiana population, it is possible for a continuous or piecewise continuous function to have points at which we cannot sketch a tangent line. Even so, it is possible for such functions to have rates of change at all other points.

To help clarify the relationship between continuous and noncontinuous functions and rates of change, consider the graphs shown in Figure 3.23. The graph in Figure 3.23a is continuous everywhere and has a rate of change at every point. You may think that the same thing is true for the graph in Figure 3.23b; however, because a tangent line drawn at P is

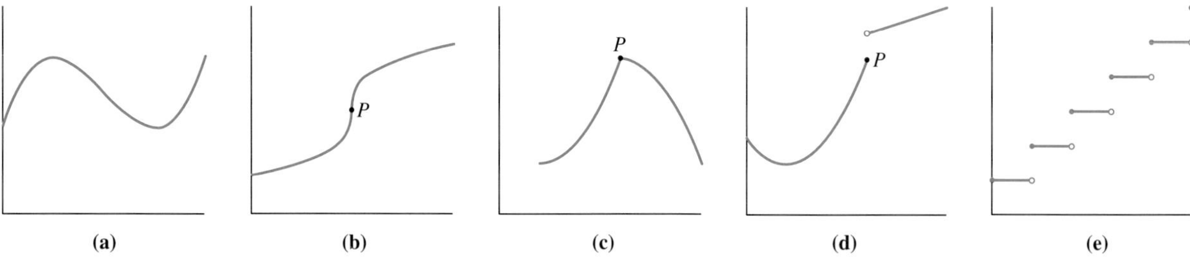

FIGURE 3.23

Functions similar to those graphed in Figures 3.23a, 3.23c, and 3.23d occur in many real-life modeling situations in this text. Graphs like the one shown in Figure 3.23b are seen in chemistry with titration data, and step functions such as the one shown in Figure 3.23e occur in finance applications and situations such as taxi cab fares, parking garage fees, and postal rates.

vertical, the run is zero. Therefore, the slope of the tangent line does not exist at that point. The graph in Figure 3.23c is also continuous; however, the graph is not smooth at P because of the sharp point there. We cannot draw a tangent line at P, because secant lines drawn with points on the right and left do not approach the same slope.

The graph in Figure 3.23d has a break at P and, therefore, is not continuous at P. This graph is similar to the Indiana population graph, and the slope does not exist at the break in the function. The slope does exist at all other points on the graph. The graph in Figure 3.23e is discontinuous at many points. There is no tangent line that can be drawn at these points, but as in Figure 3.23d, the function has an instantaneous rate of change at all other points.

The graphs in Figures 3.23d and 3.23e illustrate a general rule relating continuity and rates of change.

> If a function is not continuous at a point, then the instantaneous rate of change does not exist at that point.

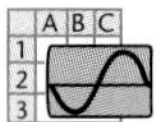

If you keep in mind the relationships among instantaneous rates of change, slopes of tangent lines, slopes of secant lines, and local linearity, then you should have little difficulty determining the times when the instantaneous rate of change does not exist.

Recall from our discussion in Chapter 1 that calculus is applied to continuous portions of functions. As we have just seen, instantaneous rates of change do not exist where a function is not continuous. Modeling is a tool by which we transform discrete data into a continuous function. Discrete data are useful for finding change, percentage change, and average rates of change, but a continuous or piecewise continuous function is necessary in order to find instantaneous rates of change. You should be starting to understand the importance of continuous and piecewise continuous models in calculus.

Approximating Instantaneous Rates of Change

Although it is possible to omit a discussion of symmetric difference quotients, many activities throughout the text ask students to use this technique.

Although a continuous or piecewise continuous model is necessary to calculate an instantaneous rate of change, it is possible to use a **symmetric difference quotient** to *estimate* a rate of change from a set of data without taking the time to construct a model. Quick approximations are sometimes as helpful as a more involved calculation.

TABLE 3.10

Year	Population (thousands)
1980	5490
1985	5459
1990	5544
1992	5648
1994	5742
1995	5788
1996	5828
1997	5864
1998	5899

Consider the data in Table 3.10, which show the population of Indiana[20] between 1980 and 1998. To estimate how quickly the population was changing in 1996 using only the data, simply find the slope of the line between the points on either side of 1996.

$$\frac{5864 - 5788}{1997 - 1995} = \frac{76 \text{ thousand people}}{2 \text{ years}} = 38{,}000 \text{ people per year}$$

In 1996, the population of Indiana was growing at a rate of approximately 38,000 people per year.

A scatter plot of the data and the secant line whose slope is our symmetric difference quotient are shown in Figure 3.24.

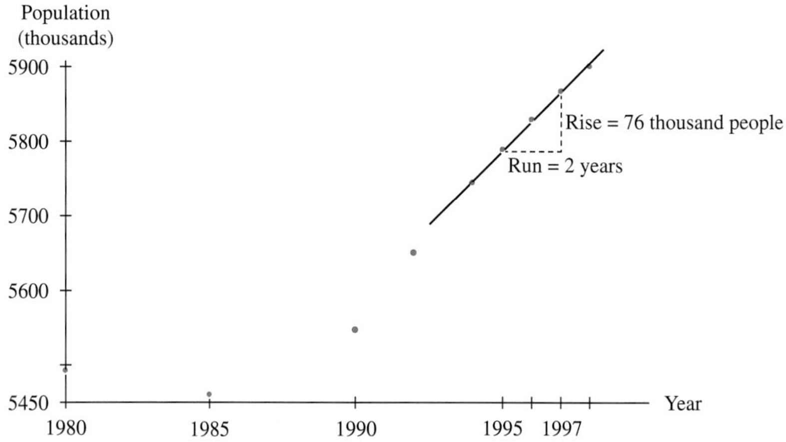

FIGURE 3.24

If you picture a smooth curve through these data points, then the slope of the secant line in Figure 3.24 should be close to the slope of the curve at 1996.

Although *symmetric difference quotient* is a cumbersome name, each word in the phrase is descriptive and should help you remember how to calculate a symmetric difference quotient. We use the term *symmetric* because we choose the closest point on either side of our point of interest and the same distance away. (This technique should be used with care if the data are not equally spaced around the point at which we desire to estimate the rate of change. Do not automatically use the closest point on either side of the point of interest.) We use the term *difference quotient* because we find the difference in the inputs and outputs of the two points and then divide to find the slope that is used to estimate the rate of change.

Again, consider the population data in Table 3.10 and the task of using a symmetric difference quotient to estimate the instantaneous rate of change in 1990. Note that the points on either side of 1990 are not equal distances away from 1990. In order to use a symmetric difference quotient, you would have to calculate the slope of the secant line using the years 1985 and 1995:

$$\text{Slope} = \frac{5788 - 5459}{1995 - 1985} = \frac{329 \text{ thousand people}}{10 \text{ years}} = 32.9 \text{ thousand people per year}$$

To summarize, symmetric difference quotients are useful if you want a fast estimate, if you cannot construct a model for a given data set, or if you need to estimate a rate of change that cannot be calculated from a given model.

20. *Statistical Abstract*, 1998.

Using Symmetric Difference Quotients

A symmetric difference quotient may be used to approximate an instantaneous rate of change if

- You have a set of data but do not wish to take the time to find a model for the data.
- You have a set of data that cannot be modeled by one of the functions we have studied or whose model does not have a tangent line at the point at which you need to know the instantaneous rate of change.

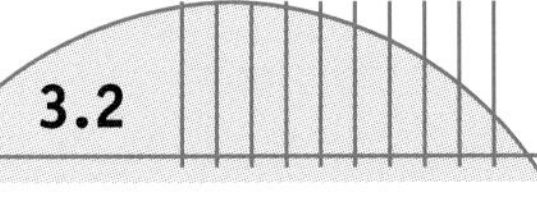

3.2 Concept Inventory

- *Instantaneous rate of change*
- *Local linearity*
- *Tangent line*
- *Slope of tangent line = instantaneous rate of change*
- *Slope and steepness of a tangent line*
- *The tangent line is the limiting position of secant lines*
- *Tangent lines lie beneath a concave-up graph*
- *Tangent lines lie above a concave-down graph*
- *Situations in which the instantaneous rate of change does not exist*
- *Symmetric difference quotient*

3.2 Activities

1. In your own words, describe the difference between
   a. discrete and continuous.
   b. average rate of change and instantaneous rate of change.
   c. secant lines and tangent lines.
2. What are some advantages of using a continuous model instead of discrete data? What are some disadvantages?
3. How are average rates of change and instantaneous rates of change measured graphically?
4. Explain in your own words how to tell visually whether a line is tangent to a smooth graph.
5. Using Table 3.9, the time/mileage table given in this section, verify that the average speed of the car from mile marker 0 to mile marker 19 is 67.1 mph.
6. a. Using Table 3.9, determine the average speed (in mph) from:
      i. milepost 66 to milepost 80.
      ii. milepost 80 to milepost 105.
   b. What might account for the difference in speed?
7. a. At each labeled point on the graph in Figure 3.2.1, determine whether the instantaneous rate of change is positive, negative, or zero.

FIGURE 3.2.1

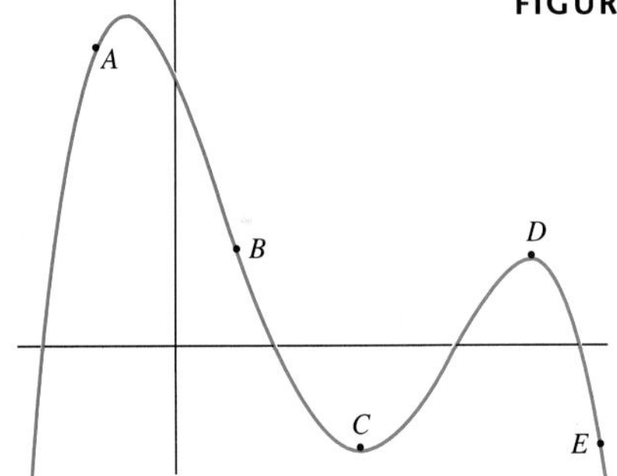

   b. Is the graph steeper at point A or at point B?
8. At each labeled point on the graph in Figure 3.2.2, determine whether the slope is positive, negative, or zero.

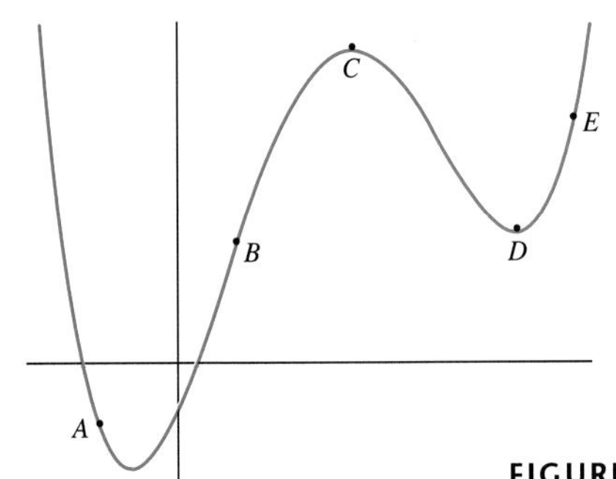

FIGURE 3.2.2

9. a. In graph *c* of Activity 10, estimate where the slope is greatest. Mark that point on the graph.

   b. In the graph in Figure 3.2.3, estimate where the output is falling most rapidly. Mark that point on the graph.

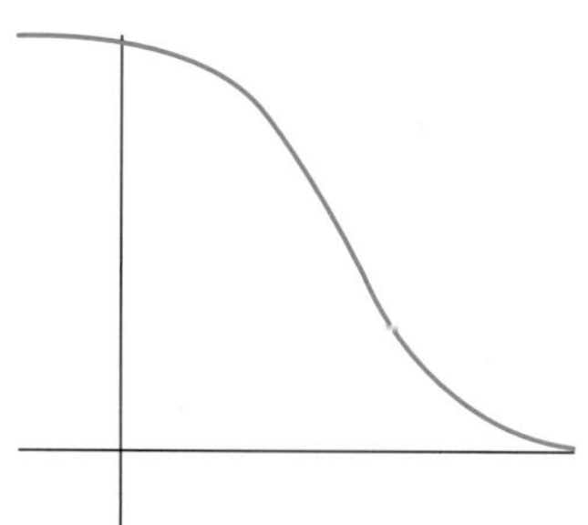

FIGURE 3.2.3

10. Discuss the slopes of the following graphs.

a.

b.

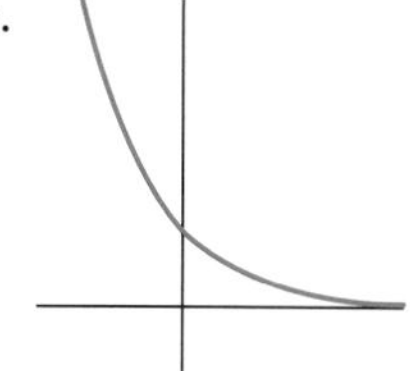

c.

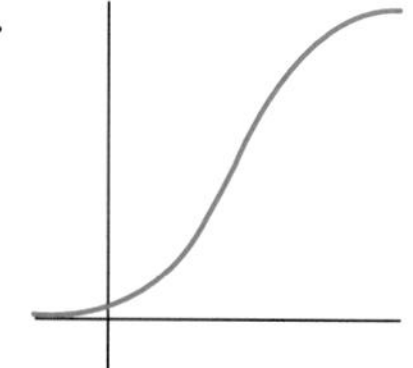

11. Which of the lines drawn on the graph in Figure 3.2.5 are *not* tangent lines?

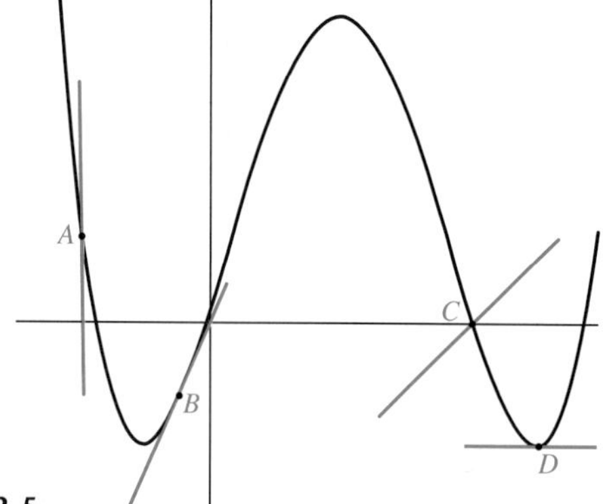

FIGURE 3.2.5

12. On Figure 3.2.6, draw secant lines through P_1 and Q, P_2 and Q, and P_3 and Q. Repeat for the points P_4 and Q, P_5 and Q, and P_6 and Q. Then draw the tangent line at Q.

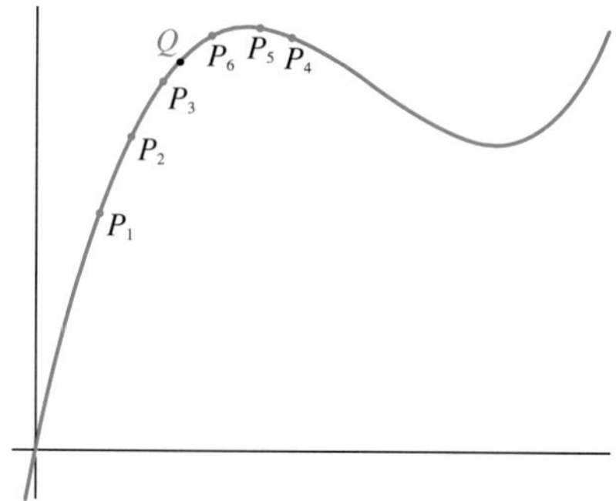

FIGURE 3.2.6

13. Draw secant lines through P_1 and Q, P_2 and Q, and P_3 and Q on the graph in Figure 3.2.7. Repeat for the points P_4 and Q, P_5 and Q, and P_6 and Q. Then draw the tangent line at Q.

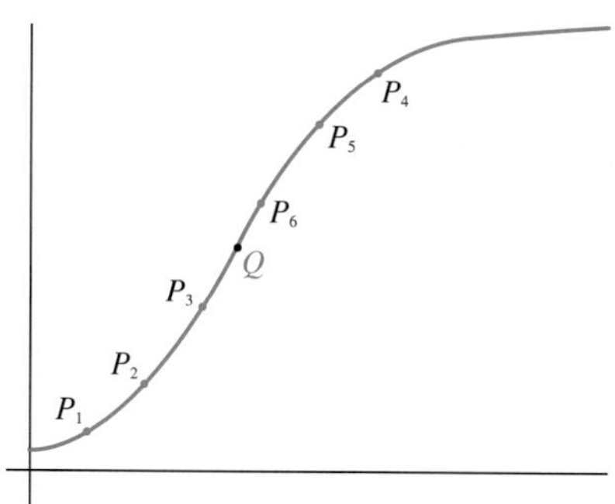

FIGURE 3.2.7

14. Explain using your own words the relationship between secant lines and tangent lines.

15. a. Is the graph shown in Figure 3.2.8 concave up, concave down, or neither (an inflection point) at A, B, C, and D?

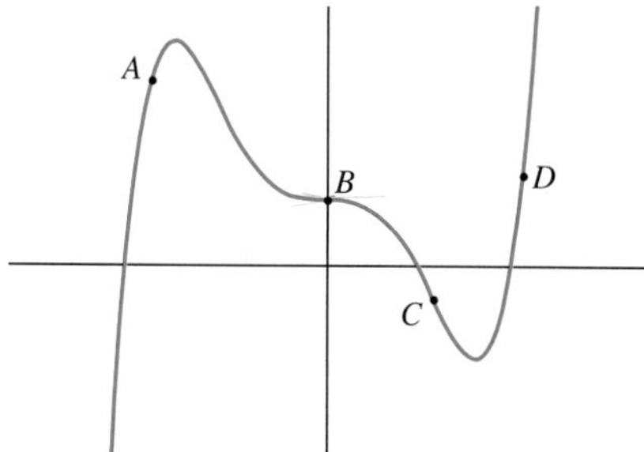

FIGURE 3.2.8

b. Should the tangent lines lie above or below the curve at each of the indicated points?

c. Carefully draw tangent lines at the labeled points in Figure 3.2.8.

d. At which of the labeled points is the slope of the tangent line positive? At which of the labeled points is the slope of the tangent line negative?

16. a. Is the graph shown in Figure 3.2.9 concave up, concave down, or neither at A, B, C, and D?

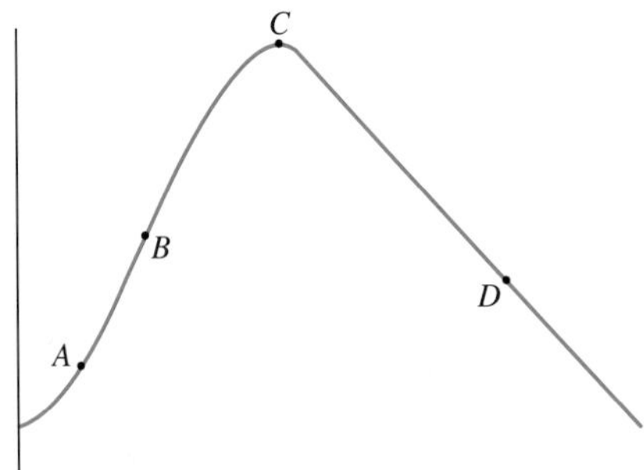

FIGURE 3.2.9

b. Should the tangent lines lie above or below the curve at each of the indicated points?

c. Carefully draw tangent lines at the labeled points.

d. At which of the labeled points is the slope of the curve positive? At which of the labeled points is the slope of the curve negative? Do any of the labeled points appear to be inflection points?

Use carefully drawn tangent lines to estimate the slopes at the labeled points in Activities 17 through 20.

17.

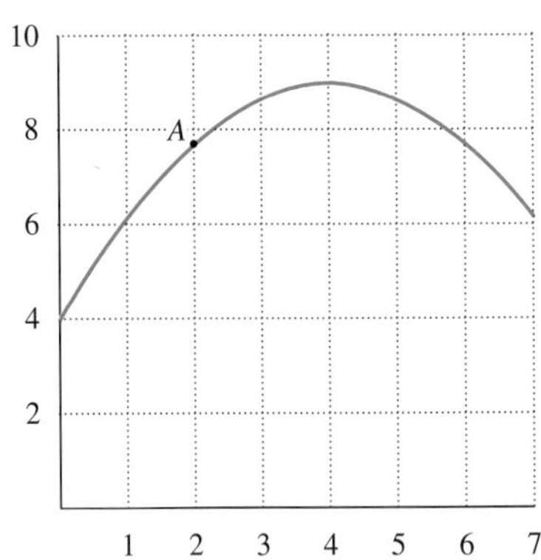

18.

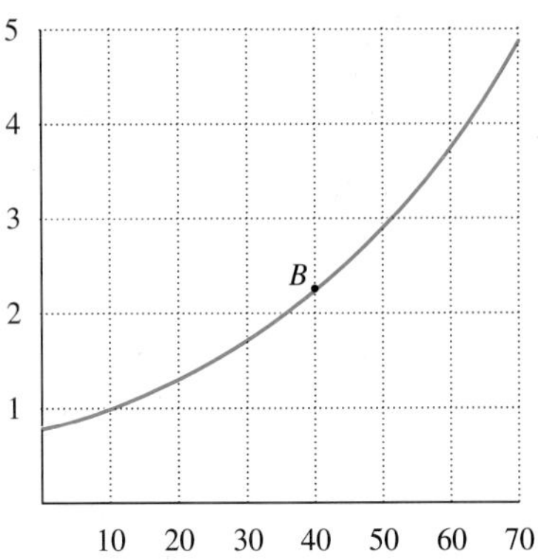

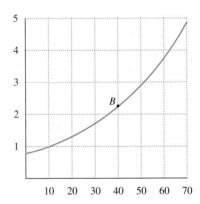

19.

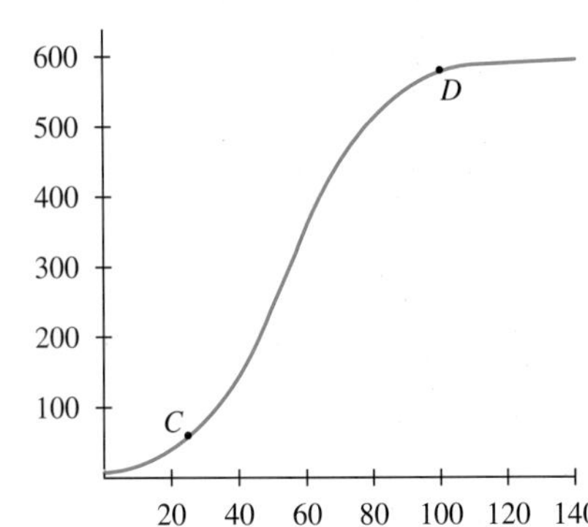

20.

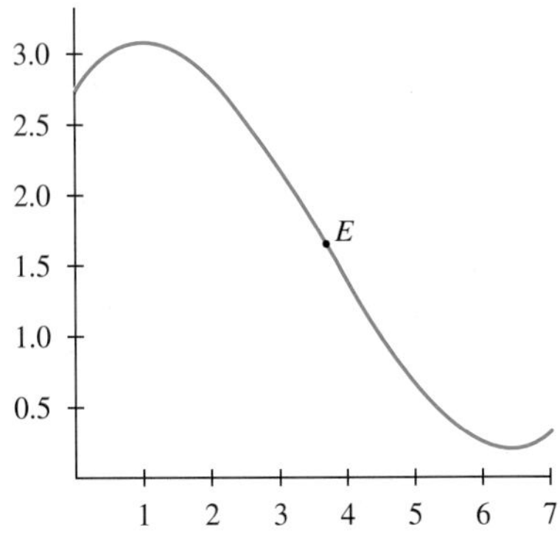

21. The graph in Figure 3.2.14 shows the total number of cellular phone subscribers[21] since 1990. The slope of the graph at point A is 11.3.

21. The Cellular Telecommunications Industry Association.

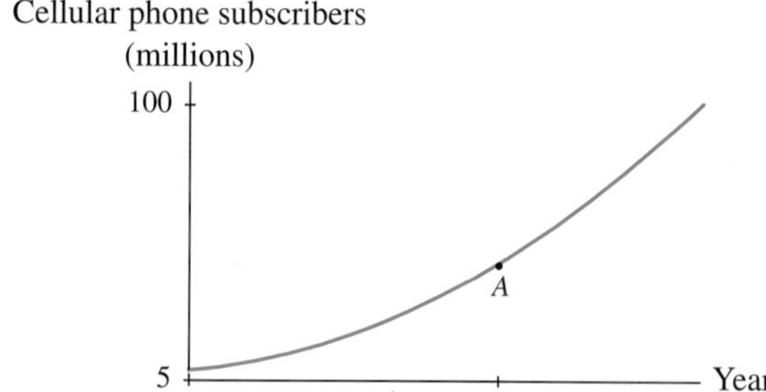

FIGURE 3.2.14

a. What are the units on the slope at point *A*?
b. How rapidly was the number of subscribers growing in 1996?
c. What is the slope of the tangent line at *A*?
d. What was the instantaneous rate of change of the number of cell phone subscribers in 1996?

22. The graph in Figure 3.2.15 shows the average monthly cellular phone bill since 1990. The slope of the curve at point *A* is −4.68.

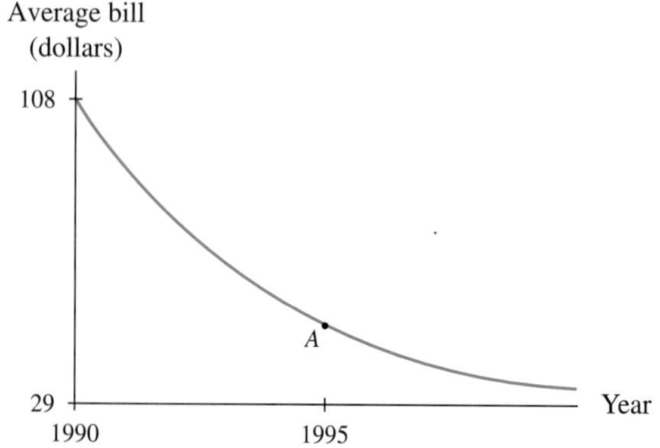

FIGURE 3.2.15

a. What should be the units on the slope at point *A*?
b. How rapidly was the dollar amount growing in 1991?
c. What is the slope of the tangent line at point *A*?
d. What is the instantaneous rate of change of the amount at point *A*?

23. The growth of a pea seedling[22] as a function of time can be modeled by two quadratic functions as shown in Figure 3.2.16. The slopes at the labeled points are (in ascending order) −4.2, 1.3, and 5.9.

FIGURE 3.2.16

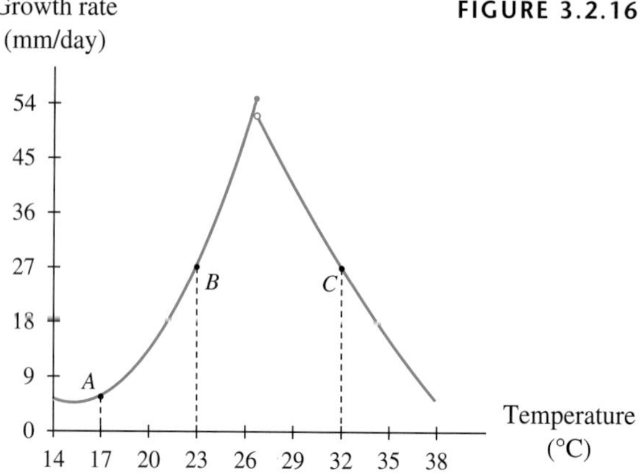

a. Match the slopes with the points, *A*, *B*, and *C*.
b. What are the units on the slopes for each of these points?
c. How quickly is the growth rate changing with respect to temperature at 23°C?
d. What is the slope of the tangent line at 32°C?
e. What is the instantaneous rate of change of the growth rate of pea seedlings at 17°C?

24. The graph in Figure 3.2.17 shows the survival rate (percentage surviving) of three stages in the development of a flour beetle (egg, pupa, and larva) as a function of the relative humidity.[23]

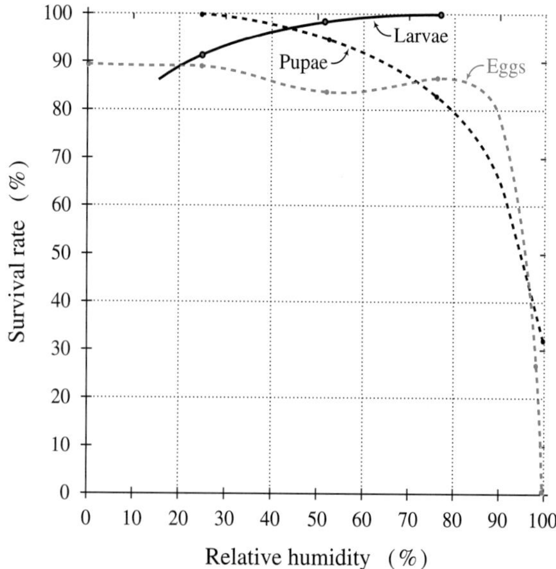

FIGURE 3.2.17

22. Based on data in George L. Clarke, *Elements of Ecology* (New York: Wiley, 1954).

23. Chapman, *Animal Ecology* (New York: McGraw-Hill, 1931).

Fill in each of the following blanks with the appropriate stage (eggs, pupae, or larvae).

a. At 60% relative humidity, the instantaneous rate of change of the survival rate of __________ is approximately zero.

b. An increase in relative humidity improves the survival rate of __________ and reduces the survival rate of __________.

c. At 97% relative humidity, the survival rate of __________ is declining faster than that of __________.

d. Any tangent lines drawn on the survival curve for __________ will have negative slope.

e. Any tangent lines drawn on the survival curve for __________ will have positive slope.

f. At 30% relative humidity, the survival rates for __________ and __________ are changing at approximately the same rate.

g. At 65% relative humidity, the survival curves for __________ and __________ have approximately the same slope.

25. Figure 3.2.18 shows a graph of the declination of the sun[24] (the angle of the sun from the equator) throughout the year.

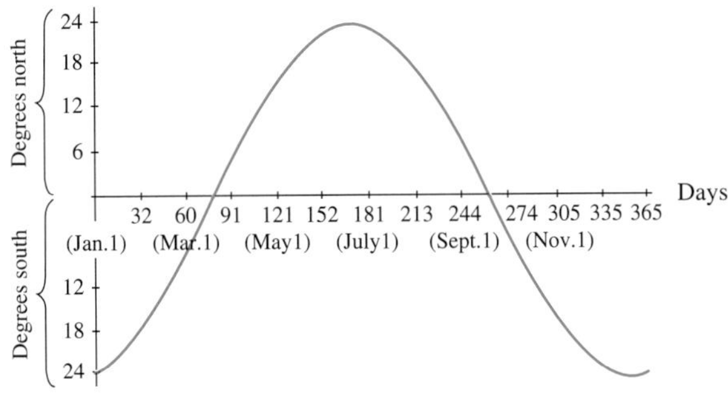

FIGURE 3.2.18

a. A *solstice* is a time when the angle of the sun from the equator is greatest. Identify the summer and winter solstices on the graph. What is the slope of the graph at these points?

b. Identify the two steepest points on the graph. Sketch tangent lines at these two points, and estimate their slopes. What is the significance of a negative slope in this context?

c. An *equinox* is a time when the sun crosses the equator, resulting in a day and night of equal length. Identify the points on the graph that correspond to the spring and fall equinoxes.

24. *The Mathematics Teacher*, March 1997, p. 238.

26. The effects of temperature on the percentage of grasshoppers' eggs from West Australia that hatch[25] is shown in the graph in Figure 3.2.19.

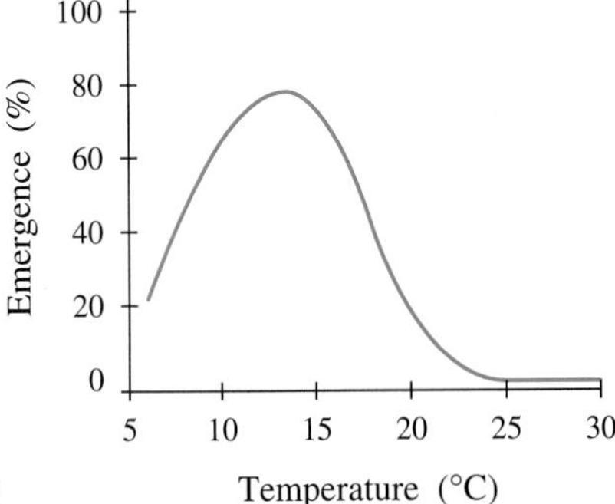

FIGURE 3.2.19

a. What is the optimum hatching temperature?

b. What is the slope of the tangent line at the optimum temperature?

c. Sketch tangent lines at 10°C, 17°C, and 22°C, and estimate the slopes at these points.

d. Where does the inflection point appear to be on this graph?

27. Predictions for the U.S. resident population from 1997 through 2050, as reported by *Statistical Abstract* for 1994, can be approximated by the model

$$p(t) = 2370.15t + 39{,}789.957 \text{ thousand people}$$

where t is the number of years since 1900. A graph of p is shown in Figure 3.2.20.

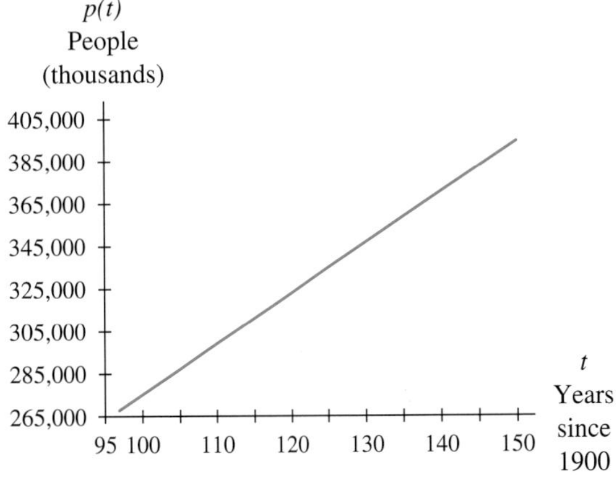

FIGURE 3.2.20

a. Sketch a tangent line at $t = 120$, and find its slope.

25. Figure 3.2.19 adapted from George L. Clarke, *Elements of Ecology* (New York: Wiley, 1954).

b. What is true about any line tangent to the graph of the function p?

c. What is the slope of any line tangent to this graph?

d. What is the slope at every point on the graph of this model?

e. According to the model, what is the instantaneous rate of change of the predicted population in any year from 1997 through 2050?

f. How rapidly does the model predict the population will be changing in 2020?

28. Predictions for the U.S. resident population from 1998 through 2050, as reported by the *Statistical Abstract* for 1998, can be approximated by the model

$$P(t) = 2393.49t + 35{,}182.931 \text{ thousand people}$$

where t is the number of years since 1900.

a. Compare the model in this activity with the one in Activity 27.

b. With the additional information about population available in *Statistical Abstract* for 1998, were the population projections adjusted up or down?

c. Was the growth rate adjusted up or down?

d. Find the slope of the graph of P at $t = 120$.

e. Describe the tangent line at $t = 120$.

29. The number of Houghton Mifflin Company employees[26] from 1989 through 1998 can be modeled by the graph in Figure 3.2.21.

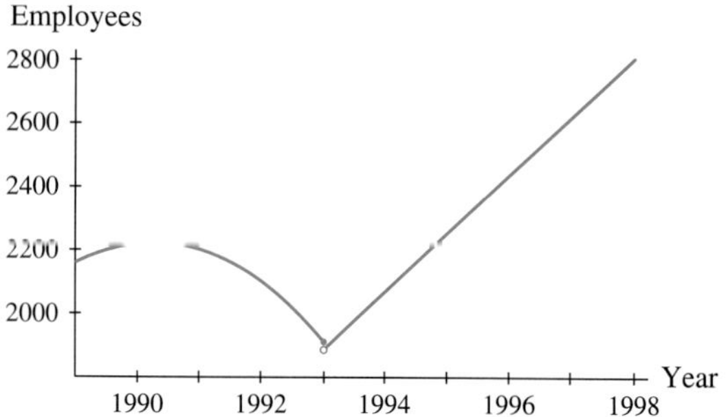

FIGURE 3.2.21

Draw tangent lines, if possible, to estimate how quickly the number of employees was changing in the indicated years. If it is impossible to do so, explain why.

a. 1990 b. 1993 c. 1997

30. The graph in Figure 3.2.22 shows employment[27] in Slovakia from 1948 through 1988.

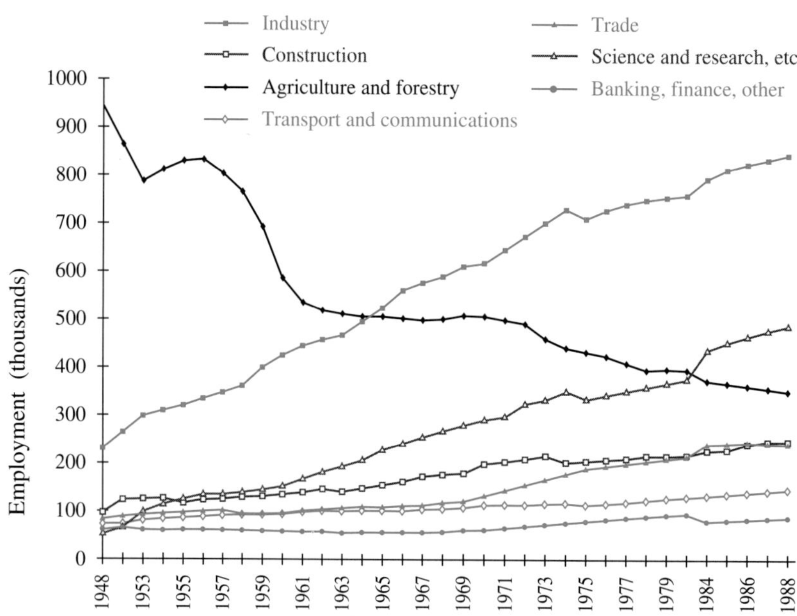

FIGURE 3.2.22

26. Hoover's Online Company Capsules.

27. Figure 3.2.22 from A. Smith, "From Convergence to Fragmentation," *Environment and Planning*, vol. 28, 1996. Pion Limited, London. Reprinted by permission. Data elaborated from *Statisticka rŏcenta* SUSR, various dates.

a. Estimate how rapidly employment in agriculture and forestry was declining in 1958.

b. Estimate the instantaneous rate of change in industry employment in 1962.

c. Why is it not possible to sketch a tangent line to the industry graph at 1974?

31. The revenues from cellular phone roaming charges[28] between 1993 and 1998 are as shown in Table 3.11.

TABLE 3.11

Year	1993	1994	1995	1996	1997	1998
Roaming revenues (millions of dollars)	1.4	1.8	2.5	2.8	3.0	3.5

a. Use a symmetric difference quotient to estimate the instantaneous rate of change of revenue in 1995.

b. Figure 3.2.23 shows the graph of an equation for the revenue data in Table 3.11.

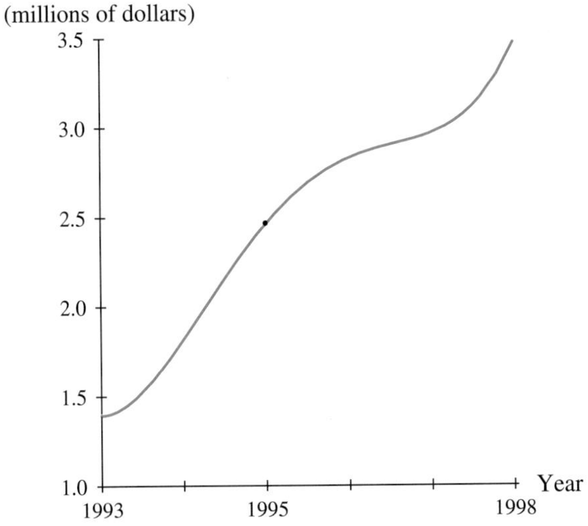

FIGURE 3.2.23

Roamer revenues (millions of dollars)

Use the graph and a tangent line to estimate the rate of change of revenue in 1995. Compare your answer to the estimate in part *a*.

32. Values of cumulative capital investment in the cellular phone industry,[29] beginning in 1985, are shown in Table 3.12.

TABLE 3.12

Year	Cumulative capital investment (millions of dollars)
1985	0.9
1987	2.2
1989	4.5
1991	8.7
1993	14.0
1995	24.1
1997	46.1
1998	60.5

a. Use a symmetric difference quotient to estimate the instantaneous rate of change of capital investment in the cellular phone industry in 1995.

b. A graph of an equation for the data in Table 3.12 is shown in Figure 3.2.24. Use the graph and a tangent line to estimate the rate of change of capital investment in 1995. Compare your answer to the estimate in part *a*.

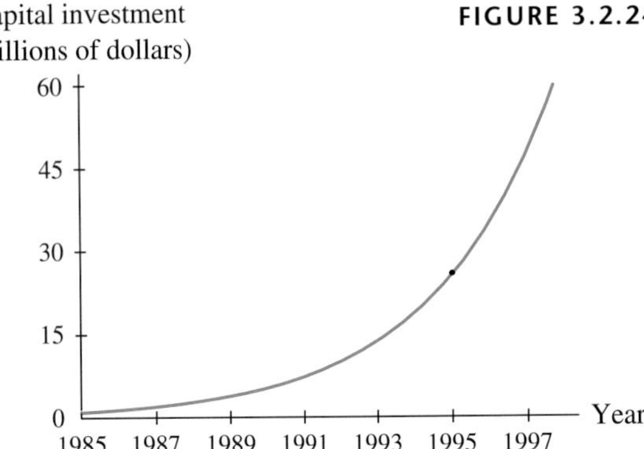

FIGURE 3.2.24

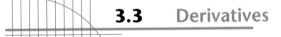

3.3 Derivatives

Derivative Terminology and Notation

By now, you should be comfortable with the concepts of average rate of change and instantaneous rate of change. Let's summarize the differences between these two rates of change.

28. The Cellular Telecommunications Industry Association.

29. Ibid.

Average Rates of Change	**Instantaneous Rates of Change**
○ measure how rapidly (on average) a quantity changes over an interval	○ measure how rapidly a quantity is changing at a point
○ can be obtained by calculating the slope of the secant line between two points	○ can be obtained by calculating the slope of the tangent line at a single point
○ require discrete data points, a continuous curve, or a piecewise continuous curve	○ require a continuous or piecewise continuous curve to calculate

Because instantaneous rates of change are so important in calculus, we commonly refer to them simply as **rates of change**. The calculus term for instantaneous rate of change is **derivative.** It is important to understand that the following phrases are equivalent.

We use these phrases interchangeably. It is easy to get in the habit of using your favorite term to the exclusion of the others, but your students will benefit from hearing all the terms in class and seeing them on tests and quizzes.

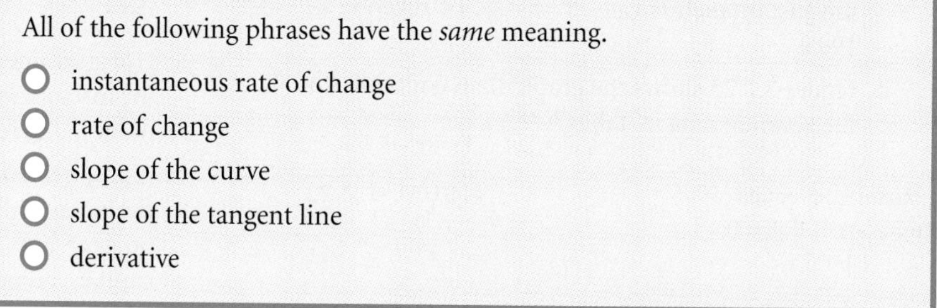
All of the following phrases have the *same* meaning.

- ○ instantaneous rate of change
- ○ rate of change
- ○ slope of the curve
- ○ slope of the tangent line
- ○ derivative

Even though we consider all these phrases synonymous, we must keep in mind that the last three phrases have specific mathematical definitions and so may not exist at a point on a function. However, the rate of change of the underlying situation does have an interpretation at that point in context. In such cases, we will have to estimate the rate of change by using a symmetric difference quotient, as we saw in Section 3.2, or by using some other estimation technique.

There are also several symbolic notations that are commonly used to represent the rate of change of a continuous function G with input t. In this book, we use three different, but equivalent, symbolic notations:

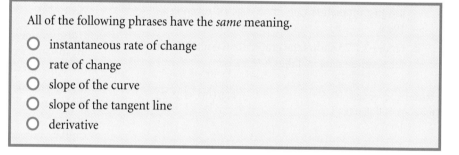

$\frac{dG}{dt}$ This is read, *"dee G–dee t," "the rate of change of G with respect to t," or "the derivative of G with respect to t."*

(or)

$G'(t)$ This is read, *"G prime of t," or "the rate of change of G with respect to t," or "the derivative of G with respect to t."*

(or)

$\frac{d}{dt}[G(t)]$ This is read, *"dee-dee-t of G of t," "the rate of change of G with respect to t," or "the derivative of G with respect to t."*

Suppose that $G(t)$ is your grade out of 100 points on the next calculus test when you study t hours during the week before the test. The graph of G may look like that shown in Figure 3.25.

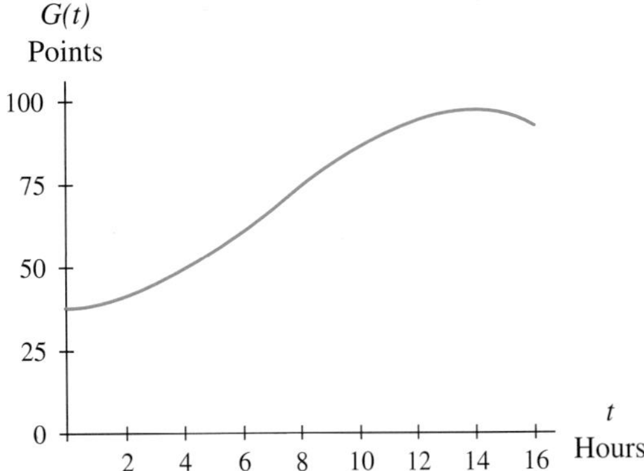

FIGURE 3.25

Note how the grade changes as your studying time increases. The grade slowly improves during the first 2 hours. The longer you study, the more rapidly the grade improves until you have studied approximately 7 hours. After 7 hours, the grade improves at a slower rate. Your grade peaks after 14 hours of studying and then actually declines. What might explain the decline?

Let us compare the rates at which your grade is increasing when $t = 1$ hour and when $t = 4$ hours of study. Tangent lines at $t = 1$ and $t = 4$ are shown in Figure 3.26. After 1 hour of studying, your grade is increasing at a rate of approximately 1.7 points per hour. (We will show you how to calculate, not estimate, this rate in a later section.) This value is the slope of the curve when $t = 1$ hour. After 4 hours of studying, your grade is increasing at a rate of approximately 5.2 points per hour. Can you see that the graph is steeper when $t = 4$ than when $t = 1$? The grade is improving more rapidly after 4 hours than it is after 1 hour. In other words, a small amount of additional study is more beneficial if you have already studied 4 hours than it is when you have studied only 1 hour.

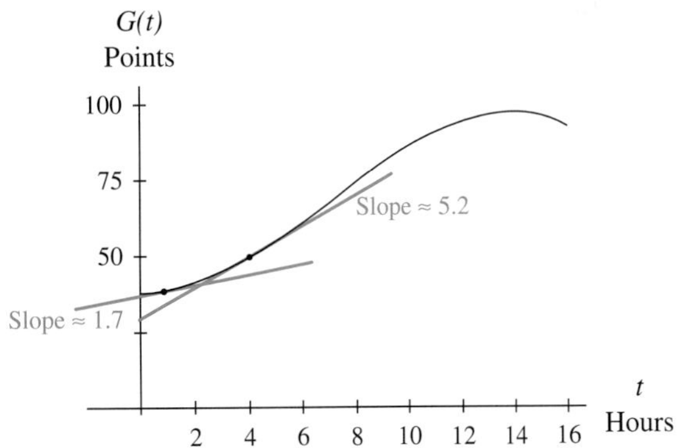

FIGURE 3.26
Tangent lines at $t = 1$ and $t = 4$

These rates can be summarized with the following notation:

$$\frac{dG}{dt} = 1.7 \text{ points per hour when } t = 1 \text{ hour, or}$$

$$G'(1) = 1.7 \text{ points per hour, and}$$

$$\frac{dG}{dt} = 5.2 \text{ points per hour when } t = 4 \text{ hours, or}$$

$$G'(4) = 5.2 \text{ points per hour.}$$

Interpreting Derivatives

As already mentioned, mathematical results are not very useful in real-world settings unless they are stated in a form that anyone can understand. For this reason, an **interpretation** of a result should be stated using a simple, nontechnical sentence. As in the case of interpreting descriptions of change discussed in Section 3.1, you should answer the questions *when, what, how,* and *by how much* when interpreting a rate of change.

Again, consider your score $G(t)$ out of 100 points on the next calculus exam as a function of the number of hours t that you have studied for the exam. Which of the following is a valid interpretation of $\frac{dG}{dt} = 5.2$ points per hour when $t = 4$ hours?

a. The rate of change of my grade after 4 hours is 5.2 points per hour.

b. The slope of the line tangent to the grade curve at $t = 4$ is 5.2.

c. My grade increased 5.2 points after I studied 4 more hours.

d. When I have studied for 4 hours, my grade is improving by 5.2 points per hour.

Choice *a* only restates the mathematical symbols in words. It does not give the meaning of the derivative in the real-life context. Choice *b* is a correct statement, but it uses technical words that a person who has not studied calculus probably would not understand. Also, the symbol t is used with no meaning attached to it, and units are not included with the value 5.2. The use of the word *increased* in choice *c* refers to an interval of time, not change at a point in time. It is an incorrect statement. Choice *d* is the only valid interpretation.

You can expect your students to complain about this distinction between verb tenses. As part of helping your students become effective communicators, you should be unrelenting in requiring them to use the progressive tense when interpreting instantaneous rates of change.

You should note that because a rate of change is measured at a point, it describes something that is in the process of changing. Therefore, we must use the progressive tense (verbs ending with *ing*) to refer to rates of change. For example, we say that "after 1 hour of studying, your grade *is increasing* by approximately 17 points per hour." It is incorrect to say that your grade *increased* or *increases* at a specific point. These verbs refer to change over an interval rather than at a point.

EXAMPLE 1 *Interpreting Derivatives*

Interpret the following four mathematical statements in the context of studying time and grades.

a. $\frac{dG}{dt} = 6.4$ points per hour when $t = 7$ hours.

b. $G'(12) = 3.0$ points per hour.

c. The derivative of G with respect to t is 0 points per hour when $t = 14$ hours.

d. The slope of the tangent line when $t = 15$ hours is approximately -2 points per hour.

Solution

a. The first statement says that when you have studied 7 hours, your grade is improving by 6.4 points per hour. As we later learn, this is the point of greatest slope—that is, the time when a small amount of additional study will benefit you the most.

b. The second statement says that after 12 hours of study, your grade is improving by 3.0 points per hour. Does this mean that at 12 hours of study, your grade is less than at 7 hours of study? No! It simply means that a small amount of additional study time beyond 12 hours may not result in as many extra points on your test as the same amount of time produces after you have studied only 7 hours.

c. The third statement says that after you have studied 14 hours, your grade will no longer be improving. A glance back at Figure 3.26 shows that you have reached your best possible score; more study will not improve your grade.

d. The fourth statement tells you that after 15 hours of study, your grade is actually declining by 2 points per hour. Additional study will only hurt your grade.

Make sure that you understand that these statements tell you nothing about what your grade is—they tell you only how quickly it is changing.

EXAMPLE 2 *Drug Concentration*

$C(h)$ is the average concentration (in nanograms per milliliter, ng/mL) of a drug in the blood stream h hours after the administration of a dose of 360 mg. On the basis of the following information, sketch a graph of C.

$C(0) = 125$ ng/mL $\quad C'(0) = 0$ ng/mL per hour

$C(4) = 215$ ng/mL $\quad C'(4) = 37$ ng/mL per hour

The concentration of the drug is increasing most rapidly after 4 hours. The maximum concentration, 380 ng/mL, occurs after 10 hours. Between $h = 10$ and $h = 24$, the concentration declines at a constant rate of 16 ng/mL per hour. The concentration after 24 hours is 31 ng/mL higher than it was when the dose was administered.

Solution The information about $C(h)$ at various values of h simply locates points on the graph of C. Plot the points (0, 125), (4, 215), (10, 380), and (24, 156).

Because $C'(0) = 0$, the curve has a horizontal tangent at (0, 125). The point of most rapid increase, (4, 215), is an inflection point. The graph is concave up to the left of that point and concave down to the right. The maximum concentration occurs after 10 hours, so the highest point on the graph of C is (10, 380). Concentration declining at a constant rate between $h = 10$ and $h = 24$ means that that portion of C is a line with slope = -16.

One possible graph is shown in Figure 3.27. Compare each statement about $C(h)$ to the graph.

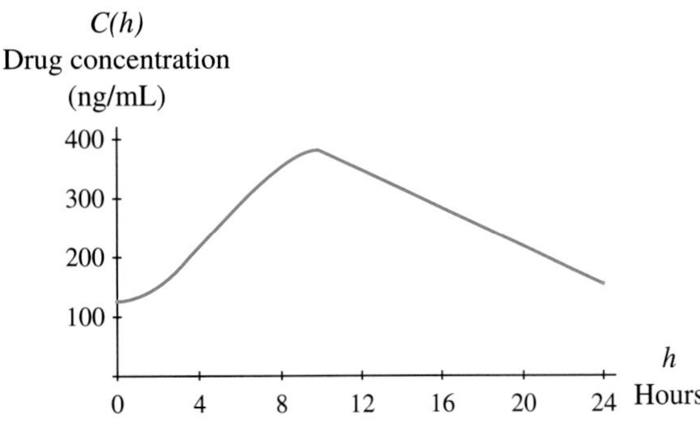

FIGURE 3.27

Note in Figure 3.27 that we cannot assign a value to $C'(10)$. However, the maximum concentration occurs at 10 hours, and on the basis of that, we can estimate that the rate of change of the concentration at that time is zero (even though there is not a horizontal tangent line at $h = 10$ on the graph in Figure 3.27).

Approximating with Derivatives

The topic of approximating change (differentials) is discussed in greater detail in Section 5.1. This topic is used again in Section 9.4 when we consider compensating for change in multivariable functions. (Chapter 9 appears only in the two-semester version of the text.)

Remember that derivatives are simply slopes of tangent lines. Return to Figure 3.25, and consider the point of the grade function graph between 8 and 15 hours studied and the tangent line at $t = 11$ hours (see Figure 3.28). If we magnify the boxed-in portion of this graph (between $t = 11$ and $t = 12$ hours), as seen in Figure 3.29, then we obtain the view of the grade function graph and tangent line at $t = 11$ shown in Figure 3.30.

It so happens that after 11 hours, the derivative (slope) is 4.2 points per hour, and the grade is 90.5 points. What is the grade after 12 hours? It is tempting to reason that if the grade after 11 hours is 90.5 and is increasing by 4.2 points per hour, then after 1 more hour of study, the grade would be $90.5 + 4.2 = 94.7$ points. However, this is not correct, because as Figure 3.30 shows, the grade after 12 hours is 94.2 points. It is the tangent line, not the grade graph, that reaches 94.7 points at 12 hours.

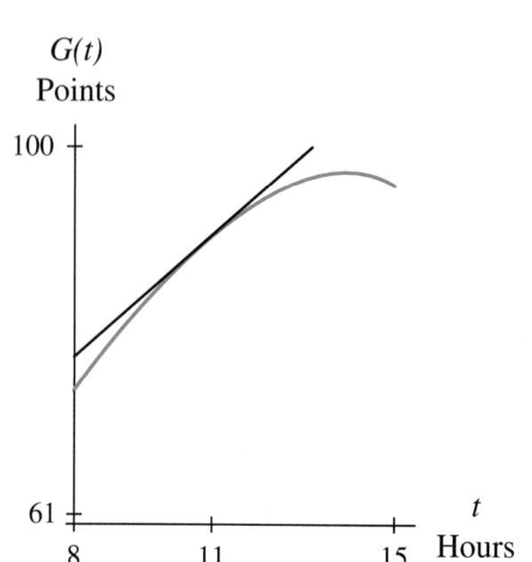

FIGURE 3.28

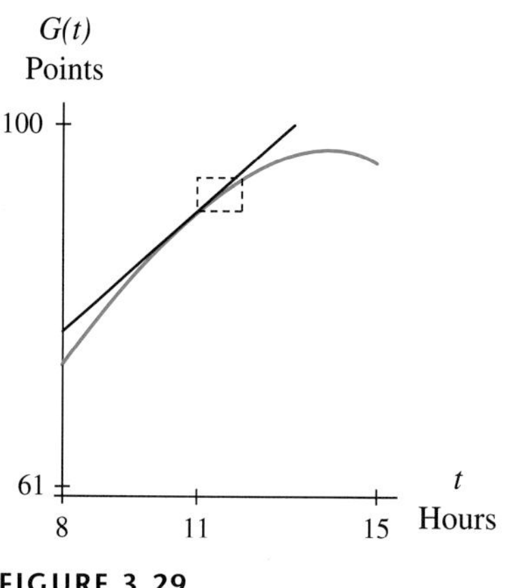

FIGURE 3.29

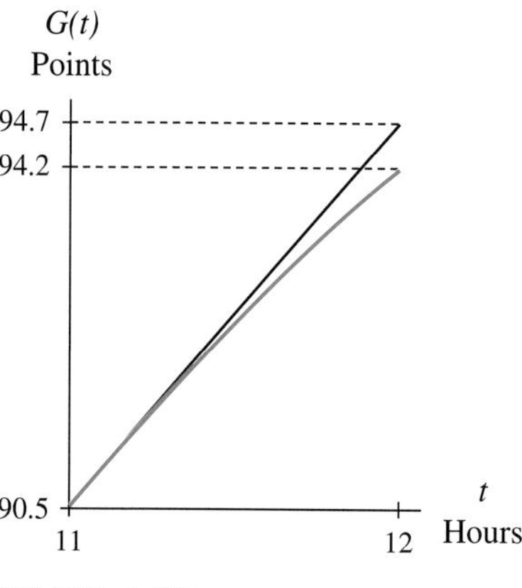

FIGURE 3.30

It is common practice to use tangent lines to estimate function outputs, but this must be done carefully. For example, it is certainly proper to say that, on the basis of a score of 90.5 points and a slope of 4.2 points per hour at 11 hours, the grade will be *approximately* $90.5 + 4.2 = 94.7$ points at 12 hours. You may also say that at 11.5 hours, the grade will be *approximately* $90.5 + 4.2(\frac{1}{2}) = 92.6$ points. But you must also be certain, in making such statements, to make it clear that the values are only *approximations* and are not exact values. As you can see from the graph in Figure 3.30, such approximations should be made only at points that are relatively close to the point of tangency. We have more to say about approximating with derivatives in a later chapter.

EXAMPLE 3 *Changing Revenue*

Suppose that $R(x)$ is a company's revenue (in thousands of dollars) when $\$x$ thousand is spent on advertising. Interpret $R(5) = 1250$ and $R'(5) = 42$.

Solution When $5000 is spent on advertising, revenue is $1,250,000 and is increasing by $42 thousand per thousand advertising dollars. It is also correct to say that when $5000 is spent on advertising, an additional $1000 of advertising will increase the company's revenue by *approximately* $42,000—from $1,250,000 to about $1,292,000.

EXAMPLE 4 *Iowa Population*

Figure 3.31 shows the graph of a model for the population of Iowa[30] during the 1990s.

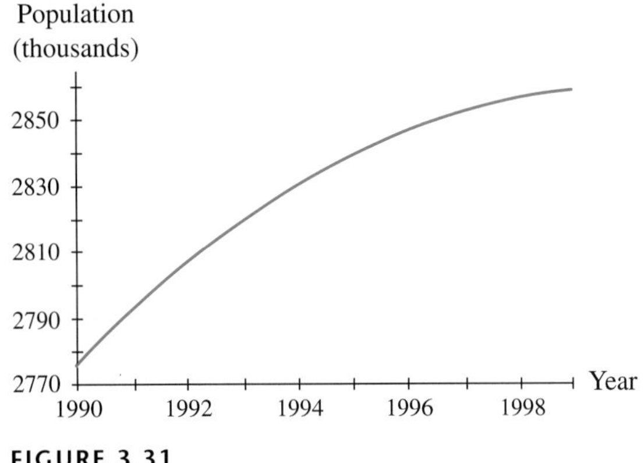

FIGURE 3.31

a. Sketch the tangent line at the point corresponding to 1996, and estimate the slope.

b. Estimate the population of Iowa in 1996.

c. Using only your answers to parts *a* and *b*, estimate the population in 1997.

Solution

a.

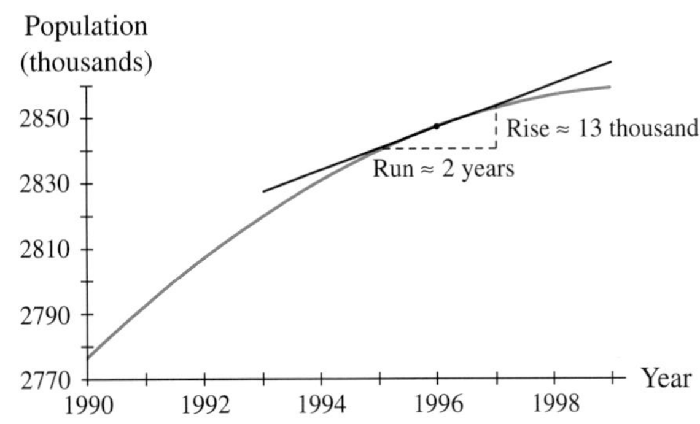

FIGURE 3.32

$$\text{Slope} = \frac{\text{rise}}{\text{run}} \approx \frac{13 \text{ thousand people}}{2 \text{ years}} = 6.5 \text{ thousand people per year}$$

30. Based on data from *Statistical Abstract*, 1998.

b. From the graph, it appears that the population was approximately 2848 thousand people in 1996.

c. The slope at 1996 is approximately 6.5 thousand people per year, so between 1996 and 1997, the population rose by approximately 6500 people.

$$2848 + 6.5 = 2853.5 \text{ thousand people}$$

In 1997, the population was approximately 2,853,500 people.

You should be aware that in part *c* of Example 4, we were using the tangent line at the point corresponding to 1996 to estimate the population in 1997. Figure 3.33 shows a close-up of the graph and tangent line near the point of tangency. Because the tangent line lies above the graph, the approximation that the tangent line yields overestimates the actual value. The population, according to the graph, was close to 2,852,900.

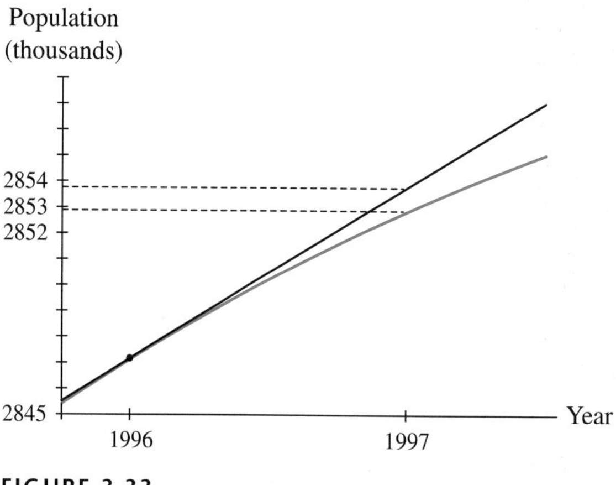

FIGURE 3.33

Does *Instantaneous* Refer to Time?

Newspaper and television reporters sometimes incorrectly use the phrase "high rate of speed." Because *rate* and *speed* describe the same concept, *rate of speed* technically describes acceleration.

We saw that the instantaneous rate of change at a point P on the graph of a continuous function is the slope of the tangent line to the graph at P. If the function inputs are measured in units of time, then it is certainly natural to use the word *instantaneous* when describing rates of change, because each point P on the graph of the function corresponds to a particular instant in time. In fact, the use of the word *instantaneous* in this context arose precisely from the historical need to understand how rapidly distance traveled changes as a function of time. Today, we are accustomed to referring to the rate of change of distance traveled as a function of time as *speed*.

You should be aware, however, that the use of the word *instantaneous* in connection with rates of change does not necessarily mean that time units are involved. For example, suppose that a graph depicts profit (in dollars) resulting from the sale of a certain number of used cars. In this case, the slope of the tangent line at any particular point (the instantaneous rate of change) expresses how rapidly profit is changing per car. The units of change are dollars per car; no time units are involved.

You should also remember that units for instantaneous rates of change, like average rates of change, are always expressed in output units per input unit. Without proper units, a number that purports to describe a rate of change is meaningless.

EXAMPLE 5 Temperature in the Polar Night Region

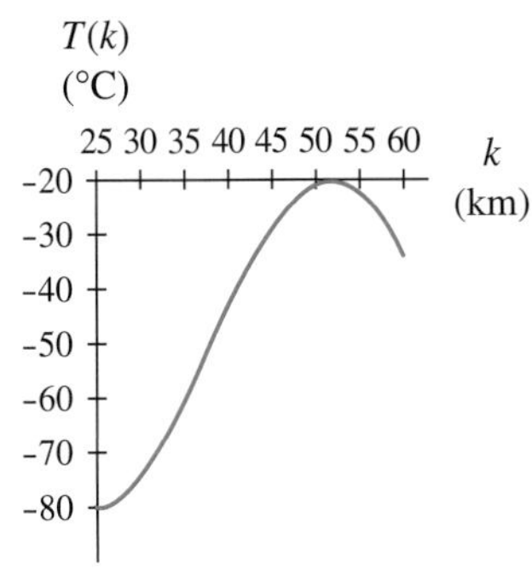

FIGURE 3.34

The graph in Figure 3.34[31] shows the temperature, $T(k)$, of the polar night region (in °C) as a function of k, the number of kilometers above sea level.

a. Sketch the tangent line at 45 km, and estimate its slope.

b. What is the derivative notation for the slope of a line tangent to the graph of T?

c. Write a sentence interpreting in context the meaning of the slope found in part *a*.

Solution

a. From Figure 3.35 we calculate

$$\text{Slope} = \frac{\text{rise}}{\text{run}} \approx \frac{23°\text{C}}{10 \text{ km}} = 2.3°\text{C per kilometer}$$

b. Correct derivative notations include $\frac{dT}{dk}$, $\frac{d}{dk}[T(k)]$, and $T'(k)$.

c. At 45 km above sea level, the temperature of the atmosphere is increasing 2.3°C per kilometer. In other words, the temperature rises by approximately 2.3°C between 45 and 46 km above sea level.

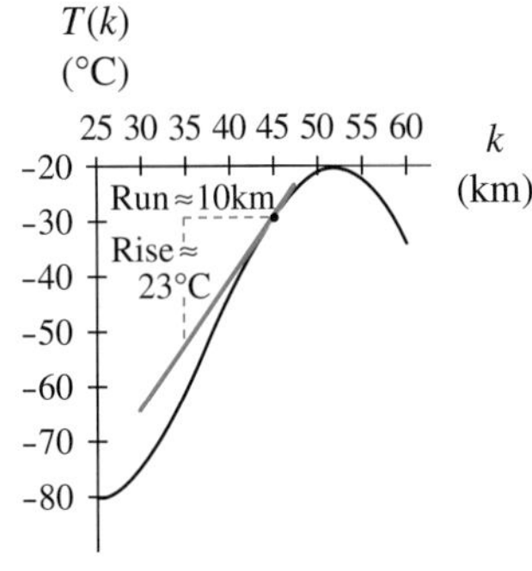

FIGURE 3.35

Percentage Rate of Change

Recall from Section 3.1 that percentage change is found by dividing change over an interval by the output at the beginning of the interval and multiplying by 100. Similarly, **percentage rate of change** can be found by dividing the rate of change at a point by the function value at the same point and multiplying by 100. The units of a percentage rate of change are percent per input unit.

Percentage rate-of-change units are the most difficult for students to identify. Encourage your students to keep units attached to all values used during calculations. (Also be sure to do this in all your calculations in class.)

Percentage Rate of Change

$$\text{Percentage rate of change} = \frac{\text{rate of change at a point}}{\text{value of the function at that point}} \cdot 100\%$$

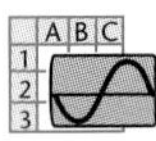

Percentage rates of change are useful in describing the relative magnitude of a rate of change. For example, suppose you are a city planner and estimate that the city's population is increasing at a rate of 50,000 people per year. Any growth in population will affect your planning activities, but just how significant is a growth rate of 50,000 people per year? If the current population is 200,000 people, then the percentage rate of change of the population is $\frac{50{,}000 \text{ people per year}}{200{,}000 \text{ people}} \cdot 100\% = 25\%$ per year. Growth of 25% per year in population is fast growth. However, if the current population is 2 million, then the percentage rate of change of the population is

31. "Atmospheric Exchange Processes and the Ozone Problem," in *The Ozone Layer*, ed. Asit K. Biswas, Institute for Environmental Studies, Toronto. Published for the United Nations Environment Program by Pergamon Press, Oxford, 1979.

$\frac{50{,}000 \text{ people per year}}{2{,}000{,}000 \text{ people}} \cdot 100\% = 2.5\%$ per year. The steps that you as a city planner must take to accommodate growth if the city is growing by 25% per year are different from the steps you must take if the city is growing by 2.5% per year. Expressing a rate of change as a percentage puts the rate in the context of the current size and adds more meaning to the interpretation of a rate of change.

EXAMPLE 6 *Sales*

The graph in Figure 3.36 shows sales (in thousands of dollars) for a small business from 1995 through 2003.

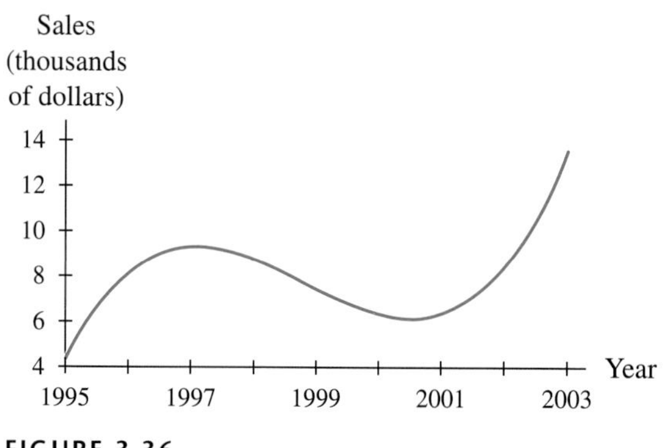

FIGURE 3.36

a. Estimate the rate of change of sales in 1999.

b. Estimate the percentage rate of change of sales in 1999.

Solution

a. A tangent line is drawn at 1999 as shown in Figure 3.37.

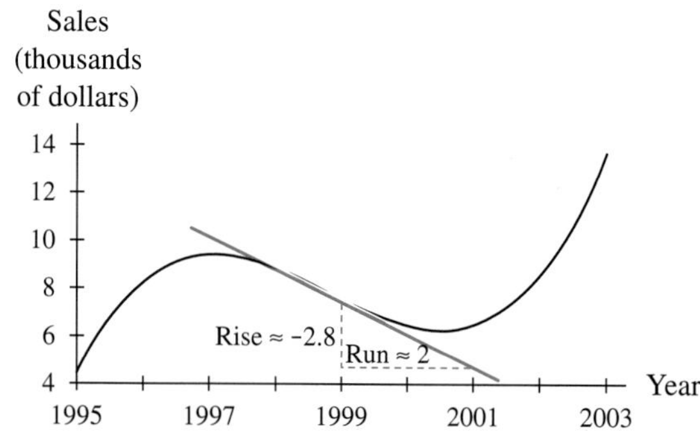

FIGURE 3.37

The slope is estimated to be

$$\frac{-\$2.8 \text{ thousand}}{2 \text{ years}} = -\$1.4 \text{ thousand per year} = -\$1400 \text{ per year}$$

In 1999, sales were falling at a rate of approximately \$1400 per year.

b. We can express this rate of change as a percentage rate of change if we divide by the sales in 1999. It appears from the graph that the sales in 1999 were approximately \$7.5 thousand dollars, or \$7500. Therefore, the percentage rate of change in 1999 was approximately

$$\frac{-\$1400 \text{ per year}}{\$7500} \cdot 100\% \approx -0.187 \cdot 100\% \text{ per year} = -18.7\% \text{ per year}$$

In 1999, sales were falling approximately 18.7% per year. Expressing the rate of change of sales as a percentage of sales gives a much clearer picture of the impact of the decline in sales. The business was experiencing a reduction in sales of nearly 20% per year in 1999.

3.3 Concept Inventory

- *Derivative = rate of change = slope of tangent line*
- *Derivative of f with respect to x* $= \frac{df}{dx} = f'(x) = \frac{d}{dx}[f(x)]$
- *Interpreting derivatives*
- *Approximating with derivatives*
- *Percentage rate of change*

3.3 Activities

1. Suppose that $P(t)$ is the number of miles from an airport that a plane has flown after t hours.
   a. What are the units on $\frac{d}{dt}[P(t)]$?
   b. What common word do we use for $\frac{d}{dt}[P(t)]$?
2. Let $B(t)$ be the balance, in dollars, in a mutual fund t years after the initial investment. Assume that no deposits or withdrawals are made during the investment period.
   a. What are the units on $\frac{dB}{dt}$?
   b. What is the financial interpretation of $\frac{dB}{dt}$?
3. Let $W(t)$ be the number of words per minute (wpm) that a student in a typing class can type after t weeks in the course.
   a. Is it possible for $W(t)$ to be negative? Explain.
   b. What are the units on $W'(t)$?
   c. Is it possible for $W'(t)$ to be negative? Explain.
4. Imagine that $C(f)$ is the number of bushels of corn produced on a tract of farm land when f pounds of fertilizer are used.
   a. What are the units on $C'(f)$?
   b. Is it possible for $C'(f)$ to be negative? Explain.
   c. Is it possible for $C(f)$ to be negative? Explain.
5. Suppose that $F(p)$ is the weekly profit (in thousands of dollars) that an airline makes on its Boston to Washington D.C. flights when the ticket price is p dollars. Interpret the following:
   a. $F(65) = 15$
   b. $F'(65) = 1.5$
   c. $\frac{dF}{dp} = -2$ when $p = 90$
6. Let $T(p)$ be the number of tickets from Boston to Washington D.C. that a certain airline sells in 1 week when the price of each ticket is p dollars. Interpret the following:
   a. $T(115) = 1750$
   b. $T'(115) = -20$
   c. $\frac{dT}{dp} = -2$ when $p = 125$
7. On the basis of the following information, sketch a possible graph of t with input x.
   - $t(3) = 7$
   - $t(4.4) = t(8) = 0$
   - $\frac{dt}{dx} = 0$ at $x = 6.2$
   - The graph of t has no concavity changes.
8. Using the information that follows, sketch a possible graph of m with input t.
   - $m(0) = 3$
   - $\frac{d}{dt}[m(t)] = 0.34$
9. Suppose that $W(t)$ is your weight t weeks after you begin a diet. Interpret the following:
   a. $W(0) = 167$
   b. $W(12) = 142$
   c. $\frac{dW}{dt} = -2$ when $t = 1$
   d. $\frac{dW}{dt} = -1$ when $t = 9$
   e. $W'(12) = 0$
   f. $W'(15) = 0.25$
   g. On the basis of the information in parts *a* through *f*, sketch a possible graph of W.

10. Suppose that $G(v)$ equals the fuel efficiency in miles per gallon (mpg) of a car going v miles per hour. Give the practical meaning of the following statements.

    a. $G(55) = 32.5$

    b. $\frac{dG}{dv} = -0.25$ when $v = 55$

    c. $G'(45) = 0.15$

    d. $\frac{d}{dv}[G(51)] = 0$

11. $P(b)$ is the percentage of all births to single mothers in the United States in year b from 1940 through 1995. Using the following information,[32] sketch a graph of P.

    - $P(1940) \approx 4\%$
    - $P'(b)$ is never zero.
    - $P(b) \approx 12\%$ when $b = 1970$
    - $P(1995)$ is about 20 percentage points more than $P(1970)$.
    - The average rate of change of P between 1970 and 1980 is 0.6 percentage points per year.
    - Lines tangent to the graph of P lie below the graph at all points between 1940 and 1990 and above the graph between 1990 and 1995.

12. $E(x)$ is the public secondary school enrollment (in millions of students) in the United States x years after 1940. Use the following information[33] to sketch a graph of E.

    - $E(40) = 13.2$
    - The graph of E is always concave down.
    - Between 1980 and 1990, enrollment declined at an average rate of 0.19 million students per year.
    - The projected enrollment for 2008 is 14,700,000 students.
    - It is not possible to draw a line tangent to the graph of E at $x = 50$.

13. Let $P(x)$ be the profit in dollars that a fraternity makes selling x T-shirts.

    a. Is it possible for $P(x)$ to be negative? Explain.

    b. Is it possible for $P'(x)$ to be negative? Explain.

    c. If $P'(200) = -1.5$, is the fraternity losing money? Explain.

14. Let $M(t)$ be the number of members in a political organization t years after its founding. What are the units on $\frac{d}{dt}[M(t)]$?

15. Let $D(r)$ be the time in years that it takes for an investment to double if interest is continuously compounded at an annual rate of $r\%$. (Here r is expressed as a percentage, not a decimal.)

    a. What are the units on $\frac{dD}{dr}$?

    b. Why does it make sense that $\frac{dD}{dr}$ is always negative?

    c. Give the practical interpretation of the following:

       i. $D(9) = 7.7$

       ii. $\frac{dD}{dr} = -2.77$ when $r = 5$

       iii. $\frac{dD}{dr} = -0.48$ when $r = 12$

16. Let $U(t)$ be the number of people unemployed in a country t months after the election of a new president.

    a. Draw and label an input/output diagram for U.

    b. Is U a function? Why or why not?

    c. Interpret the following facts about $U(t)$ in statements describing the unemployment situation.

       i. $U(0) = 3{,}000{,}000$

       ii. $U(12) = 2{,}800{,}000$

       iii. $U'(24) = 0$

       iv. $\frac{dU}{dt} = 100{,}000$ when $t = 36$

       v. $\frac{dU}{dt} = -200{,}000$ when $t = 48$

    d. On the basis of the information in part c, sketch a possible graph of the number of people unemployed during the first 48 months of the president's term. Label numbers and units on the axes.

17. The graph in Figure 3.3.1 shows the terminal speed (in meters per second) of a raindrop[34] as a function of the size of the drop measured in terms of its diameter.

    a. Sketch a secant line connecting the points for diameters of 1 mm and 5 mm, and estimate its slope. What information does this secant line slope give?

    b. Sketch a line tangent to the curve at a diameter of 4 mm. What information does the slope of this line give?

    c. Estimate the derivative of the speed for a diameter of 4 mm. Interpret your answer.

32. Based on data from L. Usdansky, "Single Motherhood: Stereotypes vs. Statistics," *The New York Times*, February 11, 1996, Section 4, page E4; and on data from *Statistical Abstract*, 1998.

33. Based on data appearing in *Datapedia of the United States*, Bernan Press, 1994; and in *Statistical Abstract*, 1998.

34. R. R. Rogers and M. K. Yau, *A Short Course in Cloud Physics* (White Plains, NY: Elsevier Science, 1989).

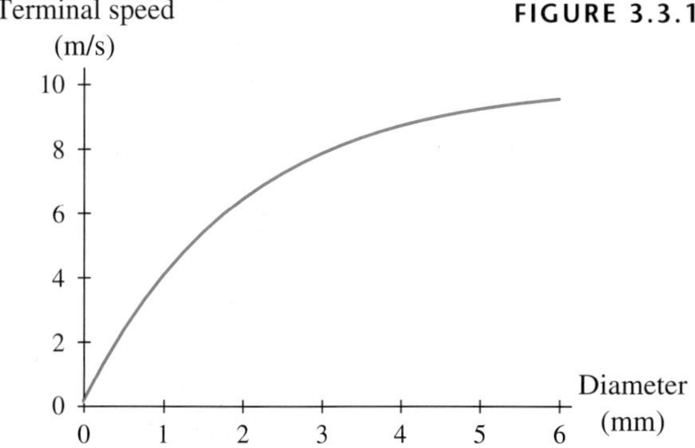

FIGURE 3.3.1

d. Estimate the rate at which the speed is rising for a raindrop with diameter 2 mm.

e. Find and interpret the percentage rate of change of speed for a raindrop with diameter 2 mm.

18. The scatter plot and graph in Figure 3.3.2 depict the number of customers that a fast-food restaurant serves each hour on a typical weekday.

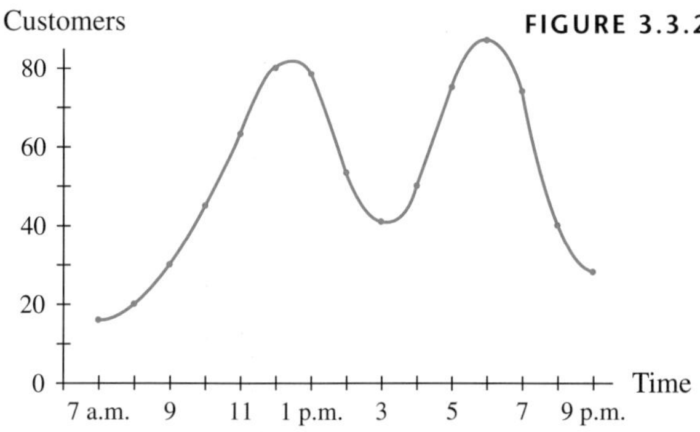

FIGURE 3.3.2

a. Estimate the average rate of change of the number of customers between 7 a.m. and 11 a.m. Interpret your answer.

b. Estimate the instantaneous rate of change and percentage rate of change of the number of customers at 4:00 p.m. Interpret your answer.

c. List the factors that might affect the accuracy of your answers to parts *a* and *b*.

19. Refer once more to the function G, your grade out of 100 points on the next calculus test when you study t hours during the week before the test. The graph of G is shown in Figure 3.3.3.

a. Carefully draw tangent lines at 4 hours and 11 hours. Estimate the slope of each tangent line.

b. Compare your answers with the slopes given on pages 221 and 224. How accurate are your estimates?

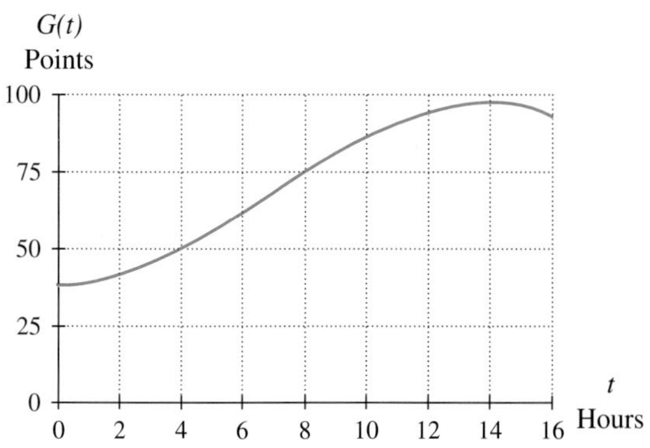

FIGURE 3.3.3

c. Estimate the average rate of change between 4 hours and 10 hours. Interpret your answer.

d. Estimate the percentage rate of change of the grade after 4 hours of study. Interpret your answer.

20.

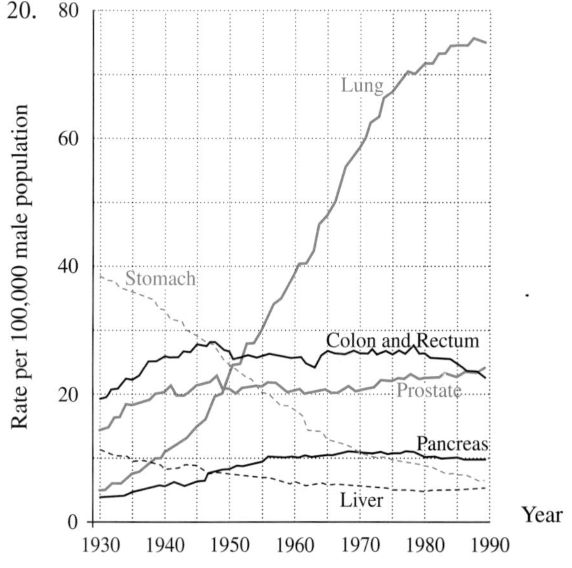

Cancer death rates by site, males, United States, 1930–89
(Rates are adjusted to the 1970 U.S. census population)

FIGURE 3.3.4

a. Figure 3.3.4 is a graph of rates of death from cancer among U.S. males.[35]

i. Estimate how rapidly the number of deaths due to lung cancer was increasing in 1970 and in 1980.

35. Figure 3.3.4 courtesy of the American Cancer Society, Inc.

ii. Estimate the percentage rate of change of deaths due to liver cancer in 1980.

iii. Estimate the slope of the stomach cancer curve in 1960.

b. Describe in detail the behavior of the lung cancer curve from 1930 to 1990. Explain why the lung cancer curve differs so radically from the other curves shown.

c. List as many factors as you can that might affect a cancer death rate curve.

21.

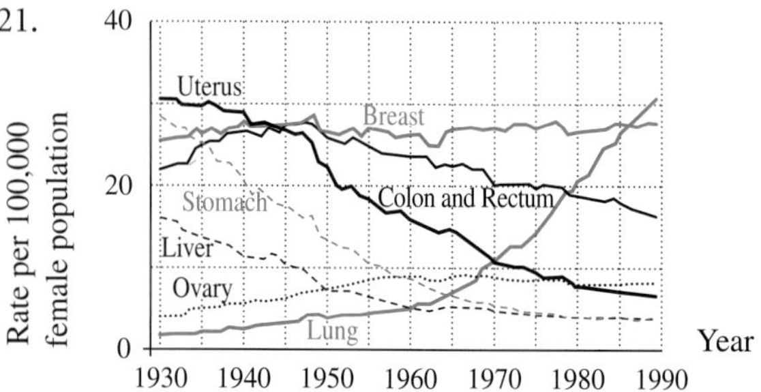

FIGURE 3.3.5

Cancer death rates by site, females, United States, 1930–89

(Rates are adjusted to the 1970 U.S. census population)

a. Figure 3.3.5 is a graph of rates of death from cancer among U.S. females.[36] Estimate how rapidly the number of deaths due to lung cancer was increasing in 1970 and 1980. Estimate the percentage rates of change in each year.

b. Compare the lung cancer death rate curves for males and females. What do you think will happen to the two curves during the next 30 years?

22.

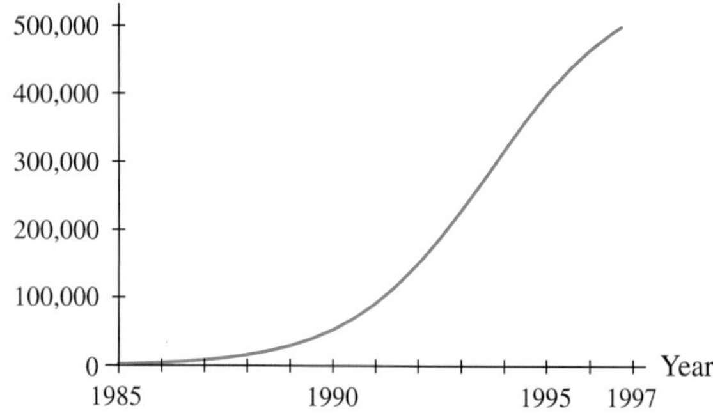

FIGURE 3.3.6

36. Ibid.

The function C gives the cumulative number of AIDS cases since 1984 diagnosed at the end of year t as represented by the curve shown in Figure 3.3.6.

a. Estimate and interpret $\frac{dC}{dt}$ at the end of 1993.

b. Estimate and interpret $C'(1996)$.

c. Estimate and interpret $C(1996)$.

d. Estimate the percentage rate of change of the cumulative number of AIDS cases diagnosed by the end of 1996.

e. Estimate the number of AIDS cases diagnosed in 1996. (*Hint:* The graph shows *cumulative* cases since 1984.)

f. Estimate the derivative of C at the end of 1990.

g. Estimate the average rate of change in the cumulative number of AIDS cases between 1985 and 1990.

23. Table 3.13 shows the cumulative totals since 1978 for deaths[37] due to alcohol-related causes.

TABLE 3.13

Year	Cumulative deaths (thousands)
1979	19.0
1981	57.5
1983	92.4
1985	127.7
1987	162.9
1989	201.6
1991	140.6
1993	279.8
1995	320.2
1997	359.6

Use a symmetric difference quotient to estimate the rate of change in the cumulative number of alcohol-related deaths in 1991. Interpret your answer.

24. Table 3.14 shows the death rates (deaths per 100,000 people) for alcohol-related deaths.[38]

a. Use a symmetric difference quotient to estimate the rates of change in the death rate in 1983 and 1993. Interpret your answers.

37. Centers for Disease Control.

38. Ibid.

TABLE 3.14

Year	Death rate (deaths per 100,000)	Year	Death rate (deaths per 100,000)
1979	8.4	1989	8.0
1981	8.1	1991	7.6
1983	7.4	1993	7.6
1985	7.5	1995	7.7
1987	7.4	1997	7.3

b. Does a negative rate of change in the death rate indicate that the number of alcohol-related deaths was declining? Explain.

25. What is the meaning of the word *derivative* in the world's financial markets? To find out, see the article entitled "Derivatives? What Are Derivatives?" in *Newsweek*, March 13, 1995, page 50.

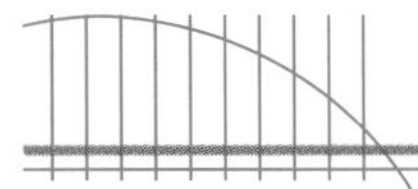

3.4 Drawing Rate-of-Change Graphs

We expect students to be able to sketch slope graphs in our discussion of the Fundamental Theorem of Calculus in Section 6.4.

In Section 1.3 we considered ways to construct a new function from two or more functions by addition/subtraction, multiplication/division, and composition. We also saw how to construct the inverse of a function by reversing the input and output of the function. We now consider another constructed function, the rate-of-change function. This function plays a central role in calculus and will be the subject of Chapters 4 and 5.

Every smooth, continuous curve with no vertical tangent lines has a slope associated with each point on the curve. When these slopes are plotted, they also form a smooth, continuous curve. We call the resulting curve a **slope graph, rate-of-change graph,** or **derivative graph.**

An effective teaching technique when beginning this discussion is to place a see-through ruler on a transparency of Figure 3.38 (or use a ruler with a curve drawn on the chalkboard) and have students discuss the sign and relative magnitude of the slopes as you move the ruler along the curve.

What do we know about the slopes of the graph shown in Figure 3.38? Sketch lines tangent to the curve at the points where $x = A$ and $x = C$ and at several other points on the curve, as shown in Figure 3.39.

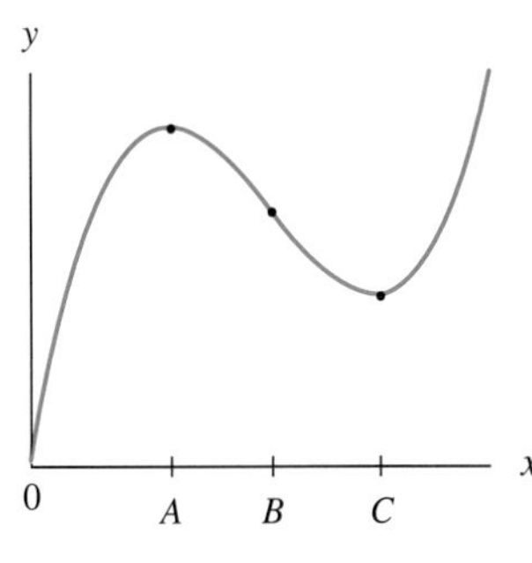

FIGURE 3.38

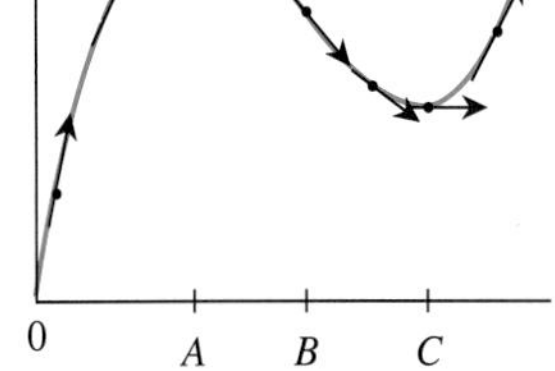

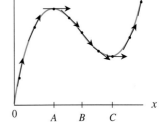

FIGURE 3.39

We deduce the following facts:

- The tangent lines at the points with inputs A and C are horizontal, so the slope is zero at those points.
- Between 0 and A, the graph is increasing, so the slopes are positive. The tangent lines become less steep moving from 0 to A, so the slopes start off large and become smaller as we approach A from the left.
- Between A and C, the graph is decreasing, so the slopes are negative.
- At B the graph has an inflection point. This is the point at which the graph is decreasing most rapidly—that is, the point at which the slope is most negative.

Because the vertical axis units of the slope graph are different from those of the function graph, we do not draw the slope graph on the same set of axes as the original graph. Doing so may be convenient, but it is sloppy mathematics.

- To the right of C, the graph is again increasing, so the slopes are positive. The tangent lines become steeper as we move to the right of C, so the slopes become larger as the input increases beyond C.

We record this information as indicated in Figure 3.40. On the basis of this information about the slopes, we sketch the shape of the slope graph, which appears in Figure 3.41.

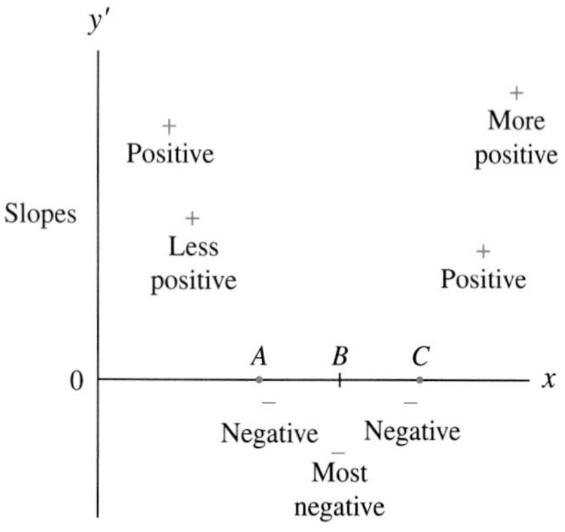

FIGURE 3.40

FIGURE 3.41

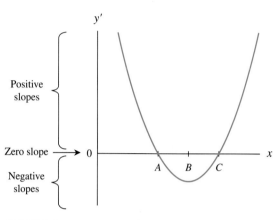

We do not know the specifics of the slope graph—how far below the horizontal axis it dips, where it crosses the vertical axis, how steeply it rises to the right of C, and so on. However, we do know its basic shape.

EXAMPLE 1 *Slope Graph of a Logistic Curve*

This example is another case where, when you are discussing the slopes, moving a ruler along the curve enhances the students' understanding. This demonstration is most effectively done before you begin sketching the slope graph.

The graph in Figure 3.42 is a logistic curve. Sketch its slope graph.

Solution First we note that the logistic curve in Figure 3.42 is always increasing. Thus its slope graph is always positive (above the horizontal axis). We also note that even though there is no relative maximum or relative minimum, the logistic curve does level off at both ends. Thus its slope graph will be near zero at both ends. (See Figure 3.43.) Finally, we note that the logistic curve has its steepest slope at A because this is the location of the inflection point. Therefore, the slope graph is greatest (has a maximum) at this point. (See Figure 3.44.) We sketch the slope graph as shown in Figure 3.45.

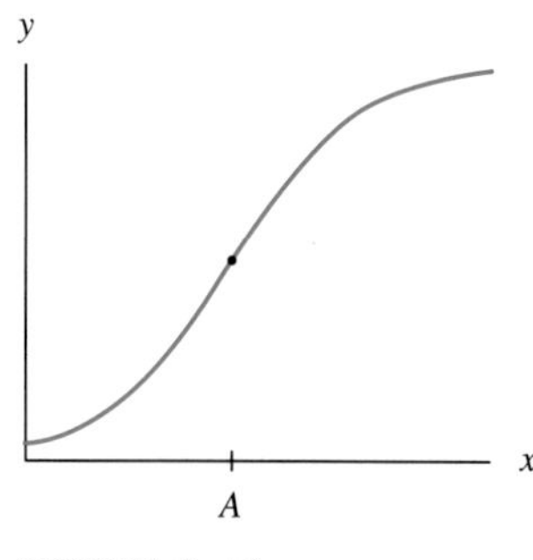

FIGURE 3.42

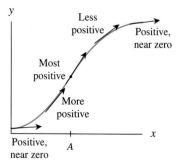

FIGURE 3.43

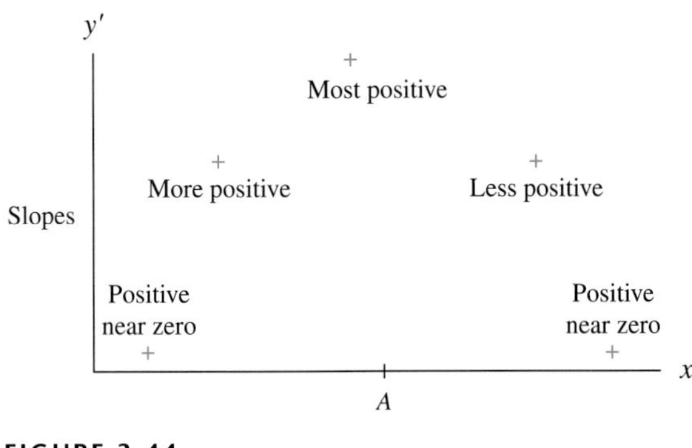

FIGURE 3.44

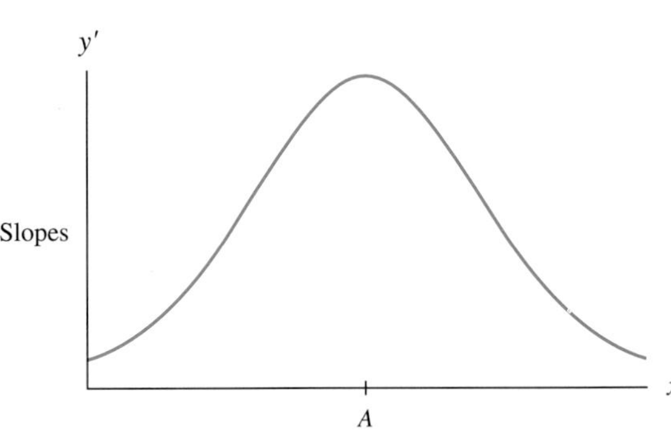

FIGURE 3.45

EXAMPLE 2 *More Slope Graphs*

a. Sketch a graph that is always decreasing but has slopes that are always increasing. Also sketch its slope graph.

b. Sketch a graph that is always decreasing with slopes that are always decreasing. Also sketch its slope graph.

Solution

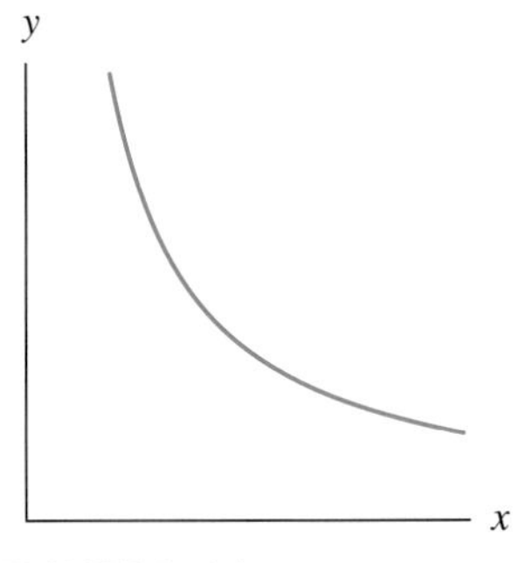

FIGURE 3.46

a. If a graph is decreasing, it is falling from left to right and has negative slopes. If the negative slopes are increasing, they are becoming less and less negative (moving toward zero) as the input increases. That means that the graph is becoming less steep. Such a graph must be concave up, as shown in Figure 3.46.

The slopes of this graph are always negative, so a slope graph must lie completely below the input axis. As Figure 3.47 shows, the slopes are increasing. This means that they are rising closer to zero, but because they will never be positive, the graph must curve downward. A slope graph with these characteristics is shown in Figure 3.48.

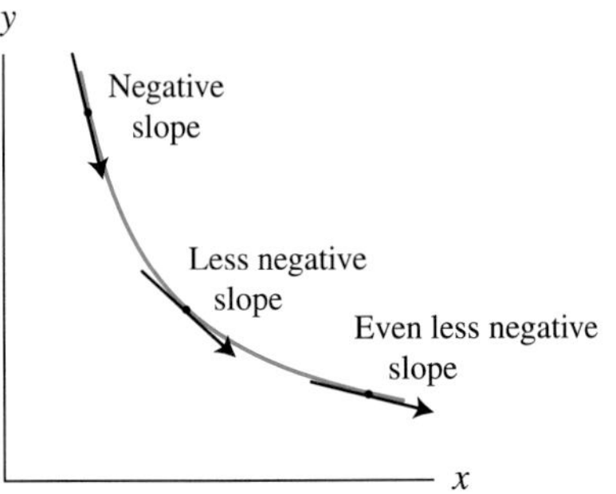

FIGURE 3.47

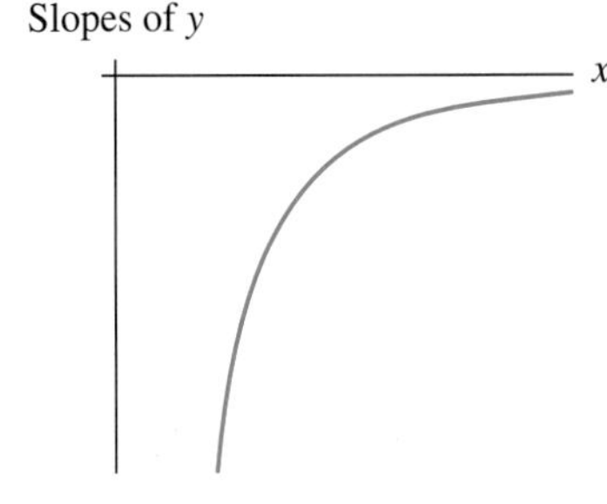

FIGURE 3.48

b. Like the graph in part *a*, this graph is falling from left to right and has negative slopes. However, the slopes are decreasing. When negative numbers decrease, they become more and more negative. This means the graph becomes more steep

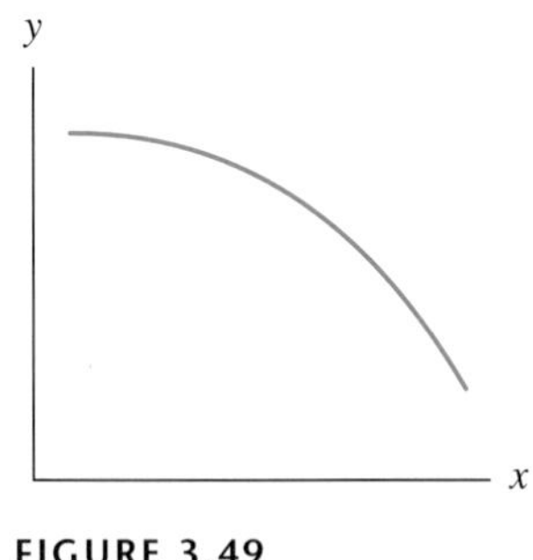

FIGURE 3.49

as input increases. A declining graph that becomes increasingly steep looks like the one shown in Figure 3.49.

Again, the slopes are always negative, so the slope graph will lie completely below the input axis. Instead of rising toward the input axis, this slope graph will fall away from the input axis as the slopes of the original graph become more and more negative. See Figure 3.50. The slope graph is shown in Figure 3.51.

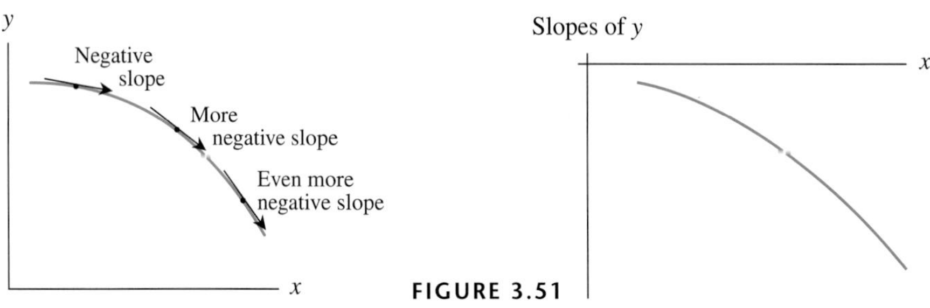

FIGURE 3.50

FIGURE 3.51

In the preceding discussion and example, the graphs we used had no labeled tick marks on the horizontal or vertical axes. In such cases, it is not possible to estimate the value of the slope of the graph at any given point. Instead, we sketch the general shape of the slope graph by observing the important points and general behavior of the original graph, such as

1. Points at which a tangent line is horizontal
2. The regions over which the graph is increasing or decreasing
3. Points of inflection
4. Places at which the graph appears to be horizontal or leveling off

As the previous examples indicate, sketching lines tangent to a curve helps us determine the relative magnitude of the slopes. As this process becomes more familiar, you should be able to visualize the tangent lines and consider the steepness of the curve itself. This technique is illustrated in Example 3.

EXAMPLE 3 *Using Relative Magnitudes to Sketch Slope Graphs*

The height of a plant often follows the general trend shown in Figure 3.52. Draw a graph depicting the growth rate of the plant.

Solution The slopes at A, B, and C are all positive. Is the slope at A smaller or larger than that at B? It is larger, so the slope graph at A should be higher than it is at B. The graph at C is not as steep as it is at either A or B, so the slope graph should be lower at C than at B. (See Figure 3.53.)

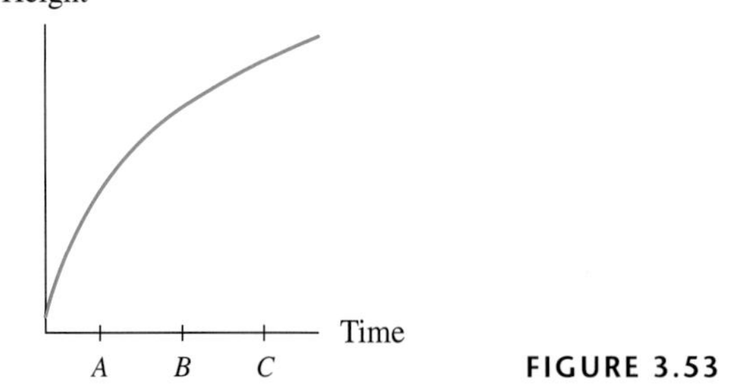

FIGURE 3.52

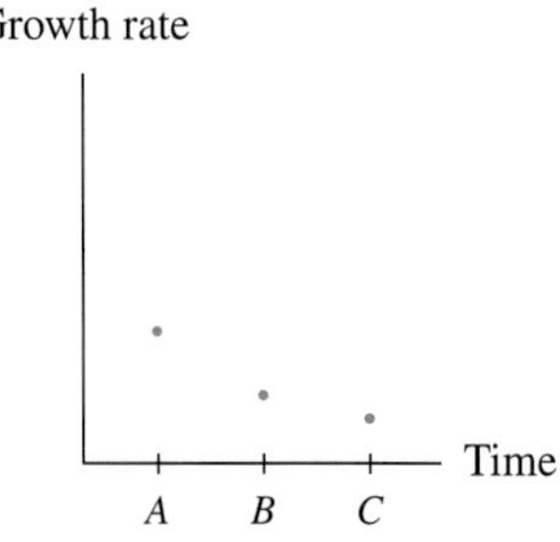

FIGURE 3.53

Where on the graph is the slope steepest? The answer is at the left endpoint. Where is the graph least steep? You can see that it is at the right endpoint. Add these observations to your plot. (See Figure 3.54.)

Now sketch the slope graph according to your plot, and be sure to include the appropriate labels on the horizontal and vertical axes. (See Figure 3.55.)

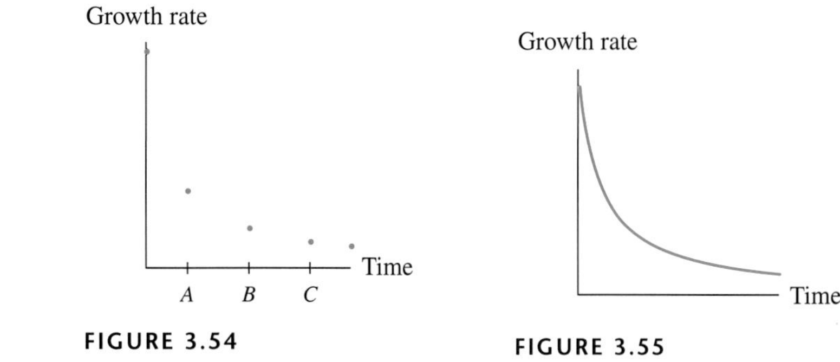

FIGURE 3.54

FIGURE 3.55

A Detailed Look at the Slope Graph

When a graph has labeled tick marks on both the horizontal and the vertical axes or an equation for the graph is known, it is possible to estimate the values of slopes at certain points on the graph. However, it would be tedious to calculate the slope graphically or numerically for every point on the graph. In fact, because there are infinitely many points on a continuous curve, it is not possible. Instead, we calculate the slope at a few special points, such as inflection points, in order to obtain a more accurate slope graph.

Consider Figure 3.42, which we saw at the beginning of this section. It is shown again in Figure 3.56, but this time the graph has labeled tick marks on both the horizontal and the vertical axes.

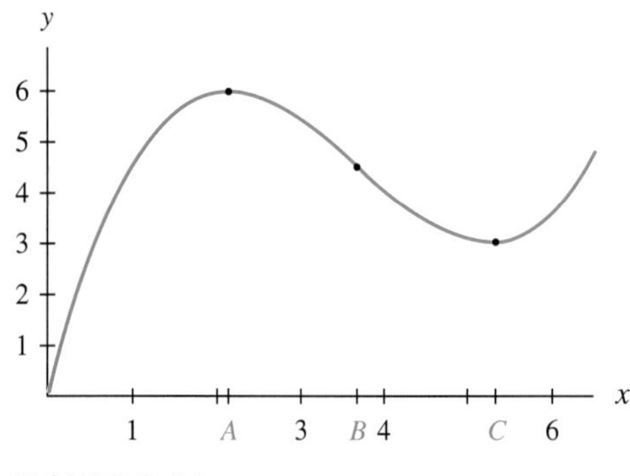

FIGURE 3.56

We know that the slope graph crosses the horizontal axis at *A* and *C* and that a minimum occurs on the slope graph below the horizontal axis at *B*. Before, we did not know how far below the axis to draw this minimum. Now that there is a numerical scale on the axes, we can graphically estimate the slope at the inflection point and use that estimate to help us sketch the slope graph.

By drawing the tangent line at *B* and estimating its slope, we find that the minimum of the slope graph is approximately 1.4 units below the horizontal axis. (See Figure 3.57.)

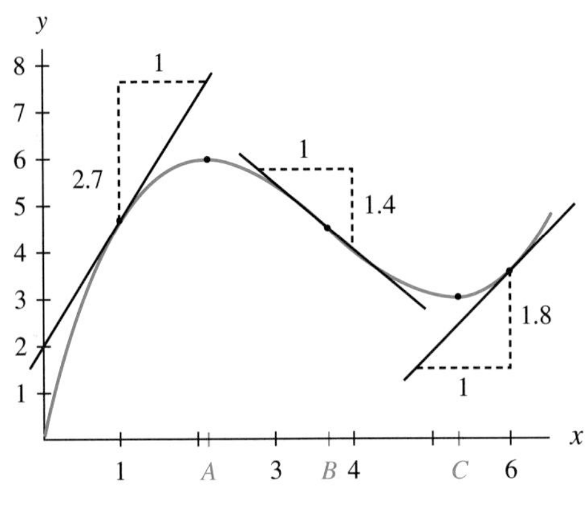

FIGURE 3.57

If we estimate the slopes at two additional points, say at $x = 1$ and $x = 6$, then we can produce a fairly accurate sketch of the slope graph. Table 3.15 shows a list of estimated slope values.

TABLE 3.15

x	1	A	B	C	6
Slope	2.7	0	-1.4	0	1.8

FIGURE 3.58

Plotting these points and sketching the slope graph give us the graph in Figure 3.58.

If we do not have a continuous curve but have a scatter plot, then we can sketch a rate-of-change graph by first sketching a smooth curve that fits the scatter plot. Then we can draw the slope graph of that smooth curve as shown in Example 4.

EXAMPLE 4 *Population of Cleveland, Ohio*

In 1797, the Lorenzo Carter family built a cabin on Lake Erie where today the city of Cleveland, Ohio, is located. Table 3.16 gives population data[39] for Cleveland from 1810 through 1990.

a. Sketch a smooth curve representing population. Your curve should have no more inflection points than the number suggested by the scatter plot.

b. Sketch a graph representing the rate of change of population.

Solution

a. Draw a scatter plot of the population data, and sketch the smooth curve. (See Figure 3.59.)

b. The population graph is fairly level in the early 1800s, so the slope graph will begin near zero. The smooth sketch increases during the 1800s and early 1900s until it peaks in the 1940s. Thus the slope graph will be positive until the mid-1940s, at which time it will cross the horizontal axis and become negative. Population decreased from the mid-1940s onward.

   There appear to be two inflection points. The point of most rapid growth appears around 1910, and the point of most rapid decline appears near 1975.

39. U.S. Department of Commerce, Bureau of the Census.

TABLE 3.16

Year	Population
1810	57
1820	606
1830	1076
1840	6071
1850	17,034
1860	43,417
1870	92,829
1880	160,146
1890	261,353
1900	381,768
1910	560,663
1920	796,841
1930	900,429
1940	878,336
1950	914,808
1960	876,050
1970	750,879
1980	573,822
1990	505,616

FIGURE 3.59

These are the years in which the slope graph will be at its maximum and at its minimum, respectively. By drawing tangent lines at 1910 and at 1975 and estimating their slopes, we find that population was increasing by approximately 22,500 people per year in 1910 and was decreasing by about 12,500 people per year in 1975. See Figure 3.60.

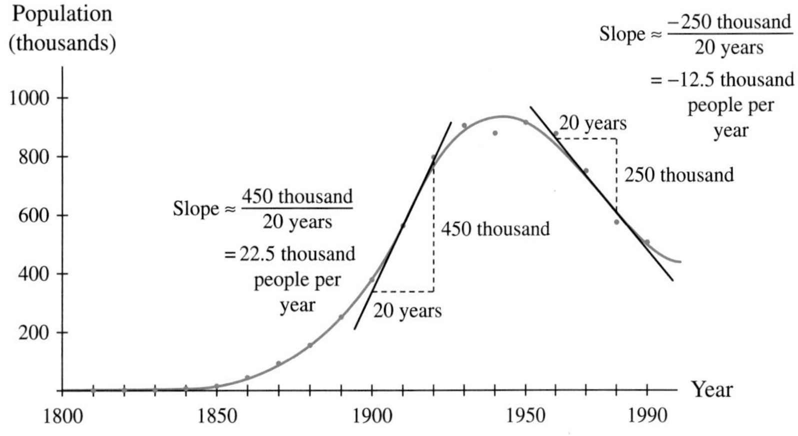

FIGURE 3.60

Now we use all the information from this analysis to sketch the slope graph shown in Figure 3.61.

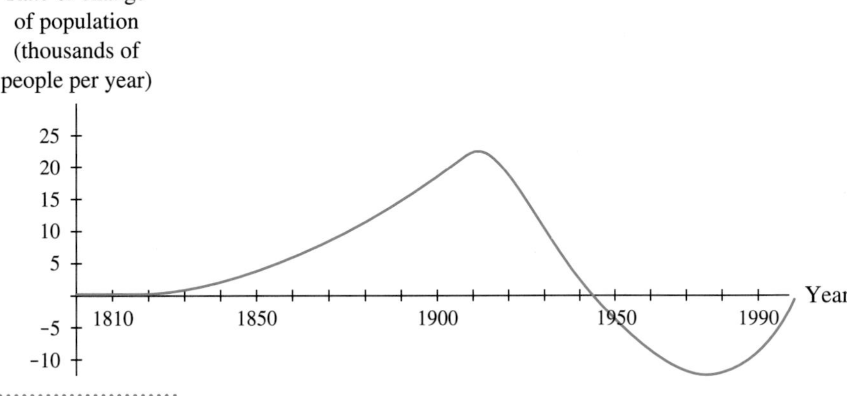

FIGURE 3.61

Points of Undefined Slope

It is possible for the graph of a function to have a point at which the slope does not exist. Remember that the slope of a tangent line is the limit of slopes of approximating secant lines and that we should be able to use secant lines through points either to the left or to the right of the point at which we are estimating the slope to find this limit.[40] If the limits from the left and from the right are not the same, then we say that the derivative does not exist at that point. We depict the nonexistence of the derivative at such points on the slope graph by drawing an open dot on each piece of the slope graph. This is illustrated in Figure 3.62a.

It is possible for there to be a point on a graph at which the derivative does not exist although the slope from the left and the slope from the right are the same. This occurs when the function is not continuous at that point (see Figure 3.62b). If a function is not continuous at a point, then its slope is undefined at that point.

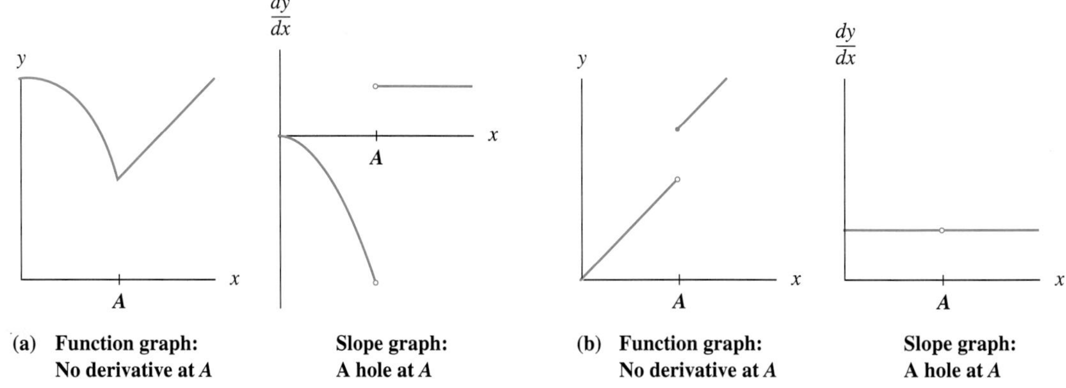

(a) **Function graph: No derivative at *A*** **Slope graph: A hole at *A*** (b) **Function graph: No derivative at *A*** **Slope graph: A hole at *A***

FIGURE 3.62

Students often need extra help in understanding that they should use open circles on a graph to indicate the nonexistence of the derivative at the break point of piecewise continous functions.

Also, points at which the tangent line is vertical (that is, the slope calculation results in a zero in the denominator) are considered to have an undefined slope. The graph of one such function is shown in Figure 3.63.

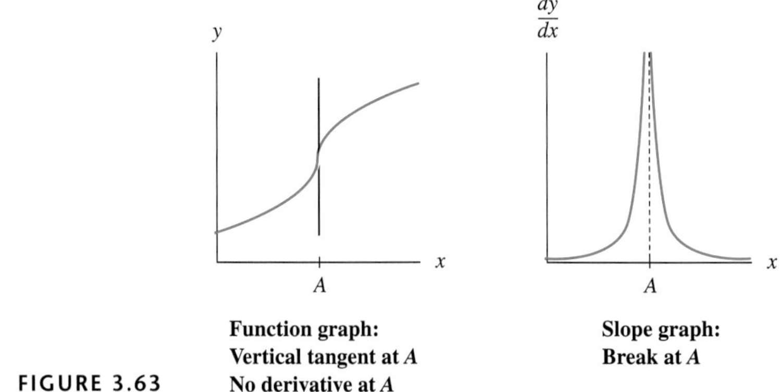

FIGURE 3.63 **Function graph: Vertical tangent at *A* No derivative at *A*** **Slope graph: Break at *A***

Most of the time when there is a break in the slope graph, it is because the original function is piecewise as shown in Figure 3.62. You should be careful when drawing slope graphs of piecewise continuous functions. Figure 3.63 shows a smooth, continuous function with a point at which the derivative is undefined. We do not often encounter such phenomena in real-life applications, but they could happen.

40. It is worth noting that in the case in which a function is defined only to the right or only to the left of a point, then the derivative at that point can be found using secant lines through points only to the right or left.

3.4 Concept Inventory

- *Slope graph = rate-of-change graph = derivative graph*
- *Increasing function ⇒ positive slopes*
- *Decreasing function ⇒ negative slopes*
- *Maximum or minimum of function ⇒ zero slope*
- *Inflection point ⇒ maximum or minimum point on slope graph or point of undefined slope*
- *Points of undefined slope*

3.4 Activities

In Activities 1 through 10, list as many facts as you can about the slopes of the graphs. Then, on the basis of those facts, sketch the slope graph of each function.

1. 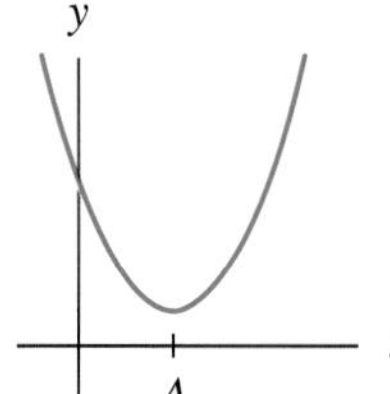

2. 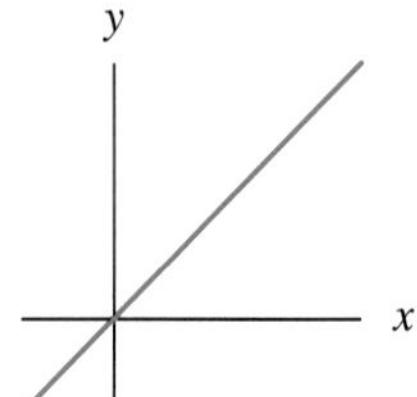

3. 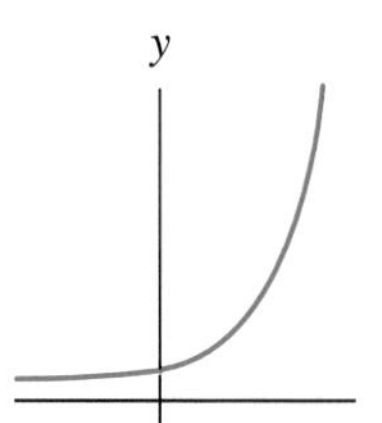

4. 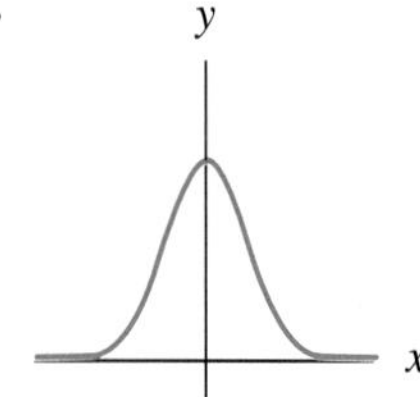

5. 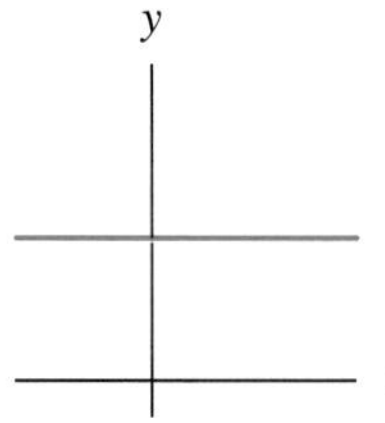

6. 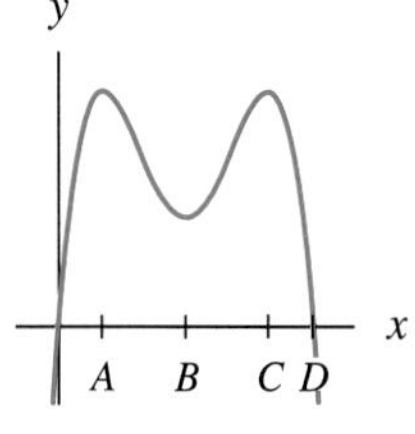

7. 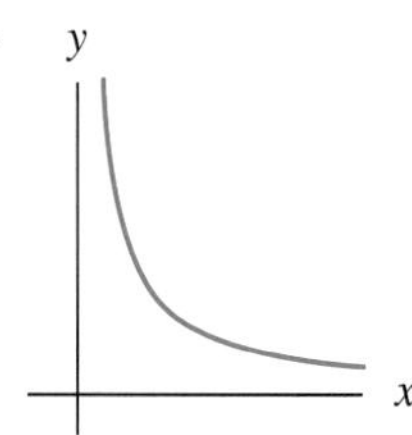

8. 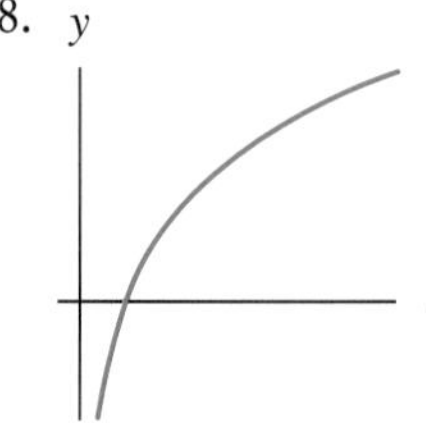

9. 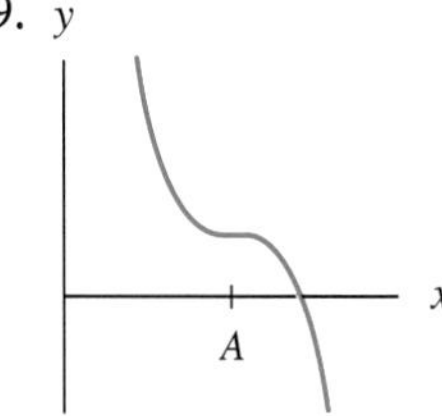

10. 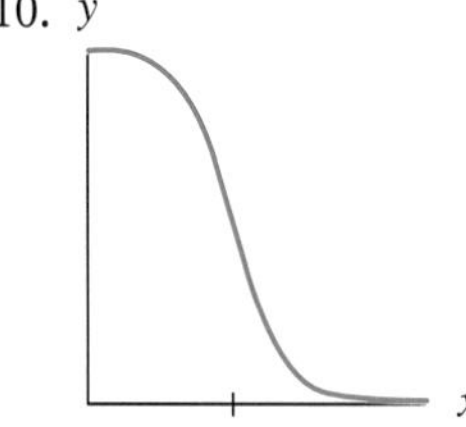

11. a. Sketch a graph that is increasing with increasing slopes. Also sketch its slope graph.
    b. Sketch a graph that is increasing with decreasing slopes. Also sketch its slope graph.

12. The graph in Figure 3.4.11 shows the cumulative amount of capital invested in the cellular phone industry[41] since 1984.

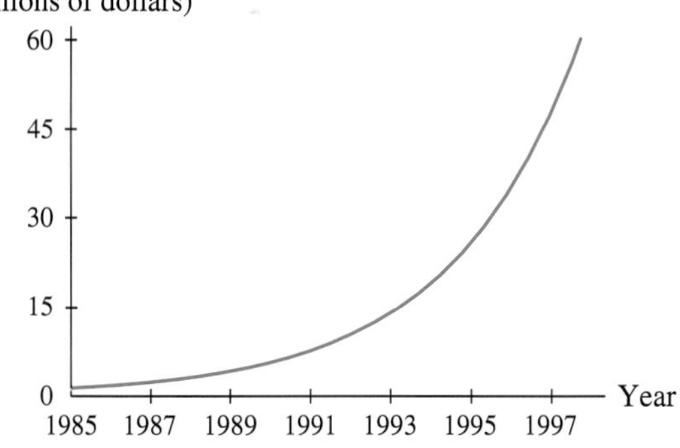

FIGURE 3.4.11

a. Sketch tangent lines for the input values shown in Table 3.17, and estimate their slopes. Record the slopes in Table 3.17.

TABLE 3.17

Year	1985	1989	1993	1997
Slope of tangent line				

41. The Cellular Telecommunications Industry Association.

b. Use the information in Table 3.17 to sketch an accurate rate-of-change graph for the cumulative amount of capital invested in the cellular phone industry. Label the axes with units as well as values.

13. The graph in Figure 3.4.12 shows the average monthly cell phone bill[42] between 1990 and 1998.

FIGURE 3.4.12

Average bill (dollars)

110
90
70
50
30
1990 1992 1994 1996 1998 Year

a. Sketch tangent lines for the input values shown in Table 3.18, and estimate their slopes. Record the slopes in Table 3.18.

TABLE 3.18

Year	1990	1992	1994	1996	1998
Slope of tangent line					

b. Use the information in Table 3.18 to sketch an accurate rate-of-change graph for the average monthly cell phone bill. Label the axes with units as well as values.

14. The graph in Figure 3.4.13 shows the population of Iowa[43] between 1990 and 1998.

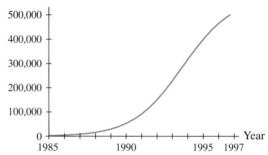

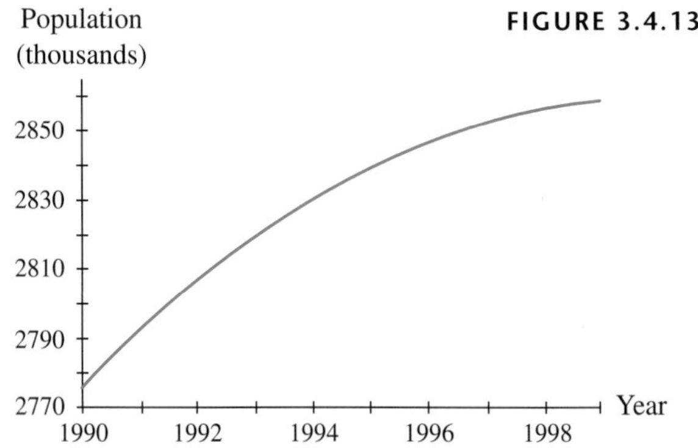

FIGURE 3.4.13

42. Ibid.
43. U.S. Bureau of the Census.

a. Sketch tangent lines for the input values shown in Table 3.19, and estimate their slopes. Record the slopes in Table 3.19.

TABLE 3.19

Year	1990	1992	1994	1996	1998
Slope of tangent line					

b. Use the information in Table 3.19 to sketch an accurate rate-of-change graph for the population of Iowa. Label the axes with units as well as values.

15. The graph in Figure 3.4.14 shows the cumulative number of AIDS cases[44] diagnosed in the United States since 1984.

Cumulative AIDS cases since 1984

500,000
400,000
300,000
200,000
100,000
0
1985 1990 1995 1997 Year

FIGURE 3.4.14

a. Sketch tangent lines for the input values shown in Table 3.20, and estimate their slopes. Record the slopes in Table 3.20.

TABLE 3.20

Year	1985	1988	1991	1994	1997
Slope of tangent line					

b. Use the information in Table 3.20 to sketch an accurate rate-of-change graph for the cumulative number of AIDS cases diagnosed in the United States. Label the axes with units as well as values.

16. The graph in Figure 3.4.15 shows the average annual fuel consumption[45] of vehicles in the United States between 1970 and 1995.

44. Centers for Disease Control.
45. *Statistical Abstract*, 1998.

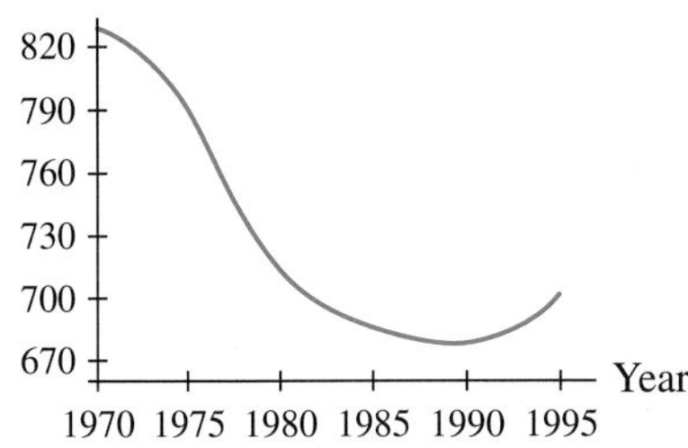

FIGURE 3.4.15

a. Sketch tangent lines for the input values shown in Table 3.21, and estimate their slopes. Record the slopes in Table 3.21.

TABLE 3.21

Year	1970	1975	1980	1985	1990	1995
Slope of tangent line						

b. Use the information in Table 3.21 to sketch an accurate rate-of-change graph for the average annual fuel consumption. Label the axes with units as well as values.

17. The graph in Figure 3.4.16 gives the membership in a campus organization during its first year.

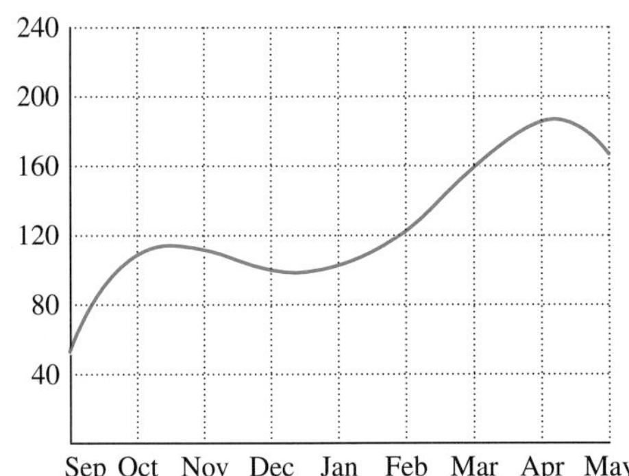

FIGURE 3.4.16

a. Estimate the average rate of change in the membership during the academic year.

b. Estimate the instantaneous rates of change in September, November, February, and April.

c. On the basis of your answers to part *b*, sketch a rate-of-change graph. Label the units on the axes.

d. The membership of the organization was growing most rapidly in September. Not including that month, when was the membership growing most rapidly? What is this point on the membership graph called?

e. Why was the result of the calculation in part *a* of no use in part *c*?

18. The scatter plot in Figure 3.4.17 depicts the number of calls placed each hour since 2 a.m. to a sheriff's department.[46]

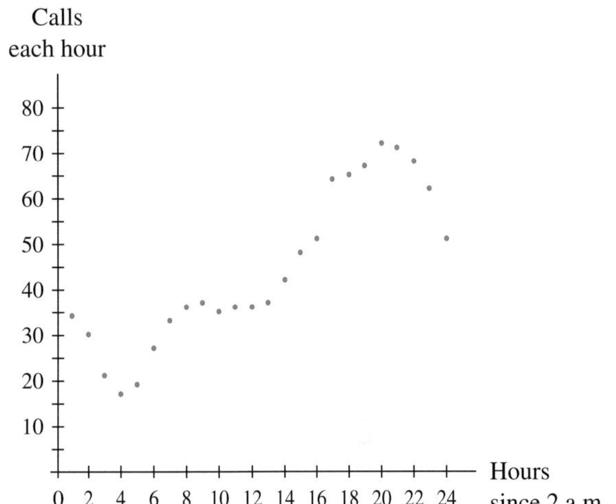

FIGURE 3.4.17

a. Sketch a smooth curve through the scatter plot with no more inflection points than the number suggested by the scatter plot.

b. At what time(s) is the number of calls a minimum? a maximum?

c. Are there any other times when the graph appears to have a zero slope? If so, when?

d. Estimate the slope of your smooth curve at any inflection points.

46. Sheriff's Department of Greenville County, South Carolina.

e. Use the information in parts a through d to sketch a graph depicting the rate of change of calls placed each hour. Label the units on both axes of the rate-of-change graph.

19. The capacity of jails in a southwestern state has been increasing since 1990. The average daily population of one jail[47] can be modeled by

$$j(t) = \begin{cases} 8.101t^3 - 55.53t^2 + 128.8t + 626.8 \text{ inmates} & \text{when } 0 \le t \le 5 \\ 18.8t + 800.6 \text{ inmates} & \text{when } t > 5 \end{cases}$$

where t is the number of years since 1990. Figure 3.4.18 shows a graph of j.

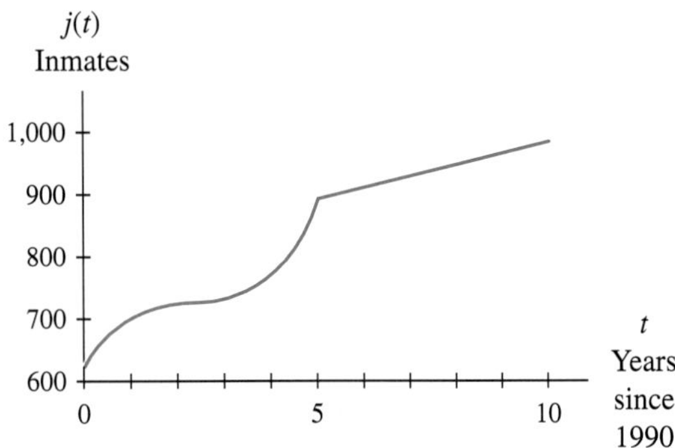

FIGURE 3.4.18

Sketch the slope graph of j. (*Hint:* Numerically estimate $j'(t)$ at $t = 0$ and $t = 4.5$ in order to sketch the slope graph accurately.) Label both the horizontal and the vertical axes.

20. Cattle prices[48] (for choice 450-pound steer calves) from October 1994 through May 1995 can be modeled by

$$p(m) = \begin{cases} -0.0025m^2 + 0.0305m + 0.8405 \text{ dollars per pound} & \text{when } 0 \le m < 3 \\ -0.028m + 0.996 \text{ dollars per pound} & \text{when } 3 \le m \le 7 \end{cases}$$

where m is the number of months since October 1994. A graph of p is given in Figure 3.4.19.

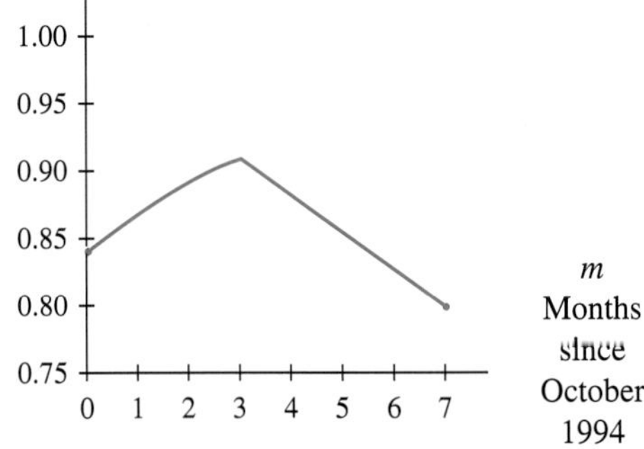

FIGURE 3.4.19

Sketch a slope graph of p. (*Hint:* Numerically estimate $p'(m)$ at $m = 0$ and $m = 2.5$ in order to sketch the slope graph accurately.) Label the horizontal and vertical axes.

21. A graph depicting the average monthly profit for Slim's Used Car Sales for the previous year is shown in Figure 3.4.20.

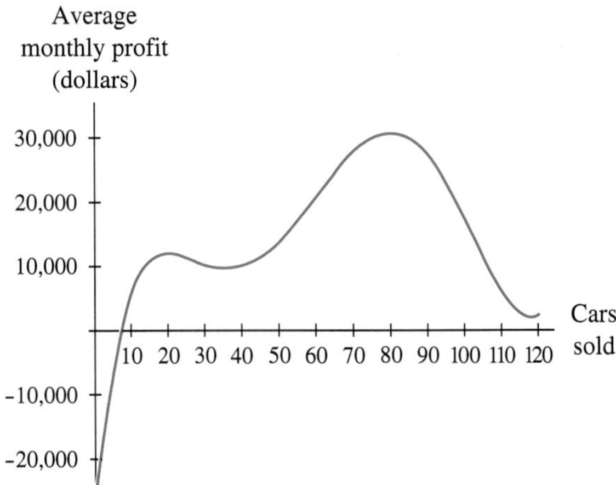

FIGURE 3.4.20

a. Estimate the average rate of change in Slim's average monthly profit if the number of cars he sells increases from 40 to 70 cars.

b. Estimate the instantaneous rates of change at 20, 40, 60, 80, and 100 cars.

c. On the basis of your answers to part b, sketch a rate-of-change graph. Label the units on the axes.

47. Based on data from Washoe County Jail, Reno, Nevada.
48. Based on data from the National Cattleman's Association.

d. For what number of cars sold between 20 and 100 is average monthly profit increasing most rapidly? For what number of cars sold is average monthly profit decreasing most rapidly? What is the mathematical term for these points?

e. Why was the result of the calculation in part *a* of no use in part *c*?

22. Using the graph in Figure 3.3.4 on page 231, carefully estimate the rate of change in deaths due to lung cancer in 1940, 1960, and 1980. Use this information to sketch an accurate rate-of-change graph for deaths due to lung cancer. Label the units on both axes of the derivative graph.

In Activities 23 through 26, indicate the input values for which the graph has no derivative. Explain why the derivative does not exist at those points. Sketch a derivative graph for each of the function graphs.

23.

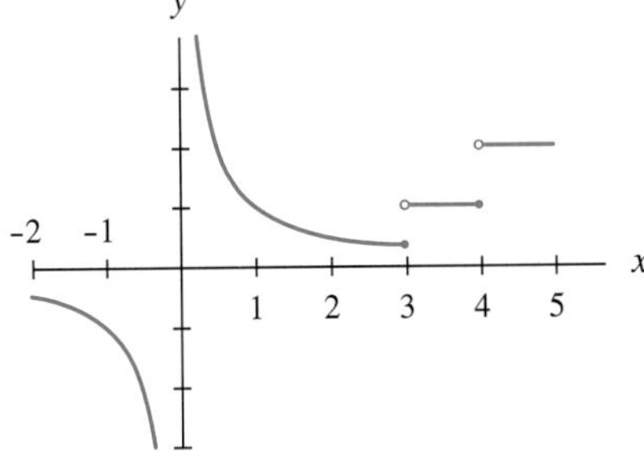

24.

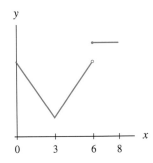

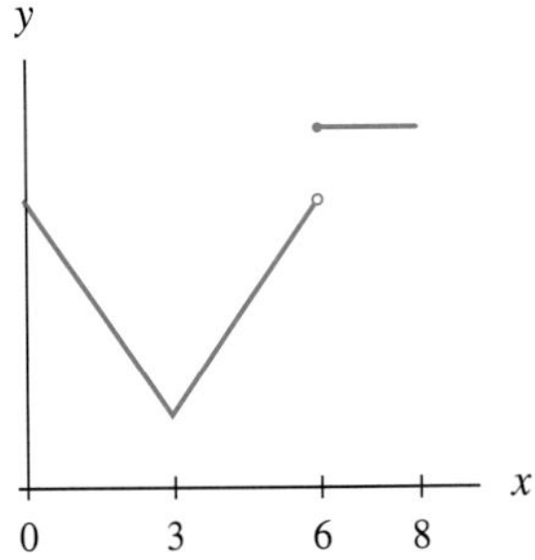

25.

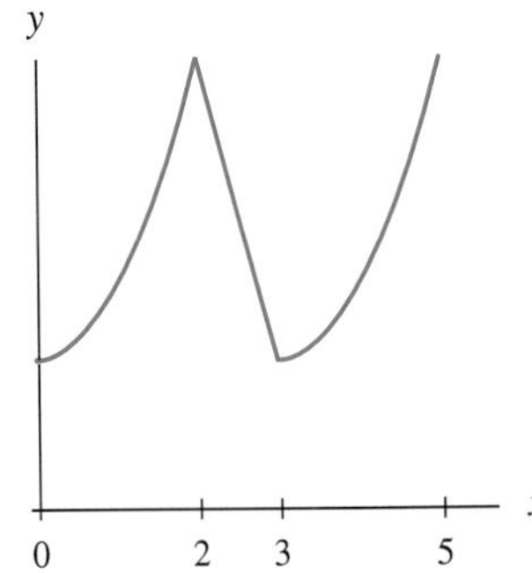

26.

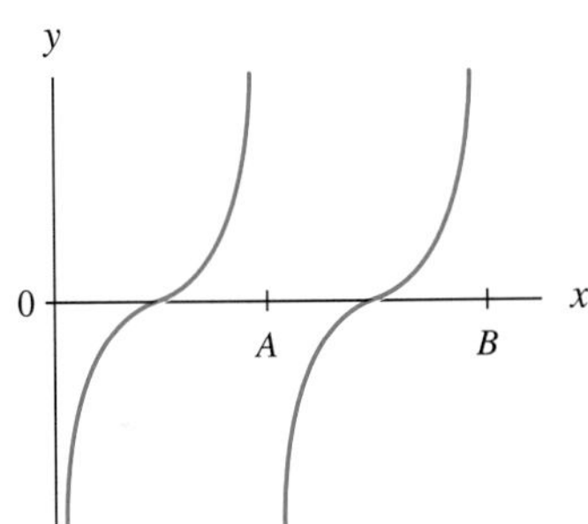

Summary

This chapter is devoted to describing change: the underlying concepts, the language, and proper interpretations.

Change, Average Rate of Change, and Percentage Change

The change in a quantity over an interval is a difference of output values. Apart from describing the actual change in a quantity that occurs over an interval, change can be described as the average rate of change over an interval or as a percentage change. In the reporting of average rates of change, the word *average* must be used, the interval must be stated, and a numerical value expressed in terms of output units per input unit must be included. When interpreting these descriptions of change, you should write a nontechnical sentence that answers the questions when, what, how, and by how much.

The numerical description of an average rate of change has an associated geometric interpretation—namely, the slope of the secant line joining two points on the graph that correspond to the two endpoints of the interval. Such slopes can be determined from discrete scatter plots, continuous graphs, or continuous models.

Instantaneous Rates of Change

Whereas average rates of change indicate how rapidly a quantity changes (on average) over an interval, instantaneous rates of change indicate how rapidly a quantity is changing at a point. The instantaneous rate of change at a point on a graph is simply the slope of the line tangent to the graph at that point. It describes how quickly the output is increasing or decreasing at that point. Given this geometric perspective, it is clear that the use of the word *instantaneous* does not necessarily mean that time units are involved (that is, we may not be speaking of change at an instant in time.)

Tangent Lines

If we zoom in closer and closer to a point on a smooth, continuous graph, we will observe the curve to look more and more like a straight line. This phenomenon is known as local linearity; that is, any graph looks like a line if you are close enough. A line tangent to a graph at a point is the line you see when you zoom in closer and closer to that point.

The line tangent to a graph at a point P can also be thought of as the limiting position of nearby secant lines—that is, secant lines through P and nearby points on the graph. With the exceptions of the line tangent to a continuous graph at an inflection point and the line tangent to each point on a line, lines tangent to a graph do not cut through the graph at the point of tangency. Rather, they include the point of tangency and lie completely on one side of the graph near the point of tangency. A tangent line reflects the tilt, or slope, of the graph at the point of tangency.

By estimating the rise and run of two points on the line, we can use tangent lines to approximate the rate of change of most points on continuous graphs, the exceptions being at sharp points and at points where the tangent line is vertical. Tangent lines can also be used to measure rates of change for piecewise continuous functions, though not usually at the break points.

Derivatives and Percentage Rate of Change

Derivative is the calculus term for (instantaneous) rate of change. Accordingly, all of the following terms are synonymous: derivative, instantaneous rate of change, rate of change, slope of the curve, and slope of the line tangent to the curve.

Three common ways of symbolically referring to the derivative of a function G with respect to x are $\frac{dG}{dx}$, $\frac{d}{dx}[G(x)]$, and $G'(x)$. You should remember to include the proper units on derivatives: output units per input unit. Without the correct units, a derivative is a meaningless number. Rates of change also can be expressed as percentages. A percentage rate of change describes the relative magnitude of the rate.

A continuous or piecewise continuous function is needed to calculate values of derivatives. However, if we are given a data set, then we can quickly estimate a rate of change by finding the slope between the nearest points an equal distance away from, and on either side of, our point of interest. This slope, which is called a symmetric difference quotient, usually gives a good approximation to the rate of change.

Drawing Slope Graphs

The smooth, continuous graphs that we use to model real-life data have slopes (derivatives) at every point on the graph except at points that have vertical tangent lines. When these slopes (derivatives) are plotted, they usually form a smooth, continuous graph—the slope graph (rate-of-change graph, or derivative graph) of the original graph. We can also obtain slope graphs for piecewise continuous models, although the slope graph is usually not defined at the points where the model is divided. Slope graphs tell us a great deal about the change that is occurring on the original graph.

Concept Check

Can you	To practice, try	
• Find and interpret change, percentage change, and average rates of change		
using data?	Section 3.1	Activity 13
using graphs?	Section 3.1	Activity 21
using equations?	Section 3.1	Activity 17
• Find the slope of a secant line?	Section 3.1	Activity 19
• Understand the relationship between secant lines and tangent lines?	Section 3.2	Activities 1, 3
• Accurately sketch tangent lines?	Section 3.2	Activity 19
• Use tangent lines to estimate rates of change?	Section 3.2	Activity 31
• Use symmetric difference quotients to estimate rates of change?	Section 3.2	Activity 31
• Understand derivative notation?	Section 3.3	Activity 9
• Correctly interpret derivatives?	Section 3.3	Activity 15
• Find and interpret percentage rate of change?	Section 3.3	Activity 21
• Sketch a general rate-of-change graph?	Section 3.4	Activities 3, 9
• Use tangent lines to sketch an accurate slope graph?	Section 3.4	Activity 15

Review Test

1. Answer the following questions about the graph shown in Figure 3.64.

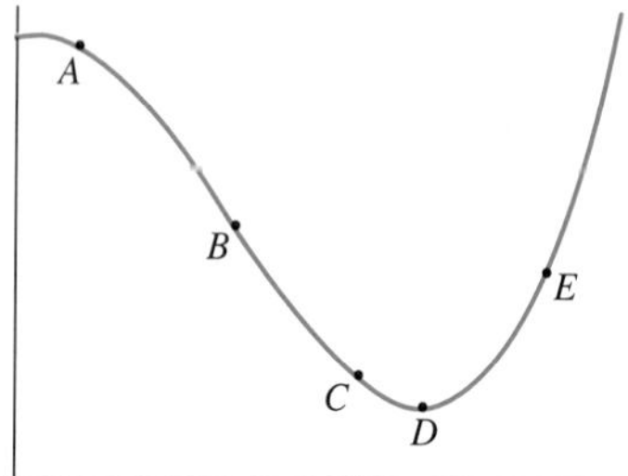

FIGURE 3.64

a. List the labeled points at which the slope appears to be (i) negative, (ii) positive, and (iii) zero.

b. If B is the inflection point, what is the relationship between the steepness at B and the steepness at the points A, C, and D on the graph?

c. For each of the labeled points, will a tangent line at the point lie above or below the graph?

d. Sketch tangent lines at points A, B, and E.

e. Suppose the graph represents the speed of a roller coaster (in feet per second) as a function of the number of seconds after the roller coaster reached the bottom of the first hill.

i. What are the units on the slopes of tangent lines? What common word is used to describe the quantity measured by the slope in this context?

ii. When, according to the graph, was the roller coaster slowing down?

iii. When, according to the graph, was the roller coaster speeding up?

iv. When was the roller coaster's speed the slowest?

v. When was the roller coaster slowing down most rapidly?

f. Sketch a slope graph of the graph in Figure 3.64.

2. The numbers of Visa cards worldwide[49] for 1995 through 1998 are shown in Table 3.22.

TABLE 3.22

Year	1995	1996	1997	1998
Visa cards (millions)	442	510	600	656

a. Find the average rate of change in the number of Visa cards between 1995 and 1998. Interpret your answer.

b. Find the percentage change in the number of Visa cards between 1997 and 1998. Interpret your answer.

3. Figure 3.65 shows the graph of a model for the number of Dell Computer Corporation[50] employees between 1992 and 1999.

a. Sketch a secant line between the points with inputs 1993 and 1997. Describe the information that the slope of this line provides.

b. Sketch a tangent line at the point corresponding to 1998. Describe the information that the slope of this line provides.

c. Estimate the rate of change and percentage rate of change in the number of employees in 1998. Interpret your answers.

d. Estimate the average rate of change in the number of employees between 1993 and 1997. Interpret your answer.

4. $T(x)$ is the number of seconds that it takes an average athlete to swim 100 meters free style at age x years.

a. Write sentences interpreting $T(22) = 49$ and $T'(22) = -0.5$.

b. What does a negative derivative indicate about a swimmer's time?

49. *Visa Press Release,* **www.visa.com.** (Access date: 11/23/99).

50. Hoover's Online Company Capsules.

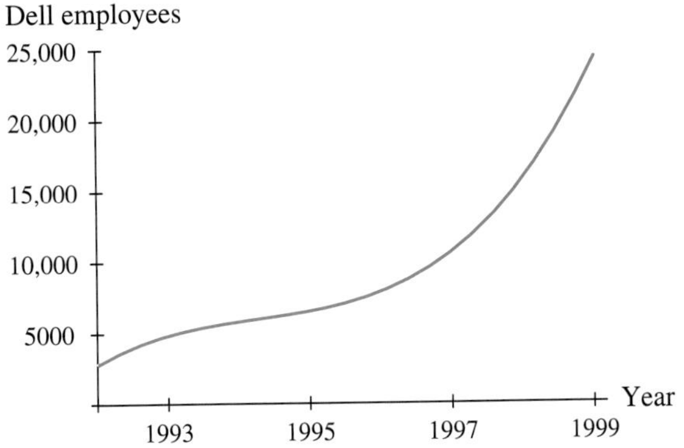

FIGURE 3.65

5. The amount of time that parents spend with their children each week,[51] based on data between 1969 and 1996, can be modeled by

$$T(x) = -0.852x + 1837.296 \text{ hours in year } x$$

   a. What is the rate of change of T?

   b. Find the change, average rate of change, and percentage change in the time parents spend with their children, as described by the model, between 1980 and 2000.

   c. Assess the validity of the following statements.

      i. The rate of change of a linear model is always constant.

      ii. The percentage rate of change of a linear model is always constant.

      iii. The average rate of change of a linear model is always constant.

51. *Time*, June 21, 1999, p. 86.

Project 3.1 Fee-Refund Schedules

Setting

Some students at many colleges and universities enroll in courses and then later withdraw from them. Such students may have part-time status upon withdrawing. Part-time students have begun questioning the fee-refund policy, and a public debate is taking place. Recently, the student senate passed a resolution condemning the current fee-refund schedule. Then, the associate vice president issued a statement claiming that further erosion of the university's ability to retain student fees would reduce course offerings. The Higher Education Commission has scheduled hearings on the issue. The Board of Trustees has hired your firm as consultants to help them prepare their presentation.

Tasks

1. Examine the current fee-refund schedule for your college or university. Present a graph and formula for the current refund schedule. Critique the refund schedule.

   Create alternative fee-refund schedules that include at least two quadratic plans (one concave up and one concave down), an exponential plan, a logistic plan, a no-refund plan, and a complete-refund plan. (Hint: Linear models have constant first differences. What is true about quadratic and exponential models?) For each plan, present the refund schedule in a table, in a graph, and with an equation. Critique each plan from the students' viewpoint and from that of the administration.

   Select the nonlinear plan that you believe to be the best choice from both the students' and the administration's perspective. Outline the reasons for your choice.

2. Estimate the rate of change of your selected equation for withdrawals after 1 week, 3 weeks, and 5 weeks. Include any other times that are indicated by your school's schedule. Interpret the rates of change in this context. How might the rate of change influence the administration's view of the model you chose? Would the administration consider a different model more advantageous? If so, why? Why did you not propose it as your model of choice?

Reporting

1. Prepare a written report of your results for the Board of Trustees. Include scatter plots, models, and graphs. Include in an appendix the reasoning that you used to develop each of your models.
2. Prepare a press release for the college or university to use when it announces the adoption of your plan. The press release should be succinct and should answer the questions *who, what, when, where,* and *why*. Include the press release in your report to the Board.
3. (Optional) Prepare a brief (15-minute) presentation on your work. You will be presenting it to members of the Board of Trustees of your college or university.

Project 3.2 Doubling Time

Setting

Doubling time is defined as the time it takes for an investment to double. Doubling time is calculated by using the compound interest formula $A = P\left(1 + \frac{r}{n}\right)^{nt}$ or the continuously compounded interest formula $A = Pe^{rt}$. An approximation to doubling time can be found by dividing 72 by $100r$. This approximating technique is known as the **Rule of 72.**

Dr. C. G. Bilkins, a nationally known financial guru, has been criticized for giving false information about doubling time and the Rule of 72 in seminars. Your team has been hired to provide mathematically correct information for Dr. Bilkins to use in future seminar presentations.

Tasks

1. Construct a table of doubling times for interest rates from 2% to 20% (in increments of 2%) when interest is compounded annually, semiannually, quarterly, monthly, and daily. Construct a table of doubling-time approximations for interest rates of 2% through 20% when using the Rule of 72. Devise similar rules for 71, 70, and 69. Then construct tables for these rules. Examine the tables and determine the best approximating rule for interest compounded semiannually, quarterly, monthly, and daily. Justify your choices. For each interest compounding listed above, compare percent errors when using the Rule of 72 and when using the rule you choose. Percent error is $\frac{\text{estimate} - \text{true value}}{\text{true value}} \cdot 100\%$. Comment on when the rules overestimate, when they underestimate, and which is preferable.
2. Dr. Bilkins is interested in knowing how sensitive doubling time is to changes in interest rates. Estimate rates of change of doubling times at 2%, 8%, 14%, and 20% when interest is compounded quarterly. Interpret your answers in a way that would be meaningful to Dr. Bilkins, who knows nothing about calculus.

Reporting

1. Prepare a written report for Dr. Bilkins in which you discuss your results in Tasks 1 and 2. Be sure to discuss whether Dr. Bilkins should continue to present the Rule of 72 or present other rules that depend on the number of times interest is compounded.
2. Prepare a document for Dr. Bilkins's speechwriter to insert into the seminar presentation (which Dr. Bilkins reads from a Teleprompter). It should include (a) a brief summary of how to estimate doubling time using an approximation rule and (b) a statement about the error involved in using the approximation. Keep in mind that Dr. Bilkins wishes to avoid further allegations of misinforming seminar participants. Also include a brief statement summarizing the sensitivity of doubling time to fluctuations in interest rates. Include the document in your written report.
3. (Optional) Prepare a brief (15-minute) presentation of your study. You will be presenting it to Dr. Bilkins and the speechwriter. Your presentation should be only a summary, but you need to be prepared to answer any technical questions that may arise.

4

Determining Change: Derivatives

Photodisc

Concept Application

The aging of the American population may be the demographic change that has the greatest impact on our society over the next several decades. Given a model for the projected number of senior Americans (65 years of age or older), the function and its derivative can be used to answer questions such as

- What is the projected number of senior Americans in 2030?
- How rapidly will that number be changing in 2030?
- What is the estimated percentage rate of change in the number of senior Americans in 2030?

You will be able to answer these questions using the model given in Activity 23 of Section 4.3 and the derivative rules presented in this chapter.

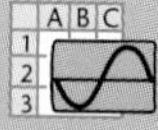

This symbol directs you to the supplemental technology guides. The different guides contain step-by-step instructions for using certain graphing calculators or spreadsheet software to work the particular example or discussion marked by the technology symbol.

This symbol directs you to the Calculus Concepts *web site, where you will find algebra review, links to data updates, additional practice problems, or some other helpful resource material.*

Concept Objectives

This chapter will help you understand the concepts of

- The slope of a tangent line as the limiting value of slopes of secant lines
- Slopes of piecewise continuous functions
- Rules for finding rate-of-change formulas

and you will learn to

- Numerically estimate rates of change
- Use limits to find rate-of-change formulas algebraically
- Apply derivative rules to many types of functions to find rate-of-change formulas quickly

We have described change in terms of rates: average rates, instantaneous rates, and percentage rates. Of these three, instantaneous rates are the most important in our study of calculus.

In this chapter, we turn our attention to determining instantaneous rates of change. Because instantaneous rates of change are slopes of tangent lines, we first consider the numerical estimation of these slopes. Next we generalize the numerical method to an algebraic method that gives us an analytic description—a formula of sorts—for the derivative of an arbitrary function.

Finally, we consider some rules for derivatives: the Sum Rule, the Chain Rule, and the Product Rule. These rules provide the foundation needed to work with more complicated functions that we often encounter in the course of real-life investigations of change.

4.1 Numerically Finding Slopes

Finding Slopes by the Numerical Method

By now you should have a firm graphical understanding of rates of change. However, sketching tangent lines is an imprecise method of determining rates of change. Although approximations are often sufficient, there are times when we need to find an exact answer.

Consider the relatively simple problem of finding the slope of the graph of $f(x) = 2\sqrt{x}$ at $x = 1$. Part of the graph of $f(x) = 2\sqrt{x}$ is shown in Figure 4.1a. Take a few moments to sketch carefully a line tangent to the graph at $x = 1$ and estimate its slope. You should find that the tangent line at $x = 1$ has slope approximately 1. See Figure 4.1b.

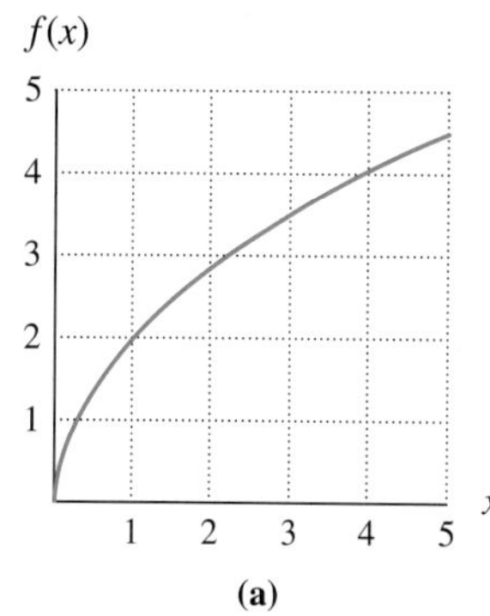

(a)

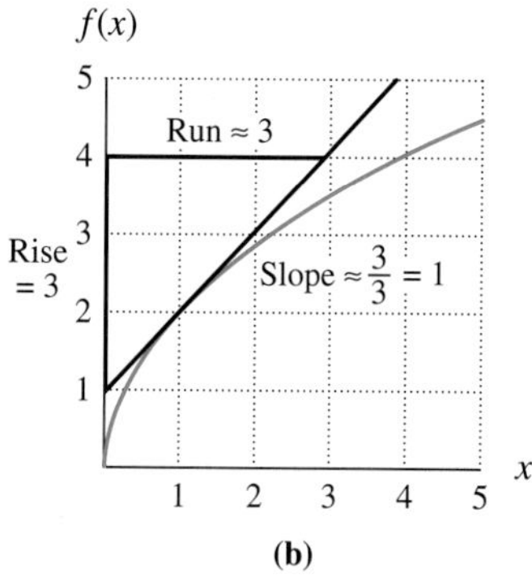

(b)

FIGURE 4.1

Another, more precise, method of estimating the slope of the graph of $f(x) = 2\sqrt{x}$ at $x = 1$ uses a technique introduced in Section 3.2. Recall that the tangent line at a point is the limiting position of secant lines through the point of tangency and other increasingly close points. In other words, *the slope of the tangent line is the limiting value of the slopes of nearby secant lines.*

To illustrate, we begin by finding the slope of the secant line on the graph of $f(x) = 2\sqrt{x}$ through $x = 1$ and $x = 1.5$. (Note that $x = 1.5$ is an arbitrarily chosen value that is close to $x = 1$.) A graph of the secant line is shown in Figure 4.2. Its slope is calculated as follows:

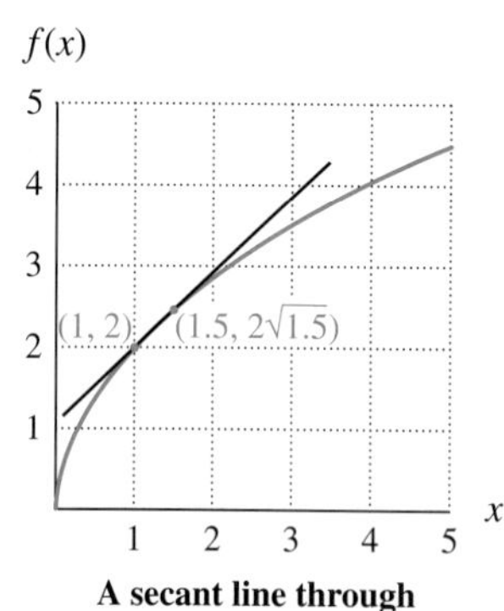

A secant line through $x = 1$ and $x = 1.5$

FIGURE 4.2

Point at $x = 1$: $\left(1, 2\sqrt{1}\right) = (1, 2)$

Point at $x = 1.5$: $\left(1.5, 2\sqrt{1.5}\right) \approx (1.5, 2.449489743)$

$$\text{Slope} \approx \frac{2.449489743 - 2}{1.5 - 1} = 0.8989794856$$

This value is an approximation to the slope of the tangent line at $x = 1$. To obtain a better approximation, we must choose a point closer to $x = 1$ than $x = 1.5$, say $x = 1.1$. (This is also an arbitrary choice.) See Figure 4.3.

Point at $x = 1$: $\left(1, 2\sqrt{1}\right) = (1, 2)$

Point at $x = 1.1$: $\left(1.1, 2\sqrt{1.1}\right) \approx (1.1, 2.097617696)$

$$\text{Slope} \approx \frac{2.097617696 - 2}{1.1 - 1} = 0.9761769634$$

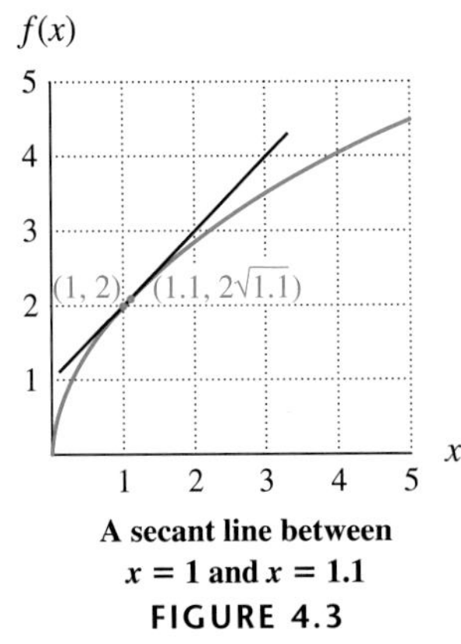

A secant line between $x = 1$ and $x = 1.1$
FIGURE 4.3

This is a better approximation to the slope of the tangent line at $x = 1$ than the one from the previous calculation. To get an even better approximation, we need only choose a closer point, such as $x = 1.01$.

$$\text{Point at } x = 1: \quad \left(1, 2\sqrt{1}\right) = (1, 2)$$

$$\text{Point at } x = 1.01: \quad \left(1.01, 2\sqrt{1.01}\right) \approx (1.01, 2.009975124)$$

$$\text{Slope} \approx \frac{2.009975124 - 2}{1.01 - 1} = 0.9975124224$$

We also use $x = 1.001$.

$$\text{Point at } x = 1: \quad \left(1, 2\sqrt{1}\right) = (1, 2)$$

$$\text{Point at } x = 1.001: \quad \left(1.001, 2\sqrt{1.001}\right) \approx (1.001, 2.00099975)$$

$$\text{Slope} \approx \frac{2.00099975 - 2}{1.001 - 1} = 0.9997501248$$

As we choose points increasingly close to $x = 1$, what do you observe about the slopes of the secant lines shown in Figure 4.4?

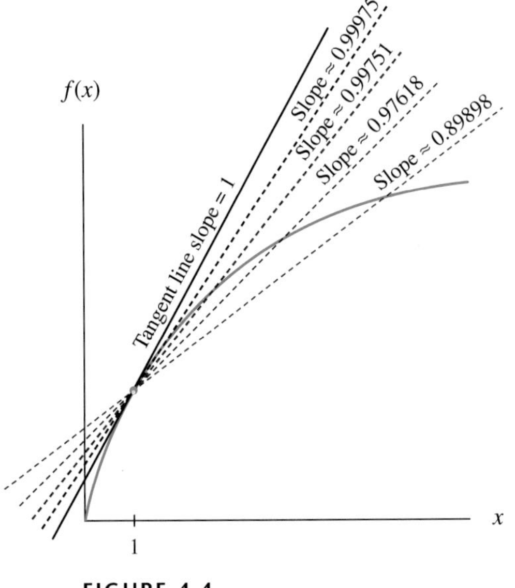

FIGURE 4.4

$x = 1.5$	Secant line slope ≈ 0.8989794856
$x = 1.1$	Secant line slope ≈ 0.9761769634
$x = 1.01$	Secant line slope ≈ 0.9975124224
$x = 1.001$	Secant line slope ≈ 0.9997501248

Following the pattern in the slope values, we conclude that for $x = 1.0001$, the slope to six decimal places will be 0.999975, for $x = 1.00001$, the slope to seven decimal places will be 0.9999975, and so on. Thus the slopes of the secant lines when we use points to the right of $x = 1$ appear to be approaching 1. You may have noticed by now that we are numerically estimating the limit of the slopes of secant lines. Recall from Section 1.4 that a limit exists only if the limit from the left and the limit from the right are equal. For this reason, in order to conclude that the slope of the tangent line is 1, we must also consider the limit of the slopes of secant lines using points to the left of $x = 1$. Choosing the x-values 0.5, 0.9, 0.99, and 0.999, we obtain the following secant line slopes:

$x = 0.5$	Secant line slope ≈ 1.171572875
$x = 0.9$	Secant line slope ≈ 1.026334039
$x = 0.99$	Secant line slope ≈ 1.002512579
$x = 0.999$	Secant line slope ≈ 1.000250125

Again, note the pattern in the slope values. For $x = 0.9999$, we can conclude that the slope to six decimal places will be 1.000025. For $x = 0.99999$, the slope to seven decimal places will be 1.0000025. Thus the slopes of secant lines using points to the left of $x = 1$ appear to be approaching 1. Because the limit of slopes using points to the left of $x = 1$ appears to be the same as the limit using points to the right of $x = 1$, we estimate that the slope of the line tangent to the graph of $f(x) = 2\sqrt{x}$ at $x = 1$ is 1. We will use algebraic methods in the next section to verify that $f'(1) = 1$.

In this case, the graphical and numerical methods for estimating the slope of the tangent line yield similar results. However, Example 1 shows that calculating the slopes of nearby secant lines generally yields a much more precise result than sketching a tangent line and estimating its slope.

EXAMPLE 1 *The 1949 Polio Epidemic*

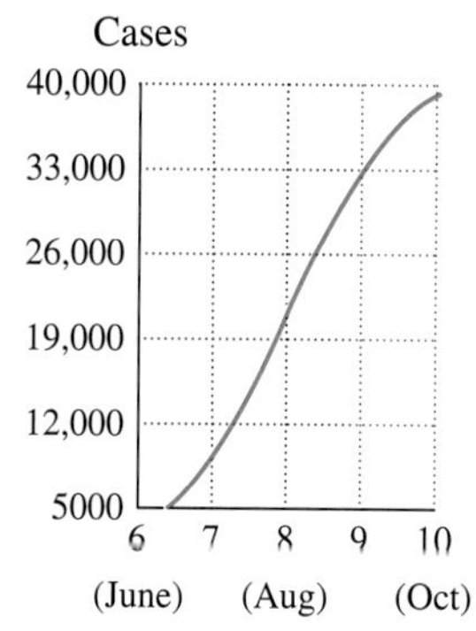

FIGURE 4.5

The number of polio cases in 1949 can be modeled[1] by the equation

$$\text{Cases} = \frac{42{,}183.911}{1 + 21{,}484.253e^{-1.248911t}}$$

where $t = 1$ on January 31, 1949, $t = 2$ on February 28, 1949, etc. How rapidly was the number of polio cases growing at the end of August 1949? Report the answer to the nearest whole number.

Solution First, note that the question asks for the slope of the tangent line at $t = 8$. One method of approximating the slope of the tangent line is to sketch the tangent line and estimate its slope. On the graph in Figure 4.5, draw the tangent line at $t = 8$, and estimate the slope. If you accurately sketched the tangent line and were careful in reading two points off that line, you should have found that the slope of the tangent line is somewhere between 12,000 and 14,000 cases per month.

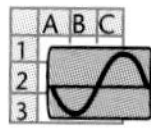

To obtain a more precise estimate of the slope of the tangent line, we calculate slopes of nearby secant lines. We choose increasingly close points to both the right and the left of the point where $t = 8$: (8, 21,262.9281). We calculate slopes until they remain constant to one decimal place beyond the desired accuracy for two or three calculations. You should verify each of the following computations.

Notice that pertinent digits in the table are underlined. We underline the digits that give the slope to one decimal position beyond our desired accuracy once we see these digits do not change. Once we have two or three underlined values in each column, we stop calculating the left and right slopes and report an answer.

Points to the left	Points to the right
Point at $t = 7.9$: (7.9, 19,946.95986) $\text{Slope} = \frac{19{,}946.95986 - 21{,}262.9281}{7.9 - 8}$ Slope ≈ 13,159.68	Point at $t = 8.1$: (8.1, 22,577.56655) $\text{Slope} = \frac{22{,}577.56655 - 21{,}262.9281}{8.1 - 8}$ Slope ≈ 13,146.38
Point at $t = 7.99$: (7.99, 21,131.22196) $\text{Slope} = \frac{21{,}131.22196 - 21{,}262.9281}{7.99 - 8}$ Slope ≈ 13,170.62	Point at $t = 8.01$: (8.01, 21,394.62098) $\text{Slope} = \frac{21{,}394.62098 - 21{,}262.9281}{8.01 - 8}$ Slope ≈ 13,169.28
Point at $t = 7.999$: (7.999, 21,249.75795) $\text{Slope} = \frac{21{,}249.75795 - 21{,}262.9281}{7.999 - 8}$ Slope ≈ <u>13,170.1</u>9	Point at $t = 8.001$: (8.001, 21,276.09819) $\text{Slope} = \frac{21{,}276.09819 - 21{,}262.9281}{8.001 - 8}$ Slope ≈ 13,170.05
Point at $t = 7.9999$: (7.9999, 21,261.61112) $\text{Slope} = \frac{21{,}261.61112 - 21{,}262.9281}{7.9999 - 8}$ Slope ≈ <u>13,170.1</u>3	Point at $t = 8.0001$: (8.0001, 21,264.24514) $\text{Slope} = \frac{21{,}264.24514 - 21{,}262.9281}{8.0001 - 8}$ Slope ≈ <u>13,170.1</u>2
Point at $t = 7.99999$: (7.99999, 21,262.79643) $\text{Slope} = \frac{21{,}262.79643 - 21{,}262.9281}{7.99999 - 8}$ Slope ≈ <u>13,170.1</u>2 ≈ 13,170	Point at $t = 8.00001$: (8.00001, 21,263.05983) $\text{Slope} = \frac{21{,}263.05983 - 21{,}262.9281}{8.00001 - 8}$ Slope ≈ <u>13,170.1</u>2 ≈ 13,170

1. Based on data from *Twelfth Annual Report*, The National Foundation for Infantile Paralysis, 1949.

We reported this answer to the nearest case because those were the specified instructions. However, this instruction was not given because the output is measured in polio cases.

There is no hard-and-fast rule for the accuracy used in reporting rates of change. It is common to report rates of change to greater accuracy than the accuracy of the output. If an accurate enough numerical investigation were conducted, a slope of 13,170.1 cases per month (or 13,170.12 cases per month) would make sense.

Whether points to the left or right of $t = 8$ are chosen, it seems clear that the slopes are approaching approximately 13,170. That is, the limit of slopes of secant lines using points to the left of $t = 8$ is approximately 13,170, and the limit of slopes of secant lines using points to the right of $t = 8$ is approximately 13,170. Thus we conclude that the slope of the line tangent to the graph at $t = 8$ is approximately 13,170 cases per month. (The actual slope correct to three decimal places is 13,170.122.)

Note that this numerical method of calculating slopes of nearby secant lines in order to estimate the slope of the tangent line at $t = 8$ gives a much more precise answer than graphically estimating the slope of the tangent line. (Be certain that you keep all decimal places in your calculations and enough decimal places in your recorded slope values to be able to see the trend.) We conclude that at the end of August 1949, the number of polio cases was growing at the rate of 13,170 cases per month.

Finding Slopes of Piecewise Functions

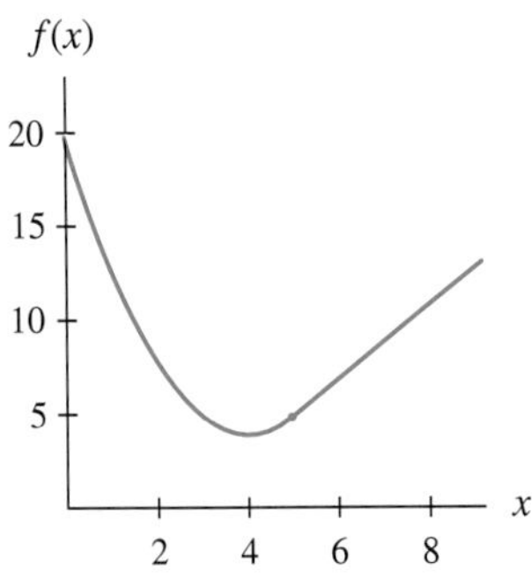

FIGURE 4.6

Remember that when we model data piecewise, we usually end up with one or more points where the graph of the function is not continuous. Even if a piecewise function is continuous, it probably will not be smooth because the pieces of the function generally do not have the same slope where they join. The derivative of the function does not exist at a point where the function is not continuous or at a point where the function is continuous but is not smooth. In order for the derivative to exist at a point where the function is continuous, the slopes that we find by approximating with secant lines through points to the left and through points to the right must be equal. For example, the function represented in Figure 4.6 is defined by

$$f(x) = \begin{cases} x^2 - 8x + 20 & \text{when } 0 \le x \le 5 \\ 2x - 5 & \text{when } x > 5 \end{cases}$$

To determine whether the rate of change of f exists at $x = 5$, we must first determine whether the function is continuous at $x = 5$. We do this by evaluating the two pieces of the function at $x = 5$ in order to determine $\lim_{x \to 5^-} f(x)$ and $\lim_{x \to 5^+} f(x)$:

You may wish to review the definition of a continuous function that is given in Section 1.4.

$$\lim_{x \to 5^-} f(x) = (5)^2 - 8(5) + 20 = 5$$

$$\lim_{x \to 5^+} f(x) = 2(5) - 5 = 5$$

Because the limits as x approaches 5 from the left and from the right are equal and are the same as the function value at $x = 5$, we see that both pieces of the function join at (5, 5), and the function is continuous.

Next we must determine if the function is smooth at $x = 5$. A continuous function is **smooth** if it has no sharp points. A sharp point occurs when two pieces of a function have different slopes at the point where they meet. In order for a function to be smooth at a point, the slopes of the left and right portions of the function must be the same at the point where the pieces join. In the case of the function f, we must check the limiting value of the slopes of secant lines for points to the left of $x = 5$ and the limiting value of the slopes of secant lines for points to the right of $x = 5$ to see whether these limiting values are equal. The equation to the left of $x = 5$ is quadratic, and we use this in our approximations from the left. The limit from the left appears to be 2.

Close point to left	Slope
(4.9, 4.81)	1.9
(4.99, 4.9801)	1.99
(4.999, 4.998001)	1.999

Now we determine the slope from the right of $x = 5$ by using the equation $f(x) = 2x - 5$. This portion of the function is a line whose slope is 2.

The limit from the left appears to be 2, which is the same as the limit from the right, so our numerical investigation suggests that the derivative is 2 at $x = 5$. In this case, even though the function is defined as piecewise, it is continuous and smooth where the split occurs. The slope of a tangent line drawn at $x = 5$ is 2. In Example 2, we see a function that is not continuous where the two pieces of the function are split.

EXAMPLE 2 *Comcast Employees*

At the end of the 1990s, Comcast Corporation was the third largest U.S. cable communications and broadcast operator. The numbers of Comcast employees[2] between 1991 and 1998 are given in Table 4.1.

TABLE 4.1

Year	1991	1992	1993	1994	1995	1996	1997	1998
Thousands of employees	3.7	5.3	5.4	6.7	12.2	16.4	17.6	17.0

a. Align the input data as the number of years since 1990, and find a piecewise model for the data. Split the data in 1994.

b. Sketch a graph of the piecewise model.

c. Do the two pieces of the graph meet in 1994?

d. Does the derivative of the piecewise model exist at $x = 4$?

e. Estimate the rate of change of the number of employees in 1994.

f. Numerically estimate the rate of change in 1992.

Again, it is important to point out to your students that in part *e* we are not asking them to estimate the rate of change of the function *E* (because this quantity does not exist). Instead, we are asking for an estimate of rate of change of the underlying real-life situation (the number of Comcast employees), which does exist.

Solution

a. The number of Comcast employees can be modeled by the equation

$$E(x) = \begin{cases} 0.45x^3 - 3.45x^2 + 8.8x - 2.1 \text{ thousand people} & \text{when } 1 \le x \le 4 \\ -1.086x^2 + 15.629x - 38.534 \text{ thousand people} & \text{when } 4 < x \le 8 \end{cases}$$

where x represents the number of years since 1990.

b.

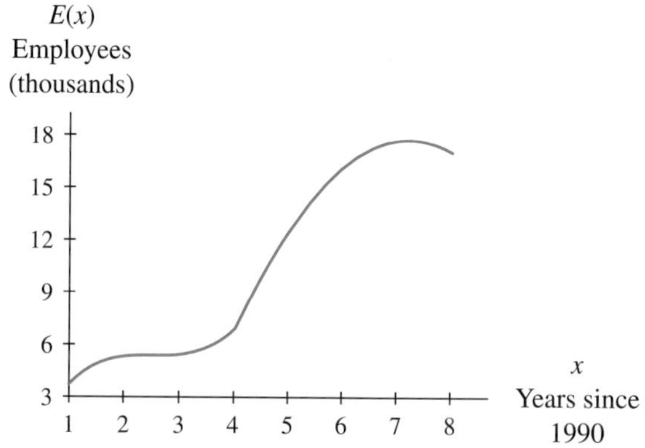

$E(x)$
Employees
(thousands)
18
15
12
9
6
3
1 2 3 4 5 6 7 8
x
Years since
1990

FIGURE 4.7

2. Hoover's Online Historical Financials.

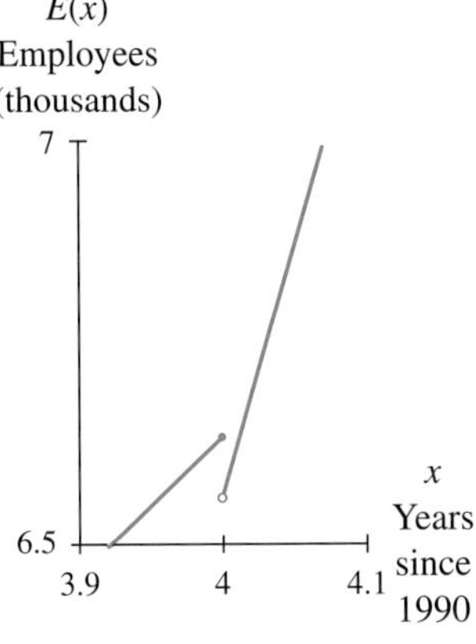

FIGURE 4.8

c. If we look at a magnified view of the graph with input values from $x = 3.9$ to $x = 4.1$ and output values from 6.5 to 7.0, we see how the graphs of the two pieces of the equation behave near $x = 4$. From Figure 4.8, we see that the two pieces of the graph do not meet at $x = 4$. The graph suggests that the function is not continuous at $x = 4$.

We algebraically verify that the function is not continuous at $x = 4$ by finding the left and right limits of f as x approaches 4.

$$\lim_{x \to 4^-} f(x) = 0.45(4)^3 - 3.45(4)^2 + 8.8(4) - 2.1 = 6.7$$
$$\lim_{x \to 4^+} f(x) = -1.086(4)^2 + 15.629(4) - 38.534 \approx 6.61$$

Thus the limit as x approaches 4 does not exist, and the function is not continuous at $x = 4$.

d. Because the two pieces of this graph do not join at $x = 4$, the function is not continuous at $x = 4$. Hence the derivative of the function does not exist at that point.

e. Even though the derivative does not exist at $x = 4$ because our piecewise model is not continuous at that point, we can still estimate the rate of change in the number of employees in 1994 by using a symmetric difference quotient (introduced in Section 3.2). We use the slope of the line through the data points (1993, 5.4) and (1995, 12.2) to estimate the rate of change in 1994:

$$\text{Slope} = \frac{12.2 - 5.4 \text{ thousand people}}{1995 - 1993} = 3.4 \text{ thousand people per year}$$

This result tells us that the number of employees was increasing by approximately 3.4 thousand people per year in 1994.

It is important that you understand that this value is not an estimate of the rate of change of the function, because that quantity does not exist. The symmetric difference quotient is an estimate of the rate of change of the underlying situation—in this case, the number of employees.

f. Because the model is continuous in 1992, we use the cubic portion of the graph to estimate the rate of change numerically. We use the point (2, 5.3) on the cubic graph and the points shown in Table 4.2 to calculate secant line slopes.

In Table 4.2 we have not underlined any digits because these slope values exhibit a pattern. Your students should be able to guess that the next secant line slope on the left in Table 4.2 would be 0.4000075 and the next value on the right would be 0.3999925. This observation should demonstrate that they see the pattern in the slope values and allow them to conclude that the limit is 0.4 thousand employees per year.

TABLE 4.2

Points to the left	Secant line slope	Points to the right	Secant line slope
(1.99, 5.2959)	0.4075	(2.01, 5.303925)	0.3925
(1.999, 5.299599)	0.40075	(2.001, 5.300399)	0.39925
(1.9999, 5.29995999)	0.400075	(2.0001, 5.30003999)	0.399925

It appears that from the left and right, the slopes of secant lines are approaching 0.4. We therefore estimate that in 1992 the number of employees was growing at a rate of 0.4 thousand employees per year, or 400 employees per year.

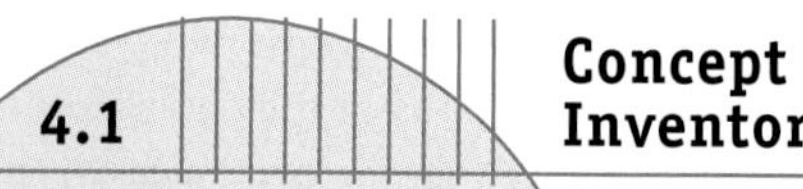

4.1 Concept Inventory

- *Slope of a tangent line = the limiting value of slopes of secant lines*
- *Numerical method of estimating the slope of a tangent line*
- *Slopes of piecewise functions*
- *The derivative does not exist at a point if the function is not continuous at that point or if the function is continuous but not smooth; that is, if the limiting value of slopes of nearby secant lines from the left does not equal the limiting value of slopes of secant lines from the right.*

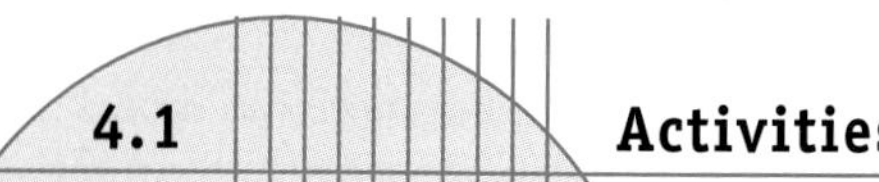

4.1 Activities

1. a. Sketch a line tangent to the graph shown in Figure 4.1.1 at the point corresponding to $x = 2$, and estimate its slope.

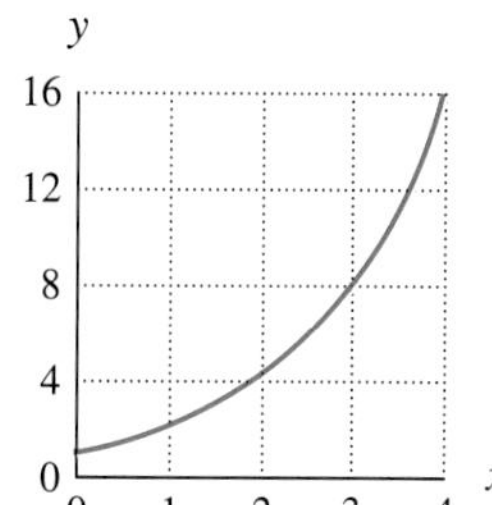

FIGURE 4.1.1

b. The equation of the graph in Figure 4.1.1 is $y = 2^x$. Use the equation to estimate numerically the slope of the line tangent to the graph at $x = 2$.

2. a. Sketch a line tangent to the graph shown in Figure 4.1.2 at the point corresponding to $x = 3$, and estimate its slope.

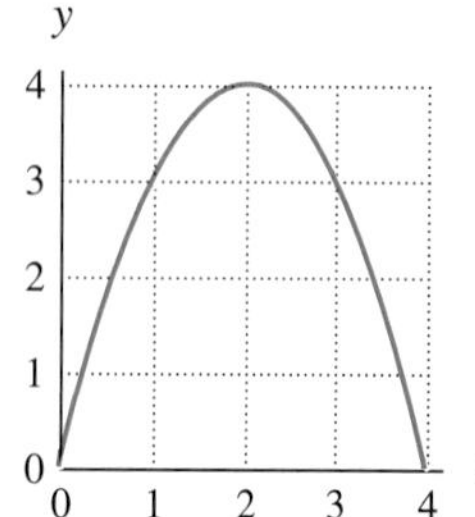

FIGURE 4.1.2

b. The equation of the graph in Figure 4.1.2 is $y = -x^2 + 4x$. Use the equation to estimate numerically the slope of the line tangent to the graph at $x = 3$.

You may wish to specify a desired accuracy in the numerical estimates of tangent line slopes in these activities. Otherwise, be prepared to accept any reasonable answer.

3. a. Numerically estimate the limit of slopes of secant lines on the graph of $g(x) = x^3 + 2x^2 + 1$ between the point corresponding to $x = 2$ and close points to the left of $x = 2$.

b. Numerically estimate the limit of slopes of secant lines on the graph of $h(x) = x^2 + 16x$ between the point corresponding to $x = 2$ and close points to the right of $x = 2$.

c. Combine the functions g and h to form the piecewise continuous function

$$f(x) = \begin{cases} x^3 + 2x^2 + 1 & \text{when } x \le 2 \\ x^2 + 16x & \text{when } x > 2 \end{cases}$$

Is the graph of f continuous at $x = 2$? Is the graph of f smooth at $x = 2$?

d. Is it possible to sketch a line tangent to the graph of f at $x = 2$? If so, what is its slope? If not, why not?

4. a. What is the slope of $f(t) = 4t + 15$ at $t = 3$?

b. Numerically estimate the limit of slopes of secant lines on the graph of $h(t) = 8(1.5^t)$ between the point corresponding to $t = 3$ and close points to the right of $t = 3$.

c. Combine the functions f and h to form the piecewise continuous function

$$g(t) = \begin{cases} 4t + 15 & \text{when } t \le 3 \\ 8(1.5^t) & \text{when } t > 3 \end{cases}$$

Is the graph of g continuous at $t = 3$? Is the graph of g smooth at $t = 3$?

d. Is it possible to sketch a line tangent to the graph of g at $t = 3$? If so, what is its slope? If not, why not?

5. The graph[3] in Figure 4.1.3 depicts the number of AIDS cases diagnosed between 1993 and 1997. The equation for the graph is Cases $= -1.85x^3 + 30.31x^2 - 168.17x + 385.05$ thousand where x is the number of years since 1990.

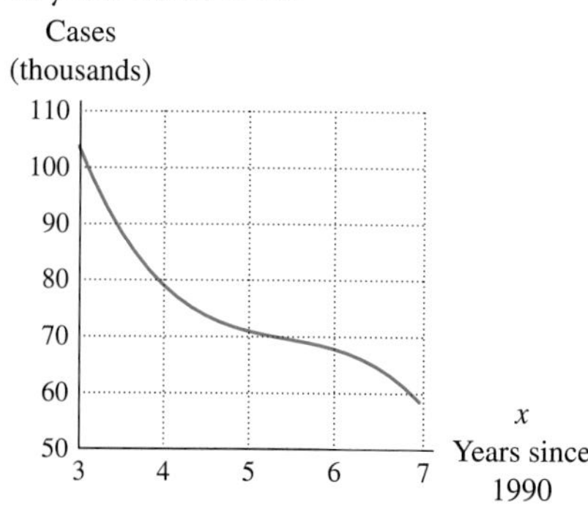

FIGURE 4.1.3

3. Based on data from *Statistical Abstract*, 1998.

a. Sketch the tangent line at $x = 4$, and estimate its slope.

b. Numerically estimate the slope at $x = 4$ using at least three carefully chosen, increasingly close points on either side of $x = 4$.

c. Interpret your answer to part *b* as a rate of change.

6. The balance in a savings account is shown in the graph in Figure 4.1.4 and is given by the equation Balance $= 1500(1.0407^t)$ dollars, where t is the number of years since the principal was invested.

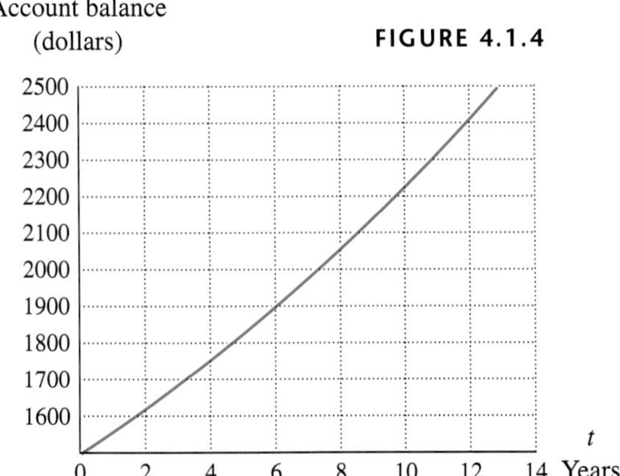

FIGURE 4.1.4

a. Using only the graph, estimate how rapidly the balance is growing after 10 years.

b. Use the equation to investigate numerically the rate of change of the balance when $t = 10$.

c. Which of the two methods (part *a* or part *b*) is more accurate? Support your answer by listing the different estimations that had to be made during each method.

7. The percentage of households between 1980 and 1996 with VCRs can be modeled[4] by the equation

$$P(t) = \frac{77.9}{1 + 39.5e^{-0.4984t}} \text{ percent}$$

where t is the number of years since 1980. A graph of the function is shown in Figure 4.1.5.

a. Use the graph to estimate $\frac{dP}{dt}$ when $t = 7$.

b. Use the equation to investigate $P'(7)$ numerically.

c. Interpret your answer to part *b*.

d. Discuss the advantages and disadvantages of using the two methods (in parts *a* and *b*) for finding derivatives.

4. Based on data from *Statistical Abstract*, 1998.

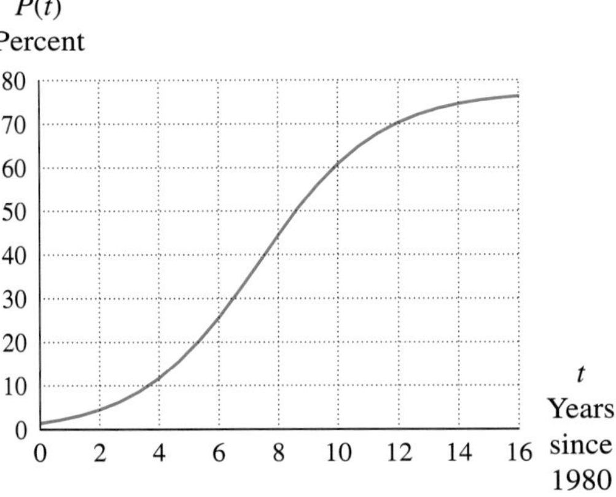

FIGURE 4.1.5

8. The time it takes an average athlete to swim 100 meters freestyle at age x years can be modeled[5] by the equation $T(x) = 0.181x^2 - 8.463x + 147.376$ seconds.

a. Use the numerical method to find the rate of change of the time for a 13-year-old swimmer to swim 100 meters freestyle.

b. Is a 13-year-old swimmer's time improving or getting worse as the swimmer gets older?

9. Annual U.S. factory sales (in billions of dollars) of electronics from 1990 through 1996 can be modeled[6] by the equation

$$S(t) = -0.23t^3 + 2.257t^2 - 1.51t + 42.8 \text{ billion dollars}$$

where t is the number of years since 1990.

a. Estimate the derivative of S when $t = 4$.

b. Interpret your answer to part *a*.

10. Table 4.3 on page 262 gives the number of cotton gins[7] in operation in the early 1900s.

a. Use a symmetric difference quotient to estimate the rate of change in the number of cotton gins in 1915.

b. Find a quadratic model for the data.

c. Use the equation to estimate numerically the derivative in 1915.

d. Compare your answers to parts *a* and *c*, and discuss whether part *a* or part *c* gives a more accurate representation of the change that was occurring in 1915.

5. Based on data from *Swim World*, August 1992.
6. Based on data from *Statistical Abstract*, 1998.
7. *Cotton Ginnings in the United States Crop of 1970*, U.S. Department of Commerce, Economics and Statistics Administration, A30–90.

TABLE 4.3

Year	Cotton gins	Year	Cotton gins
1900	29,214	1910	26,234
1901	29,254	1911	26,349
1902	30,948	1912	26,279
1903	30,218	1913	24,749
1904	30,337	1914	24,547
1905	29,038	1915	23,162
1906	28,709	1916	21,624
1907	27,592	1917	20,351
1908	27,598	1918	19,259
1909	26,669	1919	18,815

11. The per capita disposable personal income (in thousands of 1992 dollars) in the United States[8] is given for the years shown in Table 4.4.

TABLE 4.4

Year	Per capita DPI (thousands of 1992 dollars)
1981	15.0
1983	15.3
1985	16.7
1987	17.2
1989	17.9
1991	17.8
1993	18.2
1995	18.9
1997	19.5

a. Use a symmetric difference quotient to estimate the rate of change of per capita disposable personal income in 1995.

b. Find a linear model for this data set. What is the slope of the model? Compare this slope to your answer in part *a*, and discuss the information each contributes to your understanding of the change in per capita disposable personal income in 1995.

12. Table 4.5 gives the number of students per computer in U.S. public schools[9] from 1985 through 1997.

TABLE 4.5

Year	Students per computer	Year	Students per computer
1985	75	1992	18
1986	50	1993	16
1987	37	1994	14
1988	32	1995	11
1989	25	1996	10
1990	22	1997	8
1991	20		

a. Use a symmetric difference quotient to estimate how rapidly the number of students per computer was declining in 1990.

b. Find an exponential model for the data. (*Hint:* Vertically shift the data.)

c. Use the equation to estimate numerically how rapidly the number of students per computer was declining in 1990.

d. Compare your answers to parts *a* and *c*. Which answer is more accurate? Support your choice.

13. The consumer price index[10] (CPI) for college tuition between 1990 and 1997 is shown in Table 4.6 (1982–1984 = 100).

a. Use a symmetric difference quotient to approximate the rate of change of the consumer price index for college tuition in 1995.

b. Find quadratic and log models for the data.

c. Use both models and at least three carefully chosen close points from each direction to estimate numerically the rate of change of the consumer price index for college tuition in 1995.

d. Compare and contrast the ease and accuracy of the two methods you used in parts *a* and *c*.

TABLE 4.6

Year	CPI
1990	175.0
1991	192.8
1992	213.5
1993	233.5
1994	249.8
1995	264.8
1996	279.8
1997	294.1

14. Discuss the advantages and disadvantages of finding rates of change graphically and numerically. Include in your discussion a brief description of when each method might be appropriate to use.

8. *Statistical Abstract*, 1998.
9. *World Almanac and Book of Facts*, ed. Robert Famighetti (Mahwah, NJ: PRIMEDIA Reference Inc., 1999).
10. *Statistical Abstract*, 1998.

15. Cattle prices (for choice 450-pound steer calves) from April 1998 through December 1998 can be modeled[11] by the equation

$$p(m) = \begin{cases} 0.3172m^3 - 2.0820m^2 - 1.7895m + 98.6398 \text{ dollars per 100 pounds} & \text{when } 0 \le m \le 5 \\ 1.669m + 68.904 \text{ dollars per 100 pounds} & \text{when } 5 < m \le 8 \end{cases}$$

where m is the number of months since April 1998.

a. What is the limiting value of slopes of secant lines of the graph of p from the left of $m = 5$?

b. What is the slope of the portion of the graph of p to the right of $m = 5$?

c. What do your answers to parts *a* and *b* tell you about the derivative of p at $m = 5$?

d. Estimate the rate of change of cattle prices in September of 1998.

16. The capacity of jails in a southwestern state has been increasing since 1990. The average daily population of one of the jails can be modeled[12] by

$$j(t) = \begin{cases} 8.101t^3 - 55.53t^2 + 128.8t + 626.8 \text{ inmates} & \text{when } 0 \le t \le 5 \\ 18.8t + 800.6 \text{ inmates} & \text{when } t > 5 \end{cases}$$

where t is the number of years since 1990.

a. What is the limiting value of slopes of secant lines of the graph of j from the left of $t = 5$?

b. What is the slope of the portion of the graph of j to the right of $t = 5$?

c. What do your answers to parts *a* and *b* tell you about the derivative of j at $t = 5$?

d. Estimate the rate of change of jail population in 1995.

17. Let $P(x) = 1.02^x$ deutsche marks be the profit from the sale of x mountain bikes. On January 5, 2001, P deutsche marks were worth $D(P) = \frac{P}{0.4899}$ American dollars. Assume that this conversion applies today.

a. Write a function for profit in dollars from the sale of x mountain bikes.

b. What is the profit in deutsche marks and in dollars from the sale of 400 mountain bikes?

c. How quickly is profit (in dollars) changing when 400 mountain bikes are sold?

18. Refer to the functions P and D in Activity 17.

a. Write a function giving average profit per mountain bike for the sale of x mountain bikes in deutsche marks.

b. Write a function for average profit in dollars.

c. How quickly is average profit (in dollars) changing when 400 mountain bikes are sold?

4.2 Algebraically Finding Slopes

When we have enough carefully chosen close points to observe the trend in the secant line slopes, the numerical method is a fairly good way to find a slope to a specified accuracy. For this reason, the process of numerically estimating slope is valuable, but keep in mind that it gives only an estimation of the actual slope. However, we can generalize the numerical method to develop an algebraic method that will give the exact slope of a tangent line at a point.

Finding Slopes Using the Algebraic Method

Consider finding the slope of $f(x) = \frac{1}{7}x^2 + 3x$ at $x = 2$. We begin with the point $\left(2, 6\frac{4}{7}\right)$. Instead of choosing close x-values such as 1.9, 2.1, 2.01, and so on, we simply call the close value $x = 2 + h$. (Note that if $h = 0.1$, then $x = 2.1$; if $h = 0.01$,

11. Based on data from *USDA: Cattle and Beef Industry Statistics*, March 1999.
12. Based on data from Washoe County Jail, Reno, Nevada.

$x = 2.01$; if $h = -0.01$, $x = 1.99$, and so on.) The output value that corresponds to $x = 2 + h$ is

$$f(2 + h) = \tfrac{1}{7}(2 + h)^2 + 3(2 + h)$$
$$= \tfrac{1}{7}(4 + 4h + h^2) + 6 + 3h = 6\tfrac{4}{7} + 3\tfrac{4}{7}h + \tfrac{1}{7}h^2$$

Next we find the slope of the secant line between the point of tangency $\left(2, 6\tfrac{4}{7}\right)$ and the close point $\left(2 + h, 6\tfrac{4}{7} + 3\tfrac{4}{7}h + \tfrac{1}{7}h^2\right)$.

$$\text{Slope of secant line} = \frac{\left(6\tfrac{4}{7} + 3\tfrac{4}{7}h + \tfrac{1}{7}h^2\right) - 6\tfrac{4}{7}}{(2 + h) - 2} = \frac{3\tfrac{4}{7}h + \tfrac{1}{7}h^2}{h}$$

We now have a formula for the slope of the secant line through the points at $x = 2$ and $x = 2 + h$. We can apply the formula to obtain slopes at points increasingly close to $x = 2$.

Close point to left	h	Slope $= \dfrac{3\frac{4}{7}h + \frac{1}{7}h^2}{h}$	Close point to right	h	Slope $= \dfrac{3\frac{4}{7}h + \frac{1}{7}h^2}{h}$
1.9	-0.1	3.55714	2.1	0.1	3.58571
1.99	-0.01	3.57	2.01	0.01	3.57286
1.999	-0.001	3.57129	2.001	0.001	3.57157
1.9999	-0.0001	3.5714 1	2.0001	0.0001	3.5714 4
1.99999	-0.00001	3.5714 3	2.00001	0.00001	3.5714 3

We arbitrarily chose 3-decimal place accuracy for the answer in this illustration. We underline the digits that give one decimal position beyond our desired accuracy once we see that these digits do not change. Once we have two or three underlined values (in this table, 2 values) in each column, we stop calculating the left and right slopes and report an answer.

The slopes become increasingly close to approximately 3.571.

Numerically evaluating the secant line slope formula for smaller and smaller values of h gives us a good picture of where the slopes are headed, but it does not give us the exact answer. By noting that h approaches zero as the close point approaches 2, we can find the limit of the secant line slope formula as h approaches zero:

$$\lim_{h \to 0} \frac{3\tfrac{4}{7}h + \tfrac{1}{7}h^2}{h}$$

The secant line slope formula is not continuous at $h = 0$. However, recall the Cancellation Rule for limits presented in Section 1.4:

> If the numerator and denominator of a rational function share a common factor, then the new function obtained by algebraically canceling the common factor has all limits identical to those of the original function.

If we factor h out of the numerator and cancel this common factor, then we can find the limit by evaluating the new expression at $h = 0$ because the new expression $\left(3\tfrac{4}{7} + \tfrac{1}{7}h\right)$ is continuous at $h = 0$.

$$\lim_{h \to 0} \frac{h\left(3\tfrac{4}{7} + \tfrac{1}{7}h\right)}{h} = \lim_{h \to 0}\left(3\tfrac{4}{7} + \tfrac{1}{7}h\right) = 3\tfrac{4}{7} + \tfrac{1}{7}(0) = 3\tfrac{4}{7}$$

On the basis of this limit calculation, we state that the slope of the line tangent to $f(x) = \tfrac{1}{7}x^2 + 3x$ at $x = 2$ is exactly $3\tfrac{4}{7}$.

This method of finding a formula for the slope of a secant line in terms of h and then determining the limiting value of the formula as h approaches zero is important because it always yields the *exact* slope of the tangent line. Unfortunately, the method is primarily for polynomial functions and not for exponential, logarithmic, or logistic functions.

EXAMPLE 1 *Foreign-Born U.S. Residents*

The percentage of people residing in the United States who were born abroad can be modeled[13] by the equation

$$f(x) = 0.00005x^3 - 0.0034x^2 - 0.14x + 14.86 \text{ percent}$$

where x is the number of years since 1910. (*Note:* The model is based on data from 1910 through 1997.)

a. Find $\frac{df}{dx}$ at $x = 80$ by writing an expression for the slope of a secant line in terms of h and then evaluating the limit as h approaches 0.

b. Interpret $\frac{df}{dx}$ at $x = 80$ in the context given.

Solution

a. First, we find the output of f when $x = 80$.

$$f(80) = 7.5$$

Second, we write an expression for $f(x)$ when $x = 80 + h$.

$$f(80 + h) = 0.00005(80 + h)^3 - 0.0034(80 + h)^2 - 0.14\,(80 + h) + 14.86$$

algebra

We simplify this expression as much as possible, using the facts that

$$(80 + h)^2 = 6400 + 160h + h^2$$

and

$$(80 + h)^3 = (80 + h)^2\,(80 + h) = 512{,}000 + 19{,}200h + 240h^2 + h^3$$

Thus

$$f(80 + h) = 7.5 + 0.276h + 0.0086h^2 + 0.00005h^3$$

Third, we find the slope of the secant line through the two close points $(80, f(80))$ and $(80 + h, f(80 + h))$ as

$$\text{Slope of secant} = \frac{f(80 + h) - f(80)}{(80 + h) - 80}$$
$$= \frac{(7.5 + 0.276h + 0.0086h^2 + 0.00005h^3) - 7.5}{(80 + h) - 80}$$

Again, simplify this as much as possible.

$$\text{Slope of secant line (continued)} = \frac{0.276h + 0.0086h^2 + 0.00005h^3}{h}$$
$$= \frac{h(0.276 + 0.0086h + 0.00005h^2)}{h}$$

Note that we do not cancel the hs in the slope formula until we perform the limit calculation because canceling the hs changes the secant line slope function (it fills in a hole). However, we can cancel the hs in the process of finding the limit because a hole in a graph does not affect the value of any limits.

Finally, we find the slope of the tangent line (the derivative) at $x = 80$ by evaluating the limit of the slope of the secant line as h approaches 0.

13. Based on data from *World Almanac and Book of Facts*, ed. Robert Famighetti (Mahwah, NJ: PRIMEDIA Reference Inc., 1999).

$$\frac{df}{dx} = \text{slope of tangent line} = \lim_{h\to 0} \frac{h(0.276 + 0.0086h + 0.00005h^2)}{h}$$
$$= \lim_{h\to 0} (0.276 + 0.0086h + 0.00005h^2)$$
$$= 0.276$$

b. In 1990, the percentage of the United States population who were foreign-born was increasing at a rate of 0.276 percentage point per year.

A General Formula for Derivatives

This algebraic method of finding rates of change may seem tedious, but its real value is not in finding a slope at a particular point. With a simple modification, we can apply this technique to find general formulas for derivatives. These rate-of-change (or slope) formulas can be used to find rates of change for many input values.

To illustrate, consider $y = x^2$. Because we desire a general equation for any x-value, we use (x, x^2) as the point of tangency. This is the same idea as the algebraic method in Example 1 and the preceding discussion, but there we worked with a numerical value for x. Next we choose a close point. We use $x + h$ as the x-value of the close point and find the y-value by substituting $x + h$ into the function: $y = (x + h)^2 = x^2 + 2xh + h^2$. Thus the original point is (x, x^2), and a close point is $(x + h, x^2 + 2xh + h^2)$. Now we find the slope of the secant line between these two points.

$$\text{Slope of the secant line} = \frac{(x^2 + 2xh + h^2) - x^2}{(x + h) - x} = \frac{2xh + h^2}{h}$$

Finally, we determine the limiting value of the secant line slope as h approaches 0.

$$\lim_{h\to 0} \frac{2xh + h^2}{h} = \lim_{h\to 0}(2x + h) = 2x + 0 = 2x$$

Therefore, the slope formula for $y = x^2$ is $\frac{dy}{dx} = 2x$. Using this formula we find that the slope of the graph of $y = x^2$ is 6 at $x = 3$, -12 at $x = -6$, 0 at $x = 0$, 1 at $x = 0.5$, and so on. This formula also gives an equation for the slope graph of $y = x^2$.

This method can be generalized to obtain a formula for the derivative of an arbitrary function.

Four-Step Method to Find $f'(x)$

Given a function f, the equation for the derivative with respect to x can be found as follows:

1. Begin with a typical point $(x, f(x))$.
2. Choose a close point $(x + h, f(x + h))$.
3. Write a formula for the slope of the secant line between the two points.

$$\text{Slope} = \frac{f(x + h) - f(x)}{(x + h) - x} = \frac{f(x + h) - f(x)}{h}$$

It is important at this step to simplify the slope formula.

4. Evaluate the limit of the slope as h approaches 0.

$$\lim_{h\to 0} \frac{f(x + h) - f(x)}{h}$$

This limiting value is the derivative formula at each input where the limit exists.

Thus we have the following derivative formula (slope formula, rate-of-change formula) for an arbitrary function:

Derivative Formula

If $y = f(x)$, then the derivative $\frac{dy}{dx}$ is given by the formula

$$\frac{dy}{dx} = \lim_{h \to 0} \frac{f(x + h) - f(x)}{h}$$

provided the limit exists.

EXAMPLE 2 *Women in the Civilian Labor Force*

The number of women in the United States civilian labor force[14] from 1930 through 1990 can be modeled as

$$W(t) = 0.011t^2 + 0.101t + 10.817 \text{ million women}$$

where t is the number of years since 1930.

a. Use the limit definition of the derivative (the Four-Step Method) to develop a formula for the rate of change of the number of women in the labor force.

b. How quickly was the number of women in the U.S. civilian labor force growing in 1975?

Solution

a. ***Step 1.*** A typical point is $(t, 0.011t^2 + 0.101t + 10.817)$.

Step 2. A close point is $(t + h, 0.011(t + h)^2 + 0.101(t + h) + 10.817)$. We rewrite the output of the close point before we proceed:

$$\begin{aligned} &0.011(t + h)^2 + 0.101(t + h) + 10.817 \\ &= 0.011(t^2 + 2th + h^2) + 0.101(t + h) + 10.817 \\ &= 0.011t^2 + 0.022th + 0.011h^2 + 0.101t + 0.101h + 10.817 \end{aligned}$$

Step 3. Write the slope of the secant line between the two points from Steps 1 and 2.

$$\begin{aligned} \text{Slope} &= [(0.011t^2 + 0.022th + 0.011h^2 + 0.101t + 0.101h + 10.817) - \\ &\quad (0.011t^2 + 0.101t + 10.817)] \div [(t + h) - t] \\ &= \frac{0.022th + 0.011h^2 + 0.101h}{h} \\ &= \frac{h(0.022t + 0.011h + 0.101)}{h} \end{aligned}$$

Step 4. Evaluate the limit of the secant line slope as h approaches 0.

$$\begin{aligned} W'(t) &= \lim_{h \to 0} \frac{h(0.022t + 0.011h + 0.101)}{h} = \lim_{h \to 0}(0.022t + 0.011h + 0.101) \\ &= 0.022t + 0.101 \end{aligned}$$

14. *Information Please Almanac, Atlas, and Yearbook, 1996* (Boston: Houghton Mifflin, 1996).

Thus the number of women in the civilian labor force was increasing by

$$W'(t) = 0.022t + 0.101 \text{ million women per year}$$

t years after 1930.

b. We find that $W'(45) \approx 1.1$ million women per year. Thus the number of women in the U.S. civilian labor force was growing by approximately 1.1 million women per year in 1975.

The definition of the derivative of a function gives us a formula for the slope graph of the function, which enables us to calculate exact rates of change quickly.

EXAMPLE 3 *Equation of a Slope Graph*

If you are de-emphasizing an algebraic approach, you may wish to omit Example 3 and Activities 16–19.

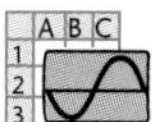

Consider the function $f(x) = 2\sqrt{x}$ that we used at the beginning of Section 4.1.

a. Sketch a slope graph for this function.

b. Use the Four-Step Method to find a formula for the slope graph.

c. Use the formula from part b to find the slope of the graph of f at $x = 1$. Compare this answer with the one found by numerical estimation in Section 4.1.

Solution

a. Examine the graph of f in Figure 4.9. Note that the slopes are always positive but that, because the graph of f becomes less steep as x becomes greater, the positive slopes are declining. A possible slope graph is given in Figure 4.10.

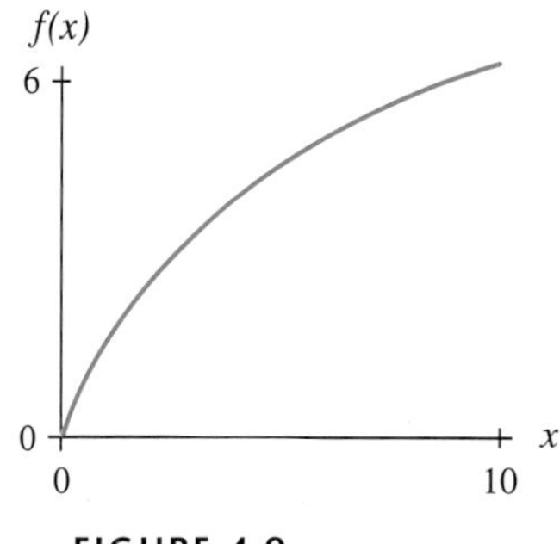

FIGURE 4.9

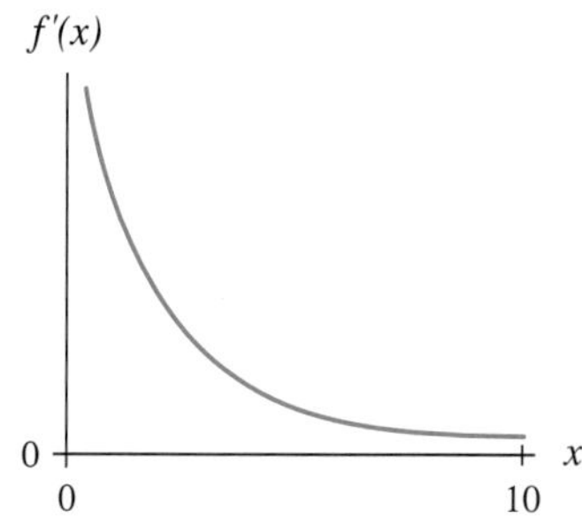

FIGURE 4.10

b. To find a formula for the slope graph, begin with a general point $(x, 2\sqrt{x})$ and a close point $(x + h, 2\sqrt{x + h})$. Next find the slope between these two points.

$$\text{Secant line slope} = \frac{2\sqrt{x+h} - 2\sqrt{x}}{x + h - x} = \frac{2\sqrt{x+h} - 2\sqrt{x}}{h}$$

Now find the limit of this formula as h approaches zero:

$$\lim_{h \to 0} \frac{2\sqrt{x+h} - 2\sqrt{x}}{h}$$

Unlike the case for polynomial functions that always contain h as a common factor, we cannot cancel the h here without rewriting the numerator. The key to this cancellation is to multiply the numerator and denominator by the term $2\sqrt{x+h}+2\sqrt{x}$. Observe how this multiplication enables us to cancel the h term:

$$\lim_{h\to 0}\frac{2\sqrt{x+h}-2\sqrt{x}}{h}\cdot\frac{2\sqrt{x+h}+2\sqrt{x}}{2\sqrt{x+h}+2\sqrt{x}}$$

$$=\lim_{h\to 0}\frac{4(x+h)-4x}{h(2\sqrt{x+h}+2\sqrt{x})}$$

$$=\lim_{h\to 0}\frac{4h}{h(2\sqrt{x+h}+2\sqrt{x})}$$

$$=\lim_{h\to 0}\frac{4}{2\sqrt{x+h}+2\sqrt{x}}$$

$$=\frac{4}{2\sqrt{x+0}+2\sqrt{x}}=\frac{4}{4\sqrt{x}}=\frac{1}{\sqrt{x}}$$

Thus the formula for the slope graph shown in Figure 4.10 is $f'(x)=\frac{1}{\sqrt{x}}$.

c. Using this slope formula, we find that $f'(1)=\frac{1}{\sqrt{1}}=1$. That is, the slope of the line tangent to the graph of $f(x)=2\sqrt{x}$ at $x=1$ is 1. This calculation confirms that the numerical estimate in Section 4.1 is correct.

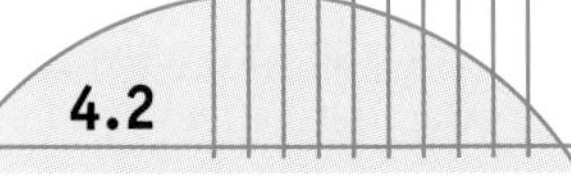

4.2 Concept Inventory

- *Using the algebraic method to find the slope of a graph at a given point*
- *Using the Four-Step Method to find a rate-of-change formula*
- *Limit definition of derivative*

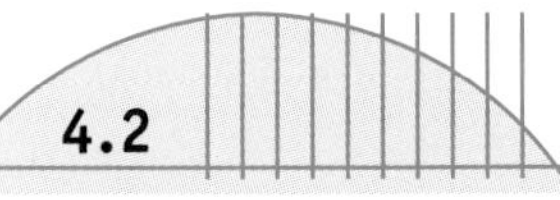

4.2 Activities

1. The time it takes an average athlete to swim 100 meters freestyle at age x years can be modeled by the equation $T(x) = 0.181x^2 - 8.463x + 147.376$ seconds.
   a. Find the swim time when $x = 13$.
   b. Write a formula for the average swim time when $x = 13 + h$.
   c. Write a simplified formula for the slope of the secant line connecting the points at $x = 13$ and $x = 13 + h$.
   d. What is the limiting value of the slope formula in part c as h approaches 0? Interpret your answer.

2. The number of AIDS cases diagnosed between 1993 and 1997 can be modeled[15] by the equation

   $$C(x) = -1.85x^3 + 30.31x^2 - 168.17x + 385.05$$

   thousand cases where x is the number of years since 1990.
   a. Find the number of AIDS cases when $x = 4$.
   b. Write an expression for the number of cases when $x = 4 + h$.
   c. Write a simplified formula for the slope of the secant line connecting the points at $x = 4$ and $x = 4 + h$.
   d. What is the limiting value of the slope formula in part c as h approaches 0? Interpret your answer.
   e. How do your answers to parts *a* and *b* of Activity 5 in Section 4.1 and part *d* of this activity compare? Which method is the most accurate and why?

3. The CPI for college tuition between 1990 and 1997 can be modeled[16] by the equation

   $$c(t) = -0.566t^2 + 21.092t + 173.996$$

   where t is the number of years since 1990.

15. Based on data from *Statistical Abstract*, 1998.
16. Based on data from *Statistical Abstract*, 1998.

a. Find the consumer price index for college tuition in 1995.

b. Write a formula in terms of h for the consumer price index of college tuition a little after 1995.

c. Write a simplified formula for the slope of the secant line connecting the points at 1995 and a little after 1995.

d. What is the limiting value for the slope formula as h approaches 0?

e. Interpret your answer to part *d*.

f. How do your answers to part *c* of Activity 13 in Section 4.1 and part *d* of this activity compare? Which method should you use in each of the following situations?

   i. You are concerned most with accuracy.

   ii. You want a quick, rough estimate.

   iii. You want a fairly good estimate without taking much time.

4. Discuss the advantages and disadvantages of finding rates of change graphically, numerically, and algebraically. Include in your discussion a brief description of when each method might be appropriate to use.

5. Consider the function $g(t) = -6t^2 + 7$. Use the algebraic method to find $\frac{dg}{dt}$ at $t = 4$ by evaluating the limit of an expression for the slope of a secant line.

6. Consider the function $m(p) = 4p + p^2$. Use the algebraic method to find $\frac{dm}{dp}$ at $p = -2$ by evaluating the limit of an expression for the slope of a secant line.

7. An object is dropped off a building. Ignoring air resistance, we know from physics that its height above the ground t seconds after being dropped is given by

$$\text{Height} = -16t^2 + 100 \text{ feet}$$

a. Use the Four-Step Method to find a rate-of-change equation for the height.

b. Use your answer to part *a* to determine how rapidly the object is falling after 1 second.

8. Clinton County, Michigan, is mostly flat farmland partitioned by straight roads (often gravel) that run either north/south or east/west. A tractor driven north on Lowell Road from the Schafer's farm is

$$d(m) = 0.28m + 0.6 \text{ miles}$$

north of Howe Road m minutes after leaving the farm's drive.

a. How far is the Schafer's drive from Howe Road?

b. Use the Four-Step Method to show that the tractor is moving at a constant speed.

c. How quickly (in miles per hour) is the tractor moving?

9. The number of licensed drivers[17] between the ages of 16 and 21 in 1997 is given in Table 4.7.

TABLE 4.7

Age (years)	Number of drivers (millions)
16	0.85
17	1.24
18	1.41
19	1.47
20	1.54
21	1.51

a. Find a quadratic model for the number of licensed drivers as a function of age. Round the coefficients of the equation to three decimal places.

b. Use the limit definition of the derivative to develop a formula for the derivative of the rounded model.

c. Use the derivative formula in part *b* to find the rate of change of the function in part *a* for an age of 20 years. Interpret your answer.

10. The data in Table 4.8 give the percentage of females of a certain age who were licensed drivers[18] in 1997.

TABLE 4.8

Age (years)	Licensed drivers (percent)
15	0.4
16	43.4
17	61.9
18	72.7
19	73.8

a. Find a quadratic model for these data. Round the coefficients in the equation to three decimal places.

b. Use the limit definition of the derivative to develop the derivative formula for the rounded equation.

17. U.S. Department of Labor and Transportation.
18. Ibid.

c. Use the derivative formula in part *b* to find the rate of change of the equation in part *a* when the input is 16 years of age. Interpret your answer.

In Activities 11 through 15, use the Four-Step Method outlined in this section to show that each statement is true.

11. The derivative of $y = 3x - 2$ is $\frac{dy}{dx} = 3$.
12. The derivative of $y = 15x + 32$ is $\frac{dy}{dx} = 15$.
13. The derivative of $y = 3x^2$ is $\frac{dy}{dx} = 6x$.
14. The derivative of $y = -3x^2 - 5x$ is $\frac{dy}{dx} = -6x - 5$.
15. The derivative of $y = x^3$ is $\frac{dy}{dx} = 3x^2$. (*Hint:* $(x + h)^3 = x^3 + 3x^2h + 3xh^2 + h^3$)
16. Figure 4.2.1 shows a graph of the function $g(x) = x^3 - x^2$.

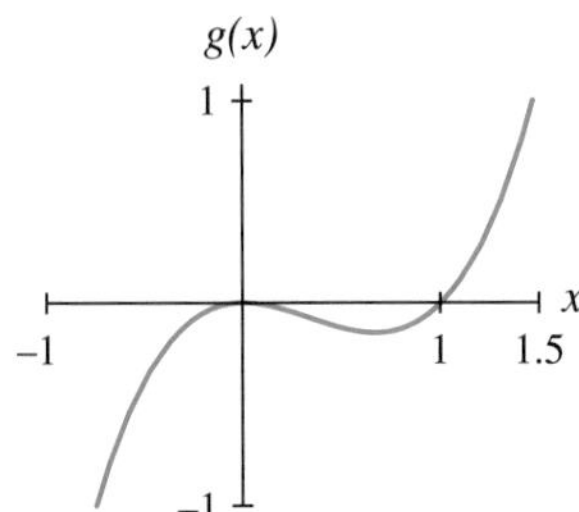

FIGURE 4.2.1

a. Sketch a slope graph of *g*.

b. Find a formula for the slope graph. Compare a graph of this formula with the graph you sketched in part *a*.

17. Figure 4.2.2 shows a graph of the function $q(t) = \frac{1}{t}$.

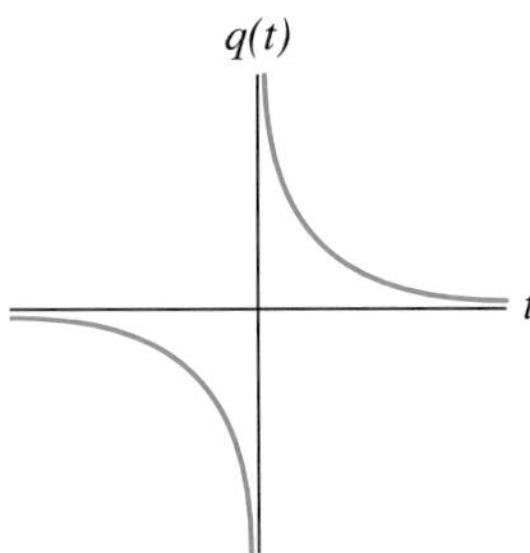

FIGURE 4.2.2

a. Sketch a slope graph of *q*.

b. Find a formula for the slope graph. [*Hint:* Multiply the numerator and denominator of the secant line slope formula by $t(t + h)$.] Compare a graph of this formula with the graph you sketched in part *a*.

18. Figure 4.2.3 shows a graph of the function $k(x) = x + \frac{1}{x}$.

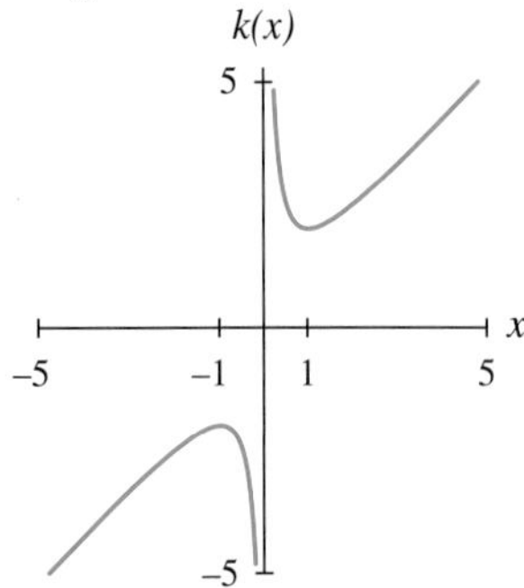

FIGURE 4.2.3

a. Sketch a slope graph of *k*.

b. Find a formula for the slope graph. Use the same hint as in Activity 17. Compare a graph of this formula with the graph you sketched in part *a*.

19. Figure 4.2.4 shows a graph of the function $p(m) = m + \sqrt{m}$.

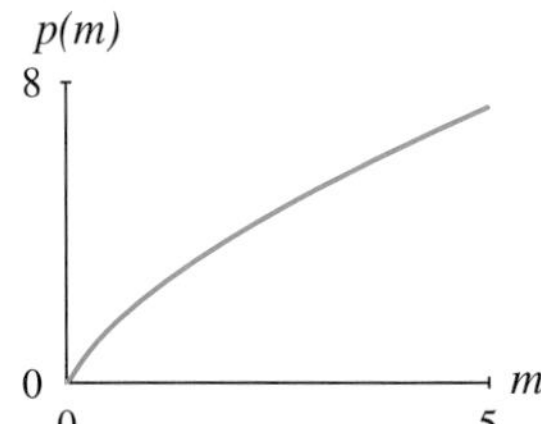

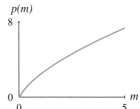

FIGURE 4.2.4

a. Sketch a slope graph of *p*.

b. Find a formula for the slope graph. Compare a graph of this formula with the graph you sketched in part *a*. (*Hint:* Multiply the numerator and denominator of the secant line slope formula by $\sqrt{m + h} + \sqrt{m}$.)

20. Recall from Example 2 the model for the number of women in the U.S. civilian labor force between 1930 and 1990 and the derivative formula for the model:

$$W(t) = 0.011t^2 + 0.101t + 10.817 \text{ million women}$$
$$W'(t) = 0.022t + 0.101 \text{ million women per year}$$

t years after 1930. Using more current data, the number of women in the U.S. civilian labor force between 1970 and 1997 can be modeled by the equation

$$C(x) = -0.0132x^2 + 1.5127x + 31.5596 \text{ million women}$$

x years after 1970.

a. Use the function *W* and its derivative to estimate the number of women in the civilian labor force in the years 1945, 1995, and 2000 and to estimate the rates of change in those years.

Activity 20 would make an effective short project. We recommend assigning it to pairs of students.

b. Use the Four-Step Method to find the derivative formula for the function C.

c. Use the function C and its derivative to estimate the number of women in the U.S. civilian labor force in the years 1945, 1995, and 2000 and to estimate the rates of change in those years.

d. Compare the answers to parts a and c. Which answers do you believe are more accurate? Why?

e. Find the most recent data available for the number of women in the U.S. civilian labor force. Use all the available data to construct a model. Round the coefficients in the equation to three decimal places.

f. Use the Four-Step Method to find the derivative formula for the function in part e.

g. Use the model in part e and its derivative to estimate the number of women in the U.S. civilian labor force in 1945, 1995, and 2000 and the rates of change in those years. Compare your answers to those you obtained in parts a and c. Do you believe the answers based on the model in part e are more reliable than the ones in parts a and c? Explain.

4.3 Simple Rate-of-Change Formulas

Even if you are de-emphasizing algebraic skill, you should not skip or rush over derivative formulas unless all of your students have technology that will find symbolic derivatives, and you are willing to allow them to use it in this manner.

By now you should have a thorough understanding of the concept of instantaneous rate of change as the slope of a line tangent to a curve at a point, defined as a limiting value of slopes of secant lines. As you discovered in the previous section, calculating rates of change as the limiting values of the slopes of secant lines is tedious and slow. We now move away from our emphasis on conceptual understanding in order to present formulas for rapid calculation of rates of change, but we urge you not to forget the graphical and numerical roots of the algebraic work we are about to begin.

You already know two rate-of-change formulas from our study of linear functions in Chapter 1. A horizontal line has slope zero, and a nonhorizontal line of the form $y = ax + b$ has slope a. We know that the rate of change, or derivative, of a line is its slope, so we can state the following derivative formulas:

Derivative of a Linear Function

If $y = ax + b$, then $\dfrac{dy}{dx} = a$.

Constant Rule

If $y = b$, then $\dfrac{dy}{dx} = 0$.

EXAMPLE 1 *Cricket's Chirping*

The frequency of a cricket's chirp is affected by air temperature and can be modeled by

$C(t) = 0.212t - 0.309$ chirps per second

when the temperature is t°F. Write a formula for the rate of change of a cricket's chirping speed with respect to a change in temperature.

Emphasize to your students that derivative formulas are models and should be labeled with all the elements of a model discussed in Section 1.5. Because the input of a derivative is the same as that of the function from which it was derived, we sometimes do not restate the input and its units when the derivative is used together with the original function. However, when the two functions are used separately, the input statement should be repeated.

Solution The frequency of a cricket's chirp is changing by

$C'(t) = 0.212$ chirps per second per degree Fahrenheit

no matter what the temperature is.

The Simple Power Rule

Next, consider quadratic and cubic functions. In Section 4.2, we proved that the rate-of-change formula for $y = x^2$ is $\frac{dy}{dx} = 2x$. In Activity 15 of that same section, you were asked to prove that the rate-of-change formula for $y = x^3$ is $\frac{dy}{dx} = 3x^2$. Note that in each instance, the power on x in the derivative formula is one less than the power on x in the original function. Also, the x-term in the derivative formula is multiplied by the power on x in the original function. This is no coincidence. These are two special cases of one of the most important rules that we use for quickly finding derivative formulas, the **Simple Power Rule**.

Simple Power Rule for Derivatives

If $y = x^n$, then $\dfrac{dy}{dx} = nx^{n-1}$

where n is any nonzero real number.

The Constant Multiplier and Sum Rules

In order to find derivative formulas for any polynomial, we need two rules in addition to the Simple Power Rule. Also, to prove these rules, we need to recall two properties of limits that were stated in Section 1.4:

Constant Multiplier Rule for Limits: If k and c are constants and f is a function of x, then

$$\lim_{x \to c} kf(x) = k \lim_{x \to c} f(x)$$

Sum Rule for Limits: If c is a constant and f and g are functions of x, then

$$\lim_{x \to c} [f(x) + g(x)] = \lim_{x \to c} f(x) + \lim_{x \to c} g(x)$$

If you are de-emphasizing algebraic rigor, you may wish to omit the proofs of the Constant Multiplier and Sum Rules as well as Activities 19 and 20.

The first derivative rule we prove is the **Constant Multiplier Rule**.

Constant Multiplier Rule for Derivatives

If $y = kf(x)$, then $\dfrac{dy}{dx} = kf'(x)$.

Proof

Let $y = kf(x)$. According to the definition of the derivative,

$$\frac{dy}{dx} = \lim_{h \to 0} \frac{kf(x + h) - kf(x)}{h}$$

Factoring k out the two terms in the numerator gives

$$\frac{dy}{dx} = \lim_{h \to 0} k\left[\frac{f(x+h) - f(x)}{h}\right]$$

The Constant Multiplier Rule for Limits allows us to move k in front of the limit.

$$\frac{dy}{dx} = k\left[\lim_{h \to 0} \frac{f(x+h) - f(x)}{h}\right]$$

You should recognize the limit as the derivative of f. We have the result

$$\frac{dy}{dx} = kf'(x)$$

This rule allows us to calculate quickly the rate-of-change formula for a function such as $y = 5x^4$. Leave the 5 alone, and apply the Simple Power Rule to x^4. This process gives $\frac{dy}{dx} = 5(4x^3) = 20x^3$.

The final rule needed for rapid calculation of the rate-of-change formula for any polynomial is the **Sum Rule**. We prove the rule using the definition of derivative and the Sum Rule for Limits.

The Sum Rule for Derivatives

If $j(x) = f(x) + g(x)$, then $\frac{dj}{dx} = \frac{df}{dx} + \frac{dg}{dx}$.

Proof

Let $j(x) = f(x) + g(x)$. According to the definition of the derivative,

$$\frac{dj}{dx} = \lim_{h \to 0} \frac{f(x+h) + g(x+h) - [f(x) + g(x)]}{h}$$

Rewrite the numerator by distributing the negative one and rearranging the terms:

$$\frac{dj}{dx} = \lim_{h \to 0} \frac{f(x+h) - f(x) + g(x+h) - g(x)}{h}$$

Now use the fact that $\frac{a+b}{c} = \frac{a}{c} + \frac{b}{c}$ to rewrite the fraction as the sum of two fractions:

$$\frac{dj}{dx} = \lim_{h \to 0} \left[\frac{f(x+h) - f(x)}{h} + \frac{g(x+h) - g(x)}{h}\right]$$

Use the Sum Rule for Limits to write the preceding limit as the sum of two limits:

$$\frac{dj}{dx} = \lim_{h \to 0} \frac{f(x+h) - f(x)}{h} + \lim_{h \to 0} \frac{g(x+h) - g(x)}{h}$$

You should recognize these two limits as derivatives. We have the result

$$\frac{dj}{dx} = \frac{df}{dx} + \frac{dg}{dx}$$

This rule can also be written as

$$j'(x) = f'(x) + g'(x)$$

In the Activities, you will be asked to prove the Difference Rule: The derivative of the difference of two functions is the difference of the two derivatives. The Sum and Difference Rules also apply to sums of more than two functions. With the rules we now have, we can find the rate-of-change formula for any polynomial function. For example, if $p(x) = 3.22x^3 - 0.15x^2 + 9.98x - 30$, we use the Constant Multiplier and Simple Power Rules to find the derivative formula for each term, and then we combine the terms using the Sum Rule. The rate-of-change formula for p is

$$p'(x) = 3.22(3x^2) - 0.15(2x) + 9.98 - 0 = 9.66x^2 - 0.3x + 9.98$$

EXAMPLE 2 *Vehicle Maintenance Costs*

Table 4.9 gives the average yearly maintenance costs per vehicle[19] in the United States from 1990 through 1997.

TABLE 4.9

Year	1990	1992	1993	1994	1995	1996	1997
Maintenance costs (cents per mile per vehicle)	2.1	2.2	2.4	2.5	2.6	2.8	2.8

Find a cubic model for the data, and use it to approximate how rapidly maintenance costs were increasing in 1995.

Solution A cubic model for the data is

$$g(t) = -0.0039t^3 + 0.043t^2 - 0.0078t + 2.098 \text{ cents per mile per vehicle}$$

where t is the number of years after 1990. Applying the Sum, Power, Constant Multiplier, and Constant Rules, we find that the derivative of g with respect to t is

$$\frac{dg}{dt} = 3(-0.0039)t^2 + 2(0.043)t - 0.0078 + 0$$

so

$$\frac{dg}{dt} = -0.0117t^2 + 0.086t - 0.0078 \text{ cents per mile per vehicle per year}$$

t years after 1990.

Evaluating the derivative at $t = 5$ gives 0.1297 cent per mile per vehicle per year as the rate of change of maintenance costs in 1995. Thus we estimate that in 1995 the average maintenance cost per vehicle was increasing at a rate of approximately 0.13 cent per mile per year.

Derivatives of Piecewise Functions

Our final example illustrates using the rules presented in this section to find a derivative formula for a piecewise continuous function. If a piecewise continuous function is discontinuous, then the derivative does not exist at the input value corresponding to the discontinuity. When you are writing the derivative formula, it is important to indicate that the derivative is not defined for this input value. If you need to estimate a rate of change of the underlying output quantity at that input value, you could use a symmetric difference quotient or some other estimation method.

19. *Statistical Abstract*, 1998.

EXAMPLE 3 *Population of Iowa*

On the basis of data from the U.S. Census Bureau for years between 1980 and 1997, the population of Iowa can be modeled by the equation

$$I(x) = \begin{cases} -2.1x^2 - 6.3x + 2914 \text{ thousand people} & \text{when } 0 \le x \le 7 \\ -0.1979x^3 + 7.254x^2 - 74.98x + 3002.28 \text{ thousand people} & \text{when } 7 < x \le 17 \end{cases}$$

where x is the number of years since 1980.

a. Write the derivative equation for the population model.

b. How quickly was the population of Iowa growing or declining in 1985? in 1995?

c. What is the slope of a line tangent to the graph of I at $x = 7$?

d. Does your answer to part *c* imply that the rate of change of the population of Iowa did not exist in 1987? Explain.

e. Estimate how quickly the population of Iowa was growing or declining in 1987.

Solution

a. The function is not continuous at $x = 7$ because the quadratic part of the function has a value of 2767 when $x = 7$, and the cubic part of the function has a value of 2764.9863 when $x = 7$. Therefore, the derivative formula must be undefined when $x = 7$. To find the formula, simply apply the derivative rules to each function individually.

$$I'(x) = \begin{cases} -4.2x - 6.3 \text{ thousand people per year} & \text{when } 0 \le x < 7 \\ -0.5937x^2 + 14.508x - 74.98 \text{ thousand people per year} & \text{when } 7 < x \le 17 \end{cases}$$

where x is the number of years since 1980. Note that the inequality $0 \le x \le 7$ in the top function becomes $0 \le x < 7$ in the function's derivative. Because there is no equals sign indicating $x = 7$ in either portion of the $I'(x)$ formula, the derivative is not defined at $x = 7$. Graphs of I and I' are shown in Figures 4.11a and b, respectively.

FIGURE 4.11

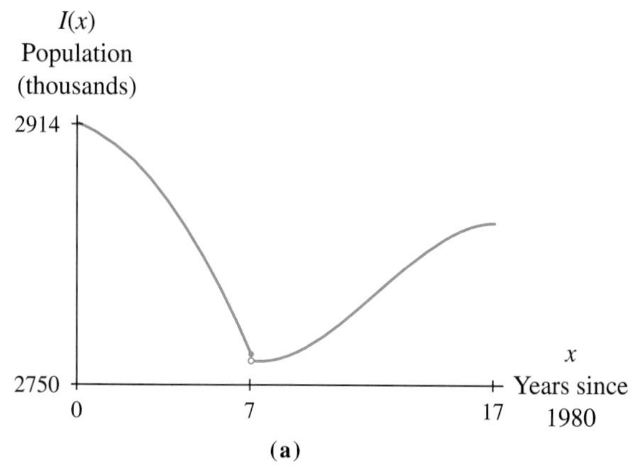

(a)

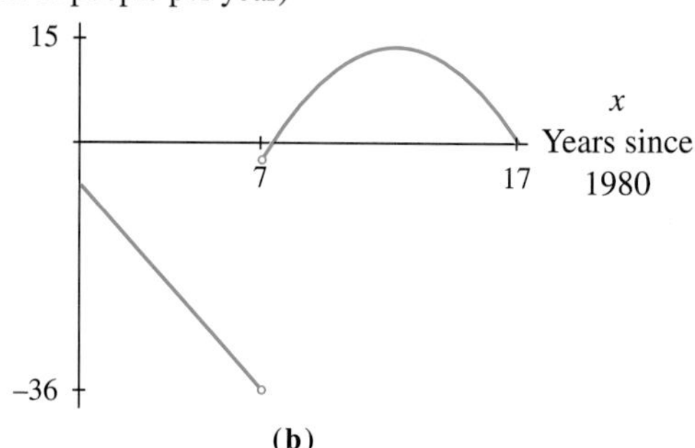

(b)

b. The rate of change of the population of Iowa in 1985 was $I'(5) = -4.2(5) - 6.3 = -27.3$ thousand people per year. In other words, the population of Iowa was declining by approximately 27,300 people per year in 1985.

In 1995, the rate of change of the population was $I'(15) \approx 9.06$ thousand people per year. The population in 1995 was increasing by approximately 9000 people per year.

c. It is impossible to draw a tangent line on the graph of I at the point corresponding to $x = 7$ because the function is not continuous there.

d. The fact that our model does not have a derivative for $x = 7$ does not mean that the Iowa population did not have a rate of change in 1987. It is, however, an indication that our model is not a good description of the behavior of the population of Iowa around 1987 and cannot be used to calculate the rate of change in 1987.

e. There are several ways in which we could estimate the rate of change of the population of Iowa in 1987:

1. Find the data from which this model was constructed and either:

   i. Use the 1986 and 1988 populations of Iowa to calculate a symmetric difference quotient, or

   ii. Use some of the data to construct, for the population of Iowa, a different model that *is* continuous and smooth at the point corresponding to 1987. Then use the derivative of that model to estimate the rate of change in 1987.

2. Examine the graph in Figure 4.11a, and note that the population declines until 1987, when it begins rising again. Because the 1987 population appears to be a minimum, we could estimate that the rate of change of the population of Iowa in 1987 was approximately 0 thousand people per year.

3. Use the model that was given to estimate the population of Iowa in 1986 and 1988. Then use those two population estimates to calculate an average rate of change (symmetric difference quotient). The model estimates the 1986 population as 2800.6 thousand people and the 1988 population as 2765.4 thousand people. The average rate of change between these two years is

$$\frac{2765.4 - 2800.6}{1988 - 1986} \approx \frac{-35.2 \text{ thousand people}}{2 \text{ years}} \approx -17.6 \text{ thousand people per year}$$

Be sure you understand that these estimates are not estimates of the derivative of the function I for $x = 7$. These are estimates of the rate of change of the underlying situation—in this case, the population of Iowa.

We have purposely kept the number of drill problems in the activities to a minimum because we want the focus of this book to be conceptual understanding rather than algebraic manipulation. However, we realize that there are times when additional practice is necessary. Additional derivative formula drill problems can be found on the *Calculus Concepts* web site.

practice

In summary, here is a list of the rate-of-change formulas you should know. Commit these formulas to memory through extensive practice. The Activities provide many opportunities for such practice.

Simple Derivative Rules

Rule Name	*Function*	*Derivative*
Constant Rule	$y = b$	$\frac{dy}{dx} = 0$
Linear Function Rule	$y = ax + b$	$\frac{dy}{dx} = a$
Power Rule	$y = x^n$	$\frac{dy}{dx} = nx^{n-1}$
Constant Multiplier Rule	$y = kf(x)$	$\frac{dy}{dx} = kf'(x)$
Sum Rule	$y = f(x) + g(x)$	$\frac{dy}{dx} = f'(x) + g'(x)$

4.3 Concept Inventory

- ***Derivative formulas***
  For constants a, b, and n,
  If $y = b$, then $y' = 0$.
  If $y = ax + b$, then $y' = a$.
  If $y = x^n$, then $y' = nx^{n-1}$.
  If $y = kf(x)$, then $y' = kf'(x)$.
  If $y = f(x) \pm g(x)$, then $y' = f'(x) \pm g'(x)$.

4.3 Activities

For each of the functions whose graphs are given in Activities 1–6, first sketch the slope graph and then give the slope equation.

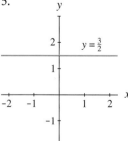

1. 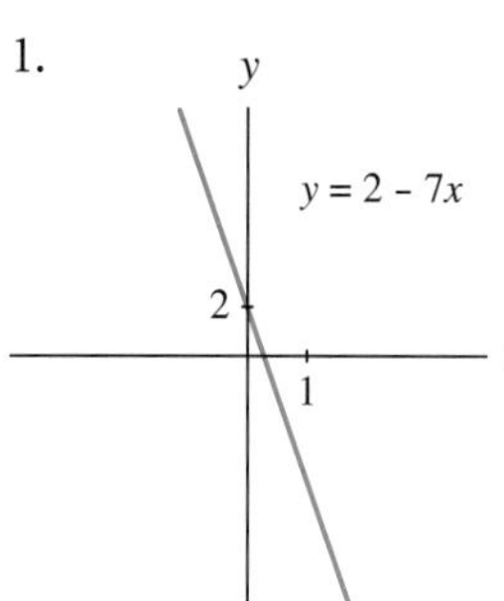

2. 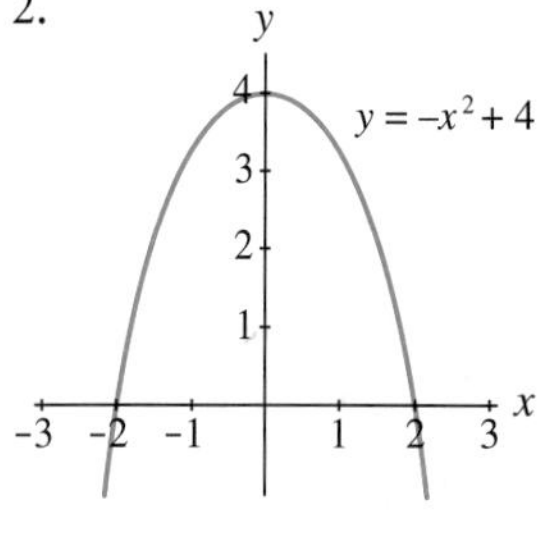

3. 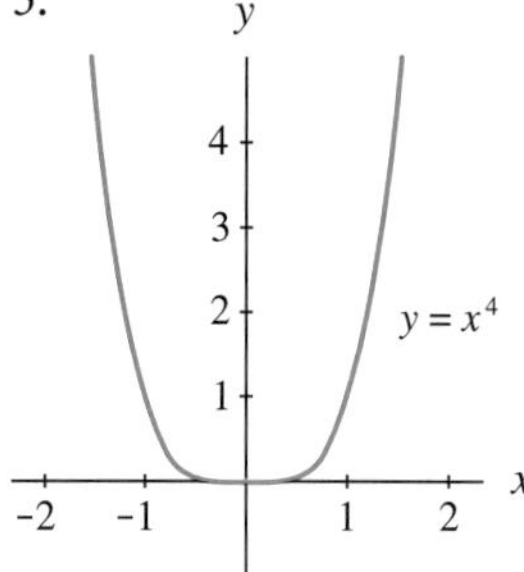

4. 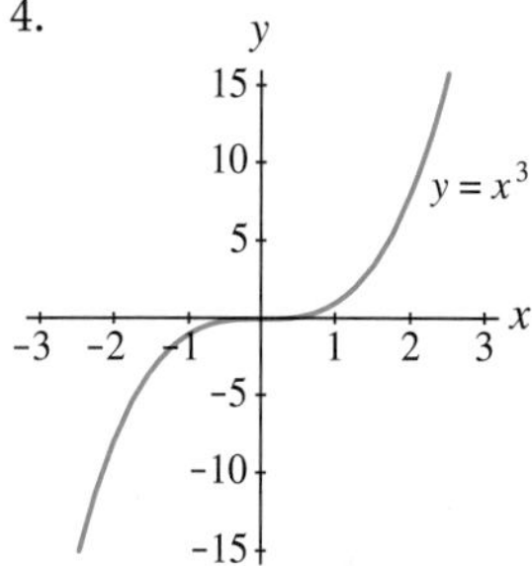

5. 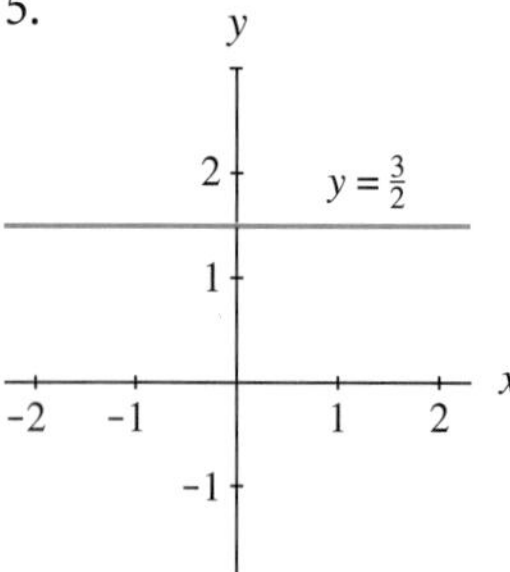

6. 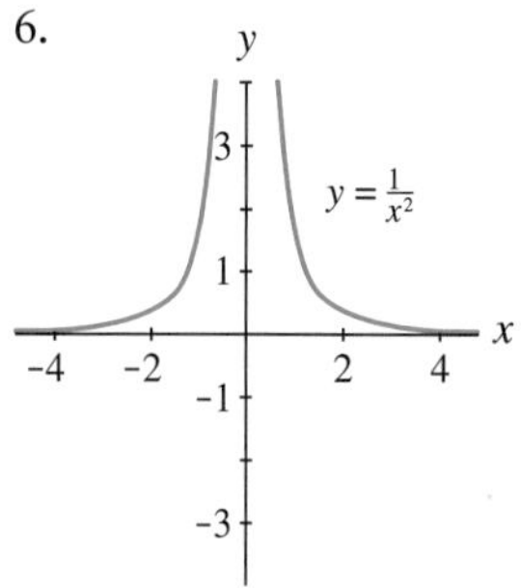

Give the derivative formulas for the functions in Activities 7 through 14.

7. $y = 7x^2 - 12x + 13$
8. $f(x) = 0.0127x^3 + 9.4861x^2 - 0.2649x + 128.98$
9. $y = 5x^3 + 3x^2 - 2x - 5$
10. $f(x) = \sqrt{x}$ (*Hint:* Rewrite as $x^{1/2}$.)
11. $y = \dfrac{-9}{x^2}$
12. $f(x) = 3x^{-2}$
13. $h(x) = 17.463 - 0.049\sqrt{x}$
14. $j(x) = \dfrac{3.694x^2 + 19.42x + 1.06}{x}$ (*Hint:* Rewrite as three separate terms.)
15. The average ATM transaction fee charged by U.S. banks between 1996 and 1999 can be modeled[20] by the equation $A(t) = 0.1333t + 0.17$ dollars t years after 1990.
    a. Write the derivative formula for A.
    b. Use the model to estimate the transaction fee in 2000.
    c. How quickly was the transaction fee changing in 1999?
16. Table 4.10 shows the metabolic rate[21] of a typical 18- to 30-year-old male according to his weight.

TABLE 4.10

Weight (pounds)	Metabolic rate (kilocalories per day)
88	1291
110	1444
125	1551
140	1658
155	1750
170	1857
185	1964
200	2071

    a. Find a formula for a typical man's metabolic rate.

20. Based on data from the U.S. Public Interest Research Group.
21. L. Smolin and M. Grosvenor, *Nutrition: Science and Applications* (Philadelphia, PA: Saunders College Publishing, 1994).

b. Write the derivative of the formula you found in part *a*.

c. What does the derivative in part *b* tell you about a man's metabolic rate if that man weighs 110 pounds? if he weighs 185 pounds?

17. The population of Hawaii between 1970 and 1990 can be modeled[22] by

$$P(t) = 15.48t + 485.4 \text{ thousand people}$$

t years after 1950.

a. Write the formula for P'.

b. How many people lived in Hawaii in 1970?

c. How quickly was Hawaii's population changing in 1990?

18. Show that the slope formula for a linear function of the form $y = ax$ is a special case of the Simple Power and Constant Multiplier Rules.

19. Following the example of the Sum Rule proof given earlier, prove that if $f(x) = g(x) - k(x)$, then $f'(x) = g'(x) - k'(x)$.

20. Following the example of the Sum Rule proof, prove that if $y = f(x) + g(x) + j(x)$, then $\frac{dy}{dx} = f'(x) + g'(x) + j'(x)$.

21. The graph in Figure 4.3.1 shows the temperature values (°F) on a typical May day in a certain midwestern city.

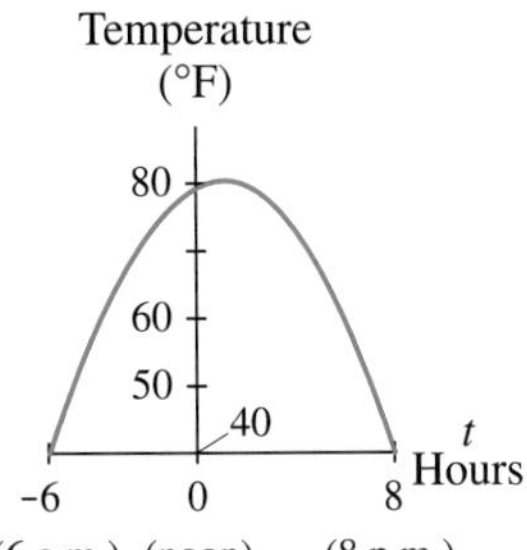

FIGURE 4.3.1

The equation of the graph is

$$\text{Temperature} = -0.8t^2 + 2t + 79 \text{ °F}$$

where t is the number of hours since noon. Use the derivative formula to verify each of the following statements.

a. The graph is not as steep at 1:30 p.m. as it is at 7 a.m.

b. The slope of the tangent line at 7 a.m. is 10°F per hour.

c. The instantaneous rate of change of the temperature at noon is 2°F per hour.

d. At 4 p.m. the temperature is falling by 4.4°F per hour.

22. A graph representing a test grade (out of 100 points) as a function of hours studied is given in Figure 3.25 on page 221. The equation of the graph is

$$G(t) = -0.044t^3 + 0.918t^2 + 38.001 \text{ points}$$

after t hours of study. Use the equation to verify each of the following statements.

a. $\frac{dG}{dt} = 1.704$ points per hour when $t = 1$ hour.

b. $G'(4) = 5.232$ points per hour.

c. The slope of the tangent line when $t = 15$ hours is approximately -2 points per hour.

23. The projected number of Americans age 65 or older for the years 1995 through 2030 can be modeled[23] by the equation

$$N(x) = 0.03x^2 + 0.315x + 34.23 \text{ million people}$$

where x is the number of years after 2000.

a. What is the projected number of Americans 65 years of age and older in 1995? in 2030?

b. What is the rate of change of the projected number in 1995? in 2030?

c. Find the percentage rate of change in the projected number in 2030.

d. The Census Bureau predicts that in 2030, 20.1% of the U.S. population will be 65 years of age or older. Use this prediction and one of your answers to part *a* to estimate the total U.S. population in 2030.

24. The number of military veterans[24] serving in the U.S. House of Representatives between 1975 and 1997 can be modeled by the equation

$$V(t) = -7.736t + 291.786 \text{ veterans}$$

where t is the number of years after 1975.

a. Was the number of veterans increasing or decreasing during the time period modeled by the equation?

22. Based on data from George T. Kurian, *Datapedia of the United States*, 1790–2000 (Latham, MD: Bernan Press, 1994).

23. Based on data from John Greenwald, "Elder Care: Making the Right Choice," *Time*, August 30, 1999, p. 52.

24. Dan Hoover, "Military Service No Longer Politically Necessary," *The Greenville News*, June 29, 1998, p. 1A.

b. How rapidly was the number of veterans in the House of Representatives changing in 1975? in 1997?

c. What was the average rate of change of the number of veterans in the House between 1975 and 1995?

25. The number of live births to U.S. women 45 years and older between 1950 and 1995 can be modeled[25] by the equation

$$B(x) = 0.3246x^3 - 18.945x^2 + 141.66x + 5300.9 \text{ births}$$

where x is the number of years since 1950.

a. Was the number of live births rising or falling in 1970? in 1990?

b. How rapidly was the number of live births rising or falling in 1970? in 1990?

26. An artisan makes hand-crafted painted benches to sell at a craft mall. Her weekly revenue and costs (not including labor) are given in Table 4.11.

TABLE 4.11

Number of benches sold each week	Weekly revenue (dollars)	Weekly cost (dollars)
1	300	57
3	875	85
5	1375	107
7	1750	121
9	1975	143
11	1950	185
13	1700	213

a. Find models for revenue and cost.

b. Construct a model for profit from the revenue and cost models in part *a*.

c. Write the derivative formula for profit.

d. Find and interpret the rate of change of profit when the artisan sells 6 benches.

e. Repeat part *d* for 9 benches and 10 benches.

f. What does the information in part *e* tell you about the number of benches the artisan should produce each week?

27. Table 4.12 gives the number of imported passenger cars[26] (in thousands) sold in the United States from 1990 through 1996.

TABLE 4.12

Year	Number of imported passenger cars sold (thousands)
1990	2403
1991	2038
1992	1937
1993	1776
1994	1735
1995	1506
1996	1273

a. Find a model for the data.

b. Write the formula for the derivative of the equation in part *a*.

c. Use your model to find how rapidly the number of imported-car sales was changing in 1996.

d. Use the model to estimate the number of imported cars sold in 1996.

e. What was the percentage rate of change in the number of imported cars sold in 1992?

28. Production costs (in dollars per hour) for a company to produce between 10 and 90 units per hour are given in Table 4.13.

TABLE 4.13

Units	Cost (dollars per hour)
10	150
20	200
30	250
40	400
50	750
60	1400
70	2400
80	3850
90	5850

a. Graph a scatter plot of the data, and discuss its curvature.

b. Consider the cost for producing 0 units to be \$0. Include (0, 0) in the data, and find the first differences of the cost data. Convert each of these differences to average rates of change. Use the average rates of change to argue that this data set has an inflection point.

25. Based on data from *The 1998 Information Please Almanac.*
26. *Statistical Abstract*, 1998.

c. Find a cubic model for production costs. Include the point (0, 0).

d. Convert the model in part *c* to one for the average cost per unit produced.

e. Find the slope formula for average cost.

f. How rapidly is the average cost changing when 15 units are being produced? 35 units? 85 units? Interpret your answers.

29. The managers of Windolux, Inc. have modeled some cost data and found that if they produce *x* storm windows each hour, the cost (in dollars) to produce one window is given by the function

$$C(x) = 0.0146x^2 - 0.7823x + 46.9125 + \frac{49.6032}{x}$$

Windolux sells its storm windows for $175 each. (You may assume that every window made will be sold.)

a. Write the formula for the profit made from the sale of one storm window when Windolux is producing *x* windows each hour.

b. Write the formula for the rate of change of profit.

c. What is the profit made from the sale of a window when Windolux is producing 80 windows each hour?

d. How rapidly is profit from the sale of a window changing when 80 windows are produced each hour? Interpret your answer.

30. A publishing company estimates that when a new book by a best-selling American author first hits the market, its sales can be predicted by the equation $n(x) = 68{,}952.921\sqrt{x}$, where $n(x)$ represents the total number of copies of the book sold in the United States by the end of the *x*th week. The number of copies of the book sold abroad by the end of the *x*th week can be modeled by

$$a(x) = -0.039x^2 + 125.783x \text{ copies of the book}$$

a. Write the formula for the total number of copies of the book sold in the United States and abroad by the end of the *x*th week.

b. Write the formula for the rate of change of the total number of books sold.

c. How many books will be sold by the end of the first year (that is, after 52 weeks)?

d. How rapidly are books selling at the end of the first year? Interpret your answer.

31. The joint population of the United States and Canada for the period from 1980 through 1994 can be modeled[27] by the equation

$$n(x) = 0.0426x^2 + 2.049x + 251.796 \text{ million people}$$

x years after 1980. The population of Mexico for the same period can be modeled[28] by the equation

$$m(x) = 68.738(1.0213^x) \text{ million people}$$

x years after 1980.

a. Find the formula for the combined population of the United States, Canada, and Mexico from 1980 through 1994.

b. Find the formula for the rate of change of the combined population of the United States, Canada, and Mexico.

c. According to the models, what was the combined population of the United States, Canada, and Mexico in 1994?

d. How rapidly was the combined population of the United States, Canada, and Mexico changing in 1994?

32. When working with sums of models, why is it important to make sure that the input values of the models correspond? (That is, why must you align all models in the sum the same way?)

33. The amount of fish produced for food by fisheries in the United States between 1970 and 1990 can be modeled[29] by the equation

$$f(x) = \begin{cases} 22.204x^2 - 108.431x - 2538.603 \text{ million lb} & \text{when } 0 \le x \le 10 \\ 85.622x^2 - 2253.95x + 17{,}772.39 \text{ million lb} & \text{when } 10 < x \le 20 \end{cases}$$

x years after 1970.

a. Write a formula for $\frac{df}{dx}$.

b. Find the rates of change of the amount of fish produced in 1970, 1980, and 1990.

c. What was the percentage rate of change in 1990?

34. The atmospheric release of chlorofluorocarbons (CFCs) between 1958 and 1978 can be modeled[30]

27. Based on data from *Statistical Abstract*, 1994.
28. Based on data from SPP and INEGI, Mexican Censuses of Population 1921 through 1990 as reported in Pick and Butler, *The Mexico Handbook* (Westview Press, 1994).
29. Based on data from *Statistical Abstract*, 1994.
30. Based on data from Ronald Bailey, ed., *The True State of the Planet* (New York: The Free Press for the Competitive Enterprise Institute, 1995).

by the equation

$$C(t) = \begin{cases} 0.708t^2 - 71.401t + 1832.4275 \text{ million kg} & \text{when } 58 \leq t < 74 \\ -2.6125t^2 + 377.775t - 13{,}230.7 \text{ million kg} & \text{when } 74 \leq t \leq 78 \end{cases}$$

where t is the number of years since 1900.

a. Write the derivative formula for C.

b. What was the atmospheric release of CFCs in 1968, and how quickly was that amount changing?

c. Estimate the rate of change in the atmospheric release of CFCs in 1974.

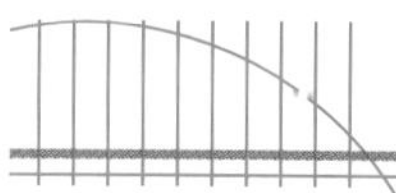

4.4 More Simple Rate-of-Change Formulas

In Section 4.3 we used the limit definition of a derivative to develop some general rules for rate-of-change formulas. In this section we will continue using those rules and develop others.

Exponential Rules

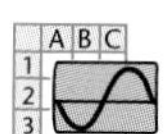

Our next formulas involve exponential functions. Because the proof of these rules is beyond the scope of this book, we will discover the rules by observing tables of slope values calculated via technology. We begin our exploration with the function $y = e^x$. Table 4.14 shows function values and slope values (rounded to three decimal places) for several inputs.

TABLE 4.14

x	-3	-2	-1	0	1	2	3
$y = e^x$	0.050	0.135	0.368	1.000	2.718	7.389	20.086
$y' = \frac{dy}{dx}$	0.050	0.135	0.368	1.000	2.718	7.389	20.086

This function is surprising in that the rate-of-change values are precisely the same as the function values. *This function is its own derivative!* In other words, if $y = e^x$, then $\frac{dy}{dx} = y' = e^x$. The slope graph of $y = e^x$ coincides with the graph of the original function.

Derivative of e^x

If $y = e^x$, then $\dfrac{dy}{dx} = e^x$

TABLE 4.15

x	$y = 2^x$	$y' = \frac{dy}{dx}$
-3	0.125	0.08664
-2	0.25	0.17329
-1	0.5	0.34657
0	1	0.69315
1	2	1.38629
2	4	2.77259
3	8	5.54518

Does this rule apply to exponential functions that have bases different from e? In other words, if $y = b^x$, is $\frac{dy}{dx} = b^x$? We begin by exploring the derivative of $y = 2^x$ with a table of values (rounded to five decimal places for convenience). See Table 4.15. It is obvious from the table that the derivative of $y = 2^x$ is *not* $y' = 2^x$.

If we use the definition of derivative for the function $f(x) = 2^x$, we obtain

$$f'(x) = \lim_{h \to 0} \frac{2^{x+h} - 2^x}{h}$$

You may recall that $2^{x+h} = 2^x 2^h$. Using this fact, we rewrite the derivative formula as

$$f'(x) = \lim_{h\to 0} \frac{2^x 2^h - 2^x}{h} = \lim_{h\to 0}\left[2^x\left(\frac{2^h - 1}{h}\right)\right]$$

Because the term 2^x is not affected by h approaching zero, we treat it as a constant:

$$f'(x) = 2^x\left[\lim_{h\to 0}\frac{2^h - 1}{h}\right]$$

Evaluating the limit term is beyond the scope of this text, but this formula indicates that the derivative of $y = 2^x$ is 2^x times a constant. If we divide each derivative value in Table 4.15 by the corresponding 2^x output value, we have the last column in Table 4.16. It may seem that the multiplier 0.693147 is arbitrary because it is not a familiar number, but that is not the case. You should verify that $0.693147 \approx \ln 2$. In fact, it can be proved that the limit term in the derivative formula for $f(x) = 2^x$ is ln 2. We state the formula for the derivative of $y = 2^x$ as $y' = (\ln 2)2^x$. A similar exploration suggests that the formula for the derivative of $y = 3^x$ is $y' = (\ln 3)3^x$.

TABLE 4.16

x	$y = 2^x$	$y' = \frac{dy}{dx}$	$\frac{y'}{y}$
-3	0.125	0.08664	0.693147
-2	0.25	0.17329	0.693147
-1	0.5	0.34657	0.693147
0	1	0.69315	0.693147
1	2	1.38629	0.693147
2	4	2.77259	0.693147
3	8	5.54518	0.693147

The two derivative formulas $\frac{d}{dx}(2^x) = (\ln 2)2^x$ and $\frac{d}{dx}(3^x) = (\ln 3)3^x$ are special cases of the general derivative formula for exponential functions. The derivative of $y = b^x$ is $y' = (\ln b)b^x$ if $b > 0$. In fact, the rule $\frac{d}{dx}(e^x) = e^x$ is also a special case of this formula. You will be asked to verify this fact in the Activities.

Derivative of b^x

If $y = b^x$ with the real number $b > 0$, then $\frac{dy}{dx} = (\ln b)b^x$.

EXAMPLE 1 *Credit Card Payment*

If credit card purchases are not paid off by the due date on the credit card statement, finance charges are applied to the remaining balance. In July 2001, one major credit card company had a daily finance charge of 0.05425% on unpaid balances. Assume that the unpaid balance is \$1 and that no new purchases are made.

a. Find an exponential function for the credit balance d days after the due date.

b. How much is owed after 30 days?

c. Write the derivative formula for the function from part *a*.

d. How quickly is the balance changing after 30 days?

e. Repeat parts *a* through *d*, assuming that the unpaid balance is \$2000.

Solution

a. Recall that the constant b in an exponential function $f(x) = ab^x$ is (1 + percentage growth) and that the constant a is the value of $f(0)$. Thus we use the function

$$f(d) = 1(1.0005425^d) = 1.0005425^d \text{ dollars}$$

to represent the balance due d days after the due date.

b. Thirty days after the due date, the balance is $f(30) \approx \$1.02$.

c. According to the derivative rules we saw in this chapter, the derivative of $y = b^x$ is $y' = (\ln b)b^x$. Thus the derivative formula for our function f is

$$f'(d) = (\ln 1.0005425)1.0005425^d \text{ dollars per day}$$

after d days.

d. We use the derivative formula evaluated at $d = 30$ to find the rate of change of the balance. After 30 days, the balance is increasing at a rate of 0.0006 dollar per day.

e. If the unpaid balance is \$2000, the balance-due function is

$$f(d) = 2000(1.0005425^d) \text{ dollars}$$

after d days. The amount due after 30 days is $f(30) = \$2032.81$.

According to the Constant Multiplier and Exponential Rules, the derivative of the balance function is

$$f'(d) = 2000(\ln 1.0005425)1.0005425^d \text{ dollars per day}$$

After 30 days, the credit balance is increasing at a rate of $f'(30) = \$1.10$ per day.

Natural Logarithm Rule

As with exponential functions, we motivate the derivative rule for the natural log function by using a table of technology-generated slope values. Table 4.17 shows derivative values (rounded to three decimal places) for the natural log function.

TABLE 4.17

x	1	2	3	4	5	6
Derivative of $y = \ln x$	1.000	0.500	0.333	0.250	0.200	0.167

Unlike the case for the exponential function, whose derivative at a certain input value is dependent on the corresponding output value, these derivatives are dependent on the input values. Note that each derivative value is the reciprocal of the input value—that is, it is 1 divided by the input value. The derivative of $y = \ln x$ at $x = 2$ is $\frac{1}{2}$, the derivative at $x = 3$ is $\frac{1}{3}$, the derivative at $x = 4$ is $\frac{1}{4}$, and so on.

Derivative of ln x

If $y = \ln x$, then $\dfrac{dy}{dx} = \dfrac{1}{x}$ for positive x values.

Our next example illustrates the use of all of the rules presented in this section and the previous section.

EXAMPLE 2 *Finding Derivative Formulas*

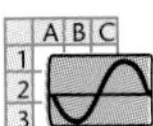

Find the derivatives of the following functions:

a. $f(x) = 12.36 + 6.2 \ln x$ b. $g(t) = 4e^t + 19$

c. $m(r) = \frac{8}{r} - 12\sqrt{r}$ d. $j(y) = 17{,}000\left(1 + \frac{0.025}{12}\right)^{12y}$

Solution

a. Apply the Constant Rule to the first term and the Constant Multiplier and Natural Log Rules to the second term. Use the Sum Rule to add the two derivatives together.

$$\frac{df}{dx} = f'(x) = 0 + 6.2\left(\frac{1}{x}\right) = \frac{6.2}{x}$$

b. Apply the Constant Multiplier and e^x Rules to the first term and the Constant Rule to the second term. Again, use the Sum Rule to add the two derivatives.

$$\frac{dg}{dt} = g'(t) = 4e^t + 0 = 4e^t$$

c. The key to finding the derivative formula for this function is rewriting the two terms using algebra rules for exponents. Recall that a negative exponent is used to indicate that a term is in the denominator and that an exponent of $\frac{1}{2}$ indicates a square root. Using these facts, rewrite m as

$$m(r) = 8r^{-1} - 12r^{1/2}$$

Now apply the Constant Multiplier, Power, and Sum Rules to obtain

$$\frac{dm}{dr} = m'(r) = -1(8)r^{-1-1} - \frac{1}{2}(12)r^{1/2-1} = -8r^{-2} - 6r^{-1/2} = \frac{-8}{r^2} - \frac{6}{\sqrt{r}}$$

d. Begin by calculating the number $\left(1 + \frac{0.025}{12}\right)^{12}$ in order to rewrite the formula in the form ab^x. We have rounded this number to six decimal places, but in using the derivative to calculate rates of change, you should keep all decimal places stored by your calculator.

$$j(y) = 17{,}000\left(1 + \frac{0.025}{12}\right)^{12y} \approx 17{,}000(1.025288^y)$$

Now apply the Constant Multiplier and Exponential Rules.

$$\frac{dj}{dy} = j'(y) \approx 17{,}000(\ln 1.025288)(1.025288^y) \approx 424.5579(1.025288^y)$$

In summary, we present a list of the rate-of-change formulas you should know. The list on page 286 includes the formulas from Section 4.3. The formulas are best learned through practice. We urge you to work as many of the Activities in this section as possible. Although you may need to refer to this table for some of the beginning Activities, you should attempt to work most of them without looking at this list.

Simple Derivative Rules

Rule Name	*Function*	*Derivative*
Constant Rule	$y = b$	$\frac{dy}{dx} = 0$
Linear Function Rule	$y = ax + b$	$\frac{dy}{dx} = a$
Power Rule	$y = x^n$	$\frac{dy}{dx} = nx^{n-1}$
Exponential Rule	$y = b^x, b > 0$	$\frac{dy}{dx} = (\ln b)b^x$
e^x Rule	$y = e^x$	$\frac{dy}{dx} = e^x$
Natural Log Rule	$y = \ln x, x > 0$	$\frac{dy}{dx} = \frac{1}{x}$
Constant Multiplier Rule	$y = kf(x)$	$\frac{dy}{dx} = kf'(x)$
Sum Rule	$y = f(x) + g(x)$	$\frac{dy}{dx} = f'(x) + g'(x)$

4.4 Concept Inventory

- *Derivative formulas*

  For constants a, b, and n,

  If $y = b$, then $y' = 0$.
  If $y = ax + b$, then $y' = a$.
  If $y = x^n$, then $y' = nx^{n-1}$.
  If $y = b^x$, then $y' = (\ln b)b^x$.
  If $y = e^x$, then $y' = e^x$.
  If $y = \ln x$, then $y' = \frac{1}{x}$.
  If $y = kf(x)$, then $y' = kf'(x)$.
  If $y = f(x) \pm g(x)$, then $y' = f'(x) \pm g'(x)$.

4.4 Activities

For each of the functions whose graphs are given in Activities 1–6, first sketch the slope graph and then give the slope equation.

1.

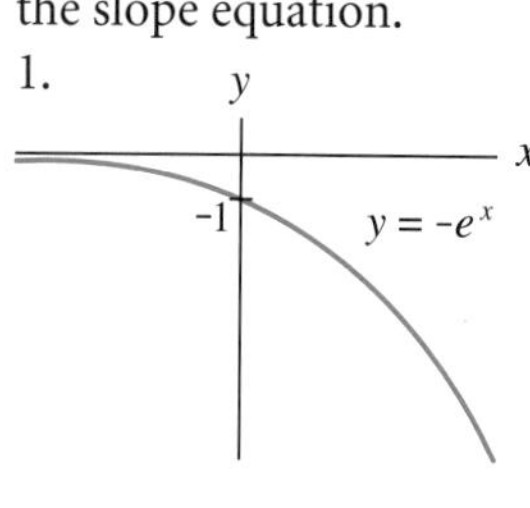

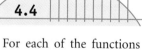

2.

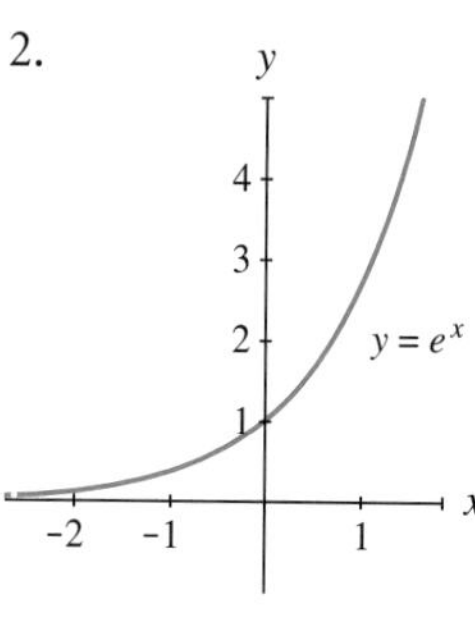

3.

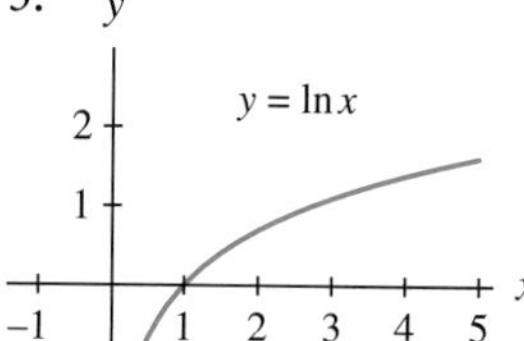

4.

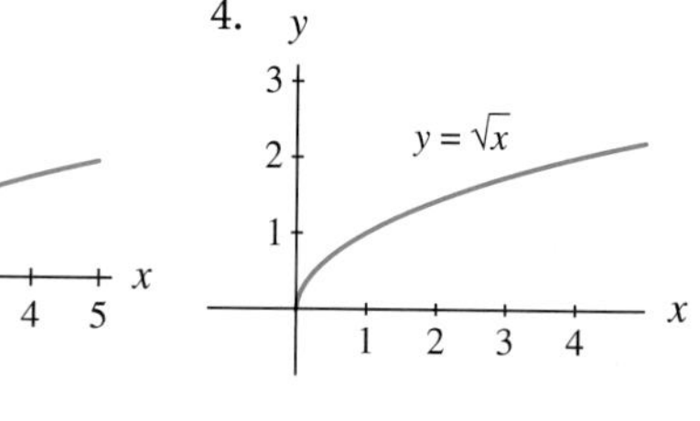

5.

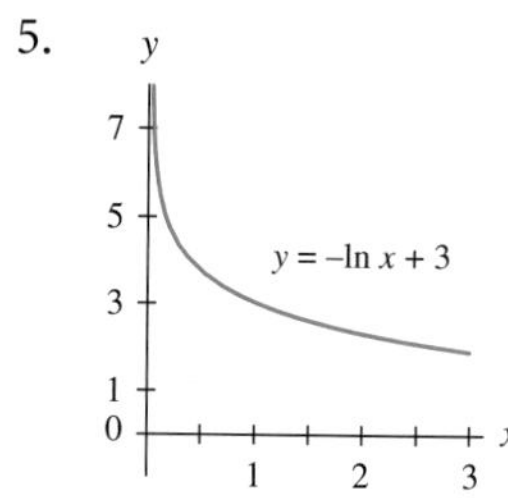

6.

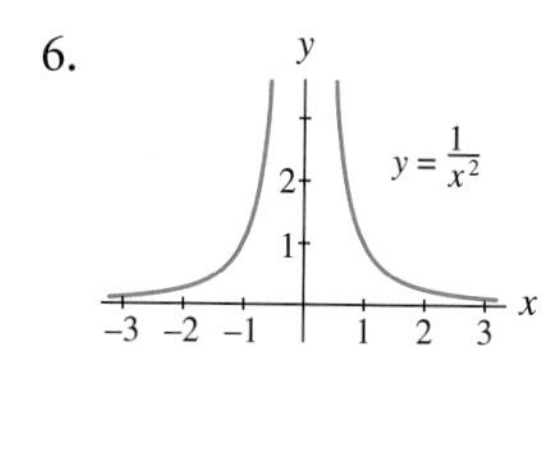

Give the derivative formulas for the functions in Activities 7 through 14.

7. $h(x) = 7x^2 + 13 \ln x$
8. $y = 1 + 4.9876e^x$
9. $g(x) = 17(4.962^x)$
10. $y = \frac{-9}{x}$
11. $f(x) = 100{,}000\left(1 + \frac{0.05}{12}\right)^{12x}$ (*Hint:* Rewrite as ab^x.)
12. $j(x) = 4.2(0.8^x) + \frac{1.6}{x}$

13. $y = 9 - 4.2 \ln x + 3.3(2.9^x)$

14. $k(x) = 3\sqrt{x} + 9e^x$

15. For the first couple of hours after yeast dough has been kneaded, it approximately doubles in volume every 42 minutes. If we prepare 1 quart of yeast dough and let it rise in a warm room, then its growth can be modeled by the function

$$V = e^h \text{ quarts}$$

where h is the number of hours the dough has been allowed to rise.

a. How many minutes will it take the dough to attain a volume of 2.5 quarts?

b. Write a formula for the rate of growth of the yeast dough.

c. How quickly is the dough expanding after 24 minutes, after 42 minutes, and after 55 minutes? Report your answers in quarts per minute.

16. The value of a \$1000 investment in an account with 4.3% interest compounded continuously can be modeled as

$$A = e^{0.043t} \text{ thousand dollars after } t \text{ years}$$

a. Write the rate-of-change formula for the value of the investment. (*Hint:* Let $b = e^{0.043}$, and use the rule for $y = b^t$.)

b. How much is the investment worth after 5 years?

c. How quickly is the investment growing after 5 years?

d. What is the percentage rate of growth after 5 years?

17. We have seen that the derivative of $y = b^x$ is $\frac{dy}{dx} = (\ln b)b^x$ as long as $b > 0$. We have also seen that the derivative of $y = e^x$ is $\frac{dy}{dx} = e^x$.

a. Show that the derivative formula for $y = e^x$ is a special case of the derivative formula for $y = b^x$ by applying the formula for $y = b^x$ to $y = e^x$ and then reconciling the result with the known derivative formula for $y = e^x$.

b. Use the derivative formula for $y = b^x$ to find a formula for the derivative of an exponential function of the form $y = e^{kx}$, where k is some known constant.

18. Dairy company managers have found that it costs them approximately $c(u) = 3250 + 75 \ln u$ dollars to produce u units of dairy products each week. They also know that it costs them approximately $s(u) = 50u + 1500$ dollars to ship u units. Assume that the company ships its products once each week.

a. Write the formula for the total weekly cost of producing and shipping u units.

b. Write the formula for the rate of change of the total weekly cost of producing and shipping u units.

c. How much does it cost the company to produce and ship 5000 units in 1 week?

d. What is the rate of change of total production and shipping costs at 5000 units? Interpret your answer.

19. The weight of a laboratory mouse between 3 and 11 weeks of age can be modeled by the equation

$$w(t) = 11.3 + 7.37 \ln t \text{ grams}$$

where the age of the mouse is $(t + 2)$ weeks after birth (thus for a 3-week-old mouse, $t = 1$).

a. What is the weight of a 9-week-old mouse, and how rapidly is its weight changing?

b. How rapidly on average does the mouse grow between ages 6 and 11 weeks?

c. What happens to the rate at which the mouse is growing as it gets older? Explain.

20. The projected number of homes with access to the Internet via cable television between 1998 and 2005 can be modeled[31] by the equation

$$I(x) = -138.27 + 76.29 \ln x \text{ million homes}$$

x years after 1990.

a. Give the rate-of-change formula for the projected number of such homes.

b. How many homes are projected to have Internet access via a cable TV system in 2005, and how rapidly is that number projected to be growing?

21. The consumer price index for college tuition[32] between 1990 and 1997 is shown in Table 4.18.

TABLE 4.18

Year	CPI	Year	CPI	Year	CPI
1990	175.0	1993	233.5	1996	279.8
1991	192.8	1994	249.8	1997	294.1
1992	213.5	1995	264.8		

31. Based on data from Paul Kagen Associates, Inc., Cable Television Technology.
32. *Statistical Abstract*, 1998.

a. Align the data as the number of years since 1980, and find a log model for the CPI.

b. Use only the data to estimate the rate of change of the CPI in 1995.

c. Use the model to find the rate of change of the CPI in 1995.

22. An individual has $45,000 to invest: $32,000 will be put into a low-risk mutual fund averaging 6.2% interest compounded monthly, and the remainder will be invested in a bond fund averaging 9.7% interest compounded continuously.

a. Find an equation for the total amount in the two investments.

b. Give the rate-of-change equation for the combined amount.

c. How rapidly is the combined amount of the investments growing after 6 months? after 15 months?

23. The Bureau of the Census reports the median family income since 1947 as shown in Table 4.19. (Median income means that half of American families make more than this value and half make less.)

TABLE 4.19

Year	Median family income (constant 1997 dollars)
1947	20,102
1957	26,133
1967	35,076
1977	40,656
1987	43,756
1997	44,568

a. Find a model for the data.

b. Find a formula for the rate of change of the median family income.

c. Find the rates of change and percentage rates of change of the median family income in 1972, 1980, 1984, 1992, and 1996.

d. Do you think the above rates of change and percentage rates of change affected the reelection campaigns of Presidents Nixon (1972), Carter (1980), Reagan (1984), Bush (1992), and Clinton (1996)?

24. The data in Table 4.20 give the value of a $2000 investment after 2 years in an account whose annual percentage yield is $100r$%.

TABLE 4.20

r APY	Value (dollars)	r APY	Value (dollars)
0.02	2080.80	0.08	2332.80
0.03	2121.80	0.09	2376.20
0.04	2163.20	0.10	2420.00
0.05	2205.00	0.11	2464.20
0.06	2247.20	0.12	2508.80
0.07	2289.80		

a. Find a model for the value of a $2000 investment after two years as a function of the interest rate (in decimals).

b. Find the rate of change of the value of the investment when the annual percentage yield is 4%. Interpret your answer.

c. Repeat part *a*; however, instead of entering the interest rate in decimals, enter the rate in whole numbers so that $r = 2$ at 2%, $r = 3$ at 3%, etc. Compare this model with the model in part *a*.

d. Use the model from part *c* to find the rate of change of the value of the investment when the annual percentage yield is 4%. Compare this answer to that of part *b*.

25. The value of a $1000 investment after 10 years in an account whose interest rate is $100r$% compounded continuously is

$$A(r) = 1000e^{10r} \text{ dollars}$$

a. Write the rate-of-change function for the value of the investment.

b. Determine the rate of change of the value of the investment at 7% interest. Discuss why the rate of change is so large.

c. If the interest rate is input as a percentage instead of a decimal, the function for the value of a $1000 investment after 10 years in an account with interest compounded continuously is

$$A(r) = 1000e^{0.1r} \text{ dollars}$$

where $r = 1.00$ when the interest rate is 1%, $r = 1.25$ when the interest rate is 1.25%, etc. Write the rate-of-change function for the value of the investment. Compare this rate of change function to that in part *a*.

d. Determine the rate of change of the value of the investment at 7% interest. Compare this answer to that of part *b*.

26. The percentage of households[33] with TVs that also have VCRs from 1990 through 1996 is shown in Table 4.21.

TABLE 4.21

Year	Households (percent)	Year	Households (percent)
1990	68.6	1994	79.0
1991	71.9	1995	81.0
1992	75.0	1996	82.2
1993	77.1		

a. Align the input data as the number of years since 1985, and find a log model for the data.

b. Write the rate-of-change formula for the model in part *a*.

c. According to the model, what was the percentage of households with TVs that also had VCRs in 1997? How rapidly was the percentage growing in that year? Interpret your anwers.

27. The population of Aurora, a Nevada ghost town, can be modeled[34] as

$$p(t) = \begin{cases} -7.91t^3 + 120.96t^2 + 193.92t - 123.21 \text{ people} \\ \qquad \text{when } 0.7 \le t \le 13 \\ 45{,}544(0.8474^t) \text{ people} \\ \qquad \text{when } 13 < t \le 55 \end{cases}$$

where t is the number of years since the beginning of 1860.

a. Write the derivative equation for the population model.

b. How quickly was the population growing or declining in the beginning of the years 1870, 1873, and 1900?

28. An alternative model[35] for the population of Iowa in the 1980s and early 1990s is

$$p(t) = \begin{cases} -2.1t^2 - 63t + 2914 \text{ thousand people} \\ \qquad \text{when } 0 \le t \le 7 \\ 0.435(1.45479^t) \text{ thousand people} \\ \qquad \text{when } 7 < t \le 13 \end{cases}$$

where t is the number of years since 1980.

a. Write the derivative equation for the population model.

b. How quickly was the population growing or declining in 1985? in 1987? in 1989?

c. Can the model be used to answer reasonably the question "How quickly was the population of Iowa growing or declining in 1995?"

4.5 The Chain Rule

We now present derivative formulas for more complicated functions than those considered in Sections 4.3 and 4.4. In particular, we introduce the method used for finding derivative formulas for composite functions.

The First Form of the Chain Rule

It is well known that high levels of carbon dioxide (CO_2) in the atmosphere are linked to increasing populations in highly industrialized societies. This is because large urban environments consume enormous amounts of energy and CO_2 is a natural by-product of the (often incomplete) consumption of that energy.

Imagine that in a certain large city, the level of CO_2 in the air is linked to the size of the population by the equation $C(p) = \sqrt{p}$, where the units of $C(p)$ are parts per million (ppm) and p is the population. Also suppose that the population is growing quadratically and can be modeled by the equation $p(t) = 400t^2 + 2500$ people, where

You may wish to introduce this section with a simpler example if you feel your students would benefit from a less involved, algebraic example of the Chain Rule before an applied illustration is given.

33. *Statistical Abstract*, 1998.
34. Based on data from Don Ashbaugh, *Nevada's Turbulent Yesterday: A Study in Ghost Towns* (Los Angeles: Westernlore Press, 1963).
35. Based on data from *Statistical Abstract*, 1994.

t is the number of years since 1990. Note that C is a function of p, and p is a function of t. Thus indirectly, C is also a function of t. Suppose we want to know the rate of change of the CO_2 concentration *with respect to time* in 2003—that is, how rapidly the CO_2 concentration is rising or falling in 2003. The notation for this rate of change is $\frac{dC}{dt}$, and the units are ppm per year.

The derivative of C is $\frac{dC}{dp} = \frac{1}{2\sqrt{p}}$ ppm per person. But this is not the rate of change that we want because $\frac{dC}{dp}$ is the rate of change *with respect to population*, not time. The question now becomes "How do we transform ppm per person to ppm per year?" If we knew the rate of change of population with respect to time (people per year), then we could multiply as indicated to get the desired units:

$$\left(\frac{\text{ppm}}{\text{person}}\right)\left(\frac{\text{people}}{\text{year}}\right) = \frac{\text{ppm}}{\text{year}}$$

Population is given as a function of time, so its derivative is the rate of change that we need: $\frac{dp}{dt} = 800t$ people per year. This motivates

$$\left(\frac{dC}{dp}\right)\left(\frac{dp}{dt}\right) = \frac{dC}{dt} \quad \text{or}$$

$$\left(\frac{1}{2\sqrt{p}}\,\frac{\text{ppm}}{\text{person}}\right)\left(800t\,\frac{\text{people}}{\text{year}}\right) = \frac{dC}{dt}\,\frac{\text{ppm}}{\text{year}}$$

Because $\frac{dC}{dt}$ is a rate of change *with respect to time*, it is standard procedure to write the derivative formula in terms of t. Recall that $p(t) = 400t^2 + 2500$ people, where t is the number of years since 1990. Substituting $400t^2 + 2500$ for p in the equation for $\frac{dC}{dt}$, we have

$$\frac{dC}{dt} = \frac{1}{2\sqrt{400t^2 + 2500}}(800t) \text{ ppm/year}$$

Now we substitute $t = 13$ (for 2003) to obtain our desired result:

$$\frac{dC}{dt} = \frac{1}{2\sqrt{400(13)^2 + 2500}}[800(13)] \approx 19.64 \text{ ppm/year}$$

In 2003, the CO_2 concentration was increasing by approximately 19.64 ppm per year.

The method used to find $\frac{dC}{dt}$ in the situation above is called the **Chain Rule** because it links together the derivatives of two functions to obtain the derivative of their composite function.

The Chain Rule (Form 1)

If C is a function of p and p is a function of t, then

$$\frac{dC}{dt} = \left(\frac{dC}{dp}\right)\left(\frac{dp}{dt}\right)$$

EXAMPLE 1 *Violin Production*

Let $A(v)$ denote the average cost to produce a student violin when v violins are produced, and let $v(t)$ represent the number (in thousands) of student violins produced t years after 2000. Suppose that 10 thousand student violins are produced in 2008 and that the average cost to produce a violin at that time is \$142.10. Also, suppose that in

2008 the production of violins is increasing by 100 violins per year and the average cost of production is decreasing by 15 cents per violin.

a. Describe the meaning and give the value of each of the following in 2008:

   i. $v(t)$ ii. $v'(t)$ iii. $A(v)$ iv. $A'(v)$

b. Calculate the rate of change with respect to time of the average cost for student violins in 2008.

Solution

a. i. There are $v(8) = 10$ thousand violins produced in 2008.
   ii. The rate of change of violin production in 2008 is $v'(8) = 0.1$ thousand violins per year. That is, $\frac{dv}{dt} = 100$ violins per year.
   iii. The average cost to produce a violin is $A(10) = \$142.10$ when 10 thousand violins are produced.
   iv. When 10 thousand violins are produced, the average cost is changing at a rate of $A'(10) = -\$0.15$ per violin. That is, $\frac{dA}{dv} = -\$0.15$ per violin.

b. The rate of change with respect to time of the average cost to produce a student violin in 2008 is

$$\begin{aligned}\frac{dA}{dt} &= \frac{dA}{dv}\cdot\frac{dv}{dt}\\ &= (-\$0.15 \text{ per violin})(100 \text{ violins per year}) = -\$15 \text{ per year}\end{aligned}$$

In 2008 the average cost to produce a violin is declining by \$15 per year.

The Second Form of the Chain Rule

Recall the discussion at the beginning of this section concerning CO_2 pollution and population. We were given two functions—p with input t and C, whose input corresponds to the output of p—and then asked to find the derivative $\frac{dC}{dt}$. You may wonder why we did not substitute the expression for population into the CO_2 equation before finding the derivative:

$$C(p) = \sqrt{p} \text{ with } p(t) = 400t^2 + 2500, \text{ so } C(p(t)) = \sqrt{400t^2 + 2500}$$

This process, called function composition (see Section 1.3), allows us to express C directly as a function of t. If we now take the derivative, we get $\frac{dC}{dt}$, which is exactly what we needed all along! The reason we did not do this before is that we did not know a formula for finding the derivative of a composite function. However, we can now use the Chain Rule to obtain a formula. First, we review some terminology from Section 1.3.

Because $p(t)$ was substituted into the formula for C to create the composite function $C \circ p$, we call p the inside function and C the outside function. Next, recall the Chain Rule process:

$$\begin{aligned}\frac{dC}{dt} &= \left(\frac{dC}{dp}\right)\left(\frac{dp}{dt}\right)\\ &= \left(\frac{1}{2\sqrt{p}}\right)(800t)\\ &= \left(\frac{1}{2\sqrt{400t^2 + 2500}}\right)800t\end{aligned}$$

The first term, $\frac{1}{2\sqrt{400t^2 + 2500}}$, is simply the derivative of $\sqrt{p}$ with p replaced by $400t^2 + 2500$. This is the derivative of the outside function with the inside function substituted for p. The second term, $800t$, is the derivative with respect to t of p, the inside function. This leads us to a second form of the Chain Rule. If a function is expressed as a result of function composition (that is, it is a combination of an inside function and an outside function), then its slope formula can be found as follows:

We have found that some students find this verbal statement of the Chain Rule easier to remember and apply than a symbolic statement.

$$\text{Slope formula of composite function} = \left(\begin{array}{c}\text{derivative of the}\\ \text{outside function}\\ \text{with the inside}\\ \text{function untouched}\end{array}\right)\left(\begin{array}{c}\text{derivative}\\ \text{of the inside}\\ \text{function}\end{array}\right)$$

Mathematically, we state this form of the Chain Rule as follows:

The Chain Rule (Form 2)

If a function f can be expressed as the composition of two functions h and g—that is, if

$$f(x) = (h \circ g)(x) = h(g(x))$$

then its slope formula is

$$\frac{df}{dx} = f'(x) = h'(g(x)) \cdot g'(x)$$

In Example 2, we consider three somewhat different forms of composite functions, identify the inside function and the outside function for each, and use the Chain Rule to find formulas for the derivatives.

EXAMPLE 2 *Using the Chain Rule*

Write the derivatives (with respect to x) for the following three functions.

a. $y = e^{x^2}$ b. $y = (x^3 + 2x^2 + 4)^{1/2}$ c. $y = \frac{3}{4 - 2x^2}$

Solution

a. We can think of $y = e^{x^2}$ as composed of an outside function $y = e^p$ and an inside function $p = x^2$. The derivative of the outside function is e^p. (This exponential function is its own derivative.) Form 2 of the Chain Rule instructs us to leave the inside function untouched (that is, in its original form) so instead of e^p appearing in the derivative, the first expression in the slope formula is e^{x^2}. The second expression in the slope formula is the derivative of the inside function, $2x$. The final answer is the product of these two derivatives.

$$\frac{dy}{dx} = (e^{x^2})(2x) = 2xe^{x^2}$$

b. The inside function of $y = (x^3 + 2x^2 + 4)^{1/2}$ is $p = x^3 + 2x^2 + 4$, and the outside function is $y = p^{1/2}$. The derivative of the outside function is $\frac{1}{2}p^{-1/2}$; with p

untouched, this becomes $\frac{1}{2}(x^3 + 2x^2 + 4)^{-1/2}$. The derivative of the inside function is $3x^2 + 4x$. Thus the Chain Rule gives

$$\frac{dy}{dx} = \frac{1}{2}(x^3 + 2x^2 + 4)^{-1/2}(3x^2 + 4x)$$

c. The function $y = \frac{3}{4 - 2x^2}$ can be thought of as the composition of the outside function $y = \frac{3}{p}$ and the inside function $p = 4 - 2x^2$. The derivative of the outside function is $\frac{-3}{p^2}$ or $\frac{-3}{(4 - 2x^2)^2}$. The derivative of the inside function is $-4x$. The derivative of the composite function is then

$$\frac{dy}{dx} = \left(\frac{-3}{(4 - 2x^2)^2}\right)(-4x) = \frac{12x}{(4 - 2x^2)^2}$$

As illustrated in Example 3, one common use of the Chain Rule is to find the derivative of a logistic function.

EXAMPLE 3 *VCRs*

The percentage of households between 1980 and 1996 with VCRs can be modeled[36] by

$$P(t) = \frac{77.9}{1 + 39.5e^{-0.4984t}} \text{ percent}$$

where t is the number of years since 1980. Find the rate-of-change formula for P with respect to t.

Solution The function P can be rewritten as

$$P(t) = 77.9(1 + 39.5e^{-0.4984t})^{-1}$$

In this form, it is easy to see that $p = 77.9u^{-1}$ is the outside function and that the inside function is $u = 1 + 39.5e^{-0.4984t}$. Further, we can split u into an outside and inside function with $u = 1 + 39.5e^{v}$ as the outside function and $v = -0.4984t$ as the inside function.

Now, the derivative of P with respect to t is

$$\begin{aligned} P'(t) &= (\text{derivative of } 77.9u^{-1})(\text{derivative of } u) \\ &= (\text{derivative of } 77.9u^{-1})[(\text{derivative of } 1 + 39.5e^{v})(\text{derivative of } v)] \\ &= (-77.9u^{-2})[(39.5e^{v})(-0.4984)] \\ &= \frac{-77.9(39.5e^{v})(-0.4984)}{u^2} \\ &\approx \frac{1533.602e^{v}}{u^2} \end{aligned}$$

Next, substitute $u = 1 + 39.5e^{-0.4984t}$ and $v = -0.4984t$ back into the expression to obtain the derivative in terms of t.

$$\frac{dP}{dt} \approx \frac{1533.602e^{-0.4984t}}{(1 + 39.5e^{-0.4984t})^2} \text{ percentage points per year}$$

where t is the number of years since 1980.

36. Based on data from *Statistical Abstract*, 1998.

4.5 Concept Inventory

- *Function composition*
- *Inside and outside functions*
- *Chain Rule:*

$$\frac{dC}{dt} = \frac{dC}{dp} \cdot \frac{dp}{dt} \quad \text{(Form 1)}$$

$$C'(t) = C'(p(t)) \cdot p'(t) \quad \text{(Form 2)}$$

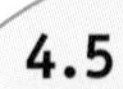

4.5 Activities

1. Let x be a function of t, and let f be a function whose input corresponds to the output of x. If $x(2) = 6$, $f(6) = 140$, $x'(2) = 1.3$, and $f'(6) = -27$, give the values of

   a. $f(x(2))$ b. $\frac{df}{dx}$ when $x = 6$

   c. $\frac{dx}{dt}$ when $t = 2$ d. $\frac{df}{dt}$ when $t = 2$

2. Let v be a function of x, and let g be a function whose input corresponds to the output of v. If $v(88) = 17$, $v'(88) = 1.6$, $g(17) = 0.04$, and $g'(17) = 0.005$, give the values of

   a. $g(v(88))$ b. $\frac{dv}{dx}$ when $x = 88$

   c. $\frac{dg}{dv}$ when $x = 88$ d. $\frac{dg}{dx}$ when $x = 88$

3. An investor has been buying gold at a constant rate of 0.2 troy ounce per day. The investor currently owns 400 troy ounces of gold. If gold is currently worth \$395.70 per troy ounce, how quickly is the value of the investor's gold increasing per day?

4. A gas station owner is unaware that one of the underground gasoline tanks is leaking. The leaking tank currently contains 600 gallons of gas and is losing 3.5 gallons per day. If the value of the gasoline is \$1.51 per gallon, how much potential revenue is the station losing per day?

5. Let $R(x)$ be the revenue in deutsche marks from the sale of x units of a commodity, and let $D(r)$ be the dollar value of r deutsche marks. On January 5, 2001, 10,000 deutsche marks were worth \$20,412, and the rate of change of the dollar value was \$2.0412 per deutsche mark. On the same day, sales were 476 units, producing a revenue of 10,000 deutsche marks, and revenue was increasing by 2.6 deutsche marks per unit. Identify the following values on January 5, 2001, and write a sentence interpreting each value.

   a. $R(476)$ b. $D(10{,}000)$ c. $\frac{dR}{dx}$

   d. $\frac{dD}{dr}$ e. $\frac{dD}{dx}$

6. Suppose that $V(t)$ is the volume of mail (in thousands of pieces) processed at a post office on the tth day of the current year and that $E(v)$ is the number of employee-hours needed to process v thousand pieces of mail. On January 1 of this year, 150 thousand pieces of mail were processed, and that number was decreasing by 200 items per day. The rate of change of the number of employee-hours is a constant 12 hours per thousand pieces of mail. Identify the following quantities on January 1 of this year, and write a sentence interpreting each value.

   a. $V(1)$ b. $\frac{dV}{dt}$

   c. $\frac{dE}{dv}$ d. $\frac{dE}{dt}$

7. The population of a city in the northeast is given by $p(t) = \frac{130}{1 + 12e^{-0.02t}}$ thousand people, where t is the number of years since 1995. The number of garbage trucks needed by the city can be modeled by the equation $g(p) = 2p - 0.001p^3$, where p is the population in thousands. Find the value of the following in 2005:

   a. $p(t)$ b. $g(p)$ c. $\frac{dp}{dt}$

   d. $\frac{dg}{dp}$ e. $\frac{dg}{dt}$

   f. Interpret the answers to parts *a–e*.

8. Let $p(x) = 1.019^x$ deutsche marks be the profit from the sale of x mountain bikes. On January 5, 2001, p deutsche marks were worth $D(p) = \frac{p}{0.4899}$ American dollars. On the same day, sales were 476 mountain bikes. Identify the following quantities on January 5, 2001:

   a. $p(x)$ b. $D(p)$ c. $\frac{dp}{dx}$

   d. $\frac{dD}{dp}$ e. $\frac{dD}{dx}$

   f. Interpret the answers to parts *a–e*.

Rewrite each pair of functions in Activities 9 through 16 as a single composite function, and then find the derivative of the composite function.

9. $c(x) = 3x^2 - 2$ $\quad x(t) = 4 - 6t$

10. $f(t) = 3e^t$ $\quad t(p) = 4p^2$

11. $h(p) = \frac{4}{p}$ $\quad p(t) = 1 + 3e^{-0.5t}$

12. $g(x) = \sqrt{7 + 5x}$ $\quad x(w) = 4e^w$

13. $k(t) = 4.3t^3 - 2t^2 + 4t - 12$ $\quad t(x) = \ln x$

14. $f(x) = \ln x$ $\quad x(t) = 5t + 11$

15. $p(t) = 7.9(1.046^t)$ $\quad t(k) = 14k^3 - 12k^2$

16. $r(m) = \frac{9.1}{m^2} + 3m$ $\quad m(f) = f^4 - f^2$

practice

For each of the composite functions in Activities 17 through 37, identify an outside function and an inside function, and find the derivative with respect to x of the composite function.

17. $f(x) = (3.2x + 5.7)^5$

18. $f(x) = (5x^2 + 3x + 7)^{-1}$

19. $f(x) = \sqrt{x^2 - 3x}$

20. $f(x) = \sqrt[3]{x^2 + 5x}$

21. $f(x) = \ln(35x)$

22. $f(x) = (\ln 6x)^2$

23. $f(x) = \ln(16x^2 + 37x)$

24. $f(x) = e^{3.785x}$

25. $f(x) = 72.378e^{0.695x}$

26. $f(x) = e^{4x^2}$

27. $f(x) = 1 + 58.32e^{0.0856x}$

28. $f(x) = \frac{8}{(x - 1)^3}$

29. $f(x) = \frac{350}{4x + 7}$

30. $f(x) = \frac{112}{1 + 18.370e^{0.695x}} + 7.39$

31. $f(x) = \frac{3706.5}{1 + 8.976e^{-1.243x}} + 89{,}070$

32. $f(x) = \left(\sqrt{x} - 3x\right)^3$

33. $f(x) = 3^{\sqrt{x} - 3x}$

34. $f(x) = 2^{\ln x}$

35. $f(x) = \ln(2^x)$

36. $f(x) = Ae^{-Bx}$

37. $f(x) = \frac{L}{1 + Ae^{-Bx}}$

38. The percentage of children living with their grandparents between 1970 and 1997 can be modeled[37] by the equation

$$p(t) = 3 + 0.214e^{0.09496t} \text{ percent}$$

t years after 1970.

a. Write the rate-of-change formula for p.

b. How rapidly was the percentage of children living with their grandparents growing in 1975?

c. How rapidly on average did the percentage of children living with their grandparents grow between 1970 and 1980?

d. Geometrically illustrate the answers to parts *b* and *c*.

39. Imagine that you invest \$1500 in a savings account at 4% annual interest compounded continuously.

a. Write an equation for the balance in the account after t years.

b. Write an equation for the rate of change of the balance.

c. What is the rate of change of the balance at the end of 1 year? 2 years?

d. Do the rates of change in part *c* tell you how much interest your account will earn over the next year? Explain.

40. The tuition at a private 4-year college from 1995 through 2005 is given in Table 4.22.

TABLE 4.22

Year	Tuition (dollars)	Year	Tuition (dollars)
1995	14,057	2001	16,918
1996	14,434	2002	17,561
1997	14,847	2003	18,264
1998	15,298	2004	19,033
1999	15,790	2005	19,873
2000	16,329		

a. Find an exponential equation of the form $f(x) = ab^x$ to fit the data.

b. Convert the equation you found in part *a* to one of the form $f(x) = ae^{kx}$.

c. Find rate-of-change formulas for both equations.

d. Use both equations to find the rate of change in 2005. How do your answers compare?

37. Based on data from the U.S. Census Bureau.

41. In October of 1999, iGo Corp. offered 5 million shares of public stock at \$9 per share. Revenue for the two years preceding the stock offering can be modeled[38] by the equation

$$R(q) = 2.9 + 0.0314e^{0.62285q} \text{ million dollars}$$

q quarters after the beginning of 1998.

a. Write the rate-of-change formula of R.

b. Complete Table 4.23.

TABLE 4.23

Quarter ending	June 1998	June 1999	June 2000
Revenue			
Rate of change of revenue			
Percentage rate of change of revenue			

42. The percentage of households with TVs that subscribe to cable from 1970 through 1996 can be modeled[39] by the logistic equation

$$P(t) = 6 + \frac{59.5}{1 + 65.1e^{-0.3049t}} \text{ percent}$$

where t is the number of years since 1970.

a. Write the rate-of-change formula for the percentage of households with TVs who subscribe to cable.

b. How rapidly was the percentage growing in 1995?

c. According to the model, what will happen to the percentage of cable subscribers in the long run? Do you believe that the model is a correct predictor of the long-term behavior? Explain.

43. Dispatchers at a sheriff's office[40] record the total number of calls received since 5 a.m. in 3-hour intervals. Total calls for a typical day are given in Table 4.24.

a. Is a cubic or a logistic model more appropriate for this data set? Explain.

b. Find the more appropriate model for the data.

c. Find the rate-of-change formula for the model.

d. Evaluate the rate of change at noon, 10 p.m., midnight, and 4 a.m. Interpret the rates of change.

TABLE 4.24

Time	Total calls since 5 a.m.
8 a.m.	81
11 a.m.	167
2 p.m.	301
5 p.m.	495
8 p.m.	738
11 p.m.	1020
2 a.m.	1180
5 a.m.	1225

e. Discuss how rates of change can help a sheriff's office schedule dispatchers for work each day.

44. A model[41] for the number of states associated with the national P.T.A. organization is

$$m(x) = \frac{49}{1 + 36.0660e^{-0.206743x}} \text{ states}$$

x years after 1895.

a. Write the derivative of m.

b. How many states had national P.T.A. membership in 1902?

c. How rapidly was the number of states joining the P.T.A. growing in 1890? in 1915? in 1927?

45. Civilian deaths due to the influenza epidemic in 1918 can be modeled[42] as

$$c(t) = \frac{93{,}700}{1 + 5095.9634e^{-1.097175t}} \text{ deaths}$$

t weeks after August 31, 1918.

a. How rapidly was the number of deaths growing on September 28, 1918?

b. What percentage increase does the answer to part a represent?

c. Repeat parts a and b for October 26, 1918.

d. Why is the percentage change for parts b and c decreasing even though the rate of change is increasing?

46. A manufacturing company has found that it can stock no more than 1 week's worth of perishable raw material for its manufacturing process. When

38. Based on data from the Securities and Exchange Commission.
39. Based on data from *Statistical Abstract*, 1998.
40. Greenville County, South Carolina, Sheriff's Office.
41. Based on data from Hamblin, Jacobsen, and Miller, *A Mathematical Theory of Social Change* (New York: Wiley, 1973).
42. Based on data from A.W. Crosby, Jr., *Epidemic and Peace 1918* (Westport, CT: Greenwood Press, 1976).

purchasing this material, however, the company receives a discount based on the size of the order. Company managers have modeled the cost data as $C(t) = 196.25 + 44.45 \ln t$ dollars to produce t units per week. Each quarter, improvements are made to the automated machinery to help enhance production. The company has kept a record of the average units per week that were produced in each quarter since January 2000. These data are given in Table 4.25.

TABLE 4.25

Quarter	Units per week
Jan–Mar 2000	2000
Apr–June 2000	2070
July–Sept 2000	2160
Oct–Dec 2000	2260
Jan–Mar 2001	2380
Apr–June 2001	2510
July–Sept 2001	2660
Oct–Dec 2001	2820
Jan–Mar 2002	3000
Apr–June 2002	3200
July–Sept 2002	3410
Oct–Dec 2002	3620
Jan–Mar 2003	3880
Apr–June 2003	4130
July–Sept 2003	4410
Oct–Dec 2003	4690

a. Find an appropriate model for production per week x quarters after January 2000.

b. Use the company cost model along with your production model to write an expression modeling cost per unit as a function of the number of quarters since January 2000.

c. Use your model to predict the company's cost per week for each quarter of 2004.

d. Carefully study a graph of the function in part *b* from January 2000 to January 2005. According to this graph, will cost ever decrease?

e. Find an expression for the rate of change of the cost function in part *b*. Look at the graph of the rate-of-change function. According to this graph, will cost ever decrease?

47. A dairy company's records reveal that it costs the company about $C(u) = 3250.23 + 74.95 \ln u$ dollars per week to produce u units each week. Consumer demand has been increasing, so the company has been increasing production to keep up with demand. Table 4.26 indicates the production of the company, in units per week, from 1990 through 2003.

TABLE 4.26

Year	Production (units per week)
1990	5915
1991	5750
1992	5940
1993	6485
1994	7385
1995	8635
1996	10,245
1997	12,210
1998	14,530
1999	17,200
2000	20,230
2001	23,610
2002	27,345
2003	31,440

a. Describe the curvature of the scatter plot of the data in Table 4.26. What types of equations could be used to fit these data?

b. Find the most appropriate model.

c. Use the company's cost model along with your production model to write an expression modeling cost per week as a function of the number of years since 1990.

d. Write the rate-of-change function of the cost function you found in part *c*.

e. Use your model to estimate the company's cost per week in 2002, 2003, 2004, and 2005. Also, estimate the rates of change for those same years.

f. Carefully study the cost graph from 1990 to 2010. According to this graph, will cost ever decrease? Why or why not?

g. Look at the slope graph of the cost function from 1990 to 2010. According to this graph, will cost ever decrease? Why or why not?

48. The marketing division of a large firm has found that it can model the effectiveness of an advertising campaign by $S(u) = 0.75\sqrt{u} + 1.8$, where $S(u)$ represents sales in millions of dollars when the firm invests u thousands of dollars in advertising. The firm plans to invest $u(x) = -2.3x^2 + 53.2x + 249.8$ thousand dollars each year x years from now.
   a. Write the formula for predicted sales x years from now.
   b. Write the formula for the rate of change of predicted sales x years from now.
   c. What will be the rate of change of sales in 2007?

49. When you are composing functions, why is it important to make sure that the output of the inside function agrees with the input of the outside function?

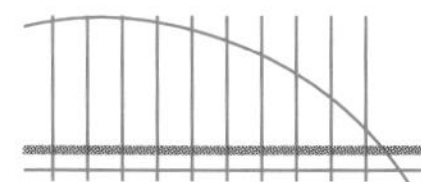

4.6 The Product Rule

You may wish to introduce this section with a simpler example if you feel your students would benefit from a less involved, algebraic example of the Product Rule before an applied illustration is given.

It is fairly common to construct a new function by multiplying two functions. For example, revenue is the number of units sold multiplied by price. If both units sold and price are given by functions, then revenue is given by the product of the two functions. How to find rates of change for product functions is the topic of this section.

Suppose that the enrollment in a university is given by a function E and the percentage (expressed as a decimal) of students who are from out of state is given by a function P. In both functions, the input is t, the year corresponding to the beginning of the school year (that is, for the 2003–04 school year, t is 2003) because school enrollment figures are stated for the beginning of the fall term. Note that the product function $N(t) = E(t) \cdot P(t)$ gives the number of out-of-state students in year t. For example, if in the current year, enrollment began at 17,000 students with 30% of those from out of state, then the number of out-of-state students is calculated as $(17{,}000)(0.30) = 5100$.

Suppose that, in addition to enrollment being 17,000 with 30% from out of state, the enrollment is decreasing at a rate of 1420 students per year $\left(\frac{dE}{dt} = -1420\right)$, and the percentage of out-of-state students is increasing at a rate of 1.5 percentage points per year $\left(\frac{dP}{dt} = 0.015\right)$. How rapidly is the number of out-of-state students changing?

Because $N(t)$, the number of out-of-state students, is the product of $E(t)$ and $P(t)$, there are two rates to consider. First, of the 1420 students per year by which enrollment is declining, 30% of the students are from out of state. The product (-1420 students per year)(0.30) = -426 gives the amount by which the number of out-of-state students is decreasing. Thus, as a consequence of the decline in enrollment, the number of out-of-state students is declining by 426 students per year.

On the other hand, of the 17,000 students enrolled, the percentage of out-of-state students is growing by 1.5% per year. Thus the product (17,000 students) · (0.015 per year) = 255 gives the amount by which the number of out-of-state students is increasing. The increasing percentage of out-of-state students results in a rate of increase of 255 out-of-state students per year.

To find the overall rate of change in the number of out-of-state students, we add the rate of change due to decline in enrollment and the rate of change due to increase in percentage.

$$\underset{\text{per year}}{-426 \text{ students}} + \underset{\text{per year}}{255 \text{ students}} = \underset{\text{per year}}{-171 \text{ students}}$$

We interpret this result as follows: As a result of the declining enrollment and the increasing percentage of students from out of state, the number of out-of-state students is declining by 171 per year.

The steps to obtain this rate of change can be summarized in the equation

$$\begin{aligned}\text{Rate of change of out-of-state students} &= (-1420 \text{ students per year})(0.30) \\ &\quad + (17{,}000 \text{ students})(0.015 \text{ per year}) \\ &= -171 \text{ out-of-state students per year}\end{aligned}$$

Expressed in terms of the functions N, E, and P, this equation can be written as

$$\frac{dN}{dt} = \left(\frac{dE}{dt}\right)P(t) + E(t)\left(\frac{dP}{dt}\right)$$

This result is known as the **Product Rule** and can be stated as follows: If a function is the product of two functions, that is,

$$\text{Product function} = \begin{pmatrix}\text{first}\\ \text{function}\end{pmatrix}\begin{pmatrix}\text{second}\\ \text{function}\end{pmatrix}$$

then

$$\begin{matrix}\text{Derivative}\\ \text{of product}\\ \text{function}\end{matrix} = \begin{pmatrix}\text{derivative}\\ \text{of first}\\ \text{function}\end{pmatrix}\begin{pmatrix}\text{second}\\ \text{function}\end{pmatrix} + \begin{pmatrix}\text{first}\\ \text{function}\end{pmatrix}\begin{pmatrix}\text{derivative}\\ \text{of second}\\ \text{function}\end{pmatrix}$$

The Product Rule

If $f(x) = g(x) \cdot h(x)$, then

$$\frac{df}{dx} = \left(\frac{dg}{dx}\right)h(x) + g(x)\left(\frac{dh}{dx}\right)$$

EXAMPLE 1 *Egg Production*

The industrialization of chicken (and egg) farming brought improvements to the production rate of eggs. Consider a chicken farm that has 1000 laying hens, each of which lays an average of 24 eggs each month. By selling or buying hens, the farmer can decrease or increase production. Also, by selective breeding and genetic research, it is possible that over a period of time the farmer can increase the average number of eggs that each hen lays.

a. How many eggs does the farm produce in a month?

b. Suppose the farmer increases the number of hens by 12 hens per month and increases the average number of eggs laid by each hen by 1 egg per month. By how much will the farmer's production be increasing?

Solution

a. The monthly egg production is the product of h, the number of hens, and l, the number of eggs each hen lays in one month. Currently, $h = 1000$ hens and $l = 24$

eggs per month. The farmer's current monthly production is

$$h \cdot l = (1000 \text{ hens})(24 \text{ eggs per hen}) = 24{,}000 \text{ eggs}$$

b. Let t be the number of months from now. We are told that $\frac{dh}{dt} = 12$ hens per month and that $\frac{dl}{dt} = 1$ egg per hen per month. Applying the Product Rule yields

$$\frac{d(hl)}{dt} = \frac{dh}{dt}l + h\frac{dl}{dt}$$
$$= \left(12\frac{\text{hens}}{\text{month}}\right)\left(24\frac{\text{eggs}}{\text{hen}}\right) + (1000 \text{ hens})\left(1\frac{\text{egg/hen}}{\text{month}}\right)$$
$$= 1288\frac{\text{eggs}}{\text{month}}$$

The farmer's egg production will be increasing by 1288 eggs per month.

As is the case with many calculations in this book, determining units of measure on a final answer is not difficult if the units are attached to the values used in the calculation. Using units of measure with calculations is essential for answering Activities 7–9.

EXAMPLE 2 *Revenue from the Sale of Compact Discs*

A music store has determined from a customer survey that when the price of each CD is $\$x$, the number of CDs sold monthly can be modeled by the function

$$N(x) = 6250\,(0.92985^x) \text{ CDs}$$

Find and interpret the rates of change of revenue when CDs are priced at \$10, \$12, \$13.75, and \$15.

Solution Revenue is the number of units sold times the selling price. In this case, the monthly revenue $R(x)$ is given by

$$R(x) = N(x) \cdot x = 6250(0.92985^x) \cdot x \text{ dollars}$$

where x dollars is the selling price. Using the Product Rule, we find that the rate of change equation is

$$\frac{dR}{dx} = \left[\frac{d}{dx}N(x)\right]x + N(x)\left[\frac{d}{dx}(x)\right]$$
$$= 6250(\ln 0.92985)(0.92985^x) \cdot x + 6250(0.92985^x)(1) \text{ dollars of revenue per dollar of price}$$

where x dollars is the selling price.

Evaluating $\frac{dR}{dx}$ at the indicated values of x yields Table 4.27. At \$10, revenue is increasing by \$823 per \$1 of CD price. In other words, increasing the price results in an increase in revenue. Similarly, at \$12, revenue is increasing by \$332 per \$1 of CD price. At \$13.75, revenue is neither increasing nor decreasing. This is the price at which revenue has reached its peak. Finally, at \$15, revenue is declining by \$191 per \$1 of CD price.

The graph of the revenue function is shown in Figure 4.12. Review the statements above about how the revenue is changing as they are related to the graph.

TABLE 4.27

Price	Rate of change of revenue (to nearest dollar)
\$10.00	823
\$12.00	332
\$13.75	0
\$15.00	-191

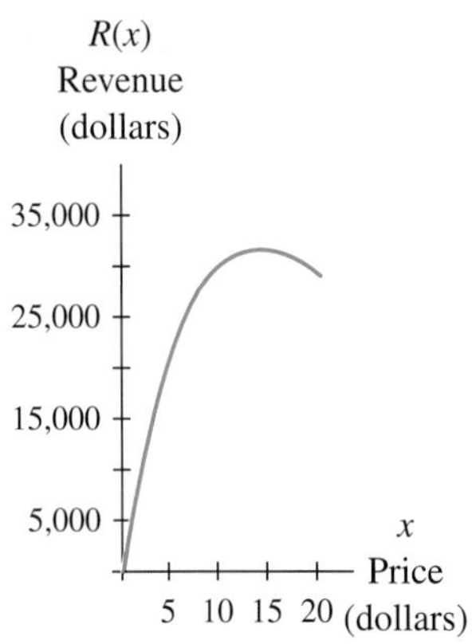

FIGURE 4.12

Often product functions are formed by multiplying a quantity function by a function that indicates the proportion of that quantity for which a certain statement is true. This is illustrated in Example 3.

EXAMPLE 3 *European Tourists*

We have not presented the Quotient Rule because we do not want our focus to be unnecessary formula memorization and the Product and Chain Rules suffice. If you feel that your students would benefit from such a rule, include it.

The number of overseas international tourists who traveled to the United States between 1990 and 1996 can be modeled[43] by the equation $f(t) = 0.069t^3 - 0.5619t^2 + 2.1639t + 14.9534$ million tourists, where t is the number of years since 1990. Suppose that during the same time period, the proportion (percentage expressed as a decimal) of foreign travelers to the United States who were from Europe is given by

$$p(t) = (9.859 \cdot 10^{-4})t^3 - 0.01122t^2 + 0.0293t + 0.4404$$

where t is the number of years since 1990.

a. Find a formula for the number of European tourists to the United States.

b. Find the derivative of the formula in part *a*.

c. Find the number of European tourists to the United States in 1995, and determine how rapidly that number was changing in that year.

Solution

a. The number $N(t)$ of European tourists to the United States is given by the product function $N(t) = f(t) \cdot p(t)$.

$$N(t) = (0.069t^3 - 0.5619t^2 + 2.1639t + 14.9534)[(9.859 \cdot 10^{-4})t^3 - 0.01122t^2 + 0.0293t + 0.4404] \text{ million European tourists}$$

t years after 1990.

b. To use the Product Rule, we need the derivatives f' and p':

$$f'(t) = 0.207t^2 - 1.1238t + 2.1639 \text{ million tourists per year, and}$$
$$p'(t) = 0.0029577t^2 - 0.02244t + 0.0293$$

(Note that we did not label $p'(t)$ with units. If $p(t)$ had been expressed as a percentage, then the units would be percentage points per year. Expressed as a decimal, $p(t)$ is actually a proportion which is a unitless number. While it is possible to label $p'(t)$ as hundredths of a percentage point per year, we choose to state the derivative of p without a label.)

Thus, by the Product Rule,

$$N'(t) = (0.207t^2 - 1.1238t + 2.1639)[(9.859 \cdot 10^{-4})t^3 - 0.01122t^2 + 0.0293t + 0.4404] + (0.069t^3 - 0.5619t^2 + 2.1639t + 14.9534)(0.0029577t^2 - 0.02244t + 0.0293)$$

European tourists per year t years after 1990.

c. The number of European tourists in 1995 is $N(5)$.

$$N(5) = (20.3504)(0.4296375) \approx 8.74 \text{ million tourists}$$

The rate of change of the number of European tourists in 1995 is $N'(5)$.

$$\begin{aligned} N'(5) &\approx (1.7199)(0.4296375) + (20.3504)(-0.0089575) \\ &\approx 0.7389 - 0.1823 \\ &\approx 0.56 \text{ million European tourists per year} \end{aligned}$$

If you are using the two-semester text and integrating Chapter 8 into Chapters 1–7, present Section 8.3 after this section. Section 8.3 covers topics from Chapters 3 and 4.

In 1995, there were approximately 8.74 million tourists from Europe, and that number was growing by about 0.56 million tourists per year.

43. Based on data from *Statistical Abstract*, 1998.

Be forewarned that some of the activities in this section are very time consuming. Be realistic in assigning activities for homework.

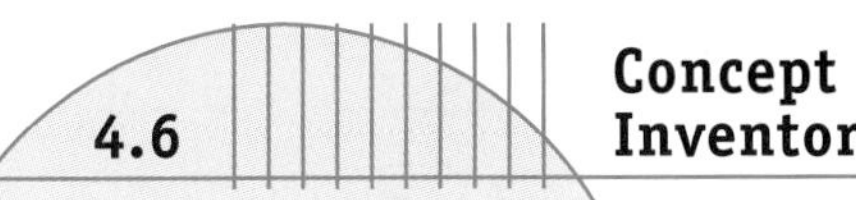

4.6 Concept Inventory

- *Product Rule*

If $f(x) = g(x) \cdot h(x)$, then $\dfrac{df}{dx} = \dfrac{dg}{dx} \cdot h(x) + g(x) \cdot \dfrac{dh}{dx}$

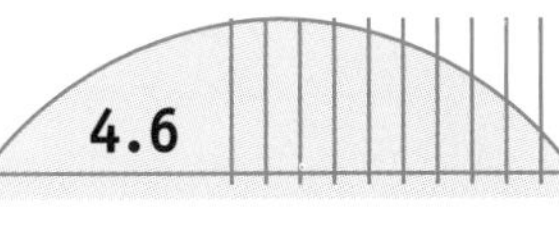

4.6 Activities

1. Find $h'(2)$ if $h(x) = f(x) \cdot g(x)$, $f(2) = 6$, $f'(2) = -1.5$, $g(2) = 4$, and $g'(2) = 3$.

2. Find $r'(100)$ if $r(t) = p(t) \cdot q(t)$, $p(100) = 4.65$, $p'(100) = 0.5$, $q(100) = 160$, and $q'(100) = 12$.

3. Let $h(t)$ be the number of households in a city, and let $c(t)$ be the proportion (expressed as a decimal) of households in that city that own a computer. In both functions, t is the number of years since 1995.
   a. Write sentences interpreting the following mathematical statements.
      i. $h(2) = 75{,}000$ ii. $h'(2) = -1200$
      iii. $c(2) = 0.52$ iv. $c'(2) = 0.05$
   b. If $N(t) = h(t) \cdot c(t)$, what are the input and output of N?
   c. Find the values of $N(2)$ and $N'(2)$. Interpret your answers.

4. Let $D(x)$ be the demand (in units) for a new product when the price is x dollars.
   a. Write sentences interpreting the following mathematical statements.
      i. $D(6.25) = 1000$ ii. $D'(6.25) = -50$
   b. Give a formula for the revenue $R(x)$ generated from the sale of the product when the price is x dollars. (Assume demand = number sold.)
   c. Find $R'(x)$ when $x = 6.25$. Interpret your answer.

5. The value of one share of a company's stock is given by $S(x) = 15 + \frac{2.6}{x+1}$ dollars x weeks after it is first offered. An investor buys some of the stock each week and owns $N(x) = 100 + 0.25x^2$ shares after x weeks. The value of the investor's stock after x weeks is given by $V(x) = S(x) \cdot N(x)$.
   a. Find and interpret the following:
      i. $S(10)$ and $S'(10)$
      ii. $N(10)$ and $N'(10)$
      iii. $V(10)$ and $V'(10)$
   b. Give a formula for $V'(x)$.

6. The number of students in an elementary school t years after 1996 is given by $S(t) = 100 \ln(t + 5)$ students. The yearly cost to educate one student can be modeled by $C(t) = 1500(1.05^t)$ dollars per student.
   a. What are the input and output of the function $F(t) = S(t) \cdot C(t)$?
   b. Find and interpret the following:
      i. $S(3)$ ii. $S'(3)$
      iii. $C(3)$ iv. $C'(3)$
      v. $F(3)$ vi. $F'(3)$
   c. Find a formula for $F'(t)$.

7. A wheat farmer is converting to corn because he believes that corn is a more lucrative crop. It is not feasible for him to convert all his acreage to corn at once. In the current year, he is farming 500 acres of corn and is increasing that number by 50 acres per year. As he becomes more experienced in growing corn, his output increases. He currently harvests 130 bushels of corn per acre, but the yield is increasing by 5 bushels per acre per year. When both the increasing acreage and the increasing yield are considered, how rapidly is the total number of bushels of corn increasing per year?

8. A point guard for an NBA team averages 15 free-throw opportunities per game. He currently hits 72% of his free throws. As he improves, the number of free-throw opportunities decreases by 1 free throw per game, while his percentage of hits increases by 0.5 percentage point per game. When his decreasing free throws and increasing percentage are taken into account, what is the rate of change in the number of free-throw points that this point guard makes per game?

9. Two candidates are running for mayor in a small town. The campaign committee for candidate A has been taking weekly telephone polls to assess the progress of the campaign. Currently there are 17,000 registered voters, 48% of whom are planning to vote. Of those planning to vote, 57% will vote for candidate A. Candidate B has begun some serious mud slinging, which has resulted in increasing public interest in the election and decreasing support for candidate A. Polls show that the percentage of people who plan to vote is increasing by 7 percentage points per week, while the percentage who will vote for candidate A is declining by 3 percentage points per week.
   a. If the election were held today, how many people would vote?

b. How many of those would vote for candidate A?

c. How rapidly is the number of votes that candidate A will receive changing?

practice

Find derivative formulas for the functions in Activities 10 through 26.

10. $f(x) = (x + 5)e^x$

11. $f(x) = (3x^2 + 15x + 7)(32x^3 + 49)$

12. $f(x) = 2.5(0.9^x)(\ln x)$

13. $f(x) = (12.8893x^2 + 3.7885x + 1.2548)[29.685(1.7584^x)]$

14. $f(x) = (5x + 29)^5(15x + 8)$

15. $f(x) = (5.7x^2 + 3.5x + 2.9)^3(3.8x^2 + 5.2x + 7)^{-2}$

16. $f(x) = \dfrac{2.97x^3 + 3.05}{2.71x + 15.29}$

17. $f(x) = \dfrac{12.624(14.831^x)}{x^2}$

18. $f(x) = (8x^2 + 13)\dfrac{39}{1 + 15.29e^{-0.0954x}}$

19. $f(x) = (79.32x)\left(\dfrac{1984.32}{1 + 7.68e^{-0.859347x}} + 1568\right)$

20. $f(x) = \ln(15.7x^3)(e^{15.7x^3})$

21. $f(x) = \dfrac{430(0.62^x)}{6.42 + 3.3(1.46^x)}$

22. $f(x) = (19 + 12\ln 2x)(17 - 3\ln 4x)$

23. $f(x) = 4x\sqrt{3x + 2} + 93$

24. $f(x) = \dfrac{4(3^x)}{\sqrt{x}}$

25. $f(x) = \dfrac{14{,}000x}{1 + 12.6e^{-0.73x}}$

26. $f(x) = \dfrac{1}{(x - 2)^2}(3x^2 - 17x + 4)$

27. During the first 8 months of last year, a grocery store raised the price of a certain brand of tissue paper from \$1.19 to \$1.54. Consequently, sales declined. The price and number sold each month are shown in Table 4.28.

a. Find models for price and number sold as functions of the month.

b. From the models in part *a*, construct an equation for revenue.

TABLE 4.28

Month	Price	Number sold
Jan	\$1.19	279
Feb	\$1.25	277
Mar	\$1.29	272
Apr	\$1.34	266
May	\$1.38	257
June	\$1.45	247
July	\$1.48	236
Aug	\$1.54	221

c. Use the equation to find the revenue in August and the projected revenue in September.

d. Would you expect the rate of change of revenue to be positive or negative in August? Why?

e. Give the rate-of-change formula for revenue.

f. How rapidly was revenue changing in February, August, and September?

28. A music store has determined that the number of CDs sold monthly is approximately

$$\text{Number} = 6250(0.9286^x) \text{ CDs}$$

where x is the price in dollars.

a. Give an equation for revenue as a function of price.

b. If each CD costs the store \$7.50, find an equation for profit as a function of price.

c. Find formulas for the rates of change of revenue and profit.

d. Complete Table 4.29.

TABLE 4.29

Price	Rate of change of revenue	Rate of change of profit
\$13		
\$14		
\$20		
\$21		
\$22		

e. What does the chart tell the store manager about the price corresponding to the highest revenue?

f. What is the price corresponding to the highest profit?

29. The population[44] (in millions) of the United States as a function of the year is given in Table 4.30.

TABLE 4.30

Year	Population (millions)
1970	205.1
1980	227.7
1985	238.5
1990	249.9
1995	263.0
1997	267.9

a. Determine the best model for the data.

b. The percentage of people in the United States who live in the midwest can be modeled[45] by the equation $m(t) = 7.667(0.9688^t)$ percent, where t is the number of years since 1970. Write an expression for the number of people who live in the midwest t years after 1970.

c. Find an expression for the rate of change of the population of the midwest.

d. According to the model, how rapidly was the population of the midwest changing in 1990, 1995, and 2000?

30. Costs for a company to produce between 10 and 90 units per hour are given in Table 4.31.

TABLE 4.31

Units	Cost (dollars)
10	150
20	200
30	250
40	400
50	750
60	1400
70	2400
80	3850
90	5850

a. Find an exponential model for production costs.

b. Find the slope formula for production costs.

c. Convert the model in part *a* to one for the average cost per unit produced.

d. Find the slope formula for average cost.

e. How rapidly is the average cost changing when 15 units are being produced? 35 units? 85 units?

f. Examine the slope graph for average cost. Is there a range of production levels for which average cost is decreasing?

g. Determine the production level at which average cost begins to increase. (That is, find the point at which the rate of change of average cost changes from negative to positive.) Explain how you found this value.

31. Table 4.32 gives the number of men 65 or older in the United States and the percentage of men age 65 or older living below the poverty level.[46]

TABLE 4.32

Year	Men 65 years or older (millions)	Percentage below poverty level
1970	8.3	20.2
1980	10.3	11.1
1985	11.0	8.7
1990	12.5	7.8
1997	14.0	6.8

a. Determine the best model for each set of data.

b. Write an expression for the number of men 65 or older who are living below the poverty level.

c. How rapidly was the number of male senior citizens living below the poverty level changing in 1990 and in 2000?

32. The number of households[47] with TVs in the United States is given in Table 4.33.

TABLE 4.33

Year	Households (millions)	Year	Households (millions)
1970	59	1985	85
1975	69	1990	92
1980	76	1995	95

44. *Statistical Abstract*, 1998.
45. Based on data from *Statistical Abstract*, 1998.
46. *Statistical Abstract*, 1994 and 1998.
47. *Statistical Abstract*, 1998.

a. Find a model for the data.

b. Table 4.34 gives, for 1978 through 1995, the percentages of households[48] with TVs that also have VCRs.

TABLE 4.34

Year	Percentage	Year	Percentage
1978	0.3	1990	71.9
1980	1.1	1995	81.0
1985	20.8		

Align the data so that the input values correspond with those in the model for part *a* (that is, if 1980 is $x = 10$ in part *a*, then you want 1980 to be $x = 10$ here also). Find a logistic model for the data.

c. Find a model for the number of households with VCRs.

d. Find the derivative equation for the number of households with VCRs.

e. How rapidly was the number of households with VCRs growing in 1980? in 1985? in 1990?

33. On the basis of data[49] from a study conducted by the University of Colorado School of Medicine at Denver, the percentage of women receiving regional analgesia (epidural pain relief) during childbirth at small hospitals between 1981 and 1997 can be modeled by the equation

$$p(x) = 0.73(1.2912^x) + 8 \text{ percent}$$

x years after 1980.

Suppose that a small hospital in southern Arizona has seen the yearly number of women giving birth decline as described by the equation

$$b(x) = -0.026x^2 - 3.842x + 538.868$$

women giving birth x years after 1980.

a. Give the equation and its derivative for the number of women receiving regional analgesia while giving birth at the Arizona hospital.

b. Was the percentage of women who received regional analgesia while giving birth increasing or decreasing in 1997?

c. Was the number of women who gave birth at the Arizona hospital increasing or decreasing in 1997?

d. Was the number of women who received regional analgesia during childbirth at the Arizona hospital increasing or decreasing in 1997?

e. If the Arizona hospital made a profit of \$57 per woman for the use of regional analgesia, what was the profit for the hospital from this method of pain relief during childbirth in 1997?

34. The amount[50] of money spent on agricultural research and services from 1990 through 1998 in the United States is given in Table 4.35.

TABLE 4.35

Year	Amount (billions of dollars)
1990	2.197
1993	2.643
1994	2.695
1995	2.758
1996	2.682
1997	2.760
1998	2.840

The purchasing power of the dollar, as measured by producer prices[51] from 1988 through 1997, is given in Table 4.36. (In 1982, one dollar was worth \$1.00.)

TABLE 4.36

Year	Purchasing power of \$1
1988	0.93
1990	0.84
1992	0.81
1994	0.80
1996	0.76
1997	0.76

a. Find models for both sets of data. (Remember to align such that both models have the same input values.)

b. Use these models to determine a new model for the amount, measured in constant 1982 dollars, spent on agricultural research and services.

c. Use your new model to find the rates of change and the percentage rates of change of the amount spent on agricultural research and services in 1992 and 1998.

48. *Statistical Abstract*, 1998.
49. "Healthfile," *Reno Gazette–Journal*, Oct. 19, 1999, p. 4.
50. *Statistical Abstract*, 1998.
51. *Statistical Abstract*, 1998.

d. Why might it be of interest to consider an expenditure problem in constant dollars?

35. Table 4.37 shows the number of students[52] enrolled in the ninth through twelfth grades and the number of dropouts from those same grades in South Carolina for each school year from 1980–1981 through 1989–1990.

TABLE 4.37

School year	Enrollment	Dropouts
1980–81	194,072	11,651
1981–82	190,372	10,599
1982–83	185,248	9314
1983–84	182,661	9659
1984–85	181,949	8605
1985–86	182,787	8048
1986–87	185,131	7466
1987–88	183,930	7740
1988–89	178,094	7466
1989–90	172,372	5768

a. Find a model for enrollment and a cubic model for the number of dropouts.

b. Use the two models that you found in part *a* to construct an equation for the percentage of high school students who dropped out each year.

c. Find the rate-of-change formula of the percentage of high school students who dropped out each year.

d. Look at the rate of change for each school year from 1980–1981 through 1989–1990. In which school year was the rate of change smallest? When was it greatest?

e. Are the rates of change positive or negative? What does this say about high school attrition in South Carolina during the 1980s?

36. A house painter has found that the number of jobs that he has per year is decreasing in inverse proportion to the number of years he has been in business. That is, the number of jobs he has each year can be modeled by $j(x) = \frac{104.25}{x}$, where x is the number of years since 1990. He has also kept a ledger of how much, on average, he was paid for each job. His income per job is presented in Table 4.38.

TABLE 4.38

Year	Income per job (dollars)
1990	430
1991	559
1992	727
1993	945
1994	1228
1995	1597
1996	2075

a. Find an exponential model for his income per job.

b. Write the formula for the painter's total income per year.

c. Write the formula for the rate of change of the painter's income each year.

d. What was the painter's total income in 1996?

e. How rapidly was the painter's income changing in 1996?

37. When you are working with products of models, why is it important to make sure that the input values of the two models correspond? (That is, why must you align both models in the same way?)

38. We have discussed three ways to find rates of change: graphically, numerically, and algebraically. Discuss the advantages and disadvantages of each method. Explain when it would be appropriate to use each method.

52. Compiled from *Rankings of the Counties and School Districts of South Carolina.*

Summary

Numerically Finding Slopes

Instantaneous rates of change are slopes of tangent lines. It is important to remember that when we are given a graph, sketching a tangent line and estimating its slope will give only an approximation to the desired rate of change. If we have no equation to associate with the graph, then such an approximation is the best that we can do, and we should therefore be as precise as possible in our work. When we have an equation $y = f(x)$ to associate with the curve, we can improve our graphical approximations with numerical calculations.

Because the tangent line at a point is the limiting position of secant lines through the point of tangency and other increasingly close points, the slope of the tangent line is the limiting value of the slopes of nearby secant lines. If (x_0, y_0) is the point of tangency on the graph of f, then we can use the equation $y = f(x)$ and the slope formula to calculate numerically the slopes of secant lines through the point (x_0, y_0) and nearby points $(x_0 + h, f(x_0 + h))$.

Algebraically Finding Slopes

The method of numerically estimating slopes can be generalized to provide a valuable algebraic method for finding exact slopes at points, as well as formulas for slopes at any input value. We call this method the Four-Step Method of finding derivatives. This method yields the formal definition of a derivative: If $y = f(x)$, then

$$\frac{dy}{dx} = f'(x) = \lim_{h \to 0} \frac{f(x + h) - f(x)}{h}$$

provided that the limit exists.

Slope Formulas

The Four-Step Method of finding derivatives is invaluable, but it is also cumbersome. For this reason, we desire formulas for derivatives of the most common functions we encounter. The formulas for derivatives of linear functions follow from our understanding of slopes of lines. We stated the Simple Power Rule without proof but proved the Sum and Constant Multiplier Rules. The rules for exponential and log functions were discovered by examining tables of slope values.

Here is a list of slope (derivative) formulas that you should know. Remember, if $y = f(x)$, then the symbolic representation for the derivative is $\frac{dy}{dx}$ or $y' = f'(x)$.

Function	*Derivative*
$y = b$	$\frac{dy}{dx} = 0$
$y = ax + b$	$\frac{dy}{dx} = a$
$y = x^n$	$\frac{dy}{dx} = nx^{n-1}$
$y = e^x$	$\frac{dy}{dx} = e^x$
$y = b^x$	$\frac{dy}{dx} = (\ln b)b^x$
$y = \ln x$	$\frac{dy}{dx} = \frac{1}{x}$
$y = kf(x)$	$\frac{dy}{dx} = kf'(x)$
$y = f(x) \pm g(x)$	$\frac{dy}{dx} = f'(x) \pm g'(x)$

The Chain Rule

The Chain Rule tells us how to calculate rates of change for a composite function. We presented it in two forms. Form 1 of the Chain Rule is most useful when we are given the two functions to compose: If C is a function of p and p is a function of t, then C can be regarded as a function of t, and

$$\frac{dC}{dt} = \left(\frac{dC}{dp}\right)\left(\frac{dp}{dt}\right)$$

If you keep track of the units on each of the two component rates of change, then this form of the Chain Rule is easy to use.

Form 2 of the Chain Rule is most useful when you are presented a function that you can recognize as the composition of an inside function p and an outside function C.

$$\frac{dC}{dt} = \begin{pmatrix}\text{the rate of change of} \\ \text{the outside function} \\ \text{with the inside} \\ \text{function untouched}\end{pmatrix}\begin{pmatrix}\text{the rate of} \\ \text{change of the} \\ \text{inside function}\end{pmatrix} = C'(p(t))\, p'(t)$$

The Product Rule

The Product Rule tells us how to calculate rates of change for a product function. If $f(x) = g(x) \cdot h(x)$, then

$$\frac{df}{dx} = \left(\frac{dg}{dx}\right)h(x) + g(x)\left(\frac{dh}{dx}\right)$$

If you need to calculate a rate of change for a general quotient function, say $h(x) = \frac{f(x)}{g(x)}$, simply view the quotient as a product $h(x) = f(x)\left(\frac{1}{g(x)}\right) = f(x)[g(x)]^{-1}$ and apply the Product Rule.

Slopes of Piecewise Continuous Functions

In order for a derivative to exist at the break point of a piecewise continuous function, the function must be both continuous and smooth at that point. If the break point of the function f occurs at $x = a$, determine if the function is continuous at $x = a$ by answering the question:

$$\text{Is } \lim_{x \to a^-} f(x) = \lim_{x \to a^+} f(x) = f(a)?$$

If the answer is no, then the function is not continuous at $x = a$ and the derivative does not exist at $x = a$. If the function is continuous at $x = a$, determine if the function is smooth by answering the question:

Is the limiting value of slopes of secant lines from the left of $x = a$ equal to the limiting value of slopes of secant lines from the right of $x = a$?

If the answer to this question is no, then the function is not smooth at $x = a$ and the derivative at $x = a$ does not exist. If the function is both continuous and smooth at $x = a$ (that is, the answer to both questions is yes), then the derivative exists and is equal to the limiting value of slopes of secant lines.

There are other formulas for derivatives that we have not given. However, we are providing the formulas that are most useful for the functions encountered in everyday situations associated with business, economics, finance, management, and the social and life sciences. You can look up other formulas (if you should ever need them) in a calculus book that emphasizes applications in science or engineering.

Concept Check

Can you	To practice, try	
• Determine if a function is continuous and/or smooth?	Section 4.1	Activity 3
• Graphically and numerically estimate rates of change?	Section 4.1	Activity 7
• Use the algebraic method to find a rate of change at a point?	Section 4.2	Activity 5
• Use the Four-Step Method to find a rate-of-change formula?	Section 4.2	Activity 9
• Apply simple derivative rules?	Sections 4.3 and 4.4	Any of Activities 7–20
• Use the Chain Rule?	Section 4.5	Activity 7 and any of 17–37
• Use the Product Rule?	Section 4.6	Activity 5 and any of 10–26

Review Test

1. The yearly high stock price[53] for Microsoft Corporation for selected years between 1989 and 1998 is shown in Table 4.39.

   a. Find an exponential model for the data.

   b. Numerically investigate the rate of change of the stock price in 1997. Choose at least three increasingly close points. In Table 4.40, record the close points, the slopes with four decimal places, and the limiting value with two decimal places.

TABLE 4.39

Year	Price per share (dollars)	Year	Price per share (dollars)
1989	0.74	1996	15.02
1990	2.11	1997	31.59
1992	4.38	1998	54.19
1994	6.44		

53. Hoover's Online Guide.

TABLE 4.40

Close point on left	Slope	Close point on right	Slope
Limiting value =		Limiting value =	

c. Interpret the limiting value in part *b*.

d. Give the formula for the derivative of the equation in part *a*. Evaluate the derivative in 1997.

2. The total outstanding mortgage debts[54] in the United States (for 1–4-unit family homes) for selected years between 1980 and 1997 are shown in Table 4.41. Also shown are a model for the data and a graph of the model. (See Figure 4.13).

$A(t) = 0.2578t^4 - 9.09t^3 + 100.655t^2 - 125.2t +$ 1464.215 billion dollars t years after 1980

a. Use only the data to estimate the rate of change in the mortgage debt amounts in 1996.

b. Use the graph to estimate how quickly the mortgage debt amount was changing in 1996.

c. Find the derivative of the equation in 1996. Interpret your answer.

FIGURE 4.13

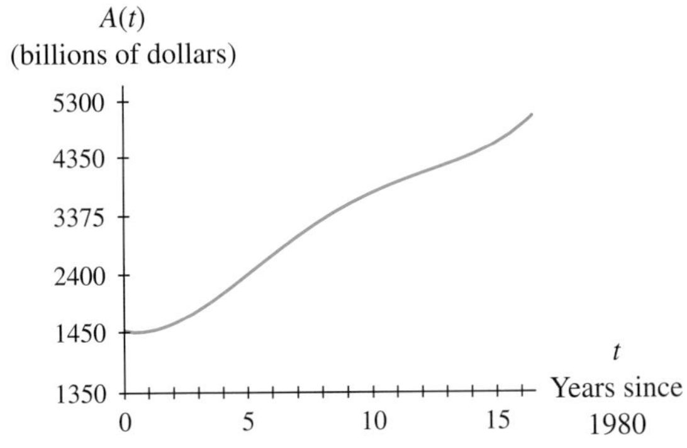

$A(t)$ (billions of dollars)

5300, 4350, 3375, 2400, 1450, 1350

0, 5, 10, 15

t Years since 1980

TABLE 4.41

Year	1980	1985	1990	1992	1994	1995	1996	1997
Total outstanding debt (billions of dollars)	1465	2374	3794	4063	4392	4606	4929	5277

54. *Statistical Abstract*, 1998.

3. The average annual per capita consumption of turkey in the United States between 1980 and 1999 can be modeled[55] by the equation

$$D(t) = \frac{7.978}{1 + 172.8e^{-0.773t}} + 10 \text{ pounds per person}$$

t years after 1980.

a. Find a formula for $\frac{dD}{dt}$.

b. Find the value of $D'(10)$. Interpret your answer.

c. How quickly was the consumption of turkey growing in 1999?

4. If $N(t)$ is the total amount (in billions of dollars) of new mortgages t years after 1980, we find the percentage of outstanding mortgage debt represented by new mortgages each year by dividing $N(t)$ by the equation in Question 2 and multiplying by 100.

$$P(t) = \frac{N(t)}{A(t)} \cdot 100 \text{ percent } t \text{ years after 1980}$$

a. Discuss how you find a formula for the rate of change of the percentage of total mortgage debt represented by new mortgages each year.

b. What are the input and output units of P'?

5. In your own words, outline the Four-Step Method for calculating derivatives. Illustrate the method for the function $f(x) = 7x + 3$.

6. Consider the function

$$g(x) = \begin{cases} 2x + 3 & \text{when } x < -1 \\ 3x + 4 & \text{when } x \geq -1 \end{cases}$$

a. Is g continuous at $x = -1$? Why or why not?

b. Is g smooth at $x = -1$? Why or why not?

c. Give the formula for $\frac{dg}{dx}$.

55. Based on data from *USDA: Cattle and Beef Industry Statistics*, March 1999.

Project 4.1 Fertility Rates

Setting

The *Statistical Abstract of the United States* (1998 edition) reports fertility rates for whites in the United States as shown in Table 4.42.

TABLE 4.42

Year	Fertility rate
1970	2385
1972	1907
1974	1749
1976	1652
1978	1668
1980	1773
1982	1767
1984	1749
1986	1776
1988	1857
1990	2003
1992	1994
1994	1985
1996	2019

Tasks

1. Find a table of fertility rate data in a current edition of the *Statistical Abstract*, and summarize in your own words the meaning of "fertility rate." Assign units to the fertility rate column in Table 4.42.

data

2. Find a piecewise model that fits the given fertility rate data. Write the derivative function for this model. Construct a table of fertility rates and the rate of change of fertility rates for all years from 1970 through 1996. Discuss any points at which the derivative of your model does not exist (and why it does not exist), and explain how you estimated the rates of change at these points.
3. Using the data in a current edition of the *Statistical Abstract*, complete the same analysis as in Task 2 for the fertility rates for blacks and other races in the United States.
4. Add any other recent data for the fertility rate of whites. (*Note:* Recent editions of the *Statistical Abstract* occasionally update older data points. Therefore, you should check the given data and change any updated values so that your data agree with the most recent *Statistical Abstract.*) Find a piecewise model for the updated data. Use this new model to calculate the rates of change that occurred in the years since 1996.

Reporting

Write a report discussing your findings and their demographical impact. Include your mathematical computations as an appendix.

Project 4.2

Superhighway

Setting

The European Communities have decided to build a new superhighway that will run from Berlin through Paris and Madrid and end in Lisbon. This superhighway, like some others in Europe, will have no posted maximum speed so that motorists may drive as fast as they wish. There will be three toll stations installed, one at each border. Because these stations will be so far apart and motorists may not anticipate their need to stop, there has been widespread concern about the possibility of high speed collisions at these stations. There have been editorials protesting the installation of the toll stations, and a private-interest group has lobbied to delay the building of the new superhighway. In response to the concern for safety, the Committee on Transportation has determined that flashing warning lights should be installed at an appropriate distance before each toll station. Your firm has been contracted to study known stopping distances and to develop a model for predicting where the warning lights should be installed.

Tasks

1. Find data that give stopping distances as a function of speed and cite the source for the data. (*Hint:* Drivers' handbooks and Department of Transportation documents are possible sources of data on stopping distance.) Present the data in a table and as a graph. Find a model to fit the data. Justify your choice. Before using your model to extrapolate, consult someone who could be considered an authority to determine whether the model holds outside of the data range. Consult a reliable source to determine probable speeds driven on such a highway. On the basis of your model, make a recommendation about where the warning lights should be placed. Justify your recommendation. Keep in mind that the Committee on Transportation does not wish to post any speed-limit signs. Suggest what other precautions could be taken to avoid accidents at the toll stations.

   data

2. Find rates of change of your model for at least three speeds, one of which should be the speed that you believe to be most likely. Interpret these rates of change in this context. Would underestimating the most likely speed have a serious adverse affect? Support your answer.

Reporting

(Bear in mind that you are reporting to Europeans who wish to see all results in the metric system. However, because you work for an American-based company, you must also have the English equivalent.)

1. Prepare a written report of your study for the Committee on Transportation.
2. Prepare a press release for the Committee on Transportation to use when it announces the implementation of your safety precautions. The press release should be succinct and should answer the questions Who, What, When, Where, and Why.
3. (Optional) Prepare a brief (15-minute) presentation of your results. You will be presenting it to members of the Committee on Transportation.

5

Analyzing Change: Applications of Derivatives

Concept Application

Michael Rosenfeld/Tony Stone Images

Businesses often experience growth or decline as a result of changes in the economy, turnover in management, introduction of new products, or even the whims of consumers. A company measures its performance by measuring revenue, profit, productivity, stock prices, and other quantities. In analyzing its performance, a company may examine the past behavior of quantities such as these and seek the answers to such questions as

- When did the quantity exhibit highs and lows?
- Was there a time when the trend of the quantity changed?
- Is it possible to identify when the rate of change of the quantity was greatest?

Such analysis may be helpful as the company seeks to improve its performance. You will find the tools in this chapter useful in answering questions such as these. One such exploration for the Russell Corporation's revenue is found in Activity 26 of Section 5.4.

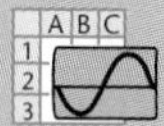

This symbol directs you to the supplemental technology guides. The different guides contain step-by-step instructions for using graphing calculators or spreadsheet software to work the particular example or discussion marked by the technology symbol.

This symbol directs you to the Calculus Concepts *web site, where you will find algebra review, links to data updates, additional practice problems, and other helpful resource material.*

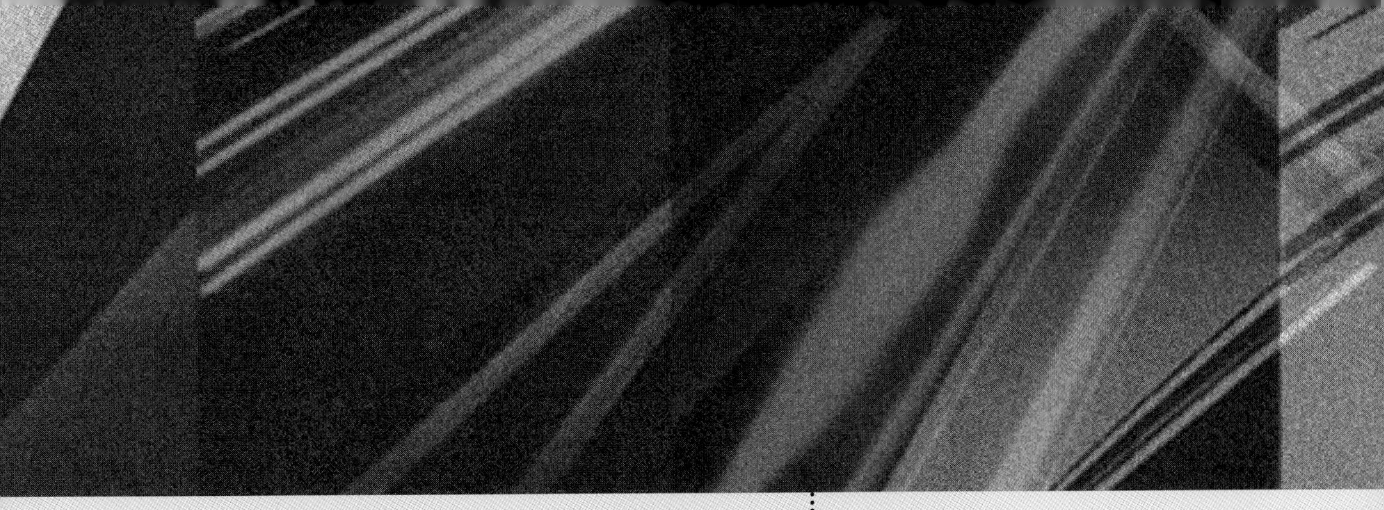

Concept Objectives

This chapter will help you understand the concepts of

- Using tangent lines to approximate change
- Marginal analysis
- Absolute and relative extreme points
- Inflection points
- Second derivative
- Dependent and independent variables
- Related-rates equations

and you will learn to

- Approximate change
- Find relative extreme points
- Find absolute extreme points
- Set up and solve optimization problems
- Find second derivatives
- Use second derivatives to determine concavity
- Find and interpret inflection points
- Find and solve related-rates equations

This chapter is devoted to exploring the ways in which rate-of-change information can be used to help analyze change.

We begin by examining how a rate of change can be used to approximate change in a function's output. We discuss marginal analysis: the use of rates of change to approximate the actual change in the economic quantities of cost, revenue, and profit.

Next, we consider the importance of those places at which the rate of change is zero. In many cases, these points correspond to maximum or minimum function values. If we are searching for maximum and minimum points on a derivative graph, then these points correspond to points of most rapid or least rapid change on the original graph.

Finally, we consider situations in which we find a rate of change of a function with respect to a variable when the variable does not appear in the equation defining the function. The resulting equation, which is called a related-rates equation, is useful in describing how rates of change are interrelated.

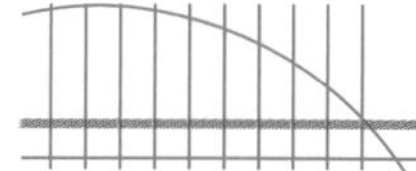

5.1 Approximating Change

The subject of approximating change comes up again in Section 9.4 in the two-semester version of this text.

Recall from our discussion of linear models that the slope of a line $y = ax + b$ can be thought of as how much y changes when x changes by 1 unit. For example, the amount spent on pollution control in the United States during the 1980s can be described by the equation[1]

$$\text{Amount} = 3788.65t - 252{,}216.11 \text{ million dollars}$$

where t is the number of years since 1900. The slope is \$3788.65 million per year, so we can say that each year during the 1980s an *additional* \$3788.65 million was spent on pollution control. It follows that every 2 years spending increased by (2)(\$3788.65) million, every 3 years spending increased by (3)(\$3788.65) million, and so on.

Because rates of change are slopes, we can apply a similar type of reasoning to functions other than lines. However, we must be careful in doing this, for although the slope of a line is constant, the slopes of other functions may change at every point. For example, if you are driving a car, your speed is constantly fluctuating. However, if you know that your speed at a certain moment is 60 mph (1 mile per minute), then you can estimate that over the next minute you will travel approximately 1 mile. Over the next 30 seconds you will travel approximately half a mile, and over the next 2 minutes you will travel approximately 2 miles. Note that we are using the rate of change of distance to estimate by how much the distance traveled changes over a short time period.

The assumption underlying these approximations is that the speed remains constant during a short interval. If the speed deviated dramatically from 60 mph, then the approximations would be invalid. For instance, in the next 30 seconds, you could encounter a stop light and have to stop the car.

As another example, consider the average retail price (cost to the consumer) of a pound of salted, grade AA butter during the 1990s, which can be modeled by the equation[2]

$$p(t) = 0.0517t^2 - 0.2872t + 1.9487 \text{ dollars}$$

where t is the number of years since the end of 1990. We calculate how rapidly the average price was increasing at the end of 1998 as follows:

$$\frac{dp}{dt} = 2(0.0517)t - 0.2872 \text{ dollars per year}$$

When $t = 8$,

$$\frac{dp}{dt} = 2(0.0517)(8) - 0.2872 = \$0.54 \text{ per year}$$

Thus, at the end of 1998, the price of butter was rising by approximately \$0.54 per year. On the basis of this rate of change and the fact that the average price at the end of 1998 was approximately $p(8) = \$2.96$, we formulate the following approximations:

- During the following year (1999), the price increased by approximately \$0.54 to a price of \$3.50.

1. Based on data from *Statistical Abstract*, 1992.
2. Based on data from *Statistical Abstract*, 1998.

- During the first 6 months of 1999, the price increased by approximately $\frac{1}{2}(\$0.54) = \$0.27$ to a price of \$3.23.
- During the first quarter of 1999, the price increased by approximately $\frac{1}{4}(0.54) = \$0.14$ to a price of \$3.10.

Again, we used the rate of change of price to estimate the change in price. Compare these approximations, including those for time periods of 9 months and 2 years, to the actual values given by the model, which are listed in Table 5.1.

TABLE 5.1

Time from end of 1998	Approximated price	Price from equation	Difference between equation value and approximation
2 years	\$2.96 + 2(\$0.54) = \$4.04	$p(10)$ = \$4.25	\$0.21
1 year	\$2.96 + 1(\$0.54) = \$3.50	$p(9)$ = \$3.55	\$0.05
9 months	\$2.96 + 0.75(\$0.54) = \$3.37	$p(8.75)$ = \$3.39	\$0.02
6 months	\$2.96 + 0.5(\$0.54) = \$3.23	$p(8.5)$ = \$3.24	\$0.01
3 months	\$2.96 + 0.25(\$0.54) = \$3.10	$p(8.25)$ = \$3.10	\$0.00

Note that the shorter the time period, the closer the approximated price is to the price given by the model. This is no coincidence, because over a short time period, the rate of change is more likely to be nearly constant than over a longer period of time. Rates of change can often be used to approximate changes in a function, and they generally give good approximations over small intervals.

The justification for using rates of change to approximate changes in a function is best understood from a graphical point of view. We know that the rate of change of a function at a point is the slope of the line tangent to the graph of the function at that point. When we use the rate of change to make a prediction about the function, we are actually making a statement about what the tangent line will do and are assuming that the function will exhibit similar behavior. Indeed, over a small enough interval, the tangent line and function graph are very much alike. (Recall the local linearity discussion in Section 3.2.) Consider the model for butter prices from $t = 8$ through $t = 9$ and the line tangent to the graph at $t = 8$. These graphs are shown in Figure 5.1.

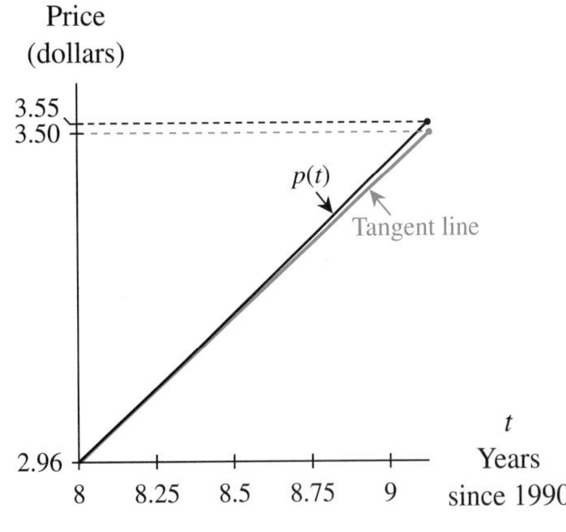

FIGURE 5.1

We used the slope of the line tangent to the graph of p at $t = 8$ to approximate values of $p(t)$ at $t = 10, t = 9, t = 8.75, t = 8.5$, and $t = 8.25$. The approximate values that we found are actually output values on the line tangent to the graph of p at $t = 8$. Note in Figure 5.1 that the closer we are to $t = 8$, the closer the tangent line is to the graph of p. Similarly, the farther we move from $t = 8$, the more the tangent line deviates from the graph. This is generally true, and it is the reason we restrict ourselves to small intervals when using a rate of change to approximate change in a function.

To summarize, consider the following statements:

- The change in a function from x to $x + h$ can be approximated by the change in the line tangent to the graph of the function at x from x to $x + h$ when h is a small number.
- The change in the tangent line from x to $x + h$ is

$$(\text{slope of the tangent line at } x) \cdot h$$

The mathematical notation for the statement "the change in a function from x to $x + h$ is approximately the slope of the tangent line at x times h" is

$$f(x + h) - f(x) \approx f'(x) \cdot h$$

for small values of h. We illustrate this statement geometrically in Figure 5.2.

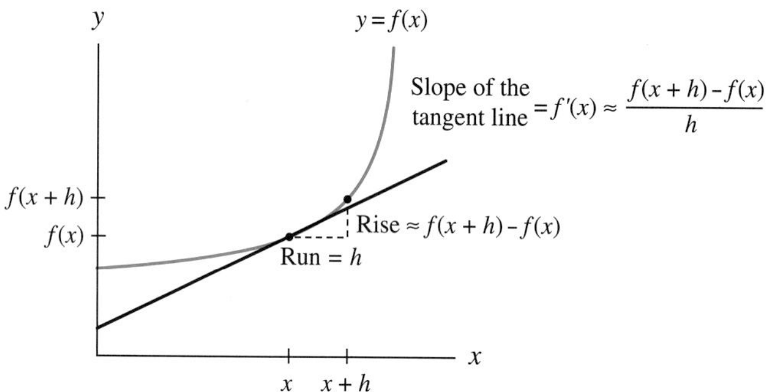

FIGURE 5.2

Recall that the derivative of a function f is defined as $\lim_{h\to 0} \frac{f(x + h) - f(x)}{h}$. For small values of h, $\frac{f(x + h) - f(x)}{h}$ is a good approximation to $f'(x)$. This concept is what motivated our numerical estimation of derivatives in Chapter 4. Multiplying both sides of the expression $\frac{f(x + h) - f(x)}{h} \approx f'(x)$ by h gives the formula for approximation of change: $f(x + h) - f(x) \approx f'(x) \cdot h$.

Approximating Change

The approximate change in f is the rate of change of f times a small change in x. That is,

$$f(x + h) - f(x) \approx f'(x) \cdot h$$

where h represents the small change in x.

It follows from this formula for approximating change that we can approximate the function value $f(x + h)$ by adding the approximate change to $f(x)$.

Approximating the Result of Change

When x changes by a small amount to $x + h$, the output of f at $x + h$ is approximately the value of f at x plus the approximate change in f.

$$f(x + h) \approx f(x) + f'(x) \cdot h$$

It was this formula that we used to obtain the approximate price column in Table 5.1 in the butter price example.

EXAMPLE 1 *Population of California*

At the end of June 1997, the population of California[3] was approximately 32,268,000 people and was increasing at a rate of about 517,000 people per year.

a. Estimate the change in the population of California during the third quarter of 1997.

b. Estimate the population of California at the end of 1997.

c. What assumptions underlie these estimations?

Solution

a.
$$\begin{aligned}\text{Change in population} &\approx (517{,}000 \text{ people per year})\left(\tfrac{1}{4}\text{ year}\right)\\ &= 129{,}250 \text{ people}\end{aligned}$$

The population increased by approximately 129,250 people during the third quarter of 1997.

b.
$$\begin{aligned}\text{Population at the end of 1997} &\approx \begin{pmatrix}\text{population at end of}\\ \text{June 1997}\end{pmatrix} + \begin{pmatrix}\text{approximate change}\\ \text{in population}\end{pmatrix}\\ &= 32{,}268{,}000 \text{ people} + (517{,}000 \text{ people per year})\left(\tfrac{1}{2}\text{ year}\right)\\ &= 32{,}268{,}000 \text{ people} + 258{,}500 \text{ people}\\ &= 32{,}526{,}500 \text{ people}\end{aligned}$$

The population at the end of 1997 was approximately 32,526,500 people.

c. In making these estimations, we assume that the growth rate remained essentially constant during the second half of 1997.

EXAMPLE 2 *Temperature*

The temperature during and after a thunderstorm can be modeled by

$$T(h) = 2.37h^4 - 5.163h^3 + 8.69h^2 - 9.87h + 78 \text{ degrees Fahrenheit}$$

where h is the number of hours since the storm began.

a. Use the rate of change of $T(h)$ at $h = 0.25$ to estimate by how much the temperature changed between 15 and 20 minutes after the storm began.

3. Based on data from *Statistical Abstract*, 1998.

b. Find the temperature and rate of change of temperature at $h = 1.5$ hours.

c. Using only the answers to part *b*, estimate the temperature 1 hour and 40 minutes after the storm began.

d. Sketch the graph of T from $h = 0$ to $h = 1.75$ with lines tangent to the graph at $h = 0.25$ and $h = 1.5$. On the basis of the graph, determine whether the answers to parts *a* and *c* are overestimates or underestimates of the temperature given by the model.

Solution

a. $$T'(0.25) = 9.48(0.25)^3 - 15.489(0.25)^2 + 17.38(0.25) - 9.87 \approx -6.34°\text{F per hour}$$

The change in the temperature between 15 and 20 minutes after the storm began is approximately $(-6.34°\text{F/hr})\left(\frac{1}{12}\text{ hr}\right) = -0.53°\text{F}$. The temperature fell approximately half a degree.

b. $T(1.5) \approx 77.3°\text{F}$ and $T'(1.5) \approx 13.3°\text{F}$ per hour

c. Note that 40 minutes $= \frac{2}{3}$ hour.

$$\text{Temperature at 1 hour and 40 minutes} \approx \begin{pmatrix}\text{temperature at}\\ 1\frac{1}{2}\text{ hours}\end{pmatrix} + \begin{pmatrix}\text{approximate change}\\ \text{in temperature}\end{pmatrix}$$

$$T\left(1 + \tfrac{2}{3}\right) \approx T(1.5) + T'(1.5)\left(\tfrac{1}{6}\right)$$

$$T\left(1 + \tfrac{2}{3}\right) \approx 77.3°\text{F} + (13.3°\text{F per hour})\left(\tfrac{1}{6}\text{ hour}\right)$$

$$\approx 77.3°\text{F} + 2.2°\text{F}$$

$$= 79.5°\text{F}$$

d.

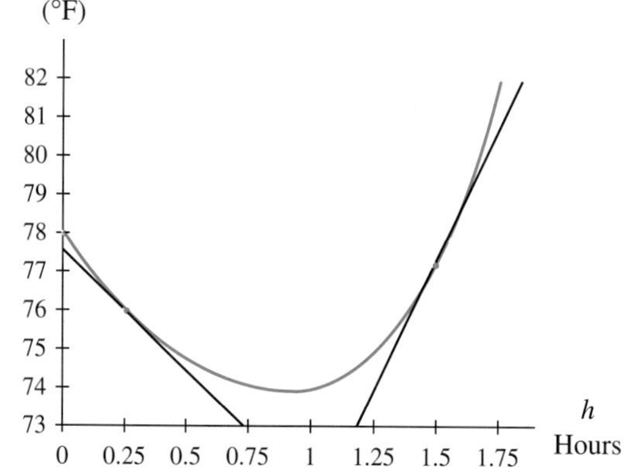

FIGURE 5.3

You may wish to make the connection between the concavity of the curve at the point of tangency and whether the approximation is an overestimate or an underestimate. Figure 3.16 in Section 3.2 is a good reference.

Because the tangent line at $h = 0.25$ hour is steeper than the graph is between $h = 0.25$ (15 minutes) and $h \approx 0.33$ (20 minutes), the approximate change in temperature overestimates the actual change (see Figure 5.3). Thus, using the approximate change to estimate the temperature at $h \approx 0.33$ gives a temperature that underestimates the temperature given by the model at $h \approx 0.33$. The tangent line at $h = 1.5$ is not as steep as the graph to the right of $h = 1.5$, so our temperature approximation for $h = 1.67$ is an underestimate of the temperature given by the model.

You may be wondering why you would want to use a tangent line to estimate a function output value. Sometimes, as in Example 1, it is necessary because we do not have enough information available to develop a model. Other times, the tangent line approximation is used for short-term extrapolation. This is illustrated in Example 3.

EXAMPLE 3 *Population of California*

The population of California from 1990 through 1997 is shown in Table 5.2. Population figures are for July 1 of each year.[4]

TABLE 5.2

Year	Population (thousands)
1990	29,786
1991	30,413
1992	30,892
1993	31,183
1994	31,369
1995	31,558
1996	31,858
1997	32,268

a. Use the last two data points given to estimate the population on July 1, 1998.

b. Find a model for the data. Use it to estimate the population on July 1, 1998.

c. Find the value of the model and the derivative of the model on July 1, 1997. Use these values to estimate the population on July 1, 1998.

d. Discuss the assumptions one makes when using each estimation technique in parts *a*, *b*, and *c*.

Solution

a. The average rate of change between 1996 and 1997 is

$$\frac{32{,}268 - 31{,}858 \text{ thousand people}}{1 \text{ year}} = 410 \text{ thousand people per year}$$

Adding this value to the population on July 1, 1997, we obtain an estimate of the July 1, 1998, population:

$$32{,}268{,}000 + 410{,}000 = 32{,}678{,}000 \text{ people}$$

b. The data can be modeled by

$$P(t) = 12.447t^3 - 151.294t^2 + 805.345t + 29{,}775.015 \text{ thousand people}$$

t years after July 1, 1990. The model estimates the population on July 1, 1998, as

$$P(8) \approx 32{,}908{,}000 \text{ people}$$

c. The derivative of P is

$$P'(t) = 37.341t^2 - 302.588t + 805.345 \text{ thousand people per year}$$

t years after July 1, 1990. On July 1, 1997, the rate of change of the population was approximately

$$P'(7) \approx 516.9 \text{ thousand people per year}$$

and the population according to the model was 32,268,000. Thus

$$\begin{aligned} P(8) &\approx P(7) + P'(7) \\ &\approx 32{,}268{,}000 + 517{,}000 \\ &= 32{,}785{,}000 \text{ people} \end{aligned}$$

4. *Statistical Abstract*, 1998.

d. To estimate using only the last two data points (part *a*) is to assume that the average rate of change in the population from July 1, 1997, through July 1, 1998, will be approximately the same as it was from July 1, 1996, through July 1, 1997. To estimate using only the model (part *b*) is to assume that future growth will continue in the manner of the cubic model. To estimate using the derivative of the model (part *c*) is to assume that the rate of change on July 1, 1997, (as estimated by the model), is a good predictor of the change in the population during the following year.

All three of these estimates are valid. If it were July 1, 1997, and someone needed an estimate for the population a year later, that person might use any one of these or many other techniques to make the prediction.

Marginal Analysis

You may omit the discussion of marginal analysis if no students in your class are majoring in a business-related field. See the *Instructor's Resource Guide* for a list of activities related to marginal analysis.

In economics, it is customary to refer to the rates of change of cost, revenue, and profit with respect to the number of units produced or sold as **marginal cost, marginal revenue,** and **marginal profit.** These rates are used to approximate actual change in cost, revenue, or profit when the number of units produced or sold is increased by one. The term **marginal analysis** is often applied to this type of approximation.

EXAMPLE 4 *Marginal Cost*

a. Suppose a manufacturer of toaster ovens currently produces 250 ovens per day with a total production cost of \$12,000 and a marginal cost of \$24 per oven. What information does the marginal cost value give the manufacturer?

b. If $C(x)$ is the cost to produce x toaster ovens, what is the notation for marginal cost?

Solution

a. The marginal cost is the rate of change of cost with respect to the number of units produced. It is the approximate increase in cost (\$24) that will result if production is increased from 250 ovens per day to 251 ovens per day.

b. The marginal cost is $C'(x)$ or $\frac{dC}{dx}$. In this example, $C'(250) = \$24$ per oven.

We know that profit = revenue − cost. If $p(x)$, $r(x)$, and $c(x)$ are the profit, revenue, and cost associated with x units, then we have the relationship $p(x) = r(x) - c(x)$. If we take the derivative of this expression, we have $p'(x) = r'(x) - c'(x)$ or

Marginal profit = marginal revenue − marginal cost

From this equation, we see that if marginal profit is to be positive, so that increased sales will increase profit, then marginal revenue must be greater than marginal cost. Example 5 explores these relationships.

EXAMPLE 5 *Marginal Revenue*

A seafood restaurant has been keeping track of the price of its Monday night all-you-can-eat buffet and the corresponding number of nightly customers. These data are given in Table 5.3.

TABLE 5.3

Number of customers	Buffet price (dollars)
86	7.70
83	7.90
80	8.20
78	8.30
76	8.60
73	8.80
70	8.90
68	9.10

a. Find a model for the data, and convert it to a model for revenue.

b. If the cost to the restaurant is \$4.50 per person regardless of the number of customers, find models for cost and profit.

c. Find the marginal revenue, cost, and profit values for 50, 92, and 100 customers. Interpret the values in context.

Solution

a. A linear model for these buffet price data is

$$B(x) = -0.078702x + 14.477879 \text{ dollars}$$

where x is the number of customers. Because revenue is equal to (price)(number of customers), the revenue is given by the equation

$$R(x) = -0.078702x^2 + 14.477879x \text{ dollars}$$

where x is the number of customers.

b. The cost model is $c(x) = 4.5x$ dollars for x customers. The profit model is

$$\begin{aligned} p(x) &= R(x) - c(x) = -0.078702x^2 + 14.477879x - 4.5x \\ &= -0.078702x^2 + 9.977879x \text{ dollars} \end{aligned}$$

where x is the number of customers.

c. The derivatives of revenue, cost, and profit are, respectively,

$$\begin{aligned} R'(x) &= -0.157404x + 14.477879 \text{ dollars per customer} \\ c'(x) &= \$4.50 \text{ per customer} \\ p'(x) &= -0.157404x + 9.977879 \text{ dollars per customer} \end{aligned}$$

where x is the number of customers. Evaluating the derivatives at 50, 92 and 100 customers gives the marginal values shown in Table 5.4.

TABLE 5.4

Demand (number of customers)	Marginal revenue (dollars per customer)	Marginal cost (dollars per customer)	Marginal profit (dollars per customer)
50	6.61	4.50	2.11
92	0.00	4.50	-4.50
100	-1.26	4.50	-5.76

What do these marginals tell us? If the buffet price is set on the basis of 50 customers, then revenue is increasing by \$6.61 per customer. Because this value is greater than marginal cost, we see a positive marginal profit. In other words, increasing the number of customers to 51 (by lowering the price) will increase nightly revenue by approximately \$6.61 and nightly profit by \$2.11. It would benefit the restaurant to increase the number of customers by lowering price.

Similarly, we estimate that if the number of customers is increased from 92 to 93, then revenue will not change significantly and profit will, therefore, decline. With 92 customers, stimulating sales by lowering the price will not benefit the restaurant.

Finally, note that when price is set so that the restaurant expects 100 customers, the marginal revenue and profit are negative. Increasing the number of customers (by decreasing price) to 101 will result in an approximate decrease in nightly revenue of \$1.26 and a decrease of nightly profit of \$5.76. That is clearly undesirable.

Knowing marginal cost, revenue, or profit can be a valuable tool when making business decisions.

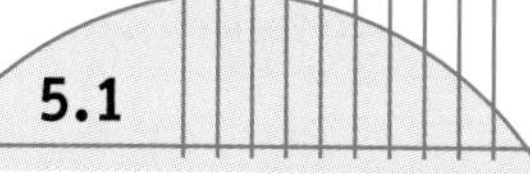

5.1 Concept Inventory

- *Change in function ≈ change in tangent line close to the point of tangency*
  $f(x + h) - f(x) \approx f'(x) \cdot h$
- $f(x + h) \approx f(x) + f'(x) \cdot h$ *for small values of h*
- *Marginal cost, marginal profit, marginal revenue*

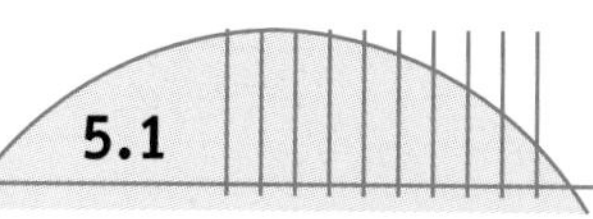

5.1 Activities

1. If the humidity is currently 32% and is falling at a rate of 4 percentage points per hour, estimate the humidity 20 minutes from now.

2. If an airplane is flying 300 mph and is accelerating at a rate of 200 mph per hour, estimate the airplane's speed in 5 minutes.

3. If $f(3) = 17$ and $f'(3) = 4.6$, estimate $f(3.5)$.

4. If $g(7) = 4$ and $g'(7) = -12.9$, estimate $g(7.25)$.

5. Interpret the following statements.
   a. At a production level of 500 units, marginal cost is \$17 per unit.
   b. When weekly sales are 150 units, marginal profit is \$4.75 per unit.

6. A fraternity currently realizes a profit of \$400 selling T-shirts at the opening baseball game of the season. If its marginal profit is -\$4 per shirt, what action should the fraternity consider taking to improve its profit?

7. A graph showing the annual premium for a one-million-dollar term life insurance policy as a function of the age of the insured person is given in Figure 5.1.1.

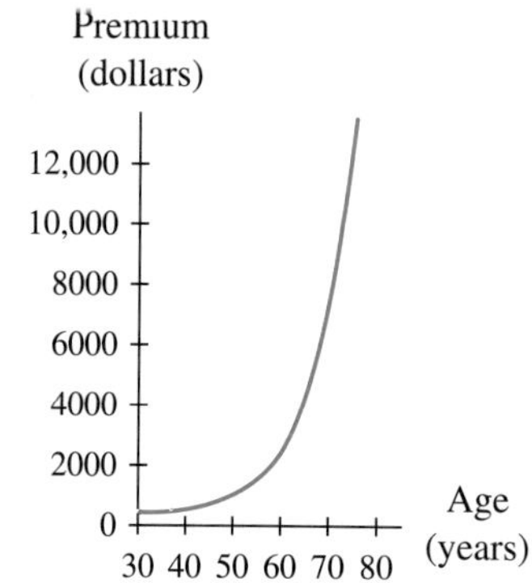

FIGURE 5.1.1

Sketch a tangent line at 70 years of age, and use it to predict the premium for a 72-year-old person.

8. A graph showing world life expectancy[5] as a function of the number of decades since 1900 is given in Figure 5.1.2.

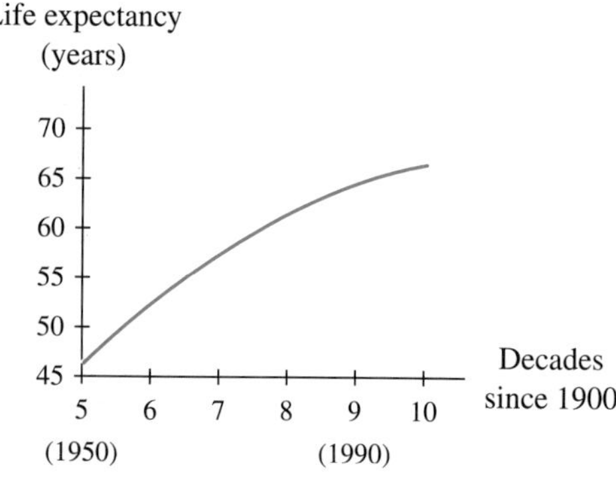

FIGURE 5.1.2

Sketch a tangent line at 1990, and use it to estimate the world life expectancy in 2000.

9. A graph of the population of Cleveland, Ohio,[6] as a function of the number of decades since 1900 is shown in Figure 5.1.3.

5. Based on data in *The True State of the Planet*, ed. Ronald Bailey (New York: The Free Press for the Competitive Enterprise Institute, 1995).
6. Based on data from U.S. Bureau of the Census.

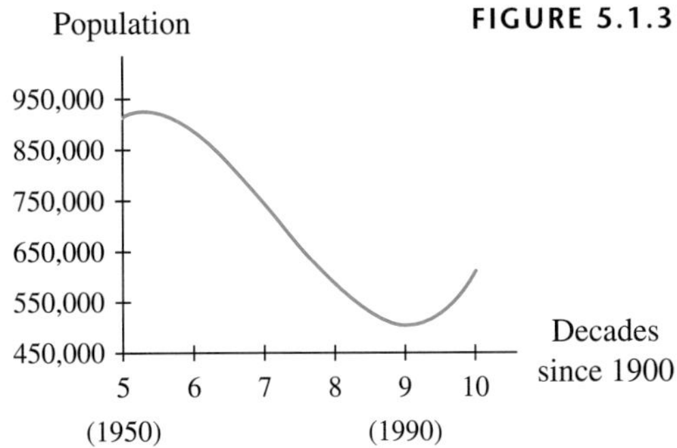

FIGURE 5.1.3

a. Sketch a tangent line at 1990, and use it to estimate the population of Cleveland in 2000.

b. The model graphed in Figure 5.1.3 is

$$P(t) = 16{,}272t^3 - 349{,}625t^2 + 2{,}335{,}380t - 4{,}057{,}270 \text{ people}$$

t decades after 1900. Note that the model applies for years after 1950. What does the model estimate as the population in 2000?

c. Which estimate do you believe is the more valid one? Why?

10. In 1987, because of concern that CFCs have a detrimental effect on the stratospheric ozone layer, the Montreal Protocol calling for phasing out all chlorofluorocarbon (CFC) production was ratified. The graph in Figure 5.1.4 shows estimated releases of CFC-11, one of the two most prominent CFCs.[7]

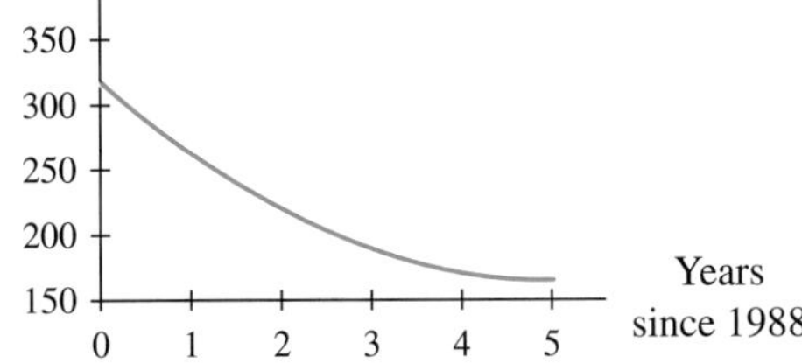

FIGURE 5.1.4

a. Sketch a tangent line at 1992, and use it to estimate CFC-11 releases in 1993.

b. The model whose graph is shown in Figure 5.1.4 is

$$C(x) = 6.107x^2 - 60.799x + 315.994 \text{ million kilograms of CFC-11}$$

where x is the number of years since 1988. What estimate does the model give for CFC-11 releases in 1993?

c. The actual amount of CFCs released into the atmosphere in 1993 was 149 million kilograms. Which estimate is the more accurate one?

11. The population of South Carolina between 1790 and 1990 can be modeled[8] by

$$\text{Population} = 0.07767x^2 - 227.8377x + 248{,}720.649 \text{ thousand people in year } x$$

a. Find the rate of change of the population of South Carolina in 1990.

b. On the basis of your answer to part *a*, approximate how much the population changed between 1990 and 1992.

c. Write an explanation of the procedure you used to find the approximate change in the population between 1990 and 1992.

12. A pizza parlor has been lowering the price of a large one-topping pizza to promote sales. The average revenues from the sale of large one-topping pizzas on a Friday night (5 p.m. to midnight) are given in Table 5.5.

TABLE 5.5

Price (dollars)	9.25	10.50	11.75	13.00	14.25
Revenue (dollars)	1202.50	1228.50	1210.25	1131.00	1054.50

a. Find a model for the data.

b. Find and interpret the rate of change of revenue at a price of \$9.25.

c. Approximate the change in revenue if the price is increased from \$9.25 to \$10.25.

d. Find and interpret the rate of change of revenue at a price of \$11.50.

e. Approximate the change in revenue if the price is increased from \$11.50 to \$12.50.

f. Explain why the approximate change is an overestimate of the change from \$9.25 to \$10.25 but an underestimate of the change from \$11.50 to \$12.50.

13. The population of Mexico between 1921 and 1990 can be modeled[9] by

$$P(t) = 12.921e^{0.026578t} \text{ million people}$$

where t is the number of years since 1921.

7. Based on data in *The True State of the Planet*, 1995.

8. Based on data from *South Carolina Statistical Abstract*, 1994.

9. SPP and INEGI, Mexican Censuses of Population 1921 through 1990 as reported in Pick and Butler, *The Mexico Handbook* (Boulder, CO: Westview Press, 1994).

a. How rapidly was the population growing in 1985?

b. On the basis of your answer to part *a*, determine by approximately how much the population of Mexico should have increased between 1985 and 1986.

14. Suppose the graph in Figure 5.1.5 shows your grade out of 100 points as a function of the time that you spend studying.

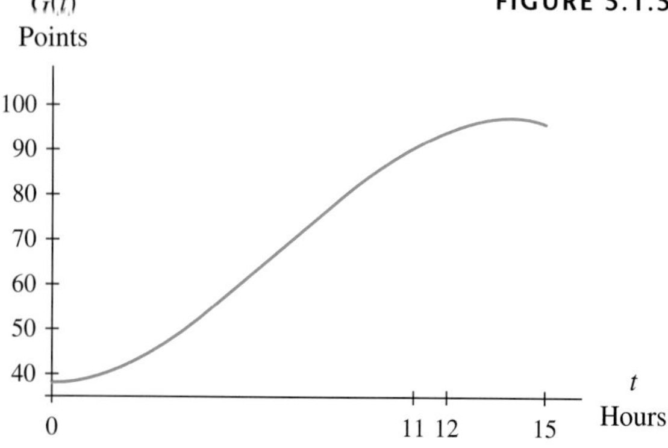

FIGURE 5.1.5

The equation for this graph is

$$G(t) = -0.044t^3 + 0.918t^2 + 38.001 \text{ points}$$

where t is the number of hours spent studying.

a. Confirm the following assertions:

i. After 11 hours of study, the slope is 4.2 points per hour, and the grade is 90.5 points.

ii. The grade after 12 hours of study is 94.2 points.

b. Use the information in the first part of *a* to estimate the grade after 12 hours. Is this an overestimate or an underestimate of the grade given by the model? Explain why.

15. A model[10] for the percentage of households with TVs that also have VCRs is

$$p(x) = 40.818 + 17.388 \ln x \text{ percent}$$

where x is the number of years since 1985. This model applies for the years 1990 through 1996.

a. How rapidly was the percentage of households with TVs and VCRs growing in 1995?

b. On the basis of the rate of change in 1995, what approximate increase would you expect between 1995 and 1996?

c. On the basis of the equation, what was the actual increase between 1995 and 1996?

d. According to the 1998 *Statistical Abstract*, the actual percentages for 1995 and 1996 were 81.0% and 82.2%, respectively. What was the actual 1995 through 1996 increase?

e. Compare the increases given by the derivative in part *b*, by the model in part *c*, and by the data in part *d*. Which of the three answers is most accurate? Explain.

16. Production costs for various hourly production levels of television sets are given in Table 5.6.

TABLE 5.6

Hourly production	Cost (dollars)
5	740
10	1060
15	1210
20	1320
25	1420
30	1580
35	1900

a. Find a model for the data.

b. Find and interpret marginal cost at production levels of 5, 20, and 30 units.

c. Find the cost to produce the 6th unit, the 21st unit, and the 31st unit.

d. Why is the cost less than the marginal cost to produce the 6th unit but greater than the marginal cost to produce the 21st and 31st units?

e. Find a model for average cost.

f. Find and interpret the rate of change of average cost at production levels of 5, 20, and 30 units.

17. A concession stand owner finds that if he prices hot dogs so as to sell a certain number at each sporting event, then the corresponding revenues are those given in Table 5.7.

a. Find a model for the data.

b. Each hot dog costs the owner of the concession stand $0.50. Use this fact and the model from part *a* to write models for cost and profit. Assume there are no fixed costs.

c. Find marginal revenue, cost, and profit for sales levels of 200, 800, 1100, and 1400 hot dogs. Interpret your answers.

10. Based on data from *Statistical Abstract*, 1998.

d. Graph revenue, profit, and cost for sales values between 200 and 1400 hot dogs. How are the marginal values in part *c* related to the graphs?

TABLE 5.7

Number of hot dogs sold	Revenue (dollars)
100	195
400	620
700	875
1000	1000
1200	1020
1500	975

18. A golf ball manufacturer knows that the cost associated with various hourly production levels are as shown in Table 5.8.

TABLE 5.8

Balls produced each hour (hundreds)	Cost (dollars)
2	248
5	356
8	432
11	499
14	532
17	567
20	625

a. Find a model for the data.

b. If 1000 balls are currently being produced each hour, find and interpret the marginal cost.

c. Repeat part *b* for 300 golf balls and for 2100 golf balls.

d. Convert the model in part *a* to one for average cost.

e. Find and interpret the rate of change of average cost for production levels of 300 and 1700 golf balls.

19. Rise in consumer prices is often used as a measure of inflation rate. Table 5.9 shows the CPI[11] during the 1980s for several different countries.

a. Find the best models for the data for the United States, Canada, Peru, and Brazil.

b. How rapidly were consumer prices rising in each of those four countries in 1987?

c. On the basis of your answers to part *b*, what would you expect the CPI to have been in the four countries in 1988?

20. The distribution of wealth is an important issue in our society today. The percentages[12] of all wealth held by the poorest fifth of the U.S. population in selected years from 1947 through 1997 are shown in Table 5.10.

TABLE 5.10

Year	Percentage	Year	Percentage
1947	5.0	1977	5.2
1957	5.1	1987	4.7
1967	5.5	1997	4.5*

* Estimate

TABLE 5.9

Country	Year							
	1980	1981	1982	1983	1984	1985	1986	1987
United States	100	110.4	117.2	120.9	126.1	130.5	133.1	137.9
Canada	100	112.4	124.6	131.8	137.5	143.0	148.9	155.4
Mexico	100	127.9	203.3	410.2	679.0	1071.2	1994.9	4624.7
Japan	100	104.9	107.8	109.9	112.3	114.6	115.3	115.4
Israel	100	217	478	1174	5560	22,498	33,330	39,937
Peru	100	175.4	288.4	609	1280	3372	5999	11,150
Brazil	100	206	407	984	2924	9556	23,436	77,258
Argentina	100	204	541	2403	17,462	134,833	256,308	592,900

11. *International Marketing Data and Statistics*, 1988/89.
12. *Statistical Abstract*, 1998.

a. Find a model for the data. Why did you choose this model?

b. What percentage of all wealth should the poorest fifth of Americans have in 2007 according to your model? Is this reasonable?

c. Estimate how rapidly the percentage of wealth held by the poorest fifth of Americans was changing in 1967.

d. Use the derivative to approximate the change in percentage of wealth between 1957 and 1958.

21. Three hundred dollars is invested in an account that compounds 6.5% APR monthly.

a. Find an equation for the balance in the account after t years.

b. Rewrite the equation in part a to be of the form $A = Pb^t$.

c. How much is in the account after 2 years?

d. How rapidly is the value of the account growing after 2 years?

e. Use the answer in part d to approximate how much the value of the account changes during the first quarter of the third year.

22. The amount in an investment after t years is given by

$$A(t) = 120{,}000(1.12682503^t) \text{ dollars}$$

a. Give the rate-of-change formula for the amount.

b. Find the rate of change of the amount after 10 years. Write a sentence interpreting the answer.

c. On the basis of your answer to part b, determine by approximately how much the investment will grow during the first half of the 11th year.

d. Find the percentage rate of change after 10 years. Given that A is exponential, what is the significance of your answer?

23. A car dealership keeps track of how much it spends on advertising each month and of its monthly revenue. From this information, the list of advertising expenditures and probable associated revenues shown in Table 5.11 was compiled.

TABLE 5.11

Advertising (thousands of dollars)	Revenue (thousands of dollars)
5	150
7	200
9	250
11	325
13	400
15	450
17	500
19	525

a. Find a model for the data.

b. Find and interpret the rate of change of revenue both as a rate of change and as an approximate change when $10,000 is spent on advertising.

c. Repeat part b for $18,000.

24. Write a brief essay that explains why, when rates of change are used to approximate change in a function, approximations over shorter time intervals generally give better answers than approximations over longer time intervals. Include graphical illustrations in your discussion.

25. Write a brief essay that explains why, when rates of change are used to approximate the change in a concave-up portion of a function, the approximation is an underestimate and when rates of change are used to approximate change in a concave-down portion of a function, the approximation is an overestimate. Include graphical illustrations in your discussion.

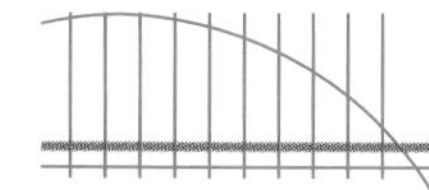

5.2 Relative and Absolute Extreme Points

We use the terms *extreme points* and *optimal points* to refer to maxima and minima (either relative or absolute, depending on the context) but not to inflection points.

In this section, we turn our attention to finding high points (maxima) and low points (minima) on the graph of a function. Points at which maximum or minimum outputs occur are called **extreme points,** and the process of **optimization** involves techniques for finding them. Maxima and minima often can be found using derivatives, and they have important applications to the world in which we live.

We begin by examining a model for the population of Kentucky[13] from 1980 through 1993:

$$\text{Population} = p(x) = 0.395x^3 - 6.674x^2 + 30.257x + 3661.147 \text{ thousand people}$$

where x is the number of years since the end of 1980.

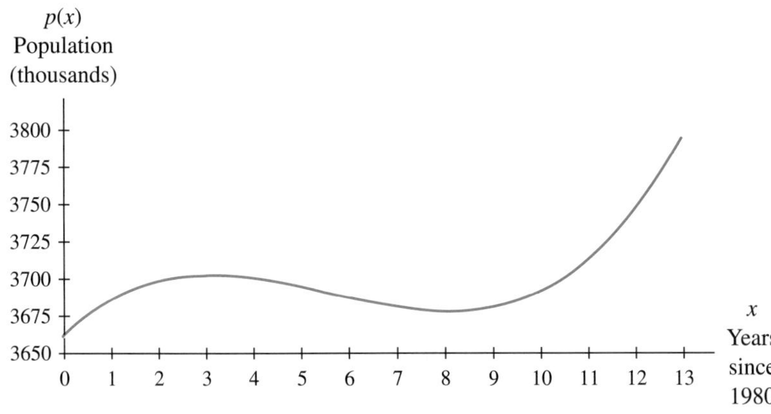

FIGURE 5.4

It is evident from the graph in Figure 5.4 that between 1980 and 1993, the population indicated by the model was smallest in 1980 (3661 thousand people) and greatest in 1993 (3794 thousand people). However, there are two other points of interest on the graph. Sometime near 1983, the population reached a peak. We call the peak a **relative** (or **local**) **maximum**. It does not represent the highest overall point, but it is a point to which the population rises and after which it declines. Similarly, near 1988, the population reached a **relative** (or **local**) **minimum**. There are lower points on the graph, but in the region around this relative minimum, the population decreases and then increases as time increases.

It should be intuitively clear from the discussions in Chapter 3 that at a point where the graph of a smooth, continuous function reaches a relative maximum or minimum, the tangent line is horizontal and the slope is 0. We can consider this important link between such a function and its derivative in more detail by examining the relationship between the Kentucky population function and its slope graph. Horizontal tangent lines on the population function graph correspond to the points at which the slope graph crosses the horizontal axis. Figure 5.5 shows the graphs of the population function and its derivative.

Note that near $x = 3$, the slope graph crosses the x-axis. This is the x-value for which $p(x)$ is a local maximum. Likewise, the slope graph crosses the x-axis at the x-value near 8 for which $p(x)$ is a local minimum. The connection between the graphical and symbolic views of this situation is a key feature of optimization techniques.

> Finding where the slope graph of a function crosses or touches the input axis is the same as finding *where the derivative of the function is zero.*

The derivative of the population function is

$$\frac{dp}{dx} = 1.185x^2 - 13.348x + 30.257 \text{ thousand people per year}$$

Students often confuse extreme values with the inputs at which the extrema occur or with the extreme point. Emphasize that the extreme value is always an output value.

where x is the number of years since 1980. Setting this expression equal to zero and solving for x results in two solutions: $x \approx 3.14$ and $x \approx 8.12$. This information, together with the graph of p shown in Figure 5.5, tells us that, according to the model, the population peaked in early 1984 at approximately $p(3.14) = 3703$ thousand people. We also conclude that the population declined to a local minimum in early 1989. The population at that time was approximately $p(8.12) = 3678$ thousand people.

13. Based on data from *Statistical Abstract*, 1994.

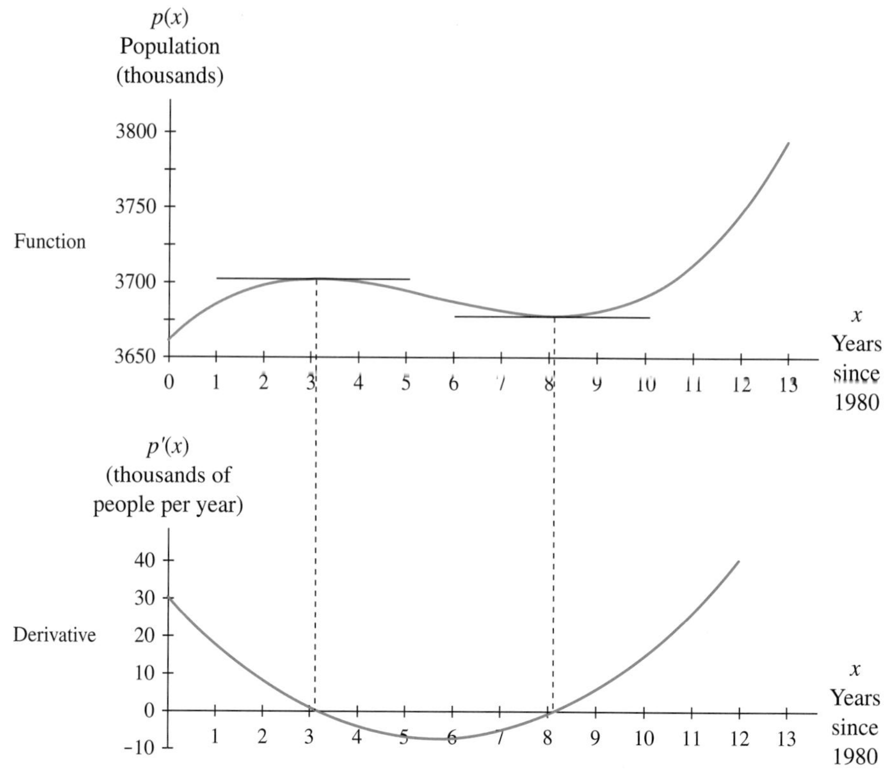

FIGURE 5.5

You may wish to point out to your students that although there is only one absolute maximum or minimum, this value may occur at more than one input. A sine function is a good example of this situation.

Recall the statement at the beginning of this discussion: "It is evident from the graph in Figure 5.4 that the population from 1980 through 1993 was smallest in 1980 (3661 thousand people)." We call this output of the population function an **absolute minimum** and refer to 3661 thousand people as the approximate **minimum value.** Also, because the 1993 population (3794 thousand people) is more than the population at the local maximum (3703 thousand people), the **absolute maximum** occurs in 1993. The **maximum value** of the population function is approximately 3794 thousand people. A function can have several different local maxima (or minima) in a given interval. However there can be only one absolute maximum value and one absolute minimum value for that interval. Any absolute maximum (or minimum) in the given interval occurs either at a local maximum (or minimum) or at a point corresponding to an endpoint of the given input interval.

Point out to your students that endpoints can never be relative extreme points but can be absolute extreme points. Also stress that an interior point can be both a relative and an absolute extreme point.

It is important to notice that relative extrema do not occur at the endpoints of an interval. A relative maximum is an output value that is larger than all other output values in some interval *around* the maximum. Similarly, a relative minimum is an output value that is smaller than all other output values in some interval *around* the minimum.

EXAMPLE 1 *Baggage Complaints*

The number of consumer complaints about baggage[14] on U.S. airlines between 1989 and 1997 can be modeled by the function

$$B(x) = 41.018x^2 - 1166.514x + 8838.325 \text{ complaints}$$

where x is the number of years after 1980.

14. Based on data from *Statistical Abstract*, 1998.

a. Consider the function out of its modeling context. Find the absolute maximum and the absolute minimum of B on the interval $9 \le x \le 17$.

Tell your students that the terms *least* and *greatest* refer to the absolute minimum and absolute maximum, respectively.

b. Find the year between 1989 and 1997 in which the number of baggage complaints was greatest and the year in which it was least.

c. Graph the function and its derivative. On each graph, clearly mark the input value that corresponds to the absolute minimum found in part *a*.

Solution

a. Because the graph of B is a concave-up parabola, we know that the minimum occurs where the derivative is zero. As long as the minimum occurs in the interval $9 \le x \le 17$, it is the absolute minimum we seek. Thus we set the derivative equal to zero and solve for x:

$$B'(x) = 82.036x - 1166.514 = 0$$
$$x \approx 14.2$$

The value of B at $x \approx 14.2$ is approximately 544.7.

To find the absolute maximum, we also must determine the value of B at the endpoints of the given interval: $B(9) \approx 1662.2$ and $B(17) \approx 861.8$. From our knowledge of the general shape of the graph of B and from these calculations, we conclude that the absolute maximum of B is approximately 1662.2 and occurs at the left endpoint when $x = 9$ and that the absolute minimum of B is approximately 544.7 and occurs at $x \approx 14.2$.

b. The left endpoint of the interval corresponds to 1989, so the number of baggage complaints was greatest in 1989 when there were approximately 1662 complaints. Because the model gives yearly complaint totals, it must be interpreted discretely. Thus the minimum number of complaints occurred in either 1994 ($x = 14$) or 1995 ($x = 15$). Checking the value of the function in each of these years, we find that the least number of complaints was approximately 547 in 1994.

c.

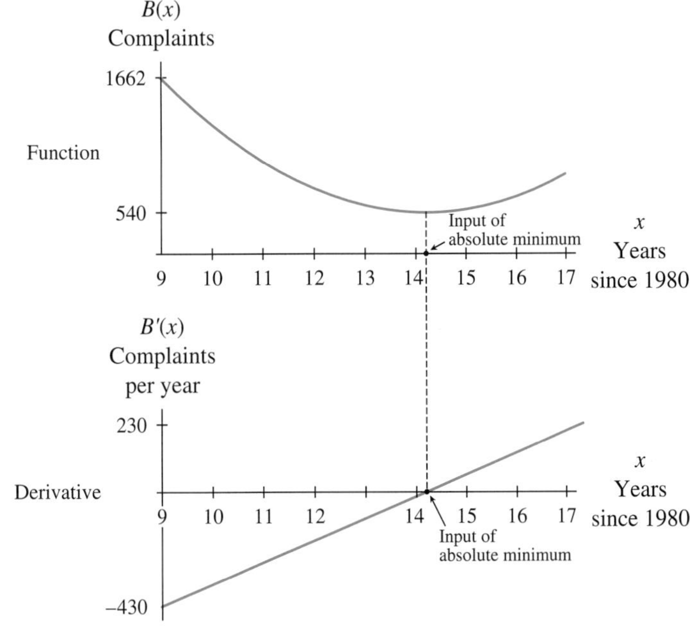

FIGURE 5.6

We see in Figure 5.6 that a relative minimum on the function graph corresponds to the point at which the derivative graph crosses the x-axis.

We have just seen how derivatives can be used to locate relative maxima and minima. *You should use caution*, however, and not automatically assume that just because the derivative is zero at a point, there is a relative maximum or relative minimum at that point. This is evident from the graphs in Figures 5.7c and 5.7d.

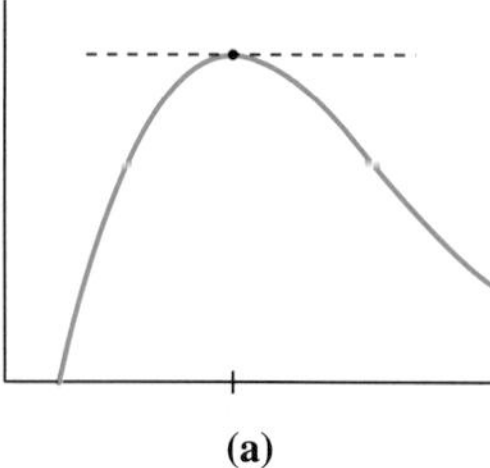

(a)
Relative maximum:
To the left of input a, slopes are positive, and to the right of a, slopes are negative.

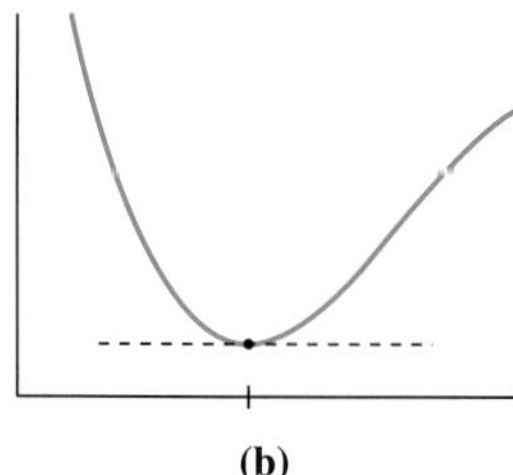

(b)
Relative minimum:
To the left of input a, slopes are negative, and to the right of a, slopes are positive.

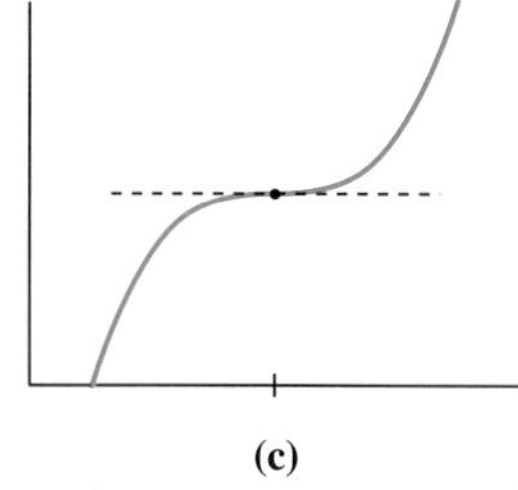

(c)
Inflection point:
Slopes are positive to the left and right of input a.

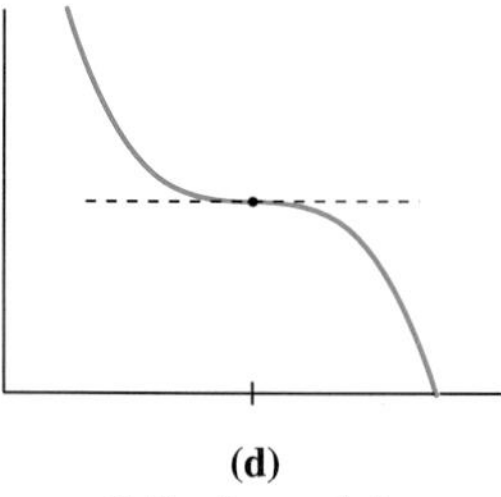

(d)
Inflection point:
Slopes are negative to the left and right of input a.

FIGURE 5.7

When there is a horizontal tangent line on the graph of a function (that is, when the derivative of the function is zero), one of the four situations depicted in Figure 5.7 occurs. It is therefore important that you graph the function when using derivatives to locate extreme points.

Because some students have difficulty finding a good view of a graph using their technology, you may wish to introduce the idea of checking the sign of the derivative on either side of a relative extremum in order to identify the extremum as a maximum, minimum, or inflection point as alluded to in Figure 5.7.

> Always begin the process of finding extreme points by graphing the function to see whether there are any relative maxima or minima before proceeding to work with derivatives.

Let us further investigate the graph shown in Figure 5.7d. Figure 5.8 shows the graph of the function that appears in Figure 5.7d and its slope (derivative) graph. If you carefully examine the slope graph, you see that it touches the x-axis but does not cross it. Thus f does not have a relative maximum or minimum at a. You may notice that the derivative graph reaches its maximum at a as it touches the x-axis. Do not confuse maxima and minima on the derivative graph with maxima and minima of the original function. In Section 5.4, we will see that maxima and minima of the derivative graph have other important interpretations in terms of the original function.

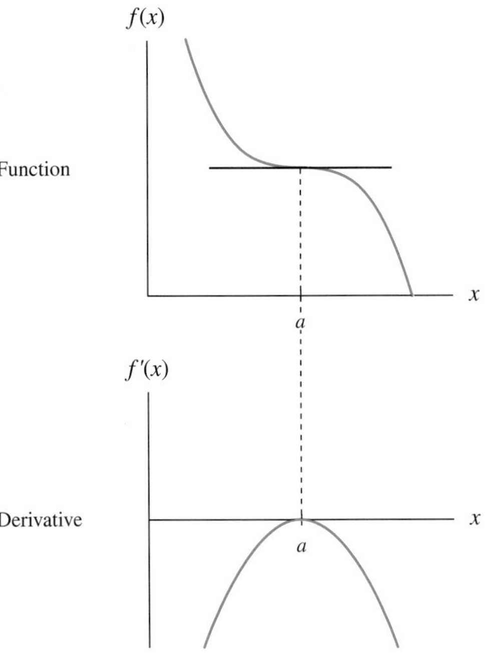

FIGURE 5.8 **$f'(a) = 0$ but the graph of f has no relative maximum or minimum at a.**

EXAMPLE 2 *Cable Company Revenue*

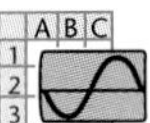

Acme Cable Company actively promoted sales in a town that previously had no cable service. Once Acme saturated the market, it introduced a new 50-channel system and raised rates. As the company began to offer its expanded system, a different company, Bigtime Cable, began offering satellite service with more channels than Acme and at a lower price. A model for Acme's revenue for the 26 weeks after it began its sales campaign follows, and the graph of the model is shown in Figure 5.9.

$$R(x) = -3x^4 + 160x^3 - 3000x^2 + 24{,}000x \text{ dollars}$$

where x is the number of weeks since Acme began sales.

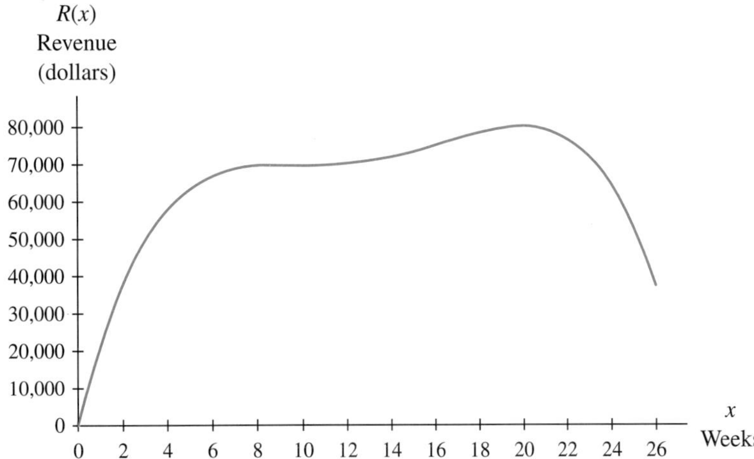

FIGURE 5.9

Determine when Acme's revenue was greatest during this 26-week interval.

Solution We know that revenue is greatest at the absolute maximum. An examination of the graph of R locates this point near 20 weeks. Solving the equation

$$R'(x) = -12x^3 + 480x^2 - 6000x + 24{,}000 = 0$$

gives two solutions, $x = 10$ weeks and $x = 20$ weeks. Revenue is greatest at 20 weeks, with a value of $R(20) = \$80{,}000$. This appears to correspond to the time immediately prior to when Bigtime Cable's sales began negatively affecting the Acme company.

The fact that the rate-of-change equation is zero at two places indicates that there are two places on the graph with horizontal tangent lines. Indeed, at $x = 10$ weeks, the line tangent to the curve is horizontal because the curve has leveled off. This corresponds to the time when Acme had saturated the market. However, no local maximum occurs at this point, because the slope graph only touches—it does not cross—the input axis at $x = 10$. Note the relationships between the rate-of-change graph and the revenue graph shown in Figure 5.10 and how they connect to the slope descriptions given in Figure 5.7.

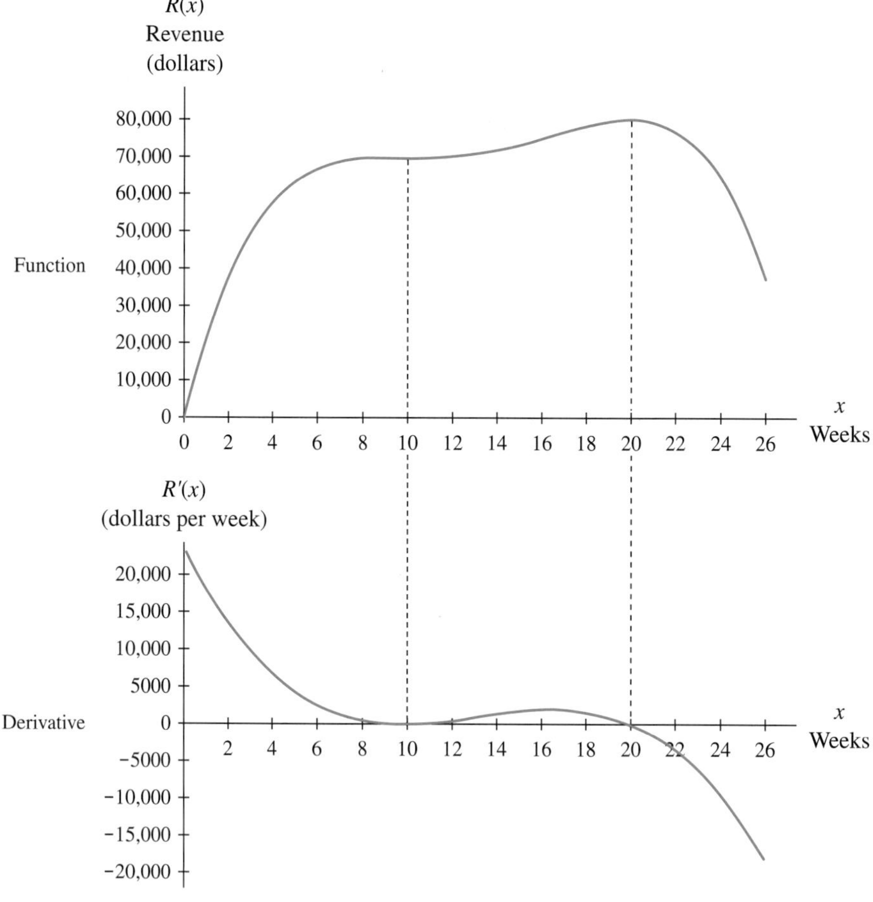

FIGURE 5.10

The maximum occurs where the derivative graph crosses the x-axis. The leveling-off point occurs where the derivative touches, but does not cross, the x-axis.

What if we do not have a specified input interval and are asked to find the absolute extrema? Let us consider the function in Example 2 apart from its context and determine the absolute extrema of the function $R(x) = -3x^4 + 160x^3 - 3000x^2 + 24{,}000x$ over all real number inputs. A careful numerical investigation indicates that $\lim_{x \to \pm\infty} R(x) \to -\infty$; that is, the graph continues to decrease infinitely in both directions. Thus there is no absolute minimum. We know that the absolute maximum is the relative maximum because of the end behavior of the function. Therefore, the greatest output value is 80,000 at $x = 20$. This value is the absolute maximum of the function over all real number inputs.

Is it true for every continuous function that relative maxima and minima occur only where the derivative crosses the input axis? Although this seems to be the case for most of the functions we use, consider the piecewise continuous function describing the average concentration (in nanograms per milliliter) of a 360-mg dose of a blood pressure drug in a patient's blood during the 24 hours after the drug is given:

$$C(h) = \begin{cases} -0.51h^3 + 7.65h^2 + 125 \text{ ng/mL} & \text{when } 0 \le h \le 10 \\ -16.07143\text{h} + 540.71430 \text{ ng/mL} & \text{when } 10 < h \le 24 \end{cases}$$

where h is the number of hours since the drug was given.

By calculating the left and right limits as h approaches 10 and comparing them to the function value at $h = 10$, we determine that C is continuous for all input values between 0 and 24. The two portions of the function join at the peak shown in Figure 5.11.

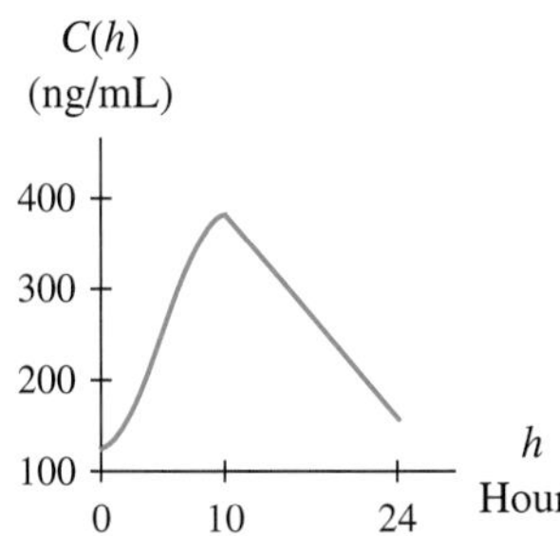

FIGURE 5.11

It is evident from the graph that the highest concentration of the drug occurs 10 hours after the patient receives the initial dose. Thus $C(10) = 380$ ng/ml is the maximum concentration. However, is $C'(10) = 0$? Does the slope graph for C cross the horizontal axis at $h = 10$? The slope at $h = 10$ exists only if the graph is smooth—that is, only if the slopes of the two portions of the graph of C are the same at $h = 10$. The slope of the graph on the left is zero at $h = 10$, but the slope of the graph on the right is -16.07143 at all points. These differing slopes cause the sharp point on the graph of C, resulting in a graph that is continuous but not smooth at $h = 10$. Thus $C'(10)$ does not exist. There is no line tangent to the graph of C at $h = 10$. That $C'(10)$ does not exist is illustrated in the graph of C' shown in Figure 5.12.

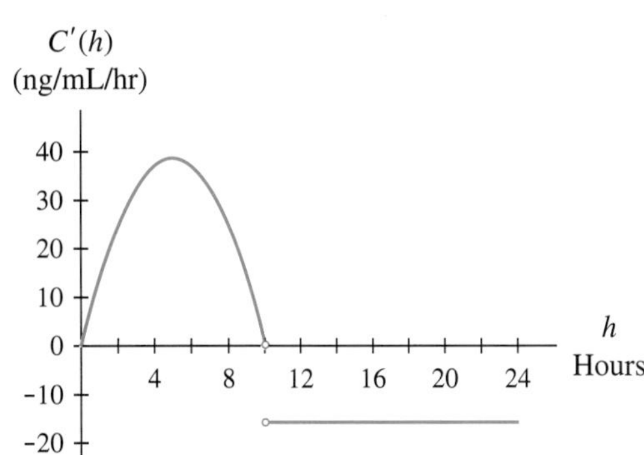

FIGURE 5.12

It is therefore possible for an extreme point to occur where the derivative of the function does not exist as long as the function has a value at that point.

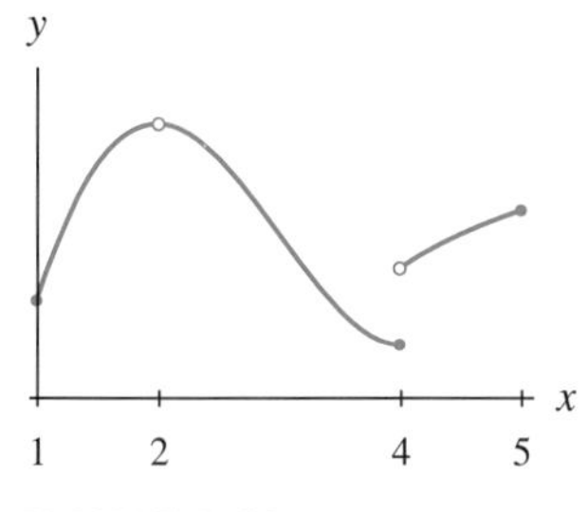

FIGURE 5.13

Our conclusion concerning the maximum drug concentration would have been the same if the function had not been continuous at $h = 0$, as long as the point with the greatest output occurred at $h = 0$. For example, consider Figure 5.13.

The absolute minimum occurs at $x = 4$, although the function is not continuous there. There is no absolute maximum of this function, because the point that corresponds to $x = 2$ is missing. An absolute extreme value cannot occur at an input for which the function is not defined.

The results of these investigations are summarized in the following statements:

Conditions When Extreme Points Exist

1. If a function has a relative maximum or relative minimum at an input value c, then either $f'(c) = 0$ or $f'(c)$ does not exist but $f(c)$ *does* exist.
2. If $f'(c) = 0$ and the derivative crosses (not just touches) the input axis at c, then a relative maximum or relative minimum exists at c.
3. An absolute maximum or absolute minimum occurs either at a relative maximum or relative minimum or at a point whose input is an endpoint of a specified interval.

In order to help you be organized as you practice the concepts presented in this section, we conclude by outlining the steps for finding relative and absolute extrema.

Finding Extrema

To find the relative maxima and minima of a function f,

Step 1: Determine the input values for which $f' = 0$ or f' is undefined.

Step 2: Examine a graph of f to determine which input values found in Step 1 correspond to relative maxima or relative minima.

To find the absolute maximum and minimum of a function f on an interval from a to b,

Step 1: Find all relative extrema of f in the interval.

Step 2: Compare the relative extreme values in the interval with $f(a)$ and $f(b)$, the output values at the endpoints of the interval. The largest of these values is the absolute maximum, and the smallest of these values is the absolute minimum.

To find the absolute maximum and minimum of a continuous function f without a specified input interval,

Step 1: Find all relative extrema of f.

Step 2: Determine the end behavior of the function in both directions in order to consider a complete view of the function. The absolute extrema either do not exist or are among the relative extrema.

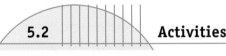

5.2 Concept Inventory

- *Relative (local) maximum*
- *Relative (local) minimum*
- *Absolute maximum*
- *Absolute minimum*
- *An extreme point occurs at an input value*
- *The extreme value is an output value*
- *Conditions under which extreme points exist*

5.2 Activities

1. Which of the six basic models discussed thus far in this book could have relative maxima or minima?
2. Discuss in detail all of the options you have available for finding the relative maxima and relative minima of a function.

In Activities 3 through 9, mark the location of all relative maxima and minima with an X and all absolute maxima and minima with an O. For each extreme point that is not an endpoint, indicate whether the derivative at that point is zero or does not exist.

3. 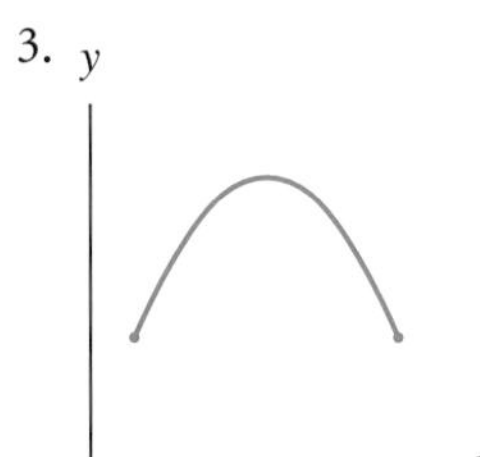

4. 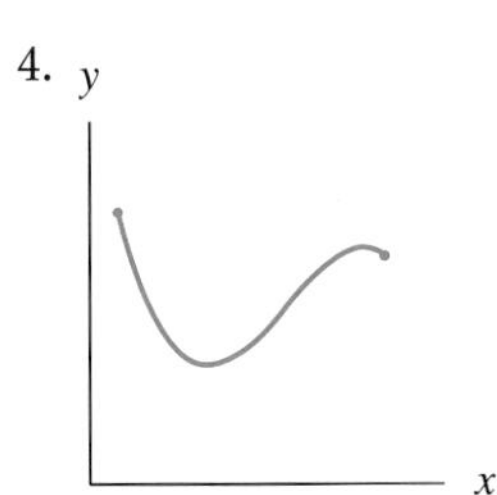

5. 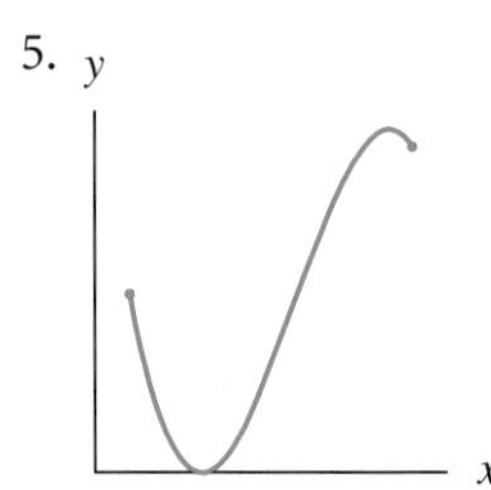

6. 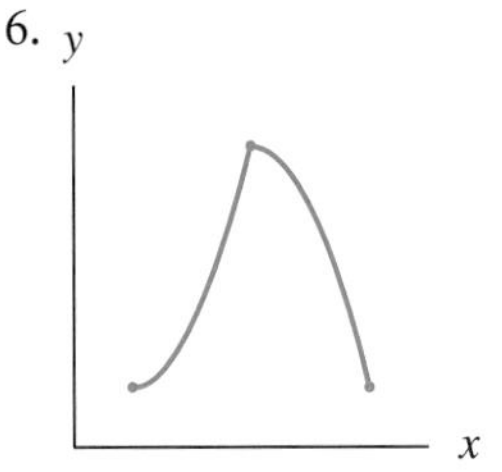

7. 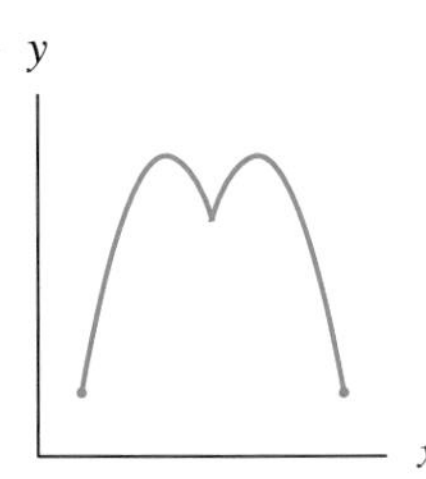

8. 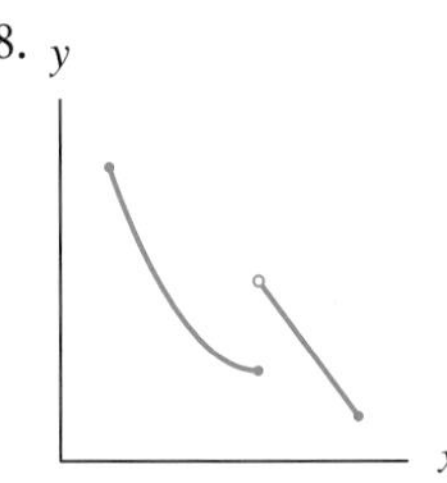

9. 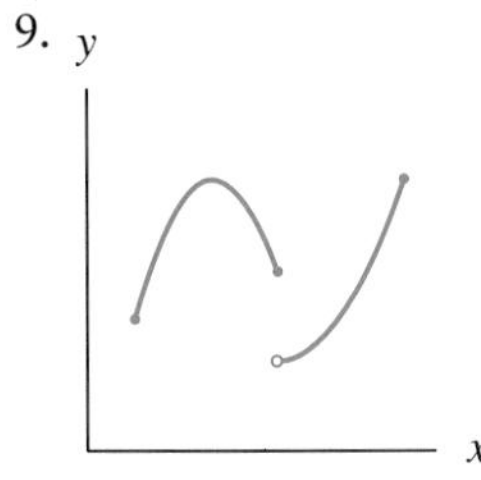

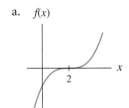

10. Sketch the graph of a function with a relative minimum at a point at which the derivative does not exist.
11. Sketch a graph of a function whose derivative is zero at a point but that does not have a relative maximum or minimum at that point.
12. Identify for which of the graphs *a* through *d* all of the following statements are true. For the other graphs, identify which statements are not true.

$$f'(x) > 0 \quad \text{for } x < 2$$
$$f'(x) < 0 \quad \text{for } x > 2$$
$$f'(x) = 0 \quad \text{for } x = 2$$

a. 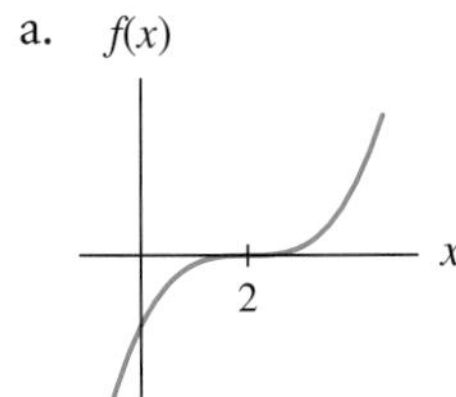

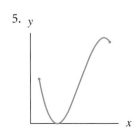

b. 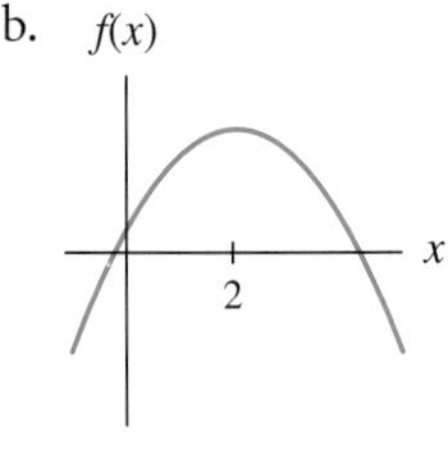

c. 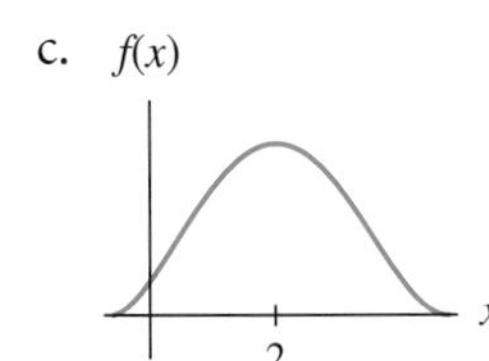

d. 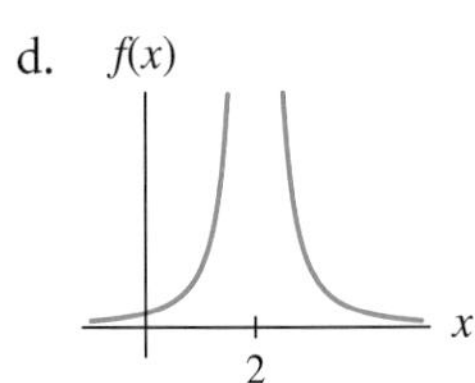

13. Identify for which of the graphs *a* through *d* all of the following statements are true. For the other graphs, identify which statements are not true.

$$f'(x) > 0 \quad \text{for } x > 2$$
$$f'(x) > 0 \quad \text{for } x < 2$$
$$f'(x) = 0 \quad \text{for } x = 2$$

a. 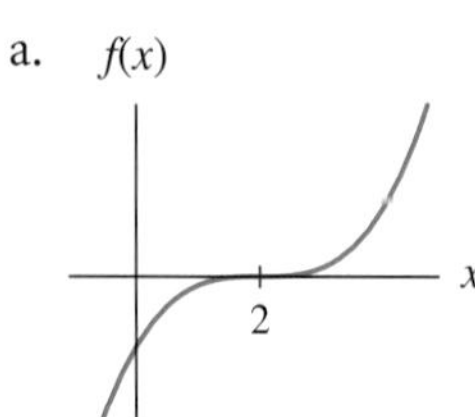

b. 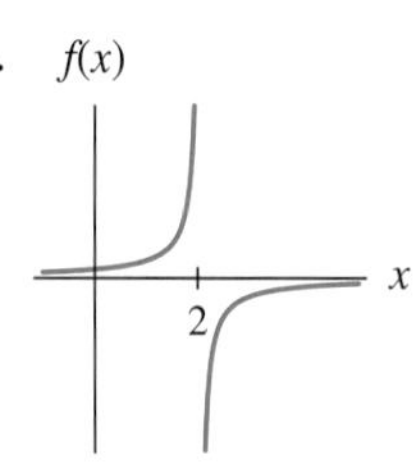

c. 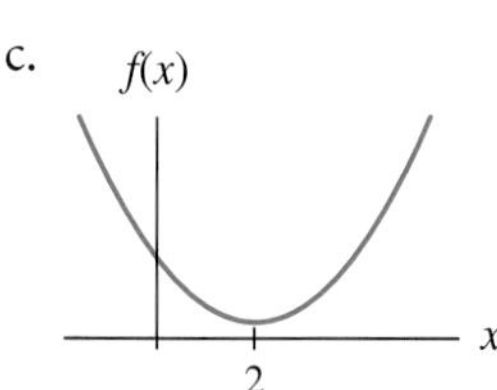

d. 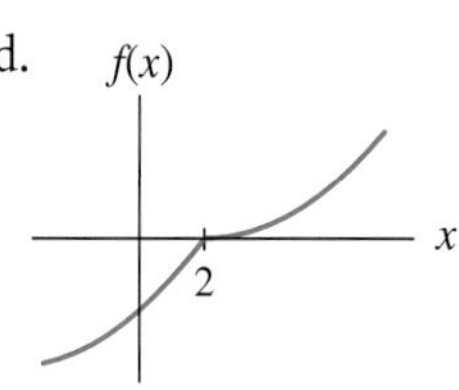

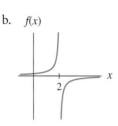

14. Sketch the graph of a function *f* such that

$$f'(x) < 0 \quad \text{for } x < -1$$
$$f'(x) > 0 \quad \text{for } x > -1$$
$$f'(-1) = 0$$

15. Sketch the graph of a function *f* such that

$$f'(x) < 0 \quad \text{for } x < -1$$
$$f'(x) > 0 \quad \text{for } x > -1$$
$$f'(-1) \text{ does not exist}$$

16. Sketch the graph of a function *f* such that

$$f'(x) > 0 \quad \text{for } x < -1 \text{ and } x > 3$$
$$f'(x) < 0 \quad \text{for } -1 < x < 3$$
$$f'(-1) = 0 \quad \text{and} \quad f'(3) = 0$$

17. Sketch the graph of a function *f* such that

f has a relative minimum at $x = 3$
f has a relative maximum at $x = -1$

$$f'(x) > 0 \quad \text{for } x < -1 \text{ and } x > 3$$
$$f'(x) < 0 \quad \text{for } -1 < x < 3$$
$$f'(-1) = 0 \quad \text{and} \quad f'(3) = 0$$

18. Consider the function $f(x) = -0.6x^2 + 4.2x + 12.8$.
   a. Find the highest and lowest points of the function between $x = 0$ and $x = 8$.
   b. Graph the function and its derivative. Indicate the relationship between the highest point on the function and the corresponding point on the derivative.

19. Consider the function

$$g(x) = 0.04x^3 - 0.88x^2 + 4.81x + 12.11$$

   a. Find the relative maximum and relative minimum of *g*.
   b. Find the absolute maximum and the absolute minimum of *g* between $x = 0$ and $x = 14.5$.
   c. Graph the function and its derivative. Indicate the relationship between the relative maximum and minimum of *g* and the corresponding points on the derivative.

20. The U.S. Bureau of the Census' prediction for the percentage of the population 65 to 74 years old from 2000 through 2050 can be modeled[15] by

$$p(x) = (1.619 \cdot 10^{-5})x^4 - (1.675 \cdot 10^{-3})x^3 + 0.050x^2 - 0.308x + 6.693 \text{ percent}$$

where *x* is the number of years after 2000. Find the absolute maximum and the absolute minimum percentages between 2000 and 2050. Give the years and the corresponding percentages.

21. The percentage of southern Australian grasshopper eggs that hatch as a function of temperature (for temperatures between 7°C and 25°C) can be modeled[16] by

$$P(t) = -0.00645t^4 + 0.488t^3 - 12.991t^2 + 136.560t - 395.154 \text{ percent}$$

where *t* is the temperature in °C.

   a. Find the temperature between 7°C and 25°C that corresponds to the greatest percentage of eggs hatching.
   b. Use the equation $°\text{F} = \frac{9}{5}(°\text{C}) + 32$ to convert your answer to °F.

22. Lake Tahoe lies on the California/Nevada border, and its level is regulated by a 17-gate concrete dam at the lake's outlet. By Federal Court decree, the lake level must never be higher than 6229.1 feet above sea level. The lake level is monitored every midnight.

15. Based on data from *Statistical Abstract*, 1998.
16. Based on information in George L. Clarke, *Elements of Ecology* (New York: Wiley, 1954), p. 170.

On the basis of this lake level data,[17] the level of the lake from October 1, 1995, through July 31, 1996, can be modeled by

$$L(d) = (-5.345\cdot 10^{-7})d^3 + (2.543\cdot 10^{-4})d^2 - 0.0192d + 6226.192 \text{ feet above sea level}$$

d days after September 30, 1995

According to the model, did the lake remain below the federally mandated level between October 1, 1995, and July 31, 1996?

23. The gender ratio[18] for the United States between 1900 and 1990 can be modeled by the equation

$$m(x) = (5.556\cdot 10^{-5})x^3 - 0.00736x^2 + 0.1029x + 104.955 \text{ males per 100 females}$$

x years after 1900.

a. Find the absolute extreme points of m. Consider the function out of its modeling context.

b. Find when between 1900 and 1990 the gender ratio was greatest and when it was least according to the model, and give the corresponding ratios.

c. Graph m and its derivative between 1900 and 2000. Label the points on each graph that correspond to the absolute extrema of m found in part *a*.

24. The amount of fish produced for human food by fisheries in the United States from 1970 through 1996 can be modeled[19] by

$$F(t) = \begin{cases} 22.204t^2 - 108.431t + 2538.603 \text{ million lb} & \text{when } 0 \le t \le 10 \\ 53.398t^2 - 1400.814t + 12{,}363.733 \text{ million lb} & \text{when } 10 < t < 23 \\ -248.6t + 13{,}913.7 \text{ million lb} & \text{when } 23 \le t \le 26 \end{cases}$$

where t is the number of years since 1970.

a. Determine all relative and absolute extrema of F between $t = 0$ and $t = 26$. Consider the function out of its modeling context.

b. F must be interpreted discretely because it gives yearly totals. With this restriction in mind, determine when production was greatest and when it was least from 1970 through 1996.

25. The flow rate (in cubic feet per second, cfs) of a river in the 24 hours after the beginning of a severe thunderstorm can be modeled by

$$C(h) = \begin{cases} -0.865h^3 + 12.045h^2 - 8.952h + 123.02 \text{ cfs} & \text{when } 0 \le h \le 10 \\ -16.643h + 539.429 \text{ cfs} & \text{when } 10 < h \le 24 \end{cases}$$

where h is the number of hours after the storm began.

a. What were the flow rates for $h = 0$ and $h = 24$?

b. Determine the absolute maximum and minimum flow rates between $h = 0$ and $h = 24$.

26. The average price (per 1000 cubic feet) of natural gas for residential use[20] from 1980 through 1997 is modeled by the equation

$$p(x) = 0.004265x^3 - 0.11816x^2 + 0.9674x + 3.681 \text{ dollars}$$

where x is the number of years since 1980.

a. Graph the function p, and find the relative maximum and the relative minimum. Consider the function out of its modeling context.

b. Graph the derivative of p, and find the x-intercepts of the derivative. Interpret your answers.

c. When (according to the model) was natural gas the most and least expensive between 1980 and 1997?

d. Repeat part *c* for the interval 1984 through 1994.

27. *Swimming World* (August 1992) lists the time in seconds that an average athlete takes to swim 100 meters free style at age x years. The data are given in Table 5.12.

TABLE 5.12

Age x (years)	Time (seconds)	Age x (years)	Time (seconds)
8	92	22	50
10	84	24	49
12	70	26	51
14	60	28	53
16	58	30	57
18	54	32	60
20	51		

a. Find the best model for the data.

17. From the Federal Watermaster, U.S. Department of the Interior.
18. Based on data from U.S. Bureau of the Census.
19. Based on data from *Statistical Abstract*, 1998.
20. Based on data from *Statistical Abstract*, 1998.

b. Using the model, find the age at which the minimum swim time occurs. Also find the minimum swim time.

c. Compare the table values with the values in part *b*.

28. Table 5.13 lists the number of live births[21] in the United States between the years 1950 and 1995 to women 45 years of age and older.

TABLE 5.13

Year	Number of births	Year	Number of births
1950	5322	1975	1628
1955	5430	1980	1200
1960	5182	1985	1162
1965	4614	1990	1638
1970	3146	1995	2727

a. Use the data values to estimate when the number of births was greatest.

b. Use the data values to estimate when the number of births was least.

c. Find a model for the data, and use the equation to answer parts *a* and *b*. (The model must be discretely interpreted because it describes yearly totals.)

d. How do the model answers differ from the table values? Which estimates do you believe are more accurate?

29. *Consumer expenditure* and *revenue* are terms for the same thing from two perspectives. Consumer expenditure is the amount of money that consumers spend on a product, and revenue is the amount of money that businesses take in by selling the product. A street vendor constructs Table 5.14 on the basis of sales data.

a. Find a model for quantity sold.

b. Convert the equation in part *a* to an equation for consumer expenditure.

c. What price should the street vendor charge to maximize consumer expenditure?

d. If each dozen roses costs $6, what price should the street vendor charge to maximize profit?

e. Why do the derivatives of the revenue and profit equations in this activity not give marginal revenue and marginal profit?

TABLE 5.14

Price of a dozen roses (dollars)	Number of dozens sold per week
10	190
15	145
20	110
25	86
30	65
35	52

30. An apartment complex has an exercise room and sauna, and tenants will be charged a yearly fee for the use of these facilities. A survey of tenants results in the demand/price data shown in Table 5.15.

TABLE 5.15

Quantity demanded	Price (dollars)
5	250
15	170
25	100
35	50
45	20
55	5

a. Find a model for price as a function of demand.

b. On the basis of the price model, give the equation for revenue.

c. Find the maximum point on the revenue model. What price and what demand give the highest revenue? What is the marginal revenue at the maximum point?

31. The yearly amount of garbage (in millions of tons) taken to a landfill outside a city during selected years from 1975 through 2005 is given in Table 5.16.

a. Find a model for the data.

b. Give the slope formula for the model.

c. How rapidly was the amount of garbage taken to the landfill increasing in 2005?

d. Graph the derivative of your model, and determine whether your model has a relative maximum and/or minimum. Explain how you reached your conclusion.

21. *The 1998 Information Please Almanac* (Boston: Houghton Mifflin).

TABLE 5.16

Year	Amount (millions of tons)
1975	81
1980	99
1985	117
1990	122
1995	132
2000	145
2005	180

32. A company analyzes the production costs for one of its products and determines the hourly operating costs when x units are produced each hour. The results are given in Table 5.17.

TABLE 5.17

Production level, x (units per hour)	Hourly cost (dollars)
1	210
7	480
13	650
19	760
25	810
31	845
37	880
43	950
49	1070
55	1280
61	1590

a. Find a model for hourly cost in terms of production level.

b. Find and interpret the marginal cost when 40 units are produced.

c. On the basis of the model in part *a*, what is the equation for the average hourly cost per unit when x units are produced each hour?

d. Find the production level that minimizes average hourly cost. Give the average hourly cost and total cost at that level.

33. Table 5.18 gives the amount spent on national defense[22] in billions of dollars (measured in constant 1987 dollars).

TABLE 5.18

Year	Amount (billions of dollars)
1970	262.9
1972	219.7
1974	185.3
1976	177.8
1978	177.2
1980	187.1
1982	214.3
1984	241.7
1986	276.4
1988	283.3
1990	272.5

a. Find a model for the data.

b. Give the slope formula for the model.

c. Find the relative maximum and relative minimum points of the function in part *a*. Consider the function out of its modeling context.

d. A model for defense spending must be discretely interpreted because it provides yearly totals. With this restriction in mind, determine when, according to the model, defense spending was greatest and when it was least between 1970 and 1990.

e. If you found the answers to part *c* by using a graphing calculator or computer, discuss how you could find the answers without using that technology.

34. Table 5.19 gives the yearly emissions[23] in millions of metric tons of nitrogen oxides, NO_x, in the United States from 1940 through 1990.

TABLE 5.19

Year	NO_x (millions of metric tons)
1940	6.9
1950	9.4
1960	13.0
1970	18.5
1980	20.9
1990	19.6

22. *Statistical Abstract*, 1994.
23. *Statistical Abstract*, 1992.

a. Find a model for the data.

b. Give the slope formula for the model.

c. Determine the year in which emissions of NO_x were greatest between 1940 and 1990. What was the amount of emissions that year? (The model must be discretely interpreted.)

d. In 1994, NO_x emissions were 21.5 million metric tons. What does this information indicate about the accuracy of the model in part *a* for extrapolation? What does this information indicate about the accuracy of your answer to part *c*?

35. Imagine that you have been hired as director of a performing arts center for a mid-sized community. The community orchestra gives monthly concerts in the 400-seat auditorium. To promote attendance, the former director lowered the ticket price every 2 months. The ticket prices and corresponding average attendances are given in Table 5.20.

TABLE 5.20

Price (dollars)	Average attendance
35	165
30	200
25	240
20	280
15	335
10	400

a. Find quadratic and exponential models for the data. Which model better reflects the probable attendance beyond a \$35 ticket price? Explain.

b. On the basis of the model that you believe is more appropriate, give the equation for revenue.

c. Find the maximum revenue and the corresponding ticket price and average attendance.

d. What other things besides the maximum revenue should you consider in setting price?

36. Refer to the Kentucky population model given at the beginning of this section.

a. Graph the equation given in the text along with the data[24] in Table 5.21. Align the input data as the number of years since 1980. How does the behavior of the graph of the model between 1993 and 1997 compare with the behavior of the scatter plot of the data?

TABLE 5.21

Year	Kentucky population (thousands)	Year	Kentucky population (thousands)
1993	3793	1996	3882
1994	3824	1997	3908
1995	3856		

b. Find a model for the data in Table 5.21. Use it to write a piecewise continuous function for the population of Kentucky between 1980 and 1997.

c. Graph the piecewise continuous model in part *b* and its derivative.

d. Find all relative extrema of the population model from 1980 through 1997 and the absolute maximum and absolute minimum of the population model during that time.

e. Find the most recent population value available for Kentucky. Does this value follow the trend indicated by the model in part *b*?

37. If they exist, find the absolute maxima and absolute minima of $y = (2 - 3x + x^2)(3.5 + x)^2$ over all real number inputs. If an absolute maxima or absolute minima does not exist, explain why not.

38. If they exist, find the absolute maxima and absolute minima of $y = \frac{2x^2 - x + 3}{x^2 + 2}$ over all real number inputs. If an absolute maxima or absolute minima does not exist, explain why not.

5.3 Derivatives in Action

By now you should have a solid understanding of the basic method of finding both relative and absolute maximum and minimum values of a function. Whether you use technology or algebra, relative extrema are found by identifying points at which the graph of

24. *Statistical Abstract*, 1998.

the derivative of a function crosses the input axis or points at which the derivative is not defined. Absolute extrema of a function on an interval are found by comparing the output values of any local extrema to the output values at the ends of the interval.

Students will find setting up the equation the most difficult part of this section.

We now extend the techniques of optimization to situations in which you must first set up an equation to optimize and then apply optimization techniques to that equation. Many of the problems presented in this section are different from most of those in this book because no data or equation is given. You must find an equation from a word description. We begin by outlining some steps that are helpful in setting up equations and optimizing their outputs. Refer to this list often until these steps become second nature to you.

Step 1: Read the question carefully, and identify the quantity to be maximized or minimized (the output variable) and the quantity or quantities on which the output quantity depends (the input variable). This information is often found in the final sentence of the problem statement. Because this step is the most crucial one for correctly directing your thinking as you approach the problem, we present four questions in Example 1 to give you some practice in identifying input and output quantities.

EXAMPLE 1 *Determining Output and Input*

For each of the following statements, determine the quantity to be maximized or minimized and the corresponding input quantity or quantities.

a. Find the order size that will minimize cost.

b. Determine the dimensions that will result in a box of greatest volume.

c. For what size dose is the sensitivity to the drug greatest?

d. What is the most economical way to lay the pipe?

Solution

a. Output quantity to be minimized: cost
Input quantity: order size

b. Output quantity to be maximized: volume
Input quantity: dimensions of box (height, length, and width)

c. Output quantity to be maximized: sensitivity
Input quantity: dosage

d. Output quantity to be minimized: cost
Input quantity: Not enough information is given in this sentence, but we can conclude that the input is probably some measure of distance related to how the pipe is laid.

Step 2: Sketch a picture if the problem situation is geometric in nature—that is, if the problem involves constructing containers, laying cable or wire, building enclosures, or the like. The picture should be labeled with appropriate variables representing distances or sizes. You should find that at least one of the variables represents the input quantity identified in Step 1.

Refer your students to the *Calculus Concepts* web site for a list of geometric formulas used in this section.

Step 3: Begin building the equation. From the results of Step 1, you will know the quantity for which you need an equation, such as cost, volume, or drug sensitivity. If the quantity is a geometric one (such as volume of a right circular cylinder, area of a

trapezoid, hypotenuse of a right triangle) you may need to look up the formula in some reference source.

Step 4: If it is not already in this form, rewrite the equation as a function of only one variable. This step may require the use of other, secondary equations relating the input variables in the equation in Step 3. For example, if the equation in Step 3 is for volume of a box as a function of length and width, you need to find an equation relating length and width. You will then use this equation to solve for one variable in terms of the other and substitute for that variable in the equation in Step 3.

Step 5: Determine any limitations on the input variable. State the limitation as an interval in which the input must lie.

Students should not assume when there is one relative extremum that this value is automatically the answer to the question, because the answer could occur at one of the endpoints. Encouraging your students to graph the function as a means of visually checking their answers will help them avoid this mistake. Instead of graphing the function, a student could check the sign of the first derivative on either side of the relative extreme point.

Step 6: Apply optimization techniques to the final equation. This step includes finding all relative extrema and determining all absolute extrema over the interval identified in Step 5. This step should also include graphing the equation or some other method of verifying that the answer you found is indeed a maximum or minimum on the interval.

Step 7: Read the problem again to be sure you have answered the question asked. Sometimes the question asks you to find the output quantity, sometimes the input quantity, sometimes both, and occasionally some other quantity that can be calculated using the input and output values that correspond to the extreme point.

Before demonstrating this multistep problem-solving method with three examples, we summarize the seven steps.

A good short quiz in the class following your discussion of this section is to have your students list these seven steps in their own words.

Problem-Solving Strategy for Optimization Problems

Step 1: Identify the quantity to be maximized or minimized (the output) and the quantity or quantities on which the output quantity depends (the input).

Step 2: If the problem is geometric in nature, sketch and label a picture.

Step 3: Build a model for the quantity that is to be maximized or minimized.

Step 4: If necessary, rewrite the equation in Step 3 in terms of only one input variable.

Step 5: Identify the input interval.

Step 6: Apply optimization techniques to the final equation.

Step 7: Answer the question or questions posed in the problem.

EXAMPLE 2 *Ranching*

A rancher removed 200 feet of wire fencing from a field on his ranch. He wishes to reuse the fencing to create a rectangular corral into which he will build a 6-foot wide wooden gate. What dimensions will result in a corral with the greatest possible area? What is the greatest area?

Solution

Step 1: Output variable to maximize: area of rectangular corral
Input variables: length and width of the corral

Step 2: A sketch of the corral is shown in Figure 5.14. (Note that you can put the gate on any side of the corral.)

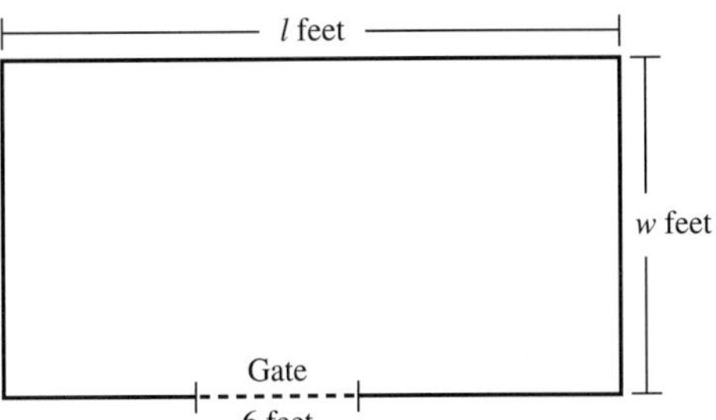

FIGURE 5.14

Step 3: Recall that the area of a rectangle is $A = lw$ square feet, where l is the length of the rectangle in feet and w is its width in feet. This equation is the one we seek to maximize.

Step 4: Because the equation in Step 3 has two input variables, we need to find a second equation relating the length and the width. A clue to this equation is found in the statement "A rancher removed 200 feet of wire fencing." If the area of the corral is to be a maximum, it makes sense that the rancher will use all of the fencing. Referring to Figure 5.14, we see that if we calculate the distance around the corral and subtract the 6-foot-wide gate, we should obtain 200 feet. This leads to the equation

$$2w + 2l - 6 = 200$$

Solving this equation for one of the variables (it doesn't matter which one) results in the equation

$$l = \frac{206 - 2w}{2} = 103 - w$$

Substituting this expression for l into the area equation gives

$$A = lw = (103 - w)w = 103w - w^2 \text{ square feet}$$

where w is the width of the corral in feet.

Step 5: Look again at Figure 5.14. If the length of the corral were 0, then the width would be 100 feet. We also know that width cannot be zero or negative. Thus we have the input interval $0 < w < 100$.

Step 6: Note that the graph of A is a concave-down parabola. We find the maximum point on the parabola by determining where the derivative of A is zero (that is, where its graph crosses the w-axis).

$$\frac{dA}{dw} = 103 - 2w = 0$$

$$w = 51.5$$

This value lies in the interval in Step 5. Because the interval is an open interval, we have no endpoints to check. Thus the corral with maximum area should be 51.5 feet wide.

Step 7: The statement of the problem asked for the dimensions of the corral of greatest area as well as for the area. We know the width and need to determine the length of the corral. To do so, we use the equation from Step 4 that expresses length in terms of width.

$$l = 103 - w$$

$$l = 103 - 51.5 = 51.5 \text{ feet}$$

This calculation tells us that the corral with maximum area is a square with sides that are 51.5 feet long. To determine the corresponding area, simply square 51.5 feet.

$$A = (51.5 \text{ feet})(51.5 \text{ feet}) = 2652.25 \text{ square feet}$$

The next example shows the solution to a more involved geometric problem.

EXAMPLE 3 *Popcorn Tins*

In an effort to be environmentally responsible, a confectionery company is rethinking the dimensions of the tins in which it packages popcorn. Each cylindrical tin is to hold 3.5 gallons. The bottom and lid are both circular, but the lid must have an additional $1\frac{1}{8}$ inch around it in order to form a lip. (Consider the amount of metal needed to create a seam on the side and to join the side to the bottom to be negligible.) What are the dimensions of a tin that meets these specifications but uses the least amount of metal possible?

Solution

Step 1: Output quantity to be minimized: amount of metal
Input quantities: dimensions of the tin

Step 2: Figure 5.15 shows the tin and its components. We have chosen to label the height of the tin h and the radius of the bottom of the tin r. (It is also possible to use the diameter, instead of the radius, of the top or bottom as the other variable.)

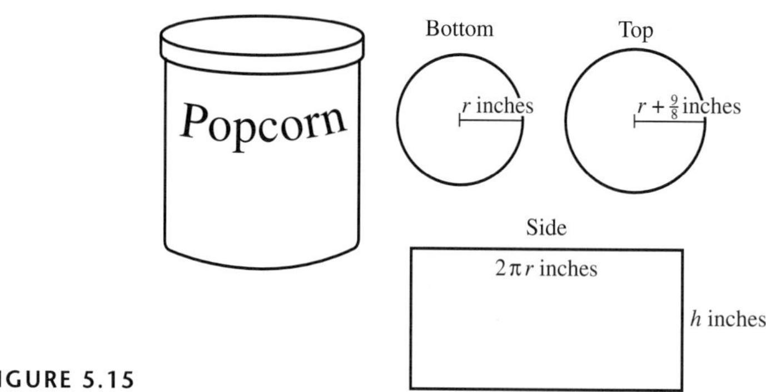

FIGURE 5.15

Step 3: The amount of metal used to construct the tin is measured by finding the combined surface area of the side, top, and bottom. Thus the total surface area is determined by the equation

$$S = \text{area of side} + \text{area of bottom} + \text{area of top}$$

The area of the side is $2\pi rh$. The area of the bottom is πr^2. The radius of the circle used to make the top is $r + \frac{9}{8}$; thus the area of the top is $\pi\left(r + \frac{9}{8}\right)^2$. Using these formulas, we have the model for the surface area of the container:

$$S = 2\pi rh + \pi r^2 + \pi\left(r + \frac{9}{8}\right)^2 \text{ square inches}$$

where r is the radius of the bottom of the container in inches and h is the height of the container in inches.

Step 4: We need to rewrite the formula for surface area in terms of either radius or height. To do so, we need an equation relating these two quantities. Recall from the statement of the problem that the volume of the container needs to be 3.5 gallons. This information allows us to set up a second equation. The volume of a cylinder is given by the equation $V = \pi r^2 h$. Because we will be using dimensions in inches, we must express the volume in cubic inches. One gallon is 231 cubic inches (the authors found this value in a dictionary), so 3.5 gallons is $3.5(231) = 808.5$ cubic inches. Thus we have the equation

$$V = \pi r^2 h = 808.5$$

We must solve for one variable in terms of the other. It makes sense to choose the variable for which this can be done most simply. In this case, we solve for h in terms of r.

$$h = \frac{808.5}{\pi r^2} \quad \text{for } r \neq 0$$

Next, we substitute this expression for h into the surface area equation in Step 3.

$$S = 2\pi r\left(\frac{808.5}{\pi r^2}\right) + \pi r^2 + \pi\left(r + \tfrac{9}{8}\right)^2 = \frac{1617}{r} + \pi r^2 + \pi\left(r + \tfrac{9}{8}\right)^2 \text{ square inches}$$

Step 5: Because we are given no indication that the dimensions are restricted (except by the volume of the container, which was taken into account in Step 4), our only restriction is that the radius must be positive: $r > 0$.

Step 6: Next we find the minimum of S by determining where a graph of $\frac{dS}{dr}$ crosses the horizontal axis. To do so, we find the derivative of S with respect to r.

$$\frac{dS}{dr} = \frac{-1617}{r^2} + 2\pi r + 2\pi\left(r + \tfrac{9}{8}\right)(1) = \frac{-1617}{r^2} + 4\pi r + \frac{9\pi}{4}$$

Setting this expression equal to zero and solving for r gives $r \approx 4.87$ inches, which lies in the interval in Step 5. To confirm that this value of r does indeed give a minimum surface area, we examine a graph of the surface area function. See Figure 5.16.

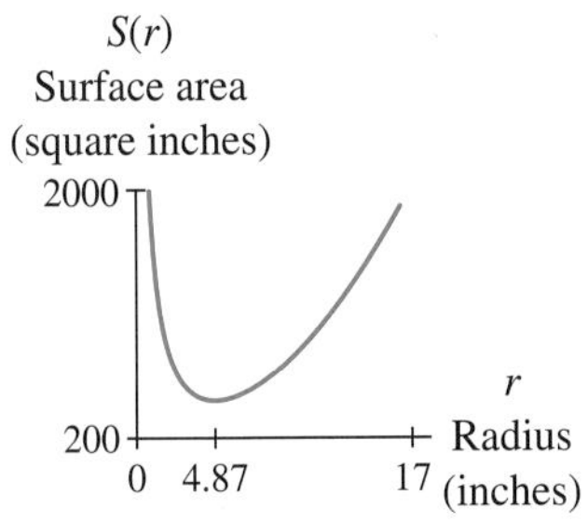

FIGURE 5.16

Step 7: Our final step is to answer the question posed. In this case, we must provide the dimensions (height and radius) of the tin with minimum surface area. We know the radius. In order to find the height, we substitute the unrounded value of the radius into the equation that we found in Step 4 for height in terms of radius.

$$h \approx \frac{808.5}{\pi(4.87)^2} \approx 10.86 \text{ inches}$$

In order to use the least amount of metal possible, the popcorn tins should be constructed with height approximately 10.86 inches and radius approximately 4.87 inches.[25]

Our final example illustrates optimization in a nongeometric setting.

25. The authors were surprised to find that the popcorn tin they purchased as a reference tool for this example did have dimensions that matched those of a tin with minimum surface area.

EXAMPLE 4 *Cruise to Cancun*

A travel agency offers spring-break cruise packages. The agency advertises a cruise to Cancun, Mexico, for \$1200 per person. In order to promote the cruise among student organizations on campus, the agency offers a discount for student groups selling the cruise to over 50 of their members. The price per student will be discounted by \$10 for each student in excess of 50. For example, if an organization had 55 members go on the cruise, each of those 55 students would pay \$1200 − 5(\$10) = \$1150.

a. What size group will produce the largest revenue for the travel agency, and what is the largest possible revenue?

b. If the travel agent limits each organization to 75 tickets, what is the agent's maximum revenue for each organization?

c. If you were the travel agent, would you set a limit on the number of students per organization?

Solution

a. ***Step 1:*** Output quantity to be maximized: travel agency's revenue from student group

Input quantity: size of the group

Step 2: Because this scenario is not geometric, there is no need for a picture.

Step 3: We need an equation for the travel agency's revenue from the students. In this example, revenue is the number of students traveling on the cruise multiplied by the price each student pays:

$$\text{Revenue} = (\text{number of students})(\text{price per student})$$

In Step 1, we determined that the input quantity is the size of the group. Because the factor affecting price is the number of students in excess of 50, we choose to use a variable to represent that quantity. (It is also possible to use a variable representing the total number of students.) If s is the number of students in excess of 50, then the total number of students is $50 + s$. The price is \$1200 minus \$10 for each student in excess of 50. Converting this statement into math symbols, we have the price per student as $1200 - 10s$ dollars. Thus

$$\text{Revenue} = R(s) = (50 + s)(1200 - 10s) = 60{,}000 + 700s - 10s^2 \text{ dollars}$$

when there are $50 + s$ students on the cruise.

Step 4: Because the revenue function has only one input variable, this step is not necessary.

Step 5: We are not given any restriction concerning the number of students. Our interval is $s > 0$.

Step 6: The revenue equation is a concave-down parabola, so we can be confident that it has a maximum. We simply set the derivative of the revenue function equal to zero and solve for s.

$$R'(s) = 700 - 20s = 0$$
$$s = 35$$

Revenue is maximized when $s = 35$.

Step 7: The question asks for the number of students and the corresponding revenue. Recall that the variable s represents the number of students in excess of 50.

Thus the total number of students is $50 + 35 = 85$. The price per student is $\$1200 - 35(10) = \$850$. The revenue is (85 students)($850 per student) = $72,250. Revenue is maximized at $72,250 for a campus group of 85 students.

b. If the number of students is limited to 75, then the interval in Step 5 is $0 < s \le 25$. The solution to part *a* is no longer valid because it does not lie in this interval. In this case, the maximum revenue occurs at the endpoint of the interval (when there are 75 students). Examine a graph of *R*, and observe that this is true. The price each student pays is $1200 - 25(10) = \$950$. The travel agent's revenue in this case is (75 students)($950 per student) = $71,250.

c. In part *a*, we found that when 85 students bought cruise tickets, the travel agent's revenue was a maximum. If more than 85 students bought tickets, the agent's revenue would actually decline because of the low price. (Recall that the revenue function graph was a concave-down parabola.) Therefore, it would make sense for the agent to limit the number of students per organization to 85.

Optimization problems such as those discussed in this section are often challenging, but as long as you approach them using the seven steps outlined here, you should be able to solve them—or at least identify the point at which you need help. As we have said before, the more problems you work, the more skilled and confident you will become.

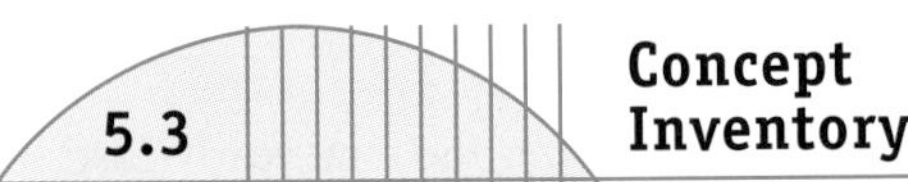

5.3 Concept Inventory

- *Seven-step problem-solving strategy*

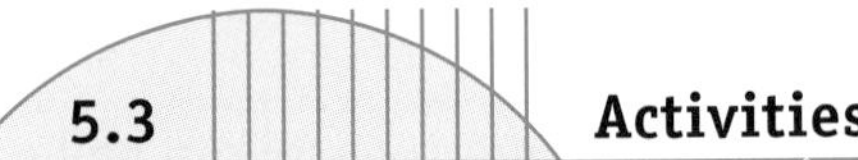

5.3 Activities

1. Refer to Example 3, and repeat the solution using the diameter of the bottom of the tin, instead of the radius, as an input variable. Show that the dimensions you obtain for minimum surface area are equivalent to those found in Example 3.

2. Refer to Example 4, and repeat the solution using the total number of students as the input variable. Show that the maximum revenue and the corresponding number of students and price are equivalent to those found in Example 4.

3. A rectangular-shaped garden has one side along the side of a house. The other three sides are to be enclosed with 60 feet of fencing. What is the largest possible area of such a garden?

4. A florist uses wire frames to support flower arrangements displayed at weddings. Each frame is constructed from a wire of length 9 feet that is cut into six pieces. The vertical edges of the frame consist of four of the pieces of wire that are each 12 inches long. One of the remaining pieces is bent into a square to form the base of the frame; the final piece is bent into a circle to form the top of the frame. See Figure 5.3.1.

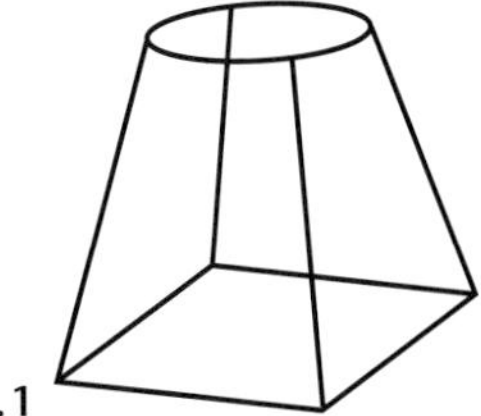

FIGURE 5.3.1

   a. How should the florist cut the wire of length 9 feet in order to minimize the combined area of the circular top and the square base of the frame?

   b. Verify that the answer to part *a* minimizes the combined area.

   c. What is the answer to part *a* if the frame must be constructed so that the area enclosed by the square is twice the area enclosed by the circle?

5. You have been selected to appear on the *Race for the Money* television program. Before the program is

aired on television, you and three other contestants are sent to Myrtle Beach, SC. Each of you is given a piece of cardboard that measures 8 inches by 10 inches, a pair of scissors, a ruler, and a roll of tape. The contestant who carries away the most sand from the beach is eligible to win the grand prize of $250,000.

There are two restrictions in the race. The sand must be carried away in an open box that each person makes from the piece of cardboard by cutting equal squares from each corner and turning up the sides. A second restriction is that the sand cannot be piled higher than the sides of the box.

a. What length should the corner cuts be for a contestant to carry away the most sand? What is the largest volume of sand that can be carried away in such a box?

b. What length should the cuts be if there is a third restriction that the box can hold no more than 50 cubic inches of sand?

c. Do you feel you would have a better chance of winning the sand race with the two restrictions given in part *a* or with all three restrictions on the amount of sand that can be carried away?

6. A portion of the shoreline of a Caribbean island is in the shape of the curve $y = 2\sqrt{x}$. A hut is located at point C, as shown in Figure 5.3.2. You wish to deliver a load of supplies to the shoreline and then hire someone to help you carry the supplies to your hut. The helper will charge you $10 per mile. At what point (x, y) on the shoreline should you land in order to minimize the amount you must pay to have the supplies carried to the hut? How much currency should you have on hand to pay the helper?

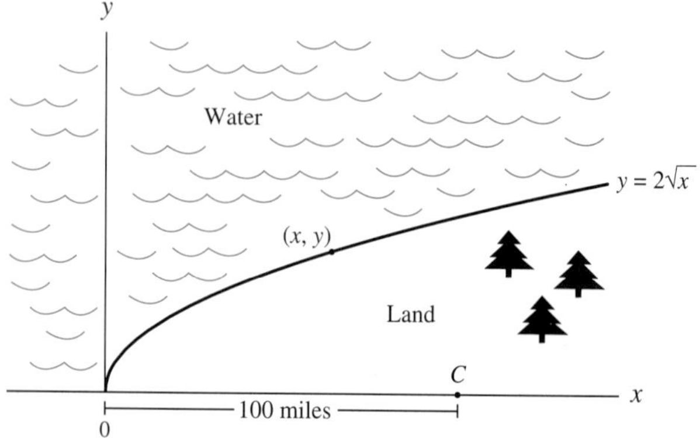

FIGURE 5.3.2

7. During one calendar year, a year-round elementary school cafeteria uses 42,000 Styrofoam plates and packets, each containing a fork, spoon, and napkin. The smallest amount the cafeteria can order from the supplier is one case containing 1000 plates and packets. Each order costs $12, and the cost to store a case for the whole year is approximately $4. In order to find the optimal order size that will minimize costs, the cafeteria manager must balance the ordering costs incurred when many small orders are placed with the storage costs incurred when many cases are ordered at once.

a. How many cases does the cafeteria need over the course of 12 months?

b. If x is the number of cases in each order, how many times will the manager need to order during one calendar year?

c. What is the total cost of ordering for one calendar year?

d. Assume that the average number of cases stored throughout the year is half the number of cases in each order. What is the total storage cost for one year?

e. Write a model for the combined ordering and storage costs for one year.

f. What order size minimizes the total yearly cost? (Note that only full cases may be ordered.) How many times a year should the manager order? What will the minimum total ordering and storage costs be for the year?

8. A situation similar to the order/storage cost problem in Activity 7 occurs when a company uses a machine to produce different types of items. For example, the same machine may be used to produce popcorn tins and 30-gallon storage drums. The machine is set up to produce a quantity of one item and then reconfigured to produce a quantity of another item.

The plant produces a run and then ships the tins out at a constant rate so that the warehouse is empty for storing the next run. Assume that the number of tins stored on average during one year is half of the number of tins produced in each run. A plant manager must take into account the cost to reset the machine (similar to the ordering cost in Activity 7) and the cost to store inventory. Although it might otherwise make sense to produce an entire year's inventory of popcorn tins at once, the cost of storing all the tins over a year's time would be prohibitive.

Suppose a company needs to produce 1.7 million popcorn tins over the course of a year. The cost to set up the machine for production is \$1300, and the cost to store one tin for a year is approximately \$1. What size production run will minimize set-up and storage costs? How many runs are needed during one year, and how often will the plant manager need to schedule a run of popcorn tins?

9. Frozen juice cans are constructed with cardboard sides and metal top and bottom. A typical can holds 12 fluid ounces. If the metal costs twice as much per square inch as the cardboard, what dimensions will minimize the cost of a 12-fluid-ounce can?

10. As a sales representative, you are required to travel to trade shows and display your company's products. You need to design a display booth for such shows. Because your company generally must pay for the amount of square footage your booth requires, you want to limit the floor size to 300 square feet. The booth is to be 6 feet tall and three-sided, with the back of the booth a display board and the two sides of the booth made of gathered fabric. The display board for the back of the booth costs \$30 per square foot. The fabric costs \$2 per square foot and needs to be twice the length of the side to allow for gathering.
    a. Find the minimum cost of constructing a booth according to these specifications. What should be the dimensions of the booth?
    b. In order to accommodate your company's display, the back of the booth must be at least 15 feet wide. Does this restriction change the dimensions necessary to minimize cost? If so, what are the optimal dimensions now?

11. A dog kennel owner needs to build a dog run adjacent to one of the kennel cages. See Figure 5.3.3.

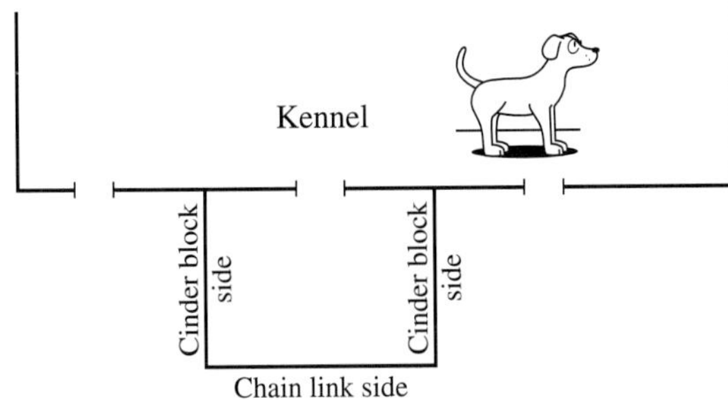

FIGURE 5.3.3

The ends of the run are to be cinder block, which costs 50 cents per square foot. The side is to be chain link, which costs \$2.75 per square foot. Local regulations require that the walls and fence be at least 7 feet high and that the area measures no less than 120 square feet.
    a. Determine the dimensions of the dog run that will minimize cost but still meet the regulatory standards.
    b. Repeat part *a* for conditions where two identical dog runs are to be built side-by-side, sharing one cinder block wall.
    c. Do the answers to parts *a* and *b* change if there is a mandatory minimum width or length of 6 feet?

12. A set of 24 duplexes (48 units) was sold to a new owner. The previous owner charged \$700 per month rent and had a 100% occupancy rate. The new owner, aware that this rental amount was well below other rental prices in the area, immediately raised the rent to \$750. Over the next 6 months, an average of 47 units were occupied. The owner raised the rent to \$800 and found that 2 of the units were unoccupied during the next 6 months.
    a. What was the monthly rental income made by the previous owner?
    b. After the first rent increase, what was the monthly income?
    c. After the second increase, how much did the new owner collect in rent each month?
    d. Find an equation for the monthly rental income as a function of the number of \$50 increases in the rent.
    e. If the occupancy rate continues to decrease as the rent increases in the same manner as it did for the first two increases, what rent amount will maximize the owner's rental income from these duplexes? How much will the owner collect in rent each month at this rent amount?
    f. What other considerations besides rental income should the owner take into account when determining an optimal rent amount?

13. A sorority plans a bus trip to The Great Mall of America during Thanksgiving break. The bus they charter seats 44 and charges a flat rate of \$350 plus \$35 per person. However, for every empty seat, the charge per person is increased by \$2.00. There is a minimum of 10 passengers. The sorority leadership

decides that each person going on the trip will pay \$35. The sorority itself will pay the flat rate and the additional amount above \$35 per person. For example, if there are only 40 passengers, the fare per passenger is \$43. Each of the 40 passengers will pay \$35, and the sorority will pay $\$350 + 40(\$8) = \$670$.

a. Find a model for the revenue made by the bus company as a function of the number of passengers.

b. Find a model for the amount the sorority pays as a function of the number of passengers.

c. For what number of passengers will the bus company's revenue be greatest? For what number of passengers will the bus company's revenue be least?

d. For what number of passengers will the amount the sorority pays be greatest? For what number of passengers will the amount the sorority pays be least?

14. A cable television company needs to run a cable line from its main line ending at point P in Figure 5.3.4 to point H at the corner of the house. The county owns the roads shown in the figure, and it costs the cable company \$25 per foot to run the line on poles along the county roads. The area bounded by the house and roads is a privately owned field, and the cable company must pay for an easement to run lines underground in the field. It is also more costly for the company to run lines underground than to run them on poles. The total cost to run the lines underground across the field is \$52 per foot. The cable company has the choice of running the line along the roads or cutting across the field.

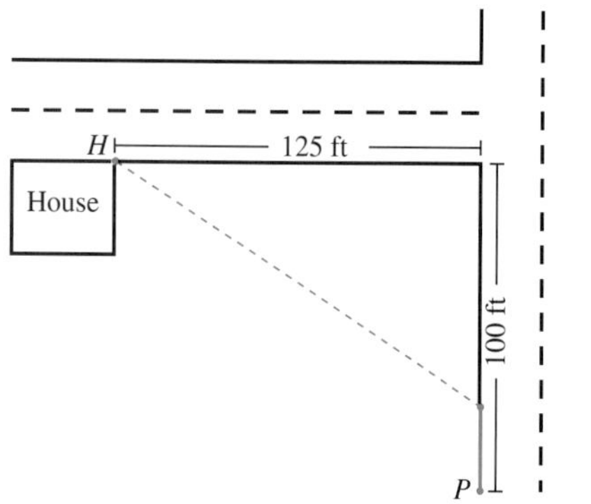

FIGURE 5.3.4

a. Calculate the cost for the company to run the line along the roads to point H.

b. Calculate the cost to run the line directly across the field from point P to point H.

c. Calculate the cost to run the line along the road for 50 feet from point P and then cut across the field.

d. Set up an equation for the cost to run the line along the road a distance of x feet from point P and then cut across the field. If $x = 0$, the line will cut directly across the field, and the value of the cost function should match your answer to part a.

e. Using calculus and your equation from part d, determine whether it is less costly for the company to cut across the field. If so, at what distance from point P should the company begin laying the line through the field?

15. A trucking company wishes to determine the recommended highway speed for its truckers to drive in order to minimize the combined cost of driver's wages and fuel required for a trip. The average wage for the truckers is \$15.50 per hour, and average fuel efficiencies for their trucks as a function of the speed at which the truck is driven are shown in Table 5.22.

Activities 15 and 16 are time consuming. You may wish to assign one or both of them as projects.

TABLE 5.22

Speed (mph)	50	55	60	65	70
Fuel consumption (mpg)	5.11	4.81	4.54	4.09	3.62

a. Find a model for fuel consumption as a function of the speed driven.

b. For a 400-mile trip, find formulas for the following quantities in terms of speed driven:

i. Driving time required

ii. Wages paid to the driver

iii. Gallons of fuel used

iv. Total cost of fuel (use a reasonable price per gallon based on current fuel prices)

v. Combined cost of wages and fuel

c. Using equation v in part b, find the speed that should be driven in order to minimize cost.

d. Repeat parts b and c for 700-mile and 2100-mile trips. What happens to the optimal speed as the trip mileage increases?

e. Repeat parts *b* and *c* for a 400-mile trip, increasing the cost per gallon of fuel by 20, 40, and 60 cents. What happens to the optimal speed as the cost of fuel increases?

f. Repeat parts *b* and *c* for a 400-mile trip, increasing the driver's wages by \$2, \$5, and \$10 per hour. What happens to the optimal speed as the wages increase?

16. A standard 3.5-inch computer diskette stores data on a circular sheet of mylar that is 3.36 inches in diameter. The disk is formatted by magnetic markers being placed on the disk to divide it into tracks (thin, circular rings). For design reasons, each track must contain the same number of bytes as the innermost track. If the innermost track is close to the center, more tracks will fit on the disk, but each track will contain a relatively small number of bytes. See Figure 5.3.5a. If the innermost track is near the edge of the disk, fewer tracks will fit on the disk, but each track will contain more bytes. See Figure 5.3.5b. For a standard double-sided, high-density 3.5-inch disk, there are 135 tracks per inch between the innermost track and the edge of the disk, and about 1400 bytes per inch of track. This type of disk holds 1,440,000 bytes of information when formatted.

a. For the disk in Figure 5.3.5a, calculate the following quantities. The answers to parts *ii* and *iii* must be positive integers, and the rounded values should be used in the remaining calculations.

i. Distance in inches around the innermost track

ii. Number of bytes in the innermost track

iii. Total number of tracks

iv. Total number of bytes on one side of the disk

v. Total number of bytes on both sides of the disk

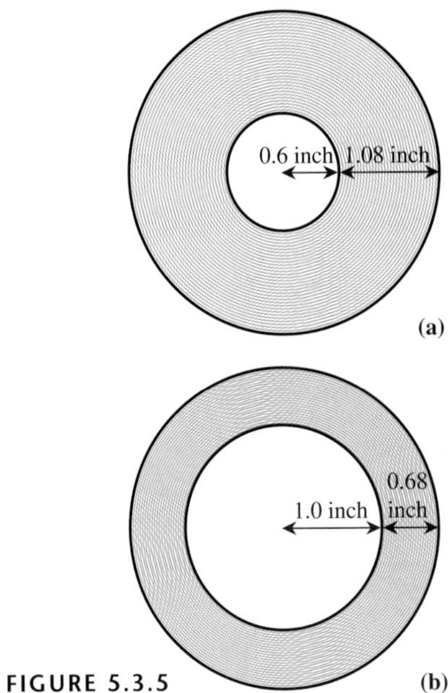

FIGURE 5.3.5

b. Calculate the quantities in part *a* for the disk shown in Figure 5.3.5b.

c. Repeat part *a* for a disk whose innermost track begins *r* inches from the center of the disk. Use the optimization techniques presented in this chapter to find the optimal distance the innermost track should be from the center of the disk in order to maximize the number of bytes stored on the disk. What is the corresponding number of bytes stored? How does this number compare to the total number of bytes stored by a standard double-sided, high-density 3.5-inch diskette?

d. If you are willing to destroy a used disk for the sake of curiosity, open a 3.5-inch disk and measure from the center to where the tracks begin. How does this measurement compare with the optimal distance you found in part *c*?

5.4 Inflection Points

Unlike relative extrema, which are output values, inflection points are coordinate points.

In Section 5.2, we discussed some important points that may occur on a graph. These were called maxima and minima (or extreme points). Another important point is an **inflection point.**

Recall from our earlier work with cubic and logistic functions that an inflection point is a point where a graph changes concavity. On a smooth, continuous graph, the

You may wish to point out to your students that whereas instructions to "find the greatest/least" refer to finding extrema, instructions to "find the point of greatest/least slope" or to "find where the function is increasing/decreasing most rapidly" refer to finding inflection points. Note also that an inflection point is a point of greatest or least slope in an interval around that point. An inflection point may not be the point of greatest or least slope over all possible input values. This concept is analogous to relative vs. absolute extrema.

inflection point can also be thought of as the point of greatest or least slope in a region around the inflection point. In real-life applications, this point is interpreted as *the point of most rapid change or least rapid change.* (See Figure 5.17.)

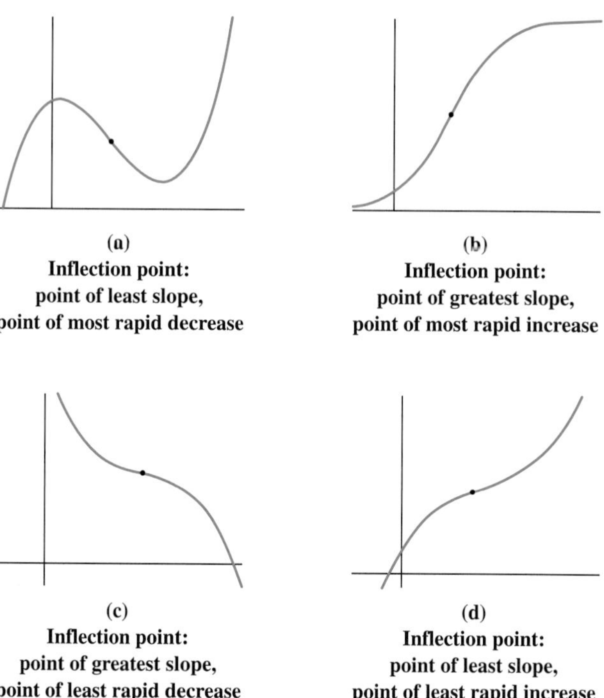

FIGURE 5.17

Relative maxima and minima on a smooth, continuous graph can be found by locating the points at which the derivative graph crosses the horizontal axis. These points are among those where the original graph has horizontal tangent lines. Inflection points also can be found by examining the derivative graph and its relation to the function graph. To find the inflection point on a smooth, continuous graph, we must find the point where the slope (derivative) graph has a relative maximum or minimum. That is, we apply the method for finding maxima and minima to the *derivative graph.*

The Second Derivative

Consider, from the discussion in the previous section, the model for the population of Kentucky from 1980 through 1993:

$$p(x) = 0.395x^3 - 6.674x^2 + 30.257x + 3661.147 \text{ thousand people}$$

where x is the number of years since the end of 1980. Graphs of the function and its derivative are shown in Figure 5.18.

We wish to determine where the inflection point occurs—that is, where the population was declining most rapidly. It appears that p' has a minimum when p has an inflection point. In fact, this is exactly the case, so we can find the inflection point of p by finding the minimum of p'. To find the minimum of p' for this smooth, continuous function p, we must find where *its* derivative crosses the x-axis. The derivative of

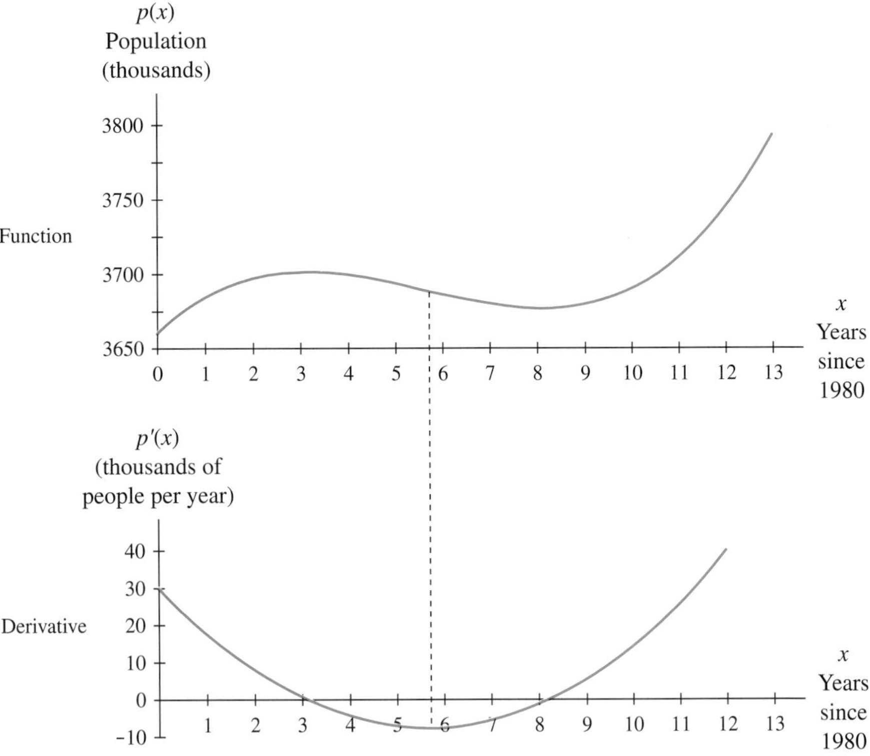

FIGURE 5.18

p' is called the **second derivative** of p, because it is the derivative of a derivative. The second derivative of p is denoted[26] p''. In this case, the second derivative is given by

$$p''(x) = 2.37x - 13.348 \text{ thousand people per year per year}$$

where $x = 0$ in 1980.

Because the second derivative represents the rate of change of the first derivative, the output units of p'' are

$$\frac{\text{output units of } p'}{\text{input units of } p'}$$

The input/output diagram for this second derivative is shown in Figure 5.19.

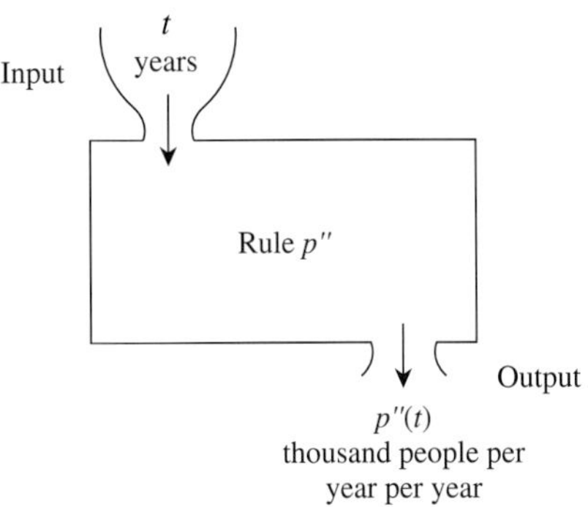

FIGURE 5.19

A graph of the second derivative (Figure 5.20) reveals that it crosses the x-axis where the graph of p has an inflection point. Note that this identifies the minimum point on the graph of the derivative of p where the tangent line is horizontal.

Setting the second derivative equal to zero and solving for x gives $x \approx 5.63$. According to the model, the population was declining most rapidly in mid-1986 at a rate of approximately $p'(5.63) \approx$ -7.3 thousand people per year. At that time, the population was approximately $p(5.63) \approx 3690$ thousand people.

26. Other acceptable notations for the second derivative of p with respect to x include $\frac{d^2p}{dx^2}$ and $\frac{d^2}{dx^2}[p(x)]$.

Function

$p(x)$ Population (thousands)

3800, 3750, 3700, 3650

x Years since 1980

0 1 2 3 4 5 6 7 8 9 10 11 12 13

Derivative

$p'(x)$ (thousands of people per year)

40, 30, 20, 10, 0, -10

x Years since 1980

1 2 3 4 5 6 7 8 9 10 11 12 13

Second derivative

$p''(x)$ (thousands of people per year per year)

15, 10, 5, 0, -5, -10, -15

x Years since 1980

1 2 3 4 5 6 7 8 9 10 11 12 13

FIGURE 5.20

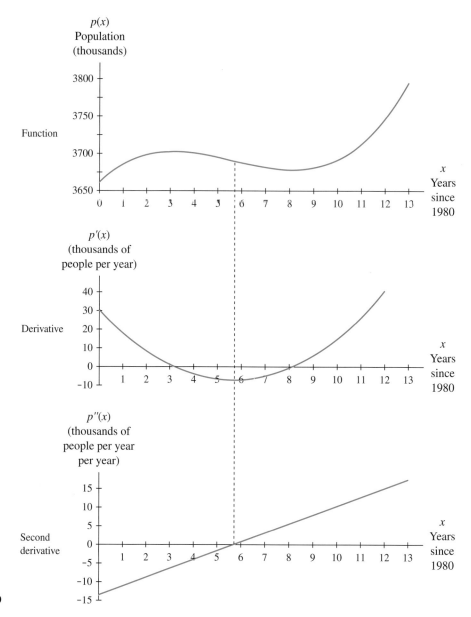

EXAMPLE 1 *Post-secondary Education*

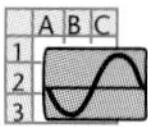

Consider a model for the percentage[27] of students graduating from high school in South Carolina from 1982 through 1990 who entered post-secondary institutions:

$$f(x) = -0.1057x^3 + 1.355x^2 - 3.672x + 50.792 \text{ percent}$$

where x is the number of years since 1982.

a. Find the inflection point of the function.

b. Determine the year between 1982 and 1990 in which the percentage was increasing most rapidly.

27. Based on data in *South Carolina Statistical Abstract*, 1992.

c. Determine the year between 1982 and 1990 in which the percentage was decreasing most rapidly.

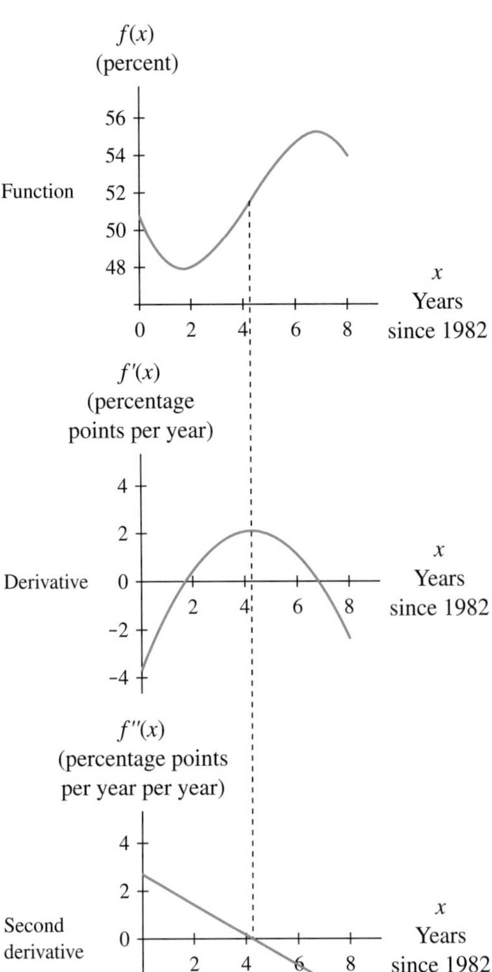

FIGURE 5.21

Connect the solution to part c with the process of checking endpoints to find extrema in Section 5.2. We find points of greatest and least slope over an interval by comparing the slopes at inflection points with the slopes at the endpoints.

Solution

a. Consider the point(s) at which the second derivative is zero. The rate-of-change formula for this function is

$$f'(x) = -0.3171x^2 + 2.71x - 3.672 \text{ percentage points per year}$$

where x is the number of years since 1982. The second derivative is

$$f''(x) = -0.6342x + 2.71 \text{ percentage points per year}$$

where x is the number of years since 1982. The second derivative is zero when $x \approx 4.27$ years after 1982. Next, look at the graph of f shown in Figure 5.21. It does appear that $x = 4.27$ is the approximate input of the inflection point. The output is $f(4.27) \approx 51.6\%$, and the rate of change at that point is $f'(4.27) \approx 2.1$ percentage points per year.

b. Although f is a continuous function, it can be interpreted only at integer values of x because educational data such as these are reported for the fall of each year. Thus, to determine which valid input actually has the greatest rate of change, we evaluate f' at the integer values on either side of $x = 4.27$.

The rate of change of the model in 1986 is $f'(4) \approx 2.09$ percentage points per year. The rate of change in 1987 is $f'(5) \approx 1.95$ percentage points per year. Thus we can say that according to the model, the percentage of South Carolina high school graduates who enter post-secondary institutions was increasing most rapidly in 1986. The percentage of graduates going on for post-secondary education in 1986 was approximately $f(4) = 51.0\%$. The percentage was increasing by about $f'(4) = 2.1$ percentage points per year at that time.

Figure 5.21 shows the function, its derivative, and its second derivative. Note again the relationship among the points at which the second derivative crosses the x-axis, the derivative has a maximum, and the function has an inflection point.

c. Observe from the graph of f shown in Figure 5.21 that the most rapid decrease occurs at one of the endpoints. Evaluate the derivative of f at both endpoints.

$$f'(0) = -3.672 \text{ percentage points per year}$$

$$f'(8) = -2.2864 \text{ percentage points per year}$$

The percentage was declining most rapidly in 1982 at a rate of approximately 3.7 percentage points per year.

You have just seen two examples of how the second derivative of a function can be used to find an inflection point. It is important to use the second derivative whenever possible, because it gives an exact answer. Sometimes, however, finding the second derivative of a function can be tedious. In such cases, you will have to decide how important extreme accuracy is. If a close approximation will suffice (as is often the case in real-world modeling), then you may wish to find the first derivative only and use appropriate technology to estimate where its maximum (or minimum) occurs.

EXAMPLE 2 *The 1949 Polio Epidemic*

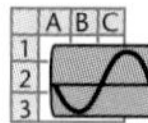

Consider the following model for the number of polio cases in the United States in 1949.

$$C(t) = \frac{42{,}183.911}{1 + 21{,}484.253e^{-1.248911t}} \text{ polio cases}$$

where $t = 1$ at the end of January, $t = 2$ at the end of February, and so forth. Find when the number of polio cases was increasing most rapidly, the rate of change of polio cases at the time, and the number of cases at that time.

Solution The graphs of C, C', and C'' are shown in Figure 5.22. We seek the inflection point on the graph of C that corresponds to the maximum point on the graph of C' that corresponds to the point at which the graph of C'' crosses the t-axis.

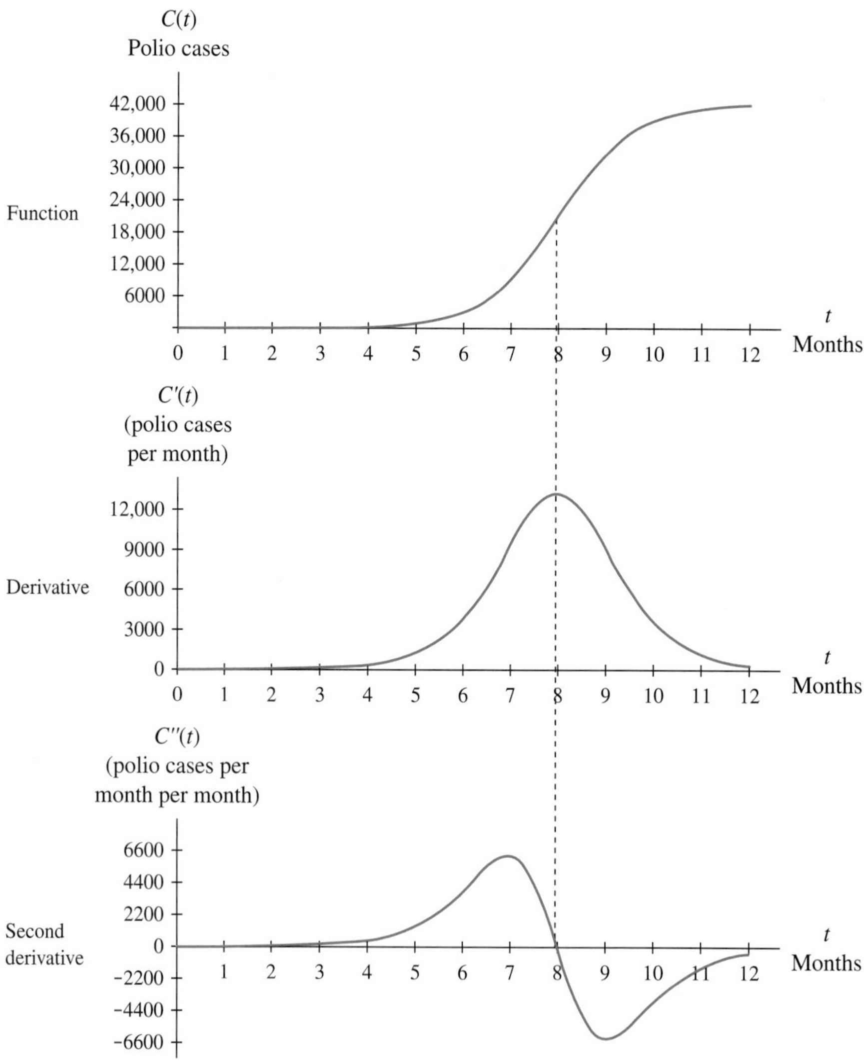

FIGURE 5.22

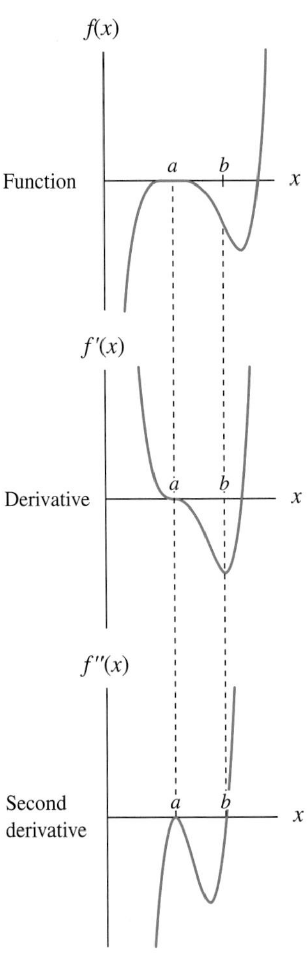

FIGURE 5.23

We choose to use technology to estimate the maximum point on the derivative graph. It occurs at $t \approx 8$ where the output is approximately 13,171. It is important to understand what these numbers represent. The t-value tells us the month in which the greatest increase occurred: $t = 8$ corresponds to the end of August. The output value is a value on the *derivative* graph. *This is the slope of the graph of C at the inflection point.* We can therefore say that polio was spreading most rapidly at the end of August 1949 at a rate of approximately 13,171 cases per month. To find the number of polio cases at the inflection point, substitute the unrounded t-value of the point into the original function to obtain approximately 21,092 cases. Note that to the right of the inflection point, the *number* of polio cases was increasing whereas the *rate* at which polio cases appeared was declining.

We saw in Section 5.2 that for a smooth, continuous function, a relative maximum or minimum occurs where the derivative graph *crosses* the horizontal axis, but not where the derivative graph touches the horizontal axis without crossing it. A similar statement can be made about inflection points. If the second derivative graph *crosses* the horizontal axis, then an inflection point occurs on the graph of the function. The graphs in Figure 5.23 of a function, its derivative, and its second derivative illustrate this issue. Note that the point at which the second derivative graph touches, but does not cross, the horizontal axis actually corresponds to a relative maximum on the function graph, not to an inflection point.

Two other situations that could occur are illustrated in Figure 5.24.

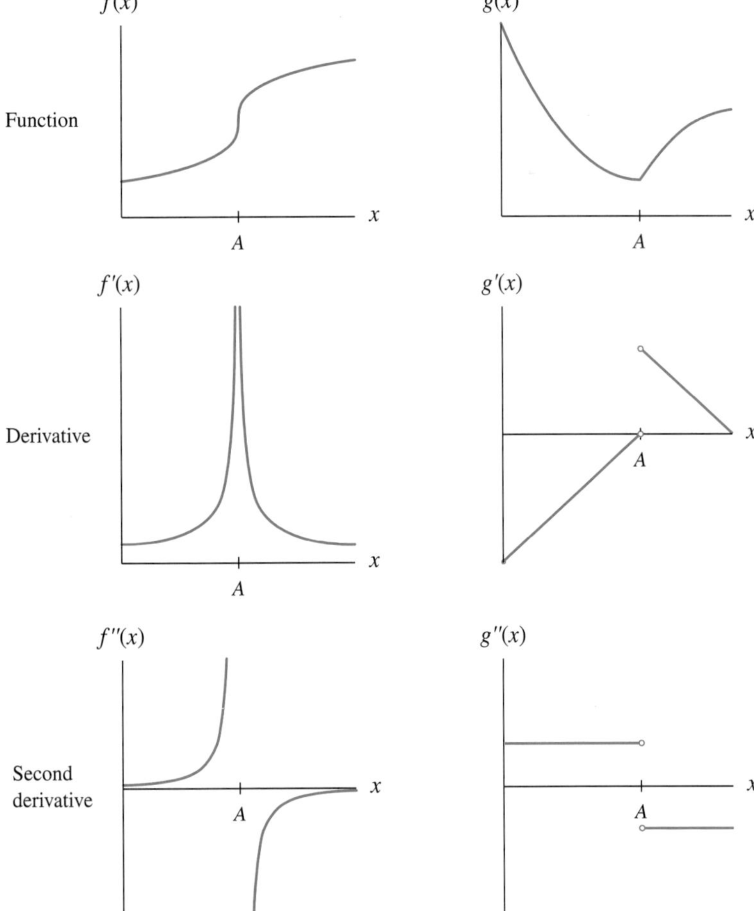

FIGURE 5.24

Note that the graphs of both f and g have inflection points at $x = A$ because they change concavity at that point. However, the second derivatives of f and g never cross the horizontal axis. In fact, in each case, the second derivative does not exist at $x = A$, because the first derivative does not exist there. The function f has a vertical tangent line at $x = A$, and the function g is not smooth at $x = A$. Even though such situations as this do not often occur in real-world applications, you should be aware that they could happen. Keep in mind the following result:

> At a point of inflection on the graph of a function, the second derivative is zero or does not exist. If the second derivative graph *crosses* the horizontal axis, then an inflection point of the function graph occurs at that input value.

In some applications, the inflection point can be regarded as the **point of diminishing returns.** Consider the college student who studies for 8 hours without a break before a major exam. The percentage of new material that the student will retain after studying for t hours can be modeled as

$$P(t) = \frac{45}{1 + 5.94e^{-0.969125t}} \text{ percent}$$

This function has an inflection point at $t \approx 1.8$. That is, after approximately 1 hour and 48 minutes, the rate at which the student is retaining new material begins to diminish. Studying beyond that point will improve the student's knowledge, but not as quickly. This is the idea behind diminishing returns: Beyond the inflection point, you gain fewer percentage points per hour than you gain at the inflection point; that is, output increases at a decreasing rate. The existence of this point of diminishing returns is one factor that has led many educators and counselors to suggest studying in 2-hour increments with breaks in between.

Concavity and the Second Derivative

An understanding of the relationship between concavity and the second derivative will help students sketch accumulation function graphs in Section 6.3.

Because the derivative of a function is simply the slope of the graph of that function, we know that a positive derivative indicates that the function output is increasing and a negative derivative indicates that the function output is decreasing. The second derivative provides similar information about where a function graph is concave up and where it is concave down.

In particular, if the second derivative is negative, it means that the first derivative graph is declining, which means that the original function graph is concave down. Similarly, a positive second derivative indicates that the first derivative is increasing, which means that the original function graph is concave up. And, as we have already seen, where the second derivative changes from positive to negative or from negative to positive, the function graph has an inflection point.

Consider, for example, the following information about the second derivative of a function f and the information it provides about the concavity of a graph of f. We use "ccu" to indicate concave up and "ccd" to indicate concave down.

x	0	1	2	3	4	5	6
$f''(x)$	15	0	-1	0	-1	0	15
Concavity of f	ccu	infl. pt.	ccd	not infl. pt.	ccd	infl. pt.	ccu

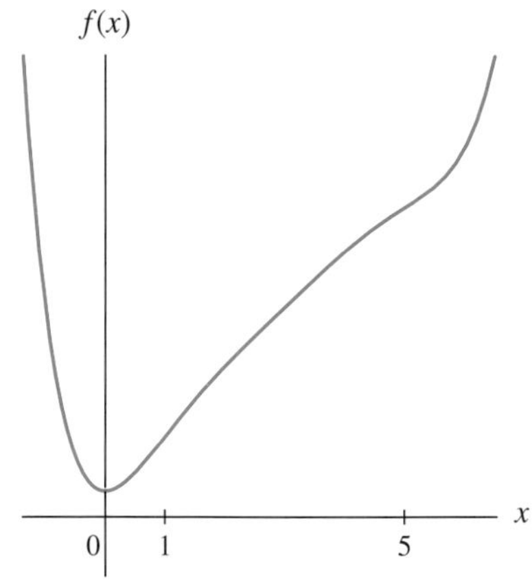

FIGURE 5.25

We can conclude from this table that a graph of the function f has two inflection points at $x = 1$ and $x = 5$. Although the second derivative is zero at $x = 3$, the function graph does not change concavity at that value (it is concave down on both sides of $x = 3$), so the corresponding point is not an inflection point.

Next, consider some values of the derivative of this same function f. We use the symbol ↗ to indicate increasing and the symbol ↘ to indicate decreasing.

x	-1	0	1	2	3	4	5	6
f′(x)	-35.6	0	5.6	4.6	4.2	3.8	2.8	8.4
f(x)	↘	local min.	↗	↗	↗	↗	↗	↗

We conclude that the graph of the function f decreases to a local minimum at zero and then increases to the right of zero. The graph is concave down between $x = 1$ and $x = 5$ and concave up to the left of $x = 1$ and to the right of $x = 5$. A possible graph of f, based on this information, is shown in Figure 5.25.

If you are using the two-semester text and integrating Chapter 8 into Chapters 1 through 7, you should cover Section 8.4 before proceeding to Chapter 6.

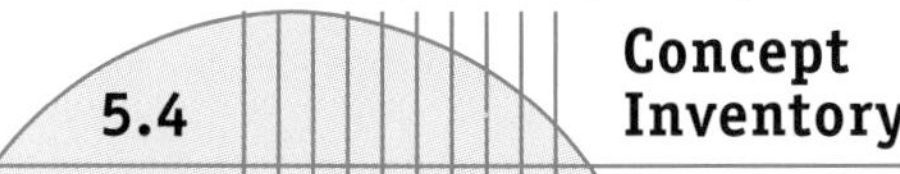

5.4 Concept Inventory

- *Inflection point*
- *Second derivative*
- *Point of diminishing returns*
- *Conditions under which inflection points exist*

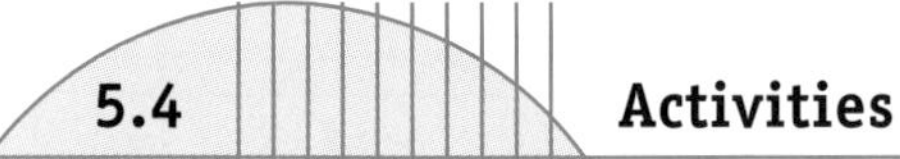

5.4 Activities

1. Which of the six basic models discussed thus far in this book could have inflection points?

2. Discuss in detail all of the options that are available for finding inflection points of a function.

3. The graph in Figure 5.4.1 shows an estimate of the ultimate crude oil production recoverable from Earth.[28]

   a. Estimate the two inflection points on the graph.

   b. Explain the meaning of the inflection points in the context of crude oil production.

FIGURE 5.4.1

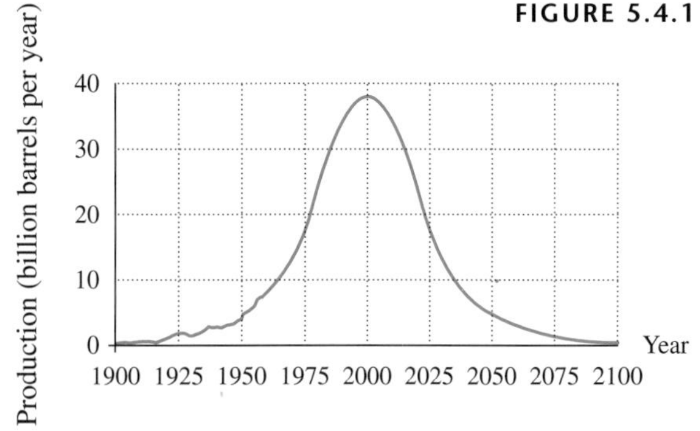

4. The graph in Figure 5.4.2 shows sales (in thousands of dollars) for a business as a function of the amount spent on advertising (in hundreds of dollars).

FIGURE 5.4.2

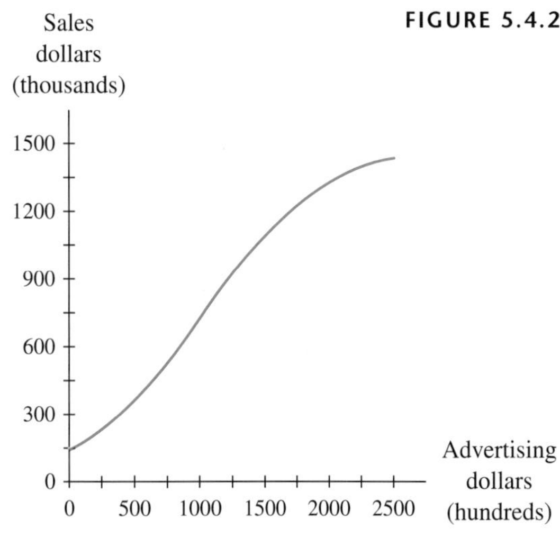

28. Figure 5.4.1 adapted from François Ramade, *Ecology of Natural Resources* (New York: Wiley, 1984). Copyright 1984 by John Wiley & Sons, Inc. Reprinted by permission of the publisher.

a. Mark the approximate location of the inflection point on the graph.

b. Explain the meaning of the inflection point in the context of this business.

c. Explain how knowledge of the inflection point might affect decisions made by the managers of this business.

5. Consider the function

$$g(x) = 0.04x^3 - 0.88x^2 + 4.81x + 12.11$$

a. Graph g, g', and g'' between $x = 0$ and $x = 15$. Indicate the relationships among points on the three graphs that correspond to maxima, minima, and inflection points.

b. Find the inflection point on the graph of g. Is it a point of most rapid decline or least rapid decline?

6. Consider the function $f(x) = \dfrac{20}{1 + 19e^{-0.5x}}$.

a. Graph f, f', and f'' between $x = 0$ and $x = 15$. Indicate the points on the graphs of f' and f'' that correspond to the inflection point on the graph of f.

b. Find the inflection point on the graph of f. Is it a point of most rapid or least rapid increase?

Each of Activities 7 through 10 presents graphs of a function, its derivative, and its second derivative. (Assume that the input is on the horizontal axis.) In each case, identify each graph as the function, the derivative, or the second derivative. Give reasons for your choice.

7. a.

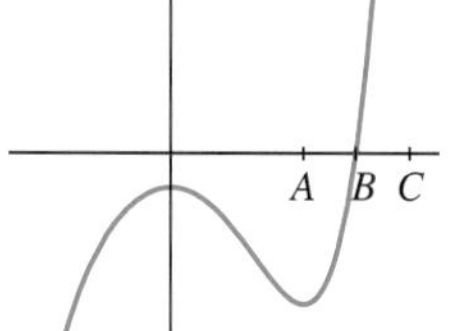

b.

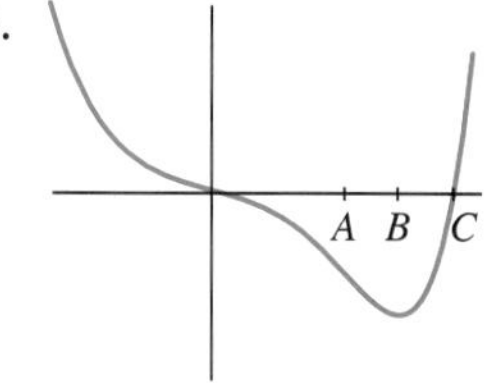

c.

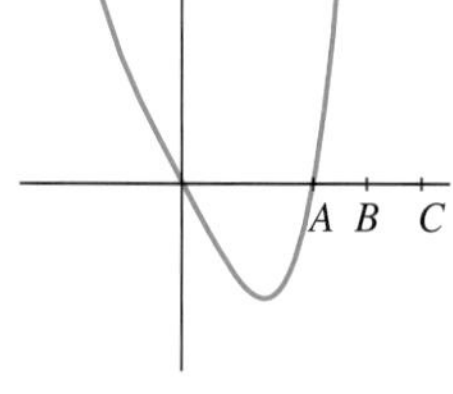

8. a.

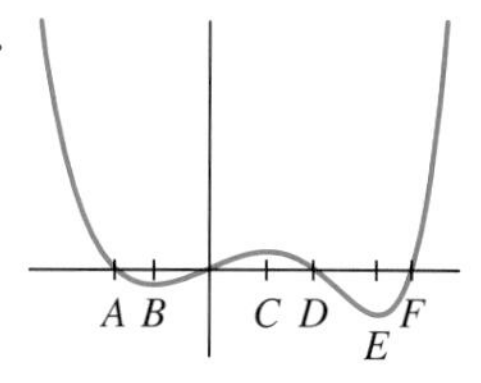

b.

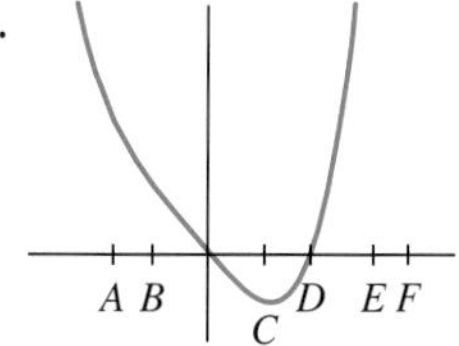

c.

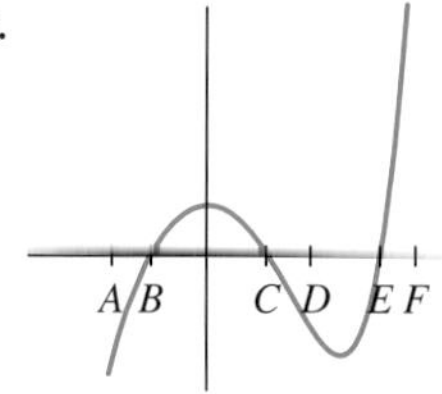

9. a.

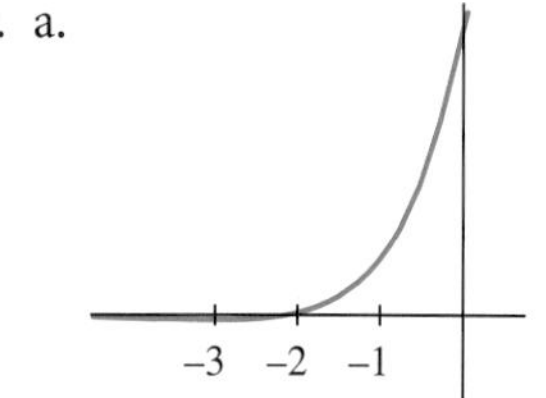

b.

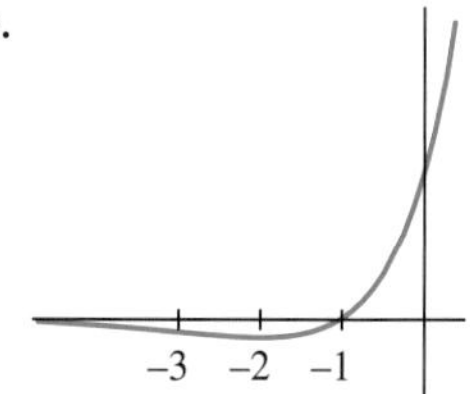

c.

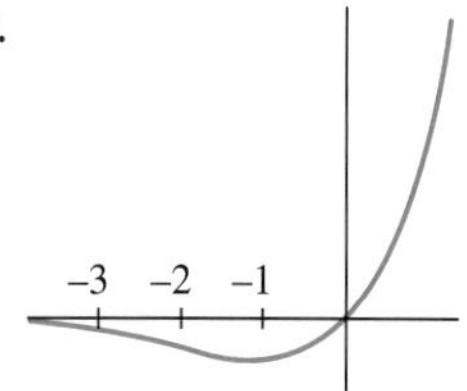

10. a. b.

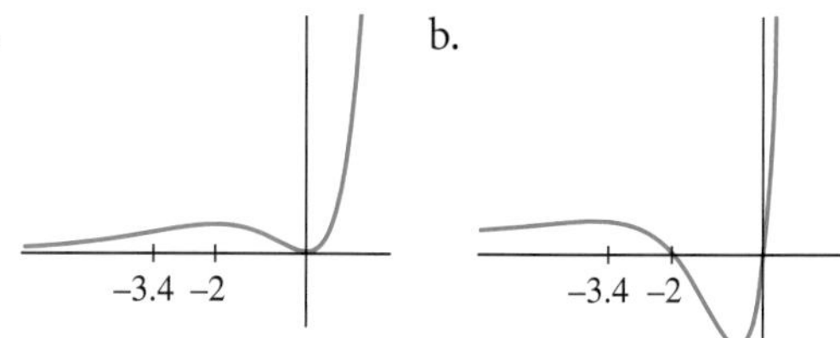

c.

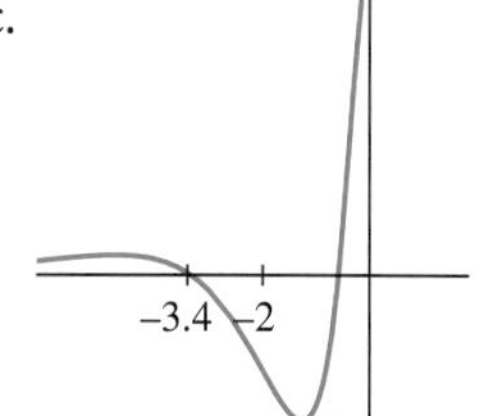

11. The percentage of new material that a student will retain after studying t hours without a break can be modeled by

$$P(t) = \frac{45}{1 + 5.94e^{-0.969125t}} \text{ percent}$$

a. Find the inflection point on the graph of P, and interpret the answer.

b. Compare your answer with that given in the discussion at the end of this section.

12. The U.S. Bureau of the Census' prediction[29] for the percentage of the population that is 65 to 74 years old from 2000 through 2050 can be modeled by

$$p(x) = (1.619 \cdot 10^{-5})x^4 - (1.675 \cdot 10^{-3})x^3 + 0.050x^2 - 0.308x + 6.693 \text{ percent}$$

where x is the number of years after 2000.

a. Determine the year between 2000 and 2050 in which the percentage is predicted to be increasing most rapidly, the percentage at that time, and the rate of change of the percentage at that time.

b. Repeat part *a* for the most rapid decrease.

13. The percentage of southern Australian grasshopper eggs that hatch as a function of temperature (for temperatures between 7°C and 25°C) can be modeled[30] by

$$P(t) = -0.00645t^4 + 0.488t^3 - 12.991t^2 + 136.560t - 395.154 \text{ percent}$$

where t is the temperature in °C.

a. Graph P, P', and P''.

b. Find the point of most rapid decrease on the graph of P. Interpret your answer.

14. The median size of a new house built in the United States between 1985 and 1998 can be modeled[31] by the equation

$$H(x) = 0.573x^3 - 23.348x^2 + 320.655x + 451.429 \text{ square feet}$$

x years after 1980.

a. Determine the time between 1985 and 1998 when the median house size was increasing least rapidly. Find the corresponding house size and rate of change in house size.

b. Graph H, H', and H'', indicating the relationships between the inflection point on the graph of H and the corresponding points on the graphs of H' and H''.

c. Determine the time between 1985 and 1998 when the median house size was increasing the most rapidly.

15. The average price (per 1000 cubic feet) of natural gas for residential use[32] from 1980 through 1997 is given by

$$p(x) = 0.004265x^3 - 0.11816x^2 + 0.9674x + 3.681 \text{ dollars}$$

where x is the number of years since 1980.

a. Sketch the graphs of p and its first and second derivatives. Label the vertical axes appropriately. Which point on the derivative graph corresponds to the inflection point of the original function graph? Which point on the second derivative graph corresponds to the inflection point of the original function graph?

b. Find the x-intercept of the second derivative graph, and interpret its meaning in context.

c. Determine when, according to the model, the average natural gas price was declining most rapidly between 1980 and 1997.

d. Determine when between 1980 and 1997 the natural gas price was increasing most rapidly.

16. On the basis of data from the 1998 *Statistical Abstract* for years from 1970 through 1996, the percentage of households with TVs that subscribed to cable can be modeled by

$$P(x) = 6 + \frac{59.3}{1 + 65.1e^{-0.3049x}} \text{ percent}$$

where x is the number of years after 1970. When was the percentage of households with TVs that subscribed to cable increasing the most rapidly? What were the percentage and the rate of change of the percentage at that time?

17. The number of people who donated to an organization supporting athletics at a certain university in the southeast from 1975 through 1992 can be modeled[33] by

$$D(t) = -10.247t^3 + 208.114t^2 - 168.805t + 9775.035 \text{ donors}$$

t years after 1975.

a. Find any relative maxima or minima that occur on a graph of the function.

29. Based on data from the *Statistical Abstract*, 1998.
30. Based on information in George L. Clarke, *Elements of Ecology* (New York: Wiley, 1954), p. 170.
31. Based on data from the National Association of Home Builders Economics Division.
32. Based on data from *Statistical Abstract*, 1998.
33. Based on data from IPTAY Association at Clemson University.

b. Find the inflection point(s).

c. How do the following events in the history of football at that college correspond with the curvature of the graph?

i. In 1981, the team won the National Championship.

ii. In 1988, Coach F was released and Coach H was hired.

18. The amount spent on cable television per person per year from 1984 through 1992 can be modeled[34] by

$$A(x) = -0.126x^3 + 1.596x^2 + 1.802x + 40.930 \text{ dollars}$$

where x is the number of years since 1984.

a. Find the inflection point on the graph of A and the corresponding points on graphs of A' and A''.

b. Find the year between 1984 and 1992 in which the average amount spent per person per year on cable television was increasing most rapidly. What was the rate of change of the amount spent per person in that year?

19. A college student works for 8 hours without a break assembling mechanical components. The cumulative number of components she has assembled after h hours can be modeled by

$$N(h) = \frac{62}{1 + 11.49e^{-0.654h}} \text{ components}$$

a. Determine when the rate at which she was working was greatest.

b. How might her employer use the information in part a to increase the student's productivity?

20. The lake level of Lake Tahoe from October 1, 1995, through July 31, 1996, can be modeled[35] by

$$L(d) = (-5.345 \cdot 10^{-7})d^3 + (2.543 \cdot 10^{-4})d^2 - 0.0192d + 6226.192 \text{ feet above sea level}$$

d days after September 30, 1995.

a. Determine when the lake level was rising most rapidly between October 1, 1995, and July 31, 1996.

b. What factors may have caused the inflection point to occur at the time you found in part a?

c. Would you expect the most rapid rise to occur at approximately the same time each year? Explain.

21. A chemical reaction begins when a certain mixture of chemicals reaches 95°C. The reaction activity is measured in Units (U) per 100 microliters (100μL) of the mixture. Measurements[36] at 4-minute intervals during the first 18 minutes after the mixture reaches 95°C are listed in Table 5.23.

TABLE 5.23

Time (minutes)	2	6	10	14	18
Activity (U/100μL)	0.10	0.60	1.40	1.75	1.95

a. According to Table 5.23, during what time interval was the activity increasing most rapidly?

b. Find a model for the data, and use the equation to find the inflection point. Interpret the inflection point in context.

22. Table 5.24 lists the number of live births[37] in the United States between the years 1950 and 1995 to women 45 years of age and older.

TABLE 5.24

Year	Births	Year	Births
1950	5322	1975	1628
1955	5430	1980	1200
1960	5182	1985	1162
1965	4614	1990	1638
1970	3146	1995	2727

a. Use the data to estimate when the number of births was declining most rapidly between 1950 and 1995.

b. Find a model for the data, and use the equation to obtain the answer to part a. Note that the model must be interpreted discretely.

c. How does the answer to part b differ from the answer to part a? Which do you believe is the better estimate?

34. Based on data from *Statistical Abstract*, 1994.
35. Based on data from the Federal Watermaster, U.S. Department of Interior.
36. David E. Birch et al., "Simplified Hot Start PCR," *Nature*, vol. 381 (May 30, 1996), p. 445.
37. *The 1998 Information Please Almanac* (Boston: Houghton Mifflin).

d. Write a careful explanation of the procedure that you used with your model to find the point of most rapid decline.

23. The yearly amount of garbage (in millions of tons) taken to a landfill outside a city during selected years from 1970 through 2000 is given in Table 5.25.

TABLE 5.25

Year	Amount (millions of tons)
1970	81
1975	99
1980	117
1985	122
1990	132
1995	145
2000	180

a. Using the table values only, determine during which 5-year period the amount of garbage showed the slowest increase. What was the average rate of change during that 5-year period?

b. Find a model for the data.

c. Give the second derivative formula for the equation.

d. Use the second derivative to find the point of slowest increase on a graph of the equation.

e. Graph the first and second derivatives, and explain how they support your answers to part *d*.

f. In what year was the rate of change of the yearly amount of garbage the smallest? What were the rate of increase and the amount of garbage in that year?

24. A business owner's sole means of advertising is to put fliers on cars in a nearby shopping mall parking lot. Table 5.26 shows the number of labor hours per month spent handing out fliers and the corresponding profit.

a. Find a model for profit.

b. For what number of labor hours is profit increasing most rapidly? Give the number of labor hours, the profit, and the rate of change of profit at that number.

c. In this context, the inflection point can be thought of as the point of diminishing returns. Discuss how knowing the point of diminishing returns could help the business owner make decisions relative to employee tasks.

TABLE 5.26

Labor hours each month	Profit (dollars)
0	2000
10	3500
20	8500
30	19,000
40	32,000
50	43,000
60	48,500
70	55,500
80	56,500
90	57,000

25. Table 5.27 gives the amount spent[38] on national defense (in billions of dollars measured in constant 1987 dollars) between 1970 and 1990.

TABLE 5.27

Year	Amount (billions of dollars)
1970	262.9
1972	219.7
1974	185.3
1976	177.8
1978	177.2
1980	187.1
1982	214.3
1984	241.7
1986	276.4
1988	283.3
1990	272.5

a. Find a model for the data.

b. Give the second derivative formula for the equation.

c. Find the point between 1970 and 1990 at which spending on national defense showed the most

38. *Statistical Abstract*, 1994.

rapid increase. Give the year, the amount spent, and how rapidly the amount was changing at that point.

d. If you found the answer to part *c* by using a graphing calculator or computer, discuss how you could find the answer without using such a tool.

e. Find the year between 1970 and 1990 when spending on national defense was declining most rapidly.

26. The revenue[39] (in millions of dollars) of the Russell Corporation from 1989 through 1998 is given in Table 5.28.

TABLE 5.28

Year	Revenue (millions of dollars)
1989	688.0
1990	713.8
1991	804.6
1992	899.1
1993	930.8
1994	1098.3
1995	1152.6
1996	1244.2
1997	1228.2
1998	1180.1

a. Use the data to estimate the year in which revenue was growing most rapidly.

b. Find a model for the data.

c. Find the first and second derivatives of the model in part *b*.

d. Determine the year in which revenue was growing most rapidly. Find the revenue and the rate of change of revenue in that year.

27. Total yearly emissions of lead into the atmosphere between 1970 and 1995 can be modeled[40] by the equation

$$L(x) = \frac{251.3}{1 + 0.1376e^{0.2854x}} \text{ million tons}$$

x years after 1970.

a. Give the slope formula for *L*.

b. How quickly were emissions decreasing in 1970 and 1995?

c. Find the year in which lead emissions were declining most rapidly. Give the emissions and the rate of change of the emissions in that year.

28. Table 5.29 gives the total emissions[41] in millions of metric tons of nitrogen oxides, NO_x, in the United States from 1940 through 1990.

TABLE 5.29

Year	NO_x (millions of metric tons)
1940	6.9
1950	9.4
1960	13.0
1970	18.5
1980	20.9
1990	19.6

a. Find a cubic model for the data.

b. Give the slope formula for the equation.

c. Determine when emissions were increasing most rapidly between 1940 and 1990. Give the year, the amount of emissions, and how rapidly they were increasing.

29. The personnel manager for a construction company keeps track of the total number of labor hours spent on a construction job each week during the construction. Some of the weeks and the corresponding labor hours are given in Table 5.30.

TABLE 5.30

Weeks after the start of a project	Cumulative labor hours
1	25
4	158
7	1254
10	5633
13	9280
16	10,010
19	10,100

Activity 29 is especially challenging if done algebraically, because it involves finding the third derivative of a logistic function. It is our intent that this activity be done using technology.

39. Hoover's Online Company Capsules.

40. Based on data from *Statistical Abstract*, 1998.

41. *Statistical Abstract*, 1992.

a. Find a logistic model for the data.

b. Find the derivative of the model. What are the units on the derivative?

c. Graph the derivative, and discuss what information it gives the manager.

d. When is the maximum number of labor hours per week needed? How many labor hours are needed in that week?

e. Find the point of most rapid increase in the number of labor hours per week. How many weeks into the job does this occur? How rapidly is the number of labor hours per week increasing at this point?

f. Find the point of most rapid decrease in the number of labor hours per week. How many weeks into the job does this occur? How rapidly is the number of labor hours per week decreasing at this point?

g. Carefully explain how the exact values for the points in parts *e* and *f* can be obtained.

h. If the company has a second job requiring the same amount of time and number of labor hours, a good manager will schedule the second job to begin so that the time when the number of labor hours per week for the first job is declining most rapidly corresponds to the time when the number of labor hours per week for the second job is increasing most rapidly. How many weeks into the first job should the second job begin?

30. Table 5.31 shows the number of males per 100 females in the United States based on census data. The number is referred to as the gender ratio.

a. Find a cubic model for the data.

b. Write the second derivative of the equation you found in part *a*.

c. In what year does the output of the model exhibit the most rapid growth? What was the gender ratio in that year, and how rapidly was it changing?

TABLE 5.31

Year	Males per 100 females
1900	104.6
1910	106.2
1920	104.1
1930	102.6
1940	100.8
1950	98.7
1960	97.1
1970	94.8
1980	94.5
1990	95.1

31. For a function f, $f''(x) > 0$ for all real number input values. Describe the concavity of a graph of f, and sketch a function for which this condition is true.

32. Draw a graph of a function g such that $g''(x) = 0$ for all real number input values.

33. For a function f, the following statements are true:

$$f''(x) > 0 \quad \text{when } 0 \le x < 2$$
$$f''(x) = 0 \quad \text{when } x = 2$$
$$f''(x) < 0 \quad \text{when } 2 < x \le 4$$

a. Describe the concavity of a graph of f between $x = 0$ and $x = 4$.

b. Draw two completely different graphs that satisfy these second-derivative conditions.

34. For a function h, the following statements are true:

$$h''(x) > 0 \quad \text{when } 0 < x < 2$$
$$h''(x) = 0 \quad \text{when } x = 2 \text{ and } x = 0$$
$$h''(x) < 0 \quad \text{when } x < 0 \text{ and } x > 2$$

a. Describe the concavity of a graph of h.

b. Draw a graph that satisfies all three of these second-derivative conditions.

5.5 Interconnected Change: Related Rates

We have seen that many situations in the world around us can be modeled mathematically and that because of this ability to model, we can use calculus to analyze changes taking place. So far we have considered how the rate of change of the output

If you wish, you may omit this section or, if you are using the two-semester text, cover it at the beginning of Chapter 11. Related rates make an excellent introduction to differential equations, because they involve equations expressed in terms of derivatives.

variable of a function is affected by a change in the input variable. We now consider the interaction of the rates of change of input and output variables with respect to a third variable. We shall also see that the interconnection of the input and output variables often reflects an interaction between their rates of change with respect to a third variable.

Today's news often brings us stories of environmental pollution in one form or another. One serious form of pollution is groundwater contamination. If a hazardous chemical is introduced into the ground, it can contaminate the groundwater and make the water source unusable. The contaminant will be carried downstream via the flow of the groundwater. This movement of contaminant as a consequence of the flow of the groundwater is known as *avection*. As a result of *diffusion*, the chemical will also spread out perpendicularly to the direction of flow. The region that the contamination covers is known as a chemical *plume*.

Hydrologists who study groundwater contamination are sometimes able to model the area of a plume as a function of the distance away from the source that the chemical has traveled by avection. That is, they develop an equation showing the interconnection between A, the area of the plume, and r, the distance from the source that the chemical has spread because of the flow of the groundwater.

If A is a function of r, then the rate of change $\frac{dA}{dr}$ describes how quickly the area of the plume is increasing as the chemical travels farther from the source. What we have not investigated is how to determine and interpret the rates of change of A and r with respect to a third variable, say time. We study the interconnection of such rates in this section. However, before continuing with the groundwater example, let us first look at some simple examples of interconnected change.

Interconnected-Change Equations

In Section 1.1 we said that the terms *independent variable* and *dependent variable* have the same meaning as the terms *input* and *output*. Because the functions in this section define the relationships between more than one output quantity and are not expressed in terms of the independent variable, we are using these terms in a slightly different way than we did in Chapter 1.

An equation relating the volume V of a basketball to its radius r is $V = \frac{4}{3}\pi r^3$. When the ball is inflated, its volume increases as the radius gets larger. It is also true that both the volume and the radius increase according to how rapidly (that is, with respect to time) air is pumped into the ball. Note that even though V and r depend on the time t, the variable t does not appear in the volume equation. In such applications, we refer to both volume V and radius r as **dependent variables** because their changes depend on time t. We refer to time, the "with respect to" variable, as the **independent variable**.

In order to develop an equation relating the rate of change of the ball's volume with respect to time and the rate of change of its radius with respect to time, we differentiate both sides of the equation with respect to time t. We use the Chain Rule to differentiate the right side of the volume equation, considering $\frac{4}{3}\pi r^3$ as the outside function and r as the inside function. The derivative of the inside function with respect to t is $\frac{dr}{dt}$.

You may wish to review the Chain Rule from Section 4.5 before proceeding with this section.

$$\frac{dV}{dt} = \left[\frac{4}{3}\pi(3r^2)\right]\frac{dr}{dt}$$

$$= 4\pi r^2\frac{dr}{dt}$$

This equation shows how the rates of change of the volume and the radius of a sphere with respect to time are interconnected and can be used to answer questions about those rates. Such an equation is referred to as a **related-rates equation**. Example 1 demonstrates how to develop related-rates equations that show how rates of change are interconnected.

EXAMPLE 1 *Related-Rates Equations*

Using each of the given equations, find an equation relating the indicated rates.

a. $p = 39x + 4$; relate $\dfrac{dp}{dt}$ with $\dfrac{dx}{dt}$.

b. $a = 4 \ln t$; relate $\dfrac{da}{dx}$ with $\dfrac{dt}{dx}$.

c. $s = \pi r\sqrt{r^2 + h^2}$; relate $\dfrac{ds}{dx}$ with $\dfrac{dr}{dx}$, assuming that h is constant.

d. $v = \pi r^2 h$; relate $\dfrac{dr}{dt}$ with $\dfrac{dh}{dt}$, assuming that v is constant.

Solution

a. Differentiating the left side of $p = 39x + 4$ with respect to t gives $\frac{dp}{dt}$. We use the Chain Rule to differentiate the right side of the equation with respect to t, considering $39x + 4$ to be the outside function and x to be the inside function.

$$\frac{dp}{dt} = 39\frac{dx}{dt}$$

b. Applying the Chain Rule yields

$$\frac{da}{dx} = 4\left(\frac{1}{t}\right)\frac{dt}{dx} = \frac{4}{t}\frac{dt}{dx}$$

c. Differentiating the right side of $s = \pi r\sqrt{r^2 + h^2}$ requires both the Chain Rule and the Product Rule. Consider first the application of the Product Rule:

$$\frac{ds}{dt} = (\pi r)\left(\text{derivative of } \sqrt{r^2 + h^2} \text{ with respect to } x\right) + \\ (\text{derivative of } \pi r \text{ with respect to } x)\left(\sqrt{r^2 + h^2}\right)$$

In order to calculate the two derivatives with respect to x, remember to apply the Chain Rule.

$$\frac{ds}{dx} = (\pi r)\left[\frac{1}{2}(r^2 + h^2)^{-1/2}(2r)\left(\frac{dr}{dx}\right)\right] + \left(\pi\frac{dr}{dx}\right)\left(\sqrt{r^2 + h^2}\right)$$

$$\frac{ds}{dx} = \pi\left(\frac{r}{\sqrt{r^2 + h^2}} + \sqrt{r^2 + h^2}\right)\frac{dr}{dx}$$

d. In order to develop a related-rates equation from $v = \pi r^2 h$ showing the interconnection between $\frac{dr}{dt}$ and $\frac{dh}{dt}$, we can isolate either r or h on one side and then find the derivative with respect to t, or we can find the derivative of both sides of the equation with respect to t and then solve for $\frac{dr}{dt}$. (In the Activities you will be asked to show that these methods are equivalent.) We choose the second method because it leads to the removal of v from the equation.

Differentiating the left side of the equation with respect to t gives zero because v is constant. Applying the Product Rule and the Chain Rule to the right side of the equation gives

$$0 = (\pi r^2)\left(\frac{dh}{dt}\right) + 2\pi r\left(\frac{dr}{dt}\right)(h)$$

Now solve for $\frac{dr}{dt}$:

$$-2\pi rh\frac{dr}{dt} = \pi r^2\frac{dh}{dt}$$

$$-2rh\frac{dr}{dt} = r^2\frac{dh}{dt}$$

$$\frac{dr}{dt} = \frac{r^2}{-2rh}\frac{dh}{dt}$$

$$\frac{dr}{dt} = \frac{r}{-2h}\frac{dh}{dt}$$

Having derived related-rates equations in Example 1, we now consider how they can be used in the context of groundwater contamination.

EXAMPLE 2 *Groundwater Contamination*

In a certain part of Michigan, a hazardous chemical leaked from an underground storage facility. Because of the terrain surrounding the storage facility, the groundwater was flowing almost due south at a rate of approximately 2 feet per day. Hydrologists studying this plume drilled wells in order to sample the groundwater in the area and determine the extent of the plume. They found that the shape of the plume was fairly easy to predict and that the area of the plume could be modeled as

$$A = 0.9604r^2 + 1.960r + 1.124 - \ln(0.980r + 1) \text{ square feet}$$

when the chemical had spread r feet south of the storage facility.

a. How quickly was the area of the plume growing when the chemical had traveled 3 miles south of the storage facility?

b. How much area had the plume covered when the chemical had spread 3 miles south?

Solution

You may wish to point out to your students that when information is given in the form of output units per input unit, it refers to a rate of change. The students need to read the problem statement carefully to determine which variable the rate is describing. For instance, after reading through the problem in Example 2, the student should return to the statement "the groundwater was flowing almost due south at a rate of approximately 2 feet per day" and conclude that $\frac{dr}{dt} = 2$ feet per day.

a. First, note that the question posed asks for the rate of change of area with respect to time. However, the equation does not contain a variable representing time. Second, note that we are given the rate of change of r with respect to time. If we represent the time in days since the leak began as t, then we know that $\frac{dr}{dt} = 2$ feet per day and we are trying to find $\frac{dA}{dt}$. We also have an equation that shows the interconnection between the dependent variables A and r. Therefore, we differentiate with respect to t both sides of the equation that relates A and r.

$$\frac{dA}{dt} = 0.9604(2r)\frac{dr}{dt} + 1.960\frac{dr}{dt} + 0 - \left(\frac{1}{0.980r + 1}\right)\left(0.980\frac{dr}{dt}\right)$$

$$\frac{dA}{dt} = \left(1.9208r + 1.960 - \frac{0.980}{0.980r + 1}\right)\frac{dr}{dt}$$

Next, we substitute the known values $r = 3$ miles $= 15{,}840$ feet and $\frac{dr}{dt} = 2$ feet per day and then solve for the unknown rate.

$$\frac{dA}{dt} = \left(1.9208(15{,}840) + 1.960 - \frac{0.980}{0.980(15{,}840) + 1}\right)(2)$$

$$\approx 60{,}855 \text{ square feet per day}$$

Unlike other problems in which we keep track of units in calculations in order to know the answer units, in these problems we determine answer units on the basis of the rate of change we are asked to find. For instance, in part *a*, the question is to find $\frac{dA}{dt}$, where A is measured in square feet and t is measured in days. Thus the units on $\frac{dA}{dt}$ will be square feet per day. This method of determining units is easier than carrying units through all the calculations.

When the chemical has spread 3 miles south, the plume is growing at a rate of 60,855 square feet per day.

b. The area of the plume is

$$A = 0.9604(15{,}840)^2 + 1.960(15{,}840) + 1.124 - \ln[0.980(15{,}840) + 1]$$
$$\approx 241{,}000{,}776 \text{ square feet}$$
$$\approx 8.6 \text{ square miles}$$

When the contamination has spread 3 miles south, the total contaminated area is approximately 8.6 square miles.

Note the method we used to answer the question posed in Example 2. First, we determined which variables were involved. Second, we identified an equation that connected the dependent variables. Third, we determined which rates of change were needed to relate to one another and took the derivative of each side of the equation with respect to the independent variable. Finally, we substituted given quantities and rates into the related-rates equation and solved for the unknown rate. We summarize this method:

Method of Related Rates

Step 1: Carefully read the problem and determine what variables are involved. Identify the independent variable (the "with respect to" variable) and all dependent variables.

Step 2: Use the given equation or find an equation relating the dependent variables. The independent variable may or may not appear in the equation.

Step 3: Differentiate both sides of the equation in Step 2 with respect to the independent variable to produce a related-rates equation. The Chain and/or Product Rule(s) may be needed.

Step 4: Substitute the known quantities and rates into the related-rates equation, and solve for the unknown rate.

Step 5: If appropriate, interpret in context the solution found in Step 4.

We illustrate this process in Example 3.

EXAMPLE 3 *Baseball Photography*

A baseball diamond is a square with each side measuring 90 feet. A baseball team is participating in a publicity photo session, and a photographer at second base wants to photograph runners when they are halfway to first base. Suppose that the average speed at which a baseball player runs from home plate to first base is 20 feet per second. The photographer needs to set the shutter speed in terms of how fast the distance between the runner and the camera is changing. At what rate is the distance between the runner and second base changing when the runner is halfway to first base?

Solution

Step 1: The three variables involved in this problem are time, the distance between the runner and first base, and the distance between the runner and the photographer at second base. Because speed is the rate of change of distance with respect to time, the independent variable is time, and the two distances are the dependent variables.

Step 2: We need an equation that relates the distance between the runner and first base and the distance between the runner and the photographer at second base. A diagram can help us better understand the relationship between these distances. See Figure 5.26.

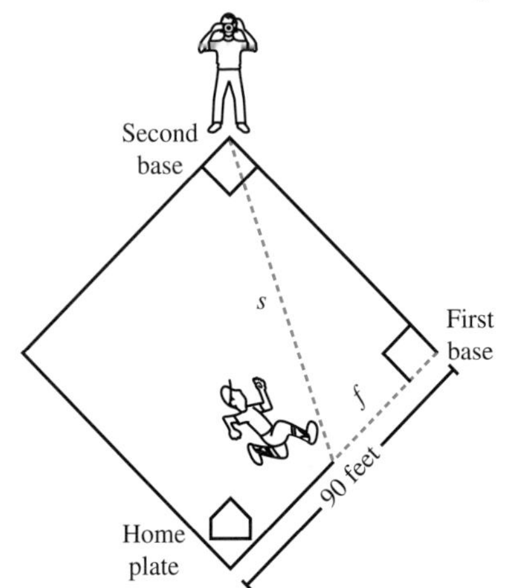

FIGURE 5.26

We consider the right triangle formed by the runner, first base, and second base. Using the Pythagorean Theorem, we know that the relationship between f and s in Figure 5.26 is

$$f^2 + 90^2 = s^2$$

Step 3: Differentiating with respect to t gives

$$2f\frac{df}{dt} + 0 = 2s\frac{ds}{dt}$$

$$\frac{ds}{dt} = \frac{f}{s}\frac{df}{dt}$$

Step 4: Note that the rate we need to find is $\frac{ds}{dt}$. We are told that $f = 45$ feet (half of the way to first base) and $\frac{df}{dt} = -20$ feet per second (negative because the distance between first base and the runner is decreasing). Substitute these values into the related-rates equation to obtain $\frac{ds}{dt} = \frac{45}{s}(-20)$. Find the value of s using the equation $f^2 + 90^2 = s^2$ and the fact that $f = 45$ feet. Thus $s = \sqrt{10{,}125} \approx 100.62$ feet, and we have

$$\frac{ds}{dt} = \frac{45 \text{ feet}}{\sqrt{10{,}125} \text{ feet}}(-20 \text{ feet per second}) \approx -8.94 \text{ feet per second}$$

Step 5: When a runner is halfway to first base, the distance between the runner and the photographer is decreasing by about 8.9 feet per second.

Understanding that the change in one variable is connected to the change in another variable is important in a variety of applications. The independent variable is often—but not always—time. Being able to work with related rates can help you solve problems that occur in many real-world applications.

5.5 Concept Inventory

- *Dependent and independent variables*
- *Related-rates equation*
- *Method of related rates*

5.5 Activities

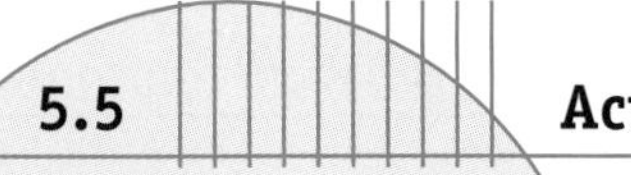

For each of the equations in Activities 1 through 14, write the indicated related-rates equation.

1. $f = 3x$; relate $\frac{df}{dt}$ and $\frac{dx}{dt}$.

2. $p^2 = 5s + 2$; relate $\frac{dp}{dx}$ and $\frac{ds}{dx}$.

3. $k = 6x^2 + 7$; relate $\frac{dk}{dy}$ and $\frac{dx}{dy}$.

4. $y = 9x^3 + 12x^2 + 4x + 3$; relate $\frac{dy}{dt}$ and $\frac{dx}{dt}$.

5. $g = e^{3x}$; relate $\frac{dg}{dt}$ and $\frac{dx}{dt}$.

6. $g = e^{15x^2}$; relate $\frac{dg}{dt}$ and $\frac{dx}{dt}$.

7. $f = 62(1.02^x)$; relate $\frac{df}{dt}$ and $\frac{dx}{dt}$.

8. $p = 5\ln(7 + s)$; relate $\frac{dp}{dx}$ and $\frac{ds}{dx}$.

9. $h = 6a \ln a$; relate $\frac{dh}{dy}$ and $\frac{da}{dy}$.

10. $v = \pi hw(x + w)$; relate $\frac{dv}{dt}$ and $\frac{dw}{dt}$, assuming that h and x are constant.

11. $s = \pi r\sqrt{r^2 + h^2}$; relate $\frac{ds}{dt}$ and $\frac{dh}{dt}$, assuming that r is constant.

12. $v = \frac{1}{3}\pi r^2 h$; relate $\frac{dh}{dt}$ and $\frac{dr}{dt}$, assuming that v is constant.

13. $s = \pi r\sqrt{r^2 + h^2}$; relate $\frac{dh}{dt}$ and $\frac{dr}{dt}$, assuming that s is constant.

14. $v = \pi hw(x + w)$; relate $\frac{dh}{dt}$ and $\frac{dw}{dt}$, assuming that v and x are constant.

15. Trees do a lot more than provide oxygen and shade. They also help pump water out of the ground and transpire it into the atmosphere. The amount of water an oak tree can remove from the ground is related to the tree's size. Suppose that a tree transpires

$$w = 31.54 + 12.97 \ln g \text{ gallons of water per day}$$

where g is the girth in feet of the tree trunk measured 5 feet above the ground. A tree is currently 5 feet in girth and is gaining 2 inches of girth per year.

a. How much water does the tree currently transpire each day?

b. If t is the time in years, find and interpret $\frac{dw}{dt}$.

16. The body-mass index[42] of an individual who weighs w pounds and is h inches tall is given as

$$B = \frac{0.45w}{0.00064516h^2} \text{ points}$$

a. Write an equation showing the relationship between the body-mass index and weight of a woman who is 5 feet 8 inches tall.

b. Find a related-rates equation showing the interconnection between the rates of change with respect to time of the weight and the body-mass index.

c. Consider a woman who weighs 160 pounds and is 5 feet 8 inches tall. If $\frac{dw}{dt} = 1$ pound per month, find and interpret $\frac{dB}{dt}$.

d. Suppose a woman who is 5 feet 8 inches tall has a body-mass index of 24 points. If her body-mass index is decreasing by 0.1 point per month, at what rate is her weight changing?

17. Refer to the body-mass index equation in Activity 16.

a. Write an equation showing the relationship between the body-mass index and the height of a young teenager who weighs 100 pounds.

b. Find a related-rates equation showing the interconnection between the rates of change with respect to time of the body-mass index and the height.

c. If the weight of the teenager who is 5 feet 3 inches tall remains constant at 100 pounds while she is growing at a rate of half an inch per year, how quickly is her body-mass index changing?

18. The apparent temperature[43] A in degrees Fahrenheit is related to the actual temperature t°F and the humidity 100h% by the equation

$$A = 2.70 + 0.885t - 78.7h + 1.20th \text{ degrees Fahrenheit}$$

a. If the humidity remains constant at 53% and the actual temperature is increasing from 80°F at a rate of 2°F per hour, what is the apparent temperature and how quickly is it changing with respect to time?

b. If the actual temperature remains constant at 100°F and the relative humidity is 30% but is dropping by 2 percentage points per hour, what is the apparent temperature and how quickly is it changing with respect to time?

42. *New England Journal of Medicine*, September 14, 1995.

43. W. Bosch and L. G. Cobb, "Temperature Humidity Indices," UMAP Module 691, *The UMAP Journal*, vol. 10, no. 3 (Fall 1989), pp. 237–256.

19. The lumber industry is interested in being able to calculate the volume of wood in a tree trunk. The volume of wood contained in the trunk of a certain fir[44] has been modeled as

$$V = 0.002198d^{1.739925}h^{1.133187} \text{ cubic feet}$$

where d is the diameter in feet of the tree, measured 5 feet above the ground, and h is the height of the tree in feet.

a. If the height of a tree is 32 feet and its diameter is 10 inches, how quickly is the volume of the wood changing when the tree's height is increasing by half a foot per year? (Assume that the tree's diameter remains constant.)

b. If the tree's diameter is 12 inches and its height is 34 feet, how quickly is the volume of the wood changing when the tree's diameter is increasing by 2 inches per year? (Assume that the tree's height remains constant.)

20. The carrying capacity of a crop is measured in the number of people for which it will provide. The carrying capacity of a certain wheat crop[45] has been modeled as

$$K = \frac{11.56P}{D} \text{ people per hectare}$$

where P is the number of kilograms of wheat produced per hectare per year and D is the yearly energy requirement for one person in megajoules per person.

a. Write an equation showing the yearly energy requirement of one person as a function of the production of the crop.

b. With time t as the independent variable, write a related-rates equation using your result from part *a*.

c. If the crop currently produces 10 kilograms of wheat per hectare per year and the yearly energy requirement for one person is increasing by 2 megajoules per year, find and interpret $\frac{dP}{dt}$.

21. A Cobb-Douglas function for the production of mattresses is

$$M = 48.10352L^{0.6}K^{0.4} \text{ mattresses produced}$$

where L is measured in thousands of worker hours and K is the capital investment in thousands of dollars.

a. Write an equation showing labor as a function of capital.

b. Write the related-rates equation for the equation in part *a*, using time as the independent variable and assuming that mattress production remains constant.

c. If there are currently 8000 worker hours and if the capital investment is \$47,000 and is increasing by \$500 per year, how quickly must the number of worker hours be changing in order for mattress production to remain constant?

22. A ladder 15 feet long leans against a tall stone wall. If the bottom of the ladder slides away from the building at a rate of 3 feet per second, how quickly is the ladder sliding down the wall when the top of the ladder is 6 feet from the ground? At what speed is the top of the ladder moving when it hits the ground?

23. A hot-air balloon is taking off from the end zone of a football field. An observer is sitting at the other end of the field 100 yards away from the balloon. If the balloon is rising vertically at a rate of 2 feet per second, at what rate is the distance between the balloon and the observer changing when the balloon is 500 yards off the ground? How far is the balloon from the observer at this time?

24. A girl flying a kite holds the string 4 feet above ground level and lets out string at a rate of 2 feet per second as the kite moves horizontally at an altitude of 84 feet. Find the rate at which the kite is moving horizontally when 100 feet of string has been let out.

25. A softball diamond is a square with each side measuring 60 feet. Suppose a player is running from second base to third base at a rate of 22 feet per second. At what rate is the distance between the runner and home plate changing when the runner is halfway to third base? How far is the runner from home plate at this time?

26. Helium gas is being pumped into a spherical balloon at a rate of 5 cubic feet per minute. The pressure in the balloon remains constant.

44. J. L. Clutter et al., *Timber Management: A Quantitative Approach* (New York: Wiley, 1983).
45. R. S. Loomis and D. J. Connor, *Crop Ecology: Productivity and Management in Agricultural Systems* (Cambridge, England: Cambridge University Press, 1982).

a. What is the volume of the balloon when its diameter is 20 inches?

b. At what rate is the radius of the balloon changing when the diameter is 20 inches?

27. A spherical snowball is melting, and its radius is decreasing at a constant rate. Its diameter decreased from 24 centimeters to 16 centimeters in 30 minutes.

a. What is the volume of the snowball when its radius is 10 centimeters?

b. How quickly is the volume of the snowball changing when its radius is 10 centimeters?

28. A leaking container of salt is sitting on a shelf in a kitchen cupboard. As salt leaks out of a hole in the side of the container, it forms a conical pile on the counter below. As the salt falls onto the pile, it slides down the sides of the pile so that the pile's radius is always equal to its height. If the height of the pile is increasing at a rate of 0.2 inch per day, how quickly is the salt leaking out of the container when the pile is 2 inches tall? How much salt has leaked out of the container by this time?

29. Soft-serve frozen yogurt is being dispensed into a waffle cone at a rate of 1 tablespoon per second. If the waffle cone has height $h = 15$ centimeters and radius $r = 2.5$ centimeters at the top, how quickly is the height of the yogurt in the cone rising when the height of the yogurt is 6 centimeters? (*Hint:* 1 cubic centimeter = 0.06 tablespoon and $r = \frac{h}{6}$.)

30. Boyle's Law for gases states that when the mass of a gas remains constant, the pressure p and the volume v of the gas are related by the equation $pv = c$, where c is a constant whose value depends on the gas. Assume that at a certain instant the volume of a gas is 75 cubic inches and its pressure is 30 pounds per square inch. Because of compression of volume, the pressure of the gas is increasing by 2 pounds per square inch every minute. At what rate is the volume changing at this instant?

31. Demonstrate that the two solution methods referred to in part *d* of Example 1 yield equivalent related-rates equations for the equation given in that part of the example.

Summary

This chapter is devoted to analyzing change. The principal topics are approximating change, optimization, inflection points, and related rates.

Approximating Change

One of the most useful approximations of change in a function is to use the behavior of a tangent line to approximate the behavior of the function. Because of the principle of local linearity, we know that tangent line approximations are quite accurate over small intervals. We estimate the output $f(x + h)$ as $f(x) + f'(x) \cdot h$, where h represents the small change in input.

When cost, revenue, and profit are approximated by using these techniques, we call the approximation marginal analysis. The rates of change of cost, revenue, and profit with respect to the number of units produced or sold are commonly called marginal cost, marginal revenue, and marginal profit. In marginal analysis, the small change in input is one unit.

Optimization

The word optimization (as we used it in Section 5.2) refers to locating relative or absolute extreme points. A relative maximum is a point to which the graph rises and after which the graph falls. Similarly, a relative minimum is a point to which the graph falls and after which the graph rises. There may be several relative maxima and relative minima on a graph. The highest and lowest points on a graph over an interval or over all possible input values are called the absolute maximum and absolute minimum points. These points may coincide with a relative maximum or minimum, or they may occur at the endpoints of a given interval.

If a smooth, continuous graph has a relative maximum or relative minimum at a point, then the line tangent to the graph at that point is horizontal and has slope zero. However, it is not the case that every horizontal tangent line occurs at a relative maximum or relative minimum, because horizontal tangent lines sometimes occur at points of inflection. The key to locating relative maxima and relative minima on smooth, continuous graphs is to locate the input values where the derivative actually *crosses* the horizontal axis. If the derivative merely *touches* the horizontal axis to produce a horizontal tangent line, then we do not have a relative maximum or relative minimum.

Some functions may have extreme points where the function has a value but the derivative does not exist; that is, there is no line tangent to the graph at those points. These functions may or may not be continuous. Some piecewise functions are examples of functions of this type. It is always wise to begin any optimization investigation with a graph of the function before using derivatives.

Inflection Points

Geometrically, inflection points are simply points where the concavity of the graph changes from concave up to concave down, or vice versa. Their importance, however, is that they identify the points of most rapid change or least rapid change in a region around the point.

Because the change along a graph is measured by its derivative, we often locate inflection points by finding the relative maxima and relative minima of the derivative graph. Analytically, we look among the points where the second derivative (the derivative of the derivative) crosses the horizontal axis.

There are functions with inflection points where the second derivative fails to exist. These points can be found by determining where the second derivative is not defined and looking at the graph and its concavity on either side of these points. We should always begin with a graph of the function and its derivative before attempting to use the second derivative to locate inflection points. In addition to locating input values of inflection points, the second derivative of a function can also be used to determine the concavity of the function.

Related Rates

When the changes in one or more variables (called dependent variables) depend on a third variable (called the independent variable), a related-rates equation can be developed to show how the rates of change of these variables are interconnected. The Chain Rule plays an important role in the development of a related-rates equation, because the independent variable (which is often time) is not always expressed in the equation that relates the dependent variables.

The method of related rates involves (1) determining what variables are involved, (2) finding an equation that relates the dependent variables, (3) differentiating both sides of the equation with respect to the independent variable, (4) substituting the given information into the related-rates equation and solving for the unknown rate, and (5) interpreting the solutionin context.

Concept Check

Can you	To practice, try	
• Use derivatives to approximate change?	Section 5.1	Activity 15
• Understand marginal analysis?	Section 5.1	Activities 5, 17
• Find relative and absolute extreme points?	Section 5.2	Activity 23
• Set up and solve applied optimization problems?	Section 5.3	Activities 3, 9, 13
• Find and interpret inflection points?	Section 5.4	Activities 3, 25
• Understand the relationship between second derivatives and concavity?	Section 5.4	Activity 33
• Set up and solve related-rates equations?	Section 5.5	Activities 19, 25

Review Test

1. The number of tourists who visited Tahiti each year between 1988 and 1994 can be modeled[46] by

$$T(x) = -0.4804x^4 + 6.635x^3 - 26.126x^2 + 26.981x + 134.848 \text{ thousand tourists}$$

x years after 1988.

a. Find any relative maxima and minima of $T(x)$ between $x = 0$ and $x = 6$. Explain how you found the value(s).

b. Find any inflection points of the graph of T between $x = 0$ and $x = 6$. Explain how you found the value(s).

c. Graph T, T', and T''. Clearly label on each graph the points corresponding to your answers to parts a and b.

d. Between 1988 and 1994, when was the number of tourists the greatest and when was it the least? What were the corresponding numbers of tourists in those years?

e. Between 1988 and 1994, when was the number of tourists increasing the most rapidly, and when was it declining the most rapidly? Give the rates of change in each of those years.

2. Let $M(t)$ represent the population of French Polynesia (of which Tahiti is a part) at the end of year t. If $M(2000) = 246.4$ thousand people, and if $M'(2000) = 4.9$ thousand people per year, estimate the following:

a. How much did the population of French Polynesia increase during the first quarter of 2001?

b. What was the population in the middle of 2001?

3. A natural gas company needs to run pipe from the point on shore marked A in Figure 5.27 to the point marked B on the island. The island is 3.2 miles down shore from point A and 1.6 miles out to sea. It will cost the gas company \$27 per foot to lay pipe underground and \$143 per foot to lay pipe underwater. The company must decide the most economical distance to run the pipe underground before cutting across the water. Determine the optimal distance x in Figure 5.27 and the corresponding total cost of laying the pipe to the island.

46. Stephen J. Page, "The Pacific Islands," *EIU International Reports*, vol. 1 (1996), page 91.

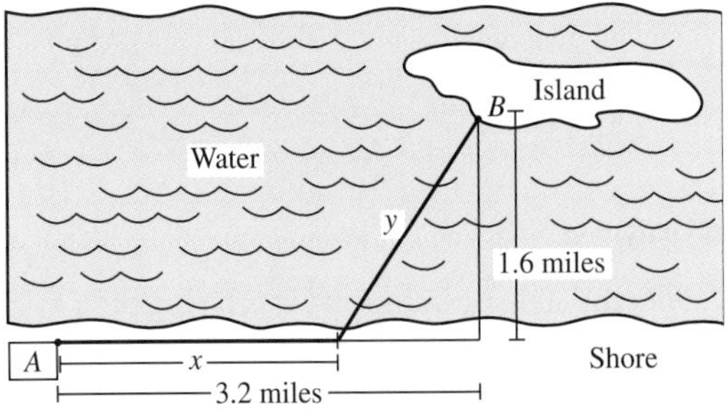

FIGURE 5.27

4. The graph of the derivative of a function h is shown in Figure 5.28.

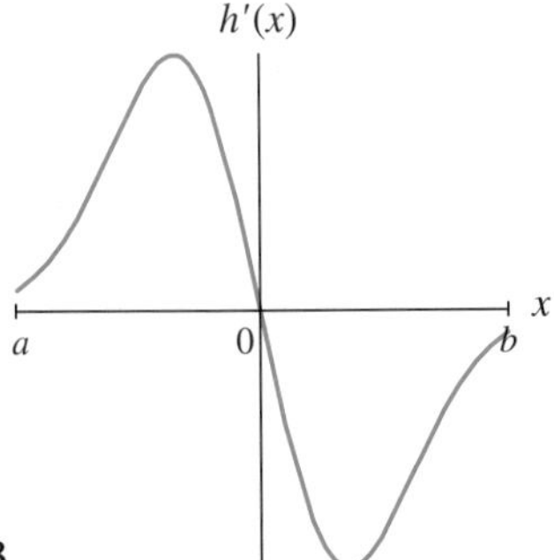

FIGURE 5.28

a. Does the graph of h have a relative maximum and/or minimum between a and b? If so, mark on the derivative graph the input location of each extreme point, and classify the type of extreme point at that location. If not, explain why not.

b. Does the graph of h have one or more inflection points between a and b? If so, mark and label on the derivative graph the input location(s) of the inflection point(s). If not, explain why not.

5. The length of skid marks made on an asphalt road when a vehicle's brakes are applied quickly is given by

$$S = 0.000013wv^2 \text{ feet}$$

where the vehicle weighs w pounds and is traveling v mph when the brakes are applied. How quickly is the length of the skid marks changing when the velocity of a 4000-pound vehicle traveling at a speed of 60 mph is decreasing by 5 mph per second when the brakes are applied? (Assume that the vehicle's weight remains constant.)

Project 5.1 Hunting License Fees

Setting

In 1986, the state of California was trying to make a decision about raising the fee for a deer hunting license. Five hundred hunters were asked how much they would be willing to pay in excess of the current fee to hunt deer. The percentage of hunters to agree to a fee increase of $x is given by the logistic model[47]

$$\text{Percentage} = \frac{1.221}{1 + 0.221e^{0.0116x}}$$

Suppose that in 1986 the license fee was $100, and 75,000 licenses were sold. Suppose that you are part of the 1986 Natural Resources Team presenting a proposed increase in the hunting license fee to the head of the California Department of Natural Resources.

47. Based on information in *Journal of Environmental Economics and Management*, vol. 24, no. 1 (January 1993).

Tasks

1. Illustrate how the model can be used by answering the following questions.
   a. What was the hunting license revenue in 1986?
   b. Suppose that in 1987 the fee increased to $150.
      i. What percentage of the 1986 hunters would buy another license?
      ii. How many hunters is that?
      iii. What would be the 1987 revenue?
   c. Repeat the questions in part *b* if the fee were increased to $300.

2. If the fee increase for 1987 were x dollars, find formulas for the following:
   a. The percentage of hunters willing to pay the new fee
   b. The number of the 75,000 hunters willing to pay the new fee
   c. The new fee
   d. The 1987 revenue

3. Use your formulas to determine the optimal license fee. Also, find the optimal fee increase, the number of hunters who will buy licenses at the new fee, and the optimal revenue.

Reporting

1. Prepare a letter to the head of the California Depeartment of Natural Resources. Your letter should address the fee increase and expectations for revenue. You should not make it technical but should give some support to back up your conclusions.

2. Prepare a technical written report outlining your findings as well as the mathematical methods you used to arrive at your conclusions.

Project 5.2 Fund-Raising Campaign

Setting

In order to raise funds, the mathematics department in your college or university is planning to sell T-shirts before next year's football game against the school's biggest rival. Your team has volunteered to conduct the fund raiser. Because several other student groups have also volunteered to head this project, your team is to present its proposal for the fund drive, as well as your predictions about its outcome, to a panel of mathematics faculty.

Task A

Follow the tasks for Project 2.2 on pages 184–185. You will find a partial price listing for the T-shirt company in Table 2.99 on page 185.

Task B

1. Review your work for Task A. If you wish to make any changes in your marketing scheme, you should do so now. If you decide to make any changes, make sure that the polling that was done is still applicable (for example, you will not be able to change your target market). Change (if necessary) any models from Task A to reflect any changes in your marketing scheme.

2. Use the models of demand, revenue, total cost, and profit developed in Task A to proceed with this section.

   Determine the selling price that generates maximum revenue. What is maximum revenue? Is the selling price that generates maximum revenue the same as the price that generates maximum profit? What is maximum profit? Which should you consider (maximum revenue or maximum profit) in order to get the best picture of the effectiveness of the drive? Re-evaluate the number of shirts you may wish to sell. Will this affect the cost you determined above? If so, change your revenue, total cost, and profit functions to reflect this adjustment and re-analyze optimal values. Show and explain the mathematics that underlies your reasoning.

   Discuss the sensitivity of the demand function to changes in price (check rates of change for \$20, \$14, and \$8). Does the demand function have an inflection point? If so, find it. Find the rate of change of demand with respect to price at this point, and interpret its meaning and impact in this context. How would the sensitivity of the demand curve affect your decisions about raising or lowering your selling price?

   On the basis of your findings, predict the optimal selling price, the number of T-shirts you intend to print, the costs involved, the number of T-shirts you expect to sell before realizing a profit (that is, the break-even point), and the expected profit.

Reporting

1. Write a report for the mathematics department concerning your proposed campaign. They will be interested in the business interpretation as well as in an accurate description of the mathematics involved. Make sure that you include graphical as well as mathematical representations of your demand, revenue, cost, and profit functions. (Include graphs of any functions and derivatives that you use. Include your calculations and your survey as appendices.) Do not forget to cover Task A in this report as well.

2. Make your proposal and present your findings to a panel of mathematics professors in a 15-minute presentation. Your presentation should be restricted to the business interpretation, and you should use overhead transparencies of graphs and equations of all models and derivatives as well as any other visual aids that you consider appropriate.

6

Accumulating Change: Limits of Sums and the Definite Integral

Concept Application

Tony Freeman/PhotoEdit

If crude oil is flowing into a holding tank through a pipe, the rate at which the oil is flowing determines how quickly the holding tank will fill. It is possible that the rate varies with time and can be mathematically modeled. Such a model could then be used to answer questions such as

- What is the change in the amount of oil in the tank during the first 10 minutes the oil is flowing into the tank?
- If there were 5000 cubic feet of oil in the tank before the oil began flowing into the tank, how much oil is in the tank after 10 minutes?
- How long can oil flow into the tank before the tank is full?

Questions such as these can be answered using definite integrals. Examples of this type of problem appear in Questions 1 and 4 of the Chapter 6 Review Test.

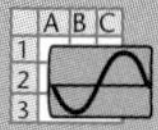

This symbol directs you to the supplemental technology guides. The different guides contain step-by-step instructions for using graphing calculators or spreadsheet software to work the particular example or discussion marked by the technology symbol.

This symbol directs you to the Calculus Concepts *web site, where you will find algebra review, links to data updates, additional practice problems, and other helpful resource material.*

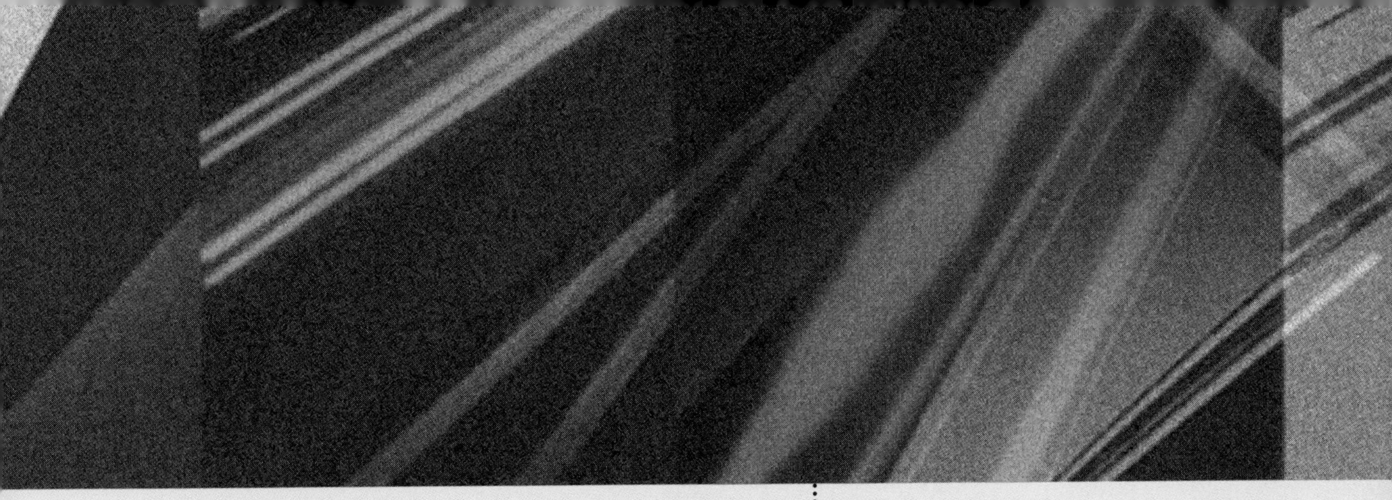

Concept Objectives

This chapter will help you understand the concepts of

- Accumulated change and area under a curve
- Area approximation techniques
- Definite integral as a limiting value
- Accumulation function
- Fundamental Theorem of Calculus
- Antiderivative
- Improper integral

and you will learn to

- Approximate area using rectangles
- Interpret the area between a graph and the horizontal axis
- Approximate area using a limiting value of sums of areas of rectangles
- Sketch and interpret accumulation function graphs
- Find simple general antiderivatives
- Find and interpret specific antiderivatives
- Recover a function from its rate-of-change equation
- Evaluate and interpret definite integrals
- Determine whether an approximation technique is necessary in order to estimate a definite integral
- Evaluate and interpret improper integrals

Chapters 1 through 5 focused on the derivative, one of the two fundamental concepts of calculus. Now we begin a study of the second fundamental concept in calculus, the integral. As before, our approach is through the mathematics of change.

We start by analyzing the accumulated change in a quantity and how it is related to areas of regions between the graph of the rate-of-change function for that quantity and the horizontal axis. As we refine our thinking about area, we are led to consider limits of sums, which, in turn, show us how to account for the results of change in terms of integrals. Integrals, as we shall see, are intimately connected to derivatives by the Fundamental Theorem of Calculus.

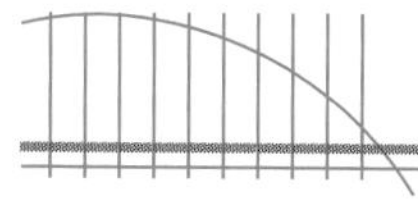

6.1 Results of Change and Area Approximations

See the *Instructor's Resource Guide* for a suggested approach to introducing the concept of accumulated change.

In our study of calculus so far, we have concentrated on finding rates of change. We now consider the results of change.

Accumulated Change

Suppose that you have been driving on an interstate highway for 2 hours at a constant speed of 60 miles per hour. Because velocity, v, is the rate of change of distance traveled, s, with respect to time, we write this mathematically as $v(t) = s'(t) = 60$ mph, where t is the time in hours $(0 \le t \le 2)$. A graph of this rate function over a 2-hour period of time appears in Figure 6.1.

FIGURE 6.1

Recall that distance = (rate)(time), so at this constant rate, the distance traveled during a time period of t hours is (60 miles per hour)(t hours) = $60t$ miles. Geometrically, we view this multiplication as giving the area of the region between the rate-of-change graph and the horizontal axis over any time period of length t hours. Figure 6.2 illustrates this fact for the 1-hour time period between 0.5 hour and 1.5 hours, the 15-minute time period between 1 hour and 45 minutes and 2 hours, and the first t hours of the trip. In each case, the change in distance traveled during a time period is given by the area of a region between the rate-of-change graph and the horizontal axis from the beginning to the end of the time period.

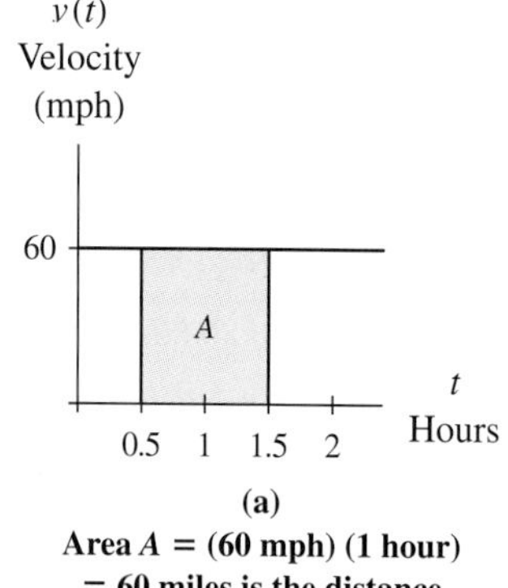

(a)
Area A = (60 mph) (1 hour) = 60 miles is the distance traveled during 1 hour

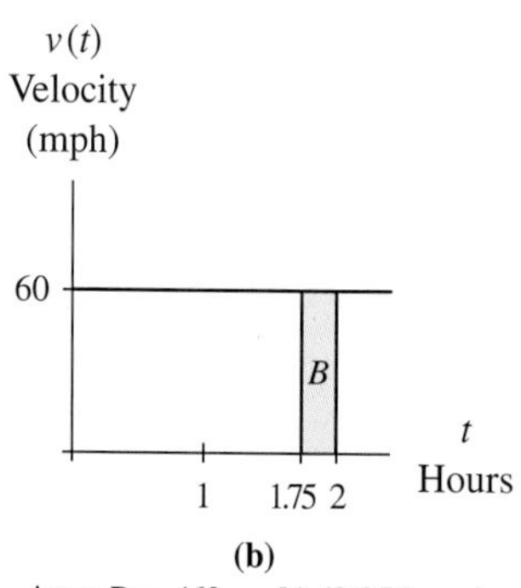

(b)
Area B = (60 mph) (0.25 hour) = 15 miles is the distance traveled during the last 15 minutes

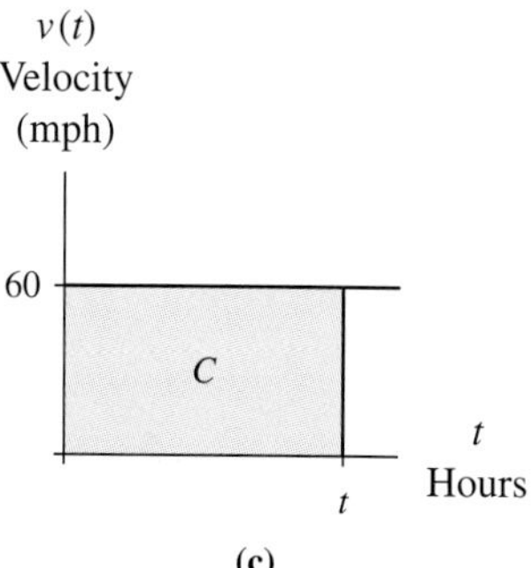

(c)
Area C = $60t$ miles is the distance traveled during t hours

FIGURE 6.2

Now, imagine that after a 2-hour drive at a constant speed of 60 mph, you increase your speed at a constant rate to 75 mph over a 10-minute interval and then maintain that constant 75-mph speed during the next half-hour. A graph of your speed appears in Figure 6.3a.

We know that the distance traveled during the times when speed is constant is the speed multiplied by the amount of time driven at that speed. But how can we calculate the distance driven between 2 hours and 2 hours 10 minutes when the speed is increasing linearly? If we knew the average speed over the 10-minute interval, then we could multiply that average speed by $\frac{1}{6}$ of an hour to obtain the distance traveled.

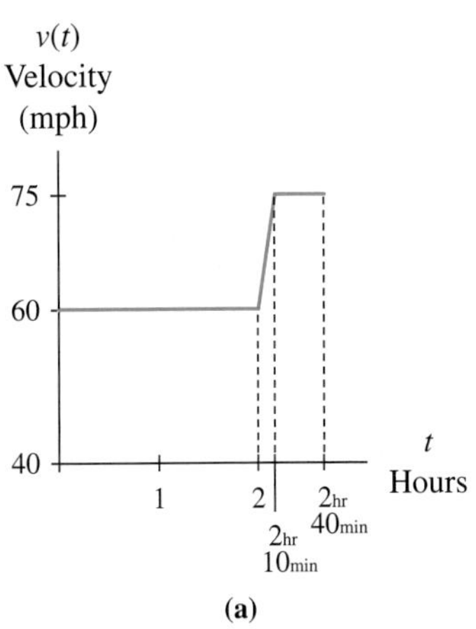

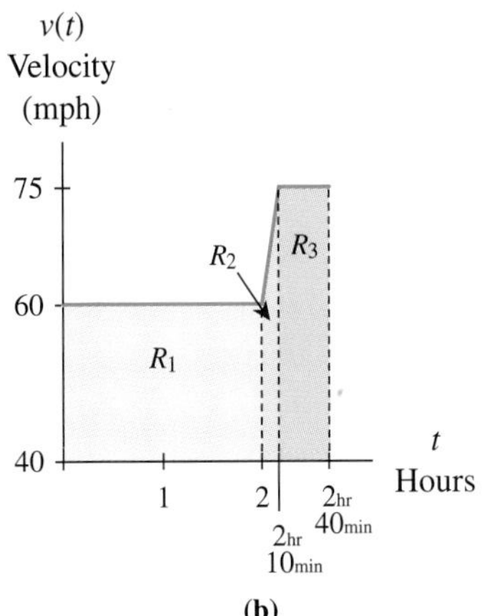

FIGURE 6.3

In this case, the average speed is simply the average of the beginning and ending speeds during the 10-minute interval. Thus we have

$$\begin{array}{c}\text{Distance traveled}\\ \text{between 2 hours}\\ \text{and 2 hours}\\ \text{10 minutes}\end{array} = \begin{pmatrix}\text{average}\\ \text{speed}\end{pmatrix}(\text{time}) = \left(\frac{60 \text{ mph} + 75 \text{ mph}}{2}\right)\left(\frac{1}{6}\text{ hour}\right)$$

This distance is, in fact, the area of the trapezoid $\left[\text{Area} = (\text{base})\left(\frac{\text{side 1 + side 2}}{2}\right)\right]$ labeled R_2 in Figure 6.3b. Thus, the distance traveled during the 2-hour 40-minute interval is the area of the region beneath the velocity graph and above the time axis between the specified inputs, calculated by dividing the region into three smaller regions as shown in Figure 6.3b.

$$\begin{aligned}\begin{array}{c}\text{Distance}\\ \text{traveled}\end{array} &= \text{area of region } R_1 + \text{area of region } R_2 + \text{area of region } R_3\\ &= (60 \text{ mph})(2 \text{ hr}) + \left(\frac{60 \text{ mph} + 75 \text{ mph}}{2}\right)\left(\frac{1}{6}\text{ hr}\right) + (75 \text{ mph})\left(\frac{1}{2}\text{ hr}\right)\\ &= 120 \text{ miles} + 11.25 \text{ miles} + 37.5 \text{ miles}\\ &= 168.75 \text{ miles}\end{aligned}$$

Once more, the change in the distance traveled is given by the area of the region between the rate-of-change graph and the horizontal axis.

EXAMPLE 1 *Draining Water*

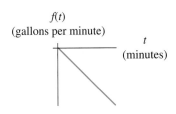

A water tank drains at a rate of $r(t) = -2t$ gallons per minute t minutes after the water began draining. The graph of the rate-of-change function is shown in Figure 6.4.

$f(t)$ (gallons per minute)

t (minutes)

FIGURE 6.4

a. What are the units on height, width, and area of the region between the time axis and the rate graph?

b. Determine the change in the volume of water in the tank during the first 4 minutes the tank was draining.

c. How much water is in the tank at $t = 4$ minutes?

Solution

a. The height corresponds to the output units, which are gallons per minute. The width is time in minutes. In the calculation of area, the height and width are multiplied, giving area in gallons.

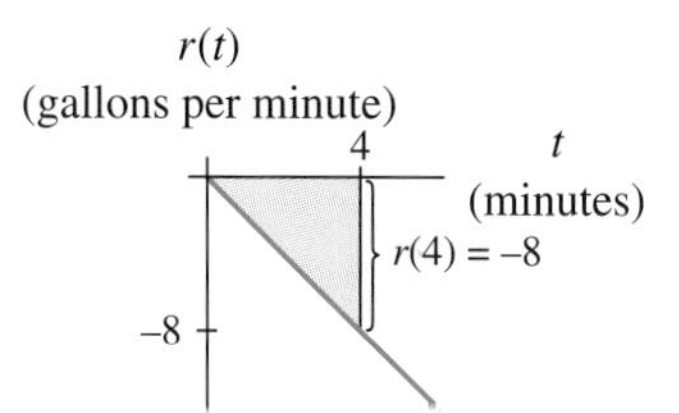

FIGURE 6.5

By *signed area* we mean the area of the region with a negative sign in front of the area value to indicate that the region lies below the input axis.

Emphasize to your students that these area calculations give changes in output rather than output values.

b. To find the change in the volume of the water, we find the area of the region between the time axis and the rate graph. This area is shaded in Figure 6.5. The region is a triangle with base of length 4. The height is determined by the function value at $t = 4$. Note that the function value is negative: $r(4) = -8$. The negative sign is simply an indication that the rate-of-change graph lies below the horizontal axis. Although we know that height is a positive measure, we use this signed value for height to remind us that the region lies beneath the horizontal axis and represents a decrease in the amount of water in the tank. The signed area of the region is

$$\frac{1}{2}(4 \text{ minutes})(-8 \text{ gallons per minute}) = -16 \text{ gallons}$$

Thus, during the first 4 minutes the tank was draining, the volume changed by -16 gallons. In other words, the tank lost 16 gallons of water.

c. Because we were not told how much water was in the tank when it began draining, we do not know how much remains after 4 minutes. However, if we knew the initial amount of water, we could subtract the 16 gallons of water calculated in part *b* from the initial amount to determine the amount remaining in the tank after 4 minutes.

The previous velocity examples illustrate the following fundamental principle:

Results of Change

The accumulated change in a quantity is represented as the area or signed area of a region between the rate-of-change function for that quantity and the horizontal axis provided the function does not cross the horizontal axis.

We will verify this principle later in this chapter by using the Fundamental Theorem of Calculus.

Left- and Right-Rectangle Approximations

The preceding velocity examples were carefully chosen so that their rate-of-change graphs were easy to obtain and the areas of the desired regions were easy to calculate. But most real-life situations are not so simple. Indeed, there are two issues that we must face:

1. Obtaining the rate-of-change function for the quantity of interest
2. Calculating the area of the desired region between the rate-of-change function and the horizontal axis

In most cases, we must resort to approximating both the rate-of-change function and the desired area. Consider the example of a store manager of a large department store who wishes to estimate the number of customers who came to a Saturday sale from 9 a.m. to 9 p.m. The manager stands by the entrance for 1-minute intervals at different times throughout the day and counts the number of people entering the store. He uses these data as an estimate of the number of customers who enter the store each minute. The manager's data may look something like Table 6.1.

TABLE 6.1

Time	Number of customers per minute	Time	Number of customers per minute
9:00 a.m.	1	2:30 p.m.	5
9:45 a.m.	2	3:10 p.m.	5
10:15 a.m.	3	4:00 p.m.	4
11:00 a.m.	4	5:15 p.m.	4
11:45 a.m.	4	6:30 p.m.	3
12:15 p.m.	5	7:30 p.m.	2
1:15 p.m.	5	8:15 p.m.	2

The number of customers who attended the sale can be calculated by summing the number of customers who entered the store during each hour for every hour of the 12-hour sale. We do not have enough information to determine the exact number of customers who entered the store each hour; however, we can estimate the number by using a continuous model for the customers-per-minute data. To build a model for the rate-of-change data, we choose to convert each of the above observation times to minutes after 9:00 a.m. Thus $m = 0$ at 9:00 a.m. and $m = 675$ at 8:15 p.m. A model for the customers-per-minute data is

$$c(m) = (4.58904 \cdot 10^{-8})m^3 - (7.78127 \cdot 10^{-5})m^2 + 0.03303m + 0.88763 \text{ customers per minute}$$

where m is the number of minutes after 9:00 a.m. The graph of this model is shown in Figure 6.6. Note that c is a continuous function modeling a discrete situation.

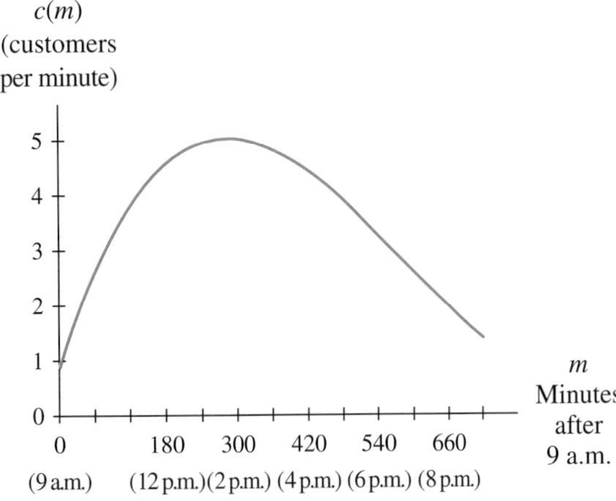

FIGURE 6.6

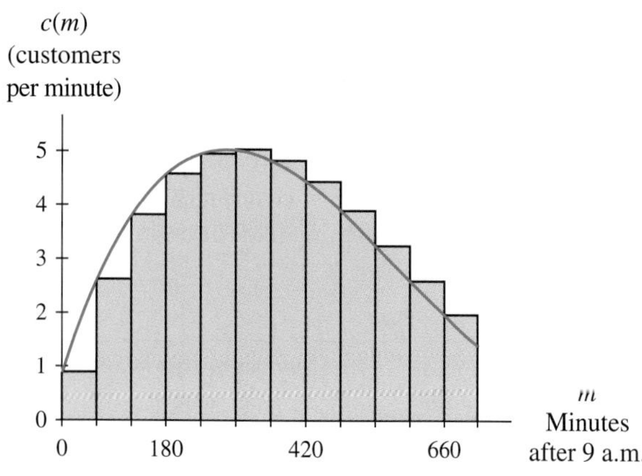

FIGURE 6.7

To estimate the number of customers who attended the sale, we use the model to estimate the number of customers per minute entering the store at the beginning of each hour and multiply by 60 minutes to estimate the number of customers entering the store during that hour. Summing the estimates for each of the 12 hours results in the estimate we desire for the total number of customers. This process is the same as drawing a set of 12 rectangles under the graph of c, one for each hour, and using the sum of their areas to estimate the total number of customers who came to the sale. (See Figure 6.7.)

Note that the area of the rectangles is an estimate of the area of the region between the graph of c and the horizontal axis. This is precisely the point made in the principle stated after Example 1 concerning results of change. The total number of customers is approximately the area of the region beneath a graph of the rate-of-change function. Here we approximate the area using 12 rectangles.

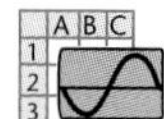

In Figure 6.7, the height of each rectangle is the function c evaluated at the left endpoint of the base of the rectangle. It is for this reason that we call these rectangles **left rectangles**. Because we are using 12 rectangles of equal width to span the 12-hour (720-minute) sale, the width of each rectangle is $720 \div 12 = 60$ minutes. We use Table 6.2 to keep track of the areas that we are summing.

TABLE 6.2

Left endpoint of rectangle m	Height of rectangle $c(m)$ (customers per minute)	Width of rectangle (minutes)	Area of rectangle[1] = height·width [(customers/min)(min) → customers]
9 a.m.	$c(0) \approx 0.9$	60	53.3
10 a.m.	$c(60) \approx 2.6$	60	156.0
11 a.m.	$c(120) \approx 3.8$	60	228.6
noon	$c(180) \approx 4.6$	60	274.8
1 p.m.	$c(240) \approx 5.0$	60	298.1
2 p.m.	$c(300) \approx 5.0$	60	302.0
3 p.m.	$c(360) \approx 4.8$	60	290.2
4 p.m.	$c(420) \approx 4.4$	60	266.1
5 p.m.	$c(480) \approx 3.9$	60	233.4
6 p.m.	$c(540) \approx 3.3$	60	195.7
7 p.m.	$c(600) \approx 2.6$	60	156.4
8 p.m.	$c(660) \approx 2.0$	60	119.2
			Total area of rectangles $\approx$ 2574 customers

Thus, using 12 rectangles, we estimate that 2574 customers came to the Saturday sale.

Note the importance in this example of measuring time in minutes. If time is measured in hours, then the area calculated by multiplying height with units cus-

1. These values were obtained using the unrounded model.

tomers per minute by width with units hours is not the number of customers. Make sure the units correspond so that their multiplication results in the desired units. Also note that our choice of using time intervals of 1 hour was arbitrary. In fact, using 30-minute or 15-minute intervals may give a more accurate estimate.

The previous discussion illustrates a way to approximate the number of customers attending a Saturday sale by using left-rectangle areas. In some situations, choosing rectangles whose heights are measured at the right endpoint of the base of each rectangle may give more reasonable area approximations. Such rectangles are called **right rectangles.** The use of such rectangles is illustrated in the following example.

EXAMPLE 2 *Drug Absorption*

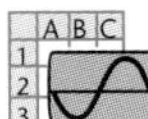

A pharmaceutical company has tested the absorption rate of a drug that is given in 20-milligram (mg) doses for 20 days. Researchers have compiled data showing the rate of change of the concentration of the drug, measured in micrograms per milliliter per day (μg/mL/day), in the bloodstream. Assume the rates are measured at the end of each day. The data and scatter plot are shown in Table 6.3 and Figure 6.8, respectively.

TABLE 6.3

Day	Concentration rate of change (μg/mL/day)
1	1.50
5	0.75
9	0.33
13	0.20
17	0.10
21	-1.10
25	-0.60
29	-0.15

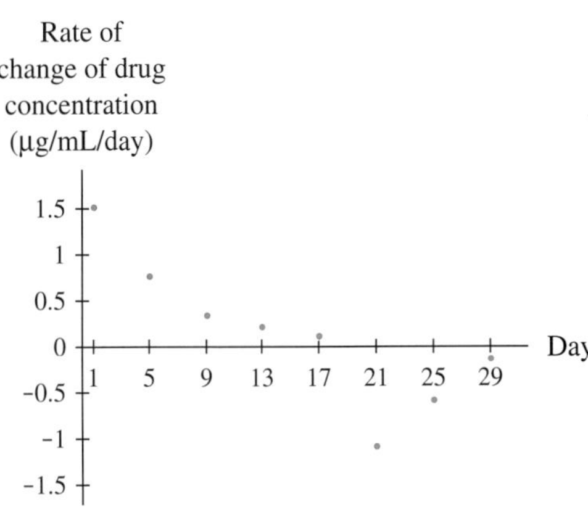

FIGURE 6.8

a. Use the data and right rectangles of width 4 days to estimate the change in the concentration of the drug from the end of day 1 through the end of day 29.

b. Find a model for the data.

c. Use the model and right rectangles of width 2 days to estimate the change in the drug concentration from the beginning of day 1 through the end of day 20.

d. Use the model to estimate the change in the drug concentration from the beginning of day 21 through the end of day 29.

e. Combine your answers to parts *c* and *d* to estimate the change in the drug concentration from the beginning of day 1 through the end of day 29.

Solution

a. Figure 6.9 shows the right rectangles constructed on the scatter plot. The height of each rectangle is simply the output of the corresponding data point. The width of each rectangle is 4 days.

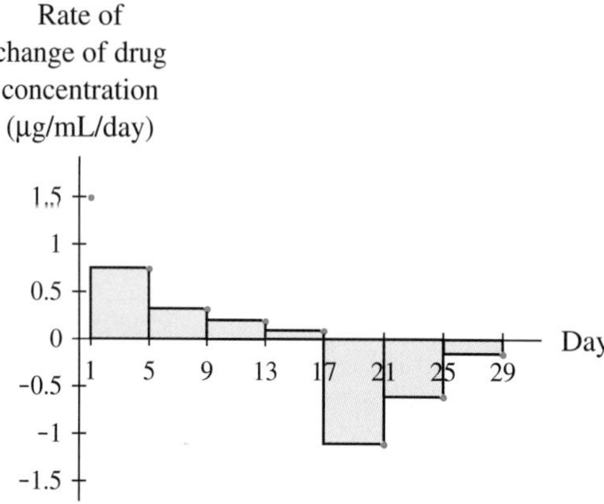

FIGURE 6.9

The sum of the areas of the rectangles above the horizontal axis is

$$\begin{aligned}\text{Area} &= (0.75\ \mu\text{g/mL/day})(4\text{ days}) + (0.33\ \mu\text{g/mL/day})(4\text{ days}) \\ &\quad + (0.20\ \mu\text{g/mL/day})(4\text{ days}) + (0.10\ \mu\text{g/mL/day})(4\text{ days}) \\ &= 3\ \mu\text{g/mL} + 1.32\ \mu\text{g/mL} + 0.8\ \mu\text{g/mL} + 0.4\ \mu\text{g/mL} \\ &= 5.52\ \mu\text{g/mL}\end{aligned}$$

The drug concentration increased by approximately 5.5 μg/mL during the first 16 days. The sum of the signed areas of the rectangles below the horizontal axis is

$$\begin{aligned}\text{Signed area} &= (-1.10\ \mu\text{g/mL/day})(4\text{ days}) + (-0.60\ \mu\text{g/mL/day})(4\text{ days}) \\ &\quad + (-0.15\ \mu\text{g/mL/day})(4\text{ days}) \\ &= -4.4\ \mu\text{g/mL} + (-2.4\ \mu\text{g/mL}) + (-0.6\ \mu\text{g/mL}) \\ &= -7.4\ \mu\text{g/mL}\end{aligned}$$

The drug concentration changed by approximately -7.4 μg/mL between the end of day 17 and the end of day 29. In other words, the drug concentration decreased by about 7.4 μg/mL during this time.

Thus between days 1 and 29, the drug concentration first increased by 5.52 μg/mL and then decreased by 7.4 μg/mL. Combining these results, we have that the total change was approximately 5.52 μg/mL − 7.4 μg/mL = -1.88 μg/mL. The drug concentration was approximately 1.88 μg/mL less at the end of day 29 than it was at the end of day 1.

b. It is clear that a piecewise model is appropriate, with one function modeling the data through the end of day 20 and a second function modeling the data after the drug is discontinued. A possible model is

$$r(x) = \begin{cases} 1.708(0.845^x)\ \mu\text{g/mL/day} & \text{when } 0 \le x < 20 \\ 0.11875x - 3.5854\ \mu\text{g/mL/day} & \text{when } 20 < x \le 29 \end{cases}$$

where x is the number of days after the drug is first administered.

c. To determine the change in the drug concentration from the beginning of day 1 ($x = 0$) through the end of day 20 ($x = 20$), we use the exponential portion of the model and 10 right rectangles as shown in Figure 6.10 and Table 6.4.

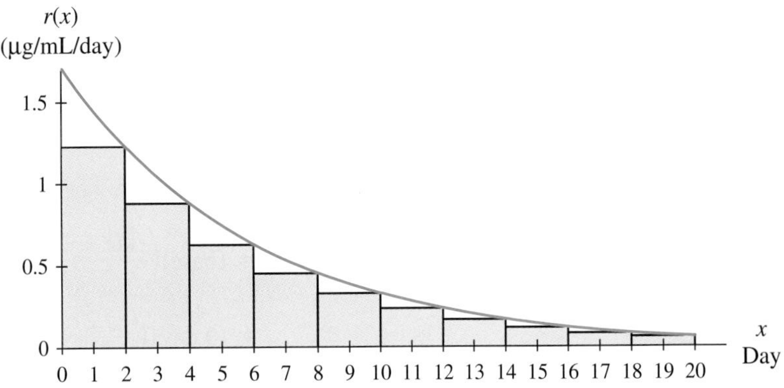

FIGURE 6.10

TABLE 6.4

Right endpoint of rectangle x	Height of rectangle $r(x)$ (μg/mL/day)	Area of rectangle = height·width [(μg/mL/day)(days)→μg/mL]
2	1.220	2.439
4	0.871	1.742
6	0.622	1.244
8	0.444	0.888
10	0.317	0.634
12	0.226	0.453
14	0.162	0.323
16	0.115	0.231
18	0.082	0.165
20	0.059	0.118
Sum of areas of rectangles ≈ 8.235 μg/mL Change in the drug concentration ≈ 8.235 μg/mL		

From the beginning of day 1 through the end of day 20, the drug concentration increased by approximately 8.235 micrograms per milliliter.

d. To determine the change in the concentration from the beginning of day 21 through the end of day 29, we use the linear portion of the model. We calculate the shaded area in Figure 6.11 using the formula for the area of a trapezoid. The signed length of the left side is the linear portion of r evaluated at $t = 20$. Although the linear portion of the function does not apply in context for $t = 20$, it is useful in giving the length of the left side of the trapezoid. The signed length of the right side is the linear portion of r evaluated at $t = 29$.

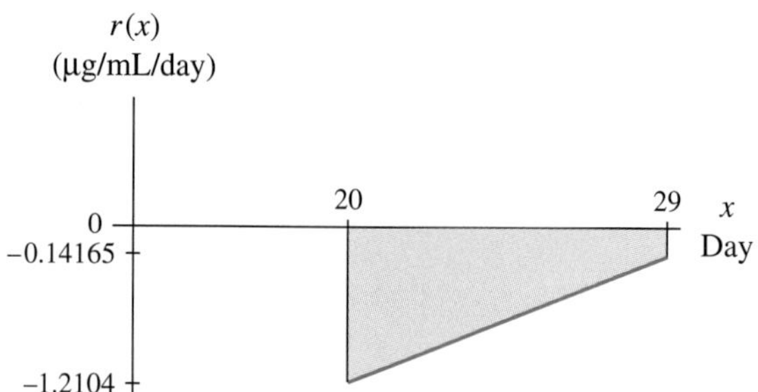

FIGURE 6.11

$$\text{Trapezoid area} = (\text{base})\left(\frac{\text{side 1} + \text{side 2}}{2}\right)$$
$$= (9 \text{ days})\left(\frac{-1.2104 + (-0.14165)}{2} \;\mu\text{g/mL/day}\right)$$
$$= (9 \text{ days})(-0.676025 \;\mu\text{g/mL/day})$$
$$\approx -6.084 \;\mu\text{g/mL}$$

Thus the change in the drug concentration is -6.084 µg/mL. In other words, from the beginning of day 21 through the end of day 29, the concentration of the drug in the bloodstream declined by approximately 6.084 µg/mL.

e. To determine the change in concentration from the beginning of day 1 through the end of day 29, we need only subtract the amount of decline from the amount of increase.

$$8.235 \;\mu\text{g/mL} - 6.084 \;\mu\text{g/mL} = 2.151 \;\mu\text{g/mL}$$

The drug concentration increased by approximately 2.151 µg/mL from the beginning of day 1 through the end of day 29.

Midpoint-Rectangle Approximation

It is often true that if a left-rectangle approximation is an over-approximation, then a right-rectangle area will be an under-approximation, and vice versa. For example, consider using four left and four right rectangles to approximate the area of the region between the function $f(x) = \sqrt{1 - x^2}$ and the x-axis between $x = 0$ and $x = 1$. Figure 6.12 shows the rectangles and the approximate areas.

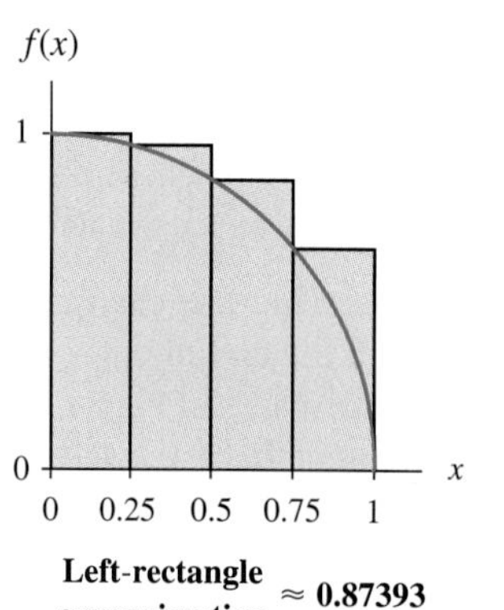

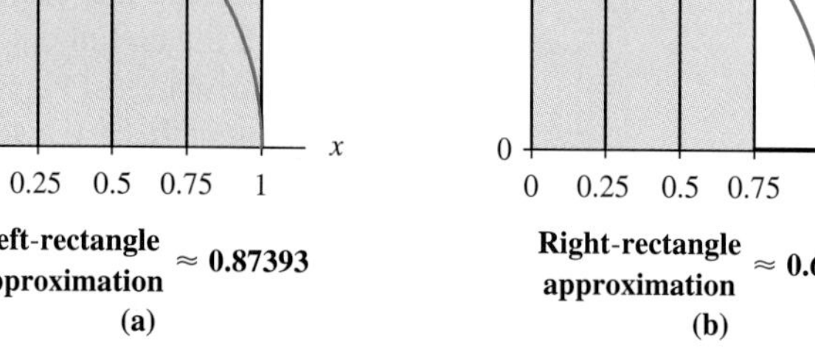

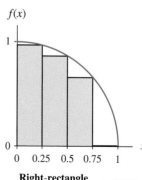

FIGURE 6.12

Left-rectangle approximation ≈ **0.87393**
(a)

Right-rectangle approximation ≈ **0.62393**
(b)

Note that the left-rectangle approximation is an overestimate and the right-rectangle approximation is an underestimate.

We next consider approximating area using a third type of rectangle. The **midpoint-rectangle approximation** uses rectangles whose heights are calculated at the midpoints of the subintervals (see Figure 6.13). Table 6.5 shows the calculations for the areas of the midpoint rectangles shown in Figure 6.13.

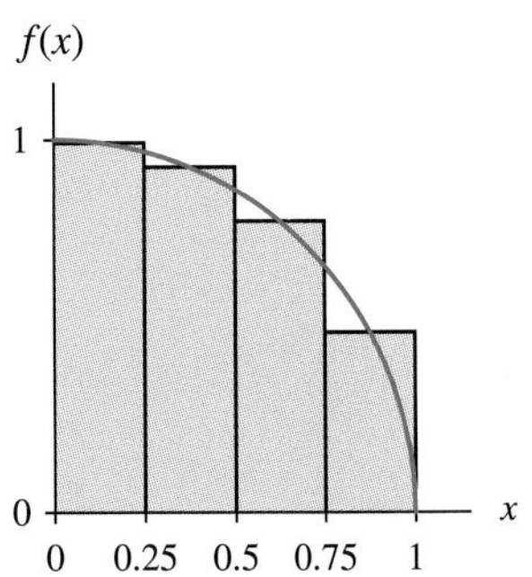

FIGURE 6.13

TABLE 6.5

Rectangle number	Midpoint of interval	Height of rectangle	Width of rectangle	Area = height·width
1	0.125	$f(0.125) \approx 0.99216$	0.25	0.24804
2	0.375	$f(0.375) \approx 0.92702$	0.25	0.23176
3	0.625	$f(0.625) \approx 0.78062$	0.25	0.19516
4	0.875	$f(0.875) \approx 0.48412$	0.25	0.12103
			Total midpoint area	$\approx$ 0.79598

The region whose area we are approximating is the interior of a quarter-circle with radius 1, so the true area is $\frac{1}{4}(\pi \cdot \text{radius}^2) = \frac{\pi}{4}(1)^2 \approx 0.78540$. The midpoint-rectangle approximation is much closer to the actual area than are the two other approximations. This is often the case.

EXAMPLE 3 *Rising River*

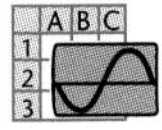

For 20 hours following a heavy rainstorm in the spring of 1996, the flow rate for the west fork of the Carson River was measured periodically. The flow rates[2] in cubic feet per second (cfs) are shown in Table 6.6.

TABLE 6.6

Time	Flow rate (cfs)
11:45 a.m. Wednesday	785
3:45 p.m. Wednesday	753
7:45 p.m. Wednesday	834
11:45 p.m. Wednesday	1070
3:45 a.m. Thursday	1470
7:45 a.m. Thursday	2090

a. If a flow-rate model were found with input in hours since 11:45 a.m. Wednesday and output in cfs, what would be the units for the area of rectangles under the graph of the function?

b. Convert the flow rates to cubic feet per hour. How does this conversion change the units on the area of the region under the graph of an equation for the converted data?

c. Find a model for the converted data.

d. Use 10 intervals and midpoint rectangles to estimate the area under the graph of the function in part *c* from 11:45 a.m. Wednesday to 7:45 a.m. Thursday.

e. What does this area represent in the context of the Carson River?

2. As reported in the *Reno Gazette–Journal*, May 17, 1996, p. 4A.

Solution

a. The heights of the rectangles will be measured in cfs and the widths of the rectangles will be in hours. Multiplying height by width results in units of (cubic feet per second)(hours). Without conversion, these are not meaningful units.

b. There are 3600 seconds each hour (60 minutes per hour times 60 seconds per minute). Multiplying cubic feet per second by seconds per hour gives cubic feet per hour. The data become as shown in Table 6.7. The area of the region beneath a graph of an equation for the data has units of (cubic feet per hour)(hour). The area is measured in cubic feet.

TABLE 6.7

Hours since 11:45 a.m. Wednesday	Flow rate (cubic feet per hour)
0	2,826,000
4	2,710,800
8	3,002,400
12	3,852,000
16	5,292,000
20	7,524,000

c. A quadratic model for the converted data is

$$f(h) = 18{,}225h^2 - 135{,}334.3h + 2{,}881{,}542.9 \text{ cubic feet per hour}$$

h hours after 11:45 a.m. Wednesday.

d. Using 10 intervals to span a 20-hour period requires that each interval be 2 hours long. Table 6.8 shows the calculation for the midpoint-rectangle approximation.

TABLE 6.8

Midpoint	Height of midpoint	Width of rectangle	Area of rectangle
1	$f(1) \approx 2{,}764{,}433.6$	2	5,528,867.1
3	$f(3) \approx 2{,}639{,}565.0$	2	5,279,130.0
5	$f(5) \approx 2{,}660{,}496.4$	2	5,320,992.9
7	$f(7) \approx 2{,}827{,}227.9$	2	5,654,455.7
9	$f(9) \approx 3{,}139{,}759.3$	2	6,279,518.6
11	$f(11) \approx 3{,}598{,}090.7$	2	7,196,181.4
13	$f(13) \approx 4{,}202{,}222.1$	2	8,404,444.3
15	$f(15) \approx 4{,}952{,}153.6$	2	9,904,307.1
17	$f(17) \approx 5{,}847{,}885.0$	2	11,695,770.0
19	$f(19) \approx 6{,}889{,}416.4$	2	13,778,832.9
		Total midpoint area $\approx$	79,042,500 ft^3

e. The area which we estimate as 79,000,000 cubic feet represents the approximate amount of water that flowed through the Carson River (at the measuring point) from 11:45 a.m. Wednesday to 7:45 a.m. Thursday.

Working with Count Data

Frequently, rate-of-change data are not available for a quantity; instead, we have **count data** (that is, totals reported at the end of a period of time) pertaining to that quantity. Table 6.9 shows the number of aluminum cans that were recycled[3] each year from 1978 through 1988.

3. Data from the Aluminum Association, Inc.

TABLE 6.9

Year	Cans each year (billions)	Year	Cans each year (billions)
1978	8.0	1984	31.9
1979	8.5	1985	33.1
1980	14.8	1986	33.3
1981	24.9	1987	36.6
1982	28.3	1988	42.0
1983	29.4		

The number of cans recycled during 1980 can be interpreted geometrically as the area of a rectangle. We use a right rectangle (one whose height is determined at the right side of the rectangle) because the data are reported at the end of each year. (See Figure 6.14.)

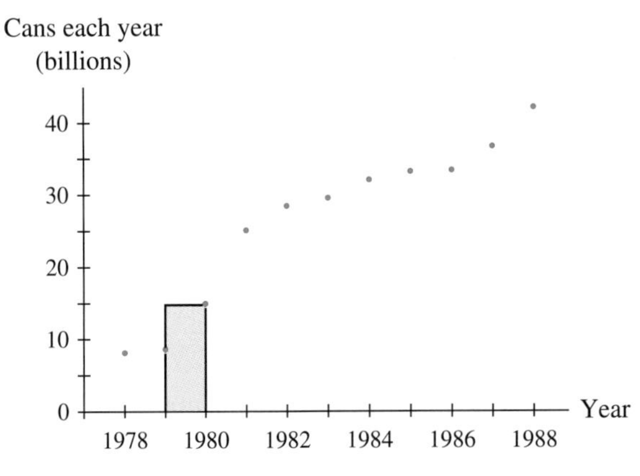

FIGURE 6.14

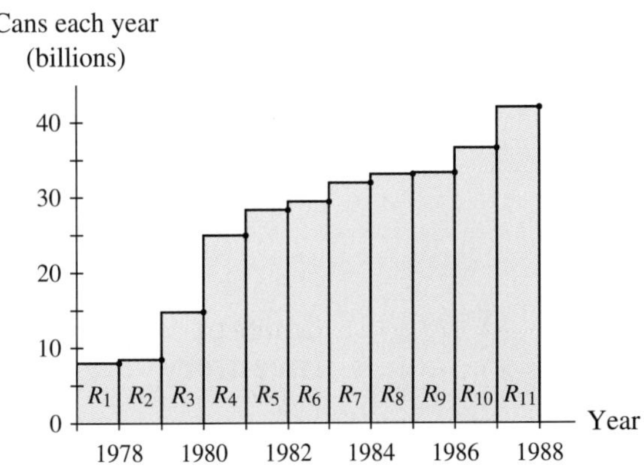

FIGURE 6.15

In other words, the number of cans recycled during 1980 is (height)(width) = (14.8 billion cans each year)(1 year) = 14.8 billion cans. If we consider the number of cans recycled during a given year as the area of a rectangle, then the number of recycled aluminum cans from the beginning of 1978 through the end of 1988 is the sum of the areas of the rectangles shown in Figure 6.15.

$$\begin{aligned}\text{Number of cans recycled} &= \text{the sum of the areas of the rectangles}\\ &= 8.0 + 8.5 + 14.8 + 24.9 + 28.3 + 29.4 +\\ &\quad\ 31.9 + 33.1 + 33.3 + 36.6 + 42.0\\ &= 290.8 \text{ billion cans}\end{aligned}$$

Note that because each rectangle has width 1 year, the area is simply the sum of the output data.

What would we do if some of the data were missing? For instance, suppose that we had data for only even-numbered years from 1978 through 1988. We could use rectangles with width 2 years, or we could find a model for the data and then construct estimates by using values determined from the equation and rectangles with width 1 year.

Students sometimes mistakenly think that more sophisticated mathematics gives more accurate results. This is not always the case, however, and count data offer a good illustration of this fact. Make sure your students understand that summing count data is more accurate than integrating a model to find accumulated change.

Most of the data recorded in statistical tables and abstracts is count data. As the aluminum can illustration indicates, calculus is not necessary to find the accumulated change in the recorded quantity. In fact, although calculus could be used to approximate an accumulated change, it is a sum of discrete data values that gives the actual change.

Count Data

Do not confuse count data with rate-of-change data. When working with rate-of-change data, accumulated change is calculated using areas. When working with count data, accumulated change is calculated by summing the output data.

The most important concept in this section is that accumulated change in a quantity can be found by finding areas between rate-of-change graphs and the horizontal axis. Sometimes we can calculate areas using geometric formulas. At other times, we approximate area using sums of areas of rectangles. The numerical calculations are often tedious and result in only approximations, but later in this chapter we will discover a valuable calculus tool that enables us to find area and determine the accumulated change in a quantity accurately and quickly.

6.1 Concept Inventory

- *Area or signed area of a region between a rate-of-change function and the horizontal axis between a and b = accumulated change in the amount function between a and b*
- *Left-rectangle approximation*
- *Right-rectangle approximation*
- *Midpoint-rectangle approximation*
- *Count data*

6.1 Activities

1. The growth rate of bacteria (in thousands per hour) in milk at room temperature is $B(t)$, where t is the number of hours that the milk has been at room temperature. We wish to use rectangles to estimate the area of the region between a graph of B and the t-axis. What are the units on
   a. the heights of the rectangles?
   b. the widths of the rectangles?
   c. the areas of the rectangles?
   d. the area of the region between the graph of B and the t-axis?
   e. the accumulated change in the number of bacteria in the milk during the first hour that the milk is at room temperature?
2. The acceleration of a car (in feet per second per second) during a test conducted by a car manufacturer is given by $A(t)$, where t is the number of seconds since the beginning of the test.
   a. What does the area of the region between the portion of the graph of A lying above the t-axis and the t-axis tell us about the car?
   b. What are the units on
      i. the heights and widths of rectangles used to estimate area?
      ii. the area of the region between the graph of A and the t-axis?
3. The distance required for a car to stop is a function of the speed of the car when the brakes are applied. The rate of change of the stopping distance could be expressed in feet per mile per hour where the input is the speed of the car, in miles per hour, when the brakes are applied.

a. What does the area of the region between the rate-of-change graph and the input axis from 40 mph to 60 mph tell us about the car?

b. What are the units on

  i. the heights and widths of rectangles used to estimate the area in part *a*?

  ii. the area in part *a*?

4. The atmospheric concentration of CO_2 is growing exponentially. If the growth rate in ppm per year is $C(t)$, where t is the number of years since 1980, what are the units on

  a. the area of the region between the graph of C and the t-axis from $t = 0$ to $t = 20$?

  b. the heights and widths of rectangles used to estimate the area in part *a*?

  c. the change in the CO_2 concentration from 1980 through 2000?

5. Figure 6.1.1 shows the graph of a function g.

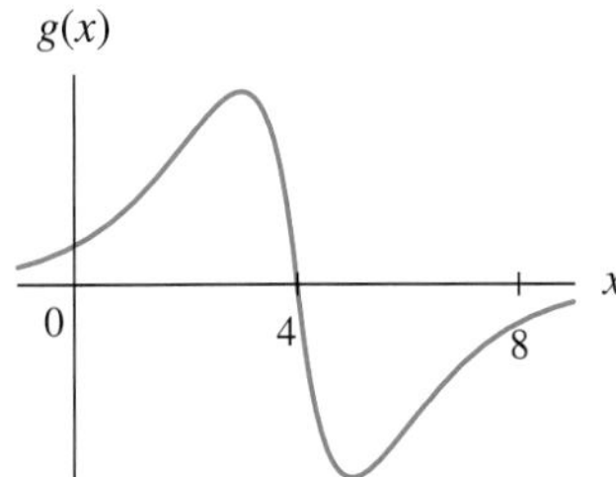

FIGURE 6.1.1

  a. Discuss how to approximate the area of the region between the graph of g and the x-axis between $x = 0$ and $x = 8$ with eight right rectangles that have the same width. Copy the graph, and draw the rectangles on the figure.

  b. Repeat part *a* using eight left rectangles.

6. The graph in Figure 6.1.2 shows the energy usage in megawatts for one day for a large university campus. The daily energy consumption for the campus is measured in megawatt-hours and is found by calculating the area of the region between the graph and the horizontal axis.

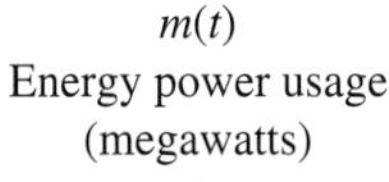

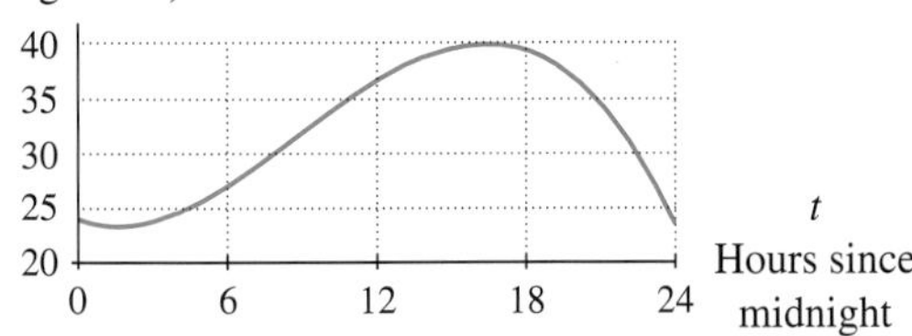

FIGURE 6.1.2

  a. Estimate the daily energy consumption using eight left rectangles.

  b. Estimate the daily energy consumption using eight right rectangles.

  c. Discuss whether the estimates in parts *b* and *c* are overestimates or underestimates of the actual daily energy consumption.

7. The graph of a function f is shown in Figure 6.1.3.

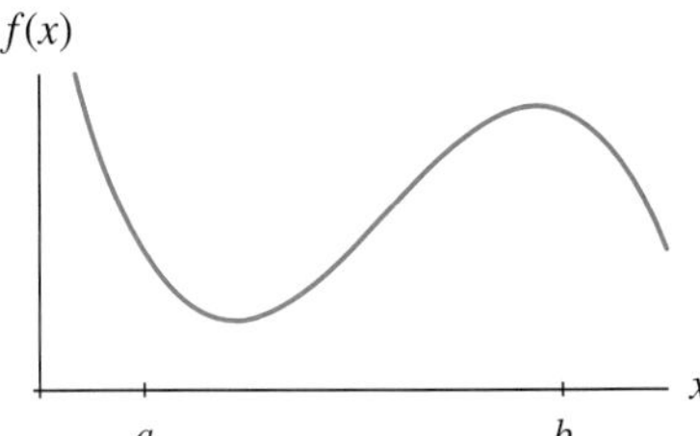

FIGURE 6.1.3

Discuss how to approximate the area of the region beneath the graph of f from $x = a$ to $x = b$ with four midpoint rectangles that have the same width. Draw the rectangles.

8. Approximate the area of the region beneath the graph of $f(x) = e^{-x^2}$ from $x = -1$ to $x = 1$ using four left rectangles, right rectangles, and midpoint rectangles.

  a. In each case,

    i. sketch the graph of f from $x = -1$ to $x = 1$.

    ii. label the points on the x-axis, and draw the rectangles.

    iii. calculate the approximating areas.

  b. Proceed as in part *a* to approximate the area of the region beneath the graph of f from $x = -1$ to $x = 1$ using eight left rectangles, eight right rectangles, and eight midpoint rectangles.

  c. The area, to nine decimal places, of the region beneath the graph of $f(x) = e^{-x^2}$ is 1.493648266. Which approximation is the most accurate?

9. The graph in Figure 6.1.4 appears in *Ecology of Natural Resources* by François Ramade (New York: Wiley, 1984).[4] It shows two estimates, labeled A and B, of oil production rates (in billions of barrels per year).

4. Figure 6.1.4 adapted from *Ecology of Naural Resources* by François Remade. Copyright 1984 by John Wiley & Sons, Inc. Reprinted by permission of the publisher.

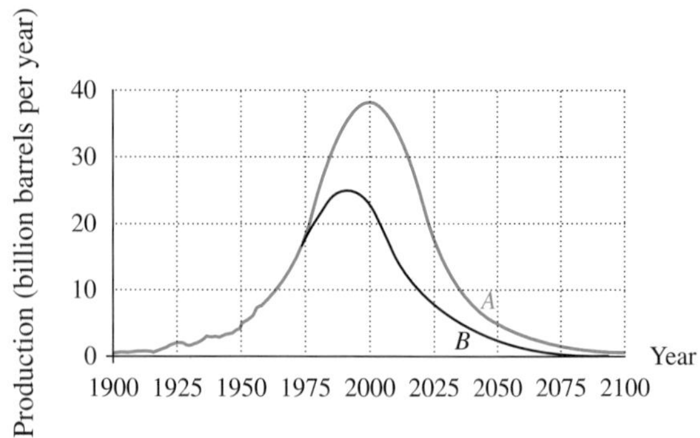

FIGURE 6.1.4

a. Use midpoint rectangles of width 25 years to estimate the total amount of oil produced from 1900 through 2100 using graph *A*.

b. Repeat part *a* for graph *B*.

c. On page 31 of *Ecology of Natural Resources*, the total oil production is estimated from graph *A* to be 2100 billion barrels and from graph *B* to be 1350 billion barrels. How close were your estimates?

d. What can you conclude from the graph about the future of oil production?

10. Scientists have long been interested in studying global climatological changes and the effect of such changes on many aspects of the environment. From carefully controlled experiments, two scientists[5] constructed a model to simulate daily snow depth in a region of the Northwest Territories in Canada. Rates of change (in equivalent centimeters of water per day) estimated from their model are shown as a scatter plot in Figure 6.1.5. Appropriate models have been sketched on the scatter plot. Note that both vertical and horizontal scales change after June 9.

   a. What does the figure indicate occurred between June 9 and June 11?

   b. Estimate the area of the region beneath the curve from April 1 through June 9. Interpret your answer.

   c. Use four midpoint rectangles to estimate the area of the region from June 11 through June 15. Interpret your answer.

11. A mechanical engineering graduate student designed a robot and is testing the ability of the robot to accelerate, decelerate, and maintain speed. The robot takes 1 minute to accelerate to 10 miles per hour (880 feet per minute). The robot maintains that speed for 2 minutes and then takes half a minute to come to a complete stop. Assume that this robot's acceleration and deceleration are constant.

   a. Draw a graph of the robot's speed during the experiment.

   b. Find the area of the region between the graph in part *a* and the horizontal axis.

   c. What is the practical interpretation of the area found in part *b*?

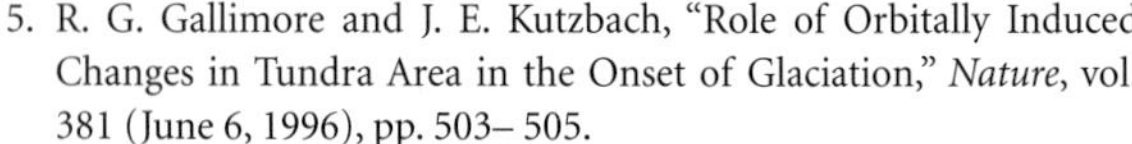

5. R. G. Gallimore and J. E. Kutzbach, "Role of Orbitally Induced Changes in Tundra Area in the Onset of Glaciation," *Nature*, vol. 381 (June 6, 1996), pp. 503–505.

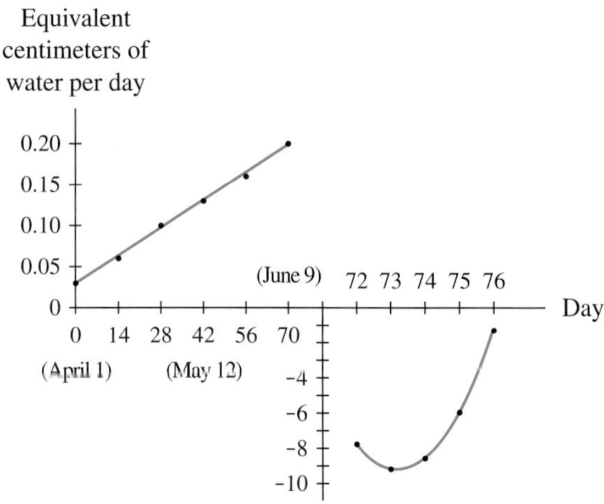

FIGURE 6.1.5

12. A certain gas expands as it is heated. Figure 6.1.6 shows the rate of expansion of the gas measured at several temperatures.

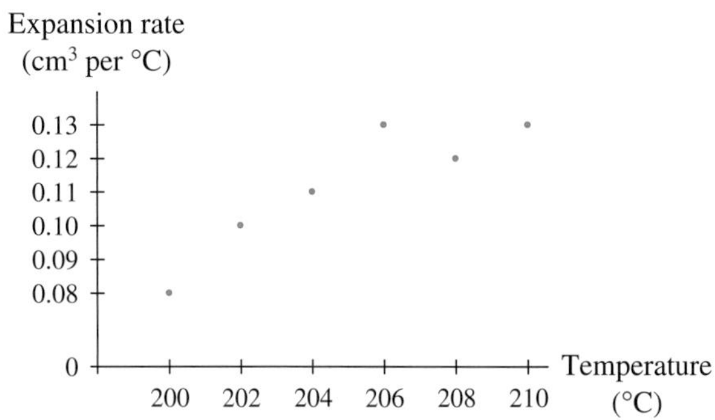

FIGURE 6.1.6

   a. Use only the data and left rectangles to estimate by how much the gas expanded as it was heated from 200°C to 210°C. Sketch the rectangles on the scatter plot. Do you believe your approximation is an overestimate or an underestimate? Explain.

b. Repeat part *a* using right rectangles.

13. An athlete is training for a mountain bike race. Her coach clocks her speed every 10 minutes during an hour-long ride. The speeds are shown in Table 6.10.

TABLE 6.10

Time (minutes)	0	10	20	30	40	50	60
Speed (mph)	22	18	20	23	15	17	12

a. Use only the data and six left rectangles to estimate the distance traveled by the cyclist during the 1-hour ride. Sketch a scatter plot of the data, and draw the rectangles on the scatter plot.

b. Repeat part *a* using right rectangles.

14. On January 4, 2000, DuPont stock was worth \$65 per share.

a. Write a function for the value of x shares of stock. Graph this function. Note that this is a continuous model representing a discrete situation.

b. Write the function for the rate of change of the continuous model for the value of DuPont stock with respect to the number of shares held. Graph this rate-of-change function.

c. Find the change in the value of stock held if the number of shares held is increased from 250 to 300 shares. Depict this change as the area of a region on the rate-of-change graph.

15. The rate of change of the population[6] of North Dakota from 1970 through 1990 can be modeled as

$$p(t) = \begin{cases} \text{3.87 thousand people per year} & \text{when } 0 \le t < 15 \\ \text{-7.39 thousand people per year} & \text{when } 15 < t \le 20 \end{cases}$$

where t represents the number of years since 1970.

a. Sketch a graph of the rate-of-change function.

b. Find the area of the region between the graph of p and the horizontal axis from 0 to 15. Interpret your answer.

c. Find the area of the region between the graph of p and the horizontal axis from 15 to 20. Interpret your answer.

d. Was the population of North Dakota in 1990 greater or less than the population in 1970? By how much did the population change between 1970 and 1990?

e. What information would you need to determine the population of North Dakota in 1990?

16. The rate of change of the per capita consumption of cottage cheese[7] in the United States between 1980 and 1996 can be modeled by the function

$$c(t) = \begin{cases} -0.10t - 0.058 \text{ pounds per person per year} & \text{when } 0 \le t < 13 \\ -0.1 \text{ pound per person per year} & \text{when } 13 < t \le 16 \end{cases}$$

where t is the number of years since 1980.

a. Sketch a graph of the rate-of-change function.

b. Find the area of the region between the graph and the horizontal axis between $t = 0$ and $t = 16$.

c. Interpret the area in part *b* in the context of cottage cheese consumption.

d. Can you determine the per capita cottage cheese consumption in 1996? Why or why not?

17. The rate of change of the length of the average hospital stay[8] between 1980 and 1996 can be modeled by the equation

$$s(t) = \begin{cases} 0.028t - 0.23 \text{ days per year} & \text{when } 0 \le t < 10 \\ -0.0408t + 0.30833 \text{ days per year} & \text{when } 10 < t \le 16 \end{cases}$$

where t is the number of years since 1980.

a. Graph s for the years between 1980 and 1996.

b. Judging on the basis of the graph, was the length of the average hospital stay increasing or decreasing between 1980 and 1996?

c. Find the area of the region lying above the axis between the graph and the t-axis.

d. Find the area of the region lying below the axis between the graph and the t-axis.

6. Based on data from *Statistical Abstract*, 1994.

7. Based on data from *Statistical Abstract*, 1998.

8. Ibid.

e. Judging on the basis of your answers to parts *c* and *d*, by how much did the average hospital stay change between 1980 and 1996? Can you determine the average stay in 1996? Why or why not?

18. The data in Table 6.11 show the marginal cost for compact disc production at the indicated hourly production levels:

TABLE 6.11

Production (CDs per hour)	Marginal cost
100	$5
150	$3.50
200	$2.50
250	$2
300	$1.60

a. What are the units on the marginal cost?

b. Use four left rectangles to approximate the change in cost when production is increased from 100 to 300 CDs per hour.

c. Find a model for the data, and sketch a graph of the model for production levels from 100 to 300 CDs per hour.

d. Repeat part *b* using eight left rectangles. Why is a model necessary when you are using eight rectangles?

e. Sketch the rectangles used in part *d* on a graph of the model. Is the approximation larger or smaller than the area of the region between the model and the horizontal axis from 100 to 300?

19. The number of live births[9] each year from 1950 through 1995 in the United States to women 45 years of age and older is given in Table 6.12.

TABLE 6.12

Year	Births	Year	Births
1950	5322	1985	1162
1955	5430	1986	1257
1960	5182	1987	1375
1965	4614	1988	1427
1970	3146	1990	1638
1975	1628	1995	2727
1980	1200		

9. *Statistical Abstract*, 1998.

a. Draw a scatter plot of the data, and sketch right rectangles depicting the total number of live births from the beginning of 1985 through 1988. Use the data to find the total number of live births for this period. Is this number an approximation or exact? Explain.

b. Find a model for the data. Sketch a graph of this function.

c. Estimate the number of such births from the beginning of 1965 through the end of 1995 by using right rectangles with a width of 1 year. How many rectangles are needed?

d. What would be required to find the exact total number of such births from the beginning of 1965 through 1995?

20. During a summer thunderstorm, the temperature drops and then rises again. The rate of change of the temperature during the hour and a half after the storm began is given by

$$T(h) = 9.48h^3 - 15.49h^2 + 17.38h - 9.87 \text{ °F per hour}$$

where h is the number of hours since the storm began.

a. Graph the function T from $h = 0$ to $h = 1.5$. Find the point at which the graph crosses the horizontal axis.

b. Consider the portion of the graph of T lying below the horizontal axis. What does the area of the region between this portion of the graph of T and the horizontal axis represent?

c. What does the area of the region lying above the axis represent?

d. Consider the graph between 0 and 1.5 hours. Use seven right rectangles to approximate the area of the region lying below the axis from 0 to 1.5 hours.

e. Repeat part *d* for the region lying above the axis.

f. According to the model, one and a half hours after the storm began, was the temperature higher, lower, or the same as the temperature at the beginning of the storm? If it was higher or lower, by how many degrees?

21. Table 6.13 shows air speed recorded from a Cessna 172. Air speed is the speed at which the air flows into a Pitot tube located on the nose or wing of an aircraft and is measured in nautical miles per hour (knots). The data were recorded for 25 seconds after the plane began to taxi for takeoff.

TABLE 6.13

Time (seconds)	Air speed (knots)
0	0
10	40
15	60
20	70
25	80

a. Convert knots to miles per second, using the fact that 1 knot $\approx$ 1.15 mph.

b. Find a model for the converted data.

c. Estimate the area of the region between a graph of the function and the horizontal axis from 0 seconds to 18 seconds.

d. If takeoff occurred at 18 seconds, interpret your estimate in part *c* in the context of the Cessna 172 data.

22. As the 76 million Americans born between 1946 and 1964 (the "baby boomers") continue to age, the United States will see an increasing proportion of Americans who are at a retirement age of 65. Table 6.14 shows past and projected rates of change for the number of U.S. citizens 65 years of age or older. These are instantaneous rates of change measured at the end of each year based on data from the U.S. Census Bureau.

TABLE 6.14

Year	Rate of change of 65+ population (millions per year)
1940	0.21
1960	0.32
1980	0.50
2000	0.76
2010	0.94
2020	1.17
2040	1.45

a. Find a model for the rate-of-change data.

b. Use the equation and ten midpoint rectangles to estimate the change in the population 65 years of age and older from the end of 2000 through the end of 2005.

23. Using data from the 1998 *Statistical Abstract*, we model the rate of change of the percentage of the U.S. population living in New England from 1970 through 1997 as

$$P(t) = (-1.970 \cdot 10^{-4})t^2 + 0.0043t - 0.0484$$
percentage points per year

where t is the number of years since 1970.

a. Sketch a graph of P from $t = 0$ to $t = 27$.

b. What does the fact that the graph of P lies below the t-axis from $t = 0$ to $t = 27$ tell you about the percentage of the population living in New England?

c. Use five midpoint rectangles to estimate the area of the region between the graph of P and the t-axis from $t = 10$ to $t = 20$. Interpret your answer.

d. What was the population of New England in 1997?

24. The rate of change of the projected total assets in the Social Security trust fund[10] for the years 2000 through 2033 can be modeled by the equation

$$S(x) = -0.4569x^2 + 1.8528x + 108.7241$$
billion dollars per year

x years after 2000.

a. Graph S between 2000 and 2033.

b. According to the graph of S, when will the trust fund assets be growing and when will they be declining?

c. Find the point on the graph of S that corresponds to the time when the amount in the trust fund will be greatest.

d. Estimate using 10 intervals the area lying above the axis and below the graph of S. Interpret your answer.

e. Estimate using 10 intervals the area lying below the axis and above the graph of S. Interpret your answer.

f. By how much will the trust fund amount change between 2000 and 2033? What information do we need to determine how much money is projected to be in the trust fund in 2033?

10. Based on data from the Social Security Administration.

25. Life expectancies in the United States are always rising because of advances in health care, increased education, and other factors. The rate of change (measured at the end of each year) of life expectancies[11] for women in the United States between 1970 and 2010 are shown in Table 6.15.

TABLE 6.15

Year	Years of life expectancy per year
1970	0.36
1975	0.26
1980	0.18
1985	0.13
1990	0.09
1995	0.08
2000	0.08*
2005	0.11*
2010	0.16*

*Projected

a. Look at a scatter plot of the data. Does the fact that the data are declining from 1970 through 1995 contradict the statement that life expectancies are always rising? Explain.

b. Find a model for the data.

c. Use eight midpoint rectangles to estimate the change in the life expectancy for women from 1970 through 2010.

26. Accelerations for a vehicle during a road test are approximated in Table 6.16.

TABLE 6.16

Time (seconds)	Acceleration (feet per second squared)
0	22.6
2	18.2
4	14.5
6	11.4
8	8.9
10	7.1
12	5.9

a. Find a model for the data.

b. Use ten midpoint rectangles to estimate the area of the region between the graph of your model and the input axis from 0 to 13.5 seconds. Interpret your answer.

c. Convert your answer in part b to miles per hour. The actual speed of the car after 13.5 seconds was 107 mph. How close is your answer to this speed?

27. The rate of change of the level of Lake Tahoe d days after September 30, 1995, can be modeled[12] by

$$r(d) = (-1.6035 \cdot 10^{-6})d^2 + (5.086 \cdot 10^{-4})d - 0.0192$$
feet per day

The graph of the model is shown in Figure 6.1.7.

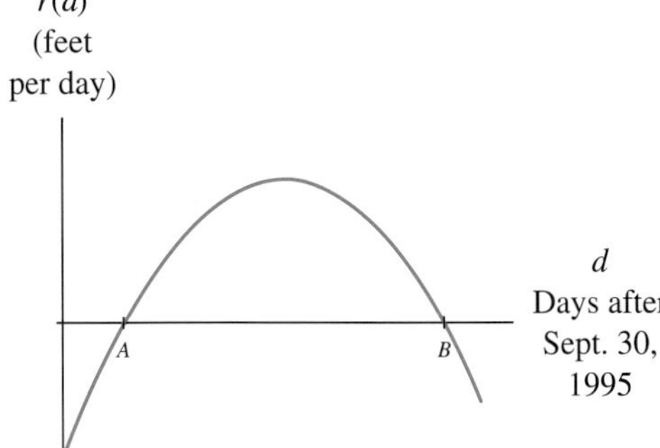

FIGURE 6.1.7

a. Find the points labeled A and B on the graph.

b. Use five midpoint rectangles to estimate the area between the graph of r and the horizontal axis from 0 to A. Interpret your answer.

c. Use ten midpoint rectangles to estimate the area between the graph of r and the horizontal axis from A to B. Interpret your answer.

d. Was the lake level higher or lower B days after September 30, 1995 than it was on that date? How much higher or lower was the lake level B days after September 30, 1995?

28. The rate of change in the percentage of southern Australian grasshopper eggs that hatch as a function of temperature (for temperatures between 7°C and 25°C) can be modeled[13] by the equation

$$p(t) = -0.0258t^3 + 1.464t^2 - 25.982t + 136.560$$
percentage points per degree

where t is the temperature in °C. A graph of p is shown in Figure 6.1.8.

11. Based on data from *Statistical Abstract*, 1998.

12. Based on data from the Federal Watermaster, U.S. Department of the Interior.

13. Based on information in George L. Clark, *Elements of Ecology* (New York: Wiley, 1954), p. 170.

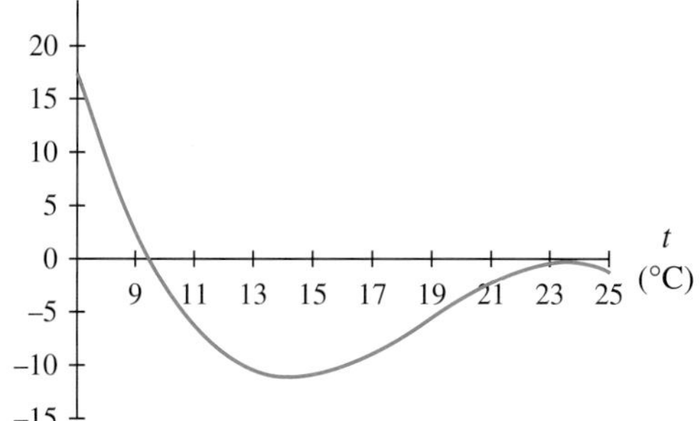

FIGURE 6.1.8

a. Find the point at which the graph of p crosses the horizontal axis.

b. Use four midpoint rectangles to estimate the area of the region between the graph of p and the horizontal axis from $t = 7$ to the point at which the curve crosses the horizontal axis. Interpret your answer.

c. Use eight midpoint rectangles to estimate the area of the region between the graph of p and the horizontal axis from the point at which the curve crosses the horizontal axis to $t = 25$. Interpret your answer.

d. Estimate the difference between the percentage of grasshopper eggs that hatch at 25°C and the percentage that hatch at 7°C.

29. Table 6.17 shows the number of new female Ph.D.s in computer science[14] in selected years from 1970 through 1993.

TABLE 6.17

Year	New female Ph.D.s	Year	New female Ph.D.s
1970	1	1987	51
1973	7	1988	60
1976	14	1989	87
1979	24	1990	97
1982	27	1991	113
1985	32	1992	108
1986	50	1993	126

a. Draw a scatter plot of the data, and sketch rectangles depicting the number of new female Ph.D.s in computer science from the beginning of 1985 through 1993. Find the total number of new female Ph.D.s from the beginning of 1985 through 1993. Is this number an approximation or exact?

b. Find a model for the data.

c. Use the model to estimate the total number of new female Ph.D.s in computer science from the beginning of 1970 through 1993.

d. The total number of new female Ph.D.s in computer science from the beginning of 1970 through 1993 was 987. How does your estimate from part *c* compare to this number?

30. The 1998 *Statistical Abstract* gives the number of international telephone calls (in millions) billed in the United States as shown in Table 6.18.

TABLE 6.18

Year	Calls (millions)	Year	Calls (millions)
1970	23	1992	1643
1980	200	1993	1926
1985	411	1994	2313
1989	835	1995	2821
1990	984	1996	3485
1991	1371		

a. Draw a scatter plot of the data, and sketch rectangles of width 1 year depicting the number of overseas calls from the beginning of 1989 through 1996. Find the total number of overseas calls from the beginning of 1989 through 1996. Is this an approximation or the exact number?

b. Find a model for the data.

c. Use your model to estimate the number of overseas calls from the beginning of 1970 through 1996.

31. In essay form (using pictures as necessary), explain why the approximation to the area of a region under a curve and above the horizontal axis generally improves when twice as many rectangles are used.

14. *Computing Research News*, January 1994.

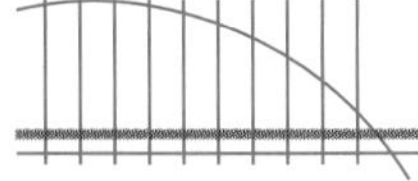

6.2 The Definite Integral as a Limit of Sums

In Section 6.1 we saw that the accumulated change in a quantity can be interpreted in terms of areas of regions between the graph of a rate-of-change function for that quantity and the horizontal axis. We approximated areas of regions between a curve and the horizontal axis by using rectangles. In the Activities, you should have noticed that the approximations became closer to the actual area of the region when twice as many intervals were used. What would you expect to happen if you were to use four, eight, or even one hundred times as many intervals?

We return to Example 2 in Section 6.1, where we considered the rate of change of the concentration of a drug in the bloodstream. Recall that we modeled the rate of change of the drug concentration for the first 20 days as

$$r(x) = 1.708(0.845^x) \text{ µg/mL/day}$$

where x is the number of days after the drug was first administered. We saw that we could estimate the change in the concentration of the drug between day 0 and day 20 as the area between the graph of r and the horizontal axis from $x = 0$ to $x = 20$. We then used ten right rectangles to estimate this area (see Figure 6.16) to obtain an estimate of 8.24 µg/mL.

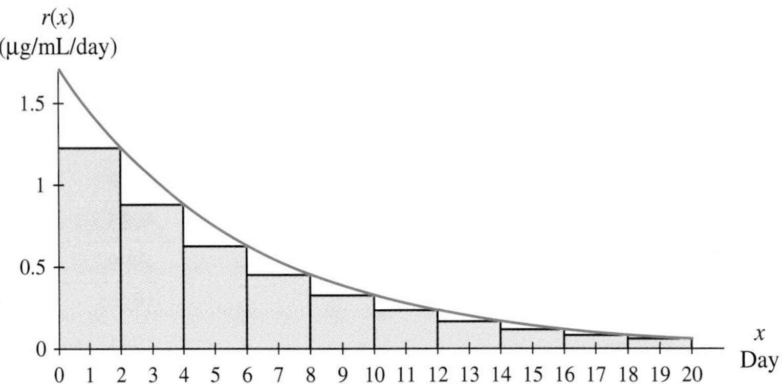

FIGURE 6.16

We know that we obtain a better estimate of this area by using midpoint rectangles, as shown in Figure 6.17.

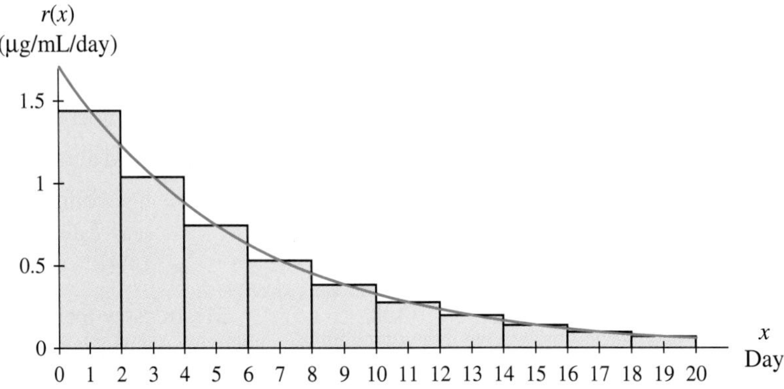

FIGURE 6.17

> From now on, whenever we speak of approximating areas using rectangles, we will use midpoint rectangles because they generally give the best approximations.

An estimate of the area between the graph of r and the horizontal axis from $x = 0$ to $x = 20$ derived by using ten midpoint rectangles is approximately 9.746 μg/mL. From this estimate of the area we can say that the drug concentration increased by approximately 9.75 μg/mL during the first 20 days of the test.

Finding a Trend

We can improve this estimate of area by using 20 rectangles instead of 10. What do you think will happen to the accuracy of the approximations if we were to use even more rectangles? Figures 6.18 through 6.21 show approximations with 10, 20, 40, and 80 rectangles, respectively.

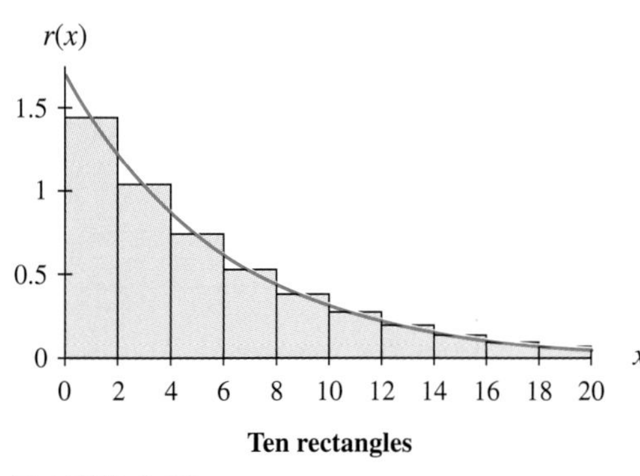

FIGURE 6.18

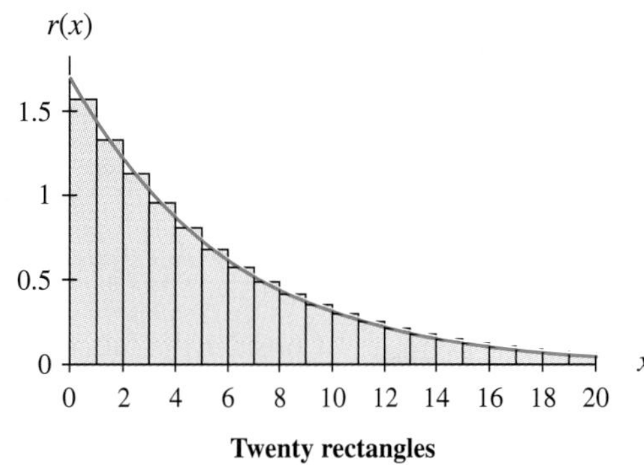

FIGURE 6.19

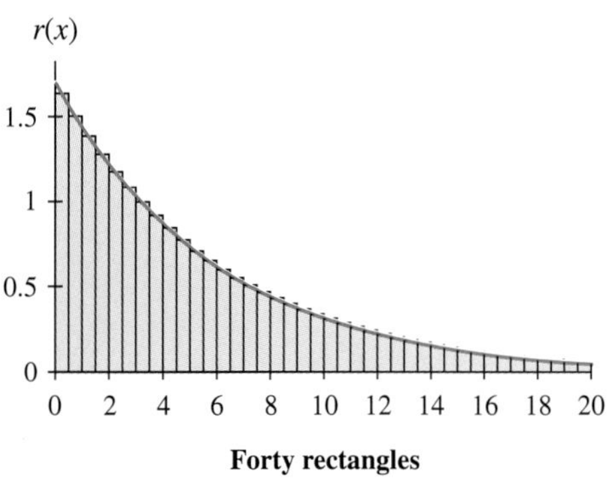

FIGURE 6.20

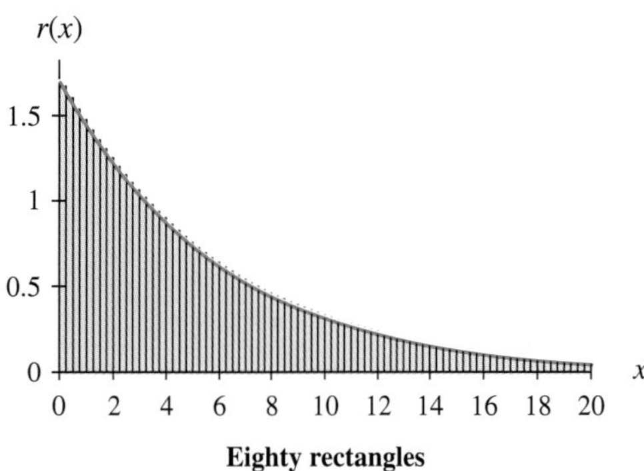

FIGURE 6.21

By carefully inspecting the graphs in these figures, you can see that as we use more rectangles, the shaded region accounted for by the rectangles more closely approximates the region of interest. This fact will help us approximate the area. If we make a table of the approximations, we may be able to recognize a *trend*—that is, a value to which the approximations seem to be getting closer and closer as the number of rectangles becomes larger and larger.

As we did in the Section 4.1 slope tables, we have underlined some of the digits in Table 6.19. We underline the digits that give one decimal position beyond our desired accuracy once we see that these digits do not change. Once we have three underlined area values, we stop calculating area approximations and report a trend. (You may wish to establish a different rule for your students.)

TABLE 6.19

Number of rectangles	Approximation of area
10	9.7459
20	9.7805
40	9.7892
80	9.7913
160	9.7919
320	9.7920
640	9.7921
1280	9.7921
Trend ≈ 9.79	

Students often forget to give units when approximating areas by using a limit of sums of rectangle areas.

The trend evident from Table 6.19 indicates that the area under consideration is approximately 9.79. Interpreting this area in context tells us that the drug concentration increased by approximately 9.79 μg/mL during the first 20 days of the test. It is important to note that we cannot conclude that the drug concentration after 20 days was 9.79 μg/mL. This is true only if the initial drug concentration was zero.

EXAMPLE 1 *A More Accurate Area Approximation*

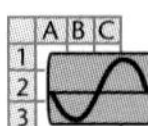

Consider the function $f(x) = \sqrt{1 - x^2}$ from $x = 0$ to $x = 1$. Approximate to three decimal places the area between this curve and the horizontal axis between $x = 0$ and $x = 1$, beginning with four midpoint rectangles and doubling the number each time until a trend is observed.

Solution Numerically, we observe the midpoint-rectangle approximations approaching a trend in Table 6.20.

TABLE 6.20

Number of rectangles	Approximation of area
4	0.79598
8	0.78917
16	0.78674
32	0.78587
64	0.78557
128	0.78546
256	0.78542
512	0.78541
Trend ≈ 0.785	

We can be fairly confident that the trend to three decimal places is 0.785, because the value through the fourth decimal position has remained constant in several approximations.

Because the region is a quarter-circle with radius 1, the exact area of this region is $\frac{\pi}{4}$, which is approximately 0.7853982. When $n = 512$, the difference in the true and approximate values is about 0.0000074.

Area Beneath a Curve

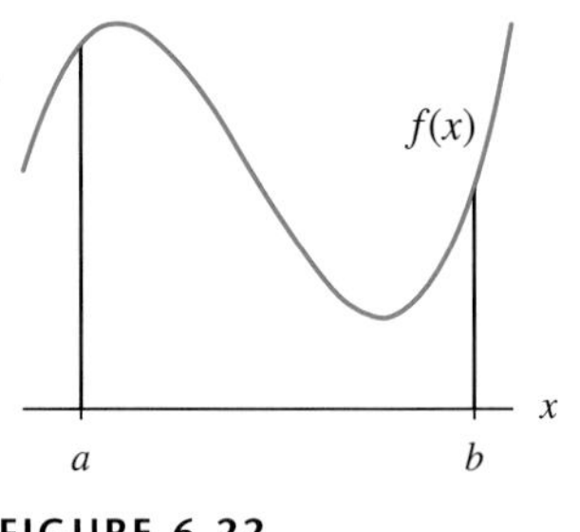

FIGURE 6.22

Because the more rectangles we use to approximate area, the better we expect the approximation to be, we are led to consider area as the limiting value of the sums of areas of approximating midpoint rectangles as the number of rectangles increases without bound. Let f be a function that is continuous and non-negative over the interval from a to b. (See Figure 6.22.) Partition the interval from a to b into n subintervals of equal length $\Delta x = \frac{b-a}{n}$, and on each subinterval construct a rectangle of width Δx whose height is given by the value of f at the midpoint of the subinterval. Figures 6.23 through 6.26 show the rectangles when $n = 4, 8, 16$, and 32.

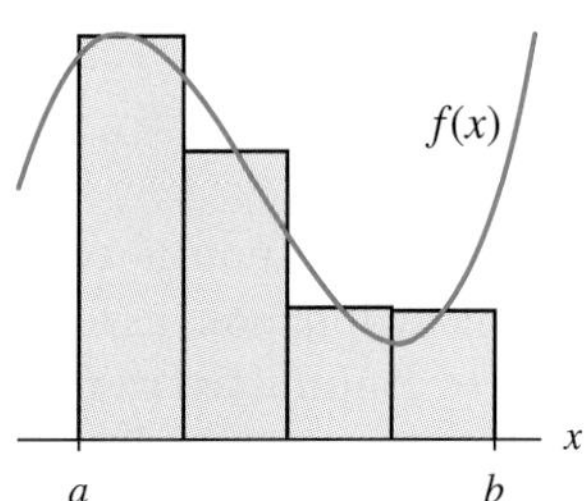

Four rectangles

FIGURE 6.23

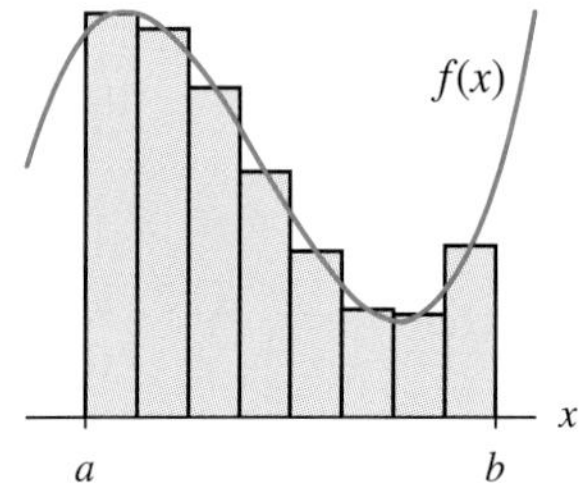

Eight rectangles

FIGURE 6.24

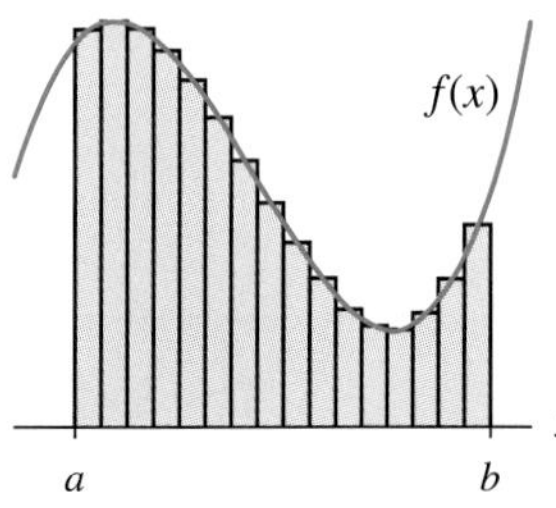

Sixteen rectangles

FIGURE 6.25

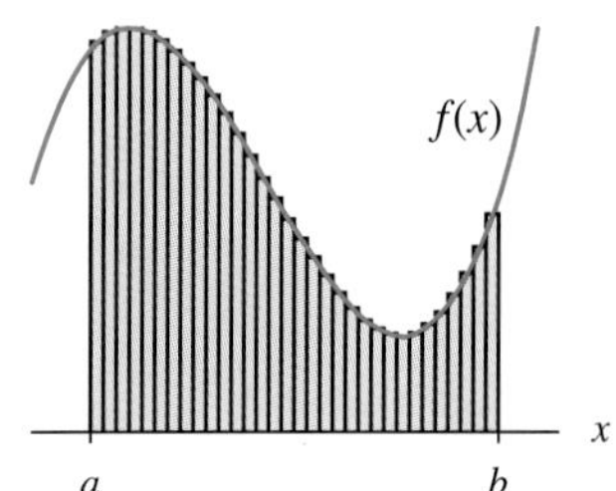

Thirty-two rectangles

FIGURE 6.26

The heights of the rectangles are given by the values

$$f(x_1), f(x_2), \ldots, f(x_n)$$

where $x_1, x_2, \ldots, x_n$ are the midpoints of the subintervals. Each rectangle has width Δx, so the areas of the rectangles are given by the values

$$f(x_1)\Delta x, f(x_2)\Delta x, \ldots, f(x_n)\Delta x$$

and the sum $[f(x_1) + f(x_2) + \cdots + f(x_n)]\Delta x$ is an approximation to the area of the region between the the graph of f and the x-axis from a to b. As our examples have shown, the approximations generally improve as n increases. In mathematical terms, the area of the region between the graph of f and the x-axis from a to b is given by a limit of sums as n gets larger and larger:

$$\text{Area} = \lim_{n\to\infty}[f(x_1) + f(x_2) + \cdots + f(x_n)]\Delta x$$

Area Beneath a Curve

Let f be a continuous or a piecewise continuous non-negative function from a to b. The area of the region between the graph of f and the x-axis from a to b is given by the limit

$$\text{Area} = \lim_{n\to\infty}[f(x_1) + f(x_2) + \cdots + f(x_n)]\Delta x$$

where $x_1, x_2, \ldots, x_n$ are the midpoints of n subintervals of length $\Delta x = \frac{b-a}{n}$ between a and b.

The Definite Integral

The limit in the area definition is central to the development of calculus and is applied to more general functions as follows: Given an arbitrary function that is continuous or piecewise continuous over an interval from a to b, partition the interval into n subintervals of length $\Delta x = \frac{b-a}{n}$ and consider the sum

$$[f(x_1) + f(x_2) + \cdots + f(x_n)]\Delta x$$

Here, the values $x_1, x_2, \ldots, x_n$ are the midpoints of the n subintervals. The limit

$$\lim_{n\to\infty}[f(x_1) + f(x_2) + \cdots + f(x_n)]\Delta x$$

is called the **definite integral** of f from a to b and is represented by

$$\int_a^b f(x)\,dx$$

The Definite Integral

Let f be a continuous or piecewise continuous function from a to b. The definite integral of f from a to b is given by the limit

$$\int_a^b f(x)\,dx = \lim_{n\to\infty}[f(x_1) + f(x_2) + \cdots + f(x_n)]\Delta x$$

where $x_1, x_2, \ldots, x_n$ are the midpoints of n subintervals of length $\Delta x = \frac{b-a}{n}$ between a and b.

The integral sign $\int$ resembles an elongated S and reminds us that we are taking a limit of sums. The values a and b identify the input interval, f is the function, and the symbol dx reminds us of the width Δx of each subinterval. If a graph of f lies above the x-axis between a and b, then the definite integral is the area of the region between the graph and the x-axis from a to b.

EXAMPLE 2 *Rising River*

In Example 3 of Section 6.1, we saw flow rates of the Carson River modeled by the equation

$$f(x) = 18{,}225x^2 - 135{,}334.3x + 2{,}881{,}542.9 \text{ ft}^3\text{/hr}$$

where x is the number of hours after 11:45 a.m. Wednesday. Use the idea of a limit of sums to estimate to the nearest ten thousand cubic feet the amount of water that flowed through the Carson River (past the point where the measurements were taken) from 11:45 a.m. Wednesday to 7:45 a.m. Thursday.

Solution The amount of water that flowed through the Carson River from 11:45 a.m. Wednesday to 7:45 a.m. Thursday can be estimated by the area between the graph of f and the horizontal axis from $x = 0$ to $x = 20$. We begin with 10 midpoint rectangles, doubling the number each time until we are confident that we know the limiting value.

TABLE 6.21

Number of rectangles	Approximation of area
10	79,042,498
20	79,133,623
40	79,156,404
80	79,162,100
160	79,163,523
320	79,163,879
640	79,163,968
Trend ≈ 79,160,000	

It appears that, according to the model, to the nearest ten thousand cubic feet the amount of water flowing through the Carson River during the 20 hours after 11:45 a.m. Wednesday was 79,160,000 ft^3. Using definite integral notation, we write

$$\int_0^{20} (18{,}225x^2 - 135{,}334.3x + 2{,}881{,}542.9)\, dx \approx 79{,}160{,}000 \text{ ft}^3$$

Interpretation of $\int_a^b f(x)\, dx$

When f is continuous or piecewise continuous and non-negative from a to b, then the integral $\int_a^b f(x)\, dx$ can be interpreted as the area of the region R between the graph of f and the x-axis from a to b. (See Figure 6.27.)

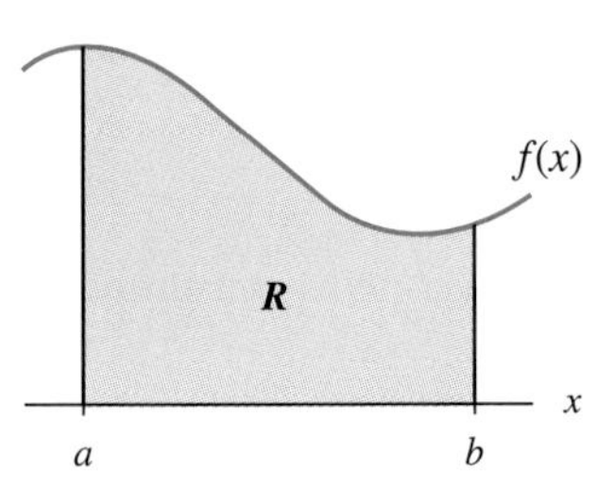

FIGURE 6.27

What interpretation can we give to the integral $\int_a^b f(x)\,dx$ if f is negative from a to b? In this case, the integral represents the signed area of the region R between the graph of f and the horizontal axis from a to b with the negative sign indicating that the region lies below the horizontal axis.

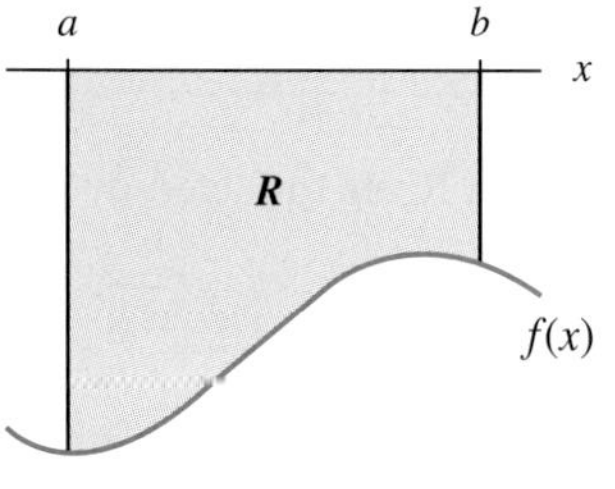

Signed area of $R = \int_a^b f(x)\,dx$

FIGURE 6.28

> When $f(x)$ is negative from a to b, the integral $\int_a^b f(x)\,dx$ is the *negative* of the area of region R. (See Figure 6.28.)

If the function f is sometimes positive and sometimes negative between a and b, then the integral $\int_a^b f(x)\,dx$ is equal to the area of the region lying under the graph of f and above the x-axis minus the area of the region lying above the graph of f and below the x-axis. (See Figure 6.29.)

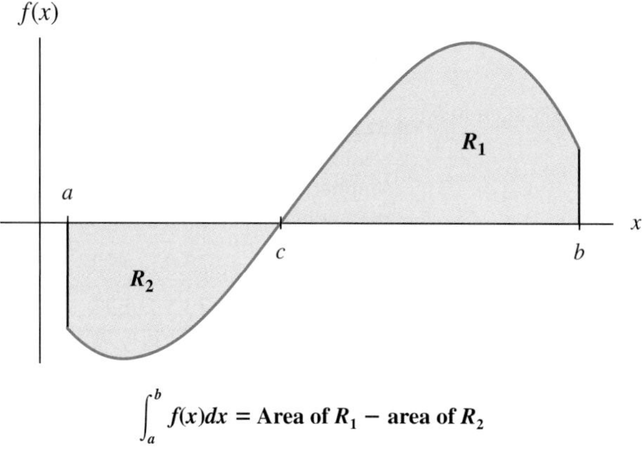

$\int_a^b f(x)dx$ **= Area of R_1 − area of R_2**

FIGURE 6.29

EXAMPLE 3 *Wine Consumption*

The rate of change of the per capita consumption of wine in the United States from 1970 through 1990 can be modeled[15] as

$$W(x) = (1.243 \cdot 10^{-4})x^3 - 0.0314x^2 + 2.6174x - 71.977 \text{ gallons per person per year}$$

where x is the number of years since the end of 1900. A graph of the function is shown in Figure 6.30.

a. Find the input value of the point labeled A.

b. From 1970 through 1990, according to the model, when was wine consumption increasing and when was it decreasing?

15. Based on data from *Statistical Abstract*, 1994.

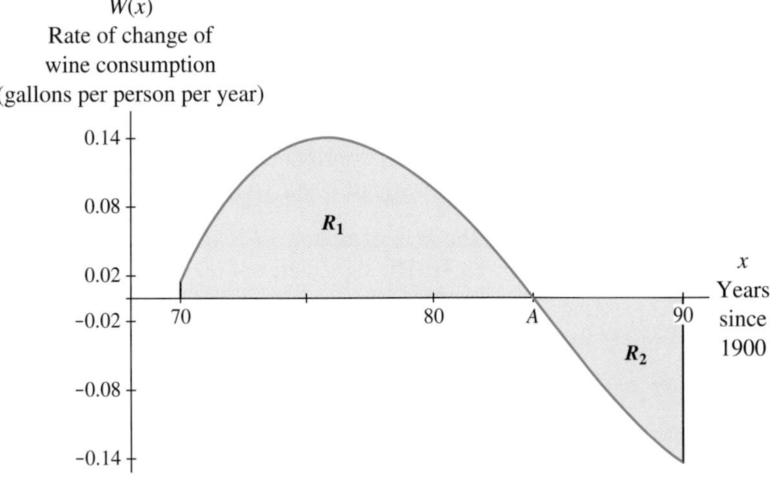

FIGURE 6.30

c. Use a limiting value of sums to estimate the areas, to two decimal places, of the regions labeled R_1 and R_2. Interpret your answers.

d. According to the model, what was the change in the per capita consumption of wine from the end of 1970 through 1990?

e. Find the values of $\int_{70}^{83.97} W(x)\,dx$, $\int_{83.97}^{90} W(x)\,dx$, and $\int_{70}^{90} W(x)\,dx$.

f. What was the per capita wine consumption in 1990?

TABLE 6.22

Number of rectangles	Approximation of area of region R_1
5	1.35966
10	1.34130
20	1.33671
40	1.33556
80	1.33527
160	1.33520
Trend ≈ 1.34	

TABLE 6.23

Number of rectangles	Approximation of signed area of region R_2
5	-0.44927
10	-0.44870
20	-0.44856
40	-0.44852
Trend ≈ -0.45	

Solution

a. Solving $W(x) = 0$ gives $A \approx 83.97$, corresponding to the end of 1984.

b. Wine consumption was increasing where the rate-of-change graph is positive—from 1970 through 1984 ($x \approx 83.97$)—and was decreasing from 1984 through 1990, where the rate-of-change graph is negative.

c. In order to find the areas of the regions R_1 and R_2, we must know the lower limit and upper limit (that is, the endpoints) of each region. Because $W(x) = 0$ when $x \approx 83.97$, we use lower limit $x = 70$ and upper limit $x \approx 83.97$ for R_1 and lower limit $x \approx 83.97$ and upper limit $x = 90$ for R_2.

The area of region R_1 is determined by examining sums of areas of midpoint rectangles for an increasing number of subintervals until a trend is observed, as shown in Table 6.22. The area of region R_1 is approximately 1.34 gallons per person. This indicates that wine consumption increased by approximately 1.34 gallons per person from the end of 1970 through 1984 ($x \approx 83.97$).

We calculate the signed area of region R_2 in a similar way as shown in Table 6.23. The area of region R_2 is approximately 0.45 gallon per person. This is an estimate for the decrease in per capita wine consumption from 1984 ($x \approx 83.97$) through 1990.

d. To determine the net change in the per capita consumption of wine from the end of 1970 through 1990, we subtract the decrease (area of R_2) from the increase (area of R_1).

Net change $= 1.34 - 0.45 = 0.89$ gallon per person

We see an approximate net increase of 0.89 gallon per person from the end of 1970 through 1990.

e. The area of region R_1 is the value of the definite integral of $W(x)$ from 70 to 83.97. The signed area of region R_2 is the value of the definite integral from 83.97 to 90, and the answer to part *d* is the value of the definite integral from 70 to 90.

$$\int_{70}^{83.97} W(x)\,dx \approx 1.34 \text{ gallons per person}$$

$$\int_{83.97}^{90} W(x)\,dx \approx -0.45 \text{ gallon per person}$$

$$\int_{70}^{90} W(x)\,dx \approx 0.89 \text{ gallon per person}$$

f. The definite integral tells us the change in a quantity over an interval, not the value of the quantity at the endpoint of the interval. Thus we do not know the per capita wine consumption in 1990.

Although finding the limiting value of sums of areas of rectangles is an invaluable tool for finding accumulated change in a quantity, there are times when it is not an appropriate technique. For example, at the end of Section 6.1 we saw count data for recycled aluminum cans. Because the data gave yearly totals, it would not be appropriate to use rectangles of width any smaller than 1 year. A limiting value of sums is most useful in truly continuous or piecewise continuous situations in which a model is defined for all input values in a certain interval.

6.2 Concept Inventory

- *Area = limiting value of sums of areas of midpoint rectangles*
- $\int_a^b f(x)\,dx$ *= definite integral*
- $\int_a^b f(x)\,dx$ *= area or signed area of the region between f and the x-axis if f does not cross the x-axis between a and b*
- $\int_a^b f(x)\,dx \neq$ *area of the region between f and the x-axis if f crosses the x-axis between a and b*

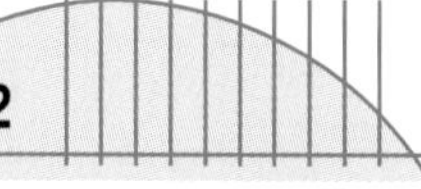

6.2 Activities

1. The rate of change of the population of a country, in thousands of people per year, is modeled by the function P with input t, where t is the number of years since 1995. What are the units on
   a. the area of the region between the graph of P and the t-axis from $t = 0$ to $t = 10$?
   b. $\int_{10}^{20} P(t)\,dt$?
   c. the change in the population from 1995 through 2000?

2. During the spring thaw a mountain lake rises by $L(d)$ feet per day, where d is the number of days since April 15. What are the units on
   a. the area of the region between the graph of L and the d-axis from $d = 0$ to $d = 15$?
   b. $\int_{16}^{31} L(d)\,dd$?
   c. the amount by which the lake rose from May 15 to May 31?

3. When warm water is released into a river from a source such as a power plant, the increased temperature of the water causes some algae to grow and other algae to die. In particular, blue-green algae that can be toxic to some aquatic life thrive. If $A(c)$ is the growth rate of blue-green algae (in organisms per °C) and c is the temperature of the water in °C, interpret the following in context:
   a. $\int_{25}^{35} A(c)\,dc$
   b. The area of the region between the graph of A and the c-axis from $c = 30$°C to $c = 40$°C

4. The value of a stock portfolio is growing by $V(t)$ dollars per day, where t is the number of days since the beginning of the year. Interpret the following in context:
   a. The area of the region between the graph of V and the t-axis from $t = 0$ to $t = 120$
   b. $\int_{120}^{240} V(t)\,dt$

5. The graph in Figure 6.2.1 shows the rate of change of profit at various production levels for a pencil manufacturer. Fill in the blanks in the following discussion of the profit. If it is not possible to determine a value, write NA in the corresponding blank.

Rate of change of profit (dollars per box)

100 200 500

Daily production level (boxes)

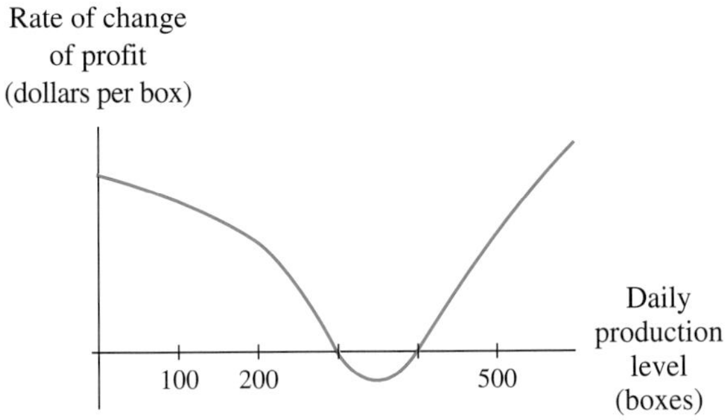

FIGURE 6.2.1

Profit is increasing when between (a) ______ and (b) ______ boxes of pencils are produced each day. The profit when 500 boxes of pencils are produced each day is (c) ______ dollars. Profit is higher than nearby profits at a production level of (d) ______ boxes each day, and it is lower than nearby profits at a production level of (e) ______ boxes of pencils each day. The profit is decreasing most rapidly when (f) ______ boxes are produced each day. The area between the rate-of-change-of-profit function and the production-level axis between production levels of 100 and 200 boxes each day has units (g) ______. If $p'(b)$ represents the rate of change of profit (in dollars per box) at a daily production level of b boxes, would $\int_{300}^{400} p'(b)\,db$ be more than, less than, or the same value as $\int_{100}^{200} p'(b)\,db$? (h) ______

6. The graph in Figure 6.2.2 shows the rate of change of cost for an orchard in Florida at various production levels during grapefruit season. Fill in the blanks in the following cost function discussion. If it is not possible to determine a value, write NA in the corresponding blank.

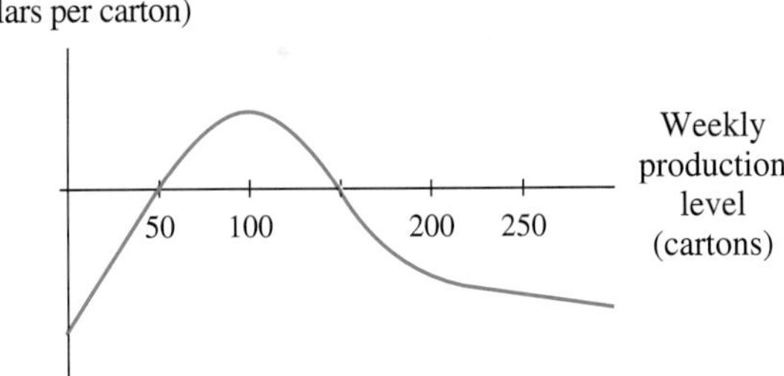

FIGURE 6.2.2

Cost is increasing when between (a) ______ and (b) ______ cartons of grapefruit are harvested each week. The cost to produce 100 cartons of grapefruit each week is (c) ______ dollars. The cost is lower than nearby costs at a production level of (d) ______ cartons, and it is higher than nearby costs at a production level of (e) ______ cartons of grapefruit each week. The cost is increasing most rapidly when (f) ______ cartons are produced each week. The area between the rate-of-change-of-cost function and the production-level axis between production levels of 50 and 150 cartons each week has units (g) ______. If $c'(p)$ represents the rate of change of cost (in dollars per carton) at weekly production level of p cartons of grapefruit, would $\int_{200}^{250} c'(p)\,dp$ be greater than, less than, or the same value as $\int_{50}^{100} c'(p)\,dp$? (h) ______

7. The rate of change of the weight of a laboratory mouse can be modeled by the equation

$$w(t) = \frac{13.785}{t} \text{ grams per week}$$

where t is the age of the mouse in weeks and $1 \le t \le 15$.

a. Use the idea of a limit of sums to estimate the value of $\int_3^{11} w(t)\,dt$.

b. Label units on the answer to part *a*. Interpret your answer.

c. If the mouse weighed 4 grams at 3 weeks, what was its weight at 11 weeks of age?

8. The rate of change of annual U.S. factory sales (in billions of dollars per year) of electronics from 1990 through 1996 can be modeled[16] by the equation

$$s(t) = -0.23x^3 + 2.257x^2 - 1.51x + 42.8 \text{ billions of dollars per year}$$

where t is the number of years since 1990.

a. Use the idea of a limit of sums to estimate the change in factory sales from 1990 through 1996.

b. Write the definite integral symbol for this limit of sums.

c. If factory sales were \$43.0 billion in 1990, what were they in 1996?

9. On the basis of data obtained from a preliminary report by a geological survey team, it is estimated that for the first ten years of production, a certain oil well can be expected to produce oil at the rate of $r(t) = 3.93546t^{3.55}e^{-1.35135t}$ thousand barrels per year t years after production begins.

a. Use the idea of a limit of sums to estimate the yield from this oil field during the first 5 years of production.

b. Use the idea of a limit of sums to estimate the yield during the first 10 years of production.

c. Write the definite integral symbols representing the limits of sums in parts *a* and *b*.

d. Estimate the percentage of the first 10 years' production that your answer to part *a* represents.

10. The rate of change of the temperature during the hour and a half after a thunderstorm began is modeled by the equation

$$T(h) = 9.48h^3 - 15.49h^2 + 17.38h - 9.87 \text{ °F per hour}$$

where h is the number of hours since the storm began. A graph of T is shown in Figure 6.2.3.

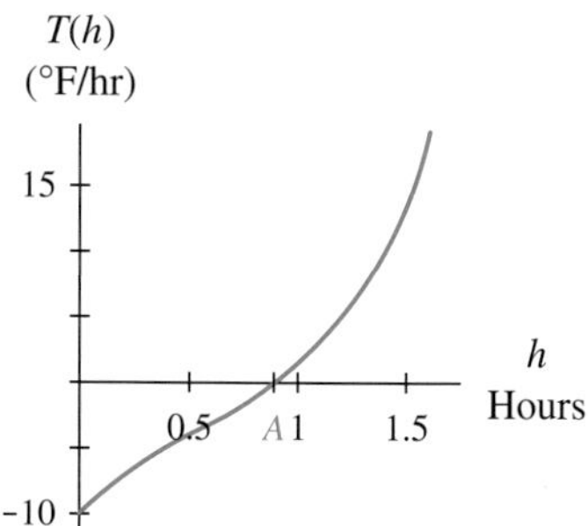

FIGURE 6.2.3

a. Determine the value of A.

b. Use a limit of sums to estimate $\int_0^A T(h)\,dh$. Interpret your answer.

c. Use a limit of sums to estimate $\int_A^{1.5} T(h)\,dh$. Interpret your answer.

d. Estimate $\int_0^{1.5} T(h)\,dh$. Interpret your answer.

e. What information is needed to determine the temperature when $h = 1.5$?

11. The acceleration of a race car during the first 35 seconds of a road test is modeled by the equation

$$a(t) = 0.024t^2 - 1.72t + 22.58 \text{ ft/sec}^2$$

where t is the number of seconds since the test began. A graph of this acceleration function is shown in Figure 6.2.4.

a. Find the time at which the acceleration curve becomes negative. This time is denoted A in the figure.

b. Use a limit of sums to estimate $\int_0^A a(t)\,dt$. Interpret your answer.

16. Based on data from *Statistical Abstract*, 1998.

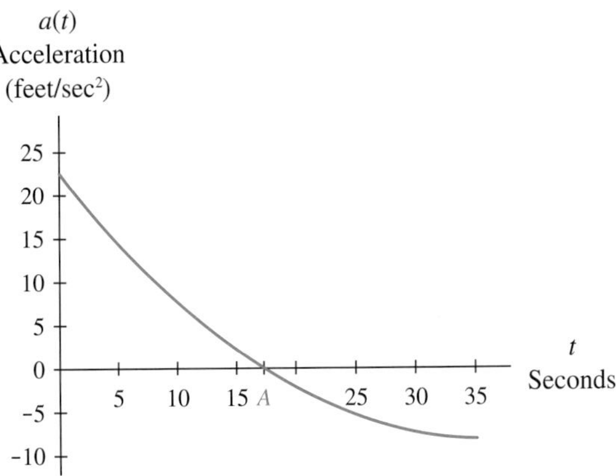

FIGURE 6.2.4

c. Use a limit of sums to estimate $\int_A^{35} a(t)\,dt$. Interpret your answer.

d. Estimate the change that occurred in the velocity of the car during the first 35 seconds of the road test.

12. Museums carefully monitor temperature and humidity. Suppose a museum has a gauge that monitors temperature as well as how rapidly temperature is changing. During a junior high school field trip to the museum, the rates of change of temperature shown in Table 6.24 are recorded.

TABLE 6.24

Time of day	Rate of change (°F/hr)
8:30 a.m.	0
8:45 a.m.	2.2
9:00 a.m.	2.4
9:15 a.m.	1.1
9:30 a.m.	-0.8
9:45 a.m.	-2.3
10:00 a.m.	-2.7
10:15 a.m.	-1.0

a. Basing your explanation on the rate-of-change data, discuss what happens to the temperature from 8:30 a.m. to 10:15 a.m.

b. According to the data, when is the temperature greatest?

c. Find a model for the data.

d. Use the model and a limit of sums to estimate the change in temperature between 8:30 a.m. and 9:30 a.m.

e. Use a limit of sums to estimate the change in temperature between 9:30 a.m. and 10:15 a.m.

f. What are the definite integral notations for your answers to parts *d* and *e*?

g. What information is needed to find the temperature at 10:15 a.m.?

13. An article in the May 23, 1996, issue of *Nature* addresses the interest some physicists have in studying cracks in order to answer the question "How fast do things break, and why?" As stated in *Nature*, "In 1991, Fineberg, Gross, Swinney and others developed a method of looking at the motion of cracks, . . . [making it] possible to measure the velocity of a crack on timescales much shorter than a millionth of a second, hundreds of thousands of times in succession."

Data estimated from a graph in this article, showing the velocity of a crack during a 60-microsecond experiment, are given in Table 6.25.

TABLE 6.25

Time (microseconds)	Velocity (meters per second)
10	148.2
20	159.3
30	169.5
40	180.7
50	189.8
60	200

a. In order to determine the distance the crack traveled during the 60-microsecond experiment, we wish to determine the area beneath the velocity curve from 0 to 60. What are the units on the heights and widths of rectangles under this curve?

b. Convert the data to millimeters per microsecond (there are 1000 millimeters in a meter and 1,000,000 microseconds in a second). Using the converted data, identify the area units.

c. Find a quadratic model for the converted data.

d. Use a limit of sums to determine how far the crack traveled during the experiment.

e. Give the definite integral notation for your answer to part *d*.

14. Table 6.26 records the volume of sales (in thousands) of a popular movie for selected months the first 18 months after it was released on video cassette.

TABLE 6.26

Months after release	Number of cassettes sold each month (thousands)
2	565
4	467
5	321
7	204
10	61
11	31
12	17
16	3
18	2

a. Find a logistic model for the data.

b. Use 5, 10, and 15 right rectangles to estimate the number of cassettes sold during the first 15 months after release.

c. Which of the following would give the most accurate value of the number of cassettes sold during the first 15 months after release?

i. The answer to part *b* for 15 rectangles.

ii. The limiting value of the sums of midpoint rectangles using the model in part *a*.

iii. The sum of actual sales figures for the first 15 months.

15. The personnel manager for a large construction company keeps records of the number of labor hours per week spent on typical construction jobs handled by the company. He has developed the following model for a labor-power curve:

$$m(x) = \frac{6{,}608{,}830e^{-0.706x}}{(1 + 925e^{-0.706x})^2} \text{ labor hours per week}$$

after the xth week of the construction job.

a. Use 5, 10, and 20 right rectangles to approximate the number of labor hours spent during the first 20 weeks of a typical construction job.

b. If the number of labor hours spent on a particular job exactly coincides with the model, which of the following would give the most accurate value of the number of labor hours spent during the first 20 weeks of the job?

i. The 20-right-rectangle sum found in part *a*

ii. The sum of 20 midpoint rectangles

iii. The limiting value of the sums of midpoint rectangles

16. Though all companies receive revenue in a discrete fashion, if a company is large enough, then its revenue can be thought of as flowing in at a continuous rate. Suppose that it is possible to measure the rate of flow of revenue for a company and that the flow rates measured at the end of the year can be modeled[17] by the equation

$$r(x) = 9.907x^2 - 40.769x + 58.492 \text{ million dollars per year}$$

x years after the end of 1987.

a. Use the idea of a limit of sums to estimate the value of $\int_0^6 r(x)\,dx$.

b. Interpret your answer to part *a*.

c. The company's revenue from 1987 through 1993 was reported to be 543 million dollars. How close is this value to your estimate from part *a*?

17. In Section 6.1 we considered a department store manager's effort to estimate the number of people who attended a Saturday sale. Recall that after taking samples of the number of customers who arrived during certain 1-minute times throughout the sale, the manager estimated the number of customers who entered the store each minute with the model

$$c(m) = (4.58904 \cdot 10^{-8})m^3 - (7.78127 \cdot 10^{-5})m^2 + 0.03303m + 0.88763 \text{ customers per minute}$$

where m is the number of minutes after 9:00 a.m.

a. Use a limiting value of sums to estimate the value of $\int_0^{720} c(m)\,dm$.

b. Interpret your answer to part *a* in context.

18. Table 6.27 gives rates of change of the amount in an interest-bearing account for which interest is compounded continuously.

17. Based on data for Timberland in *Allen Financial Advisors*.

TABLE 6.27

At end of year	Rate of change (dollars per day)
1	2.06
3	2.37
5	2.72
7	3.13
9	3.60

a. Convert the input to days. Disregard leap years. Why is this conversion important for a definite integral calculation?

b. Find an exponential model for the converted data.

c. Use a limiting value of sums to estimate the change in the balance of the account from the day the money was invested to the last day of the ninth year after the investment was made. Again, disregard leap years.

d. Give the definite integral notation for your answer to part *c*.

e. What is the balance in the account at the end of 9 years? Note that Activities 18*e* and 19*g* cannot be answered with the information given.

19. Blood pressure (BP) varies for individuals throughout the course of a day, typically being lowest at night and highest from late morning to early afternoon. The estimated rate of change in diastolic blood pressure for a patient with untreated hypertension is shown in Table 6.28.

a. During which time intervals was the patient's diastolic blood pressure rising? falling?

TABLE 6.28

Time	Rate of change of diastolic BP (mm Hg per hour)
8 a.m.	3.0
10 a.m.	1.8
12 p.m.	0.7
2 p.m.	-0.1
4 p.m.	-0.7
6 p.m.	-1.1
8 p.m.	-1.3
10 p.m.	-1.1
12 a.m.	-0.7
2 a.m.	0.1
4 a.m.	0.8
6 a.m.	1.9

b. Estimate the times when diastolic blood pressure was rising and falling most rapidly.

c. Find a model for the data.

d. Find the times at which the output of the model is zero. Of what significance are these times in the context of blood pressure?

e. Use the idea of a limiting value of sums to estimate by how much the diastolic blood pressure changed from 8 a.m. to 8 p.m.

f. Write the definite integral notation for your answer to part *e*.

g. What was this patient's blood pressure at 8 a.m.?

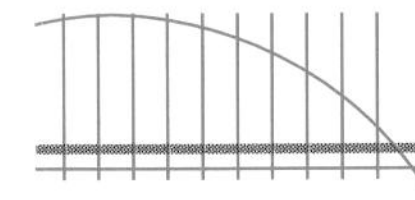

6.3 Accumulation Functions

The topic of accumulation functions is not commonly covered in applied calculus books, but we believe that this section is one of the most important in contributing to students' understanding of the concept of integration and in motivating the Fundamental Theorem of Calculus. It is worth any time and energy you spend to become comfortable with our presentation before introducing this section to your class.

In the last two sections, we saw that when we have a rate-of-change function for a certain quantity, we approximate the accumulation of change in that quantity between two values of the input variable using the area between the rate-of-change curve and the horizontal axis. We approximated area with sums of areas of rectangles and found that when we used more rectangles, the approximation usually improved. That discovery led us to view the area of a region between a rate-of-change function f and the x-axis as a limit of sums:

$$\lim_{n\to\infty} [f(x_1) + f(x_2) + \cdots + f(x_n)]\Delta x$$

This limit, when applied to an arbitrary, continuous function over an interval from a to b, is called the definite integral of f from a to b and is denoted by the symbol

$\int_a^b f(x)\,dx$. Note that a definite integral is a number that tells by how much a quantity has changed over a specific input interval. We now turn our attention to the more general situation of finding a function that gives the amount by which a quantity has changed from a specific starting point to an arbitrary ending point.

Using Estimated Areas to Sketch Accumulation Graphs

Consider the graph shown in Figure 6.31 of the speed of a vehicle during a 10-second drive. We wish to use the information in this graph to sketch a graph of the accumulated distance that the vehicle has traveled from the starting point at any time during the 10-second interval. Recall from Section 6.1 that distance is measured by the area of the region between a graph of the vehicle's speed and the horizontal axis. Thus the graph of accumulated distance can be thought of as a graph of accumulated area. We could use the methods of Section 6.1 and Section 6.2 to obtain numerical estimates of the distance the vehicle has traveled between the starting point and specified end points during the 10-second drive. This technique would give us points on the accumulated distance graph. Instead, however, we obtain a quick estimate of the accumulated distance by counting boxes on the grid in Figure 6.31 to estimate the area beneath the graph.

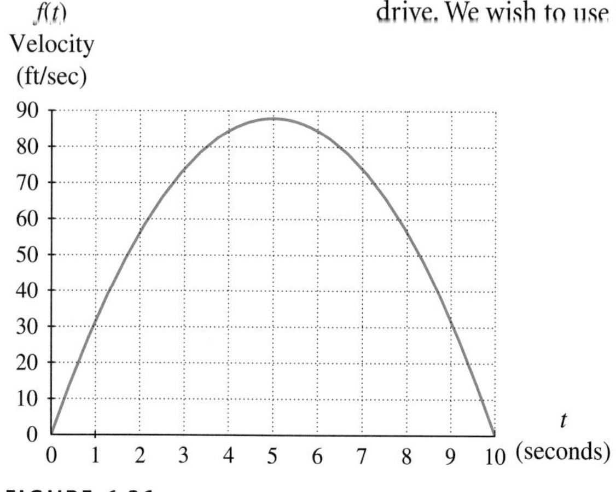

FIGURE 6.31

Figures 6.32a through 6.32e show shaded areas for time intervals between the starting point and 1, 2, 3, 4, and 5 seconds, respectively. On each graph, the lighter-shaded region represents the additional distance traveled from the previous stopping point. These areas give the distance traveled after 1, 2, 3, 4, and 5 seconds.

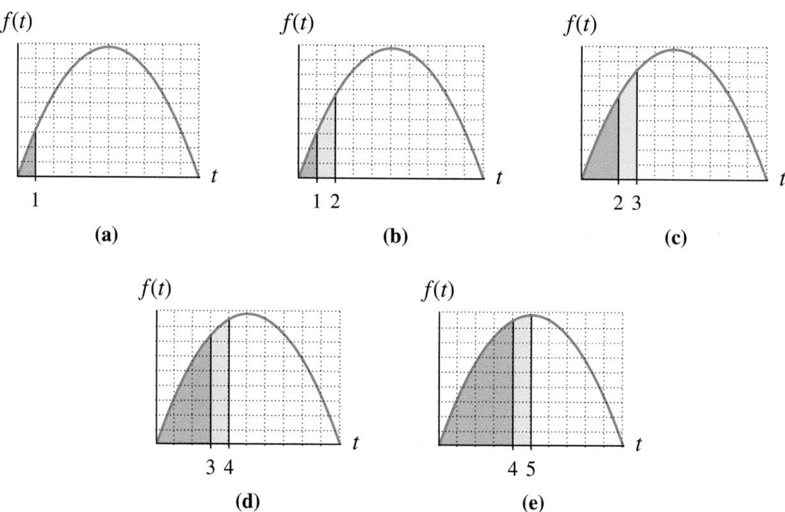

FIGURE 6.32

Note that the area of each box in the grid in Figure 6.31 is

$$\text{height} \cdot \text{width} = 10 \frac{\text{feet}}{\text{second}} \cdot 1 \text{ second} = 10 \text{ feet}$$

By counting boxes, we estimate the areas as shown in Table 6.29.

TABLE 6.29

Time (seconds)	1	2	3	4	5
Estimated number of boxes	1.6	6.1	12.7	20.7	29.3
Accumulated area (feet)	16	61	127	207	293

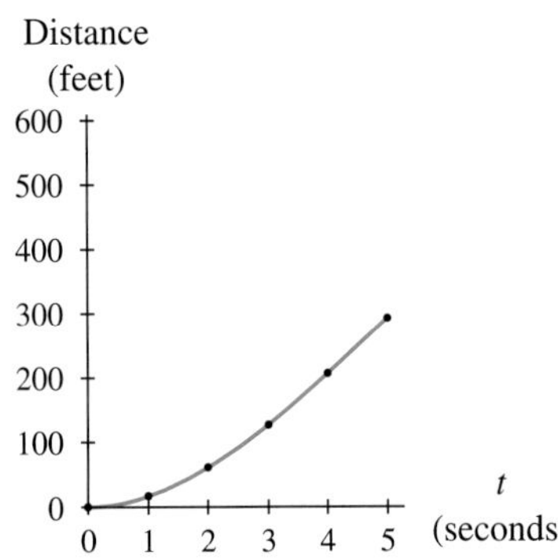

FIGURE 6.33

The accumulated area, which is distance traveled, has units

$$(\text{number of boxes})(10 \text{ feet per box}) \rightarrow \text{feet}$$

Note that with each additional second, the area grows by a greater and greater amount. When we plot these accumulated distances and imagine the points connected by a smooth curve, we see that the curve is increasing and concave up. (See Figure 6.33.)

We continue counting boxes to estimate the total distance traveled for times between 5 and 10 seconds. Figures 6.34a through 6.34e show accumulated areas for times of 6, 7, 8, 9, and 10 seconds, and Table 6.30 shows estimated areas corresponding to total distance traveled.

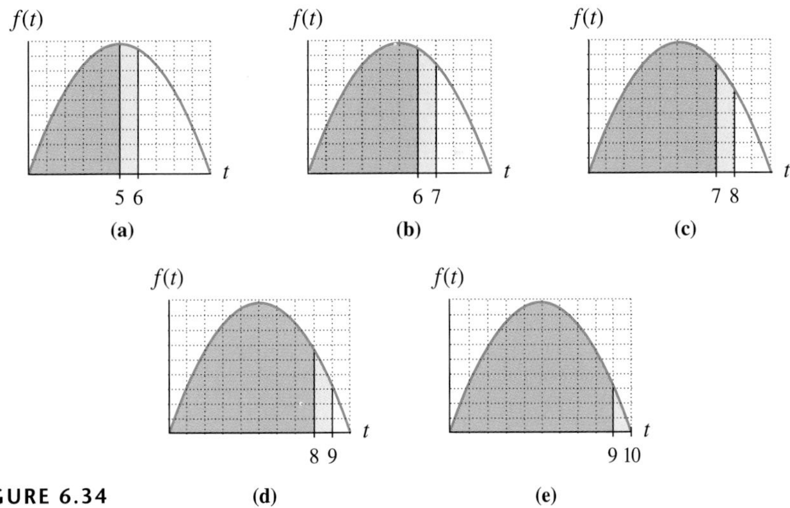

FIGURE 6.34

TABLE 6.30

Time (seconds)	6	7	8	9	10
Estimated number of boxes	38.0	46.0	52.6	57.0	58.7
Accumulated area (feet)	380	460	526	570	587

For the times shown in Table 6.30, each additional second still results in the area growing, but by a smaller and smaller amount. When we plot these accumulated distance points and imagine a smooth curve through them, we see that the curve is increasing and concave down. (See Figure 6.35.)

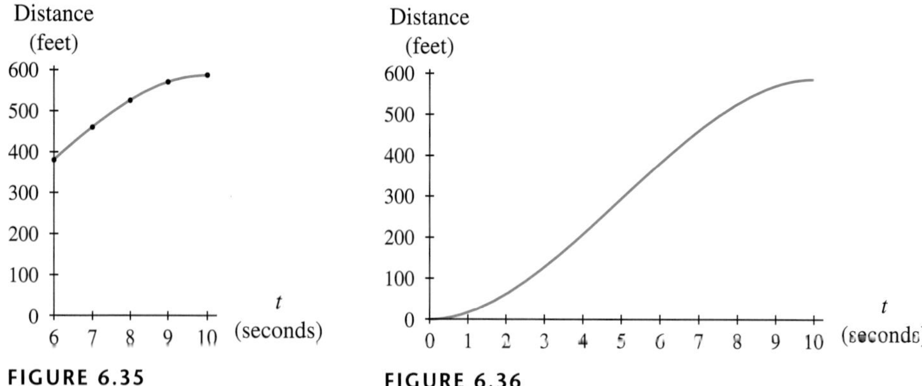

FIGURE 6.35

FIGURE 6.36

Combining the portions of the distance graphs in Figure 6.33 and Figure 6.35, we obtain the graph shown in Figure 6.36.

When we repeat this process with an arbitrary ending value, say $t = x$ instead of $t = 10$, we call a graph such as the one in Figure 6.36 a graph of the **accumulation function of f from 0 to x** and use the notation $A(x) = \int_0^x f(t)\,dt$. A definite integral of f between two specified inputs simply represents an output of an accumulation function of f. For example, we used (10, 587) as a point on the accumulation function graph in Figure 6.36. That is, we estimated the distance traveled after 10 seconds to be 587 feet. Thus $\int_0^{10} f(t)\,dt \approx 587$ feet. Similarly, $\int_0^4 f(t)\,dt \approx 207$ feet.

EXAMPLE 1 *Plant Growth*

The graph in Figure 6.37 shows $g(t)$, the growth rate of a plant in millimeters per day, t days after germination.

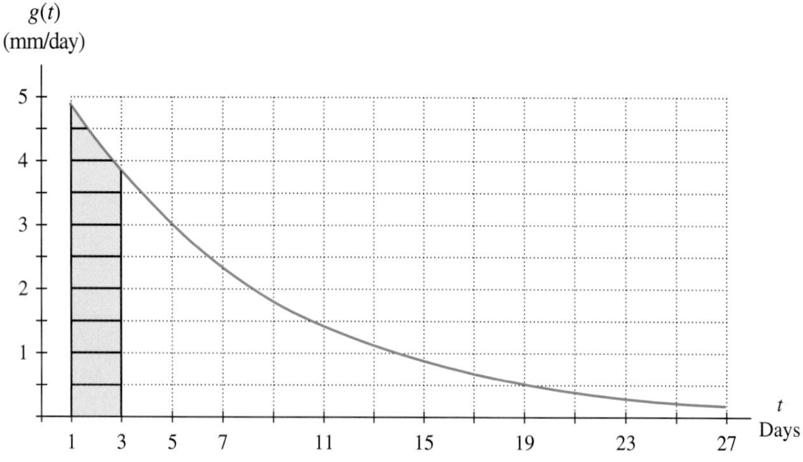

FIGURE 6.37

a. What does the area of the region between the graph of g and the t-axis between $t = 1$ and $t = x$ represent for x between 1 and 27?

b. Estimate the growth of the plant between day 1 and each of the days between 3 and 27 that are labeled in Figure 6.37.

c. Sketch a graph of the accumulation function $G(x) = \int_1^x g(t)\,dt$ for values of x between 1 and 27.

Solution

a. The area of the region between the graph of g and the t-axis between 1 and x represents the growth of the plant (its change in height) between day 1 and day x.

b. Begin by noting that each box on the grid has height 0.5 mm/day and width 2 days. Thus, each box has area (0.5mm/day)(2 days) = 1 mm. The shaded area in Figure 6.37 represents by how much the plant grew from day 1 to day 3. The shaded area is approximately $8\frac{2}{3}$ boxes, representing $8\frac{2}{3}$ mm of growth. This means that the point $\left(3, 8\frac{2}{3}\right)$ is a point on the accumulation function $G(x) = \int_1^x g(t)\,dt$; that is, $G(3) = 8\frac{2}{3}$.

Table 6.31 shows estimated areas obtained by counting boxes from 1 to x. You should confirm these numbers by finding your own estimate of area by counting boxes. Your estimates probably won't agree exactly with these, but they should be close.

TABLE 6.31

x	3	5	7	11	15	19	23	27
Accumulated number of boxes = millimeters of growth	$8\frac{2}{3}$	$15\frac{1}{2}$	21	$28\frac{1}{3}$	$32\frac{2}{3}$	$35\frac{2}{3}$	37	38

c. The values in Table 6.31 correspond to points on the accumulation function graph. We can add the point (1, 0) because we are measuring the growth in the plant after day 1. (In fact, an accumulation function is always zero at its starting value because no accumulation has occurred at the starting point.) Plotting these points and sketching a smooth curve through them result in the accumulation function graph shown in Figure 6.38.

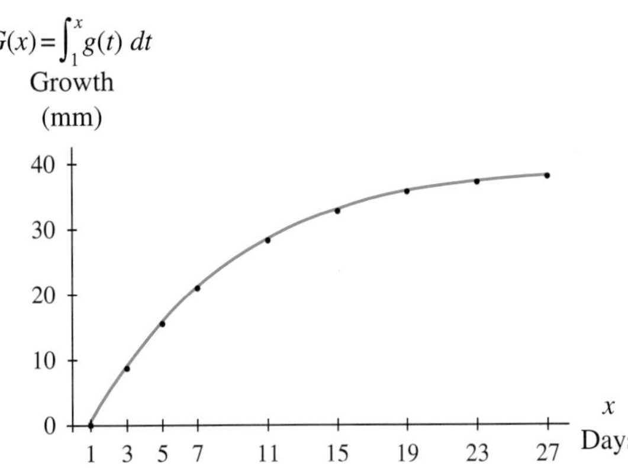

FIGURE 6.38

How would the accumulation graph in Figure 6.38 change if we started our accumulation at day 5 instead of day 1? In other words, consider a graph of the accumulated height of the plant after day 5. The first observation we make is that at day 5, the accumulation is zero. As noted before, an accumulation function is always zero at its starting point. The points on the accumulation graph then become the values in Table 6.31 minus 15.5, the estimated growth of the plant between days 1 and 5. The graph looks like the one shown in Figure 6.39.

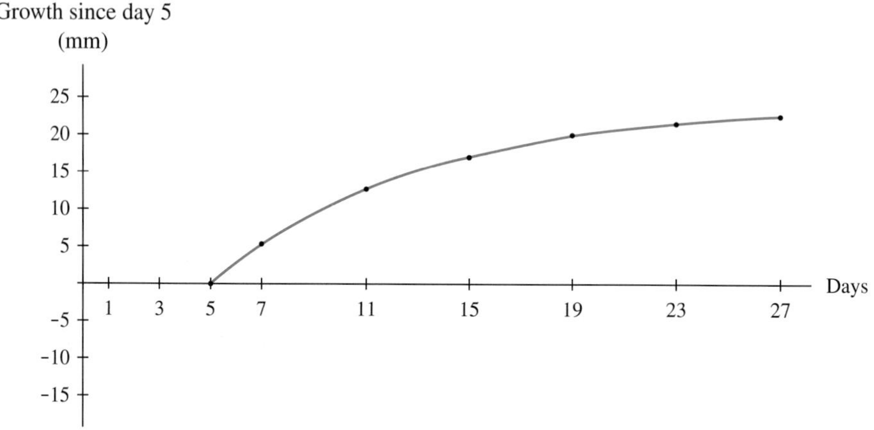

FIGURE 6.39

TABLE 6.32

x	3	1
Negative of the accumulated number of boxes = −(millimeters of growth)	$-6\frac{3}{4}$	$-15\frac{1}{2}$

We could also consider growth between day 5 and day 3. Because we chose day 5 as our starting value, we are looking backward in time to day 3. We indicate this backward direction by making the growth values negative. Estimates of these values for days 3 and 1 are given in Table 6.32.

Adding these points to the graph in Figure 6.39 and continuing the smooth curve, we obtain the graph of the accumulation function $G(x) = \int_5^x g(t)\,dt$ for days 1 through 27. (See Figure 6.40.)

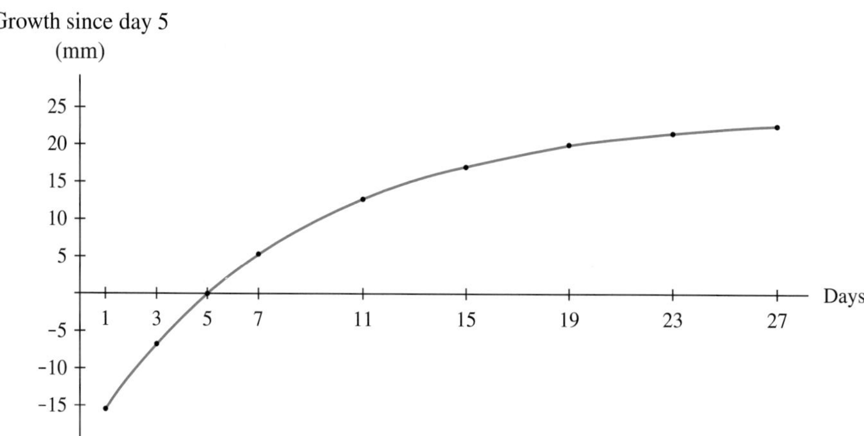

FIGURE 6.40

Compare Figure 6.38 with Figure 6.40. The two accumulation graphs have exactly the same shape. The graph in Figure 6.40 (considering day 5 as the starting point) is the graph in Figure 6.38 (considering day 1 as the starting point) shifted down 15.5 units. It is always the case that changing the starting point of an accumulation graph has the effect of vertically shifting the graph so that the graph is zero at the starting point.

These observations lead us to establish an important rule of thumb regarding sketching accumulation graphs:

> **Sketching Accumulation Graphs**
>
> Given the graph of a function, consider accumulation beginning at the far left, regardless of the location of the specified starting point. Sketch the accumulation graph starting at the far left, and then shift the graph up or down so that the output value at the specified starting point is zero.

So far, our examples have dealt with sketching accumulation graphs for functions whose graphs are positive. When a portion of a graph is negative, the area below the horizontal axis indicates a decrease in the accumulation. Example 2 illustrates how to sketch accumulation graphs for graphs that go below the horizontal axis and how to apply the rule of thumb stated above.

EXAMPLE 2 *Sketching an Accumulation Function Graph*

Consider the graph of f shown in Figure 6.41.

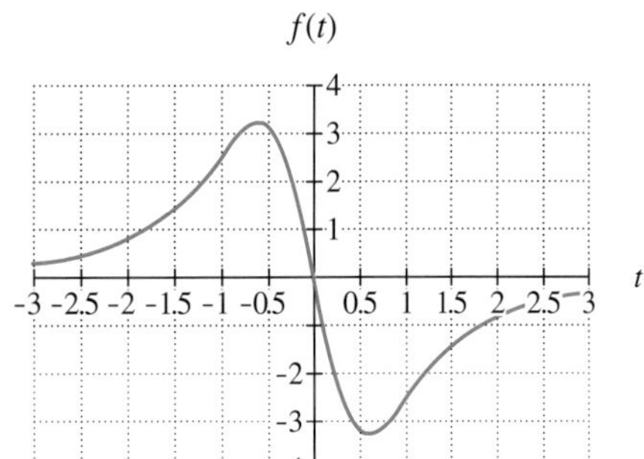

FIGURE 6.41

a. Construct a table of accumulation function values for $x = -3, -2.5, \ldots, 2.5, 3$.

b. Sketch a scatter plot and continuous graph of the accumulation function $A(x) = \int_{-3}^{x} f(t)\,dt$.

c. Use the graph in part b to sketch a graph of the accumulation function $B(x) = \int_{0}^{x} f(t)\,dt$.

Solution

a. Begin estimating accumulated area from the far left side of the graph by counting the boxes between the graph of f and the horizontal axis from -3 to x. You should obtain values similar to those in the second column of Table 6.33. Note that the boxes have height 1 unit and width 0.5 unit; thus the area of each box is $(1)(0.5) = 0.5$ unit^2. The third column of Table 6.33 is the number of boxes multiplied by 0.5 to obtain the accumulated area value.

TABLE 6.33

x	Accumulated number of boxes	Accumulation function value	x	Accumulated number of boxes	Accumulation function value
-3	0	0	0.5	$9 - 2 = 7$	3.5
-2.5	0.5	0.25	1	$7 - 3 = 4$	2
2	1	0.5	1.5	$4 - 2 = 2$	1
-1.5	2	1	2	$2 - 1 = 1$	0.5
-1	4	2	2.5	$1 - 0.5 = 0.5$	0.25
-0.5	7	3.5	3	$0.5 - 0.5 = 0$	0
0	9	4.5			

For values of x greater than zero, the boxes lie below the horizontal axis. The areas of these boxes should be subtracted from the accumulated area of boxes above the horizontal axis in order to obtain the net number of accumulated boxes. For example, the number of boxes from -3 to 0 is approximately 9. The number of boxes from 0 to 0.5 is approximately 2, and these boxes lie below the horizontal axis. To determine the accumulated number of boxes from -3 to 0.5, we subtract the number of boxes lying below the axis from the number lying above the axis: $9 - 2 = 7$. Because the number of boxes from -3 to 0 is the same as the number of boxes from 0 to 3, the net result of accumulated area from -3 to 3 is zero.

b. A scatter plot and continuous graph for $A(x) = \int_{-3}^{x} f(t)\,dt$ are shown in Figure 6.42.

c. In order to sketch the accumulation graph with 0 as the starting point, we vertically shift the graph in Figure 6.42 so that the function value at the starting point is zero. In this example, we shift the graph down 4.5 units so that the peak of the graph is the point (0, 0). (See Figure 6.43.)

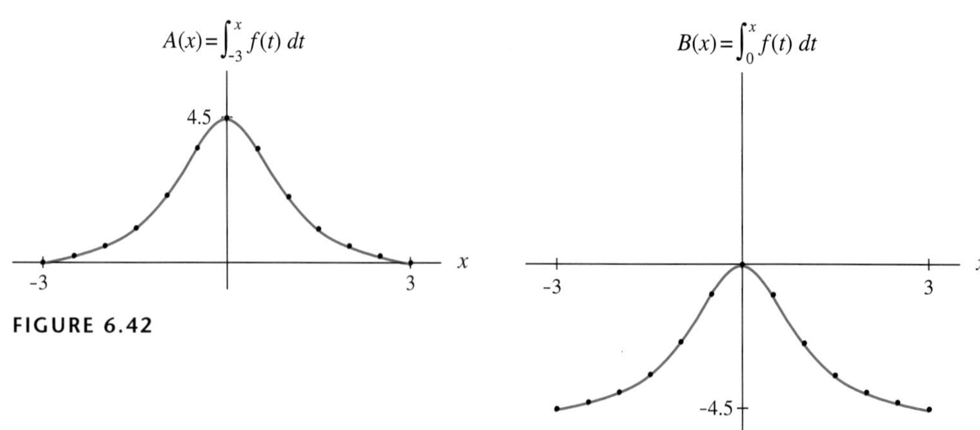

FIGURE 6.42

FIGURE 6.43

Sketching General Accumulation Function Graphs

By now, you should have a good idea of what we mean by an accumulation function. We now formally define an accumulation function and then turn our attention to the more difficult task of sketching accumulation function graphs from a graph without a grid or any numerical labels.

You should help your students visualize the accumulation of area with some graphical presentation, such as shading regions on a graph of the original function as the area accumulates. The most important tool you can give your students to help them understand accumulation functions is a clear mental picture of area accumulating between a graph and the horizontal axis.

Accumulation Function

The accumulation function of a function f, denoted by $A(x) = \int_a^x f(t)\,dt$, gives the accumulation of the area between the horizontal axis and the graph of f from a to x. The constant a is referred to as the *starting value* of the accumulation.

To sketch an accumulation function with starting value a given a graph of f,

Step 1: Choose as an initial starting value the input value at the left end of the axis shown in the graph.

Step 2: Sketch the accumulation of area from the far left such that

- Regions between the graph of f and the horizontal axis that lie above the axis contribute positive accumulation equal to the area of the region, and
- Regions that lie beneath the horizontal axis contribute negative accumulation equal to the negative of the area of the region.

Step 3: Vertically shift the accumulation graph in Step 2 so that the graph is zero at the actual starting value a.

It is important to remember that unlike the definite integral, which is a number, the accumulation function is a function whose input is the upper limit of the integral. The output of an accumulation function at a certain input value is the value of the definite integral from the given starting point to that value.

We have already sketched an accumulation graph for a portion of a concave-down parabola. We now consider a similar parabola, but this time we will sketch the accumulation graph without the aid of a grid, and we will consider the accumulation graph over all real number inputs. The graph we consider is $f(t) = -t^2 + 4$ (see Figure 6.44) with -2 as the starting point, so we seek a graph of the function $A(x) = \int_{-2}^{x} (-t^2 + 4)\,dt$.

Although the starting value of the accumulation function we seek is -2, we begin by considering the initial starting value to be the input value at the left end of the axis shown in Figure 6.44. At the end of this process, we will adjust the graph so that the starting value is -2.

We begin on the horizontal axis at the far left. As x begins moving to the right, the region is below the axis, so the accumulation is negative. The accumulation becomes more and more negative (that is, the area increases below the horizontal axis) as we

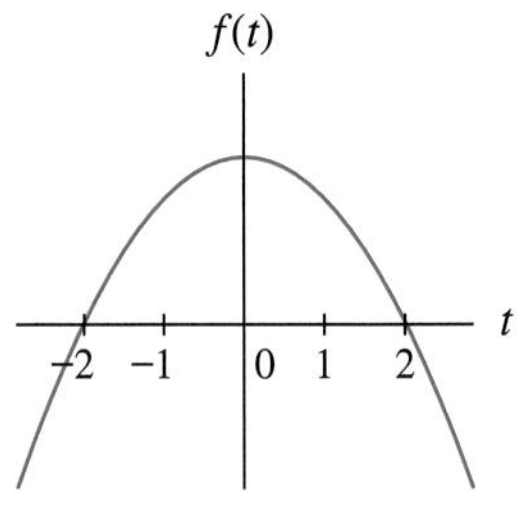

FIGURE 6.44

A mistake students commonly make when sketching accumulation function graphs is to reason that because the original function decreases, the accumulation function graph must decrease. Emphasize that students should consider only the accumulating area between the original function and the horizontal axis, *not* whether the original function increases or decreases.

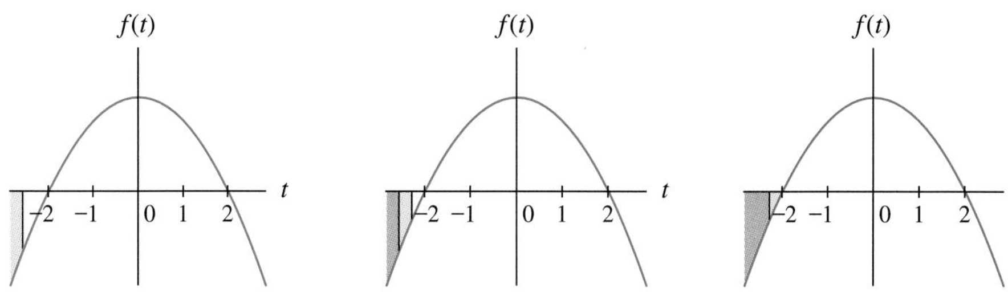

Accumulation of area is negative and growing more and more slowly.

FIGURE 6.45

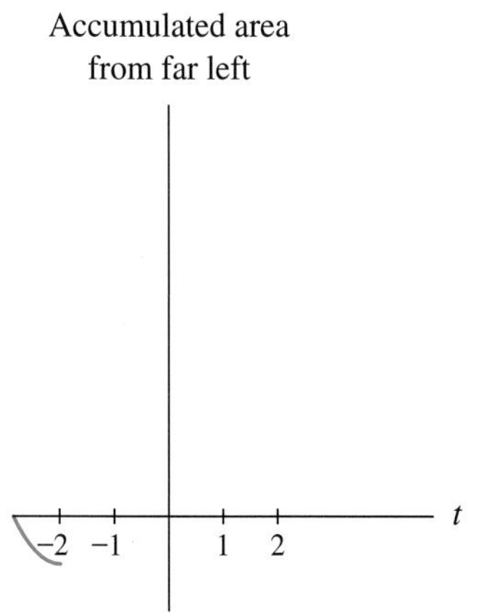

Accumulation function to the left of -2 is decreasing and concave up.

FIGURE 6.46

move toward -2, as shown in Figure 6.45. However, each time we move toward -2, the additional portion of area is smaller and smaller, indicating that the accumulation occurs more and more slowly. We draw the accumulation graph below the axis and concave up as shown in Figure 6.46.

Between -2 and 0, the accumulation function is increasing (because the new accumulated area is above the axis) and is concave up (because the accumulated area is getting larger more and more rapidly). See Figure 6.47. It appears that the area below the axis and the area above the axis are approximately equal at $x = -1$. Thus we expect the accumulation function to be zero near $x = -1$. See Figure 6.48.

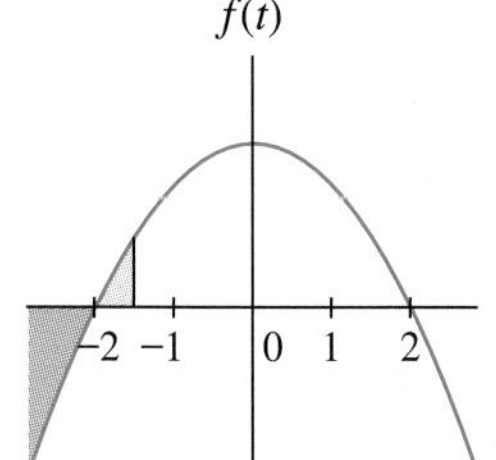

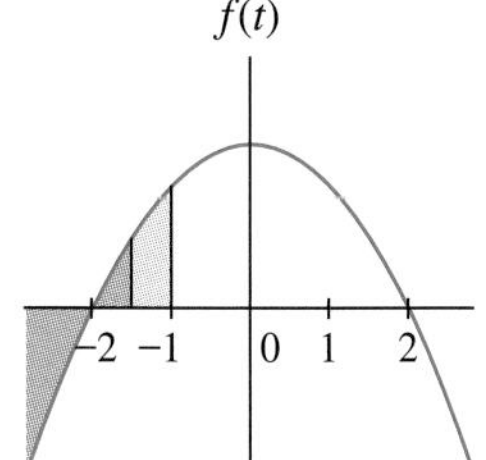

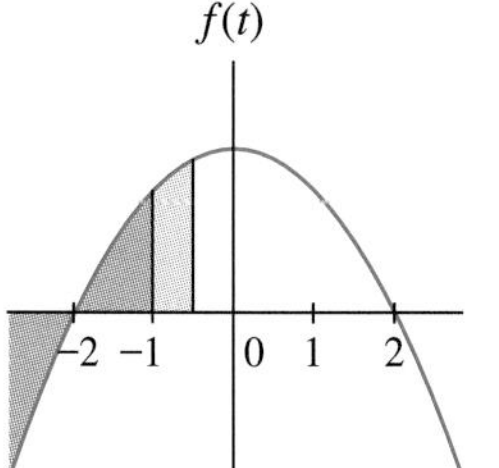

Accumulation of area is growing more and more rapidly.

FIGURE 6.47

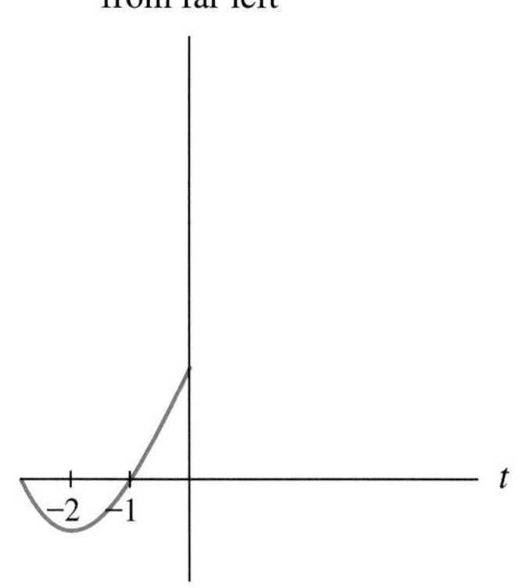

Accumulation function from -2 to 0 is increasing and concave up.

FIGURE 6.48

As x increases to the right of 0, the accumulated area continues to become larger (more positive), but it does so more and more slowly. (See Figure 6.49.)

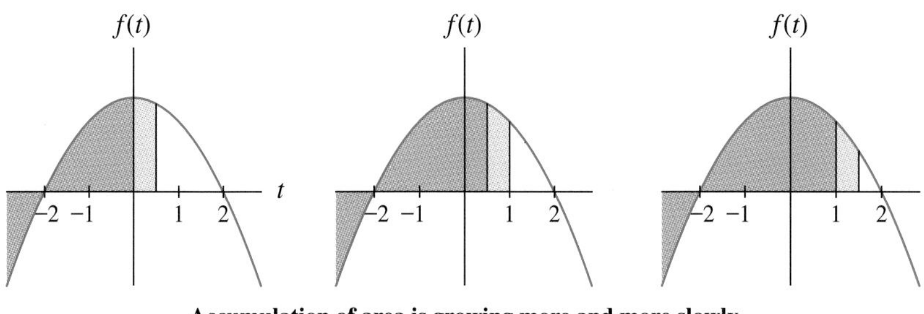

Accumulation of area is growing more and more slowly.

FIGURE 6.49

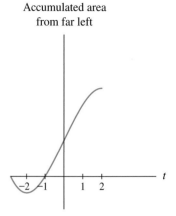

Accumulated area from far left

Accumulation function from 0 to 2 is increasing and concave down.

FIGURE 6.50

That means that from 0 to 2, the accumulation function continues to increase but is concave down, as shown in Figure 6.50.

As x moves to the right of 2, $f(t)$ is again negative, so the accumulation function begins to decline as the area below the axis is subtracted from the area that has accumulated above. (Refer to the second statement under Step 2 in the definition of the accumulation of area.) Also, the decline occurs more and more rapidly (resulting in the accumulation function being concave down) until the area below the axis is equal to the area above the axis. At this input value (B in Figure 6.51), the accumulation of area is again 0.

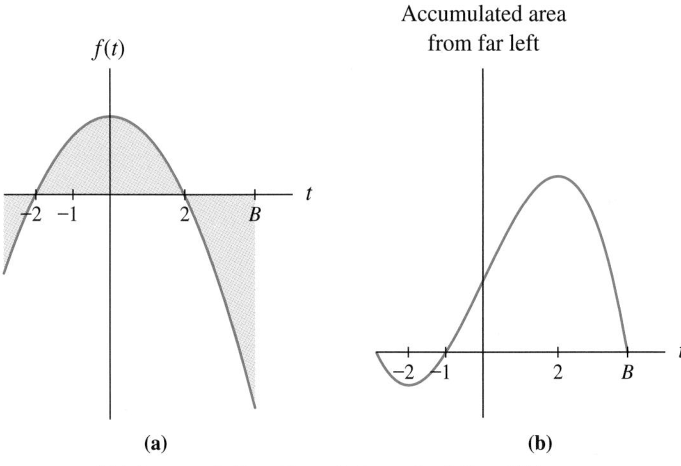

(a) (b)

Total accumulation of area between far left and B is zero.

FIGURE 6.51

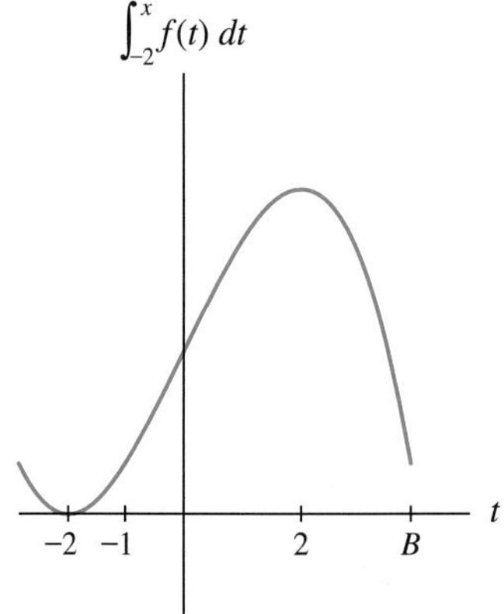

The accumulation function of $f(t) = -t^2 + 4$ from -2 to x

FIGURE 6.52

Our final step is to shift the graph in Figure 6.51b vertically so that the graph is zero at the starting value $a = -2$. The result of the vertical shift is shown in Figure 6.52.

Concavity and Accumulation

An important tool for accurately sketching accumulation function graphs is an understanding of how the increasing or decreasing nature, as well as the concavity, of the graph of an accumulation function describe the accumulation of area. Increasing accumulation graphs describe rising accumulation associated with accumulation of area above the horizontal axis, whereas decreasing accumulation graphs describe declining accumulation associated with accumulation of area below the horizontal axis. Along with the increasing or decreasing nature of an accumulation graph, we also consider how the concavity of an accumulation graph describes the accumulation of area.

Warn your students that they should not try to memorize these statements relating concavity to accumulation. As long as students picture the systematic accumulation of area, they will sketch these graphs correctly.

Concavity and Accumulation

A concave-up, increasing shape describes faster and faster positive accumulation. A concave-down, increasing shape indicates slower and slower positive accumulation. A concave-down, decreasing shape indicates faster and faster decline (negative accumulation). A concave-up, decreasing shape indicates slower and slower decline.

Note that the direction of the graph of the accumulation function A depends on whether the graph of f is positive or negative, not on whether the graph of f is increasing or decreasing. To determine whether the accumulation is faster and faster or slower and slower, picture systematically filling in area as illustrated in Figures 6.45, 6.47, and 6.49. If each additional portion is larger than the previous portion, then the accumulation is faster and faster. If each additional portion is smaller than the previous one, then the accumulation is slower and slower.

Example 3 shows how we can apply these concepts to sketching a general accumulation function graph.

EXAMPLE 3 *Sketching a General Accumulation Function Graph*

Consider the graph shown in Figure 6.53.

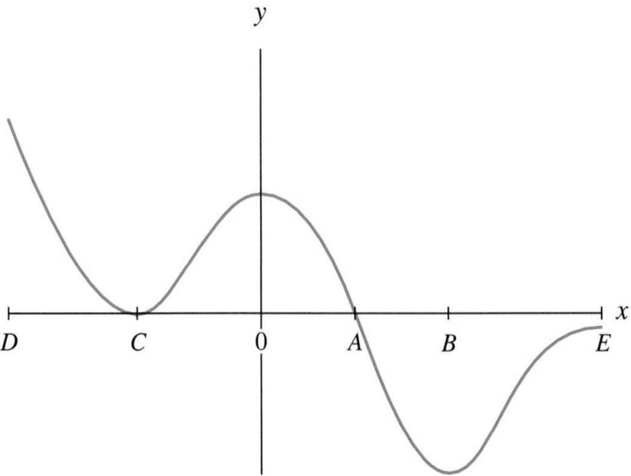

FIGURE 6.53

a. Consider each portion of the graph between the points marked on the horizontal axis. Label these regions as contributing positive or negative accumulated area, and indicate whether each area is accumulated faster and faster or slower and slower.

b. On the basis of the labels in part *a*, sketch the accumulation function with zero as the starting point.

Solution

a. Figure 6.54 shows the correct labels.

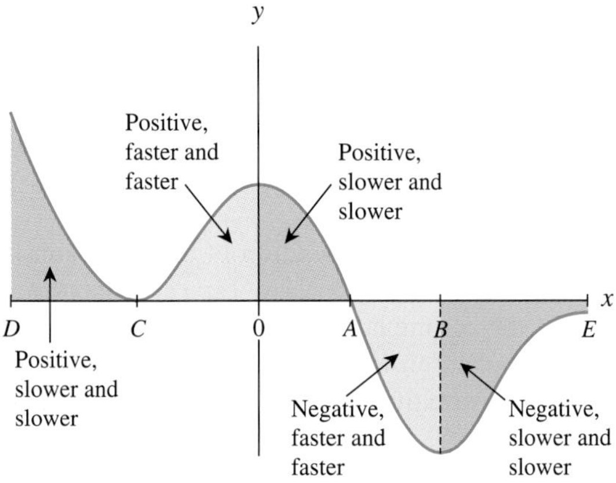

FIGURE 6.54

b. We begin at the far left with the accumulation function output at zero. On the basis of the labels in part *a*, we conclude that the accumulation function graph is increasing and concave down until *C* and is increasing and concave up between *C* and 0. Between 0 and *A*, the accumulation function graph continues to increase but is concave down. Once the graph goes below the horizontal axis, the accumulated area begins decreasing. The accumulation function graph is decreasing and concave down between *A* and *B* and is decreasing and concave up to the right of *B*.

A graph illustrating these properties is shown in Figure 6.55.

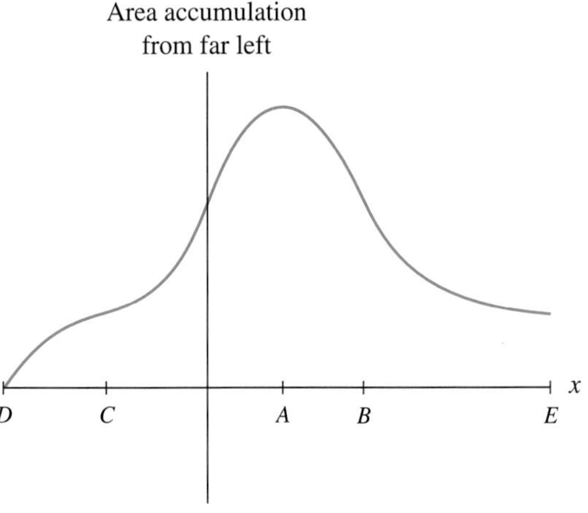

FIGURE 6.55

To obtain the accumulation function graph with zero as the starting point, shift the graph down so that it passes through (0, 0). See Figure 6.56.

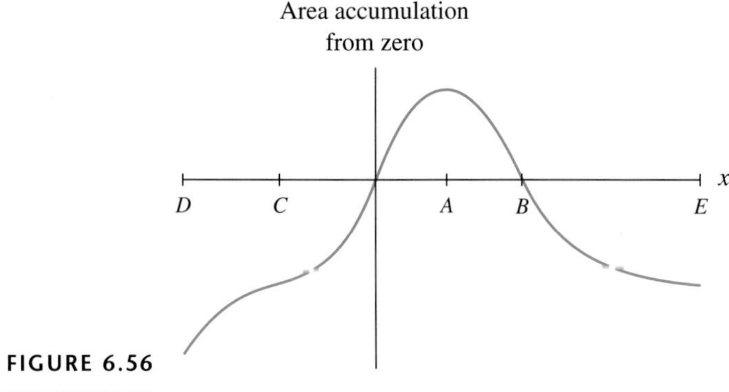

FIGURE 6.56

Formulas for Accumulation Functions

Our final consideration of accumulation functions involves those accumulation functions for which we can calculate formulas. This calculation is possible when the accumulated area forms a geometric shape for which we know an area formula. Consider, for example, the function $f(t) = 3$ and the accumulation of area between the graph of f and the t-axis, beginning at $t = 0$. (See Figure 6.57.)

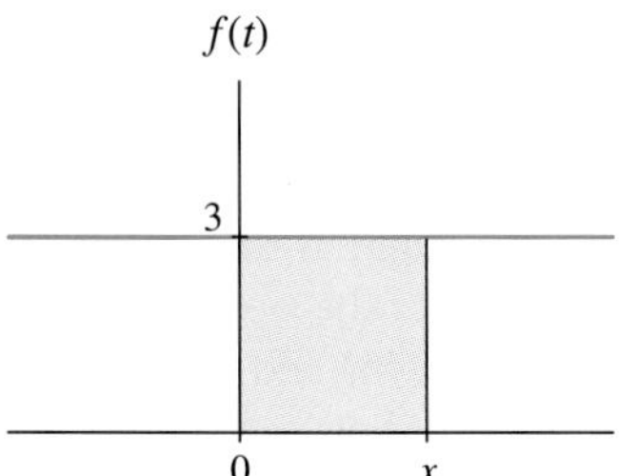

FIGURE 6.57

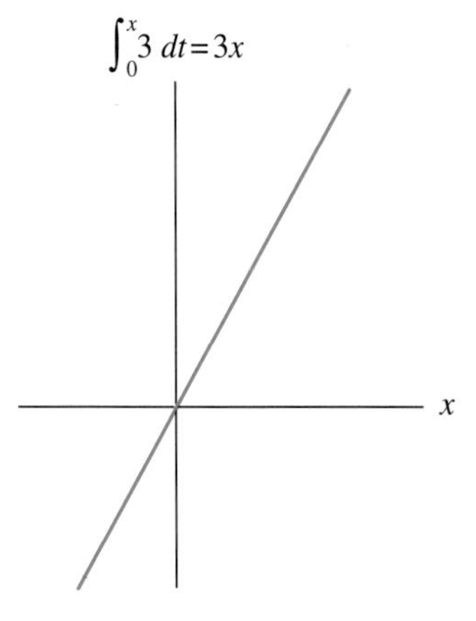

FIGURE 6.58

We can determine the formula for the accumulation function $A(x) = \int_0^x 3\,dt$ by noticing that for x to the right of 0, the region is a rectangle with height 3 and width x. Thus for any value of x to the right of 0, the accumulated area is given by the formula $3x$. For x to the left of 0, the argument is slightly more involved but results in the same conclusion: The accumulation formula for $f(t) = 3$ beginning at 0 is $\int_0^x 3\,dt = 3x$. The graph of this accumulation function is shown in Figure 6.58.

For the cases we consider here, the formula for x to the left of the starting point will be the same as that for x to the right of the starting point, so we consider only the second case when determining an accumulation function formula.

EXAMPLE 4 *Finding Accumulation Function Formulas*

Determine the following accumulation functions for $f(t) = 2t - 4$ and $x > 2$.

a. $\int_2^x f(t)\,dt$ b. $\int_{-1}^x f(t)\,dt$ c. $\int_6^x f(t)\,dt$

d. Use the formulas you obtained in parts a, b, and c to find the areas of the regions between $f(t) = 2t - 4$ and the horizontal axis from 2 to 5 and from 6 to 12.

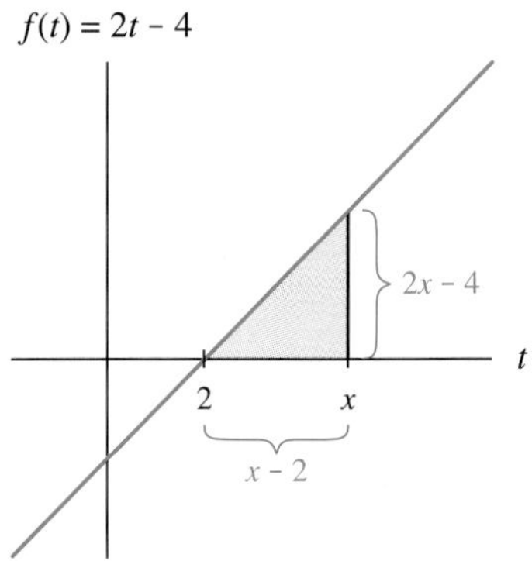

FIGURE 6.59

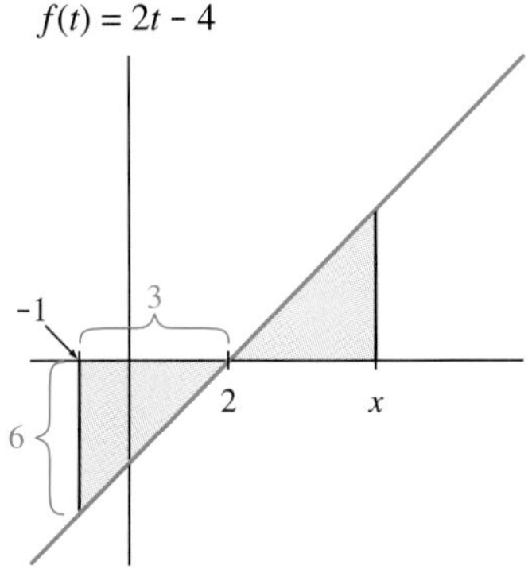

FIGURE 6.60

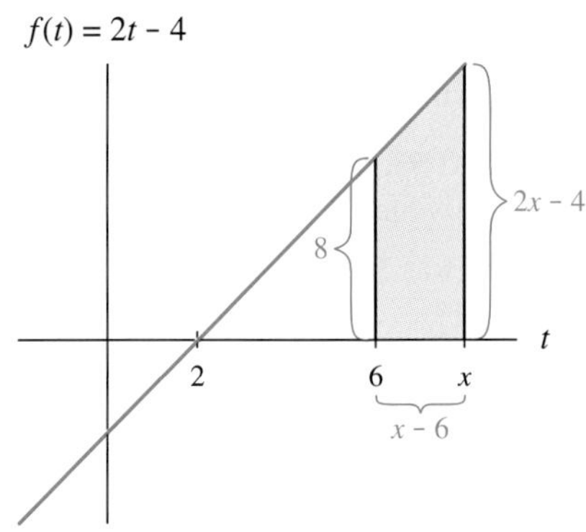

FIGURE 6.61

Solution

a. The graph in Figure 6.59 shows that the formula we desire is the area of a triangle with base $x - 2$ and height $2x - 4$.

$$\text{Hence, } \int_2^x (2t - 4)\,dt = \tfrac{1}{2}(x - 2)(2x - 4)$$
$$= x^2 - 4x + 4.$$

b. If we change the starting point to -1, we now must consider two triangles, with the one above the horizontal axis contributing its area (which we found in part *a*) and the one beneath the axis contributing the negative of its area. (See Figure 6.60.)

The base of the triangle below the axis is 3 and the height is 6 (determined by substituting -1 for t in the formula $2t - 4$ and noting that heights are always positive measurements). Thus the area we must subtract is $\frac{1}{2}(3)(6) = 9$. Therefore,

$$\int_{-1}^x (2t - 4)\,dt = \text{area of triangle above axis} - \text{area of triangle below axis}$$
$$= (x^2 - 4x + 4) - 9$$
$$= x^2 - 4x - 5$$

c. Using 6 as the starting value, we redraw Figure 6.60 with x to the right of 6 (for convenience) to obtain the graph in Figure 6.61. Now, instead of a triangle, we have a trapezoid. Its base is $x - 6$, and its heights are determined by evaluating $f(t) = 2t - 4$ at $t = 6$ and $t = x$. Thus the accumulation function we desire is

$$\int_6^x (2t - 4)\,dt = (x - 6)\left(\frac{8 + (2x - 4)}{2}\right)$$
$$= (x - 6)(x + 2)$$
$$= x^2 - 4x - 12$$

d. The area from 2 to 5 (using the formula found in part *a*) is

$$\int_2^5 (2t - 4)\,dt = 5^2 - 4(5) + 4 = 9$$

The area from 6 to 12 (using the formula found in part *c*) is

$$\int_6^{12} (2t - 4)\,dt = 12^2 - 4(12) - 12 = 84$$

Be sure you understand that we found formulas only for $x > 2$ and x to the right of the starting point. The formulas we found do hold for all values of x, but we have not proven that.

This discussion of accumulation functions has been largely theoretical, but we will see in the next section that accumulation functions play an important role in providing us with an algebraic method for calculating areas. Indeed, accumulation functions are the link between what we have examined thus far in our study of calculus and what is to come in future chapters.

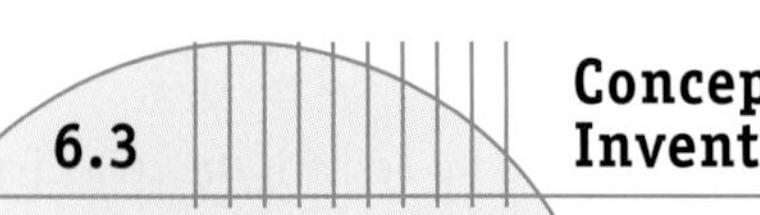

6.3 Concept Inventory

- *Accumulation and area*
- *Accumulation function:*
  - Sketching graphs
  - Interpreting
  - Finding simple formulas
- *Accumulation and concavity*

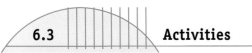

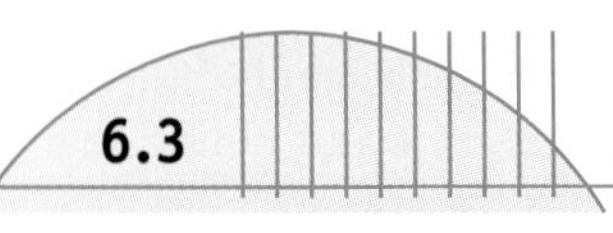

6.3 Activities

1. Refer to the velocity graph shown in Figure 6.31 on page 414.
   a. Sketch the accumulation function with 5 seconds as the starting point.
   b. Give the mathematical notation for the function you sketched in part *a*.
   c. In the context of the moving vehicle, what is the interpretation of the output values of the accumulation function in part *a*?
2. Repeat Activity 1 using 10 seconds as the starting point.
3. Refer to the plant growth rate function graph shown in Figure 6.37 on page 416.
   a. Sketch the accumulation function with day 13 as the starting point.
   b. Give the mathematical notation for the function you sketched in part *a*.
   c. In the context of plant growth, what is the interpretation of the output values of the accumulation function in part *a*?
4. Repeat Activity 3 using day 27 as the starting point.
5. Consider the graph shown in Figure 6.3.1 of the rate of change in the price of a certain technology stock during the first 55 trading days of 2003.
   a. What does the area of the region between the graph and the horizontal axis between days 0 and 18 represent?
   b. What does the area of the region between the graph and the horizontal axis between days 18 and 47 represent?
   c. Was the stock price higher or lower on day 47 than it was on day 0? How much higher or lower?

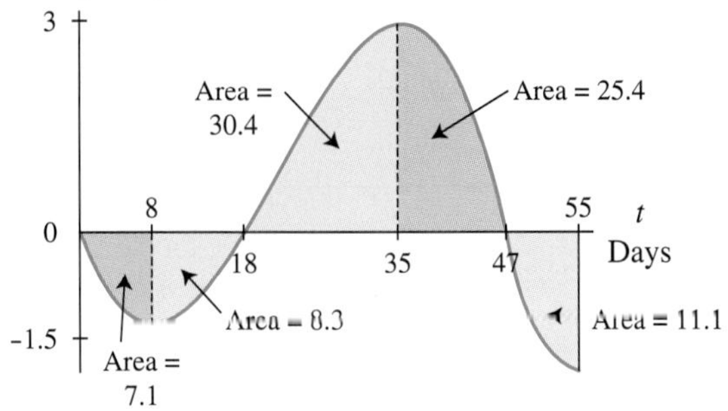

FIGURE 6.3.1

   d. Was the stock price higher or lower on day 55 than it was on day 47? How much higher or lower?
   e. Using the information presented in the graph, fill in the accumulation function values in Table 6.34.

TABLE 6.34

x	0	8	18	35	47	55
$\int_0^x r(t)\,dt$						

   f. Graph the function $R(x) = \int_0^x r(t)\,dt$ for values of x between 0 and 55, labeling the vertical axis as accurately as possible.
   g. If the stock price was $127 on day 0, what was the price on day 55?
6. Refer again to Figure 6.3.1.
   a. Label each of the shaded areas as representing positive or negative change in price. Also label each region as describing faster and faster or slower and slower change in stock price.
   b. On the basis of your answers to part *a*, sketch a graph of the accumulation function $P(x) = \int_{18}^x r(t)\,dt$.
   c. Sketch a graph of the accumulation function $Q(x) = \int_{35}^x r(t)\,dt$.
   d. What differences do you notice among the three accumulation functions in part *f* of Activity 5 and parts *b* and *c* of this activity?

7. The graph in Figure 6.3.2 shows the rate of change of the number of subscribers to an Internet service provider during its first year of business.

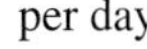

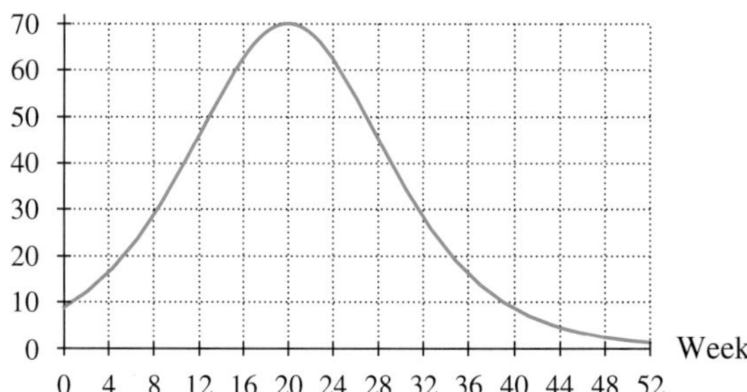

FIGURE 6.3.2

a. According to the graph, did the number of subscribers ever decline during the first year?

b. What is the significance of the peak in the graph at 20 weeks?

c. If $n(x)$ is the number of subscribers per day at the end of the xth day of the year, what does the accumulation function $N(t) = \int_{140}^{t} n(x)\,dx$ describe?

d. How many subscribers does each box in Figure 6.3.2 represent?

e. Use the grid and graph in Figure 6.3.2 to estimate the accumulation function values in Table 6.35.

TABLE 6.35

Week	t (days)	$\int_0^t n(x)\,dx$	Week	t (days)	$\int_0^t n(x)\,dx$
4	28		28	196	
8	56		36	252	
12	84		44	308	
16	112		52	364	
20	140				

f. Sketch a graph of the accumulation function with 140 days as the starting value.

8. Refer again to Figure 6.3.2 and Table 6.35. Sketch a graph of the accumulation function of n with zero as the starting point.

9. The graph shown in Figure 6.3.3 is a model of the rate of change of profit for a new business during its first year. The input is the number of weeks since the business opened, and the output units are thousands of dollars per week.

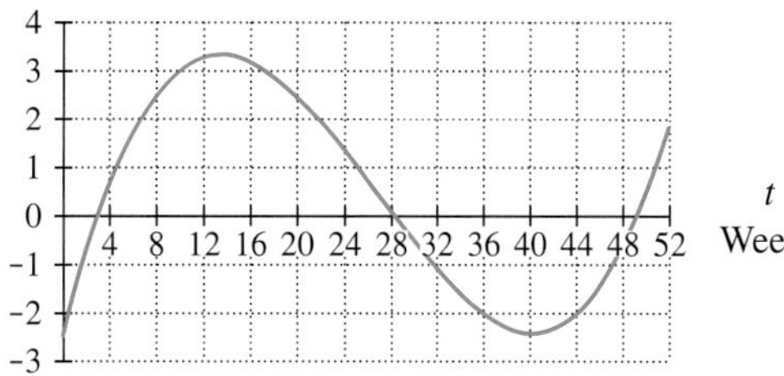

FIGURE 6.3.3

a. What does the area of each box in the grid represent?

b. What is the interpretation, in context, of the accumulation function $P(x) = \int_0^x p(t)\,dt$?

c. Count boxes to estimate accumulation function values from 0 to x for the values of x given in Table 6.36.

TABLE 6.36

x	Accumulation function value	x	Accumulation function value
0		28	
4		32	
8		36	
12		40	
16		44	
20		48	
24		52	

d. Use the data in part c to sketch an accurate graph of the accumulation function $P(x) = \int_0^x p(t)\,dt$. Label units and values on the horizontal and vertical axes.

10. After observing the growth of a certain bacterium, a microbiologist models the growth rate $b(t)$, measured in bacteria per hour, as a function of t, the number of hours since 8 a.m. on the day she began observing. A graph of the model is shown in Figure 6.3.4.

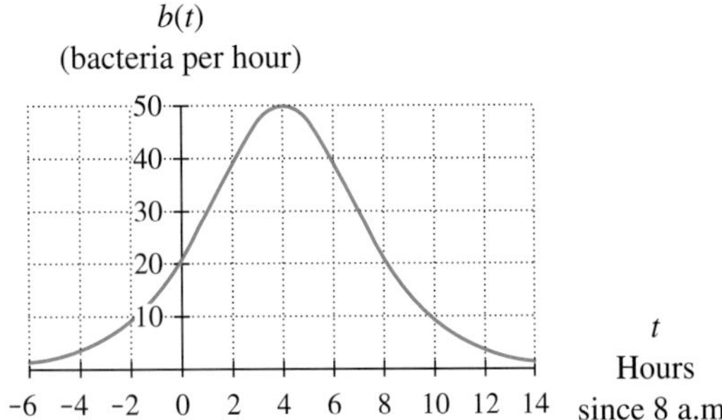

FIGURE 6.3.4

a. What does the area of each box on the grid represent?

b. Estimate the area under the curve by counting boxes from 0 to x for the values of x given in Table 6.37.

TABLE 6.37

x	Accumulation function value	x	Accumulation function value
0		12	
2		14	
4		-2	
6		-4	
8		-6	
10			

c. What is the interpretation of the accumulation function $B(x) = \int_0^x b(t)\,dt$ in the context of bacteria growth?

d. Use the data in part b to sketch an accurate graph of B from 0 to x. Label units and values on the horizontal and vertical axes.

11. The graph shown in Figure 6.3.5 represents $r(t)$, the rate of change of rainfall in Florida during a severe thunderstorm t hours after the rain began falling. Draw a graph of the total amount of rain that fell during this storm, using the facts that

a. the rain first started falling at noon and did not stop until 6 p.m.

b. 3 inches of rain fell between noon and 3 p.m.

c. the total amount of rain that fell during the storm was 5.5 inches.

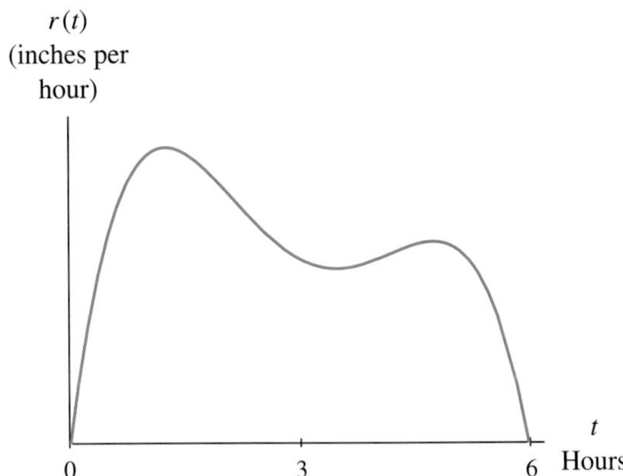

FIGURE 6.3.5

12. The Brazilian government has established a program to protect a certain species of endangered bird that lives in the Amazon rain forest. The program is to be phased out gradually by the year 2020. An environmental group believes that the government's program is destined to fail and has projected that the rate of change in the bird population between 2000 and 2050 will be as shown in Figure 6.3.6.

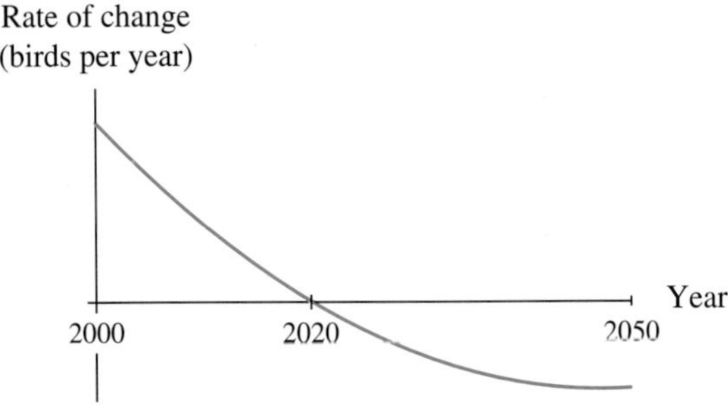

FIGURE 6.3.6

Draw a graph of the bird population between 2000 and 2050, using the facts that (1) at the beginning of 2000 there were 1.3 million birds in existence, and (2) the birds will be extinct by 2050.

In Activities 13 through 18, sketch the indicated accumulation function graphs.

13. a. $\int_A^x f(t)\,dt$

b. $\int_B^x f(t)\,dt$

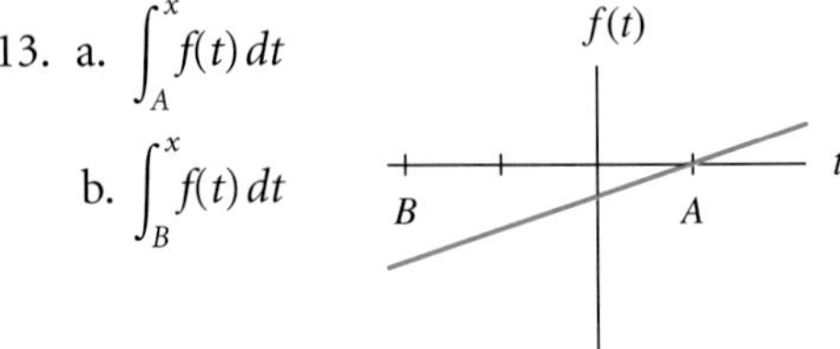

14. a. $\int_0^x f(t)\,dt$

b. $\int_A^x f(t)\,dt$

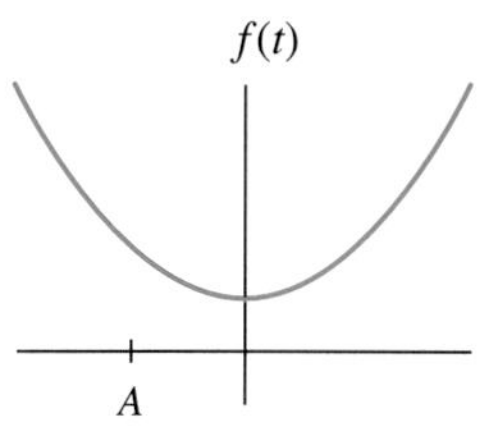

15. a. $\int_0^x f(t)\,dt$

b. $\int_A^x f(t)\,dt$

c. $\int_B^x f(t)\,dt$

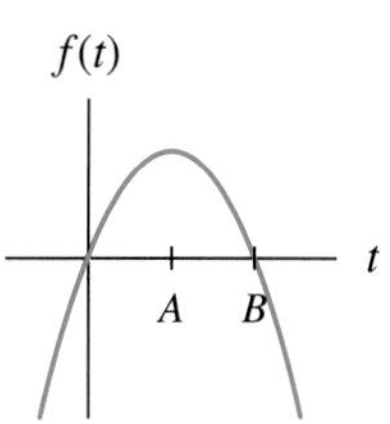

16. $\int_0^x f(t)\,dt$

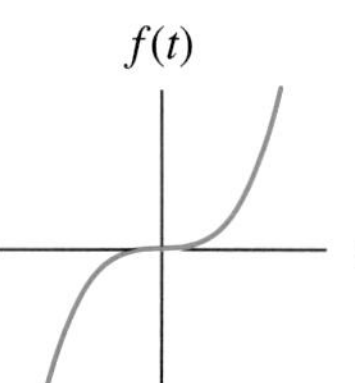

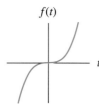

17. $\int_0^x f(t)\,dt$

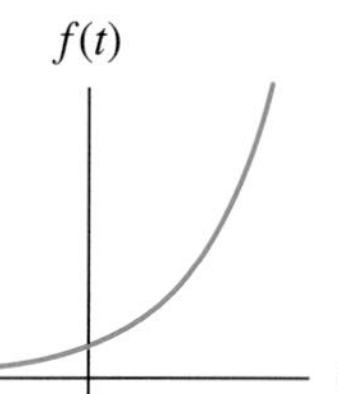

18. $\int_A^x f(t)\,dt$

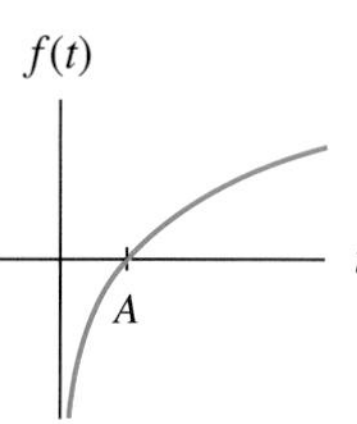

Find formulas for the accumulation functions in Activities 19 through 24. In each case, draw a graph, shading a region whose area the formula represents.

19. a. $\int_0^x 6\,dt$ b. $\int_0^x -2\,dt$ c. $\int_0^x \frac{1}{3}\,dt$

20. a. $\int_{-4}^x 6\,dt$ b. $\int_2^x 6\,dt$ c. $\int_{2000}^x 6\,dt$

21. a. $\int_{-4}^x -2\,dt$ b. $\int_2^x -2\,dt$ c. $\int_{2000}^x -2\,dt$

22. a. $\int_0^x 3t\,dt$ b. $\int_0^x -\frac{1}{2}t\,dt$ c. $\int_0^x 100t\,dt$

23. a. $\int_5^x 3t\,dt$ b. $\int_5^x -\frac{1}{2}t\,dt$ c. $\int_5^x 100t\,dt$

24. a. $\int_{-1}^x (2t+2)\,dt$ b. $\int_0^x (2t+2)\,dt$

c. $\int_1^x (2t+2)\,dt$ d. $\int_{-2}^x (2t+2)\,dt$

On the basis of your answers to Activities 19 through 24, speculate about formulas for the accumulation functions in Activities 25 through 30.

25. $\int_0^x k\,dt$

26. $\int_a^x k\,dt$ for $k > 0$

27. $\int_a^x k\,dt$ for $k < 0$

28. $\int_0^x kt\,dt$

29. $\int_5^x kt\,dt$

30. $\int_a^x (2t+2)\,dt$

31. a. Use the formula you found in Activity 29 to find the area of the region between $f(t) = 3t$ and the horizontal axis from 5 to 19.

b. If $f(t)$ is the rate of change of the revenue of an airline, in hundreds of dollars per day, and t is the number of days since December 31 of last year, what does the answer to part *a* represent?

32. a. Use the formula you found in Activity 28 to find the area of the region between $f(t) = 7t$ and the horizontal axis from 0 to 9.

b. If $f(t)$ is the rate of change in the temperature of an oven, in degrees per minute, and t is the number of minutes since the oven was turned on, what does your answer to part *a* represent?

In each of Activities 33 through 36, a graph is given. Identify, from graphs *a* through *f*, the derivative graph and the accumulation graph (with 0 as the starting point) of the given graph. Graphs *a* through *f* on page 432 may be used more than once.

33.

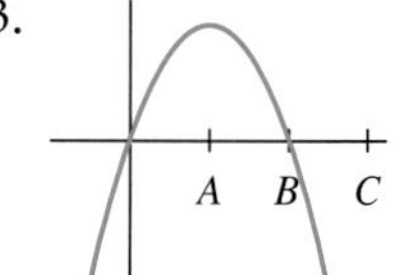

34.

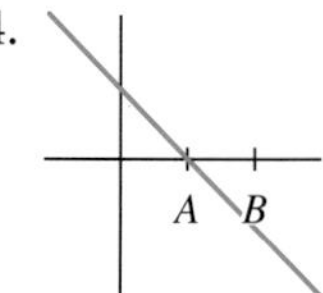

35.

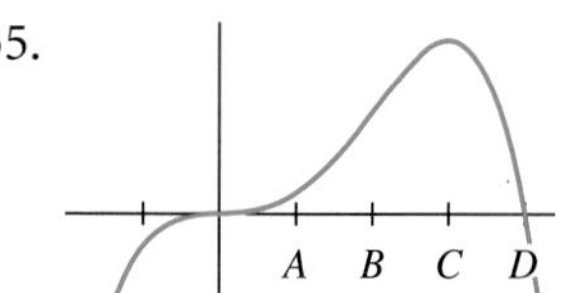

36.

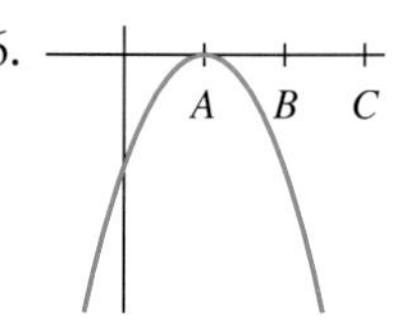

a.

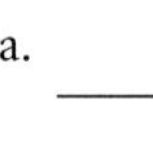

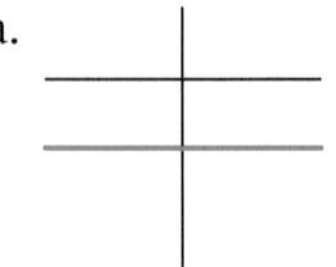

b.

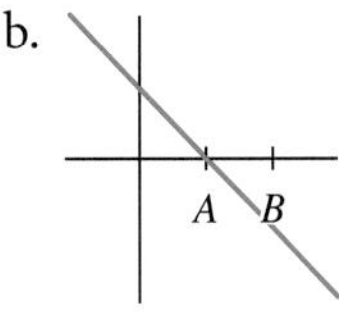

c.

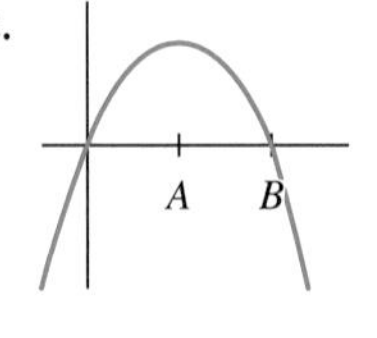

d.

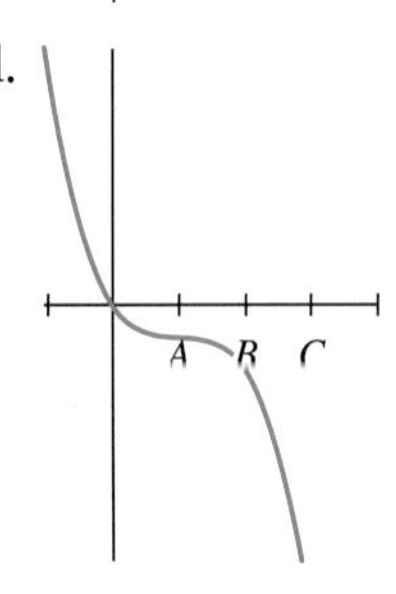

e.

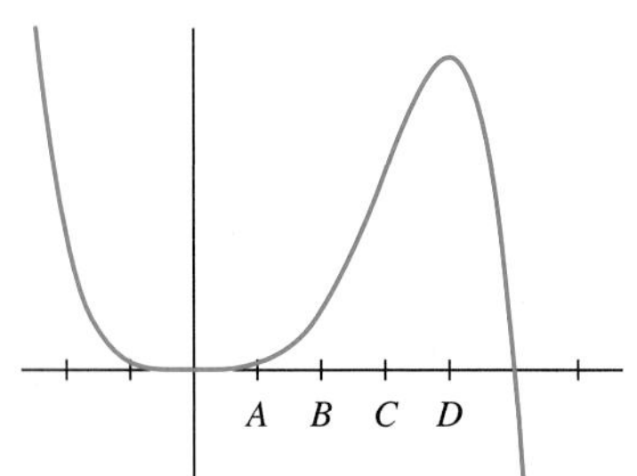

f.

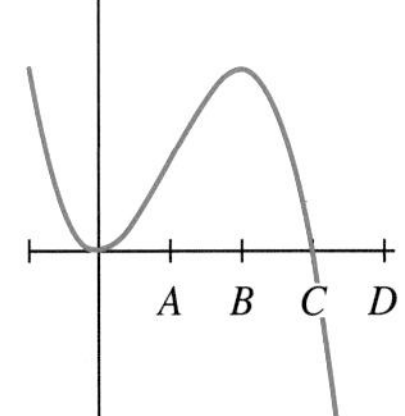

In each of Activities 37 and 38, a table of selected values for a function is given. Also shown are tables of values for the derivative and the accumulation function with 0 as the starting point. Determine which table contains values of the derivative and which contains values of the accumulation function. Justify your choice.

37.

t	f(t)
0	4
1	3
2	0
3	-5
4	-12

Input	Output
0	0
1	-2
2	-4
3	-6
4	-8

Input	Output
0	0
1	3.667
2	5.333
3	3
4	-5.333

38.

m	p(m)
0	0
1	-8
2	-16
3	0
4	64
5	200

Input	Output
0	0
1	-12
2	0
3	36
4	96
5	180

Input	Output
0	0
1	-3
2	-16
3	-27
4	0
5	125

6.4 The Fundamental Theorem

In Section 6.3 we developed accumulation formulas for linear functions and found that these formulas vary slightly depending on the location at which we begin accumulating. Some of the formulas from the examples and activities in the last section are shown below:

$f(t) = 3$ $\qquad \int_0^x 3\,dt = 3x$

$f(t) = -2$ $\qquad \int_0^x -2\,dt = -2x$

$$\int_{2000}^x -2\,dt = -2x + 4000$$

$f(t) = 2t - 4$ $\qquad \int_2^x (2t - 4)\,dt = x^2 - 4x + 4$

$$\int_{-1}^x (2t - 4)\,dt = x^2 - 4x - 5$$

Note that if you take the derivative with respect to x of each of the accumulation formulas, you obtain the original function in terms of x. For example,

$$\frac{d}{dx}\int_2^x (2t - 4)dt = \frac{d}{dx}(x^2 - 4x + 4) = 2x - 4$$

which differs from $2t - 4$ only in the input variable. You may think this relationship is a special case for lines, but consider some of the more complicated accumulation functions that we sketched. In Section 6.3 we began with the function in Figure 6.62a and drew the accumulation function graph shown in Figure 6.62b.

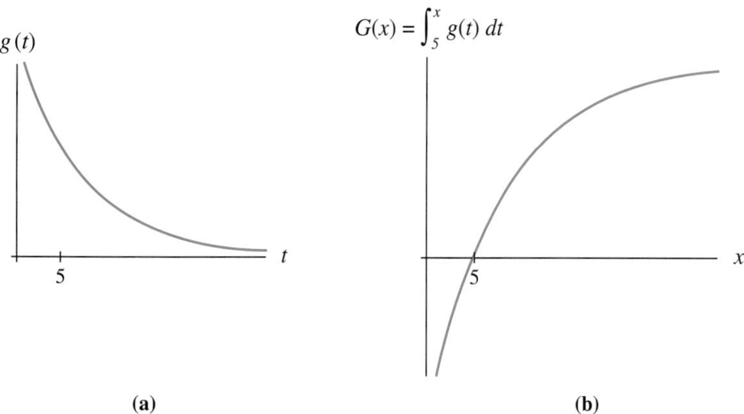

FIGURE 6.62 (a) (b)

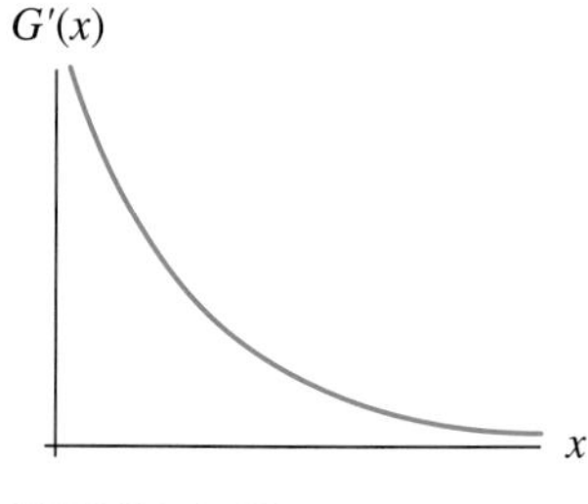

FIGURE 6.63

Now let us sketch the slope graph (derivative graph) of the accumulation function in Figure 6.62b. Note that the slopes are positive everywhere but seem to be getting smaller and smaller as x gets larger. Also, close to the vertical axis, the graph is very steep, so the slopes are very large. The slope graph appears in Figure 6.63. The slope graph is exactly the graph with which we began in Figure 6.62a (with the input variable labeled x instead of t).

In Example 2 in Section 6.3, we began with the graph in Figure 6.64a and sketched the accumulation function shown in Figure 6.64b.

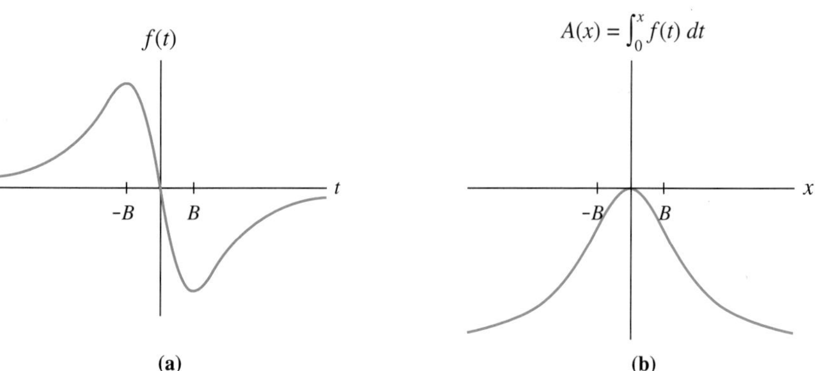

FIGURE 6.64 (a) (b)

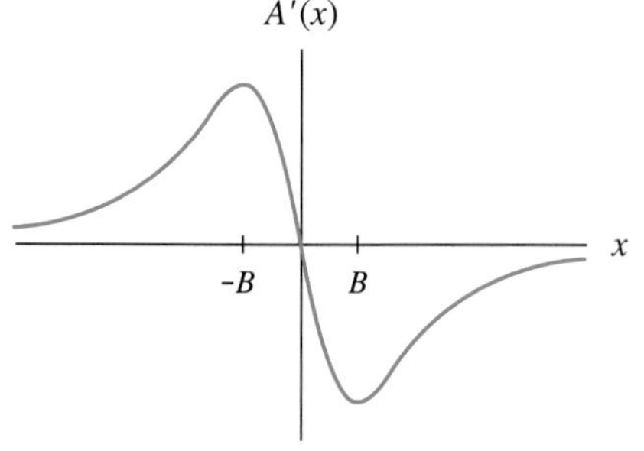

FIGURE 6.65

Again, let us sketch the slope graph of the accumulation function. To the left of zero, the graph has positive slopes. The slopes are near zero to the far left, and the graph becomes steeper until $-B$. Between $-B$ and 0, the slopes are still positive but are approaching 0 as the accumulation function approaches its maximum. At $x = 0$, the slope is zero. To the right of $x = 0$, the slopes are negative. They become more and more negative until B. To the right of B, the slopes are still negative but are getting closer and closer to zero as the graph levels off. The slope graph is shown in Figure 6.65. Again, this is exactly the graph with which we began in Figure 6.64a.

You are probably beginning to see that if we begin with a function f with input t, graph or find a formula for the accumulation function $A(x) = \int_a^x f(t)\,dt$, and then take the derivative or draw the slope graph, we get f, the function with which we began but in terms of x. In order to explore this connection between accumulation functions and derivatives, consider the following argument:

Let A be the accumulation function of f from a to x. The graph in Figure 6.66a shows the function f and the area representing the accumulation function value from a to x. Figure 6.66b shows the region whose area is the accumulation value from a to $x + h$.

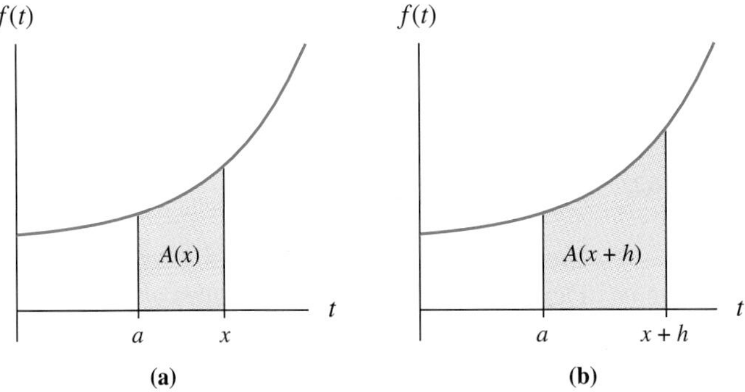

FIGURE 6.66

Next, consider the difference between the two areas. The small region with this area is shown in Figure 6.67a.

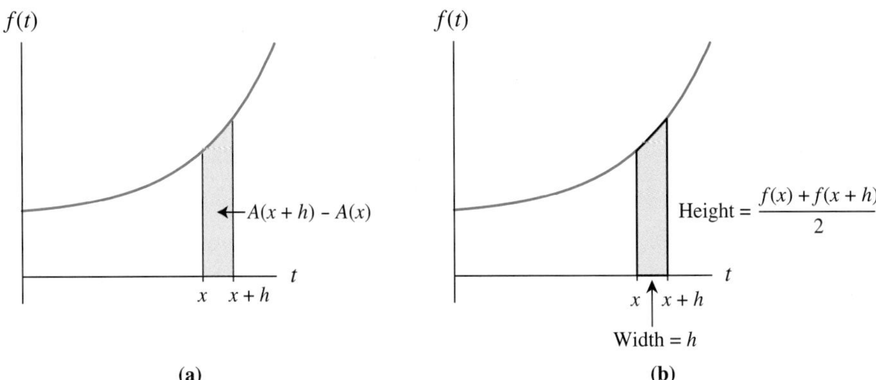

FIGURE 6.67

We can approximate the area of this region by using a trapezoid. Recall that the height of a trapezoid constructed over an interval is the average of the function value at the two endpoints. Figure 6.67b shows this trapezoid and its height and width.

The true area of the shaded region in Figure 6.67a is $A(x + h) - A(x)$, and it can be approximated by the area of the trapezoid. Thus we have

$$A(x + h) - A(x) \approx \left[\frac{f(x) + f(x + h)}{2}\right]h$$

We now divide each side of the expression by h. The reason for this division will be evident later.

$$\frac{A(x+h)-A(x)}{h} \approx \frac{f(x)+f(x+h)}{2}$$

Consider what happens as h becomes smaller and smaller. In other words, what happens when we take the limit of the above expression as h approaches zero?

$$\lim_{h\to 0}\frac{A(x+h)-A(x)}{h} \approx \lim_{h\to 0}\frac{f(x)+f(x+h)}{2}$$

You should recognize the term on the left as the derivative of A. As h approaches 0, $f(x+h)$ gets closer and closer to $f(x)$, so the term on the right approaches $\frac{f(x)+f(x)}{2} = \frac{2f(x)}{2} = f(x)$. Thus we have

$$\frac{dA}{dx} \approx f(x)$$

In fact, we can replace the inequality with an equality (although a discussion of why this is the case is beyond the scope of this book):

$$\frac{dA}{dx} = f(x)$$

This expression says that the derivative of the accumulation function is the original function. This is actually a surprising result—accumulation functions and areas are related to derivatives! In fact, this is such an important result that it is called the **Fundamental Theorem of Calculus**.

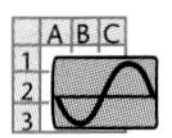

You may wish to ask your students to translate the symbolic statement of the Fundamental Theorem into graphical terms: "The slope graph of an accumulation function graph is the original graph with a different input variable."

The Fundamental Theorem of Calculus

For any continuous function f with input t, the derivative of an accumulation function of f is the function f in terms of x. In symbols, we write

$$\frac{d}{dx}\left(\int_a^x f(t)\,dt\right) = f(x)$$

We can infer from the Fundamental Theorem that to find an accumulation formula, we need only reverse the process of finding a derivative. For this reason, we call the accumulation function $F(x) = \int_a^x f(t)\,dt$ an **antiderivative** of the function f.

Antiderivative

Let f be a function of x. A function F is called an antiderivative of f if $F'(x) = f(x)$; that is, the derivative of F is f.

Our motivation for developing accumulation functions (antiderivatives) is not only to have a formula for accumulated change but also (and more importantly) to develop a function for a quantity if we know a function for that quantity's rate of change. In Example 3 of Section 6.1, we were interested in finding the total volume of water that ran through the Carson River in the 20 hours after 11:45 a.m. on a Wednesday but were given values for only the flow rate in cubic feet per hour. If we could find an antiderivative of the flow rate equation, we would have a model for the volume of the water flowing through the river and could use the model to answer questions about the volume rather than summing areas of rectangles as we did in Section 6.1.

Similarly, in Example 2 of Section 6.1, we found a model for the rate of change of the concentration of a drug and summed areas of rectangles to find a change in the concentration. If we could find an antiderivative of the rate-of-change function, we would then have a function modeling the concentration of the drug and could use it to answer the questions posed in Section 6.1.

Recovering a Function

Recovering a function is the phrase we use for the process of beginning with a rate-of-change function for a quantity and finding its antiderivative to obtain a function for the quantity. An important part of recovering a function from its rate of change is recovering the units of the function from the units of its rate of change. If $\frac{dM}{dt}$ is the rate-of-change function of the amount of insulin in a patient's body t hours after an injection, and if it is reported in milliliters per hour, then we can recover the units of the amount function M by recalling that the rate of change of a function is a slope of a tangent line. Slope is calculated as $\frac{\text{rise}}{\text{run}}$, where the units are $\frac{\text{output units}}{\text{input units}}$. In the case of $\frac{dM}{dt}$, the units milliliters per hour can be rewritten as $\frac{\text{milliliters}}{\text{hour}}$. Now we can see that the output units of M are milliliters and the input units are hours. Figure 6.68 shows input/output diagrams for $\frac{dM}{dt}$ and M.

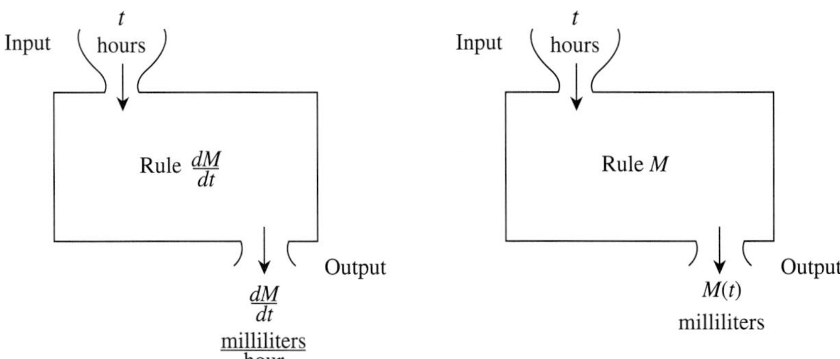

FIGURE 6.68

Occasionally, units of a rate of change are expressed in terms of a squared unit. For example, acceleration is often expressed in feet per second squared. Suppose $A(t)$ is acceleration of a vehicle in feet per second squared, where t is the number of seconds since the vehicle began accelerating. Again, this is obviously the rate of change of some function. In fact, it is the rate of change of velocity. However, the input and output units of that velocity function may not be immediately apparent because of the use of the word *squared*. Rewrite feet per second squared as

Point out to your students that squared units in the denominator can be written with two "per" statements.

$$\frac{\text{feet}}{(\text{second})^2} = \frac{\text{feet}}{(\text{second})(\text{second})} = \frac{\text{feet}}{\text{second}} \div (\text{second})$$
$$= (\text{feet per second}) \text{ per second}$$

The output units of the velocity function are now identifiable as feet per second. Input/output diagrams for these functions are shown in Figure 6.69.

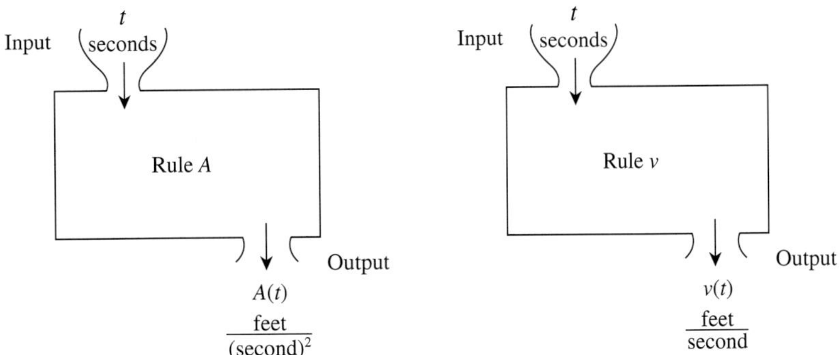

FIGURE 6.69

We have seen how to recover the quantity units from rate-of-change units, and we have noted that the Fundamental Theorem of Calculus implies that we have a powerful algebraic tool for recovering the accumulated-quantity formula from its rate-of-change formula. However, we have yet to develop the specific algebraic rules that will enable us actually to use this tool. We now turn our attention to developing those rules.

Antiderivative Formulas

Given a function, how do we find an antiderivative? As we have seen, we must reverse the process of differentiation. Antidifferentiation starts with the known derivative and finds the unknown function. For example, consider the constant function $f(x) = 3$. To find an antiderivative of f, we need to find a function of x whose derivative is 3. One such function is $F(x) = 3x$. Other functions whose derivatives are 3 include $F(x) = 3x + 7$ and $F(x) = 3x - 24.9$.

In fact, having found one antiderivative F for a given function f, we can obtain infinitely many antiderivatives for that function by adding an arbitrary constant C to F. Thus we call $y = F(x) + C$ the **general antiderivative of *f*.** We use the notation

$$\int f(x)\,dx = F(x) + C$$

Help your students understand that a general antiderivative is a formula for all possible accumulation functions (that is, for all possible starting points). Also point out that graphically, the *C*-term in the general antiderivative shifts the function up or down, accounting for various accumulation function starting points.

for the general antiderivative. The general antiderivative is a group of infinitely many functions. (A particular accumulation function is one specific function from that group.) Note that the integral sign has no upper and lower limits for general antiderivative notation. The dx in this notation is to remind us that we are finding the general antiderivative with respect to x, so our antiderivative formula will be in terms of x. For example, we say that the general antiderivative for $f(x) = 3$ is $F(x) = 3x + C$, and we write $\int 3\,dx = 3x + C$.

As we saw in Section 6.2, an antiderivative (accumulation formula) of any constant function will be a line because lines have constant derivatives. We can write this general rule as

$$\int k\,dx = kx + C$$

where k and C are any constants.

Now consider finding the general antiderivative of $f(x) = 2x$. We are seeking a function whose derivative is $2x$. The function is $F(x) = x^2 + C$, and we write

$$\int 2x\,dx = x^2 + C$$

It is more difficult to reverse the derivative process for $f(x) = x^2$. Recall that the power rule for derivatives $\left[\frac{d}{dx}(x^n) = nx^{n-1}\right]$ instructs us to

- multiply by the power
- subtract 1 from the power to get the new power

To reverse the process for antiderivatives, we

- *add* 1 to the power to get the new power
- *divide* by that new power

This formula is known as the **Simple Power Rule** for antiderivatives.

Simple Power Rule for Antiderivatives

$$\int x^n\,dx = \frac{x^{n+1}}{n+1} + C \quad \text{for any } n \neq -1$$

This rule requires that $n \neq -1$, because otherwise, we would be dividing by zero.

In the case of $f(x) = x^2$, the general antiderivative is $F(x) = \frac{x^{2+1}}{2+1} + C = \frac{x^3}{3} + C$:

$$\int x^2\,dx = \frac{x^3}{3} + C$$

EXAMPLE 1 *Finding Antiderivatives*

Find the following general antiderivatives and their appropriate units.

a. $\int -7\,dx$ degrees per hour squared where x is in hours

b. $\int h^{0.5}\,dh$ parts per million per day where h is in days

Solution

a. $\int -7\,dx - -7x + C$ degrees per hour

b. $\int h^{0.5}\,dh = \frac{h^{1.5}}{1.5} + C$ parts per million

Recall the Constant Multiplier Rule for derivatives:

If $g(x) = kf(x)$, then $g'(x) = kf'(x)$, where k is a constant.

A similar rule applies for antiderivatives:

Constant Multiplier Rule for Antiderivatives

$$\int kf(x)\,dx = k\int f(x)\,dx$$

Thus $\int 12x^6\,dx = 12\int x^6\,dx = 12\left(\frac{x^7}{7}\right) + C = \frac{12x^7}{7} + C$.

Another property of antiderivatives that can be easily deduced from a similar property for derivatives is the Sum Rule.

Sum Rule

$$\int [f(x) \pm g(x)]dx = \int f(x)\,dx \pm \int g(x)\,dx$$

The Sum Rule lets us find an antiderivative for a sum (or difference) of functions by operating on each function independently. For example,

$$\begin{aligned}\int (7x^3 + x)\,dx &= \int 7x^3\,dx + \int x\,dx \\ &= \left(\frac{7x^4}{4} + C_1\right) + \left(\frac{x^2}{2} + C_2\right) \\ &= \frac{7x^4}{4} + \frac{x^2}{2} + C \quad \text{(Combine } C_1 \text{ and } C_2 \text{ into one constant } C.)\end{aligned}$$

Repeated applications of the Simple Power Rule, the Constant Multiplier Rule, and the Sum Rule enable us to find an antiderivative of any polynomial function. We now have the tools we need to begin with a simple polynomial rate-of-change function for a quantity and recover an amount function for that quantity.

EXAMPLE 2 *Birth Rate*

An African country has an increasing population but a declining birth rate, a situation that results in the number of babies born each year increasing but at a slower rate. The rate of change in the number of babies born each year is given by

$$b(t) = 87{,}000 - 1600t \text{ births per year}$$

t years from the end of this year. Also, the number of babies born in the current year is 1,185,800.

a. Find a function describing the number of births each year t years from now.

b. Use the function in part *a* to estimate the number of babies born next year.

Solution

a. A function B describing the number of births each year is found as

$$B(t) = \int b(t)\,dt = 87{,}000t - \frac{1600}{2}t^2 + C = 87{,}000t - 800t^2 + C \text{ births}$$

t years from now.

We also know that $B(0) = 1{,}185{,}800$, so $1{,}185{,}800 = 87{,}000(0) - 800(0)^2 + C$, which gives $C = 1{,}185{,}800$. Thus we have

$$B(t) = 87{,}000t - 800t^2 + 1{,}185{,}800 \text{ births each year}$$

t years from now.

b. The number of babies born next year is estimated as $B(1) = 1{,}272{,}000$ babies.

We have just presented and applied three antiderivative rules: the Simple Power Rule, the Constant Multiplier Rule, and the Sum Rule. Now let us look at four more rules for finding antiderivatives.

Refer to the Simple Power Rule, and note that it did not apply for $n = -1$. The case where $n = -1$ is special. This results in the antiderivative $\int x^{-1}\,dx = \int \frac{1}{x}\,dx$. Recall that $\frac{d}{dx}(\ln x) = \frac{1}{x}$. This is valid only for $x > 0$, because $\ln x$ is not defined for $x < 0$. When $x < 0$, we use $\ln(-x)$ because $\frac{d}{dx}[\ln(-x)] = \frac{-1}{-x} = \frac{1}{x}$. Both cases ($x > 0$ and $x < 0$) can be handled very simply by using $\ln|x|$.

You may wish to motivate the Natural Log Rule by sketching the accumulation function of $f(x) = \frac{1}{x}$. Your students should recognize the result as a log function.

Natural Log Rule

$$\int \frac{1}{x}\,dx = \ln|x| + C$$

The two final antiderivative formulas that we consider are for exponential functions. Recall that the derivative of $f(x) = e^x$ is e^x. Similarly, the general antiderivative of $f(x) = e^x$ is also e^x plus a constant.

e^x Rule

$$\int e^x\,dx = e^x + C$$

The other exponential function that we have encountered is $f(x) = b^x$. Its derivative was found by multiplying b^x by $\ln b$. To find the general antiderivative, we divide b^x by $\ln b$ and add a constant:

Exponential Rule

$$\int b^x\,dx = \frac{b^x}{\ln b} + C$$

Note that the e^x Rule is a special case of this Exponential Rule with $b = e$.

Because we often encounter functions of the form $f(x) = e^{ax}$, it is helpful to know this function's derivative formula. Recall that the Chain Rule for derivatives applied to $f(x) = e^{ax}$ gives $f'(x) = ae^{ax}$. Thus, to reverse the derivative process, we leave the e^{ax}-term and divide, rather than multiply, by a.

e^{ax} Rule

$$\int e^{ax}\,dx = \frac{1}{a}e^{ax} + C \quad \text{for any } a \neq 0$$

In summary, we now have the following antiderivative formulas, where k, n, b, and a are constants.

Antiderivative Formulas

	Function: f	General antiderivative: $\int f(x)\,dx$
Constant Rule	k	$kx + C$
Simple Power Rule	$x^n, n \neq -1$	$\left(\frac{1}{n+1}\right)x^{n+1} + C$
Natural Log Rule	$\frac{1}{x}$	$\ln\lvert x\rvert + C$
Exponential Rule	$b^x, b > 0$	$\left(\frac{1}{\ln b}\right)b^x + C$
e^x Rule	e^x	$e^x + C$
e^{ax} Rule	$e^{ax}, a \neq 0$	$\frac{1}{a}e^{ax} + C$
Constant Multiplier Rule	$kg(x)$	$k\int g(x)\,dx$
Sum Rule	$g(x) \pm h(x)$	$\int g(x)\,dx \pm \int h(x)\,dx$

EXAMPLE 3 More Antiderivatives

Find the following general antiderivatives.

a. $\int\left(3^x - 7e^x + \frac{5}{x}\right)dx$ quarts per hour where x is measured in hours

b. $\int\left(4\sqrt{x} + 100e^{0.06x} + 0.46\right)dx$ mpg per mph where x is measured in mph

Solution

a. $$\int\left(3^x - 7e^x + \frac{5}{x}\right)dx = \int 3^x\,dx - 7\int e^x\,dx + 5\int \frac{1}{x}\,dx$$
$$= \frac{3^x}{\ln 3} - 7e^x + 5\ln\lvert x\rvert + C \text{ quarts}$$

b. We must first rewrite $\sqrt{x}$ as $x^{1/2}$ and then apply the appropriate rules:

$$\int\left(4\sqrt{x} + 100e^{0.06x} + 0.46\right)dx = \int [4x^{1/2} + 100e^{0.06x} + 0.46]\,dx$$
$$= 4\int x^{1/2}\,dx + 100\int e^{0.06x}\,dx + \int 0.46\,dx$$
$$= 4\frac{x^{3/2}}{3/2} + 100\frac{e^{0.06x}}{0.06} + 0.46x + C$$
$$= \frac{8}{3}x^{3/2} + \frac{100}{0.06}e^{0.06x} + 0.46x + C \text{ mpg}$$

....

Specific Antiderivatives

Point out to your students that a specific antiderivative is the formula for a particular accumulation function where the given condition specifies a point (though not necessarily the starting point) on this accumulation function graph.

We have seen that any function has infinitely many antiderivatives, each differing by a constant. When we seek an antiderivative with a particular constant, we call the resulting function a **specific antiderivative.** An example of a specific antiderivative is the function in Example 2 for the number of births in an African country. Using the information that in the current year the number of births is 1,185,800, we were able to solve for the constant to obtain the specific antiderivative $B(t) = 87{,}000t - 800t^2 + 1{,}185{,}800$ births each year t years from now.

In general, to find a specific antiderivative, you must be given additional information about the output quantity that the antiderivative describes. After finding a general antiderivative, you simply substitute the given input and corresponding output into the general antiderivative and solve for the constant. Then you replace the constant in the general antiderivative formula with the value you found to obtain the specific antiderivative. Example 4 illustrates this process.

EXAMPLE 4 *Marginal Cost*

Suppose that a manufacturer of small toaster ovens has collected the data given in Table 6.38, which shows, at various production levels, the approximate cost to produce one more oven. Recall from Section 5.1 that this is marginal cost and can be interpreted as the rate of change of cost.

TABLE 6.38

Production level (ovens per day)	200	300	400	500	600	700
Cost to produce an additional oven	\$29	\$20	\$15	\$11	\$9	\$7

The manufacturer also knows that the total cost to produce 250 ovens is \$12,000.

a. Find a model for the marginal cost data.

b. Recover a cost model from the model you found in part *a*.

c. Estimate the cost of producing 500 ovens.

Solution

a. Either a quadratic or an exponential model is a good fit to the data. We choose an exponential model:

$$C'(x) = 47.638(0.9972^x) \text{ dollars per oven}$$

where x is the number of ovens produced each day.

b. To recover a model for cost, we need an antiderivative of C' satisfying the known condition that the cost to produce 250 ovens is \$12,000. The general antiderivative is

$$C(x) = \frac{47.638(0.9972^x)}{\ln 0.9972} + K \approx -16{,}991.852(0.9972^x) + K$$

where K is a constant.

Using the fact that $C(250) = \$12{,}000$, we substitute 250 for x, set the antiderivative equal to 12,000, and solve for K.

$$-16{,}991.852(0.9972^{250}) + K = 12{,}000$$
$$-8430.309 + K = 12{,}000$$
$$K \approx 20{,}430.309$$

Thus the specific antiderivative giving the approximate cost of producing x toaster ovens is

$$C(x) = -16{,}991.852(0.9972^x) + 20{,}430.309 \text{ dollars}$$

c. You can readily verify that the cost of producing 500 toaster ovens is estimated by $C(500) \approx \$16{,}248$.

It is sometimes necessary to find an antiderivative twice in order to obtain the appropriate accumulation formula. For example, to obtain distance from acceleration, you must determine the specific antiderivative of the acceleration function to obtain a velocity function and then determine the specific antiderivative of the velocity function to obtain a function for distance traveled.

EXAMPLE 5 *Falling Pianos*

A mathematically inclined cartoonist wants to make sure his animated cartoons accurately portray the laws of physics. In a particular cartoon he is creating, a grand piano falls from the top of a 10-story building. How many seconds should he allow the piano to fall before it hits the ground? Assume that 1 story equals 12 feet and that acceleration due to gravity is -32 feet per second squared.

Solution We begin with the equation for acceleration, $a(t) = -32$ feet per second squared t seconds after the piano falls. The antiderivative of acceleration gives an equation for velocity.

$$\int a(t)\,dt = v(t) = -32t + C \text{ feet per second after } t \text{ seconds of fall}$$

To find the specific velocity equation for this example, we need information about the velocity at a specific time. The fact that the piano falls (rather than being pushed with an initial force) tells us that the velocity is zero when the time is zero. Substituting zero for both t and v and solving for C give $C = 0$. Thus the specific antiderivative describing velocity is $v(t) = -32t$ feet per second after t seconds of fall.

From this velocity equation, we can derive an equation for position (the distance the piano is above the ground) by finding the antiderivative. In this case, the general antiderivative of v is

$$\int v(t)\,dt = s(t) = -16t^2 + K \text{ feet after } t \text{ seconds of fall}$$

Again, we find the specific position equation by using information given about the position of the piano at a certain time. In this example we know that when time is zero, the piano is 10 stories, or 120 feet, above the ground. We substitute this information into the position equation and solve for K, obtaining $K = 120$ feet. Substituting this value for K in the position equation yields the specific position equation

$$s(t) = -16t^2 + 120 \text{ feet after } t \text{ seconds of fall}$$

Now that we have equations for acceleration, velocity, and position, we can answer the question posed: How long until the piano hits the ground? Let us phrase this question mathematically: When position is zero, what is time? To answer this question, we set the position equation equal to zero and solve for t.

$$0 = -16t^2 + 120$$
$$16t^2 = 120$$
$$t^2 = \frac{120}{16}$$
$$t \approx \pm 2.7 \text{ seconds}$$

Because negative time doesn't make sense in the context of the question, we conclude that the cartoonist should allow approximately 2.7 seconds for the piano to fall.

Let us summarize what we have learned thus far about integrals. The definite integral $\int_a^b f(x)\,dx$ is a limiting value of sums of areas of rectangles and gives us the area of the region between the graph of f and the x-axis if the graph lies above the horizontal axis from a to b. When the graph of f lies below the horizontal axis from a to b, the definite integral is the negative of the area inscribed. If $f(x)$ is a rate of change of some quantity, then $\int_a^b f(x)\,dx$ is the change in the quantity from a to b.

Make certain your students can articulate the differences among the three symbols in Table 6.39. You may wish to discuss with your students which types of questions are associated with each type of integral. See the *Instructor's Resource Guide* for some examples.

The accumulation function $A(x) = \int_a^x f(t)\,dt$ is a formula in terms of x for the accumulated change in $f(t)$ from a to x. We use the integral symbol without the upper and lower limits, $\int f(x)\,dx$, to represent the general antiderivative of f. Although these three symbols are similar, it is important that you have a clear understanding of what each one represents. Their interpretations are summarized in Table 6.39.

TABLE 6.39

Symbol	Name	Interpretation
$\int_a^b f(x)\,dx$	definite integral	a number that can be thought of in terms of area
$\int_a^x f(t)\,dt$	accumulation function, specific antiderivative	a formula for an accumulated amount
$\int f(x)\,dx$	general antiderivative	a formula whose derivative is f

The Fundamental Theorem of Calculus tells us that accumulation functions are specific antiderivatives. As we shall see in Section 6.5, antiderivatives enable us to find areas algebraically by using accumulation formulas rather than numerically as limiting values of sums of areas of rectangles.

Apart from helping us find areas, antiderivatives are useful in allowing us to recover functions from rates of change. We have seen several examples of that in this section. It may seem difficult to reverse your thinking from finding derivatives to finding antiderivatives, but with practice you will soon be proficient at both.

6.4 Concept Inventory

- *Fundamental Theorem of Calculus*
- *Antiderivative*
- *Recovering a function from its rate of change*
- $\int f(x)\,dx$ = *general antiderivative*
- *Antiderivative formulas*
- *Specific antiderivative*

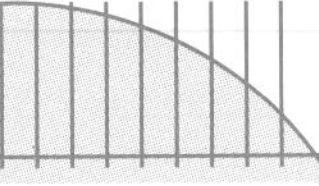

6.4 Activities

In Activities 1 through 12, a and b are constants and x and t are variables. In these activities, label each notation as always representing

a. A function of x

b. A function of t

c. A number

1. $f'(t)$
2. $\dfrac{df}{dx}$
3. $f'(3)$
4. $\int f(t)\,dt$
5. $\int f(x)\,dx$
6. $\int_a^b f(t)\,dt$
7. $\int_a^b f(x)\,dx$
8. $\int_a^x f(t)\,dt$
9. $\int_b^t f(x)\,dx$
10. $\frac{d}{dx}\int_a^x f(t)\,dt$
11. $\frac{d}{dt}\int_a^t f(x)\,dx$
12. $\frac{d}{dx}\int_a^a f(t)\,dt$

For each of the rates of change in Activities 13 through 16:

a. Write the units of the rate of change as a fraction.

b. Draw an input/output diagram for the recovered function.

c. Interpret the recovered function in a sentence.

13. When m thousand dollars are being spent on advertising, the annual revenue of a corporation is changing by $\frac{dR}{dm}$ million dollars per thousand dollars spent on advertising.

14. The percentage of households with washing machines was changing by $\frac{dW}{dt}$ percentage points per year, where t is the number of years since 1950.

15. The concentration of a drug in the bloodstream of a patient is changing by $\frac{dc}{dh}$ milligrams per liter per hour h hours after the drug was given.

16. The level of production at a tire manufacturer h hours after production began is increasing by $P'(h)$ tires per hour squared.

practice

Find the general antiderivative as indicated in Activities 17 through 22. Check each of your antiderivatives by taking its derivative.

17. $\int 19.436(1.07^x)\,dx$
18. $\int 39.24e^{3.9x}\,dx$
19. $\int [6e^x + 4(2^x)]\,dx$
20. $\int (32.685x^3 + 3.296x - 15.067)\,dx$
21. $\int \left(10^x + 4\sqrt{x} + 8\right)dx$
22. $\int \left[\frac{1}{2}x + \frac{1}{2x} + \left(\frac{1}{2}\right)^x\right]dx$

We have purposely kept the number of drill problems in the activities to a minimum, because we want this book to focus on conceptual understanding rather than algebraic manipulation. However, we realize that there are times when additional practice is necessary. Additional antiderivative formula drill problems can be found on the *Calculus Concepts* web site.

For each of the rate-of-change functions in Activities 23 through 27, find the general antiderivative, and label the units on the antiderivative.

23. $s(m) = 6250(0.92985^m)$ CDs per month m months since the beginning of the year

24. $p(x) = 0.03731x^2 - 0.4841x + 1.4069$ dollars per 1000 cubic feet per year x years since 1989

25. $j(x) = \dfrac{15.29}{x} + 7.95$ units per dollar where x is the price (in dollars)

26. $c(x) = \dfrac{0.7925}{x} + 0.3292(0.009324^x)$ dollars per unit squared when x units are produced

27. $p(t) = 1.724928e^{0.0256t}$ millions of people per year t years after 1990

In Activities 28 through 31, find F, the specific antiderivative of f.

28. $f(x) = 5x + 3$; $F(1) = 3$

29. $f(t) = t^2 + 2t$; $F(12) = 700$

30. $f(u) = \frac{2}{u} + u$; $F(1) = 5$

31. $f(z) = \frac{1}{z^2} + e^z$; $F(2) = 1$

32. The rate of change of the average yield of short-term German bonds can be described by the equation $G(t) = \frac{0.5696}{t}$ percentage points per year for a bond with a maturity time of t years. The average 10-year bond has a yield of 4.95%. Find the specific antiderivative describing the average yield of short-term German bonds. How is this specific antiderivative related to an accumulation function of G?

33. The rate of change of the weight of a laboratory mouse can be modeled by the equation $W(t) = \frac{7.372}{t}$ grams per week, where t is the age of the mouse, in weeks, beyond 2 weeks. At an age of 9 weeks, the mouse weighed 26 grams. Find the specific antiderivative describing the weight of the mouse. How is this specific antiderivative related to an accumulation function of W?

34. The rate of change of the average annual fuel consumption of passenger vehicles, buses, and trucks from 1970 through 1995 can be modeled by the equation $f(t) = 0.85t - 16.431$ gallons per vehicle per year t years after 1970. The average annual fuel consumption was 712 gallons per vehicle in 1980. Find the specific antiderivative giving the average annual fuel consumption. How is this specific antiderivative related to an accumulation function of f?

35. The rate of change of the gender ratio for the United States during the twentieth century can be modeled as $g(t) = (1.667 \cdot 10^{-4})t^2 - 0.01472t - 0.103$ males per 100 females per year t years after 1900. In 1970 the gender ratio was 94.8 males per 100 females. Find a specific antiderivative giving the gender ratio. How is this specific antiderivative related to an accumulation function of g?

36. On the basis of data in the 1998 *Statistical Abstract*, between 1980 and 1995, the amount of outstanding consumer credit grew at a rate described by the function $c(x) = 27.375(1.0904^x)$ billion dollars per year x years after 1980.

   a. Using the fact that outstanding consumer credit was \$794.6 billion in 1990, find an equation for the amount of outstanding consumer credit. For what years can this equation be used with confidence?

   b. By how much did consumer credit grow between 1985 and 1995?

37. On the basis of data from the Cellular Telecommunications Industry Association, the rate of change in the number of cellular telephone subscribers in the United States from 1988 through 1995 can be modeled by the equation $p(x) = 891.6(1.5^x)$ thousand subscribers per year x years after 1988.

   a. Find a model for the number of cellular phone subscribers in the United States. Use the fact that there were 2.069 million subscribers in 1988.

   b. Find the change in the number of cellular phone subscribers in the United States between 1990 and 1994.

38. During the 1990s the civilian labor force in the Trident region of South Carolina was expanding[18] at a rate of $p(x) = -2915x + 19{,}433$ people each year x years after 1989. The civilian labor force was 227,120 people in 1989.

   a. Find a model for the size of the civilian labor force in the 1990s.

   b. Determine, according to the model, how many people were in the labor force by 1992.

   c. According to the model, how many people were added to the labor force between 1989 and 1991?

39. For the first 2 months after sweet corn is planted (given optimal growing conditions) the rate of change of its growth rate can be modeled by

$$g'(d) = -0.00036d - 0.018 \text{ inches per day squared}$$

where d is the number of days after sprouting.

   a. Find a function for the growth rate of sweet corn d days after it sprouts. The corn was growing by 1.95 inches per day on the day it sprouted.

   b. Find a function for the height of sweet corn d days after it sprouts.

   c. How tall is the sweet corn 60 days after sprouting?

40. An investment worth \$1 million in 2000 has been growing at a rate of $f(t) = 0.1397619(1.15^t)$ million dollars per year t years after 2000.

   a. Recover the amount function, and determine the current value of the investment and its projected value next year.

   b. Determine how much the investment has grown since 2000 and how much it is projected to grow over the next year.

18. Based on information in *1993 Trident Economic Forecast*.

McDonald's study referred to in Activity 45 was based on observations of veterinarians who treated cats that had fallen from buildings in New York City. None of the cats' falls were deliberately caused by the researchers.

41. The amount[19] of poultry produced in the United States annually from 1990 through 1998 can be modeled as $p(x) = -0.019x^3 + 0.191x^2 + 0.927x + 23.539$ billion pounds x years after 1990.
    a. Write the general antiderivative of p.
    b. According to the model, how much poultry was produced in the United States from 1990 through 1998?
    c. Find a model for the rate of change in annual poultry production.
    d. How do p, the answer to part a, and the answer to part c differ?

42. As a person's age increases, the amount she or he would have to pay as a first-year premium on a new life insurance policy also increases. According to a recent report from an insurance company, the first-year premium (per \$1000) on an annual renewable term life insurance policy for nonsmokers between the ages of 30 and 60 increases at a rate of $p'(x) = 0.017714(1.158887^x)$ dollars per year of age, where $x + 30$ is the age of the insured in years.
    a. What value is x if the person is 45 years old?
    b. Find the function for the premium per \$1000. The premium per \$1000 for a 40-year-old is \$1.49.
    c. Determine the first-year premium for a 30-year-old who is purchasing a \$500,000 policy.

43. The Washington Monument, located at one end of the Federal Mall in Washington, DC, is the world's tallest obelisk at 555 feet. Suppose that a tourist drops a penny from the observation deck atop the monument. Let us assume that the penny falls from a height of 540 feet.
    a. Recover the velocity function for the penny using the facts that
        i. Acceleration due to gravity near the surface of the earth is -32 feet per second squared.
        ii. Because the penny is dropped, velocity is 0 when time is 0.
    b. Recover the distance function for the penny using the velocity function from part a and the fact that distance is 540 feet when time is 0.
    c. When will the penny hit the ground?
    d. What is the impact velocity of the penny in miles per hour?

44. According to the *Guinness Book of Records*, the world's record high dive from a diving board is 176 feet, 10 inches. It was made by Olivier Favre (Switzerland) in 1987. Ignoring air resistance, approximate Favre's impact velocity in miles per hour from a height of 176 feet, 10 inches.

45. a. What is the impact velocity (in feet per second and miles per hour) of a cat that accidentally falls off a building from a height of 66 feet $\left(5\frac{1}{2}\text{ stories}\right)$?
    b. In the 1960s, Donald McDonald claimed in an article in *The New Scientist* that plummeting cats never fall faster than 40 mph. What accounts for the difference between your answer to part a and McDonald's claim (assuming McDonald's claim is accurate)?

46. Table 6.40 gives the increase or decrease in the number of donors to a college athletics support organization for selected years.

TABLE 6.40

Year	Rate of change in donors (donors per year)
1985	-169
1988	803
1991	1222
1994	1087
1997	399
2000	-842

    a. Find a model for the rate of change in the number of donors.
    b. Find a model for the number of donors. Use the fact that in 1990 there were 10,706 donors.
    c. Estimate the number of donors in 2002.

47. From 1997 through 2002, an Internet company was hiring new employees at a rate of $n(x) = \frac{593}{x} + 138$ new employees per year, where x represents the number of years since 1996. By 2001 the company had hired 896 employees.
    a. Write the function that gives the number of employees who had been hired by the xth year after 1996.
    b. For what years will the function in part a apply?

19. Based on data from *Statistical Abstract*, 1998.

c. Find the total number of employees the company had hired between 1997 and 2002. Would this figure necessarily be the same as the number of employees the company had at the end of 2002? Explain.

48. Table 6.41 shows the rate of change in the purchasing power of one dollar at various times since 1985.

TABLE 6.41

Year	Rate of change in purchasing power of $1 (dollars per year)
1985	-0.014
1987	-0.021
1989	-0.024
1991	-0.023
1993	-0.018
1995	-0.008
1997	0.006

a. Examine a scatter plot of the data, and find a model for the rate of change over the period from 1985 through 1997.

b. Use your model to derive a function that models the purchasing power of one dollar as a function of time. Write the function in terms of constant 1990 dollars. (That is, one dollar was worth one dollar in 1990.)

c. Rewrite the function in terms of constant 1980 dollars.

d. Rewrite the function in terms of the current value of a dollar.

e. What is the difference in the three functions from parts *b*, *c*, and *d*?

49. Super Bowl I was held on January 15, 1967. From that time through Super Bowl XXX held in 1997, the price of a Super Bowl ticket[20] was growing at a rate of $p(x) = 0.338e^{0.17422x}$ dollars per year x years after 1966.

a. Find the specific antiderivative of p, using the information that the ticket price for Super Bowl I was $10.

b. Estimate the price of a ticket for the Super Bowl in 2000. Compare your estimate to the actual price of $325.

6.5 The Definite Integral

So far, we have been finding general and specific antiderivatives to recover functions from their rate-of-change functions. The Fundamental Theorem tells us that accumulation functions are antiderivatives. Consider finding the accumulation function for $f(t) = 6t^2 - 2t + 3$ from 4 to x, $A(x) = \int_4^x (6t^2 - 2t + 3)\,dt$. We know that the answer is an antiderivative of the form $y = 2x^3 - x^2 + 3x + C$, and we also know that when $x = 4$, the accumulation must be zero: $A(4) = \int_4^4 (6t^2 - 2t + 3)\,dt = 0$. Substituting $x = 4$ into the general antiderivative, setting it equal to zero, and solving for C gives the value of C in the accumulation function with starting point $x = 4$.

$$\begin{aligned} 2(4)^3 - (4)^2 + 3(4) + C &= 0 \\ 128 - 16 + 12 + C &= 0 \\ 124 + C &= 0 \\ C &= -124 \end{aligned}$$

Thus we have

$$\int_4^x (6t^2 - 2t + 3)\,dt = 2x^3 - x^2 + 3x - 124$$

20. Based on data in *Sports Magazine*, February 1999.

Because $f(t)$ is greater than zero for $t \geq 4$, we use this formula to find the area of the region between the graph of $f(t) = 6t^2 - 2t + 3$ and the horizontal axis from 4 to 7 by substituting 7 into the accumulation formula: $2(7)^3 - (7)^2 + 3(7) - 124 = 534$.

EXAMPLE 1 *Finding Accumulation Formulas*

a. Find a formula for $\int 69.7966(1.07229^t)\,dt$.

b. Find a formula for the accumulation function $A(x) = \int_1^x 69.7966(1.07229^t)\,dt$.

c. Use the accumulation function to find the area of the region between the graph of $f(t) = 69.7966(1.07229^t)$ and the horizontal axis from 1 to 4.

d. If $f(t) = 69.7966(1.07229^t)$ is the rate of change of the balance in a savings account given in dollars per year and t is the number of years since the savings account was opened, what does your answer to the question in part *c* represent?

Solution

a. $$\int 69.7966(1.07229^t)\,dt = \frac{69.7966(1.07229^t)}{\ln 1.07229} + C$$

b. The accumulation function is the specific antiderivative (in terms of x) for which the antiderivative is zero when $x = 1$.

$$A(1) = \frac{69.7966(1.07229^1)}{\ln 1.07229} + C = 0$$

$$C \approx -1072.2908$$

Thus we have

$$A(x) = \int_1^x 69.7966(1.07229^t)\,dt \approx \frac{69.7966(1.07229^x)}{\ln 1.07229} - 1072.2908$$

c. Substituting 4 for x in the formula above gives the area between 1 and 4 as approximately 249.7637.

d. The answer to part *c* represents the change in the amount in the savings account between 1 and 4 years. The amount grew by \$249.76.

The Fundamental Theorem of Calculus gives us a method, as illustrated in Example 1, for finding an accumulation function and using it to determine the area of the region between the graph of a function and the horizontal axis. In general, we know that $\int_a^x f(t)\,dt = F(x) + C$, where F is an antiderivative of f. We also know that when $x = a$, the accumulation function is zero.

$$\int_a^a f(t)\,dt = F(a) + C = 0$$

This tells us that $C = -F(a)$. Thus we have

$$\int_a^x f(t)\,dt = F(x) - F(a)$$

To find the value of the accumulation function from a to b, we simply substitute b for x.

$$\int_a^b f(t)\,dt = F(b) - F(a)$$

We now have an efficient algebraic method for evaluating definite integrals. We summarize the discussion as follows:

Evaluating a Definite Integral

If f is a continuous function from a to b and F is any antiderivative of f, then

$$\int_a^b f(x)\,dx = F(b) - F(a)$$

Recall from Section 6.2 that we define the definite integral, $\int_a^b f(x)\,dx$, as the limiting value of sums of areas of rectangles. That is,

$$\int_a^b f(x)\,dx = \lim_{n\to\infty} [f(x_1) + f(x_2) + \cdots + f(x_n)]\Delta x$$

The antiderivative definition for a definite integral, $\int_a^b f(x)\,dx = F(b) - F(a)$, gives us another method for evaluating a definite integral, so for many functions, we no longer have to rely on the sometimes tedious process of finding limiting values of sums of areas of rectangles. The fact that we can find accumulated change using a reverse derivative process should continue to surprise you. The connection between accumulation and derivatives is a remarkable one.

To illustrate the ease with which we can calculate areas of regions between graphs of functions and the horizontal axis, consider the region between the graph of $f(x) = x^2 + 2$ and the x-axis from -2 to 4. (See Figure 6.70.)

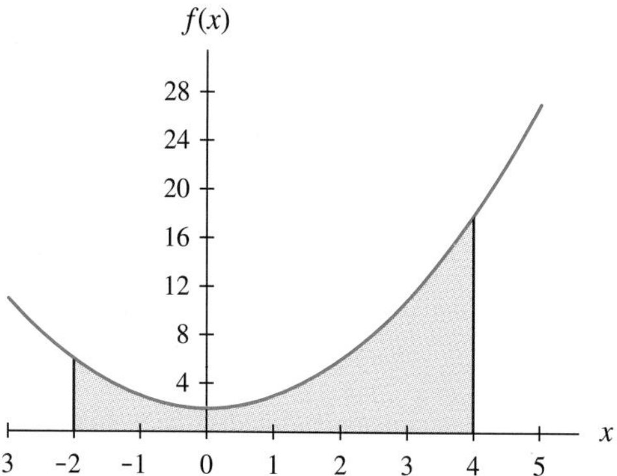

FIGURE 6.70

To find the area of this region in Sections 6.1 and 6.2, we would have used sums of areas of rectangles. Now all we need to do is calculate the value of the definite integral, $\int_{-2}^4 (x^2 + 2)\,dx$, by simply finding an antiderivative and subtracting the value of the antiderivative at -2 from the value at 4.

$$\int_{-2}^{4} (x^2 + 2)\,dx = F(4) - F(-2)$$

where F is an antiderivative of $x^2 + 2$. Here are the details of this process:

You may wish to point out to your students that in the box "Evaluating a Definite Integral," the function F is described as *any* antiderivative. The word *any* implies that we can choose any value for the constant C in the antiderivative. In this illustration, we choose $C = 0$. You may wish to show your students that the result is the same if you choose a different value of C. The role of the constant C in definite integral calculations is further discussed following Example 2.

1. Find an antiderivative:

$$\int_{-2}^{4} (x^2 + 2)\,dx = \left(\frac{x^3}{3} + 2x\right)\Bigg|_{-2}^{4}$$

← This notation is used to indicate that we have found an antiderivative and now must evaluate it at 4 and -2 and then subtract the results.

2. Evaluate at upper and lower limits and then subtract:

$$= \left[\frac{4^3}{3} + 2(4)\right] - \left[\frac{(-2)^3}{3} + 2(-2)\right]$$

$$= 29\tfrac{1}{3} - \left(-6\tfrac{2}{3}\right) \approx 29.333 - (-6.667)$$

$$= 36$$

Thus the area of the region depicted in Figure 6.70 is exactly 36.

EXAMPLE 2 *Plant Growth*

Suppose a plant is 9 mm tall at the beginning of an experiment and is growing at a rate of $f(t) = 6t$ mm per day t days after the beginning of the experiment.

a. Find the specific antiderivative of f that gives the plant's height t days after the experiment began.

b. Find $\int_0^2 f(t)\,dt$. Interpret your answer.

Solution

a. The general antiderivative we seek is

$$F(t) = \frac{6t^2}{2} + C = 3t^2 + C$$

Using the fact that when $t = 0$ the plant is 9 mm tall, we have

$$F(0) = 9$$
$$3(0)^2 + C = 9$$
$$C = 9$$

Thus the specific antiderivative giving the plant's height is

$$F(t) = 3t^2 + 9 \text{ mm}$$

t days after the beginning of the experiment.

b.
$$\int_0^2 f(t)\,dt = F(t)\Big|_0^2$$
$$= (3t^2 + 9)\Big|_0^2$$
$$= [3(2)^2 + 9] - [3(0)^2 + 9]$$
$$= (12 + 9) - 9$$
$$= 12 \text{ mm}$$

This result is the area of the region between the graph of $f(t) = 6t$ and the horizontal axis from 0 to 2. It is the change in the plant's height during the first 2 days of the experiment. That is, the plant grew 12 mm during the first 2 days of the experiment.

Now consider the result of Example 2 if the plant had been 0.5 mm tall at the beginning of the experiment. The function giving the plant's height would then be $F(t) = 3t^2 + 0.5$ mm t days into the experiment. The definite integral calculation would be

$$\begin{aligned}\int_0^2 f(t)\,dt &= F(t)\Big|_0^2 \\ &= (3t^2 + 0.5)\Big|_0^2 \\ &= [3(2)^2 + 0.5] - [3(0)^2 + 0.5] \\ &= (12 + 0.5) - 0.5 \\ &= 12 \text{ mm}\end{aligned}$$

Note that the initial height of the plant made no difference in the definite integral calculation. This is not a coincidence. If F_1 and F_2 are any two antiderivatives of f, they differ only by a constant: $F_1(x) = F_2(x) + C$. Thus the change in $F_1(x)$ from a to b is

$$F_1(b) - F_1(a) = [F_2(b) + C] - [F_2(a) + C] = F_2(b) - F_2(a)$$

the same as the change in $F_2(x)$ from a to b. The constant term C always cancels out during the definite integral calculation.

> The constant term in an antiderivative does not affect definite integral calculations. If you are concerned only with finding change in a quantity, finding the constant in the antiderivative is not necessary.

We return to Example 4 in Section 6.4. Recall that we modeled the marginal cost for toaster ovens using the exponential function

$$C'(x) = 47.638(0.9972^x) \text{ dollars per oven}$$

where x is the number of ovens produced per day. Suppose that the current production level is 300 ovens per day and that the manufacturer wishes to increase production to 500 ovens per day. The definite integral $\int_{300}^{500} C'(x)\,dx = C(500) - C(300)$ gives the change in cost as a result of this increase. Finding the change requires two steps. First, find an antiderivative of C'. Then evaluate the antiderivative at the two production levels and subtract the value at the lower limit from the value at the upper limit of the integral.

In Section 6.4 (using an unrounded model), we found the general antiderivative for C' to be

$$C(x) = \int C'(x)\,dx = -16{,}991.852(0.9972^x) + K$$

The constant K will not affect our calculations of change, so we set K to be 0. (Note that if we need to answer other questions about cost, we should use the information that the cost to produce 250 ovens is \$12,000 to find the proper value of K, as was illustrated in Example 4.)

The definite integral is

$$\int_{300}^{500} C'(x)\,dx = C(x)\Big|_{300}^{500}$$
$$= C(500) - C(300)$$
$$= -16{,}991.852(0.9972^{500}) - [-16{,}991.852(0.9972^{300})]$$
$$\approx \$3145$$

We conclude that when production is increased from 300 to 500 ovens per day, cost increases by approximately \$3145.

Sums of Definite Integrals

Definite integral calculations for piecewise continuous functions require special care. In Example 3 of Section 6.1, we saw flow rates for a river during heavy rains; however, only some of the data were given. The complete available data[21] and a scatter plot are shown in Table 6.42 and Figure 6.71, respectively.

TABLE 6.42

Hours since 11:45 a.m. Wednesday	Flow rate (cubic feet per hour)
0	2,826,000
4	2,710,800
8	3,002,400
12	3,852,000
16	5,292,000
20	7,524,000
23	6,624,000
27	5,760,000

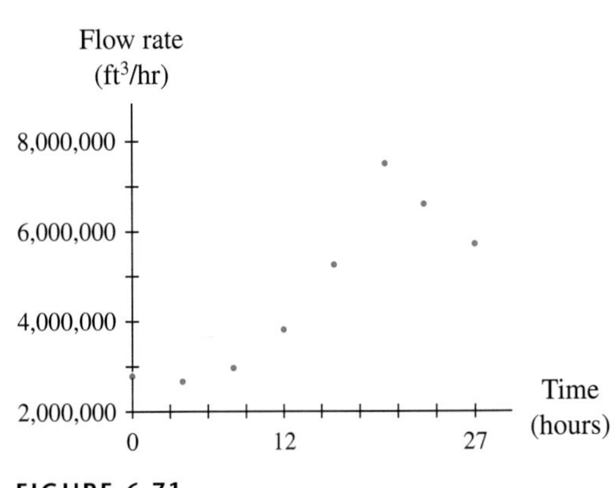

FIGURE 6.71

A piecewise continuous function is appropriate. We choose to divide the data at 20 hours and obtain the model[22]

$$f(h) = \begin{cases} 18{,}225h^2 - 135{,}334.3h + 2{,}881{,}542.9 \text{ ft}^3/\text{hr} & \text{when } 0 \le h \le 20 \\ 12{,}000h^2 - 816{,}000h + 19{,}044{,}000 \text{ ft}^3/\text{hr} & \text{when } 20 < h \le 27 \end{cases}$$

where h is the number of hours since 11:45 a.m. Wednesday.

To estimate the amount of water that flowed through the river from 11:45 a.m. Wednesday to 2:45 p.m. Thursday, 27 hours later, we calculate the value of the definite integral, $\int_0^{27} f(h)\,dh$. Note that the point of division for the model occurs in the interval from 0 to 27. For this reason, we cannot calculate the value of the definite integral simply by evaluating an antiderivative of f at 27 and 0 and subtracting.

Note that the area of the region from a to b shaded in Figure 6.72 is equal to the sum of the area of R_1 and the area of R_2.

21. As reported in the *Reno Gazette–Journal*, May 17, 1996, p. 4a.
22. When the entire data set is viewed, the three points on the right appear to follow a linear pattern. However, when viewed in isolation, the points are distinctively concave up. For this reason, we chose quadratic models for both portions of the data.

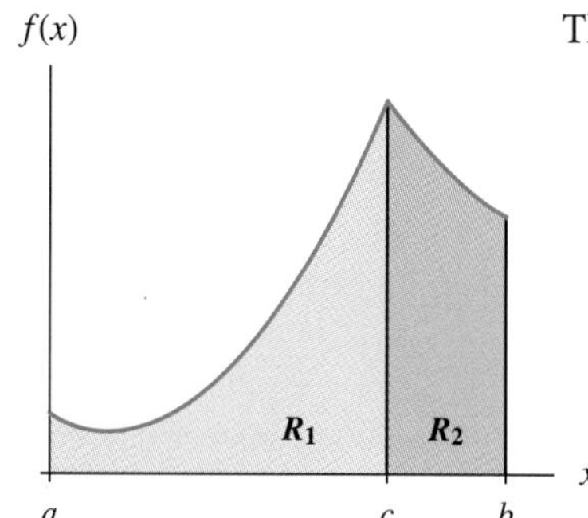

FIGURE 6.72

This figure illustrates the following property of integrals:

Sum Property of Integrals

$$\int_a^b f(x)\,dx = \int_a^c f(x)\,dx + \int_c^b f(x)\,dx$$

where c is between a and b.

It is this property that enables us to calculate definite integrals for piecewise continuous functions.

Although we state the Sum Property of Integrals for $a \le c \le b$, it is true for any a, b, and c. You may wish to discuss with your students that $\int_c^b f(x)\,dx = -\int_b^c f(x)\,dx$.

Returning to the river flow function, in order to calculate $\int_0^{27} f(h)\,dh$, we divide the integral into two pieces at the point where the model changes and sum the results.

$$\begin{aligned}\int_0^{27} f(h)\,dh &= \int_0^{20} f(h)\,dh + \int_{20}^{27} f(h)\,dh \\ &= \int_0^{20} (18{,}225h^2 - 135{,}334.3h + 2{,}881{,}542.9)\,dh + \\ &\quad \int_{20}^{27} (12{,}000h^2 - 816{,}000h + 19{,}044{,}000)\,dh \\ &= \left(\frac{18{,}225}{3}h^3 - \frac{135{,}334.3}{2}h^2 + 2{,}881{,}542.9h\right)\Bigg|_0^{20} + \\ &\quad \left(\frac{12{,}000}{3}h^3 - \frac{816{,}000}{2}h^2 + 19{,}044{,}000h\right)\Bigg|_{20}^{27} \\ &= (79{,}164{,}000 - 0) + (295{,}488{,}000 - 249{,}680{,}000) \\ &= 79{,}164{,}000 + 45{,}808{,}000 \\ &= 124{,}972{,}000 \text{ cubic feet}\end{aligned}$$

We estimate that during the first 20 hours, 79,164,000 ft^3 of water flowed through the river. Between 20 and 27 hours, the volume of water was about 45,808,000 ft^3. Summing these two values, we estimate that from 11:45 a.m. Wednesday to 2:45 p.m. Thursday, 124,972,000 ft^3 of water flowed through the river.

In Example 3 in Section 6.2, we saw a rate-of-change function for per capita wine consumption whose graph was both above and below the horizontal axis. In order to calculate the change in per capita wine consumption from the end of 1970 through 1990, we calculated the increase in wine consumption from the end of 1970 through 1984 (actually through $x \approx 83.97$) and subtracted the decrease in wine consumption from 1984 ($x \approx 83.97$) through 1990. In other words, we calculated the definite integral from 1970 ($x = 70$) through 1990 ($x = 90$) as

$$\int_{70}^{83.97} W(x)\,dx + \int_{83.97}^{90} W(x)\,dx \approx 1.34 - 0.45 = 0.89 \text{ gallon per person}$$

where the value of the integral from 83.97 to 90 is negative because the graph of W is below the horizontal axis. Because W is continuous from 70 to 90 (not piecewise continuous), we can calculate $\int_{70}^{90} W(x)\,dx$ by finding an antiderivative of W, evaluating it at 90 and 70, and subtracting.

$$\begin{aligned}\int_{70}^{90} W(x)\,dx &= \int_{70}^{90} [(1.243 \cdot 10^{-4})x^3 - 0.0314x^2 + 2.6174x - 71.977]\,dx \\ &= [(3.1075 \cdot 10^{-5})x^4 - 0.01046667x^3 + 1.3087x^2 - 71.977x]\Big|_{70}^{90} \\ &\approx -1468.83 - (-1469.72) = 0.89 \text{ gallon per person}\end{aligned}$$

EXAMPLE 3 *Changing Sea Levels*

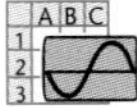

Scientists believe that the average sea level is dropping and has been for some 4000 years. They also believe that was not always the case. Estimated rates of change in the average sea level[23] in meters per year during the past 7000 years are given in Table 6.43.

TABLE 6.43

Time, t (thousands of years before present)	Rate of change of average sea level, $r(t)$ (meters/year)
-7	3.8
-6	2.6
-5	1.0
-4	0.1
-3	-0.6
-2	-0.9
-1	-1.0

a. Find a model for the data.

b. Find the areas of the regions above and below the t-axis from $t = -7$ to $t = 0$. Interpret the areas in the context of sea level.

c. Find $\int_{-7}^{0} r(t)\,dt$, and interpret your answer.

Solution

a. A quadratic model for the data is

$$r(t) = 0.14762t^2 + 0.35952t - 0.8 \text{ meters per year}$$

t thousand years from the present (past years are represented by negative numbers).

b. A graph of r is shown in Figure 6.73. The graph crosses the t-axis at $t \approx -3.845$ thousand years.

23. Estimated from information in François Ramade, *Ecology of Natural Resources* (New York: Wiley, 1981).

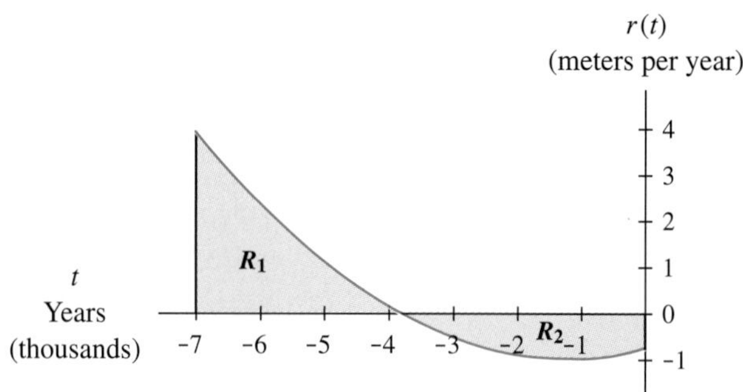

FIGURE 6.73

The area of region R_1 (above the t-axis) is

$$\int_{-7}^{-3.845} (0.14762t^2 + 0.35952t - 0.8)\,dt = \left(\frac{0.14762}{3}t^3 + \frac{0.35952}{2}t^2 - 0.8t\right)\Bigg|_{-7}^{-3.845}$$
$$\approx 2.9365 - (-2.4694) \approx 5.4 \text{ meters}$$

The area of region R_2 (below the t-axis) is

$$-\int_{-3.845}^{0} (0.14762t^2 + 0.35952t - 0.8)\,dt = -\left(\frac{0.14762}{3}t^3 + \frac{0.35952}{2}t^2 - 0.8t\right)\Bigg|_{-3.845}^{0}$$
$$\approx -(0 - 2.9365) \approx 2.9 \text{ meters}$$

From 7000 years ago through 3845 years ago, the average sea level rose by approximately 5.4 meters. From 3845 years ago to the present, the average sea level fell by approximately 2.9 meters.

c. $$\int_{-7}^{0} r(t)\,dt = \int_{-7}^{0} (0.14762t^2 + 0.35952t - 0.8)\,dt$$
$$= \left(\frac{0.14762}{3}t^3 + \frac{0.35952}{2}t^2 - 0.8t\right)\Bigg|_{-7}^{0}$$
$$\approx 0 - (-2.4694) \approx 2.5 \text{ meters}$$

From 7000 years ago to the present, the average sea level has risen approximately 2.5 meters. This result is the same as that obtained by subtracting the amount that sea level has fallen from the amount that it has risen:

5.4 meters − 2.9 meters = 2.5 meters

The Fundamental Theorem gives us a technique for evaluating definite integrals using antiderivatives. However, we have seen rules for finding antiderivatives for very few classifications of functions. There are many other rules for antiderivatives, and there are some functions to which no antiderivative rule applies.

Consider the function that models the rate of worldwide[24] oil production:

$p(x) = 0.018e^{-0.001x^2}$ billion barrels per year

where x is the number of years since 2000. Consider using this model to estimate how much oil will be produced from the year 2000 through the year 2010.

24. Estimated from information in François Ramade, *Ecology of Natural Resources* (New York: Wiley, 1981).

TABLE 6.44

n	Approximation of area
10	0.174189
20	0.174179
40	0.174177
80	<u>0.174176</u>
160	<u>0.174176</u>
320	<u>0.174176</u>
Trend ≈ 0.17418	

To find this estimate, we evaluate $\int_0^{10} 0.018e^{-0.001x^2}\,dx$. This is a function for which there is not an antiderivative rule. However, recall that the definite integral was first defined as a limiting value of sums. We rely on that definition whenever an algebraic formula for the antiderivative of a function is not known.

We investigate the limiting value of sums beginning with 10 midpoint rectangles as shown in Table 6.44 and conclude that approximately 174,180,000 barrels of oil will be produced from the year 2000 through the year 2010.

It is important for you to understand that if you are ever unable to find an algebraic formula for the antiderivative of a function, you can still estimate the value of a definite integral of that function by using the limiting value of sums of areas of rectangles.

If you are using the two-semester text and are integrating Chapter 8 throughout Chapters 1 through 7, you may cover Section 8.5 now. However, be aware that Section 8.5 contains a discussion of average value that is better understood after your students have seen Section 7.4. See the *Instructor's Resource Guide* for more detailed suggestions.

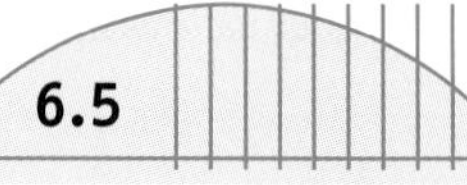

Concept Inventory

- $\int_a^b f(x)\,dx = F(b) - F(a)$, where F is an antiderivative of f
- $\int_a^b f(x)\,dx = \int_a^c f(x)\,dx + \int_c^b f(x)\,dx$

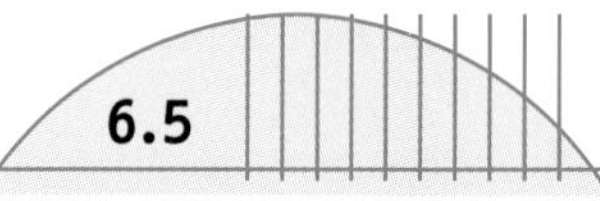

Activities

For each of Activities 1 through 7, determine which of the following processes you would use when answering the question posed. Note that a and b are constants.

a. Find a derivative.

b. Find a general antiderivative (with unknown constant).

c. Find a specific antiderivative (solve for the constant).

1. Given a rate-of-change function for population and the population in a given year, find the population in year t.

2. Given a velocity function, determine the distance traveled from time a to time b.

3. Given a function, find its accumulation function from a to x.

4. Given a velocity function, determine acceleration at time t.

5. Given a rate-of-change function for population, find the change in population from year a to year b.

6. Given a function, find the area of the region between the function and the horizontal axis from a to b.

7. Given a function, find the slope of the tangent line at input a.

8. a. Find the accumulation function $A(x) = \int_0^x 2e^t\,dt$.

   b. Use part a to find the value of $A(3) = \int_0^3 2e^t\,dt$.

9. a. Find the accumulation function $G(x) = \int_1^x (t^3 - 2t + 3)\,dt$.

   b. Use part a to find the value of $G(2)$.

10. a. Find a formula for $\int_2^x (4.63t^2 - 9.2t + 6)\,dt$.

    b. Use the formula from part a to find the area of the region between the graph of $f(t) = 4.63t^2 - 9.2t + 6$ and the horizontal axis from 2 to 3.

    c. If $f(t)$ is the rate of change of the population of a state in thousands of people per year and t is the number of years since 1975, what is the interpretation of the answer to part b?

11. a. Find the accumulation function from 0 to x for $f(t) = 18{,}000(0.974^t) + 1500$.

    b. Use the formula in part a to find the area of the region between between the graph of f and the horizontal axis from 0 to 10.

c. If $f(t)$ is the rate of increase in a piece of real estate in dollars per year, where t is the number of years since the current owner purchased the property, what is the meaning of the answer to part *b*?

Evaluate the definite integrals in Activities 12 to 17.

12. $\int_{-1}^{4} 3\,dx$

13. $\int_{1}^{2} (3x - 7)\,dx$

14. $\int_{3}^{10} 4.298(1.036^x)\,dx$

15. $\int_{1}^{2} \left(1 + \frac{1}{x} + \frac{1}{x^3}\right) dx$

16. $\int_{0}^{4.5} (6.27e^{-1.3296x} - 0.324x^3 + 1.59)\,dx$

17. $\int_{10}^{40} [427.705(1.043^x) - 413.226e^{-0.4132x}]\,dx$

In Activities 18 through 21:

a. Graph the function f from a to b.

b. Find the area of the region between the graph of f and the x-axis from a to b.

c. Is the area in part b equal to $\int_{a}^{b} f(x)\,dx$? Explain.

18. $f(x) = -4x^{-2}$; $a = 1, b = 4$

19. $f(x) = -1.3x^3 + 0.93x^2 + 0.49$; $a = -1, b = 2$

20. $f(x) = \frac{9.295}{x} - 1.472$; $a = 5, b = 10$

21. $f(x) = -965.27(1.079^x)$; $a = 0.5, b = 3.5$

22. The air speed of a small airplane during the first 25 seconds of takeoff and flight can be modeled by

$$v(t) = -940{,}602t^2 + 19{,}269.3t - 0.3 \text{ mph}$$

t hours after takeoff.

a. Find the value of $\int_{0}^{0.005} v(t)\,dt$.

b. Interpret your answer in context.

23. From data taken from the 1998 *Statistical Abstract*, we model the rate of change of the percentage of the population of the United States living in New England from 1970 through 1997 as

$$P(t) = (-1.970 \cdot 10^{-4})t^2 + 0.00493t - 0.0484$$
percentage points per year

where t is the number of years since 1970. Find and interpret the value of $\int_{10}^{20} P(t)\,dt$.

24. The rate of change of the weight of a laboratory mouse t weeks (for $1 \le t \le 15$) after the beginning of an experiment can be modeled by the equation

$$w(t) = \frac{13.785}{t} \text{ grams per week}$$

Evaluate $\int_{3}^{9} w(t)\,dt$, and interpret your answer.

25. A corporation's revenue flow rate can be modeled by

$$r(x) = 9.907x^2 - 40.769x + 58.492$$
million dollars per year

x years after the end of 1987. Evaluate $\int_{0}^{5} r(x)\,dx$, and interpret your answer.

26. A dam develops a leak that ultimately results in a major fissure. The rate at which water is flowing out of the leak is given by

$$l(m) = e^{0.0062m^2} \text{ cubic feet per minute}$$

m minutes after the leak develops.

Evaluate $\int_{0}^{30} e^{0.0062m^2}\,dm$, and interpret your answer.

27. The production rate of a certain oil well can be modeled by

$$r(t) = 3.93546t^{3.55}\,e^{-1.35135t} \text{ thousand barrels per year}$$

t years after production begins.

Evaluate $\int_{0}^{10} r(t)\,dt$, and interpret your answer.

28. In Section 6.1 we saw the rate of change in the concentration of a drug modeled by

$$r(x) = \begin{cases} 1.708(0.845^x) & \text{when } 0 \le x < 20 \\ \quad \mu\text{g/mL/day} & \\ 0.11875x - 3.5854 & \text{when } 20 < x \le 29 \\ \quad \mu\text{g/mL/day} & \end{cases}$$

where x is the number of days after the drug was administered. Determine the values of the following definite integrals, and interpret your answers.

a. $\int_{0}^{20} r(x)\,dx$

b. $\int_{20}^{29} r(x)\,dx$

c. $\int_{0}^{29} r(x)\,dx$

29. The rate of change of the snow pack in an area in the Northwest Territories in Canada can be modeled by

$$s(t) = \begin{cases} 0.00241t + 0.02905 \text{ cm of water per day} & \text{when } 0 \le t < 70 \\ 1.011t^2 - 147.971t + 5406.578 \text{ cm of water per day} & \text{when } 72 < t \le 76 \end{cases}$$

where t is the number of days since April 1.

a. Evaluate $\int_0^{70} s(t)\,dt$, and interpret your answer.

b. Evaluate $\int_{72}^{76} s(t)\,dt$, and interpret your answer.

c. Explain why it is not possible to find the value of $\int_0^{76} s(t)\,dt$.

30. The rate of change of the temperature during the hour and a half after the beginning of a thunderstorm is given by

$$T(h) = 9.48h^3 - 15.49h^2 + 17.38h - 9.87 \text{ °F per hour}$$

where h is the number of hours since the storm began.

a. Graph the function T from $h = 0$ to $h = 1.5$.

b. Calculate the value of $\int_0^{1.5} T(h)\,dh$. Interpret your answer.

31. The rate of change of the temperature in a museum during a junior high school field trip can be modeled by

$$T(h) = 9.07h^3 - 24.69h^2 + 14.87h - 0.03 \text{ °F per hour}$$

h hours after 8:30 a.m.

a. Find the area of the region that lies above the axis between the graph of T and the h-axis between 8:30 a.m. and 10:15 a.m. Interpret the answer.

b. Find the area of the region that lies below the axis between the graph of T and the h-axis between 8:30 a.m. and 10:15 a.m. Interpret the answer.

c. There are items in the museum that should not be exposed to temperatures greater than 73°F. If the temperature at 8:30 a.m. was 71°F, did the temperature exceed 73°F between 8:30 a.m. and 10:15 a.m.?

32. The acceleration of a race car during the first 35 seconds of a road test is modeled by

$$a(t) = 0.024t^2 - 1.72t + 22.58 \text{ ft/sec}^2$$

where t is the number of seconds since the test began.

a. Graph the function a from $t = 0$ to $t = 35$.

b. Write the definite integral notation representing the amount by which the car's speed increased during the road test. Calculate the value of the definite integral.

33. The estimated production rate[25] of marketed natural gas, in trillion cubic feet per year, in the United States (excluding Alaska) from 1900 through 1960 is shown in Table 6.45.

TABLE 6.45

Year	Estimated production rate (trillions of cubic feet per year)
1900	0.1
1910	0.5
1920	0.8
1930	2.0
1940	2.3
1950	6.0
1960	12.7

a. Find a model for the data in Table 6.45.

b. Use the model to estimate the total production of natural gas from 1940 through 1960.

c. Give the definite integral notation for your answer to part *b*.

34. Many businesses spend money each year on advertising in order to stimulate sales of their products. The data given in Table 6.46 show the approximate increase in sales (in thousands of dollars) that an additional $100 spent on advertising, at various levels, can be expected to generate.

TABLE 6.46

Advertising expenditures (hundreds of dollars)	Revenue increase due to an extra $100 advertising (thousands of dollars)
25	5
50	60
75	95
100	105
125	104
150	79
175	34

25. From information in *Resources and Man*, National Academy of Sciences, 1969, p. 165.

a. Find a model for these data.

b. Use the model in part *a* to determine a model for the total sales revenue $R(x)$ as a function of the amount x spent on advertising. Use the fact that revenue is approximately 877 thousand dollars when \$5000 is spent on advertising.

c. Find the point where returns begin to diminish for sales revenue.

d. The managers of the business are considering an increase in advertising expenditures from the current level of \$8000 to \$13,000. What effect could this decision have on sales revenue?

35. Table 6.47 shows the marginal cost to produce one more compact disc, given various hourly production levels.

a. Find an appropriate model for the data.

TABLE 6.47

Production (CDs per hour)	Cost of an additional CD
100	\$5
150	\$3.50
200	\$2.50
250	\$2
300	\$1.60

b. Use your model from part *a* to derive an equation that specifies production cost $C(x)$ as a function of the number x of CDs produced. Use the fact that it costs approximately \$750 to produce 150 CDs in a 1-hour period.

c. Calculate the value of $\int_{200}^{300} C'(x)\,dx$. Interpret your answer.

6.6 Additional Definite Integral Topics

You may omit this section; however, the material on improper integrals is used in Section 7.3 and in Section 7.5 (which appears in the two-semester version).

We have thus far examined situations in which it is useful to determine accumulated change in a quantity by finding or estimating the area of a region between the graph of a rate-of-change function and the horizontal axis. We estimated such areas by summing areas of rectangles until we discovered that the Fundamental Theorem of Calculus allows us to find areas using antiderivatives. However, as noted at the end of the previous section, there are limitations on our ability to find antiderivatives for functions. In this section, we further explore these limitations. We also examine a definite integral whose upper or lower limit is not a number.

Antiderivative Limitations

There are many relatively simple functions for which we are unable to find an antiderivative formula either because one does not exist or because the methods needed to determine a formula for the antiderivative are beyond the scope of this book. It is important for you to be able to identify such functions. In order to understand situations in which you cannot find antiderivatives, it is essential for you to keep in mind the processes involved in finding derivatives. Recall the Product Rule: Given a function that is the product of two functions, the derivative is found by summing two terms. In particular,

$$\text{If } f(x) = g(x)h(x), \text{ then } f'(x) = g(x)h'(x) + g'(x)h(x).$$

With this rule in mind, consider the process of finding the general antiderivative $\int xe^x\,dx$. Seeing a product of two functions, a typical student who has forgotten the Product Rule for derivatives may reason that the antiderivative is simply the product of the antiderivatives of the two terms x and e^x and will thereby incorrectly conclude that $y = \left(\frac{x^2}{2}\right)(e^x) + C$ is the general antiderivative. A quick derivative calculation, however, reveals otherwise: $\frac{dy}{dx} = \left(\frac{x^2}{2}\right)e^x + xe^x \neq xe^x$. There is a method[26] for determining the general antiderivative of xe^x, but it is beyond the scope of this text. If we need to calculate a definite integral in the form $\int_a^b xe^x\,dx$, we choose to rely on approximation methods (using technology) to estimate the value of the integral. Example 1 illustrates antiderivatives that involve products.

EXAMPLE 1 *Antiderivatives and Products*

algebra

Determine whether the following general antiderivatives can be found using the techniques presented in this book. If so, find the antiderivative.

a. $\int x\sqrt{x}\,dx$

b. $\int x\sqrt{x+1}\,dx$

c. $\int e^{3x}(e^{4x} + 4)\,dx$

d. $\int e^{3x}(3x + 1)\,dx$

e. $\int e^{3x}(4x + 4)\,dx$

Solution

a. Rewrite $x\sqrt{x}$ as $x(x^{1/2}) = x^{3/2}$. Then apply the Power Rule for antiderivatives to obtain $\int x\sqrt{x}\,dx = \frac{2x^{5/2}}{5} + C$.

b. Unlike the expression in part *a*, it is not possible to rewrite $x\sqrt{x+1}$ in such a way that we can easily find its antiderivative.

c. Multiply the two terms to rewrite as $e^{7x} + 4e^{3x}$ and apply antiderivative rules to obtain

$$\int e^{3x}(e^{4x} + 4)\,dx = \int (e^{7x} + 4e^{3x})\,dx = \frac{e^{7x}}{7} + \frac{4e^{3x}}{3} + C$$

d. Although this question is similar to part *c*, the x-term complicates the solution. If we multiply the terms, we obtain $3xe^{3x} + e^{3x}$. You may recognize this expression as a result of the application of the Product Rule. The function $y = xe^{3x}$ has derivative $\frac{dy}{dx} = x(3e^{3x}) + (1)e^{3x} = 3xe^{3x} + e^{3x} = e^{3x}(3x + 1)$. Recognizing the expression as a result of the Product Rule is the key to discovering the antiderivative. Thus $\int e^{3x}(3x + 1)\,dx = xe^{3x} + C$. If you did not see the connection to the Product Rule and needed this antiderivative for a definite integral calculation, you could use technology to estimate the value of a definite integral involving this product.

e. This question is similar to part *d*, and a reasonable approach would be to consider $y = 4xe^{3x}$ as a candidate for the antiderivative. Finding the derivative of $y = 4xe^{3x}$ using the Product Rule, we have $\frac{dy}{dx} = 4x(3e^{3x}) + 4e^{3x} = e^{3x}(12x + 4)$. Although this answer is close to $e^{3x}(4x + 4)$, it is not the same. Therefore, the antiderivative is not $y = 4xe^{3x} + C$. This is another example of an antiderivative

26. You can find a discussion of *integration by parts* in most other calculus books.

that can be found using methods not discussed in this text. If we needed to evaluate a definite integral of $e^{3x}(4x + 4)$, we would use technology to approximate the numerical answer.

In addition to keeping the Product Rule in mind when finding antiderivatives, you should also consider the Chain Rule. If a function is a composite function, you may not have the ability to find its antiderivative with the rules discussed in this book. Consider the general antiderivative $\int (x^2 + 2x)^4 dx$. A common mistake is to conclude that this general antiderivative is $y = \frac{(x^2 + 2x)^5}{5} + C$. However, if you take the derivative of y to check this answer, you find by applying the Chain Rule that $\frac{dy}{dx} = (x^2 + 2x)^4 (2x + 2)$. Therefore, the Fundamental Theorem of Calculus tells us that $y = \frac{(x^2 + 2x)^5}{5} + C$ is not the general antiderivative $\int (x^2 + 2x)^4 dx$. Note that it is possible to expand $(x^2 + 2x)^4$ by repeated multiplication to obtain $x^8 + 8x^7 + 24x^6 + 32x^5 + 16x^4$. Thus

$$\int (x^2 + 2x)^4 dx = \frac{x^9}{9} + x^8 + \frac{24x^7}{7} + \frac{16x^6}{3} + \frac{16x^5}{5} + C$$

EXAMPLE 2 *Antiderivatives and Composite Functions*

For the following definite integrals, find the exact answer using antiderivative formulas if possible. If finding the exact answer is not possible, use technology to estimate the answer.

a. $\int_1^6 (3x^2 - 7)^2 dx$

b. $\int_4^5 \sqrt{3x^2 - 7}\, dx$

c. $\int_0^2 6xe^{3x^2 - 7} dx$

d. $\int_1^2 \frac{3}{1 + 7e^{-3x}} dx$

Solution

a. Rewrite the integral as $\int_1^6 (3x^2 - 7)^2 dx = \int_1^6 (9x^4 - 42x^2 + 49)\, dx$. Then apply the Sum Rule, Power Rule, and Constant Multiplier Rule for antiderivatives:

$$\int_1^6 (9x^4 - 42x^2 + 49)\, dx = \left(\tfrac{9}{5}x^5 - 14x^3 + 49x\right)\Big|_1^6$$

$$= 11{,}266.8 - 36.8 = 11{,}230$$

b. Rewrite the integral as $\int_4^5 (3x^2 - 7)^{\frac{1}{2}} dx$. Unlike part *a*, there is no way to rewrite this integral in order to apply the antiderivative rules we have learned. The value of the integral can be approximated as 7.329.

c. The function in this integral is the combination of a product and a composite function. The first term, $6x$, is the derivative of the inside function, $3x^2 - 7$. This function is the result of applying the Chain Rule to the function $e^{3x^2 - 7}$. Thus the exact solution of this definite integral is $\int_0^2 6xe^{3x^2 - 7} dx = e^{3x^2 - 7}\Big|_0^2 = e^5 - e^{-7}$.

d. You should recognize this function as a logistic function. We can rewrite it as $\int_1^2 3(1 + 7e^{-3x})^{-1}\,dx$. Doing so makes it easy to see that $u = 1 + 7e^{-3x}$ is the inside function and $3u^{-1}$ is the outside function. Because the derivative of the inside function does not appear in the integral, the function in the integral cannot be thought of as the result of the Chain Rule, and its antiderivative cannot be found using the methods we have studied. The value of the integral can be approximated via technology as 2.718.

When finding antiderivatives, it is always a good idea to take the derivative of your answer in order to determine whether it is correct. Finding antiderivatives is one problem for which there is a simple way to check the answer. If finding an antiderivative is needed to evaluate a definite integral, then it is also wise to use technology to estimate the integral as a means of checking your answer.

Perpetual Accumulation

Definite integrals have specific numbers for both the upper limit and the lower limit. We now consider what happens to the accumulation of change when one or both of the limits of the integral is infinite. That is, we wish to evaluate integrals of the form $\int_a^{\infty} f(x)\,dx$, $\int_{-\infty}^{b} f(x)\,dx$, or $\int_{-\infty}^{\infty} f(x)\,dx$. We call integrals of this form **improper integrals.** Improper integrals play a role in economics and statistics as well as in other fields of study.

Consider evaluating the improper integral $\int_2^{\infty} 4.3e^{-0.06x}\,dx$. We can interpret this integral as the area of the region between the graph of $y = 4.3e^{-0.06x}$ and the x-axis from 2 to infinity. See Figure 6.74. One way to estimate this area is to consider the area between 2 and some large value. In Table 6.48 we show several area calculations for increasingly large values.

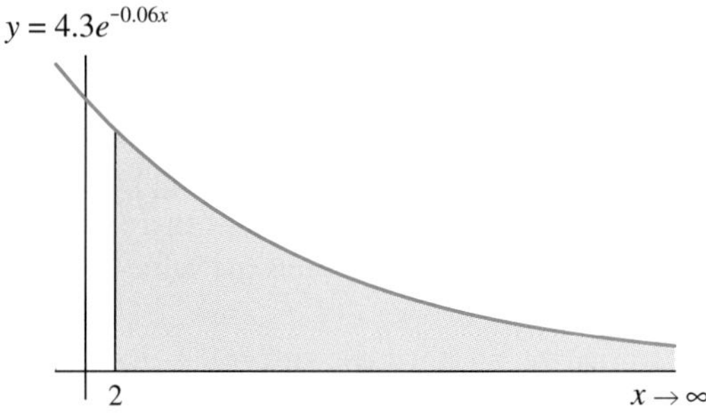

FIGURE 6.74

TABLE 6.48

N	$\int_2^N 4.3e^{-0.06x}\,dx$
50	59.994558
100	63.384987
200	63.562191
400	63.562631
800	63.562631
1600	63.562631
Trend $\approx$ 63.56263	

Note that the area between 2 and 400 is not significantly different from the area between 2 and 1600. The difference is smaller than can be shown by the technology that we used. However, the limiting value seen in the table is still just an estimate of the integral $\int_2^{\infty} 4.3e^{-0.06x}\,dx$.

You should recognize that we are numerically investigating a limit in Table 6.48. The limit we are investigating is $\lim\limits_{N\to\infty}\int_2^N 4.3e^{-0.06x}\,dx$. Because of our knowledge of limits, we can calculate this limit algebraically in order to obtain the exact answer. (Before continuing, you may wish to review the discussion in Section 1.4 of limits describing end behavior.) We begin by finding the general antiderivative of $4.3e^{-0.06x}$, evaluating it at 2 and N, and subtracting the results.

$$\begin{aligned}\int_2^N 4.3e^{-0.06x}\,dx &= \frac{4.3}{-0.06}e^{-0.06x}\Big|_2^N\\ &= \frac{4.3}{-0.06}e^{-0.06N} - \left(\frac{4.3}{-0.06}e^{-0.06(2)}\right)\\ &= \frac{4.3}{-0.06}e^{-0.06N} + \frac{4.3}{0.06}e^{-0.12}\end{aligned}$$

Next we find the limit of this expression as N becomes infinitely large.

$$\begin{aligned}&\lim_{N\to\infty}\left(\frac{4.3}{-0.06}e^{-0.06N} + \frac{4.3}{0.06}e^{-0.12}\right)\\ &\quad= \lim_{N\to\infty}\frac{4.3}{-0.06}e^{-0.06N} + \lim_{N\to\infty}\frac{4.3}{0.06}e^{-0.12}\end{aligned}$$

The first term is a decreasing exponential, so we know that as N approaches infinity, it approaches zero. The second term is a constant that is not affected by the value of N.

Rather than stating for your students a rule for finding limits of exponential functions or having them estimate the limits numerically, we suggest that you draw on their knowledge of the behavior of the graphs of increasing and decreasing exponential functions.

$$\begin{aligned}&\lim_{N\to\infty}\frac{4.3}{-0.06}e^{-0.06N} + \lim_{N\to\infty}\frac{4.3}{0.06}e^{-0.12}\\ &\quad= 0 + \frac{4.3}{0.06}e^{-0.12}\\ &\quad= \frac{4.3}{0.06}e^{-0.12}\end{aligned}$$

We see that this answer confirms our former numerical estimate, because

$$\frac{4.3}{0.06}e^{-0.12} \approx 63.56263$$

It is important for you to understand, however, that the answer $\frac{4.3}{0.06}e^{-0.12}$ is exact, whereas the answer 63.56263 is not. Although an answer accurate to the fifth decimal place is sufficient for most applications, there are situations in which greater precision is necessary. Perfect precision is the value of an exact answer.

To summarize, an improper integral $\int_a^\infty f(x)\,dx$ is evaluated by replacing infinity with a variable, say N, and evaluating the limit of the integral $\int_a^N f(x)\,dx$ as N approaches infinity. That is, provided the limits exist,

$$\int_a^\infty f(x)\,dx = \lim_{N\to\infty}\int_a^N f(x)\,dx = [\lim_{N\to\infty}F(N)] - F(a)$$

$$\int_{-\infty}^b f(x)\,dx = \lim_{N\to-\infty}\int_N^b f(x)\,dx = F(b) - \lim_{N\to-\infty}F(N)$$

where F is an antiderivative of f. We now have the tools we need to apply improper integrals to some real-world problems.

EXAMPLE 3 *Carbon-14*

Carbon-14 dating methods are sometimes used by archeologists to determine the age of an artifact. The rate at which 100 milligrams of ^{14}C is decaying can be modeled by

$$r(t) = -0.01209(0.999879^t) \text{ milligrams per year}$$

where t is the number of years since the 100 milligrams began to decay.

a. How much of the ^{14}C will have decayed after 1000 years?

b. How much of the ^{14}C will eventually decay?

Solution

a. The amount of ^{14}C to decay during the first 1000 years is

$$\int_0^{1000} r(t)\,dt = \int_0^{1000} -0.01209(0.999879^t)\,dt \approx -11.4 \text{ milligrams}$$

Thus approximately 11.4 milligrams will decay during the first 1000 years. Note that -11.4 milligrams is the signed area of the shaded region in Figure 6.75.

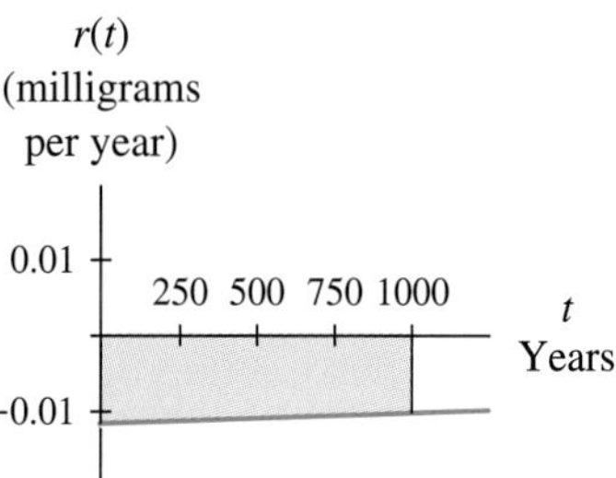

FIGURE 6.75

b. In the long run, the amount that will decay is

$$\begin{aligned}\int_0^{\infty} r(t)\,dt &= \lim_{N\to\infty} \int_0^{N} -0.01209(0.999879^t)\,dt \\ &= \lim_{N\to\infty} \left[\frac{-0.01209(0.999879^t)}{\ln 0.999879}\right]\Bigg|_0^N \\ &= \lim_{N\to\infty} \left(\frac{-0.01209(0.999879^N)}{\ln 0.999879} - \frac{-0.01209(0.999879^0)}{\ln 0.999879}\right) \\ &\approx 99.91131\left[\lim_{N\to\infty}(0.999879^N)\right] - 99.91131 \\ &= 99.91131(0) - 99.91131 \approx -100 \text{ milligrams}\end{aligned}$$

Explain to your students that the answer to part *b* is not exactly 100 milligrams because of rounding of the values in the function *r*.

Eventually all of the ^{14}C will ultimately decay. In terms of the graph shown in Figure 6.76, the area of the region between the graph of the function r and the horizontal axis gets closer and closer to 99.91131 as t gets larger and larger. Because the parameters in the equation are rounded, the area is getting closer and closer to 99.91131 rather than to 100, which is the amount that must ultimately decay.

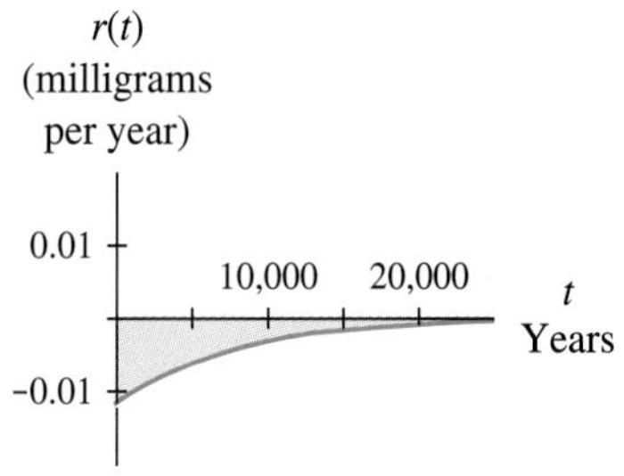

FIGURE 6.76

Divergence

It is possible that when we are evaluating an improper integral, the limit does not exist. (For example, as we numerically approximate the limit, the limit estimates become increasingly large.) In this case, we say that the improper integral **diverges.** Example 4 illustrates this situation.

EXAMPLE 4 *Diverging Integrals*

If possible, determine the value of $\int_1^{\infty} \frac{1}{x}\,dx$.

Solution

We begin by replacing ∞ with the variable N and finding the limit as $N \to \infty$.

$$\int_1^{\infty} \frac{1}{x}\,dx = \lim_{N\to\infty} \ln x \Big|_1^N$$
$$= \lim_{N\to\infty} (\ln N - \ln 1)$$
$$= \lim_{N\to\infty} \ln N - \lim_{N\to\infty} \ln 1$$

To evaluate these limits, recall the shape of the graph of $y = \ln x$ (see Figure 6.77). We see from the graph that $\ln 1 = 0$ and that as the input becomes increasingly large, the output also becomes increasingly large. Thus, $\lim_{N\to\infty} \ln N \to \infty$, $\int_1^{\infty} \frac{1}{x}\,dx \to \infty$, and we say that this improper integral diverges.

$f(x) = \ln x$

FIGURE 6.77

6.6 Concept Inventory

- *Antiderivative formula limitations*
- *Improper integrals*
- *Divergence*

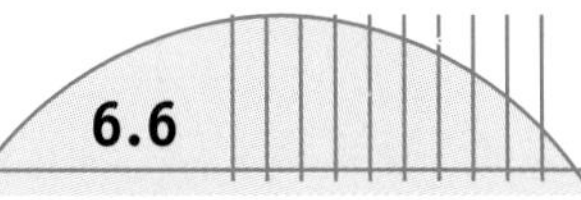

6.6 Activities

In Activities 1 through 8, find the general antiderivative if possible.

1. $\int 2e^{2x}\,dx$
2. $\int 2xe^{x^2}\,dx$
3. $\int x^2e^{x^2}\,dx$
4. $\int 3(\ln 2)2^x(1 + 2^x)^3\,dx$
5. $\int (1 + e^x)^2\,dx$
6. $\int \sqrt{1 + e^x}\,dx$
7. $\int \frac{1}{2^x + 2}\,dx$
8. $\int \frac{e^x}{e^x + 2}\,dx$

practice

In Activities 9 through 20, find the exact value of the integral by using antiderivative formulas if possible. If not possible, use technology to estimate the answer. In either case, state whether your answer is exact or an approximation.

9. $\int_1^4 \ln x\,dx$
10. $\int_1^4 x\ln x\,dx$
11. $\int_2^5 \frac{\ln x}{x}\,dx$
12. $\int_2^5 \frac{5(\ln x)^4}{x}\,dx$
13. $\int_1^2 2x\ln(x^2 + 1)\,dx$
14. $\int_1^2 \ln(x^2 + 1)\,dx$
15. $\int_3^4 \frac{2x}{x^2 + 1}\,dx$
16. $\int_3^4 \frac{1}{x^2 + 1}\,dx$
17. $\int_1^6 \frac{2x^2}{x^2 + 1}\,dx$
18. $\int_3^4 \frac{x^2 + 1}{2x}\,dx$
19. $\int_3^4 \frac{x^2 + 1}{x^2}\,dx$
20. $\int_0^1 -x\sqrt{x^2 + 1}\,dx$

For Activities 21 through 31, evaluate the indicated improper integral.

21. $\int_0^\infty 3e^{-0.2t}\,dt$

22. $\int_{15}^\infty 5(0.36^t)\,dt$

23. $\int_{10}^\infty 3x^{-2}\,dx$

24. $\int_{-\infty}^3 7e^{7x}\,dx$

25. $\int_{-\infty}^{-10} 4x^{-3}\,dx$

26. $\int_2^\infty \frac{1}{\sqrt{x}}\,dx$

27. $\int_{0.36}^\infty 9.6x^{-0.432}\,dx$

28. $\int_{-\infty}^{-2} \left(\frac{3}{x^2} + 1\right) dx$

29. $\int_2^\infty \frac{2x}{x^2 + 1}\,dx$

30. $\int_5^\infty [5(0.36^x) + 5]\,dx$

31. $\int_a^\infty [f(x) + k]\,dx$, where $\int_a^\infty f(x)\,dx = b$ and a, b, and k are constants

32. The rate at which 15 grams of ^{14}C is decaying can be modeled by

$$r(t) = -0.027205(0.998188^t) \text{ grams per year}$$

where t is the number of years since the 15 grams began decaying.

a. How much of the ^{14}C will decay during the first 1000 years? during the fourth 1000 years?

b. How much of the ^{14}C will eventually decay?

33. An isotope of uranium, ^{238}U, is commonly used in atomic weapons and nuclear power generators. Because of its radioactive nature, the United States government is concerned with safe ways of storing used uranium. The rate at which 100 milligrams of ^{238}U is decaying can be modeled by

$$r(t) = -1.55(0.9999999845^t) \cdot 10^{-6} \text{ milligrams per year}$$

where t is the number of years since the 100 milligrams began decaying.

a. How much of the ^{238}U will decay during the first 100 years? during the first 1000 years?

b. How much of the ^{238}U will eventually decay?

The following information is used in Activities 34 and 35:

In the study of markets, economists define *consumers' willingness and ability to spend* as the maximum amount that consumers are willing and able to spend for a specific quantity of goods or services. If some consumers will purchase a product or service regardless of its price, then the consumers' willingness and ability to spend is defined by

$$C = qp_0 + \int_{p_0}^\infty D(p)\,dp$$

where q is a specific quantity, p_0 is the price associated with quantity q, and $D(p)$ is the demand for the commodity when the price is p. (Section 7.3 discusses these concepts in greater detail.)

34. The weekly demand for a dozen roses is given by $D(p) = 316.765(0.949^p)$ dozen roses, where $\$p$ is the price per dozen.

a. Find the price that corresponds to a weekly demand of 80 dozen roses.

b. Use the price (p_0) found in part *a* to calculate how much consumers are willing and able to spend for 80 dozen roses per week.

35. The yearly demand for a certain hardback science fiction novel is

$$D(p) = 499.589(0.958^p) \text{ thousand books}$$

where $\$p$ is the price per book.

a. Find the price that corresponds to a yearly demand of 150,000 books.

b. Use the price (p_0) found in part *a* to calculate how much consumers are willing and able to spend for 150,000 books each year.

36. The work required to propel a 10-ton rocket an unlimited distance from the surface of Earth into space is defined in terms of force and is given by the improper integral

$$W = \int_{4000}^\infty \frac{160{,}000{,}000}{x^2}\,dx$$

The expression $\frac{160{,}000{,}000}{x^2}$ is force in tons. The variable x is the distance between the rocket and the center of Earth measured in miles.

a. What are the units on work in this context?

b. Calculate the work to propel this rocket infinitely into space.

The following information is used in Activities 37 and 38:

A *probability density function* is defined as a non-negative function f with the property that $\int_{-\infty}^\infty f(x)\,dx = 1$.

37. Consider the function

$$f(x) = \begin{cases} 0.1e^{-0.1x} & \text{when } x \geq 0 \\ 0 & \text{when } x < 0 \end{cases}$$

Prove that f is a probability density function.

38. Consider the function

$$g(x) = \begin{cases} \dfrac{1}{x^2} & \text{when } x \geq 1 \\ 0 & \text{when } x < 1 \end{cases}$$

Prove that g is a probability density function.

Summary

Approximating Results of Change

The accumulated results of change are best understood in geometric terms: Positive accumulation is the area of a region between the graph of a positive rate-of-change function and the horizontal axis, and negative accumulation is the signed area of a region between the graph of a negative rate-of-change function and the horizontal axis. We can approximate the areas of regions of interest by summing areas of rectangular regions using left rectangles, right rectangles, or midpoint rectangles. Of these three, midpoint rectangles usually produce the most accurate approximations.

Limits of Sums

If we agree to estimate the accumulated change in a quantity by calculating areas of regions with sums of midpoint-rectangle areas, then it is evident that the estimates generally improve as we increase the number of approximating rectangles. If we imagine that there is no bound on the number of rectangles, then we see that the area of a region between the graph of a continuous, non-negative function f and the horizontal axis from a to b is given by a limit of sums:

$$\text{Area} = \lim_{n\to\infty} [f(x_1) + f(x_2) + \cdots + f(x_n)]\,\Delta x.$$

Here, the points $x_1, x_2, \ldots, x_n$ are the midpoints of n rectangles of width $\Delta x = \frac{b-a}{n}$ between a and b.

More generally, we consider the limit applied to an arbitrary continuous function f over the interval from a to b and call this limit the definite integral of f from a to b. In symbols, we write

$$\int_a^b f(x)\,dx = \lim_{n\to\infty} [f(x_1) + f(x_2) + \cdots + f(x_n)]\,\Delta x$$

Accumulation Functions

An accumulation function is an integral of the form $\int_a^x f(t)\,dt$ where the upper limit x is a variable. The value of this integral depends on the variable x, so the accumulation function is actually a function of x: $A(x) = \int_a^x f(t)\,dt$. This function gives us a formula for calculating accumulated change in a quantity.

The Fundamental Theorem of Calculus

The Fundamental Theorem sets forth the fundamental connection between the two main concepts of calculus, the derivative and the integral. It tells us that for any continuous function f,

$$\frac{d}{dx}\int_a^x f(t)\,dt = f(x)$$

In other words, the derivative of an accumulation function of $y = f(t)$ is precisely $y = f(x)$. If we imagine differentiation and integration as processes, then the process of integration (from a fixed value a to a variable x) followed by the process of differentiation takes us back to where we began.

If we reverse the order of these two processes and begin by differentiating first, then we obtain the starting function plus a constant.

$$\int_a^x f'(t)\,dt = f(x) + C$$

Antiderivatives

A function F is an antiderivative of f if $F'(x) = f(x)$. Because the derivative of $y = \int_a^x f(t)\,dt$ is $y' = f(x)$, we see that $y = \int_a^x f(t)\,dt$ is an antiderivative of $y = f(x)$. Each continuous function has infinitely many antiderivatives, but any two differ by only a constant. The Fundamental Theorem enables us to find accumulation function formulas by finding antiderivatives.

The Symbol $\int$

We use the symbol $\int$ in three ways. The definite integral $\int_a^b f(t)\,dt$ is a number that can be interpreted in terms of the area(s) of the region(s) between the graph of f and the horizontal axis from a to b. The accumulation function $A(x) = \int_a^x f(t)\,dt$ is a function of x; for each input value, it gives the accumulated change from a to x in the quantity for which $f(t)$ is a rate of change. Finally, the general antiderivative of a function f is written $\int f(x)\,dx$; it is a set of functions whose derivatives are f and involves an arbitrary constant C. By the Fundamental Theorem of Calculus, the accumulation function $A(x) = \int_a^x f(t)\,dt$ is an antiderivative of f.

Recovering Functions from Rates of Change

The formulas for antiderivatives that were developed in Section 6.4 enable us to recover an unknown function f quickly from its rate of change f' because

$$\int f'(x)\,dx = f(x) + C$$

In real-life situations, the constant C must be determined from information about a known value of $f(x)$.

An important aspect of recovering a function from its rate of change is to recover the units used to describe the function from the units associated with its rate of change. Ordinarily, this is not difficult; $f'(x)$ is usually expressed in terms of (output units) per (input unit), where the input and output units are those of f.

The Definite Integral

We initially found the value of a definite integral by examining the trend in a sequence of sums of the form $f(x_1)\Delta x + \cdots + f(x_n)\Delta x$ as n increases. But the Fundamental Theorem of Calculus enabled us to show that when f is a smooth, continuous function, the definite integral $\int_a^b f(x)\,dx$ can be evaluated by

$$\int_a^b f(x)\,dx = F(b) - F(a)$$

where F is any antiderivative of f.

The Fundamental Theorem of Calculus ensures that each continuous function does indeed have an antiderivative. Thus, to the extent that we can actually obtain an algebraic expression for an antiderivative, we can easily evaluate a definite integral. In situations where an antiderivative cannot be found, we use one of the approximation techniques discussed in Sections 6.1 and 6.2, allowing technology to perform the calculations.

Improper Integrals

Improper integrals of the forms $\int_a^\infty f(x)\,dx$, $\int_{-\infty}^a f(x)\,dx$, and $\int_{-\infty}^\infty f(x)\,dx$ can be evaluated by substituting a constant for each infinity symbol, finding an antiderivative and evaluating it to determine an expression in terms of x and the constant(s), and then determining the limit of the resulting expression as the constant approaches infinity or negative infinity. If the limit does not exist, we say the integral diverges. We will see further applications of definite integrals and improper integrals in Chapter 7.

Concept Check

Can you	To practice, try	
• Interpret accumulated change and area?	Section 6.1	Activity 3
• Approximate areas using rectangles?	Section 6.1	Activities 11, 13
• Interpret definite integrals?	Section 6.2	Activities 3, 5
• Approximate area using a limiting value?	Section 6.2	Activities 9, 11
• Sketch and interpret accumulation functions?	Section 6.3	Activities 5, 15
• Find general antiderivatives?	Section 6.4	Activities 17–27
• Find and interpret specific antiderivatives?	Section 6.4	Activities 33, 35, 37
• Recover a function from its rate-of-change equation?	Section 6.4	Activity 43
• Use the Fundamental Theorem to evaluate definite integrals?	Section 6.5	Activities 15, 23, 29
• Determine whether an approximation technique is necessary in order to estimate the value of a definite integral?	Section 6.6	Activities 13, 15
• Evaluate improper integrals?	Section 6.6	Activities 21, 27

Review Test

1. The rate at which crude oil flows through a pipe into a holding tank can be modeled by

$$r(t) = 10(-3.2t^2 + 93.3t + 50.7) \text{ ft}^3\text{/minute}$$

where t is the number of minutes the oil has been flowing into the tank.

a. Sketch a graph of r for t between 0 and 25 minutes.

b. Use five midpoint rectangles to estimate the area of the region between the graph of r and the t-axis from 0 to 25 minutes. Sketch the rectangles on the graph you drew in part *a*.

c. Interpret your answer to part *b*.

2. A hurricane is 300 miles off the east coast of Florida at 1 a.m. The speed at which the hurricane is moving toward Florida is measured each hour. Speeds between 1 a.m. and 5 a.m. are recorded in Table 6.49.

TABLE 6.49

Time	Speed (mph)	Time	Speed (mph)
1 a.m.	15	4 a.m.	38
2 a.m.	25	5 a.m.	40
3 a.m.	35		

a. Find a model for the data.

b. Use a limiting value of sums of areas of midpoint rectangles to estimate how far (to the nearest tenth of a mile) the hurricane traveled between 1 a.m. and 5 a.m. Begin with five rectangles, doubling the number each time until you are confident that you know the limiting value.

3. The graph shown in Figure 6.78 depicts the rate of change in the weight of someone who diets for 20 weeks.

a. What does the area of the shaded region beneath the horizontal axis represent?

b. What does the area of the shaded region above the horizontal axis represent?

c. Is this person's weight at 30 weeks more or less than it was at 0 weeks? How much more or less?

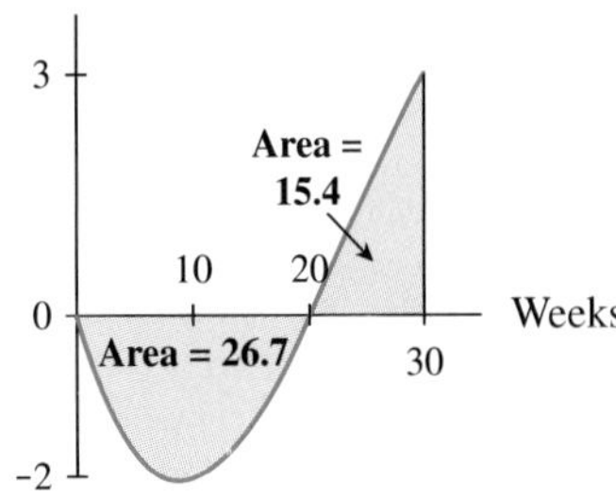

FIGURE 6.78

d. If w is the function shown in Figure 6.78, sketch a graph of $W(t) = \int_0^x w(t)\,dt$. Label units and values on both axes of your graph.

e. What does the graph in part *d* represent?

4. Consider again the model for the flow rate of crude oil into a holding tank.

$$r(t) = 10(-3.2t^2 + 93.3t + 50.7) \text{ ft}^3\text{/minute}$$

after t minutes.

a. If the holding tank contains 5000 ft^3 of oil when $t = 0$, find a model for the amount of oil in the tank after t minutes.

b. Use your model in part *a* to find how much oil flowed into the tank during the first 10 minutes.

c. If the capacity of the tank is 150,000 ft^3, according to the model, how long can the oil flow into the tank before the tank is full?

5. Ten thousand dollars invested in a mutual fund is growing at a rate of

$$a(x) = 840(1.08763^x) \text{ dollars per year}$$

x years after it was invested.

a. Determine the value of $\int_0^{2.75} a(x)\,dx$.

b. Interpret your answer to part *a*.

6. Evaluate the improper integral $\int_{-\infty}^{-4} \frac{13}{x^3}\,dx$.

Project 6.1 Acceleration, Velocity, and Distance

Setting

According to tests conducted by *Road and Track*, a 1993 Toyota Supra Turbo accelerates from 0 to 30 mph in 2.2 seconds and travels 1.4 miles (1320 feet) in 13.5 seconds, reaching a speed of 107 mph. *Road and Track* reported the data given in Table 6.50.

TABLE 6.50

Time (seconds)	Speed reached from rest (mph)
0	0
2.2	30
2.9	40
4.0	50
5.0	60
6.5	70
8.0	80
9.0	90
11.8	100

Tasks

1. Convert the speed data to feet per second, and find a quadratic model for velocity (in feet per second) as a function of time (in seconds). Discuss how close your model comes to predicting the 107 mph reached after 13.5 seconds.
2. Add the data point for 13.5 seconds, and find a quadratic model for velocity.
3. Use four rectangles and your model from Task 2 to estimate the distance traveled during acceleration from rest to a speed of 50 mph and the distance traveled during acceleration from a speed of 50 mph to a speed of 100 mph. Repeat the estimate using twice as many rectangles.
4. Use nine rectangles to approximate the distance traveled during the first 13.5 seconds. How close is your estimate to the reported value?
5. Find the distances traveled during
   a. Acceleration from rest to a speed of 50 mph
   b. Acceleration from a speed of 50 mph to a speed of 100 mph
   c. The first 13.5 seconds of acceleration

   Compare these answers to your estimates in Tasks 3 and 4. Explain how estimating with areas of rectangles is related to calculating the definite integral.

Reporting

Prepare a written report of your work. Include scatter plots, models, graphs, and discussions of each of the above tasks.

Project 6.2 Estimating Growth

Setting

Table 6.51 lists the rate of growth[27] of a typical male from birth to 18 years.

TABLE 6.51

Age (years)	Rate of growth (centimeters per year)
0	18.0
1	16.0
2	12.0
3	6.5
4	6.0
5	6.25
6	6.5
7	6.0
8	5.75
9	5.25
10	5.0
11	4.5
12	5.0
13	6.75
14	9.0
15	7.0
16	2.0
17	0.75
18	0.5

27. Based on data collected in the Berkeley Growth Study.

Tasks

1. Use the data and right rectangles to approximate the height of a typical 18-year-old male.
2. Sketch a smooth, continuous curve over a scatter plot of the data. Remember that such a curve should exhibit only the curvature implied by the data. Find a piecewise model for the data. Use no more than three pieces.
3. Use your piecewise model and limits of sums to approximate the height of a typical 18-year-old male. Convert centimeters to feet and inches, and compare your answer to the estimate you obtained using right rectangles. Which is likely to be the more accurate approximation? Why?
4. Use your piecewise model and what you know about definite integrals to find the height of a typical 18-year-old male in feet and inches. Compare your answer with the better of the approximations you obtained in Task 3.
5. Randomly choose ten 18-year-old male students, and determine their heights. (Include your data—names are not necessary, only the heights.) Discuss your selection process and why you feel that it is random. Find the average height of the 18-year-old males in your sample. Compare this average height with your answer to Task 4. Discuss your results.
6. Refer to your sketch of the rate-of-growth graph in Task 2, and draw a possible graph of the height of a typical 18-year-old male from birth to age 18.

Reporting

Prepare a report that presents your findings in Tasks 1 through 6. Explain the different methods that you used, and discuss why these methods should all give similar results. Attach your mathematical work as an appendix to your report.

7

Analyzing Accumulated Change: Integrals in Action

Concept Application

Mark Richards/PhotoEdit

The CEO of a large corporation must concern himself or herself with many facets of the economy and the corporation's relationship to it. For example, the CEO may be interested in answering questions as diverse as

- What was the total U.S. trade gap over the past 2 years?
- What is the 5-year future value of an income stream that the corporation invests?
- What is the market equilibrium price for a particular product produced by the corporation?
- Is the average level of carbon monoxide in the air likely to be a threat to employee health?

In this chapter you will learn how to answer questions such as these.

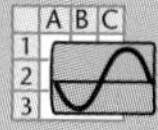

This symbol directs you to the supplemental technology guides. The different guides contain step-by-step instructions for using graphing calculators or spreadsheet software to work the particular example or discussion marked by the technology symbol.

This symbol directs you to the Calculus Concepts *web site, where you will find algebra review, links to data updates, additional practice problems, and other helpful resource material.*

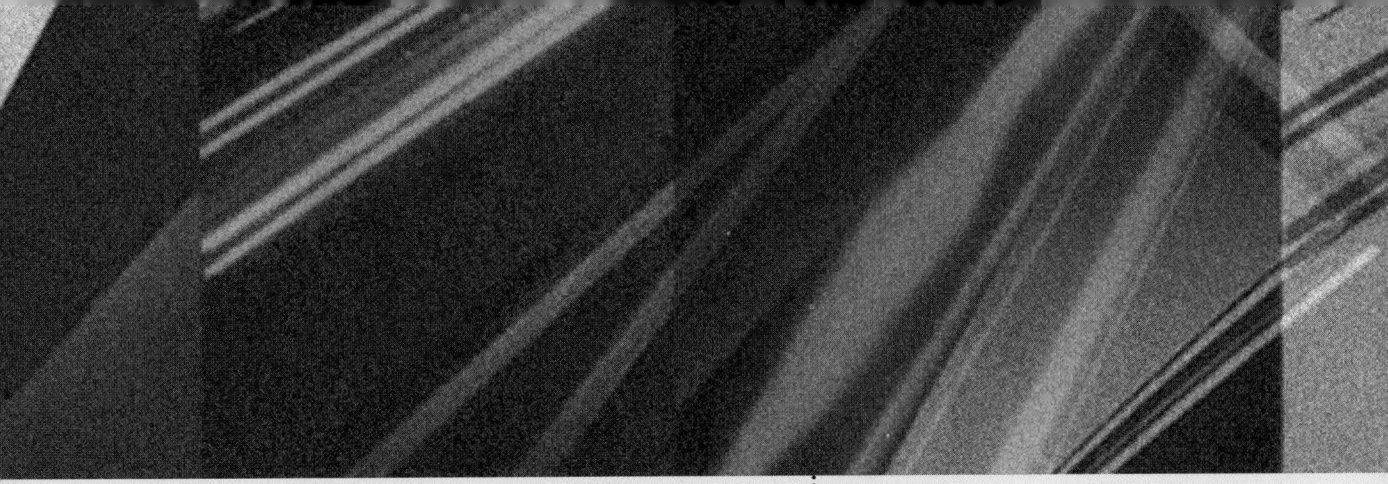

Concept Objectives

This chapter will help you understand the concepts of

- Differences in accumulated change
- Discrete and continuous income streams
- Present and future value
- Supply and demand
- Average value
- Probability distributions and density functions

and you will learn to

- Calculate and interpret the area between two curves
- Calculate and interpret present and futures values for discrete and continuous income streams
- Calculate and interpret the following economic quantities:
  consumers' willingness and ability to spend, expenditure, and surplus; suppliers' willingness and ability to receive, revenue, and surplus; market equilibrium point; total social gain
- Calculate and interpret the average value of a function and average rates of change
- Calculate and interpret probabilities, means, and standard deviations
- Solve problems using probability density functions
- Solve problems using cumulative density functions

Chapter 6 established that the accumulated results of change are limiting values of approximating sums, known as definite integrals whose magnitudes can be expressed as areas of regions between a rate-of-change function and the horizontal axis. The Fundamental Theorem of Calculus provides a simple method for evaluating definite integrals using antiderivatives.

In Chapter 7 we present several applications of integration. We begin by considering the difference of two accumulated changes. We also use integrals to calculate present and future values of income streams and future values of biological streams. We discuss how integrals can be applied to economics topics and used to calculate economic quantities of interest to consumers and producers. We conclude by using integrals to calculate averages and probabilities.

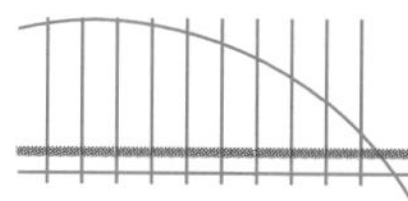

7.1 Differences of Accumulated Changes

You should feel free to omit any sections of Chapter 7 that are not appropriate for your students. For those using the complete version of the text: No portions of this chapter are prerequisite for the remainder of the text.

The idea of accumulated change was introduced in Chapter 6. We saw that when given a continuous or piecewise continuous rate-of-change function, we obtain the accumulation of change over an interval by finding the limiting value of sums of areas of rectangles. In Section 6.2 we defined this limiting value as the definite integral

$$\int_a^b f(x)\,dx = \lim_{n\to\infty}[f(x_1) + f(x_2) + \cdots + f(x_n)]\Delta x$$

We used accumulation functions to present the Fundamental Theorem of Calculus, which introduced the concept of an antiderivative. We saw that if F is an antiderivative of a smooth, continuous function f, then

$$\int_a^b f(x)\,dx = F(b) - F(a)$$

As we proceeded through Chapter 6, we considered many applications of the definite integral.

Now we turn our attention to the difference of two accumulated changes. This difference can often be thought of as the area of a region between two curves. For example, suppose the number of patients admitted to a large inner-city hospital is changing by

$$a(h) = 0.0145h^3 - 0.549h^2 + 4.85h + 8.00 \text{ patients per hour}$$

h hours after 3 a.m. We find the approximate number of patients admitted to the hospital between 7 a.m. ($h = 4$) and 10 a.m. ($h = 7$) as

$$\begin{aligned}\int_4^7 a(h)\,dh &= \int_4^7 (0.0145h^3 - 0.549h^2 + 4.85h + 8.00)\,dh \\ &= (0.003625h^4 - 0.183h^3 + 2.425h^2 + 8.00h)\Big|_4^7 \\ &\approx 120.760 - 60.016 \approx 61 \text{ patients}\end{aligned}$$

Graphically, this value is the area of the region between the graph of a and the horizontal axis from 4 to 7. (See Figure 7.1.)

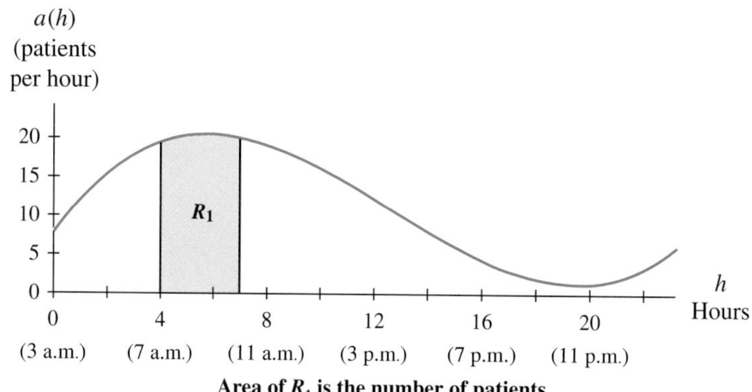

Area of R_1 is the number of patients admitted between 7 a.m. and 10 a.m.

FIGURE 7.1

Now suppose that the rate at which patients are discharged is modeled by the equation

$$y(h) = \begin{cases} -0.028h^3 + 0.528h^2 + 0.056h - 1.5 \text{ patients/hour} & \text{when } 4 \le h \le 17 \\ 0 \text{ patients/hour} & \text{when } 0 \le h < 4 \text{ and } 17 < h \le 24 \end{cases}$$

where h is the number of hours after 3 a.m. The approximate number of patients discharged between 7 a.m. and 10 a.m. is calculated as

$$\begin{aligned}\int_4^7 y(h)\,dh &= \int_4^7 (-0.028h^3 + 0.528h^2 + 0.056h - 1.5)\,dh \\ &= (-0.007h^4 + 0.176h^3 + 0.028h^2 - 1.5h)\Big|_4^7 \\ &= 34.433 - 3.92 \approx 31 \text{ patients}\end{aligned}$$

Graphically, this value is the area of the region between the graph of y and the horizontal axis from 4 to 7. (See Figure 7.2.)

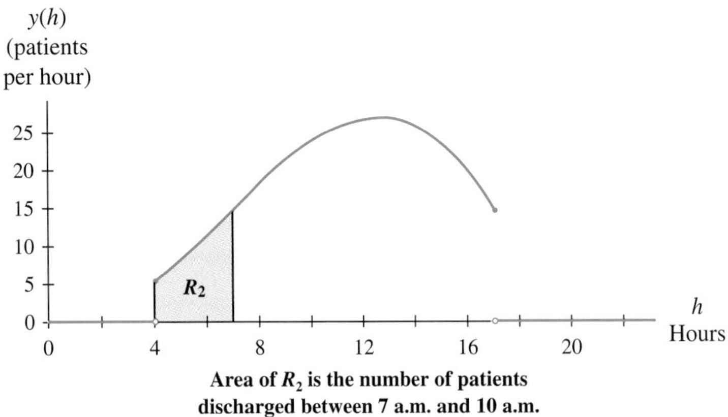

FIGURE 7.2

The net change in the number of patients at the hospital from 7 a.m. to 10 a.m. is the difference between the number of patients admitted and the number discharged between 7 a.m. and 10 a.m. That is,

$$\begin{aligned}\text{Change in the number of patients from 7 a.m. to 10 a.m.} &= \int_4^7 a(h)\,dh - \int_4^7 y(h)\,dh \\ &\approx 60.744 - 30.513 \\ &\approx 30 \text{ patients}\end{aligned}$$

Geometrically, we represent this value as the area of the region below the graph of a and above the graph of y from 4 to 7. (See Figure 7.3.)

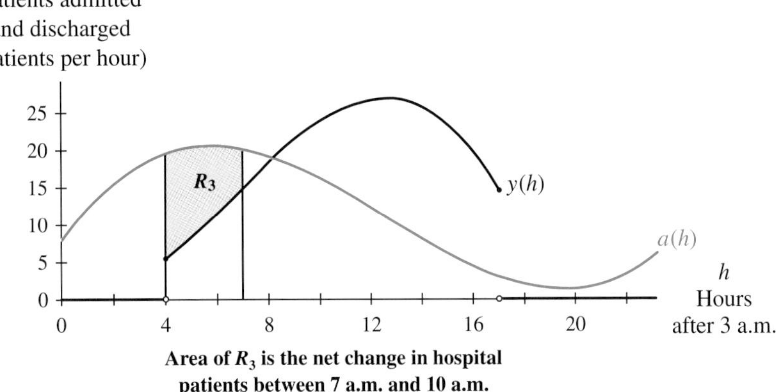

FIGURE 7.3

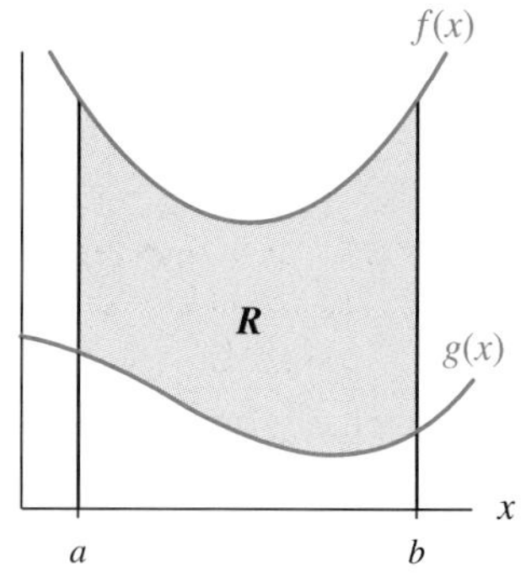

FIGURE 7.4

In general, when we want to find the area of a region that lies below one curve f and above another curve g from a to b (as in Figure 7.4), we calculate it as

$$\text{Area of the region between the graphs of } f \text{ and } g = \text{area beneath the graph of } f - \text{area beneath the graph of } g = \int_a^b f(x)\,dx - \int_a^b g(x)\,dx$$

Using the Sum Rule for antiderivatives, we obtain

$$\text{Area of the region between the graphs of } f \text{ and } g = \int_a^b [f(x) - g(x)]\,dx$$

Note that when f and g are obtained by fitting equations to data, the input variables of the functions must represent the same quantity measured in the same units.

Tell your students that if they obtain the negative of the answer they expect, then it is very likely that they have interchanged the positions of the functions in the integrand.

Area Between Two Curves

If the graph of f lies above the graph of g from a to b, then the area of the region between the two curves from a to b is given by

$$\int_a^b [f(x) - g(x)]\,dx$$

EXAMPLE 1 *Tire Manufacturers*

A major European tire manufacturer has seen its sales from tires skyrocket since 1989. A model for the rate of change of sales (in U.S. dollars) accumulated since 1989 is

$$s(t) = 3.7(1.19376^t) \text{ million dollars per year}$$

where t is the number of years since the end of 1989.

At the same time, an American tire manufacturer's rate of change of sales accumulated since 1989 can be modeled by

$$a(t) = 0.04t^3 - 0.54t^2 + 2.5t + 4.47 \text{ million dollars per year}$$

where t is the number of years since the end of 1989. These models apply through the year 2000.

a. Assuming the patterns indicated by the models remain valid through 2003, by how much did the amount of accumulated sales differ for these two companies from the end of 1995 through 2003?

b. Interpret the answer to part *a* geometrically.

Solution

a. First, determine whether one graph lies above the other on the interval in the question. The two rate-of-change functions are graphed in Figure 7.5.

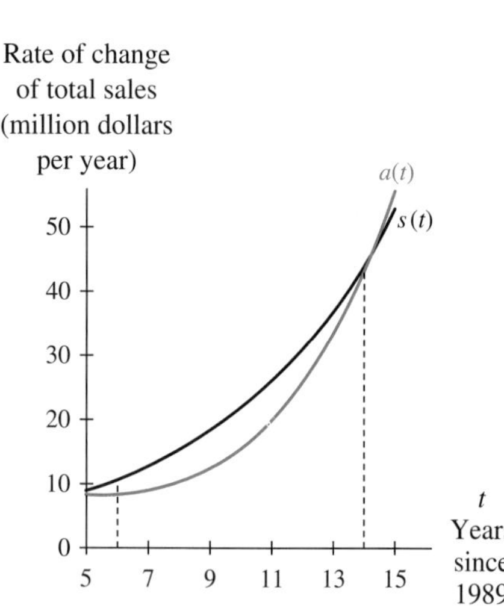

FIGURE 7.5

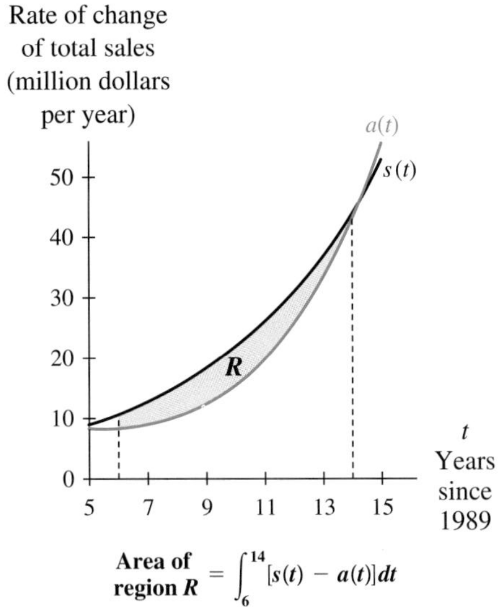

The graph of s is above the graph of a for all input values between 6 and 14, so the difference in the two companies' accumulated sales is

$$\begin{aligned}
\text{Difference in total sales} &= \left(\begin{array}{c}\text{accumulated sales for}\\ \text{the European company}\end{array}\right) - \left(\begin{array}{c}\text{accumulated sales for}\\ \text{the American company}\end{array}\right)\\
&= \int_6^{14} s(t)\,dt - \int_6^{14} a(t)\,dt\\
&= \int_6^{14} [s(t) - a(t)]\,dt\\
&= \int_6^{14} [3.7(1.19376^t) - (0.04t^3 - 0.54t^2 + 2.5t + 4.47)]\,dt\\
&= \left[\frac{3.7(1.19376^t)}{\ln 1.19376} - 0.01t^4 + 0.18t^3 - 1.25t^2 - 4.47t\right]\Bigg|_6^{14}\\
&\approx 36.9660 \text{ million dollars}
\end{aligned}$$

The European company's accumulated sales were approximately \$37,000,000 more than the American company's sales between 1995 and 2003.

Area of region $R = \int_6^{14} [s(t) - a(t)]dt$

FIGURE 7.6

b. The difference in the accumulated sales for the two companies is represented as the area of the region between the s curve and the a curve from $t = 6$ to $t = 14$. (See Figure 7.6.) As we saw in part a, this area is approximately \$37 million.

If you think of finding the area between two graphs as a limit of sums of areas of rectangles, then you should have no problem seeing that the heights of areas of rectangles for the shaded region in Figure 7.4 are given by $f(x) - g(x)$. But how are the heights of rectangles determined if one or both of the graphs lie below the horizontal axis?

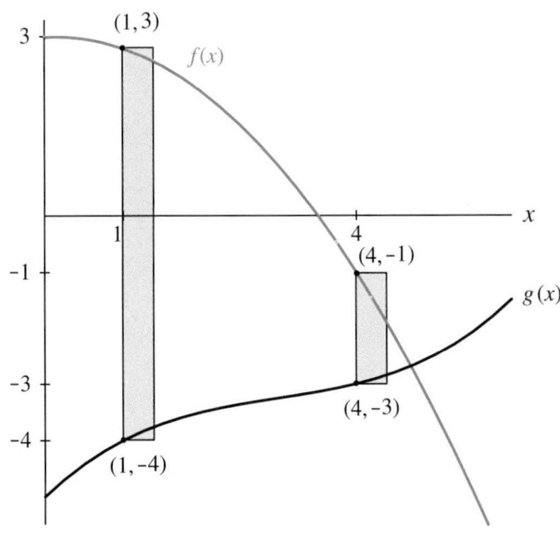

FIGURE 7.7

Consider the graphs shown in Figure 7.7. Examine the rectangle whose height is determined by the function values at $x = 1$, where the function value on top is positive but the function value on the bottom is negative. If we subtract the value of the bottom function from the value of the top function, we have $f(x) - g(x) = 3 - (-4) = 3 + 4 = 7$. The difference of the function values is the height of the rectangle.

Now consider the rectangles whose height is determined by the function values at $x = 4$. At $x = 4$, both function graphs lie below the x-axis. If we subtract the value of the bottom function from the value of the top function, we have $f(x) - g(x) = -1 - (-3) = -1 + 3 = 2$. Again, this difference is the height of the rectangle. The statement in the box on page 478 concerning the area between two curves is true whether the functions are both positive, both negative, or one of each. The important thing is to make sure that the graph of the first function in the subtraction is always on top of the graph of the second function. Care must be taken whenever the two graphs intersect.

Let us consider again the two tire companies from Example 1. Assuming that the models remain valid through 2009, what is the difference in the accumulated sales for the two companies from 1999 through 2009? As shown in Figure 7.8, the graphs of the two rate-of-change functions s and a cross near $t = 5$ and cross again near $t = 14$.

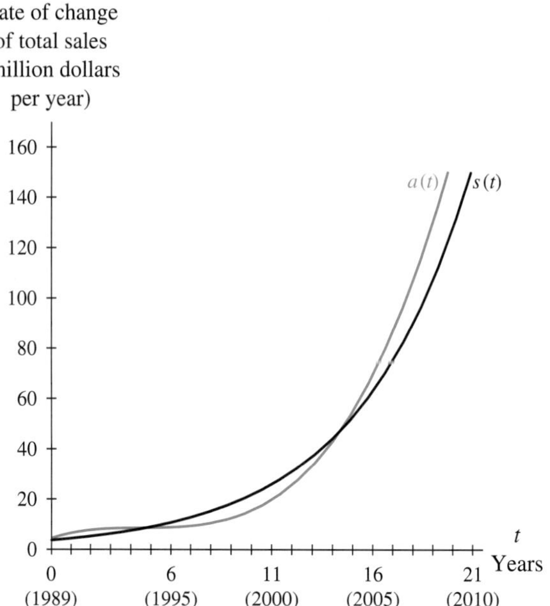

FIGURE 7.8

If we set $s(t) = 3.7(1.19376^t)$ equal to $a(t) = 0.04t^3 - 0.54t^2 + 2.5t + 4.47$ and solve for t, we find that the two functions intersect when $t \approx 4.657$ (in 1994) and when $t \approx 14.242$ (in 2004), as well as when $t \approx -0.372$. Accumulated sales were greater for the European company than for the American company from $t \approx 4.657$ to $t \approx 14.242$. After $t \approx 14.242$, the American company saw greater accumulated sales than the European company.

From the beginning of 2000 ($t = 10$) through most of the first quarter of 2004 ($t \approx 14.242$), the European company accumulated approximately

$$\int_{10}^{14.242} [s(t) - a(t)]dt = 18.5383 \text{ million dollars}$$

more in sales than the American company. This is the area of region R_1 in Figure 7.9.

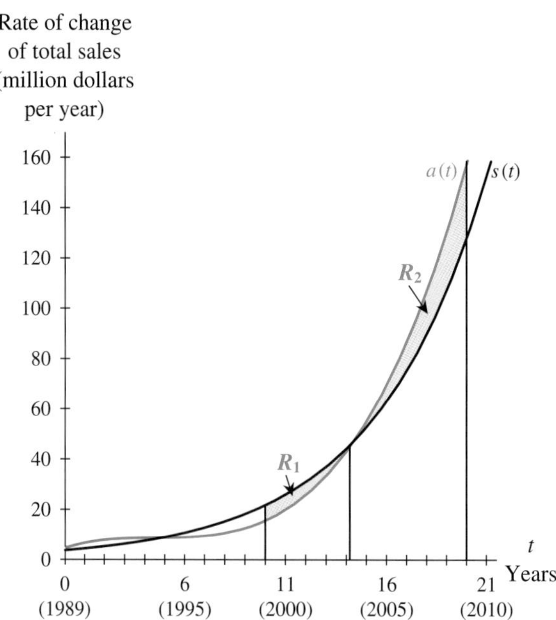

FIGURE 7.9

From close to the end of the first quarter of 2004 ($t \approx 14.242$) through 2009 ($t = 20$), the American company accumulated approximately

$$\int_{14.242}^{20} [a(t) - s(t)]\,dt = 79.4076 \text{ million dollars}$$

more in sales than the European company. This is the area of region R_2 in Figure 7.9.

In order to calculate the estimated difference in accumulated sales between the two companies from 1999 through 2009, we subtract the portion where the American company's accumulated sales were greater from the portion where the European company's accumulated sales were greater. That is,

$$\begin{aligned}\text{Difference in accumulated sales} &= \int_{10}^{14.242} [s(t) - a(t)]\,dt - \int_{14.242}^{20} [a(t) - s(t)]\,dt \\ &\approx 18.5383 \text{ million dollars} - 79.4076 \text{ million dollars} \\ &\approx -60.869 \text{ million dollars}\end{aligned}$$

The European company's accumulated sales were nearly 61 million dollars less than the American company's accumulated sales over the years considered.

If we use the Constant Multiplier Rule and the Sum Rule for antiderivatives, we see that we did not need to split the interval from 10 to 20 into two intervals.

$$\begin{aligned}\text{Difference in accumulated sales} &= \int_{10}^{14.242} [s(t) - a(t)]\,dt - \int_{14.242}^{20} [a(t) - s(t)]\,dt \\ &= \int_{10}^{14.242} [s(t) - a(t)]\,dt + \int_{14.242}^{20} [-a(t) + s(t)]\,dt \\ &= \int_{10}^{14.242} [s(t) - a(t)]\,dt + \int_{14.242}^{20} [s(t) - a(t)]\,dt \\ &= \int_{10}^{20} [s(t) - a(t)]\,dt\end{aligned}$$

This is true in general—whenever you wish to find the difference between accumulated change for two continuous rate-of-change functions, you can calculate the definite integral of the difference of the functions, regardless of where the functions intersect.

Difference of Two Accumulated Changes

If f and g are two continuous rate-of-change functions, then the difference between the accumulated change of f from a to b and the accumulated change of g from a to b is

$$\int_a^b [f(x) - g(x)]\,dx$$

It is important to note, however, that the integral $\int_a^b [f(x) - g(x)]\,dx$ may not represent the area between the graphs of f and g from a to b. Note that in the tire sales example, the area of the regions between the two curves from 10 to 20 is $18.538 + 79.408 = 97.946$, whereas the difference in accumulated sales is $18.538 - 79.408 = -60.869$.

> If two rate-of-change functions intersect in the interval from a to b, then the difference between their accumulated changes is *not* the same as the area of the regions between the two curves.

EXAMPLE 2 *Area Between Two Curves*

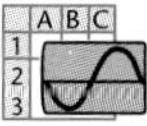

The functions $f(x) = 0.3x^3 - 3.3x^2 + 9.6x - 5.1$ and $g(x) = -0.15x^2 + 2.03x - 5.1$ are shown in Figure 7.10.

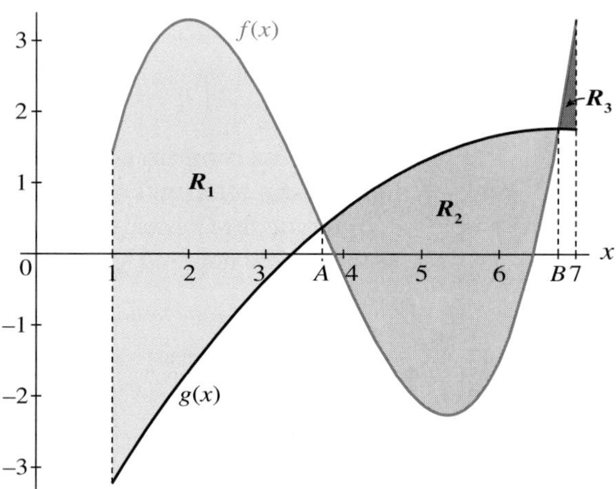

FIGURE 7.10

a. Determine the input values marked A and B.

b. Find the areas of regions R_1, R_2, and R_3.

c. Find the value of $\int_1^7 [f(x) - g(x)]\,dx$.

d. What is the area of the regions between the two functions from 1 to 7?

e. If $f(x)$ is the rate of change of the revenue of a small business and $g(x)$ is the rate of change of the costs of the business x years after its establishment (both quantities measured in thousands of dollars per year), interpret the areas in part *b* and the value of the definite integral in part *c*.

Solution

a. By solving the equation $f(x) = g(x)$ for x, we find that $A \approx 3.724$ and $B \approx 6.776$. (The functions also intersect at $x = 0$).

b. Area $R_1 = \int_1^A [f(x) - g(x)]\,dx \approx 9.878$

Area $R_2 = \int_A^B [g(x) - f(x)]\,dx \approx 7.465$

Area $R_3 = \int_B^7 [f(x) - g(x)]\,dx \approx 0.167$

c. $\int_1^7 [f(x) - g(x)]\,dx = 2.58$

d. Adding the three areas from part *b*, we find the total area to be approximately

$$9.878 + 7.465 + 0.167 \approx 17.510$$

e. The areas of the regions between the rate-of-change-of-revenue and the rate-of-change-of-cost curves represent the accumulated change in profit. The change in profit is positive when the rate-of-change-of-revenue curve is on top and negative when the rate-of-change-of-revenue curve is on the bottom. Thus, between the end of the first year and approximately 3.75 years (the end of the third quarter of the fourth year), profit increased by approximately \$9.9 thousand. Profit decreased by approximately \$7.5 thousand between $x \approx 3.724$ (near the end of the third quarter of the fourth year) and $x \approx 6.776$ (near the beginning of the fourth quarter of the seventh year). In the final quarter of the seventh year (between $x \approx 6.776$ and $x = 7$), the profit increased by approximately \$0.2 thousand.

The definite integral in part *c* gives the change in profit from the end of the first year through the end of the seventh year. The business's profits at the end of the seventh year was about \$2.6 thousand more than its profits at the end of the first year.

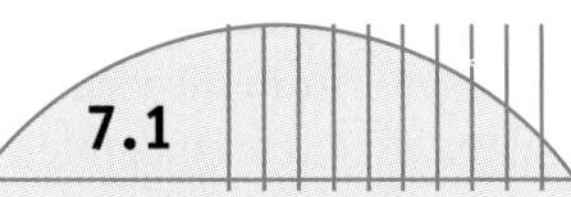

Concept Inventory

- *Area(s) of region(s) between two curves*
- *Differences of accumulated changes*

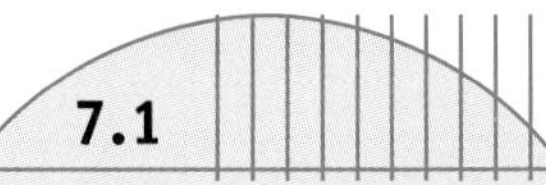

Activities

For Activities 1 through 4:

a. Sketch graphs of the functions f and g on the same axes.

b. Shade the region between the graphs of f and g from a to b.

c. Calculate the area of the shaded region.

1. $f(x) = 0.25x^2 + 2$ $\quad a = 0$
   $g(x) = -2x + 24$ $\quad b = 6$

2. $f(x) = 10(0.85^x)$ $\quad a = 2$
   $g(x) = 6(0.75^x)$ $\quad b = 10$

3. $f(x) = x^2 - 4x + 10$ $\quad a = 1$
   $g(x) = 2x^2 - 12x + 14$ $\quad b = 7$

4. $f(x) = e^x$ $\quad a = 0$
   $g(x) = e^{-x}$ $\quad b = 2$

For Activities 5 through 8:

a. Sketch graphs of the functions f and g on the same axes.

b. Find the input value(s) at which the graphs of f and g intersect.

c. Shade the region(s) between the graphs of f and g from a to b.

d. Calculate the difference in the area of the region between the graph of f and the horizontal axis and the area of the region between the graph of g and the horizontal axis from a to b.

e. Calculate the total area of the shaded region(s).

5. $f(x) = e^{0.5x}$ $\quad a = 0.5$
   $g(x) = \dfrac{2}{x}$ $\quad b = 3$

6. $f(x) = 0.25x - 3$ $\quad a = 15$
   $g(x) = 14(0.93^x)$ $\quad b = 50$

7. $f(x) = x + 1.5$ $\quad a = -5$
   $g(x) = 2(1.25^x)$ $\quad b = 8$

8. $f(x) = -0.25x^2 + 1.9x + 1.8$ $\quad a = 0$
   $g(x) = -0.08x^3 + 0.5x^2 + 0.3x + 2.0$ $\quad b = 10$

9. Figure 7.1.1 shows graphs of $r'(x)$, the rate of change of revenue, and $c'(x)$, the rate of change of costs (both in thousands of dollars per thousand dollars of capital investment) associated with the production of solid wood furniture as functions of x, the amount (in thousands of dollars) invested in capital. The area of the shaded region is 13.29.

   a. Interpret the area in the context of furniture manufacturing.

   b. Write an equation for the area of the shaded region.

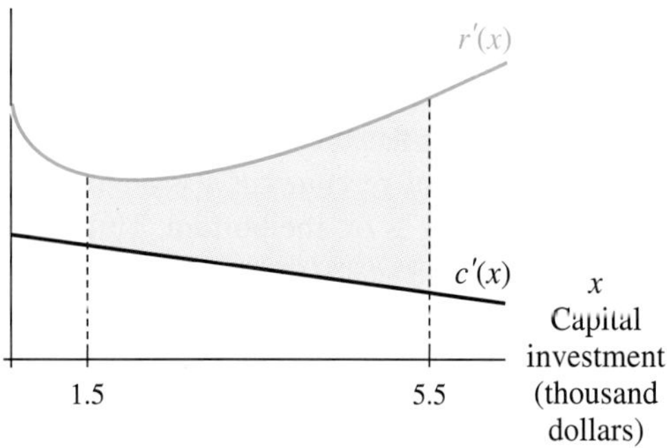

FIGURE 7.1.1

10. Figure 7.1.2 depicts graphs of the rate of change of total revenue $R'(x)$ (in billions of dollars per year) and the rate of change of total cost $C'(x)$ (in billions of dollars per year) of a company in year x. The area of the shaded region is 126.5.

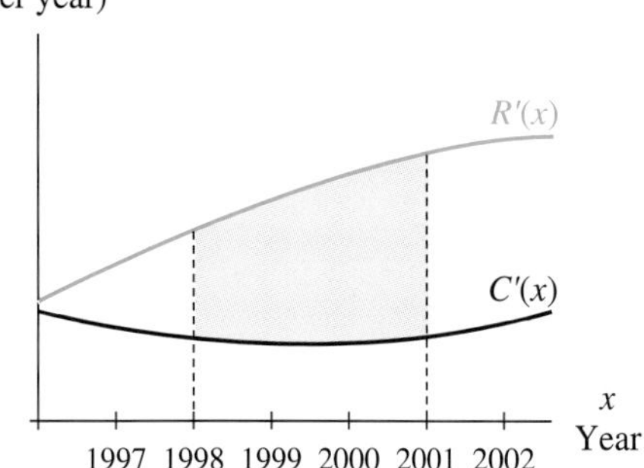

FIGURE 7.1.2

a. Interpret the area in context.

b. Write an equation for the area of the shaded region.

11. A country is in a state of civil war. As a consequence of deaths and people fleeing the country, its population is decreasing at a rate of $D(x)$ people per month. The rate of increase of the population as a result of births and immigration is $I(x)$ people per month. The variable x is the number of months since the beginning of the year. Graphs of D and I are shown in Figure 7.1.3. Region R_1 has area 3690, and region R_2 has area 9720.

a. Interpret the area of R_1 in context.

b. Interpret the area of R_2 in context.

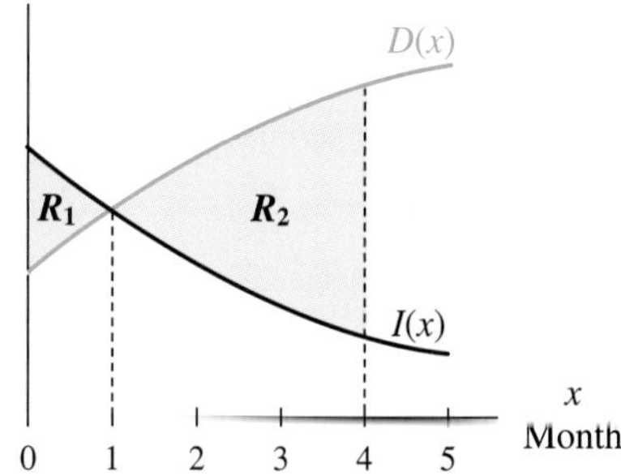

FIGURE 7.1.3

c. Find the change in population from the beginning of January through the end of April.

d. Explain why the answer to part *c* is not the sum of the areas of the two regions.

12. Figure 7.1.4 depicts graphs of $c(t)$, the rate at which people contract a virus during an epidemic, and $r(t)$, the rate at which people recover from the virus, where t is the number of days after the epidemic begins.

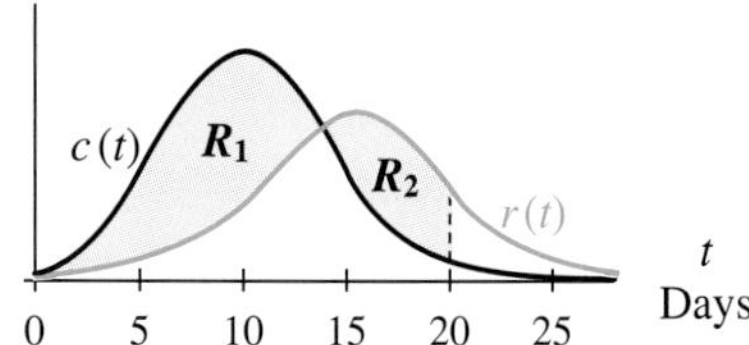

FIGURE 7.1.4

a. Interpret the area of region R_1 in the context of the epidemic.

b. Interpret the area of region R_2 in the context of the epidemic.

c. Explain how you could use a definite integral to find the number of people who contracted the virus since day 0 but have not recovered by day 20.

13. Table 7.1 shows the time it takes for a Toyota Supra and a Porsche 911 Carerra to accelerate from 0 mph to the speeds given.[1]

a. Find models for the speed of each car, given the number of seconds after starting from 0 mph. (*Hint*: Add the point (0, 0), and convert miles per hour to feet per second before modeling.)

1. *Road and Track.*

In Activity 17, the models give *monthly* values, thus we can consider the output units to be billion dollars per month. Similarly, Activity 18 gives *annual* earnings, so output units can be thought of as thousand dollars per year. In both of these activities, the models are based on count data. Thus both integrating and summing a discrete number of output values give approximations. Summing the actual data would give the most accurate answer, but we have not provided those data here.

TABLE 7.1

Toyota Supra		Porsche 911 Carerra	
Time (seconds)	Speed (mph)	Time (seconds)	Speed (mph)
2.2	30	1.9	30
2.9	40	3.0	40
4.0	50	4.1	50
5.0	60	5.2	60
6.5	70	6.8	70
8.0	80	8.6	80
9.9	90	10.7	90
11.8	100	13.3	100

b. How much farther than a Porsche 911 Carerra does a Toyota Supra travel during the first 10 seconds, assuming that both cars begin from a standing start?

c. How much farther than a Porsche 911 does a Toyota Supra travel between 5 seconds and 10 seconds of acceleration?

Table 7.2 shows the approximate rates of change of revenue[2] for the United States Postal Service (USPS), Federal Express (FedEx), and United Parcel Service (UPS). The information in the table is used in Activities 14 and 15.

TABLE 7.2

Year	USPS (billions of dollars per year)	FedEx (billions of dollars per year)	UPS (billions of dollars per year)
1993	2.7	0.2	1.4
1994	2.7	1.7	1.4
1995	2.6	2.2	1.3
1996	2.4	2.4	1.3
1997	2.2	2.2	1.2
1998	2.0	2.4	1.1
1999	1.7	2.2	1.1

14. a. Using the data in Table 7.2, find and graph equations for the rates of change of revenue for USPS and UPS.

b. Find and interpret the area of the region between the graphs of the two equations in part *a* from 1993 through 1999.

15. a. Using the data in Table 7.2, find and graph equations for the rates of change of revenue for FedEx and UPS.

b. Find and interpret the areas of the two regions bounded by the graphs in part *a* between 1993 and 1999.

c. Find the definite integral of the difference of the two equations in part *a* between 1993 and 1999. Interpret your answer.

16. The rate of change of the value of goods exported from the United States between 1990 and 1997 can be modeled[3] as

$$E'(t) = 7.342t + 16.49 \text{ billion dollars per year}$$

t years after the end of 1990. Likewise, the rate of change of the value of goods imported into the United States during those years can be modeled[4] as

$$I'(t) = -5.23t^2 + 45.71t - 21.75 \text{ billion dollars per year}$$

t years after the end of 1990.

a. Find the difference between the accumulated value of imports and the accumulated value of exports from the end of 1990 through 1997.

b. Is your answer from part *a* the same as the area of the region(s) between the graphs of E' and I'? Explain.

17. Based on data from the Department of Commerce between November 1998 and November 1999, the value of monthly exports of U.S. goods can be modeled by the equation

$$E(m) = -0.01107m^3 + 0.62213m^2 - 10.83633m + 137.86264 \text{ billion dollars}$$

Based on data from the same time period, the value of monthly imports can be modeled by the equation

$$I(m) = -0.02637m^3 + 1.38031m^2 - 21.8869m + 202.64835 \text{ billion dollars}$$

In both equations, m is the number of months since the beginning of 1998. The *trade deficit* is calculated as exports minus imports.

a. Use a definite integral and the equations given to estimate the trade deficit in 1999.

b. Use a definite integral and the equations given to estimate the trade deficit in 1998.

c. On January 21, 2000, an Associated Press article titled "Trade gap widens to record $26.5 billion"

2. Based on data from Hoover's Online Company Capsules.

3. Based on data from *Statistical Abstract*, 1995.

4. Ibid.

was published in the *Reno Gazette–Journal.* This article states that ". . . [in 1999] the U.S. trade deficit was running at an annual rate of \$266.8 billion, far surpassing last year's record deficit of \$164.3 billion." Compare your answers to parts *a* and *b* with the values given in the article for the 1998 and 1999 total trade deficits. Explain why one estimate is close but the other is not.

18. Based on data provided by the Census Bureau for the years 1980 through 1988, the full-time average annual earnings of men and women in the United States can be modeled by the following equations:

    Men: $m(t) = 0.0625t^2 - 10.38t + 466.8075$ thousand dollars

    Women: $w(t) = -0.03125t^2 + 5.695t - 234.89875$ thousand dollars

    where t is the number of years since 1900. From the beginning of 1980 and through the end of 1988, by how much did the 9-year earnings of a man who earned the average wage exceed those of a woman who earned the average wage?

19. The amount of water in a reservoir is controlled by a dam. During a brief but severe storm, water begins pouring into the reservoir from streams and runoff at a rate given by the equation

    $I(m) = (5.6972 \cdot 10^{-6})m^4 - 0.00137m^3 + 0.08176m^2 + 0.0338m + 0.877828$ thousand cubic feet per minute

    where m is the number of minutes after 6 a.m. At 7:54 a.m., the runoff stops. At 6:30 a.m., the dam is opened to release 20,000 cubic feet of water per minute. Before 6:30 a.m., the dam releases no water. Figure 7.1.5 shows a graph of I, as well as a graph of the rate at which water flows out of the dam.

    a. Express the following quantities in terms of the areas of the regions labeled in Figure 7.1.5:
        i. The excess amount of water in the reservoir from the storm before the dam was opened
        ii. The excess amount of water in the reservoir from the storm at the time the runoff ended
        iii. The amount of water drained through the dam from the time it was opened until the runoff had ended
        iv. The total amount of water that ran into the reservoir as a result of the storm runoff

    b. Determine the values of the quantities in part *a.*

    c. How long did the dam need to remain open after the runoff ended in order to return the reservoir to its level before the storm began?

20. When modeling populations, biologists consider many factors that affect mortality. Some mortality factors (such as those that may be weather-related) are not dependent on the size of the population; that is, the proportion of the population killed by such factors remains constant regardless of the size of the population. Other mortality factors are dependent on the size of the population. In these cases, the proportion of a population killed by a certain mortality factor increases (or decreases) as the size of the population increases. Such factors are called *density-dependent mortality factors.*

    Varley and Gradwell studied the population size of a winter moth[5] in a wooded area between 1950 and 1968. They found that predatory beetles represented the only density-dependent mortality factor in the life cycle of the winter moth. When the population of moths was small, the beetles ate few moths, searching elsewhere for food, but when the

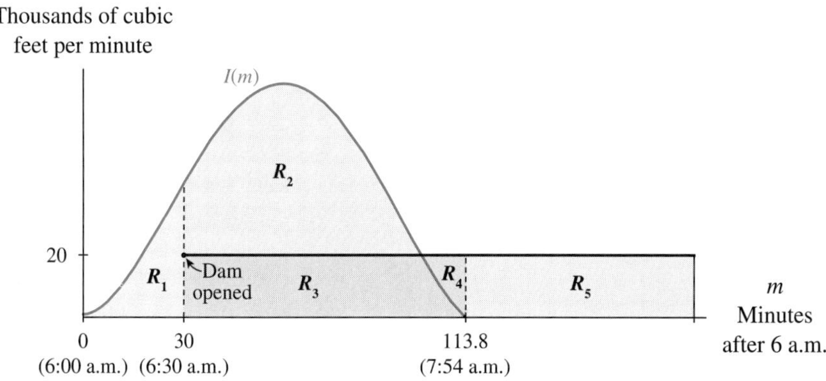

FIGURE 7.1.5

population was large, the beetles assembled in large clusters in the area of the moth population and laid eggs, thus increasing the proportion of moths eaten by the beetles.

Suppose the annual number of winter moth larvae in Varley and Gradwell's study that survived winter kill and parasitism each year between 1961 and 1968 can be modeled by the equation

$$l(t) = -0.0505 + 1.516 \ln t \text{ hundred moths per square meter of tree canopy}$$

and that the number of pupae surviving the predatory beetles each year during the same time period can be modeled by the equation

$$p(t) = 0.251 + 0.794 \ln t \text{ hundred moths per square meter of tree canopy}$$

In both equations, t is the number of years since 1960. The area of the region below the graph of l and above the graph of p is referred to as the *accumulated density-dependent mortality* of pupae by predatory beetles. Use an integral to estimate this value between the years 1962 and 1965. Interpret your answer.

21. In response to EPA regulations, a factory that produces carbon emissions plants 22 hectares of forest[6] in 1990. The trees absorb carbon dioxide as they grow, thus reducing the carbon level in the atmosphere. The EPA requires that the trees absorb as much carbon in 20 years as the factory produces during that time. The trees absorb no carbon until they are 5 years old. Between 5 and 20 years of age, the trees absorb carbon at the rates indicated in Table 7.3.

Activity 21 makes an effective short project. We recommend assigning it to pairs of students.

TABLE 7.3

Tree age (years)	5	10	15	20
Carbon absorption (tons per hectare per year)	0.2	6.0	14.0	22.0

a. Find a model for the rate (in tons per year) at which carbon is absorbed by the 22 hectares of trees between 1990 and 2010.

b. The factory produced carbon at a constant rate of 246 tons per year between 1990 and 1997. In 1997, the factory made some equipment changes that reduced the emissions to 190 tons per year. Graph the rate of emissions produced by the factory together with the model in part *a* between 1990 and 2010. Find and label the time in which the absorption rate equals the production rate.

c. Label the regions of the graph in part *b* whose areas correspond to the following quantities:

   i. The carbon emissions produced by the factory but not absorbed by the trees

   ii. The carbon emissions produced by the factory and absorbed by the trees

   iii. The carbon emissions absorbed by the trees from sources other than the factory

d. Determine the values of the three quantities in part *c*.

e. After 20 years, will the amount of carbon absorbed by the trees be at least as much as the amount produced by the factory during that time period, as required by the EPA?

7.2 Streams in Business and Biology

Picture a stream flowing into a pond. You have probably just created a mental picture of water that is flowing continuously into the pond. We can also imagine moneys that are "flowing" continuously into an investment or new individuals that are "flowing" continuously into an existing population.

It is not unreasonable to consider the income of large financial institutions and major corporations as being received continuously over time in varying amounts.

5. Adapted from P.J. denBoer and J. Reddingius, *Regulation and Stabilization Paradigms in Population Ecology* (London: Chapman and Hall), 1996.
6. Adapted from A. R. Ennos and S. E. R. Bailey, *Problem Solving in Environmental Biology* (Harlow, Essex, England: Longman House), 1995.

For instance, consider utility companies that receive payments at varying times throughout each month. Furthermore, with electronic transfer of funds, these payments can be made at any time during the day or night. Such a flow of money is called a **continuous income stream**. When you make payments to a bank or to some other financial institution for the purpose of investing money or repaying a loan, your payments are usually for the same fixed amount and are made at regular times that are separated by a specified interval. Such a flow of money is called a **discrete income stream**. Whether continuous or discrete, an income stream is usually described as a rate $R(t)$ that varies with time t.

Determining Income Streams

Consider a business that currently posts a yearly profit of \$4.3 million. The business allocates 5% of its profits in a continuous stream among several investments. There are several situations that could determine the flow rate into these investments. If the company's profits remain constant, then the function that describes the income stream flowing into the investments is $R(t) = 0.05(\$4.3 \text{ million per year}) = \$0.215$ million per year.

Suppose, however, that the company's profits increase by \$0.2 million each year. In this case, the company's profit is linear, beginning with \$4.3 million dollars the first year and increasing by a constant \$0.2 million each year: $4.3 + 0.2t$ million dollars per year after t years. The investment stream is 5% of the profit, so $R(t) = 0.05(4.3 + 0.2t)$ million dollars per year t years after the company posted a profit of \$4.3 million.

It is also possible that the company's profits could increase by a constant 7% each year. Recall that constant percentage change is modeled by an exponential equation. In this case, the function that describes the flow rate of the investment stream is

$$R(t) = 0.05[4.3(1.07^t)] \text{ million dollars per year}$$

t years after the company posted a profit of \$4.3 million. Determining the rate at which income flows into an investment is the first step in answering questions about the present and future values of the invested income stream.

EXAMPLE 1 *Starting a Business*

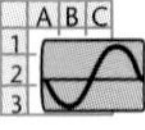

After you graduate from college, you start a small business that immediately becomes successful. When you establish the business, you determine that 10% of your profits will be invested each year. In the first year you post a profit of \$579,000. Determine the income stream flow rate for your investments over the next several years if

a. The business's profit remains constant.

b. The profit grows by \$50,000 each year.

c. The profit increases by 17% each year.

d. The profits for the first six years are as shown in Table 7.4 and are expected to follow the trend indicated by the data.

TABLE 7.4

Year	1	2	3	4	5	6
Profit (thousands of dollars)	579	600	610	618	623	627

Solution

a. If the profit remains constant, then the flow rate of the investment stream is constant, calculated as 10% of \$579,000. Thus $R(t) = \$57{,}900$ per year.

b. Profit increasing by a constant amount each year indicates linear growth. In this case, the amount of profit that is invested is described by the linear function $R(t) = (0.10)(579{,}000 + 50{,}000t)$ dollars per year t years after the first year.

c. An increase in profit of 17% each year indicates exponential growth with a constant percentage change of 17%. In this case, the flow rate of the investment stream is described by the equation $R(t) = 0.10[570{,}000(1.17^t)]$ dollars per year t years after the first year of business.

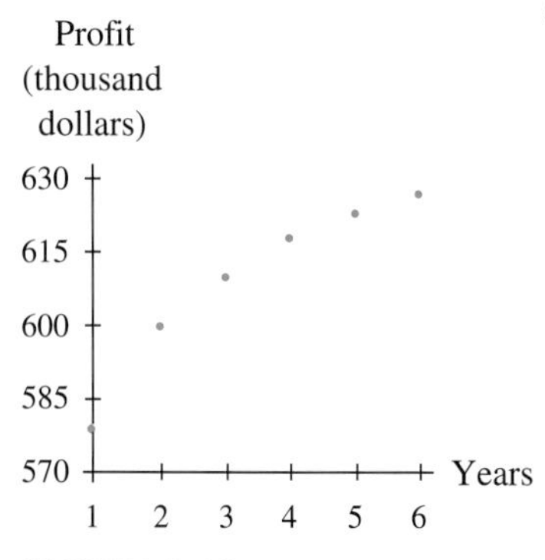

FIGURE 7.11

d. A scatter plot of the data in Table 7.4 indicates an increasing, concave-down shape. (See Figure 7.11.) A quadratic model is a reasonably good fit, although the model indicates declining profits beyond the years given in the table. A better model for profit is the log model $P(t) = 580.117 + 26.7 \ln t$ thousand dollars per year after t years of business. Thus the investment flow rate is $R(t) = 0.10P(t) = 58.0117 + 2.67 \ln t$. Note that in parts *a* through *c*, the first year of business corresponds to an input value of 0. In this log model, the first year of business corresponds to an input value of 1.

Future Value of a Continuous Stream

We calculated future and present values of single payments in Section 2.2. Future and present values of a stream of payments also have meaningful applications.

The **future value** of a continuous stream is the total accumulated value of the income stream and its earned interest. Suppose that an income stream flows continuously into an interest-bearing account at the rate of $R(t)$ dollars per year where t is measured in years and the account earns interest at the annual rate of $100r\%$ compounded continuously. What is the future value of the account at the end of T years?

To answer this question, we begin by imagining the time interval from 0 to T years as being divided into n subintervals, each of length Δt.

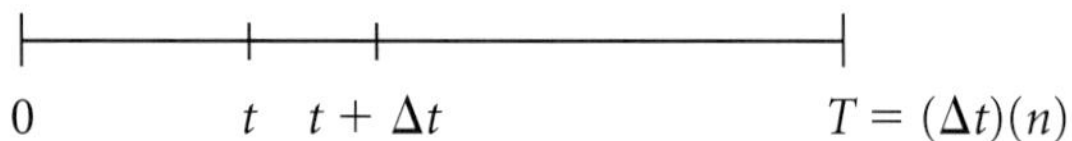

We regard Δt as being small—so small that over a typical subinterval $[t, t + \Delta t]$, the rate $R(t)$ can be considered constant. Then the amount paid into the account during this subinterval can be approximated by

$$\begin{aligned}\text{Amount paid in} &\approx [R(t) \text{ dollars per year}](\Delta t \text{ years})\\ &\approx R(t)\Delta t \text{ dollars}\end{aligned}$$

We consider this amount as being paid in at t, the beginning of the interval, and earning interest continuously for $(T - t)$ years. Using the continuously compounded interest formula ($A = Pe^{rt}$), we see that the amount grows to

$$[R(t)\,\Delta t]e^{r(T-t)} = R(t)e^{r(T-t)}\Delta t \text{ dollars}$$

at the end of T years. Summing over the n subintervals, we have the approximation

$$\text{Future value} \approx [R(t_1)e^{r(T-t_1)} + R(t_2)e^{r(T-t_2)} + \cdots + R(t_n)e^{r(T-t_n)}]\Delta t \text{ dollars}$$

where $t_1, t_2, \ldots, t_n$ are the left endpoints of the n subintervals. This sum should look familiar to you. If we simplify the expression by letting $f(t) = R(t)e^{r(T-t)}$ and rewrite the sum as $[f(t_1) + f(t_2) + \cdots + f(t_n)]\Delta t$ then you should recognize it as the type of sum we used in Sections 6.1 and 6.2.

Because we are considering the income as a continuous stream and interest as being compounded continuously, we let the time interval Δt become extremely small ($\Delta t \to 0$). That is, we use an infinite number of intervals ($n \to \infty$). Thus

$$\begin{aligned}\text{Future value} &= \lim_{n\to\infty} [f(t_1) + f(t_2) + \cdots + f(t_n)]\Delta t \\ &= \int_0^T f(t)\,dt \\ &= \int_0^T R(t)e^{r(T-t)}\,dt \text{ dollars}\end{aligned}$$

Future Value of a Continuous Income Stream

Suppose that an income stream flows continuously into an interest-bearing account at the rate of $R(t)$ dollars per year where t is measured in years and the account earns interest at the annual rate of $100r\%$ compounded continuously. The future value of the account at the end of T years is

$$\text{Future value} = \int_0^T R(t)e^{r(T-t)}\,dt \text{ dollars}$$

Using the Fundamental Theorem, we can find the rate-of-change function for future value:

$$\begin{aligned}\text{Rate of change of future value} &= \frac{d}{dx}\int_0^x R(t)e^{r(T-t)}\,dt \quad \text{for } 0 \le x \le T \\ &= R(x)e^{r(T-x)} \text{ dollars per year}\end{aligned}$$

Thus the function $f(t) = R(t)e^{r(T-t)}$ gives the rate of change after t years of the future value (in T years) of an income stream whose income is flowing continuously in at a rate of $R(t)$ dollars per year. It is the rate-of-change function $f(t) = R(t)e^{r(T-t)}$, not the flow rate of the income stream $R(t)$, that we graph when depicting future value as the area of a region beneath a rate-of-change function.

EXAMPLE 2 *Airline Expansion*

The owners of a small airline are making big plans. They hope to be able to buy out a larger airline 10 years from now by investing into an account returning 9.4% APR. Assume a continuous income stream and continuous compounding of interest.

a. The owners have determined that they can afford to invest \$3.3 million each year. How much will these investments be worth 10 years from now?

b. If the airline's profits increase so that the amount the owners invest each year increases by 8% per year, how much will their investments be worth in 10 years?

Solution

a. The flow rate of the income stream is $R(t) = 3.3$ million per year with $r = 0.094$ and $T = 10$ years. The value of these investments in 10 years is calculated as

$$\begin{aligned}
\text{Future value} &= \int_0^{10} 3.3e^{0.094(10-t)}\,dt \\
&= \int_0^{10} 3.3e^{0.94}e^{-0.094t}\,dt \\
&= \frac{3.3e^{0.94}}{-0.094}e^{-0.094(10)} - \frac{3.3e^{0.94}}{-0.094}e^{-0.094(0)} \\
&\approx -35.106 + 89.872 \\
&\approx \$54.8 \text{ million}
\end{aligned}$$

b. The function modeling exponential growth of 8% per year in the investment stream is $R(t) = 3.3(1.08^t)$ million dollars per year after t years. The future value is calculated (using technology) as $\int_0^{10} 3.3(1.08^t)e^{0.094(10-t)}\,dt \approx \$77.7$ million.

Present Value of a Continuous Stream

The **present value** of a continuous income stream is the amount P that would have to be invested now in an interest-bearing account in order for the amount to grow to a given future value. Because P dollars earning continuously compounded interest would grow to a future value of Pe^{rT} dollars in T years, we have

$$Pe^{rT} = \int_0^T R(t)e^{r(T-t)}\,dt = \int_0^T R(t)e^{rT}e^{-rt}\,dt = e^{rT}\int_0^T R(t)e^{-rt}\,dt$$

Solving for P, we obtain

$$\text{Present value} = P = \int_0^T R(t)e^{-rt}\,dt$$

Present Value of a Continuous Income Stream

Suppose that an income stream flows continuously into an interest-bearing account at the rate of $R(t)$ dollars per year where t is measured in years and that the account earns interest at the annual rate of $100r\%$ compounded continuously. The present value of the account is

$$\text{Present value} = \int_0^T R(t)e^{-rt}\,dt \text{ dollars}$$

It is worth noting that once you have calculated future value, it is easy to calculate the associated present value by solving for P in the equation

$$Pe^{rt} = \text{Future value}$$

EXAMPLE 3 *Revenue Expansion*

Last year, profit for the HiTech Corporation was \$17.2 million. Assuming that HiTech's profits increase continuously for the next 5 years at a rate of \$1.3 million per year, what are the future and present values of the corporation's 5-year profits? Assume an interest rate of 12% compounded continuously.

Solution We note that the rate of the stream is $R(t) = 17.2 + 1.3t$ million dollars per year in year t. In order to calculate the future value of this stream, we evaluate $\int_0^5 (17.2 + 1.3t)e^{0.12(5-t)}\,dt$. We have not developed a method for finding the antiderivative of $f(t) = (17.2 + 1.3t)e^{0.12(5-t)}$, so we numerically estimate the definite integral using a limiting value of sums or use technology to evaluate the integral.

$$\text{Future value} = \int_0^5 (17.2 + 1.3t)e^{0.12(5-t)}\,dt \approx \$137.9 \text{ million}$$

The invested revenue will be worth approximately \$138 million in 5 years. Again, numerically estimate a limiting value of sums or use technology to find the present value.

$$\text{Present value} = \int_0^5 (17.2 + 1.3t)e^{-0.12t}\,dt \approx \$75.7 \text{ million}$$

This amount is the lump sum (\$75.7 million) that would have to be invested at 12% compounded continuously in order to earn \$137.9 million (the future value) in 5 years.

We could also use the future value (\$137.9 million) to calculate the present value:

$$Pe^{(0.12)(5)} \approx \$137.9 \text{ million, so } P \approx \$75.7 \text{ million}$$

The integral definition of the present value is most useful in situations in which you do not know the future value.

Discrete Income Streams

As was the case with count data in Section 6.1, the calculation of future value and present value of a discrete income stream is an example of a case in which more sophisticated mathematics (integration) produces an approximation, whereas less sophisticated mathematics (summing) gives the exact answer.

The assumptions that income is flowing continuously and that interest is compounded continuously make it possible to use calculus and are often imposed by economists. Unfortunately, they do not generally hold in the real world of business. It is much more realistic to consider an income stream that flows monthly into an account with monthly compounding of interest or a stream flowing quarterly with quarterly compoundings.

The process of finding the future value for discrete income streams begins in a similar way to that for continuous streams: Determine the rate-of-flow function for the income stream, and multiply by a term that accounts for compounding interest. In the discrete case, we base our interest calculations on the formula $A = P\left(1 + \frac{r}{n}\right)^{nt}$ and, instead of integrating the resulting function, we sum a series of values.

Consider a small business that begins investing 7.5% of its monthly profit into an account that pays 10.3% annual interest compounded monthly. When the company begins investing, the monthly profit is \$12,000 and is growing by \$500 each month. We wish to determine the 2-year future value of the company's investment (assuming that profit continues to grow in the manner described).

The company will make a total of 24 deposits during the 2-year period. The first deposit is (0.075)(\$12,000) = \$900. This deposit earns interest each month for 24 months. The future value of this first deposit is $\$900\left(1+\frac{0.103}{12}\right)^{24} = \$1104.91$.

The second deposit is (0.075)(\$12,000 + \$500) = \$937.50. This deposit earns interest each month for 23 months. The future value of the second deposit is $\$937.50\left(1+\frac{0.103}{12}\right)^{23} = \$1141.15$.

The third deposit is (0.075)(\$12,000 + \$1000) = \$975. This deposit earns interest each month for 22 months. The future value of the third deposit is $\$975\left(1+\frac{0.103}{12}\right)^{22} = \$1176.60$.

By now you should be able to see a pattern in the future values of these successive monthly deposits. (See Table 7.5.)

TABLE 7.5

Time *d* (months after first deposit)	Future value of monthly deposit *F*(*d*)
0	$F(0) = \$900.00\left(1+\frac{0.103}{12}\right)^{24} = \$1104.91$
1	$F(1) = \$937.50\left(1+\frac{0.103}{12}\right)^{23} = \$1141.15$
2	$F(2) = \$975.00\left(1+\frac{0.103}{12}\right)^{22} = \$1176.60$
⋮	⋮
22	$F(22) = \$1725.00\left(1+\frac{0.103}{12}\right)^{2} = \$1754.74$
23	$F(23) = \$1762.50\left(1+\frac{0.103}{12}\right)^{1} = \$1777.63$
	Future value = sum of 24 values = \$35,204.03

Because the first deposit occurs at time $d=0$, the last (Dth) deposit occurs at time $d=D-1$.

The amount deposited each month is given by $R(d) = (0.075)(\$12{,}000 + \$500d)$, where d is the number of deposits after the first one. The deposit associated with input d accrues interest for $24-d$ months. The future value of each month's deposit is given by the formula

$$F(d) = R(d)\left(1+\frac{0.103}{12}\right)^{24-d} = (0.075)(12{,}000 + 500d)\left(1+\frac{0.103}{12}\right)^{24-d}$$

To determine the 2-year future value, we add the future values of each month's deposit beginning with 0 (month 1) and ending with 23 (month 24). Using summation notation, we write

$$\text{Future value} = \sum_{d=0}^{23} F(d) = \sum_{d=0}^{23} (0.075)(12{,}000 + 500d)\left(1+\frac{0.103}{12}\right)^{24-d}$$

We realize that many students in this course have difficulty with mathematical symbols. We suggest taking the time to explain completely the meaning of the summation symbol.

The symbol $\sum_{d=0}^{23} F(d)$ is the notation used for the sum $F(0) + F(1) + F(2) + \cdots + F(23)$. In this example, the future value of the deposits is \$35,204.03.

Your students will benefit from a comparison of the formulas for the future values of continuous and discrete income streams. What the formulas have in common is the flow rate function R and the exponent ($T - t$ and $D - d$). The formulas differ in the upper limit (T and $D-1$) and in how the interest is determined $\left(e^r \text{ and } 1+\frac{r}{n}\right)$. Connect the difference in the interest term to the formulas for continuously and discretely compound interest.

We generalize the definition of future value for discrete income streams as follows:

Future Value of a Discrete Income Stream

Suppose that a deposit is made into an interest-bearing account at n equally spaced times throughout a year. The value of the dth deposit is $R(d)$ dollars per period, and it earns interest at an annual percentage rate of $100r\%$ compounded once in each deposit period. The future value of the deposits at the end of D deposit periods is

$$\text{Future value} = \sum_{d=0}^{D-1} R(d)\left(1 + \frac{r}{n}\right)^{D-d} \text{ dollars}$$

Once you have determined future value, present value can be found by solving for P in the equation

$$P\left(1 + \frac{r}{n}\right)^{D} = \text{Future value}$$

EXAMPLE 4 *Saving for the Future*

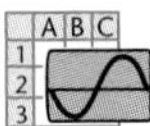

When you graduate from college (say, in 3 years), you would like to purchase a car. You have a job and can put \$75 into savings each month for this purchase. You choose a money market account that offers an APR of 6.2% compounded quarterly.

a. How much money will you have deposited in 3 years?

b. What will be the value of your savings in 3 years?

c. How much money would you have to deposit now (in one lump sum) to achieve the same future value in 3 years?

d. You are considering a second money market account that pays monthly interest of 0.5%. Will this account result in a greater future value than that calculated in part *b*?

Solution

a. The total amount deposited is (36)(\$75) = \$2700.

b. Because interest is compounded quarterly ($n = 4$), the monthly deposits each quarter do not earn interest until the end of the quarter in which they are deposited. We therefore consider three \$75 monthly deposits equivalent to one \$225 quarterly deposit, and this gives $R(d) = \$75(3) = \$225$ per quarter d quarters after the first one. We assume that the first deposit is made at the beginning of a quarter. During the 3 years, there will be 12 quarters to consider ($D = 12$). Table 7.6 shows the pattern of the future values of each quarter's deposit.

The 3-year future value is given by

$$\text{Future value} = \sum_{d=0}^{12-1} R(d)\left(1 + \frac{0.062}{4}\right)^{12-d}$$

$$= \sum_{d=0}^{11} \$225\left(1 + \frac{0.062}{4}\right)^{12-d} = \$2988.10$$

TABLE 7.6

Time (quarters after first deposit)	Future value of quarterly deposit
0	$\$225\left(1+\frac{0.062}{4}\right)^{12} = \$270.61$
1	$\$225\left(1+\frac{0.062}{4}\right)^{11} = \$266.48$
2	$\$225\left(1+\frac{0.062}{4}\right)^{10} = \$262.41$
⋮	⋮
10	$\$225\left(1+\frac{0.062}{4}\right)^{2} = \$232.03$
11	$\$225\left(1+\frac{0.062}{4}\right)^{1} = \$228.48$
Future value = sum of 12 values = \$2988.10	

c. Because we know the future value, we can solve $P\left(1+\frac{0.062}{4}\right)^{12} \approx 2988.10$ for the present value P to obtain $P \approx \$2484.48$. This is the amount that you would need to deposit now to have \$2988.10 in 3 years.

d. Because the account pays interest monthly, we have the values $n = 12$ and $D = 3(12) = 36$. Also, because interest is compounded monthly, we consider the deposits being made monthly, so that $R(d) = \$75$. In this case, the interest rate is given in terms of a monthly rather than a yearly rate. In other words, we are given the term $\frac{r}{n} = \frac{r}{12} = 0.005$. Thus the 3-year future value of \$75 monthly deposits into this money market account is

$$\text{Future value} = \sum_{d=0}^{35} \$75(1.005)^{36-d} = \$2964.96$$

This is a smaller future value than for the account that pays 6.2% interest quarterly.

Streams in Biology

Biology and other fields involve situations very similar to income streams. An example of this is the growth of populations of animals. As of 1978, there were approximately 1.5 million sperm whales[7] in the world's oceans. Each year approximately 0.06 million sperm whales are added to the population. Also each year, 4% of the sperm whale population either die of natural causes or are killed by hunters. Assuming that these rates (and percentage rates) have remained constant since 1978, we estimate the sperm whale population in 1998 using the same procedure as when determining future value of continuous income streams.

There are two aspects of the population that we must consider when estimating the population of sperm whales in 1998. First, we must determine the number of whales that were living in 1978 that will still be living in 1998. Because 4% of the

7. Delphine Haley, *Marine Mammals* (Seattle: Pacific Search Press, 1978).

sperm whales die each year, we calculate the number of whales that have survived the entire 20 years as $1.5(0.96^{20}) \approx 0.663$ million whales.

The second aspect that we must consider is the impact on the population made by the birth of new whales. We are told that 0.06 million whales per year are added to the population and that 96% of those survive each year. Therefore, the growth rate of the population of sperm whales associated with those that were born t years after 1978 is

$$f(t) = 0.06(0.96)^{20-t} \text{ million whales per year}$$

Thus the sperm whale population in 1998 is calculated as

$$\begin{aligned}
\text{Whale population} &= 1.5(0.96^{20}) + \int_0^{20} 0.06(0.96)^{20-t}\,dt \\
&\approx 0.663 + \int_0^{20} 0.06(0.96^{20})(0.96)^{-t}\,dt \\
&= 0.663 + 0.06(0.96^{20}) \int_0^{20} (0.96^{-1})^t\,dt \\
&= 0.663 + \left.\frac{0.06(0.96^{20})(0.96^{-1})^t}{\ln(0.96^{-1})}\right|_0^{20} \\
&\approx 1.48 \text{ million sperm whales}
\end{aligned}$$

Functions that model such biological streams, in which new individuals are added to the population and the rate of survival of the individuals is known, are referred to as *survival* and *renewal functions.*

You may wish to tell your students that the first term in the formula for the future value of a population gives the number of the initial population that will survive. The second term indicates how many of the new members of the population (introduced throughout the time period) will survive until the end of the time period.

Future Value of a Biological Stream

The future value (in b years) of a biological stream with initial population size P, survival rate $100s\%$, and renewal rate $r(t)$, where t is the number of years, is

$$\text{Future value} \approx Ps^b + \int_0^b r(t)s^{b-t}\,dt$$

In the whale example, the initial population is $P = 1.5$ million. The survival rate is 96% per year, so $s = 0.96$. The renewal rate is $r(t) = 0.06$ million whales per year.

EXAMPLE 5 *Flea Population*

An example of a stream in entomology is the growth of a flea population. In cooler areas of the country, adult fleas die before winter, but flea eggs survive and hatch the following spring when temperatures again reach 70°F. Not all the eggs hatch at the same time, so part of the growth in the flea population is due to the hatching of the original eggs. Another part of the growth in the flea population is due to propagation. Suppose fleas propagate at the rate of 134% per day and that the original set of fleas (from the dormant eggs) become reproducing adults at the rate of 600 fleas per day. What will the flea population be 10 days after the first 600 fleas begin reproducing? Assume that none of the fleas die during the 10-day period and that all fleas become reproducing adults 24 hours after hatching and propogate every day thereafter at the rate of 134% per day.

Solution We first note that because we begin counting when the first 600 fleas have become mature adults, we consider the initial population to be $P = 600$ fleas. The renewal rate is also 600 fleas per day, so $r(t) = 600$.

Because, in this case, the renewal rate function r does not account for renewal due to propagation, we must incorporate the propagation rate of 134% into the survival rate of 100%. Thus the survival/propagation rate is $s = 2.34$.

Because the renewal rate and survival/propagation rate are given in days, we let t be the input variable measured in days. The flea population will grow over 10 days to

This solution can be calculated discretely as $\sum_{t=0}^{10} 600(2.34)^{10-t} \approx$ 5.2 million fleas. See the *Instructor's Resource Guide* for a more detailed discussion.

$$\text{Flea population} \approx Ps^{10} + \int_0^{10} r(t)s^{10-t}dt = 600(2.34^{10}) + \int_0^{10} 600(2.34)^{10-t}dt$$

$$\approx 2{,}953{,}315 + 3{,}473{,}166 \approx 6.4 \text{ million fleas}$$

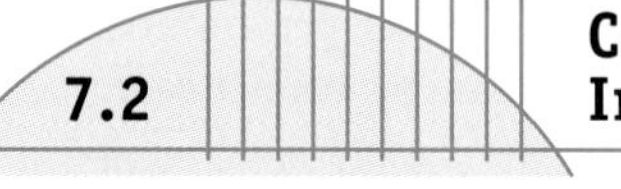

7.2 Concept Inventory

- *Income stream*
- *Flow rate of a stream*
- *Future and present value of a continuous stream*
- *Future and present value of a discrete stream*
- *Biological stream*
- *Future value of a biological stream*

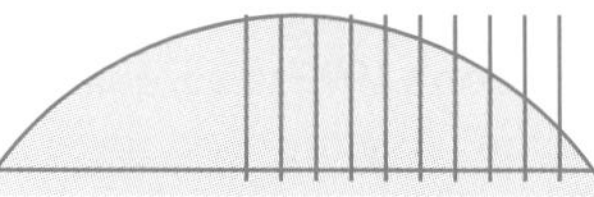

Activities

1. Suppose that after you graduate, you are hired by a company in the San Francisco Bay area. Housing prices in that region are the highest in the nation; however, you are determined to buy a house within 5 years of beginning your new job. Your starting salary is $47,000. After talking with fellow employees, you consider three possibilities for what might happen to your salary over the next 5 years:

   i. Your salary remains at your starting level.

   ii. Your salary increases by $100 a month.

   iii. Your salary increases by 0.5% each month.

   You have decided to save 20% of your salary each month for a down payment on a house. Give the function describing your monthly investments for each of the three salary possibilities.

2. A company is hoping to expand its facilities but needs capital to do so. In an effort to position itself for expansion in 3 years, the company will direct half of its profits into investments in a continuous manner. The company's profits for the past 5 years are shown in Table 7.7.

TABLE 7.7

Years ago	5	4	3	2	1
Profit (thousands of dollars)	860	890	930	990	1050

   The company's current yearly profit is $1,130,000. Find the function that describes the flow of the company's investments for each of the following profit scenarios:

   a. The profit for the next 3 years follows the trend shown in the table.

   b. The profit increases each year for the next 3 years by the same percentage that it increased in the current year.

   c. The profit remains constant at the current year's level.

   d. The profit increases each year for the next 3 years by the same fixed amount that it increased this year.

3. Consider again the situation described in Activity 1. You estimate that you will need $60,000 for a down payment, and you are unwilling to accept any

investment risk. You will be investing your money monthly in a bank savings account that pays an annual interest rate of 5% compounded monthly. For which salary possibilities will the total amount saved in 5 years be at least $60,000?

4. Consider again the situation described in Activity 2. If the company's investments can earn 16.4% annual interest compounded continuously, how much capital will it have saved after 3 years of investing for each of the profit scenarios given?

5. For the year ending December 31, 1998, the revenue of the Sara Lee Corporation[8] was $20.011 billion. Assume that Sara Lee's revenue will increase by 5% per year and that beginning on January 1, 1999, 12.5% of the revenue is invested each year (continuously) at an APR of 9% compounded continuously. What is the future value of the investment at the end of the year 2006?

6. A high school student is trying to save money to help pay for her first-year college tuition. She plans to invest $300 each quarter for 3 years into an account that pays interest at an APR of 6% compounded quarterly. Her parents decide to lend her the money so that she can devote more time to her studies while in high school. How much should they lend her so that she can invest the loaned amount now, as one lump sum, into the account and accumulate the same amount as if she had made the quarterly deposits for 3 years?

7. The revenue of General Motors Acceptance Corporation[9] (GMAC) in December 1998 was $20.165 million. Assume that GMAC's revenue remains constant and that 3% of the revenue is invested continuously throughout each year beginning at the end of December 1998 into an account that pays interest at a rate of 12.8% compounded continuously.
   a. Find the value of the account in December 2008.
   b. How much would GMAC have had to invest at the end of December 1998, in one lump sum, into this account in order to build the same 10-year future value as the one found in part *a*?

8. For the year ending December 31, 1998, the General Electric Company's revenue[10] was $99.82 billion. Assume that the revenue increases by 8% per year and that General Electric will (continuously) invest 10% of its profits each year at an APR of 11.5% compounded continuously for a period of 9 years beginning at the end of December of 1998. What is the present value of this 9-year investment?

9. To save for the purchase of your first home (in 6 years), suppose you begin investing $500 per month in an account with a fixed rate of return of 8.34%.
   a. Assuming a continuous stream, what will the account be worth at the end of 6 years?
   b. Assuming monthly activity (deposits and interest compounding), what will the account be worth at the end of 6 years?
   c. Is the answer to part *a* or part *b* more likely to be the actual future value of the account? Explain.

10. In preparing for your retirement (in 40 years), suppose you plan to invest 14% of your salary each month in an annuity with a fixed rate of return of 9.2%. You currently make $2800 per month and expect your income to increase by 4% per year.
    a. Assuming a continuous stream, what will the annuity be worth at the end of 40 years?
    b. Assuming monthly activity (deposits and interest compounding), what will the annuity be worth at the end of 40 years?
    c. Is the answer to part *a* or part *b* more likely to be the actual future value of the annuity? Explain.

11. Compaq Computer Corporation's 1999 third-quarter profits[11] were $2.1 billion. Assume that these profits will increase by 5% per quarter and that Compaq will invest 15% of its quarterly profits in an investment with a quarterly return of 13%.
    a. Write a function for the rate at which money flows into this investment each quarter.
    b. Write a function for the rate at which the 4-year future value of this investment is changing.
    c. Find the value of this investment at the end of the year 2003. (Assume a quarterly stream beginning on January 1, 2000 with the investment of 4th quarter 1999 profits.)

12. For the 1999 fiscal year, Abercrombie and Fitch Company[12] reported an annual net income of

8. Hoover's Online Company Capsules.
9. Ibid.
10. Ibid.
11. Ibid.
12. Ibid.

$102,100,000. Assume the income can be reinvested continuously at an annual rate of return of 10% compounded continuously. Also assume that Abercrombie and Fitch will maintain this annual net income for the next 5 years.

a. What is the future value of its 5-year net income?

b. What is the present value of its 5-year net income?

13. In 1993, PepsiCo installed a new soccer scoreboard for Alma College in Alma, MI. The terms of the installation were that Pepsi would have sole vending rights at Alma College for the next 7 years. It is estimated that in the 3 years after the scoreboard was installed, Pepsi sold 36.4 thousand liters of Pepsi products to Alma College students, faculty, staff, and visitors. Suppose that the average yearly sales and associated revenue remained constant and that the revenue from Alma College sales was reinvested at 4.5% APR. Also assume that PepsiCo makes a revenue of $0.80 per liter of Pepsi.

a. The vending of Pepsi products on campus can be considered a continuous process. Assuming that the revenue was invested in a continuous stream and that interest on that investment was compounded continuously, how much did Pepsi make from its 7 years of sales at Alma College?

b. Still assuming a continuous stream, find how much Pepsi would have had to invest in 1993 to create the same 7-year future value.

14. Refer to Activity 10. How much would you have to invest now, in one lump sum instead of in a continuous stream, in order to build to the same future (40-year) value?

15. Refer to Activity 9.

a. How much would you have to invest now, in one lump sum instead of in a continuous stream, in order to build to the same future (6-year) value? Assume that interest is compounded continuously.

b. How much would you have to invest now, in one lump sum instead of in a monthly stream, in order to build to the same future (6-year) value? Assume monthly compounding of interest.

c. Is the answer to part *a* or to part *b* more likely to be the actual present value of the annuity? Explain.

16. Between 1994 and 1998, the revenue of Sears, the second largest retailer in the United States, can be modeled[13] as

$$R(t) = \frac{12.574}{1 + 3.845e^{-1.002t}} + 30 \text{ billion dollars per year}$$

t years after 1994. Assume that the revenue can be reinvested at 12% compounded continuously.

a. How much is Sears' revenue invested since 1994 worth in 2000?

b. How much is this accumulated investment worth in 1994?

17. In 1956, AT&T laid its first underwater phone line. By 1996, AT&T Submarine Systems, the division of AT&T that installs and maintains undersea communication lines, had seven cable ships and 1000 workers. On October 5, 1996, AT&T[14] announced that it was seeking a buyer for its Submarine Systems division. The Submarine Systems division of AT&T was posting a profit of $850 million per year.

a. If AT&T assumed that the Submarine Systems division's annual profit would remain constant and could be reinvested at an annual return of 15%, what would AT&T have considered to be the 20-year present value of its Submarine Systems division? (Assume a continuous stream.)

b. If prospective bidder A considered that the annual profits of this division would remain constant and could be reinvested at an annual return of 13%, what would bidder A consider to be the 20-year present value of AT&T's Submarine Systems? (Assume a continuous stream.)

c. If prospective bidder B considered that over a 20-year period, profits of the division would grow by 10% per year (after which it would be obsolete) and that profits could be reinvested at an annual return of 14%, what would bidder B consider to be the 20-year present value of AT&T's Submarine Systems? (Assume a continuous stream.)

18. On October 4, 1996, Tenet Healthcare Corporation, the second-largest hospital company in the United States at that time, announced that it would buy Ornda Healthcorp.[15]

13. Based on data from the *Sears 1998 Financial Report.*

14. "AT&T Seeking a Buyer for Cable-Ship Business," *Wall Street Journal*, October 5, 1996.

15. "Tenet to Acquire Ornda," *Wall Street Journal*, October 5, 1996.

a. If Tenet Healthcare Corporation assumed that Ornda's annual revenue of \$0.273 billion would increase by 10% per year and that the revenues could be continuously reinvested at an annual return of 13%, what would Tenet Healthcare Corporation consider to be the 15-year present value of Ornda Healthcorp at the time of the buyout?

b. If Ornda Healthcorp's forecast for its financial future was that its \$0.273 billion annual revenue would remain constant and that revenues could be continuously reinvested at an annual return of 15%, what would Ornda Healthcorp consider its 15-year present value to be at the time of the buyout?

c. Tenet Healthcare Corporation bought Ornda Healthcorp for \$1.82 billion in stock. If the sale price was the 15-year present value, did either of the companies have to compromise on what it believed to be the value of Ornda Healthcorp?

19. CSX Corporation, a railway company, announced in October of 1996 its intention to buy Conrail Inc. for \$8.1 billion.[16] The combined company, CSX-Conrail, would control 29,000 miles of track and have an annual revenue of \$14 billion the first year after the merger, making it one of the largest railway companies in the country.

a. If Conrail assumed that its \$2 billion annual revenue would decrease by 5% each year for the next 10 years but that the annual revenue could be reinvested at an annual return of 20%, what would Conrail consider to be its 10-year present value at the time of CSX's offer? Is this more or less than the amount CSX offered?

b. CSX Corporation forecast that its Conrail acquisition would add \$1.2 billion to its annual revenue the first year and that this added annual revenue would increase by 2% each year. Suppose CSX is able to reinvest that revenue at an annual return of 20%. What would CSX Corporation have considered to be the 10-year present value of the Conrail acquisition in October of 1996?

c. Why might CSX Corporation have forecast an increase in annual revenue when Conrail forecast a decrease?

16. "Seeking Concessions from CSX-Conrail Is Seen as Most Likely Move by Norfolk," *Wall Street Journal*, October 5, 1996.

20. Company A is attempting to negotiate a buyout of Company B. Company B accountants project an annual income of 2.8 million dollars per year. Accountants for Company A project that with Company B's assets, Company A could produce an income starting at 1.4 million dollars per year and growing at a rate of 5% per year. The discount rate (the rate at which income can be reinvested) is 8% for both companies. Suppose that both companies consider their incomes over a 10-year period. Company A's top offer is equal to the present value of its projected income, and Company B's bottom price is equal to the present value of its projected income. Will the two companies come to an agreement for the buyout? Explain.

21. A company involved in videotape reproduction has just reported \$1.2 million net income during its first year of operation. Projections are that net income will grow over the next 5 years at the rate of 6% per year. The *capital value* (present sales value) of the company has been set as its present value over the next 5 years. If the rate of return on reinvested income can be compounded continuously for the next 5 years at 12% per year, what is the capital value of this company?

22. There were once more than 1 million elephants in West Africa.[17] Now, however, the elephant population has dwindled to 19,000. Each year 17.8% of West Africa elephants die or are killed by hunters. At the same time, elephant births are decreasing by 13% per year.

a. How many of the current population of 19,000 elephants will still be alive 30 years from now?

b. Considering that 47 elephants were born in the wild this year, write a function for the number of elephants that will be born *t* years from now and will still be alive 30 years from now.

c. Estimate the elephant population of West Africa 30 years from now.

23. In 1979 there were 12 million sooty terns (a bird) in the world.[18] Assume that the percentage of terns that survive from year to year has stayed constant at 83% and that approximately 2.04 million terns hatch each year.

17. Douglas Chawick, *The Fate of the Elephant* (Sierra Club Books, 1992).

18. Bryan Nelson, *Seabirds: Their Biology and Ecology* (New York: Hamlyn Publishing Group, 1979).

a. How many of the terns that were alive in 1979 are still alive?

b. Write a function for the number of terns that hatched t years after 1979 and are still alive.

c. Estimate the present population of sooty terns.

24. From 1936 through 1957, a population of 15,000 muskrats in Iowa[19] bred at a rate of 468 new muskrats per year and had a survival rate of 75%.

a. How many of the muskrats alive in 1936 were still alive in 1957?

b. Write a function for the number of muskrats that were born t years after 1936 and were still alive in 1957.

c. Estimate the muskrat population in 1957.

25. There are approximately 200 thousand northern fur seals.[20] Suppose the population is being renewed at a rate of $r(t) = 60 - 0.5t$ thousand seals per year and that the survival rate is 67%.

a. How many of the current population of 200 thousand seals will still be alive 50 years from now?

b. Write a function for the number of seals that will be born t years from now and will still be alive 50 years from now.

c. Estimate the northern fur seal population 50 years from now.

26. Explain, using related examples, the difference between a continuous income stream and a discrete income stream.

7.3 Integrals in Economics

Note that this section uses improper integrals, which were discussed in Section 6.6.

When you purchase an item in a store, you ordinarily have no control over the price that you pay. Your only choice is whether to buy or not to buy the item at the current price. In general, consumers hold to the view that price is a variable to which they can only respond. As the price per unit increases, consumers usually respond by purchasing (demanding) less. The typical relation between the price per unit (as input) and the quantity in demand (as output) is shown in Figure 7.12a.

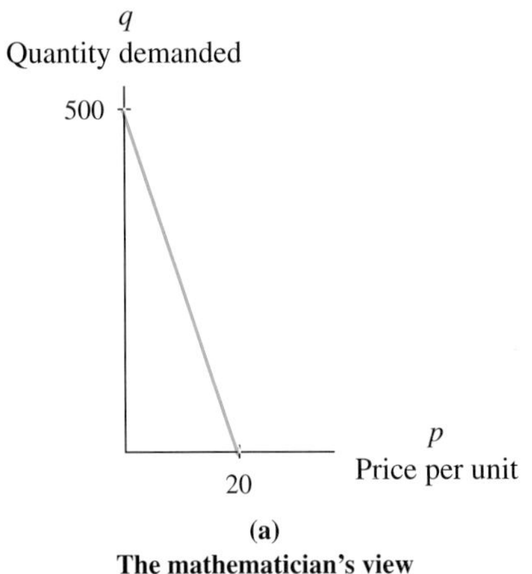

(a)
The mathematician's view

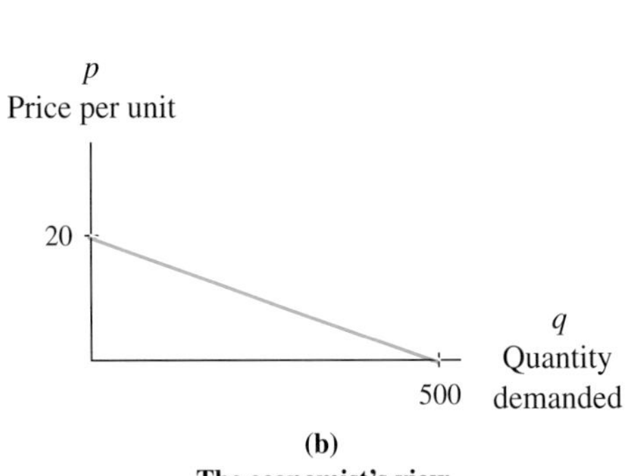

(b)
The economist's view

FIGURE 7.12

The traditional approach to graphing in economic theory is to put the price per unit along the vertical axis and the quantity in demand along the horizontal axis. (See

19. Paul L. Errington, *Muskrat Population* (Ames, IA: Iowa State University Press, 1963).
20. Delphine Haley, *Marine Mammals* (Seattle: Pacific Search Press, 1978).

Figure 7.12b) We choose not to graph the economists' way but instead use price per unit as input along the horizontal axis. This will help us understand price as input and visualize the definite integrals used later in this section.

Demand Curves

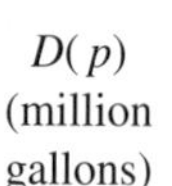

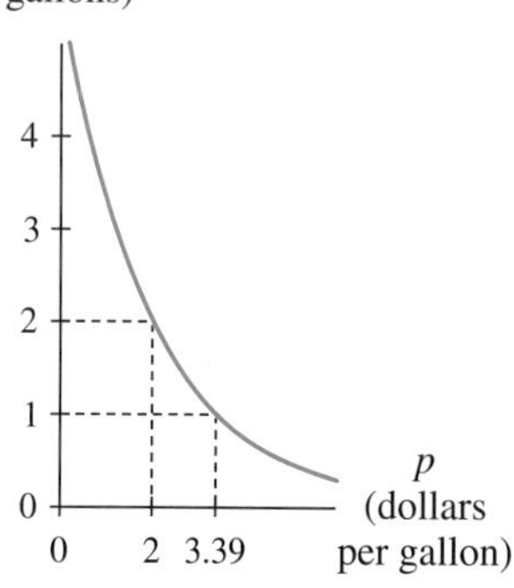

FIGURE 7.13

The graph relating quantity in demand q to price per unit p is called a **demand curve.** In economic theory, demand is actually a function that has several input variables, such as price per unit, consumers' ability to buy, consumers' need, and so on. The demand curve we consider here is a simplified version. We assume that all the possible input variables are constant except price. We denote this demand function as D with input p.

Even though the demand function is not a rate-of-change function, there are economic interpretations for the areas of certain regions lying beneath the demand curve. In order to interpret the area of these regions, you must understand how to interpret the information the demand curve represents.

For instance, suppose the graph in Figure 7.13 represents the weekly demand for regular unleaded gasoline in a California city. A point on the demand curve indicates the quantity that consumers will purchase at a given price. For instance, at \$3.39 per gallon, consumers will purchase 1 million gallons of gas. At \$2.00 per gallon, consumers will purchase 2 million gallons of gas.

Consumers' Willingness and Ability to Spend

Even though points on the demand curve tell us how much consumers will actually purchase at certain prices, consumers are willing and able to pay more than this amount for the quantity they purchase. For instance, consumers are willing and able to spend approximately \$3.39 million for the first million gallons of regular unleaded gasoline, but they are willing and able to spend only approximately \$2.00 million for the second million gallons. Thus, in total, consumers are willing and able to spend approximately \$5.39 million for 2 million gallons of gas.

If the price of gas is \$1.19 a gallon, consumers are willing and able to buy the third million gallons. That is, consumers are willing and able to spend approximately

$$(1 \text{ million gallons})(\$3.39 \text{ per gallon}) + (1 \text{ million gallons})(\$2.00 \text{ per gallon}) \\ + (1 \text{ million gallons})(\$1.19 \text{ per gallon}) = \$6.58 \text{ million}$$

for 3 million gallons of gas, even though in actuality they spend only

$$(3 \text{ million gallons})(\$1.19 \text{ per gallon}) = \$3.57 \text{ million}$$

Consumers' willingness and ability to spend can be approximated graphically as the areas of stacked horizontal rectangles. The amount that consumers actually spend is depicted as the area of a single vertical rectangle. (See Figures 7.14a and b.)

You should have noticed that the amount that consumers are willing and able to spend for 3 million gallons was given as *approximately* \$6.58 million. We can make this approximation better by considering smaller increments for price. If we were to approximate consumers' willingness and ability to spend using price increments of \$0.5 per gallon, \$0.25 per gallon, \$0.125 per gallon, etc., we would see that the areas of the stacked rectangles representing these approximations would become closer to being the true area depicted in Figure 7.15.

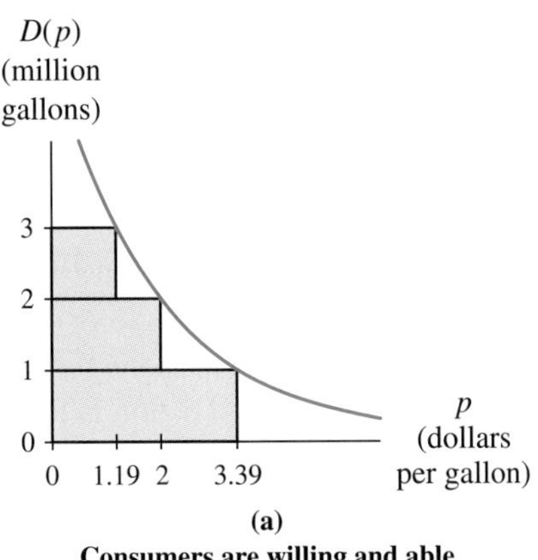

(a)
Consumers are willing and able to spend about \$6.58 million for 3 million gallons of gas.

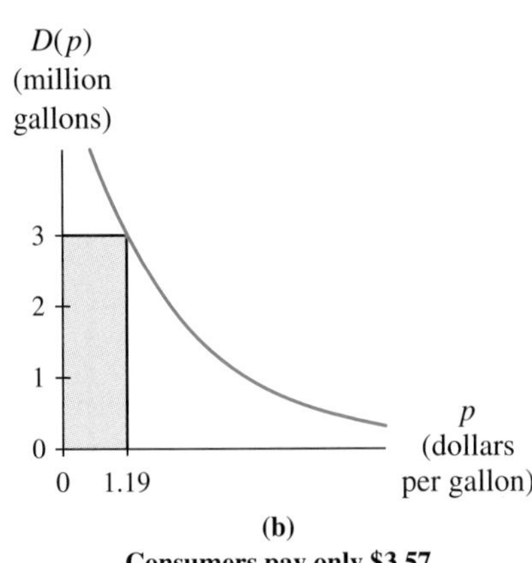

(b)
Consumers pay only \$3.57 for 3 million gallons of gas.

FIGURE 7.14

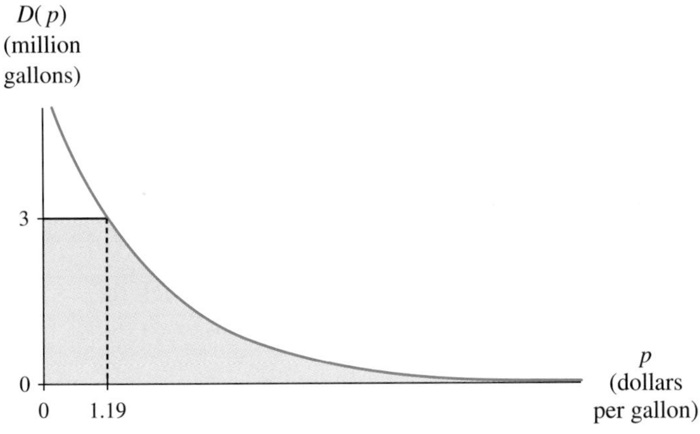

FIGURE 7.15

Thus consumers' willingness and ability to purchase 3 million gallons of gas can be visually represented by the sum of the area of the rectangle with width 1.19 under the horizontal line $D(p) = 3$ and the area under the demand curve from 1.19 to P, where P is the price above which consumers cannot and will not purchase any gas. We calculate the consumers' willingness and ability to spend as

$$3(1.19) + \int_{1.19}^{P} D(p)\,dp \text{ million dollars}$$

Suppose the demand for gas can be modeled by

$$D(p) = 5.43(0.607^p) \text{ million gallons}$$

where p dollars is the price per gallon. The only piece of information we still need is P, the price above which no gas will be purchased. You should notice that the demand function approaches 0 as p becomes large; however, it will never be exactly 0 for any p. Hence, we let P approach ∞. (You may wish to review the discussion of improper integrals in Section 6.6.) This is true for most demand functions in economics—some people will always want the product or service, regardless of the price. In this case, we consider the area under the demand curve as P becomes infinitely large. That is,

$$3(1.19) + \int_{1.19}^{\infty} 5.43(0.607^p)\,dp$$

$$= 3.57 + \lim_{P\to\infty} \int_{1.19}^{P} 5.43(0.607^p)\,dp$$

$$= 3.57 + \lim_{P\to\infty} \left(\frac{5.43(0.607^p)}{\ln 0.607} \right)\Bigg|_{1.19}^{P}$$

$$= 3.57 + \left(\lim_{P\to\infty} \frac{5.43(0.607^P)}{\ln 0.607} \right) - \frac{5.43(0.607^{1.19})}{\ln 0.607}$$

$$\approx 3.57 + 0 + 6.00$$

$$\approx \$9.57 \text{ million}$$

Remind your students to view this limit graphically as the limit of a decreasing exponential function.

Thus consumers are willing and able to spend \$9.57 million in order to purchase 3 million gallons of gas.

In general, we make the following definition:

Consumers' Willingness and Ability to Spend

For a continuous demand function $q = D(p)$, the maximum amount that consumers are willing and able to spend for a certain quantity q_0 of goods or services is the area of the shaded region in Figure 7.16.

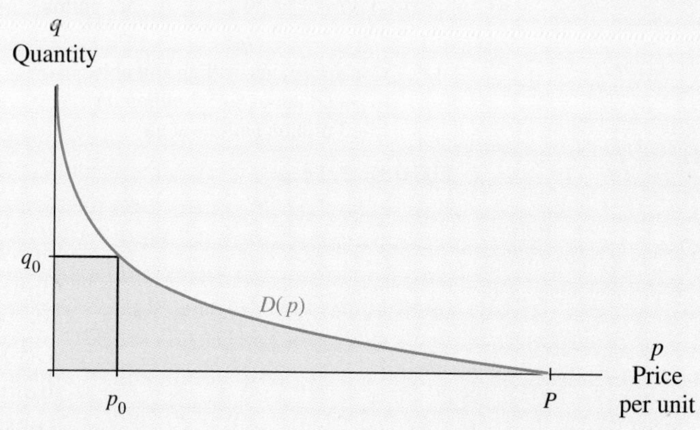

FIGURE 7.16

where p_0 is the market price at which q_0 units are in demand and P is the price above which consumers will purchase none of the goods or services. This area is calculated as

$$p_0 q_0 + \int_{p_0}^{P} D(p)\,dp$$

(Note that ∞ is used as the upper limit on the integral if the demand function approaches, but does not cross, the input axis.)

It is likely that some of your students will cover this same topic in an economics course near the time that you are discussing it in your calculus course. If so, the economics text probably contains graphs with price per unit on the vertical axis. You might want to warn these students that most of the regions we show graphically will be in different locations than they are in the graphs in their economics text.

We strongly urge you to encourage your students not to memorize the formulas in this section but rather to view them graphically as areas of rectangles and/or areas under curves.

EXAMPLE 1 *Cellular Phones*

Suppose the average weekly demand of a certain brand of cellular phone can be modeled by the equation $D(p) = 1952(0.959^p)$ phones, where p is the wholesale price per phone in dollars. How much money are consumers willing and able to spend each week on the wholesale market for 300 such cellular phones?

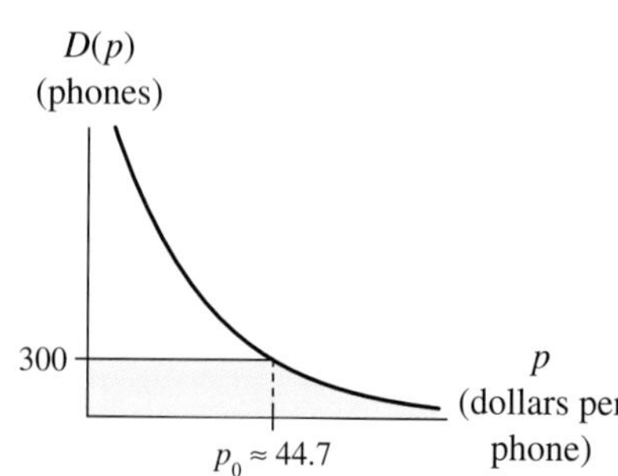

FIGURE 7.17

Solution The amount that consumers are willing and able to spend on the wholesale market for 300 cellular telephones each week is given by the area of the region below the line $q_0 = 300$ and the graph of D. (See Figure 7.17.)

We use $D(p) = 300$ to solve for $p_0 \approx 44.73577$ and note that $D(p)$ is never 0. However, it approaches 0 as p increases, so consumers are willing and able to spend

Your students will be tempted to use a rounded value for p_0 and/or P when calculating willingness and ability to spend (as well as in calculations of other quantities in this section). Encourage your students to store these values in the memories of their technologies rather than rounding them. As previously mentioned, we, of necessity, show rounded values but use unrounded values in intermediate calculations.

$$(300)(44.73577) + \int_{44.73577}^{\infty} 1952(0.959^p)\,dp$$

$$= (300)(44.73577) + \lim_{P\to\infty}\left(\frac{1952(0.959^p)}{\ln 0.959}\right)\Bigg|_{44.73577}^{P}$$

$$\approx 13{,}420.73019 + \left(\lim_{P\to\infty}\frac{1952(0.959^P)}{\ln 0.959}\right) - (-7166.026596)$$

$$= 13{,}420.73019 + 0 + 7166.026596$$

$$\approx \$20{,}587$$

According to the demand model, consumers are willing and able to spend \$20,587 for 300 cellular phones each week.

Consumers' Expenditure and Surplus

Now that we have considered what consumers are willing and able to spend for a certain quantity of a product, let us turn our attention to calculating what consumers actually spend for that quantity. We return to the discussion of gasoline demand. As was previously mentioned, if the market price for gas is \$1.19 per gallon, consumers will purchase 3 million gallons. The actual amount spent by consumers is (3 million gallons)(\$1.19 per gallon) = \$3.57 million, even though they are willing and able to spend much more. This actual amount spent is (price)(quantity), which is the area of the rectangular region from the vertical axis to $p = 1.19$ with height 3 as shown in Figure 7.18. This amount is known as the **consumers' expenditure.** The amount that consumers are willing and able to spend but do not actually spend is known as the **consumers' surplus.**

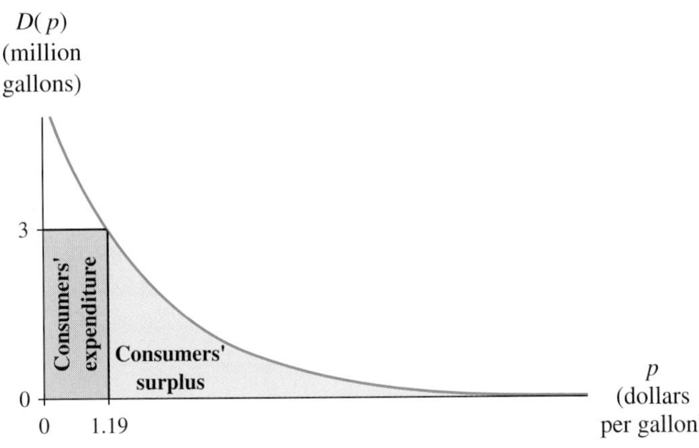

FIGURE 7.18

Earlier we found that consumers are willing and able to spend \$9.57 million to purchase 3 million gallons of gas, so the consumers' surplus from buying 3 million gallons of gas at \$1.19 per gallon is

$$\$9.57 \text{ million} - \$3.57 \text{ million} = \$6 \text{ million}$$

Consumers' surplus can also be computed as the area between the demand function and the horizontal axis as

$$\text{Consumers' surplus} = \int_{1.19}^{\infty} 5.43(0.607^p)\,dp \approx \$6 \text{ million}$$

In general, we make the following definitions:

Consumers' Expenditure and Surplus

For a continuous demand function $q = D(p)$, the amount that consumers spend at a certain market price is called consumers' expenditure; it is represented by the rectangular area in Figure 7.19. Furthermore, the amount that consumers are willing and able to spend, but do not spend, for q_0 items at a market price p_0 is called consumers' surplus; it is represented by the area of the nonrectangular shaded region in Figure 7.19. The value P is the price above which consumers will purchase none of the goods or services.

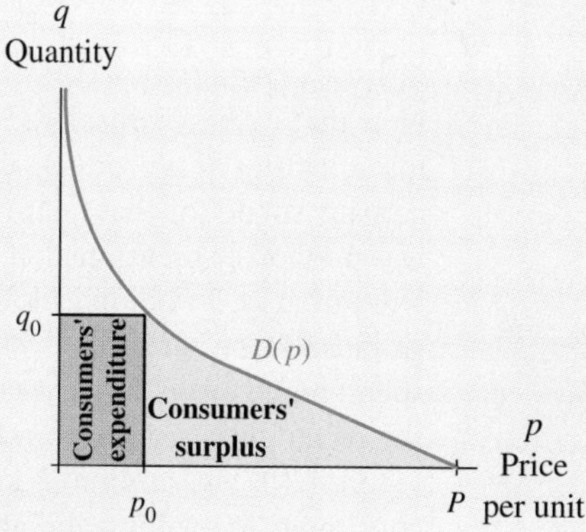

FIGURE 7.19

These areas are calculated as follows:

$$\text{Consumers' expenditure} = p_0 q_0$$

$$\text{Consumers' surplus} = \int_{p_0}^{P} D(p)\,dp$$

(Note that ∞ is used as the upper limit on the integral if the demand function approaches, but does not cross, the input axis.)

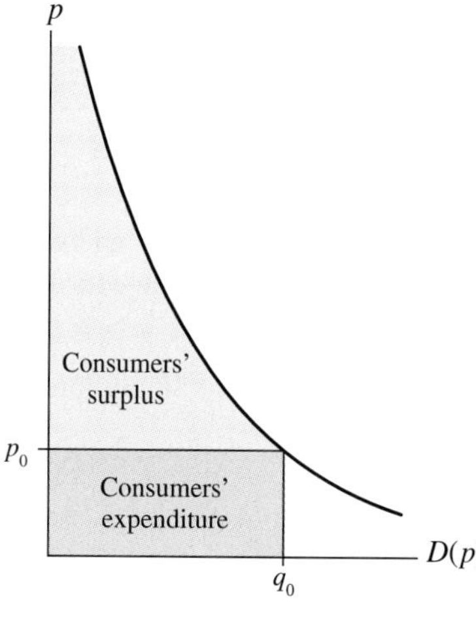

FIGURE 7.20 The economist's view

It is worth noting again that because economists graph with the input on the vertical axis, graphs showing consumers' expenditure and surplus appear in economics books with the rectangle that represents consumers' expenditure lying below the area that represents consumers' surplus, as shown in Figure 7.20.

EXAMPLE 2 *Minivans*

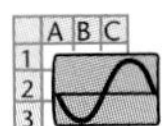

Suppose the demand for a certain model of minivan in the United States can be described as

$$D(p) = 14.12(0.933^p) - 0.25 \text{ million minivans}$$

when the market price is p thousand dollars per minivan.

a. At what price per minivan will consumers purchase 2.5 million minivans?

b. What is the consumers' expenditure when purchasing 2.5 million minivans?

c. Does the model indicate a possible price above which consumers will purchase no minivans? If so, what is this price?

d. When 2.5 million minivans are purchased, what is the consumers' surplus?

e. What is the total amount consumers are willing and able to spend on 2.5 million minivans?

Solution

a. We solve $D(p) = 2.5$ to find the market price at which consumers will purchase 2.5 million minivans. The equation

$$14.12(0.933^p) - 0.25 = 2.5$$

is satisfied when $p \approx 23.59033$. That is, at a market price p_0 of approximately \$23,600 per minivan, consumers will purchase $q_0 = 2.5$ million minivans.

b. When they purchase 2.5 million minivans, consumers' expenditure will be

$$\begin{aligned} p_0q_0 &\approx (23.59033 \text{ thousand dollars per minivan})(2.5 \text{ million minivans}) \\ &\approx \$59.0 \text{ billion} \end{aligned}$$

c. If the demand function approaches but does not cross the horizontal axis as price per unit increases without bound, then there is no price above which consumers will not purchase minivans. However, in this case, the demand function crosses the horizontal axis near $p = 58.16701$ (found by solving $D(p) = 0$). According to the model, the price above which consumers will purchase no minivans is approximately $p =$ \$58.2 thousand per minivan.

d. Consumers' surplus is the area of the region shaded in Figure 7.21, calculated as

$$\begin{aligned} \int_{p_0}^{P} D(p)\,dp &\approx \int_{23.59033}^{58.16701} [14.12(0.933^p) - 0.25]\,dp \\ &\approx 27.40482 \end{aligned}$$

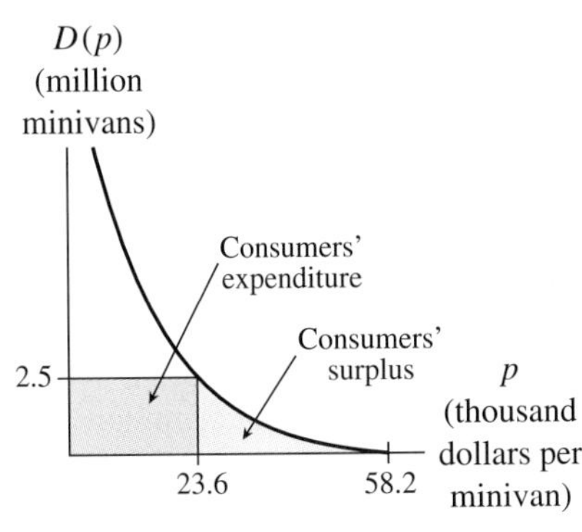

FIGURE 7.21

To determine the appropriate units for consumers' surplus, remember that we are finding the area of a region whose width is measured in thousand dollars per minivan and whose height is measured in million minivans. Thus the units on consumers' surplus are (thousand dollars per minivan)(million minivans) which simplify to billion dollars. Therefore, we estimate the consumers' surplus when purchasing 2.5 million minivans to be \$27.4 billion.

e. The amount that consumers are willing and able to spend for 2.5 million minivans is the combined area of the two shaded regions in Figure 7.21. This area is approximately $59.0 + 27.4 =$ \$86.4 billion.

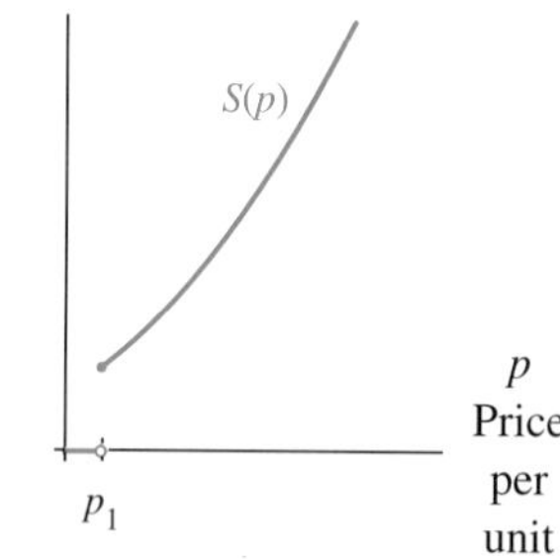

FIGURE 7.22

Supply Curves

We have seen that when prices go up, consumers usually respond by demanding less. However, manufacturers and producers respond to higher prices by supplying more. Thus a typical curve that relates the quantity supplied, S, to price per unit, p, is usually increasing and appears as shown in Figure 7.22.

The graph that expresses the quantity supplied in terms of the price per unit is called a **supply curve.** You should note from Figure 7.22 that there is a price p_1 below which producers are not willing or able to supply any quantity of the product. The point $(p_1, S(p_1))$ is known in economics as the **shutdown point.** If the market price (and the corresponding quantity) fall below this point, producers will shut down their production.

The supply function has an interpretation very similar to that of the demand function. Suppose the quantity of regular unleaded gasoline that producers will supply is modeled as

$$S(p) = \begin{cases} 0 \text{ million gallons} & \text{when } p < 1 \\ 0.792p^2 - 0.433p + 0.314 \text{ million gallons} & \text{when } p \geq 1 \end{cases}$$

where the market price of gas is p dollars per gallon. This function is graphed in blue in Figure 7.23. At \$1.24 a gallon, producers will supply 1 million gallons of gas. At \$1.78 per gallon, producers will supply 2 million gallons, and if the price is \$2.14 a gallon, producers will supply 3 million gallons.

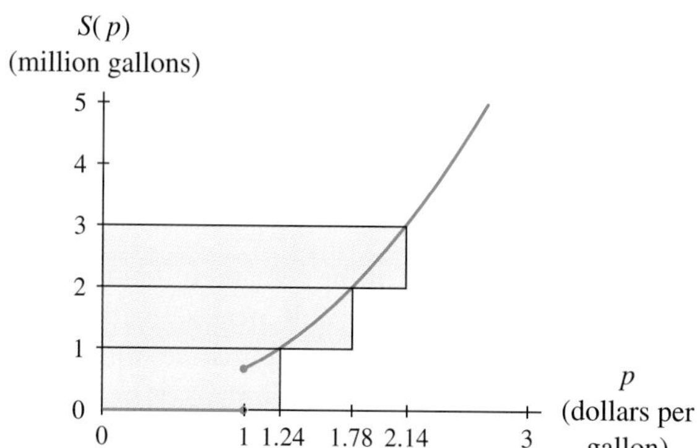

FIGURE 7.23

Producers' Willingness and Ability to Receive

We approximate the minimum that producers are willing and able to receive for 3 million gallons of gas by summing the amount they are willing and able to receive for the first million gallons, for the second million gallons, and for the third million gallons:

$$\begin{aligned} &(\$1.24 \text{ per gallon})(\text{first 1 million gallons}) + (\$1.78 \text{ per gallon})(\text{second} \\ &\text{1 million gallons}) + (\$2.14 \text{ per gallon})(\text{third 1 million gallons}) \\ = \; &\$1{,}240{,}000 + \$1{,}780{,}000 + \$2{,}140{,}000 = \$5{,}160{,}000 \end{aligned}$$

Thus the minimum that producers are willing and able to receive for 3 million gallons of gas is approximately \$5,160,000. This amount can be thought of as the sum of the areas of the three stacked rectangles shaded in Figure 7.23.

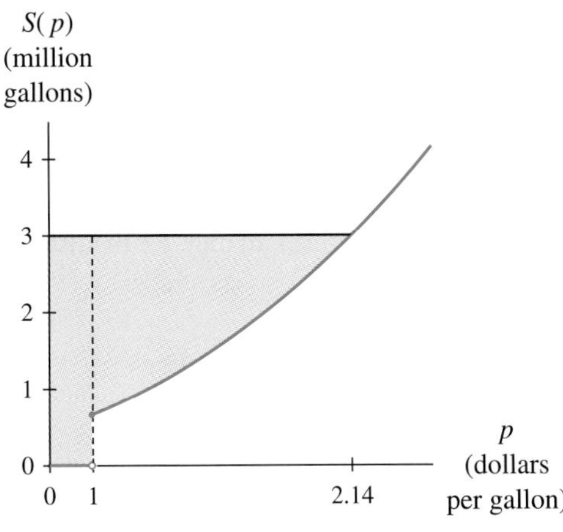

FIGURE 7.24

As we use more intervals, the area of stacked rectangles comes closer to the true area of the region below the $q = 3$ line and above the $q = S(p)$ curve shown in Figure 7.24. Because a portion of $S(p)$ is zero, we find the total area by dividing the region into a rectangular region and the region below $q = 3$ and above $q = S(p)$ to the right of the shutdown point.

Therefore, the minimum amount that suppliers are willing and able to receive is calculated as

$$3(1) + \int_1^{2.14} [3 - S(p)]\,dp$$

$$= 3 + \int_1^{2.14} (-0.792p^2 + 0.433p + 2.686)\,dp$$

$$\approx \$4.5 \text{ million}$$

According to the supply model S, suppliers are willing and able to receive no less than \$4.5 million for 3 million gallons of gas.

We make the following general definition:

Producers' Willingness and Ability to Receive

For a continuous supply function $q = S(p)$, the minimum amount that producers are willing and able to receive for a certain quantity q_0 of goods or services is the area of the shaded region in Figure 7.25

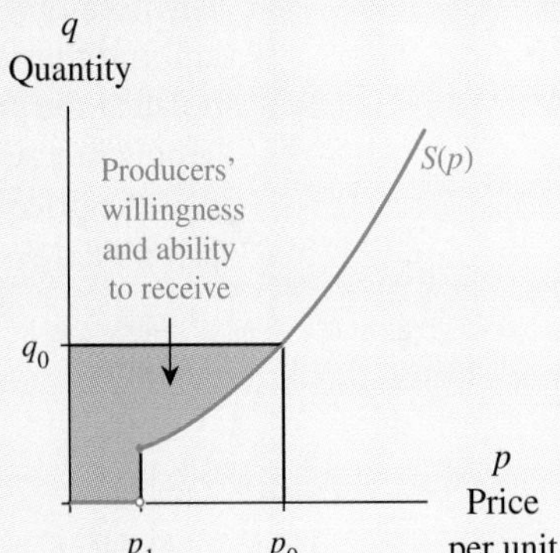

FIGURE 7.25

where p_0 is the market price at which q_0 units are supplied and p_1 is the shutdown price. This area is calculated as

$$p_1 q_0 + \int_{p_1}^{p_0} [q_0 - S(p)]\,dp$$

If there is no shutdown price, then $p_1 = 0$.

Producers' Revenue and Surplus

The market price that will lead to the supply of 3 million gallons of gas is $p \approx \$2.14$ per gallon. The **producers' revenue** is (price)(quantity), which is the area of the

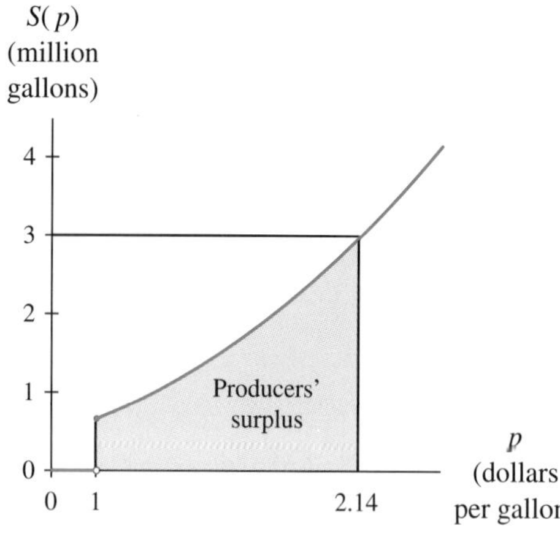

The area of the rectangle is producers' revenue.

FIGURE 7.26

rectangle shown in Figure 7.26: (\$2.14 per gallon)(3 million gallons) $\approx$ \$6.4 million. Producers will therefore receive \$6.4 million $-$ \$4.5 million $\approx$ \$1.9 million in excess of the minimum they are willing and able to receive. This excess amount is known as the **producers' surplus** and is the area of the shaded region in Figure 7.26.

We calculate the producers' surplus from the sale of 3 million gallons of gas at the market price of approximately \$2.14 directly from the supply function as follows:

$$\int_{1}^{2.14} S(p)\,dp = \int_{1}^{2.14} (0.792p^2 - 0.433p + 0.314)\,dp$$
$$\approx \$1.9 \text{ million}$$

In general, we find the producers' total revenue and the producers' surplus as follows:

Producers' Revenue and Surplus

For a continuous or piecewise continuous supply function $q = S(p)$, the amount that producers receive at a certain market price is called the producers' revenue and is the area of the shaded rectangle in Figure 7.27a. Furthermore, the amount that producers receive above the minimum amount they are willing and able to receive for q_0 items at a market price p_0 is called the producers' surplus and is the area of the region between the supply function and the horizontal axis as shown in Figure 7.27b. The value p_1 is the price below which production shuts down. (If there is no shutdown price, then $p_1 = 0$.)

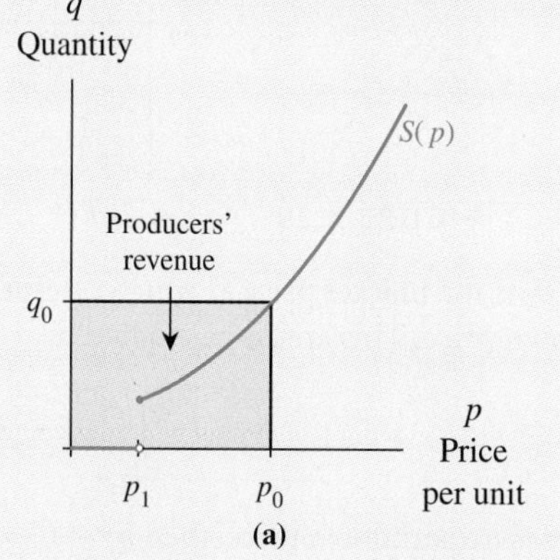

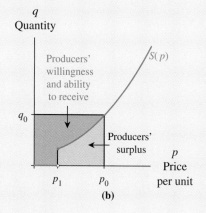

FIGURE 7.27

These areas are calculated as follows:

$$\text{Producers' revenue} = p_0 q_0$$

$$\text{Producers' surplus} = \int_{p_1}^{p_0} S(p)\,dp$$

EXAMPLE 3 *Cellular Phone Supply*

Suppose the function for the average weekly supply of a certain brand of cellular phone can be modeled by the equation

$$S(p) = \begin{cases} 0 \text{ phones} & \text{when } p < 15 \\ 0.047p^2 + 9.38p + 150 \text{ phones} & \text{when } p \geq 15 \end{cases}$$

where p is the market price in dollars per phone.

a. How many phones (on average) will producers supply at a market price of \$45.95?

b. What is the least amount that producers are willing and able to receive for the quantity of phones that corresponds to a market price of \$45.95?

c. What is the producers' revenue when the market price is \$45.95?

d. What is the producers' surplus when the market price is \$45.95?

Solution

a. When the market price is \$45.95, producers will supply an average of $S(45.95) \approx$ 680 phones each week.

b. The minimum amount that producers are willing and able to receive is the area of the labeled region in Figure 7.28 and is calculated as

$$(15)(680.247) + \int_{15}^{45.95} [680.247 - S(p)]\,dp$$
$$= 10{,}203.704 + (-0.0157p^3 - 4.69p^2 + 530.247p)\Big|_{15}^{45.95}$$
$$\approx \$16{,}300.53$$

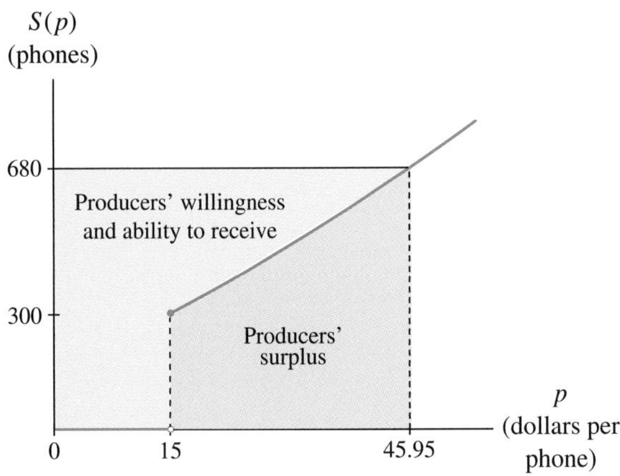

FIGURE 7.28

c. When the market price is \$45.95, the producers' revenue is

(Quantity supplied at \$45.95)(\$45.95 per phone)

$\approx$ (680.247 phones)(\$45.95 per phone)

$\approx$ \$31,257.35

Graphically, the producers' revenue is the area of the rectangle in Figure 7.28.

d. When the market price is \$45.95, the producers' surplus (see Figure 7.28) is

$$\int_{15}^{45.95} S(p)\,dp = (0.0157p^3 + 4.69p^2 + 150p)\Big|_{15}^{45.95} \approx \$14{,}956.82$$

Note that the producers' surplus plus the minimum amount that the producers are willing and able to receive is equal to the producers' revenue.

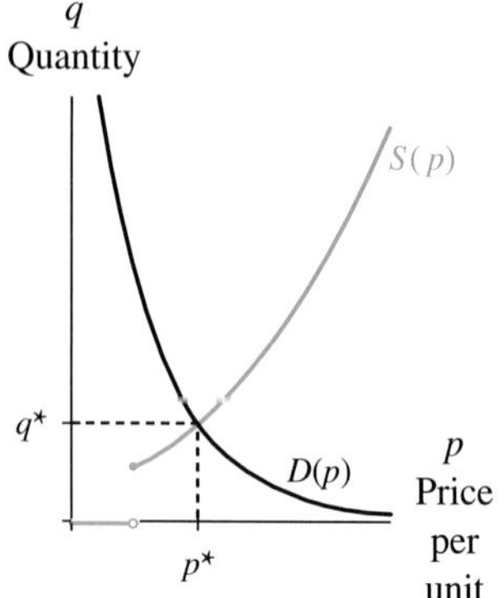

Market equilibrium (p^*, q^*) occurs when demand is equal to supply.

FIGURE 7.29

Social Gain

Consider the economic market for a particular item for which the demand and supply curves are shown in Figure 7.29. The point (p^*, q^*) where the demand curve and supply curve cross is called the **equilibrium point.** At the equilibrium price p^*, the quantity demanded by consumers coincides with the quantity supplied by producers. This quantity is q^*.

Economists consider that society is benefited whenever consumers and/or producers have surplus funds. When the market price of a product is the equilibrium price for that product, the total benefit to society is the consumers' surplus plus the producers' surplus. This amount is known as the **total social gain.**

Market Equilibrium and Social Gain

Market equilibrium occurs when the supply of a product is equal to the demand for that product. If $q = D(p)$ is the demand function and $q = S(p)$ is the supply function, then the equilibrium point is the point (p^*, q^*), where p^* is the price that satisfies the equation $D(p) = S(p)$ and $q^* = D(p^*) = S(p^*)$.

The total social gain for a product is the sum of the producers' surplus and the consumers' surplus. When q^* units are produced and sold at a market price of p^*, the total social gain is the area of the entire shaded region in Figure 7.30. The value p_1 is the price below which production shuts down, and P is the price above which consumers will not purchase.

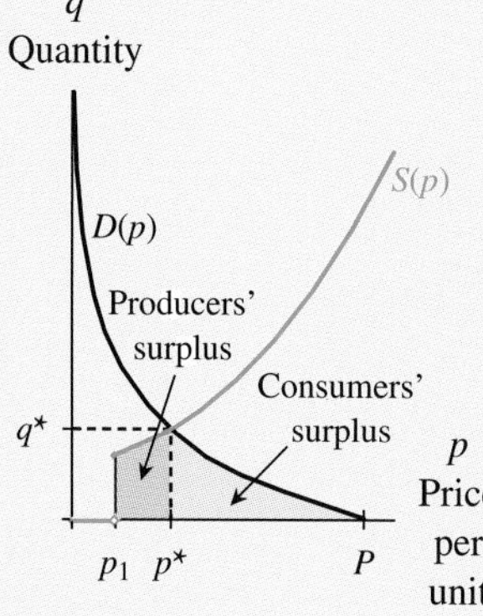

FIGURE 7.30

The combined area is calculated as

$$\text{Total social gain} = \text{producers' surplus} + \text{consumers' surplus}$$

$$= \int_{p_1}^{p^*} S(p)\,dp + \int_{p^*}^{P} D(p)\,dp$$

When we solve equations for input values and then use those values to calculate, we must make a choice: We must either abide by our convention that we never round the results of intermediate calculations or take into consideration the interpretation of the input value in the problem context and round as appropriate. For example, in the gasoline scenario, we know that gas is priced in dollars and cents per gallon. A student might argue that it makes sense to round the equilibrium price to \$1.83 per gallon and use that rounded value in calculating social gain. The student has a valid point. You must decide whether to instruct your students never to round intermediate calculation results or to accept rounded intermediate values when doing so makes sense in context. We have chosen the former option.

For the gasoline example, we solve $D(p) = S(p)$ to find the price that results in market equilibrium. The solution to $5.43(0.607^p) = 0.792p^2 - 0.433p + 0.314$ is found using technology as $p^* \approx \$1.83$ per gallon; the corresponding demand and supply are given by $q^* = D(p^*) = S(p^*) \approx 2.2$ million gallons of gas. Thus market equilibrium occurs at a price of \$1.83 per gallon. At this price, 2.2 million gallons of gas will be sold and supplied.

Recall that for the gasoline supply function, the shutdown price is \$1.00 per gallon and that for the demand function, there is no value above which consumers will no longer purchase any gas. Using these two facts and the market equilibrium price, we calculate total social gain as

$$\text{Total social gain} \approx \int_{1.00}^{1.8311} (0.792p^2 - 0.433p + 0.314)\,dp + \int_{1.8311}^{\infty} 5.43(0.607^p)\,dp$$
$$\approx 1.108 + 4.360 \approx \$5.5 \text{ million}$$

At the market equilibrium price of \$1.83 per gallon, the total social gain is approximately \$5.5 million.

EXAMPLE 4 *Minivans*

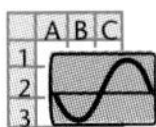

Suppose the demand and supply functions for a certain model of minivans are

$$\text{Demand} = D(p) = 14.12(0.933^p) - 0.25 \text{ million minivans}$$

and

$$\text{Supply} = S(p) = \begin{cases} 0 \text{ million minivans} & \text{when } p < 15 \\ 0.25p - 3.75 \text{ million minivans} & \text{when } p \geq 15 \end{cases}$$

where p is the market price in thousand dollars per minivan.

a. Find the market equilibrium point for minivans.

b. Find the total social gain when minivans are sold at the market equilibrium price.

Solution

a. Solving

$$14.12(0.933^p) - 0.25 = 0.25p - 3.75$$

for p yields $p^* \approx \$24.39963$ thousand. At this market price, $q^* \approx 2.34991$ million minivans will be purchased. [*Note:* q^* can be found as either $D(p^*)$ or $S(p^*)$.]

b. The total social gain at market equilibrium is the area of the shaded regions in Figure 7.31. We must find p_1, the shutdown price, and P, the price beyond which consumers will purchase no minivans, before we can proceed.

The shutdown price is given in the statement of the supply function as $p_1 = 15$. The price beyond which consumers will not purchase can be found by solving $D(p) = 0$. In this case, $P \approx 58.16701$.

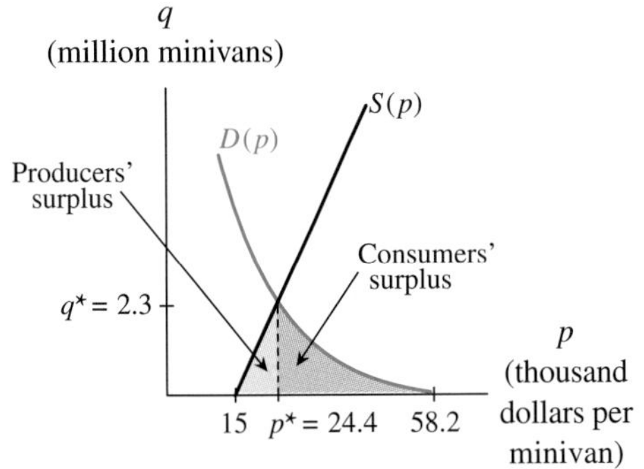

FIGURE 7.31

Now we proceed with our calculation of total social gain:

$$\begin{aligned}\text{Total social gain} &\approx \int_{15}^{24.39963} (0.25p - 3.75)\,dp + \int_{24.39963}^{58.16701} [14.12(0.933^p) - 0.25]\,dp \\ &\approx 11.04413 + 25.44287 \\ &\approx 36.487\end{aligned}$$

Hence the total social gain is approximately $36.5 billion. See Example 2, solution *d* for an explanation of these units. [Note that producers' surplus is the area of a triangle and that it can be calculated as $\frac{1}{2}(24.39963 - 15)(2.34991) \approx 11.04$.]

7.3 Concept Inventory

- *Market price*
- *Demand curve*
- *Consumers' willingness and ability to spend*
- *Consumers' expenditure*
- *Consumers' surplus*
- *Supply curve*
- *Shutdown point*
- *Producers' revenue*
- *Producers' surplus*
- *Producers' willingness and ability to receive*
- *Market equilibrium*
- *Total social gain*

7.3 Activities

1. Give the economic name by which we call each of the following:
   a. The function relating the number of items the consumer will purchase at a certain price and the price per item
   b. The function relating the quantity of items the supplier of the items will sell at a certain price and the price per item
   c. The area of the region below the supply curve between the shutdown price and the market price
   d. The area of the region below the demand curve between the market price and the price above which consumers will cease to purchase
2. For each of the following amounts:
   i. Describe the region whose area gives the specified amount.
   ii. Illustrate that region by sketching an example.
   a. The maximum amount that consumers are willing and able to spend
   b. The minimum amount that producers are willing and able to receive
   c. The consumers' expenditure
   d. The total social gain at market equilibrium
3. Explain how to find each of the following:
   a. The price P above which consumers will purchase none of the goods or services
   b. The shutdown point
   c. The point of market equilibrium
4. The demand for wooden chairs can be modeled as
   $$D(p) = -0.01p + 5.55 \text{ million chairs}$$
   where p is the price (in dollars) of a chair.
   a. According to the model, at what price will consumers no longer purchase chairs? Is this price guaranteed to be the highest price any consumer will pay for a wooden chair? Explain.
   b. Find the quantity of wooden chairs that consumers will purchase if the market price is $99.95.
   c. Determine the amount that consumers are willing and able to spend to purchase 3 million wooden chairs.
   d. Find the consumers' surplus when consumers purchase 3 million wooden chairs.
5. The demand for ceiling fans can be modeled as
   $$D(p) = 25.92(0.996^p) \text{ thousand ceiling fans}$$
   where p is the price (in dollars) of a ceiling fan.
   a. According to the model, is there a price above which consumers will no longer purchase fans? If so, what is it? If not, explain why not.

b. Find the amount that consumers are willing and able to spend to purchase 18 thousand ceiling fans.

c. Find the quantity of fans consumers will purchase if the market price is $100.

d. Find the consumers' surplus when the market price is $100.

6. The demand for a 12-ounce bottle of sparkling water is given in Table 7.8.

TABLE 7.8

Price (dollars per bottle)	Demand (million bottles)
2.29	25
2.69	9
3.09	3
3.49	2
3.89	1
4.29	0.5

a. Find a model for demand as a function of price.

b. Does your model indicate a price above which consumers will purchase no bottles of water? If so, what is it? If not, explain.

c. Find the quantity of water that consumers will purchase if the market price is $2.59.

d. Find the amount that consumers are willing and able to spend to purchase the quantity you found in part c.

e. Find the consumers' surplus when the market price is $2.59.

7. The average daily demand for a new type of kerosene lantern in a certain hardware store is as shown in Table 7.9.

TABLE 7.9

Price (dollars per lantern)	Average quantity demanded (lanterns)
21.52	1
17.11	3
14.00	5
11.45	7
9.23	9
7.25	11

a. Find a model giving the average quantity demanded as a function of the price.

b. How much are consumers willing and able to spend each day for these lanterns if the market price is $12.34 per lantern?

c. Find the consumers' surplus if the equilibrium price for these lanterns is $12.34 per lantern.

8. The willingness of saddle producers to supply saddles can be modeled by the following function:

$$S(p) = \begin{cases} 0 \text{ thousand saddles} & \text{if } p < 5 \\ 2.194(1.295^p) \text{ thousand saddles} & \text{if } p \geq 5 \end{cases}$$

when saddles are sold for p thousand dollars.

a. How many saddles will producers supply if the market price is $4000? $8000?

b. At what price will producers supply 10 thousand saddles?

c. Find the producers' revenue if the market price is $7500.

d. Find the producers' surplus if the market price is $7500.

9. The willingness of answering machine producers to supply can be modeled by the following function:

$$S(p) = \begin{cases} 0 \text{ thousand answering machines} & \text{if } p < 20 \\ 0.024p^2 - 2p + 60 \text{ thousand answering machines} & \text{if } p \geq 20 \end{cases}$$

when answering machines are sold for p dollars.

a. How many answering machines will producers supply if the market price is $40? $150?

b. Find the producers' revenue and the producers' surplus if the market price is $99.95.

10. Table 7.10 shows the number of CDs that producers will supply at the given prices.

TABLE 7.10

Price per CD (dollars)	CDs supplied (millions)
5.00	1
7.50	1.5
10.00	2
15.00	3
20.00	4
25.00	5

a. Find a model giving the quantity supplied as a function of the price per CD. *Note:* Producers will not supply CDs if the market price falls below \$4.99.

b. How many CDs will producers supply if the market price is \$15.98?

c. At what price will producers supply 2.3 million CDs?

d. Find the producers' revenue and producers' surplus if the market price is \$19.99.

11. Table 7.11 shows the average number of prints of a famous painting that producers will supply at the given prices.

TABLE 7.11

Price per print (hundred dollars)	Prints supplied (hundreds)
5	2
6	2.2
7	3
8	4.3
9	6.3
10	8.9

a. Find a model giving the quantity supplied as a function of the price per print. *Note:* Producers will not supply prints if the market price falls below \$500.

b. At what price will producers supply 5 hundred prints?

c. Find the producers' revenue and producers' surplus if the market price is \$630.

12. The daily demand for beef can be modeled by

$$D(p) = \frac{40.007}{1 + 0.033e^{0.35382p}} \text{ million pounds}$$

when the price for beef is p dollars per pound. Likewise, the supply for beef can be modeled by

$$S(p) = \begin{cases} 0 \text{ million pounds} & \text{if } p < 0.5 \\ \dfrac{51}{1 + 53.98e^{-0.3949p}} \text{ million pounds} & \text{if } p \geq 0.5 \end{cases}$$

when the price for beef is p dollars per pound.

a. How much beef is supplied when the price is \$1.50 per pound? Will supply exceed demand at this quantity?

b. Find the point of market equilibrium.

13. The average quantity of sculptures that consumers will demand can be modeled as $D(p) = -1.003p^2 - 20.689p + 850.375$ sculptures, and the average quantity that producers will supply can be modeled as

$$S(p) = \begin{cases} 0 \text{ sculptures} & \text{when } p < 4.5 \\ 0.256p^2 + 8.132p \\ \quad + 250.097 \text{ sculptures} & \text{when } p \geq 4.5 \end{cases}$$

where the market price is p hundred dollars per sculpture.

a. How much are consumers willing and able to spend for 20 sculptures?

b. How many sculptures will producers supply at \$500 per sculpture? Will supply exceed demand at this quantity?

c. Determine the total social gain when sculptures are sold at the equilibrium price.

14. A florist constructs Table 7.12 on the basis of sales data for roses.

TABLE 7.12

Price of 1 dozen roses (dollars)	Dozens sold per week
10	190
15	145
20	110
25	86
30	65
35	52

a. Find a model for the quantity demanded.

b. Determine how much money consumers will be willing and able to spend for 80 dozen roses each week.

c. If the actual market price of the roses is \$22 per dozen, find the consumers' surplus.

Suppose the suppliers of roses collect the data shown in Table 7.13.

d. Find an equation that models the supply data. Suppliers will supply no roses for prices below \$5.00 per dozen.

e. What is the producers' surplus when the market price is \$17 per dozen?

f. For what price will roses be sold at the equilibrium point?

g. What is the total social gain from the sale of roses at market equilibrium?

TABLE 7.13

Price of 1 dozen roses (dollars)	Dozens supplied per week
20	200
18	150
14	100
11	80
8	60
5	50

15. Table 7.14 gives both number of copies of a hardback science fiction novel in demand and the number supplied at certain prices.
    a. Find an exponential model for demand given the price per book.
    b. Find a model for supply given the price per book. *Note:* Producers are not willing to supply any books when the market price is less than \$18.97.
    c. At what price will market equilibrium occur? How many books will be supplied and demanded at this price?
    d. Find the total social gain from the sale of a hardback science fiction novel at the market equilibrium price.
16. Table 7.15 shows both the number of a certain type of graphing calculator in demand and the number supplied at certain prices.
    a. Find a model for demand given the price per calculator.
    b. Find a model for supply given the price per calculator. *Note:* Producers are not willing to supply any of these graphing calculators when the market price is less than \$47.50.
    c. At what price will market equilibrium occur? How many calculators will be supplied and demanded at this price?
    d. Find the producers' surplus at market equilibrium.
    e. Estimate the consumers' surplus at market equilibrium.
    f. Estimate the total social gain from the sale of this type of graphing calculator at the market equilibrium price.

TABLE 7.14

Price (dollars per book)	20	23	25	28	30	32
Books demanded (thousands)	214	186	170	150	138	128
Books supplied (thousands)	120	130	140	160	190	210

TABLE 7.15

Price (dollars per calculator)	60	90	120	150	180	210
Calculators demanded (millions)	35	31	15	5	4	3
Calculators supplied (millions)	10	32	50	80	100	120

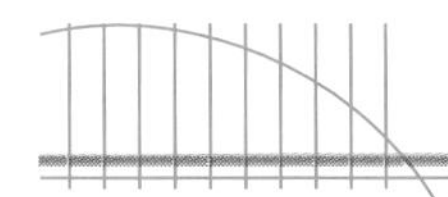

7.4 Average Value and Average Rate of Change

Several times throughout our consideration of calculus concepts, we have discussed the differences between discrete and continuous situations. Generally, we can answer mathematical questions about discrete situations without the use of calculus, but we often need calculus to describe continuous situations accurately. In this section we consider averages—another mathematical description that requires simple arithmetic in discrete cases and calculus in continuous cases.

As a student, you are intimately acquainted with grade averages and the method of calculating averages by adding grades and dividing by the number of them. Grades are discrete data. Let us consider a situation in which it may be important to know an average but discrete averaging is not practical.

Suppose that a student in your calculus class has a medical condition that exhibits itself in a racing heartbeat. His doctor has warned this student that his average resting heart rate over any given hour should not be allowed to exceed 101 beats per minute. The student has a history of math anxiety and is certain that during math tests, his heart rate averages over 101 beats per minute. If he can prove that his average heart rate during a test is greater than 101 beats per minute, he will be allowed to test with methods other than traditional in-class tests.

In order to determine his average heart rate during a math test, a nurse accompanies the student to one of the tests with a machine that automatically measures his heart rate every 10 minutes. The nurse records each measurement. The data are shown in Table 7.16.

TABLE 7.16

Time into test (minutes)	Heart rate (beats per minute)
0	95
10	105
20	100
30	94
40	101
50	110

The nurse averages the 6 heart rate values (by adding and dividing by 6) to determine that his average heart rate is 100.8 beats per minute. The student argues that this average is not accurate because his heart rate is constantly changing and hence the 6 measurements are not a complete picture of his heart rate during the test time.

The student convinces the doctor to obtain a more accurate estimate of his heart rate by having the nurse record the heart rate every 5 minutes. Averaging the resulting 11 heart rates should provide a more accurate estimate. This is done as he takes the next test, and the results are shown in Table 7.17.

TABLE 7.17

Time into test (minutes)	Heart rate (beats per minute)	Time into test (minutes)	Heart rate (beats per minute)
0	95	30	94
5	102	35	97
10	105	40	101
15	104	45	105
20	100	50	110
25	95		

Convinced that these data will prove that his average heart rate during a math test is above 101, he calculates the average by summing the heart rates and dividing by 11. The result is an average of 100.73 beats per minute.

The student again argues that this average is not accurate because it uses a discrete number of heart rates, whereas he knows his heart rate is constantly changing. He requests that his heart rate be measured every minute, but the doctor refuses, saying that he is convinced that the patient's heart rate is averaging below 101. Frustrated but not defeated, the student turns to mathematics for help. He realizes that his actual heart rate is calculated as the number of times his heart beats during the test divided by the time. He also realizes that it is impractical to count the number of heartbeats during a 50-minute period.

Having learned (despite his math anxiety!) about integrals, he reasons that if he could model with a continuous function his heart rate during a test and integrate that function over the 50-minute interval, then he would have a close approximation for the total number of times his heart beat during the test. Dividing this value by 50 minutes would give a more accurate average heart rate than the method of summing discrete data values.

Applying this reasoning, he models the data in Table 7.17. The result is the equation

$$H(t) = (-4.80186 \cdot 10^{-5})t^4 + 0.00599t^3 - 0.229476t^2 + 2.813403t + 94.370629 \text{ beats per minute}$$

where t is the number of minutes since the test began. Integrating from 0 to 50 minutes gives the result $\int_0^{50} H(t)\,dt \approx 5033$ beats. Dividing this total number of beats by 50 minutes, the student obtains an average heart rate of 100.66 beats per minute. Defeated, this time by his own reasoning, he goes to see his math instructor to find out whether there was a flaw in his logic. With amazement in her voice, she informs him that his reasoning is essentially correct. She reminds him that using a continuous model for a discrete situation must be done with care but that in this case, there are so many beats during the 50-minute interval (nearly 2 per second) that it is reasonable to model the heart rate with a continuous function. She is so impressed by his solution to the problem that she exempts him from taking the final exam!

Average Value of a Function

If $y = f(x)$ is a continuous function, then we can approximate the average value (or average) of the function over an interval from $x = a$ to $x = b$ by dividing the interval into n equally spaced subintervals, evaluating the function at a point in each subinterval, summing the function values, and dividing by n.

$$\text{Average value} \approx \frac{f(x_1) + f(x_2) + \cdots + f(x_{n-1}) + f(x_n)}{n}$$

We denote the length of each subinterval by Δx and calculate the length as $\Delta x = \frac{b-a}{n}$. Rewrite the average value estimate by multiplying the top and bottom terms by Δx:

$$\begin{aligned}\text{Average value} &\approx \frac{[f(x_1) + f(x_2) + \cdots + f(x_{n-1}) + f(x_n)]\,\Delta x}{n\Delta x} \\ &= \frac{[f(x_1) + f(x_2) + \cdots + f(x_{n-1}) + f(x_n)]\,\Delta x}{b-a}\end{aligned}$$

As in the heart rate example, the estimate improves as the number of intervals increases. Thus we obtain the exact average value by finding the limit of the estimate as n approaches infinity:

$$\text{Average value} = \lim_{n\to\infty} \frac{[f(x_1) + f(x_2) + \cdots + f(x_{n-1}) + f(x_n)]\,\Delta x}{b - a}$$

which can be written as

$$\text{Average value} = \frac{\int_a^b f(x)\,dx}{b - a}$$

Thus we have

A good visual illustration of the average value concept is water sloshing in a bathtub. The water levels out at its average value.

Average Value

If $y = f(x)$ is a smooth, continuous function from a to b, then the average value of $f(x)$ from a to b is

$$\text{Average value of } f(x) \text{ from } a \text{ to } b = \frac{\int_a^b f(x)\,dx}{b - a}$$

The average value of a function over an interval can be graphically interpreted as the height (or signed height) of a rectangle whose area equals the area between the function and the horizontal axis over the interval. In Figure 7.32, the heart rate function is shown together with the rectangle whose height is the average value. (Note that we do not show the vertical axis to zero; however, it is true that the area of the rectangle and the area between the graph and the line representing the horizontal axis shown in Figure 7.32 are equal. You will have an opportunity to justify this statement in the Activities.)

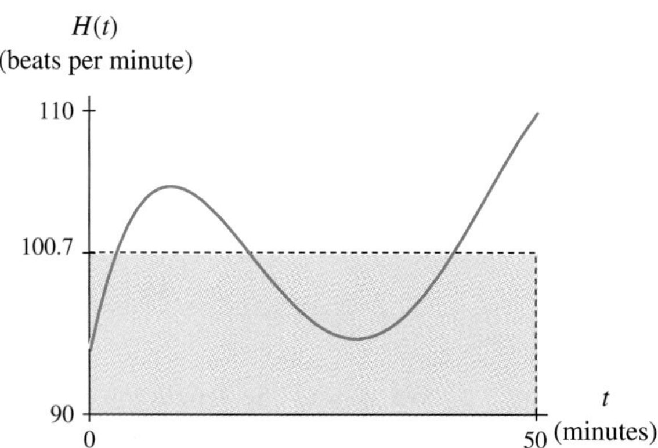

FIGURE 7.32

EXAMPLE 1 *Temperature*

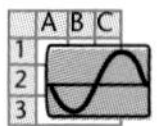

Suppose that the hourly temperatures shown in Table 7.18 were recorded from 7 a.m. to 7 p.m. one day in September.

TABLE 7.18

Time	Temperature (°F)	Time	Temperature (°F)
7 a.m.	49	2 p.m.	80
8 a.m.	54	3 p.m.	80
9 a.m.	58	4 p.m.	78
10 a.m.	66	5 p.m.	74
11 a.m.	72	6 p.m.	69
noon	76	7 p.m.	62
1 p.m.	79		

a. Find a cubic model for this set of data.

b. Calculate the average temperature between 9 a.m. and 6 p.m.

c. Graph the equation together with the rectangle whose upper edge is determined by the average value.

d. Calculate the average rate of change of temperature from 9 a.m. to 6 p.m.

Solution

a. The temperature on this particular day can be modeled as

$$t(h) = -0.03526h^3 + 0.71816h^2 + 1.584h + 13.689 \text{ degrees Fahrenheit}$$

h hours after midnight. This model applies only from $h = 7$ (7 a.m.) to $h = 19$ (7 p.m.).

b. The average temperature between 9 a.m. ($h = 9$) and 6 p.m. ($h = 18$) is

$$\text{Average temperature} = \frac{\int_9^{18} t(h)\,dh}{18 - 9} \approx 74.4\text{ °F}$$

c.

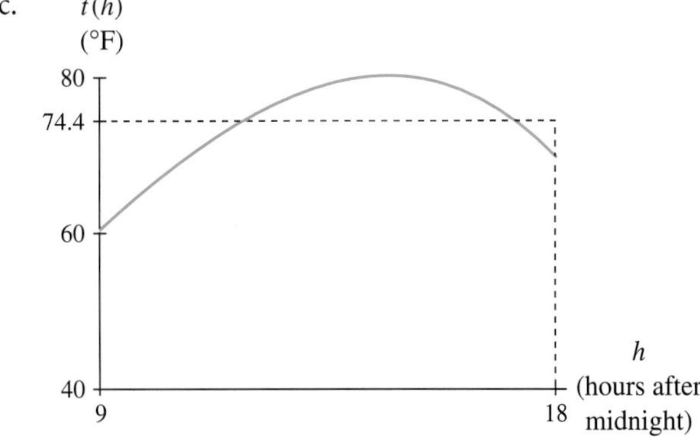

FIGURE 7.33

d. Recall from Section 3.1 that the average rate of change of a function on an interval is the change in output divided by the change in input. The average rate of change of temperature from 9 a.m. to 6 p.m. is

$$\text{Average rate of change of temperature} = \frac{t(18) - t(9)}{18 - 9} \approx 0.98\text{°F per hour}$$

Average Rate of Change

The preceding example asked for an average rate of change. We know from Section 3.1 that the average rate of change of a continuous function $y = f(x)$ from $x = a$ to $x = b$ is calculated as $\frac{f(b) - f(a)}{b - a}$. However, consider what happens when you have a function that describes the rate of change of a quantity (that is, you have $y = f'(x)$) and you need to find the average rate of change of the quantity $f(x)$. In this case, you do not use the average rate-of-change formula but should instead use an integral to find the average value of the rate-of-change function. Note that we use the terms *average rate of change* and *average value of the rate of change* interchangeably.

The Average Value of the Rate of Change

If $y = f'(x)$ is a smooth, continuous rate-of-change function from a to b, then the average value of $f'(x)$ from a to b is

$$\text{Average value of the rate of change of } f(x) \text{ from } a \text{ to } b = \frac{\int_a^b f'(x)\,dx}{b - a} = \frac{f(b) - f(a)}{b - a}$$

where $f(x)$ is an antiderivative of $f'(x)$.

EXAMPLE 2 *Population Growth*

The growth rate of the population of South Carolina can be modeled[21] as

$p'(t) = 0.1552t + 0.223$ thousand people per year

where t is the number of years since 1790. The population of South Carolina in 1990 was 3486 thousand people.

a. What was the average rate of change in population from 1995 through 2000?

b. What was the average size of the population from 1995 through 2000?

Solution

Point out to your students that keeping units of measure attached to values will help them correctly calculate and label average values.

a. The average rate of change in population between 1995 and 2000 is calculated directly from the rate-of-change function as

$$\frac{\int_{205}^{210} p'(t)\,dt \text{ (thousand people/year) (years)}}{210 - 205 \text{ years}} \approx 32.4 \text{ thousand people per year}$$

b. In order to calculate the average population, we must have a function for population. That is, we need an antiderivative of the rate-of-change function:

$$p(t) = \int p'(t)\,dt$$
$$= 0.0776t^2 + 0.223t + C \text{ thousand people}$$

21. Based on data from *Statistical Abstract*, 1998.

We know that the population in 1990 was 3486 thousand people. Using this fact, we solve for C, so the function for population is

$$p(t) = 0.0776t^2 + 0.223t + 337.4 \text{ thousand people}$$

where t is the number of years since 1790. Now we calculate the average population between 1995 and 2000 as

$$\frac{\int_{205}^{210} p(t)\,dt}{210 - 205} \approx 3725 \text{ thousand people}$$

Note in Example 2 that the average value of the population was found by integrating the population function, whereas the average rate of change was found by integrating the population rate-of-change function. This example illustrates an important principle:

> In using integrals to find average values, integrate the function whose output is the quantity you wish to average.

Note also that in part *a* of Example 2, we could have calculated the average rate of change by using the population function and the formula

$$\frac{p(210) - p(205)}{210 - 205} = \frac{162.135 \text{ thousand people}}{5 \text{ years}} \approx 32.4 \text{ thousand people per year}$$

We summarize this discussion as follows:

Average Values and Average Rates of Change

If $y = f(x)$ is a continuous or piecewise continuous function describing a quantity from $x = a$ to $x = b$, then the average value of the quantity from a to b is calculated by using the quantity function and the formula

$$\text{Average value of } f(x) = \frac{\int_a^b f(x)\,dx}{b - a}$$

The average value has units the same as the output of the function f.

The average rate of change of the quantity, also called the average value of the rate of change, can be calculated from the quantity function as

$$\text{Average rate of change} = \frac{f(b) - f(a)}{b - a}$$

or from the rate-of-change function as

$$\text{Average rate of change} = \frac{\int_a^b f'(x)\,dx}{b - a}$$

The average rate of change has the same units as the rate of change of f.

EXAMPLE 3 Carbon-14

Scientists estimate that 100 milligrams of the isotope ^{14}C used in carbon dating methods decay at a rate of

$$r(t) = -0.0121(0.999879^t) \text{ milligrams per year}$$

where t is the number of years since the 100 milligrams of isotope began to decay. The amount of the 100 milligrams remaining after t years of decay is

$$a(t) = 100(0.999879^t) \text{ milligrams}$$

a. What is the average amount of the remaining isotope during the first 1000 years?

b. What is the average rate of decay during the first 1000 years?

c. Graphically illustrate the answers to parts *a* and *b*.

Solution

a. We calculate the average amount remaining during the first thousand years as

$$\frac{\int_0^{1000} a(t)\,dt}{1000 - 0} \approx 94.2 \text{ milligrams}$$

b. We find the average rate of decay during the first 1000 years as

$$\frac{\int_0^{1000} r(t)\,dt}{1000 - 0} \approx -0.0114 \text{ milligram per year}$$

In other words, the amount of ^{14}C decreased by an average of 0.0114 milligram per year during the first 1000 years. Note that this average rate of change can also be calculated by using the amount function:

$$\frac{a(1000) - a(0)}{1000 - 0} \approx \frac{88.6 - 100 \text{ milligrams}}{1000 \text{ years}} = -0.0114 \text{ milligram per year}$$

c. The average amount determines the top of the rectangle shown in Figure 7.34a. The average rate of decay determines the bottom of the rectangle shown in Figure 7.34b. The average decay rate also can be graphically illustrated as the slope of a secant line through two points on the amount function.

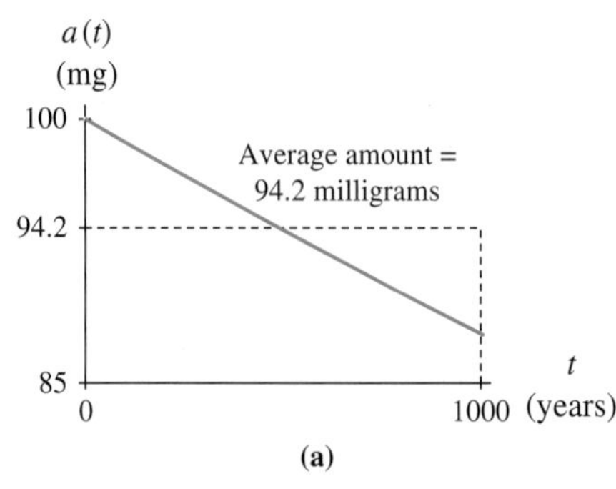

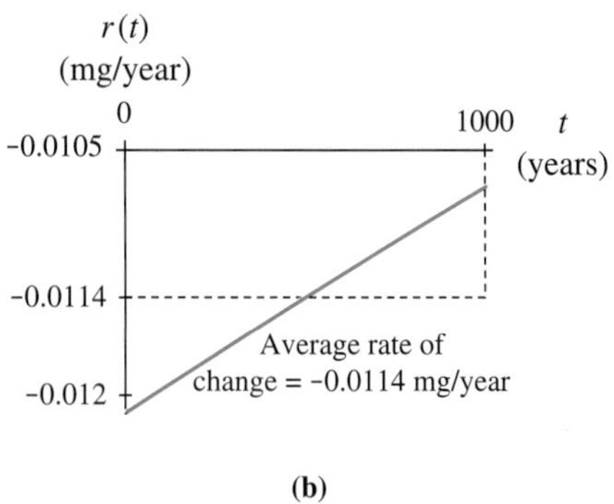

FIGURE 7.34

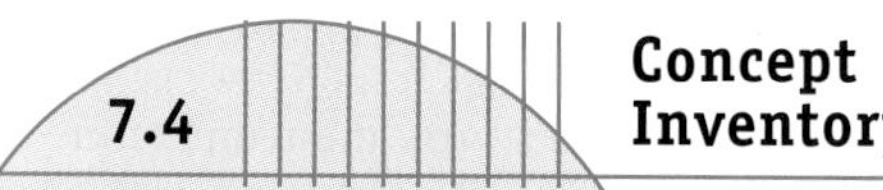

7.4 Concept Inventory

- *Average value of a function*
- *Graphical illustration of average value*
- *Average rate of change of a function*
- *Average value of a rate-of-change function*

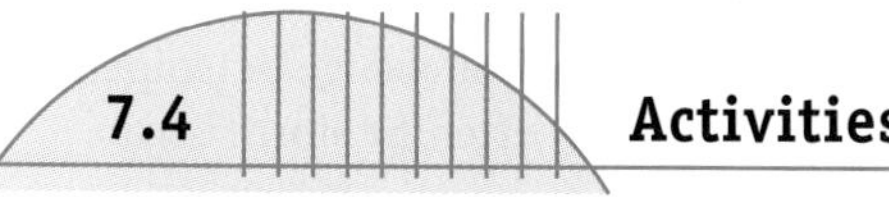

7.4 Activities

1. The Highway Department is concerned about the high speed of traffic during the weekday afternoon rush hours from 4 p.m. to 7 p.m. on a newly widened stretch of interstate highway that is just inside the city limits of a certain city. The Office of Traffic Studies has collected the data given in Table 7.19, which show typical weekday speeds during the 4 p.m. to 7 p.m. rush hours.

TABLE 7.19

Time	Speed (mph)	Time	Speed (mph)
4:00	60	5:45	72.25
4:15	61	6:00	74
4:30	62.5	6:15	74.5
4:45	64	6:30	75
5:00	66.25	6:45	74.25
5:15	67.5	7:00	73
5:30	70		

   a. Find a model for the data.
   b. Use the equation to approximate the average weekday rush-hour speed from 4 p.m. to 7 p.m.
   c. Use the equation to approximate the average weekday rush-hour speed from 5 p.m. to 7 p.m.

2. U.S. factory sales of electronics from 1990 through 1996 can be modeled[22] by the equation

$$\text{Sales} = -0.23x^3 + 2.257x^2 - 1.51x + 42.8 \text{ billion dollars}$$

   where x is the number of years since 1990.

   a. Use a definite integral to approximate the average annual value of U.S. factory sales of electronics over the 6-year period from 1990 through 1996.
   b. Sketch the graph of sales from 1990 through 1996, and draw the horizontal line representing the average value.

3. The most expensive rates (in dollars per minute) for a 2-minute telephone call using a long-distance carrier are listed in Table 7.20.

TABLE 7.20

Year	Rate (dollars per minute)
1982	1.32
1984	1.24
1985	1.14
1986	1.01
1987	0.83
1988	0.77
1989	0.65
1990	0.65
1995	0.40
2000	0.20

   a. Find a model for the data.
   b. Use a definite integral to estimate the average of the most expensive rates from 1982 through 1990.
   c. Use a definite integral to estimate the average of the most expensive rates from 1982 through 2000.

4. Table 7.21 gives the price (in dollars) of a round-trip flight from Denver to Chicago on a certain airline and the corresponding monthly profit (in millions of dollars) for that airline for that route.

TABLE 7.21

Ticket price (dollars)	Profit (millions of dollars)
200	3.08
250	3.52
300	3.76
350	3.82
400	3.70
450	3.38

   a. Find a model for the data.
   b. Determine the average profit for ticket prices from \$325 to \$450.

22. Based on data from *Statistical Abstract*, 1998.

c. Determine the average rate of change of profit when the ticket price rises from \$325 to \$450.

d. Graphically illustrate the answers to parts *b* and *c*.

5. The population of Mexico between 1921 and 1997 is given by the model[23]

$$\text{Population} = 7.391(1.02695^t) \text{ million people}$$

where t is number of years since the end of 1900.

a. What was the average population of Mexico from the beginning of 1980 through the end of 1989?

b. In what year was the population of Mexico equal to its 1980s average?

c. What was the average rate of change of the population of Mexico during the 1980s?

6. The number of AIDS cases diagnosed from 1994 through 1997 can be modeled[24] by

$$C(x) = -x^3 + 16.1x^2 - 90.7x + 247.5 \text{ thousand cases}$$

where x is the number of years since the end of 1990.

a. Use a definite integral to estimate the average number of cases diagnosed each year between the end of 1994 and the end of 1997.

b. Find the average rate of change in cases diagnosed from the end of 1994 through 1997.

c. In which year was the number of cases diagnosed closest to the average number of cases diagnosed from the end of 1994 through 1997?

7. The number of general-aviation aircraft accidents from 1975 through 1997 can be modeled[25] by

$$\begin{matrix}\text{Number of} \\ \text{accidents}\end{matrix} = -100.6118x + 3967.5572 \text{ accidents}$$

where x is the number of years since 1975.

a. Calculate the average rate of change in the yearly number of accidents from 1976 through 1997.

b. Use a definite integral to estimate the average number of accidents that occurred each year from 1976 through 1997.

c. Graphically illustrate the answers to parts *a* and *b*.

8. During a summer thunderstorm, the temperature drops and then rises again. The rate of change of the temperature during the hour and a half after the storm began is given by

$$T(h) = 9.48h^3 - 15.49h^2 + 17.38h - 9.87 \text{ °F per hour}$$

where h is the number of hours since the storm began.

a. Calculate the average rate of change of temperature from 0 to 1.5 hours after the storm began.

b. If the temperature was 85°F at the time the storm began, find the average temperature during the first 1.5 hours of the storm.

9. The acceleration of a race car during the first 35 seconds of a road test is modeled by

$$a(t) = 0.024t^2 - 1.72t + 22.58 \text{ ft/sec}^2$$

where t is the number of seconds since the test began. Assume that velocity and distance were both 0 at the beginning of the road test.

a. Calculate the average acceleration during the first 35 seconds of the road test.

b. Calculate the average velocity during the first 35 seconds of the road test.

c. Calculate the distance traveled during the first 35 seconds of the road test.

d. If the car had been traveling at its average velocity throughout the 35 seconds, how far would the car have traveled during that 35 seconds?

e. Graphically illustrate the answers to parts *a* and *b*. Explain how the answer to part *c* relates to the graphical illustration of the part *b* answer.

10. On the basis of data obtained from a preliminary report by a geological survey team, it is estimated that for the first 10 years of production, a certain oil well in Texas can be expected to produce oil at the rate of $r(t) = 3.93546t^{3.55}e^{-1.35135t}$ thousand barrels per year, t years after production begins. Estimate the average annual yield from this oil well during the first 10 years of production.

11. An article in the May 23, 1996, issue of *Nature* addresses the interest some physicists have in studying cracks in order to answer the question "How fast do things break, and why?" Data estimated from a graph in this article showing velocity of a

23. Based on data from SPP and INEGI, Mexican Censuses of Population 1921 through 1990 as reported by Pick and Butler, *The Mexico Handbook* (Boulder, CO: Westview Press, 1994) and *Statistical Abstract*, 1998.
24. Based on data from *Statistical Abstract*, 1998.
25. Based on data from *Statistical Abstract*, 1994 and 1998.

crack during a 60-microsecond experiment are shown in Table 7.22.

TABLE 7.22

Time (microseconds)	Velocity (meters per second)
10	148.2
20	159.3
30	169.5
40	180.7
50	189.8
60	200.0

a. Find a model for the data.

b. Determine the average speed at which a crack travels between 10 and 60 microseconds.

12. The amount[26] of poultry produced in the United States annually from 1960 through 1993 can be modeled as $p(x) = 6.178983e^{0.047509x}$ million pounds per year x years after 1960.

a. Estimate the average amount of poultry produced annually from 1960 through 1990.

b. In what year was poultry production closest to the average annual production from 1960 through 1990?

c. Graphically illustrate the answer to part *a*.

13. Blood pressure varies for individuals throughout the course of a day, typically being lowest at night and highest from late morning to early afternoon. The estimated rate of change in diastolic blood pressure for a patient with untreated hypertension is shown in Table 7.23 in the next column.

a. Find a model for the data.

b. Estimate the average rate of change in diastolic blood pressure from 8 a.m. to 8 p.m.

c. Assuming that diastolic blood pressure was 95mm Hg at 12 p.m., estimate the average diastolic blood pressure between 8 a.m. and 8 p.m.

14. The air speed of a small airplane during the first 25 seconds of takeoff and flight can be modeled by

$$v(t) = -940{,}602t^2 + 19{,}269.3t - 0.3 \text{ mph}$$

t hours after takeoff.

TABLE 7.23

Time	Diastolic BP (mm Hg per hour)
8 a.m.	3.0
10 a.m.	1.8
12 p.m.	0.7
2 p.m.	-0.1
4 p.m.	-0.7
6 p.m.	-1.1
8 p.m.	-1.3
10 p.m.	-1.1
12 a.m.	-0.7
2 a.m.	0.1
4 a.m.	0.8
6 a.m.	1.9

a. Find the average air speed during the first 25 seconds of takeoff and flight.

b. Find the average acceleration during the first 25 seconds of takeoff and flight.

15. From data taken from the 1994 *Statistical Abstract*, we model the rate of change of the percentage of the population of the United States living in the Great Lakes region from 1960 through 1990 as

$$P(t) = (9.371 \cdot 10^{-4})t^2 - 0.141t + 5.083$$
percentage points per year

where t is the number of years since 1900. Find the average rate of change of the percentage of the population of the United States living in the Great Lakes region from 1960 through 1990.

16. The federal government sets standards for toxic substances in the air. Often these standards are stated in the form of average pollutant levels over a period of time on the basis of the reasoning that exposure to high levels of toxic substances is harmful, but prolonged exposure to moderate levels is equally harmful.[27] For example, carbon monoxide (CO) levels may not exceed 35 ppm (parts per million) at any time, but they also must not exceed 9 ppm averaged over any 8-hour period.

26. Based on data from National Agricultural Statistics Service.

27. Douglas J. Crawford-Brown, *Theoretical and Mathematical Foundations of Human Health Risk Analysis* (Boston: Kluwer Academic Publishers, 1997).

The concentration of carbon monoxide in the air in a certain metropolitan area is measured and modeled as

$$c(h) = -0.004h^4 + 0.05h^3 - 0.27h^2 + 2.05h + 3.1 \text{ ppm}$$

h hours after 7 a.m.

a. Did the city exceed the 35-ppm maximum in the 8 hours between 7 a.m. and 3 p.m.?

b. Did the city exceed the 9-ppm maximum average between 7 a.m. and 3 p.m.?

17. Refer to the discussion in Activity 16. Table 7.24 shows measured concentrations of carbon monoxide in the air of a city on a certain day between 6 a.m. and 10 p.m.

TABLE 7.24

Time (hours since 6 a.m.)	CO concentration (ppm)
0	3
2	12
4	22
6	18
8	16
10	20
12	28
14	16
16	6

a. Sketch a scatter plot of the data and determine (by examination of the data) over which 8-hour period the average CO concentration was greatest.

b. Model the data. Use the equation to calculate the average CO concentration during the 8-hour period determined in part *a*.

c. Use the equation in part *b* to estimate the average CO concentration in this city between 6 a.m. and 10 p.m. The city issues air quality warnings based on the daily average CO concentration of the previous day between 6 a.m. and 10 p.m. The warnings are as follows:

Average concentration	*Warning*
$0 < \text{CO} \le 9$	None
$9 < \text{CO} \le 12$	Moderate pollution. People with asthma and other respiratory problems should remain indoors if possible.
$12 < \text{CO} \le 16$	Serious pollution. Ban on all single-passenger vehicles. Everyone is encouraged to stay indoors.
$\text{CO} > 16$	Severe pollution. Mandatory school and business closures.

Judging on the basis of the data in Table 7.24 and your answer to part *c*, which warning do you believe should be posted?

18. Aurora, Nevada, was a mining boom town in the 1860s and 1870s. Its population can be modeled[28] by the function

$$p(t) = \begin{cases} -7.91t^3 + 120.96t^2 + 193.92t - 123.21 \text{ people} & \text{when } 0.7 \le t \le 13 \\ 45{,}544(0.8474^t) \text{ people} & \text{when } 13 < t \le 55 \end{cases}$$

with rate-of-change function

$$p'(t) = \begin{cases} -23.73t^2 + 241.92t + 193.92 \text{ people per year} & \text{when } 0.7 \le t < 13 \\ -7541.287(0.8474^t) \text{ people per year} & \text{when } 13 < t \le 55 \end{cases}$$

In both functions, t is the number of years since 1860.

a. What was the average population of Aurora between 1861 and 1871? between 1871 and 1881?

b. Demonstrate two methods for calculating the average rate of change of the population of Aurora between 1861 and 1871.

19. We know that the area of the region between the graph of a function and the horizontal axis from $x = b$ to $x = c$ is equal to the area of the rectangle whose height is the average value of the function from $x = b$ to $x = c$ and whose width is $c - b$. The graphs presented in this section show only a portion of the vertical axis. That is, in each graph, the vertical axis does not extend all the way to zero. Instead, the vertical axis is shown above (or below) a line $y = k$. However, on the interval from $x = b$ to $x = c$, the area of the region between the graph and the line $y = k$ and the area of the rectangle between the average value and $y = k$ are equal. Explain, illustrating with graphs, why this is true.

28. Based on data from Don Ashbaugh, *Nevada's Turbulent Yesterday: A Study in Ghost Towns* (Los Angeles: Westernlore Press, 1963).

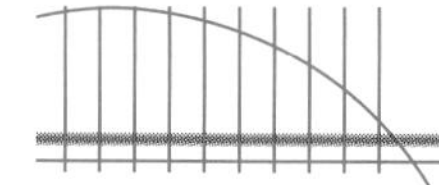

7.5 Probability Distributions and Density Functions

The depth of coverage of this section depends on whether your students are required to take a probability and/or statistics course and, if so, on whether that course is taken before or after this calculus course.

We have used derivatives to measure rates of change and integrals to measure accumulation of change. This section discusses another application of integrals—to measure the likelihood that certain events will occur in situations that involve some degree of chance or uncertainty.

Most of the applications in this text contain numerical values studied in a particular context; that is, they incorporate situations involving data. As we saw in previous chapters, understanding how data change is important in making decisions and giving predictions about the variable being studied. When all outcomes of a particular situation are considered, the pattern indicated by the variability in data is called the **distribution** of the quantity being studied. Consider, for instance, the variability in scores on the SAT I: Reasoning Test. Table 7.25 gives a distribution of scores for 1996 college-bound seniors[29] on the mathematics portion of the SAT I. This table shows the proportion of the 1,084,725 students taking the test whose math scores fell in each indicated interval. Scores are based on the recentered scale. (See Activity 15 at the end of this section.)

TABLE 7.25

Math SAT score x (points)	Number of students in score interval	Proportion of students in score interval
$200 \le x < 300$	26,967	0.02486
$300 \le x < 400$	146,161	0.13474
$400 \le x < 500$	337,267	0.31092
$500 \le x < 600$	327,932	0.30232
$600 \le x < 700$	188,024	0.17334
$700 \le x \le 800$	58,374	0.05381

This distribution of the SAT math scores can be viewed with a graph called a **histogram** that is composed of rectangles. We choose to construct the histogram so that the area of each rectangle is the proportion of students in the corresponding score interval. The intervals must be such that no score is in more than one interval and all possible scores are included. The height of each rectangle is calculated by dividing the area of the rectangle by the width of the score interval.

TABLE 7.26

Math SAT score x (points)	Width of rectangle (points)	Area of rectangle (proportion of students)	Height of rectangle (proportion of students per point of score)
$200 \le x < 300$	100	0.02486	0.00025
$300 \le x < 400$	100	0.13474	0.00135
$400 \le x < 500$	100	0.31092	0.00311
$500 \le x < 600$	100	0.30232	0.00302
$600 \le x < 700$	100	0.17334	0.00173
$700 \le x \le 800$	100	0.05381	0.00054

29. "SAT/SAT I Scores for College-Bound Seniors, 1996," College Entrance Examination Board and Educational Testing Service.

A histogram based on the SAT score groupings in Table 7.26 is shown in Figure 7.35.

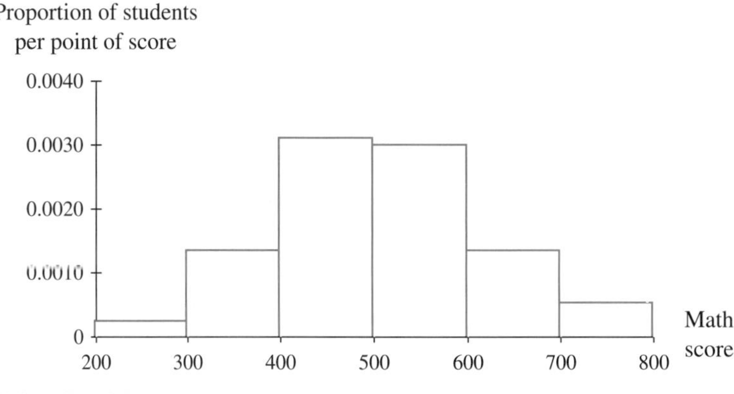

FIGURE 7.35

EXAMPLE 1 *Using Histograms*

Refer to Table 7.26 and the histogram of math SAT scores in Figure 7.35.

a. Describe the distribution of the scores.

b. Find the total area enclosed by the rectangles in the histogram.

c. What proportion of scores is between 300 and 600?

Solution

a. The distribution peaks near 500, and most students taking the test had math scores that ranged between 300 and 700.

b. The total area enclosed by the rectangles in the histogram equals 1:

$$\text{Area} = 0.02486 + 0.13474 + 0.31092 + 0.30232 + 0.17334 + 0.05381 = 1$$

c. From the third column in Table 7.26, we find that the proportion of scores between 300 and 600 is $0.13474 + 0.31092 + 0.30232 \approx 0.75$.

Area as Probability

The **probability** of an event is a measure of how likely it is to happen. Because there is no uncertainty associated with a known outcome, we assume that all variables for which probabilities are determined are **random variables**—that is, variables whose numerical values are determined by the results of an experiment involving chance. We intuitively consider the probability of an event to be the proportion of times the event occurs when the experiment associated with the event is repeated under similar conditions a large number of times.

To illustrate this concept, we return to the distribution of SAT math scores given numerically in Table 7.26 and graphically in the histogram in Figure 7.35. What is the probability that a 1996 college-bound senior made a score between 400 and 600 on the mathematics portion of the SAT? To answer this question, we first need to understand the terms involved and how they are related to probability.

- Taking the math portion of the SAT is the experiment that was repeated, under similar conditions, a large number of times (1,084,725 students took the test).

- Assuming that the 1996 college-bound senior in the question is any of those taking the test and that we do not know any student's individual score, the random variable is the SAT math score.
- The event being considered is the value of the SAT math score lying in the interval of values between 400 and 600.

We can now compute the probability that a student's math score is between 400 and 600 by finding the proportion of scores that are in that score interval. From the information in Table 7.26, the proportion of students whose scores are between 400 and 600 is 0.61324, so the probability that a student's math score is between 400 and 600 is 0.61324.

Probabilities are proportions, so they are real numbers between 0 and 1. Suppose we want to know the probability that a student makes a math score between 200 and 800. Because all SAT math scores lie between 200 and 800, this event is known as a **certain** or **sure event,** so its probability should be 1. This was verified in part *b* of Example 1.

What is the likelihood that a student makes an SAT math score of 950? The highest score is 800, so the probability that a student makes an SAT math score of 950 is 0. An event that has a probability of zero is known as an **impossible event.**

The histograms used in this section have been designed so that all rectangle heights are non-negative values, and the total area enclosed by the rectangles in each histogram is 1. When expressed as a proportion, the area of each rectangle in the histogram gives the probability that the value of the random variable under discussion is in the interval that forms the base of the rectangle. For that reason, we also call these histograms **probability histograms.**

The above discussions suggest the following statements:

Probability

The probability that any event occurs is a real number between 0 and 1. The probability of a certain or sure event is 1, and the probability of an impossible event is 0. Even though probabilities are real numbers between 0 and 1, they are often referred to as percentages between 0% and 100%.

When the distribution of a random variable is represented by a probability histogram, the probability that the value of the variable lies in certain intervals is given by the sum of the areas of the rectangles whose bases are those intervals.

If there are a large number of values very close together, then it often simplifies matters to approximate their behavior with a continuous function. Such an approximating function, shown in Figure 7.36, is overdrawn on a more deailed SAT math scores histogram. This continuous function is the familiar bell-shaped curve commonly called the **normal distribution.**

Instead of summing areas of rectangles in a histogram to find probabilities, we can either compute or estimate probabilities by finding areas under curves like the one shown in Figure 7.36. Such functions are called **probability density functions.** We do not discuss the interpretation of output values of probability density functions. Instead, we focus on the interpretation of areas of regions beneath these functions. The remainder of this section discusses probability density functions and the computation of probabilities for random variables whose distributions are described

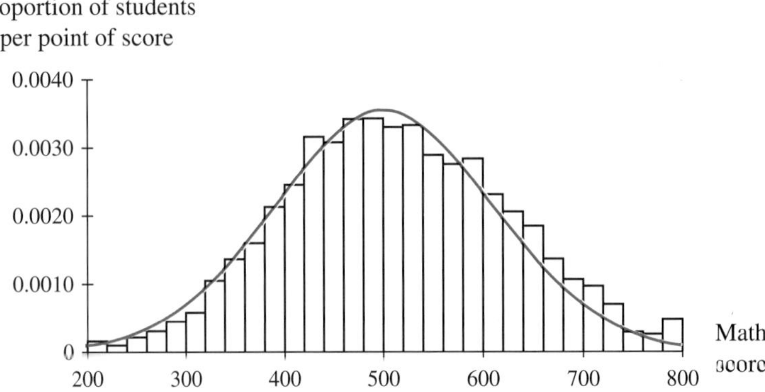

FIGURE 7.36

by these functions. Because probability density functions determine how probabilities are distributed over various intervals associated with random variables, they are also called **probability distributions.**

> **Probability Density Function**
>
> A probability density function is a continuous function or a piecewise continuous function with input consisting of some interval of real numbers and with output satisfying the following two conditions:
>
> 1. Each output value is greater than or equal to 0.
> 2. The area of the region between the function and the horizontal axis is 1.

Before proceeding further, we introduce a notation to simplify writing statements involving probability. We write the statement "the probability that the value of x is between a and b" using the notation $P(a \leq x \leq b)$. For instance, if x is the SAT math score, then $P(500 \leq x \leq 800)$ is read "the probability that the value of the SAT math score is between 500 and 800." We also consider "the probability that the SAT math score is between 500 and 800" to have the same meaning. Using these notations, we restate the definition of a probability density function and define probability for random variables having such density functions:

> **Probability and Probability Density Functions**
>
> A probability density function $y = f(x)$ for a random variable x is a continuous function or a piecewise continuous function such that
>
> 1. $f(x) \geq 0$ for each real number x, and
> 2. $\int_{-\infty}^{\infty} f(x)\,dx = 1$
>
> The probability that a value of x lies in an interval with endpoints a and b, where $a \leq b$, is given by
>
> $$P(a \leq x \leq b) = \int_a^b f(x)\,dx$$
>
> It is possible for the interval from a through b to be the entire set of real numbers.

Probability density functions are constructed from experimental data and/or statistical theory using techniques you will probably study if you take a course in statistics. This section is not intended to discuss those statistical techniques. Rather, we are illustrating another use of integrals to find area under a curve.

EXAMPLE 2 *Recovery Times*

Suppose that the proportion of patients who recover from mild dehydration x hours after receiving treatment is given by

$$f(x) = \begin{cases} 12x^2 - 12x^3 & \text{when } 0 \leq x \leq 1 \\ 0 & \text{when } x < 0 \text{ or } x > 1 \end{cases}$$

a. What is the random variable in this situation?

b. Verify that f is a probability density function for this random variable.

c. Find the probability that the recovery time is between 42 minutes and 48 minutes. Interpret your answer.

d. Find the probability that recovery takes at least half an hour.

Solution

a. The random variable x is the number of hours until recovery. After 1 hour all patients will have recovered.

b. To verify that f is a probability density function for this random variable, we must show that the two properties of a probability density function are true.
   1. As the graph in Figure 7.37 shows, all values of $f(x)$ are non-negative.
   2. The area under the graph of f is 1 because

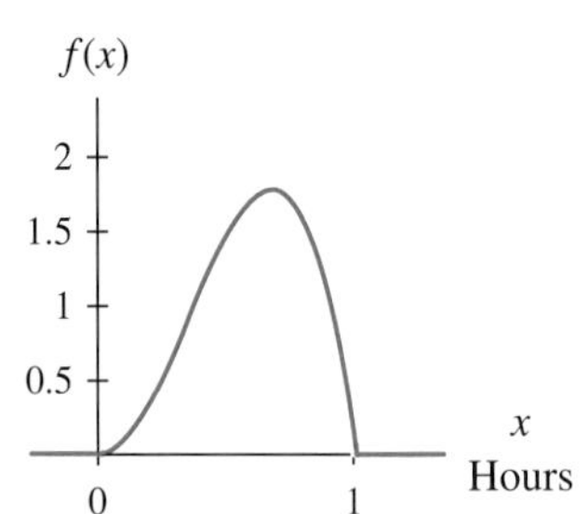

FIGURE 7.37

$$\begin{aligned}\int_{-\infty}^{\infty} f(x)\,dx &= \int_{-\infty}^{0} 0\,dx + \int_{0}^{1} (12x^2 - 12x^3)\,dx + \int_{1}^{\infty} 0\,dx \\ &= 0 + (4x^3 - 3x^4)\Big|_0^1 + 0 \\ &= 4 - 3 = 1\end{aligned}$$

c. Because 42 minutes is 0.7 hours and 48 minutes is 0.8 hours, we find the probability that recovery time is between 42 minutes and 48 minutes by computing

$$\begin{aligned}P(0.7 \leq x \leq 0.8) &= \int_{0.7}^{0.8} (12x^2 - 12x^3)\,dx \\ &= (4x^3 - 3x^4)\Big|_{0.7}^{0.8} \\ &= (2.048 - 1.2288) - (1.372 - 0.7203) \\ &= 0.1675\end{aligned}$$

There is a 16.75% chance that the recovery time for a patient will be between 42 minutes and 48 minutes. (Note that 0.1675 is the area between the graph of f and the horizontal axis from $x = 0.7$ to $x = 0.8$.)

d. $$\begin{aligned}P(x \geq 0.5) &= \int_{0.5}^{\infty} f(x)\,dx = \int_{0.5}^{1} (12x^2 - 12x^3)\,dx + \int_{1}^{\infty} 0\,dx \\ &= (4x^3 - 3x^4)\Big|_{0.5}^{1} \\ &= 0.6875\end{aligned}$$

There is a 68.75% chance that recovery takes at least a half hour.

It is important that you note the difference between the meanings of $f(x)$ and $P(a \le x \le b)$. To see more clearly that the output of a probability density function is *not* a probability, look again at the graph of the density function in Figure 7.37. There are values of $f(x)$ that are obviously larger than 1, and we know that probabilities can never be greater than 1. Always keep in mind that whereas f is a function that shows how the probabilities associated with values of a random variable x are distributed over the input interval, $P(a \le x \le b)$ is a real number between 0 and 1. Figure 7.38 illustrates these ideas for the probability density function f.

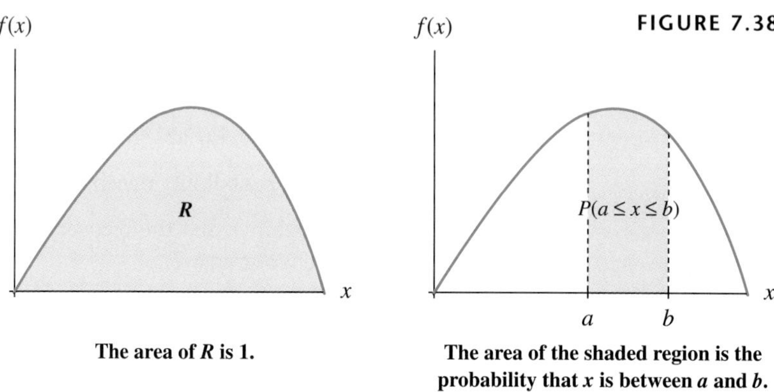

FIGURE 7.38

The area of R is 1.

The area of the shaded region is the probability that x is between a and b.

The Uniform Probability Distribution

Even though probability density functions have a variety of shapes, some naturally occur so often that they have names. Perhaps the simplest density function is one that assumes a constant value between two specified inputs and is zero elsewhere. This probability distribution is called the **uniform density function** and has the equation

$$u(x) = \begin{cases} \dfrac{1}{b-a} & \text{when } a \le x \le b \\ 0 & \text{when } x < a \text{ or } x > b \end{cases}$$

A graph of this piecewise continuous function is shown in Figure 7.39.

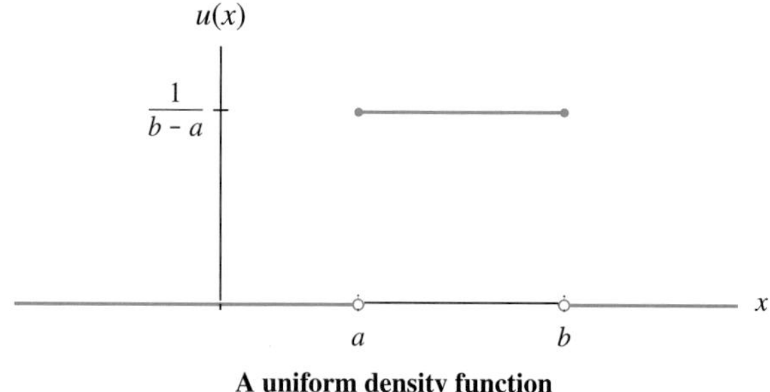

A uniform density function

FIGURE 7.39

The uniform density function provides a good model for random variables that are evenly distributed over an interval. For instance, the distribution of winning numbers picked since the beginning of a state lottery is approximated by a uniform density function. The probability distribution describing how tires wear (in terms of the remaining tread) on a properly aligned and balanced set of wheels is a uniform density function.

EXAMPLE 3 *Campus Transportation*

Buses that transport students from one location to another on a large campus arrive at the student parking lot every 15 minutes between 7:30 a.m. and 4:30 p.m. If t is the number of minutes before the next bus arrives at the lot, then the distribution of waiting times is modeled by the density function

$$u(t) = \begin{cases} \dfrac{1}{15} & \text{when } 0 \leq t \leq 15 \\ 0 & \text{when } t > 15 \text{ or } t < 0 \end{cases}$$

a. Explain, in the context of this situation, why it makes sense for $u(t)$ to equal 0 when $t > 15$ or $t < 0$.

b. Represent on a graph of u the probability that a student arriving at the parking lot will have to wait more than 5 minutes for the next bus.

c. Find $P(t \geq 5)$. Interpret the result.

Solution

a. If buses arrive every 15 minutes at the parking lot, no one will have to wait more than 15 minutes for a bus. Because time cannot be negative, the waiting time distribution is zero for $t < 0$. It therefore makes sense that $u(t) = 0$ when t is not in the interval with endpoints 0 and 15.

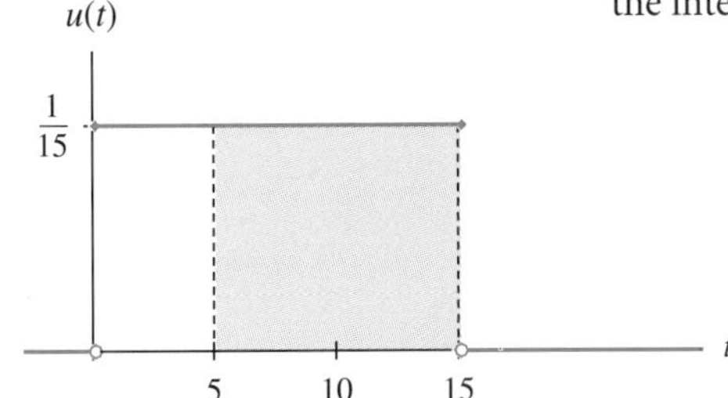

$P(t > 5)$ is the area of the shaded region.

FIGURE 7.40

b. $P(t > 5)$ is the area of the region between the graph of u and the t-axis to the right of $t = 5$. This region is shaded on the graph shown in Figure 7.40.

c. Because $P(t = 5) = 0, P(t \geq 5) = P(t > 5)$. This probability is represented by the area of the rectangle with height $\frac{1}{15}$ and width 10. (See Figure 7.40.) Thus

$$P(t \geq 5) = \frac{10}{15} \approx 0.667$$

Of course, this probability could have also been found using integrals:

$$P(t \geq 5) = \int_5^{\infty} u(t)\,dt = \int_5^{15} \frac{1}{15}\,dt + \int_{15}^{\infty} 0\,dt = \frac{1}{15}(15 - 5) = \frac{10}{15} \approx 0.667$$

When a student arrives at the parking lot between 7:30 a.m. and 4:30 p.m., there is about a 66.7% chance that he or she will have to wait at least 5 minutes for a bus to arrive.

Measures of Center and Variability

One important characteristic of a density function is its central value. Familiar examples of such central values include the average life span of a light bulb, the expected waiting time at a grocery store checkout counter, and the typical income of American families. The measure of central value that we consider is the mean[30] (commonly referred to as the *average*). The **mean** is also called the *expected value* and is geometrically associated with the "balance point" of the region between the density function and the horizontal axis. The Greek letter μ is used to denote the mean.

30. Another common measure of the center of a distribution is the median. The *median* is the input value such that half of the area under the density function lies to the left and half to the right of it.

Distributions can have the same mean but a completely different spread, or variability, about that center. One measure of how closely the values of the distribution cluster about its mean is the **standard deviation.** If most of the values of the input variable are close to the mean, then the standard deviation is small. On the other hand, if it is likely that the input values are widely scattered about the mean, the standard deviation is large. The Greek letter σ is used to denote the standard deviation.

EXAMPLE 4 *Mean and Standard Deviation of SAT Math Scores*

Refer to the continuous function overdrawn on the histogram of the 1996 college-bound seniors' SAT math scores in Figure 7.36. The mean math SAT score is 507.95 points, and the standard deviation is 112.356 points.[31] Locate the mean and standard deviation on the graph of this continuous function that approximates the distribution of the SAT math scores.

Solution Because the mean is an input value for this continuous function, μ is located on the horizontal axis. (See Figure 7.41.) The standard deviation tells us how the math scores are spread out around the mean. For instance, one standard deviation to the right of the mean would be $\mu + \sigma = 507.950 + 112.356 = 620.306$ points.

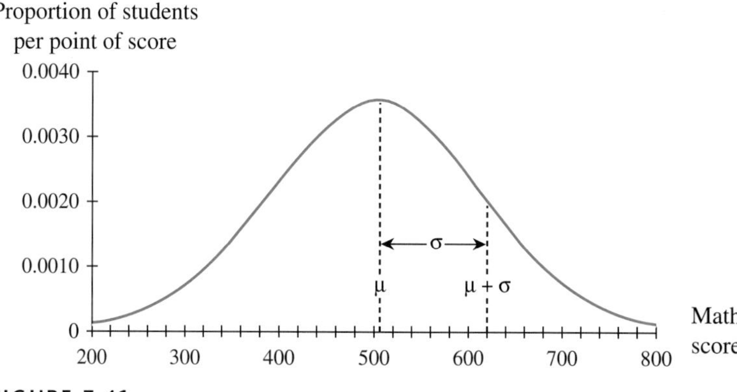

FIGURE 7.41

The mean and standard deviation of a known density function are determined using the following definitions:

Mean and Standard Deviation

For the density function $y = f(x)$ with x defined on the interval of real numbers,

- the mean is $\mu = \int_{-\infty}^{\infty} xf(x)\,dx$
- the standard deviation is $\sigma = \sqrt{\int_{-\infty}^{\infty} (x - \mu)^2 f(x)\,dx}$

provided the integrals exist.

31. "SAT/SAT I Scores for College-Bound Seniors, 1996," College Entrance Examination Board and Educational Testing Service.

EXAMPLE 5 *Computing a Mean and Standard Deviation*

Refer to the distribution of waiting times for campus transportation in Example 3:

$$u(t) = \begin{cases} \dfrac{1}{15} & \text{when } 0 \le t \le 15 \\ 0 & \text{when } t > 15 \text{ or } t < 0 \end{cases}$$

where t is the time in minutes until the next bus arrives.

a. Find the mean waiting time.

b. Find the standard deviation of the waiting times.

Solution

a. $$\mu = \int_{-\infty}^{\infty} t \cdot u(t)\,dt = \int_{-\infty}^{0} t \cdot u(t)\,dt + \int_{0}^{15} t \cdot u(t)\,dt + \int_{15}^{\infty} t \cdot u(t)\,dt$$
$$= \int_{-\infty}^{0} t \cdot 0\,dt + \int_{0}^{15} \frac{1}{15} t\,dt + \int_{15}^{\infty} t \cdot 0\,dt$$
$$= 0 + \frac{1}{30} t^2 \Big|_0^{15} + 0 = \frac{1}{30}(225 - 0) = 7.5$$

The average waiting time for a bus is 7.5 minutes.

b. $$\sigma = \sqrt{\int_{-\infty}^{\infty} (t - \mu)^2 u(t)\,dt}$$
$$= \sqrt{\int_{-\infty}^{0} (t - 7.5)^2 \cdot 0\,dt + \int_{0}^{15} (t - 7.5)^2 \cdot \frac{1}{15}\,dt + \int_{15}^{\infty} (t - 7.5)^2 \cdot 0\,dt}$$
$$= \sqrt{0 + \frac{1}{15} \int_{0}^{15} (t - 7.5)^2\,dt + 0}$$
$$= \sqrt{\frac{1}{15} \cdot \frac{(t - 7.5)^3}{3} \Big|_0^{15}} = \sqrt{\frac{1}{15}\left(\frac{(15 - 7.5)^3}{3} - \frac{(0 - 7.5)^3}{3}\right)}$$
$$= \sqrt{18.75} \approx 4.33$$

The standard deviation of waiting times is about 4.33 minutes.

As illustrated in Example 5, a definite integral involving a density function f has a value of 0 if the height of the density function is 0. In these cases, we are finding the area of a rectangle with height 0 and width determined by the upper and lower limits on the integral:

Whenever $f(x) = 0$, area = (height)(width) = (0)(width) = 0

Therefore, from this point on, we do not show the integrals involved in probability, mean, or standard deviation calculations for the portions of the horizontal axis where the value of a density function is 0.

The Exponential Probability Distribution

The amount of time it takes to learn a task, the duration of a phone call, the time you wait for service at a bank teller's window, and the time between arrivals at the drive-through station of a fast-food restaurant are examples of random variables that are more likely to be small than large. With events such as these, we are interested in the time elapsed or space between any two occurrences of an event rather than in the number of times the event happens. The likelihood of encountering certain intervals of time or space between consecutive occurrences of an event can be modeled by an **exponential density function.** The general formula for the exponential probability distribution is

$$e(x) = \begin{cases} ke^{-kx} & \text{when } x \geq 0 \\ 0 & \text{when } x < 0 \end{cases}$$

where k is some positive constant. The graph of the general exponential density function is shown in Figure 7.42.

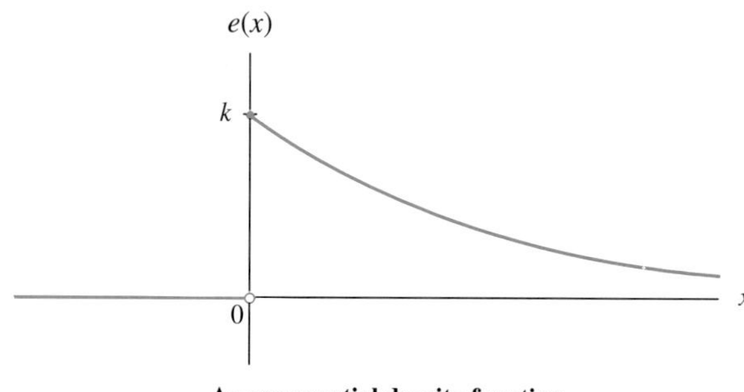

An exponential density function

FIGURE 7.42

An exponential distribution is appropriate in situations that involve the waiting time between successive events where that waiting time is more likely to be small than large. The mean of the exponential distribution is $1/k$, which can be interpreted in these situations as the mean time interval between consecutive events. When, on average, an event occurs at a rate of k arrivals per unit of time, the average gap between consecutive arrivals is $1/k$ time units.

For instance, suppose that you arrive at your local Burger King sometime between 11:30 a.m. and 12:30 p.m. and that the time you have to wait for service while in line at the drive-through window is an exponential random variable with a mean of 2.5 minutes. This means that there are 2.5 minutes between cars coming into the line, with cars arriving at the rate of 0.4 car per minute (that is, 4 cars every 10 minutes). When applying this distribution, remember that the values of the random variable x and the value of k must be in the same time units.

EXAMPLE 6 *Emergency Room Arrivals*

The distribution of the time between successive arrivals at an emergency room of a large city hospital on Saturday nights can be approximated by an exponential density function. Two patients arrive at the emergency room every 10 minutes.

a. What is the equation for this exponential density function?

b. What is the probability that the time between successive arrivals will be more than 1 minute? Interpret this result.

Solution

a. The mean of the distribution is the time between consecutive events. Thus the mean is $\mu = \frac{1}{k} = \frac{10}{2} = 5$ minutes per arrival. The reciprocal of the mean gives the average number of arrivals per minute, thus $k = \frac{1}{5}$. The equation of the exponential density function with $k = \frac{1}{5}$ is

$$e(t) = \frac{1}{5}e^{-t/5} = 0.2e^{-0.2t} \quad \text{when } t \geq 0$$

where t is the time, in minutes, between successive arrivals. (In the context of time, it is assumed that $e(t) = 0$ when $t < 0$.)

b. $P(t > 1) = \int_1^{\infty} 0.2e^{-0.2t}\,dt = \lim_{T\to\infty} \int_1^{T} 0.2e^{-0.2t}\,dt$

$= \lim_{T\to\infty} [-e^{-0.2T} - (-e^{(-0.2)(1)})]$

$= 0 + e^{-0.2} \approx 0.819$

There is an 81.9% chance that the time between successive arrivals will be more than 1 minute. This event is likely to occur.

The Normal Distribution

We now turn our attention to the single most important density function in statistics, the **normal density function**, which is also called the *normal distribution*. The equation of the normal density function with mean μ and standard deviation σ is

$$f(x) = \frac{1}{\sigma\sqrt{2\pi}}e^{\frac{-(x-\mu)^2}{2\sigma^2}} \quad \text{where } -\infty < x < \infty$$

The graph of a normal distribution is called a *normal curve*. The mean controls the location of a normal curve, and the standard deviation controls its spread. Figure 7.43 gives us a good idea of some of the features of any normal distribution.

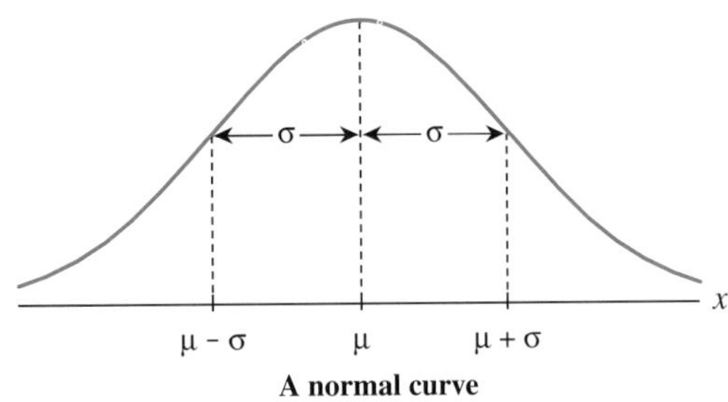

FIGURE 7.43

Note these important properties of any normal curve:

- The curve is bell-shaped with the absolute maximum occurring at the mean μ.
- The curve is symmetric about a vertical line through μ.
- The curve approaches the horizontal axis but never touches or crosses it.
- The inflection points occur at $\mu - \sigma$ and $\mu + \sigma$.

EXAMPLE 7 *Comparing Normal Curves*

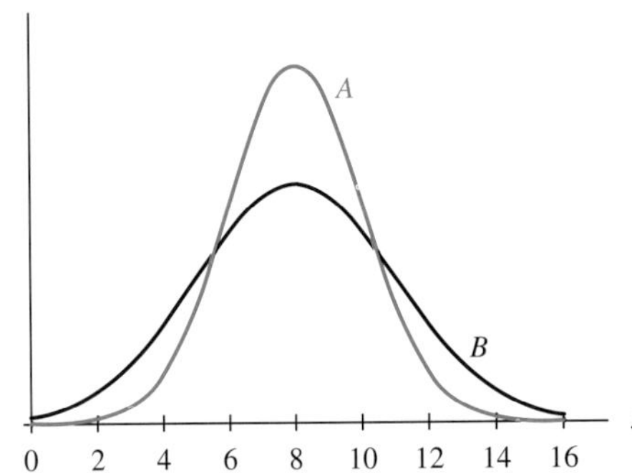

FIGURE 7.44

Consider the two normal curves shown in Figure 7.44.

a. Compare the means of the two normal distributions.

b. One of the curves has $\sigma = 3$, and the other curve has an inflection point at $x = 10$. Compare the spreads of the two curves.

c. What percentage of the total area under each normal curve lies to the right of 8?

Solution

a. The means of the two density functions are the same, because each curve has a maximum at $x = 8$.

b. From Figure 7.44, we see that the rightmost inflection point on normal curve A occurs around 10. Thus $8 + \sigma = 10$, and the standard deviation of curve A is approximately 2. Curve B is more spread out from its center than curve A, so it has the larger standard deviation, $\sigma = 3$.

c. Because a normal curve is the graph of a density function, the area between each normal curve and the horizontal axis must be 1. Each normal curve is symmetric about its mean, so 50% of the total area is to the right of μ for each normal curve.

The normal curve rises gradually to an absolute maximum and then decreases in a symmetric manner. Although it is not unique in exhibiting this form, it has been found to provide a reasonable approximation to certain other distributions that occur in many real-life situations. Also, many naturally occurring phenomena, as well as various mental and physical characteristics of human beings, can be described by normal distributions.

EXAMPLE 8 *Life of Light Bulbs*

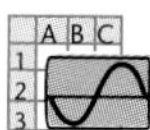

A manufacturer of light bulbs advertises that the average life of these bulbs is 900 hours with a standard deviation of 100 hours. Suppose the distribution of the length of life of these light bulbs, with the life span measured in hundreds of hours, is modeled by a normal density function.

a. Write the definite integral that represents the probability that a light bulb lasts between 900 and 1000 hours.

b. Approximate the value of the integral in part *a*, and interpret the result in context.

Solution

a. Note that the input for the normal density function is measured in hundreds of hours. The integral that gives $P(9 \le x \le 10)$ is $\int_9^{10} f(x)\,dx$, where f is the normal density function with $\mu = 9$ and $\sigma = 1$. That is,

$$P(9 \le x \le 10) = \int_9^{10} \frac{1}{\sqrt{2\pi}} e^{\frac{-(x-9)^2}{2}}\,dx$$

b. There is no antiderivative formula that can be used to find the exact value of an integral of the normal density function. However the value of the definite integral can be numerically approximated using technology:

$$\int_9^{10} \frac{1}{\sqrt{2\pi}} e^{-0.5(x-9)^2}\,dx \approx 0.34$$

Thus the chance that any one of these light bulbs will last between 900 and 1000 hours is about 34%.

Cumulative Density Functions

Suppose that the length of time a student waits for a computer terminal with which to preregister for next term's classes is described by the uniform density function

$$f(x) = \begin{cases} \frac{1}{25} & \text{when } 0 \leq x \leq 25 \\ 0 & \text{when } x > 25 \end{cases}$$

where the value of the random variable x is measured in minutes. The probability that the waiting time is less than or equal to 5 minutes is found by computing the area of the shaded region in Figure 7.45.

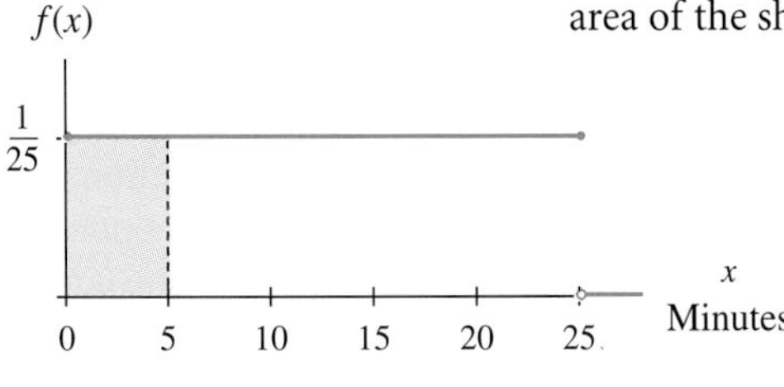

FIGURE 7.45

Because the event that the waiting time is more than 25 minutes is impossible by the definition of this density function, $P(x > 25) = 0$. Also, $P(0 \leq x \leq a)$, when a is a waiting time between 0 and 25 minutes, is computed by finding the area of a rectangle with height $\frac{1}{25}$ and width a. Table 7.27 gives the results of some of these probability computations.

TABLE 7.27

Time interval	Probability that waiting time is in time interval
Between 0 and 2 minutes	$P(0 \leq x \leq 2) = 0.08$
Between 0 and 5 minutes	$P(0 \leq x \leq 5) = 0.20$
Between 0 and 12.5 minutes	$P(0 \leq x \leq 12.5) = 0.5$
Between 0 and 20.38 minutes	$P(0 \leq x \leq 20.38) = 0.8152$
Between 0 and 25 minutes	$P(0 \leq x \leq 25) = 1$
Between 0 and 33 minutes	$P(0 \leq x \leq 33) = P(0 \leq x \leq 25) + P(25 < x \leq 33) = 1 + 0 = 1$
Between 0 and 40.2 minutes	$P(0 \leq x \leq 40.2) = P(0 \leq x \leq 25) + P(25 < x \leq 40.2) = 1 + 0 = 1$

Note that when the waiting time x is between 0 and 25 minutes, the probability that a student waits between 0 minutes and t minutes seems to be a function of the upper endpoint of the interval. Also, the probability that a student waits between 0 minutes and t minutes, where $t > 25$, is 1 because 25 minutes is the maximum waiting time. A scatterplot of the points $(t, P(0 \leq x \leq t))$ that appear in Table 7.27 is shown in Figure 7.46a.

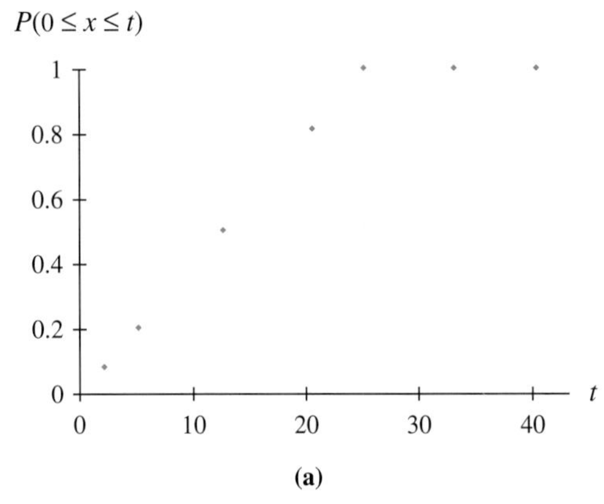

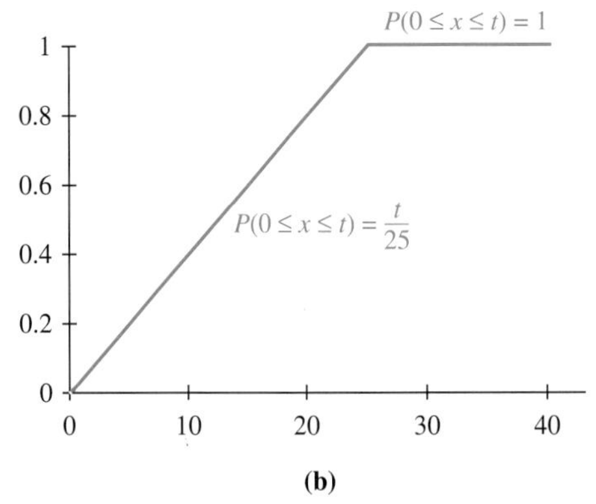

FIGURE 7.46

If all probabilities $P(0 < x < t)$, where $0 < t < \infty$, were computed and plotted, then the graph shown in Figure 7.46b would result.

A function that shows how probabilities accumulate as the value of the random variable increases is called a **cumulative density function**. For instance, if $y = F(x)$ is a cumulative density function for waiting times where x is measured in minutes, then the output $F(5) = 0.2$ represents the probability that the waiting time is less than or equal to 5 minutes. That is, $F(5) = P(x \le 5)$. Similarly, $F(12.5) = P(x \le 12.5)$, and so forth. In general, for a random variable x with probability density function f, where c is some constant, the cumulative density function of f at c is

$$F(c) = P(x \le c) = \int_{-\infty}^{c} f(x)\,dx$$

The left end behavior of any cumulative density function will be 0, corresponding to impossible events, and the right end behavior of any cumulative density function will be 1, corresponding to sure events. Cumulative density functions are always nondecreasing. Outputs of cumulative density functions are the areas between the corresponding probability density functions and the horizontal axis.

Does the link between probability density functions and cumulative density functions sound familiar? It should—it is the same as the relationship between functions and their accumulation functions, which we discussed in Chapter 6.

Cumulative Density Function

The cumulative density function for a random variable x defined on the interval of real numbers with probability density function f is

$$F(x) = \int_{-\infty}^{x} f(t)\,dt \quad \text{for all real numbers } x$$

For any value of the random variable x, say c, $F(c) = P(x \le c)$. The cumulative density function F is an accumulation function of the probability density function f.

EXAMPLE 9 *Graphing a Cumulative Density Function*

The graph in Figure 7.47 shows the probability density function for a random variable x.

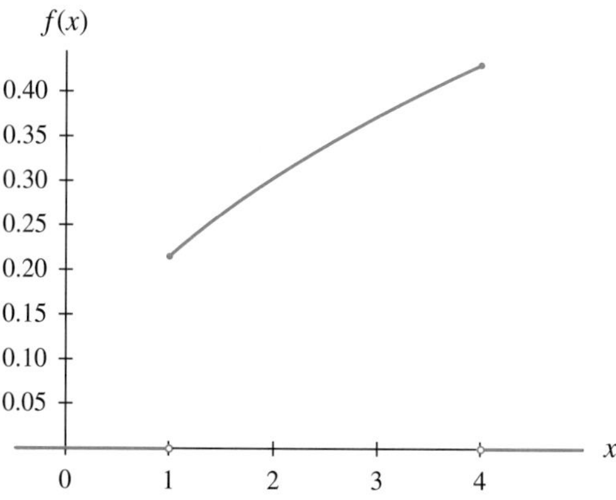

FIGURE 7.47

a. Sketch the graph of the cumulative density function for x.

b. Express $P(2 \leq x < 3.4)$ using the probability density function f.

c. Express $P(2 \leq x < 3.4)$ using the cumulative density function.

Solution

a. Because $f(x) = 0$ for $x < 1$, no area is accumulated when $x < 1$. Thus $F(x) = 0$ when $x < 1$. Note too that the accumulation of nonzero area under the graph of f begins at $x = 1$. Also, all accumulated area is positive because the graph of f is above the horizontal axis. As x moves to the right of the starting point ($x = 1$), area is accumulating faster and faster. Thus the cumulative density function is increasing and concave up until $x = 4$. At this point, no more area is accumulated, because the value of the density function is once again 0. Because f is a probability density function, the total accumulated area must be 1. Thus $F(4) = \int_{-\infty}^{4} f(x)\,dx = 1$. No area is added or subtracted past $x = 4$, so $F(x) = 1$ for $x > 4$. Combining all of this information, we can draw a possible graph of the cumulative density function F. (See Figure 7.48.)

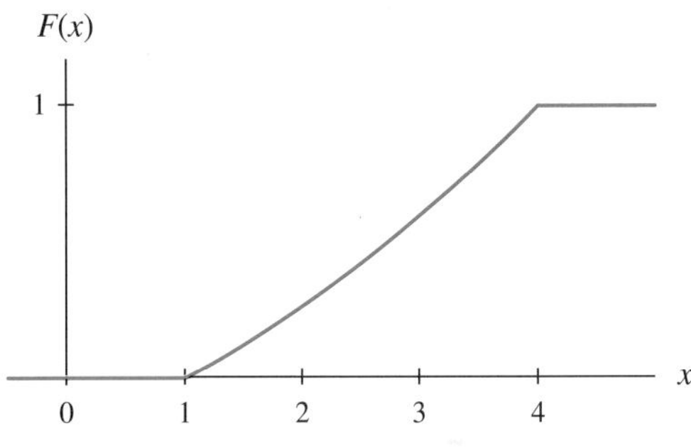

FIGURE 7.48

b. $P(2 \leq x < 3.4) = \int_{2}^{3.4} f(x)\,dx$

c. $P(2 \leq x < 3.4) = F(3.4) - F(2)$

The Fundamental Theorem guarantees the equivalence of the results of parts b and c of Example 9. It also tells us that the derivative of a cumulative density function is the corresponding probability density function at all points where the derivative exists. That is, $\frac{d}{dx}[F(x)] = f(x)$.

EXAMPLE 10 A Temperature Distribution

The graph in Figure 7.49 shows the cumulative density function for the distribution of temperatures in a Southwestern city during a 24-hour period in May. The random variable x measures the temperature recorded in degrees Fahrenheit (beginning at midnight), and the output $T(x)$ is the proportion of the time that the temperature is less than or equal to x°F.

a. What proportion of the time do you expect the temperature to be at most 80°F? Interpret this result in a probability context.

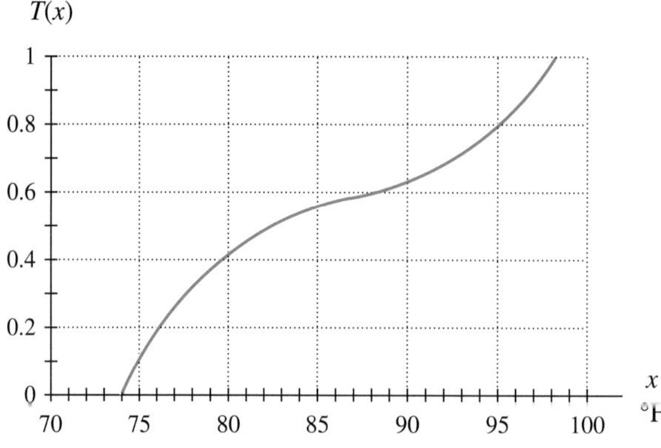

FIGURE 7.49

b. Estimate the high and low temperatures.

c. Estimate the probability that the temperature will be above 90°F.

d. Sketch a graph of the probability density function for the distribution of May temperatures in this location.

Solution

a. The proportion of the time that the temperature will be at most (less than or equal to) 80°F is $P(x \leq 80) = T(80) \approx 0.41$. In a Southwestern city during any 24-hour period in May, the temperature will be less than or equal to 80°F about 41% of the time.

b. $T(x)$ appears to be 0 at approximately 74°F. Because $T(x)$ is a proportion, it cannot be negative. Thus, the cumulative proportion to the left of 74°F must also be 0, and the minimum temperature is 74°F. The cumulative probability has a maximum of 1 at a temperature of about 98°F, so the maximum temperature is approximately 98°F.

c. The temperature on any day must be either less than 90°F or greater than or equal to 90°F. Thus the probability that the temperature is less than 90°F added to the probability that the temperature is 90°F or greater must equal 1.

$$\begin{aligned} T(90) + P(x \geq 90) &= 1 \\ P(x \geq 90) &= 1 - T(90) \\ &\approx 1 - 0.63 \\ &= 0.37 \end{aligned}$$

d. We know that the probability density function is the slope function of the cumulative density function. Note that because the graph of T is always increasing for temperatures between 74°F and 98°F, its slope graph is positive over this interval. (Recall that no output of a density function can be negative.) Also, the graph of T appears to have an inflection point (point of least slope) located at approximately 87°F, so there is a minimum on the slope graph at this temperature.

Using 74°F as the minimum and 98°F as the maximum temperatures, we know that $P(x < 74)$ and $P(x > 98)$ are both 0. Thus the value of the probability density function for temperatures less than 74°F and greater than 98°F is 0.

A possible graph of the probability density function is shown in Figure 7.50.

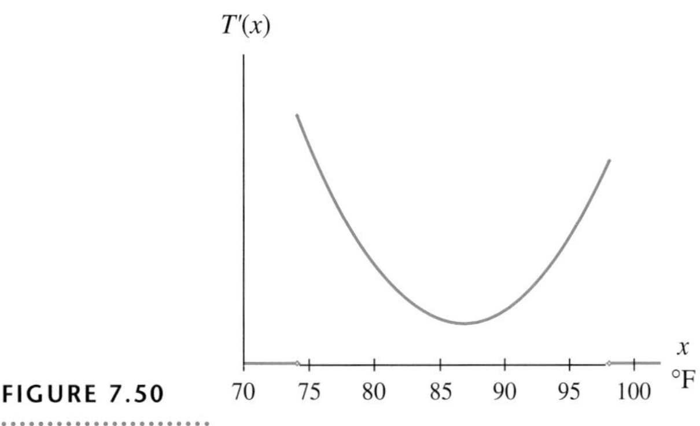

FIGURE 7.50

The connection between probability density functions and cumulative density functions lies at the very heart of calculus, for it is the relationship between a function and its accumulation function.

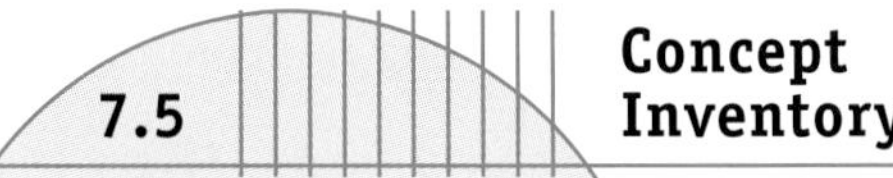

7.5 Concept Inventory

- ***Probability***
  - Histogram
  - Random variable
  - Notation, computation, and interpretation
- ***Probability density function***
- ***Probability distribution***
  - Uniform
  - Exponential
  - Normal
  - Mean and standard deviation
- ***Cumulative density function***

7.5 Activities

1. Interpret each of the following probability statements in the context of the given situation.
   a. Suppose the random variable x is the length, in minutes, of a telephone call made on a computer software technical support line. Interpret $P(x \geq 5) = 0.46$.
   b. Suppose the random variable x is the distance in feet between cars on a certain two-lane highway. Interpret $P(x < 7) \approx 0.25$.
   c. The probability that New Orleans will receive between 2 and 4 inches of rain during the month of March is 0.15.

2. Figure 7.5.1a through Figure 7.5.1d show the shapes of some probability distributions. Match each given situation to a possible graph of its density function. Give reasons for each of your choices.

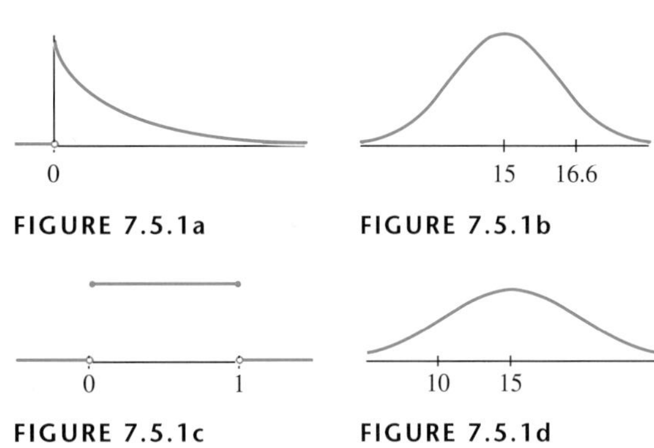

FIGURE 7.5.1a FIGURE 7.5.1b FIGURE 7.5.1c FIGURE 7.5.1d

   a. The amount of cereal put into a box is normally distributed with $\mu = 15$ ounces and $\sigma = 1.6$ ounces.
   b. The height of a certain species of plant is normally distributed with a mean of 15 inches and a standard deviation of 5 inches.
   c. You use a random number generator to choose a number between 0 and 1. The random variable x is the number that is chosen.
   d. The time, in minutes, that a customer waits to pay for items at a department store is exponentially distributed.

3. Which of the following could be probability density functions? Explain.

a. $f(x) = \begin{cases} 1.5(1 - x^2) & \text{when } 0 \le x \le 1 \\ 0 & \text{when } x < 0 \text{ or } x > 1 \end{cases}$

b. $h(x) = \begin{cases} 6(x - x^2) & \text{when } 0 \le x \le 1 \\ 0 & \text{when } x < 0 \text{ or } x > 1 \end{cases}$

c. $r(t)$

1.2

0 0.5 1 1.5 2 t

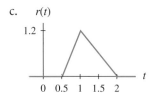

d. $s(c)$

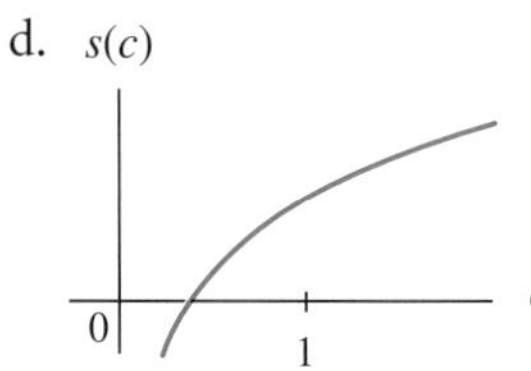

4. Which of the following could be probability density functions? Give reasons.

a. $g(x) = \begin{cases} 3x(1 - x^2) & \text{when } 0 \le x \le 1 \\ 0 & \text{when } x < 0 \text{ or } x > 1 \end{cases}$

b. $h(y) = \begin{cases} 0.625e^{-1.6y} & \text{when } y > 0 \\ 0 & \text{when } y \le 0 \end{cases}$

c. $g(t)$

(2, 1)

1
0.75
0.50
0.25 (1, 0.50) (3, 0.25)

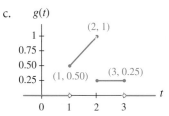

0 1 2 3 t

d. $f(x)$

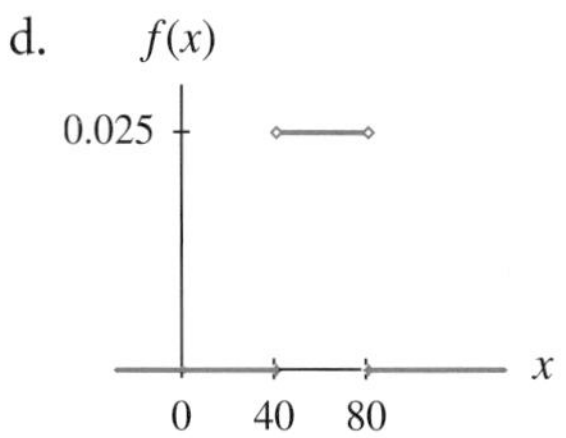

5. Let x represent the amount of frozen yogurt (in hundreds of gallons) sold by the G&T restaurant on any day during the summer. Storage limitations dictate that the maximum amount of frozen yogurt that can be kept at G&T on any given day is 250 gallons. Records of past sales indicate that the probability density function for x is approximated by $y(x) = 0.32x$ for $0 \le x \le 2.5$.

   a. What is the probability that, on some summer day, G&T will sell less than 100 gallons of frozen yogurt?

   b. What is the mean number of gallons of frozen yogurt that G&T expects to sell on a summer day?

   c. Sketch a graph of y, and locate the mean on the graph. Also, shade the region that the answer to part *a* represents.

6. Suppose that a traffic light on your campus remains red for 30 seconds at a time. You arrive at that light and find it red. Assume that your waiting time t (in seconds) at the light follows a uniform density function $y = u(t)$.

   a. Why does it make sense that $u(t) = 0$ for $t < 0$ and $t > 30$?

   b. Sketch a graph of u.

   c. Find your chances of waiting at least 10 seconds at the red light. What does this value represent on the graph of u?

   d. Find the probability of waiting no more than 20 seconds at the red light.

   e. What is the average time you would expect to wait at the light?

7. Explain, using the definition of a density function, why the definite integral calculation of a probability must result in a value between 0 and 1.

8. At a certain grocery checkout counter, the average waiting time is 2.5 minutes. Suppose the waiting times follow an exponential density function.

   a. Why would customers prefer the waiting times at the grocery checkout to follow an exponential distribution rather than a uniform distribution with the same mean?

   b. Write the equation for the exponential distribution of waiting times. Graph the equation and locate the mean waiting time on the graph. Does $\mu = 2.5$ seem reasonable for the graph?

   c. What is the likelihood that a customer waits less than 2 minutes to check out?

   d. What is the probability of waiting between 2 and 4 minutes to check out?

   e. What is the probability of waiting more than 5 minutes to check out?

9. Consider the exponential density function discussed in Example 6:

$$e(t) = 0.2e^{-0.2t} \text{ when } t \ge 0$$

where t is the time in minutes between successive arrivals at an emergency room in a large hospital on Saturday nights.

a. Find the probability that successive arrivals are between 20 and 30 minutes apart.

b. Find the probability that 10 minutes or less elapses between successive arrivals.

c. Find the probability that successive arrivals will be more than 15 minutes apart.

10. a. Graph the exponential density function for $k = 0.5, 1, 1.5, 2,$ and 4.

b. Comment on how changing the value of k affects the shape of the graph.

c. How does the mean of the exponential density function change as k increases?

11. The manufacturer of a new board game believes that the time it takes a child between the ages of 8 and 10 to learn the rules of its new board game has the probability density function

$$P(t) = \begin{cases} \frac{3}{32}(4t - t^2) & \text{when } 0 \le t \le 4 \\ 0 & \text{when } t > 4 \end{cases}$$

where t is time measured in minutes.

a. Find the mean time that it takes a child age 8 to 10 to learn the rules of the game.

b. Find the standard deviation of the learning times.

c. Find $P(0 \le t \le 1.5)$. Interpret this result.

d. Find $P(t \ge 3)$. Interpret this result.

12. Suppose the weight of pieces of passenger luggage for domestic airline flights follows a normal distribution with $\mu = 40$ pounds and $\sigma = 10.63$ pounds.

a. Find the probability that a piece of luggage weighs less than 45 pounds.

b. Find the probability that the total weight of the luggage for 80 passengers on a particular flight is between 1200 and 2400 pounds. (Assume each passenger has one piece of luggage.)

c. Find the probability that the total weight of the luggage for 125 passengers on a particular flight is more than 5600 pounds. (Assume each passenger has one piece of luggage.)

d. Find where the probability density function for the weight of passenger luggage is decreasing most rapidly.

13. The number of customers served daily by the ATM machines for a certain bank follows a normal distribution with a mean of 167 customers and a standard deviation of 30 customers.

a. Find where the probability density function for the number of customers who require daily ATM service at this bank is increasing the fastest.

b. Give two specific reasons why it would benefit a bank to know the probability distribution of its customers who are served daily by the ATM machines.

c. Sketch a graph of this normal distribution. In parts *i*, *ii*, and *iii*, shade on the graph the area representing the probability. Remembering that the area to the left (or right) of the mean of any normal distribution is 0.5, find the likelihood that, on a particular day,

i. between 150 and 200 customers require service at the ATM machines.

ii. fewer than 220 customers require service.

iii. more than 235 customers require service.

14. Figure 7.5.2a is a probability histogram of math SAT scores for male 1996 college-bound seniors, and Figure 7.5.2b is a probability histogram for

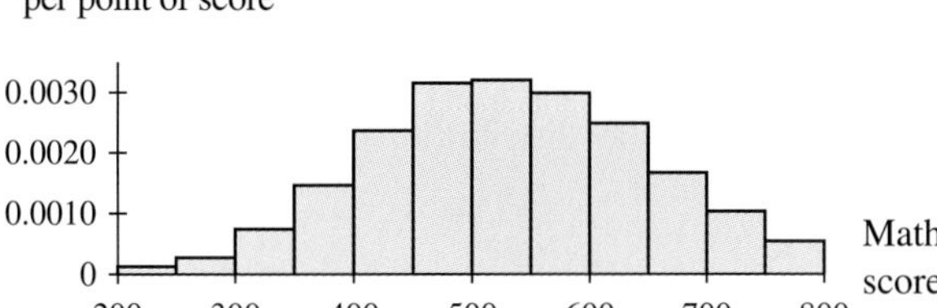

FIGURE 7.5.2a

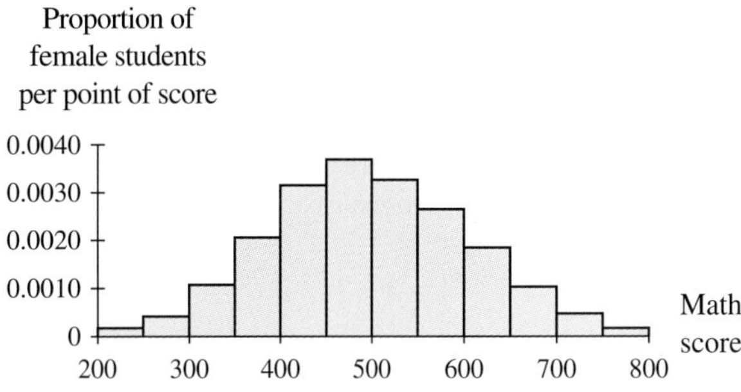

FIGURE 7.5.2b

math SAT scores for female 1996 college-bound seniors. All scores are based on the recentered scale. (See Activity 15.) Overdraw, on the histograms, continuous curves representing normal density functions that approximate the distributions of scores. Discuss any similarities and/or differences between the two normal distributions.

15. For all test dates on or after April 1, 1995, SAT I: Reasoning Test scores have been reported on a new, "recentered" scale. Over the years, the average score on the math portion of the SAT I moved away from 500, the midpoint of the original 200-to-800 scale. This change reestablished the average score near the midpoint of the scale and realigned the verbal and math scores so that a student with a score of 450 on each test can conclude that his or her math and verbal scores are equal. The previous scales showed the average verbal score to be about 425 and the average math score to be about 475, which made comparison between the two difficult.[32]

    a. If the interval between 200 and 800 included all scores within three standard deviations of the mean score on the original scale, what was the standard deviation of the original math SAT I distribution?

    b. Is the realigned mean score for verbal scores more or less than 425? Is the realigned mean score for math scores more or less than 475? Explain.

    c. Most standardized test scores follow a normal distribution. Using the fact that the probability of a score falling in a particular interval is the same as the percentage of students expected to score in that interval, determine what percentage of students were expected to make a math score of at least 475 under the "old" score scale.

    d. Assuming that the SAT I math and verbal scores follow a normal distribution, do you have enough information to draw a graph of the density function for either of the recentered SAT I scores?

    e. Do you think that recentering the SAT scores moved only the mean of the distribution or did it also change the standard deviation? Give reasons for your answer.

    f. Why do the recentered "higher" scores not translate into improved performance?

16. As we have previously noted, a quick approximation is sometimes useful when an exact answer is not required. For a distribution that is symmetric and bell-shaped (in particular, for a normal distribution), the *Empirical Rule* states that

    - Approximately 68% of the data values lie between $\mu - \sigma$ and $\mu + \sigma$.
    - Approximately 95% of the data values lie between $\mu - 2\sigma$ and $\mu + 2\sigma$.
    - Approximately 99.7% of the data values lie between $\mu - 3\sigma$ and $\mu + 3\sigma$.

    a. Verify the statements in the Empirical Rule for the normal probability density function with $\mu = 5.3$ and $\sigma = 8.372$.

    b. Estimate $P(-11.444 < x \leq 13.672)$ using the Empirical Rule if x has a normal probability distribution with $\mu = 5.3$ and $\sigma = 8.372$.

    c. Use the normal probability density function to find the probability in part *b*.

17. Scores on a final exam administered to all calculus classes at a large university are normally distributed with a mean of 72.3 and a standard deviation of 28.65. What percentage of students taking the test made

    a. a score between 60 and 80?

    b. a score of at least 90?

    c. a score that was more than one standard deviation away from the mean?

    d. At what score was the rate of change of the probability density function for the scores a maximum?

18. Another measure of the center of a probability distribution is the *median*. The median is the value m such that $\int_a^m f(x)\,dx = \int_m^b f(x)\,dx$.

    a. Refer to the distribution of waiting times in Example 5:

$$u(t) = \begin{cases} \frac{1}{15} & \text{when } 0 \leq t \leq 15 \\ 0 & \text{when } t > 15 \end{cases}$$

32. Used by permission from *Peterson's Guide to Four-Year Colleges, 1997*, 27th edition (Princeton, NJ: Peterson's Guides, Inc., 1996). © 1996 by Peterson's, Princeton, NJ.

where t is the time until the next bus. Find the median time to wait for the next bus.

b. How does the median compare to the mean for a probability distribution that is symmetric about a vertical line drawn through its mean?

c. For the density function $y = f(x)$ with $a \leq x \leq b$, explain why the statement

"The median is the value m such that

$$\int_a^m f(x)\,dx = \frac{1}{2}\text{"}$$

is equivalent to saying

"The median is the value m such that

$$\int_a^m f(x)\,dx = \int_m^b f(x)\,dx\text{"}$$

d. Verify that the median of the general exponential density function is $x = \frac{\ln 2}{k}$.

19. Verify the following statements for the uniform density function

$$u(x) = \begin{cases} \frac{1}{b-a} & \text{when } a \leq x \leq b \\ 0 & \text{when } x < a \text{ or } x > b \end{cases}$$

a. The mean is $\mu = \frac{a+b}{2}$.

b. The standard deviation is $\sigma = \frac{b-a}{\sqrt{12}}$.

c. The cumulative density function is

$$F(x) = \begin{cases} 0 & \text{when } x < 0 \\ \frac{x-a}{b-a} & \text{when } a \leq x \leq b \\ 1 & \text{when } x > b \end{cases}$$

20. The graph of a cumulative density function is shown in Figure 7.5.3. Sketch the graph of the corresponding probability density function if the input set for both functions is all real numbers.

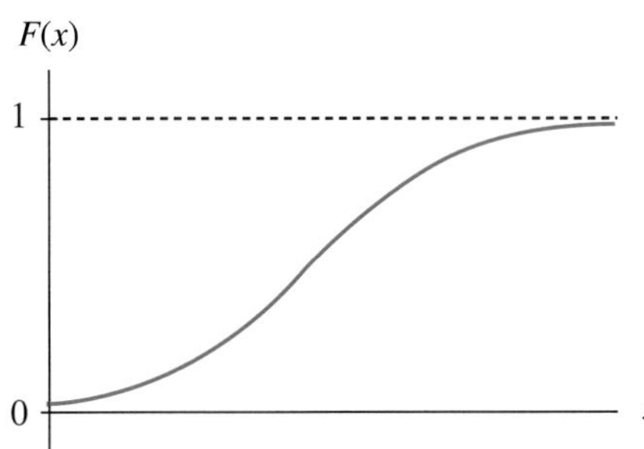

FIGURE 7.5.3

21. The graph in Figure 7.5.4 shows a probability distribution of the United States population at the time of the 1990 census.[33]

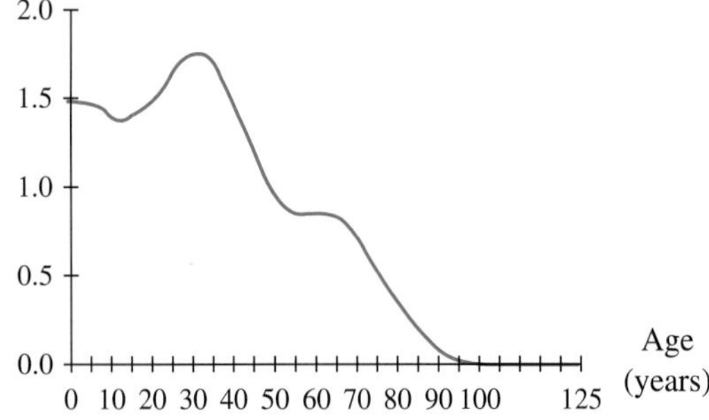

FIGURE 7.5.4

a. Sketch the graph of the corresponding cumulative density function.

b. Locate, on the graph of the cumulative density function you sketched in part *a* and on the graph of the probability density function in Figure 7.5.4, the probability that a person is between 20 and 40 years old.

22. Suppose $y = f(x)$ is the exponential density function with $k = 2$.

a. Find F, the corresponding cumulative density function.

b. Use both f and F to find the probability that $x \leq 0.35$.

c. Use F to find the probability that $x > 0.86$.

d. Sketch graphs of f and F.

23. f is the density function

$$f(x) = \begin{cases} 2x & \text{when } 0 \leq x < 1 \\ 0 & \text{when } x < 0 \text{ and } x \geq 1 \end{cases}$$

a. Find F, the corresponding cumulative density function.

b. Use both f and F to find the probability that $x < 0.67$.

c. Use F to find the probability that $x > 0.25$.

d. Sketch graphs of f and F.

33. Based on data from the U.S. Bureau of the Census.

TABLE 7.28

x (years)	F(x) (percent)	x (years)	F(x) (percent)	x (years)	F(x) (percent)
0		35	52.8	70	91.3
5	7.6	40	61.1	75	94.7
10	14.8	45	68.4	80	97.2
15	21.9	50	74.5	85	98.8
20	28.6	55	79.2	90	99.7
25	36.1	60	83.3	95	99.9
30	44.0	65	87.4	100	100.0

24. In mid-1992 the U.S. resident population was about 255 million, with only approximately 45,000 persons aged 100 or older. Table 7.28 gives the distribution of ages[34] of those persons less than 100 years old. $F(x)$ is the percentage of U.S. residents of age less than x years. (Assume that no one is more than 100 years old.)[35]
   a. Fill in the value for $F(0)$.
   b. If possible, give the units of the corresponding probability density function. Use symmetric difference quotients to estimate $F'(x)$ at $x = 10$, 25, 30, 35, 45, 50, 70, and 90 years of age.
   c. The plot in Figure 7.5.5 shows some approximate values of the continuous density function. Label the axes and draw a smooth curve through the points so that there are no more concavity changes in your curve than those indicated by the given points. What is the demographic significance of the large bump on the graph?

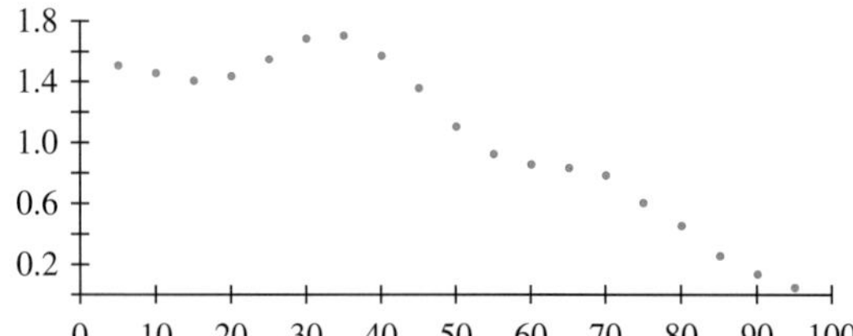

FIGURE 7.5.5

   d. Use the cumulative density function F to find the percentage and number of 1992 U.S. residents who were age 20 or more but under age 50.
   e. Use the density function F' to write an integral expression for the percentage of residents who were age 20 or more but under age 50.
   f. Use three midpoint rectangles and the appropriate $F'(x)$ estimates from part *b* to estimate the value of the integral in part *e*. Sketch the rectangles you use on the plot in Figure 7.5.5.
   g. Use the appropriate $F'(x)$ estimates calculated in part *b* and five midpoint rectangles to estimate $\int_0^{100} xF'(x)\,dx$. What does this integral represent, and what are its units?

25. Table 7.29 shows income intervals (in thousands of dollars) and the percentage of Texas households[36] with 1990 income in each interval. For instance, 9.6% of Texas households had income greater than or equal to \$5000 but less than \$10,000 in 1990.

TABLE 7.29

Income interval (thousands of dollars)	Percent of households
0–5	8.2
5–10	9.6
10–15	9.8
15–25	18.8
25–35	15.8
35–50	16.6
50–75	13.3
75–100	4.3

34. Ibid.
35. The authors wish to express their appreciation to Dr. Dan Lewis of Texas A&M University for his contributions to Example 10 and Activities 24–26.

36. U.S. Bureau of the Census.

a. Complete Table 7.30, which shows the cumulative percent $F(x)$ of households with incomes less than x thousand dollars.

TABLE 7.30

x	$F(x)$	x	$F(x)$
5		35	
10		50	
15		75	
25		100	

Comment on the value $F(100)$. Why is it not 100%?

b. Use symmetric difference quotients to estimate the values of the density function f at the midpoint of each income interval. Record your results in Table 7.31.

TABLE 7.31

x	$F(x)$	x	$F(x)$
2.5		30	
7.5		42.5	
12.5		62.5	
20		87.5	

c. Examine a plot of the data in part *b*. Does it appear that the continuous density function f would have an inflection point or limiting value?

d. An approximate model for the density function is

$$f(x) = 0.85403395x^{0.56}e^{-0.0437980796x}$$

percentage points per thousand dollars

where x is household income in thousands of dollars. Examine a graph of this model on the plot of density function estimates that you constructed in part *c*. Comment on the fit.

e. Use the equation for $f(x)$ to estimate the percentage of households with 1990 annual income less than \$100,000. Compare your result with $F(100)$ from part *a*.

f. Use the equation for $f(x)$ to estimate the mean 1990 income for Texas households with incomes less than \$100,000.

26. A study of serious accidents in British coal mines[37] focused on the time interval between successive accidents. If, for example, successive accidents were observed on August 2 and August 5, then the time interval between these accidents was recorded as 3 days. During the study period, there were 34 serious accidents separated by 33 time intervals. Table 7.32 gives some of the data. $A(x)$ is the number of time intervals that were greater than or equal to x days.

TABLE 7.32

x	$A(x)$	x	$A(x)$
1	31	13	7
4	19	16	4
7	13	19	3
10	9		

a. Find $A(0)$. Was there a day on which two accidents occurred? Was there a day on which more than three accidents occurred?

b. Explain why $F(x) = 1 - \frac{A(x)}{33}$ is the distribution of the time intervals observed between successive accidents.

c. Find an exponential model for $A(x)$.

d. Use the model from part *c* to write a model for the distribution F, and then write a model for the density function F'.

e. What does the integral $\int_0^\infty x\,F'(x)\,dx$ represent, and what are its units?

27. In Michael Crichton's novel *Jurassic Park*,[38] dinosaur clones are alive and roaming about a remote jungle island intended to be a theme park. All the dinosaurs have been cloned female so that the populations can be controlled in Jurassic Park. Ian Malcolm, a cynical mathematician who is invited to the island, finds one of the first clues that all is not well when he examines one of the graphs given in Figure 7.5.6. Both graphs show height distributions of the compy (Procompsognathid). (The park's computer that produced the graphs constructed them with straight lines connecting the data rather than using smooth curves.)

37. B. A. Maguire, E. S. Person, and A. W. Wynn, "The Time Interval Between Industrial Accidents," *Biometrika* 39 (1952), 168–180.

38. Michael Crichton, *Jurassic Park* (New York: Knopf, 1990).

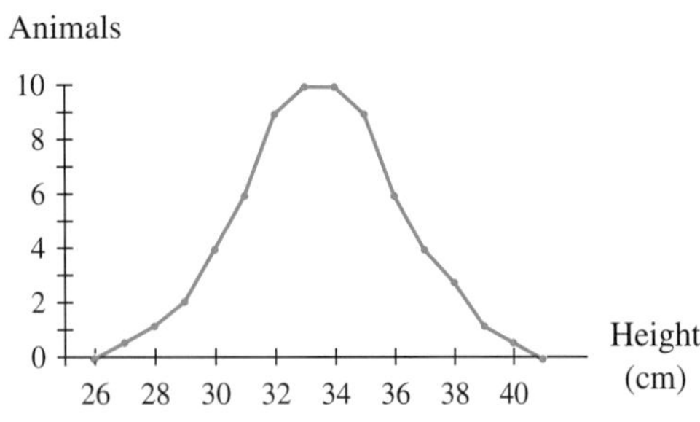

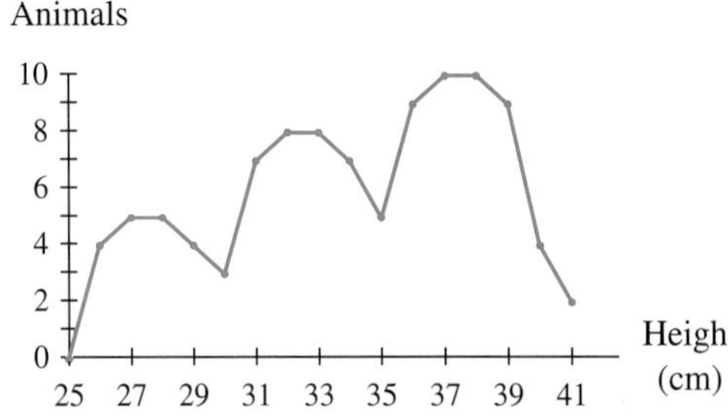

Height distribution (Procompsognathids)

FIGURE 7.5.6

a. Malcolm claims that one of the graphs in Figure 7.5.6 is characteristic of a breeding population and that the other graph is what would be expected from a controlled population in which the compys were introduced in three batches at six-month intervals. Which distribution corresponds to which population?

b. Which graph did Malcolm first see that indicated something was amiss?

c. Are the height distribution graphs in Figure 7.5.6 graphs of probability density functions? Explain.

Upon completing the Chapter 7 topics you have chosen for your course, you may do one of the following: (1) Proceed to Chapter 8 if you have not already covered it and wish to do so. (2) Skip Chapter 8 and proceed to Chapters 9 and 10. (3) Proceed to Chapter 11, returning to Chapters 9 and 10 later, if desired.

Summary

Differences of Accumulated Changes

In this chapter, we investigated situations that lend themselves to analysis by examination of the area between two curves. To compute the area between two curves, we used the fact that if the graph of f lies above the graph of g from a to b, then the integral $\int_a^b [f(x) - g(x)]\,dx$ is the area of the region between the two graphs from a to b.

We observed that when f and g are two continuous rate-of-change functions, then the difference between the accumulated change of f from a to b and the accumulated change of g from a to b is also given by $\int_a^b [f(x) - g(x)]\,dx$. However, if the two functions intersect between a and b, then the difference between the accumulated changes of the functions is *not* the same as the total area of the regions between the two rate-of-change curves.

Streams in Business and Biology

An income stream is a flow of money into an interest-bearing account over a period of time. If the stream flows continuously into an account at a rate of $R(t)$ dollars per year and the account earns annual interest at the rate of $100r\%$ compounded continuously, then the future value of the account at the end of T years is given by

$$\text{Future value} = \int_0^T R(t)e^{r(T-t)}\,dt \text{ dollars}$$

The present value of an income stream is the amount that would have to be invested now in order for the account to grow to a given future value. The present value of a continuous income stream whose future value is given by the above equation is

$$\text{Present value} = \int_0^T R(t)e^{-rt}\,dt \text{ dollars}$$

Although it may be feasible to consider money flowing as a continuous stream in very large corporations, there are many business applications for income streams that have quarterly or monthly activity. We call such a stream a discrete income stream and determine future value by summing rather than integrating:

$$\text{Future value} = \sum_{d=0}^{D-1} R(d)\left(1 + \frac{r}{n}\right)^{D-d} \text{ dollars}$$

where n is the number of deposits made each year, $R(d)$ is the value of the dth deposit, $100r\%$ is the annual interest rate, and D is the total number of deposits made. Once the future value of a discrete income stream is calculated, the present value is determined by solving for P in the formula

$$P\left(1 + \frac{r}{n}\right)^{D} = \text{Future value}$$

Streams also have applications in biology and related fields. The future value (in b years) of a biological stream with initial population size P, survival rate s (in decimals), and renewal rate $r(t)$, where t is the number of years of the stream, is

$$\text{Future value} \approx Ps^{b} + \int_{0}^{b} r(t)s^{b-t}\,dt$$

Integrals in Economics

A demand curve for a commodity is determined by economic factors. Points on the demand curve tell us how much consumers will purchase at the associated prices. However, consumers are usually willing and able to pay more for the quantity they purchase. Once a market price (the price at which the item is sold) is set, the actual amount consumers spend for a certain quantity of items is the consumers' expenditure. The amount that consumers are willing and able to spend but do not actually spend is known as the consumers' surplus. We use areas of regions under the demand curve to represent these quantities. (See Figure 7.51.)

The interaction between supply and demand usually determines the quantity of an item that is available. The producers' supply curve gives the number of units that a producer is willing and able to supply as a function of the price per unit of the item. When prices increase, consumers usually respond by demanding less, but producers ordinarily respond by supplying more. Commodities may have a price, called the shutdown price, under which producers are not willing or able to supply any of the product, so the supply curve has output 0 to the left of this price. Thus a supply curve is often the graph of a piecewise continuous function.

Points on the supply curve tell us how much producers will supply at the associated prices. However, producers are often willing and able to supply more of the commodity they produce. The amount that producers receive for supplying a certain quantity of items at a certain market price is the producers' total revenue. The amount that producers receive above the minimum amount they are willing and able to receive is the producers' surplus. (See Figure 7.52.)

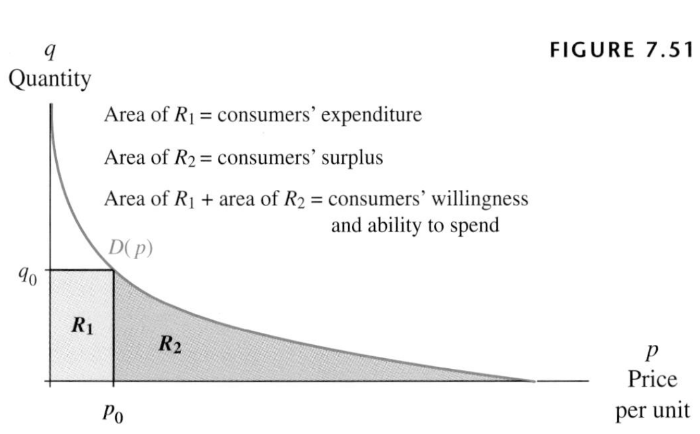

FIGURE 7.51

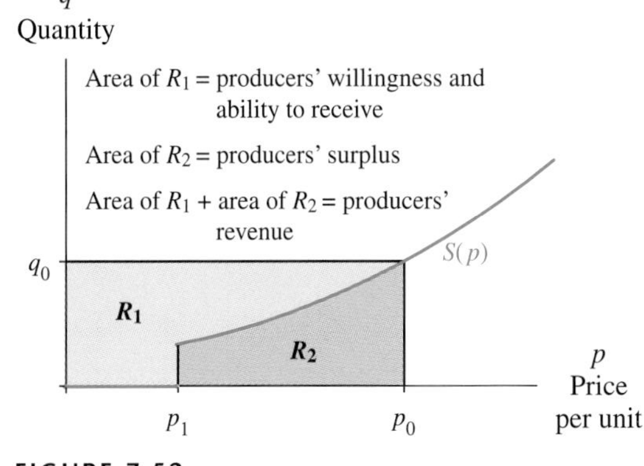

FIGURE 7.52

The point (p^*, q^*) where the supply and demand curves for a particular item cross is called the equilibrium point. At the equilibrium price p^*, the quantity demanded by consumers coincides with the quantity supplied by producers, which is q^*. When the market price of a commodity is the equilibrium price for that product, the total social gain is the sum of the consumers' surplus and the producers' surplus.

Average Values and Average Rates of Change

We use definite integrals to calculate the average value of a continuous function for a quantity:

$$\text{Average value of } f(x) \text{ from } a \text{ to } b = \frac{\int_a^b f(x)\,dx}{b-a}$$

When we are given a rate-of-change function $y = f'(t)$, the average rate of change of $f(t)$ from $t = a$ to $t = b$ is

$$\text{Average rate of change of } f(t) \text{ from } a \text{ to } b = \frac{\int_a^b f'(t)\,dt}{b-a}$$

The average rate of change of f can also be calculated by using the quantity function $y = f(t)$:

$$\text{Average rate of change of } f(t) \text{ from } a \text{ to } b = \frac{f(b) - f(a)}{b-a}$$

Probability Distributions and Density Functions

The probability $P(a \le x \le b)$ is a measure of the likelihood that an outcome of an experiment involving a random quantity will lie between a and b. Functions that describe how the probabilities associated with a random variable are distributed over various intervals of numbers are called probability density functions (or probability distributions).

A probability density function f is a continuous or piecewise continuous function that has two conditions on its output: The graph of f must lie above or on the horizontal axis, and the area between the graph of f and the horizontal axis must be 1. The input of f is the interval of real numbers. We present no interpretation of the outputs of probability density functions. Instead, we give meaning to integrals of a probability density function f, provided, of course, that the integrals exist.

- The likelihood that x is between a and b is

$$P(a \le x \le b) = \int_a^b f(x)\,dx$$

- A measure of the center of the distribution is the mean:

$$\mu = \int_{-\infty}^{\infty} x f(x)\,dx$$

- A measure of the spread of the distribution is the standard deviation:

$$\sigma = \sqrt{\int_{-\infty}^{\infty} (x-\mu)^2 f(x)\,dx}$$

In real-world applications, some situations are described by known probability distributions. Two of these are the uniform density function and the exponential density function. The uniform distribution has a constant value on a specified interval and is zero elsewhere. It is a good model for random variables that are evenly distributed on a specified interval. The exponential density function is often called the waiting time distribution and has applications for random variables that are more likely to be small than large.

A density function with numerous applications in the behavioral and social sciences is the normal distribution. A normal curve is symmetric about a vertical line through its mean, is bell-shaped, and has inflection points that are located one standard deviation on either side of the mean.

A cumulative density function is an accumulation function of a probability density function. Outputs of cumulative density functions are the areas between the corresponding probability density functions and the horizontal axis. Probabilities can be determined by using either probability density functions or cumulative density functions.

Concept Check

Can you	To practice, try	
• Find and interpret areas between two curves?	Section 7.1	Activity 17
• Determine income flow rate functions?	Section 7.2	Activity 1
• Calculate and interpret present and future values of discrete and continuous income streams?	Section 7.2	Activities 9, 15
• Find various quantities related to a demand function?	Section 7.3	Activity 5
• Find various quantities related to a supply function?	Section 7.3	Activity 9
• Find the market equilibrium point and total social gain?	Section 7.3	Activity 13
• Find average value and average rates of change?	Section 7.4	Activity 7
• Find and interpret probability, mean, and standard deviation?	Section 7.5	Activity 11
• Understand and use probability density functions?	Section 7.5	Activities 3, 5
• Work with cumulative density functions?	Section 7.5	Activity 23

Review Test

1. In preparing to start your own business (in 6 years), you plan to invest 10% of your salary each month in an account with a fixed rate of return of 8.3%. You currently make \$3000 per month and expect your income to increase by \$500 per year.
   a. Find a function for the yearly rate at which you will invest money in the account.
   b. If you start investing now, to what amount will your account grow in 6 years? (Consider a continuous stream.)
   c. How much would you have to invest now in one lump sum, instead of in a continuous stream, in order to build to the same 6-year future value?
2. A teacher is planning to retire in 8 years. To supplement her state retirement income, she plans to invest 7% of her salary each month until retirement in an annuity with a fixed rate of return of 8.2% compounded monthly. She currently makes \$3100 per month and expects her income to increase, due to consulting work, by 0.4% per month.
   a. How much will be in the annuity at the end of 8 years?
   b. How much would she have to invest now, in one lump sum, to accumulate the same amount as the 8-year future value found in part *a*?
3. Suppose a 1990 population of 10,000 foxes breeds at a rate of 500 pups per year and has a survival rate of 63%.
   a. Assuming that the survival and renewal rates remain constant, determine how many of the foxes alive in 1990 will still be alive in 2010.
   b. Write a function for the number of foxes that were born t years after 1990 and will still be alive in 2010.
   c. Estimate the fox population in the year 2010.

4. The average quantity of marble fountains that consumers will demand can be modeled as

$$D(p) = -1.0p^2 - 20.6p + 900 \text{ fountains}$$

and the average quantity that producers will supply can be modeled as

$$S(p) = \begin{cases} 0 \text{ fountains} & \text{if } p < 2 \\ 0.3p^2 + 8.1p + 300 \text{ fountains} & \text{if } p \geq 2 \end{cases}$$

when the market price is p hundred dollars per fountain.

   a. How much are consumers willing to spend for 30 fountains?

   b. How many fountains will producers supply at \$1000 per fountain? Will supply exceed demand at this quantity?

   c. Determine the total social gain when fountains are sold at the equilibrium price.

5. The projected number of households shopping online[39] between 1999 and 2004 can be modeled by the equation

$$h(y) = 15.936 + 18.753 \ln y \text{ million}$$

y years after 1998.

   a. Find the average projected number of households shopping online between 1999 and 2004.

   b. Find the average rate of change in the projected number of households shopping online between 1999 and 2004.

6. The density function for a random variable x is

$$f(x) = \begin{cases} 0.125x & \text{when } 0 \leq x \leq 4 \\ 0 & \text{when } x < 0 \text{ or } x > 4 \end{cases}$$

   a. Find the probability that x is less than 3.8. Interpret this result.

   b. Find the probability that x is between 1.3 and 5. Indicate, on a graph of f, what this answer represents.

   c. Find the mean value of x. Locate the mean on the graph you drew in part *b*.

   d. Find and graph $y = F(x)$, the cumulative density function for x.

   e. Use F to find the answer to part *b*. Show all work involved in your calculation.

39. Based on data from Forrester Research, Inc.

Project 7.1

Arch Art

Setting

A popular historical site in Missouri is the Gateway Arch. Designed by Eero Saarinen, it is located on the original riverfront town site of St. Louis and symbolizes the city's role as gateway to the West. The stainless steel Gateway Arch (also called the St. Louis Arch) is 630 feet (192 meters) high and has an equal span.

In honor of the 200th anniversary of the Louisiana Purchase, which made St. Louis a part of the United States, the city has commissioned an artist to design a work of art at the Jefferson National Expansion Memorial National Historic Site. The artist plans to construct a hill beneath the Gateway Arch, located at the Historic Site, and hang strips of mylar from the arch to the hill so as to completely fill the space. (See Figure 7.53.) The artist has asked for your help in determining the amount of mylar needed.

Tasks

1. If the hill is to be 30 feet tall at its highest point, find an equation for the height of the cross section of the hill at its peak. Refer to Figure 7.53.
2. Estimate the height of the arch in at least ten different places. Use the estimated heights to construct a model for the height of the arch. (You need not consider only the models presented in this text.)
3. Estimate the area between the arch and the hill.
4. The artist plans to use strips of mylar 60 inches wide. What is the minimum number of yards of mylar that the artist will need to purchase?
5. Repeat Task 4 for strips 30 inches wide.
6. If the 30-inch strips cost half as much as the 60-inch strips, is there any cost benefit to using one width instead of the other? If so, which width? Explain.

Reporting

Write a memo telling the artist the minimum amount of mylar necessary. Explain how you came to your conclusions. Include your mathematical work as an attachment.

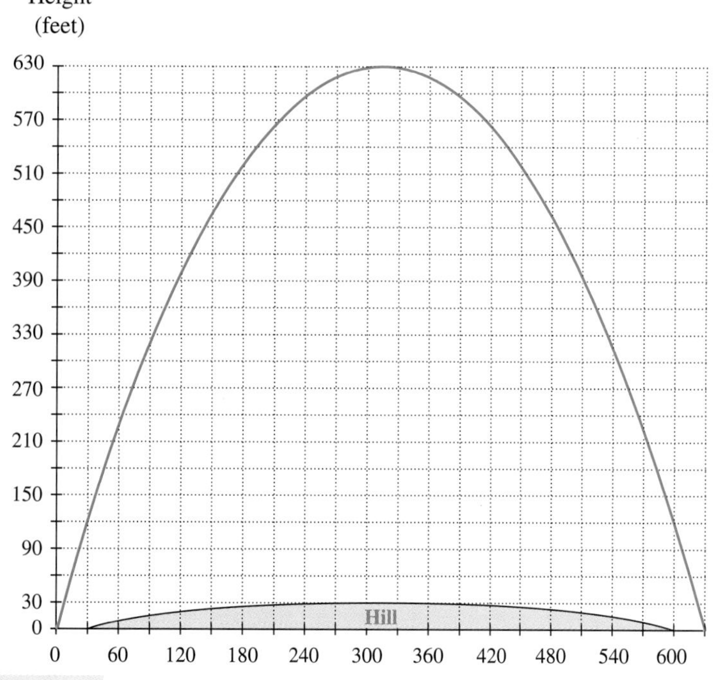

FIGURE 7.53 The Gateway Arch in St. Louis

8

Repetitive Change: Cycles and Trigonometry

Eric Simmons/Stock Boston

Concept Application

The levels of carbon dioxide (CO_2) in the atmosphere fluctuate in a periodic manner throughout the year, the lowest levels occurring in the summer when plants absorb significant amounts of CO_2. The amount of CO_2 in the atmosphere has been increasing for decades, and by using a model constructed from a trigonometric function and a quadratic function, we can model the carbon dioxide level over time. The model can then be used to answer questions such as

- How quickly is the level of carbon dioxide changing in a given month?
- What are the maximum and minimum levels of CO_2 in a given year, and when do those levels occur?
- When during a year is the amount of carbon dioxide increasing most rapidly?

You will be given an opportunity to answer questions such as these in Examples 2 and 3 of Section 8.3 and Activities 14 and 15 of Section 8.4.

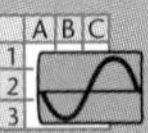

This symbol directs you to the supplemental technology guides. The different guides contain step-by-step instructions for using graphing calculators or spreadsheet software to work the particular example or discussion marked by the technology symbol.

This symbol directs you to the Calculus Concepts *web site, where you will find algebra review, links to data updates, additional practice problems, and other helpful resource material.*

Concept Objectives

This chapter will help you understand the concepts of

- Sine and cosine functions in terms of the unit circle
- Amplitude, reflection, frequency, period, and shifts of sine functions
- Derivatives and integrals of sine and cosine functions
- Expected value of a cyclic function

and you will learn to

- Calculate and interpret sine and cosine values
- Identify the amplitude, reflection, period, frequency, horizontal shift, and vertical shift of a sine function
- Use data to construct and graph a sine model without the aid of technology
- Use technology to construct a sine model
- Calculate and interpret derivatives of sine and cosine functions
- Find extrema and inflection points of sine and cosine functions
- Calculate and interpret definite integrals of sine and cosine functions

Many things change and then return to repeat the same change. This creates a repetitive pattern and forms a cycle. Patterns of repeating cycles occur in nature as well as in business applications.

Many cycles are best described with the functions from an area of mathematics called trigonometry. This term has its roots in the Greek word for "triangle measurement." Most of early trigonometry was centered on using ratios in similar triangles to determine indirect measurements. However, the trigonometric functions are now widely used to describe cyclic patterns of change.

Understanding the repetitive process in cycles and the mathematical models used to describe such cycles is the goal of the first part of this chapter. The second part of the chapter shows how calculus concepts can be applied to cyclic functions.

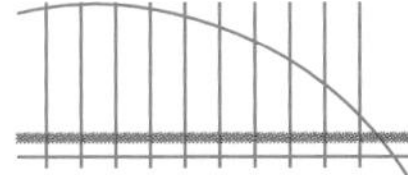

8.1 Functions of Angles: Sine and Cosine

Chapter 8 has been written so that it can be easily integrated throughout Chapters 1–7. Specifically, Section 8.1 can be presented in Chapter 1, Section 8.2 in Chapter 2, Section 8.3 in Chapter 4, Section 8.4 in Chapter 5, and Section 8.5 in either Chapter 6 or Chapter 7. See the *Instructor's Resource Guide* for more detailed information about using this approach.

It the intent of the authors that the sine function be presented as another type of model useful in situations that are cyclic in nature. As a result, we do not include a thorough trigonometry discussion. If you wish to expose your students to more trigonometry (including the relationship between radians and degrees and right-triangle trigonometry), we recommend that you supplement your coverage of Chapter 8 with the Trigonometry Appendix at the *Calculus Concepts* web site.

A study of functions would not be complete without an examination of functions that repeat a pattern. If a function repeats itself and the length between the repetitions is constant, then we say the function is **periodic**. The shortest interval over which the function repeats itself is called the **period**. Furthermore, when a function is periodic and varies continuously between two extremes, we say the function is **cyclic**.

Cyclic Function

A cyclic function is a periodic, continuous function that varies between two extremes.

The most common periodic functions are the trigonometric (or trig) functions. There are six trig functions, all of them periodic. However, only two of the trig functions are cyclic, and we restrict our study to those: the sine function (abbreviated sin) and the cosine function (abbreviated cos). Before we define these functions, we briefly discuss the angle measurement that we use in calculus.

Radian Measure

FIGURE 8.1

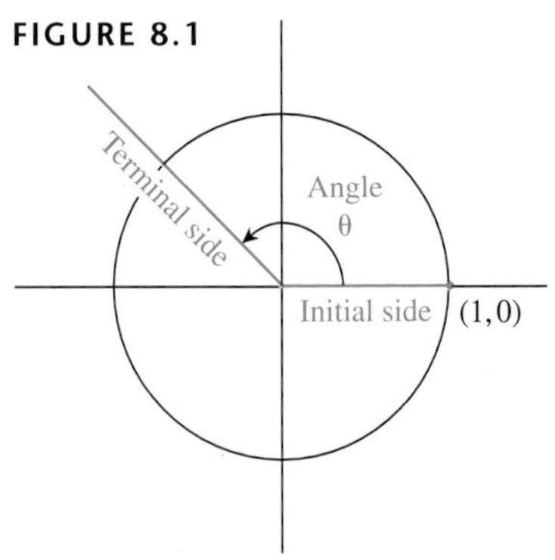

An angle drawn in standard position on a unit circle

FIGURE 8.2

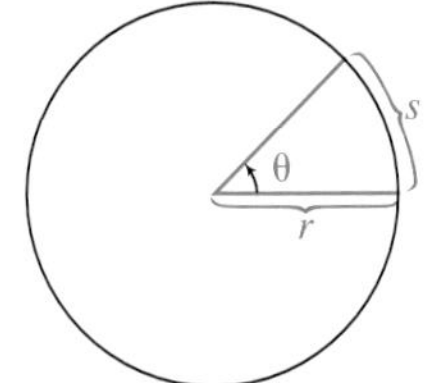

The radian measure of an angle is the ratio of the arc length to the radius.

$$\theta = \frac{s}{r}$$

We place an angle on the coordinate axes so that the vertex of the angle is at the origin and so that one side, called the **initial side**, is along the positive x-axis. The other side of the angle is called the **terminal side**. An angle drawn in this position is referred to as being in **standard position**. The size of the angle is determined by the measure of the rotation of the initial side into the terminal side. A positive angle represents counterclockwise rotation, and a negative angle represents clockwise rotation. The trig functions are defined in terms of a unit circle, so we draw angles in standard position on a unit circle—that is, a circle centered at the origin with radius 1 unit. (See Figure 8.1.)

You are accustomed to measuring angles in degrees; however, it is common in a study of calculus to measure angles in radians. We define the **radian measure** of an angle to be the ratio of the length of the arc s cut out of the circumference of a circle by the angle θ to the radius r of the circle. (See Figure 8.2.) Because the total arc length (that is, the circumference) of a circle is 2π times its radius, a full revolution has radian measure of $\frac{2\pi r}{r} = 2\pi$. Note that 2π is a real number approximately equal to 6.283. When we are using a unit circle, the radius is 1, so the radian measure is simply the corresponding arc length: $\theta = \frac{s}{r} = \frac{s}{1} = s$.

It is important to recognize that the radian measure of an angle is a real number with no units attached. In the definition of radian measure, both s and r measure length and must have the same units, which causes θ to be a unitless quantity. At first this might seem to be a strange way to measure angles, but we later find that radian measure of angles gives simple calculus results for the trigonometric functions with which we will be dealing.

Radian Measurement

2π radians $=$ 1 complete counterclockwise rotation

1 radian $= \dfrac{1}{2\pi}$ counterclockwise rotation (a little less than $\frac{1}{6}$ revolution)

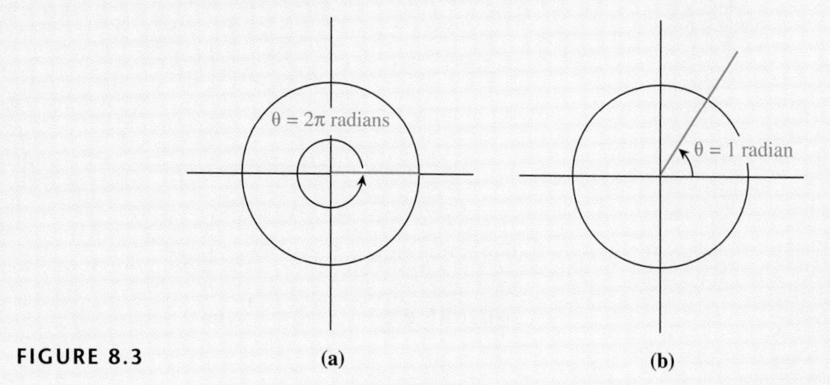

FIGURE 8.3

Figure 8.4 shows some common radian measures on a circle.

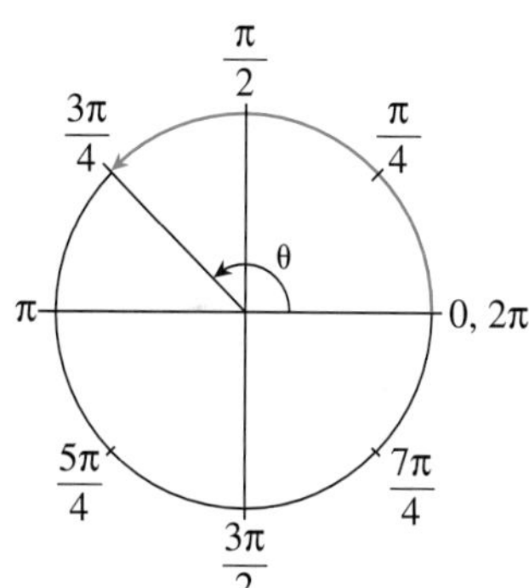

The angle θ corresponds to $\frac{3}{8}$ of a complete rotation, or $\frac{3}{8}(2\pi) = \frac{3\pi}{4}$ units around the circle. Thus its radian measure is $\frac{3\pi}{4}$.

FIGURE 8.4

EXAMPLE 1 *Radian Measurement*

a. Draw angles with radian measure $\frac{-5\pi}{4}$ and 5 on a unit circle.

b. Estimate the radian measure of the two angles shown in Figures 8.5a and b.

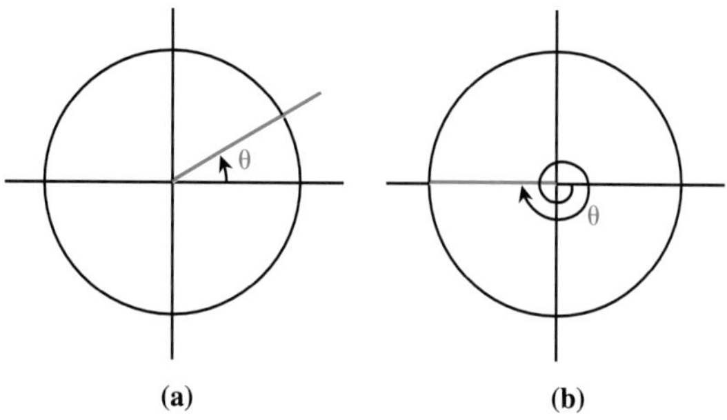

FIGURE 8.5

Solution

a. An angle of $\frac{-5\pi}{4} = -\pi - \frac{\pi}{4} = -\left[\frac{1}{2}(2\pi) + \frac{1}{8}(2\pi)\right]$ is half of a clockwise rotation plus another $\frac{1}{8}$ of a clockwise rotation. See Figure 8.6a. An angle of 5 is between $\frac{3}{4}$ of a counterclockwise rotation $\left(\frac{3\pi}{2} \approx 4.7\right)$ and a complete rotation ($2\pi \approx 6.3$). See Figure 8.6b.

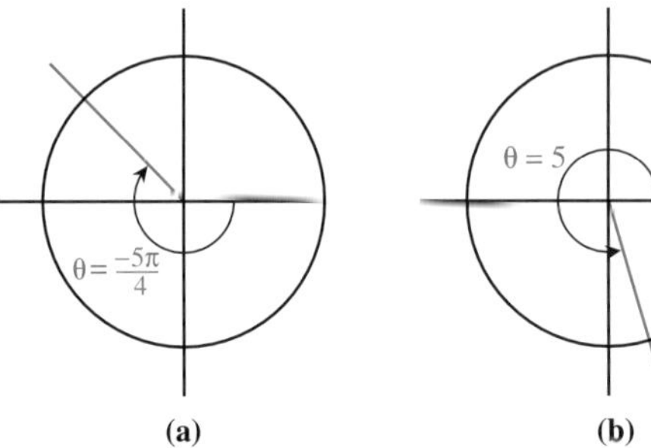

FIGURE 8.6 (a) (b)

b. The angle in Figure 8.5a appears to be approximately $\frac{1}{12}$ of a rotation, or $\frac{2\pi}{12} \approx 0.5$ radian. The angle in Figure 8.5b is -3π: one complete clockwise rotation (-2π) plus a half of a clockwise rotation ($-\pi$).

Sine and Cosine Functions

Having explained radian measure, we now define sine and cosine functions of angles in standard position on a unit circle. It is also possible to define these functions in terms of the sides of a right triangle. More information on this approach to the sine and cosine functions is found in the Trigonometry Appendix at the *Calculus Concepts* web site.

Trigonometric Values for a General Angle

Consider an angle θ in standard position. The terminal side of the angle intercepts the unit circle at a point (x, y). The sine and cosine functions are defined as follows:

$f(\theta) = \sin\theta$ is the function whose output is the y-coordinate

$g(\theta) = \cos\theta$ is the function whose output is the x-coordinate

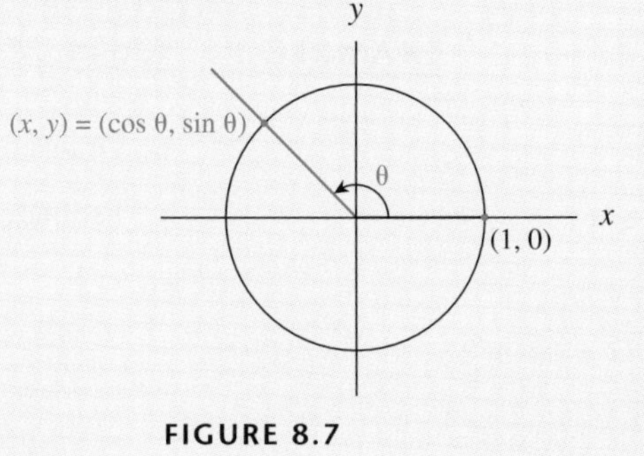

FIGURE 8.7

Because the equation of the unit circle shown in Figure 8.7 is $x^2 + y^2 = 1$, it follows that $\cos^2\theta + \sin^2\theta = 1$ for any angle θ.

It is not important in our study of cyclic functions that students memorize any angles and their corresponding points on the unit circle.

Figure 8.8 shows the four points where the unit circle intersects the axes, the corresponding angles in radians, and the values of the sine and cosine functions.

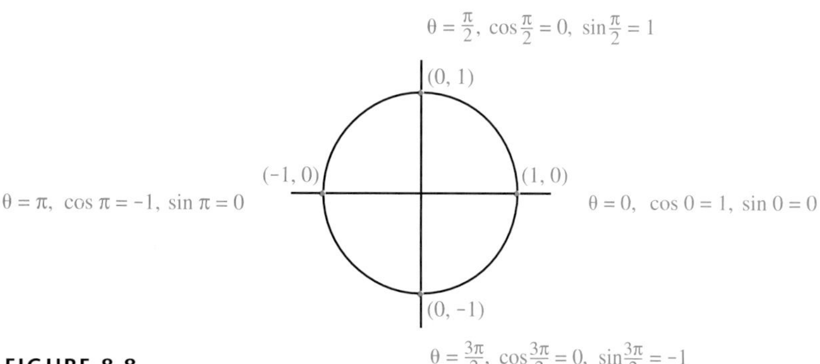

FIGURE 8.8

Any point on the unit circle has multiple corresponding angles with the same initial and terminal sides. For example, an angle of $\frac{5\pi}{2}$ is 1 complete revolution plus a quarter of a revolution $\left(2\pi + \frac{\pi}{2}\right)$, resulting in the same point on the unit circle as the angle $\frac{\pi}{2}$. Hence $\cos\frac{5\pi}{2} = \cos\frac{\pi}{2} = 0$.

The values of the sine and cosine functions are not obvious for points other than those shown in Figure 8.8. You can use a calculator or computer to find other sine and cosine values, as illustrated in Example 2. Be sure your technology is set in radian mode.

EXAMPLE 2 *Sine and Cosine Values*

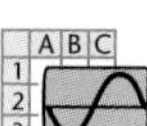

a. Find $\sin\frac{9\pi}{8}$ and $\cos\frac{9\pi}{8}$.

b. Interpret your answer to part *a* in terms of the coordinates of a point on the unit circle.

c. Find two other angles whose sine and cosine are the same as those in part *a*.

Solution

a. Using a calculator or computer you should find that $\sin\frac{9\pi}{8} \approx -0.38268$ and $\cos\frac{9\pi}{8} \approx -0.92388$.

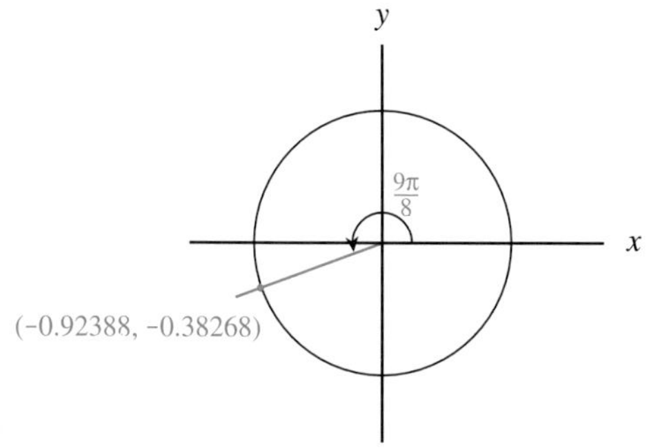

FIGURE 8.9

b. The cosine and sine values are the *x*- and *y*-coordinates of the point where the terminal side of the angle $\frac{9\pi}{8}$ intersects the unit circle. This angle and the corresponding point are shown in Figure 8.9. Note that because the *x*- and *y*-coordinates of the point are negative, both the cosine and the sine of this angle are negative.

c. We seek two other angles corresponding to the point where the terminal side of the angle shown in Figure 8.9 intersects the unit circle. One such angle is obtained by a clockwise rotation from the positive *x*-axis. This angle is $2\pi - \frac{9\pi}{8} = \frac{7\pi}{8}$ and is expressed as a negative number to denote the direction. Thus the angle is $\frac{-7\pi}{8}$, and $\sin\left(\frac{-7\pi}{8}\right) = \sin\frac{9\pi}{8}$ and $\cos\left(\frac{-7\pi}{8}\right) = \cos\frac{9\pi}{8}$.

It is also possible to reach the point under consideration by going around the circle 1 full rotation counterclockwise and then an additional $\frac{9\pi}{8}$ radians. This angle is $2\pi + \frac{9\pi}{8} = \frac{25\pi}{8}$. Thus $\sin \frac{25\pi}{8} = \sin \frac{9\pi}{8}$ and $\cos \frac{25\pi}{8} = \cos \frac{9\pi}{8}$. There are infinitely more positive and negative angles corresponding to the point indicated in Figure 8.9, all with a sine value of approximately -0.38268 and a cosine value of approximately -0.92388.

So far we have referred to $f(\theta) = \sin \theta$ and $g(\theta) = \cos \theta$ as functions, but we have not verified that this is the case. Recall that a function is a rule that assigns to each input exactly one output. To verify that $f(\theta) = \sin \theta$ is a function, we must ask, "Can $\sin \theta$ have more than one value for a particular angle θ?" Because the terminal side of the angle θ intersects the unit circle at a single point, there will be only one output $f(\theta) = \sin \theta$ corresponding to each angle θ. Thus $f(\theta) = \sin \theta$ is indeed a function. Similar reasoning confirms that $g(\theta) = \cos \theta$ is also a function.

You should be careful not to confuse inputs of these functions. Even though angles such as $\frac{\pi}{6}$ and $\frac{13\pi}{6}$ have the same initial and terminal sides and therefore the same trigonometric function outputs, $\frac{\pi}{6}$ is $\frac{1}{12}$ of a revolution, whereas $\frac{13\pi}{6}$ represents 1 complete revolution plus $\frac{1}{12}$ additional revolution. The two angles are certainly not the same. For cyclic functions, infinitely many distinct inputs correspond to the same output.

Activity 12 asks the student to graph the cosine function. Activity 14 asks the student to identify properties of the cosine graph.

To graph the cosine function, we put angle measures on the horizontal axis and put the corresponding x-coordinate on the unit circle on the vertical axis. In a graph of the sine function, the vertical axis represents the y-coordinates on the unit circle. We graph the sine function here and leave the cosine function graph for the activities.

Refer to Figure 8.8. If we plot the angles shown in that figure and the corresponding y-coordinates, we obtain the graph shown in Figure 8.10. We also add the point $(2\pi, 0)$ corresponding to 1 complete revolution.

By means of technology, more points can be added to this graph. Table 8.1 shows selected angles between 0 and 2π (corresponding to sixteenths of a revolution) and the associated y-coordinates (to five decimal places) on a unit circle.

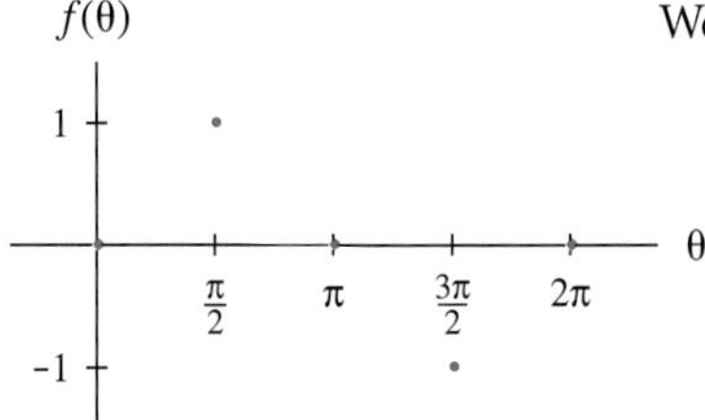

Points on the graph of $f(\theta) = \sin \theta$

FIGURE 8.10

TABLE 8.1

Angle θ	y-coordinate on unit circle $\sin \theta$	Angle θ	y-coordinate on unit circle $\sin \theta$
0	0		
$\frac{\pi}{8}$	0.38268	$\frac{9\pi}{8}$	-0.38268
$\frac{\pi}{4}$	0.70711	$\frac{5\pi}{4}$	-0.70711
$\frac{3\pi}{8}$	0.92388	$\frac{11\pi}{8}$	-0.92388
$\frac{5\pi}{8}$	0.92388	$\frac{13\pi}{8}$	-0.92388
$\frac{3\pi}{4}$	0.70711	$\frac{7\pi}{4}$	-0.70711
$\frac{7\pi}{8}$	0.38268	$\frac{15\pi}{8}$	-0.38268
π	0	2π	0

Figure 8.11 shows the points in Table 8.1 added to the graph in Figure 8.10.

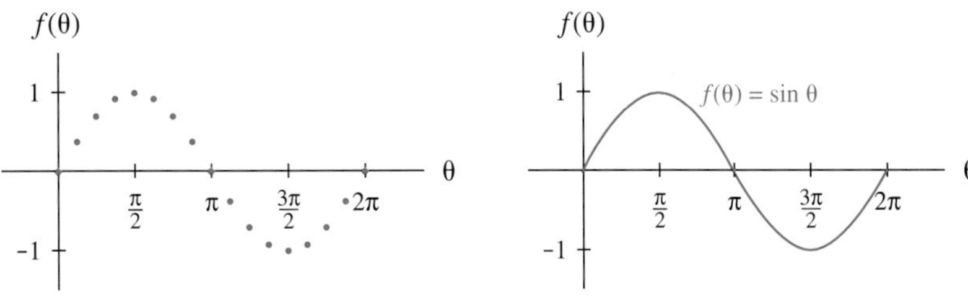

FIGURE 8.11

FIGURE 8.12

Because the y-coordinates increase, decrease, and increase again and take on all real number values between -1 and 1 as we move around the circle, we expect the sine function graph to increase and decrease in a smooth, continuous manner, taking on all values between -1 and 1. We therefore connect these points with a smooth curve to obtain the graph of the sine function shown in Figure 8.12.

Once the angle exceeds 2π, we begin retracing the unit circle, and the sine function begins repeating itself. Figure 8.13 shows a sine graph with more of the repetitions. This graph extends infinitely far in both directions.

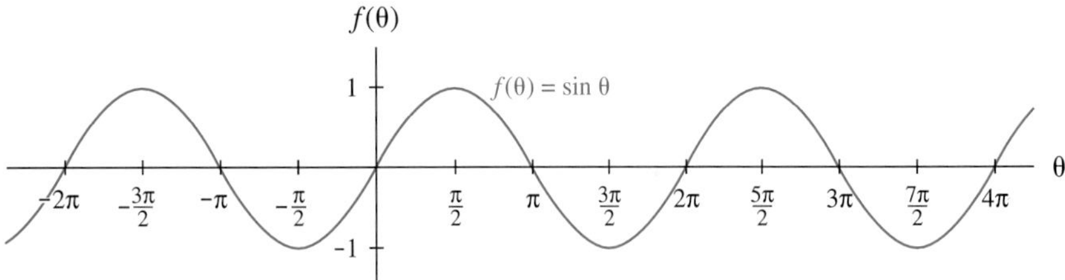

FIGURE 8.13

The sine function is periodic because it repeats itself every 2π input units. The period of the sine function is 2π. The sine function is also cyclic, because it varies continuously, alternating between -1 and 1. The portion of the sine function over one period—that is, the part that keeps repeating itself—is called a **cycle** of the function.

Although we have viewed the sine and cosine functions as having angle measure as input, we are not restricted to this interpretation. In fact, we can consider the sine and cosine functions as having any real number as the input. Because most applications to which we will apply a sine model have input that is not an angle measure, we will no longer use θ to denote the input. Instead we use the notation $y = \sin x$ and $y = \cos x$. Do not confuse x and y here with points on the unit circle. The variable x is simply the input, which can be interpreted as an angle measured in radians, and the variable y is the output, which can be interpreted as the x- or y-coordinate on the unit circle, depending on whether we are considering a cosine function or a sine function.

Variations of the Sine Function

The sine function can be modified to vary between extremes other than -1 and 1, to shift up or down, to repeat itself more or less often than 2π, and to shift right or left. The ability to modify the sine function in these ways will be valuable when we use a sine function to model data.

Multiplying the function $y = \sin x$ by a constant changes the extremes between which the output values of the function vary. In particular, the function $y = a \sin x$ varies between $-a$ and a. The parameter a is called the **amplitude** (unless a is negative, in which case the amplitude is $|a|$), and its effect is illustrated in Figure 8.14. A negative value of a indicates that the graph is reflected across the horizontal axis.

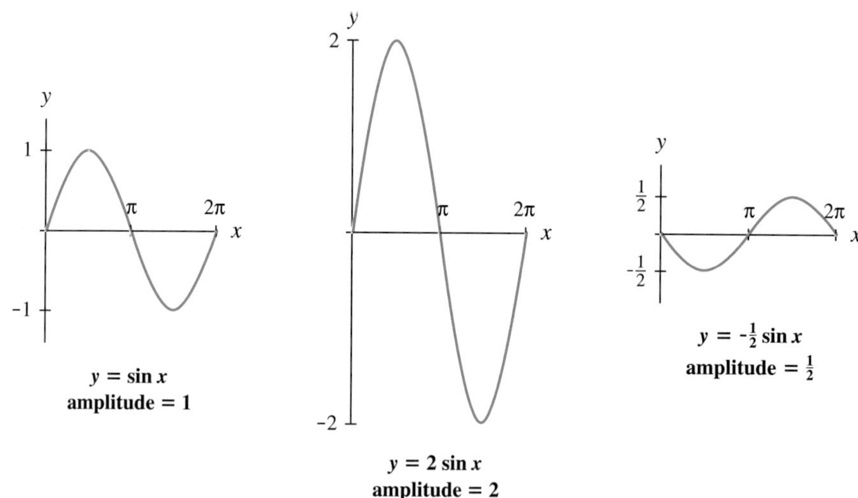

FIGURE 8.14

Point out to your students that *expected value* and *average value* have the same meaning in this context. Understanding these two terms will be important for the discussion of average value in Section 8.5. Also, point out that for a sine function, the average value of any one cycle occurs at the inflection points.

Adding a constant to, or subtracting a constant from, the sine function shifts the graph vertically up or down. This constant can be thought of as the **average** or **expected value** of one cycle. It is the midpoint between the high and low values of the function. Figure 8.15 illustrates the effect of adding or subtracting a constant to or from a sine function.

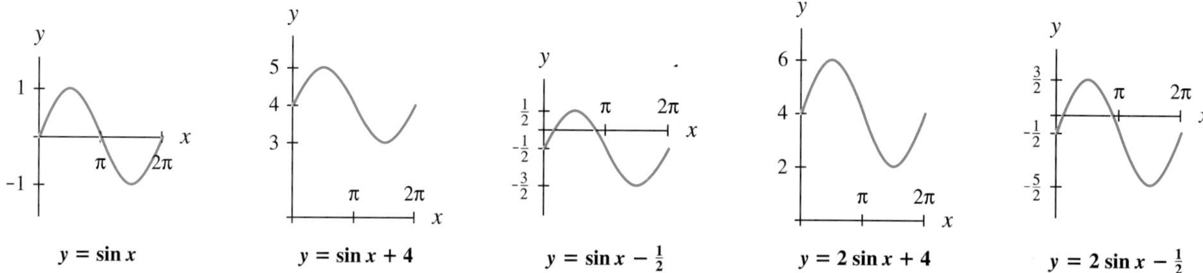

FIGURE 8.15

The **frequency** of a sine or cosine function is the number of cycles of the graph on an interval of length 2π. Thus the frequency of the function $y = \sin x$ is 1. Multiplying the input by a constant changes the frequency to that constant. In other words, if you want the function to cycle twice every 2π units, multiply x by 2. If you want to elongate the function so that one cycle stretches over 2 periods (4π), multiply by $\frac{1}{2}$.

Students may have previously heard the term *frequency* defined with simple harmonic motion as the number of cycles of a function per unit time or as the reciprocal of the period. You may need to contrast the use of the term *frequency* here with its use elsewhere.

The frequency is $\frac{1}{2}$ because only half of a cycle would cover an interval of length 2π. These examples illustrate the fact that the function $y = \sin bx$ has frequency b. (We assume that $b > 0$.) To find the period of $y = \sin bx$, we simply divide 2π by b:

$$\text{Period} = \frac{2\pi}{b}$$

Similarly, if we know the period, then we can find b by using the equation

$$b = \frac{2\pi}{\text{period}}$$

Figure 8.16 illustrates several functions of the form $y = \sin bx$.

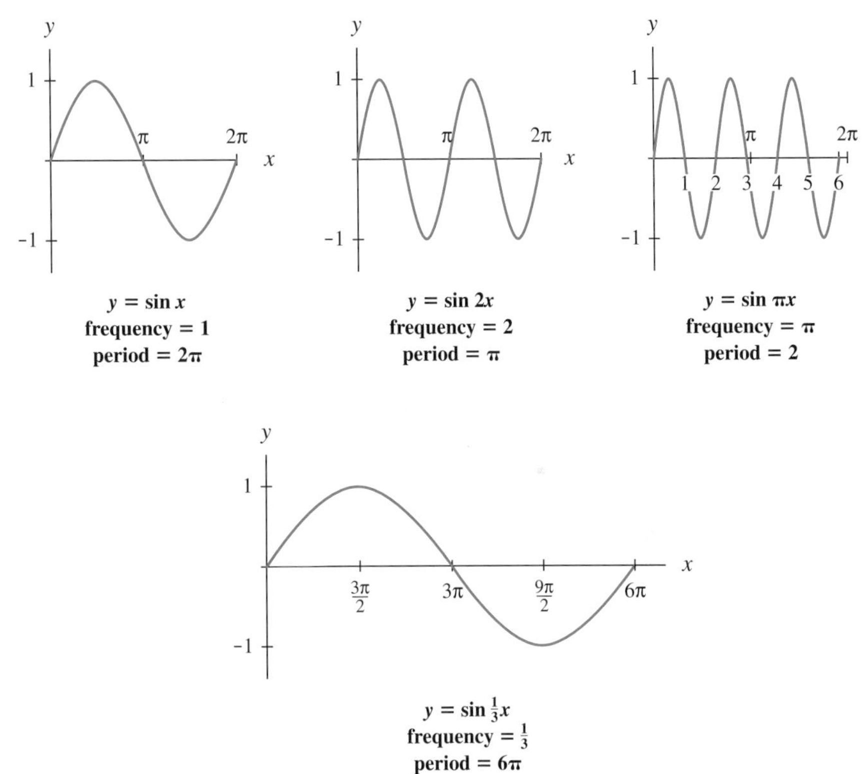

FIGURE 8.16

Finally, we can shift a sine function horizontally by adding a constant to the input. Specifically, for $y = \sin(bx + h)$, the graph is shifted horizontally by $\frac{|h|}{b}$ units. If h is negative, the shift is to the right. If h is positive, the shift is to the left. One way to identify the horizontal shift of a graph is to find the location of an inflection point at which the function is increasing. Because the function is cyclic, an inflection point at which the function is increasing is not unique. One such inflection point occurs at $x = 0$ for $y = \sin x$, and one occurs at $x = \frac{-h}{b}$ for $y = \sin(bx + h)$. Figure 8.17 illustrates several functions with horizontal shifts.

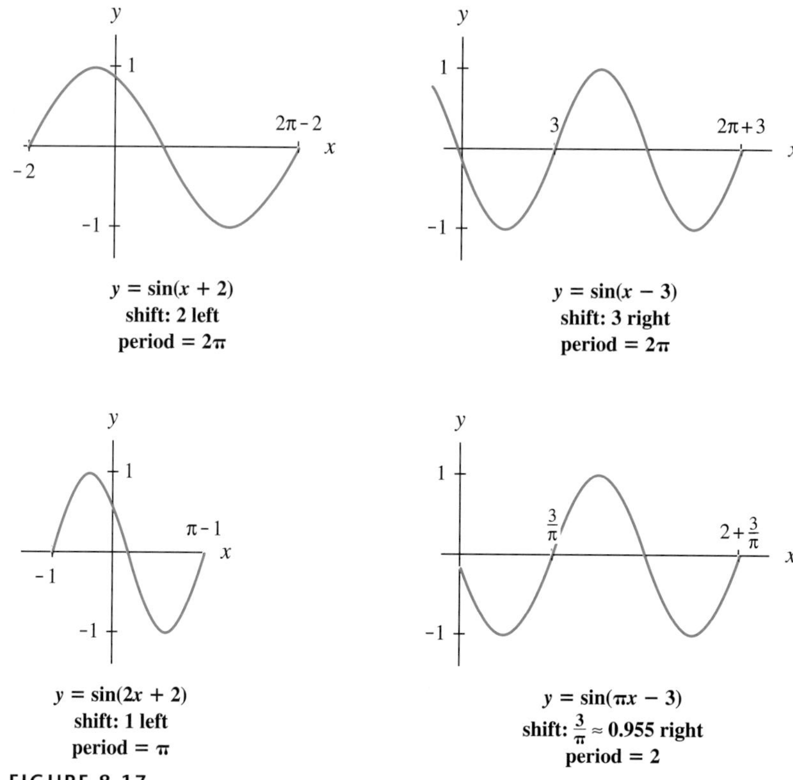

$y = \sin(x + 2)$
shift: 2 left
period $= 2\pi$

$y = \sin(x - 3)$
shift: 3 right
period $= 2\pi$

$y = \sin(2x + 2)$
shift: 1 left
period $= \pi$

$y = \sin(\pi x - 3)$
shift: $\frac{3}{\pi} \approx 0.955$ right
period $= 2$

FIGURE 8.17

There is one final variation of the sine function to consider. The functions $y = \sin(-x)$ and $y = -\sin x$ are equivalent functions. That is, $\sin(-x) = -\sin x$. The graph of each of these functions is the reflection of the graph of $y = \sin x$ across the horizontal axis. See Figure 8.18.

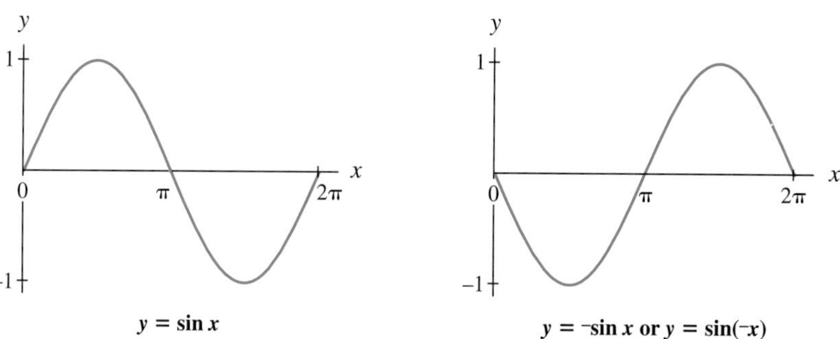

$y = \sin x$

$y = -\sin x$ or $y = \sin(-x)$

FIGURE 8.18

We earlier stated that we assume b is positive. This restriction does not limit the functions we consider because a negative value for either a or b in $y = a\sin(bx + h) + k$ indicates reflection. If we encounter a function in which b is negative, we simply factor the negative one out of the $(bx + h)$ term and use the fact that $\sin(-x) = -\sin x$ to rewrite the function with a positive b. For example, $y = -2\sin(-3x + 4) - 6$ can be rewritten as

$$y = -2\sin[-(3x - 4)] - 6 = -[-2\sin(3x - 4)] - 6 = 2\sin(3x - 4) - 6$$

Consider the cycle of the graph $y = a\sin(bx + h) + k$ that begins at $x = \frac{-h}{b}$. The output value is always k at the point with this input. If a is positive, this cycle will first increase and then decrease. If a is negative, this cycle is reflected across the line $y = k$, so it will first decrease and then increase.

We summarize the effects of these parameters:

Variations of the Sine Function

The parameters of the sine function[1]

$$f(x) = a\sin(bx + h) + k$$

are interpreted as follows (we assume $b > 0$):

$|a|$ is the amplitude where $a < 0$ denotes reflection across the horizontal axis
b is the frequency, and $\frac{2\pi}{b}$ is the period
$\frac{|h|}{b}$ is the horizontal shift, right if $h < 0$ and left if $h > 0$
k is the vertical shift, up if $k > 0$ and down if $k < 0$

EXAMPLE 3 Variations of the Sine Function

For each of the following sine functions, specify the amplitude, frequency, period, horizontal shift, and vertical shift. Graph one cycle of each function.

a. $y = 3\sin x - 4$ b. $y = \sin(-2x - 6)$ c. $y = 0.5\sin(0.2x - 1) + 2.5$

Solution

a. The parameters of $y = 3\sin x - 4$ are $a = 3$, $b = 1$, $h = 0$, and $k = -4$. Thus the amplitude of this function is 3, and the graph is shifted down 4 units. Without the shift down, the function would vary between -3 and 3. Subtracting 4 from each of these values, we see that the graph of $y = 3\sin x - 4$ varies between -7 and -1. The period is 2π, and the frequency is 1. We choose to graph the cycle that begins at 0 and ends at 2π. Because a is positive, this cycle begins with the graph increasing. The graph is shown in Figure 8.19.

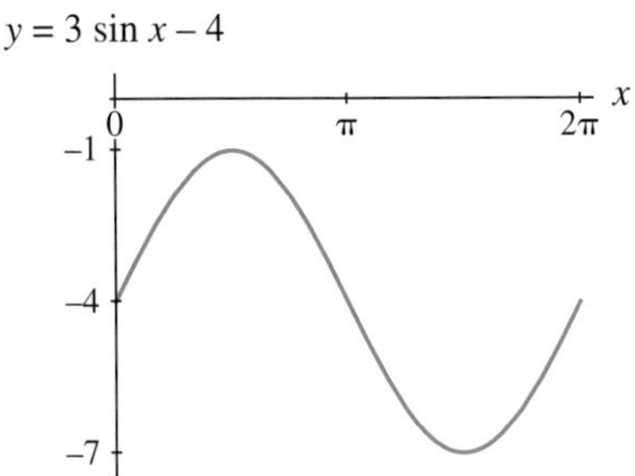

FIGURE 8.19

1. Instead of the equation $y = a\sin(bx + h) + k$, some technologies use $y = a\sin b(x - h) + k$. The advantage in this alternative form is that the quantities influenced by the parameter have direct readouts. That is, in the equation $y = a\sin b(x - h) + k$, h is precisely the horizontal shift. In the first formulation, the one used in the text, $|\frac{h}{b}|$ is the horizontal shift.

b. Because b is negative in this function, we factor out the negative one and use the property that $\sin(-x) = -\sin x$:

$$\sin(-2x - 6) = \sin[-(2x + 6)] = -\sin(2x + 6)$$

Thus we consider the function in the form $y = -\sin(2x + 6)$. For this function $a = -1$, $b = 2$, $h = 6$, and $k = 0$. The amplitude of $y = -\sin(2x + 6)$ is 1. There is no vertical shift, so the function varies between -1 and 1. The frequency is 2, and the period is $\frac{2\pi}{2} = \pi$. The graph is shifted horizontally $\frac{|6|}{2} = 3$ units, and the shift is to the left because 6 is positive. Without the horizontal shift, one cycle would begin at 0 and end at π. With a horizontal shift of 3 units to the left, one cycle begins at -3 and ends at $\pi - 3$. Because a is negative, this cycle begins by declining rather than rising because it is reflected across the horizontal axis. (See Figure 8.20.)

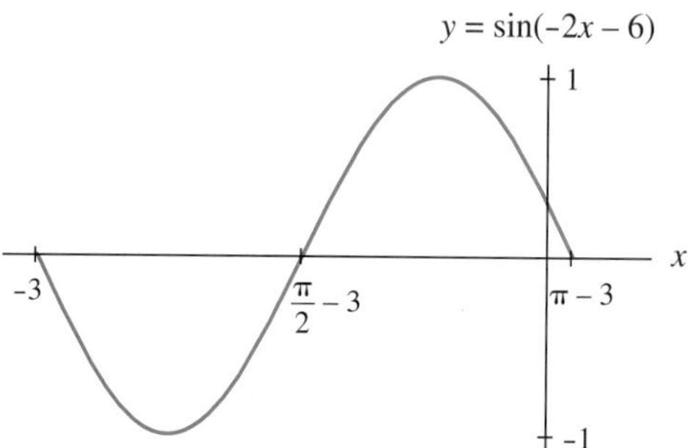

FIGURE 8.20

c. For the function $y = 0.5 \sin(0.2x - 1) + 2.5$, the parameters are $a = 0.5$, $b = 0.2$, $h = -1$, and $k = 2.5$. The graph has amplitude 0.5 and is shifted up 2.5 units. Thus the function varies between $-0.5 + 2.5 = 2$ and $0.5 + 2.5 = 3$. The frequency of the function is 0.2, and the period is $\frac{2\pi}{0.2} = 10\pi$. The graph is shifted horizontally $\frac{|-1|}{0.2} = 5$ units, and the shift is to the right because -1 is negative. Without the horizontal shift, one cycle begins at 0 and ends at 10π. With the horizontal shift 5 units to the right, one cycle begins at 5 and ends at $10\pi + 5$. Because a is positive, this cycle begins with the graph increasing. The graph is shown in Figure 8.21.

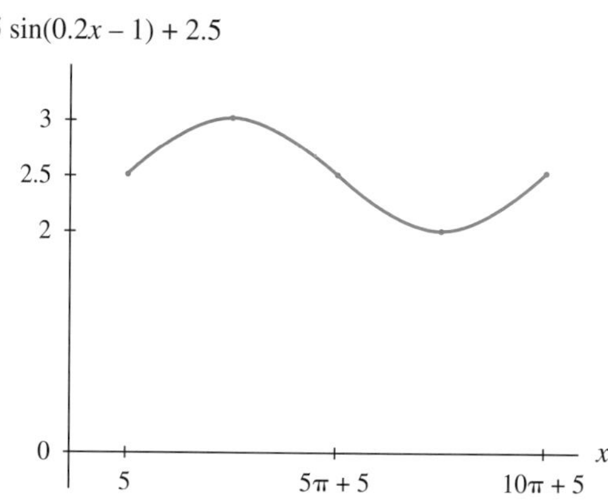

FIGURE 8.21

Now that we have seen the basic form of the sine function $f(x) = a \sin(bx + h) + k$ and understand how the parameters are related to the shape of a graph of the function, we can consider the usefulness of the sine function when modeling cyclic data. In the next section, we look at modeling various sets of cyclic data with sine functions.

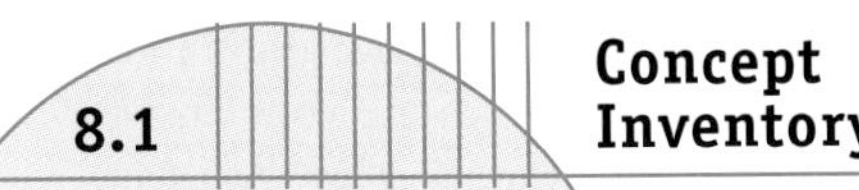

8.1 Concept Inventory

- *Periodic function*
- *Cyclic function*
- *Radian measure of angles*
- *Unit-circle definition of the sine and cosine of an angle*
- *Sine and cosine functions*
- *Sine function graph*
- *Sine function variations: $y = a\sin(bx + h) + k$*

Amplitude	Horizontal shift
Frequency	Vertical shift
Period	Expected value
Cycle	Reflection

8.1 Activities

1. a. Use the graph in Figure 8.1.1 to estimate sin 4, cos 4, $\sin\left(\frac{-\pi}{8}\right)$, and $\cos\left(\frac{-\pi}{8}\right)$.

   b. Mark the estimated locations of the angles $\theta = -1$ and $\theta = \frac{144\pi}{16}$ on Figure 8.1.1, and use the grid to estimate the values of sin(-1), cos(-1), $\sin\frac{144\pi}{16}$, and $\cos\frac{144\pi}{16}$.

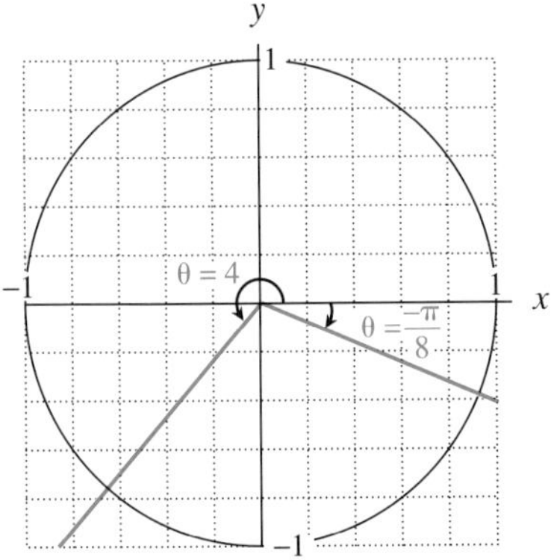

FIGURE 8.1.1

2. a. Use the graph in Figure 8.1.2 to estimate sin 2.5, cos 2.5, $\sin\left(\frac{-15\pi}{16}\right)$, and $\cos\left(\frac{-15\pi}{16}\right)$.

   b. Mark the estimated locations of the angles $\theta = \frac{-1}{2}$ and $\theta = \frac{-5\pi}{12}$ on Figure 8.1.2, and use the grid to estimate the values of $\sin\left(\frac{-1}{2}\right)$, $\cos\left(\frac{-1}{2}\right)$, $\sin\left(\frac{-5\pi}{12}\right)$, and $\cos\left(\frac{-5\pi}{12}\right)$.

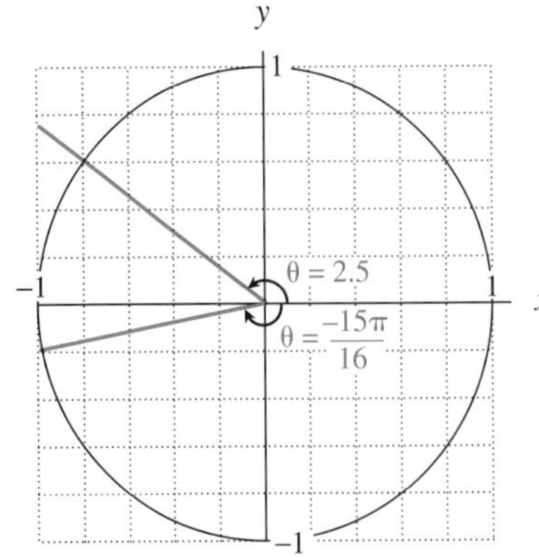

FIGURE 8.1.2

3. a. Find the values of $\sin\frac{2\pi}{3}$ and $\cos\frac{2\pi}{3}$.

   b. Interpret each answer to part *a* in terms of the coordinates of a point on the unit circle.

   c. Give two other angles whose sine and cosine values are the same as those for $\theta = \frac{2\pi}{3}$.

4. a. Find the values of sin 15 and cos 15.

   b. Interpret each answer to part *a* in terms of the coordinates of a point on the unit circle.

   c. Give two other angles whose sine and cosine values are the same as those for $\theta = 15$.

5. a. Find the values of $\sin\frac{3\pi}{2}$ and $\cos\frac{3\pi}{2}$.

   b. Interpret the answers to part *a* in terms of the unit circle.

   c. Give three other angles whose sine and cosine values are the same as those for $\theta = \frac{3\pi}{2}$.

6. a. Find the values of $\sin\frac{4\pi}{9}$ and $\cos\frac{4\pi}{9}$.

   b. Interpret the answers to part *a* in terms of the unit circle.

   c. Give three other angles whose sine and cosine values are the same as those for $\theta = \frac{4\pi}{9}$.

7. Is it possible for the trigonometric functions of an angle θ to be $\cos\theta = 0.5$ and $\sin\theta = -0.5$? Explain.

The equation $\cos^2\theta + \sin^2\theta = 1$ is found below the box on page 562.

8. Is it possible for the trigonometric functions of an angle θ to be $\cos\theta = 0.35$ and $\sin\theta = 0.82$? Explain.

9. a. Without using technology, tell whether each of the following values is positive, negative, or zero by using its interpretation as the *y*-coordinate of a point on the unit circle.

Trig value	sin 0	$\sin\frac{\pi}{4}$	$\sin\frac{\pi}{2}$	$\sin\frac{3\pi}{4}$	$\sin\pi$
Sign					
Trig value	$\sin\frac{5\pi}{4}$	$\sin\frac{3\pi}{2}$	$\sin\frac{7\pi}{4}$	$\sin 2\pi$	
Sign					

b. Use your calculator or computer to complete the following table. Compare the signs of the values to your results from part *a*.

Trig value	sin 0	sin $\frac{\pi}{4}$	sin $\frac{\pi}{2}$	sin $\frac{3\pi}{4}$	sin π
Decimal value					
Trig value	sin $\frac{5\pi}{4}$	sin $\frac{3\pi}{2}$	sin $\frac{7\pi}{4}$	sin 2π	
Decimal value					

c. Plot the values you obtained in part *b* as a function of the angle. How is this plot related to the graph of the function $f(x) = \sin x$?

10. a. Without using technology, tell whether each of the following values is positive, negative, or zero by using its interpretation as the *x*-coordinate of a point on the unit circle.

Trig value	cos 0	cos $\frac{\pi}{4}$	cos $\frac{\pi}{2}$	cos $\frac{3\pi}{4}$	cos π
Sign					
Trig value	cos $\frac{5\pi}{4}$	cos $\frac{3\pi}{2}$	cos $\frac{7\pi}{4}$	cos 2π	
Sign					

b. Use technology to complete the following table. Compare the signs of the values to your results from part *a*.

Trig value	cos 0	cos $\frac{\pi}{4}$	cos $\frac{\pi}{2}$	cos $\frac{3\pi}{4}$	cos π
Decimal value					
Trig value	cos $\frac{5\pi}{4}$	cos $\frac{3\pi}{2}$	cos $\frac{7\pi}{4}$	cos 2π	
Decimal value					

11. Construct an input/output diagram for the function $s(x) = \sin x$.

12. a. Construct an input/output diagram for the function $c(x) = \cos x$.

b. On axes like those in Figure 8.10, plot the four cosine values for the points labeled in Figure 8.8.

c. Make a table of values for the cosine function for the angles shown in Table 8.1.

d. Add the points in part *c* to the plot in part *b*. Sketch a smooth curve through the points.

e. Use technology to verify that the graph in part *d* is a graph of the cosine function.

f. Sketch a graph of $c(\theta) = \cos\theta$ for θ between -2π and 6π.

g. What could you do to the cosine graph in order to obtain the sine graph? State your answer in words and as an equation.

13. a. For *x*-values from 0 to 2π, what is the largest value that $s(x) = \sin x$ achieves? Also find the value(s) of *x* at which the maximum occurs.

b. For any real number *x*, what is the largest value that $s(x) = \sin x$ achieves? Also find the value(s) of *x* at which the maximum occurs.

c. For *x*-values from 0 to 2π, what is the smallest value that $s(x) = \sin x$ achieves? Also find the value(s) of *x* at which the minimum occurs.

d. For any real number *x*, what is the smallest value that $s(x) = \sin x$ achieves? Also find the value(s) of *x* at which the minimum occurs.

e. Relate the results of parts *a–d* to the nature of the circle on which the trigonometric function $s(\theta) = \sin\theta$ achieves its values.

14. Repeat Activity 13 for the cosine function.

15. a. Graph $c(x) = \cos x$ and $s(x) = \sin x$ on the same axes for $0 \le x \le 2\pi$.

b. Find where $c(x) = s(x)$.

c. On the basis of your answer to part *b*, what are the *x*-coordinates of all the points of intersection for the sine and cosine functions over all real numbers?

For each of the functions in Activities 16 through 20, state the amplitude, frequency, period, vertical shift, and horizontal shift. Also state whether the function is reflected across the horizontal axis and whether the graph is increasing or decreasing at $x = \frac{-h}{b}$.

16. $f(x) = 0.1\sin(2x - 4) - 0.5$

17. $g(x) = -\sin(x - \pi)$

18. $g(x) = 3.62\sin(0.22x + 4.81) + 7.32$

19. $p(x) = 235\sin(-300x + 100) - 65$

20. $p(x) = \sin(-x + 4) + 0.7$

21. Match functions *a–f* with graphs i–vi in Figure 8.1.3.

a. $y = 3\sin x + 8$
b. $y = 3\sin x - 3$
c. $y = 3\sin x + 2$
d. $y = 4\sin x + 8$
e. $y = 2\sin x - 3$
f. $y = 2\sin x + 2$

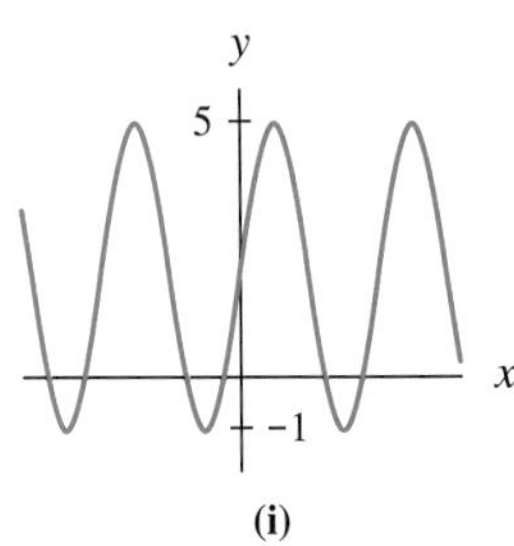

(i)

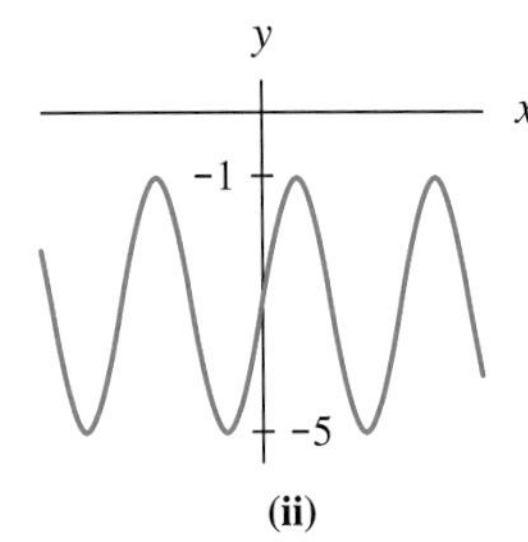

(ii)

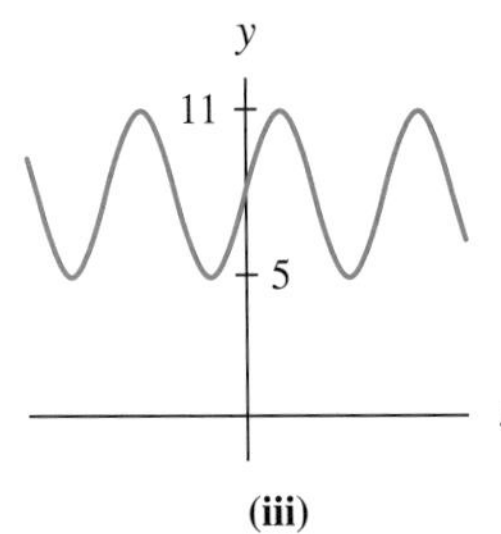

(iii)

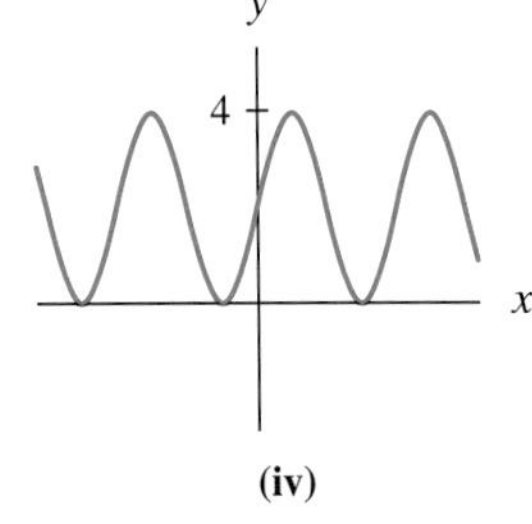

(iv)

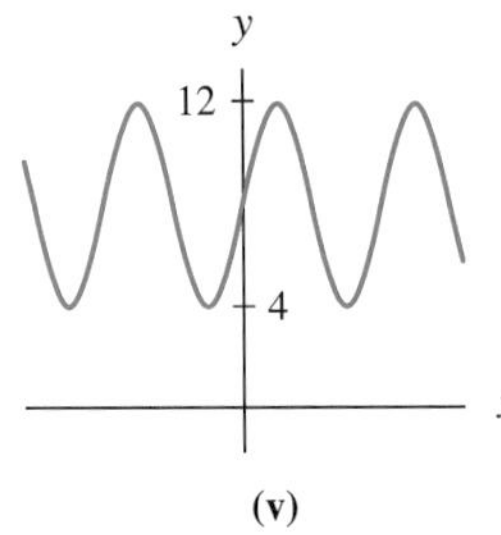

(v)

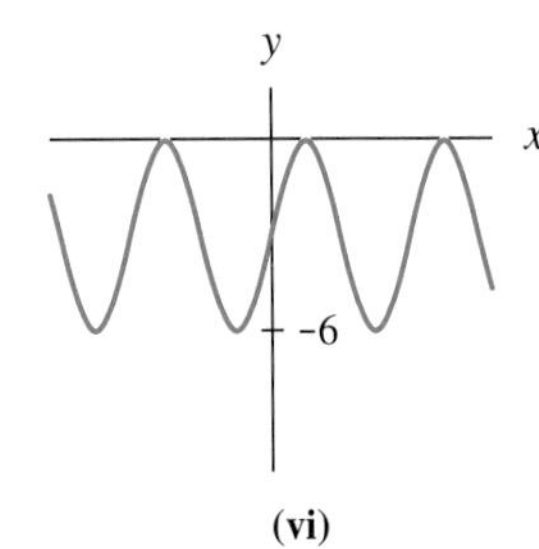

(vi)

FIGURE 8.1.3

22. a. What is the distance between two consecutive repeating maxima of the function $f(x) = \sin(2x)$? the function $g(x) = \sin(0.5x)$?

    b. For the function $h(x) = \sin\left(\frac{1}{3}x + \frac{2}{3}\right)$ and the function $j(x) = \sin\left(\frac{1}{3}x - \frac{1}{3}\right)$, estimate the locations of the points of greatest slope closest to $x = 0$.

    c. Match the four functions given in this Activity with graphs i–iv in Figure 8.1.4.

Without the aid of technology, graph one cycle of each of the functions in Activities 23 through 28.

23. $f(x) = \sin 2x + 3$
24. $f(x) = -2 \sin x + 3$
25. $f(x) = 2 \sin(x + 3)$
26. $f(x) = \sin(-2x + 3)$
27. $f(x) = 2 \sin(-2x + 3)$
28. $f(x) = -2 \sin(-2x + 3) + 3$
29. Compare the graph of $y = -2 \sin x$ with the graph of $y = 2 \sin(x - \pi)$. Also compare the graphs of $y = -0.5 \sin x$ and $y = 0.5 \sin(x + 3\pi)$. What generalization can you make from your observations?

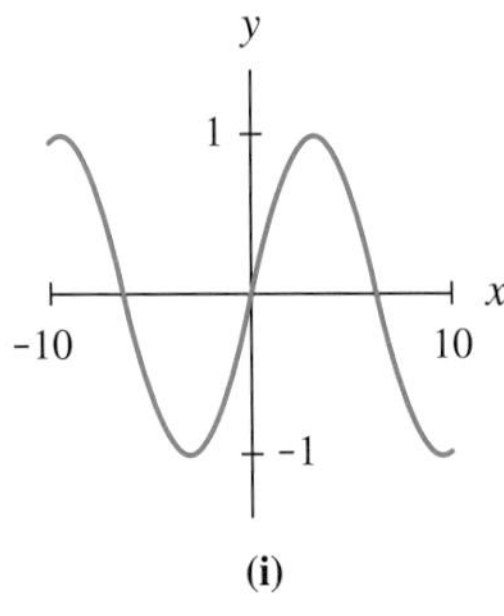

(i)

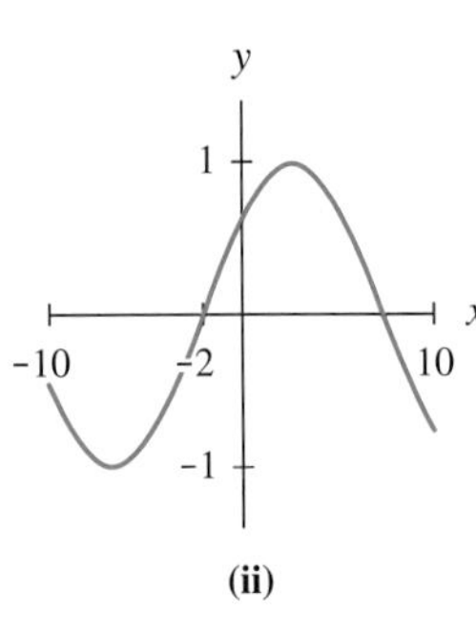

(ii)

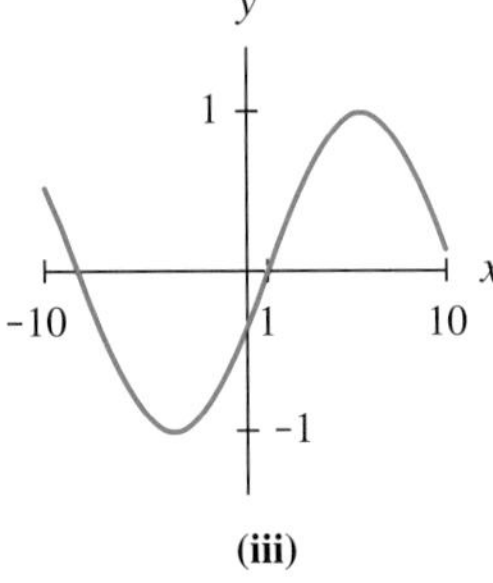

(iii)

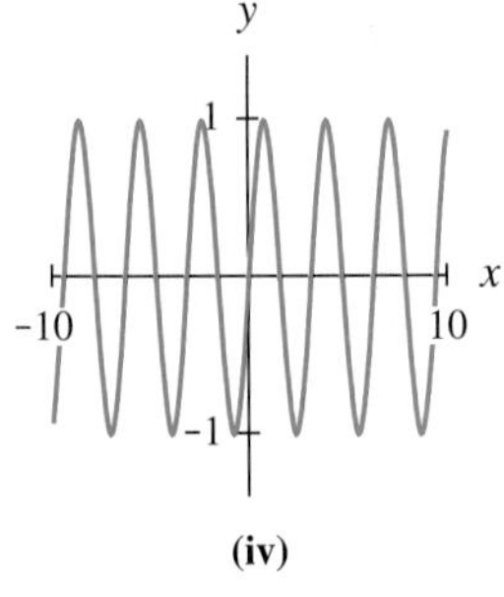

(iv)

FIGURE 8.1.4

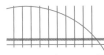

8.2 Cyclic Functions as Models

Be aware that not all technologies have a built-in sine regression routine. Students who use such technologies should obtain a sine curve-fitting program. See the technology supplements or the *Calculus Concepts* web site for more information.

The world is full of cyclic data that repeatedly return to the same maximum and minimum. Our ability to analyze such data is helpful. If we learn to model these data, then we can use calculus and the model to obtain special insight into the changing nature of the data. The sine and cosine functions are cyclic, and they are useful in modeling data that exhibit cyclic behavior. Because these two functions are similar (one can be obtained from the other by a horizontal shift), they are equally useful in

Your students may wonder why technologies use the sine function instead of the cosine function in curve fitting. Although there are reasons for this (such as the origin being a point on $y = \sin x$ and the positive cosine being the derivative), assure your students that technology programmers could have chosen the cosine function rather than the sine. One has no great advantages over the other.

modeling. However, the technologies that include a cyclic function in their regression library use the sine, rather than the cosine function. Therefore, we choose to use only the sine function when modeling cyclic data.

The Sine Model

In Section 8.1 we examined the sine function of the form $y = a \sin(bx + h) + k$. We saw that the parameter a determines the height and reflection of the curve, the parameter b is the frequency that determines the period, the parameter h influences the horizontal shift, and the parameter k is the vertical shift. We now consider using this function to model data in real-life situations.

The monthly amount[2] of ultraviolet (UV) radiation (in watts per square centimeter) received by a horizontal surface of area 1 cm^2 in Auckland, New Zealand, is given in Table 8.2, and a scatter plot of these data is shown in Figure 8.22.

TABLE 8.2

Month	Jan	Feb	Mar	Apr	May	Jun	Jul	Aug	Sep	Oct	Nov	Dec
UV radiation (watts per cm^2)	55	46	32	19	10	7	8	15	27	40	52	58

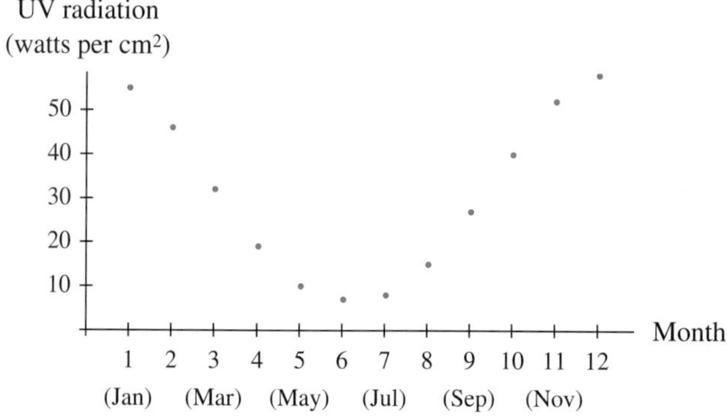

FIGURE 8.22

Explain to your students that ultraviolet radiation is based on the tilt of the earth toward or away from the sun. Physical phenomena that depend on either the earth's rotation on its axis or its revolution around the sun are probably cyclic. You may wish to ask your students to name as many such phenomena as they can. Many of them appear in the examples and activities.

We see from the data and scatter plot that the data have a minimum and a maximum. We assume that the amount of UV radiation reaches this same maximum and minimum each year; hence the data are cyclic. Because the repetitions occur on a yearly basis, the period is 12 months. Thus it makes sense to fit a sine function to the data. Before using technology to fit the function $f(x) = a \sin(bx + h) + k$, let us inspect the data more carefully to see if we can estimate the parameters a, b, h, and k.

We begin our inspection by looking for k. The parameter k is the vertical shift of the sine function. It tells us how high (or low) the middle value of the output range is. This is the average (or expected) value of the cycle. In order to determine k, we must know the maximum and the minimum output. The maximum of the data occurs in December at 58 watts per cm^2, and the minimum occurs in June at 7 watts per cm^2. To find the middle value of the output, we average the maximum and minimum values.

2. W. Rudloff, *World-Climates*, (Stuttgart: Wissenschaftliche Verlagsgesellschaft, 1981).

If you wish, you may omit the discussion of finding the parameters of a sine function from data. This omission is appropriate if you are presenting the sine function as simply another choice of model type and are not concerned with your students understanding the role each parameter plays in the function. See the *Instructor's Resource Guide* to learn which activities are affected by such an omission.

$$k = \frac{\text{max} + \text{min}}{2} = \frac{58 + 7}{2} = 32.5 \text{ watts per cm}^2$$

In other words, if we did not take into account the cyclic fluctuations in the amount of UV radiation, we would expect the average amount to be 32.5 watts per cm^2 for any given month. However, there are monthly variations in the amount. The amplitude tells us by how much the amount may vary from the expected (or average) value.

We find the amplitude a of the sine model by calculating half the difference between the maximum and minimum values.

$$a = \frac{\text{max} - \text{min}}{2} = \frac{58 - 7}{2} = 25.5 \text{ watts per cm}^2$$

That is, UV radiation varies by $\pm$ 25.5 watts per cm^2 away from the expected value of 32.5 watts per cm^2.

Because the data are periodic with a cycle that we assume to repeat each year, the period is 12 months. We calculate the parameter b (the frequency) using the equation

$$\text{Period} = \frac{2\pi}{b}$$

Thus $12 = \frac{2\pi}{b}$, or $b = \frac{\pi}{6} \approx 0.52$.

Finally, we look for the parameter h, which is an indication of the horizontal shift of the sine function. When looking for the horizontal shift of a sine function of the form $f(x) = a\sin(bx + h) + k$, we seek an input value x_0 at which the sine function f behaves the same way $g(x) = \sin x$ behaves at $x = 0$. At $x = 0$, a graph of $g(x) = \sin x$ is increasing and has an inflection point, so we are looking for an input value x_0 at which the graph of $f(x) = a\sin(bx + h) + k$ is increasing and has an inflection point. More specifically, we seek a value for x_0 such that

1. $f(x_0) = k =$ the average value of the cycle
2. The graph of f is increasing at x_0
3. The graph of f has an inflection point at x_0; that is, the graph of f changes from concave up to concave down at x_0.[3]

On a function of the form $f(x) = a\sin(bx + h) + k$, such an input value x_0 occurs exactly halfway between the x-value at which $f(x)$ is lowest and the x-value at which $f(x)$ is highest. Thus for the UV-radiation data in Table 8.2 we find that a candidate for the horizontal shift x_0 is halfway between June and December. If we align so that $x = 1$ in January and $x = 6$ in June, then $x = 12$ in December, and $x_0 = \frac{12 + 6}{2} = 9$ (September). This shift is to the right, indicating that h is negative. To calculate h from x_0, we use the equation

$$x_0 = \text{horizontal shift} = \frac{|h|}{b}$$

with $b = \frac{\pi}{6}$ as previously calculated. Thus $9 = \frac{|h|}{\pi/6}$ or $h = -\left(\frac{\pi}{6}\right)(9) \approx -4.71$.

Now we can write a model to approximate the amount of ultraviolet radiation received in Auckland, New Zealand, as

3. For a function of the form $f(x) = a\sin(bx + h) + k$, if conditions 1 and 2 are satisfied, then condition 3 is automatically satisfied. If conditions 2 and 3 are satisfied, then condition 1 is also satisfied. However, conditions 1 and 3 being satisfied does not guarantee that condition 2 is true.

$$r(m) = 25.5 \sin(0.52m - 4.71) + 32.5 \text{ watts per cm}^2$$

when $m = 1$ in January, $m = 2$ in February, and so on. Figure 8.23 shows this equation graphed on the scatter plot of the data.

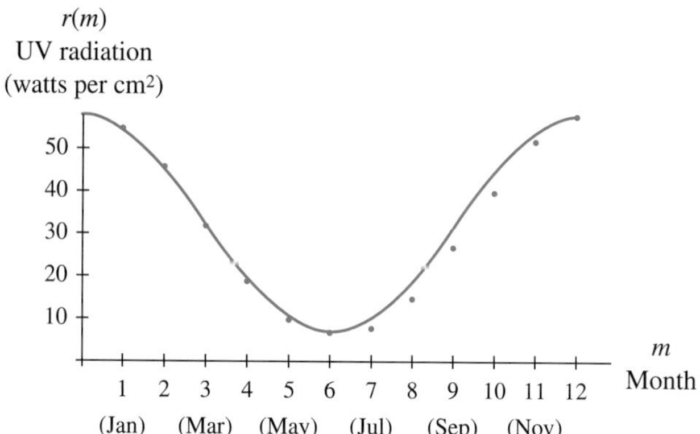

FIGURE 8.23

As we have seen, the context of a real-life situation often indicates reasonable estimates for the parameters of a sine model. Knowing the interpretation of the parameters a, b, h, and k in the general sine model $f(x) = a \sin(bx + h) + k$ allows us to calculate an equation by hand. We again illustrate these ideas in Example 1 by using the sine function to model the number of daylight hours at a location on the Arctic Circle.

EXAMPLE 1 *Hours of Daylight on the Arctic Circle*

On the Arctic Circle there are 24 hours of daylight on the summer solstice, June 21, and 24 hours of darkness on the winter solstice, December 21. We use this information, along with the fact that there are 12 hours of daylight and of darkness midway between the summer and winter solstices, to construct a model that approximates the number of hours of daylight throughout the year on the Arctic Circle. We use as input the number of days since the beginning of the year and show just over 2 years of data so that the pattern in the data can be easily seen. Note that if January 1 is day 1, then December 21 of the previous year is day -10 and December 21 of the current year is day 355. The data are given in Table 8.3, and a scatter plot of the data is shown in Figure 8.24.

TABLE 8.3

Date	12/21		6/21		12/21		6/21		12/21	
Day of the year	-10	81.5	173	264	355	446.5	538	629	720	811.5
Hours of daylight	0	12	24	12	0	12	24	12	0	12

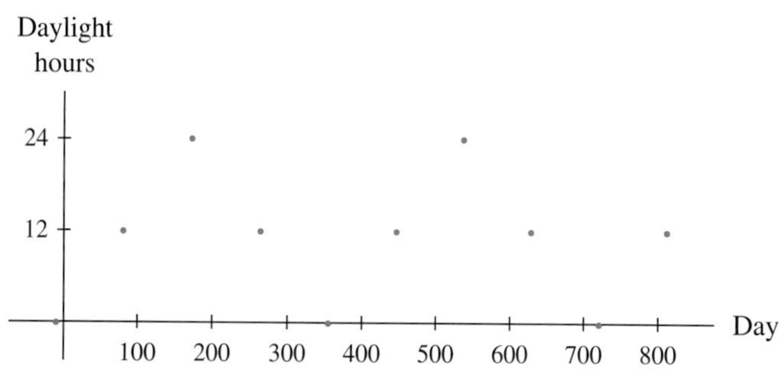

FIGURE 8.24

Use the data in Table 8.3 to calculate the parameters a, b, h, and k of a model of the form $f(x) = a \sin(bx + h) + k$.

Solution The scatter plot suggests that a sine model is appropriate. The output values of the data range from 0 to 24 hours; therefore, we expect the amplitude of our model to be $\frac{24}{2} = 12$ hours. However, without a vertical shift, the output of a sine model with amplitude 12 is between -12 and +12 hours. Because the data are between 0 and 24, we expect a vertical shift upward of 12 hours. We also expect the model to repeat itself every 365 days. To find the horizontal shift, we need to locate where the number of hours of daylight is the average value of the cycle (12 hours) and is first increasing. This occurs at 81.5 days, so 81.5 is the horizontal shift to the right. Thus in the sine function $f(x) = a \sin(bx + h) + k$, we expect

$$\text{Amplitude} = a = 12$$

$$\text{Period} = \frac{2\pi}{b} = 365, \text{ so frequency} = b \approx 0.01721$$

$$\text{Horizontal shift to the right} = \frac{|h|}{b} = 81.5, \text{ so } h \approx -1.40296$$

$$\text{Vertical shift} = k = 12$$

A model for the number of hours of daylight on the Arctic Circle is

$$h(t) = 12 \sin(0.01721t - 1.40296) + 12 \text{ hours}$$

where t is the number of days since December 31 of the previous year. Figure 8.25 shows this model graphed on the scatter plot.

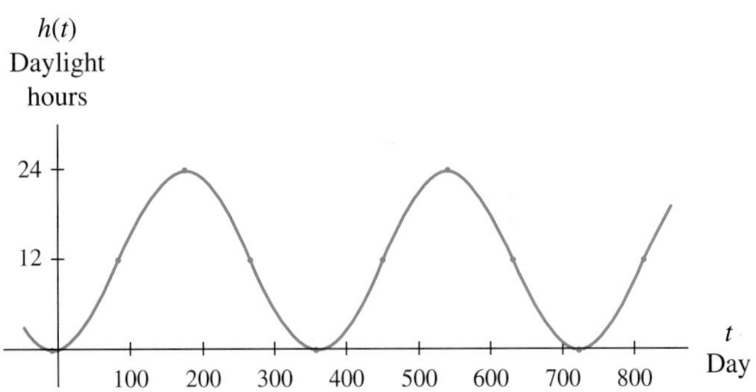

FIGURE 8.25

It may be the case that the data or information provided are not sufficient to predict accurately the maximum, minimum, or period of the cycle, or you may wish to find a model quickly. In such cases, you can use a calculator or computer to fit a sine function to the data. Keep in mind that it is still important to understand the role of the parameters a, b, h, and k of the equation so that you can interpret what the model is telling you.

At the beginning of this section we calculated a model of the data for the amount of ultraviolet radiation in Auckland, New Zealand. However, our estimates of the maximum and minimum of the function were not exact. The minimum of the function probably occurs somewhere between the June and the July data, and the maximum probably occurs somewhere between the December and the January data. A sine function fitted by technology[4] to the data in Table 8.2 is

4. In this text, all sine models found using technology were constructed on a TI-83. If you use a different technology, your model parameters may vary.

$$A(m) = 27.1 \sin(0.485m + 1.707) + 32.9 \text{ watts per cm}^2$$

when $m = 1$ in January, $m = 2$ in February, and so on.

Let us examine this model and compare the parameters to those we calculated earlier. The function A gives the average value of UV radiation as approximately 32.9 watts per cm^2 instead of the 32.5 watts per cm^2 previously calculated from the data. Ultraviolet radiation fluctuates as much as $\pm$ 27.1 watts per cm^2 above and below this average, according to the function A, giving a slightly larger amplitude than that calculated earlier. The period of the function A is $\frac{2\pi}{b} \approx \frac{2\pi}{0.485} \approx 13.0$. Thus the equation that technology fit has a period of more than 12 months.

Finally, the horizontal shift of A is $\frac{|h|}{b} \approx \frac{|1.707|}{0.485} \approx 3.5$ to the left. Because this value is outside the data interval, we add one period (approximately 13) to it in order to interpret it correctly. In this case, adding the period to the horizontal shift gives approximately 9.4. This number is within the range of our data and can be interpreted as halfway between the September data and the October data. Thus UV radiation hits its average value and has an inflection point at which the function is increasing halfway between the September data and the October data. (The horizontal shift calculated earlier was 9, or September.) This supports our analysis that the minimum probably occurs halfway between the June and the July data and that the maximum probably occurs halfway between the December and the January data.

When we compare the two models, we see that the graph of the model calculated via technology is a better fit to the data (see Figure 8.26a). However, because the period of the technology-generated model is more than 1 year (the period we expect from the context), the two models will begin to differ by more and more after the first cycle (see Figure 8.26b). Thus the technology-generated model should be used only for inputs between 0 and 12.

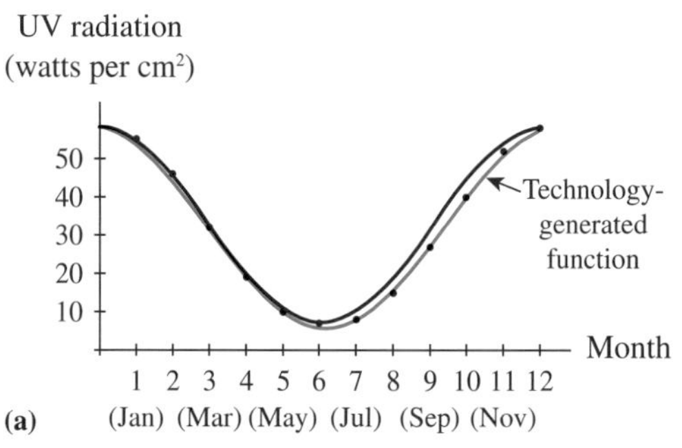

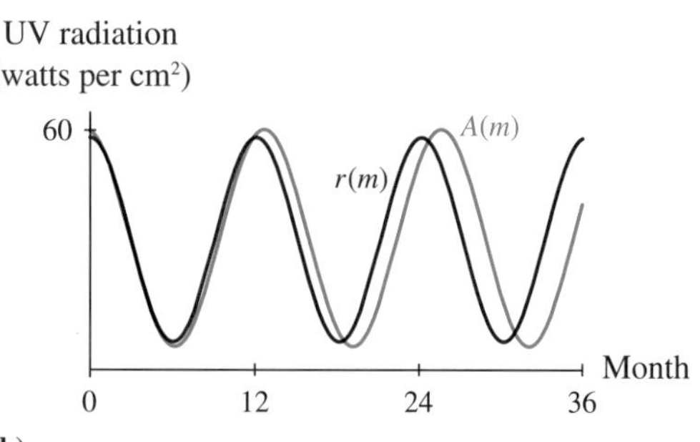

FIGURE 8.26 (a) (b)

EXAMPLE 2 *Hours of Daylight on the Arctic Circle (Revisited)*

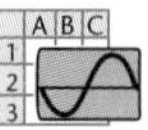

Use technology to find a sine model for the Arctic Circle hours-of-daylight data. Compare the equation with the one calculated in Example 1.

Solution A technology-determined model is

$$y = 12.00011 \sin(0.01721x - 1.40296) + 12 \text{ hours}$$

where x is the number of days since December 31 of the previous year. We see that the amplitude coefficient of this equation agrees to three decimal places with that of the equation found in Example 1. The other coefficients agree to at least the number of decimal places shown here.

We now have the ability to use a sine function to model data that exhibit a wave pattern. A sine model may also be used if the data do not show more than one complete cycle but the context indicates that the situation is cyclic in nature. It is also possible to use a sine model if a situation is not cyclic but the curvature of the data matches that of a sine function. In such cases the model would be useless for extrapolation, but it could be used for interpolation.

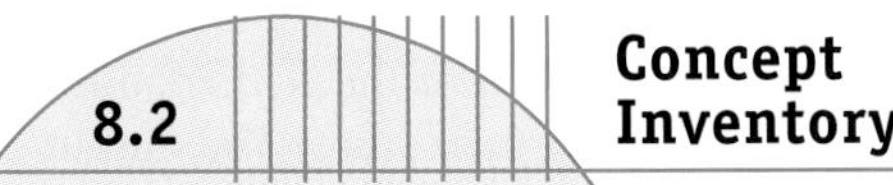

Concept Inventory

- *The sine function:* $f(x) = a \sin(bx + h) + k$
- *Determining sine function parameters from data without the aid of technology*
- *Finding a sine function using technology*

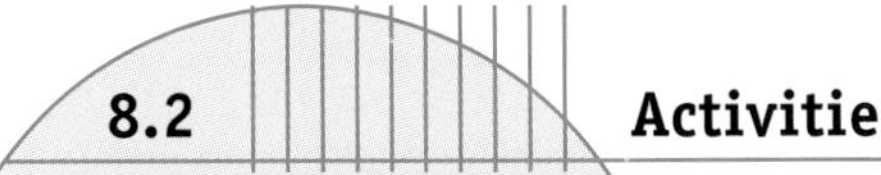

Activities

1. Table 8.4 shows the average number of gas thermal units per day used by a residential gas customer in Reno, Nevada.

TABLE 8.4

Month	Gas use (therms per day)
Nov 2000	1.2
Dec 2000	1.7
Jan 2001	3.2
Feb 2001	3.3
Mar 2001	3.3
Apr 2001	2.5
May 2001	1.5
June 2001	0.9
July 2001	0.4
Aug 2001	0.3
Sept 2001	0.3
Oct 2001	0.4
Nov 2001	1.4
Dec 2001	2.3
Jan 2002	2.9

   a. Find a sine function to fit the data in Table 8.4. State the amplitude, period, frequency, and horizontal and vertical shifts of the equation.

   b. According to the function, what is the average value (or expected value) of the number of therms per day?

   c. Assuming that a billing period has 30 days and that the cost per therm is \$0.5632, estimate the average gas bill for this customer.

2. In order to smooth out seasonal work loads, the Campbell Soup Company produces both canned soup and powdered drink mix.[5] Estimated sales for 1992 are shown in Table 8.5. The table values are in millions of reconstituted pints or millions of 16-ounce cans sold in a four-week period beginning on the date indicated.

TABLE 8.5

Date	1/5	2/2	3/1	3/29	4/26
Soup (millions of 16-oz cans)	200	215	180	160	150
Drink (millions of pints)	375	425	430	460	570

Date	5/24	6/21	7/19	8/16	9/13
Soup (millions of 16-oz cans)	120	110	100	110	130
Drink (millions of pints)	720	740	800	740	600

Date	10/11	11/8	12/6	1/3
Soup (millions of 16-oz cans)	155	200	210	200
Drink (millions of pints)	500	410	375	350

5. *Wall Street Journal*, July 7, 1993. p. B1.

a. Find a model for the soup sales.

b. Find a model for the drink sales.

c. Compare and contrast the amplitudes, periods, frequencies, and vertical and horizontal shifts of the equations in parts *a* and *b*.

d. How do you suppose management would use the two models that you have found in their production scheduling?

3. Table 8.6 gives the normal daily mean temperatures,[6] based on the 30-year period 1961 through 1990, for Omaha, Nebraska.

TABLE 8.6

Month	Daily mean temperature (°F)
January	21.1
February	26.9
March	38.6
April	51.9
May	62.4
June	72.1
July	76.9
August	74.1
September	65.1
October	53.4
November	39.0
December	25.1

a. Plot these daily mean temperatures. Explain why the data should be cyclic.

b. Use the data to determine the amplitude and the vertical shift.

c. Approximate the period and horizontal shift from the data.

d. Construct, without the aid of technology, a model of the form

Mean daily temperature $= a\sin(bx + h) + k$ °F

during the xth month of the year.

e. Graph your function on the plot of the data. Is the fit reasonable?

f. Use your function to predict the mean daily temperature in Omaha in July of this year. Discuss the reasonableness of the prediction.

Note that Activities 3 and 4, 5 and 6, and 7 and 8 are corresponding pairs and should be assigned together.

4. Use your calculator or computer to find a sine function that fits the data in Activity 3. Write the model. Compare and contrast the equation found in Activity 3 with the one found with technology in this activity.

5. Some populations of animals exhibit periodic behavior over time. During the 1960s and 1970s, two scientists carefully studied the population of the Carabid beetle in a region of the Netherlands.[7] Table 8.7 shows data from their study. The output data are log (base 10) data for the number of beetles not surviving in a given year. These values are referred to as the *reduction* in the number of beetles. The number of beetles not surviving in a given year is found by raising 10 to the reduction value in the table corresponding to that year.

TABLE 8.7

Year	log (base 10) reduction	Year	log (base 10) reduction
1965	2.7	1971	2.7
1966	2.8	1972	2.9
1967	4.5	1973	3.7
1968	4.4	1974	4.5
1969	3.3	1975	4.4
1970	2.8	1976	2.7

a. Renumber the input data as years since 1965. Using only the data in Table 8.7, estimate the amplitude and vertical shift.

b. Using only the data in Table 8.7, estimate the period, frequency, and horizontal shift.

c. On the basis of the values calculated in parts *a* and *b*, give a sine model for the reduction data. Sketch a graph of the function on a scatter plot of the data.

d. Use the equation in part *c* to construct a model for the number of beetles not surviving in a given year. Sketch a graph of this function, and compare it with the graph in part *c*.

6. a. Use technology to find a sine equation that fits the data in Activity 5. Write the model. Compare and contrast the parameters of the equation you constructed in part *c* of Activity 5 with those of the technology-obtained equation.

6. U.S. National Oceanic and Atmospheric Administration, *Climatography of the United States*, No. 81.

7. P. J. denBoer and J. Reddingius, *Regulation and Stabilization Paradigms in Population Ecology* (Chapman & Hall: London, 1996).

b. Compare the graph of the technology-obtained equation in part *a* with the graph in part *c* of Activity 5. Judging on the basis of these graphs, which model do you believe is the more appropriate one?

7. Table 8.8 shows the normal daily low temperatures,[8] based on the 30-year period 1961 through 1990, for Omaha, Nebraska.

TABLE 8.8

Month	Daily minimum temperature (°F)	Month	Daily minimum temperature (°F)
Jan	10.9	July	65.9
Feb	16.7	Aug	62.9
Mar	27.7	Sept	53.6
Apr	39.9	Oct	41.2
May	50.9	Nov	28.7
June	60.4	Dec	15.6

a. How would you expect the scatter plot and model for the normal daily low temperatures in Omaha to be related to the scatter plot and model for the normal daily mean temperatures in Omaha?

b. Use the data for the normal daily low temperatures to determine the amplitude and the vertical shift. Plot the daily low temperatures, and use the plot to approximate the period and horizontal shift. Compare these results with those found for the daily mean temperature in Activity 3. Are the findings what you expected?

c. On the basis of your answers to parts *a* and *b*, write a model for the normal daily low temperatures.

8. a. Use your calculator or computer to find a sine equation that fits the Omaha daily low-temperature data in Activity 7. Write the model. Graph this equation and the one found in Activity 7 on a plot of the data, and examine the plot and the graphs. Discuss any differences you observe.

b. What, if any, are the advantages of using technology to find this equation?

9. Table 8.9 shows the approximate numbers of large, civil-transport aircraft produced in the United States[9] from 1949 through 1963.

TABLE 8.9

Year	Aircraft	Year	Aircraft
1949	185	1957	230
1950	165	1958	235
1951	155	1959	225
1952	160	1960	210
1953	160	1961	180
1954	180	1962	160
1955	215	1963	155
1956	220		

a. Examine a scatter plot, and describe the curvature of the data. Which of the models that we have studied could be used to model these data?

b. Find a sine model for the data.

c. What does the equation give as the number of large, civil-transport aircraft produced in the United States in 1964?

d. Do you think the periodic nature of a sine model is a good description of aircraft production beyond 1963? Explain.

10. An exclusive fishing club on the Restigouche River in Canada kept detailed records on the numbers of fish caught by its members between 1880 and 1930. Approximate data[10] for the number of salmon caught per rod per day are given in Table 8.10.

TABLE 8.10

Year	Salmon per rod per day	Year	Salmon per rod per day
1880	1.0	1893	1.5
1881	0.75	1894	1.65
1882	0.85	1895	1.8
1883	1.1	1896	1.5
1884	1.4	1897	1.35
1885	1.6	1898	0.9
1886	1.6	1899	1.0
1887	1.2	1900	1.1
1888	1.1	1901	1.2
1889	1.15	1902	1.35
1890	1.1	1903	1.2
1891	1.1	1904	1.5
1892	1.2	1905	1.75

8. U.S. National Oceanic and Atmospheric Administration, *Climatography of the United States*, No. 81.

9. Data estimated from J. J. Van Duijn, *The Long Wave in Economic Life* (London: George Allen & Unwin, 1983).

10. Data estimated from E. R. Dewey and E. F. Dakin, *Cycles: The Science of Prediction* (New York: Holt, 1947).

In *Cycles: The Science of Prediction*, Dewey and Dakin assert that the salmon data are periodic with period $9\frac{2}{3}$ years. Find a sine model for the salmon data. Does your equation agree with the assertion that the period is $9\frac{2}{3}$ years?

11. Utility needs are dynamic, and electric companies have to plan for base and peak loads carefully. Forecasting is fundamental to their management operations. A regular pattern of supply from the coal and nuclear plants is pre-planned, and hydroelectric units are on standby to pick up the slack. The reason for this program is that coal and nuclear plants are slow to react, and their efficiency is lost without a regular, consistent pattern of operation. Hydroelectric facilities can come on line immediately, but their resources are limited and must be used sparingly. Pump storage may be used to restore hydroelectric resources during times of excess.

    In the summer, high temperatures affect air conditioning energy requirements; in the winter, low temperatures affect heat requirements. The weather bureau's forecast for the high and low temperatures for the next 5 days will be used as the information base for a temperature model from which the planners can anticipate power load requirements. A 5-day forecast is shown in Figure 8.2.1.

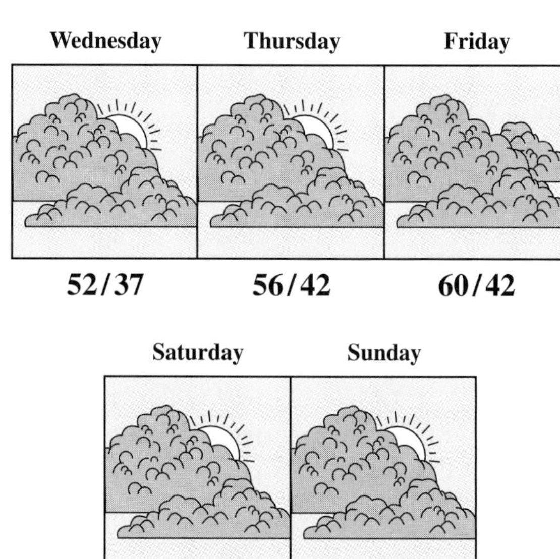

FIGURE 8.2.1

a. Construct a table using as input the number of hours since midnight on Wednesday, assuming that the high temperatures occur at 5 p.m. and the low temperatures occur at 5 a.m. Also assume that the midrange temperatures between the highs and lows occur at 11 a.m. and 11 p.m.

b. Find a sine model for the data. What is the period of the function?

c. What are the high and low temperatures given by the equation?

12. Figure 8.2.1 shows a 5-day forecast for temperatures. The actual highs and lows are given in Table 8.11.

TABLE 8.11

Weekday	High (°F)	Low (°F)
Wednesday	56	35
Thursday	59	40
Friday	71	52
Saturday	66	53
Sunday	56	33

Compare the actual high and low temperatures to the highest and lowest temperatures predicted by the equation in Activity 11. Assuming that the coal and nuclear power generators can supply enough energy for the highs and lows predicted by the model, when will the hydroelectric plant's energy be needed?

13. a. Why do electric production facilities have to operate in sync with power demands?

    b. Note your local temperature forecast, and build a power demand model for your area. During the next several days, keep a record of the daily highs and lows, and comment on how well your model forecasts the demand.

    c. Imagine limits on your hydroelectric facilities that would make them inadequate to meet the residual demands. Explain how increasing the amplitude parameters on the model could reduce the calls on the hydroelectric facilities.

14. Find local information, and build a mathematical model for the hours of daylight in the town or city where your school is located.

15. Use two different methods to build a sine model for the hours of darkness in Anchorage, Alaska, using the data in Table 8.12.

16. Locate appropriate data, and build a mathematical model for the hours of daylight at

    a. the Tropic of Capricorn

    b. the Equator

TABLE 8.12

Day of the year	Hours of darkness	Day of the year	Hours of darkness
1	18	211	7
31	16	241	9.5
61	13	271	11.5
91	10.5	301	15.5
121	7.5	331	17.5
151	5	361	18.5
181	4.5		

17. To study the rhythmic synchronization of neurons, a team of scientists exposed locusts to odors and measured the firing patterns of pairs of neurons in response to those odors. Table 8.13 gives the number of times per millisecond that the pair of neurons simultaneously fired during the experiment when exposed to cherry odor, apple odor, and a combination of the two.[11]

TABLE 8.13

Time (ms)	Cherry (firings/ms)	Apple (firings/ms)	Cherry/Apple (firings/ms)
0	14	16	7.5
5	11.5	12.5	8.5
10	10.5	12.5	3
15	4	6	6.5
20	2	2	1.5
25	1	5	1.5
30	7	5	4
35	4	14	5
40	7	17	7.5
45	19	23	7.5
50	15	15.5	10
55	17	15	8.5
60	11.5	10	6.5
65	8.5	6	4
70	5	8	3
75	5	6	3.5

The scientists claim that the period of neuron activity is not influenced by a particular odor. Find models for the data for cherry odor, apple odor, and combined odor, and report the period of each model. Do your models seem to support the scientists' claim? Explain.

18. The data in Table 8.14 describe the monthly sales for an ice cream company.

    a. Find the change in sales between January and June and between July and December.

TABLE 8.14

Month	Monthly sales (thousand dollars)
Jan	50
Feb	60
Mar	77
Apr	96
May	137
June	158
July	167
Aug	159
Sept	108
Oct	75
Nov	61
Dec	54

    b. Find a model for the ice cream data, and use it to estimate the changes in part *a*. Compare your answers with those in part *a*.

19. A model[12] fit to the normal mean air temperature at Fairbanks, Alaska, is

$$f(x) = 37 \sin[0.0172(x - 101)] + 25 \text{ degrees Fahrenheit}$$

where x is the number of days since the last day of the previous year.

    a. Estimate the highest and lowest mean daily temperatures.

    b. Find the average of the highest and lowest mean daily temperatures. Relate this value to the vertical shift of the graph of f.

20. Table 8.15 gives a patient's temperature during a 6-day illness, as recorded at noon each day.

11. Data estimated from M. Wier and G. Laurent: "Odour Encoding by Temporal Sequences of Firing in Oscillating Neural Assemblies," *Nature*, vol. 384 (November 14, 1996), pp. 162–165.

12. B. Lando and C. Lando, "Is the Curve of Temperature Variation a Sine Curve?" *The Mathematics Teacher*, vol. 7, no. 6 (September, 1977), p. 535.

TABLE 8.15

Day	Temperature (°F)
1	103.9
2	104.4
3	102.8
4	100.2
5	98.6
6	99.2

a. Use the data to find the average rate of change of the temperature between day 2 and day 5.

b. Find a sine model for these data.

c. Judging by the graph of the model, what is your prognosis for this patient? Discuss the implications of extrapolating in this situation.

Answer the questions in Activities 21 and 22 on the basis of the data in Table 8.16 of the normal daily high temperatures[13] for the 30-year period 1961 through 1990 in New Orleans, Louisiana, and Boston, Massachusetts.

TABLE 8.16

Month	New Orleans daily high temperature (°F)	Boston daily high temperature (°F)
January	60.8	35.7
February	64.1	37.5
March	71.6	45.8
April	78.5	55.9
May	84.4	66.6
June	89.2	76.3
July	90.6	81.8
August	90.2	79.8
September	86.6	72.8
October	79.4	62.7
November	71.1	52.2
December	64.3	40.4

21. a. Would you expect these normal daily high temperature data to be cyclic? Explain.

b. Use the data to find the difference between high temperatures in January and July for these two cities. What do these differences tell you about how the climates of the cities compare?

22. a. Discuss any differences and similarities you would expect in the models for the normal daily high temperature on the basis of the 30-year period 1961 through 1990 for New Orleans and Boston.

b. Find sine models for each set of data. Compare the amplitude, vertical shift, horizontal shift, and period of these functions. What do these values indicate about the climates of the two cities?

c. Use the models to estimate the differences in normal maximum daily temperatures for the cities between April 15 and May 1. What do these values indicate about the climates of the two cities during the last half of April?

23. In desert areas of the western United States, lizards and other reptiles are harvested for sale as pets. Because reptiles hibernate during the winter months, the number of reptiles gathered is periodic. Table 8.17 shows the number of lizards gathered during two consecutive years in a certain western state.[14]

TABLE 8.17

Year 1	Lizards gathered (thousands)	Year 2	Lizards gathered (thousands)
January	0	January	0
February	0	February	0
March	2	March	2
April	14	April	15
May	30	May	27
June	32	June	33
July	25	July	27
August	17	August	15
September	10	September	9
October	1	October	2
November	0	November	0
December	0	December	0

a. From an examination of the data, estimate the amplitude, period, and vertical shift.

13. U.S. National Oceanic and Atmospheric Administration, *Climatography of the United States*, No. 81.

14. Based on information from the Nevada Division of Wildlife.

b. Use technology to find a sine function that fits the 2-year data. Compare the amplitude, period, and vertical shift of the equation with those estimated in part *a*.

c. Examine a graph of the equation on a scatter plot of the data. Discuss how well the graph fits the data.

d. Remove the months in which no lizards were harvested, and renumber the remaining months from 1 to 16. Find a sine model for the modified data. Compare the amplitude, period, and vertical shift of this equation with those of the equation in part *b*. Does the graph of this equation fit the data significantly better than the one in part *b*?

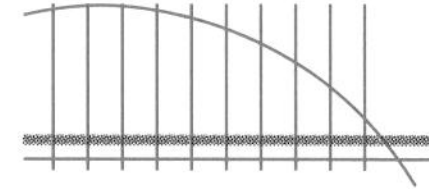

8.3 Rates of Change and Derivatives

As something moves through a cycle, we are often interested in following the change that takes place. Derivatives are one of the tools that can be used to measure how rapidly change occurs.

Derivative Formulas

You may wish to introduce this section by giving your students a graph of two or three periods of the sine function and asking them to sketch the slope graph. This is a more effective approach than sketching the slope graph for them.

As we seek a function to describe the rate of change of the sine function, we note that because the sine curve is periodic, we would expect the rate-of-change function to be periodic as well. We know that the rate of change of a sine graph is zero at the high and low points on the graph. The values in the sine curve repeat in a regular manner, so the zero slopes occur repeatedly. Thus the rate-of-change function of $y = \sin x$ has a sequence of repeating x-intercepts at $\frac{-\pi}{2}, \frac{\pi}{2}, \frac{3\pi}{2}, \frac{5\pi}{2}$, etc.

Consider the slopes between the high and low points of $y = \sin x$. Figure 8.27 indicates where the slopes are positive and where they are negative. In addition, the blue dots in the figure mark the location of the inflection points. These are points of greatest or least slope and correspond to high or low points on the derivative graph.

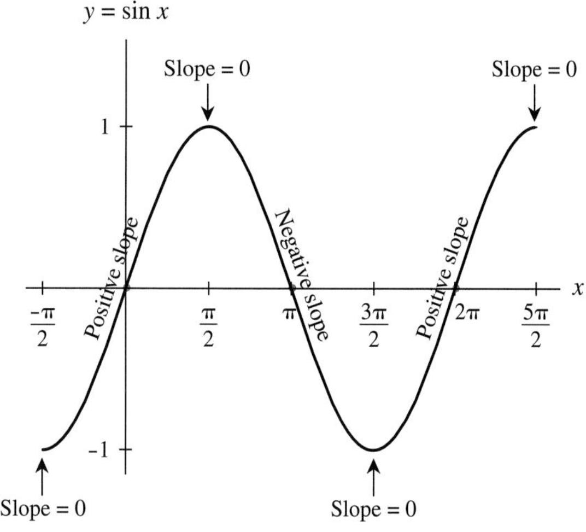

FIGURE 8.27

Sketching the slope graph results in the graph shown in Figure 8.28a.

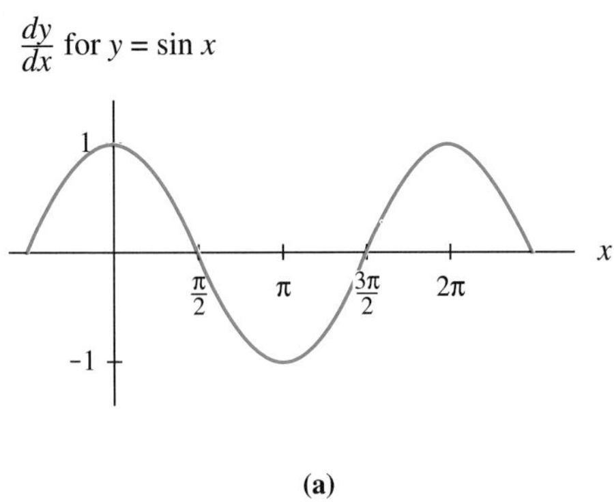

(a)

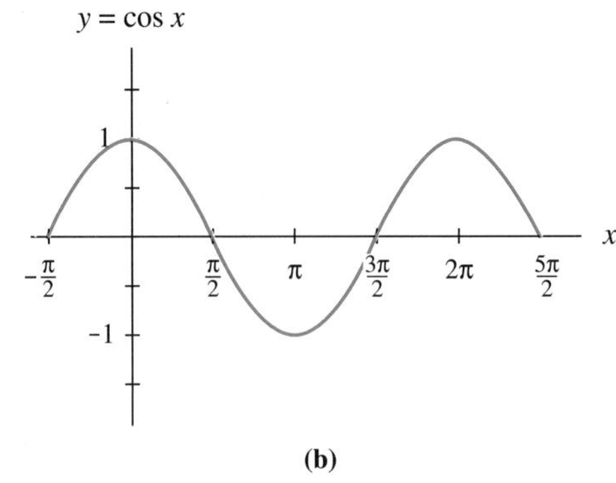

(b)

FIGURE 8.28

The rate-of-change graph of $y = \sin x$ is periodic with period 2π and amplitude 1. A careful numerical investigation reveals that the output is 1 at $x = 0$, −1 at $x = \pi$, and 0 at $x = \frac{\pi}{2}$ and at $x = \frac{3\pi}{2}$. In the Section 8.1 Activities, you were asked to graph the cosine function. This graph is shown in Figure 8.28b. Note that it has the same characteristics as those previously stated for the rate-of-change graph of the sine function. In fact, it can be shown mathematically that the derivative of the sine function, $y = \sin x$, is the cosine function, $y' = \cos x$.

Derivative of the Sine Function

When x is measured in radians, the rate of change of the sine function is the cosine function.

$$\text{If } f(x) = \sin x, \text{ then } \frac{df}{dx} = \cos x.$$

Derivative Rules

Now that we have a derivative formula for $y = \sin x$, we can apply the Constant Multiplier Rule, the Chain Rule, and the Sum Rule given in Chapter 4 to find slope formulas for functions of the form $f(x) = a \sin(bx + h) + k$.

To illustrate the use of these rules, we consider the temperatures[15] in Philadelphia on August 27, 1993, as shown in Table 8.18.

TABLE 8.18

Time	1 a.m.	2 a.m.	3 a.m.	4 a.m.	5 a.m.	6 a.m.	7 a.m.	8 a.m.	9 a.m.	10 a.m.	11 a.m.	12 p.m.
Temperature (°F)	78	77	77	76	76	75	75	77	79	83	87	90
Time	1 p.m.	2 p.m.	3 p.m.	4 p.m.	5 p.m.	6 p.m.	7 p.m.	8 p.m.	9 p.m.	10 p.m.	11 p.m.	12 a.m.
Temperature (°F)	91	93	94	95	93	92	89	86	84	83	81	79

15. *Philadelphia Inquirer*, August 28, 1993.

The data in Table 8.18 can be modeled as

$$t(x) = 9.6 \sin(0.285x - 2.93) + 84.2 \text{ °F}$$

x hours after midnight on August 26, 1993. Using the Constant Multiplier Rule (with 9.6 as the constant multiplier) and the Sum Rule (with the derivative of 84.2 being zero), we find that a rate-of-change formula for this function is

$$t'(x) = 9.6 \frac{d}{dx}[\sin(0.285x - 2.93)] + 0$$

Now, consider $u = 0.285x - 2.93$ to be an inside function with the sine function being the outside function. Using the Chain Rule, we have

$$\begin{aligned} t'(x) &= 9.6 \frac{d}{du}(\sin u) \frac{d}{dx}(0.285x - 2.93) \\ &= 9.6(\cos u)(0.285) \\ &\approx 2.736 \cos(0.285x - 2.93) \text{ °F per hour} \end{aligned}$$

x hours after midnight.

According to the model and its rate-of-change formula, the temperature at noon on August 27, 1993, was $t(12) \approx 89$°F, and it was increasing by $t'(12) \approx 2.4$°F per hour.

EXAMPLE 1 *Calls for Service*

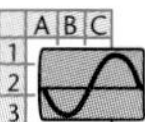

The calls for service made to a county sheriff's department in a certain rural/suburban county can be modeled as

$$c(t) = 2.8 \sin(0.262t + 2.5) + 5.38 \text{ calls}$$

during the tth hour after midnight.

a. What is the average number of calls the department receives each hour?

b. Find a formula for $c'(t)$.

c. How quickly is the number of calls received each hour changing at noon? at midnight?

Solution

a. The parameter k of a sine model $f(x) = a \sin(bx + h) + k$ tells us the average (or expected) value of the cycle. Thus the department receives an average of 5.38 calls each hour during a day.

b. A rate-of-change model for calls received each hour can be found by using the Constant Multiplier Rule, the Chain Rule, and the Sum Rule:

$$c'(t) = 2.8 \frac{d}{dt}[\sin(0.262t + 2.5)] + 0$$

Now, let $u = 0.262t + 2.5$, and proceed using the Chain Rule.

$$\begin{aligned} c'(t) &= 2.8 \frac{d}{du}(\sin u) \frac{d}{dt}(0.262t + 2.5) \\ &= 2.8(\cos u)(0.262) \\ &= 0.7336 \cos(0.262t + 2.5) \text{ calls per hour} \end{aligned}$$

t hours after midnight.

c. At noon the number of calls received each hour is changing by $c'(12) \approx 0.6$ call per hour. At midnight the number of calls received each hour is changing by $c'(0) \approx -0.6$ call per hour. [*Note*: You could use $t = 24$ for midnight. If the cycle has close to a 24-hour period, then $c'(0)$ should be approximately the same as $c'(24)$.]

The Sum Rule is useful when we are finding the derivative of a sine function that is added to another function. This type of function occurs when a phenomenon creates cycles of the same period and amplitude but the expected value of each cycle changes according to some other function. For example, the amount of soup sold each week is cyclic over a year and has approximately the same amplitude from year to year. However, we expect that overall, the amount of soup sold each year will increase. Example 2 illustrates this type of model in the context of atmospheric carbon dioxide.

EXAMPLE 2 *Atmospheric Carbon Dioxide*

Figure 8.29 appeared in a *Scientific American*[16] article in July of 1990.

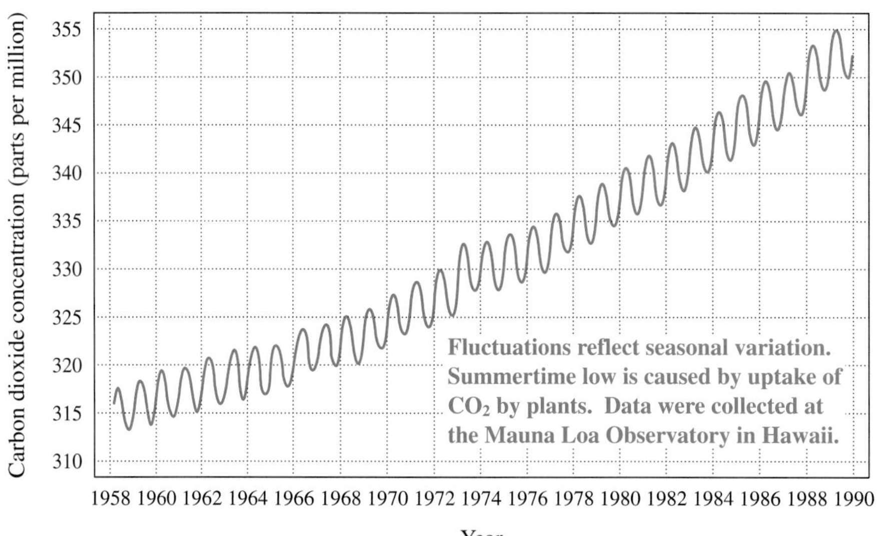

FIGURE 8.29

Our goal in including this example is not that students understand exactly how this model was created but rather that they understand how the model can be used to calculate rates of change. To create this function, we relied on data (not presented here) from the Mauna Loa Observatory.

This figure illustrates the rise in the amount of atmospheric carbon dioxide (CO_2) along with the seasonal fluctuations. We wish to find a model for the CO_2 concentration. An examination of the data from 1959 to the present (readily available on the web) reveals that the data are cyclic with period 1 year. Using the equation $\frac{2\pi}{b} = 1$, we find that $b \approx 6.283$. The data show that the approximate location of the increasing inflection point of each cycle is the beginning of each year; thus there is no horizontal shift for each cycle. On the basis of these approximations, a model for the yearly fluctuation in the amount of atmospheric CO_2 is $f(x) = a \sin(6.283x) + d$

16. Figure 8.29 from R. M. White, "The Great Climate Debate," *Scientific American*, vol. 263, no. 1 (1990), pp. 36–43. Copyright © 1990 by Scientific American, Inc. All rights reserved. Reprinted with permission.

parts per million, where x is the time in years measured from the beginning of any year,[17] a is the amplitude, and d is the vertical shift.

The amplitude of each cycle is slightly different, but using the data to estimate amplitudes for the years between 1959 and 1999, we observe that there is a general rise in amplitudes. Modeling the amplitudes gives the equation

$$a = 0.010345x + 2.105676 \text{ parts per million}$$

x years after the beginning of 1900.

We also notice that the average amount of carbon dioxide (the parameter d) is increasing each year. The average value of a cycle of a sine function that is not shifted horizontally occurs at an input of zero. By fitting a quadratic model to the CO_2 concentration in January of each year for the years between 1959 and 1999, we find that the average amount of CO_2 can be described by the equation

$$d(x) = 0.013783x^2 - 0.844566x + 316.913080 \text{ parts per million}$$

x years after the beginning of 1900. Adding this equation to the fluctuation cycle as the expected value and substituting the amplitude function for the parameter a give the model for atmospheric CO_2 concentration as

$$c(x) = (0.010345x + 2.105676)\sin(6.283x) + 0.013783x^2 - 0.844566x + 316.913080 \text{ parts per million}$$

x years after the beginning of 1900. The model applies for years between 1959 and 1999.

a. Use the model to estimate the amount of atmospheric CO_2 in July of 1989.

b. Write a formula for the rate of change of the amount of atmospheric CO_2.

c. Use the formula from part b to estimate the rate of change of the amount of atmospheric CO_2 in July of 1989.

Solution

a. There were approximately $c(89.5) \approx 351.8$ parts per million of CO_2 in the atmosphere in July of 1989.

b. The derivative of c is

$$c'(x) = (0.010345x + 2.105676)(6.283)\cos(6.283x) + 0.010345\sin(6.283x) + 2(0.013783)x - 0.844566 \text{ parts per million per year}$$

x years after the beginning of 1900.

c. $c'(89.5) \approx -17.4$ parts per million per year. The amount of atmospheric CO_2 was decreasing by approximately 17.4 parts per million per year in July of 1989.

It is effective to have your students sketch the slope graph of the cosine function as a means of motivating the derivative formula.

Finally, let us consider finding a formula for the derivative of a rate-of-change formula for the sine model. This is the second derivative of the sine model. Because the derivative of $y = \sin x$ is $y' = \cos x$, we must know the derivative of $\cos x$ in order to find the second derivative of $y = \sin x$.

17. Because we are considering the general shape of each cycle over a 40-year period rather than modeling each individual cycle, the input variable is not a particular year but, rather, applies to all years between 1959 and 1999.

If we were to sketch the slope graph of the cosine function, the result would be the solid graph shown in Figure 8.30.

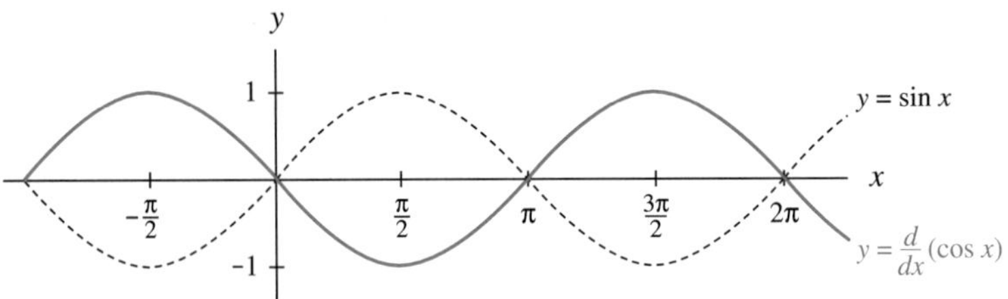

FIGURE 8.30

This solid graph appears to be a reflection of $y = \sin x$ across the horizontal axis. That is, the slope graph of $y = \cos x$ appears to be $y' = -\sin x$. In fact, using more rigorous mathematics, it can be proven that if $f(x) = \cos x$, then its derivative is $f'(x) = -\sin x$.

Derivative of the Cosine Function

When x is measured in radians, the rate of change of the cosine function is the negative of the sine function.

$$\text{If } g(x) = \cos x, \text{ then } \frac{dg}{dx} = -\sin x.$$

We use this fact when writing the second derivative of a sine function. For example, the model for Philadelphia temperatures x hours after midnight,

$$t(x) = 9.6 \sin(0.285x - 2.93) + 84.2 \text{ °F}$$

has the derivative

$$t'(x) = 2.736 \cos(0.285x - 2.93) \text{ °F per hour}$$

and its second derivative is

$$t''(x) = -0.77976 \sin(0.285x - 2.93) \text{ °F per hour per hour}$$

Note that in the first and second derivatives of t, the period and horizontal shift are the same as for t, whereas the amplitude and vertical shift are different.

Asking your students why the period and horizontal shift remain constant while the other parameters do not may result in a thoughtful class discussion.

EXAMPLE 3 *Atmospheric Carbon Dioxide*

Recall from Example 2 that a model for the amount of atmospheric carbon dioxide is

$$c(x) = (0.010345x + 2.105676) \sin(6.283x) + 0.013783x^2 - 0.844566x + 316.913080 \text{ parts per million}$$

x years after the beginning of 1900. The model applies for years between 1959 and 1999.

a. Write a formula for the second derivative of this function.

b. Find and interpret $c''(80)$.

Solution

a. The derivative of c was found in Example 2 to be

$$c'(x) = (0.010345x + 2.105676)(6.283)\cos(6.283x) + 0.010345\sin(6.283x) + 2(0.013783)x - 0.844566 \text{ parts per million per year}$$

x years after the beginning of 1900. This derivative was found using the Product Rule, and the Product Rule is also required for the second derivative.

$$\begin{aligned} c''(x) &= (0.010345x + 2.105676)(6.283)(6.283)[-\sin(6.283x)] \\ &\quad + (0.010345)(6.283)\cos(6.283x) \\ &\quad + 0.010345(6.283)\cos(6.283x) + 2(0.013783) \\ &= (0.010345x + 2.105676)(6.283)(6.283)[-\sin(6.283x)] \\ &\quad + 2(0.010345)(6.283)\cos(6.283x) \\ &\quad + 2(0.013783) \text{ parts per million per year per year} \end{aligned}$$

x years after the beginning of 1900.

b. $c''(80) \approx 1.87$ parts per million per year per year. At the beginning of 1980 the rate of change of the amount of atmospheric carbon dioxide was increasing by about 1.9 parts per million per year per year.

Even though in this section we have illustrated only some of the general derivative rules, any previously developed rule for derivatives of functions applies to the sine and cosine functions. Recall also that if you have a graph and not a function formula, then you can estimate rates of change by sketching tangent lines. This is illustrated in Example 4.

EXAMPLE 4 *Nebraska Mean Temperatures*

Figure 8.31 shows the graph of a function m giving the normal daily mean temperature,[18] based on the 30 year period 1961 through 1990, for Omaha, Nebraska.

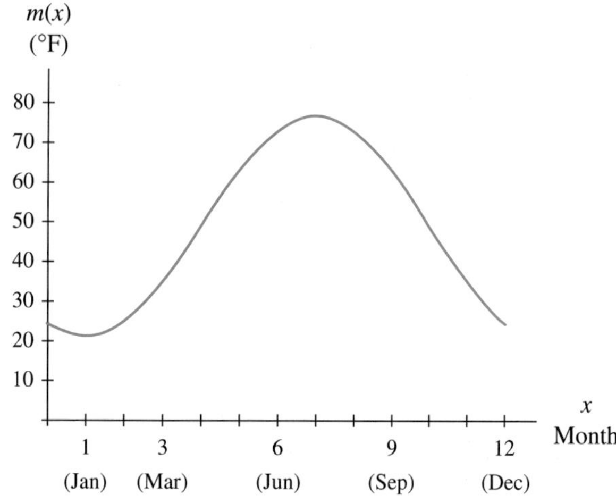

FIGURE 8.31

18. Based on data from U.S. National Oceanic and Atmospheric Administration, *Climatography of the United States*, No. 81.

a. Estimate the maximum difference in daily mean temperature. Estimate the average rate of change between the lowest daily mean temperature and the highest daily mean temperature.

b. Estimate where the daily mean temperature is increasing most rapidly.

c. Estimate the rate of change of the daily mean temperature at the point where the temperature is increasing most rapidly.

d. Interpret each of the following:

i. $m(9) - m(2) = 38.1$

ii. $m'(8) = -7.3$

Solution

a. The maximum difference in daily mean temperature is the difference between the highest daily mean temperature and the lowest daily mean temperature. According to Figure 8.31, the highest temperature is approximately 77°F (in July), and the lowest temperature is approximately 21°F (in January). The maximum difference is approximately 56°F. The average rate of change in daily mean temperature between January and July is $\frac{77 - 21}{7 - 1} \approx 9.3$°F per month.

b. The daily mean temperature is increasing most rapidly at the inflection point of this graph. This (according to Figure 8.31) occurs near $x = 4$, which corresponds to April.

c. We estimate the rate of change of daily mean temperature in April by carefully drawing the tangent line to the graph at $x = 4$ and approximating its slope. (See Figure 8.32.)

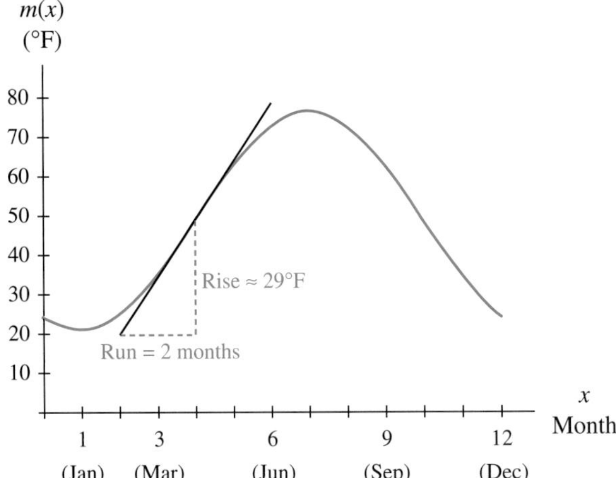

FIGURE 8.32

The daily mean temperature during April was increasing by approximately $\frac{29\text{°F}}{2 \text{ months}} = 14.5$°F per month.

d. i. The change in the daily mean temperature between February and September was 38.1°F.

ii. In August, the daily mean temperature was decreasing by 7.3°F per month.

We now have the tools we need not only to find rates of change of a sine model but also to use rates of change to locate maxima, minima, and inflection points on sine models. This is the topic of the next section.

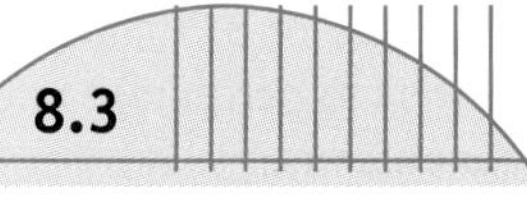

Concept Inventory

- *Derivatives:* If $y = \sin x$, then $y' = \cos x$.
  If $y = \cos x$, then $y' = -\sin x$.
- *Second derivatives:* If $y = \sin x$, then $y'' = -\sin x$.
  If $y = \cos x$, then $y'' = -\cos x$.

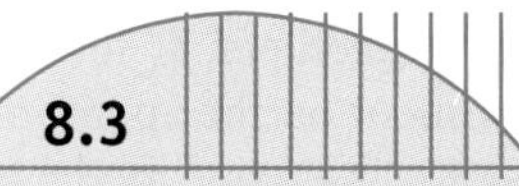

Activities

In Activities 1 through 6, find formulas for the indicated rates of change.

1. $f(x) = \sin 3x + 10.456$, $f'(x)$
2. $g(x) = \cos 0.45x - 3.763$, $\frac{dg}{dx}$
3. $t(r) = 5.2\cos(0.45r + \pi) + 80r - 6.34$, $\frac{dt}{dr}$
4. $f(t) = 3\sin(-0.45t + 3\pi) - 12.8t^3 + 4.6t$, $f''(t)$
5. $h(x) = 2.08\sin(-0.16x + 12.3) + 3.58\cos[1.35(2x + 6)] - 4.79$, $h'(x)$ and $h''(x)$
6. $y(x) = -0.023\sin(2x + 7) + 4\cos[3(0.5 - x)]$, $y'(x)$ and $y''(x)$
7. In order to smooth out seasonal work loads, the Campbell Soup Company produces both canned soup and powdered drink mix. The 1992 daily sales can be modeled[19] by

   $c(x) = 1.8961\sin(0.0186x + 1.1801) + 5.4660$ million cans (16 ounce) of soup

   $p(x) = 7.3026\cos(0.0197x - 3.7526) + 20.3728$ million (reconstituted) pints of drink mix

   where x is the day of the year. Assume that soup is sold at an average price of \$0.70 per can and drink mix is sold at \$0.20 per pint.

   a. Find a function giving the revenue from the combined 1992 daily sales as a function of the day of the year on which soup and powdered drink mix are produced.
   b. Give a function that models the rate of change of revenue from combined 1992 sales.
   c. How quickly was revenue changing on February 15, 1992?
   d. At what other times during 1992 was revenue decreasing at the same rate as that found in part *c*?
   e. Give one time during 1992 at which revenue was increasing at the same rate at which it was decreasing in part *c*.
8. A model[20] for the normal mean air temperature in Fairbanks, Alaska, is

   $f(x) = 37\sin[0.0172(x - 101)] + 25$ °F

   where x is the number of days since the last day of last year. Estimate how rapidly the normal mean temperature in Fairbanks is changing on March 15 and on December 31.
9. The normal daily high temperature in Boston can be modeled[21] by

   $B(m) = 22.926\sin(0.510m - 2.151) + 58.502$ °F

   at the end of the mth month of the year. Estimate how rapidly the normal daily high temperature in Boston is changing at the end of November and in the middle of May.
10. Ice cream cones are sold at a county fair. The total number of cones sold since the booth opened at 8 a.m. is approximated by

    $c(t) = 80.153\sin(0.511t - 1.680) + 87.564$ cones

    where t is the number of hours since 8 a.m. Estimate how rapidly the number of cones sold is changing at 12 p.m. and 8 p.m.

19. Based on information in *Wall Street Journal*, July 7, 1993, p. B1.
20. B. Lando and C. Lando, "Is the Curve of Temperature Variation a Sine Curve?" *The Mathematics Teacher*, vol. 7, no. 6 (September, 1977), p. 535.
21. Based on data from U.S. National Oceanic and Atmospheric Administration, *Climatography of the United States*, No. 81.

11. The graph in Figure 8.3.1 shows a model of the distance in kilometers that a weather satellite is north or south of the equator at time x minutes since the satellite was launched.

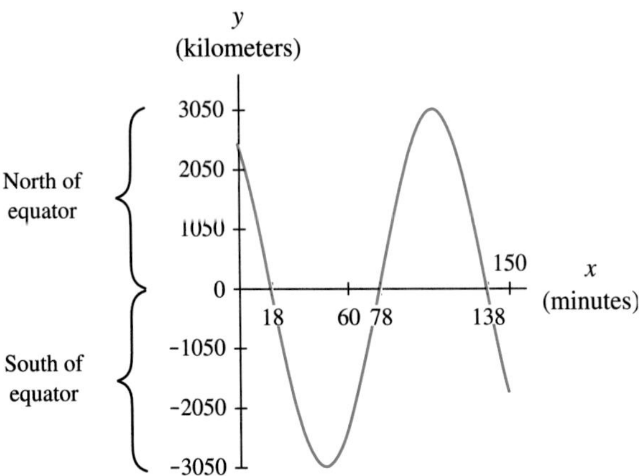

FIGURE 8.3.1

a. How long does it take the satellite to complete one orbit of the Earth?
b. What is the greatest distance south of the equator that the satellite ever reaches?
c. Carefully draw the tangent line at the time the satellite has been in orbit for 1 hour.
d. Use your tangent line to estimate the rate of change of the satellite's distance from the equator after it has been in orbit for 1 hour.
e. Is your answer to part *d* the speed of the satellite after it has been in orbit for 1 hour? Explain.

12. Figure 8.3.2 shows the power output[22] of a diesel engine in an irrigation pumping system that can be supplemented by a wind turbine when the wind speeds are sufficient.

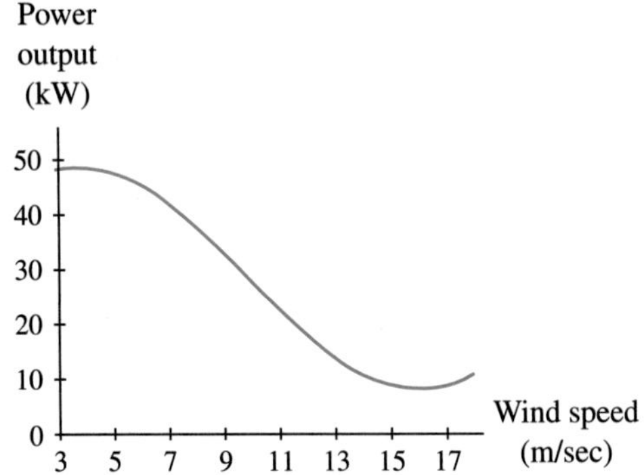

FIGURE 8.3.2

a. Estimate the change in power output when the wind speed changes from 5 m/sec to 10 m/sec and from 10 m/sec to 15 m/sec.
b. Report the estimates in part *a* as percentage changes and average rates of change.
c. Estimate the instantaneous rate of change in power output for wind speeds of 5, 10, and 15 m/sec. Convert your answers to percentage rates of change. Interpret your answers.
d. Why are the rates of change negative? What would you expect the rates of change to be for the power output of the wind turbine as a function of wind speed?
e. The graph in Figure 8.3.2 was generated from a sine function, which is an excellent fit for the power output data. Comment on the appropriateness and limitations of using a sine model in this context.

13. The power output of a diesel engine in an irrigation pumping system that can be supplemented by a wind turbine can be modeled[23] by

$$P(s) = 20.204 \sin(0.258s + 0.570) + 28.119 \text{ kilowatts}$$

when the wind speed is s meters per second.

a. Find the change in power output when the wind speed changes from 6 m/sec to 10 m/sec, from 10 m/sec to 14 m/sec, and from 14 m/sec to 18 m/sec.
b. Report the answers to part *a* as percentage changes and average rates of change.
c. Find a formula for $P'(s)$.
d. Find $P'(s)$ for $s = 6$, $s = 10$, $s = 14$, and $s = 18$. Interpret your answers.
e. Convert the answers to part *d* to percentage rates of change.

14. Figure 8.3.3 shows a temperature graph[24] for a location in the south Gobi Desert over a period of 2 years.

a. Estimate average rates of change in temperature from
   i. July through July
   ii. February through February
   iii. July through December of the first year

22. B. A. Stout, *Handbook of Energy for World Agriculture* (London: Elsevier Applied Science, 1990).

23. Ibid.

24. Neil E. West, ed., *Ecosystems of the World: Temperate Deserts and Semi-Deserts* (Amsterdam: Elsevier Scientific Publishing Company, 1983).

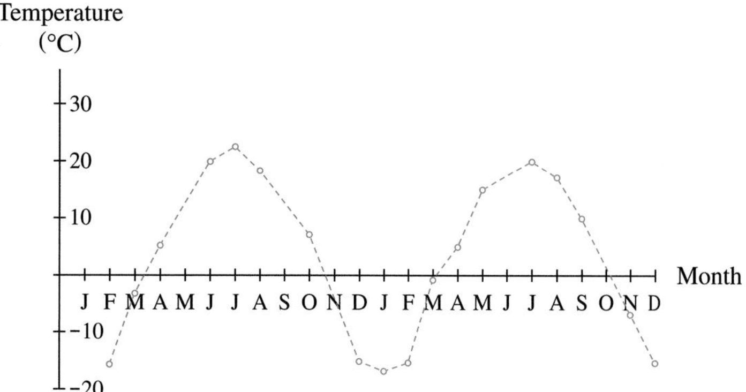

FIGURE 8.3.3

b. Estimate during which month the temperature
   i. increased most rapidly
   ii. decreased most rapidly
   iii. increased most slowly
   iv. decreased most slowly

15. A graph of the average amount of extraterrestrial radiation[25] in Amarillo, Texas, for each month of the year is shown in Figure 8.3.4. The amount is measured in equivalent millimeters of water evaporation per day. Month 1 corresponds to January, month 12 to December.

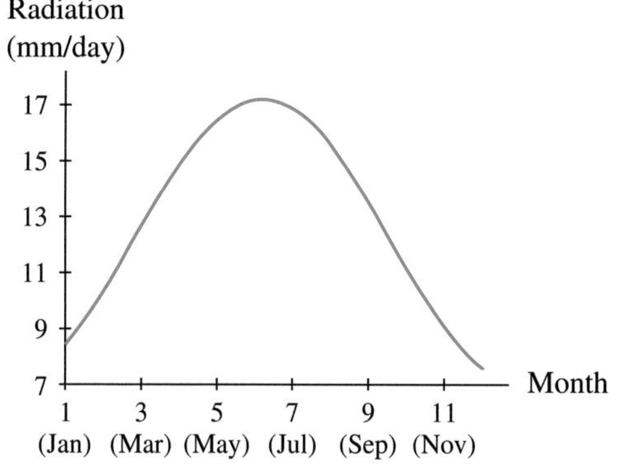

FIGURE 8.3.4

a. Use the graph to estimate the change in the amount of radiation from June through November.

b. Report the change in part *a* as a percentage change and an average rate of change.

c. Use the graph to estimate the rate of change in March.

d. Convert the rate of change in part *c* to a percentage rate of change.

25. A. A. Hanson, ed., *Practical Handbook of Agricultural Science* (Boca Raton, FL: CRC Press, 1990).

16. The average amount of extraterrestrial radiation in Amarillo, Texas, can be modeled[26] by

$$R(t) = 5.059 \sin(0.461t - 1.287) + 12.151$$
equivalent mm of water evaporation per day

where $t = 1$ in January and $t = 12$ in December.

a. By how much does the amount of extraterrestrial radiation change between April and October? What is the percentage change?

b. How quickly (on average) is the amount of extraterrestrial radiation changing from April through October? Is this value an accurate representation of how extraterrestrial radiation changed during these months? Explain.

c. Find $R'(t)$ for $t = 4$ and $t = 10$. Also express your answers as percentage rates of change. Interpret your answers.

17. Table 8.19 shows the weekly pneumonia death rate[27] (number of deaths per 100,000 people per week) in the United States from 1923 through 1925.

TABLE 8.19

Weeks since Jan. 1, 1923	Pneumonia death rate	Weeks since Jan. 1, 1923	Pneumonia death rate
0	100	80	60
10	210	90	45
20	95	100	95
30	35	110	160
40	55	120	140
50	85	130	80
60	140	140	40
70	160	150	90

26. From data in A. A. Hanson, ed., *Practical Handbook of Agricultural Science* (Boca Raton, FL: CRC Press, 1990).

27. Estimated from E. R. Dewey and E. F. Dakin, *Cycles: The Science of Prediction* (New York: Holt, 1947).

a. Find a sine model for the pneumonia death rate data. Why would such data be periodic?

b. Use the sine model to estimate the rate of change of the pneumonia death rate in the middle of 1924 and at the end of 1924.

18. The sales (in thousands of dollars) for a retail women's clothing store[28] for 2 years are given in Table 8.20.

a. Why have the December data been omitted from Table 8.20?

b. Find a model for the 2-year data set.

c. Use the model to estimate the rate of change of sales in April of both years.

19. The February 1995 tide chart for the Savannah River entrance notes the times of high and low tides.[29] A portion of the tide chart is shown in Table 8.21. Assume that the tidal variation is 4 feet; that is, it varies from -2 to +2 feet around mean sea level.

TABLE 8.20

Year 1		Year 2	
Month	Sales (thousands of dollars)	Month	Sales (thousands of dollars)
1	5.0	1	5.6
2	5.7	2	5.3
3	10.6	3	10.4
4	9.4	4	11.7
5	9.9	5	11.7
6	14.2	6	13.5
7	12.9	7	12.2
8	11.9	8	14.0
9	10.1	9	14.0
10	8.3	10	10.6
11	8.5	11	7.3

TABLE 8.21

Feb 26	Feb 27	Feb 28	Mar 1	Mar 2	Mar 3	Mar 4
HI 0522 LO 1146 HI 1739	LO 0003 HI 0619 LO 1238 HI 1833	LO 0056 HI 0708 LO 1326 HI 1922	LO 0145 HI 0753 LO 1411 HI 2006	LO 0231 HI 0835 LO 1453 HI 2047	LO 0313 HI 0915 LO 1532 HI 2128	LO 0353 HI 0955 LO 1609 HI 2209
Mar 5	**Mar 6**	**Mar 7**	**Mar 8**	**Mar 9**	**Mar 10**	**Mar 11**
LO 0431 HI 1036 LO 1646 HI 2251	LO 0509 HI 1120 LO 1724 HI 2336	LO 0549 HI 1206 LO 1805	HI 0024 LO 0634 HI 1255 LO 1853	HI 0115 LO 0728 HI 1347 LO 1950	HI 0208 LO 0831 HI 1442 LO 2053	HI 0304 LO 0934 HI 1539 LO 2154

a. On the basis of the data with mean sea level at 0 feet, what would you expect the period, amplitude, and average water height to be if the data are modeled by a sine function?

b. Use the data for March 2 through March 4 to find a sine model for the height of the water as a function of time.

c. How fast is the water falling at noon on March 2?

d. How fast is the water falling at noon on March 4?

e. If a boat needs at least 1 foot of water above mean sea level in order to float away from being grounded at low tide, at what times is it possible for this boat to float away on March 3?

20. Table 8.22 shows the monthly sales (in thousands of dollars) for an ice cream company for a 4-year time period.

a. Find a sine model for the data. Comment on how well it fits the data.

b. Use the data in the table to find the average of the high and low sales each year.

c. Find a model for the average sales found in part *b*. Use as input the first month of each year in which sales would be expected to reach the average value.

28. Shennanigans, Bellingham, WA.

29. *Sea Island Scene of Beaufort* (Beaufort, SC: Sands Publishing Co., 1995).

If you have discussed degree and radian measure with your students, they may be concerned that the data values in Table 8.23 are measured in degrees and may want to convert them to radians. Point out that these degree measures are output values, not input values. It is input values that we convert from degrees to radians.

TABLE 8.22

Month	Ice cream sales			
	Year 1	Year 2	Year 3	Year 4
January	50	52	55	58
February	60	62	65	68
March	77	79	83	86
April	96	99	102	106
May	137	140	144	149
June	158	163	165	170
July	167	174	175	180
August	159	165	166	171
September	108	114	116	120
October	75	78	81	85
November	61	62	64	67
December	54	56	58	61

d. Replace the parameter d in the function in part a with the function for average value found in part c. Comment on how well this new function fits the data in Table 8.22 compared with the function in part a.

e. Use the model in part d to estimate the rate of change of sales in September of the third year and January of the fourth year. Interpret your answers.

21. The declination of the sun is the angle of the sun as measured from the equator. Table 8.23 shows the angle of declination[30] of the sun in Greenwich, England, for selected days throughout the year. These values vary slightly from year to year.

TABLE 8.23

Day of the year	Declination of sun	Day of the year	Declination of sun
1	23.02° S	182	23.05° N
32	17.15° S	213	17.82° N
60	7.17° S	244	8.00° N
91	4.83° N	273	3.48° S
121	15.30° N	304	14.67° S
152	22.09° N	334	21.92° S

a. Considering southern declinations to be negative and northern declinations positive, find a sine model for the data. State and interpret the amplitude and period of the function.

b. Use the equation to find the rate of change of the declination of the sun at the two equinoxes (when the sun is directly on the equator).

c. What is the relationship between the rate of change of the declination of the sun and the summer and winter solstices?

The functions in Activities 22 through 27 are composite functions. For each function, determine the inside and outside function, and give the formula for the derivative.

22. $y = 2.4^{3 \sin x - 12}$

23. $y = \sin(2.4^x)$

24. $y = \sqrt{\sin(2x + 3)}$

25. $y = 4(\sin x - 7)^2 + 8(\sin x - 7) + 13$

26. $y = \sin(3x^2 + 19x)$

27. $y = \ln(\sin x)$

8.4 Extrema and Points of Inflection

When describing a cycle, we must note the height of its peaks and the depth of its valleys. The height and depth between which a cycle oscillates determine its amplitude. How often the maximum values and minimum values repeat depends on how often the cycle repeats.

We might also want to find the points at which the concavity of a cycle changes, because inflection points often have interesting interpretations in context. For example, suppose the economy is in an inflationary cycle with costs increasing by 3% per

30. 1996 *Old Farmer's Almanac.*

year in January. Economists would be interested in finding the rate of change of costs in February, rather than just the change in costs, to determine whether inflation is getting better or worse.

Inflection points are also important to product managers in business. A manager must constantly watch sales figures and try to anticipate consumer interest. If consumer interest is increasing and is continuing to exhibit an ever larger increase, then demand for the product is accelerating, and the manager should pay close attention to production so that there will be adequate product available to meet the sales needs. If customer interest is increasing but at a declining rate, then the manager may conclude that the market is maturing and that serious attention needs to be given to creating more interest in the product by discounts, advertising, and so on. Because such fundamental shifts take place at inflection points, they are important in these types of applications.

Insightful students are likely to question why we bother using derivatives to identify optimal points and inflection points of sine graphs when such points are easily located algebraically if one understands the role of the parameters in the sine function. This is a valid question. The value of knowing derivatives of the sine and cosine function lies in our ability to calculate rates of change, to identify interesting points of more complicated functions (such as product or composite functions) involving the sine or cosine, and to find antiderivatives needed in determining integrals. If you are emphasizing algebraic skill in your course, you may wish to have your students verify that the algebraically found optimal points and inflection points of a sine function are the same as those found using calculus.

Determining Extreme Points

It is often important to know the location of, and output values for, maxima and minima of cyclic functions. Recall from Section 5.2 that if a function has a relative maximum or a relative minimum at an input c, then either $f'(c) = 0$ or $f'(c)$ does not exist but $f(c)$ does exist. We also know that if the derivative is zero and the slope graph crosses (not just touches) the horizontal axis at a certain input, then a relative maximum or a relative minimum exists at that input. When we are using sine or cosine functions of the forms discussed in this chapter, the derivative exists at all input values, so in our search for extreme points, we need consider only the inputs where the derivative of the function crosses the input axis.

EXAMPLE 1 *Heating Oil Sales*

Weekly sales of heating oil are cyclic with a winter peak and a summer lull. A model for the weekly heating oil sales for a certain company is

$$f(t) = 2000 \sin(0.1208t + 1.5707) + 3000 \text{ gallons}$$

t weeks after the beginning of the year.

a. Find the period of the heating oil sales model.

b. Sketch a graph of one cycle of f, and mark the locations of maximum and minimum heating oil sales.

c. Use derivatives to find the maximum and minimum of a cycle.

Solution

a. The period of the heating oil sales model is $\frac{2\pi}{0.1208} \approx 52$ weeks.

b. A graph of the weekly heating oil sales for a 1-year period is shown in Figure 8.33.

c. The derivative of f is

$$f'(t) = 241.6 \cos(0.1208t + 1.5707) \text{ gallons per week}$$

t weeks after the beginning of the year. By setting the derivative equal to zero and solving for t, we find that $f'(t) = 0$ when $t \approx 0$ and again when $t \approx 26$. The graph

We arbitrarily chose the value $t = 0$, rather than $t = 52$, as the input corresponding to the maximum. Either input value is correct, and both correspond to the same output value.

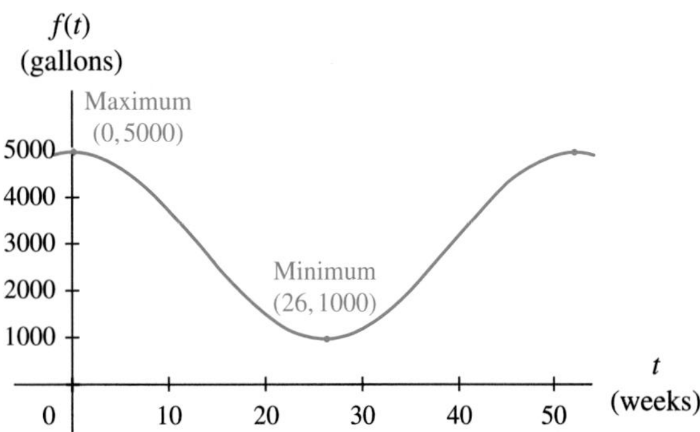

FIGURE 8.33

in Figure 8.33 shows that the second solution gives the minimum. Heating oil sales reach a maximum of $f(0) \approx 5000$ gallons at the beginning of January and reach a minimum of $f(26) \approx 1000$ gallons at the beginning of July. Sales again reach the maximum at 52 weeks.

Being able to solve for the values at which the first derivative of a sine or cosine function is zero helps us locate maxima and minima. We use a similar technique when we are looking for inflection points.

Inflection Points

The opening paragraphs in this section discussed some situations that illustrate the importance of finding the inflection points of a cyclic model. Recall from Section 5.4 that at a point of inflection on the graph of a function, the second derivative is zero or does not exist. If the second derivative graph crosses the horizontal axis, then an inflection point occurs on the graph of the function at that input value. Let us again consider the model for heating oil sales.

EXAMPLE 2 *Heating Oil Sales*

A model for weekly heating oil sales is

$$f(t) = 2000 \sin(0.1208t + 1.5707) + 3000 \text{ gallons}$$

t weeks after the beginning of the year. A graph of f is shown in Figure 8.33 in Example 1.

a. Use the graph in Figure 8.33 to estimate the locations of the inflection points.

b. Use derivatives to find the inflection points.

c. What would be the best course of action for the sales manager during the time before and after each of the inflection points?

Solution

a. The graph of weekly heating oil sales appears to be concave down from week 0 through week 13, concave up from week 13 through week 39, and concave down from week 39 through week 52. Thus we estimate that inflection points occur at the beginning of April (week 13) and the beginning of October (week 39).

b. In order for inflection points to occur on a sine or cosine graph of the forms used in this chapter, the second derivative of the function must be zero. The second derivative of the heating oil function is

$$f''(t) = -29.18528 \sin(0.1208t + 1.5707) \text{ gallons per week per week}$$

t weeks after the beginning of the year. By setting $f''(t) = 0$ and solving for t, we find that inflection points occur at $t \approx 13.0$ and $t \approx 39.0$. In both cases, 3000 gallons of heating oil are sold, so the inflection points are (13, 3000) and (39, 3000). The point of most rapid decrease occurs at the 13th week, and the point of most rapid increase occurs at the 39th week. (See Figure 8.33.)

c. Between January 1 and April 1, the oil manager should be concentrating on maintaining supply, making timely deliveries, and advertising the product. From April 1 until October 1, heating fuel demand is in the summer doldrums, and the company should be running promotion specials to move the product. Between October 1 and December 31, the manager should focus on ensuring that there is enough heating oil available to meet the demand.

Of course, in the real world there are not such idealized models as that in the previous example, but managers routinely talk about changes in the rate of increase or rate of decrease. Through those observations, they try to determine when fundamental shifts in demand occur. In mathematical terms, they are looking for the inflection points.

The connection between relative maxima and minima and inflection points of a function and the zero values of the first and second derivatives is what makes the use of calculus concepts so powerful in the analysis of cycles.

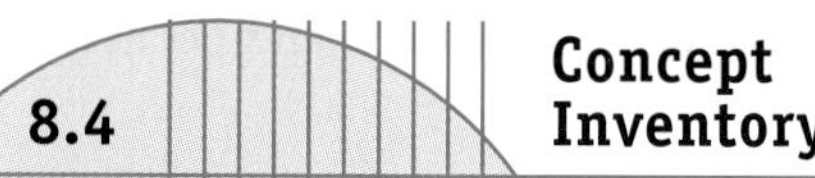

8.4 Concept Inventory

- *Maxima and minima of sine and cosine models*
- *Inflection points of sine and cosine models*

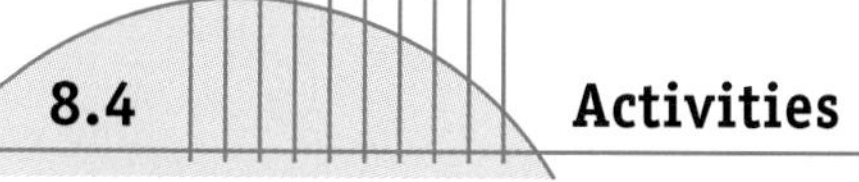

8.4 Activities

1. Table 8.24 gives the normal daily mean temperatures[31], based on the 30-year period 1961 through 1990, for Omaha, Nebraska.

   a. Use only the data to estimate the maximum and minimum daily mean temperatures for Omaha.

   b. Find a sine model for the data, and determine the highest and lowest points on the graph. Interpret these points in the context of Omaha's weather.

   c. How do the answers obtained with the equation compare with those found using the data? What relationship is there between the maximum and minimum daily mean temperatures for Omaha found in parts *a* and *b* and the actual yearly high and low temperatures in a given year?

TABLE 8.24

Month	Daily mean temperature (°F)
January	21.1
February	26.9
March	38.6
April	51.9
May	62.4
June	72.1
July	76.9
August	74.1
September	65.1
October	53.4
November	39.0
December	25.1

31. U.S. National Oceanic and Atmospheric Administration, *Climatography of the United States*, No. 81.

2. An exclusive fishing club on the Restigouche River in Canada kept detailed records of the number of fish caught by its members. A model[32] for the number of salmon caught per rod per day between 1880 and 1905 is

$$s(y) = 0.327 \sin(0.675y - 2.122) + 1.267 \text{ salmon per rod per day}$$

y years after 1880.

a. Find the nearest year after 1905 in which, according to the model, the number of salmon is expected to peak.

b. Find the nearest year after 1905 in which, according to the model, the number of salmon is expected to be at a minimum.

c. How might the manager of the fishing club have used the information in parts *a* and *b* to benefit the club?

3. The number of males per 100 females in the United States can be modeled[33] by

$$m(t) = 5.3365 \sin(0.0425t + 1.268) + 99.938 \text{ males per 100 females}$$

t years after 1900.

a. Find the maximum and minimum values of the function and the years between 1900 to 1990 in which they occur. Interpret your answers.

b. Determine when the graph is decreasing most rapidly and how rapidly it is decreasing at that time. Interpret your answers.

4. The average amount of extraterrestrial radiation[34] in Amarillo, Texas, for each month of the year is given in Table 8.25. The amount is measured in equivalent millimeters of water evaporation per day.

a. Use only the data to estimate the maximum and minimum radiation levels and when those levels occur.

b. Find a sine model for the data, and determine the highest and lowest points on the graph. Interpret these points in the context of radiation.

TABLE 8.25

Month	Radiation (mm of H_2O per day)	Month	Radiation (mm of H_2O per day)
Jan	8.3	July	16.8
Feb	10.2	Aug	15.6
Mar	12.8	Sep	13.6
Apr	15.0	Oct	11.2
May	16.5	Nov	9.0
June	17.0	Dec	7.8

c. How might knowledge of high and low radiation levels be useful to someone living in Amarillo?

5. On December 15, 1995, a team of astronomers discovered X-ray pulses emitting from what they believe to be a neutron star. The speed of the pulses (in counts per second) can be modeled[35] by

$$p(s) = 40.5 \sin(0.01345s - 1.5708) + 186.5 \text{ counts per second}$$

after s milliseconds.

a. What is the period of the pulsations?

b. Determine two times when the pulse speed was highest and two times when it was lowest. What were the pulse speeds at those times?

c. Find and interpret one inflection point where the function is increasing and one inflection point where the function is decreasing.

6. The combined power output of a diesel engine and wind turbine in an irrigation pumping system can be modeled[36] by

$$P(s) = 20.204 \sin(0.258s + 0.570) + 19.026 \sin(0.247s - 2.549) + 44.952 \text{ kilowatts}$$

when the wind speed is s meters per second. The model should be used for wind speeds between 6 and 18 meters per second, because the wind turbine produces power only if wind speeds exceed 6 meters per second.

32. Based on data estimated from E. R. Dewey and E. F. Dakin, *Cycles: The Science of Prediction* (New York: Holt, 1947).
33. Based on data from the U.S. Bureau of the Census.
34. A. A. Hanson, ed., *Practical Handbook of Agricultural Science* (Boca Raton, FL: CRC Press, 1990).
35. Based on information in M. H. Finger et al., "Discovery of Hard X-ray Pulsations from the Transient Source GRO J1744 - 28," *Nature*, vol. 381 (May 23, 1996), pp. 291–292.
36. Based on information in B. A. Stout, *Handbook of Energy for World Agriculture* (London: Elsevier Applied Science, 1990).

a. Find the wind speeds associated with the greatest and the least power output for this pumping system.

b. On the basis of your answer to part *a*, what do you believe is the benefit of having a wind turbine in this pumping system?

c. Find all inflection points of *P* between $s = 6$ and $s = 18$. Interpret your answers.

7. Table 8.26 shows the weekly pneumonia death rate[37] (number of deaths per 100,000 people per week) in the United States from 1923 through 1925.

TABLE 8.26

Weeks since Jan 1, 1923	Pneumonia death rate	Weeks since Jan 1, 1923	Pneumonia death rate
0	100	80	60
10	210	90	45
20	95	100	95
30	35	110	160
40	55	120	140
50	85	130	80
60	140	140	40
70	160	150	90

a. Find a sine function to fit the pneumonia death rate data.

b. Use the sine equation to predict the next peak in the pneumonia death rate beyond the data given. How might this information have been useful to health care providers in 1926? What factors might cause the actual peak to differ from the one found by the equation?

c. Find and interpret the first inflection point located beyond the data given.

8. Explain how you could determine without using derivatives the maximum, minimum, and inflection points (both output and input) on a graph of $f(x) = a \sin(bx + h) + k$.

9. Figure 8.4.1 shows graphs of *w*, the indexed price of wheat in the United Kingdom, and a sine function, *i*, for the index.[38]

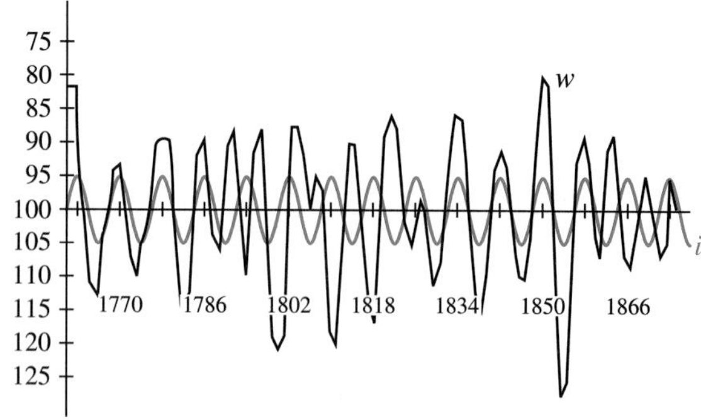

FIGURE 8.4.1

a. How many of the maxima of the sine graph are within 1 year of a relative maxima in the wheat price index? Note that the vertical axis is inverted.

b. How many of the minima of the sine graph are within 1 year of a relative minima point in the wheat price index?

c. An equation for the sine graph shown in Figure 8.4.1 is $y = 100.2 + 4.4 \sin\left(\frac{\pi}{4}t + \frac{62}{45}\right)$ where *t* is the number of years since 1761. Use this equation to find when the next maximum and minimum occur after 1874.

d. On the basis of the graph in Figure 8.4.1 and your answers to parts *a* and *b*, how accurately do you think the years you found in part *c* predict when the next maximum and minimum wheat price indices occurred? Do you think the sine model should be used to predict the next maximum and minimum price index?

10. A model[39] for the Suaerbeck index of general wholesale prices in England between 1818 and 1914 is

$$p(x) = 3.1 \sin\left(\frac{\pi}{3.69}x + \frac{23}{45}\right) + 100$$

x years after 1818. Use the model to estimate when the next high and low wholesale prices occurred after 1914.

11. The *Anchorage Daily News* in Anchorage, Alaska, includes a chart of the hours of daylight in its weather section. Apparently the amount of daylight is as much a topic of conversation in Alaska as the temperature is in the other 49 states! A copy of the newspaper chart is shown in Figure 8.4.2.[40]

37. Estimated from E. R. Dewey and E. F. Dakin, *Cycles: The Science of Prediction* (New York: Holt, 1947).
38. Figure 8.4.1 from H. L. Moore, *Generating Economic Cycles* (New York: The Macmillan Co., 1923; repr. 1967, Augustus M. Kelley Publishers), p. 75. Reprinted by permission.
39. Ibid.
40. Figure 8.4.2 used with the permission of the *Anchorage Daily News*.

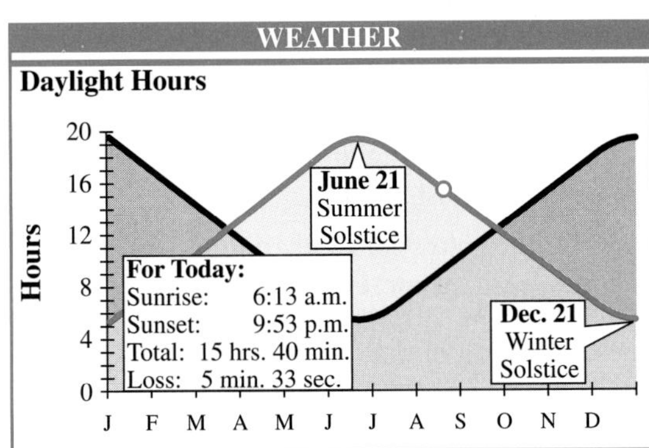

FIGURE 8.4.2

a. There are two graphs shown in Figure 8.4.2. Which graph represents daylight hours, and what does the other graph represent?

b. On the basis of the graph of daylight hours, estimate the period, amplitude, and vertical shift.

c. Reading carefully from the chart, estimate the hours of daylight at the beginning of each month, and create a scatter plot of these data. Explain why it makes sense to use a sine model for this data set.

d. Find a sine function to fit the data. Compare the period, amplitude, and vertical shift of the function to your estimates in part *b*.

e. The amount of daylight determines one's moods. In particular, less daylight has been known to cause "winter blues." Also important to mood is a person's perception of future change. For example, when the number of hours of daylight is increasing rapidly, some people may develop an optimistic outlook, whereas a pessimistic outlook may result when the number of hours of daylight is declining rapidly. Use the sine model to find when the number of hours of daylight is increasing most rapidly and when it is decreasing most rapidly in Anchorage.

f. If you were directing a public campaign, you would not want to introduce the campaign when the public was in a pessimistic winter mood. Use your answers to part *e* to estimate when politicians in Alaska should begin a new campaign.

12. Find a function (or find data and create a function) for the number of daylight hours in your hometown. Repeat Activity 11 parts *e* and *f* for your hometown.

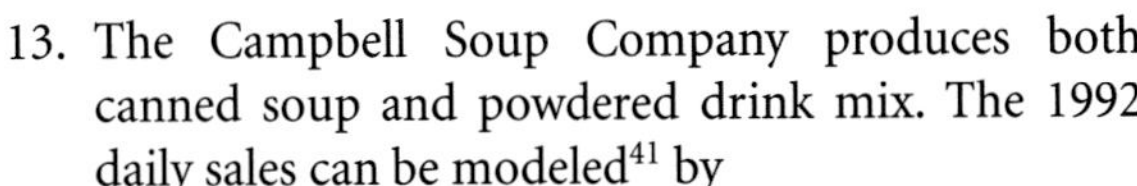

13. The Campbell Soup Company produces both canned soup and powdered drink mix. The 1992 daily sales can be modeled[41] by

$$c(x) = 1.8961 \sin(0.0186x + 1.1801) + 5.4660$$
million cans (16 ounce) of soup

$$p(x) = 7.3026 \cos(0.0197x - 3.7526) + 20.3728$$
million (reconstituted) pints of drink mix

where x is the day of the year. Assume that soup sold at an average price of 70 cents per can and that drink mix sold at an average price of 20 cents per pint.

a. Write a function giving the revenue from the combined 1992 daily sales as a function of the day of the year on which soup and powdered drink mix are produced.

b. Write a function that models the rate of change of the revenue from the combined 1992 sales.

c. Use the model and derivative to estimate the highest revenue and the day(s) on which this maximum occurred.

d. Use the model and derivative to estimate the lowest revenue and the day(s) on which this minimum occurred.

14. A sine model for the amount of carbon dioxide in the atmosphere[42] was given in Section 8.3. We could also have used the following cosine model:

$$c(x) = (0.010345x + 2.105676)\cos\left(6.283x - \frac{\pi}{2}\right) + 0.013783x^2 - 0.844566x + 316.913080$$
parts per million

x years after the beginning of 1900. The model applies for years between 1959 and 1999.

a. Graph the sine function given in Section 8.3 in Example 3 and the cosine function here. Are there any differences between these two graphs? Why can either a sine function or a cosine function be used?

b. Write a formula for the derivative of the cosine function.

c. Use c and its derivative to find the maximum and minimum carbon dioxide levels in 1960,

41. *Wall Street Journal*, July 7, 1993, p. B1.
42. R. M. White, "The Great Climate Debate," *Scientific American*, vol. 263, no. 1 (1990), pp. 36–43.

1970, 1980, and 1990. In what months do these maximum and minimum levels occur?

d. Why are the maxima different for each year?

15. Refer to the cosine model for the amount of carbon dioxide in the atmosphere given in Activity 14 to answer the following questions.

a. Write a formula for c''.

b. Determine in what months of 1960, 1970, 1980, and 1990 the carbon dioxide level was increasing most rapidly.

c. According to the function c, what was the carbon dioxide level at the time of most rapid increase in 1960, 1970, 1980, and 1990? Explain why these values are increasing from 1960 through 1990.

16. A model of the data for the amount of ultraviolet radiation[43] in Auckland, New Zealand, is

$$a(m) = 27.1 \sin(0.485m - 1.707) + 32.9 \text{ watts per cm}^2$$

when $m = 1$ in January, $m = 2$ in February, and so on.

a. Write a function for the rate of change of the amount of ultraviolet radiation.

b. Estimate the time when ultraviolet radiation is at its highest level and the time when it is at its lowest level.

c. How much ultraviolet radiation is received at the times found in part *b*?

17. Refer to the model for the amount of ultraviolet radiation in Auckland, New Zealand, given in Activity 16 to answer the following questions.

a. Write a function for the second derivative of the ultraviolet radiation model.

b. Estimate the time when ultraviolet radiation is increasing most rapidly and the time when it is decreasing most rapidly.

c. How much ultraviolet radiation is received at the times found in part *b*?

18. The declination of the sun in Greenwich, England, can be described by the equation[44]

$$D(d) = 23.677 \sin(0.0168x - 1.312) - 0.292 \text{ degrees}$$

on the xth day of the year. Negative output values correspond to southern declinations, and positive output values correspond to northern declinations.

a. Use calculus to find the inflection points and extrema over one period of the function. Interpret your answers.

b. Relate your answers in part *a* to the earth's equinoxes and solstices.

19. Based on data for 1965 through 1976, the reduction in the number of beetles in a certain geographical area in the Netherlands can be modeled by the equation[45]

$$R(x) = 10^{0.9 \sin\left(\frac{\pi}{3}x - \frac{6.5\pi}{3}\right) + 3.6} \text{ beetles}$$

x years after 1960.

a. What is the period of this function?

b. Use calculus to find the extrema of this function over one period. Interpret your answers.

8.5 Accumulation in Cycles

The first four sections of this chapter discussed sine and cosine functions and rates of change of sine and cosine functions. We now consider the accumulated change in sine and cosine functions and also investigate applications of integration using these functions.

We begin by stating general antiderivative formulas for the sine and cosine functions. Because $\frac{d}{dx}(-\cos x) = -1 \cdot \frac{d}{dx}(\cos x) = -1(-\sin x) = \sin x$, the Fundamental Theorem of Calculus tells us that a general antiderivative of the sine function is the negative of the cosine function. Also, because $\frac{d}{dx}(\sin x) = \cos x$, a general antiderivative of the cosine function is the sine function.

43. Based on data from W. Rudloff, *World-Climates* (Stuttgart: Wissenschaftliche Verlagsgesellschaft, 1981).
44. Based on data from the 1996 *Old Farmer's Almanac*.
45. Based on data from P. J. denBoer and R. Reddingius, *Regulation and Stabilization Paradigms in Population Ecology* (Chapman & Hall: London, 1996).

Antiderivatives of Sine and Cosine Functions

When x is measured in radians,

$$\int \sin x\, dx = -\cos x + C$$

$$\int \cos x\, dx = \sin x + C$$

To find the general antiderivative of the sine function $f(x) = a \sin(bx + h) + k$, we need only apply the antiderivative rules we used in Chapter 6. We begin by applying the Constant Multiplier Rule and the Constant Rule for antiderivatives:

$$\int [a \sin(bx + h) + k]\, dx = a\int \sin(bx + h)\, dx + \int k dx$$

$$= a\int \sin(bx + h)\, dx + kx + C_1$$

Now we need to find the general antiderivative of $y = \sin(bx + h)$. Recall that the antiderivative process reverses the derivative process. In using the Chain Rule to find the derivative of $y = \sin(bx + h)$, we would multiply by b, the derivative of the inside function, to obtain $y' = b \cos(bx + h)$. For the antiderivative, we *divide* by b:

$$\int [a \sin(bx + h) + k]\, dx = a\left[\frac{-\cos(bx + h)}{b}\right] + C_2 + kx + C_1$$

$$= \frac{-a}{b} \cos(bx + h) + kx + C$$

Similarly, we have

$$\int [a \cos(bx + h) + k]\, dx = a\left[\frac{\sin(bx + h)}{b}\right] + C_2 + kx + C_1$$

$$= \frac{a}{b} \sin(bx + h) + kx + C$$

Antiderivatives of Trigonometric Models

When x is measured in radians,

$$\int [a \sin(bx + h) + k]\, dx = \frac{-a}{b} \cos(bx + h) + kx + C$$

$$\int [a \cos(bx + h) + k]\, dx = \frac{a}{b} \sin(bx + h) + kx + C$$

With these formulas, we can find accumulated change for sine models. For example, tides in 1995 for the Savannah River entrance changed according to the rate-of-change model[46]

46. Based on data in *Sea Island Scene of Beaufort* (Beaufort, SC: Sands Publishing Co., 1995).

$$t(x) = 1.00934 \sin(0.50467x + 2.5633) \text{ feet per hour}$$

x hours after midnight February 7, 1995. The height of the water x hours after midnight is the accumulated change in the rate of change from 0 to x. Thus we can find a function for the height of the water x hours after midnight on February 7 by writing the general antiderivative T of the rate-of-change function t:

$$T(x) = \frac{-1.00934}{0.50467} \cos(0.50467x + 2.5633) + C \text{ feet above sea level}$$

Suppose we know that midtide occurs at 4:16 a.m. That is, $T(4.26667) \approx 0$. Using this fact, we solve for C and obtain $C \approx 0.008$. Thus the height of the water x hours after midnight on February 7 can be modeled as

$$T(x) \approx -2 \cos(0.50467x + 2.5633) + 0.008 \text{ feet above sea level}$$

(*Note*: Negative values of $T(x)$ denote tides that are lower than the average sea level.)

As illustrated in the tide example, when we have functions for rates of change, we find the accumulation of that rate of change by using the antiderivative of the function. Example 1 again illustrates this principle.

EXAMPLE 1 *Temperatures in Philadelphia*

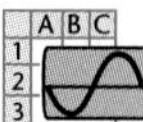

Earlier in this chapter, we saw that the rate of change of the temperature in Philadelphia on August 27, 1993, can be modeled as

$$t(x) = 2.733 \cos(0.285x - 2.93) \text{ °F per hour}$$

x hours after midnight. Use this rate-of-change model for temperature to find the accumulated change in the temperature between 9 a.m. and 3 p.m. on August 27, 1993.

Solution The accumulated change in the temperature between 9 a.m. and 3 p.m. is given by the definite integral

$$\int_9^{15} 2.733 \cos(0.285x - 2.93)\, dx$$
$$= \frac{2.733}{0.285} \sin(0.285x - 2.93)\Big|_9^{15}$$
$$= \left(\frac{2.733}{0.285} \sin[0.285(15) - 2.93]\right) - \left(\frac{2.733}{0.285} \sin[0.285(9) - 2.93]\right)$$
$$\approx 9.34606 - (-3.42296)$$
$$\approx 12.769°\text{F}$$

That is, the temperature increased by approximately 13°F between 9 a.m. and 3 p.m.

Average Values and Sine Models

Throughout this chapter, we have referred to the parameter k of the function $f(x) = a \sin(bx + h) + k$ as the average (or expected) value of a cycle. This is the value we would expect the function to take on if it were not for the fluctuations that occur. In Chapter 7 we defined the average value of a function f from c to d as

$$\text{Average value} = \frac{\int_c^d f(x)\, dx}{d - c}$$

Example 2 illustrates that the parameter k of a sine function is its average value over one cycle, according to the definition in Chapter 7.

EXAMPLE 2 *Ultraviolet Radiation*

A model for the amount of ultraviolet radiation received in Auckland, New Zealand, is

$$r(m) = 25.5\sin(0.52m - 1.57) + 32.5 \text{ watts per cm}^2$$

during the mth month of the year.

a. Calculate the average value for one cycle of this model beginning at $m = 1$.

b. Calculate the average value for one cycle of this model between two successive minima. Interpret the result.

Solution

a. The function r has period $\frac{2\pi}{0.52} \approx 12.083$ months. Thus the cycle beginning at $m = 1$ ends at $m \approx 13.083$, so the average value of the cycle can be calculated as

$$\frac{\int_1^{13.083} [25.5\sin(0.52m - 1.57) + 32.5]\,dm}{13.083 - 1}$$

$$= \frac{\left(\frac{-25.5}{0.52}\cos(0.52m - 1.57) + 32.5m\right)\Big|_1^{13.083}}{12.083}$$

$$\approx \frac{400.798 - (8.100)}{12.083}$$

$$\approx 32.5 \text{ watts per cm}^2$$

That is, discounting seasonal fluctuation, we expect the UV radiation over a period of approximately 1 year in Auckland to be 32.5 watts per cm^2. Note that this average value is the parameter k in the function r.

b. The cycle between the first two consecutive minima starts at $m \approx -0.002$ and ends at $m \approx 12.081$. The average value of the cycle is calculated as

$$\frac{\int_{-0.002}^{12.081} [25.5\sin(0.52m - 1.57) + 32.5]\,dm}{12.081 - (-0.002)}$$

$$= \frac{\left(\frac{-25.5}{0.52}\cos(0.52m - 1.57) + 32.5m\right)\Big|_{-0.002}^{12.081}}{12.083}$$

$$\approx \frac{392.6457 - (-0.0531)}{12.083}$$

$$\approx 32.5 \text{ watts per cm}^2$$

That is, discounting seasonal fluctuation, we expect the UV radiation in Auckland to be 32.5 watts per cm^2. This is, again, the parameter k of the function r.

No matter which cycle you use to compute the average value of a sine function, the average value of that cycle will always be the parameter k of that sine function.

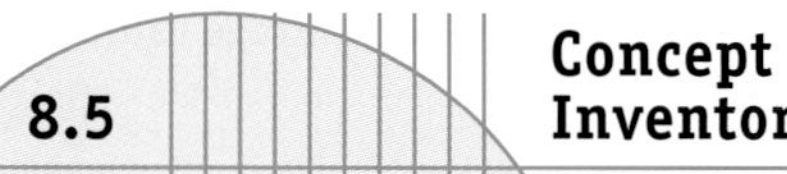

8.5 Concept Inventory

- *The antiderivative of the sine function*
- *The antiderivative of the cosine function*
- *Average value of a sine or cosine function*

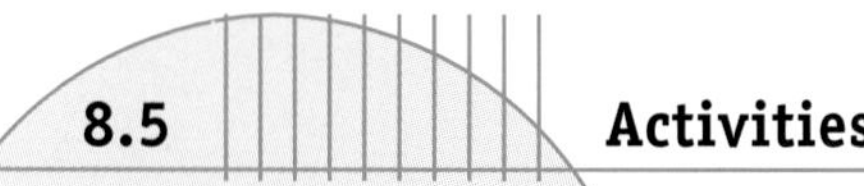

8.5 Activities

For Activities 1 and 2, sketch the indicated accumulation function graphs using the graph in Figure 8.5.1.

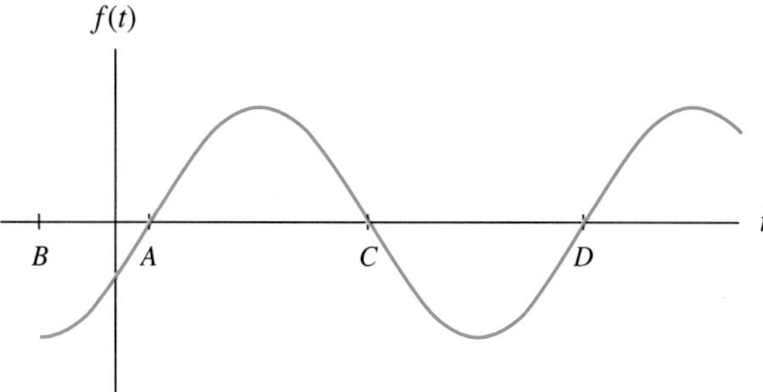

FIGURE 8.5.1

1. $\int_A^x f(t)\,dt$

2. $\int_B^x f(t)\,dt$

For Activities 3 through 6, find the indicated general antiderivative.

3. $\int (7.3 \sin x + 12)\,dx$

4. $\int 7.3 \sin(x + 12)\,dx$

5. $\int \sin(7.3x + 12)\,dx$

6. $\int [7.3 \sin(7.3x + 12) + 12]\,dx$

7. Find the general antiderivative of
$f(x) = 4.67 \sin(0.024x + 3.211) + 14.63$

8. Find the general antiderivative of
$f(t) = 49.88 \cos(3.54t - 4.86) + 7.02$

9. The rate of change of sales for a store specializing in swimming pools in the summer and ski gear in the winter can be modeled by

$$S'(t) = 23.944 \cos(0.987t + 1.276) \text{ thousand dollars per month}$$

where $t = 1$ in January, $t = 2$ in February, and so on.

a. Find a model for sales in month t if sales in January are \$54,000.

b. Use the model to estimate when peak sales occur each year. Also estimate when sales will be lowest.

10. The rate of change of the average temperature in New York from 1873 through 1923 can be modeled by

$$T'(x) = 42.059 \cos(0.524x - 2.27) \text{ °F per month}$$

where $x = 1$ in January, $x = 2$ in February, etc.

a. Find a model for the average temperature in New York. The average temperature in July is 73.5°F.

b. What does the model give as the average temperature in December?

11. The function f in Figure 8.5.2 is cyclic. The area of the shaded region is 2. Using the symmetry indicated by the graph, find each of the following:

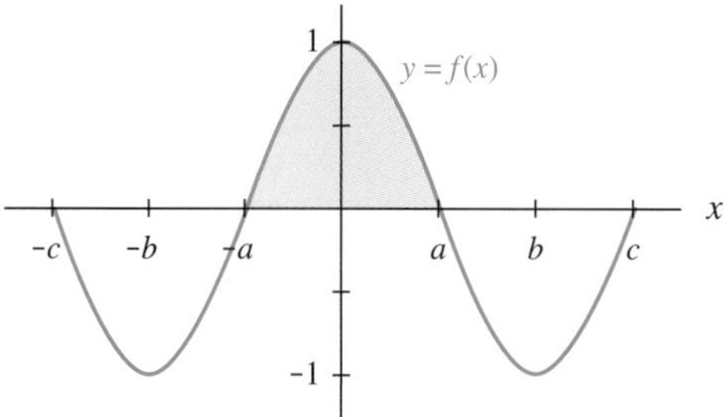

FIGURE 8.5.2

a. $\int_0^a f(x)\,dx$

b. $\int_{-a}^c f(x)\,dx$

c. $\int_{-a}^a [f(x) + 1]\,dx$

d. $\int_{-c}^a 3\,|f(x)|\,dx$

12. The rate of change of the number of hours of daylight in Honolulu, Hawaii, can be modeled by

$$d'(m) = 1.18 \cos(0.514m - 1.616) \text{ hours per month}$$

m months after the beginning of each year.

a. Find the area beneath the graph of d' that lies above the m-axis between $m = 0$ and $m = 12$. Interpret your answer.

b. Find the area above the graph of d' that lies below the m-axis between $m = 0$ and $m = 12$. Interpret your answer.

13. On December 15, 1995, a team of astronomers discovered X-ray pulses emitting from what they

believe to be a neutron star. The speed of the pulses can be modeled[47] by

$$p(s) = 40.5 \sin(0.01345s - 1.5708) + 186.5 \text{ counts per second}$$

after s milliseconds.

a. What will the units be on the heights and widths of midpoint rectangles constructed between the graph of p and the s-axis? What will the units be on the area of such midpoint rectangles?

b. Using the fact that there are 1000 milliseconds in one second, convert $p(s)$ to counts per millisecond.

c. Find and interpret the area between the graph of p and the s-axis for one period of the function, beginning at $s = 0$.

14. a. Referring to the model in Activity 10, find and interpret the value of $\int_2^8 T'(x)\,dx$.

b. Find the area of the region between the graph of T' and the x-axis from $x = 2$ to $x = 8$.

15. a. Referring to the model in Activity 9, find and interpret the value of $\int_1^6 S'(t)\,dt$.

b. Find the area of the region between the graph of S' and the t-axis from $t = 1$ to $t = 6$.

c. Find the average sales between January and June.

16. The rate of change of the ratio of males to females in the United States from 1900 through 1990 can be modeled[48] by

$$r(t) = 0.2268 \sin(0.0425t + 2.839) \text{ males per 100 females per year}$$

t years after 1900. Find and interpret the values of $\int_0^{40} r(t)\,dt$, $\int_{50}^{90} r(t)\,dt$, and $\int_0^{90} r(t)\,dt$.

17. The number of lizards legally gathered from a western state for selected months during a 2-year time period can be modeled by the equation[49]

$$L(x) = 15.388 \sin(0.787x - 1.667) + 16.346 \text{ thousand lizards per month}$$

where x is a number assigned to each month. The numbers 1 through 8 represent each of the 8 consecutive months in which lizards are harvested in the first year, and the numbers 9 through 16 represent each of the 8 consecutive harvesting months of the second year. In the months during the 2-year period not defined by the model, there were no lizards gathered.

a. Use a definite integral to estimate the number of lizards harvested during the 2-year harvesting period.

b. Because this model was formed from count data, you could also estimate the number of lizards harvested during the 2-year time period by summing each of the output values for the 16 months. Calculate this sum, and compare it to the estimate in part *a*.

c. Which method (the one in part *a* or the one in part *b*) do you prefer?

18. The rate of change of the declination of the sun for Greenwich, England, can be described by the equation[50]

$$S(d) = 0.398 \sin(0.0168d + 0.259) \text{ degrees per day}$$

on the dth day of the year. Use a definite integral to determine by how many degrees the declination changes between January 1 and March 1.

47. Based on information in M. H. Finger et al., "Discovery of Hard X-ray Pulsations from the Transient Source GRO J1744 - 28," *Nature*, vol. 381 (May 23, 1996), pp. 291–292.
48. Based on data from the U.S. Bureau of the Census.
49. Based on information from the Nevada Division of Wildlife.
50. Based on data from the 1996 *Old Farmer's Almanac*.

Summary

Trigonometry has been an important topic in mathematics for centuries. We consider two trigonometric functions in terms of a unit circle. The x- and y-coordinates, $\cos \theta$ and $\sin \theta$, respectively, are at the point where the terminal side of the angle θ intersects the unit circle. We saw that the angle θ can be measured in radians. Measuring angles in radians leads to a wider range of applications and is convenient in calculus. When we use sine functions to model data, we no longer consider the input as an angle measurement. Instead, the input is simply a real number with an interpretation determined by the context.

The Sine Model

Many types of data exhibit a cyclic pattern. Such cyclic data can be modeled by a sine function of the form $f(x) = a \sin(bx + h) + k$. The parameters of the sine function correspond to certain physical properties of the data. The parameter k is the average value of a cycle and is the function's vertical shift. The parameter a is the amplitude of the fluctuation away from that average value. A negative value of a denotes reflection across the horizontal axis. The parameter b (considered positive) is associated with the period of the cycle by the equation

$$\text{period} = \frac{2\pi}{b}$$

The parameter h is associated with the horizontal shift of the function by the equation

$$\text{horizontal shift} = \frac{|h|}{b}$$

where the shift is to the left if $h > 0$ and to the right if $h < 0$. Understanding these parameters and their interpretations in the context of the model helps us to understand the model itself.

Rates of Change

As with all the other functions that we have seen in this text, we are able to use calculus to analyze the change that is occurring in a sine or cosine model. Derivatives answer the question "How rapidly is the quantity changing?" and help us locate the maxima and minima of a sine or cosine function. Second derivatives help us locate the inflection points of these functions.

The derivative of the sine function is simply the cosine function, and using derivative rules developed in Chapter 4, we have

$$\frac{d}{dx}[a \sin(bx + h) + k] = ab \cos(bx + h)$$

We also saw that the derivative of the cosine function is the negative of the sine function, which enables us to find second derivatives for sine models.

Accumulation of Change

The antiderivative formulas for sine and cosine models are

$$\int [a \sin(bx + h) + k]\,dx = \frac{-a}{b}\cos(bx + h) + kx + C$$

$$\int [a \cos(bx + h) + k]\,dx = \frac{a}{b}\sin(bx + h) + kx + C$$

We use these antiderivatives to answer questions about cyclic models such as "How much change has accumulated?" We also used definite integrals to show that the parameter k of a sine model is indeed the average value of its cycle.

Concept Check

Can you	To practice, try	
• Find and interpret sine and cosine values in terms of a unit circle?	Section 8.1	Activity 3
• Identify the amplitude, reflection, frequency, period, vertical shift, and horizontal shift of a sine function?	Section 8.1	Activity 17
• Graph a sine function?	Section 8.1	Activity 25
• Construct a sine model with and without the use of technology?	Section 8.2	Activities 3, 4
• Calculate and interpret derivatives of sine and cosine functions?	Section 8.3	Activity 13
• Use derivatives to find extrema and inflection points of sine and cosine functions?	Section 8.4	Activity 3
• Find and interpret definite integrals of sine and cosine functions?	Section 8.5	Activity 13
• Find the expected (average) value of a cyclic function?	Section 8.5	Activity 15

Review Test

1. Lawn mower sales for a certain manufacturer are cyclic, with a low in the winter months, an increase in the spring, and a decline in the fall. Orders at this company for a 1-year period are shown in Table 8.27.

TABLE 8.27

Month	Lawn mowers	Month	Lawn mowers
Jan	97	July	1400
Feb	450	Aug	1050
Mar	800	Sept	795
Apr	1150	Oct	400
May	1406	Nov	190
June	1500	Dec	75

a. Use the data to determine the amplitude, vertical shift, period, and horizontal shift for the lawn mower orders.

b. Use your calculator or computer to find a sine function for the number of lawn mowers ordered in terms of the month of the year. How do the function parameters compare to those you gave in part *a*?

c. Use the equation to predict lawn mower orders in April of the following year. Under what conditions should this estimate be used as a prediction?

d. Use the equation to determine when lawn mower orders are greatest.

e. When is the number of lawn mowers ordered increasing most rapidly?

2. The graph in Figure 8.34 shows the Sauerbeck Index[51] of general wholesale prices between 1818 and 1913.

51. H. L. Moore, *Generating Economic Cycles* (New York: Augustus M. Kelley, 1967).

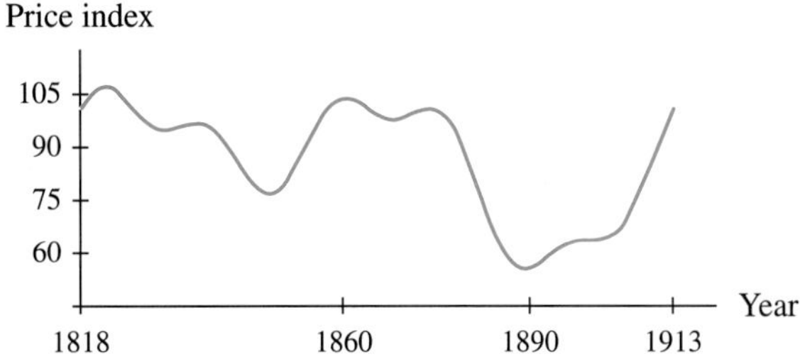

FIGURE 8.34

a. Estimate the average rate of change of the price index between 1860 and 1890.

b. Estimate how quickly the price index was changing in 1860.

3. The equation for the graph in Figure 8.34 is

$$S(t) = 88.6 + 11.2\sin\left(\frac{\pi}{48}t + \frac{19\pi}{10}\right) + 13.8\sin\left(\frac{\pi}{24}t + \frac{11\pi}{36}\right) + 5.3\sin\left(\frac{\pi}{9.6}t + \frac{13\pi}{45}\right)$$

t years after 1818.

a. Use the equation to estimate the average rate of change of the price index between 1860 and 1890.

b. Find and interpret $S'(42)$.

c. Find the average price index between 1860 and 1890.

4. The rate of change of the daily sales of powdered drink mix produced in 1992 by the Campbell Soup Company can be modeled by

$$r(t) = 0.1439\sin(0.0197t - 3.7526)$$ million (reconstituted) pints per day

t days after January 1, 1992.

a. Determine when the rate of change of the sales model was greatest in 1992.

b. Find a model for the sales if the sales were 14.4 million pints on January 1, 1992.

c. How is the answer to part *a* related to the graph of the function in part *b*?

5. Find and interpret the value of $\int_{32}^{61} r(t)\,dt$, where r is rate-of-change function in Question 4.

Project 8.1 Seasonal Sales

Setting

Some companies sell products or offer services that are seasonal in nature. Summer or winter sports equipment is one example of a seasonal product, and swimming pool cleaning is an example of a seasonal service. Sometimes companies specialize in two different products or services for which the demand peaks in different seasons. For example, a snow ski shop may sell patio furniture in the summer.

Tasks

Find a company that specializes in a seasonal product or service and that is willing to release sales data. Obtain data from the company. The data should span at least 1 year. If the company specializes in two different seasonal products, you may need to consider sales for the two products separately.

1. Find a sine function for the data. Do you believe a sine model is the most appropriate model for the data you found? Give the period, frequency, amplitude, and vertical shift of the sine model. Interpret these values in the context of the company's sales.
2. Find the maximum and minimum sales for the model. When do these sales occur? Compare the extreme points of the sine function with the highest and lowest data values and corresponding times. According to the function, when are sales increasing and decreasing most rapidly?
3. If a sine function is not a good fit for the data, find a more appropriate function. Explain your reasons for choosing a different function.

Reporting

1. Write a letter to the company from which you obtained data. If the company sells only one product or offers only one service, suggest another product or service to offset the minimum sales created by the current product or service. Suggest when the company should begin advertising each product or service.

   If the company already sells two products or offers two services, analyze whether the peak sales of one offset the low sales of the other. If you believe the company might benefit from offering a third product or service or from increasing advertising at particular times during the year, offer suggestions in your letter.
2. Write a report for your files containing all the mathematics used in performing Tasks 1 through 3 to use as reference should the sales manager for the company call to ask you questions.

Project 8.2 Lake Tahoe Levels

Setting

Lake Tahoe lies on the California–Nevada border. Its level is regulated by a 17-gate concrete dam at the lake's outlet. By federal court decree, the lake level must never be higher than 6229.1 feet above sea level. The lake level is monitored every midnight. The United States Department of the Interior reports (8/2/96 ADAPS report) that the lake level on the first day of each month from October 1995 through September 1996 was as shown in Table 8.28.

TABLE 8.28

Month	Lake level (feet above 6220 feet above sea level)
Oct 1995	6.37
Nov	5.93
Dec	5.68
Jan 1996	6.05
Feb	6.68
Mar	7.58
Apr	8.00
May	8.29
June	8.87
July	9.03
Aug (est.)	8.90
Sep (est.)	8.20

Tasks

1. Examine a scatter plot of this set of data. Discuss why a sine function might be appropriate. If a sine function were used to describe these data, estimate what would be its period, amplitude, and vertical shift. Find a sine model to fit the data. What are the function's period, amplitude, and vertical shift? Compare these answers to your expectations, and discuss any discrepancies. Rewrite your model so that its output is in feet above sea level.
2. Carefully sketch a graph of the function for the lake level. Use the sketch to estimate the lake level in January 1996 and the rate at which the level of the lake was changing at that time.
3. Use the model to estimate the lake level in January 1996 and to estimate numerically how quickly the level of the lake was changing at that time.
4. Use the derivative of the function to find the rate at which the level of the lake was changing in January 1996.
5. According to the data, when was the lake at its lowest level? When was it at its highest level? On a sketch of the function for lake level, draw lines that are tangent to the graph of the model at the model's minimum and maximum points. What is important about the tangent lines at these points?
6. Use the derivative of the function to estimate the month and day between October 1, 1995 and September 1, 1996 when the lake was at its lowest level. Also determine when the lake was at its highest level. According to the model, did the lake remain below the federally mandated level between October 1, 1995 and September 1, 1996?
7. When (between October 1, 1995 and September 1, 1996) was the level of the lake changing most rapidly? At that time, what was the level of the lake and how quickly was it changing?

8. Discuss any other types of models that could be used for these data. Find any such models for the data. Discuss how well they fit the data, and use them to perform Tasks 6 and 7.

Reporting

1. Write a report addressed to the Department of the Interior with your findings on Lake Tahoe levels. Keep in mind that this should be a nonmathematical report of your conclusions. Use graphs in your report as appropriate. Include mathematical support for your conclusions as an appendix. Refer to your appendix in the body of your report as appropriate.
2. Prepare a 10-minute presentation of your conclusions to be given to the Department of the Interior. You should be prepared to discuss the mathematics if questioned, but keep in mind that your target audience is not expecting a math talk. Use overhead transparencies and/or other visual aids to enhance your presentation.
3. Prepare a poster of your conclusions. The poster should be self-explanatory, attractive, and easily readable from 3 feet away. Show enough mathematics on the poster to support your conclusions.

9

Ingredients of Multivariable Change: Models, Graphs, Rates

Concept Application

Competition is a fundamental element of a free-market society. We are accustomed to seeing companies compete for sales and customer loyalty. Although many factors (such as economic indicators, advertising budgets, and even the weather) affect sales of competing products, the most obvious factor is the price of the product itself. If side-by-side vending machines sell competitive products, then the price of each product is a primary determiner of sales. If sales data are collected and modeled as a function of the two prices, then we can use the data and model to answer questions such as

Robert Brenner/PhotoEdit

- If one of the competitors lowers the price, what change can the other company expect in sales?
- How quickly are sales changing for two given prices?
- If one company lowers its price, how should the other company respond in order to maintain its current sales level?

You will be given the opportunity to answer questions such as these in Activity 24 of Section 9.3 and Activity 23 of Section 9.4.

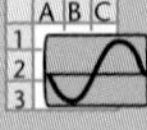
This symbol directs you to the supplemental technology guides. The different guides contain step-by-step instructions for using graphing calculators or spreadsheet software to work the particular example or discussion marked by the technology symbol.

This symbol directs you to the Calculus Concepts *web site, where you will find algebra review, links to data updates, additional practice problems, and other helpful resource material.*

Concept Objectives

This chapter will help you understand the concepts of

- Multivariable functions
- Cross-sectional models
- Contour curves and graphs
- Partial rates of change
- Compensating for change

and you will learn to

- Construct input/output diagrams for multivariable functions
- Interpret multivariable function inputs and outputs
- Find and interpret cross-sectional models given data or a function
- Draw contour graphs and answer questions using them
- Find and interpret first and second partial derivatives
- Estimate partial derivatives using tables and cross-sectional models
- Calculate the slope at a point on a contour curve
- Estimate the change needed in one variable to compensate for a change in another variable in order that the output remains constant

The first eight chapters of this text discussed the mathematics of change for functions of a single input variable. Chapters 9 and 10 discuss change for multivariable functions—that is, functions with two or more input variables.

Graphs of functions with two input variables are three-dimensional and represent surfaces in space. We consider associated two-dimensional contour graphs and how they relate to the three-dimensional surfaces. Our understanding of any function and how it changes is greatly enhanced by a visual presentation.

We investigate change in a multivariable function by considering rates of change in two directions. We also discuss compensating for change in the context of an important consideration in any manufacturing process, the allocation of available resources to produce a fixed level of output.

9.1 Cross-Sectional Models and Multivariable Functions

In Chapters 1 through 8 we considered functions with which we analyzed change. In each case, the function had only one input variable and one output variable. We now turn our attention to data and functions that have two or more input variables and one output variable. We call such functions **multivariable functions.** In fact, many of the functions we encounter in everyday situations are multivariable functions that depend on two or more input variables. A manufacturer's profit depends on sales, variable costs, fixed costs, and perhaps other input variables that are related to inventory costs, equipment failures, labor strikes, and so on. In forestry, the volume of a tree is a function of its height and diameter. Crop yield is a function of temperature, rainfall, amount of fertilizer, and so forth.

Recall that a single-variable equation is a function when each input has only one corresponding output. The input of a multivariable function is written as an n-tuple, denoted $(x_1, x_2, x_3, \ldots, x_n)$. For each input $(x_1, x_2, x_3, \ldots, x_n)$ of a multivariable function there is only one output. For instance, a packaging container in the shape of a box will hold a volume equal to the length times the width times the height of the box. Inputing $(l, w, h) = $ (2 inches, 4 inches, 7 inches) into a function giving volume produces a single output volume of 56 cubic inches.

Let us now consider another example of a multivariable function—the elevation of land above sea level. Elevation is a function whose output (distance above sea level) relies on two inputs (longitude and latitude). At one time or another, we have all seen a topographical map with elevation curves indicated. Let us consider the elevation function for a fenced tract of farmland in Missouri. For ease of calculation, instead of using the actual latitude and longitude of locations on this tract, we consider the southwest corner of the tract to be our origin of reference. We measure distance from this point in miles from the southern fence and miles from the western fence. The two input variables are e and n, where e is the distance in miles east of the western fence and n is the distance in miles north of the southern fence. The output of the elevation function is $E(e, n)$, the elevation of the tract in feet above sea level at position (e, n). Figure 9.1 shows an input/output diagram for this function.

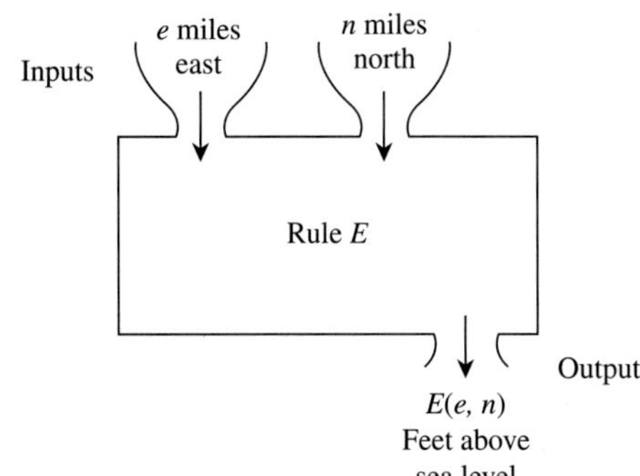

FIGURE 9.1

This function can be represented as a two-dimensional contour graph (Figure 9.2) or a three-dimensional graph (Figure 9.3).

Maps like the one in Figure 9.2 are drawn from data that surveyors gather. Data for this map are given in Table 9.1. From the table, we read the elevation at a position 0.3 mile east and 1 mile north of the southwest corner as $E(0.3, 1) = 799.7$ feet above sea level. Note that $e = 0.3$ because the position is 0.3 mile east of the western fence. Likewise, $n = 1$ because the position is 1 mile north of the southern fence.

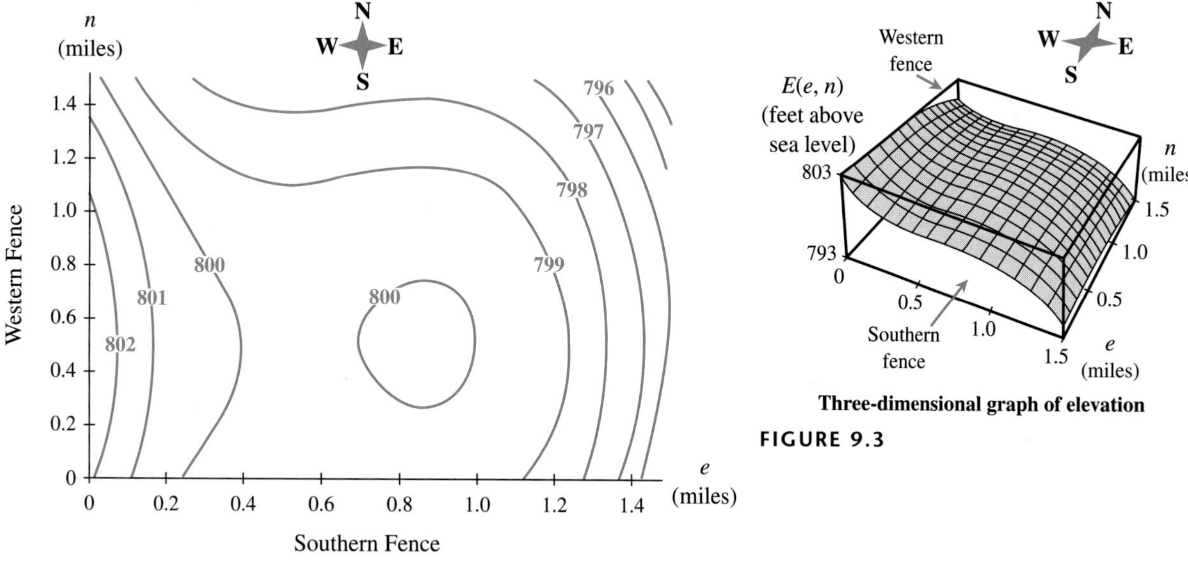

Contour graph for elevation in feet above sea level

FIGURE 9.2

Three-dimensional graph of elevation

FIGURE 9.3

We recommend choosing a point and demonstrating to the students how to locate that point on each of the three function representations. Then, on each function representation, briefly discuss moving away from the point in some direction and whether the output is increasing or decreasing. Your students probably will conclude that the table gives the most detailed description of output, whereas the three-dimensional graph is the hardest to read.

TABLE 9.1 Elevation *E*(*e*, *n*) in feet above sea level on a tract of Missouri farmland measured *e* miles east of the western fence and *n* miles north of the southern fence

e→	0	0.1	0.2	0.3	0.4	0.5	0.6	0.7	0.8	0.9	1.0	1.1	1.2	1.3	1.4	1.5
1.5	800.3	799.1	798.3	797.8	797.5	797.4	797.4	797.5	797.6	797.7	797.6	797.3	796.8	796.0	794.8	793.2
1.4	800.8	799.6	798.8	798.3	798.0	797.9	797.9	798.0	798.1	798.1	798.1	797.8	797.3	796.5	795.3	793.7
1.3	801.2	800.0	799.2	798.7	798.4	798.3	798.3	798.4	798.5	798.6	798.5	798.2	797.7	796.9	795.7	794.1
1.2	801.6	800.4	799.6	799.1	798.8	798.7	798.7	798.8	798.9	798.9	798.9	798.6	798.1	797.3	796.1	794.5
1.1	801.9	800.7	799.9	799.4	799.1	799.0	799.0	799.1	799.2	799.3	799.2	798.9	798.4	797.6	796.4	794.8
1.0	802.2	801.0	800.2	799.7	799.4	799.3	799.3	799.4	799.5	799.5	799.5	799.2	798.7	797.9	796.7	795.1
0.9	802.4	801.2	800.4	799.9	799.6	799.5	799.5	799.6	799.7	799.8	799.7	799.4	798.9	798.1	796.9	795.3
0.8	802.6	801.4	800.6	800.1	799.8	799.7	799.7	799.8	799.9	799.9	799.9	799.6	799.1	798.3	797.1	795.5
0.7	802.7	801.5	800.7	800.2	799.9	799.8	799.8	799.9	800.0	800.1	800.0	799.7	799.2	798.4	797.2	795.6
0.6	802.8	801.6	800.8	800.3	800.0	799.9	799.9	800.0	800.1	800.1	800.1	799.8	799.3	798.5	797.3	795.7
0.5	802.8	801.6	800.8	800.3	800.0	799.9	799.9	800.0	800.1	800.2	800.1	799.8	799.3	798.5	797.3	795.7
0.4	802.8	801.6	800.8	800.3	800.0	799.9	799.9	800.0	800.1	800.1	800.1	799.8	799.3	798.5	797.3	795.7
0.3	802.7	801.5	800.7	800.2	799.9	799.8	799.8	799.9	800.0	800.1	800.0	799.7	799.2	798.4	797.2	795.6
0.2	802.6	801.4	800.6	800.1	799.8	799.7	799.7	799.8	799.9	799.9	799.9	799.6	799.1	798.3	797.1	795.5
0.1	802.4	801.2	800.4	799.9	799.6	799.5	799.5	799.6	799.7	799.8	799.7	799.4	798.9	798.1	796.9	795.3
0	802.2	801.0	800.2	799.7	799.4	799.3	799.3	799.4	799.5	799.5	799.5	799.2	798.7	797.9	796.7	795.1
n↑																

Cross-Sectional Models from Data

Although the technology exists to model multivariable data, we have chosen not to make multivariable modeling the focus of our discussion. Instead, we will partially model the data by holding one input variable constant and finding an equation in terms of the other input variable. For example, we can model the elevation along a line parallel to the western fence at a distance of $e = 0.8$ mile east of that fence by looking at only that one column of elevations. (You may wish to highlight this column in Table 9.1.) The equality $e = 0.8$ mile east of the western fence is depicted on the contour graph as a line (Figure 9.4) and on the three-dimensional graph as a cross-sectional plane (Figure 9.5). The curve resulting from the intersection of the plane $e = 0.8$ and the elevation function E is called a cross section of E and has output denoted by $E(0.8, n)$. For a function with two input variables, a **cross section** is the curve that results when the three-dimensional graph of the function is intersected with a plane. A cross section of a function will always have one less dimension (variable) than the function itself. For a three-dimensional function, a cross section will always be two-dimensional.

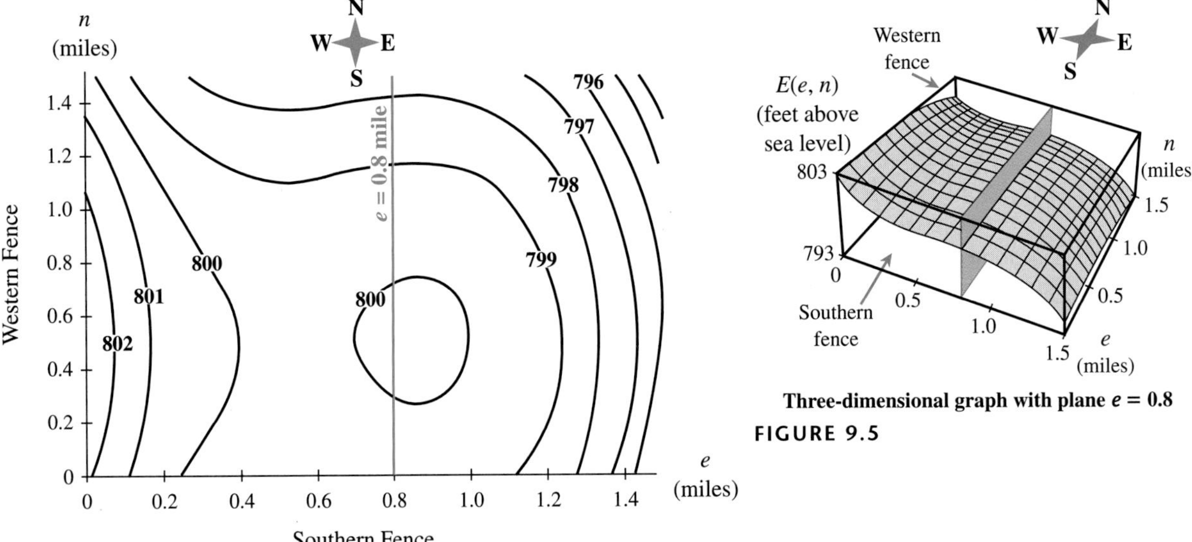

FIGURE 9.4

Three-dimensional graph with plane $e = 0.8$

FIGURE 9.5

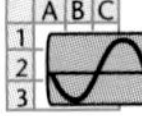

By holding the input variable e constant at 0.8 mile, we have the data shown in Table 9.2. Because the distance east of the western fence can no longer vary, the only input is n, the number of miles north of the southern fence. We can now model $E(0.8, n)$, the elevation in feet above sea level on the farmland measured at $e = 0.8$ mile east of the western fence.

$$E(0.8, n) = -2.5n^2 + 2.497n + 799.490 \text{ feet above sea level}$$

n miles north of the southern fence and 0.8 mile east of the western fence. Remember, this model is only a partial description of all of the data. It tells us nothing about the elevations when the distance east of the western fence is 0.2 mile or 0.5 mile or 1.5 miles. It gives us elevation information for only the cross section at 0.8 mile east of the western fence of the tract. For this reason, we call it a **cross-sectional model**. Figure 9.6 shows a graph of this cross section of E with output $E(0.8, n)$.

As shown in Example 1, we can also find a cross-sectional model for the elevation when distance north of the southern fence is held constant.

TABLE 9.2

n	$E(0.8, n)$	n	$E(0.8, n)$
1.5	797.6	0.7	800.0
1.4	798.1	0.6	800.1
1.3	798.5	0.5	800.1
1.2	798.9	0.4	800.1
1.1	799.2	0.3	800.0
1.0	799.5	0.2	799.9
0.9	799.7	0.1	799.7
0.8	799.9	0.0	799.5

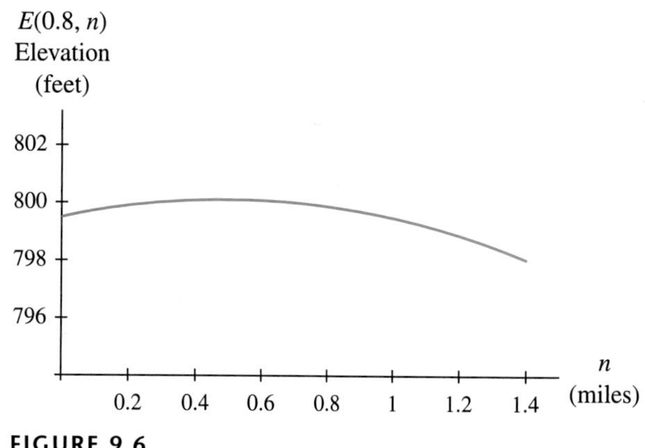

FIGURE 9.6

EXAMPLE 1 *Missouri Farmland*

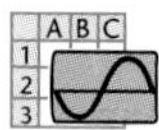

Consider the elevation of the tract of Missouri farmland as described in Table 9.1.

a. Interpret the notation $E(e, 0.6)$, where $E(e, n)$ is the elevation in feet above sea level of the land e miles east of the western fence and n miles north of the southern fence.

b. Use Table 9.1 to find a model for the cross section in part *a*.

Solution

a. The notation $E(e, 0.6)$ represents the output of the cross-sectional function for elevation where distance north of the southern fence is held constant at 0.6 mile and distance e miles east of the western fence is the input. Figure 9.7 shows the line $n = 0.6$ mile north of the southern fence on the contour graph, and Figure 9.8 illustrates the cross section formed by the intersection of the three-dimensional graph and the cross-sectional plane.

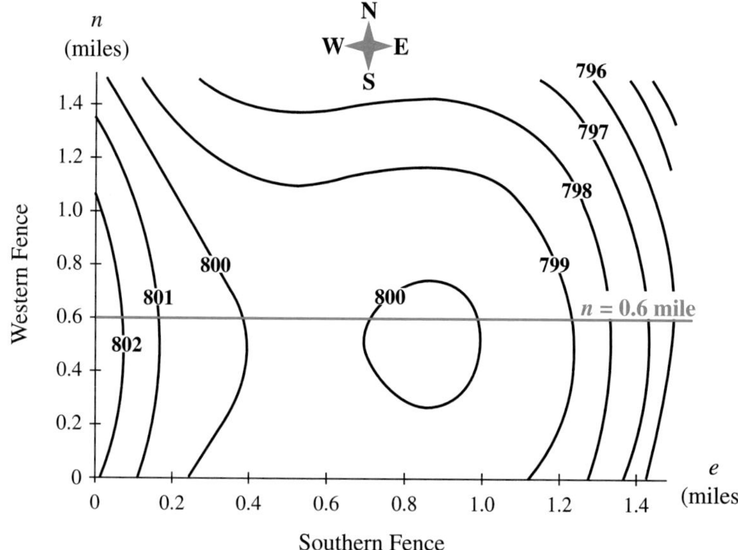

FIGURE 9.7

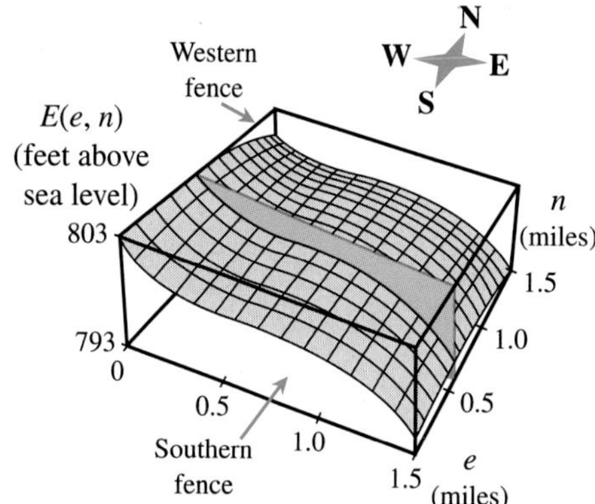

Three-dimensional graph with plane $n = 0.6$

FIGURE 9.8

b. We consider the row in Table 9.1 for $n = 0.6$ mile to obtain the data shown in Table 9.3.

TABLE 9.3

e	0	0.1	0.2	0.3	0.4	0.5	0.6	0.7
E(e, 0.6)	802.8	801.6	800.8	800.3	800.0	799.9	799.9	800.0
e	**0.8**	**0.9**	**1.0**	**1.1**	**1.2**	**1.3**	**1.4**	**1.5**
E(e, 0.6)	800.1	800.1	800.1	799.8	799.3	798.5	797.3	795.7

A model for $E(e, 0.6)$, the elevation in feet above sea level of the land measured at $n = 0.6$ mile north of the southern fence, is

$$E(e, 0.6) = -10.124e^3 + 21.347e^2 - 13.972e + 802.809 \text{ feet above sea level}$$

e miles east of the western fence and 0.6 mile north of the southern fence. Figure 9.9 shows a graph of this cross section of E with output $E(e, 0.6)$.

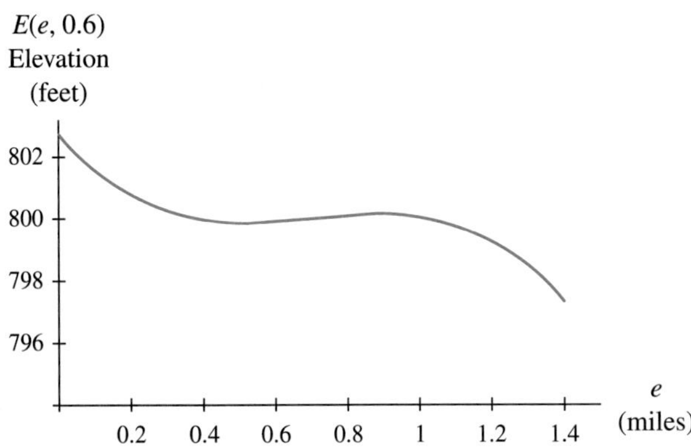

FIGURE 9.9

Cross-Sectional Models from Equations

Another example of a multivariable function is the value of an investment. There are several factors that play a role in the accumulated value of a bank account: the principal P (in dollars), the annual interest rate $100r\%$, the number n of compoundings per year, and the time t (in years) since the principal was deposited. The formula for the accumulated value as a function of these four input variables is

$$A(P, r, n, t) = P\left(1 + \frac{r}{n}\right)^{nt} \text{ dollars}$$

An input/output diagram of this function is shown in Figure 9.10.

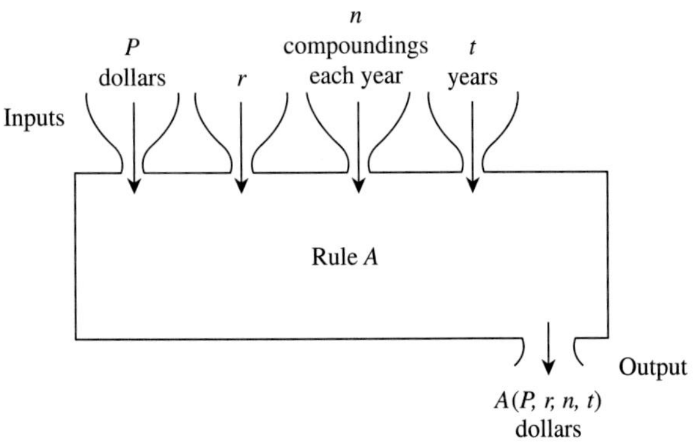

FIGURE 9.10

The notation $A(P, 0.06, 4, t)$ represents the accumulated value if the annual interest rate is 6% compounded quarterly. For simplicity, we write $A(P, t)$.

$$A(P, t) = P\left(1 + \frac{0.06}{4}\right)^{4t} \approx P(1.061364^t) \text{ dollars}$$

where P dollars are initially invested and left untouched, except for accumulating interest, for t years. $A(P, t)$ is the output of a function with two input variables P and t. Table 9.4 can be generated from this formula.

TABLE 9.4 Accumulated value (in dollars) of a bank account paying 6% interest compounded quarterly on an initial investment of *P* dollars after *t* years

t (years)	*P* (dollars) 3500	4000	4500	5000	5500	6000	6500
6	5003.26	5718.01	6432.76	7147.51	7862.27	8577.02	9291.77
8	5636.14	6441.30	7246.46	8051.62	8856.78	9661.95	10,467.11
10	6349.06	7256.07	8163.08	9070.09	9977.10	10,884.11	11,791.12
12	7152.17	8173.91	9195.65	10,217.39	11,239.13	12,260.87	13,282.61

As with the elevation data, we can form a cross-sectional function by holding one of the input variables constant. For example, if we keep the principal investment constant at $P = \$5000$, then the cross-sectional model for the accumulated value is given in terms of time t by

$$A(5000, t) \approx 5000(1.061364^t) \text{ dollars}$$

when t years have elapsed since the deposit was made. Because this function is exponential, we know that for a principal of \$5000 (or any principal, for that matter), the accumulated value has a constant percentage increase of about 6.14% per year. If we did not have a formula for $A(P, t)$, we could have approximated the cross-sectional function for a principal investment of \$5000 by modeling the column corresponding to $P = \$5000$ in Table 9.4.

EXAMPLE 2 *Value of an Investment*

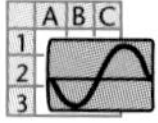

a. Find a cross-sectional model for the accumulated value $A(P, t)$ of an investment of P dollars over 10 years with annual interest rate of 6% compounded quarterly.

b. Find a model for the data in the row in Table 9.4 that corresponds to 10 years. Compare the cross-sectional function in part *a* to this function.

c. Use the answer to part *a* to find the accumulated value of an initial investment of \$5300 after 10 years.

Solution

a. We substitute 10 for t in $A(P, t) = P\left(1 + \frac{0.06}{4}\right)^{4t}$.

$$A(P, 10) = P\left(1 + \frac{0.06}{4}\right)^{40} \approx 1.814018P \text{ dollars}$$

Because this cross-sectional function is linear, we see that the accumulated value after 10 years is a constant multiple of the amount invested.

b. The row in Table 9.4 associated with 10 years can be modeled by the equation

$$A(P, 10) = 1.81402P - 0.01 \text{ dollars}$$

when P dollars is invested at 6% compounded quarterly. The function has a slightly different slope than and is shifted down from the one in part *a* because the values in the table are rounded to the nearest cent.

c. The accumulated value for a principal investment of \$5300 is calculated as

$$A(5300, 10) = 5300\left(1 + \frac{0.06}{4}\right)^{40} \approx \$9614.30$$

As we have seen with both the elevation example and the accumulated value example, we can use cross-sectional functions to help us understand the behavior of a multivariable function. As illustrated in this section, equations for cross sections can be found by using tables of data or multivariable equations.

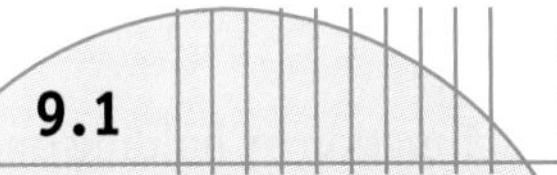

9.1 Concept Inventory

- *Multivariable function*
- *Cross-sectional model from data and from a multivariable function*
- *Input/output diagram*

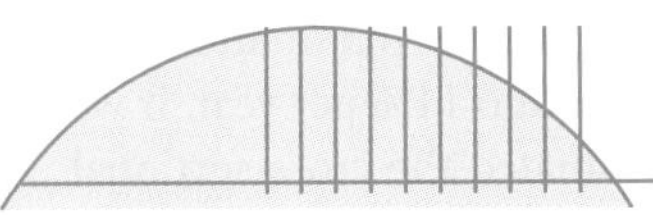

9.1 Activities

Assume that the rules in Activities 1 through 6 are functions. Draw an input/output diagram for each of the multivariable functions.

1. $W(a, h)$ is the average weight (in pounds) of a person who is a years old and h inches tall.
2. $T(i, d)$ is the federal income tax (in dollars) owed by a person who made i dollars and can claim d dependents in a given year, assuming that the person has no other deductions.
3. $T(g, t)$ is the average temperature (in °F) at longitude g degrees and latitude t degrees on a particular day.
4. $P(k, c)$ is the price (in thousands of dollars) of a diamond that weighs k karats and is of color code c.
5. $R(b, c)$ is the revenue (in hundreds of dollars) a farmer makes from the sale of b bulls and c cows.
6. $G(h, s)$ is the expected grade-point average of a typical freshman college student who had a GPA of h in high school and made a combined score of s points on the SAT.
7. Let $P(c, s)$ be the profit in dollars from the sale of 1 yard of fabric when c dollars is the production cost per yard and s dollars is the selling price per yard. Write a sentence interpreting each of the following mathematical notations.

   a. $P(1.2, s)$ b. $P(c, 4.5)$ c. $P(1.2, 4.5) = 3.0$
8. Let $T(i, d)$ dollars be the amount of income tax owed by a household claiming d dependents with adjusted income i dollars. Write a sentence interpreting each of the following notations.

   a. $T(36{,}000, d)$ b. $T(i, 4)$
   c. $T(36{,}000, 4) = 10{,}000$
9. Let $P(l, m)$ be the probability that a certain senator votes in favor of a tobacco-ban bill when the senator receives l letters supporting the bill and m million dollars is invested by the tobacco industry lobbying against the bill. Write a sentence interpreting each of the following mathematical notations.

   a. $P(100{,}000, m)$ b. $P(l, 53)$
10. Let $N(p, s)$ be the number of skiers on a Saturday at a ski resort in Utah when p dollars is the price of an all-day lift ticket and s is the number of inches of fresh snow that fell since the previous Saturday. Write a sentence interpreting each of the following mathematical notations.

    a. $N(25, s)$ b. $N(p, 6)$
11. Your comfort in summer depends on the combined effects of air temperature and humidity. Table 9.5 shows how hot it feels[1] (the apparent temperature in degrees Fahrenheit) for a given air temperature and relative humidity.

1. National Oceanic and Atmospheric Administration.

TABLE 9.5 Apparent temperature (°F)

Air temperature (°F)	Relative humidity (%) 40	45	50	55	60	65	70	75	80	85	90	95	100
110	135												
108	130	137											
106	124	130	137										
104	119	124	130	137									
102	114	119	124	130	137								
100	109	113	118	123	129	136							
98	105	108	113	117	122	128	134						
96	101	104	107	111	116	121	126	132					
94	97	100	103	106	110	114	119	124	129	135			
92	94	96	98	101	104	108	112	116	121	126	131		
90	91	92	94	97	99	102	106	109	113	117	122	126	131
88	88	89	91	93	95	97	100	103	106	109	113	117	121
86	85	86	88	89	91	93	95	97	99	102	105	108	111
84	83	84	85	86	87	89	90	92	94	96	98	100	102
82	81	82	83	83	84	85	86	87	88	90	91	93	94
80	80	80	81	81	82	82	83	83	84	85	85	86	87

TABLE 9.6 Average daily weight gain/loss for a pig (kg/day)

w Mean live weight (kg)	t, Air temperature (°C) 4.4	10	15.6	21.1	26.7	32.2	37.8
45		0.62	0.72	0.91	0.89	0.64	0.18
68	0.58	0.67	0.79	0.98	0.83	0.52	-0.09
91	0.54	0.71	0.87	1.01	0.76	0.40	-0.35
113	0.50	0.76	0.94	0.97	0.68	0.28	-0.62
136	0.46	0.80	1.02	0.93	0.62	0.16	-0.88
156	0.43	0.85	1.09	0.90	0.55	0.05	-1.15

In Activity 11b the student should reason that because 70% is in the table and 89°F is not, the desired cross-sectional model has output $A(h, 70)$. Similar situations occur throughout the Activities.

a. Find a cross-sectional model for the apparent temperature as a function of the relative humidity when the air temperature is 96°F.

b. Find an appropriate cross-sectional model, and use it to estimate the apparent temperature when the air temperature is 89°F and the relative humidity is 70%.

12. Table 9.6 gives $W(t, w)$, the average daily weight gain/loss of a pig[2] in kilograms per day as a function of its weight and the air temperature.

a. What is the average daily weight gain/loss for a 91-kg pig when the temperature is 37.8°C?

b. Find and interpret a model for $W(t, 68)$.

c. Graph the model in part *b*, and describe its behavior. What does the behavior of the graph tell you about the weight gain of a pig?

d. Find and interpret a model for $W(26.7, w)$.

e. Repeat part *c* for the model in part *d*.

2. A. A. Hanson, ed., *Practical Handbook of Agricultural Science* (Boca Raton, FL: CRC Press, 1990).

TABLE 9.7 Frequency of cloud cover over Minneapolis in January

Fraction of sky covered	Hour of the day							
	Midnight	3:00 a.m.	6:00 a.m.	9:00 a.m.	Noon	3:00 p.m.	6:00 p.m.	9:00 p.m.
overcast	0.45	0.48	0.47	0.45	0.41	0.39	0.41	0.38
≥ 9/10	0.48	0.52	0.49	0.57	0.55	0.53	0.46	0.46
≥ 8/10	0.50	0.55	0.52	0.62	0.60	0.57	0.49	0.43
≥ 7/10	0.52	0.56	0.54	0.64	0.63	0.59	0.52	0.47
≥ 6/10	0.54	0.59	0.57	0.66	0.65	0.61	0.54	0.49
≥ 5/10	0.55	0.59	0.59	0.69	0.67	0.64	0.55	0.51
≥ 4/10	0.56	0.59	0.62	0.72	0.68	0.66	0.58	0.56
≥ 3/10	0.58	0.62	0.64	0.74	0.71	0.69	0.60	0.60
≥ 2/10	0.60	0.67	0.66	0.78	0.74	0.71	0.62	0.62
≥ 1/10	0.66	0.71	0.72	0.89	0.86	0.84	0.79	0.66

TABLE 9.8 Monthly payments per $1000 of loan amount

	24 months	36 months	42 months	48 months	60 months
5%	43.87	29.97	26.00	23.03	18.87
6%	44.32	30.42	26.46	23.49	19.33
7%	44.77	30.88	26.91	23.95	19.80
8%	45.23	31.34	27.38	24.41	20.28
9%	45.68	31.80	27.84	24.89	20.76
10%	46.14	32.27	28.32	25.36	21.25
11%	46.61	32.74	28.79	25.85	21.74
12%	47.07	33.21	29.28	26.33	22.24

Graphing the function in part *a* of Activity 14 together with the one obtained in part *a* of Activity 22 over the interval between 0 and 84 months is an effective illustration of why extrapolating using a model obtained from data is dangerous.

13. Table 9.7 shows the frequency of cloud cover[3] over Minneapolis in January given the fraction of the sky covered by clouds and the hour of the day. The frequency may be interpreted as the percentage of time that a certain situation occurs.
    a. What percentage of time is at least $\frac{8}{10}$ of the sky covered by clouds at noon?
    b. Find a cross-sectional model for the frequency of cloud cover at 9 a.m.

14. Table 9.8, taken from a magazine advertisement, shows monthly payments[4] on a $1000 loan at different interest rates over different loan periods.
    a. Find an appropriate cross-sectional model, and use it to estimate the monthly payments for a 52-month loan at 9%.
    b. Find an appropriate cross-sectional model, and use it to estimate the monthly payments for a 42-month loan at 10.5%.

15. Table 9.9 shows selected values of the average yearly consumption of peaches per person, based on the price of peaches and the yearly income of the person's family.
    a. Find a cross-sectional model for a yearly income of $40,000. Use the model to estimate the consumption for someone in a family with a yearly income of $40,000 when the price of peaches is $1.55 per pound.

3. I. I. Gringorten, "Modelling Conditional Probability," *Journal of Applied Meteorology*, vol. 10, no. 4 (August 1971), pp. 646–657.
4. *Automobile*, vol. 9, no. 4 (July 1994), p. 72.

TABLE 9.9 Per capita consumption of peaches (pounds per person per year)

Yearly income (tens of thousands of dollars)	Price per pound above $1.50					
	0	0.10	0.20	0.30	0.40	0.50
1	5.0	4.9	4.8	4.7	4.7	4.6
2	6.4	6.3	6.2	6.1	6.1	6.0
3	7.2	7.1	7.0	6.9	6.9	6.8
4	7.8	7.7	7.6	7.5	7.4	7.4
5	8.2	8.1	8.0	8.0	7.9	7.8
6	8.6	8.5	8.4	8.3	8.3	8.2

TABLE 9.10 Daily sales (number of cans) of Coke products from a vending machine

Cost of Pepsi products	Cost of Coke products				
	$0.50	$0.75	$1.00	$1.25	$1.50
$0.50	157	143	123	98	65
$0.75	206	192	172	146	114
$1.00	255	241	221	195	163
$1.25	304	290	270	244	211
$1.50	353	339	319	293	260

b. Find a cross-sectional model for a price of $1.80 per pound. Use the model to estimate consumption for a person in a family with yearly income of $35,000 when the price is $1.80 per pound.

c. The data in Table 9.9 can be modeled by the equation

$$C(p, i) = 2 \ln i + 2.7183^{-p} + 4 \text{ pounds per person per year}$$

where the price of peaches is $\$(1.50 + p)$ per pound and the person lives in a family with annual income $\$10{,}000i$. Use this equation to repeat parts *a* and *b*. What might account for the discrepancies in the cross-sectional functions found by fitting equations to data and those derived from the multivariable equation?

16. Two vending machines sit side by side in a college dorm. One machine sells Coke products, and the other sells Pepsi products. Daily sales of Coke products, based on the prices of the products in the two machines, are as shown in Table 9.10.

a. Find an appropriate cross-sectional model, and use it to estimate sales of Coke products when the price of Coke products is $1.00 and the price of Pepsi products is $0.90.

b. Find an appropriate cross-sectional model, and use it to estimate sales of Coke products when the price of Coke products is $1.40 and the price of Pepsi products is $0.50.

c. The data in Table 9.10 can be modeled by the equation

$$S(c, p) = 196.42p - 50.2c^2 + 9.6c + 66.4 - 1.04cp \text{ cans of Coke products per day}$$

when each Coke product costs $\$c$ and each Pepsi product costs $\$p$. Find and interpret $S(1.30, 1.10)$.

d. Repeat parts *a* and *b* using the equation in part *c*. Compare the answers you find by using this equation to those you found by using the data.

17. The carrying capacity[5] of a particular farm system is defined as the number of animals or people that can be supported by the crop production from a given land area. Carrying capacity is defined as

$$K(k_c, k_d, G, P, A) = \frac{k_c k_d G P}{A} \text{ animals per hectare}$$

where k_c = fraction of the crop consumed, k_d = fraction of the amount consumed that is digested as useful nutrients, G = gross energy content of the crop (in megajoules per kilogram), P = net crop production (in kilograms per hectare per year), and A = annual energy requirement of the animal (in megajoules per animal).

a. Rye grass is an important grazing crop in cool climates. Annual rye grass production of 15,000 kg per hectare is common, and the gross energy of rye grass is 17 megajoules per kilogram. The annual energy requirement of a milking cow is approximately 64,000 megajoules. If $\frac{7}{10}$ of the crop is available for consumption (the remainder is used as seed) and $\frac{7}{10}$ of the amount consumed is digestible, how many milk cows can be supported by 1 hectare? How many milk cows can be supported by 1 acre?

5. R. S. Loomis and D. J. Connor, *Crop Ecology: Productivity and Management in Agricultural Systems* (Cambridge, England: Cambridge University Press, 1992).

b. Write the cross-sectional model for $k_c = 0.8$ and $k_d = 0.85$.

c. Find and interpret $K(0.7, 0.8, 17, P, A)$.

18. The amount of organic matter that one beef cow grazing on the Northern Great Plains rangeland eats each day can be modeled[6] by

$$I(s, m) = 8.61967 - 1.244s + 0.0897s^2 - 0.20988m + 0.035947m^2 + 0.21491sm$$

kilograms per day, where the cow produces m kilograms of milk per day and s is a number between -4 and 4 describing the size of the cow.

a. Find and interpret the values $I(4, 8)$ and $I(0, 4)$.

b. Find and interpret $I(4, m)$ and $I(s, 4)$.

19. Before applying for a loan, you should carefully consider how much debt you can afford. Three factors that affect the amount of the loan you can afford are the interest rate, the period of the loan, and the size of the monthly payments. The following equation gives the amount of a loan A (in dollars) given the annual interest rate $100r\%$, the period n (in months), and the monthly payments m (in dollars).

$$A(r, n, m) = \frac{12m}{r}\left[1 - \left(1 + \frac{r}{12}\right)^{-n}\right]$$

a. What is the loan amount when the APR is 10% and \$532.98 is paid monthly for 30 years?

b. Write the cross-sectional function for the loan amount for a fixed interest rate of 7.5%.

c. Interpret $A(r, 36, m)$.

d. Write the cross-sectional function for the loan amount for a 3-year loan with a fixed interest rate of 6%.

20. The monthly payment on a loan carrying 9% interest can be calculated by using the function

$$m(A, t) = \frac{0.0075A}{1 - 0.914238^t} \text{ dollars}$$

when the loan is for A dollars and is to be repaid over t years.

a. How much is the monthly payment on a loan of \$6000 for 15 years?

b. Find and interpret $m(76{,}000, 30)$.

c. If you are able to pay \$250 per month for 4 years, what loan amount can be financed?

21. Siple and Passel were two explorers who accompanied Admiral Byrd to the Antarctic and were among the first researchers to measure loss of body heat due to the wind. From their data, they obtained the equation[7]

$$H(v, t) = (10.45 + 10\sqrt{v} - v)(33 - t)$$

which gives the body's heat loss in kilogram-calories per square meter of body surface area per hour for wind speed v in meters per second when the air temperature is t°C.

a. What is the body's heat loss at -15°C when the wind speed is 15 meters per second?

b. Siple and Passel found that unprotected areas of a person's face will freeze in about 1 minute when the heat loss is 2000 kilogram-calories per square meter of body surface area per hour. If the wind is blowing at 10 meters per second, at what temperature will exposed areas of the face freeze?

22. The amount of the monthly payments on a \$1000 loan can be modeled by the equation

$$m(r, n) = \frac{1000r}{12\left[1 - \left(1 + \frac{r}{12}\right)^{-n}\right]} \text{ dollars}$$

where $100r\%$ is the interest rate and the period of the loan is n months.

a. Find $m(0.09, n)$ and compare it to the equation found in part *a* of Activity 14.

b. Find $m(r, 42)$ and compare it to the equation found in part *b* of Activity 14.

23. The apparent temperature data in Table 9.5 was constructed from the equation[8]

$$A(h, t) = -42.379 + 2.049t + 10.1433h - 0.2248ht - (6.8378 \cdot 10^{-3})t^2 - (5.4817 \cdot 10^{-2})h^2 + (1.2287 \cdot 10^{-3})t^2h + (8.5282 \cdot 10^{-4})th^2 - (1.99 \cdot 10^{-6})t^2h^2 \text{ °F}$$

for an air temperature of t°F and a relative humidity of h%.

6. E. E. Grings et al., "Efficiency of Production in Cattle of Two Growth Potentials on Northern Great Plains Rangelands During Spring-Summer Grazing," *Journal of Food Science*, vol. 74, no. 10 (1996), pp. 2317–2326.

7. W. Bosch and L. G. Cobb, "Windchill," UMAP Module 658, *The UMAP Journal*, vol. 5, no. 4 (Winter 1984), pp. 477–492.

8. Australia Bureau of Meteorology.

a. How hot does it feel when the relative humidity is 85% and the air temperature is 90°F?

b. If the apparent temperature is 100°F and the actual air temperature is 95°F, what is the relative humidity?

c. Find an equation for $A(h, 96)$. Compare it to the one in part *a* of Activity 11.

d. Find an equation for $A(70, t)$. Compare it to the one in part *b* of Activity 11.

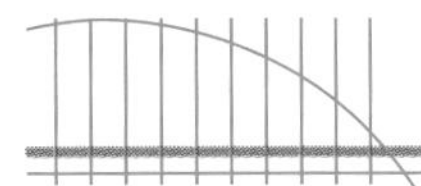

9.2 Contour Graphs

Let us return to the example of the tract of farmland in Missouri. A topographical map of this tract shows the contour of the terrain by outlining different elevations. The curves that outline the land at different but constant elevations are known as **contour curves.** In general, a contour curve for a three-dimensional function is the collection of all points (x, y) for which $f(x, y) = K$, where K is a constant. That is, a contour curve is a two-dimensional outline of a three-dimensional graph at a given output level. For a specific value of K, we sometimes call the contour curve the K contour curve.

Topographical maps are constructed from data that surveyors have gathered. Table 9.1 shows data for the elevation of the Missouri tract. We can draw contour curves on the table of data. The blue contour curves shown in Figure 9.11 outline the shape of the land at an elevation of 800 feet above sea level. The contour curves we will sketch and the ones you will be asked to sketch should be as smooth as possible—that is, with as few concavity changes as possible.

$e \rightarrow$	0	0.1	0.2	0.3	0.4	0.5	0.6	0.7	0.8	0.9	1.0	1.1	1.2	1.3	1.4	1.5
1.5	800.3	799.1	798.3	797.8	797.5	797.4	797.4	797.5	797.6	797.7	797.6	797.3	796.8	796.0	794.8	793.2
1.4	800.8	799.6	798.8	798.3	798.0	797.9	797.9	798.0	798.1	798.1	798.1	797.8	797.3	796.5	795.3	793.7
1.3	801.2	800.0	799.2	798.7	798.4	798.3	798.3	798.4	798.5	798.6	798.5	798.2	797.7	796.9	795.7	794.1
1.2	801.6	800.4	799.6	799.1	798.8	798.7	798.7	798.8	798.9	798.9	798.9	798.6	798.1	797.3	796.1	794.5
1.1	801.9	800.7	799.9	799.4	799.1	799.0	799.0	799.1	799.2	799.3	799.2	798.9	798.4	797.6	796.4	794.8
1.0	802.2	801.0	800.2	799.7	799.4	799.3	799.3	799.4	799.5	799.5	799.5	799.2	798.7	797.9	796.7	795.1
0.9	802.4	801.2	800.4	799.9	799.6	799.5	799.5	799.6	799.7	799.8	799.7	799.4	798.9	798.1	796.9	795.3
0.8	802.6	801.4	800.6	800.1	799.8	799.7	799.7	799.8	799.9	799.9	799.9	799.6	799.1	798.3	797.1	795.5
0.7	802.7	801.5	800.7	800.2	799.9	799.8	799.8	799.9	800.0	800.1	800.0	799.7	799.2	798.4	797.2	795.6
0.6	802.8	801.6	800.8	800.3	800.0	799.9	799.9	800.0	800.1	800.1	800.1	799.8	799.3	798.5	797.3	795.7
0.5	802.8	801.6	800.8	800.3	800.0	799.9	799.9	800.0	800.1	800.2	800.1	799.8	799.3	798.5	797.3	795.7
0.4	802.8	801.6	800.8	800.3	800.0	799.9	799.9	800.0	800.1	800.1	800.1	799.8	799.3	798.5	797.3	795.7
0.3	802.7	801.5	800.7	800.2	799.9	799.8	799.8	799.9	800.0	800.1	800.0	799.7	799.2	798.4	797.2	795.6
0.2	802.6	801.4	800.6	800.1	799.8	799.7	799.7	799.8	799.9	799.9	799.9	799.6	799.1	798.3	797.1	795.5
0.1	802.4	801.2	800.4	799.9	799.6	799.5	799.5	799.6	799.7	799.8	799.7	799.4	798.9	798.1	796.9	795.3
0	802.2	801.0	800.2	799.7	799.4	799.3	799.3	799.4	799.5	799.5	799.5	799.2	798.7	797.9	796.7	795.1
$n\uparrow$																

FIGURE 9.11

Note that there are two portions of the 800-feet contour curve. To the left of the contour curve and also inside the egg-shaped region, the elevations are higher than 800 feet. Elsewhere the elevations are lower than 800 feet. Figure 9.12 shows the 800-feet contour as it occurs on a plane at $E = 800$ feet cutting through the three-dimensional graph of the land.

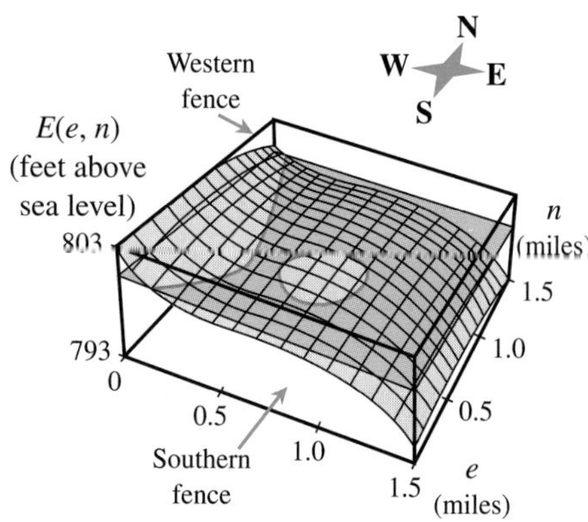

FIGURE 9.12

Figure 9.13 shows the contour curves at elevations of 796, 797, 798, 799, 800, 801, and 802 feet above sea level. The contour curves at various elevations form the topographical map, or **contour graph**, of the tract shown in Figure 9.14a. Figure 9.14b shows the contour curves drawn on the three-dimensional graph. Because each contour curve represents all the points on the surface that correspond to a particular level of output, we sometimes call contour curves *level curves*.

In general, a contour graph is a graph of contour curves $f(x, y) = K$ for more than one value of K. Usually, the values of K are equally spaced. For example, a contour graph may contain contour curves for $K = 0.25, 0.5, 0.75$, and 1 or for $K = 1000, 1020, 1040, 1060$, and 1080. When the values of K are evenly spaced, we can tell something about the steepness of the underlying three-dimensional surface by noting how close together the contour curves are. The closer the contour curves, the steeper the graph.

Contour graphs help us understand and interpret the interactions between the different input variables in a multivariable model. For instance, topographical maps may guide a hiker in finding a path that is less steep than another or help a farmer determine a pattern for plowing fields in such a way that erosion is minimized. Of course, contour graphs are also helpful in applications other than land elevation, as illustrated in Example 1.

e→	0	0.1	0.2	0.3	0.4	0.5	0.6	0.7	0.8	0.9	1.0	1.1	1.2	1.3	1.4	1.5
1.5	800.3	799.1	798.3	797.8	797.5	797.4	797.4	797.5	797.6	797.7	797.6	797.3	796.8	796.0	794.8	793.2
1.4	800.8	799.6	798.8	798.3	798.0	797.9	797.9	798.0	798.1	798.1	798.1	797.8	797.3	796.5	795.3	793.7
1.3	801.2	800.0	799.2	798.7	798.4	798.3	798.3	798.4	798.5	798.6	798.5	798.2	797.7	796.9	795.7	794.1
1.2	801.6	800.4	799.6	799.1	798.8	798.7	798.7	798.8	798.9	798.9	798.9	798.6	798.1	797.3	796.1	794.5
1.1	801.9	800.7	799.9	799.4	799.1	799.0	799.0	799.1	799.2	799.3	799.2	798.9	798.4	797.6	796.4	794.8
1.0	802.2	801.0	800.2	799.7	799.4	799.3	799.3	799.4	799.5	799.5	799.5	799.2	798.7	797.9	796.7	795.1
0.9	802.4	801.2	800.4	799.9	799.6	799.5	799.5	799.6	799.7	799.8	799.7	799.4	798.9	798.1	796.9	795.3
0.8	802.6	801.4	800.6	800.1	799.8	799.7	799.7	799.8	799.9	799.9	799.9	799.6	799.1	798.3	797.1	795.5
0.7	802.7	801.5	800.7	800.2	799.9	799.8	799.8	799.9	800.0	800.1	800.0	799.7	799.2	798.4	797.2	795.6
0.6	802.8	801.6	800.8	800.3	800.0	799.9	799.9	800.0	800.1	800.1	800.1	799.8	799.3	798.5	797.3	795.7
0.5	802.8	801.6	800.8	800.3	800.0	799.9	799.9	800.0	800.1	800.2	800.1	799.8	799.3	798.5	797.3	795.7
0.4	802.8	801.6	800.8	800.3	800.0	799.9	799.9	800.0	800.1	800.1	800.1	799.8	799.3	798.5	797.3	795.7
0.3	802.7	801.5	800.7	800.2	799.9	799.8	799.8	799.9	800.0	800.1	800.0	799.7	799.2	798.4	797.2	795.6
0.2	802.6	801.4	800.6	800.1	799.8	799.7	799.7	799.8	799.9	799.9	799.9	799.6	799.1	798.3	797.1	795.5
0.1	802.4	801.2	800.4	799.9	799.6	799.5	799.5	799.6	799.7	799.8	799.7	799.4	798.9	798.1	796.9	795.3
0	802.2	801.0	800.2	799.7	799.4	799.3	799.3	799.4	799.5	799.5	799.5	799.2	798.7	797.9	796.7	795.1
n↑	802	801	800							800		799		798 797	796	

FIGURE 9.13

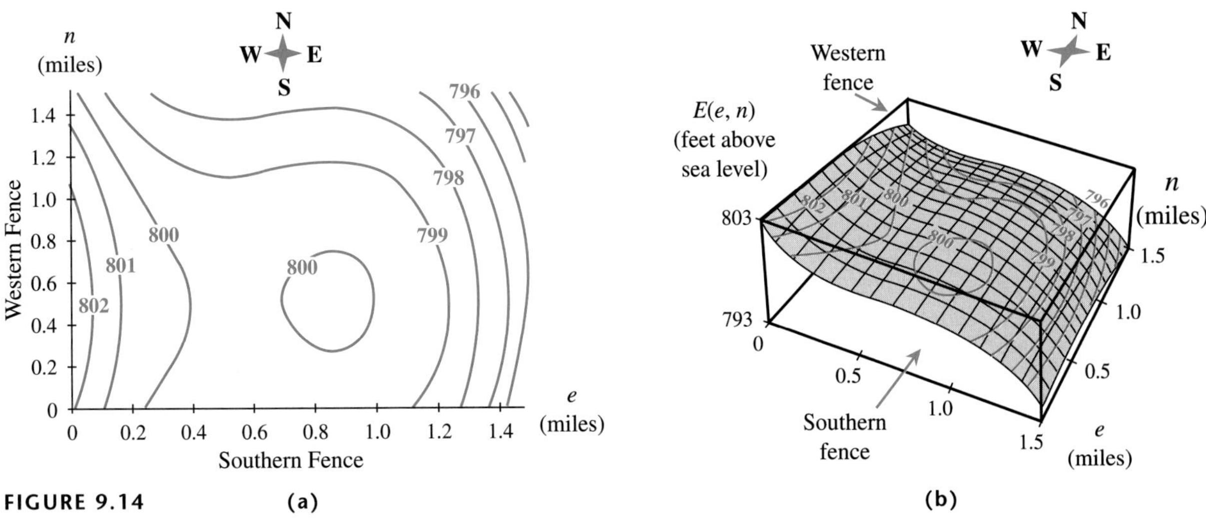

FIGURE 9.14 (a) (b)

EXAMPLE 1 *Contour Graphs and Monthly Payments on a Mortgage*

The data in the table shown in Figure 9.15 represent the monthly payments $M(t, A)$ required to pay off a mortgage of A thousand dollars over a period of t years when interest is paid on the loan at a rate of 7%.

Amount borrowed (thousand dollars)	Period of loan (years)								
	10	12.5	15	17.5	20	22.5	25	27.5	30
5	58.05	50.11	44.94	41.36	38.76	36.82	35.34	34.18	33.27
10	116.11	100.22	89.88	82.72	77.53	73.65	70.68	68.36	66.53
15	174.16	150.32	134.82	124.08	116.29	110.47	106.02	102.54	99.80
20	232.22	200.43	179.76	165.44	155.06	147.30	141.36	136.72	133.06
25	290.27	250.54	224.71	206.80	193.82	184.12	176.69	170.90	166.33
30	348.32	300.65	269.65	248.16	232.59	220.95	212.03	205.08	199.59
35	406.38	350.75	314.59	289.52	271.35	257.77	247.37	239.26	232.86
40	464.43	400.86	359.53	330.88	310.12	294.60	282.71	273.45	266.12
45	522.49	450.97	404.47	372.24	348.88	331.42	318.05	307.63	299.39
50	580.54	501.08	449.41	413.60	387.65	368.24	353.39	341.81	332.65
55	638.59	551.18	494.35	454.96	426.41	405.07	388.73	375.99	365.92
60	696.65	601.29	539.29	496.31	465.18	441.89	424.07	410.17	399.18
65	754.70	651.40	584.24	537.67	503.94	478.72	459.41	444.35	432.45
70	812.76	701.51	629.18	579.03	542.71	515.54	494.74	478.53	465.71
75	870.81	751.62	674.12	620.39	581.47	552.37	530.08	512.71	498.98
80	928.86	801.72	719.06	661.75	620.24	589.19	565.42	546.89	532.24

Contour curve for $520 payment

FIGURE 9.15

a. If a prospective house buyer can afford to pay only \$520 each month, what are the options?

b. Explain how increasing the period of the loan affects the mortgage amount if the monthly payment remains constant.

Solution

The students' natural inclination is to draw a squiggly contour curve. Encourage them to make their contour curves as smooth (that is, containing as few concavity changes) as possible.

a. We answer this question by sketching a contour curve at the \$520 level (as depicted in Figure 9.15) and identifying points on that curve. There are many available options. For instance, a buyer who wants to pay off the mortgage within 12 years can apply for a loan of around \$50,000. If a buyer wants a 20-year mortgage, then the amount borrowed can be between \$65,000 and \$70,000. However, if the buyer is willing to pay on the mortgage for 30 years, a \$75,000 mortgage can be obtained.

b. If the loan period increases and the buyer wishes to remain on the contour curve (thus keeping the monthly payment constant), then the mortgage amount also increases.

Data tables do not show every possible value for the input and output variables. For instance, in Example 1, you cannot determine from the data table the exact loan amount that can be obtained for a 20-year mortgage at 7% interest paid off with monthly payments of \$520. If, however, there is a multivariable function given for the data table, then exact values on a particular contour curve can be determined.

Sketching Contour Curves

When a formula for a multivariable function is known, we draw contour graphs for functions with two input variables by plotting points that correspond to certain levels of output. The function for the monthly payment table in Example 1 is

$$M(t, A) = \frac{5.833333A}{1 - 0.932583^t} \text{ dollars}$$

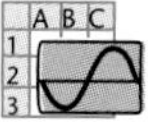

where A thousand dollars is the mortgage amount and t is the number of years the mortgage is held. We graph the \$520 contour curve by setting the function equal to \$520, picking values for one variable, solving for values of the other variable, and plotting the points generated. This three-step process is illustrated below:

Step 1. Set $M(t, A) = 520$.

$$\frac{5.833333A}{1 - 0.932583^t} = 520$$

Step 2. Let $t = 12$, and solve for A.

$$\frac{5.833333A}{1 - 0.932583^{12}} = 520, \text{ so } A \approx 50.565$$

Thus one point on the \$520 contour curve is $(12, 50.565)$. With $t = 16$, we have

$$\frac{5.833333A}{1 - 0.932583^{16}} = 520, \text{ so } A \approx 59.963$$

Stress to your students that we are finding points to help us sketch a contour graph, not fit a curve to the points. The technique of sketching contour curves when given a multivariable function formula is used again in Section 9.4.

Therefore, another point on the \$520 contour curve is (16, 59.963). If we repeat Step 2 a few more times with different values for t (we choose $t = 20, 24, 28$, and 30), then we obtain the following points on the contour curve:

$$(20, 67.071), (24, 72.448), (28, 76.515), (30, 78.160)$$

Step 3. Plot the points, and sketch[9] the curve (see Figure 9.16).

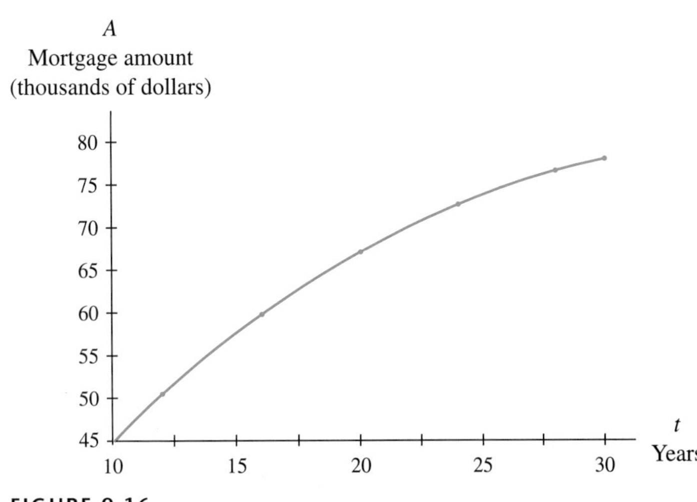

FIGURE 9.16

Why does this \$520 contour curve appear to differ from the one drawn on the data in Figure 9.15? Note that in the table, the mortgage amount is increasing from the top of the table to the bottom of the table. When we graph, it is customary to have the units on the vertical axis increase from the bottom of the graph to the top of the graph. In other words, if we flip the table over so that the mortgage amount is decreasing from the top of the table to the bottom of the table, then the curve on the table will appear to be the same as the one graphed in Figure 9.16.

EXAMPLE 2 *Windchill and Heat Loss*

Siple and Passel were two explorers who accompanied Admiral Byrd to the Antarctic and were among the first researchers to measure loss of body heat due to the wind. From their data, they obtained the equation[10]

$$H(v, t) = (10.45 + 10\sqrt{v} - v)(33 - t)$$

which gives the body's heat loss in kilogram-calories per square meter of body surface area per hour for wind speed v in meters per second when the air temperature is t°C. Table 9.11 shows data generated from the model.

a. On the table, sketch a contour curve for a heat loss of 2000 kilogram-calories per square meter of body surface area per hour.

9. The sketch of the curve should not be done by fitting a model to the points. Often, the model that should be used is not available in your calculator or computer.

10. W. Bosch and L. G. Cobb, "Windchill," UMAP Module 658, *The UMAP Journal*, vol. 5, no. 4 (Winter 1984), pp. 477–492.

TABLE 9.11 **A body's heat loss in kilogram-calories per square meter of body surface area per hour**

	Wind speed (meters per second)					
Air temperature (°F)	**0**	**5**	**10**	**15**	**20**	**25**
-20	554	1474	1700	1812	1864	1879
-25	606	1613	1860	1982	2040	2056
-30	658	1752	2021	2153	2216	2233
-35	711	1891	2181	2324	2392	2411
-40	763	2030	2341	2495	2568	2588

b. Use the equation to sketch this contour curve more accurately for wind speed values between 5 and 25 meters per second.

Solution

a. The 2000 contour curve is shown on the table in Figure 9.17.

	Wind speed (meters per second)					
Air temperature (°F)	**0**	**5**	**10**	**15**	**20**	**25**
-20	554	1474	1700	1812	1864	1879
-25	606	1613	1860	1982	2040	2056
-30	658	1752	2021	2153	2216	2233
-25	711	1891	2181	2324	2392	2411
-40	763	2030	2341	2495	2568	2588

2000

FIGURE 9.17

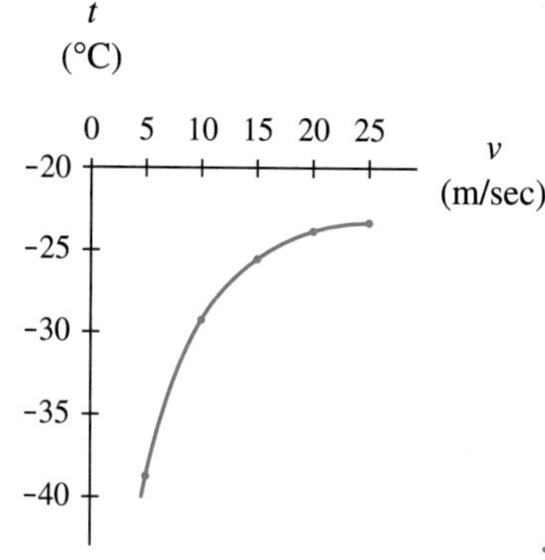

FIGURE 9.18

b. To obtain a more accurate graph of this contour curve, we choose values, v, for wind speed and solve the function $2000 = (10.45 + 10\sqrt{v} - v)(33 - t)$ for the temperature t. Let $v = 5$. Solving

$$2000 = (10.45 + 10\sqrt{5} - 5)(33 - t)$$

for t yields $t \approx$ -38.9°C. We repeat this process for $v = 10$, 15, 20, and 25 and obtain the points (5, -38.9), (10, -29.4), (15, -25.5), (20, -23.9), (25, -23.4). When we plot these points and sketch a curve through them, we obtain the contour curve shown in Figure 9.18. This curve has a similar increasing, concave-down shape to the one in Figure 9.17 but is more accurate. We can also use the equation to sketch contour curves for inputs other than those shown in the table.

Formulas for Contour Curves

We have seen how we can plot points and sketch a contour curve. Often it is helpful to be able to sketch a contour graph showing several contours. In this case, it is beneficial to find a formula to use to graph each contour curve.

Consider again the monthly payment function

$$M(t, A) = \frac{5.833333A}{1 - 0.932583^t} \text{ dollars}$$

Because contour curves are generated by choosing a constant value for output, we replace $M(t, A)$ with K, which denotes a constant:

$$K = \frac{5.833333A}{1 - 0.932583^t}$$

Then we solve for one of the variables (whichever is easier—in this case, A).

$$K(1 - 0.932583^t) = 5.833333A$$

$$A = \frac{K(1 - 0.932583^t)}{5.833333}$$

Now we have a general formula that we can use to graph contour curves by replacing K with appropriate constants.

Graphing the function for A with $K = 520$:

$$A = \frac{520(1 - 0.932583^t)}{5.833333}$$

gives the contour curve shown in Figure 9.16. Graphing additional contour curves for the equally spaced values $K = 360, 440, 600, 680, 760$, and 840 gives the contour graph in Figure 9.19.

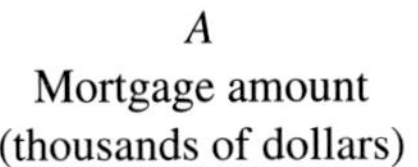

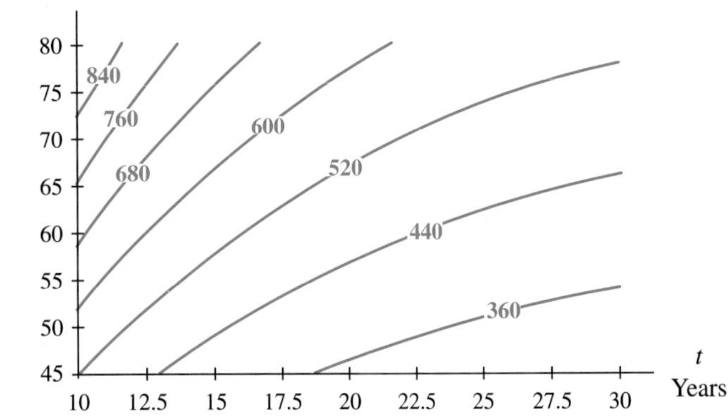

FIGURE 9.19

EXAMPLE 3 *Heat Loss*

Using the Siple and Passel heat-loss equation described in Example 2,

$$H(v, t) = (10.45 + 10\sqrt{v} - v)(33 - t)$$

find a general formula for the $H(v, t)$ contour curves, and use it to sketch a contour graph for heat-loss levels of 500, 1000, 1500, 2000, and 2500 kilogram-calories per square meter per hour.

Solution Replacing $H(v, t)$ with K and solving for t give

$$t = 33 - \frac{K}{10.45 + 10\sqrt{v} - v}$$

Substituting 500, 1000, 1500, 2000, and 2500 for K and graphing give the contour graph shown in Figure 9.20.

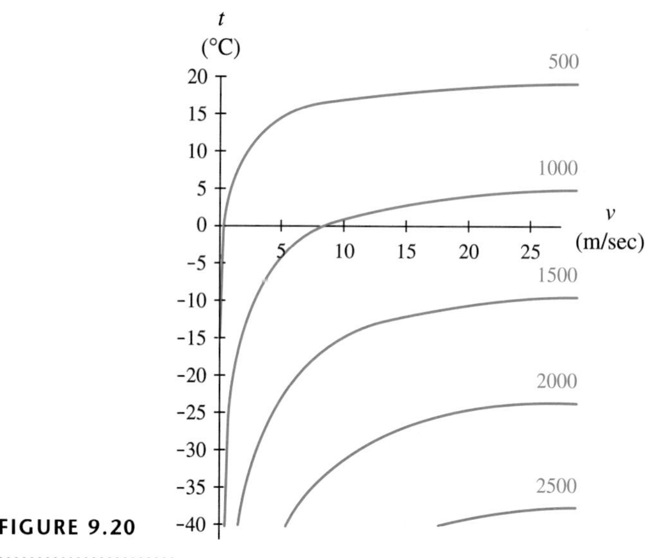

FIGURE 9.20

Estimating Output and Change in Output Using Contour Graphs

Figures 9.21 and 9.22 show a three-dimensional representation and a contour graph of the function $z = M(t, A)$, which gives the amount of the monthly payments necessary to pay off a loan of A thousand dollars over t years at a 7% interest rate.

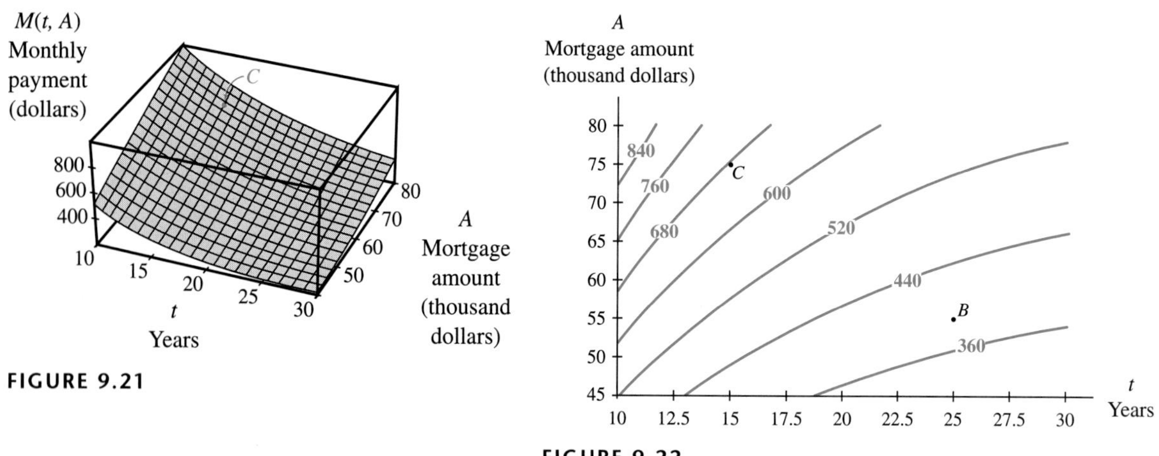

FIGURE 9.21

FIGURE 9.22

Note that in Figure 9.21, it is not difficult to estimate that the input values corresponding to the point C are $t = 15$ years and $A = 75$ thousand dollars. However, because of the orientation of the graph, it is difficult to estimate the output at point C. Looking at a contour graph may help us estimate this output. Use the contour graph and the input values $t = 15$ and $A = 75$ to locate the point C on the contour graph (see Figure 9.22). Because C is located slightly below the \$680 contour curve on the side closer to the \$600 contour curve rather than on the side closer to the \$760 contour curve, we estimate that the output at C is slightly less than \$680.

Similarly, we can use the contour graph to estimate the monthly payment corresponding to a mortgage amount of \$55,000 over 25 years. The corresponding point on the contour curve is labeled B. The monthly payment amount is between \$360 and \$440 but is closer to \$360. [The actual output is $M(25, 55) \approx \$389$.]

Consider the question "By how much does the payment on a \$55,000 mortgage change when the time is increased from 15 years to 25 years?" To use the contour graph to answer this question, first locate the points (15, 55) and (25, 55). (See Figure 9.23.)

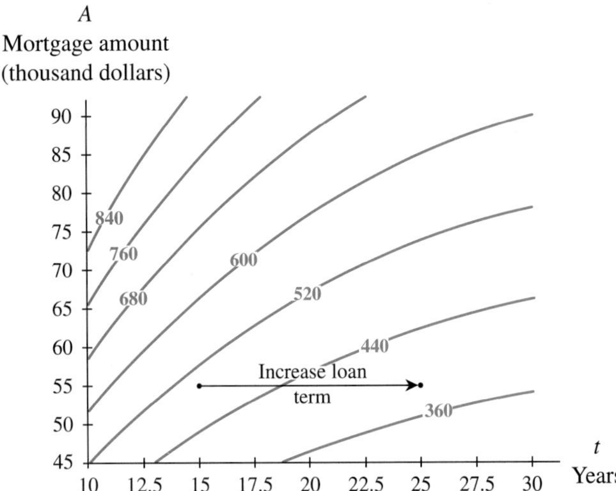

FIGURE 9.23

Next estimate the output of these two points on the basis of their location between particular contours. We estimate that the output corresponding to (15, 55) is \$490 and the output corresponding to (25, 55) is \$390. Therefore, when the term of a \$55,000 mortgage is increased from 15 years to 25 years, the output (payment) decreases by approximately $\$490 - \$390 = \$100$.

Your students' answers will vary when they estimate points on contour curves. You should allow variations in estimates, but do not allow sloppy estimates. In particular, your students should use a straightedge to help them more accurately identify points on contour graphs.

It is important to understand that when estimating the change in a multivariable function from a contour graph, you are concerned with estimating output values on the basis of nearby contour curves. Consider a \$75,000, 15-year mortgage and the question "Does the monthly payment change more when the amount of the mortgage is increased by \$5000 or when the term of the mortgage is increased by 10 years?" To answer this question, first locate the point (15, 75) on the contour graph and estimate its output. One possible estimate is \$675. (See Figure 9.24.)

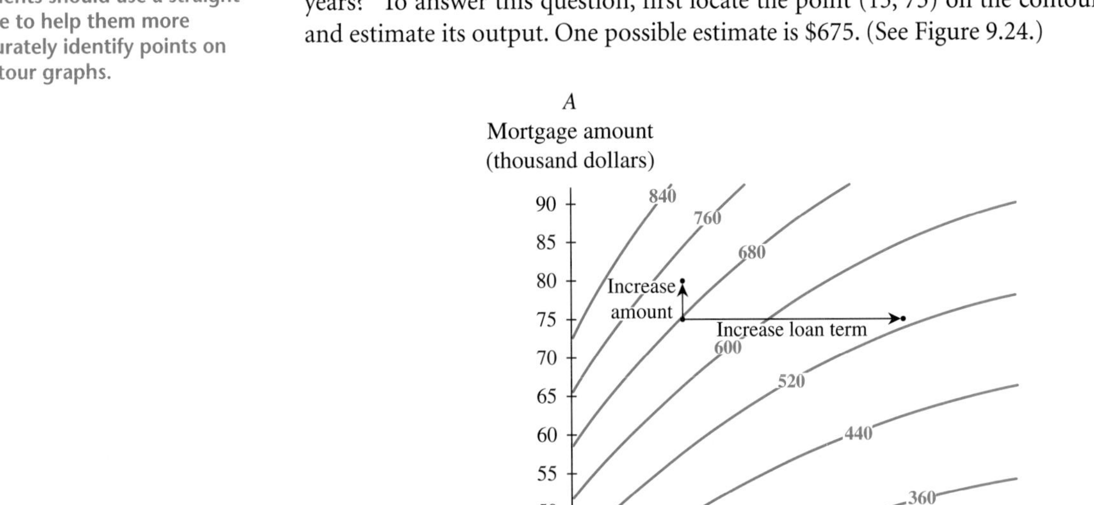

FIGURE 9.24

Increasing the amount by \$5000 corresponds to moving to the point (15, 80). Again, estimate the output on the basis of nearby contour curves. A possible estimate is \$720. Thus increasing the amount of the mortgage by \$5000 increases the monthly payments by approximately \$720 − \$675 = \$45.

Note that we distinguish between the change and the "signed" change. In comparing changes, we compare the absolute values of the changes rather than the directions of change.

Next locate the point (25, 75), which corresponds to an increase of 10 years in the term of the mortgage. The output at this point is approximately \$530, so the decrease is \$675 − \$530 = \$145, representing a change of −\$145. In determining which change is greater, we consider the magnitude (absolute value) of the changes and conclude that a decrease of \$145 is a greater change than an increase of \$45. Thus increasing the term of the mortgage by 10 years has a greater effect on the monthly payment than increasing the amount of the mortgage by \$5000.

Note that the vertical arrow in Figure 9.24 has length \$5000, whereas the horizontal arrow has length 10 years. These two lengths are not what we consider when determining which change is greater. Indeed, because these lengths have different units (thousands of dollars and years), it makes no sense to compare their lengths. The function outputs, estimated from the contours, are what we compare.

EXAMPLE 4 *Missouri Farmland*

Once again consider the elevation of the tract of Missouri farmland with graphs shown in Figures 9.25 and 9.26.

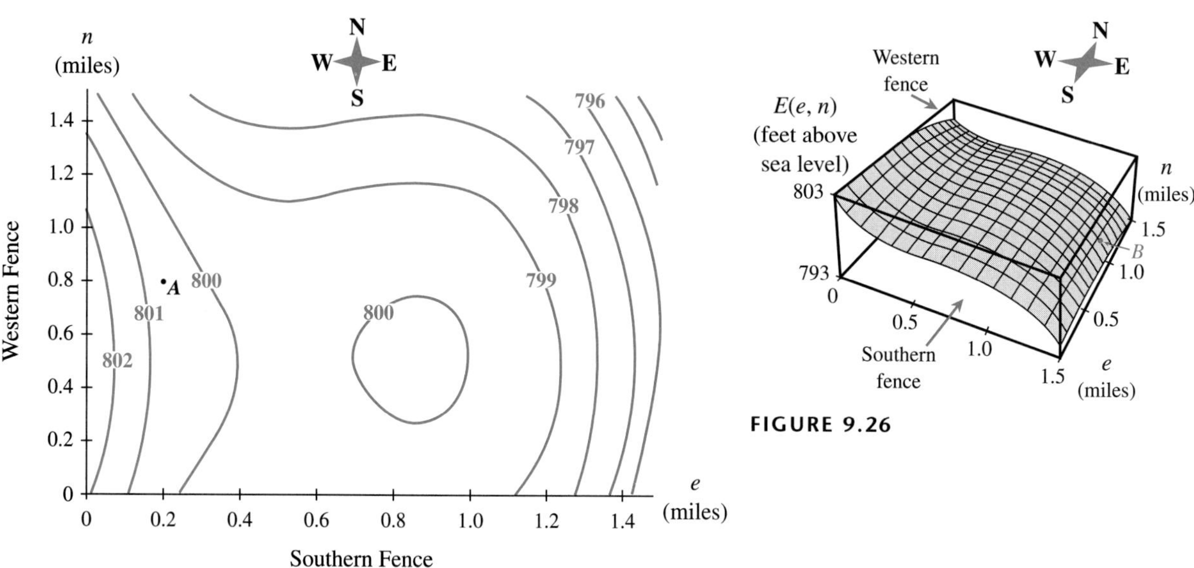

FIGURE 9.25

FIGURE 9.26

a. Estimate the input and output values associated with point *A* in Figure 9.25.

b. Estimate the input and output values associated with point *B* in Figure 9.26.

c. If you are standing at $(e, n) = (0.4, 1.0)$, will you be headed downhill or uphill if you walk 0.4 mile north? south? east? west?

d. If a tractor is at (1.0, 0.6), which direction results in the steeper descent: east 0.4 mile or north 0.4 mile?

There are two ways to determine the direction of steepest descent or ascent. You can look for the direction in which the contours are closest together, or you can look for the direction in which you first reach the next contour. If, as in part *d* of this example, you are comparing movements the same distance away from a point in two different directions, then you can determine the direction of steepest descent or ascent by comparing the change in output in each direction.

Solution

a. Point A is approximately 0.8 mile north of the southern fence ($n \approx 0.8$) and approximately 0.2 mile east of the western fence ($e \approx 0.2$). Because A lies between the 800-feet and 801-feet contour curves but closer to the 801 curve, we estimate the elevation to be $E(0.2, 0.8) \approx 800.6$ feet.

b. Looking at the three-dimensional graph, we find the input values to be $e \approx 1.4$ and $n \approx 1.0$. Transferring this point to the contour graph (see Figure 9.27), we see that it lies approximately midway between the 796-feet and 797-feet contour curves, so we estimate the elevation to be $E(1.4, 1.0) \approx 796.5$ feet.

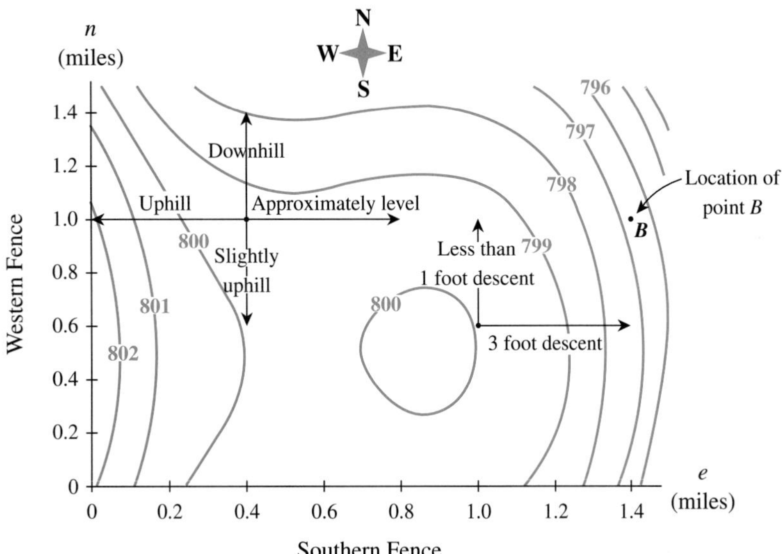

FIGURE 9.27

c. The point (0.4, 1.0) corresponds to an elevation of about 799.5 feet above sea level. Walking 0.4 mile north takes you to (0.4, 1.4) with elevation of approximately 798 feet above sea level. Therefore, you have walked downhill.

Walking 0.4 mile south from (0.4, 1.0) takes you to (0.4, 0.6) with elevation close to 800 feet above sea level, so you have walked uphill. However, if you walk farther south, you will walk uphill for approximately another 0.1 mile and then begin walking downhill.

It is not clear whether you would be walking slightly uphill, slightly downhill, or basically level when walking east from (0.4, 1.0) to (0.8, 1.0); you remain between the 799 and 800 contour curves, and this level is ambiguous. Walking west, you will walk uphill to just over 802 feet above sea level. (See Figure 9.27.)

d. Moving 0.4 mile east from (1.0, 0.6) to (1.4, 0.6) results in an approximately 3-foot descent from about 800 feet to about 797 feet, whereas moving north 0.4 mile results in less than a 1-foot descent. (See Figure 9.27.) Therefore, movement 0.4 mile to the east results in a steeper descent.

We have seen in this section how to draw contour curves and how to estimate inputs and outputs at points on contour graphs and three-dimensional graphs. Being able to visualize a multivariable function often helps us better understand the function and the relationships between the variables in the function.

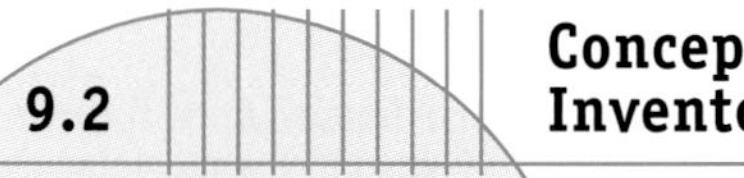

9.2 Concept Inventory

- *Contour curve*
- *Contour graph*
- *Three-dimensional graph*
- *Sketching contour curves*
- *Finding equations of contour curves*
- *Estimating output and change in output from a contour graph*
- *Estimating steeper direction*

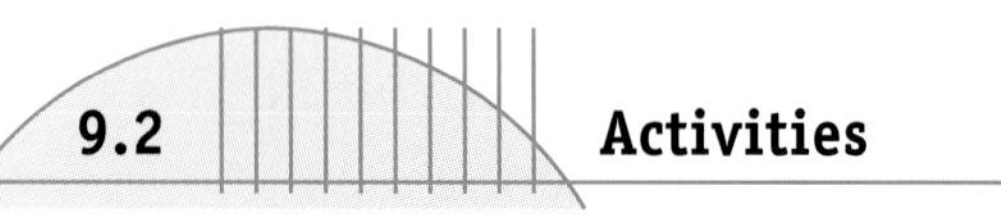

9.2 Activities

1. Your comfort in summer depends on the combined effects of air temperature and humidity. Table 9.12 shows how hot it feels[11] (the apparent temperature) for a given air temperature and relative humidity.

   The National Weather Service has established the following guidelines for the health threat on the basis of the apparent temperature:

Apparent temperature	*Health risk*
80–90°F	Fatigue possible with prolonged exposure and/or physical activity
90–104°F	Heat exhaustion or heat cramps possible
105–130°F	Heat exhaustion or heat cramps likely
above 130°F	Heat exhaustion or heat cramps highly likely

   Draw contour curves on Table 9.12 for apparent temperatures of 90°F, 105°F, and 130°F. Shade in the region on the table corresponding to apparent temperatures that are likely to cause heat exhaustion or heat cramps.

2. Table 9.13 shows the mean monthly temperature for Halley Bay, Antarctica, at noon Greenwich Mean Time at heights where the air pressure was measured at the given value during the particular month. These data were collected during balloon ascents and reported in *Report of the Royal Society, IGY Antarctic Expedition to Halley Bay*. (Temperatures have been converted to °F.)

TABLE 9.12 Apparent temperature (°F)

Air temperature (°F)	Relative humidity (%)												
	40	45	50	55	60	65	70	75	80	85	90	95	100
110	135												
108	130	137											
106	124	130	137										
104	119	124	130	137									
102	114	119	124	130	137								
100	109	113	118	123	129	136							
98	105	108	113	117	122	128	134						
96	101	104	107	111	116	121	126	132					
94	97	100	103	106	110	114	119	124	129	135			
92	94	96	98	101	104	108	112	116	121	126	131		
90	91	92	94	97	99	102	106	109	113	117	122	126	131
88	88	89	91	93	95	97	100	103	106	109	113	117	121
86	85	86	88	89	91	93	95	97	99	102	105	108	111
84	83	84	85	86	87	89	90	92	94	96	98	100	102
82	81	82	83	83	84	85	86	87	88	90	91	93	94
80	80	80	81	81	82	82	83	83	84	85	85	86	87

11. National Oceanic and Atmospheric Administration.

TABLE 9.13 Mean monthly temperature at noon (°F)

Pressure (millibars)	Month											
	Jan	Feb	Mar	Apr	May	Jun	Jul	Aug	Sep	Oct	Nov	Dec
30	12.9	8.5	2.6	-6.2	-11.7	-16.3	-18.9	-17.2	-13.9	-4.2	10.1	13.9
40	12.3	8.4	2.6	-4.9	-11.3	-15.2	-18.1	-17.5	-15.3	-7.1	7.4	12.7
50	11.6	8.4	3.3	-3.6	-11.2	-14.5	-17.3	-17.2	-15.2	-8.0	6.2	12.6
60	11.2	8.2	3.7	-3.3	-10.3	-13.7	-16.3	-16.7	-14.6	-8.4	5.2	11.4
80	10.6	8.1	4.3	-1.9	-7.9	-11.8	-14.8	-15.2	-19.6	-8.6	3.1	10.2
100	10.0	7.8	4.4	-0.5	-6.7	-10.3	-13.1	-14.2	-13.6	-9.1	-1.2	9.6
150	9.1	7.8	5.8	1.1	-3.8	-7.8	-10.7	-12.0	-11.8	-9.1	-1.4	7.2
200	8.5	7.8	6.0	1.4	-3.0	-6.9	-9.7	-10.3	-10.2	-8.0	-2.8	6.4
250	6.8	6.7	4.7	-0.4	-4.2	-5.0	-6.9	-7.9	-7.3	-5.6	-2.9	2.0
300	4.1	4.4	1.6	0.3	-2.1	-2.2	-3.5	-4.3	-3.8	-2.2	-0.6	1.1
400	8.1	8.3	6.3	5.8	3.7	4.2	3.1	2.8	3.6	4.3	6.2	6.5
500	13.4	13.7	11.9	11.3	9.2	10.0	8.8	8.8	9.8	10.1	12.3	12.3
700	22.1	21.7	20.2	19.4	17.0	17.9	17.7	16.9	18.2	19.5	20.7	20.4
850	16.8	25.2	24.3	22.2	20.4	20.6	20.5	19.9	20.3	23.4	24.9	24.9

a. Draw contour curves on the data from -10°F to 20°F in increments of 10.

b. Interpret these contours in the context of the mean monthly temperature.

c. Why are the temperatures in December and January warmer than the temperatures in June and July?

d. In general, are temperatures higher at 30 millibars or 850 millibars? Which air pressure represents elevation closer to the Earth's surface?

3. Table 9.14 on page 642 gives the number of hours of daylight[12] for a given month at a given latitude (measured in degrees away from the equator) for the northern and southern hemispheres.

   a. How many hours of daylight will there be at a location at latitude 25° north in March?

   b. How many hours of daylight will there be at a location at latitude 25° south in September?

   c. How many hours of daylight will there be in January in the region where you attend school?

   d. Draw contour curves representing 13, 14, 15, 16, 17, and 18 hours of daylight. In a different color, draw contour curves representing 12, 11, 10, and 9 hours of daylight.

4. The Hutterite Brethren live on communal farms in parts of Canada and the United States. They are a religious group that migrated from Europe to North America in the 1870s. Practically all marriages are within the group, and the data[13] in Table 9.15 on page 642 give the number of the 1236 married Dariusleut or Lehrerleut Hutterite women born between 1879 and 1936 who had *s* sons and *d* daughters. For example, there were 28 women who had no sons and 1 daughter and 39 women with 4 sons and 4 daughters.

   a. How many women had 2 daughters and 2 sons?

   b. How many women had 5 children?

   c. Draw contour curves at 34, 25, 16, and 7 women.

5. Your body-mass index (BMI) is a measure of how thin you are compared to your height. It is calculated as your weight (in kilograms) divided by your height (in meters) squared. Table 9.16 on page 643 shows BMI[14] values for people between 5 and 6 feet tall who weigh between 90 and 200 pounds.

12. A. A. Hanson, ed., *Practical Handbook of Agricultural Science* (Boca Raton, FL: CRC Press, 1990).

13. P. Guttorp, *Statistical Inference for Branching Processes* (New York: Wiley, 1991), p. 194.

14. Based on information obtained from **shapeup.org** (accessed on 6/27/00).

TABLE 9.14 Hours of daylight for a given month and latitude

Latitude	Month											
North South	Jan Jul	Feb Aug	Mar Sep	Apr Oct	May Nov	Jun Dec	Jul Jan	Aug Feb	Sep Mar	Oct Apr	Nov May	Dec Jun
0	12.12	12.12	12.12	12.12	12.12	12.12	12.12	12.10	12.11	12.12	12.12	12.12
5	11.87	11.96	12.08	12.22	12.35	12.41	12.38	12.28	12.16	12.02	11.90	11.83
10	11.61	11.81	12.06	12.35	12.57	12.70	12.64	12.45	12.17	11.91	11.67	11.55
15	11.34	11.66	12.04	12.47	12.82	13.00	12.92	12.62	12.22	11.81	11.44	11.25
20	11.07	11.50	12.01	12.60	13.07	13.32	13.22	12.81	12.26	11.70	11.20	10.94
25	10.78	11.33	11.97	12.74	13.34	13.66	13.53	13.02	12.31	11.58	10.94	10.62
30	10.45	11.14	11.97	12.88	13.65	14.05	13.88	13.23	12.35	11.47	10.67	10.26
35	10.09	10.95	11.95	13.06	13.98	14.47	14.27	13.47	12.42	11.33	10.36	9.86
40	9.68	10.71	11.91	13.25	14.36	14.96	14.71	13.76	12.48	11.18	10.00	9.39
45	9.19	10.45	11.87	13.48	14.82	15.55	15.25	14.09	12.55	11.01	9.60	8.85
50	8.61	10.13	11.84	13.78	15.38	16.29	15.91	14.48	12.66	10.80	9.07	8.17
55	7.83	9.73	11.79	14.10	16.14	17.28	16.78	14.99	12.76	10.55	8.45	7.28
60	6.79	9.21	11.74	14.62	17.10	18.70	18.01	15.67	12.92	10.22	7.60	6.04

TABLE 9.15 Number of women with *s* sons and *d* daughters

s	*d*, Daughters												
Sons	0	1	2	3	4	5	6	7	8	9	10	11	12
0		28	8	2	3	2	2	0	0	0	0	1	0
1	21	29	21	11	5	7	6	4	1	0	0	0	1
2	11	27	22	21	21	14	10	15	5	3	1	0	0
3	6	16	27	20	35	29	18	12	10	2	2	0	0
4	9	10	20	21	39	28	30	24	10	2	5	1	0
5	3	7	22	22	40	17	18	23	16	7	2	0	0
6	2	9	15	16	27	26	26	17	10	4	1	0	0
7	1	4	7	27	19	20	16	7	2	2	0	0	0
8	0	3	12	14	12	7	10	5	3	1	0	0	0
9	0	2	4	8	11	4	5	2	0	0	0	0	0
10	0	1	1	2	3	2	2	1	1	0	0	0	0
11	0	0	1	2	0	1	0	0	0	0	0	0	0
12	0	0	1	1	0	0	0	0	0	0	0	0	0

For input units in inches and pounds, the body-mass index is calculated by the equation

$$B(h, w) = \frac{0.4536w}{0.00064516h^2} \text{ points}$$

where h is the person's height in inches and w is the person's weight in pounds.

a. Estimate your BMI from Table 9.16. Calculate your BMI using the equation. Use the chart that follows to assess your health risk on the basis of your BMI.

TABLE 9.16 Body-mass index

Weight (pounds)	Height (inches)						
	60	62	64	66	68	70	72
90	17.6	16.5	15.4	14.5	13.7	12.9	12.2
100	19.5	18.3	17.2	16.1	15.2	14.3	13.6
110	21.5	20.1	18.9	17.8	16.7	15.8	14.9
120	23.4	21.9	20.6	19.4	18.2	17.2	16.3
130	25.4	23.8	22.3	21.0	19.8	18.7	17.6
140	27.3	25.6	24.0	22.6	21.3	20.1	19.0
150	29.3	27.4	25.7	24.2	22.8	21.5	20.3
160	31.2	29.3	27.5	25.8	24.3	23.0	21.7
170	33.2	31.1	29.2	27.4	25.8	24.4	23.1
180	35.2	32.9	30.9	29.1	27.4	25.8	24.4
190	37.1	34.8	32.6	30.7	28.9	27.3	25.8
200	39.1	36.6	34.3	32.3	30.4	28.7	27.1

BMI	*Health risk based solely on BMI*
19–24	Minimal
25–26	Low
27–29	Moderate
30–34	High
35–39	Very high
above 40	Extremely high

b. On the table, sketch contour curves for index values of 15, 20, 25, 30, and 35 points.

c. Find a general equation for the contour curves of the function B.

d. Use the equation in part c to sketch a contour graph for index values of 15, 20, 25, 30, and 35 points. Compare this graph with the one sketched on the table.

6. The demand for air travel between the United States and Europe between 1965 and 1978 can be modeled[15] by the function

$$D(g, p) = \frac{15.4406g^{1.905}}{p^{1.247}} \text{ thousand passengers}$$

where p is the yearly average airfare (in dollars) between New York City and London, adjusted for inflation, and g is the U.S. yearly gross national product (in billions of dollars), adjusted for inflation. Table 9.17 on page 644 shows selected values of this function.

a. On the table, sketch contour curves for demand values of 3000, 6000, 9000, 12,000, 15,000, 18,000, 21,000, and 24,000 thousand passengers.

b. Find a general equation for the contour curves of the function D.

c. Use the equation in part b to sketch a contour graph for demand values of 3000, 6000, 9000, 12,000, 15,000, 18,000, 21,000, and 24,000 thousand passengers. Compare this graph with the one sketched on the table.

7. Draw the contour curves $g(s, k) = 200$, $g(s, k) = 400$, and $g(s, k) = 600$ for the multivariable function $g(s, k) = 100k(1.093^s)$ for $0 \le s \le 60$.

8. Draw the contour curves $f(L, K) = 50$, $f(L, K) = 55$, and $f(L, K) = 60$ for the multivariable function $f(L, K) = 38.954L^{0.5}K^{0.5}$ for $0.25 \le L \le 1$.

9. Draw the contour curves $P(s, u) = 40$, $P(s, u) = 60$, and $P(s, u) = 100$ for the multivariable function $P(s, u) = 38.567s - 2.155su + 12.87u + 0.987u^2$ for $0 \le u \le 10$.

10. The function $M(t, A) = \frac{5.833333A}{1 - 0.932583^t}$ gives the amount (in dollars) of the monthly payments necessary to pay off a loan of A thousand dollars at 7% annual interest over t years. Draw the \$400 monthly payment contour curve for M for loans between 0 and 30 years.

15. J. M. Cigliano, "Price and Income Elasticities for Airline Travel: The North Atlantic Market," *Business Economics*, September 1980, pp. 17–21.

TABLE 9.17 Demand (thousands of passengers) for air travel between the United States and Europe between 1965 and 1978

Average yearly airfare (dollars)	Yearly U.S. gross national product (billions of dollars)					
	1400	1700	2000	2300	2600	2900
400	8655	12,529	17,075	22,284	28,146	34,655
500	6553	9485	12,927	16,871	21,309	26,237
600	5220	7556	10,298	13,440	16,976	20,902
700	4307	6235	8497	11,090	14,007	17,246
800	3647	5279	7194	9389	11,859	14,601
900	3148	4558	6211	8106	10,239	12,606
1000	2761	3996	5447	7108	8978	11,054
1100	2451	3549	4836	6312	7972	9816
1200	2199	3184	4339	5663	7152	8806
1300	1990	2881	3927	5125	6473	7970
1400	1815	2627	3580	4672	5902	7266

11. In 1986, Cotterill[16] developed a model for measuring the performance of supermarkets by considering their price index level. The price index level was an aggregate of 121 representative prices. The lowest-price supermarket was assigned a price index of 100. According to the Cotterill study, the price index level of an independent supermarket can be modeled by the function

$$P(c, d, p, s) = 109.168 - 0.730s + 0.027s^2 + 0.002d - 0.041p + 0.175c$$

where the supermarket has s thousand square feet of sales space and is d miles from the warehouse, and the consumer base grew by p thousand people in 10 years and had a per capita income of c thousand dollars. Assume that the distance from a supermarket to its distribution warehouse is 100 miles and that the consumer base grew by 10,000 people in 10 years.

   a. Write an equation for $P(c, s) = P(c, 100, 10, s)$.

   b. Draw the 110 contour curve for $P(c, s)$ for supermarkets between 5000 and 25,000 square feet.

12. In the timber industry, being able to predict the volume of wood in a tree stem is important, especially when one is trying to determine the number of boards a tree will yield. Several multivariable models for total-stem volume for different varieties of trees have been developed. In 1973, Brackett[17] developed the following model for predicting the total-stem (inside bark) volume for Douglas fir trees in British Columbia:

$$V(d, h) = 0.002198d^{1.739925}h^{1.133187} \text{ cubic feet}$$

where d is the diameter of the tree at breast height (4.5 feet above the ground), which is denoted by dbh and measured in inches, and h is the height of the tree in feet.

   a. Find the stem volume of a Douglas fir with a dbh of 1 foot and a height of 32 feet.

   b. Draw a contour curve representing the volume you found in part *a* for diameters between 8 dbh and 18 dbh.

   c. Explain how a change in dbh affects height if volume remains constant.

13. The percentage cooking loss in sausage can be modeled[18] by

$$P(w, s) = 10.65 + 1.13w + 1.04s - 5.83ws \text{ percent}$$

when w and s represent the proportions of whey protein and skim milk powder, respectively, used in the sausage.

16. P. G. Helmberger and J. P. Chavas, *The Economics of Agricultural Prices* (Upper Saddle River, NJ: Prentice-Hall, 1996).

17. J. L. Clutter et al., *Timber Management: A Quantitative Approach* (New York: Wiley, 1983).

18. M. R. Ellekjaer, T. Naes, and P. Baardseth, "Milk Proteins Affect Yield and Sensory Quality of Cooked Sausages," *Journal of Food Science*, vol. 61, no. 3 (1996), pp. 660–666.

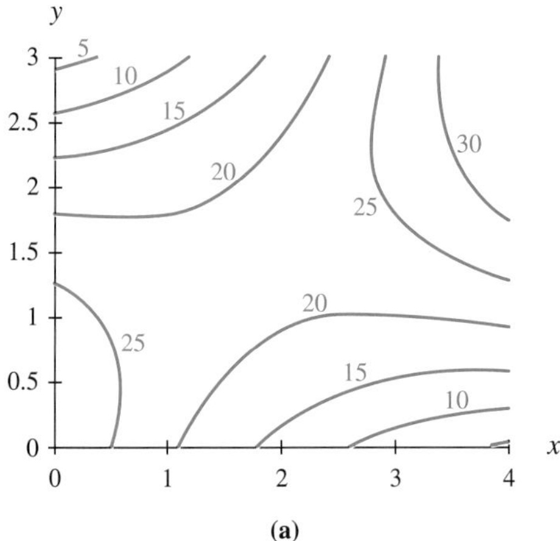

(a)

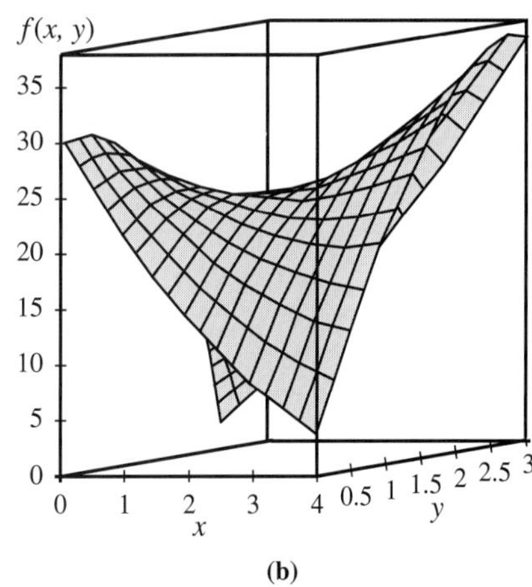

(b)

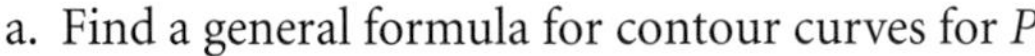

FIGURE 9.2.1

a. Find a general formula for contour curves for P.

b. Sketch a contour graph for percentages of 10.6, 10.7, 10.8, 10.9, and 11.0. Note that because w and s are proportions, they should be graphed from 0 to 1.

14. The carrying capacity of a particular farm system is defined as the number of animals or people that can be supported by the crop production from a given land area. The carrying capacity of a wheat crop can be modeled[19] by

$$K(P, D) = \frac{11.56P}{D} \text{ people per hectare}$$

where P is the number of kilograms of wheat produced per hectare per year and D is the yearly energy requirement for one person measured in megajoules per person. Sketch a contour graph for carrying capacities of 13, 15, 17, and 19 people per hectare.

15. For the contour graph and its associated three-dimensional graph in Figures 9.2.1a and 9.2.1b, answer the questions by analyzing the graphs. Explain in sentences why you answered as you did.

a. At $(1.5, 2)$, does $f(x, y)$ increase when y increases or when y decreases?

b. At $(2.5, 2.5)$, will $f(x, y)$ decrease more quickly as x decreases or as y decreases?

c. Will the change in $f(x, y)$ be greater when $(2, 2)$ shifts to $(1, 2.5)$ or when $(1, 0)$ shifts to $(4, 1)$?

16. For each of the following questions, refer to the graphs in Figures 9.2.2a and 9.2.2b on page 646, and answer the questions by analyzing the graphs. Explain in sentences why you answered as you did.

a. At $(10, 10)$, is $K(s, w)$ increasing when w increases or when s decreases?

b. At $(15, 2.5)$, will $K(s, w)$ be decreasing more rapidly when w increases or when s decreases?

c. Will the change in $K(s, w)$ be greater when the position is changed from $(5, 5)$ to $(6, 3)$ or from $(5, 5)$ to $(6, 7)$?

17. Answer each of the following questions by analyzing the contour graph in Figure 9.2.3 on page 646.

a. Describe in as much detail as possible the three-dimensional graph that this contour graph represents.

b. Is the descent greater when the position changes from $(0.7, 0.1)$ to $(0.4, 0.4)$ or when it changes from $(0.7, 0.1)$ to $(0, 0.3)$?

c. At $(0, 0.1)$, does the function output increase as x increases or as y increases?

d. Locate a point with output approximately 0.15 units greater than the output at $(-0.2, -0.3)$.

18. Answer each of the following questions by analyzing the contour graph in Figure 9.2.4 on page 646.

a. Consider any point (x, y) on the graph. What happens to the function output when x remains constant and y increases?

19. R. S. Loomis and D. J. Connor, *Crop Ecology: Productivity and Management in Agricultural Systems* (Cambridge, England: Cambridge University Press, 1992).

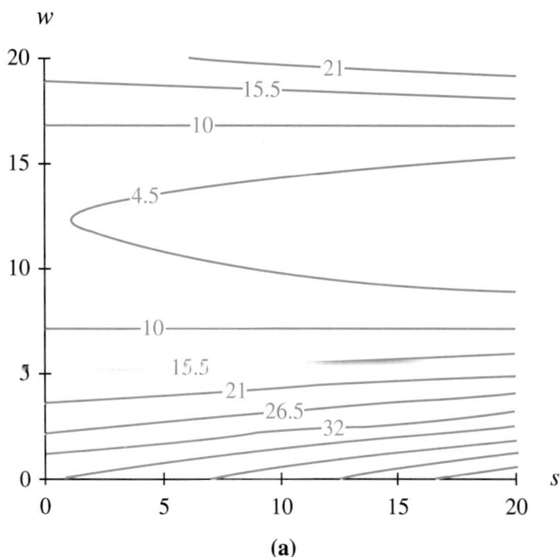

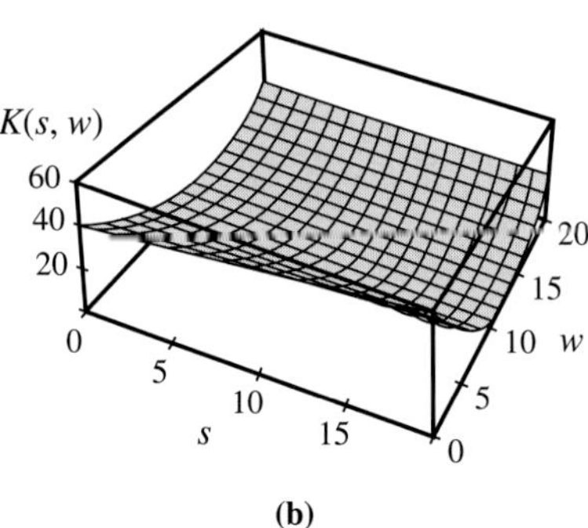

FIGURE 9.2.2 (Graphs for Activity 16)

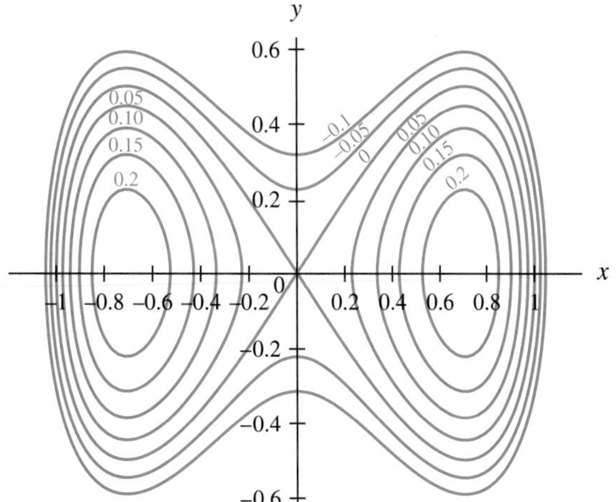

FIGURE 9.2.3 (Graph for Activity 17)

b. Consider the point (-1, 5). When x is decreased by 1/2, by what amount must y change in order for the point to remain on the same contour curve?

c. Consider the point (-2, -2). When y is increased by 1.5, by what amount must x change in order for the output to remain constant?

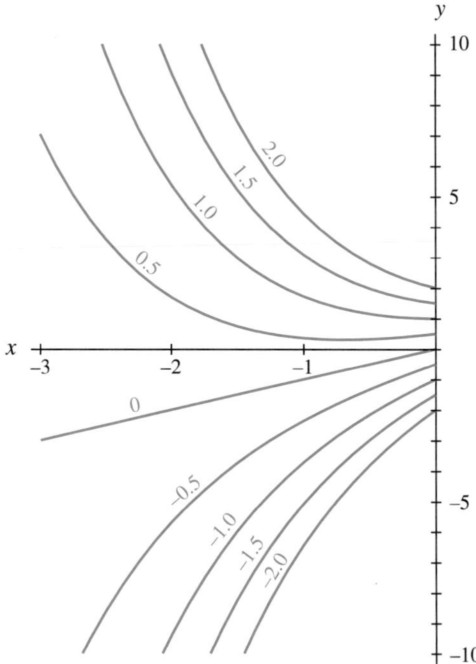

FIGURE 9.2.4 (Graphs for Activity 18)

19. The contour graph in Figure 9.2.5a and the three-dimensional graph in Figure 9.2.5b represent the revenue $R(c, r)$ in thousands of guilders when c thousand tons of 40% fat cheese and r thousand tons of regular cheese are produced.

   a. Use the contour graph to approximate the production levels that will maximize revenue. Why did you pick the production levels you chose?

   b. Use the contour graph and the three-dimensional graph to approximate the maximum revenue. Explain how you arrived at this conclusion.

   c. The function for revenue is

$$R(c, r) = -52.196r^2 + 5935.497r - 59.128cr + 10{,}299.325c - 384.386c^2$$

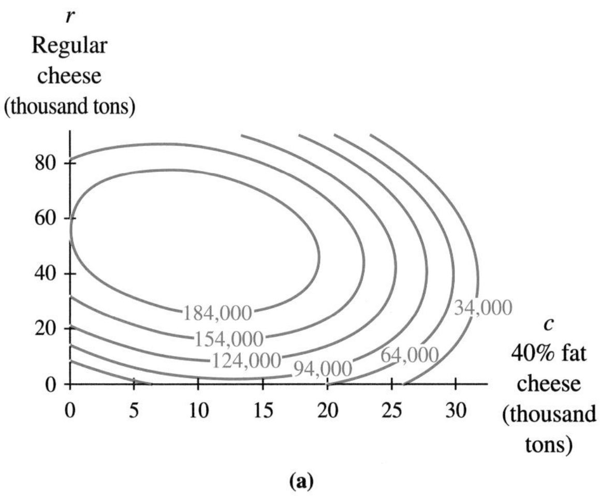

FIGURE 9.2.5 (Graphs for Activity 19)

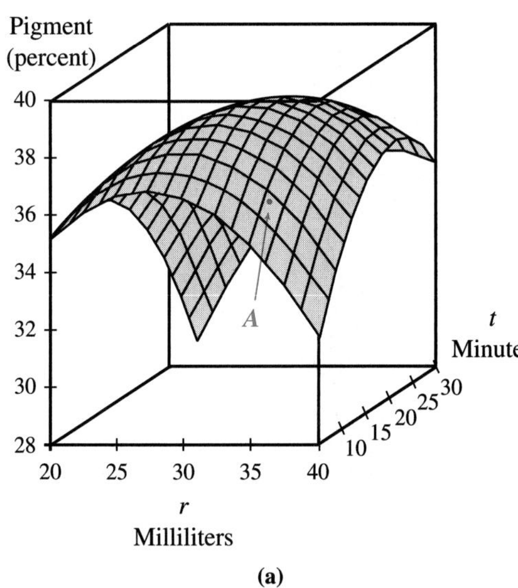

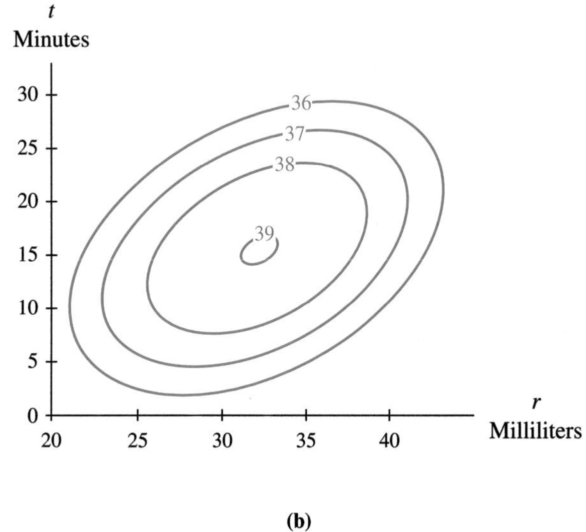

FIGURE 9.2.6 (Graphs for Activity 20)

thousand guilders where c thousand tons of 40% fat cheese and r thousand tons of regular cheese are produced. Use the revenue function to find the revenue that corresponds to the production levels from part *a*. Compare this answer with your answer to part *b*.

20. A process to extract pectin and pigment from sunflower heads involves washing the sunflower heads in heated water. Figures 9.2.6a and 9.2.6b show a three-dimensional graph and the associated contour graph of the percentage of pigment[20] that can be removed from a sunflower head by washing for t minutes in r milliliters of water per gram of sunflower heads when the water temperature is 75°C.

    a. Find and interpret the input and output values for point *A*.

    b. Estimate the percentage of pigment removed from 15 grams of sunflower heads if the sunflower heads are washed for 10 minutes in 37.5 milliliters of 75°C water.

    c. Estimate how much water would be necessary to remove 36% of the pigment from 9 grams of sunflower heads if they are washed for 15 minutes in 75°C water. Discuss why there are two answers.

20. X. Q. Shi et al., "Optimizing Water Washing Process for Sunflower Heads Before Pectin Extraction," *Journal of Food Science*, vol. 61, no. 3 (1996), pp. 608–612.

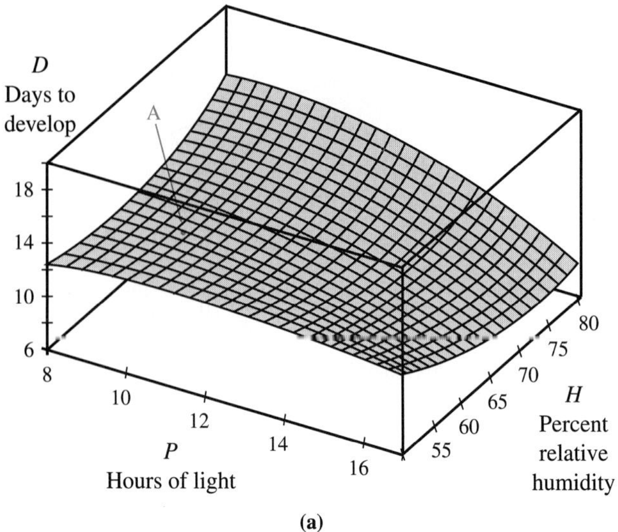

(a)

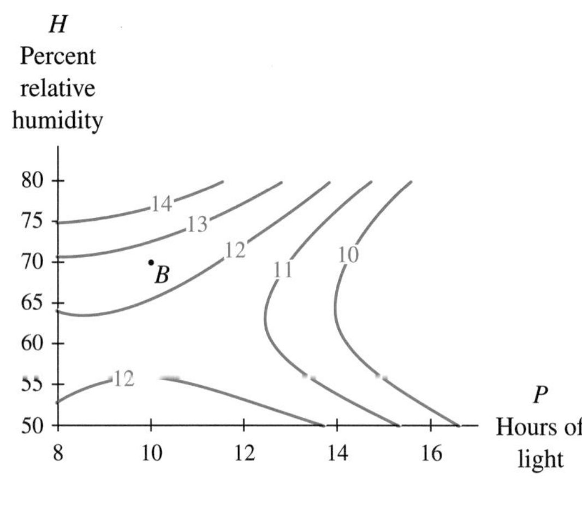

(b)

FIGURE 9.2.7

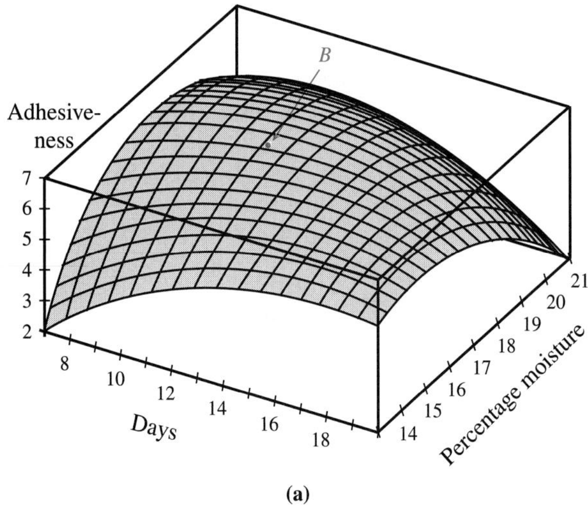

(a)

(b)

FIGURE 9.2.8

21. Boll weevils have long presented a threat to cotton crops in the southern United States. Research has been done to determine the optimal conditions for reproduction of *Catolaccus grandis*, a parasitoid that attaches to boll weevils and kills them. Figures 9.2.7a and 9.2.7b show a three-dimensional graph and an associated contour graph of the developmental time[21] (in days) of *C. grandis* as a function of the relative humidity and the number of hours of light.

    a. Estimate and interpret the input and output values at point *B*.

    b. Estimate and interpret the input and output values at point *A*.

22. The ability of honey to attach to a surface on which it is spread is known as its adhesiveness. The adhesiveness of honey relies on several factors: the percentage of glucose and maltose (two sugars), the percentage of moisture, the percentage of crystallization in the honey before it begins to set, and the number of days the honey is allowed to set at 12°C. Figures 9.2.8a and 9.2.8b show a three-dimensional graph and a contour graph of a measure of the

21. J. A. Morales-Ramos, S. M. Greenberg, and E. G. King, "Selection of Optimal Physical Conditions for Mass Propagation of *Catolaccus grandis*," *Environmental Entomology*, vol. 25, no. 1 (February 1996), pp. 165–173.

adhesiveness of honey[22] given the percentage of moisture and the number of days the honey is allowed to set when 40.9% of the honey is glucose and maltose and 12.5% of the honey was crystallized before setting began.

a. Estimate and interpret the input and output values at point *A*.

b. Estimate and interpret the input and output values at point *B*.

23. In meteorology, temperature contours are known as *isotherms*. Figure 9.2.9 shows surface temperature isotherms[23] (in °F) on February 15, 1990. The shaded regions indicate the areas of precipitation. Label the regions most likely to have received snow, ice pellets, freezing rain, and rain.

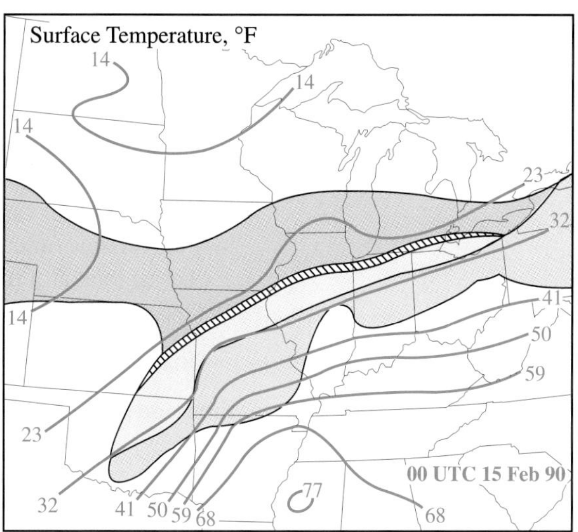

FIGURE 9.2.9

24. Wind turbines can be used to harness the power of the wind and generate electricity. An optimally placed wind machine is located where wind power will generate the most electricity. Figure 9.2.10 shows a contour map of average available wind power[24] (measured in watts per square meter) over the contiguous United States. Estimate four optimal wind machine placement sites in (or near the shore of) the United States.

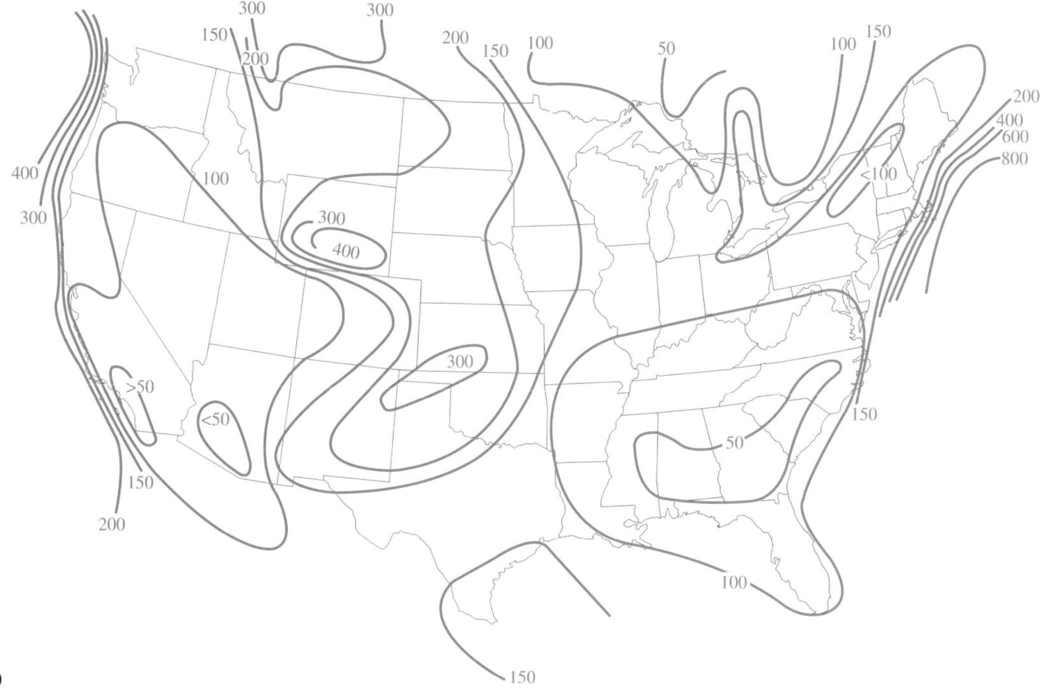

FIGURE 9.2.10

22. J. M. Shinn and S. L. Wang, "Textural Analysis of Crystallized Honey Using Response Surface Methodology," *Canadian Institute of Food Science Technology Journal*, vol. 23, nos. 4/5 (1990), pp. 178–182.

23. Figure 9.2.9 from R. R. Czys and R. W. Scott, "Forecasting Techniques: A Physically Based, Nondimensional Parameter for Discriminating Between Locations of Freezing Rain and Ice Pellets," *Weather and Forecasting*, vol. II (1996), pp. 591–598. Used by permission of the American Meteorological Society.

24. Figure 9.2.10 from D. M. Gates, *Energy and Ecology* (Sunderland, MA: Sinauer Associates, 1985). Copyright © 1985. Reprinted with permission of Sinauer Associates, Inc.

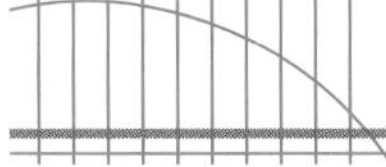

9.3 Partial Rates of Change

Calculus is the study of change. As we saw in Chapters 3 and 4, one of the aspects of calculus is the study of rates of change. In those chapters, we considered functions with a single input variable. We now apply the concepts of calculus to analyze change in multivariable functions.

Let us look at a model[25] for the revenue generated by the sale of cheeses in the Netherlands:

$$R(c, r) = -52.196r^2 + 5935.497r - 59.128cr + 10{,}299.325c - 384.386c^2$$

thousand guilders (a guilder is a Dutch monetary unit) when c thousand tons of 40% fat cheese and r thousand tons of regular cheese are sold. Suppose that sales of 40% fat cheese are 17,000 tons. Figures 9.28 and 9.29 show the graph of the cross section with output $R(17, r)$ in two and three dimensions, respectively.

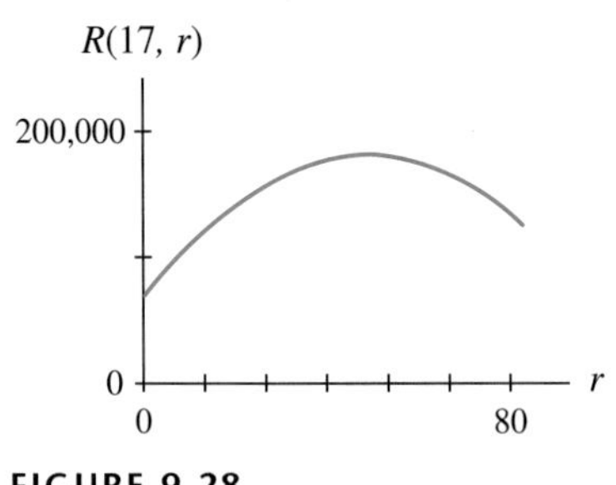

FIGURE 9.28

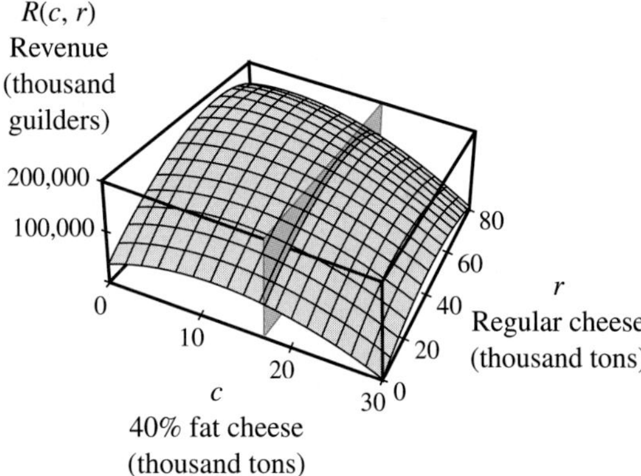

FIGURE 9.29

It would be interesting to know how quickly revenue is changing if we remain on the $c = 17$ cross section but let r (the amount of regular cheese sold) vary. In order to do this, we must find a rate-of-change formula for this cross section. To do so, we will first write a formula for the cross-sectional model. The cross-sectional model for revenue is

$$R(17, r) = -52.196r^2 + 4930.321r + 64{,}000.971 \text{ thousand guilders}$$

when 17 thousand tons of 40% fat cheese and r thousand tons of regular cheese are sold. Because this is a function of one input variable, we can find its derivative with respect to r:

When indicating derivatives of single-variable functions, your students may be more familiar with the notation y' than with the symbol $\frac{dy}{dx}$. You may wish to tell them that the input variable following the words *with respect to* is the variable that should be used in the "denominator" of the derivative symbol.

$$\frac{dR(17, r)}{dr} = -104.392r + 4930.321 \quad \frac{\text{thousand guilders per}}{\text{thousand tons of regular cheese}}$$

Evaluating this derivative at different values of r gives us the slopes of tangent lines on the cross section. For instance, evaluating $\frac{dR(17, r)}{dr}$ at $r = 40$ gives the slope of the tangent line shown in Figures 9.30a and 9.30b as 754.6 thousand guilders per thousand tons of regular cheese.

25. S. Louwes, J. Boot, and S. Wage, "A Quadratic-Programming Approach to the Problem of the Optimal Use of Milk in the Netherlands," *Journal of Farm Economics*, vol. 45, no. 2 (May 1963), pp. 309–317.

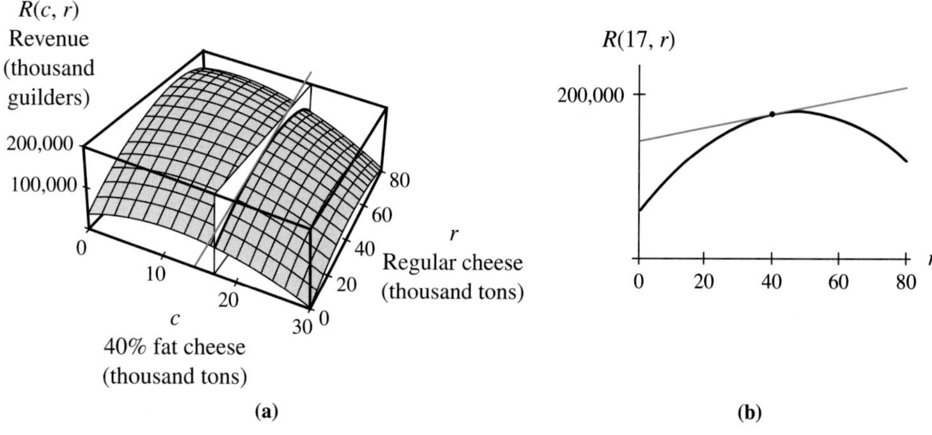

FIGURE 9.30

Note in Figure 9.30a that the tangent line is drawn on a plane cutting through the three-dimensional revenue graph at $c = 17$. The tangent line at $r = 40$ is more easily seen in Figure 9.30b on a graph of the cross-sectional function with output $R(17, r)$. This tangent line runs in the r direction, not the c direction. Thus we interpret the slope as follows:

> When 17,000 tons of 40% fat cheese is sold and 40 thousand tons of regular cheese is sold, if additional regular cheese is sold, then the revenue is increasing by 754.6 thousand guilders per thousand tons of regular cheese.

We could also determine the rates of change of sales on the $R(17, r)$ cross section at other values of r, say when 20,000 and 63,000 tons of regular cheese are sold. Note in Figure 9.30b that a tangent line drawn at $r = 20$ will be steeper than the one shown at $r = 40$. Thus we expect the derivative of $R(17, r)$ at $r = 20$ to be greater than the derivative at $r = 40$. Evaluating $\frac{dR(17, r)}{dr}$ at $r = 20$ gives 2842.5 thousand guilders per thousand tons of regular cheese.

Thus, when the sale of 40% fat cheese is held constant at a level of 17 thousand tons and 20 thousand tons of regular cheese is sold, revenue is increasing at a rate of 2842.5 thousand guilders per thousand tons of regular cheese. That is, when the sale of 40% fat cheese is constant at a level of 17 thousand tons, if sales of regular cheese increase from 20 to 21 thousand tons, then revenue will increase by approximately 2842.5 thousand guilders.

Look again at Figure 9.30b. We expect the tangent line at $r = 63$ to be negative. If we evaluate $\frac{dR(17, r)}{dr}$ at $r = 63$, we find that revenue is decreasing by 1646.4 thousand guilders per thousand tons of regular cheese when 63 thousand tons of regular cheese and 17 thousand tons of 40% fat cheese are sold.

Figure 9.31a depicts the tangent lines at (17, 20) and (17, 63) on a cross section of a three-dimensional graph of $R(c, r)$. Figure 9.31b shows the same tangent lines on the cross-sectional function with output $R(17, r)$. The partial rates of change $\frac{dR(17, r)}{dr}$ for $r = 20$ and $r = 63$ are the slopes of the tangent lines shown in Figure 9.31.

Keep in mind that these rates of change tell us only part of the story about the rate of change in revenue, because they are restricted by a sales level of 17 thousand tons of 40% fat cheese. They do not give us any information about what is happening when the sales level of 40% fat cheese is something other than 17,000 tons.

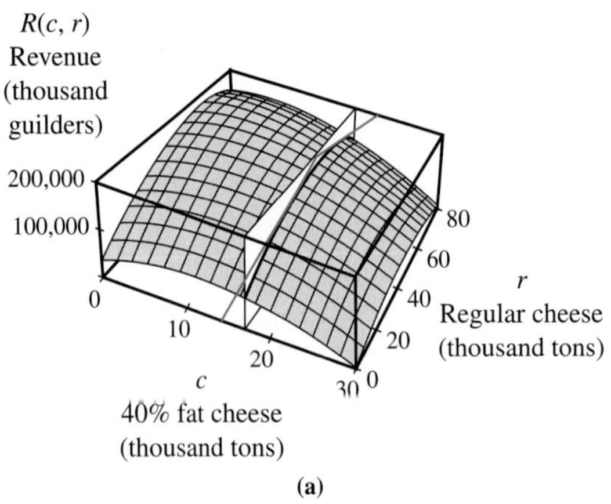

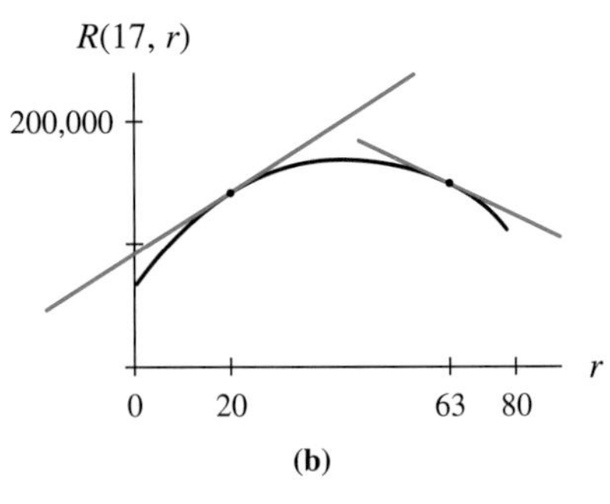

FIGURE 9.31

Partial Derivatives

Because rates of change of a cross-sectional function give us only part of the picture, we refer to them as **partial rates of change** or **partial derivatives**, and we use the notation $\frac{\partial R}{\partial r}$ instead of $\frac{dR}{dr}$ to remind ourselves that there is more than one input variable in the underlying function R. Another common notation for the partial derivative of R with respect to r is R_r.

When finding partial derivatives using a cross-sectional function, we limit ourselves to considering points on that particular cross section. In the cheese example, we are limited to the cross section that corresponds to 17 thousand tons of 40% fat cheese ($c = 17$). A more useful approach is to find a general formula for the partial derivative of revenue with respect to regular cheese sales that will enable us to choose a particular cross section and then to choose a particular point on that cross section. In this way, one formula can be used to find any partial derivative of revenue with respect to sales of regular cheese.

To find $\frac{\partial R}{\partial r}$, we treat r as a variable and c as a constant. We highlight r to help us remember that it is the variable that is changing and is not held at a constant value.

$$R(c, r) = -52.196r^2 + 5935.497r - 59.128cr + 10{,}299.325c - 384.386c^2$$

so that

$$R_r = \frac{\partial R}{\partial r} = -52.196(2r) + 5935.497(1) - 59.128c(1) + 0 - 0$$

$$= -104.392r + 5935.497 - 59.128c \quad \frac{\text{thousand guilders per}}{\text{thousand tons of regular cheese}}$$

Note that the derivatives of the first and second terms are straightforward. In the third term, c acts as a constant multiple of r, so the derivative follows the Constant Multiplier Rule. Because c acts as a constant, the last two terms are both constants, and the derivative of any constant is 0.

The partial derivative of R with respect to r $\left(R_r \text{ or } \frac{\partial R}{\partial r}\right)$ is a function of both r and c. If we evaluate $\frac{\partial R}{\partial r}$ at a particular value of c—say $c = 17$—we obtain a single-variable function with input r. This function gives the slope of any line tangent to the cross section shown in Figures 9.30b and 9.31b. We use the notation $\left.\frac{\partial R}{\partial r}\right|_{c=17}$ to represent this slope function. The end result of finding the cross-sectional function of R when $c = 17$ and then taking its derivative with respect to r is the same as that of finding the

Make sure your students understand that when they are given a function with two input variables, the following processes produce the same single-variable function:

1. Finding a partial derivative of the function with respect to one of the input variables and then evaluating the partial derivative at a particular value of the other input.
2. Evaluating the function at a particular input value and then finding the derivative with respect to the other input variable.

We use the notation $\frac{\partial R}{\partial r}\Big|_{c=17}$ or $R_r\Big|_{c=17}$ in the first case and the notation $\frac{dR(17, r)}{dr}$ in the second case. We use different notations to distinguish between the two processes, not to distinguish between the identical results.

partial derivative $\frac{\partial R}{\partial r}$ and then evaluating it at $c = 17$. In mathematical notation,[26] we say that $\frac{dR(17, r)}{dr} = \frac{\partial R}{\partial r}\Big|_{c=17}$. The advantage of the second method is that the formula for the partial derivative can be used to find rates of change at points on any cross section. The first method restricts us to points on one particular cross section.

We now use this partial derivative $\frac{\partial R}{\partial r}$ to estimate the partial rate of change of revenue with respect to regular cheese sales when 40 thousand tons of 40% fat cheese and 30 thousand tons of regular cheese are sold. Using $c = 40$ and $r = 30$, we find that

$$\begin{aligned}\frac{\partial R}{\partial r}\Big|_{(c,\,r)=(40,\,30)} &= -104.392(30) + 5935.497 - 59.128(40)\\ &\approx 438.6 \text{ thousand guilders per thousand tons of regular cheese}\end{aligned}$$

That is, when 40% fat cheese sales remain constant at 40 thousand tons and regular cheese sales increase from 30 to 31 thousand tons, revenue increases by approximately 438.6 thousand guilders. This partial rate of change is the slope of the tangent line at $r = 30$ on the graph of the cross section with output $R(40, r)$.

Note that we use the notation $\frac{\partial R}{\partial r}\Big|_{(c,\,r)=(40,\,30)} = R_r\Big|_{(c,\,r)=(40,\,30)}$ when finding the partial derivative of the two-variable function R with respect to r, treating c as a constant, and then substituting $c = 40$ and $r = 30$. If we had substituted only $c = 40$ into the partial derivative of R with respect to r, then the resulting one-variable function would have been denoted by $\frac{\partial R}{\partial r}\Big|_{c=40} = R_r\Big|_{c=40}$. This one-variable function could also be found by first substituting $c = 40$ into the formula for $R(c, r)$ and then finding the derivative with respect to r.

What would happen if the regular cheese sales were held constant and the 40% fat cheese sales were varied? In this case, we could find a general formula for the partial rate of change of revenue with respect to the quantity of 40% fat cheese sold:

$$R(c, r) = -52.196r^2 + 5935.497r - 59.128cr + 10{,}299.325c - 384.386c^2$$

so that

$$\begin{aligned}R_c = \frac{\partial R}{\partial c} &= 0 + 0 - 59.128(1)r + 10{,}299.325(1) - 384.386(2c)\\ &= -59.128r + 10{,}299.325 - 768.772c\end{aligned}$$

thousand guilders per thousand tons of 40% fat cheese when sales of regular cheese are held constant.

We now use this partial derivative $\frac{\partial R}{\partial c}$ to estimate the partial rate of change of revenue with respect to regular cheese sales when 40 thousand tons of 40% fat cheese and 30 thousand tons of regular cheese are sold. When $c = 40$ and $r = 30$,

$$\begin{aligned}\frac{\partial R}{\partial c}\Big|_{(c,\,r)=(40,\,30)} &= -59.128(30) + 10{,}299.325 - 768.772(40)\\ &\approx -22{,}225.4 \text{ thousand guilders per thousand tons of 40\% fat cheese}\end{aligned}$$

26. When writing the partial rate of change of a function $z = f(x, y)$ with respect to one variable, say x, we use the notation $\frac{\partial f}{\partial x}$ or f_x. The partial derivative of f with respect to x and evaluated at $y = b$ is denoted by $\frac{\partial f}{\partial x}\Big|_{y=b}$ or $f_x\Big|_{y=b}$. This partial derivative gives the slope of a line tangent to the cross-sectional function for which $y = b$. The partial derivative of f with respect to y and evaluated at $x = a$ is denoted by $\frac{\partial f}{\partial y}\Big|_{x=a}$ or $f_y\Big|_{x=a}$. This partial derivative gives the slope of a line tangent to the cross-sectional function for which $x = a$. We write $\frac{\partial f}{\partial x}\Big|_{(x,\,y)=(a,\,b)}$ or $f_x\Big|_{(x,\,y)=(a,\,b)}$ when evaluating the partial derivative $\frac{\partial f}{\partial x}$ at the point where $x = a$ and $y = b$, where a and b are constants.

That is, when regular cheese sales remain constant at 30 thousand tons and 40% fat cheese sales increase from 40 to 41 thousand tons, revenue decreases by approximately 22,225.4 thousand guilders. Again, this partial rate of change is the slope of the tangent line at $c = 40$ on the graph of the cross section with output $R(c, 30)$.

In general, we find the partial derivative of a multivariable function with respect to one input variable by treating all other input variables as constants and proceeding to take derivatives as functions of a single input variable. For a two-variable function, a partial rate of change can be visualized graphically as the slope of a line tangent to a cross section.

EXAMPLE 1 *Value of an Investment*

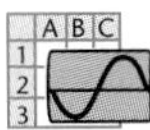

The formula for the accumulated value of an investment of P dollars over t years is

$$A(P, r, n, t) = P\left(1 + \frac{r}{n}\right)^{nt} \text{ dollars}$$

If the annual interest rate is 6% compounded quarterly, then the accumulated value formula is

$$A(P, t) = P\left(1 + \tfrac{0.06}{4}\right)^{4t} \approx P(1.061363551^t) \text{ dollars}$$

a. Find the cross-sectional models for $P = 1000$, $P = 7500$, and $P = 190{,}000$. Using appropriate notation, write the partial derivatives of these functions with respect to t.

b. Write a general formula for $\frac{\partial A}{\partial t}$. What notation is used to represent this general formula evaluated at $P = 1000$ and $t = 10$?

c. Find the cross-sectional models for $t = 2$, $t = 10$, and $t = 37$. Also find the partial derivatives of these functions with respect to P.

d. Write a general formula for $\frac{\partial A}{\partial P}$. Find and interpret $A_P\big|_{(P,\,t) = (7500,\,10)}$.

Solution

a. When $P = 1000$, $A(1000, t) = 1000(1.061363551^t)$ dollars,
and $A_t\big|_{P=1000} = \frac{\partial A}{\partial t}\big|_{P=1000} = 1000(\ln 1.061363551)(1.061363551^t)$ dollars per year.

When $P = 7500$, $A(7500, t) = 7500(1.061363551^t)$ dollars,
and $A_t\big|_{P=7500} = \frac{\partial A}{\partial t}\big|_{P=7500} = 7500(\ln 1.061363551)(1.061363551^t)$ dollars per year.

When $P = 190{,}000$, $A(190{,}000, t) = 190{,}000(1.061363551^t)$ dollars,
and $A_t\big|_{P=190{,}000} = \frac{\partial A}{\partial t}\big|_{P=190{,}000} = 190{,}000(\ln 1.061363551)(1.061363551^t)$ dollars per year.

b. Note that for each choice of principal P,

$$\frac{\partial A}{\partial t} = A_t = P(\ln 1.061363551)(1.061363551^t) \text{ dollars per year}$$

This result should not surprise you, because it follows the Constant Multiplier Rule for derivatives: When $f(x) = kg(x)$, then $f'(x) = kg'(x)$. In this case, P is the constant in the Constant Multiplier Rule. The general formula for $\frac{\partial A}{\partial t}$ evaluated at $P = 1000$ and $t = 10$ is denoted by

$$\left.\frac{\partial A}{\partial t}\right|_{(P,\,t)=(1000,\,10)} \quad \text{or} \quad A_t\Big|_{(P,\,t)=(1000,\,10)}$$

c. When $t = 2$, $A(P, 2) = P(1.06136551^2) = 1.126492587P$ dollars, and $\left.\frac{\partial A}{\partial P}\right|_{t=2} = 1.126492587 = 1.06136551^2$ dollars per dollar invested.
When $t = 10$, $A(P, 10) = P(1.06136551^{10}) = 1.814018409P$ dollars, and $\left.\frac{\partial A}{\partial P}\right|_{t=10} = 1.814018409 = 1.06136551^{10}$ dollar per dollar invested.
When $t = 37$, $A(P, 37) = P(1.06136551^{37}) = 9.056789062P$ dollars, and $\left.\frac{\partial A}{\partial P}\right|_{t=37} = 9.056789062 = 1.06136551^{37}$ dollars per dollar invested.

d. For any number of years t,

$$\frac{\partial A}{\partial P} = A_P = 1.061363551^t \text{ dollars per dollar invested}$$

$A_P\Big|_{(P,\,t)=(7500,\,10)} = 1.061363551^{10} \approx \$1.81$ per dollar invested when \$7500 is invested for 10 years at 6% compounded quarterly. The accumulated value of the investment will increase by approximately \$1.81 if an additional dollar is invested.

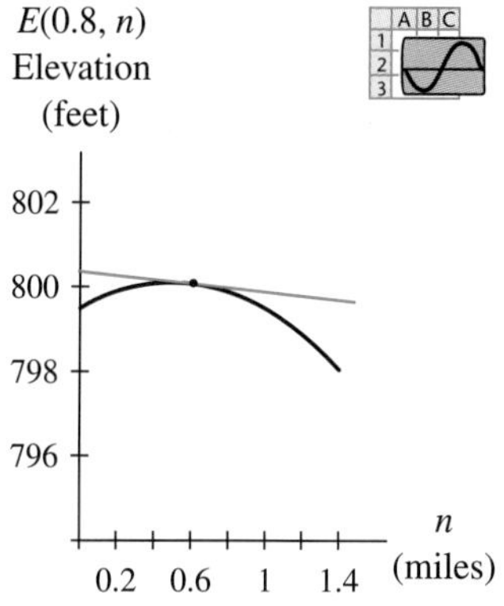

When we are given a table of two-variable data rather than a model, we approximate partial rates of change by finding rates of change of cross-sectional functions. Consider again the elevation of a tract of Missouri farmland. We use cross-sectional models to help answer questions about the elevation and the steepness (that is, the magnitude of the rate of change of elevation) of the tract. When we are 0.8 mile east of the western fence and 0.6 mile north of the southern fence, how steep is the land? In other words, how rapidly is elevation changing, or what is the rate of change of elevation? To answer this question we must first know in which direction to look: from south to north, from west to east, or possibly in some particular northeasterly direction.

Let us first look to the north. The question is "At $e = 0.8$ mile and $n = 0.6$ mile, what is the rate of change of elevation *with respect to the distance north of the southern fence*?" This partial rate of change is the slope of the tangent line that runs in the north-south direction at (0.8, 0.6). This tangent line is depicted in Figure 9.32 on the graph of the $E(0.8, n)$ cross section.

E(0.8, n) Elevation (feet)

802, 800, 798, 796

n (miles) 0.2 0.6 1 1.4

Line tangent to $E(0.8, n)$ at $n = 0.6$

FIGURE 9.32

We use the "approximately equals" symbol in the statement $\left.\frac{\partial E}{\partial n}\right|_{e=0.8} \approx \frac{dE(0.8,\,n)}{dn}$ to indicate that we are approximating the partial derivative of the underlying elevation function E with the derivative of a cross-sectional model found from data. If we had known the formula for the elevation function and had used it to find the cross-sectional function $E(0.8, n)$, then it would be appropriate to write $\left.\frac{\partial E}{\partial n}\right|_{e=0.8} = \frac{dE(0.8,\,n)}{dn}$.

The cross-sectional model for $e = 0.8$ was found in Section 9.1 to be

$$E(0.8, n) = -2.5n^2 + 2.497n + 799.490 \text{ feet above sea level}$$

where n is the number of miles north of the southern fence. To estimate the partial rate of change $E_n(0.8, n)$, we first find the derivative of the cross-sectional function with output $E(0.8, n)$ with respect to n:

$$\left.\frac{\partial E}{\partial n}\right|_{e=0.8} \approx \frac{dE(0.8,\,n)}{dn} = -5n + 2.497 \text{ feet per mile}$$

We now evaluate $\frac{dE(0.8,\,n)}{dn}$ at $n = 0.6$ to find that at 0.6 mile north of the southern fence and 0.8 mile east of the western fence, elevation is decreasing by 0.503 foot per mile toward the north. This also means that elevation is increasing by 0.503 foot per mile toward the south.

Keep in mind that this rate of change tells us only part of the story about the rate of change in elevation on the tract, because it is restricted to the north-south direction at a particular point. It does not tell us what is happening toward the east or west. In order to find the steepness in the east-west direction, we must look at the derivative of a cross-sectional function with respect to distance e miles east of the western fence.

EXAMPLE 2 *Missouri Farmland*

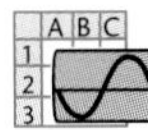

Estimate and interpret the rate of change of elevation with respect to distance from the western fence 0.8 mile east of the western fence and 0.6 mile north of the southern fence.

Solution We use the cross-sectional model that has distance e east of the western fence as input with the distance 0.6 mile north of the southern fence held constant. That is, we use the model

$$E(e, 0.6) = -10.124e^3 + 21.347e^2 - 13.972e + 802.809 \text{ feet}$$

The rate of change of $E(e, 0.6)$ with respect to distance e east of the western fence is

$$\frac{dE(e, 0.6)}{de} = -30.372e^2 + 42.694e - 13.972 \text{ feet per mile east}$$

We now evaluate $\frac{dE(e, 0.6)}{de}$ at $e = 0.8$ to estimate $\left.\frac{\partial E}{\partial e}\right|_{n=0.6}$, the partial rate of change of elevation with respect to distance east of the western fence, at 0.6 mile north of the southern fence and 0.8 mile east of the western fence. At that point on the land, elevation is increasing toward the east by 0.745 foot per mile. Figure 9.33 shows the line tangent to a graph of $E(e, 0.6)$ at $e = 0.8$.

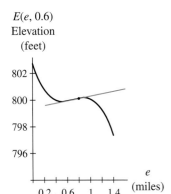

$E(e, 0.6)$ Elevation (feet)

802 800 798 796

0.2 0.6 1 1.4 e (miles)

Line tangent to $E(e\ 0.6,)$ at $e = 0.8$

FIGURE 9.33

Being able to calculate partial rates of change and evaluate these partial derivatives at various input values is important for the applications we will consider in subsequent sections.

Second Partial Derivatives

In Chapter 5 we considered the second derivative of a single-variable function and used the second derivative to locate inflection points. With multivariable functions, we can consider the second partial derivatives—that is, partial derivatives of partial derivatives. As we shall see in Chapter 10, these second partial derivatives are useful in identifying extrema and other interesting points on three-dimensional graphs. They also indicate how the partial rates of change of the function are changing.

Consider again the Example 1 investment function $A(P, t) = P(1.06136551^t)$ giving the accumulated value (in dollars) of an investment of P dollars over t years when the interest rate is 6% compounded quarterly. The two partial derivatives are

$$A_P = 1.06136551^t \text{ dollars per dollar invested}$$

$$A_t = P(\ln 1.0613551)(1.06136551^t) \text{ dollars per year}$$

What is happening to the rate of change of A with respect to t as t increases? Let us first look at this question graphically on one cross-sectional model, say $A(1000, t)$.

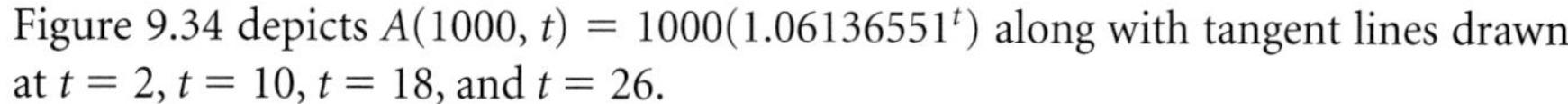

Figure 9.34 depicts $A(1000, t) = 1000(1.06136551^t)$ along with tangent lines drawn at $t = 2$, $t = 10$, $t = 18$, and $t = 26$.

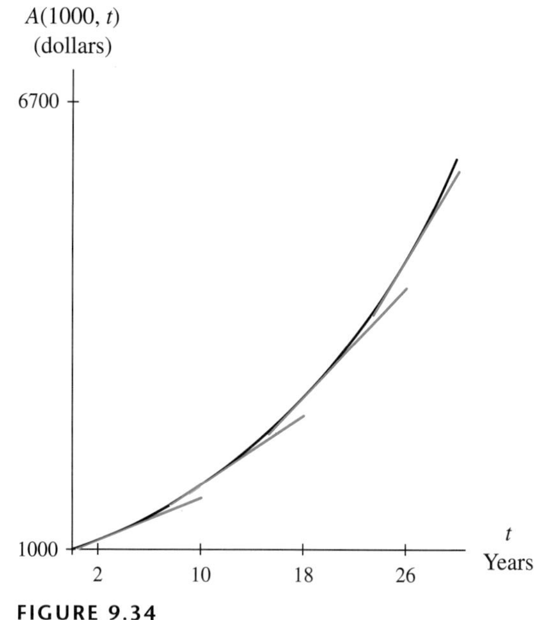

FIGURE 9.34

As t increases, the slope becomes larger. Thus the rate of change with respect to t, $A_t\big|_{P=1000}$, is increasing as t increases, so the second derivative must be positive. We can verify this by finding the derivative of $A_t\big|_{P=1000}$ with respect to t.

$$\frac{\partial}{\partial t}\left(A_t\big|_{P=1000}\right) = \frac{\partial}{\partial t}[1000(\ln 1.0613551)(1.06136551^t)]$$

$$= 1000(\ln 1.0613551)^2(1.06136551^t) \text{ dollars per year per year}$$

This second partial derivative is positive for all values of t, as Figure 9.34 suggests. Note also that because this partial derivative of the partial derivative $A_t\big|_{P=1000}$ is positive, the graph of the cross-sectional model with output $A(1000, t)$ is concave up. Before continuing with our interpretation of second partial derivatives, we find all second partial derivatives for A.

The partial derivative A_t is a function with two input variables P and t. The partial derivatives of A_t are A_{tP} (the partial derivative of A_t with respect to P) and A_{tt} (the partial derivative of A_t with respect to t). Even though the partial derivative of A_P is a function of only one input variable, it is derived from a function with two input variables. For this reason, A_P also has two second partial derivatives: A_{PP} (the partial derivative of A_P with respect to P) and A_{Pt} (the partial derivative of A_P with respect to t).[27]

The second partial derivatives for A are

$$A_{PP} = 0 \text{ dollars per dollar invested per dollar invested}$$

$$A_{Pt} = (\ln 1.0613551)(1.06136551^t) \text{ dollars per dollar invested per year}$$

$$A_{tP} = (\ln 1.0613551)(1.06136551^t) \text{ dollars per year per dollar invested}$$

$$A_{tt} = P(\ln 1.0613551)^2(1.06136551^t) \text{ dollars per year per year}$$

27. The second partial derivative notation may also be written $A_{PP} = \frac{\partial^2 A}{\partial P^2}$, $A_{Pt} = \frac{\partial^2 A}{\partial t \partial P}$, $A_{tP} = \frac{\partial^2 A}{\partial P \partial t}$, and $A_{tt} = \frac{\partial^2 A}{\partial t^2}$.

You may have noticed that $A_{Pt} = A_{tP}$. We refer to A_{Pt} and A_{tP} as *mixed second partial derivatives.* The mixed second partial derivatives of a multivariable function are equal whenever the function and its first and second partial derivatives are continuous. This is the case for all the multivariable functions in this book.

We now further consider the interpretation of the second partial derivatives of $A(P, t)$. First, $A_{PP} = 0$ tells us that the rate of change of A with respect to P does not change when P changes. This can be seen by looking at the cross-sectional models of A for different values of t. Figure 9.35 shows graphs of the linear cross-sectional functions $A(P, 2)$, $A(P, 12)$, $A(P, 18)$, and $A(P, 26)$. In A_{PP} we consider P to be changing but consider t to be constant; that is, we look at what happens to the slope of each cross section as P changes. The cross sections are all lines, and as P increases, the slope of any cross section remains constant. Thus $A_{PP} = 0$, because the slopes do not change as P changes.

Now let us consider $A_{Pt} = (\ln 1.0613551)(1.06136551^t)$. No matter what t is, A_{Pt} is positive. This tells us that the rate of change (slope) of A in the P direction increases as we move in the t direction. To see this graphically, refer again to the cross sections in Figure 9.35. Now, however, instead of considering the effect on the slope of each cross section as P increases, we consider what happens at a particular value of P as t increases—that is, as we move from one cross section to another. Figure 9.36 depicts the cross-sectional models shown in Figure 9.35 with the slopes of each cross section labeled at $P = 3000$. Note that as t increases, the slopes of the cross sections also increase. Hence, $A_{Pt} > 0$. There are similar interpretations for A_{tt} and A_{tP}.

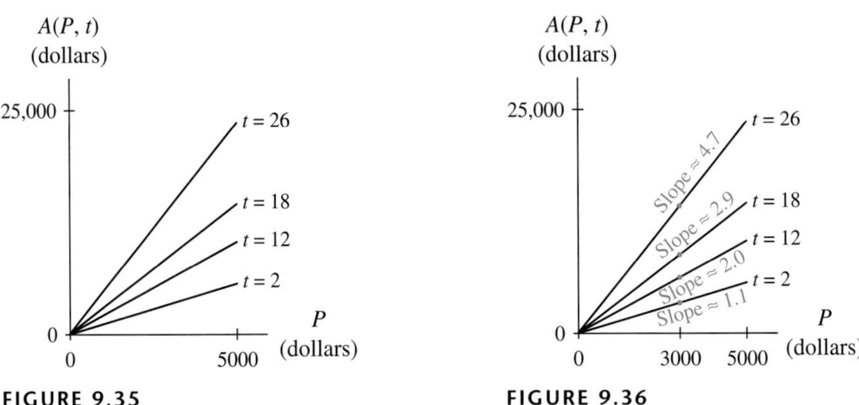

FIGURE 9.35

FIGURE 9.36

Keeping track of all the second partial derivatives for a particular function is easier with a mathematical array called a **matrix.** We write the second partial derivatives in a square matrix—that is, a table of rows and columns where the numbers of rows and columns are equal—with one of the second partial derivatives in each position. We call this the **second partials matrix,**[28] and we use this matrix in the next chapter to help identify extreme points. A second partials matrix can be formed by labeling the rows with the input variables and labeling the columns with the input variables in the same order. For example, one possible setup for a second partials matrix for a multivariable function $z = f(x, y)$ is

We use the determinant of the second partials matrix to classify critical points as maxima, minima, or saddle points in Section 10.2.

$$\begin{array}{c} \\ x \\ y \end{array}\begin{array}{c} x \quad y \\ \begin{bmatrix} & \\ & \end{bmatrix} \end{array}$$

28. This matrix is also called the *Hessian* of the function.

For convenience, we choose always to use the matrix with the input variables in the same order in which they appear in the notation used to represent the function. The labels on the rows and columns determine which second partial derivative we write in each position of the matrix. Using the function $z = f(x, y)$, we write f_{xx} in the position corresponding to the row and column labeled x, we write f_{yx} in the row labeled y and the column labeled x, and so forth.

$$\begin{array}{cc} & \begin{array}{cc} x & y \end{array} \\ \begin{array}{c} x \\ y \end{array} & \begin{bmatrix} f_{xx} & f_{xy} \\ f_{yx} & f_{yy} \end{bmatrix} \end{array}$$

For the function $A(P, t) = P(1.06136551^t)$, the second partials matrix is

$$\begin{array}{cc} & \begin{array}{cc} P & t \end{array} \\ \begin{array}{c} P \\ t \end{array} & \begin{bmatrix} 0 & (\ln 1.06136551)(1.06136551^t) \\ (\ln 1.06136551)(1.06136551^t) & P(\ln 1.06136551)^2(1.06136551^t) \end{bmatrix} \end{array}$$

Example 3 shows a second partials matrix for a function with three input variables.

EXAMPLE 3 *Cheese Spread*

A measure of the adhesiveness of cheese spread can be modeled[29] as a function of the percentage of glycerol, salt, and lactose used in preparation:

$$A(g, s, l) = 10^3(5.647 - 3.436s + 1.577s^2 + 0.375l - 1.757g + 0.141g^2 + 1.217sl - 0.533s^2l + 0.186sg)$$

where g is the percentage of glycerol, s is the percentage of salt, and l is the percentage of lactose.

a. Write the partial derivatives of A.

b. Write the second partials matrix of A.

c. Interpret A_{gg} and A_{sl}.

d. Write the second partials matrix of A when $g = 10$, $s = 2$, and $l = 6$.

e. Interpret $A_{ls}\big|_{(g, s, l) = (10, 2, 6)}$.

Solution

a. The three partial derivatives are

$$\frac{\partial A}{\partial g} = 10^3(-1.757 + 0.282g + 0.186s) \text{ units per percentage point of glycerol}$$

$$\frac{\partial A}{\partial s} = 10^3(-3.436 + 3.154s + 1.217l - 1.066sl + 0.186g) \text{ units per percentage point of salt}$$

$$\frac{\partial A}{\partial l} = 10^3(0.375 + 1.217s - 0.533s^2) \text{ units per percentage point of lactose}$$

29. E. Kombila-Moundounga and C. Lacroix, "Effet des combinaisons de chloure de sodium, de lactose et de glycerol sur les caractéristiques rhéologiques et la couleur des fromages fondus à tartiner," *Canadian Institute of Food Science and Technology Journal*, vol. 24, no. 5 (1991), pp. 239–251.

b. The second partials matrix is

$$\begin{array}{c} \\ g \\ s \\ l \end{array}\begin{array}{c} \begin{array}{ccc} g & s & l \end{array} \\ \begin{bmatrix} A_{gg} = 282 & A_{gs} = 186 & A_{gl} = 0 \\ A_{sg} = 186 & A_{ss} = 3154 - 1066l & A_{sl} = 1217 - 1066s \\ A_{lg} = 0 & A_{ls} = 1217 - 1066s & A_{ll} = 0 \end{bmatrix} \end{array}$$

c. The second partial derivative $A_{gg} = 282$ is positive. This means that as g increases, the rate of change (slope) of A with respect to g (in the g direction) is increasing. When the percentage of glycerol increases by 1 percentage point, the rate of change of adhesiveness increases by approximately 282 units per percentage point of glycerol if salt and lactose percentages are held constant.

The second partial derivative $A_{sl} = 1217 - 1066s$ is positive when $s < 1.14165$ and negative when $s > 1.14165$. Consider the percentage of glycerol as constant. If the percentage of salt is less than 1.14165%, then the rate of change of adhesiveness with respect to salt is increasing as the percentage of lactose increases. Likewise, if the percentage of salt is more than 1.14165%, then the rate of change of adhesiveness with respect to salt is decreasing as the percentage of lactose increases.

d. The second partials matrix of A at (10, 2, 6) is

$$\begin{array}{c} \\ g \\ s \\ l \end{array}\begin{array}{c} \begin{array}{ccc} g & s & l \end{array} \\ \begin{bmatrix} 282 & 186 & 0 \\ 186 & -3242 & -915 \\ 0 & -915 & 0 \end{bmatrix} \end{array}$$

e. We first note that $A_l\big|_{(g,s,l)=(10,2,6)} = 677$. This means that adhesiveness is increasing with respect to an increase in lactose above the 6% level. Because $A_{ls}\big|_{(g,s,l)=(10,2,6)} = -915$ is negative, as the percentage of salt increases above 2%, adhesiveness is increasing less rapidly (the rate of change A_l is decreasing) with respect to an increase in the percentage of lactose above the 6% level when the percentage of glycerol is held constant at 10%.

Partial rates of change and second partial derivatives will be important to us as we continue our study of multivariable functions.

9.3 Concept Inventory

- *Partial derivatives/partial rates of change*
- *Partial derivative notation*
- *Derivatives of cross-sectional models*
- *Second partial derivatives*
- *Second partials matrix*
- *Interpreting partial derivatives*

9.3 Activities

In Activities 1 through 6, write the mathematical notation for the partial rate-of-change function needed to answer the question posed. Also give the units of that rate-of-change function.

1. $W(a, h)$ is the weight (in pounds) of a person who is a years old and h inches tall. What is the rate of change of weight with respect to height?

2. $P(k, c)$ is the price of a diamond (in dollars) that weighs k karats and is of color code c. How quickly is the price of a diamond changing with respect to weight?

3. $T(g, t)$ is the mean daily temperature (in degrees Fahrenheit) at longitude g degrees and latitude t degrees. If longitude is 23° and latitude is changing, how rapidly is temperature changing?

4. $T(i, d)$ is the federal income tax (in dollars) owed by a person who made i dollars and can claim d dependents in a given year. When 4 dependents can be claimed and income is changing, how quickly is the amount of taxes owed changing?

5. $R(b, c)$ is the revenue (in dollars) a farmer makes from the sale of b bulls and c cows. When the farmer sells 2 bulls and 100 cows, how rapidly is revenue changing if the number of cows sold changes?

6. $G(h, s)$ is the expected grade-point average of a typical freshman college student who had a GPA of h in high school and made a combined score of s on the SAT. What is the rate of change of expected GPA with respect to the SAT score when the high school GPA is 3.5 and the SAT score is 1048?

7. Let $P(l, m)$ be the probability that a certain senator votes in favor of a tobacco ban bill when the senator receives l letters supporting the bill and m million dollars is invested by the tobacco industry lobbying against the bill.

   a. Interpret $\left.\frac{\partial P}{\partial m}\right|_{l=100{,}000}$.

      Would you expect $\left.\frac{\partial P}{\partial m}\right|_{l=100{,}000}$ to be positive or negative? Explain.

   b. Interpret $\left.\frac{\partial P}{\partial l}\right|_{m=53}$.

      Would you expect $\left.\frac{\partial P}{\partial l}\right|_{m=53}$ to be positive or negative? Explain.

8. Let $N(p, s)$ be the number of skiers on a Saturday at a ski resort in Utah when p dollars is the price of an all-day lift ticket and s is the number of inches of fresh snow received since the previous Saturday.

   a. Interpret $\left.\frac{\partial N}{\partial s}\right|_{p=25}$.

      Would you expect $\left.\frac{\partial N}{\partial s}\right|_{p=25}$ to be positive or negative? Explain.

   b. Interpret $\left.\frac{\partial N}{\partial p}\right|_{s=6}$.

      Would you expect $\left.\frac{\partial N}{\partial p}\right|_{s=6}$ to be positive or negative? Explain.

In Activities 9 through 14, find the indicated partial derivatives for each of the multivariable functions.

9. $f(x, y) = 0.529x^3 + 0.375x^2y^3 + 8.971xy + 14.390x + 7.982$

   a. $\frac{\partial f}{\partial x}$ b. $\frac{\partial f}{\partial y}$ c. $\left.\frac{\partial f}{\partial x}\right|_{y=2}$

10. $K(a, b) = 3.7ab^3 + 7.9(1.375^b)$

    a. $\frac{\partial K}{\partial a}$ b. $\frac{\partial K}{\partial b}$ c. $\left.\frac{\partial K}{\partial b}\right|_{a=6}$

11. $M(t, s) = s \ln t + 3.75s + 14.96$

    a. M_t b. M_s c. $\left.M_s\right|_{t=3}$

12. $g(x, y, z) = 3.2x^2yz^2 + 2.9x^y + z$

    a. g_x b. g_y c. g_z

13. $h(s, t, r) = \frac{s}{t} + \frac{t}{r} - (st - tr)^2$

    a. $\frac{\partial h}{\partial s}$ b. $\frac{\partial h}{\partial t}$ c. $\frac{\partial h}{\partial r}$ d. $\left.\frac{\partial h}{\partial r}\right|_{(s, t, r) = (1, 2, -1)}$

14. $k(x, y) = x \ln(x + y)$

    a. $\frac{\partial k}{\partial x}$ b. $\frac{\partial k}{\partial y}$ c. $\left.\frac{\partial k}{\partial y}\right|_{x=3}$ d. $\left.\frac{\partial k}{\partial y}\right|_{(x, y) = (3, 2)}$

15. Your comfort in summer depends on the combined effects of air temperature and humidity. Table 9.5 on page 625 shows how hot it feels[30] (the apparent temperature in degrees Fahrenheit) for a given air temperature and relative humidity.

    a. When the air temperature is 96°F and the relative humidity is 50%, how rapidly is the apparent temperature increasing with respect to humidity? (*Hint:* Begin by finding an appropriate cross-sectional model.)

    b. When the air temperature is 96°F and the relative humidity is 50%, how rapidly is the apparent temperature increasing with respect to air temperature?

    c. Use Table 9.5 and symmetric difference quotients to estimate the answers to parts *a* and *b*.

16. Table 9.6 on page 625 gives the average daily weight gain/loss of a pig[31] in kilograms per day as a function of its weight and the air temperature.

    a. Find a model for the average daily weight gain/loss of a 68-kg pig as a function of the air temperature.

    b. Find and interpret the rate of change of the average daily weight gain of a 68-kg pig with respect

30. National Oceanic and Atmospheric Administration.
31. A. A. Hanson, ed., *Practical Handbook of Agricultural Science* (Boca Raton, FL: CRC Press, 1990).

TABLE 9.18 Amount (in dollars) after 2 years

	Initial investment				
Rate (%)	**$10,000**	**$12,000**	**$14,000**	**$16,000**	**$18,000**
2	10,404.00	12,484.80	14,565.60	16,646.40	18,727.20
3	10,609.00	12,730.80	14,852.60	16,974.40	19,096.20
4	10,816.00	12,979.20	15,142.40	17,305.60	19,468.80
5	11,025.00	13,230.00	15,435.00	17,640.00	19,845.00
6	11,236.00	13,483.20	15,730.40	17,977.60	20,224.80
7	11,449.00	13,738.80	16,028.60	18,318.40	20,608.20
8	11,664.00	13,996.80	16,329.60	18,662.40	20,995.20
9	11,881.00	14,257.20	16,633.40	19,009.60	21,385.80
10	12,100.00	14,520.00	16,940.00	19,360.00	21,780.00
11	12,321.00	14,785.20	17,249.40	19,713.60	22,177.80
12	12,544.00	15,052.80	17,561.60	20,070.40	22,579.20

to a change in temperature when the temperature is 26.7°C.

c. Find a model for the average daily weight gain/loss of a pig as a function of the pig's weight when the air temperature is 26.7°C.

d. Find and interpret the rate of change of the average daily weight gain of a 68-kg pig with respect to the pig's weight when the air temperature is 26.7°C.

e. Use Table 9.6 and symmetric difference quotients to estimate the answers to parts *b* and *d*. Compare these answers to the ones you found in parts *b* and *d*.

17. Table 9.18 gives the value $A(P, r)$ of an investment of P dollars after 2 years in an account whose annual percentage yield is $100r\%$.

a. Find a model for the value of a $14,000 investment after 2 years in terms of r.

b. Find and interpret $\frac{dA(14{,}000,\, r)}{dr}$ when the APY is 12.7%.

c. Repeat part *a*; however, instead of entering the interest rate in decimals, enter r in whole numbers so that $r = 2$ at 2%, $r = 3$ at 3%, etc. Compare this model with the model found in part *a*.

d. Use the model from part *c* to find $\frac{dA(14{,}000,\, r)}{dr}$ when the APY is 12.7%. Compare this answer to that of part *b*.

18. Table 9.19 gives the value $A(t, r)$ of an initial investment of $1000 after t years in an account whose interest rate $100r\%$ is compounded continuously.

TABLE 9.19 Value (in dollars) of $1000 investment

	Time (years)			
Rate (%)	**5**	**7**	**10**	**12**
2	1105.17	1150.27	1221.40	1271.25
3	1161.83	1233.68	1349.86	1433.33
4	1221.40	1323.13	1491.82	1616.07
5	1284.03	1419.07	1648.72	1822.12
6	1349.86	1521.96	1822.12	2054.43
7	1419.07	1632.32	2013.75	2316.37
8	1491.82	1750.67	2225.54	2611.70
9	1568.31	1877.61	2459.60	2944.68
10	1648.72	2013.75	2718.28	3320.12
11	1733.25	2159.77	3004.17	3743.42
12	1822.12	2316.37	3320.12	4220.70

a. Find a cross-sectional model for the value of a $1000 investment in terms of r, assuming 10 years have elapsed since the initial investment.

b. Use the model in part *a* to estimate the partial rate of change with respect to the interest rate of a $1000 investment after 10 years in an account paying 7% interest. Interpret the partial rate of change as an approximate change. Discuss why the partial rate of change appears to be 100 times as large as expected.

c. Find a cross-sectional model for the value of a $1000 investment in terms of the interest rate expressed as a percentage, assuming 10 years have elapsed. Compare this equation with the one found in part *a*.

d. Use the model in part *c* to estimate the partial rate of change with respect to the interest rate of a \$1000 investment after 10 years in an account paying 7% interest. Interpret the partial rate of change as an approximate change. Compare these answers with those found in part *b*, and reconcile any differences.

19. The value $A(P, r)$ of an investment of P dollars after 2 years in an account with annual percentage yield $100r\%$ is given by the function $A(P, r) = P(1 + r)^2$ dollars.

    a. Find and interpret $A(14{,}000, r)$. Compare this function with the one in part *a* of Activity 17.

    b. Find and interpret $\frac{\partial A}{\partial r}\Big|_{(P,\, r) = (14{,}000,\, 0.127)}$. Compare this answer with the one in part *b* of Activity 17.

    c. Explain how the rate of change in part *b* is related to a graph of the cross-sectional function in part *a*. Illustrate graphically.

20. The value $A(t, r)$ of an investment of \$1000 after t years in an account for which the interest rate is $100r\%$ compounded continuously is given by the function $A(t, r) = 1000e^{rt}$ dollars.

    a. Find and interpret $A(10, r)$. Compare this answer with the one in part *a* of Activity 18.

    b. Find and interpret $\frac{\partial A}{\partial r}\Big|_{(t,\, r) = (10,\, 0.07)}$. Compare this answer with the one in part *b* of Activity 18.

    c. Explain how the rate of change in part *b* is related to a graph of the cross-sectional function in part *a*. Illustrate graphically.

21. Values of UV (ultraviolet) radiation are sometimes useful for the medical consultant, the chemist, and the technical engineer. Table 9.20 gives the amount of a certain type of UV radiation in watts per square meter per month.[32]

    a. Find cross-sectional models for the amount of UV radiation at a given latitude in March and for the amount of radiation for a given month at a latitude of 50° south of the equator.

TABLE 9.20 Amount of UVA radiation (watts/m²/month)

	Jan	Feb	Mar	Apr	May	Jun	Jul	Aug	Sep	Oct	Nov	Dec	Year
North Pole	—	—	—	17	41	52	46	27	—	—	—	—	183
80°	—	—	3	19	42	52	47	28	8	—	—	—	179
70°	—	2	11	26	43	51	47	32	16	5	—	—	233
60°	3	8	20	34	46	53	49	39	24	12	4	2	294
50°	9	16	28	40	50	55	53	44	32	20	11	7	365
40°	19	25	35	45	53	56	54	49	39	29	20	15	439
30°	26	33	41	49	54	55	54	51	44	36	28	24	495
20°	34	40	47	51	53	53	53	52	48	42	36	32	541
10°	42	46	50	51	50	50	50	50	50	47	43	40	569
Equator	49	50	52	50	47	45	46	48	50	50	49	47	583
10°	53	53	50	46	41	38	39	43	48	52	53	53	569
20°	56	54	49	42	34	31	32	38	45	51	55	56	543
30°	58	53	44	35	26	22	24	31	40	49	56	58	496
40°	57	50	39	27	18	13	15	23	34	45	55	59	435
50°	55	46	32	19	10	7	8	15	27	40	52	58	369
60°	52	40	24	11	5	1	2	7	18	33	47	56	296
70°	50	35	15	4	—	—	—	1	10	26	43	54	238
80°	49	27	8	—	—	—	—	—	3	19	41	56	203
South Pole	48	26	—	—	—	—	—	—	—	16	41	56	187

32. W. Rudloff, *World-Climates* (Stuttgart: Wissenschaftliche Verlagsgesellschaft, 1981).

TABLE 9.21 Number of cars sold, $S(m, y)$

Year	Month m											
y	Jan	Feb	Mar	Apr	May	June	July	Aug	Sept	Oct	Nov	Dec
1960	6550	8728	12,026	14,395	14,587	13,791	9498	8251	7049	9545	9364	8456
1961	7237	9374	11,837	13,784	15,926	13,821	11,143	7975	7610	10,015	12,759	8816
1962	10,677	10,947	15,200	17,010	20,900	16,205	12,143	8997	5568	11,474	12,256	10,583
1963	10,862	10,965	14,405	20,379	20,128	17,816	12,268	8642	7962	13,932	15,936	12,628
1964	12,267	12,470	18,944	21,259	22,015	18,581	15,175	10,306	10,792	14,752	13,754	11,738
1965	12,181	12,965	19,990	23,125	23,541	21,247	15,189	14,767	10,895	17,130	17,697	16,611
1966	12,674	12,760	20,249	22,135	20,677	19,933	15,388	15,113	13,401	16,135	17,562	14,720
1967	12,225	11,608	20,985	19,692	24,081	22,114	14,220	13,434	13,598	17,187	16,119	13,713
1968	13,210	14,251	20,139	21,725	26,099	21,084	18,024	16,722	14,385	21,342	17,180	14,577

b. Estimate the rate of change of UV radiation at 50° south in March with respect to latitude.

c. Estimate the rate of change of UV radiation at 50° south in March with respect to the month.

d. Graph the cross-sectional functions in part *a*. On the graphs, illustrate the rates of change in parts *b* and *c*.

22. Table 9.21 lists the number of cars sold[33] in Quebec from January 1960 through December 1968.

a. Find cross-sectional models for car sales $S(m, y)$ in January and June as functions of the number of years y since 1960.

b. Approximate the partial rates of change of car sales with respect to the year of the sales in January 1965 and in June 1965.

c. Estimate and interpret $\frac{\partial S}{\partial y}$ at $m = 1$ (January) and $y = 2$.

23. Table 9.22 shows selected values of the average yearly consumption of peaches per person, based on the price of peaches and the yearly income of the person's family.

a. Use a symmetric difference quotient to estimate the rate of change of the per capita consumption of peaches with respect to yearly income when the yearly income is \$30,000 and the price is \$1.70 per pound.

b. Use a symmetric difference quotient to estimate the rate of change of the per capita consumption of peaches with respect to price when the yearly income is \$30,000 and the price is \$1.70 per pound.

TABLE 9.22 Per capita consumption of peaches (pounds per person per year)

Yearly income (tens of thousands of dollars)	Price per pound (dollars above \$1.50)					
	0	0.10	0.20	0.30	0.40	0.50
1	5.0	4.9	4.8	4.7	4.7	4.6
2	6.4	6.3	6.2	6.1	6.1	6.0
3	7.2	7.1	7.0	6.9	6.9	6.8
4	7.8	7.7	7.6	7.5	7.4	7.4
5	8.2	8.1	8.0	8.0	7.9	7.8
6	8.6	8.5	8.4	8.3	8.3	8.2

c. Use the data in Table 9.22 and cross-sectional models to estimate the rates of change in parts *a* and *b*.

d. The data in Table 9.22 can be modeled by the equation

$$C(p, i) = 2 \ln i + 2.7183^{-p} + 4$$

pounds per person per year

where the price of peaches is $\$(1.50 + p)$ per pound and the person lives in a family with annual income $\$10{,}000i$. Use this model to find the rates of change in parts *a* and *b*.

e. Compare the answers you found using the three methods in this activity. Which method was easiest to use? Which answers do you believe are the most reliable?

33. B. Abraham and J. Ledolter, *Statistical Methods for Forecasting* (New York: Wiley, 1983).

24. Two vending machines sit side by side in a college dorm. One machine sells Coke products, and the other sells Pepsi products. Daily sales of Coke products, based on the prices of the products in the two machines, are as shown in Table 9.23.

TABLE 9.23 Daily sales (number of cans) of Coke products from a vending machine

	Cost of Coke products				
Cost of Pepsi products	**\$0.50**	**\$0.75**	**\$1.00**	**\$1.25**	**\$1.50**
\$0.50	157	143	123	98	65
\$0.75	206	192	172	146	114
\$1.00	255	241	221	195	163
\$1.25	304	290	270	244	211
\$1.50	353	339	319	293	260

a. Use a symmetric difference quotient to estimate the rate of change of the sale of Coke products with respect to the price of Coke products when Coke products cost \$0.75 and Pepsi products cost \$1.25.

b. Use a symmetric difference quotient to estimate the rate of change of the sale of Coke products with respect to the price of Pepsi products when Coke products cost \$0.75 and Pepsi products cost \$1.25.

c. Use the data in Table 9.23 and cross-sectional functions to estimate the rates of change in parts *a* and *b*.

d. The data in Table 9.23 can be modeled by the equation

$$S(c, p) = 196.42p - 50.2c^2 + 9.6c + 66.4 - 1.04cp$$
cans per day

when Coke products cost \$$c$ and Pepsi products cost \$$p$. Use this model to find the rates of change in parts *a* and *b*.

e. Compare the answers you found using the three methods in this activity. Which method was easiest to use? Which answers do you believe are the most reliable?

f. Use the equation in part *d* to find and interpret $\left.\frac{\partial S}{\partial c}\right|_{(c,\,p)=(1.30,\,1.20)}$ and $\left.\frac{\partial S}{\partial p}\right|_{(c,\,p)=(1.30,\,1.20)}$.

25. Siple and Passel developed the equation

$$H(v, t) = (10.45 + 10\sqrt{v} - v)(33 - t)$$

giving the body's heat loss in kilogram-calories per square meter of body surface area per hour for wind speed v in meters per second when the air temperature is t degrees Celsius.[34]

a. Find the function expressing the partial rate of change of heat loss with respect to air temperature and the function expressing the partial rate of change of heat loss with respect to wind speed.

b. Would you expect $\frac{\partial H}{\partial v}$ at $(v, t) = (20, 12)$ to be positive or negative? Give reasons.

c. Find the value of $\frac{\partial H}{\partial v}$ at $(v, t) = (20, 12)$. Interpret your answer.

d. Would you expect $\frac{\partial H}{\partial t}$ at $(v, t) = (20, 12)$ to be positive or negative? Give reasons.

e. Find the value of $\frac{\partial H}{\partial t}$ at $(v, t) = (20, 12)$. Interpret your answer.

26. The carrying capacity[35] of a particular farm system is defined as the number of animals or people that can be supported by the crop production from a given land area. The carrying capacity of a crop of rye grass when 70% of the crop is consumed and 80% of the amount consumed is digested as useful nutrients is

$$K(P, A) = \frac{9.52P}{A} \text{ animals per hectare}$$

where P is the net crop production (in kilograms per hectare per year) and A is the annual energy requirements of the animal (in megajoules per animal).

a. Write a function for the partial rate of change of carrying capacity with respect to net crop production.

b. How quickly is carrying capacity changing with respect to net crop production when 15,000 kg of rye grass is produced per hectare annually and the crop is used to support milking cows that require approximately 64,000 megajoules of energy per cow each year?

c. Write a function for the partial rate of change of the carrying capacity with respect to the animal's energy requirements.

d. How quickly is the carrying capacity changing with respect to the animal's energy requirements

34. W. Bosch and L. G. Cobb, "Windchill," UMAP Module 658, *The UMAP Journal*, vol. 5, no. 4 (Winter 1984), pp. 477–492.

35. R. S. Loomis and D. J. Connor, *Crop Ecology: Productivity and Management in Agricultural Systems* (Cambridge, England: Cambridge University Press, 1992).

when 15,000 kg of rye grass is produced per hectare annually and the crop is used to support milking cows that require approximately 64,000 megajoules of energy per cow each year?

e. Construct a second partials matrix for K, and interpret each of the second partial derivatives.

27. The amount of organic matter that one beef cow grazing on the Northern Great Plains rangeland eats each day can be modeled[36] by

$$I(s, m) = 8.61967 - 1.244s + 0.0897s^2 - 0.20988m + 0.035947m^2 + 0.214915sm$$

kilograms per day when the cow produces m kilograms of milk per day and s is a number between -4 and 4 that describes the size of the cow.

a. Would you expect $\frac{\partial I}{\partial s}$ and $\frac{\partial I}{\partial m}$ to be positive or negative? Explain.

b. Write and interpret functions for $\frac{\partial I}{\partial s}$ and $\frac{\partial I}{\partial m}$.

c. When a cow is of size 2 and produces 6 kg of milk per day, how quickly is the amount of organic matter consumed by the cow changing as its milk production increases?

d. When a cow is of size 2 and produces 6 kg of milk per day, how quickly is the amount of organic matter consumed by the cow changing as its size increases?

e. Construct a second partials matrix for I, and interpret each of the second partial derivatives.

28. A measure of the adhesiveness of honey that is being seeded with crystals to cause controlled crystallization can be modeled[37] by

$$A(g, m, s, h) = -151.78 + 4.26g + 5.69m + 0.67s + 2.48h - 0.05g^2 - 0.14m^2 - 0.03s^2 - 0.05h^2 - 0.07mh$$

where g is the percentage of glucose and maltose, m is the percentage of moisture, s is the percentage of seed, and h is the holding time in days.

a. Write functions for each of the partial derivatives of A.

b. Tell which partial derivative should be used to answer the question "How quickly is adhesiveness changing as the percentage of glucose and maltose changes?"

c. For which input variable(s) do you need a specific value in order to determine the actual rate at which adhesiveness is changing with respect to the percentage moisture?

d. Find a second partials matrix for A.

29. The value $A(t, r)$ of an investment of \$1000 after t years in an account for which the interest rate $100r\%$ is compounded continuously is given by the function $A(t, r) = 1000e^{rt}$ dollars.

a. Write the partial derivatives $\frac{\partial A}{\partial t}$ and $\frac{\partial A}{\partial r}$.

b. Construct a second partials matrix for A, and interpret each of the second partial derivatives for $t = 30$ and $r = 0.047$.

30. The value $A(P, r)$ of an investment of P dollars after 2 years in an account with annual percentage yield $100r\%$ is given by the function $A(P, r) = P(1 + r)^2$ dollars.

a. Find the first partials of A.

b. Construct a second partials matrix for A, and interpret each of the second partial derivatives for $P = 10{,}000$ and $r = 0.09$.

31. The value $A(r, t)$ of an investment of 1 million dollars with an annual yield of $100r\%$ is given by the function $A(r, t) = (1 + r)^t$ million dollars.

a. Write the partial derivative $\frac{\partial A}{\partial t}$. What are the units on $\frac{\partial A}{\partial t}$?

b. Write the partial derivative $\frac{\partial A}{\partial r}$. What are the units on $\frac{\partial A}{\partial r}$?

c. How quickly will the value of the investment be changing with respect to time 5 years after the investment is made if the investment yields 15% annually?

d. Illustrate the answer to part c using a graph.

32. The following equation gives the amount of a loan A (in dollars) given the interest rate $100r\%$, the period n (in months), and the monthly payments m (in dollars).

$$A(r, n, m) = \frac{12m}{r}\left[1 - \left(1 + \frac{r}{12}\right)^{-n}\right]$$

36. E. E. Grings et al., "Efficiency of Production in Cattle of Two Growth Potentials on Northern Great Plains Rangelands During Spring-Summer Grazing," *Journal of Food Science*, vol. 74, no. 10 (1996), pp. 2317–2326.

37. J. M. Shinn and S. L. Wang, "Textural Analysis of Crystallized Honey Using Response Surface Methodology," *Canadian Institute of Food Science and Technology Journal*, vol. 23, no. 4/5 (1990), pp. 178–182.

a. Write an expression for the rate of change of the loan amount with respect to the amount of the monthly payments.

b. How quickly is the loan amount changing with respect to the amount of the monthly payments when \$500 is paid monthly for 15 years on a loan with 12% interest?

c. Write the appropriate partial derivative, and calculate the rate of change of the loan amount when the interest rate and the monthly payment amount are fixed at 11% and \$250, respectively, and the period is 3 years but may vary.

d. Use graphs to illustrate the rates of change in parts *b* and *c*.

Construct a second partials matrix for each of the functions in Activities 33 through 36.

33. $f(x, y) = \frac{x}{y} - \frac{y}{x}$

34. $g(x, y) = 6(3x - y + 4)^3$

35. $h(x, y) = e^{2x-3y}$

36. $j(x, y) = y^2 \ln x$

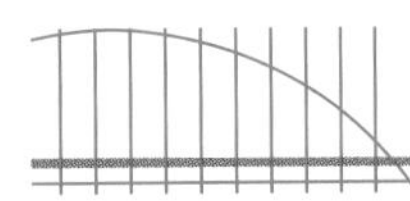

9.4 Compensating for Change

The following question often arises in situations where output depends on two input variables: "If output must remain fixed at some constant level and one of the input variables changes, how must the other input variable change in order to compensate?" In other words, given a function of two variables $z = f(x, y)$, if x increases (or decreases), how must y change in order to guarantee that the output $f(x, y)$ remains the same?

Let us consider this question in the context of a manufacturing problem where production depends on more than one input. The number of units produced during a manufacturing process can often be regarded as depending on several different input variables, two of which are the size L of the labor force and the amount K invested in capital. Economists often use the Cobb-Douglas production function

$$f(L, K) = cL^aK^{1-a}$$

as a simple model for the number of units produced. The parameters c and a in the Cobb-Douglas production function are constants that depend on the particular manufacturing process. Furthermore, the model requires that a be greater than 0 but less than 1.

The Cobb-Douglas monthly production function for a certain mattress manufacturing plant is

$$f(L, K) = 48.1L^{0.6}K^{0.4} \text{ mattresses}$$

where L represents the number of worker hours (in thousands) and K represents the amount invested in capital (in thousands of dollars). Let us suppose that currently each month the manufacturer spends \$47,000 in capital, has a labor force that works 8320 worker hours, and produces $f(8.32, 47) \approx 800$ mattresses.

A competing plant offering higher wages is hiring new employees. Management expects some of its employees to transfer to the new plant. Instead of replacing the employees, management decides to upgrade the equipment to improve efficiency. The number of lost worker hours is not yet known, so the information that is most beneficial to management is the rate of change of capital K with respect to worker hours L for the current production level. How do we find $\frac{dK}{dL}$?

To determine how to compensate for the change in the labor force, we look at the contour curve corresponding to 800 mattresses. This helps us understand the connection between the two input variables L and K. By solving

$$800 = 48.1L^{0.6}K^{0.4}$$

for K in terms of L, we obtain the equation of the contour curve shown in Figure 9.37.

$$K = \left(\frac{800}{48.1L^{0.6}}\right)^{\frac{1}{0.4}}$$
$$= \left(\frac{800L^{-0.6}}{48.1}\right)^{2.5}$$
$$= \left(\frac{800}{48.1}\right)^{2.5} L^{-1.5} \text{ thousand dollars}$$

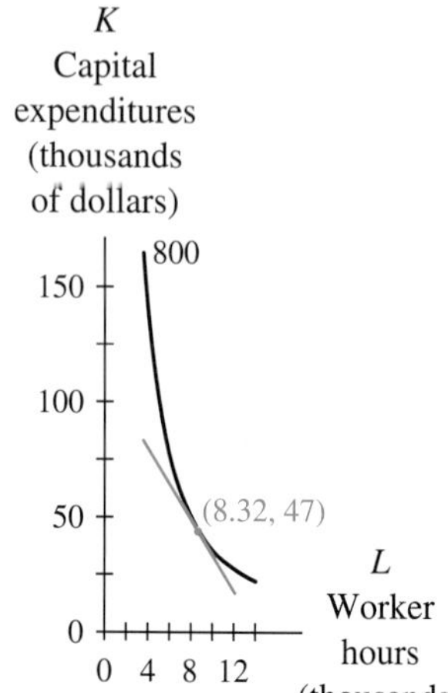

FIGURE 9.37

A tangent line is drawn on the curve in Figure 9.37 at the current amount of capital and number of worker hours.

To find $\frac{dK}{dL}$ we simply differentiate the contour curve equation with respect to L to obtain

$$\frac{dK}{dL} = -1.5\left(\frac{800}{48.1}\right)^{2.5} L^{-2.5} \text{ thousand dollars per thousand worker hours}$$

Evaluating this derivative at the current number of worker hours ($L = 8.32$) gives $\frac{dK}{dL} \approx -8.475$ thousand dollars per thousand worker hours. This value is the slope of the tangent line shown in Figure 9.37.

Recall from Chapter 5 that we use the slope of a tangent line to estimate by how much an output quantity will change if the input changes by a small amount. That is, for a function $y = f(x)$, if the slope of the tangent line is $\frac{dy}{dx}$ when $x = c$ and the input x changes by a small amount (Δx), then the corresponding change (Δy) in the output y can be approximated by

$$\Delta y \approx \frac{dy}{dx}\Delta x$$

This approximation is illustrated in Figure 9.38.

Rather than using the formula $\Delta y \approx \frac{dy}{dx}\Delta x$, some students may prefer to use $\frac{dy}{dx} \approx \frac{\Delta y}{\Delta x}$, substitute known values, and solve for the unknown values.

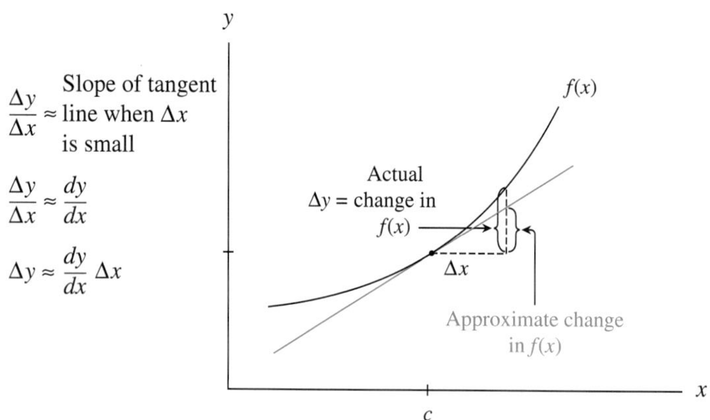

FIGURE 9.38

In the Cobb–Douglas example, the number of worker hours is currently $L = 8.32$, and $\frac{dK}{dL} \approx -8.475$ thousand dollars per thousand worker hours. That is, if the plant decreases the monthly labor force by 1000 worker hours, the amount spent on capital must increase by approximately \$8475 if the production level is to remain constant.

Suppose that the plant loses 500 worker hours as a result of employees leaving to work at the competing plant. In this case, the change in L is $\Delta L = -0.5$ thousand worker hours. Using the approximating-change equation from Chapter 5, we estimate the corresponding change in capital K as

$$\Delta K \approx \frac{dK}{dL}\Delta L$$

$$\Delta K \approx (-\$8.475 \text{ thousand/thousand worker hours})(-0.5 \text{ thousand worker hours})$$
$$\approx \$4.24 \text{ thousand}$$

Thus approximately \$4240 more will need to be spent on capital improvements in order to maintain the current production level and compensate for the loss of 500 worker hours. Figure 9.39 illustrates this approximation. As the figure shows, the approximation is an underestimate.

A check on (or an estimate of) the size and direction of ΔK is obtained from a contour graph by drawing a directional line segment representing the given change and then drawing and estimating the length of another line segment perpendicular to the first and returning back to the contour curve. This process is illustrated in Figure 9.39b.

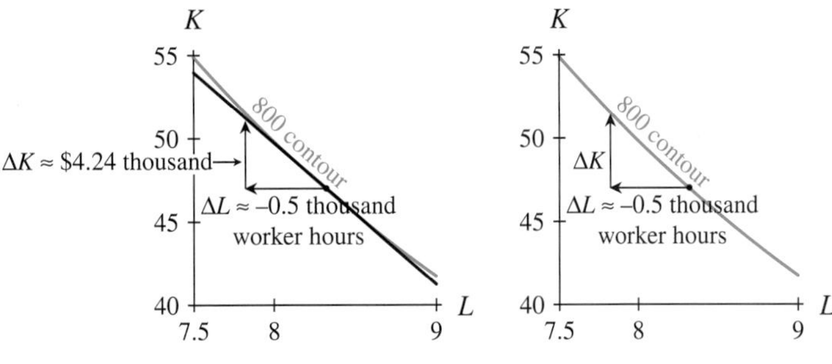

FIGURE 9.39

Similarly, if the plant loses 2500 worker hours to the competing plant, then the change in capital is approximated as

$$\Delta K \approx \frac{dK}{dL}\Delta L$$

$$\Delta K \approx (-\$8.475 \text{ thousand/thousand worker hours})(-2.5 \text{ thousand worker hours})$$
$$\approx \$21.2 \text{ thousand}$$

Thus, in order not to reduce production, an additional amount of approximately \$21,200 in capital expenditures is needed to compensate for the loss of 2500 worker hours.

A General Formula

The preceding method of finding $\frac{dK}{dL}$ works only if you can solve the contour curve equation for one of the variables. How do we find a derivative of one input variable with respect to another input variable if we are not able to solve the contour curve equation algebraically for one of the variables?

Consider a function $z = f(x, y)$ and the contour curve equation $f(x, y) = c$. Even if we cannot solve for y in terms of x, it is true that we can consider y to be a function[38] of x. We replace y with $y(x)$ and write the contour curve equation as $f(x, y(x)) = c$. Note that the function f in this form is a composite function, where f is the outside

38. When y is not a function of x, y can be broken into parts so that each part is a function of x.

function and y is the inside function. Recall from Chapter 4 the Chain Rule for the derivative of a composite function $f \circ g$. If we call g the inside function and f the outside function, then the derivative of f with respect to x is

$$\frac{df}{dx} = (\text{derivative of the outside function})(\text{derivative of the inside function})$$
$$= \left(\frac{df}{dg}\right)\left(\frac{dg}{dx}\right)$$

We use the Chain Rule as we develop the formula for the derivative of $z = f(x, y(x))$ with respect to x.

To take the derivative of $z = f(x, y(x))$ with respect to x, we consider the effect that a change in x has on the function f as well as the effect that a change in x has on y and the subsequent effect on the function f; that is,

$$\begin{matrix} \text{Rate of change of } f(x, y(x)) \\ \text{with respect to } x \end{matrix} = \begin{matrix} \text{partial rate of change} \\ \text{of } f \text{ with respect to } x \end{matrix} + \begin{pmatrix} \text{partial rate of change} \\ \text{of } f \text{ with respect to } y \end{pmatrix} \begin{pmatrix} \text{rate of change of } y \\ \text{with respect to } x \end{pmatrix}$$
$$\frac{d}{dx}[f(x, y(x))] = \frac{\partial f}{\partial x} + \left(\frac{\partial f}{\partial y}\right) \cdot \left(\frac{dy}{dx}\right)$$

You may wonder why we use both derivative and partial derivative notation in the preceding equation. We use partial derivative notation when considering f to be a function of two variables, x and y, and we use derivative notation when considering f or y to be a function of the single variable x.

With this formula, we return to the contour equation, differentiate both sides with respect to x, and solve for $\frac{dy}{dx}$.

$$f(x, y(x)) = c$$
$$\frac{d}{dx}[f(x, y(x))] = 0$$
$$\frac{\partial f}{\partial x} + \left(\frac{\partial f}{\partial y}\right)\left(\frac{dy}{dx}\right) = 0$$
$$\frac{dy}{dx} = \frac{-\left(\frac{\partial f}{\partial x}\right)}{\left(\frac{\partial f}{\partial y}\right)} = \frac{-f_x}{f_y}$$

This formula gives a method of finding a rate of change of one input variable with respect to another input variable at a point on a contour curve. It gives the slope of the tangent line at a point on a contour curve. This method is particularly useful when we cannot (or do not wish to) solve for one variable in terms of another.

The Slope of a Line Tangent to a Contour Curve

When a function of two variables $z = f(x, y)$ is held constant at a value c, the slope at any point on the contour curve $f(x, y) = c$ (that is, the slope of the line tangent to the c contour curve) is given by

$$\frac{dy}{dx} = \frac{-\left(\frac{\partial f}{\partial x}\right)}{\left(\frac{\partial f}{\partial y}\right)} = \frac{-f_x}{f_y}$$

whenever $f_y \neq 0$.

To illustrate this method, we consider the volume index for a cake that provides a measure of how much the cake rises. An index of 100 corresponds to the volume of the batter. The index can be modeled by

$$V(l, t) = -3.1l^2 + 22.4l - 0.1t^2 + 5.3t \text{ points}$$

when l grams of leavening is used and the cake is baked at 177°C for t minutes. Suppose that for a certain kind of cake, a volume index of 110 is most desirable. This volume can be achieved by using approximately 3.15 grams of leavening and baking for 26 minutes.

We are interested in investigating the change in baking time needed to compensate for an increase or decrease in the amount of leavening if the volume index is to remain constant. Figure 9.40 shows a portion of the 110 contour curve for V. The point corresponding to 3.15 grams of leavening and 26 minutes is labeled on the graph.

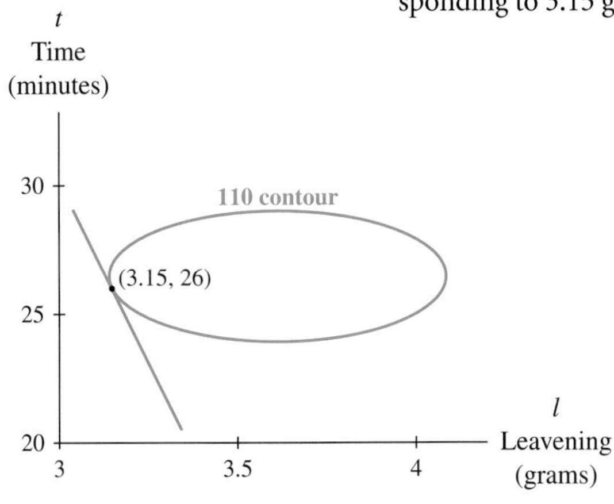

FIGURE 9.40

If the amount of leavening is increased or decreased, by how much will the baking time need to be adjusted in order for the volume index to remain constant? In other words, if the l coordinate of the point indicated in Figure 9.40 is changed, by how much will the t coordinate need to change in order for the point to return to the 110 contour curve? We approximate this amount by estimating by how much the t coordinate needs to change in order for the point to return to the tangent line shown in Figure 9.40, and we use the slope of the tangent line to calculate this estimate.

In the Cobb–Douglas example, it was fairly simple to find the equation for the contour curve and calculate the slope of the tangent line directly from that equation. In this example, the contour curve equation is not as easily found. Thus we calculate $\frac{dt}{dl}$ by using the formula for the slope of a line tangent to a contour curve:

$$\frac{dt}{dl} = \frac{-V_l}{V_t}$$

The partial derivatives of V are

$$V_l = -6.2l + 22.4 \ \frac{\text{index points}}{\text{per gram}} \quad \text{and} \quad V_t = -0.2t + 5.3 \ \frac{\text{index points}}{\text{per minute}}$$

Evaluating these partials at the point (3.15, 26) gives $V_l = 2.87$ and $V_t = 0.1$, so we have as the slope of the tangent line

$$\frac{dt}{dl} = \frac{-2.87}{0.1} = -28.7 \text{ minutes per gram of leavening}$$

You may wish to determine answer units on the basis of the slope we are asked to find rather than by keeping track of units during calculations. For instance, in this discussion, the question is to find $\frac{dt}{dl}$, where t is measured in minutes and l is measured in grams. Thus the units on $\frac{dt}{dl}$ will be minutes per gram. This method of determining units may be easier than carrying units through all the calculations.

This rate of change tells us that if the amount of leavening increases by 0.2 gram, we must reduce the baking time by approximately 5.7 minutes in order to maintain the volume index of 110. That is, the change in baking time Δt can be approximated by

$$\Delta t \approx \frac{dt}{dl}\Delta l$$

$$= \left(-28.7 \frac{\text{minutes}}{\text{gram}}\right)(0.2 \text{ gram}) = -5.74 \text{ minutes}$$

Similarly, if the amount of leavening decreases by 0.1 gram, then the baking time must be increased by approximately 2.9 minutes in order to maintain the volume index of 110.

$$\Delta t \approx \left(-28.7 \frac{\text{minutes}}{\text{gram}}\right)(-0.1 \text{ gram}) = 2.87 \text{ minutes}$$

We have seen how to approximate the change needed in one variable to compensate for a change in another variable in order to maintain a constant function output in two ways: using a contour curve equation solved for one of the input variables to directly calculate a tangent line slope or using partial derivatives to calculate the needed slope. We summarize these two methods as follows:

Compensating for Change

In order to compensate for a small change Δx in x to keep $f(x, y)$ constant at c, y must change by approximately

$$\Delta y \approx \left(\frac{dy}{dx}\right)\Delta x$$

which can also be calculated as

$$\Delta y \approx \left(\frac{-f_x}{f_y}\right)\Delta x$$

EXAMPLE 1 *Body-Mass Index*

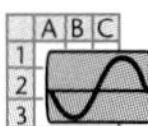

Your body-mass index is a measurement of how thin you are compared to your height. A person's body-mass index[39] is given by

$$B(h, w) = \frac{0.4536w}{0.00064516h^2} \text{ points}$$

where h is your height in inches and w is your weight in pounds. A teenage boy, in the middle of a growth spurt, has an appointment to see his doctor. He is 5 feet 7 inches tall and weighs 129 pounds. The doctor is concerned that the boy's body-mass index remain constant.

a. Find $\frac{dw}{dh}$ at the point (67, 129) on the contour curve corresponding to the boy's current body-mass index.

b. Use $\frac{dw}{dh}$ to estimate the weight change needed to compensate for growths of 0.5 inch, 1 inch, and 2 inches if the boy's body-mass index is to remain constant.

Solution

a. We begin by finding the two partial derivatives of B:

$$B_h = \frac{(-2)(0.4536w)}{0.00064516h^3} \text{ points per inch}$$

$$B_w = \frac{0.4536}{0.00064516h^2} \text{ points per pound}$$

where h is height in inches and w is weight in pounds. We evaluate these partial derivatives at $w = 129$ pounds and $h = 67$ inches and obtain

$$B_h \approx -0.60312 \text{ point per inch} \quad \text{and} \quad B_w \approx 0.15662 \text{ point per pound}$$

39. *New England Journal of Medicine*, September 14, 1995.

Thus,

$$\frac{dw}{dh} = \frac{-B_h}{B_w} \approx \frac{-(-0.60312)\text{ point per inch}}{0.15662\text{ point per pound}}$$
$$\approx 3.85\text{ pounds per inch}$$

b. We know that the change in weight needed to compensate for an increase in height can be approximated by $\Delta w \approx \frac{dw}{dh}\Delta h$. Thus,

For a 0.5-inch growth, $\Delta w \approx$ (3.85 pounds per inch)(0.5 inch) = 1.9 pounds

For a 1-inch growth, $\Delta w \approx$ (3.85 pounds per inch)(1 inch) = 3.85 pounds

For a 2-inch growth, $\Delta w \approx$ (3.85 pounds per inch)(2 inches) = 7.7 pounds

It is possible to estimate compensating change by using a table of data rather than a multivariable function. This is illustrated in Example 2.

EXAMPLE 2 *College Enrollment*

A college wishes to keep its enrollment constant at 5600 students. From experience, the trustees know that the two factors that have the greatest influence on enrollment are tuition and the amount of available financial aid. Some research has been done at similar colleges, and Table 9.24 gives enrollment figures based on that research.

TABLE 9.24 College enrollment

Average financial aid per student (thousand dollars)	Tuition (thousand dollars)								
	16	16.5	17.0	17.5	18.0	18.5	19	19.5	20.0
1.50	6358	6000	5662	5344	5043	4759	4491	4239	4000
1.75	6722	6346	5990	5656	5339	5040	4757	4492	4240
2.00	7086	6691	6318	5967	5634	5320	5024	4745	4480
2.25	7451	7037	6647	6279	5930	5600	5290	4997	4720
2.50	7815	7383	6975	6590	6226	5882	5557	5250	4960
2.75	8179	7729	7303	6902	6522	6163	5823	5503	5200
3.00	8543	8074	7631	7213	6817	6443	6089	5756	5440
3.25	8907	8420	7959	7525	7113	6724	6356	6009	5680
3.50	9272	8766	8288	7836	7409	7005	6622	6261	5920
3.75	9636	9111	8616	8148	7704	7285	6889	6514	6160
4.00	10,000	9457	8944	8459	8000	7566	7155	6767	6400

a. Using only the table values, estimate the change in the amount of financial aid the college should make available if it boosts tuition from $18.5 thousand to $19 thousand and wants to maintain an enrollment of 5600 students.

b. Use cross-sectional models to estimate the change in part *a*.

Solution

a. Reading in the column corresponding to a tuition of $18,500, we find that the enrollment of 5600 students occurs when the average financial aid is $2250 per

Average financial aid per student (thousands of dollars)	Tuition (thousands of dollars)			
	18.0	18.5	19	19.5
1.50	5043	4759	4491	4239
1.75	5339	5040	4757	4492
2.00	5634	5320	5024	4745
2.25	5930	5600	5290	4997
2.50	6226	5882	5557	5250
2.75	6522	6163	5823	5503
3.00	6817	6443	6089	5736

5600

FIGURE 9.41

student. If tuition is increased to \$19,000, we must move one column to the right. Because the college wishes to keep enrollment constant, we seek a value of 5600 students in the \$19,000 column. This is equivalent to staying on the 5600 contour curve (shown in blue in Figure 9.41) while increasing tuition to \$19,000.

A look at the table shows that there is no 5600 entry in the \$19,000 column. However, we know that 5600 occurs between 5557 (corresponding to financial aid of \$2500) and 5823 (corresponding to financial aid of \$2750). Thus the average financial aid will need to increase by a little more than \$250 per student in order for tuition to increase to \$19,000 and enrollment to remain constant.

b. We choose to represent enrollment as $E(t, a)$ students, where tuition is t thousand dollars and average financial aid is a thousand dollars per student. Note that if $t = 18.5$ and $E = 5600$, then $a = 2.25$. We seek to estimate the change in financial aid a needed to compensate for an increase in tuition t using the equation

$$\Delta a \approx \frac{da}{dt}\Delta t = \frac{-E_t}{E_a}\Delta t$$

We begin by calculating rates of change of the appropriate cross-sectional models. To estimate E_a when $t = 18.5$ and $a = 2.25$, we find a cross-sectional model for $E(18.5, a)$. This is a linear model given by

$$E(18.5, a) = 1122.87a + 3074.55 \text{ students}$$

when a thousand dollars of financial aid is available and the tuition is \$18,500. Now we find

$$\left.\frac{\partial E}{\partial a}\right|_{t=18.5} \approx \frac{dE(18.5, a)}{da} = 1122.87 \text{ students per thousand dollars of financial aid}$$

Thus if $t = 18.5$ and $a = 2.25$, then $E_a \approx 1122.87$ students per thousand dollars.

Likewise, we find the cross-sectional model $E(t, 2.25) = 46{,}275.637(0.892129^t)$ students when tuition is t thousand dollars and financial aid is \$2250. Thus

$$\left.\frac{\partial E}{\partial t}\right|_{a=2.25} \approx \frac{dE(t, 2.25)}{dt} \approx -5282.10(0.892129^t) \text{ students per thousand dollars of tuition}$$

Thus if $t = 18.5$ and $a = 2.25$, then $E_t \approx -639.3$ students per thousand dollars.

Because the change in tuition is $\Delta t = 0.5$ thousand dollars, we estimate the necessary change in financial aid as

$$\begin{aligned}\Delta a &\approx \frac{-E_t}{E_a}\Delta t \\ &\approx -\left(\frac{-639.3 \text{ students per \$1000 of financial aid}}{1122.87 \text{ students per \$1000 of tuition}}\right)(0.5 \text{ thousand tuition dollars}) \\ &\approx 0.28 \text{ thousand dollars of financial aid}\end{aligned}$$

The average financial aid available per student will need to increase by approximately \$280 in order to compensate for the increased tuition to maintain the current enrollment. (Our estimate from part a using only table values, a little more than \$250, agrees with this result.)

In this section, we have seen examples of finding rates of change on contour curves and using them to compensate for a change in one input variable. Compensating for change is a valuable tool with many real-world applications.

9.4 Concept Inventory

- *Approximating change*
- *Slope of a line tangent to a contour curve*
- *Compensating for change*

9.4 Activities

For each of the functions in Activities 1 through 4, sketch the contour curve indicated in the activity. Also sketch the tangent line indicated, and determine its slope.

Note that in Activity 1 it is possible to consider the slope of the tangent line as $\frac{dy}{dx}$ or as $\frac{dx}{dy}$, depending on which input variable is placed on the horizontal axis. A similar note applies to Activities 2 through 4 and 9 through 12.

1. $g(x, y) = x(1.0511^y)$
   $g(x, y) = 100$ for $0 \le y \le 10$
   tangent line at $y = 5$
2. $f(s, t) = s\ln(2t) + e^{-1.34t}$
   $f(s, t) = 2$ for $1 \le t \le 20$
   tangent line at $s = 1$
3. $f(a, b) = 2.846a^2b^3 - 1.834a^2 + 12.5b$
   $f(a, b) = 15$ for $0 \le a \le 10$ and $0.87 \le b \le 1.2$
   tangent line at $b = 0.9$
4. $f(m, n) = 10n(3.67 - m)^2e^{-0.2m}$
   $f(m, n) = 80$ for $1 \le m \le 10$ and $0 \le n \le 50$
   tangent lines at $n = 20$

For each of the functions in Activities 5 through 8, write a formula for the indicated rate of change.

5. $f(x, y) = 15x^2y^3$; $\frac{dx}{dy}$
6. $S(c, k) = c(39.871^k)$; $\frac{dc}{dk}$
7. $g(m, n) = 59.372\ln m + 49.283nm + 16.873$; $\frac{dm}{dn}$
8. $A(r, t) = 4983.22e^{rt}$; $\frac{dr}{dt}$

For the functions in Activities 9 through 12, determine the output associated with the input given, and then approximate the change needed in one input variable to compensate for the given change in the other input variable.

9. $f(m, n) = 3m^2 + 2mn + 5n^2$ when $m = 2, n = 1$, and $\Delta m = 0.2$
10. $f(h, k) = (32.371h^3 + 15.297h^2 - 10.290h + 47.291)(42.871k + 15.297)$ when $h = 4.2$, $k = 3.7$, and $\Delta k = 0.6$
11. $f(h, s) = 0.00091s[0.103(2.505^h) + 1]$ when $h = 3.5$, $s = 1148$, and $\Delta h = -0.5$
12. $W(r, h) = 2.79r^2(1.0792^h) + 59.78r(0.29h^2 - 3.25h + 72.41)$ when $r = 10$, $h = 60$, and $\Delta r = -1.3$
13. The cost of having specialty T-shirts made depends on the number of colors used in the T-shirt design and the number of T-shirts being ordered. A function[40] giving *A*, the cost per T-shirt (the average cost), when *c* colors are used and *n* T-shirts are ordered is

    $$A(c, n) = (-0.02c^2 + 0.35c + 0.99)(0.99897^n) + 0.46c + 2.57 \text{ dollars}$$

    a. Determine the average cost when 250 T-shirts are printed with six colors.
    b. When 250 T-shirts are printed with six colors, how quickly is the average cost changing when more T-shirts are printed?
    c. Find a formula for $\frac{dn}{dc}$. If average cost is to remain constant, would you expect $\frac{dn}{dc}$ to be positive or negative? Explain.
    d. A fraternity is planning to buy 500 four-color shirts. One of the members has proposed several alternative designs, some using more and some fewer than four colors. Use $\frac{dn}{dc}$ to estimate the change in order size needed to compensate for an increase or decrease in the number of colors if the average cost per T-shirt is to remain constant.
14. Another model[41] for the apparent temperature discussed in Activity 23 of Section 9.1 is

    $$A(t, h) = 2.70 + 0.885t - 78.7h + 1.20th \text{ °F}$$

    for an air temperature of *t* degrees Fahrenheit and a relative humidity of $100h$%.

40. Based on data compiled from 1993 prices at Tiger Town Graphics, Inc., Clemson, SC.
41. W. Bosch and L. G. Cobb. "Temperature Humidity Indices," UMAP Module 691, *The UMAP Journal*, vol. 10, no. 3 (Fall 1989), pp. 237–256.

a. How hot does it feel when the relative humidity is 85% and the air temperature is 90°F?

b. Find a formula for $\frac{dh}{dt}$. If the apparent temperature is to remain constant, would you expect $\frac{dh}{dt}$ to be positive or negative? Explain.

c. Use $\frac{dh}{dt}$ to estimate the change in the relative humidity needed to compensate for a 2°F increase in temperature if the current conditions are those stated in part *a* and the apparent temperature is to remain constant.

d. Repeat part *c* for a 3.5°F decrease in temperature.

15. A process to extract pectin and pigment from sunflower heads involves washing the sunflower heads in heated water. It has been shown[42] that the percentage of pigment that can be removed from a sunflower head by washing for 20 minutes can be modeled as

$$p(t, r) = 306.761 - 9.6544t + 1.9836r + 0.07368t^2 - 0.02958r^2 \text{ percent}$$

where r milliliters of t°C water is used for each gram of sunflower heads.

a. Draw the 53% contour curve for $20 \le r \le 45$ and $85 \le t \le 88$.

b. Draw the line(s) tangent to the 53% contour curve when the temperature is 86.5°C.

c. Write a formula for the rate of change of temperature with respect to a change in the amount of water used when 53% of the pigment is removed.

d. Use the formula from part *c* to determine the slope(s) of the tangent line(s) drawn in part *b*.

16. In 1965, Honer[43] developed a model for the total-stem volume of red pine trees in Canada. His model is

$$V(d, h) = \frac{d^2}{0.691 + 363.676h^{-1}} \text{ cubic feet}$$

where d is the diameter of the tree at breast height (4.5 feet above the ground), which is denoted by dbh and measured in inches, and the tree is h feet tall.

a. Find the volume of a 40-foot tree with a 2-foot dbh.

b. Draw a contour curve corresponding to the volume you found in part *a*.

c. On the contour curve, draw a line whose slope represents how quickly height is changing as dbh is changing if the volume does not change for a 40-foot tree with a 2-foot dbh.

d. Assuming a constant volume, find the rate at which height changes with respect to dbh for a 40-foot tree with a 2-foot dbh.

e. A logging company wishes to cut only trees that have a volume of at least 59 cubic feet. Use the answer to part *d* to estimate the height of a tree with dbh = 25.5 inches that satisfies the volume requirement. Repeat for a dbh of 23 inches.

f. How is your answer to part *d* related to the line you drew in part *c*?

17. a. Solve for w as a function of h for the body mass of the teenage boy in Example 1.

b. Use the formula in part *a* to find $\frac{dw}{dh}$ when $h = 67$ inches and $w = 129$ pounds.

c. Compare your answer in part *b* to the answer in part *a* of Example 1.

18. Refer to the body-mass function in Example 1 to answer the following questions.

a. Find the body-mass indices for an individual 5 feet 2 inches tall who weighs 130 pounds and an individual 6 feet 3 inches tall who weighs 190 pounds.

b. Approximate the height of an individual who weighs 132 pounds and has the same body-mass index as an individual 5 feet 2 inches tall who weighs 130 pounds.

c. Find the actual height of the 132-pound individual in part *b*, and compare it to your approximation.

d. Find a formula that approximates the difference in weight of two individuals who have the same body-mass index but slightly different weights.

e. Use your answer to part *d* to approximate the difference in weight of two individuals who have the same body-mass index: one individual is 5 feet 10 inches tall and weighs 165 pounds; the other is 5 feet 8 inches tall but will not reveal his weight. Approximately how much does the second individual weigh?

19. The amount of skin covering a person's body (in square feet) depends on the person's height and

42. X. Q. Shi et al., "Optimizing Water Washing Process for Sunflower Heads Before Pectin Extraction," *Journal of Food Science*, vol. 61, no. 3 (1996), pp. 608–612.

43. J. L. Clutter et al., *Timber Management: A Quantitative Approach* (New York: Wiley, 1983).

weight. One model for estimating this skin surface area is given by the multivariable function

$$A(w, h) = 0.6416w^{0.425}h^{0.725} \text{ square feet}$$

for a person weighing w pounds who is h feet tall. If a person 5 feet 11 inches tall who weighs 130 pounds grows 2 inches in height, by approximately how much must this person's weight change if the skin surface area is to remain the same?

20. The amount of a monthly payment on a loan with 6% interest compounded monthly can be calculated by using the function

$$m(A, t) = \frac{0.005A}{1 - 0.9419^t} \text{ dollars}$$

when the loan is for A dollars and is to be repaid over t years.

a. Suppose that you borrow $10,000 to buy a car and will repay the loan over a period of 5 years. What is your monthly payment?

b. If you decided to finance the car over a period of 4 years instead of 5, approximate the amount you could borrow to buy a car without increasing or decreasing the monthly payment determined in part *a*.

21. Table 9.25 shows body-mass index values as a function of height and weight. At her last doctor's visit, a teenage girl weighed 110 pounds and had a body-mass index of 20.

TABLE 9.25 Body-mass index

Weight (pounds)	Height (inches)						
	60	62	64	66	68	70	72
90	17.6	16.5	15.4	14.5	13.7	12.9	12.2
100	19.5	18.3	17.2	16.1	15.2	14.3	13.6
110	21.5	20.1	18.9	17.8	16.7	15.8	14.9
120	23.4	21.9	20.6	19.4	18.2	17.2	16.3
130	25.4	23.8	22.3	21.0	19.8	18.7	17.6
140	27.3	25.6	24.0	22.6	21.3	20.1	19.0
150	29.3	27.4	25.7	24.2	22.8	21.5	20.3
160	31.2	29.3	27.5	25.8	24.3	23.0	21.7
170	33.2	31.1	29.2	27.4	25.8	24.4	23.1

a. Sketch the 20 contour curve on the table.

b. Use only the table values to estimate by how much the girl should have grown if she gained 20 pounds but her body-mass index remained constant.

c. Use cross-sectional models to estimate the change in the girl's height that is needed to compensate for a 20-pound weight increase.

d. Use the equation given in Example 1 to determine the change in height that is needed to compensate for the girl's 20-pound weight increase.

e. Compare and contrast the three methods of estimating change used in this activity in terms of ease of use and accuracy.

22. Refer to Table 9.24 on page 673, which shows a college's enrollment as a function of tuition and financial aid. The college now has enrollment of 7000 and average financial aid of $3500 per student.

a. Sketch the 7000 contour curve, and locate the approximate point on the table corresponding to the college's current situation.

b. The board of trustees has decided to reduce the average financial aid by $500 per student in order to build a new dormitory. Use the table to estimate the corresponding decrease in tuition that will be necessary to offset the cut in financial aid and allow enrollment to remain constant.

c. Use cross-sectional models to estimate the change in part *b*.

23. Table 9.26 shows the sales of Coke products from a vending machine that is located next to a Pepsi vending machine. The current price of Coke products is $1.00, and Pepsi products cost $1.25.

TABLE 9.26 Number of cans of Coke products sold daily from a vending machine

Cost of Pepsi products	Cost of Coke products				
	$0.50	$0.75	$1.00	$1.25	$1.50
$0.50	157	143	123	98	65
$0.75	206	192	172	146	114
$1.00	255	241	221	195	163
$1.25	304	290	270	244	211
$1.50	353	339	319	293	260

a. Pepsi is planning to lower its price to $1.00. Use the table values to estimate by how much Coke would need to change its price in order to compensate for this change and retain current sales.

b. Using cross-sectional models, estimate the change in part *a*.

c. Judging on the basis of your estimates in parts *a* and *b*, what action do you think Coke should take when Pepsi lowers its prices?

24. Table 9.27 gives the per capita consumption of peaches as a function of annual family income and the price of peaches. The current average annual family income in a community is \$40,000, and the average price of peaches is \$1.99 per pound.

a. Suppose that in the next year, a company that employs several thousand people in this community goes out of business, thereby increasing unemployment and lowering the average annual family income by \$5000. Use the table to estimate the change in peach prices needed to compensate for the decrease in income.

b. Use cross-sectional models to estimate the change in part *a*.

c. The peach consumption data can be modeled by the equation

$$C(p, i) = 2 \ln i + 2.718^{-p} + 4 \text{ pounds per person per year}$$

where the price of peaches is $\$(1.50 + p)$ per pound and the person lives in a family with annual income $\$10{,}000i$. Estimate the change in part *a* using this equation.

d. Compare and contrast the three methods of estimating change used in this activity in terms of ease of use and accuracy.

TABLE 9.27 Per capita consumption of peaches (pounds per person per year)

Yearly income (tens of thousands of dollars)	Price per pound (dollars above \$1.50)					
	0	0.10	0.20	0.30	0.40	0.50
1	5.0	4.9	4.8	4.7	4.7	4.6
2	6.4	6.3	6.2	6.1	6.1	6.0
3	7.2	7.1	7.0	6.9	6.9	6.8
4	7.8	7.7	7.6	7.5	7.4	7.4
5	8.2	8.1	8.0	8.0	7.9	7.8
6	8.6	8.5	8.4	8.3	8.3	8.2

Summary

Multivariable Functions and Cross-Sectional Models

Multivariable functions occur in applications whenever output depends on two or more inputs. When a table of data describes a function with two input variables and one output variable, we obtain cross-sectional models of the function by holding one input variable constant and finding the equation that best fits the data in that row or column as a function of the other input variable. Geometrically, cross-sectional models describe the curves that result when cross-sectional planes cut a three-dimensional graph of the function of two variables parallel to one of the two input axes.

Contour Graphs

Although a multivariable function with two inputs is graphed using a three-dimensional coordinate system, it can be represented graphically in two dimensions by its contour curves. If f has two inputs, x and y, then the contour curves of f are the graphs of the equations $f(x, y) = k$ for various values of k. Contour graphs help us better understand and interpret the interactions between the different input variables in a multivariable function.

Rates of Change and Partial Derivatives

When considering the rate of change of a function with more than one input variable, we must consider the direction of movement. Rates of change of cross-sectional models describe how the function is changing when moving on the three-dimensional graph of the function in directions parallel to the two input axes. When we have a multivariable function, we can find functions for the partial rates of change or partial derivatives. If a function f has two input variables x and y, then we use the symbols $\frac{\partial f}{\partial x}$ and $\frac{\partial f}{\partial y}$ $\left(\text{rather than } \frac{df}{dx}\right.$ and $\left.\frac{df}{dy}\right)$ to remind us that there is more than one input variable and to indicate the direction of movement. Partial derivatives of multivariable functions with respect to a particular input variable are found by treating all the other input variables as constants and computing the derivative with respect to the changing variable using the rules developed in Chapter 4.

Compensating for Change

In situations with fixed output and changing input, the adjustment needed to compensate for small changes in input can be determined by using a multivariable equation, estimated by using a table and/or cross-sectional model, or estimated by using the slope of a tangent line drawn on the fixed contour curve. The slope of the line tangent to the contour curve $f(x, y) = c$ at the point (a, b) is

$$\frac{dy}{dx} = \frac{-f_x}{f_y}$$

where the partial derivatives are evaluated at $x = a$ and $y = b$.

To remain on the contour curve $f(x, y) = c$ for some constant c when x changes by a small amount Δx, y must change by approximately

$$\Delta y \approx \frac{-f_x}{f_y} \Delta x$$

The partial derivatives in this formula are evaluated at the point from which the change occurs.

Concept Check

Can you	To practice, try	
• Construct input/output diagrams for multivariable functions?	Section 9.1	Activity 5
• Interpret multivariable function inputs and outputs?	Section 9.1	Activity 9
• Find and interpret cross-sectional models?	Section 9.1	Activity 15
• Draw contour graphs on tables?	Section 9.2	Activity 3
• Draw contour graphs using equations?	Section 9.2	Activity 5
• Answer questions using contour graphs?	Section 9.2	Activity 17
• Interpret partial rates of change?	Section 9.3	Activity 7
• Estimate partial rates of change using tables and cross-sectional models?	Section 9.3	Activity 15
• Calculate first and second partial derivatives using an equation?	Section 9.3	Activities 29, 33
• Calculate the slope at a point on a contour curve?	Section 9.4	Activities 1, 15
• Estimate compensating change?	Section 9.4	Activities 7, 21

Review Test

1. The storage time of apples and the temperature at which they are blanched when making applesauce affect the consistency of the applesauce. Table 9.28 gives the consistometer values for applesauce as a function of the number of months the raw apples were stored and the temperature at which they were blanched.[44] The consistometer value is a measure of how far (in centimeters) an amount of applesauce flows down a vertical surface in 30 seconds. $C(t, s)$ denotes the consistometer value for a storage time of s months and a blanch temperature of t°C.

TABLE 9.28 Consistometer values (cm) for Rome applesauce

Storage time (months)	Temperature (Celsius)				
	35°	47°	59°	71°	83°
0	3.0	2.8	2.6	2.6	2.8
1	3.3	3.1	2.8	2.8	3.0
2	3.5	3.2	3.0	2.9	3.2
3	3.4	3.2	3.0	2.9	3.2
4	3.2	2.9	2.7	2.7	3.0

a. Draw an input/output diagram for C.

b. Find and interpret $C(71, 3)$.

c. Find an appropriate cross-sectional model, and use it to estimate the consistometer value of applesauce made from apples that were stored for 2 weeks and blanched at 35°C.

d. Interpret the notation $\left.\frac{\partial C}{\partial t}\right|_{s=4}$.

e. Use a cross-sectional model to estimate the value of the partial derivative in part d for $t = 45$°C. Interpret your answer.

f. Sketch contour curves on Table 9.28 for consistometer values of 2.8, 3.0, and 3.2.

2. A model for the number of eggs laid per female in 30 days for a certain insect[45] is

$$E(t, h) = -4191.6877 + 299.7038t + 23.1412h - 5.2210t^2 - 0.0937h^2 - 0.4023th \text{ eggs}$$

where t is the temperature in degrees Celsius and h is the percentage relative humidity.

a. Find and interpret $E(24, 60)$.

b. Find and interpret $\frac{\partial E}{\partial h}$ for a temperature of 27°C and relative humidity 77%.

c. A researcher wishes to hold the egg production constant but desires to investigate the change in humidity needed to compensate for an increase in temperature. Find a formula for the derivative that is most useful to this researcher.

d. If the researcher is currently conducting an experiment with temperature 25°C and humidity 63%, what approximate change in humidity will be needed to compensate for a decrease in temperature of 0.5°C while maintaining the current egg production level?

e. Discuss the graphical interpretation of the answer to part d.

3. Figure 9.42 shows a contour curve for the function described in Question 2. Use the contour curve to answer the questions that follow.

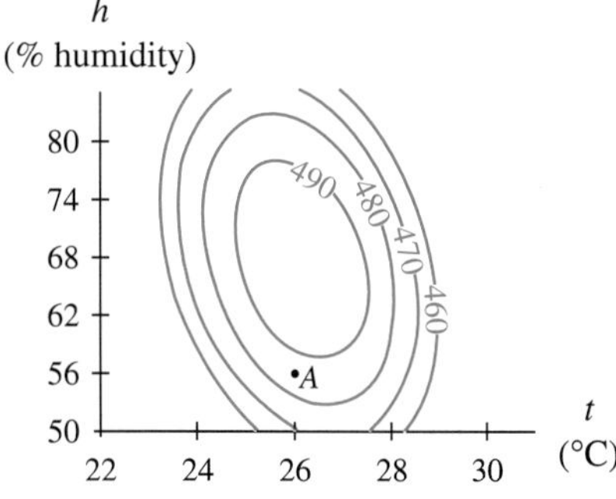

FIGURE 9.42

44. Based on information in A. M. Godfrey Usiak, M. C. Bourne, and M. A. Rao, "Blanch Temperature/Time Effects on Rheological Properties of Applesauce," *Journal of Food Science*, vol. 60, no. 6 (1995), pp. 1289–1291.

45. J. A. Morales-Ramos, S. M. Greenberg, and E. G. King, "Selection of Optimal Physical Conditions for Mass Propagation of *Catolaccus grandi*," *Environmental Entomology*, vol. 25, no. 1 (February 1996), pp. 165–173.

a. Estimate the inputs and output associated with point A. Interpret your answers.

b. At point A, which will result in a greater change in the number of eggs laid: an increase in temperature of 3°C or an increase in humidity of 3 percentage points?

c. At point A, will the number of eggs laid decrease more quickly when h decreases or when t decreases?

d. Estimate the slope of the line tangent to the contour graph at $t = 24$ and $h = 62$.

e. Calculate the answer to part d by using the equation given in Question 2. Interpret your answer.

4. Construct a second partials matrix for the function in Question 2.

Project 9.1 Competitive and Complementary Products

Setting

Competitive and complementary products play an important role in today's business markets. Two commodities are called *competitive* if an increase in demand for one results in a decrease in demand for the other. Two different brands of the same item are competitive. For example, Pepsi and Coke products in side-by-side vending machines are competitive. In this case, when the demand for Pepsi increases, the demand for Coke decreases, and when the demand for Coke increases, the demand for Pepsi decreases. It is important to note that the driving force in the change in the demands for competitive products is often the increase (or decrease) in the prices of the products. That is, when the price of Pepsi increases, the demand for Pepsi will decrease and the demand for Coke will increase. Likewise, when the price of Coke increases, the demand for Coke will decrease and the demand for Pepsi will increase.

Sometimes products are used together because they complement each other (coffee and cream, computers and floppy disks, VCRs and tapes). Two products are called *complementary* when each loses sales if one raises prices.

Tasks

1. Let $f(x, y)$ be the demand for a certain automobile sold for x dollars by Ford, and let $h(x, y)$ be the demand for a similar automobile sold for y dollars by Honda. If the total number of these two types of automobiles remains constant, explain why the automobiles are competitive if

$$\frac{\partial f}{\partial y} > 0 \quad \text{and} \quad \frac{\partial h}{\partial x} > 0$$

2. Give four examples of commodities you believe would be competitive in today's market.
3. Let $D(a, b)$ be the demand for product A which sells for a dollars, and let $F(a, b)$ be the demand for product B which sells for b dollars. If products A and B are complementary, restate the definition of complementary products in terms of their demand functions using partial derivative notation.
4. Give four examples of products you believe would be complementary in today's market.

Reporting

Prepare a written report containing detailed answers to the preceding tasks.

Project 9.2 Expert Witness

Setting

Investigators at accident scenes often wish to determine the speed at which a vehicle was traveling before the driver applied the brakes. Normally, all an investigator has to go on is the length of the skid marks and the weight of the vehicle that caused the skid marks. The investigator for an involuntary manslaughter lawsuit has called you as an expert witness to testify regarding the probable speed of the vehicle at the time of the accident.

Tasks

1. Under normal driving conditions, if a person driving a vehicle on an asphalt road slams on the brakes and skids to a stop, then the length of the skid marks can be modeled by $S(v, w) = 0.000013wv^2$ feet, where the vehicle was traveling v miles per hour when the driver slammed on the brakes. The vehicle weighs w pounds.

   Suppose the investigator measures the skid marks and weighs the vehicle that made them. He finds that the skid marks are 112 feet long and the vehicle weighs 3145 pounds. Determine the velocity at which the vehicle must have been traveling if it was empty.

2. It is possible that when it was weighed, the vehicle did not contain everything that was in it when the skid marks were created. For instance, it is extremely unlikely that the driver and passengers were still in the vehicle. Therefore, the investigator's ability to predict accurately the velocity of the vehicle is affected by the rate of change of velocity with respect to weight. Find a formula for the rate of change of the velocity of a vehicle with respect to its weight. How quickly will the velocity of the 3145-pound vehicle that made skid marks 112 feet long be changing when its weight increases?

3. The investigator knows that the vehicle at the time of the accident contained one male driver 6 feet 1 inch tall and of average build, one female passenger 5 feet 5 inches tall and of average build, one 2-year-old child described as "chubby," and a child safety seat. Discuss whether the weight of the driver and passengers would significantly affect the calculation of the velocity of the vehicle in the case under investigation. Note that the accident occurred in a 50-mph zone.

Reporting

1. Prepare a document summarizing your analysis of the vehicle's velocity to be admitted as court evidence. This report should be technical in nature with your conclusions clearly stated.

2. Prepare a witness statement to be given under oath. You will be cross-examined by the defense attorney and should be prepared to justify your assumptions and conclusions.

10

Analyzing Multivariable Change: Optimization

Jim Harrison/Stock Boston

Concept Application

Honey is composed of sugars (fructose, glucose, and maltose) as well as moisture content influenced by the floral source and variations in climate where the honey is gathered. The percentages of sugars and moisture affect the texture of the honey. Measures of the texture of honey include cohesiveness and adhesiveness. With the ability to model multivariable data, food scientists are able to answer such questions as

- What is the minimum cohesiveness possible?
- What is the maximum adhesiveness possible?
- If the FDA restricts the percentage of sugars and moisture in Grade A honey, how does this affect the smallest cohesiveness and largest adhesiveness possible if honey is to be labeled Grade A?

You will have the opportunity to answer these questions in Activities 3, 4, 10, and 11 of Section 10.3.

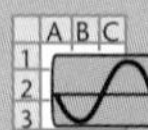

This symbol directs you to the supplemental technology guides. The different guides contain step-by-step instructions for using graphing calculators or spreadsheet software to work the particular example or discussion marked by the technology symbol.

This symbol directs you to the Calculus Concepts *web site, where you will find algebra review, links to data updates, additional practice problems, and other helpful resource material.*

Concept Objectives

This chapter will help you understand the concepts of

- Critical points of multivariable functions
- Absolute extrema of multivariable functions
- Multivariable optimization
- Constrained optimization and Lagrange multipliers
- The least-squares method

and you will learn to

- Estimate critical points using tables and contour graphs
- Estimate absolute extrema using tables and contour graphs
- Find critical points using equations
- Use the Determinant Test to classify critical points
- Use Lagrange multipliers to solve constrained optimization problems
- Interpret the Lagrange multiplier
- Use multivariable optimization to find a best-fit linear model

The optimization techniques for functions with a single input variable readily generalize to multivariable functions. In the same way that derivatives play an important role in determining critical points of a function with a single input variable, partial derivatives are used for locating critical points of multivariable functions. Critical points of functions of two input variables include maxima, minima, and saddle points.

We investigate methods for unconstrained optimization where the variables are free to change in any direction. We also consider constrained optimization where there are limits on the values of the variables involved. We conclude with an application of optimization that is related to how we began: finding the equation of the line that best fits a set of data points.

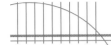

10.1 Multivariable Critical Points

In Chapter 5 we considered relative maxima and minima (that is, extreme points) and inflection points of single-variable functions. Now we consider similar points on multivariable functions—maxima, minima, and saddle points.

Although the term *relative extrema* technically refers only to the output value of an extreme point, we sometimes informally use the term to refer to the point itself. Students should be able to distinguish between and to interpret the point, the inputs, and the output.

Critical Points

Figure 10.1 shows a three-dimensional graph with a **relative maximum**—that is, an output value greater than any of those around it. Figure 10.2 shows a three-dimensional graph with a **relative minimum** (an output value smaller than any of those around it).

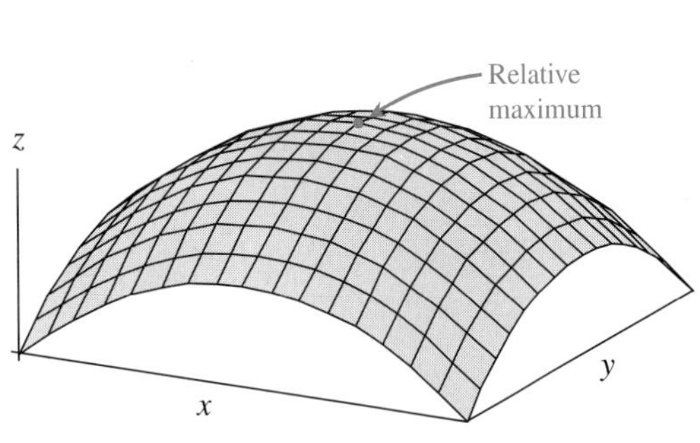

FIGURE 10.1

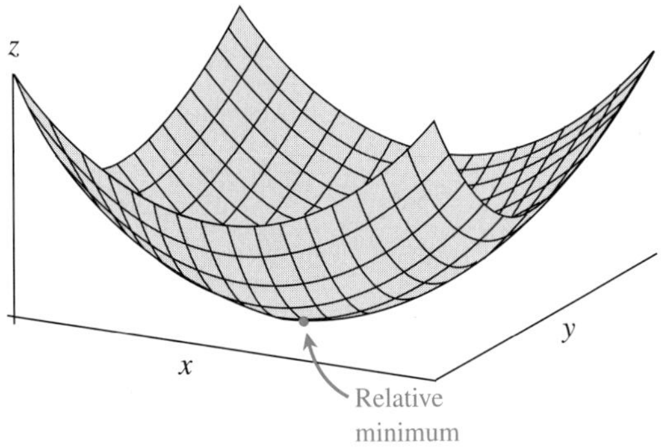

FIGURE 10.2

In labeling *B* in Figure 10.3 as an absolute maximum, we are restricting ourselves to the input region shown. If we know that the graph continues to decrease in every direction from what is shown in Figure 10.3, then point *B* can be labeled as an "overall" absolute maximum.

Some of your students may question why the high and low points on the edges of the visible region in Figure 10.4 are not labeled as absolute extreme points. We deal with absolute

If for a given input region, there are no output values greater than a certain relative maximum, then we say that the relative maximum is also an **absolute maximum** in that region. Similarly, if for a given input region there are no output values less than a certain relative minimum, then that relative minimum is called an **absolute minimum** in that region. Figure 10.3 shows a graph that contains relative maxima at points *A* and *B* and an absolute maximum at point *B*.

A **saddle point** is a point that corresponds to a relative maximum of a cross section in one direction and a relative minimum of a cross section in another. Figure 10.4 shows a function with a saddle point. Indeed, a three-dimensional graph with a saddle point often looks like a saddle in the area around that point. Note that the function in Figure 10.3 also has a saddle point in the valley between the points labeled *A* and *B*.

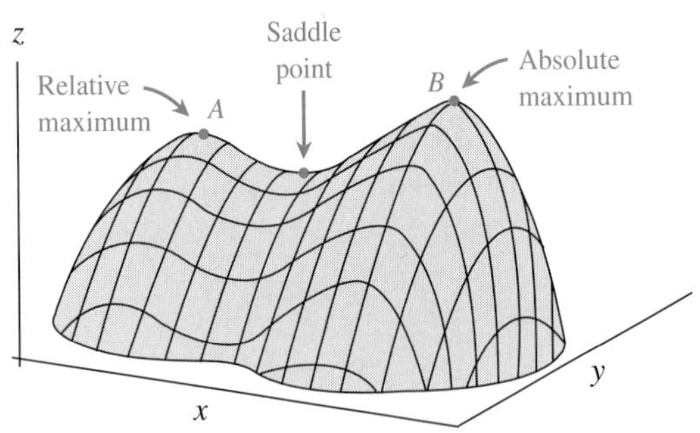

FIGURE 10.3

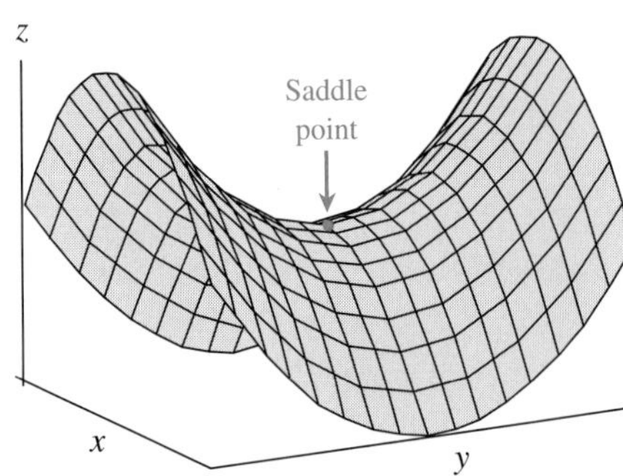

FIGURE 10.4

extrema on the edges of graphs (and tables) in the subsection "Locating Absolute Extrema Using Tables and Contour Graphs."

Saddle points and points corresponding to relative maxima and minima are called **critical points** (or stationary points). Although we call the relative extrema *critical points*, not all absolute extrema are critical points. This is because critical points are identified, as we will see in the next section, when partial rates of change are zero. Absolute extrema that occur on edges of tables or graphs may not fit this definition. Thus the term *critical point* refers only to relative maxima and minima and to saddle points.

Locating Critical Points on Contour Graphs

Contour graphs that correspond to the graphs in Figures 10.1 through 10.4 are shown in Figures 10.5 through 10.8, respectively.

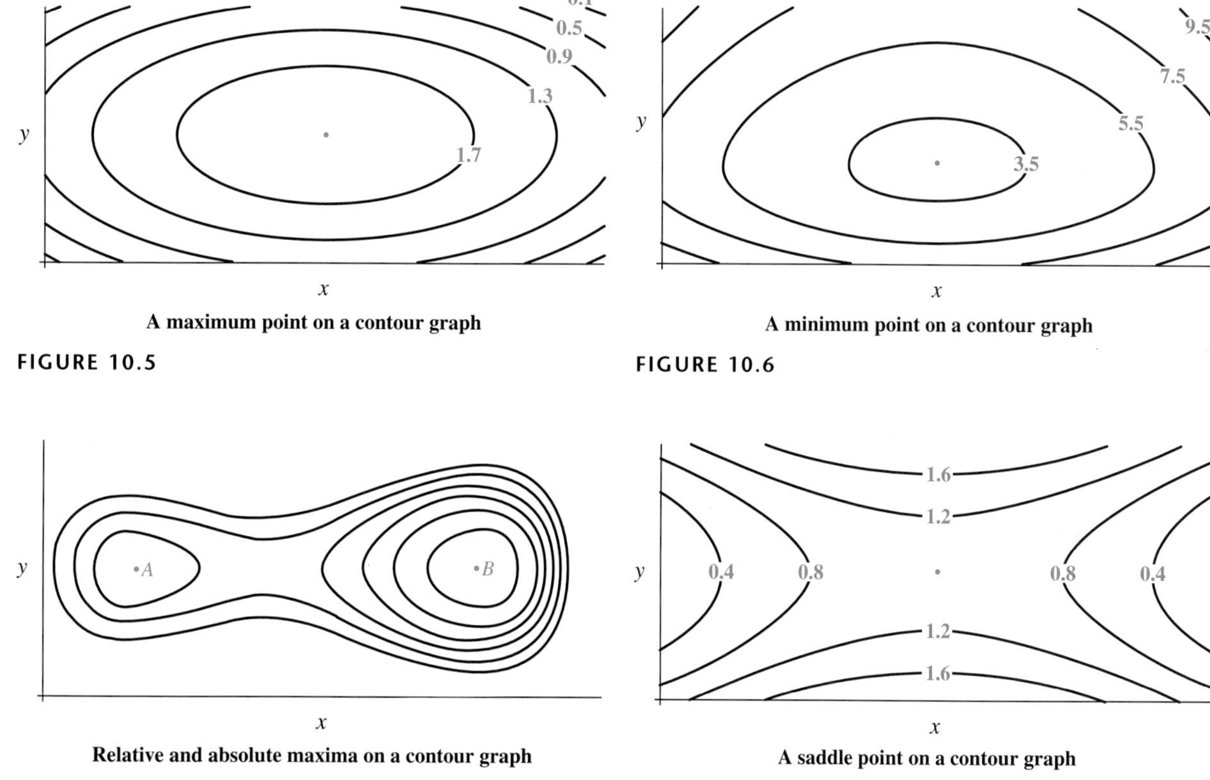

A maximum point on a contour graph

FIGURE 10.5

A minimum point on a contour graph

FIGURE 10.6

Relative and absolute maxima on a contour graph

FIGURE 10.7

A saddle point on a contour graph

FIGURE 10.8

Note that on the contour graph of a continuous function, relative maximum and minimum points lie within a *simple closed contour* (provided the contour graph is detailed enough). A simple closed contour is a contour that completely encloses a region but never crosses over itself. (See Figure 10.9.) In particular, relative maximum and minimum points will lie in the center of a group of concentric simple closed contours. The contour curves surrounding a relative maximum or minimum point often appear as a target with the relative maximum or minimum point as the bull's-eye. (See Figures 10.5, 10.6, and 10.7.)

Saddle points are also fairly easy to locate on contour graphs. The contour curves near the saddle point (but not through the saddle point) are all curved away from the point. Some of the contours decrease as you move away from the saddle point; others increase as you move away from the saddle point. (See Figure 10.8.)

Three examples of simple closed contours

Three curves that are not simple closed contours

FIGURE 10.9

To illustrate how to identify critical points, we consider an application from the field of food science. Olestra,[1] a fat substitute with no food calories, was approved by the FDA in 1996. Olestra is a simple sugar (sucrose) molecule with fatty acids attached. One method of creating olestra is to combine peanut oil and simple sugar with a catalyst to cause fatty acids from the peanut oil to attach to the sugar molecules. This process requires the mixture to be heated for several hours. The amount of olestra produced is a function of the ratio of peanut oil to sugar, the temperature to which the mixture is heated, and the amount of time the mixture is heated. The amount of olestra produced is measured as a percentage of the sugar that is changed into olestra.

Figure 10.10 shows a contour graph for the percentage conversion of sugar to olestra as a function of time and the temperature at which the mixture is heated (with a 10:1 ratio of peanut oil to sugar).

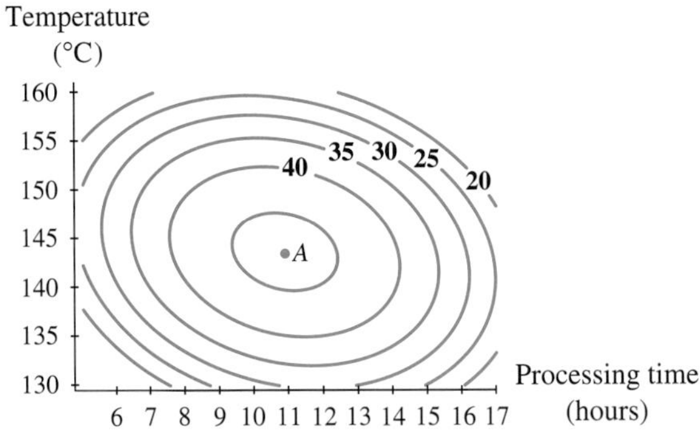

Percentage of sugar converted to olestra

FIGURE 10.10

Because point A appears to be a bull's-eye within concentric simple closed contours, it is an approximate relative maximum or minimum. The contour values decrease in every direction away from A. Thus the output value at A must be larger than any of the outputs around it, which makes A the location of a relative maximum. In order to

1. C. J. Shieh, P. E. Koehler, and C. C. Akoh, "Optimization of Sucrose Polyester Synthesis Using Response Surface Methodology," *Journal of Food Science*, vol. 61, no. 1 (1996), pp. 97–100.

estimate the output value at point A, note that the contour curves in Figure 10.10 are in increments of 5 units. We therefore assume that the unlabeled contour near A is the 45% contour and estimate that the percentage of sugar converted to olestra at the point A is slightly greater than 45%.

Locating Critical Points in Tables

When given only a table of data, we can use contour curves to identify critical points within the table. We simply sketch contour curves on the table for several equally spaced output values and then apply the methods for locating critical points on contour graphs. For example, Figure 10.11 shows selected values of the percentage of sugar converted to olestra when the ratio of peanut oil to sugar is 10:1. Contour curves for 35%, 40%, and 45% are drawn on the data.

		Processing time (hours)								
		6	7	8	9	10	11	12	13	14
Temperature (°C)	160	14.5	18.4	21.1	22.6	22.9	22.0	20.0	16.7	12.3
	155	24.5	28.8	31.9	33.8	34.5	34.0	32.4	29.5	25.5
	150	30.2	34.8	38.3	40.6	41.8	41.7	40.5	38.0	34.4
	145	31.5	36.6	40.5	43.2	44.7	45.1	44.3	42.2	39.0
	140	28.6	34.1	38.4	41.5	43.4	44.2	43.7	42.1	39.3
	135	21.3	27.2	31.9	35.5	37.8	39.0	38.9	37.7	35.3
	130	9.8	16.1	21.2	25.2	27.9	29.5	29.8	29.0	27.0

Percent conversion of sugar to olestra (with 10:1 ratio of fatty acids to sucrose)

FIGURE 10.11

We see in Figure 10.11 that the approximate maximum percentage of sugar converted to olestra is 45.1% and that it is achieved when the mixture is heated at 145°C for 11 hours. Because the value 45.1% lies in the center of concentric closed contour curves that decrease in every direction away from it, we know that this is an approximate relative maximum for the function from which the values in the table were generated. This can also be seen by noticing that 45.1% is larger than all eight values around it.

Note that 45.1% is only an approximation of the relative maximum output value because the table does not show every possible output of the function. For example, a processing time of 10 hours and 50 minutes at a temperature of 146°C may result in a percentage higher than 45.1%. Because most tables do not show all possible input and output values, both input and output of critical points can be only approximated from tables.

We now consider the percentage of sugar converted to olestra when the temperature is held constant at 150°C. Figure 10.12 shows the conversion percentages as a function of the processing time and the ratio of peanut oil to sugar. Contour curves for 20%, 25%, 30%, 35%, 40%, 45%, and 50% are drawn on the data.

The table entry corresponding to 11 processing hours and an 11:1 ratio of peanut oil to sugar appears to be a saddle point. Note that the contour curves decrease to the left and right of this value and increase above and below it. This value is a relative

Ratio of peanut oil to sugar	Processing time (hours)									
	5	6	7	8	9	10	11	12	13	14
15:1	15.7	24.4	31.9	38.2	43.3	47.3	50.1	51.6	52.0	51.2
14:1	13.8	22.0	28.9	34.7	39.2	42.6	44.8	45.9	45.7	44.3
13:1	13.7	21.2	27.6	32.8	36.8	39.6	41.3	41.7	41.0	39.1
12:1	15.2	22.2	28.0	32.6	36.1	38.3	39.4	39.3	38.0	35.5
11:1	18.3	24.8	30.0	34.1	37.0	38.7	39.2	38.5	36.6	33.6
10:1	23.2	29.0	33.7	37.2	39.5	40.7	40.6	39.4	36.9	33.3
9:1	29.7	35.0	39.1	42.0	43.8	44.3	43.7	41.9	38.9	34.7
8:1	37.8	42.6	46.1	48.5	49.7	49.7	48.5	46.1	42.5	37.8

50% 45% 40% 35% 30% 25% 20% 40% 45% 40% 35%

Percent conversion of sugar to olestra (with 150°C processing temperature)

FIGURE 10.12

maximum in the row in which it appears but is a relative minimum in its column. A table value that appears to be a relative maximum when approached from one direction and a relative minimum when approached from another direction is identified as a saddle point.

Note in Figure 10.12 that the entry 37.8 corresponding to 5 hours and a ratio of 8:1 is the largest value in the 5-hour column and the smallest value in the 8:1 row. We do not, however, classify this point as a saddle point because it lies on the edge of the table. In order to identify a saddle point visually on a table or contour graph, we must be able to identify a point that is a maximum in one direction and a minimum in another direction. We cannot visually identify relative extrema of cross sections on the edges of tables or contour graphs, because we cannot observe the edges lying within concentric closed contours. Similarly, we cannot visually identify saddle points on edges.

> Relative extreme points and saddle points cannot be visually identified on the edges of tables or contour graphs.

Sketching contour curves on a table is often the best way to identify its critical points.

Estimating Absolute Extrema Using Tables and Contour Graphs

Make sure that when you ask for absolute extrema on tests and quizzes, you either specify an input region or give your students enough information to determine the end behavior of the graph.

Recall from Chapter 5 that in order to find absolute extrema of a continuous function over a specified interval, we first find all relative extrema and then compare the outputs of the relative extrema with the function outputs at the endpoints. If we are given no interval over which to find the absolute extrema, then we must also consider the end behavior of the function. This process can be extended to find absolute extrema of a continuous function with two input variables (a three-dimensional function). We first locate all relative extrema (which lie within concentric closed contour curves), and we then compare the output values of the relative extrema with any output values on the edges of the function. Some functions are restricted on the basis of the context or the nature of the function. Some functions have no edges—that is, they extend infinitely in all directions. In this case, we must also consider the end behavior of the function.

When using tables and contour graphs to estimate absolute extrema for continuous, three-dimensional functions, we must keep in mind the following steps:

These steps can be applied to finding absolute extrema for three-dimensional graphs. For instance, if we consider all the edges of the graph in Figure 10.4 to be terminal edges, then we can identify an absolute minimum and an absolute maximum on the graph. If the edges are not terminal and there are no other critical points, then we can assume that the graph continues upward to the left and right and downward to the front and back and can conclude that there are no absolute extrema.

- Consider any restrictions on output suggested by the context.
- Sketch several contour curves on the table.
- Locate all relative extrema. In a table, relative extrema will be values that are smaller or larger than all eight values around them. Relative extrema cannot be identified on the edges of a table or contour graph, because to be visually identified, they must lie within concentric closed contour curves.
- Determine whether the edges of the table or contour graph can extend beyond what is shown. We call any edges beyond which the table or graph cannot extend **terminal edges.** Terminal edges are determined by the context of the situation that the function describes.
- Determine absolute extrema by comparing relative extrema and values on terminal edges of the table or graph and by considering the behavior of the function beyond the nonterminal edges. As with functions of one input variable, all absolute extrema will either be relative extreme points or lie on terminal edges of the graph.

We illustrate these steps by returning to the synthesis of olestra as a function of the processing temperature and time given in the table on page 689. We wish to identify the absolute maximum percentage of sugar converted to olestra. First note that the output values will never exceed 100%, although we do not know from the context whether a value of 100% is actually achieved. Next locate all critical points. The contour curves in Figure 10.11 indicate that 45.1% is a relative maximum. This value corresponds to the only critical point in the table.

To find the absolute maximum, we next determine whether the edges of the table are terminal edges. It is possible to consider processing times of less than 6 hours or more than 14 hours and temperatures of less than 130°C or more than 160°C; therefore, none of the edges in the table are terminal edges. Thus there may be other critical points outside of the table that correspond to relative maxima with output values greater than 45.1%. Therefore, we do not, in this case, have enough information to identify an absolute maximum percentage of sugar converted to olestra.

Now consider the question of estimating the absolute maximum if we are given the additional information that all of the critical points are contained within the region represented by the table. (We will prove that this is true in the next section.) With this additional information, we can conclude that the function peaks near 45.1% and decreases in all directions away from that point. Thus the relative maximum percentage of approximately 45.1% is the absolute maximum percentage.

Finally, consider the question of estimating the absolute minimum percentage of sugar converted to olestra using the information that all critical points are located within the region represented by the table. The context suggests that the absolute minimum of the function is 0% (no sugar converted to olestra) and that it occurs when the temperature is too high or too low and the processing time is too little or too long. The table and contour graph do not give us enough information to identify the place(s) where the absolute minimum occurs.

EXAMPLE 1 *Cohesiveness of Honey*

Table 10.1 shows a measure of the cohesiveness of honey as a function of the percentage moisture and the percentage of glucose and maltose.[2] All of the critical points lie within the region represented by the table.

a. Sketch a contour graph on the table, and mark any approximate critical points with an X. Use contour curves at cohesiveness values of 2.5, 2.75, 3, and 3.25. Identify each critical point as a relative maximum, a relative minimum, or a saddle point.

b. If possible, find the absolute maximum and minimum cohesiveness measures.

TABLE 10.1 A measure of the cohesiveness of honey

		Percentage of glucose and maltose								
		25	27.5	30	32.5	35	37.5	40	42.5	45
Percentage moisture	14	8.09	6.04	4.49	3.44	2.89	2.84	3.29	4.24	5.69
	14.5	7.62	5.65	4.17	3.20	2.72	2.75	3.27	4.30	5.82
	15	7.20	5.30	3.90	3.00	2.60	2.70	3.30	4.40	6.00
	15.5	6.81	4.99	3.66	2.84	2.51	2.69	3.36	4.54	6.21
	16	6.47	4.72	3.47	2.72	2.47	2.72	3.47	4.72	6.47
	16.5	6.16	4.49	3.31	2.64	2.46	2.79	3.61	4.94	6.76
	17	5.90	4.30	3.20	2.60	2.50	2.90	3.80	5.20	7.10
	17.5	5.67	4.15	3.12	2.60	2.57	3.05	4.03	5.50	7.47
	18	5.49	4.04	3.09	2.64	2.69	3.24	4.29	5.84	7.89
	18.5	5.34	3.97	3.09	2.72	2.84	3.47	4.60	6.22	8.35
	19	5.24	3.94	3.14	2.84	3.04	3.74	4.94	6.64	8.84

Solution

a.

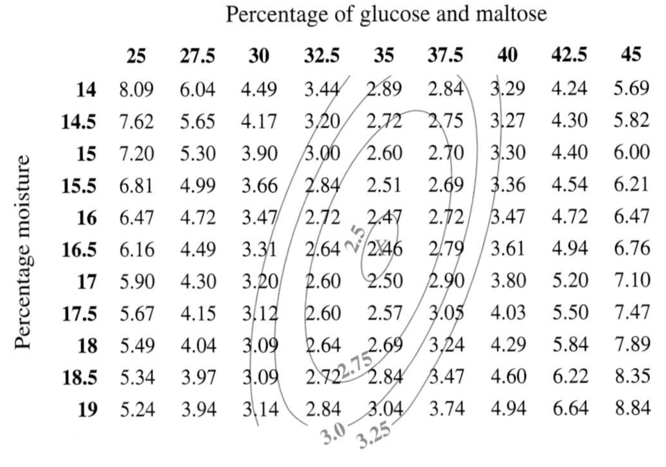

		Percentage of glucose and maltose								
		25	27.5	30	32.5	35	37.5	40	42.5	45
Percentage moisture	14	8.09	6.04	4.49	3.44	2.89	2.84	3.29	4.24	5.69
	14.5	7.62	5.65	4.17	3.20	2.72	2.75	3.27	4.30	5.82
	15	7.20	5.30	3.90	3.00	2.60	2.70	3.30	4.40	6.00
	15.5	6.81	4.99	3.66	2.84	2.51	2.69	3.36	4.54	6.21
	16	6.47	4.72	3.47	2.72	2.47	2.72	3.47	4.72	6.47
	16.5	6.16	4.49	3.31	2.64	2.46	2.79	3.61	4.94	6.76
	17	5.90	4.30	3.20	2.60	2.50	2.90	3.80	5.20	7.10
	17.5	5.67	4.15	3.12	2.60	2.57	3.05	4.03	5.50	7.47
	18	5.49	4.04	3.09	2.64	2.69	3.24	4.29	5.84	7.89
	18.5	5.34	3.97	3.09	2.72	2.84	3.47	4.60	6.22	8.35
	19	5.24	3.94	3.14	2.84	3.04	3.74	4.94	6.64	8.84

FIGURE 10.13

2. J. M. Shinn and S. L. Wang, "Textural Analysis of Crystallized Honey Using Response Surface Methodology," *Canadian Institute of Food Science Technology Journal*, vol. 23, nos. 4/5 (1990) pp. 178–182.

We begin by sketching contour curves on the table as shown in Figure 10.13. From the contour curves, we conclude that there is only one critical point. Because this point lies within concentric closed contours and the values of the contour curves increase in all directions away from it, the point is the approximate location of a relative minimum.

b. The context gives us no information concerning the greatest or least possible measures of cohesiveness, although it is probably safe to assume that the measure is a positive number. The contour graph, together with the information that all critical points lie within the region represented by the table, indicates that the function has a relative minimum of approximately 2.46 and rises in all directions away from that minimum. Thus the relative minimum is also the absolute minimum.

The context indicates that the table can extend in all directions but will not extend to negative input values or to input values over 100%. Because the relative minimum is the only critical point, and Table 10.1 does not include any terminal edges, we cannot identify either the absolute maximum cohesiveness measure or where it occurs.

EXAMPLE 2 *Cheese Spread*

Table 10.2 shows a measure of the consistency of cheese spread[3] as a function of the percent of salt and the percent of glycerol used in processing. All of the critical points lie within the region represented by the table.

TABLE 10.2 A measure of the consistency of cheese spread

	Percent glycerol										
Percent salt	**0**	**1**	**2**	**3**	**4**	**5**	**6**	**7**	**8**	**9**	**10**
0	8028	6412	5078	4026	3256	2768	2562	2638	2996	3636	4558
0.3	9153	7593	6315	5319	4604	4172	4022	4154	4568	5263	6241
0.6	9953	8448	7226	6286	5627	5251	5156	5344	5814	6565	7599
1	10513	9083	7935	7069	6485	6183	6163	6425	6969	7795	8903
1.3	10553	9179	8087	7276	6748	6502	6538	6856	7455	8337	9501
1.6	10268	8950	7913	7159	6686	6496	6588	6961	7617	8554	9774
2	9382	8138	7176	6496	6098	5982	6148	6596	7326	8338	9632
2.3	8338	7150	6243	5619	5277	5217	5439	5942	6728	7796	9146
2.6	6968	5836	4985	4417	4131	4126	4404	4963	5805	6929	8334
3	4636	3578	2802	2308	2096	2166	2518	3152	4068	5266	6746
3.3	2507	1505	785	347	191	316	724	1414	2386	3640	5175

a. Sketch contour curves on Table 10.2 for consistency values of 4000, 5000, 6000, 7000, 8000, and 9000. Mark the approximate saddle point and the largest and smallest table values.

b. Use the table and contour graph to estimate relative and absolute maximum and minimum consistency measures.

3. E. Kombila-Moundounga and C. Lacroix, "Effet des combinaisons de chloure de sodium, de lactose et de glycerol sur les caractéristiques rhéologiques et la couleur des fromages fondus à tartiner," *Canadian Institute of Food Science and Technology Journal*, vol. 24, no. 5 (1991), pp. 239–251.

Solution

a. In Figure 10.14, the requested contour curves are sketched, and boxes are drawn around the largest and smallest table values. The approximate saddle point is circled.

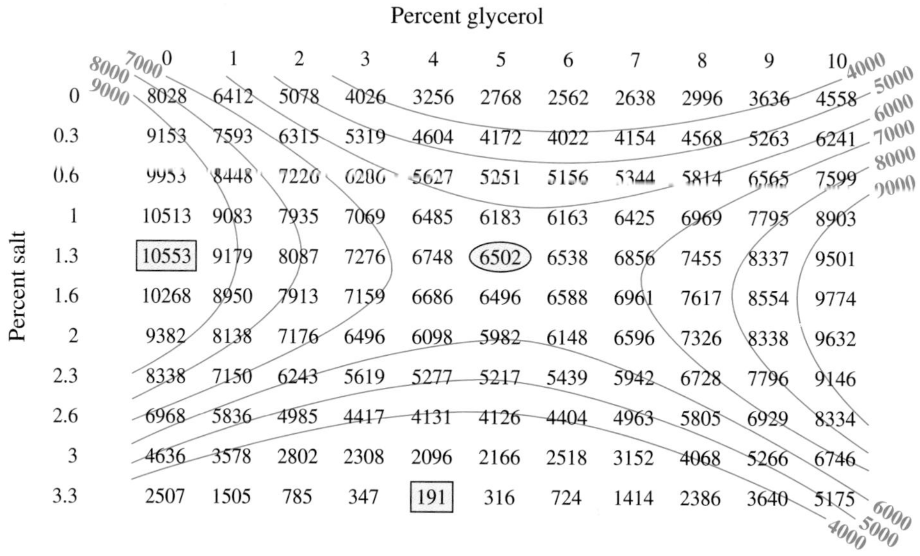

Percent glycerol

Percent salt	0	1	2	3	4	5	6	7	8	9	10
0	8028	6412	5078	4026	3256	2768	2562	2638	2996	3636	4558
0.3	9153	7593	6315	5319	4604	4172	4022	4154	4568	5263	6241
0.6	9953	8448	7220	6286	5627	5251	5156	5344	5814	6565	7599
1	10513	9083	7935	7069	6485	6183	6163	6425	6969	7795	8903
1.3	10553	9179	8087	7276	6748	6502	6538	6856	7455	8337	9501
1.6	10268	8950	7913	7159	6686	6496	6588	6961	7617	8554	9774
2	9382	8138	7176	6496	6098	5982	6148	6596	7326	8338	9632
2.3	8338	7150	6243	5619	5277	5217	5439	5942	6728	7796	9146
2.6	6968	5836	4985	4417	4131	4126	4404	4963	5805	6929	8334
3	4636	3578	2802	2308	2096	2166	2518	3152	4068	5266	6746
3.3	2507	1505	785	347	191	316	724	1414	2386	3640	5175

FIGURE 10.14

Note that the 6502 entry corresponding to 5% glycerol and 1.3% salt is an approximate saddle point because it is a maximum in the 5% glycerol column and a minimum in the 1.3% salt row.

We cannot consider the values 4558 (upper right corner) and 2507 (lower left corner) as candidates for absolute extrema, even though they lie in the terminal edges along the top row and left column of Figure 10.14, because these terminal edges can extend down and to the right. In order for a corner entry in a table to be considered in the search for absolute extrema, both the row and the column in which the entry lies must be terminal edges.

b. We are not given information concerning how large or small the consistency measure can be, although from the data it appears that the consistency measure is a positive number. Because there are no closed contour curves indicated by Figure 10.14, we conclude that the function has no relative extrema. Thus any absolute extrema must occur on terminal edges of the function. Note that the table cannot extend beyond the top and left edges because negative percentages do not make sense in this context. Thus these edges are terminal edges, and we can look in these two edges for approximate absolute extrema. We must also consider the behavior of the function beyond the lower and right edges.

The greatest value in the table is approximately 10,553, and it occurs when the percentages of glycerol and salt are approximately 0 and 1.3, respectively. This value is on a terminal edge; however, the contour curves suggest that the output values will continue to increase for glycerol percentages greater than 10%. If the table were extended far enough to the right, we would probably see an output value greater than 10,553. We conclude that we do not have enough information to identify an absolute maximum.

The smallest value in the table is 191, found in the bottom row. However, if the table were extended downward for percentages of salt greater than 3.3%, we would probably observe an absolute minimum of zero. The places where this minimum occur are difficult to estimate from the information presented in the table.

Three-dimensional graphs, contour graphs, and tables are helpful in estimating critical points and absolute extrema. However, we are limited to approximations of these points when using graphs and tables and sometimes do not have enough information to identify absolute extrema. In the next section, we will explore a more accurate method of finding critical points and absolute extrema for three-dimensional functions.

10.1 Concept Inventory

- *Critical points*
- *Relative maxima, relative minima*
- *Absolute maxima, absolute minima*
- *Saddle points*
- *Locating critical points and absolute extrema in tables and on contour graphs*
- *Terminal edges*

10.1 Activities

1. Describe in your own words how to determine whether a table of data has each of the following:

   a. relative maximum b. relative minimum
   c. saddle point d. absolute maximum
   e. absolute minimum

2. Describe in your own words (illustrating with pictures) how to determine whether a contour graph has each of the following:

   a. relative maximum b. relative minimum
   c. saddle point d. absolute maximum
   e. absolute minimum

3. Is the point in Figure 10.1.1 at which $x = 4.885$ and $y = 1.382$ a relative maximum point, a relative minimum point, or a saddle point? Explain.

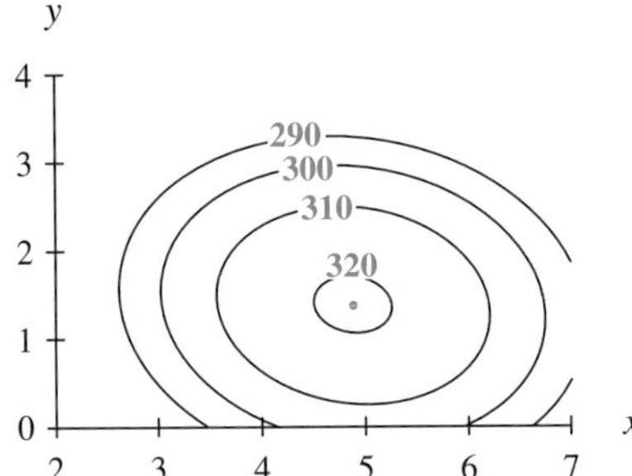

FIGURE 10.1.1

4. Figure 10.1.2 shows a contour graph and a three-dimensional graph for a function T with inputs p and f.

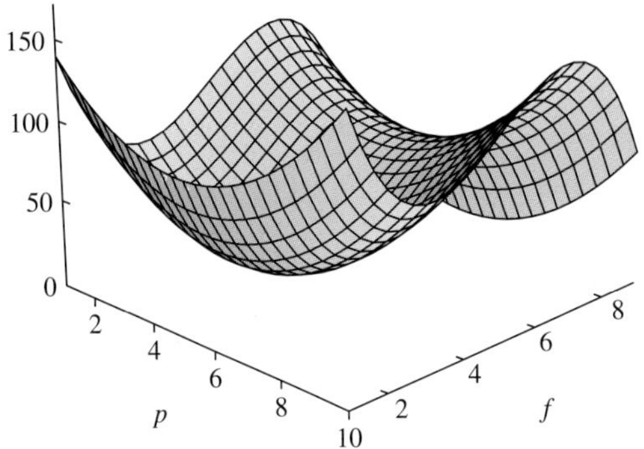

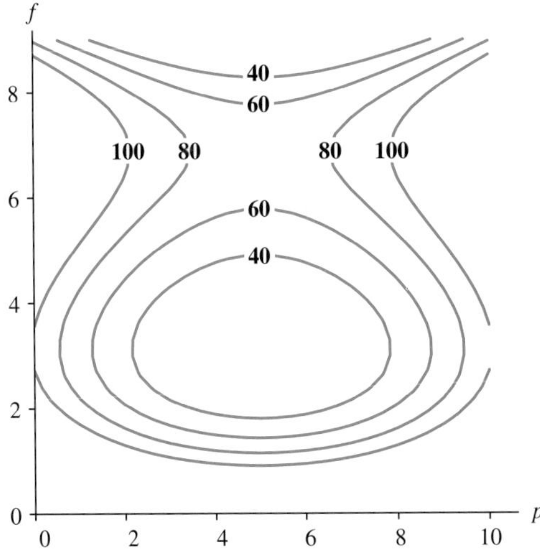

FIGURE 10.1.2

   a. Mark with large dots on the contour plot the approximate locations of all critical points. Label each point as a relative maximum point, a relative minimum point, or a saddle point.

   b. Mark on the three-dimensional graph the location of each critical point identified in part *a*.

   c. On the basis of the graphs, approximate the inputs and output of each critical point.

5. Figure 10.1.3 shows a contour graph and a three-dimensional graph for a function R with inputs g and h.

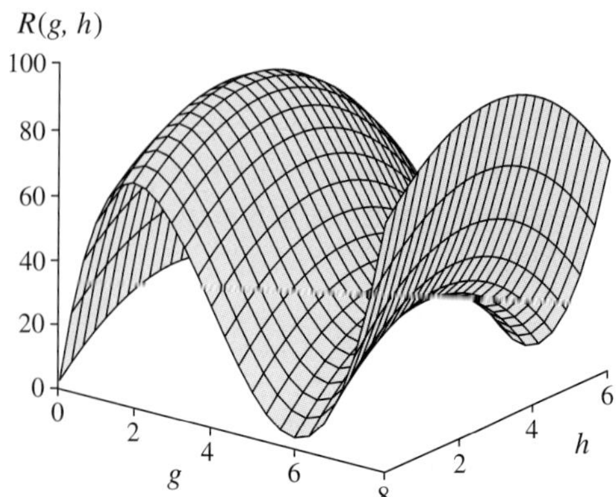

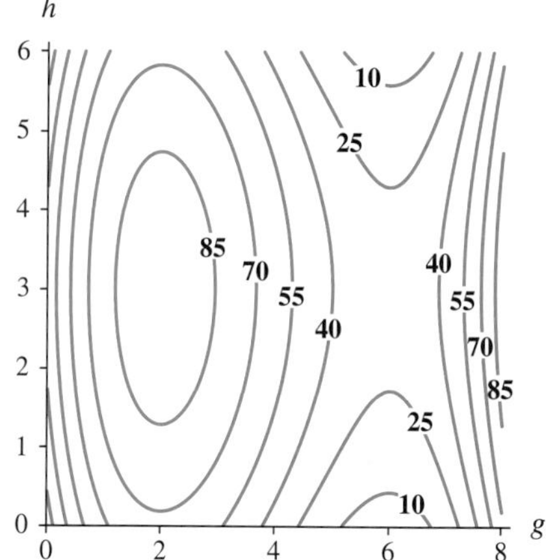

FIGURE 10.1.3

a. Mark with large dots on the contour plot the approximate locations of all critical points. Label each point as a relative maximum point, a relative minimum point, or a saddle point.

b. Mark on the three-dimensional graph the location of each critical point identified in part a.

c. On the basis of the graphs, approximate the inputs and output of each critical point.

6. Table 10.3 shows the average price of bananas[4] in cents per pound in U.S. cities for selected months and years. Locate all critical points in the table. Identify each point as a relative maximum point, a relative minimum point, or a saddle point.

TABLE 10.3 Price of bananas (cents per pound)

	May	June	July	August	September
1994	47	45	45	48	45
1995	48	49	52	51	49
1996	51	50	50	48	46
1997	48	49	49	48	46
1998	51	51	53	49	48
1999	49	50	49	49	48

7. Table 10.4 shows the average price of iceberg lettuce[5] in cents per pound in U.S. cities for selected months and years.

TABLE 10.4 Price of iceberg lettuce (cents per pound)

	February	March	April	May	June
1994	54	61	53	54	52
1995	60	61	134	145	80
1996	59	65	65	61	67
1997	59	61	67	60	59
1998	64	70	84	88	71
1999	66	77	75	69	65

Locate all critical points in the table. Identify each point as a relative maximum point, a relative minimum point, or a saddle point.

8. The contour graph in Figure 10.1.4a and the three-dimensional graph in Figure 10.1.4b represent the elevation of a tract of farmland where elevation is measured in feet above sea level. Identify each of the following input values as corresponding to a relative maximum, a relative minimum, a saddle point, or none of these.

a. $e = 0.9$ and $n = 0.5$

b. $e = 0.5$ and $n = 0.5$

c. $e = 1.5$ and $n = 1.5$

9. Consider the contour graph in Figure 10.1.5a and its associated three-dimensional graph, shown in Figure 10.1.5b. Is the point at which $x = 1.8$ and $y = 1.5$ a relative maximum point, a relative minimum point, or a saddle point? Explain.

4. Bureau of Labor Statistics.

5. Bureau of Labor Statistics.

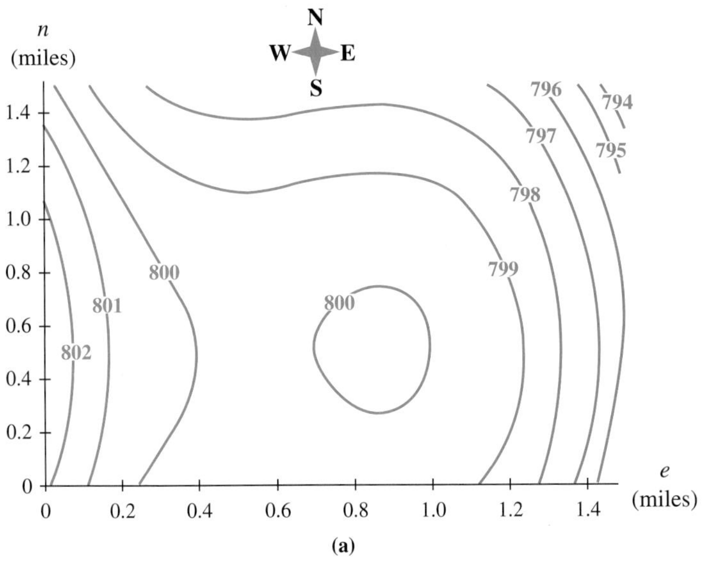

(a)

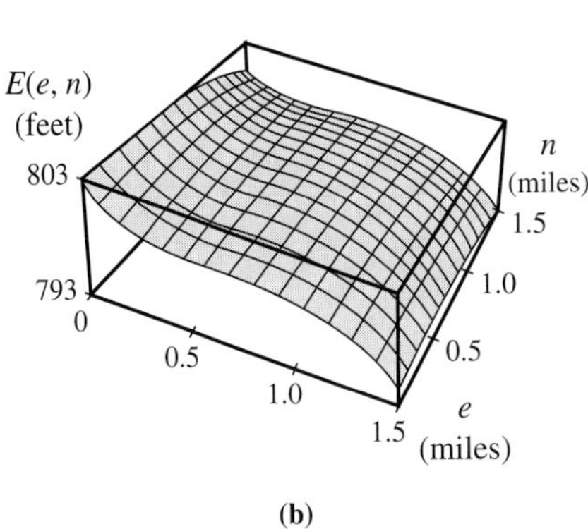

(b)

FIGURE 10.1.4

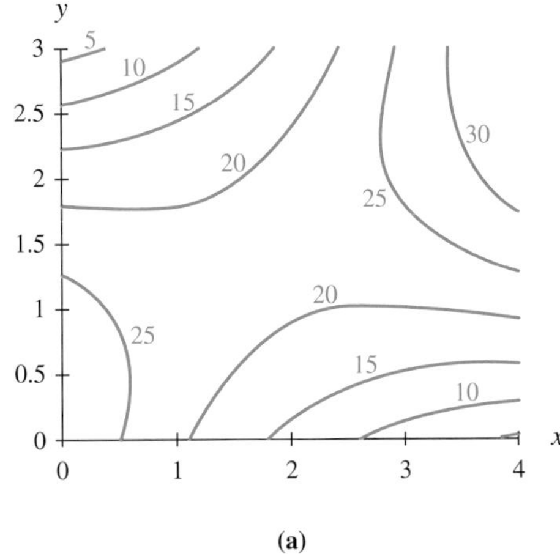

(a)

(b)

FIGURE 10.1.5

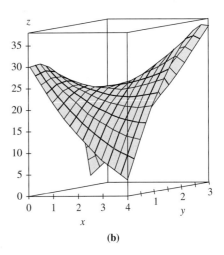

10. Table 10.5 on page 698 gives the elevation $E(e, n)$ in feet above sea level of a tract of Missouri farmland measured e miles east of the western fence and n miles north of the southern fence.

    a. Find any relative extreme points in the table.

    b. Find the saddle point in the table.

    c. Explain how you located the saddle point.

11. The monthly average global radiation[6] (measured in kilowatt-hours per square meter per day) is shown in Table 10.6 on page 698 as a function of the month and the number of degrees of latitude.

    a. Is it possible for Table 10.6 to extend in any direction? Explain.

    b. Sketch contour curves on the table for 5.2, 6.2, 7.2, 8.2, and 9.2 kW-h/m^2/day.

    c. Use the contours to locate all critical points in the table.

    d. Locate the largest and smallest values in the table. Write a sentence interpreting these values.

    e. Use the table to estimate the absolute maximum and minimum monthly average global radiation

6. W. Rudolff, *World-Climates* (Stuttgart: Wissenschaftliche Verlagsgesellschaft, 1981).

TABLE 10.5 Elevation (feet above sea level) *e* miles east of western fence and *n* miles north of southern fence

$e \rightarrow$	0	0.1	0.2	0.3	0.4	0.5	0.6	0.7	0.8	0.9	1.0	1.1	1.2	1.3	1.4	1.5
1.5	800.3	799.1	798.3	797.8	797.5	797.4	797.5	797.6	797.6	797.7	797.6	797.3	796.8	796.0	794.8	793.2
1.4	800.8	799.6	798.8	798.3	798.0	797.9	797.9	798.0	798.1	798.1	798.1	797.8	797.3	796.5	795.3	793.7
1.3	801.2	800.0	799.2	798.7	798.4	798.3	798.4	798.5	798.5	798.6	798.5	798.2	797.7	796.9	795.7	794.1
1.2	801.6	800.4	799.6	799.1	798.8	798.7	798.7	798.8	798.9	798.9	798.9	798.6	798.1	797.3	796.1	794.5
1.1	801.9	800.7	799.9	799.4	799.1	799.0	799.1	799.2	799.2	799.3	799.2	798.9	798.4	797.6	796.4	794.8
1.0	802.2	801.0	800.2	799.7	799.4	799.3	799.3	799.4	799.5	799.5	799.5	799.2	798.7	797.9	796.7	795.1
0.9	802.4	801.2	800.4	799.9	799.6	799.5	799.6	799.7	799.7	799.8	799.7	799.4	798.9	798.1	796.9	795.3
0.8	802.6	801.4	800.6	800.1	799.8	799.7	799.7	799.8	799.9	799.9	799.9	799.6	799.1	798.3	797.1	795.5
0.7	802.7	801.5	800.7	800.2	799.9	799.8	799.9	800.0	800.0	800.1	800.0	799.7	799.2	798.4	797.2	795.6
0.6	802.8	801.6	800.8	800.3	800.0	799.9	799.9	800.0	800.1	800.1	800.1	799.8	799.3	798.5	797.3	795.7
0.5	802.8	801.6	800.8	800.3	800.0	799.9	800.0	800.1	800.1	800.2	800.1	799.8	799.3	798.5	797.3	795.7
0.4	802.8	801.6	800.8	800.3	800.0	799.9	799.9	800.0	800.1	800.1	800.1	799.8	799.3	798.5	797.3	795.7
0.3	802.7	801.5	800.7	800.2	799.9	799.8	799.9	800.0	800.0	800.1	800.0	799.7	799.2	798.4	797.2	795.6
0.2	802.6	801.4	800.6	800.1	799.8	799.7	799.7	799.8	799.9	799.9	799.9	799.6	799.1	798.3	797.1	795.5
0.1	802.4	801.2	800.4	799.9	799.6	799.5	799.6	799.7	799.7	799.8	799.7	799.4	798.9	798.1	796.9	795.3
0	802.2	801.0	800.2	799.7	799.4	799.3	799.3	799.4	799.5	799.5	799.5	799.2	798.7	797.9	796.7	795.1
$n \uparrow$																

TABLE 10.6 Monthly and yearly means of the diurnal sums of the global radiation, in kW-h/m^2 per day

	Jan	Feb	Mar	Apr	May	Jun	Jul	Aug	Sep	Oct	Nov	Dec
North Pole	—	—	—	3.1	6.9	8.9	7.9	4.9	0.80	—	—	—
80°	—	—	0.77	3.4	6.9	8.8	7.8	4.8	1.55	0.13	—	—
70°	—	0.51	2.0	4.5	7.1	8.5	7.8	5.5	2.9	0.95	0.06	—
60°	0.55	1.53	3.4	5.7	7.7	8.8	8.2	6.5	4.3	2.2	0.81	0.34
50°	1.66	2.8	4.7	6.7	8.4	9.1	8.8	7.4	5.5	3.4	1.97	1.35
40°	3.0	4.2	5.9	7.5	8.8	9.3	9.0	8.1	6.5	4.8	3.4	2.6
30°	4.4	5.6	6.9	8.1	9.0	9.2	9.1	8.4	7.4	6.1	4.7	4.1
20°	5.8	6.7	7.8	8.5	8.8	8.9	8.8	8.6	8.0	7.1	6.0	5.5
10°	7.1	7.7	8.3	8.5	8.4	8.3	8.3	8.4	8.3	7.9	7.2	6.8
Equator	8.1	8.5	8.6	8.3	7.8	7.5	7.6	8.0	8.4	8.4	8.2	7.9
10°	8.9	8.8	8.4	7.7	6.9	6.4	6.5	7.2	8.1	8.6	8.8	8.8
20°	9.4	9.0	8.1	6.9	5.7	5.1	5.4	6.3	7.5	8.6	9.2	9.5
30°	9.6	8.8	7.4	5.8	4.4	3.8	4.1	5.2	6.7	8.2	9.3	9.8
40°	9.6	8.3	6.5	4.6	3.1	2.5	2.7	3.9	5.6	7.5	9.1	9.9
50°	9.3	7.6	5.4	3.3	1.84	1.25	1.49	2.6	4.5	6.6	8.7	9.7
60°	8.7	6.6	4.1	2.0	0.72	0.31	0.47	1.36	3.1	5.6	8.0	9.3
70°	8.2	5.5	2.8	0.84	—	—	—	0.38	1.78	4.3	7.2	9.1
80°	8.2	4.7	1.42	0.099	—	—	—	—	0.62	3.2	7.0	9.3
South Pole	8.1	4.6	0.60	—	—	—	—	—	—	2.9	7.0	9.4

TABLE 10.7 Car sales in Quebec

Year	Month											
	Jan	Feb	Mar	Apr	May	Jun	Jul	Aug	Sep	Oct	Nov	Dec
1960	6550	8728	12,026	14,395	14,587	13,791	9498	8251	7049	9545	9364	8456
1961	7237	9374	11,837	13,784	15,926	13,821	11,143	7975	7610	10,015	12,759	8816
1962	10,677	10,947	15,200	17,010	20,900	16,205	12,143	8997	5568	11,474	12,256	10,583
1963	10,862	10,965	14,405	20,379	20,128	17,816	12,268	8642	7962	13,932	15,936	12,628
1964	12,267	12,470	18,944	21,259	22,015	18,581	15,175	10,306	10,792	14,752	13,754	11,738
1965	12,181	12,965	19,990	23,125	23,541	21,247	15,189	14,767	10,895	17,130	17,697	16,611
1966	12,674	12,760	20,249	22,135	20,677	19,933	15,388	15,113	13,401	16,135	17,562	14,720
1967	12,225	11,608	20,985	19,692	24,081	22,114	14,220	13,434	13,598	17,187	16,119	13,713
1968	13,210	14,251	20,139	21,725	26,099	21,084	18,024	16,722	14,385	21,342	17,180	14,577

TABLE 10.8 Expected annual corn yield as a percentage of average annual yield

ΔT (°C)	ΔP (%)																		
	-100	-90	-80	-70	-60	-50	-40	-30	-20	-10	0	10	20	30	40	50	60	70	80
6	0	7	14	18	23	27	32	40	48	56	64	67	69	72	74	75	75	75	76
5.5	0	7	15	20	25	29	34	43	51	59	67	70	73	75	78	78	78	78	78
5	0	8	16	21	26	32	37	45	54	62	71	73	76	79	81	81	81	81	81
4.5	0	8	17	22	28	34	39	48	57	65	74	77	79	82	85	85	84	84	84
4	0	9	18	24	30	36	42	51	60	68	77	80	83	86	88	88	88	87	87
3.5	0	9	19	25	32	38	45	54	62	71	80	83	86	89	92	91	91	90	89
3	0	10	20	27	33	40	47	56	65	75	84	87	90	93	96	95	94	93	92
2.5	0	10	21	28	35	42	49	59	69	78	87	91	94	96	98	97	95	94	93
2	0	11	21	29	37	44	52	63	74	82	90	95	99	99	100	98	97	96	94
1.5	0	11	22	30	38	46	54	65	76	85	93	99	101	101	102	100	99	97	95
1	0	12	23	31	40	48	56	67	78	87	95	102	103	104	104	102	100	98	96
0.5	0	12	24	33	41	50	58	69	80	90	98	104	105	106	106	104	102	100	97
0	0	12	25	34	43	51	60	71	83	92	100	107	107	108	109	106	103	101	98
−0.5	0	13	26	35	43	52	61	72	84	93	101	107	108	108	107	105	102	100	97
−1	0	13	27	35	44	53	61	73	85	94	103	108	109	108	106	104	101	99	96
−1.5	0	14	28	36	45	53	62	73	84	93	102	107	108	107	105	103	100	98	95
−2	0	14	29	37	46	54	63	73	84	93	101	105	107	106	104	102	99	97	94
−2.5	0	15	29	38	46	55	63	72	82	90	98	101	103	103	103	101	98	95	93
−3	0	15	30	39	47	55	64	72	80	88	96	97	99	101	102	100	97	94	92
−3.5	0	15	30	38	46	54	62	70	77	85	92	94	95	97	98	96	94	91	89
−4	0	15	31	38	46	53	61	68	74	81	88	90	91	93	95	92	90	88	85
−4.5	0	15	31	38	45	52	59	65	72	78	85	86	88	89	91	89	87	85	82
−5	0	15	31	38	44	51	58	63	69	75	81	82	84	85	87	85	83	81	79
−5.5	0	16	31	37	44	50	56	61	67	72	77	79	80	82	83	81	80	78	76
−6	0	16	31	37	43	49	55	59	64	69	74	75	76	78	79	78	76	75	73

levels. Are these answers the same as those you found in part *d*?

f. Under what conditions will the largest and smallest values in a table also be the absolute extrema of the function represented by the table?

12. Table 10.7 on page 699 lists the number of cars sold[7] in Quebec from January 1960 through December 1968.

a. During what month and year did the absolute minimum occur in the given time period? What is the minimum value?

b. Interpret your answer to part *a*.

c. What type of critical point is May 1965?

13. Table 10.8 on page 699 shows expected annual yield of U.S. corn production[8] as a percentage of the average annual yield as a function of the percentage change in precipitation (above or below the average precipitation) and the change in temperature (in °C above or below the average temperature). Assume that no critical points lie outside of the table.

a. If a cold, wet year produces average temperatures 3°C below normal and 50% more precipitation than normal, what is the expected corn yield?

b. If a hot, dry year results in 50% less precipitation than normal and an average temperature 3°C above normal, what is the expected corn yield?

c. Sketch contour curves on Table 10.8 at 105%, 100%, 95%, and 90%.

d. Locate the absolute maximum in Table 10.8, and write a sentence interpreting the corresponding point(s).

14. Table 10.9 shows the probability of precipitation[9] in the U.S. corn belt being ΔP% above or below normal and the temperature being ΔT°C above or below normal. Assume that if the table were extended in any direction, all the additional entries would be zero.

a. Sketch contour curves on the table at 10%, 20%, 30%, 40%, and 50%.

TABLE 10.9 Probability (expressed as a percentage) of precipitation and temperature variations

ΔT (°C)	ΔP (%)								
	-40	-30	-20	-10	0	10	20	30	40
4	0	0	0	0	0	0	0	0	0
3.5	0	0	1	1	0	0	0	0	0
3	0	1	2	3	2	0	0	0	0
2.5	0	1	5	7	5	1	0	0	0
2	0	2	9	15	12	4	1	0	0
1.5	0	3	12	26	24	10	2	0	0
1	0	3	15	37	40	19	4	0	0
0.5	0	2	14	42	54	30	7	1	0
0	0	1	11	39	59	39	11	1	0
-0.5	0	1	7	30	54	42	14	2	0
-1	0	0	4	19	40	37	15	3	0
-1.5	0	0	2	10	24	26	12	3	0
-2	0	0	1	4	12	15	9	2	0
-2.5	0	0	0	1	5	7	5	1	0
-3	0	0	0	0	2	3	2	1	0
-3.5	0	0	0	0	0	1	1	0	0
-4	0	0	0	0	0	0	0	0	0

b. Use the contours to locate and estimate the absolute maximum percentage. Interpret your answer.

15. Table 10.10 shows the amount of extraterrestrial radiation[10] measured in millimeters per day equivalent evaporation in the Northern hemisphere as a function of the month and the number of degrees of latitude.

a. According to the table, approximately how much extraterrestrial radiation does your campus receive in September?

b. Locate any relative extreme points in Table 10.10. Give the month, degrees of latitude, and amount of radiation for the relative extreme points.

c. In which directions can the table extend?

d. Estimate the least and greatest amounts of extraterrestrial radiation in the Northern hemisphere. Give the corresponding months and degrees of latitude.

7. B. Abraham and J. Ledolter, *Statistical Methods for Forecasting* (New York: Wiley, 1983).
8. *Crop Yields and Climate Change to the Year 2000* (U.S. Department of Agriculture, 1980).
9. Ibid.
10. A. A. Hanson, ed., *Practical Handbook of Agricultural Science* (Boca Raton, FL: CRC Press, 1990).

TABLE 10.10 Extraterrestrial radiation expressed in equivalent evaporation in mm/day

Latitude (degrees)	Northern hemisphere											
	Jan	Feb	Mar	Apr	May	Jun	Jul	Aug	Sep	Oct	Nov	Dec
50	3.8	6.1	9.4	12.7	15.8	17.1	16.4	14.1	10.9	7.4	4.5	3.2
48	4.3	6.6	9.8	13.0	15.9	17.2	16.5	14.3	11.2	7.8	5.0	3.7
46	4.9	7.1	10.2	13.3	16.0	17.2	16.6	14.5	11.5	8.3	5.5	4.3
44	5.3	7.6	10.6	13.7	16.1	17.2	16.6	14.7	11.9	8.7	6.0	4.7
42	5.9	8.1	11.0	14.0	16.2	17.3	16.7	15.0	12.2	9.1	6.5	5.2
40	6.4	8.6	11.4	14.3	16.4	17.3	16.7	15.2	12.5	9.6	7.0	5.7
38	6.9	9.0	11.8	14.5	16.4	17.2	16.7	15.3	12.8	10.0	7.5	6.1
36	7.4	9.4	12.1	14.7	16.4	17.2	16.7	15.4	13.1	10.6	8.0	6.6
34	7.9	9.8	12.4	14.8	16.5	17.1	16.8	15.5	13.4	10.8	8.5	7.2
32	8.3	10.2	12.8	15.0	16.5	17.0	16.8	15.6	13.6	11.2	9.0	7.8
30	8.8	10.7	13.1	15.2	16.5	17.0	16.8	15.7	13.9	11.6	9.5	8.3
28	9.3	11.1	13.4	15.3	16.5	16.8	16.7	15.7	14.1	12.0	9.9	8.8
26	9.8	11.5	13.7	15.3	16.4	16.7	16.6	15.7	14.3	12.3	10.3	9.3
24	10.2	11.9	13.9	15.4	16.4	16.6	16.5	15.8	14.5	12.6	10.7	9.7
22	10.7	12.3	14.2	15.5	16.3	16.4	16.4	15.8	14.6	13.0	11.1	10.2
20	11.2	12.7	14.4	15.6	16.3	16.4	16.3	15.9	14.8	13.3	11.6	10.7
18	11.6	13.0	14.6	15.6	16.1	16.1	16.1	15.8	14.9	13.6	12.0	11.1
16	12.0	13.3	14.7	15.6	16.0	15.9	15.9	15.7	15.0	13.9	12.4	11.6
14	12.4	13.6	14.9	15.7	15.8	15.7	15.7	15.7	15.1	14.1	12.8	12.0
12	12.8	13.9	15.1	15.7	15.7	15.5	15.5	15.6	15.2	14.4	13.3	12.5
10	13.2	14.2	15.3	15.7	15.5	15.3	15.3	15.5	15.3	14.7	13.6	12.9
8	13.6	14.5	15.3	15.6	15.3	15.0	15.1	15.4	15.3	14.8	13.9	13.3
6	13.9	14.8	15.4	15.4	15.1	14.7	14.9	15.2	15.3	15.0	14.2	13.7
4	14.3	15.0	15.5	15.5	14.9	14.4	14.6	15.1	15.3	15.1	14.5	14.1
2	14.7	15.3	15.6	15.3	14.6	14.2	14.3	14.9	15.3	15.3	14.8	14.4
0	15.0	15.5	15.7	15.3	14.4	13.9	14.1	14.8	15.3	15.4	15.1	14.8

16. The average daily weight gain of a pig[11] (in kilograms per day) is shown in Table 10.11 on page 702 as a function of the air temperature and the pig's weight.

    a. Use the table to estimate relative and absolute extrema for temperatures between 4.4°C and 37.8°C and weights between 45 kg and 156 kg.

    b. Convert the temperatures to degrees Fahrenheit using the equation $°F = \frac{9}{5}(°C) + 32$.

    c. Write a sentence interpreting the absolute extrema found in part *a*.

17. Table 10.12 on page 702 gives the consistometer values for applesauce as a function of the number of months the raw apples were stored and the temperature at which they were blanched.[12] The

11. Ibid.

12. Based on information in A. M. Godfrey Usiak, M. C. Bourne, and M. A. Rao, "Blanch Temperature/Time Effects on Rheological Properties of Applesauce," *Journal of Food Science*, vol. 60, no. 6 (1995), pp. 1289–1291.

TABLE 10.11 Average daily weight gain of a pig (kg/day)

Mean live weight (kg)	Air temperature (°C)						
	4.4	10	15.6	21.1	26.7	32.2	37.8
45		0.62	0.72	0.91	0.89	0.64	0.18
68	0.58	0.67	0.79	0.98	0.83	0.52	-0.09
91	0.54	0.71	0.87	1.01	0.76	0.4	-0.35
113	0.5	0.76	0.94	0.97	0.68	0.28	-0.62
136	0.46	0.8	1.02	0.93	0.62	0.16	-0.88
156	0.43	0.85	1.09	0.9	0.55	0.05	-1.15

consistometer value is a measure of how far (in centimeters) an amount of applesauce flows down a vertical surface in 30 seconds.

TABLE 10.12 Consistometer values for Rome applesauce (cm)

Storage time (months)	Temperature (°C)				
	35°	47°	59°	71°	83°
0	3.0	2.8	2.6	2.6	2.8
1	3.3	3.1	2.8	2.8	3.0
2	3.5	3.2	3.0	2.9	3.2
3	3.4	3.2	3.0	2.9	3.2
4	3.2	2.9	2.7	2.7	3.0

a. Sketch contour curves on the table for consistometer values of 2.7, 3.0, and 3.3.

b. Locate all critical points in the table.

c. Estimate the absolute maximum and minimum consistometer values for storage times between 0 and 4 months and temperatures between 35°C and 83°C.

d. If you were an applesauce manufacturer, which of the points found in parts *b* and *c* would be of the most interest to you? Explain.

18. The Food and Drug Administration (FDA) defines Grade A applesauce as having a consistometer value of 6.5 cm or less (see Activity 17 for more information). It is common for applesauce manufacturers to add to thick applesauce enough water to thin it to a 6.5-cm consistometer value. Table 10.13 shows the percentage of water that can be added to applesauce while maintaining an FDA Grade A. This percentage of water is a function of the number of months the apples are stored and the blanching temperature.

TABLE 10.13 Percentage of water that can be added to applesauce

Storage time (months)	Blanching temperature (°C)				
	35°	47°	59°	71°	83°
0	43	48	54	53	45
0.5	38	44	48	47	40
1	35	41	45	45	38
1.5	34	40	44	44	35
2	33	39	44	44	34
2.5	33	38	43	43	33
3	34	40	45	45	35
3.5	35	43	47	47	38
4	39	45	50	50	42

a. Sketch contour curves on the table for water percentages of 35, 40, 45, and 50.

b. Locate all critical points in the table.

c. Estimate the absolute maximum and minimum percentage for storage times between 0 and 4 months and temperatures between 35°C and 83°C.

d. Why would a maximum value be of interest to an applesauce manufacturer?

e. If not all harvested apples can be used right away, how long (on the basis of the table) do you think a manufacturer should store apples before processing them?

19. The contour graph[13] in Figure 10.1.6 shows the volume index of a cake baked at 350°F, given the baking time (in minutes) and the amount of leavening used (in grams). An index of 100 corresponds to the volume of the batter. No critical points lie outside the contour graph shown in Figure 10.1.6.

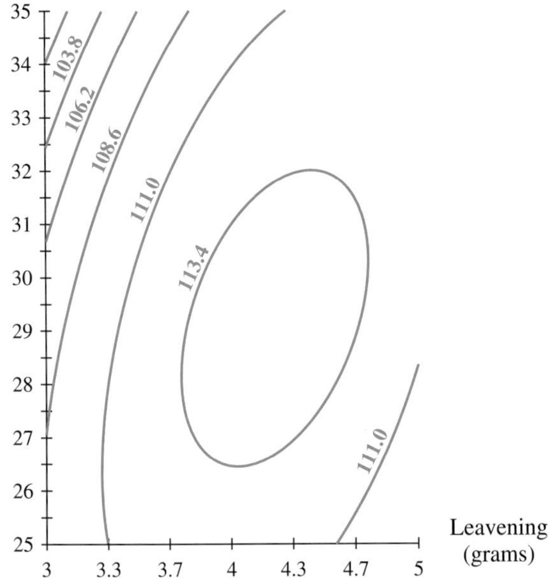

FIGURE 10.1.6

a. Would a food scientist be more interested in finding a maximum or a minimum volume index?
b. Approximately where on Figure 10.1.6 does the optimal point occur?
c. Estimate and interpret the optimal value.

20. Even in desert climates, the ground contains water that in some cases can be pumped and used. When major pumping occurs, it is possible for the land to sink. This sinking is referred to as *land subsidence.* Figure 10.1.7 shows the land subsidence[14] in feet in the Santa Clara Valley between 1934 and 1960.

a. Mark with an X on the contour graph in Figure 10.1.7 the approximate location of any relative extreme points.

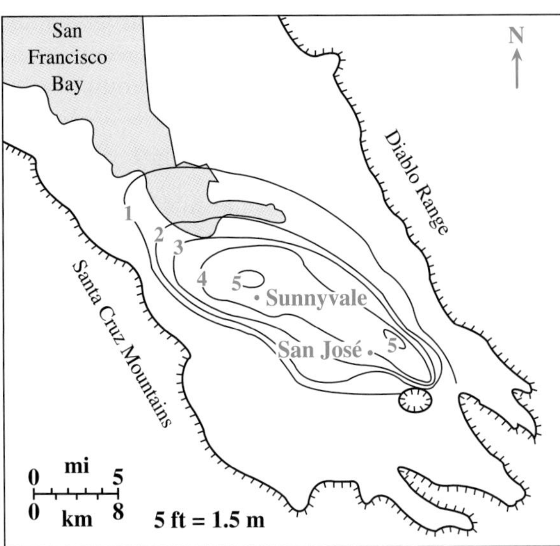

Land subsidence in feet, 1934–1960, Santa Clara Valley, California

FIGURE 10.1.7

b. Write a sentence interpreting the points in part *a* and characterizing them as maxima or minima.

21. A technique commonly used by physical therapists is the application of heat to muscles and tissues. A contour graph showing the change in the temperature of a thigh model after exposure to microwave radiation[15] is given in Figure 10.1.8.

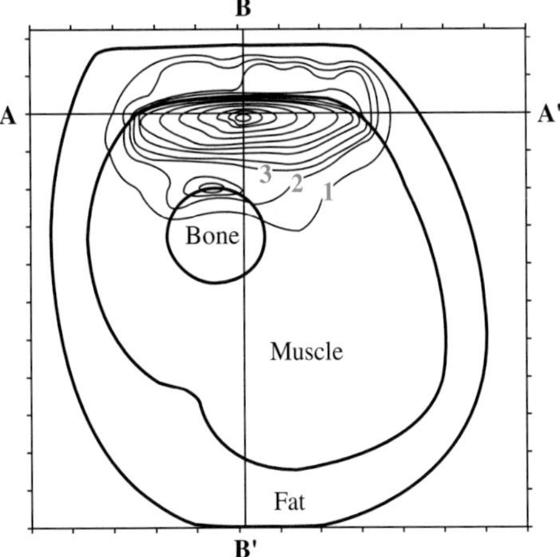

Increase in temperature (°C) of a thigh model

FIGURE 10.1.8

13. A. Li and C. E. Walker, "Cake Baking in Conventional, Impingement and Hybrid Ovens," *Journal of Food Science*, vol. 61, no. 1 (1996), pp. 188–192.
14. Figure 10.1.7 from R. A. Freeze and J. A. Cherry, *Groundwater* (Englewood Cliffs, NJ: Prentice-Hall), 1979. Reprinted by permission of Prentice-Hall, Inc. Upper Saddle River, NJ.
15. F. J. Kottke and J. F. Lehman, *Krusen's Handbook of Physical Medicine and Rehabilitation* (Philadelphia: Saunders, 1990).

a. Mark with an X on the contour graph the approximate location of any relative and absolute extrema, and estimate the output value at each point.

b. Are the points you marked maxima or minima?

22. In the summer of 1993, the midwestern United States experienced catastrophic flooding that resulted in damages estimated at over $10 billion. A particularly heavy rainfall occurred in Missouri on June 6 of that year. Figure 10.1.9 shows a contour map of the number of millimeters of rain[16] that fell on June 6, 1993.

a. Identify the Missouri and Mississippi Rivers on the map in Figure 10.1.9.

b. Mark with an X on the contour map the area of greatest rainfall. Estimate the amount of rain that fell on June 6 in the area you marked.

c. Identify any other critical points on the contour map, and classify them as relative maxima, relative minima, or saddle points.

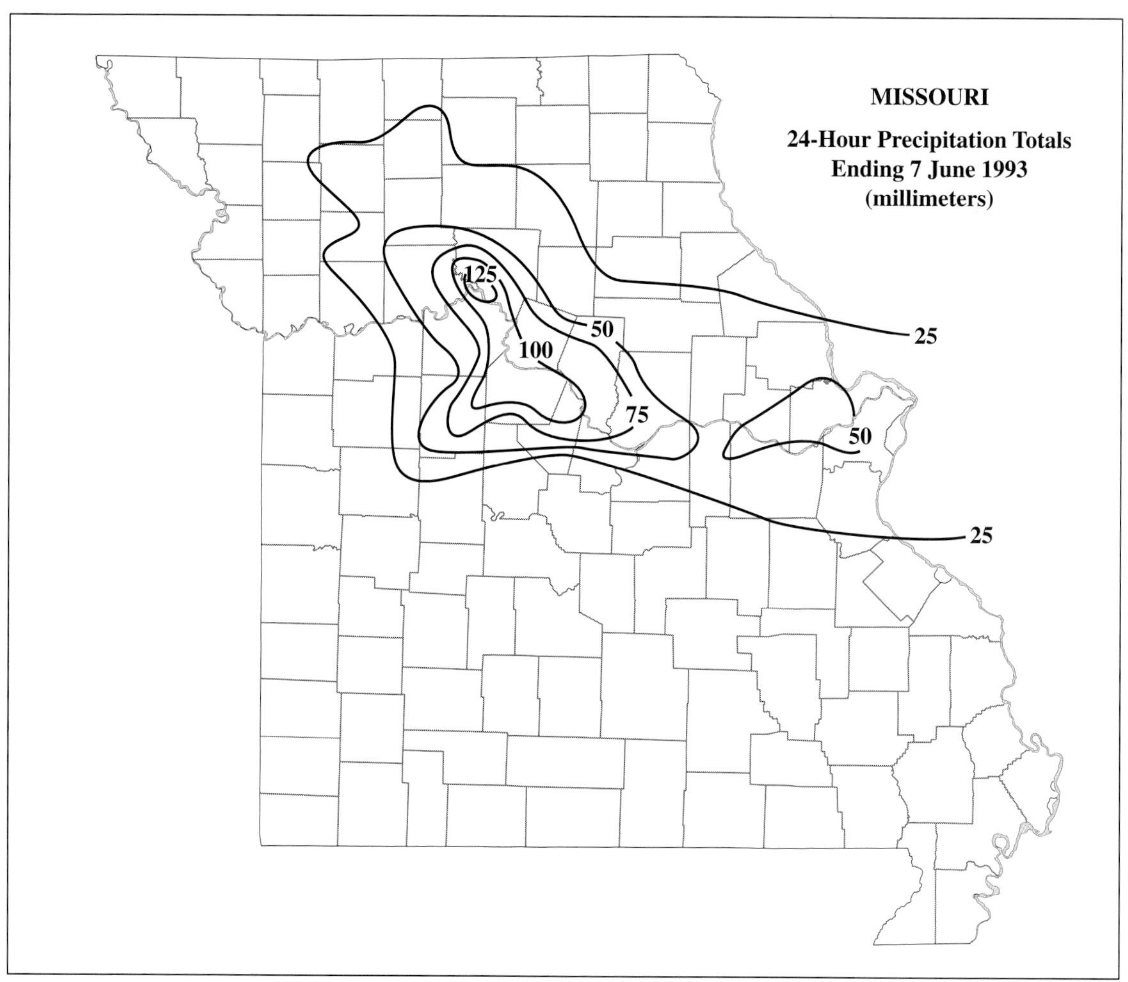

FIGURE 10.1.9

16. Figure 10.1.9 from S. M. Rochette and J. T. Moore, "Initiation of an Elevated Mesoscale Convective System Associated with Heavy Rainfall," *Weather and Forecasting*, vol. 11 (1996), p. 444. Used by permission of the American Meteorological Society.

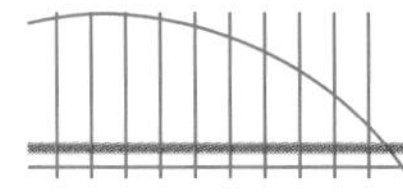

10.2 Multivariable Optimization

In Section 5.2 we saw how to locate relative extreme points (maxima or minima) on the graph of a two-dimensional function by analyzing the function's derivative. Because relative extreme points on a smooth, continuous graph can occur only where the tangent line is horizontal (that is, with slope equal to zero), we search for points where the derivative is 0. If we find such a point, we determine whether the graph has a local maximum, a local minimum, or neither by determining whether the graph is concave down, is concave up, or has an inflection point at that point.

Finding Critical Points Algebraically

We now direct our attention to looking for critical points of smooth, continuous multivariable functions. As an example, we again consider the problem of maximizing olestra production as a function of processing time and temperature. The percentage of sugar converted to olestra when the ratio of peanut oil to sugar is 10 to 1 is given by the equation

$$P(h, c) = -1924.3916 + 24.5402h + 25.5695c - 0.5927h^2 - 0.0860c^2 - 0.0810hc$$

percent where h is the processing time in hours and c is the temperature in degrees Celsius. We seek the values of h and c that cause $P(h, c)$ to be as large as possible. The three-dimensional graph in Figure 10.15 helps us visualize what is happening.

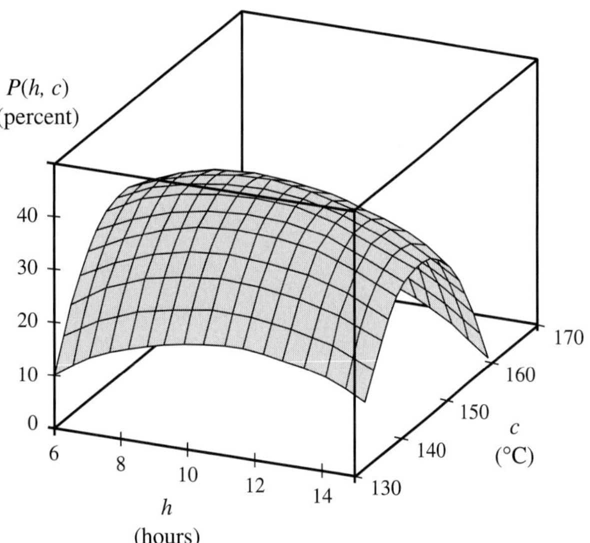

FIGURE 10.15

Let us look at the maximum from two different perspectives. First, ignore the effect of time on olestra production, and think only about the impact due to the temperature. In other words, look at a cross-sectional function of P with output $P(h_0, c)$ where h is held constant at some value h_0. Figure 10.16 shows graphs of cross-sectional models for $h = 7$ hours and $h = 10$ hours. Figure 10.17 shows the same cross sections depicted in three dimensions.

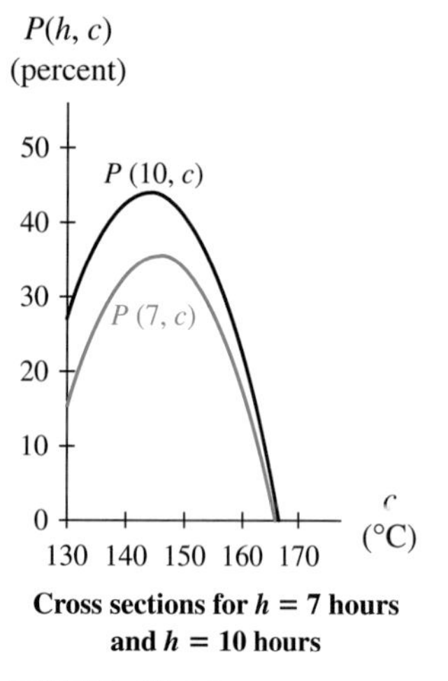

Cross sections for $h = 7$ hours and $h = 10$ hours

FIGURE 10.16

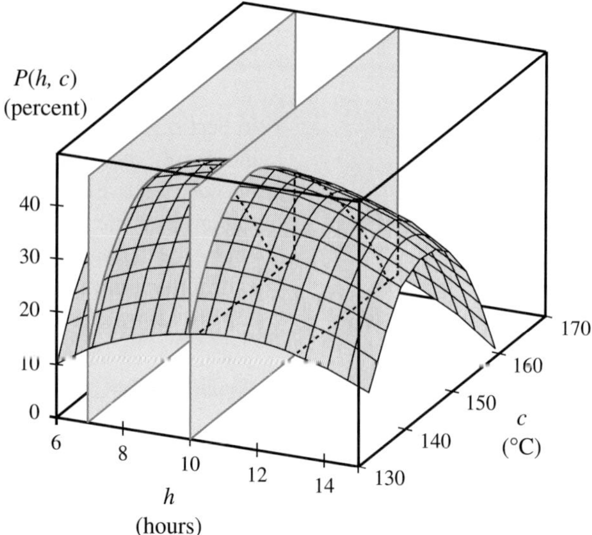

FIGURE 10.17

In order for the olestra model to have a maximum with respect to the temperature, there must be a point at which the derivative of a cross-sectional function of P with output $P(h_0, c)$ is zero. In other words, the partial derivative of olestra production with respect to the temperature must be zero; that is, $\frac{\partial P}{\partial c} = 0$.

Next, ignore the effect of temperature on olestra production, and think only about the impact due to the processing time. In other words, look at a cross-sectional function of P with output $P(h, c_0)$ where c is held constant at some value c_0. Figure 10.18 shows graphs of cross-sectional functions for $c = 135°C$ and $c = 145°C$. Figure 10.19 shows the same cross sections depicted in three dimensions.

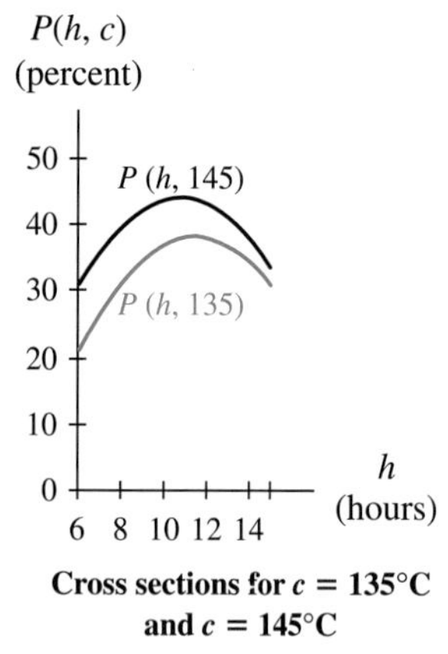

Cross sections for $c = 135°C$ and $c = 145°C$

FIGURE 10.18

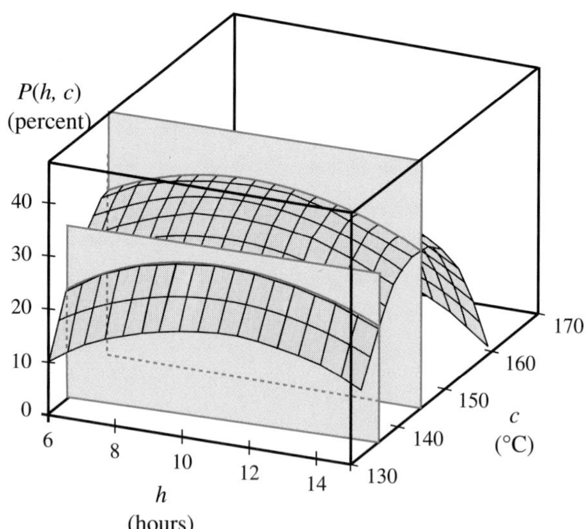

FIGURE 10.19

If we ignore the impact of temperature on olestra production and think only about the impact due to the processing time, we see that for maximum olestra production to occur, the partial derivative of the percent production with respect to time must also be zero; that is, $\frac{\partial P}{\partial h} = 0$. In order for a maximum of P to occur, both partial derivatives must be zero:

Warn your students to check their partial derivative formulas carefully. One simple mistake when calculating a partial derivative results in errors in all subsequent calculations.

$$\frac{\partial P}{\partial c} = 25.5695 - 0.172c - 0.081h = 0$$

$$\frac{\partial P}{\partial h} = 24.5402 - 1.1854h - 0.081c = 0$$

We must simultaneously solve this system of equations to find a point (h, c) that satisfies the condition that both partial derivatives are equal to zero. We write the equations without the partial derivative symbols.

$$25.5695 - 0.172c - 0.081h = 0 \quad (1)$$

$$24.5402 - 1.1854h - 0.081c = 0 \quad (2)$$

Then we rewrite one of the equations so that one of the variables is expressed in terms of the other variable. Solving equation 1 for c in terms of h yields

$$c = \frac{25.5695 - 0.081h}{0.172} \quad (3)$$

Next, we substitute this expression for c into equation 2:

$$24.5402 - 1.1854h - 0.081\left(\frac{25.5695 - 0.081h}{0.172}\right) = 0 \quad (4)$$

Solving equation 4 for h yields

$h \approx 10.9$ hours

We use this value of h to determine the value of c. Substituting the unrounded value of h in equation 3 yields

$c \approx 143.5$°C

Thus the point (h, c) where $h \approx 10.9$ hours and $c \approx 143.5$°C satisfies the condition that the partial derivatives are zero. The output corresponding to these input values is approximately 45.7%. (Compare this result with the approximation in Section 10.1 obtained from a contour graph sketched on the table of data in Figure 10.11. We estimated there that the maximum percentage was 45.1% and occurred when $h = 11$ hours and $c = 145$°C.) Are we certain that $h \approx 10.9$, $c \approx 143.5$, and $P \approx 45.7$ is a point of maximum olestra production rather than a different type of critical point? Figure 10.15 on page 705 shows that a maximum is the only possibility.

Throughout this text, our philosophy is to have students set up a problem and then let technology perform the algebraic computations. Using matrices to solve systems of linear equations is just another example of this approach. If you wish to emphasize algebraic skill, you may omit the matrix solution and solve the systems algebraically.

It is possible to solve equations 1 and 2 by forming a matrix of the coefficients. To do so, we write the equations with the variables in the same order in both equations and move the constants to the other side:

$$-0.172c - 0.081h = -25.5695$$

$$-0.081c - 1.1854h = -24.5402$$

A matrix corresponding to these equations is

$$\begin{bmatrix} -0.172 & -0.081 & -25.5695 \\ -0.081 & -1.1854 & -24.5402 \end{bmatrix}$$

The form of the matrix may vary, depending on which solution method a particular technology uses. Some technologies require a coefficient matrix and a constant matrix; others use a single matrix.

The matrix method for solving systems of equations applies only to linear systems. Make sure your students understand this, because the matrix solution method cannot be used in Activities 3, 5, 6, and 19.

Using steps that are further explained in the technology supplements (and based on mathematics not covered in this text), a calculator or computer software with matrix operations can rewrite this matrix as

$$\begin{bmatrix} 1 & 0 & 143.52934 \\ 0 & 1 & 10.894485 \end{bmatrix}$$

which corresponds to the equations

$$c = 143.52934$$

$$h = 10.894485$$

These equations give the solution to the system of equations.

Note that this matrix method of solving systems of equations works only for linear systems of equations. A **linear system of equations** is one in which all the variables occur to the first power and there are no terms in which two variables are multiplied or divided.

The algebraic method for finding the relative maximum of the percentage of sugar converted to olestra can be used to find any critical points of smooth, continuous three-dimensional functions because a critical point of such a function has the characteristic that both partial derivatives are zero. We locate critical points of any three-dimensional function as follows:

Critical Points of Three-dimensional Functions

Let f be a three-dimensional function with input (x, y). A critical point of f occurs at (a, b) if

1. $f_x(a, b) = 0$ and $f_y(a, b) = 0$

   or

2. $f_x(a, b)$ does not exist or $f_y(a, b)$ does not exist

The first case yields critical points that are relative extrema or saddle points. Because the second case does not occur often in nonscientific applications, we will restrict the remainder of our discussion of critical points to the case where both partial derivatives are zero.

How do we determine which type of critical point occurs? As seen in the olestra illustration, if we can produce either a three-dimensional graph or a contour graph of a function, then we can visually verify whether a point at which all of the first partial derivatives are zero is a maximum, a minimum, or a saddle point. However, if we cannot easily produce such a graph, then we can use the second partial derivatives to determine what type of point we have found. To do so, we begin by defining the determinant of a second partials matrix.

The Determinant Test

If a function f has two input variables x and y, then we can calculate the **determinant** D of the second partials matrix evaluated at a point for which $f_x = f_y = 0$ as

$$D = \begin{vmatrix} f_{xx} & f_{xy} \\ f_{yx} & f_{yy} \end{vmatrix} = f_{xx}f_{yy} - f_{xy}f_{yx}$$

If the determinant D is 0, it tells us nothing about the nature of the point. If $D > 0$, then the point is either a relative maximum or a relative minimum. We can determine which it is by looking at f_{xx} at the point. If $f_{xx} < 0$, then the cross section with respect to x is concave down, so the point is a relative maximum in the x direction. Also, when $f_{xx} < 0$ and $D > 0$, then f_{yy} must also be less than zero. Thus there is also a relative maximum in the y direction. Because $D > 0$ guarantees a maximum or minimum, this critical point must be a relative maximum. Similarly, if $f_{xx} > 0$, then the point is a relative minimum. If $D < 0$, then the point is a saddle point regardless of whether f_{xx} and f_{yy} are positive or negative.

Determinant Test

Let f be a continuous multivariable function with two input variables x and y. Let (a, b) be a point at which the first partial derivatives of f are both 0. The determinant of the second partials matrix evaluated at the point (a, b) is

$$D(a, b) = \begin{vmatrix} f_{xx} & f_{xy} \\ f_{yx} & f_{yy} \end{vmatrix} = f_{xx}f_{yy} - f_{xy}f_{yx}$$

If $D(a, b) > 0$ and $f_{xx} < 0$ at (a, b), then f has a relative maximum at (a, b).
If $D(a, b) > 0$ and $f_{xx} > 0$ at (a, b), then f has a relative minimum at (a, b).
If $D(a, b) < 0$, then f has a saddle point at (a, b).
If $D(a, b) = 0$, then the test does not give any information about (a, b).

In the previous olestra example, we found the point representing $h \approx 10.9$ hours and $c \approx 143.5°C$ to be an extreme point. To use the Determinant Test to verify that the point is a maximum, we first form the second partials matrix:

$$\begin{bmatrix} P_{cc} & P_{ch} \\ P_{hc} & P_{hh} \end{bmatrix} = \begin{bmatrix} -0.172 & -0.081 \\ -0.081 & -1.1854 \end{bmatrix}$$

Then we find the determinant. Note that if the matrix had contained variables, we would have substituted $h \approx 10.9$ and $c \approx 143.5$ before continuing.

$$D = (-0.172)(-1.1854) - (-0.081)^2 \approx 0.197$$

Because D is positive, we know the critical point is either a maximum or a minimum. Because $P_{cc} = -0.172$ is less than zero (and $P_{hh} = -1.1854$ is less than zero) we have a function that is concave down in both the c and h directions. Thus we know the point is a maximum.

EXAMPLE 1 *Cake Volume*

Consider the volume index that provides a measure of how much a cake rises. An index of 100 corresponds to the volume of the batter. The index can be modeled by

$$V(l, t) = -3.1l^2 + 22.4l - 0.1t^2 + 5.3t$$

when l grams of leavening is used and the cake is baked at 177°C for t minutes.

a. Find the maximum volume possible and the conditions needed to achieve that volume.

b. Verify that the point you found in part a is a maximum.

Solution

a. We begin by finding the partial derivatives of V and setting them equal to zero:

$$V_l = -6.2l + 22.4 = 0$$
$$V_t = -0.2t + 5.3 = 0$$

Solving for l and t gives $l \approx 3.6$ grams of leavening and $t = 26.5$ minutes baking time. These values result in a volume index of $V(3.6, 26.5) \approx 110.7$.

b. At $(3.6, 26.5, 110.7)$, $V_{ll} = -6.2$ and $V_{tt} = -0.2$. These second partials are both negative, so V is concave down in both the l and t directions. Figures 10.20a and 10.20b depict the cross sections of V at $l \approx 3.6$ and $t = 26.5$, respectively.

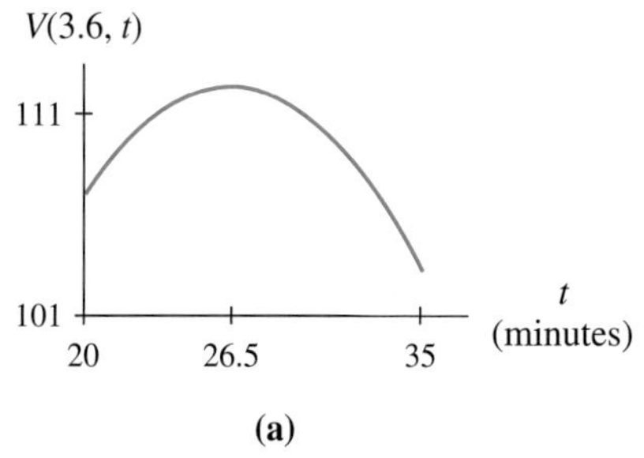

(a)

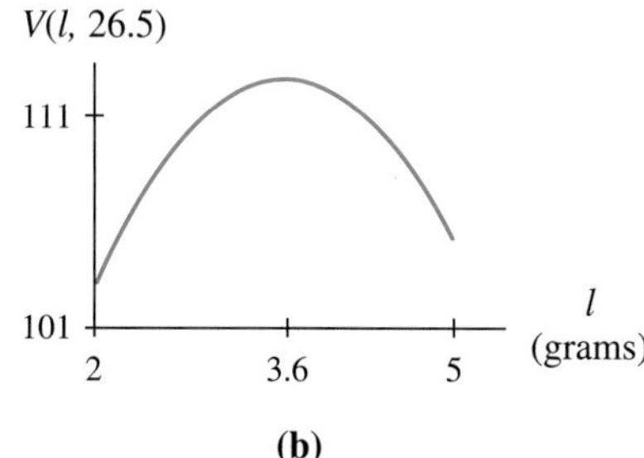

(b)

FIGURE 10.20

Figures 10.20a and 10.20b and negative V_{ll} and V_{tt} suggest that the critical point $(3.6, 26.5, 110.7)$ is a relative maximum. We confirm this by evaluating the determinant of the second partials matrix.

$$D = \begin{vmatrix} -6.2 & 0 \\ 0 & -0.2 \end{vmatrix} = (-6.2)(-0.2) - (0)(0) = 1.24$$

Because D is positive and V_{ll} is negative, the critical point is indeed a relative maximum.

Returning to the example of olestra production, consider the production model for a processing temperature of 150°C:

$$P(h, r) = 144.0344 + 6.7402h - 25.1161r - 0.5927h^2 + 0.8316r^2 + 0.5650hr$$

percent sugar converted to olestra, where h is the processing time in hours and r is the ratio of parts of peanut oil to one part sugar. We find any critical points of P by finding where its two partial derivatives are zero.

$$\frac{\partial P}{\partial h} = 6.7402 - 1.1854h + 0.5650r = 0$$

$$\frac{\partial P}{\partial r} = -25.1161 + 1.6632r + 0.5650h = 0$$

The solution for this linear system of equations is $r \approx 11.3$ and $h \approx 11.1$ hours. When we substitute the unrounded values of r and h in P, we obtain an output value of 39.1%. To determine whether the point $(11.1, 11.3, 39.1)$ is a maximum, a minimum, or a saddle point, we apply the Determinant Test. The determinant of the second partials matrix is

$$D = \begin{vmatrix} P_{hh} & P_{hr} \\ P_{rh} & P_{rr} \end{vmatrix} = \begin{vmatrix} -1.1854 & 0.5650 \\ 0.5650 & 1.6632 \end{vmatrix} = (-1.1854)(1.6632) - (0.5650)^2 \approx -2.291$$

Because this value is negative, we know that P has a saddle point at $h \approx 11.1$ hours and a ratio of $r \approx 11.3$.

EXAMPLE 2 Grazing Beef Cattle

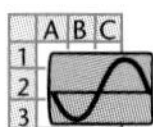

The total daily intake of organic matter required by a beef cow grazing the Northern Great Plains rangeland can be modeled[17] by

$$I(s, m) = 8.61967 - 1.244s + 0.0897s^2 - 0.20988m + 0.035947m^2 + 0.214915sm \text{ kg per day}$$

when the cow produces m kilograms of milk per day. The variable s is a number between -4 and 4 indicating the size of the cow. Find the critical point for I, and determine whether it is a maximum, a minimum, or a saddle point.

Solution We find the critical point by finding where the two partial derivatives of I are zero.

$$\frac{\partial I}{\partial s} = -1.244 + 0.1794s + 0.214915m = 0$$

$$\frac{\partial I}{\partial m} = -0.20988 + 0.071894m + 0.214915s = 0$$

Solving the linear system of equations yields $m \approx 6.90$ kg of milk per day and $s \approx -1.33$. When we substitute the unrounded values of m and s in I, we obtain a value of approximately 8.72 kg of organic matter eaten each day. At the point (-1.33, 6.90, 8.72), $I_{ss} = 0.1794$ and $I_{mm} = 0.071894$ are both positive, indicating that the cross sections of I with outputs $I(-1.33, m)$ and $I(s, 6.90)$ are both concave up. We can verify this graphically. See Figure 10.21 showing these cross sections.

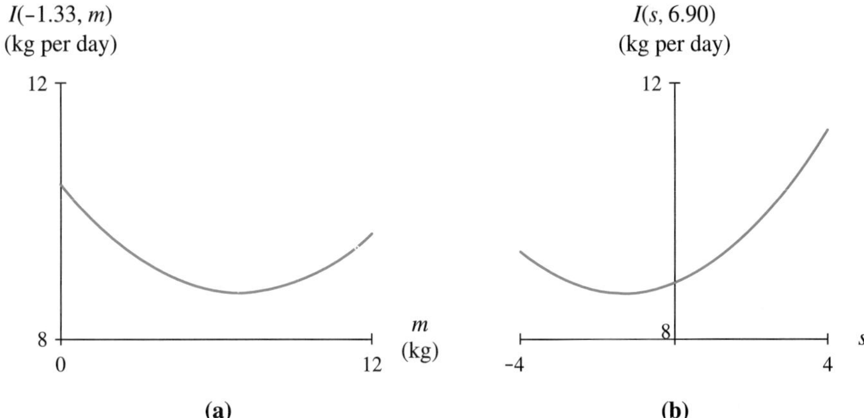

FIGURE 10.21

However, to determine whether the critical point (-1.33, 6.90, 8.72) is a minimum or a saddle point, we apply the Determinant Test. The determinant of the second partials matrix is

$$D = \begin{vmatrix} I_{ss} & I_{sm} \\ I_{ms} & I_{mm} \end{vmatrix} = \begin{vmatrix} 0.1794 & 0.214915 \\ 0.214915 & 0.071894 \end{vmatrix} = (0.1794)(0.071894) - (0.214915)^2 \approx -0.033$$

17. E. E. Grings et al., "Efficiency of Production in Cattle of Two Growth Potentials on Northern Great Plains Rangelands During Spring-Summer Grazing," *Journal of Food Science*, vol. 74, no. 10 (1996), pp. 2317–2326.

Because this value is negative, we know that I has a saddle point rather than a minimum at $s \approx -1.33$ and $m \approx 6.90$ kg of milk per day. Figure 10.22 shows the planes $s = -1.33$ and $m = 6.90$ depicted on the graph of I. Figure 10.23 shows a contour graph of I.

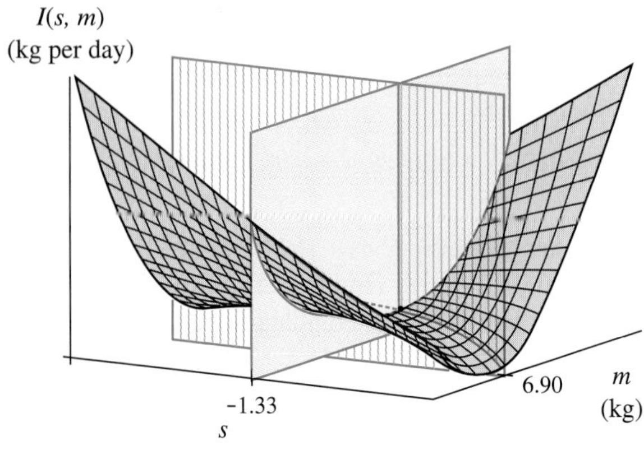

FIGURE 10.22

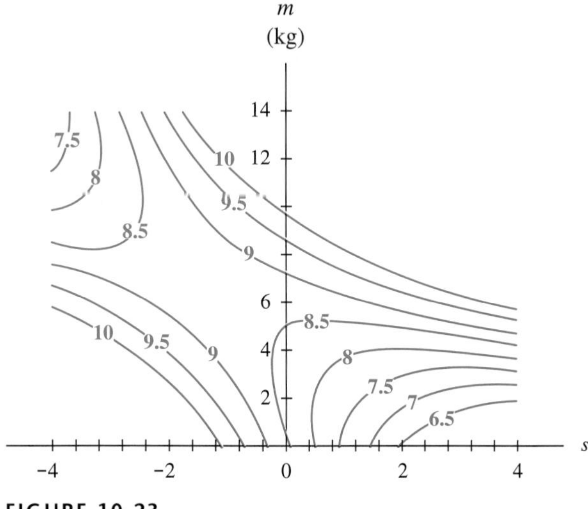

FIGURE 10.23

In general, when looking for the critical points of a three-dimensional function, we solve for the point at which the partial derivatives are zero and then determine what type of point we have found either by visually inspecting a three-dimensional graph or a contour graph or by using the Determinant Test.

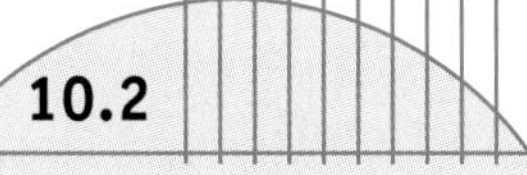

10.2 Concept Inventory

- *Extreme points: maxima and minima*
- *Saddle points*
- *Critical points*
- *Solving systems of equations to find critical points*
- *Determinant Test*

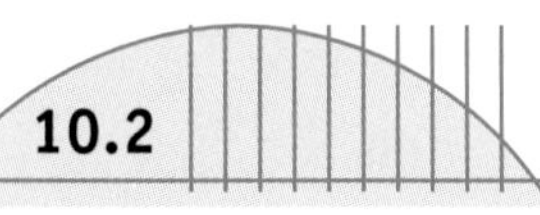

10.2 Activities

For each of the multivariable functions in Activities 1 through 8, find and classify any critical points.

1. $R(k, m) = 3k^2 - 2km - 20k + 3m^2 - 4m + 60$
2. $H(r, s) = rs + 2s^2 + r^2$
3. $G(t, p) = pe^t - 3p$
4. $f(a, b) = a^2 - 4a + b^2 - 2b - 12$
5. $h(w, z) = 0.6w^2 + 1.3z^3 - 4.7wz$
6. $R(s, t) = 1.1s^3 - 2.6s^2 + 0.9s + 6 - 3.1t^2 + 5.3t$
7. $f(x, y) = 3x^2 - x^3 + 12y^2 - 8y^3 + 60$
8. $g(x, y) = 4xy - x^4 - y^4$
9. A restaurant mixes ground beef that costs \$$b$ per pound with pork sausage that costs \$$p$ per pound to make a meat mixture that is used on the restaurant's signature pizza. The quarterly revenue in thousands of dollars from the sale of this pizza is given by the equation

$$R(b, p) = 14b - 3b^2 - bp - 2p^2 + 12p$$

   a. Find the prices at which the restaurant should try to purchase ground beef and pork sausage in order to maximize the quarterly revenue from the sale of the pizza.
   b. Explain how the Determinant Test verifies that your result gives the maximum revenue.
   c. What is the maximum quarterly revenue from the sale of the restaurant's signature pizza?

10. A nursery sells mulch by the truckload. Bark mulch sells for \$$b$ per load, and pine straw sells for \$$p$ per load. The nursery's average weekly profit from the sale of these two types of mulch can be modeled by the equation

$$P(p, b) = 144p - 3p^2 - pb - 2b^2 + 120b + 35 \text{ dollars}$$

   a. Find how much the nursery should charge for each type of mulch in order to maximize the weekly profit from the sale of mulch.
   b. Explain in detail how the Determinant Test verifies that your result in part *a* gives the maximum profit.
   c. What is the maximum weekly profit from the sales of these two types of mulch?

11. Boll weevils have long presented a threat to cotton crops in the southern United States. Research has been done to determine the optimal conditions for reproduction of *Catolaccus grandis*, a parasitoid that attaches to boll weevils and kills them. The number of eggs produced in 30 days by one female *C. grandis* under conditions of 16 hours of light each day, a constant temperature of T°C, and a relative humidity of H% can be modeled[18] by

$$E(T, H) = -4191.6877 + 299.7038T + 23.1412H - 5.2210T^2 - 0.0937H^2 - 0.4023TH \text{ eggs}$$

   a. Determine the point where the partial derivatives of E are both equal to zero.
   b. Write a sentence interpreting the point in part *a* and characterizing it as a maximum, a minimum, or a saddle point.

12. The average time for *C. grandis* eggs to develop into an adult can be modeled[19] by

$$D(H, L) = 25.6691 - 0.838H + 2.4297L + 0.0084H^2 - 0.0726L^2 - 0.0181HL \text{ days}$$

   where the relative humidity is held constant at H%, the eggs are exposed to L hours of light each day, and the temperature is held constant at 30°C.

   a. Determine the point where the partial derivatives of D are both equal to zero.
   b. Write a sentence interpreting the point in part *a* and characterizing it as a maximum, a minimum, or a saddle point.

13. The increased consumption of low-fat foods in this country has led to a surplus of milk fat. Research is being done to find new uses for milk fat such as extracting flavor from the milk fat to utilize it in other products. The process of flavor extraction entails hydrolysis—a chemical reaction in which fatty acids are liberated from the milk fat. It is desirable to maximize the rate at which the fatty acids are liberated. The initial liberation rate can be modeled[20] by

$$R(P, T) = 32.614 - 1.544P - 1.625T - 4.905P^2 - 7.053T^2 - 3PT$$

   μmoles (micromoles) of free fatty acids per minute, where $5.5 + 1.5P$ is the pH of the milk fat mixture and the temperature is $(60 + 8T)$°C. Find the pH and temperature that maximize the initial rate. Verify that the optimal point is a maximum.

14. In the chemical process described in Activity 13, it is also desirable to maximize the amount of fatty acids liberated in the hydrolysis process. The number of grams of fatty acids per 100 grams of water can be modeled by

$$F(p, t) = 25.224 - 0.129p - 3.052t - 0.446p^2 - 3.761t^2 - 1.249pt \text{ grams}$$

   where the pH of the milk fat is $5.5 + 1.5p$ and the temperature is $(60 + 8t)$°C. Find the pH and temperature that maximize the amount of fatty acid. Verify that the optimal point is a maximum.

15. The Louisiana crayfish industry produces over 38,600 tons of crayfish-processing byproducts every year. One potential use of the byproducts is the production of flavor extracts. In order to extract flavoring from the crayfish-processing byproducts, peptides from the byproducts must be drawn out. This process is a complicated one with many variables. Two of these variables are the pH and temperature maintained during the process. A model[21] for the amount of peptides released is

$$P(x, y) = 2.14 - 0.26x - 0.34y - 0.23x^2 - 0.16y^2 - 0.25xy \text{ milligrams}$$

18. J. A. Morales-Ramos, S. M. Greenberg, and E. G. King, "Selection of Optimal Physical Conditions for Mass Propagation of *Catolaccus grandis*," *Environmental Entomology*, vol. 25, no. 1 (1996), pp. 165–173.
19. Ibid.
20. M. T. Patel, R. Nagarajan, and A. Kilara, "Hydrolysis of Milk Fat by Lipase in Solvent-Free Phospholipid Reverse Micellar Media," *Journal of Food Science*, vol. 61, no. 1 (1996), pp. 33–38.
21. H. H. Baek and K. R. Cadwallader, "Enzymatic Hydrolysis of Crayfish Processing By-products," *Journal of Food Science*, vol. 60, no. 5 (1995), pp. 929–934.

where the pH is $9 + x$ and the temperature is $(70 + 5y)$°C. Figure 10.2.1 shows a contour plot of the amount of peptides released.

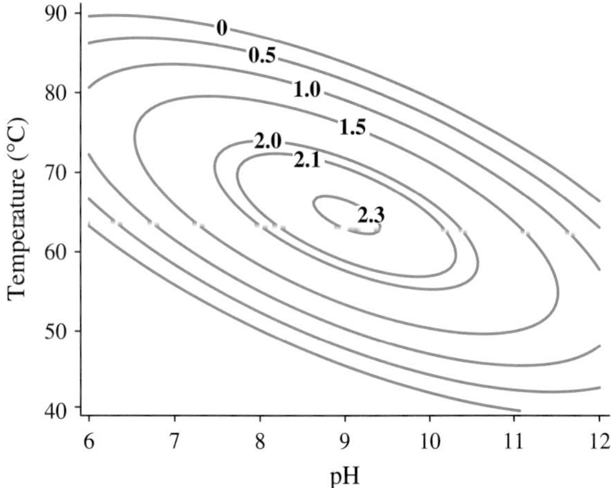

FIGURE 10.2.1

a. Use the contour plot to estimate the maximum amount of peptides and the pH and temperature needed to achieve the maximum.

b. Use the model to determine the pH and temperature that will maximize the amount of peptides produced.

16. The amount of peptides in Activity 15 can also be modeled as a function of pH and processing time:

$$P(x, t) = 2.14 - 0.26x + 0.04t - 0.23x^2 - 0.04t^2 - 0.07xt \text{ milligrams}$$

where the pH is $9 + x$ and the processing time is $2.5 + t$ hours. Figure 10.2.2 shows a contour plot of the amount of peptides.

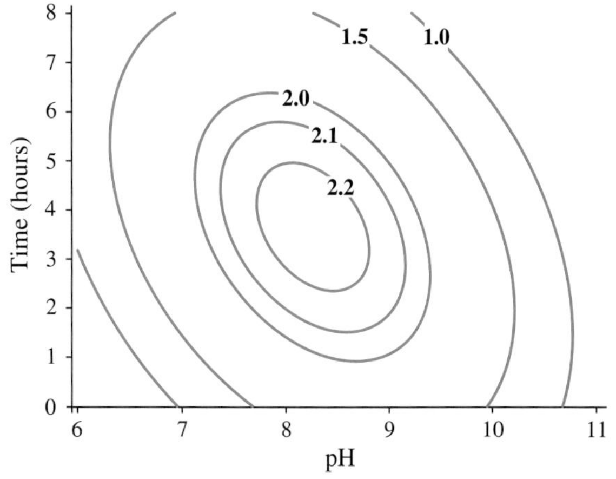

FIGURE 10.2.2

a. Use the contour plot to estimate the maximum amount of peptides as well as the pH and time needed to achieve the maximum.

b. Use the model to find the pH and time that will maximize the amount of peptides.

17. Milk proteins are sometimes added to sausage to reduce shrinking due to cooking loss and to improve the texture of the sausage. Research has shown that when sausage is prepared with three milk proteins—sodium caseinate, whey protein, and skim milk powder—the cooking loss (expressed as a percentage of initial weight) can be modeled[22] by

$$L(w, s) = 10.65 + 1.13w + 1.04s - 5.83ws \text{ percent}$$

where w is the proportion of whey protein, s is the proportion of skim milk powder, and the proportion c of sodium caseinate can be calculated as $c = 1 - w - s$. Determine whether the function L has a relative maximum, a relative minimum, or a saddle point, and find the corresponding proportions of sodium caseinate, whey protein, and skim milk powder.

18. A process to extract pectin and pigment from sunflower heads involves washing the sunflower heads in heated water. A model for the percentage of pigment[23] that can be removed from a sunflower head by washing for 20 minutes in r milliliters of water per gram of sunflower when the water temperature is T°C is

$$P(T, r) = 306.761 - 9.6544T + 1.9836r + 0.07368T^2 - 0.02958r^2 \text{ percent}$$

Find the critical point of the pigment removal function, and identify it as a maximum, a minimum, or a saddle point.

19. A model for the elevation above sea level of a tract of farmland previously discussed is

$$E(e, n) = -10.124e^3 + 21.347e^2 - 13.972e - 2.5n^2 + 2.497n + 802.2 \text{ feet above sea level}$$

where e is the distance in miles east of the western fence and n is the distance in miles north of the southern fence.

22. M.R. Ellekjaer, T. Naes, and P. Baardseth, "Milk Proteins Affect Yield and Sensory Quality of Cooked Sausages," *Journal of Food Science*, vol. 61, no. 3 (1996), pp. 660–666.

23. X. Q. Shi, et al., "Optimizing Water Washing Process for Sunflower Heads Before Pectin Extraction," *Journal of Food Science*, vol. 61, no. 3 (1996), pp. 608–612.

a. Find the two critical points of E.

b. Use the contour graph for E in Figure 10.1.4 on page 697 to identify each critical point as a maximum, a minimum, or a saddle point.

c. Use the Determinant Test to verify that the critical points are as you identified them in part b.

20. A model[24] for the revenue generated by the sale of cheeses in the Netherlands is

$$R(c, r) = -52.196r^2 + 5935.497r - 59.128cr + 10{,}299.325c - 384.386c^2 \text{ thousand guilders}$$

where c thousand tons of 40% fat cheese and r thousand tons of regular cheese are sold. Find the combination of cheeses whose sales result in maximum revenue.

21. A measure of the adhesiveness of honey[25] after processing can be modeled by

$$A(g, m, s, h) = -151.78 + 4.26g + 5.69m + 0.67s + 2.48h - 0.05g^2 - 0.14m^2 - 0.03s^2 - 0.05h^2 - 0.07mh$$

where g is the percentage of glucose and maltose, m is the percentage of moisture, s is the percentage of crystal used to seed the honey, and h is the holding time of the honey in days.

a. Find the point at which all four partial derivatives are zero. Find the corresponding adhesiveness measure at that point.

b. The Determinant Test applies only to functions of two variables. Use some other method to make a conjecture about the nature of the critical point you found. (That is, is it a maximum or a minimum or something else?)

22. A measure of the thickness of processed cheese spread can be modeled by

$$V(g, s, l) = 1000(5.647 - 3.436s + 1.577s^2 + 0.375l - 1.757g + 0.141g^2 + 1.217sl - 0.533s^2l + 0.186sg)$$

where g is the percentage of glycerol, s is the percentage of salt, and l is the percentage of lactose.

a. Find the point at which all three partial derivatives are zero. Find the corresponding thickness measure at that point.

b. The Determinant Test applies only to functions of two variables. Use some other method to make a conjecture about the nature of the critical point you found. (That is, is it a maximum or a minimum or something else?)

10.3 Optimization Under Constraints

Often it is not enough to be able to find the input that leads to a maximum (or minimum) output. In many applications there are constraints on what input values can be used. For example, when trying to optimize revenue made from the sale of cheese, the producer is constrained by the amount of cheese the company is able to produce. In the case where a manufacturer is trying to maximize production, budget constraints must be considered. Sometimes constraints are implied by a particular context.

It is common in meat production to add nonmeat proteins to meat products in order to decrease ingredient costs and increase yield. Milk proteins, in particular, reduce the amount of shrinkage due to cooking. The percentage of volume lost when cooking sausage can be modeled[26] by the equation

$$P(w, s) = 11.78w + 11.69s - 5.83ws \text{ percent}$$

24. S. Louwes, J. Boot, and S. Wage, "A Quadratic-programming Approach to the Problem of the Optimal Use of Milk in the Netherlands," *Journal of Farm Economics*, vol. 45, no. 2 (May 1963), pp. 309–317.

25. J. M. Shinn and S. L. Wang, "Textural Analysis of Crystallized Honey Using Response Surface Methodology," *Canadian Institute of Food Science Journal*, vol. 23, no. 4/5 (1990), pp. 178–182.

26. M. R. Ellekjaer, T. Naes, and P. Baardseth, "Milk Proteins Affect Yield and Sensory Quality of Cooked Sausages," *Journal of Food Science*, vol. 61, no. 3 (1996), pp. 660–672.

where w is the proportion of the added milk proteins that is whey protein, and s is the proportion of the added milk proteins that is skim milk powder. The fact that the proportions of whey protein and skim milk powder must add up to 1 gives us a constraint that affects the optimal value of the function for the percentage of volume lost in cooking. This constraint is $g(w, s) = w + s = 1$.

Figure 10.24 shows several contour curves for the function P. A graph of the constraint equation $g(w, s) = 1$ is the line that intersects different curves on the contour graph. Minimum cooking loss is determined by finding the lowest-valued contour curve that touches the constraint curve. The output value associated with this contour curve is the minimum cooking loss. Note that the lowest-valued contour curve that the constraint line $g(w, s) = 1$ appears to touch is $P(w, s) = 10.3$. Thus the minimum cooking loss subject to the constraint $w + s = 1$ is approximately 10.3%. The line $g(w, s) = 1$ touches the 10.3% contour curve but remains completely on one side of it. This description should sound familiar, for it is the description of a tangent line.

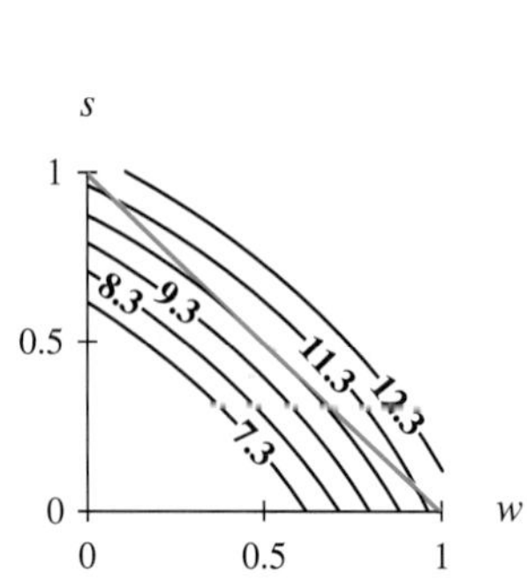

Contour graph of P with constraint $w + s = 1$

FIGURE 10.24

For other multivariable functions f, the constraint function g may not be linear. The constraint curve drawn on a contour graph may look more like the one shown in Figure 10.25.

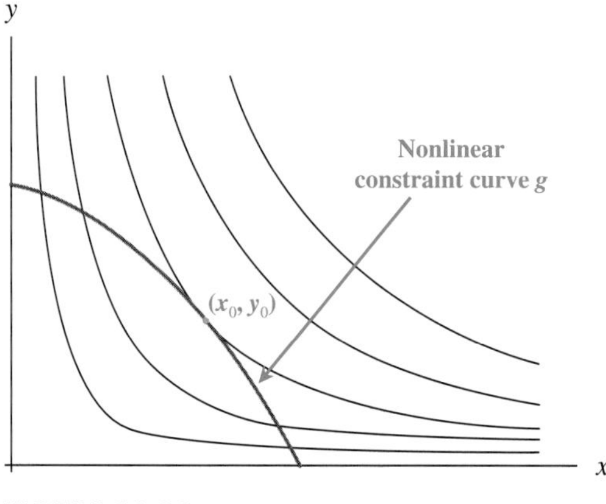

FIGURE 10.25

The Lagrange Multiplier

The point of minimum (or maximum) output of f subject to $g(x, y) = c$ can be determined by first finding the smallest (or largest) value M for which the constraint curve $g(x, y) = c$ and the contour curve $f(x, y) = M$ touch and then determining the point (x_0, y_0) where these two curves meet. Figure 10.25 illustrates that the constraint curve touches this lowest or highest possible contour curve at only one point on that contour curve and does not pass through the contour curve but remains completely on one side of it. What does this say about the lines tangent to the contour curve and the constraint curve at this point (x_0, y_0)? Compare the tangent lines shown in Figure 10.26 and 10.27. The tangent line in Figure 10.26 is the tangent line at (x_0, y_0) on the constraint curve $g(x, y) = c$, whereas the tangent line in Figure 10.27 is the tangent line at the same point but on the contour curve $f(x, y) = M$. They appear to be the same. In fact, they are the same.[27]

27. Although we use this result to develop further the optimization technique, realize that we have only graphically illustrated it in a special case. We do not give an analytic proof.

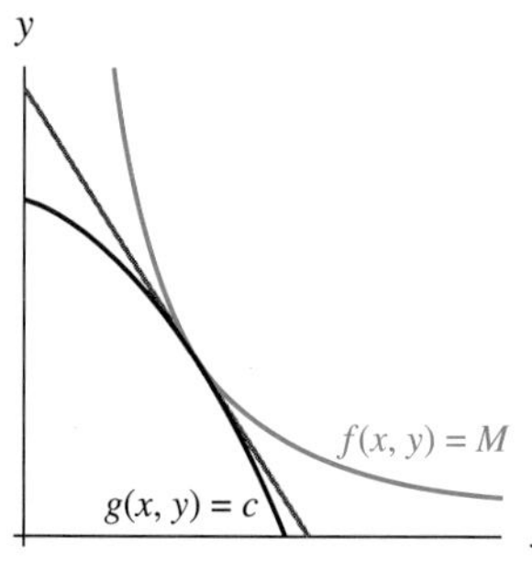

Line tangent to the constraint curve at the optimal point

FIGURE 10.26

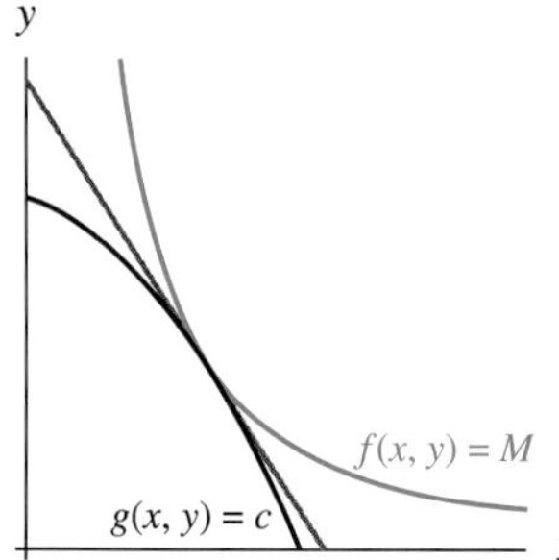

Line tangent to the contour curve at the optimal point

FIGURE 10.27

Let us now use the fact that the tangent line at (x_0, y_0) on $g(x, y) = c$ is the same line as the tangent line at the same point on $f(x, y) = M$ to develop a method for solving constrained optimization problems. The tangent line in Figure 10.26 is tangent to the curve $g(x, y) = c$, so from Chapter 9 we know that its slope is

$$\frac{dy}{dx} = \frac{-\left(\frac{\partial g}{\partial x}\right)}{\left(\frac{\partial g}{\partial y}\right)} = \frac{-g_x}{g_y}$$

The tangent line in Figure 10.27 is a tangent to the contour curve $f(x, y) = M$, so its slope is

$$\frac{dy}{dx} = \frac{-\left(\frac{\partial f}{\partial x}\right)}{\left(\frac{\partial f}{\partial y}\right)} = \frac{-f_x}{f_y}$$

Because these two rates of change are equal, we have

$$\frac{-g_x}{g_y} = \frac{-f_x}{f_y}$$

and some algebra shows that

$$\frac{f_x}{g_x} = \frac{f_y}{g_y}$$

For convenience, we denote this common ratio as λ.

We now use λ to set up a system of equations that we solve in the process of finding any extreme points. First, we know that any extreme point on the constraint curve satisfies the condition

$$\frac{f_x}{g_x} = \lambda, \quad \text{so} \quad \frac{\partial f}{\partial x} = \lambda\frac{\partial g}{\partial x}$$

The point also satisfies the condition

$$\frac{f_y}{g_y} = \lambda, \quad \text{so} \quad \frac{\partial f}{\partial y} = \lambda\frac{\partial g}{\partial y}$$

Finally, the point we are trying to find must satisfy the constraint $g(x, y) = c$. Thus we have a system of three equations with three unknown variables, x, y, and λ:

$$\frac{\partial f}{\partial x} = \lambda\frac{\partial g}{\partial x}$$
$$\frac{\partial f}{\partial y} = \lambda\frac{\partial g}{\partial y}$$
$$g(x, y) = c$$

Solving this system of equations for x and y gives a point on the constraint curve at which an extreme value may occur. In fact, Joseph Lagrange (1736–1813), an Italian-born French mathematician, showed that any extreme point for the function f subject to the constraint $g(x, y) = c$ occurs among the solutions to the above system of equations. For this reason, λ is called the **Lagrange multiplier**.

Once a solution to the system of equations is found, a contour graph of the multivariable function f and the constraint $g(x, y) = c$ may be helpful in identifying the nature of the corresponding extreme point. Testing values of $f(x, y)$ for values of x and y that satisfy the constraint and are near the extreme point may also help tell whether that point is a maximum or minimum of the function f.

We return to the sausage example where cooking loss is modeled by the equation

$$P(w, s) = 11.78w + 11.69s - 5.83ws \text{ percent}$$

subject to the constraint

$$g(w, s) = w + s = 1$$

where w is the proportion of whey protein and s is the proportion of skim milk powder. What combination of whey protein and skim milk powder will cause the minimum amount of sausage shrinkage during cooking?

The point at which the minimum shrinkage occurs must satisfy the following system of equations:

$$\frac{\partial P}{\partial w} = \lambda \frac{\partial g}{\partial w} \qquad 11.78 - 5.83s = \lambda$$

$$\frac{\partial P}{\partial s} = \lambda \frac{\partial g}{\partial s} \qquad 11.69 - 5.83w = \lambda$$

$$g(w, s) = c \qquad w + s = 1$$

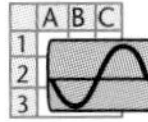

This system of equations can be solved algebraically or using matrices. A matrix for this linear system of equations is

$$\begin{bmatrix} -5.83 & 0 & -1 & -11.78 \\ 0 & -5.83 & -1 & -11.69 \\ 1 & 1 & 0 & 1 \end{bmatrix}$$

Using technology, we find that the solution to this system is $s \approx 0.508$, $w \approx 0.492$, and $\lambda = 8.82$. With these proportions of whey protein and skim milk powder, the shrinkage is $P(0.492, 0.508) \approx 10.28\%$. To verify that this value is a minimum, we look at the contour graph shown in Figure 10.24 on page 716. The contour curve to which the constraint curve is tangent has the smallest value of all contour curves that intersect or touch the constraint curve. This confirms that we have found the minimum of P subject to the constraint $w + s = 1$.

EXAMPLE 1 *Cobb–Douglas Production Function*

Consider a mattress manufacturing process for which the Cobb–Douglas production function is

$$f(L, K) = 48.1L^{0.6}K^{0.4} \text{ mattresses}$$

where L represents the number of worker hours (in thousands) and K represents the amount invested in capital (in thousands of dollars). Suppose that the plant manager has a budget of \$98,000 to be invested in capital or spent on labor and that the average wage of an employee at the mattress plant is \$8 per hour.

a. Write the budget constraint equation.

b. Use Lagrange multipliers to find the expenditures on labor and capital investment that will result in the greatest number of mattresses being produced subject to the budget constraint. Also find the maximum production level.

c. Verify that the answer to part *b* is a relative maximum.

Solution

a. The manager's budget constraint equation is

$$g(L, K) = 8L + K = 98 \text{ thousand dollars}$$

where L and K are as described in the production function.

b. The point at which the maximum production occurs must satisfy the following system of equations:

$$\frac{\partial f}{\partial L} = \lambda \frac{\partial g}{\partial L} \qquad 28.86\, L^{-0.4}K^{0.4} = \lambda(8) \qquad (1)$$

$$\frac{\partial f}{\partial K} = \lambda \frac{\partial g}{\partial K} \qquad 19.24\, L^{0.6}K^{-0.6} = \lambda(1) \qquad (2)$$

$$g(L, K) = c \qquad 8L + K = 98 \qquad (3)$$

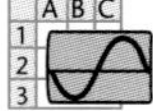

Because this system of equations is not linear, we cannot use a matrix to find the solution. To solve this system, we begin by solving equation 1 for one of the variables L, K, or λ. In this example, we solve for λ.

$$\lambda = \frac{28.86L^{-0.4}K^{0.4}}{8} = 3.6075\, L^{-0.4}K^{0.4}$$

Substituting this expression for λ into equation 2, we have

$$19.24\, L^{0.6}K^{-0.6} = 3.6075\, L^{-0.4}K^{0.4}$$

Next, we multiply both sides of the equation by $L^{0.4}K^{0.6}$ to remove the negative exponents and solve for L:

$$(19.24L^{0.6}\, K^{-0.6})\, L^{0.4}K^{0.6} = (3.6075L^{-0.4}\, K^{0.4})\, L^{0.4}\, K^{0.6}$$
$$19.24L^{0.6+0.4}\, K^{-0.6+0.6} = 3.6075L^{-0.4+0.4}\, K^{0.4+0.6}$$
$$19.24L = 3.6075K$$
$$L = 0.1875K$$

Finally, substituting $L = 0.1875K$ into equation 3 allows us to solve for K:

$$8(0.1875\, K) + K = 98$$
$$2.5\, K = 98$$
$$K = 39.2 \text{ thousand dollars}$$

Next we use $K = 39.2$ to find that $L = 0.1875(39.2) = 7.35$ thousand worker hours. Thus the labor costs are (7.35 thousand worker hours) ($8 per hour) = $58,800, and the capital investment is $39,200. With those investments of time and money, the production level is $f(7.35, 39.2) = 690.608 \approx 690$ mattresses. Note that we discount the fraction of a mattress and round down the production level. A production level of 691 mattresses would (technically) result in labor exceeding the budget constraint.

c. To verify that the values of K and L found in part b give a maximum production level subject to the constraint, we consider nearby points on the constraint curve. We choose values for L that are close to 7.35. In this case, we choose 7.3 and 7.4. Substituting these values into the constraint equation $g(L, K) = 8L + K = 98$, we obtain the corresponding values of K, so the points we check are (7.3, 39.6) and (7.4, 38.8). We find that the values $f(7.3, 39.6) \approx 690.585$ and $f(7.4, 38.8) \approx 690.584$ are both less than the extreme value $f(7.35, 39.2) \approx 690.608$, and we therefore have good evidence that 690 mattresses is a maximum production level. The manager will be able to maximize the production of mattresses by investing \$39,200 in capital and having his employees put in a total of 7350 worker hours. Notice that we could also verify that our solution is a maximum by examining the graph in Figure 10.28. The budget constraint $g(L, K) = 98$ intersects the $f(L, K) = 290, 490,$ and 690 contours. Because the contour to which the constraint is tangent is the largest of these values, the answer to part *b* is the location of a relative maximum.

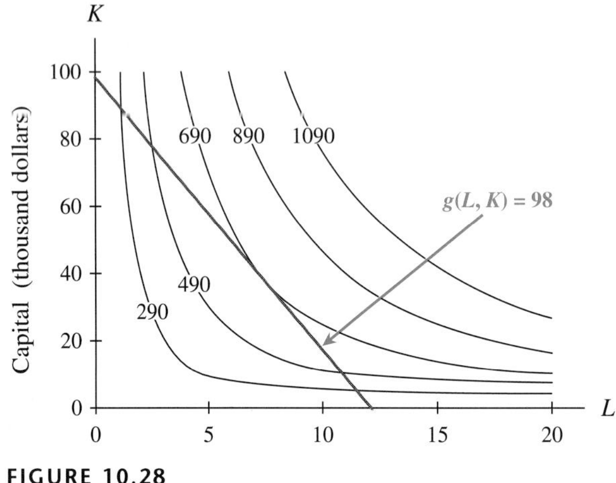

FIGURE 10.28

Interpreting λ

If the level of the constraint function g changes from c to $c + \Delta c$, it will determine a different extreme point because adding Δc to the constraint equation shifts the constraint curve (see Figure 10.29).

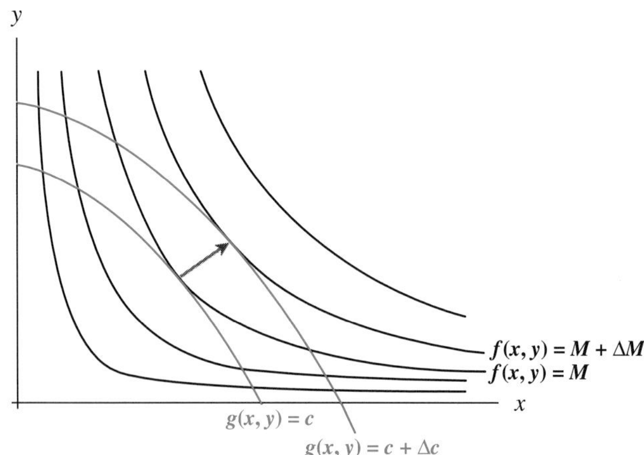

FIGURE 10.29

Note that $g(x, y) = c + \Delta c$ is tangent to the contour curve $f(x, y) = M + \Delta M$, whereas $g(x, y) = c$ is tangent to the curve $f(x, y) = M$. Therefore, we can consider the extreme value M to be a function of the level c of the constraint. In order to analyze the effect that a change in the constraint level c has on the extreme value M, we look at the rate of change of M with respect to c; that is, $\frac{dM}{dc}$. It can be shown using more rigorous mathematics that the Lagrange multiplier λ is this rate of change.

When the level of the constraint changes by a small amount Δc, the extreme value M of $f(x, y)$ changes by

Because M is an output value $f(x, y)$, we could write

$$\Delta f = \Delta c\left(\frac{df}{dc}\right) = \Delta c\lambda.$$

We use this notation in Example 2.

$$\Delta M \approx \Delta c\left(\frac{dM}{dc}\right) = \Delta c(\lambda)$$

When $f(x, y)$ represents the output of a production function, the Lagrange multiplier λ can be used to determine the approximate increase in the optimal production level effected by an additional allocation of resources. Economists often refer to the value of λ as the *marginal productivity of money*.

In the mattress production example, we can use equation 2 on page 719 to find the value of λ when $L = 7.35$ and $K = 39.2$:

$$\lambda = 19.24\,(7.35)^{0.6}\,(39.2)^{-0.6} \approx 7.047$$

Because λ can be interpreted as $\frac{dM}{dc}$, its units are mattresses per thousand budget dollars.

This value of λ can be used to approximate the maximum production level that would occur if the budget were increased. For example, if the manager's budget were \$99 thousand instead of \$98 thousand, the production level would increase by

$$\begin{aligned}\Delta M \approx \frac{dM}{dc}\Delta c &= (\lambda)\Delta c \\ &= \left(7.047\,\frac{\text{mattresses}}{\text{thousand budget dollars}}\right)(1 \text{ thousand budget dollars}) \\ &\approx 7 \text{ mattresses}\end{aligned}$$

That is, the company could produce approximately 697 mattresses, instead of 690 mattresses, if the manager's budget were \$99 thousand instead of \$98 thousand.

Interpretation of λ

If the constraint level c is increased by 1 unit, then the extreme value M of the function changes by approximately λ units.

EXAMPLE 2 *Sausage Production*

Consider the production of sausage, in which the amount of shrinkage due to cooking is reduced by the addition of three milk proteins. The percentage of volume lost in cooking sausage can be modeled[28] by the equation

$$P(c, w, s) = 10.65c + 11.78w + 11.69s - 5.83ws \text{ percent}$$

where c, w, and s are the proportions of the milk proteins sodium caseinate, whey protein, and skim milk powder, respectively.

a. Rewrite the function P in terms of w and s, using the fact that $c + w + s = 1$.

28. M. R. Ellekjaer, T. Naes, and P. Baardseth, "Milk Proteins Affect Yield and Sensory Quality of Cooked Sausages," *Journal of Food Science*, vol. 61, no. 3 (1996), pp. 660–672.

b. To determine the effect that changing k in the constraint $w + s = k$ has on the optimal cooking loss P, find the value of $\frac{dP}{dk}$ for $k = 1$.

c. Use the answer to part b to estimate the effect on the minimum cooking loss if sodium caseinate makes up 10% of the total milk proteins.

d. On the basis of your answer to part c, do you believe that a sausage producer would be wise to use sodium caseinate in addition to whey and skim milk powder? Explain.

Solution

a. Substituting $1 - w - s$ for c in the equation, we obtain

$$P(w, s) = 10.65 + 1.13w + 1.04s - 5.83ws \text{ percent}$$

b. To find $\frac{dP}{dk}$, we must find λ in the system of equations used to minimize the model in part a, subject to the constraint $g(w, s) = 1$. The system is

$$1.13 - 5.83s = \lambda$$
$$1.04 - 5.83w = \lambda$$
$$w + s = 1$$

Solving this system gives the same solution for s and w found earlier: $s \approx 0.508$ and $w \approx 0.492$ with $\lambda = -1.83$. Therefore, $\frac{dP}{dk} = -1.83$ percentage points per unit of k.

c. We can approximate the effect on optimal cooking loss using the formula

$$\Delta P \approx \frac{dP}{dk} \Delta k$$

In this case, k is decreased by 0.1 (from 1 to 0.9), so the change in minimum cooking loss is approximately $(-1.83)(-0.1) = 0.183$ percentage points. In other words, if sodium caseinate is added so that the proportions of whey and skim milk proteins sum to 0.9, minimum cooking loss will increase to approximately $10.28 + 0.183 \approx 10.46\%$.

d. Because $\frac{dP}{dk}$ is negative, any decrease in k (the combined proportion of whey and skim milk proteins) will increase P (the percentage cooking loss). Thus the addition of sodium caseinate will increase the percentage loss in cooking. Because a sausage producer desires minimum cooking loss, it probably is not beneficial for producers to use sodium caseinate. We do not know, however, what other benefits sodium caseinate may confer, such as improved texture, improved taste, or decreased processing expenses. A sausage producer would need to evaluate all contributing factors when making a decision.

Concept Inventory

- *Constrained optimization*
- *Lagrange multiplier*
- *Interpretation of* λ

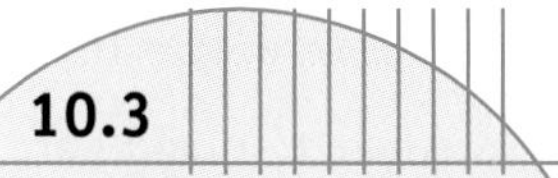

Activities

1. For the multivariable function $f(a, b) = ab$ with constraint $g(a, b) = a + b = 90$:

   a. Use the graph shown in Figure 10.3.1 of several of the contour curves of f and the constraint

Some of your students may be used to graphing lines by plotting the two axis intercepts. Warn them that the point where $G = 40$ and $M = 0$ does not lie on the horizontal axis that is shown in Figure 10.3.3 because the horizontal axis corresponds to $M = 6$. This situation also occurs in Activity 4.

$g(a, b) = 90$ to estimate the optimal point and optimal value of $f(a, b)$ subject to the constraint.

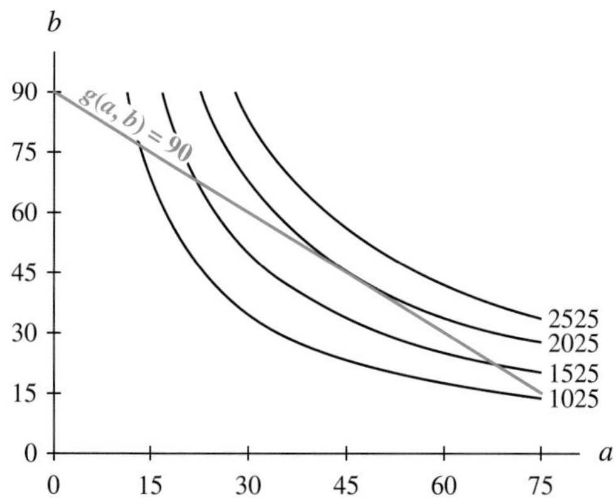

FIGURE 10.3.1

b. Classify the optimal point as a maximum or a minimum of $f(a, b)$.

c. Use the methods of this section to find the optimal point and the optimal value.

2. For the multivariable function $f(k, h) = 2k^2h$ with constraint $g(k, h) = k^2 + h = 2$:

a. Use the graph shown in Figure 10.3.2 of several of the contour curves of f and the constraint $g(k, h) = 2$ to estimate each optimal point and optimal value of $f(k, h)$ subject to the constraint.

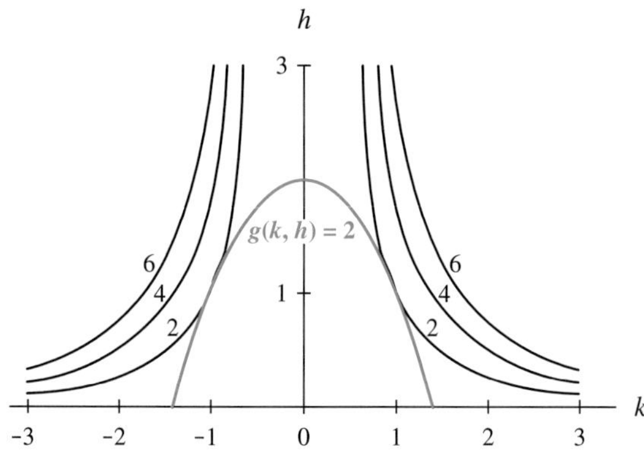

FIGURE 10.3.2

b. Classify each optimal point as a maximum or a minimum of $f(k, h)$.

c. Use the methods of this section to find the optimal points and the optimal values.

3. Figure 10.3.3 shows a contour graph of a measure of the cohesiveness of honey as a function of the percentage of glucose and maltose and the percentage of moisture.

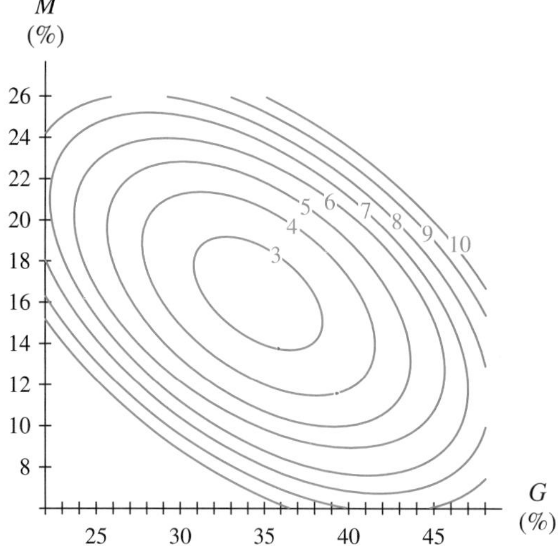

FIGURE 10.3.3

a. Estimate the critical point indicated by the contour graph. Interpret your answer.

b. Graph the constraint $g(G, M) = G + M = 40\%$ on the contour graph. Estimate the optimal point of cohesiveness subject to the constraint. Is the point the location of a maximum or a minimum?

4. Figure 10.3.4 shows a contour graph of a measure of the adhesiveness of honey as a function of G, the percentage of glucose and maltose, and M, the percentage of moisture.

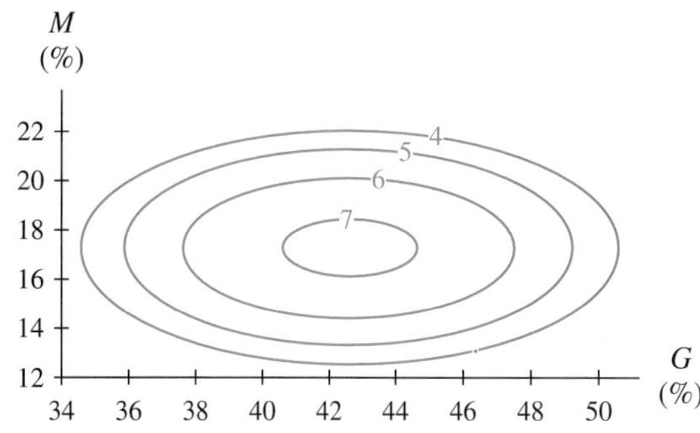

FIGURE 10.3.4

a. Estimate the critical point indicated by the contour graph. Interpret your answer.

b. Graph the constraint $g(G, M) = G + M = 55\%$ on the contour graph. Estimate the optimal point of adhesiveness subject to the constraint. Is the point the location of a maximum or a minimum?

5. Find the optimal point of

$$f(r, p) = 2r^2 + rp - p^2 + p$$

under the constraint

$$g(r, p) = 2r + 3p = 1$$

Identify the output at the point as a maximum or a minimum using:

a. A contour graph.

b. Close points.

6. Find the optimal point of

$$f(x, y) = 80x + 5y^2$$

under the constraint

$$g(x, y) = 2x + 2y = 1.4$$

Identify the output of the point as a maximum or a minimum using:

a. A contour graph.

b. Close points.

7. Use the model given in part *a* of Example 2 and the constraint $w + s = 0.9$ to find the minimum percentage cooking loss if 10% sodium caseinate is used. Compare your answer to the approximation given in part *c* of Example 2.

8. Use the Cobb–Douglas model for mattress production given in Example 1 to find maximum production if the manager's budget were increased to \$99 thousand. Compare your answer to the estimate given above the box on page 721.

9. A health club is trying to determine how to allocate funds for advertising. The manager decides to advertise on the radio and in the newspaper. Previous experience with such advertising leads the club to expect $A(r, n) = 0.1r^2n$ responses when r ads are run on the radio and n ads appear in the newspaper. Each ad on the radio costs \$12, and each newspaper ad costs \$6.

a. Find the optimal allocation for advertising to produce the maximum number of responses if the club has budgeted \$504 for advertising.

b. What is the maximum number of responses expected with the given advertising costs?

c. Suppose the manager budgeted an additional \$26 for advertising. What is the approximate change in the optimal value as a result of this change in the constraint level?

10. A measure of the adhesiveness of honey[29] can be modeled by

$$A(G, M) = -125.48 + 4.26G + 4.85M - 0.05G^2 - 0.14M^2$$

where G is the percentage of glucose and maltose and M is the percentage of moisture. A contour graph of A is shown in Figure 10.3.4 in Activity 4.

a. Find the absolute maximum of the adhesiveness of honey. Compare your answer with the estimate in part *a* of Activity 4.

b. If FDA restrictions for Grade A honey require that the combined percentages of glucose, maltose, and moisture must not exceed 55%, what is the maximum measure of adhesiveness possible?

c. Verify that the point in part *b* corresponds to a maximum.

d. Compare your answer in part *b* to the estimate in part *b* of Activity 4.

11. A measure of the cohesiveness of honey[30] can be modeled by

$$C(G, M) = 106.35 - 3.76G - 4.71M + 0.04G^2 + 0.08M^2 + 0.06GM$$

where G is the percentage of glucose and maltose and M is the percentage of moisture. A contour graph of C is shown in Figure 10.3.3 in Activity 3.

a. Find the absolute minimum cohesiveness of honey. Compare your answer with the estimate in part *a* of Activity 3.

b. If FDA restrictions for Grade A honey require that the combined percentages of glucose, maltose, and moisture must not exceed 40%, what is the minimum measure of cohesiveness possible?

c. Verify that the point in part *b* corresponds to a minimum.

d. Compare your answer in part *b* to the estimate in part *b* of Activity 3.

12. A manufacturer is designing a packaging carton for shipping. The carton will be a box with fixed volume of 8 cubic feet. The cost to construct each box is

$$C(l, w) = lw + \frac{4.8}{w} + \frac{3.2}{l} \text{ dollars}$$

29. J. M. Shinn and S. L. Wang, "Textural Analysis of Crystallized Honey Using Response Surface Methodology," *Canadian Institute of Food Science and Technology Journal*, vol. 23, no. 4/5 (1990), pp. 178–182.

30. Ibid.

where the box is l feet long and w feet wide. If shipping requirements are such that the length and width of the box are constrained by $l + 2w = 5$ feet, what size box will cost the least to produce? What is the minimum cost?

13. The daily output at a plant manufacturing transistor radios is approximated by the function

$$f(L, K) = 10.5463L^{0.3}K^{0.5} \text{ radios}$$

where L is the size of the labor force measured in hundreds of worker hours and K is the capital investment in thousands of dollars.

a. Suppose that the plant manager has a daily budget of \$15,000 to be invested in capital or spent on labor and that the average wage of an employee at the radio plant is \$7.50 per hour. What combination of worker hours and capital expenditures will yield maximum daily production?

b. Verify that the output value you obtain is a maximum.

c. Find and interpret the marginal productivity of money for this manufacturing process.

14. Refer to the model and constraint in Activity 10.

a. If $f(G, M) = k$ is the constraint equation and A is the maximum adhesiveness measure, find $\frac{dA}{dk}$ for $k = 55\%$.

b. Use the answer to part a to estimate the change in the maximum adhesiveness measure if the FDA approves the percentage of glucose, maltose, and moisture to be 56%.

c. Find the maximum measure of adhesiveness if there are no restrictions on the percentage of glucose, maltose, and moisture.

15. Refer to the model and constraint in Activity 11.

a. If $f(G, M) = k$ is the constraint equation and C is the minimum cohesiveness measure, find $\frac{dC}{dk}$ for $k = 40\%$.

b. Use the answer to part a to estimate the change in the minimum cohesiveness measure if the FDA approves the percentage of glucose, maltose, and moisture to be 42%.

c. Find the minimum measure of cohesiveness if there are no restrictions on the percentage of glucose, maltose, and moisture.

16. Refer to Activity 12 to answer the following questions.

a. If M is the minimum cost and $f(l, w) = k$ is the constraint equation, find $\frac{dM}{dk}$ for $k = 5$.

b. Use the answer to part a to estimate the minimum cost if the constraint curve equation is $l + 2w = 5.5$.

17. Refer to Activity 13 to answer the following questions.

a. If P is the maximum daily production and the budget constraint equation is $g(L, K) = c$, find $\frac{dP}{dc}$ for $c = \$15{,}000$.

b. Use the answer to part a to estimate the change in maximum production if the budget amount is increased by \$1500.

c. Estimate the maximum production if the budget amount is decreased to \$14,000.

18. In Example 2 of Section 5.3, a rancher needed to maximize the area of a corral subject to a constraint. Read through that example, and then answer the following questions.

a. What is the multivariable function to be maximized?

b. What is the constraint function?

c. Maximize the equation in part a subject to the constraint in part b, using the method of Lagrange multipliers.

d. Compare your answers with those found in Example 2 of Section 5.3.

e. Describe the solution method in the example and the solution method in part c of this activity. Which method do you think is easier?

19. In Example 3 of Section 5.3, we considered minimizing the surface area of a popcorn tin subject to a constraint. Read through that example, and then answer the following questions.

a. What is the multivariable function to be minimized?

b. What is the constraint function?

c. Minimize the equation in part a subject to the constraint in part b, using the method of Lagrange multipliers.

d. Compare your answers with those found in Example 3 of Section 5.3.

e. Describe the solution method in the example and the solution method in part c of this activity. Which method do you think is easier?

20. In Example 4 of Section 5.3, we maximized revenue subject to a constraint. Read through that example, and then answer the following questions.
    a. Write revenue as a multivariable function of the number of students in excess of 50 and the price per student.
    b. Write a constraint function in terms of the number of students in excess of 50 and the price.
    c. Maximize the equation in part *a* subject to the constraint in part *b*, using the method of Lagrange multipliers.
    d. Compare your answers with those found in Example 4 of Section 5.3.
    e. Describe the solution method in the example and the solution method in part *c* of this activity. Which method do you think is easier?

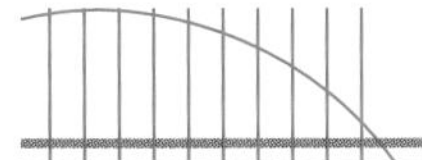

10.4 Least-Squares Optimization

The beauty of Section 10.4 is that we have come full circle. Our understanding of calculus concepts now enables us to comprehend fully the method of least squares that was introduced in Section 1.5.

We have used different models in this text to represent real-world situations, and we put those models to use in our study of the mathematics of change. We called upon technology to obtain the equations of the models so that interesting and current data could be utilized in our study of calculus concepts. We are now in a position to understand a procedure that technology uses to find linear models. This section is devoted to understanding the **least-squares method** of determining the linear model that best fits a set of data points.

In Section 1.5 we intuitively considered the least-squares method when answering the question "What is best fit?" Let us review how we decide when a particular line best fits the data in the context of an application dealing with retirement plans.

The average number of investment choices for people who invest in 401(k) plans at their place of employment is given by the data[31] in Table 10.14.

TABLE 10.14

Year	1990	1991	1992	1993
Average number of 401(k) plans	3.5	4.0	4.2	4.8

Let x represent the number of years since 1990, and let y be the average number of 401(k) plans. A scatter plot of the data and a linear model are shown in Figure 10.30.

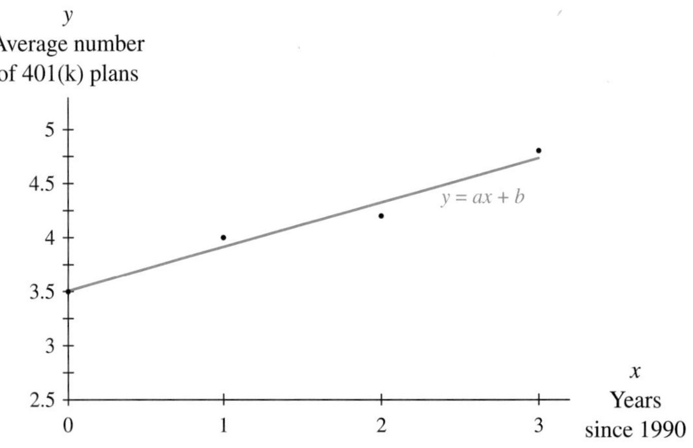

FIGURE 10.30

31. *USA Today*, November 5, 1993. Source: Foster Higgins.

A visual indication of how well the line fits the data is given by the extent to which the data points deviate from the line. We determine a numerical measure of this visual observation by calculating the amount by which the line misses the data. That is, we calculate the vertical distance (the deviation) from each data point to the line. These deviations are the lengths of the vertical segments shown in Figure 10.31.

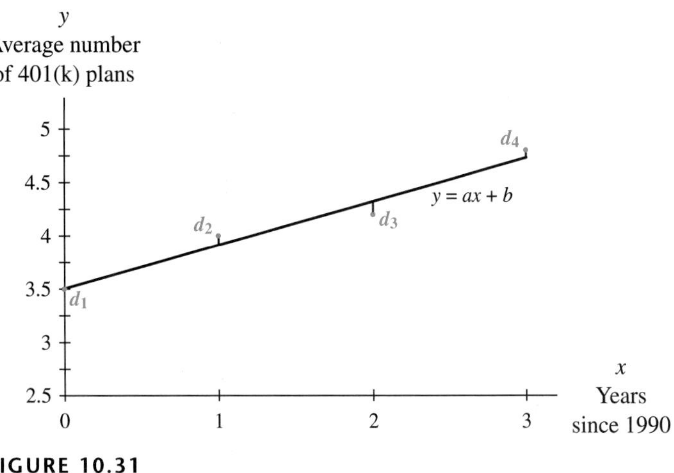

FIGURE 10.31

Each **deviation** measures the **error** between the associated data point and the line.

$$\text{Deviation} = \text{error} = y_{\text{data}} - y_{\text{line}}$$

To obtain a single numerical measure, we square the errors and sum. When there are n data points, we have

$$\begin{aligned} \text{SSE} &= \text{the sum of squared errors} \\ &= (\text{first error})^2 + (\text{second error})^2 + \cdots + (\text{last error})^2 \\ &= d_1^2 + d_2^2 + \cdots + d_n^2 \end{aligned}$$

This number, denoted by *SSE* (Sum of Squared Errors), is our overall measure of how well the line fits the data. The key idea to its correct use is that smaller values of SSE arise from lines where the errors are small, and larger values of SSE arise from lines where the errors are large. The strategy for choosing the best-fitting line is to choose the line for which the sum of squared errors (SSE) is as small as possible. Such a line is designated as the *line of best fit* (also called the *regression line*), and this procedure for choosing it is called the **method of least squares**.

The Line of Best Fit

The method of least squares is the procedure that is used to find the best-fitting line based on the criterion that the sum of squared errors is as small as possible—that is, at a minimum value. The linear model obtained by this method is called the **least-squares line**.

Thus our task is to find the values of a and b in the model $y = ax + b$ such that $\text{SSE} = d_2^2 + d_2^2 + \cdots + d_n^2$ is a minimum. Because each deviation is the

y-value of the data minus the y-value on the line, we proceed as follows using the aligned data:

$$\begin{aligned} d_1^2 + d_2^2 + d_3^2 + d_4^2 &= [3.5 - (a \cdot 0 + b)]^2 + [4 - (a \cdot 1 + b)]^2 + \\ &\quad [4.2 - (a \cdot 2 + b)]^2 + [4.8 - (a \cdot 3 + b)]^2 \\ &= (3.5 - b)^2 + (4 - a - b)^2 + (4.2 - 2a - b)^2 + \\ &\quad (4.8 - 3a - b)^2 \end{aligned}$$

This sum of squares of errors is a function of the slope, a, and the y-intercept, b, of the least-squares line. Thus it is a multivariable function with two input variables:

$$f(a, b) = (3.5 - b)^2 + (4 - a - b)^2 + (4.2 - 2a - b)^2 + (4.8 - 3a - b)^2$$

Recall that to find the minimum value of $f(a, b)$, we first find $\frac{\partial f}{\partial a}$ and $\frac{\partial f}{\partial b}$:

$$\frac{\partial f}{\partial a} = 0 + 2(4 - a - b)(-1) + 2(4.2 - 2a - b)(-2) + 2(4.8 - 3a - b)(-3)$$

$$\frac{\partial f}{\partial b} = 2(3.5 - b)(-1) + 2(4 - a - b)(-1) + 2(4.2 - 2a - b)(-1) + 2(4.8 - 3a - b)(-1)$$

Next, we set $\frac{\partial f}{\partial a} = 0$ and $\frac{\partial f}{\partial b} = 0$. Some algebraic simplification results in the equations

$$\begin{aligned} 28a + 12b &= 53.6 \\ 12a + 8b &= 33 \end{aligned}$$

An algebraic or matrix solution to this system of equations is (0.41, 3.51).

Finally, we must verify that the point (0.41, 3.51) is a minimum for the function f by considering the determinant of the second partials matrix:

$$D = \begin{vmatrix} 28 & 12 \\ 12 & 8 \end{vmatrix} = 224 - 144 = 80$$

Because $D(0.41, 3.51) > 0$ and $f_{aa} > 0$, we conclude that the sum of squared errors, SSE, is at a minimum for the linear model with slope $a = 0.41$ and y-intercept $b = 3.51$. The minimum value of SSE is $f(0.41, 3.51) = 0.027$.

The best-fitting line determined by the least-squares method for the investment data is $y = 0.41x + 3.51$ investment choices of 401(k) plans x years after 1990. You should verify that this is the equation found by using the linear regression routine in your calculator or computer.

The least-squares method is illustrated in this section with small data sets to simplify calculations. In real-world situations, you should be cautious when using only a few data points because a small error in one of the points may lead to a large error in the least-squares line.

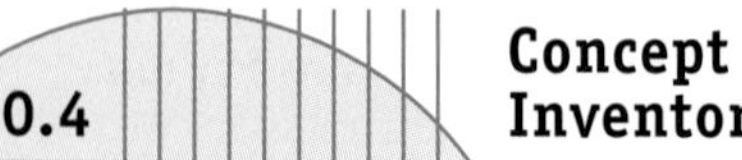

Concept Inventory

- *Least-squares method*
- *Sum of squared errors*
- *Line of best fit*

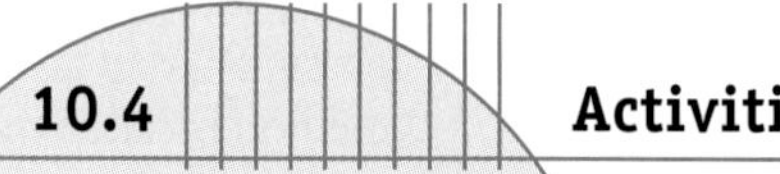

Activities

For the data given in Activities 1 and 2,

a. Express SSE as a multivariable function of a and b for the best-fitting line $y = ax + b$.

b. Find $\frac{\partial f}{\partial a}, \frac{\partial f}{\partial b}$, and the determinant of the second partials matrix.

c. Find the minimum value of $f(a, b)$, and verify that the value you obtain is a minimum.

d. Use the results of part *c* to give the linear function that best fits the data.

1.

x	1	6	12
y	7	11	19

2.

x	2	3	6	8
y	5	7	11	15

3. Homes that lack complete kitchen facilities have been rare in the United States for many years. The first census for which data were tabulated on this subject was in 1970. The percentage of housing units lacking complete kitchen facilities in the western United States[32] is shown in Table 10.15.

TABLE 10.15

Year	1970	1980	1990
% lacking	3	2	1

a. Use the method of least squares to find the multivariable function f with inputs a and b for the best-fitting line $y = ax + b$, where x is years since 1970.

b. Find the minimum value of $f(a, b)$, and verify that it is a minimum value. Interpret the meaning of this minimum value in terms of the data and the best-fitting line.

c. Give the linear model that best fits these data.

d. Use the best-fitting line to predict the percentage of housing units lacking complete kitchen facilities in the western United States this year. Is this a reasonable prediction? Why or why not?

4. Are young adults staying single longer? The percent of adults 20–24 years of age who have not yet been married[33] is given in Table 10.16.

a. Use the method of least squares to find the multivariable function f with inputs a and b for the best-fitting line $y = ax + b$, where x is the number of years since 1970.

TABLE 10.16

Year	1970	1980	1990
Percent never married	46	60	72

b. Find the minimum value of $f(a, b)$, and verify that it is a minimum value.

c. Give the linear model that best fits these data.

d. Use your model to estimate the percentage of adults 20–24 years old who had never been married in 1997.

e. The percentage of adults 20–24 years old who had never been married in 1960 was 41. Add this data point, and use technology to find a linear model. Use this model to estimate the percentage of adults 20–24 years old who had never been married in 1997.

f. Use the results of parts *d* and *e* to discuss what effect additional data may have when one is predicting by using models.

5. A factory makes 7-mm aluminum ball bearings. Company planners have determined how much it costs them to make the following numbers of cases of ball bearings in a single run. These costs are shown in Table 10.17.

TABLE 10.17

Cases of ball bearings	Cost (dollars)
1	3.10
2	4.25
6	8.95
9	12.29
14	18.45

a. Construct a scatter plot of these data.

b. Use your calculator or computer to find the best-fitting linear model, and sketch the model on the scatter plot. Interpret the slope and vertical-axis intercept in the context of this situation.

c. Find the deviation of each data point from the line, and find the sum of the squares of the deviations.

32. *Census Questionnaire Content*, 1990 CQC-25, Bureau of the Census, December 1993.

33. *Census Questionnaire Content*, 1990 CQC-6, Bureau of the Census, April 1993.

d. Without performing any computations, discuss the process you would use to find the multivariable function f with inputs a and b for the best-fitting line $y = ax + b$. How is this line related to the results of parts b and c?

6. Table 10.18 gives the number of inches of precipitation[34] that fell in New York City in early 1855.

TABLE 10.18

Month	Jan	Feb	March
Precipitation (inches)	5.50	3.25	1.35

a. Use the method of least squares to find the multivariable function f with inputs a and b for the best-fitting line $y = ax + b$, where x is 1 in January, 2 in February, and 3 in March.

b. Find the minimum value of $f(a, b)$, and verify that it is a minimum value. Interpret the meaning of this minimum value in terms of the data and the best-fitting line.

c. Give the linear model that best fits these data.

7. The number of animal experiments in England[35] declined between 1970 and 1980. The numbers for selected years are shown in Table 10.19. Use the method of least squares to find the best-fitting linear model for the data.

TABLE 10.19

Year	1970	1975	1978	1980
Experiments (millions)	5.5	5	4.8	4.6

8. The number of days that milk will keep as a function of the temperature[36] is shown in Table 10.20. Use the method of least squares to find the best-fitting linear model for the data.

TABLE 10.20

Temperature (°F)	30	38	45	50
Days	24	10	5	0.5

9. Before the technology was available to fit many kinds of models to data, researchers and others were restricted to using linear models. Because exponential data are common in many fields of study, it has always been important to be able to fit an exponential model to data. Consider Table 10.21, which shows past and predicted world population.[37]

TABLE 10.21

Year	Population (billions)
1850	1.1
1930	2.0
1975	4.0
2013	8.0

a. Construct a scatter plot of the data. Comment on the curvature.

b. Change the data so that they represent the year and the natural log of the population. Construct a scatter plot of the new data, and comment on the curvature.

c. Use the technique discussed in this section to find the best-fitting linear model for the changed data in part b. You should keep the data in the form $\ln y$ and should not round the values until the final calculation. Label input and output units on your model.

d. If a and b are the parameters of the linear function $y = ax + b$ found in part c, graph the function $y = e^b(e^a)^x$ on the scatter plot of the original data. Label the input and output units on the exponential model.

e. Use technology and an exponential regression routine to find the best exponential model for the population data. Compare it with the model in part d, and reconcile any differences.

10. Repeat Activity 9 for the data shown in Table 10.22, which gives the numbers of infants born to each of five generations in a certain family.

TABLE 10.22

Generation	1	2	3	4	5
Infants	11	7	5	3	2

34. H. Helm Clayton, ed., *World Weather Records* (Washington, DC: The Smithsonian Institution, 1927).
35. A. Rowan, *Of Mice, Models, and Men* (Albany: State University of New York Press, 1984).
36. Data taken from a milk carton produced by Model Dairy.
37. *Information Please Almanac*, 1994.

Summary

Multivariable Optimization

In this chapter, we consider optimization techniques for multivariable functions. High points and low points on the graph of a multivariable function are called extreme points, and points that appear to be a maximum in one direction and a minimum in another directon are called saddle points. Such points can be estimated from tables of data or contour plots of a multivariable function.

On a contour graph of a continuous function, relative maxima and minima lie in the center of concentric simple closed contours. Saddle points are located where contours decrease in one direction and increase in another direction as you move away from the saddle point.

Critical points are located on a table of data by first sketching several contours on the table and then using methods for locating relative maxima and minima and saddle points on contour graphs. We identify a value in a table or on a graph as an absolute extreme value by comparing relative extrema with values on terminal edges and by considering the context of the problem and the behavior of the function beyond the nonterminal edges.

While the location of critical points in tables and on contour graphs often depends on how detailed the table or contour graph is surrounding such points, extreme points and saddle points of a smooth, continuous function f with inputs x and y are located by solving the system of equations $\frac{\partial f}{\partial x} = 0$ and $\frac{\partial f}{\partial y} = 0$. We then classify any point found as a solution as a relative maximum, a relative minimum, or a saddle point by looking at a three-dimensional graph or contour plot or by using the Determinant Test.

Optimization Under Constraints

Many practical problems involve finding the maximum or minimum of a multivariable function when the input variables are restricted or constrained in some manner. The constraint equation may not always be linear; however, at the point where a multivariable function f has an extreme point subject to the constraint $g(x, y) = c$, the constant contour curve $f(x, y) = M$ and the constraint $g(x, y) = c$ have the same tangent line. The slope of this tangent line is

$$\frac{dy}{dx} = \frac{-\left(\frac{\partial f}{\partial x}\right)}{\left(\frac{\partial f}{\partial y}\right)} = \frac{-f_x}{f_y}$$

We solve the following system of equations to find extreme points, when they exist, for a multivariable function f with two inputs, x and y, under the constraint $g(x, y) = c$.

$$\frac{\partial f}{\partial x} = \lambda\frac{\partial g}{\partial x}$$

$$\frac{\partial f}{\partial y} = \lambda\frac{\partial g}{\partial y}$$

$$g(x, y) = c$$

We classify solutions to this system of equations as maxima or minima by viewing a contour graph or three-dimensional graph or checking the value of the function at nearby points on the constraint curve.

The number λ is called a Lagrange multiplier and, when M is the extreme value and c is the level of the constraint, λ equals $\frac{dM}{dc}$, the rate of change of the extreme value with respect to a change in the constraint. When the constraint level changes by a small amount, Δc, the extreme value of $f(x, y)$ changes by

$$\Delta M \approx \frac{dM}{dc}\Delta c = (\lambda)\Delta c$$

Economists often refer to the value of λ as the marginal productivity of money.

Least-Squares Optimization

Mathematical models developed for data have been used throughout this text to describe real-world events. We end the chapter with a calculus explanation for the intuitive idea of the line of best fit that was presented in Chapter 1.

We define the line of best fit (also called the regression line) to be the one for which the sum of squared errors, SSE, has a minimum value. SSE is a multivariable function with two input variables: the slope, a, and the y-intercept, b, of the line of best fit, $y = ax + b$. The optimization technique that finds the minimum SSE is called the method of least squares and is an application of the method of unconstrained optimization.

Concept Check

Can you	To practice, try	
• Identify critical points and absolute extrema on three-dimensional and contour graphs?	Section 10.1	Activities 5, 21
• Estimate critical points and absolute extrema in tables?	Section 10.1	Activities 7, 17
• Find critical points using equations?	Section 10.2	Activities 1, 5
• Use the Determinant Test?	Section 10.2	Activity 19
• Use the method of Lagrange multipliers?	Section 10.3	Activities 1, 11
• Interpret the Lagrange multiplier λ?	Section 10.3	Activities 13, 17
• Use multivariable optimization to find a line of best fit for a set of data?	Section 10.4	Activities 1, 3

Review Test

1. Figure 10.32 shows a contour graph of ozone levels[38] (in thousandths of a centimeter) as a function of the month (beginning and ending in June) and the degrees of latitude from the equator.
   a. On Figure 10.32, label relative maxima with an H, relative minima with an L, and saddle points with an S.
   b. Approximate the highest and lowest ozone concentrations. Tell when and where they occur.
2. Banana chips are a deep-fried snack food, popular in many countries. Research[39] has shown that blanching the bananas before peeling increases the crispness of the chips. Table 10.23 shows observed measures of crispness for selected blanching times and temperatures. (The larger the measure, the crisper the banana chip.)

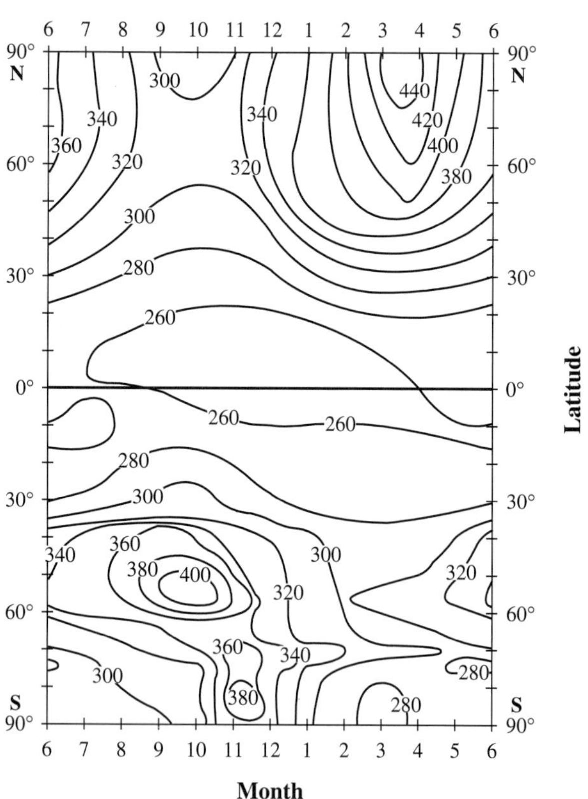

FIGURE 10.32

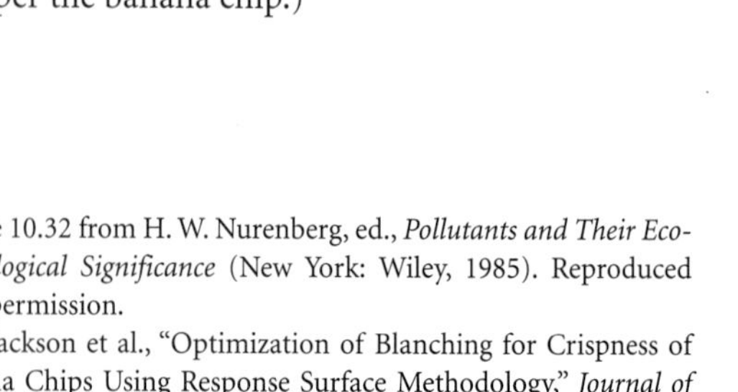

38. Figure 10.32 from H. W. Nurenberg, ed., *Pollutants and Their Ecotoxicological Significance* (New York: Wiley, 1985). Reproduced with permission.
39. J. C. Jackson et al., "Optimization of Blanching for Crispness of Banana Chips Using Response Surface Methodology," *Journal of Food Science*, vol. 61, no. 1 (1996), pp 165–166.

TABLE 10.23 Measures of crispness of banana chips

Blanching temperature (°C)	Blanching time (minutes)		
	2	15	30
50	4.2	6.8	4.2
60	4.5	4.7	7.1
70	8.2	11.4	8.6
80	7.3	6.9	3.9

The measure of crispness can be modeled by

$$C(x, y) = -5.2x^2 + 299.7x - 0.1y^2 + 23.1y - 0.4xy - 4671$$

where x is the blanching time in minutes and y is the blanching temperature in degrees Celsius.

a. Sketch contour curves on the table for crispness measures of 6, 8, and 10.

b. Using only table values, estimate the point of maximum crispness.

c. Use the equation to find the time and temperature that produce the maximum crispness. What is the crispness measure at that point?

d. Verify that the point found in part *c* is a relative maximum. Is it an absolute maximum?

3. A company sells Superbowl T-shirts and hats each year. The company has modeled data from previous years and has obtained a model for profit as a function of s, the number of Superbowl T-shirts sold, and h, the number of Superbowl hats sold. Both s and h are measured in thousands. The graph in Figure 10.33 shows contour curves for values of profit (in dollars).

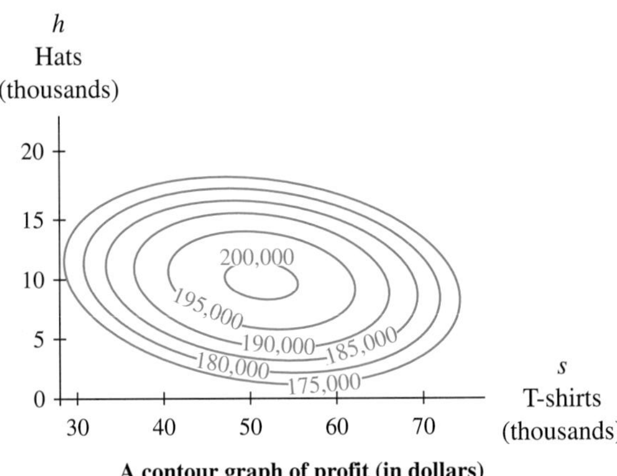

A contour graph of profit (in dollars)

FIGURE 10.33

a. If there are no restrictions on sales, estimate the maximum profit possible and the corresponding numbers of shirts and hats sold.

b. T-shirts cost the company \$4 each, and hats cost \$1.25 each. The budget for the cost of producing next year's Superbowl shirts and hats is \$150 thousand. Write this constraint as a mathematical equation.

c. Graph the constraint function on the contour graph in Figure 10.33. Estimate the maximum profit subject to the constraint, and give the corresponding shirt and hat sales.

4. The model for the profit function in Question 3 is

$$P(s, h) = -52.2s^2 + 5935.5s - 59.1sh + 10{,}299.3h - 384.3h^2 \text{ dollars}$$

when s thousand T-shirts and h thousand hats are sold.

a. Set up the system of equations used to find the optimal profit if the constraint in part *b* of Question 3 applies.

b. Solve the system of equations in part *a*. Find the profit corresponding to the solution.

c. Find and interpret the value of λ.

d. Approximate the change in the maximum profit if the cost budget in part *b* of Question 3 were increased to \$152 thousand.

5. The data in Table 10.24 give the average body weight[40] of an 8-week-old pig who is kept in an environment with a fixed temperature.

TABLE 10.24

Temperature (°C)	10	15	20
Body weight (kg)	29.9	33.4	37.5

a. Use the method of least squares to find the multivariable function f with inputs a and b for the best-fitting line $y = ax + b$, where x is the temperature.

b. Find and interpret the minimum value of $f(a, b)$, and verify that it is a minimum value.

c. Give the linear model that best fits these data.

40. A. A. Hanson, ed., *Practical Handbook of Agricultural Science* (Boca Raton, FL: CRC Press, 1990).

Project 10.1 Snow Cover

Setting

Often contour maps of a geographical region are not topographical maps showing land elevation. Contour maps could indicate other features such as depth of the water table, richness of mineral deposits, population density, precipitation levels, or atmospheric conditions. You are probably most familiar with weather maps where contours called *isotherms* indicate temperature or contours called *isobars* indicate air pressure. The contour graph[41] in Figure 10.34 shows the probability of snow cover in January in the northern hemisphere.

Tasks

1. Carefully examine the contour graph in Figure 10.34. You may wish to make several copies of the graph for use in this project. An enlarged copy may also be helpful. Begin by marking on a copy of the graph all the relative high and low percentages indicated in the graph. Use Hs and Ls to distinguish between high and low percentages.
2. Find a map or atlas of the world that indicates special topographical features. Choose at least five of the relative extreme points in Figure 10.34, and determine the precise topographical characteristic of the planet at each location that results in the corresponding relative extreme point.
3. What would you expect the snow cover probability on Mt. Everest to be in January? Locate Mt. Everest on the contour map. Does the graph indicate the probability you expected? If not, why not?

Reporting

Prepare a poster to present the contour map in such a way as to highlight the relative extrema and the corresponding topography at each point. Give some information about each point (such as latitude, longitude, elevation, and topography). Include on the poster the location of and information about Mt. Everest. Make the poster attractive and readable from a distance of 3 feet.

41. Figure 10.34 is taken from W. Rudloff, *World-Climates* (Stuttgart: Wissenschaftliche Verlagsgesellschaft, 1981), p. 100. Reproduced by permission of the publisher.

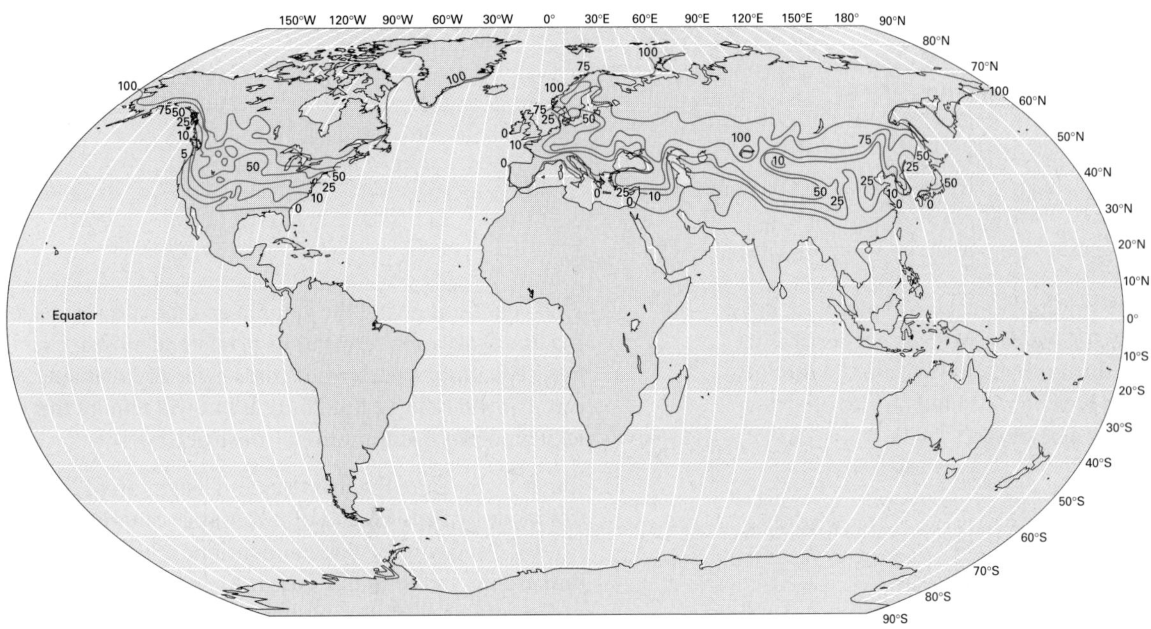

——— Snow cover probability in January

FIGURE 10.34

Project 10.2 Carbonated Beverage Packaging

Setting

A new firm that wants to compete with Pepsi and Coca-Cola has enlisted your services to design the optimal can shape for its product. The can is to hold 12 fluid ounces of a carbonated beverage.

Tasks

1. Find equations giving the volume and the surface area of a cylindrical can in terms of its height and radius. Use these equations to determine the height and diameter of a can that will hold 12 fluid ounces but will require the least amount of aluminum to construct.
2. Carefully measure the diameter and height of a 12-ounce can made by PepsiCo and of a 12-ounce can made by Coca-Cola. Compare these dimensions to the optimal dimensions you found in Task 1. Are there any beverage cans on the market that conform to the optimal dimensions?
3. Repeat Tasks 1 and 2, considering that it will cost twice as much to form the ends of the can as it does to form the cylindrical surface.

Reporting

1. Write a letter to the company directors explaining what dimensions you recommend that they use for their cans. Include your findings on the optimal dimensions as well as the dimensions of cans produced by competitors. Attach your mathematical work as an appendix to support your conclusions.
2. Prepare an oral presentation to be given in front of the company directors. You should prepare visual aids in order to enhance your presentation. Your goal is to persuade the directors to construct cans according to the dimensions you suggest. Keep in mind that the company directors are probably not as concerned with all of the mathematical details as they are with your analysis of your findings. However, be prepared to answer mathematical questions in case any of the directors asks you to elaborate.

Answers to Odd Activities

CHAPTER 1

Section 1.1

1. Input: weight of a letter — Output: first-class postal rate
   Input variable: w — Output variable: $R(w)$
   Input units: ounces — Output units: cents
   R is a function of w.
3. Input: day of the year — Output: number of students with birthday on day x
   Input variable: x — Output variable: $B(x)$
   Input units: days — Output units: students
   B is a function of x.
5. Input: day of the week — Output: amount spent on lunch
   Input variable: m — Output variable: $A(m)$
   Input units: none — Output units: dollars
   A is not a function of m unless you always spend the same amount on lunch every Monday, Tuesday, and so on, or unless the input is the days in only 1 week.
7. a, b. Input: years since investment — Output: amount in account
   Input variable: t — Output variable: $B(t)$, $A(t)$
   Input units: years — Output units: dollars
   A and B are both functions of t.
9. Input: name of movie starlet — Output: hair color in 2002
   Input variable: x — Output variable: $P(x)$
   Input units: none — Output units: none
   P is not a function of x.
11. This is a function.
13. This is not a function.
15. $P(\text{Honolulu}) = 307$
    $P(\text{Springfield, MA}) = 107$
    $P(\text{Portland, OR}) = 152$
17. Graphs b and c are functions.
19. a. \$60
    b. 3 CDs
    c. 7 CDs
    d. The average price of 3 CDs is \$12 each.
    The average price of 6 CDs is \$10 each.
21. a. Approximately \$9000
    b. Approximately \$340
    c. Approximately \$105
    d. The graph would pass through (0,0) but would lie below the graph in Figure 1.1.8 because the same monthly payment would pay for a smaller loan amount.
23. a. 3.0%
    b. The cost-of-living increase was greatest in 1990 at 5.4%.
    c. 1994
    d. Benefits increased, but the percentage by which they increased decreased.
25.

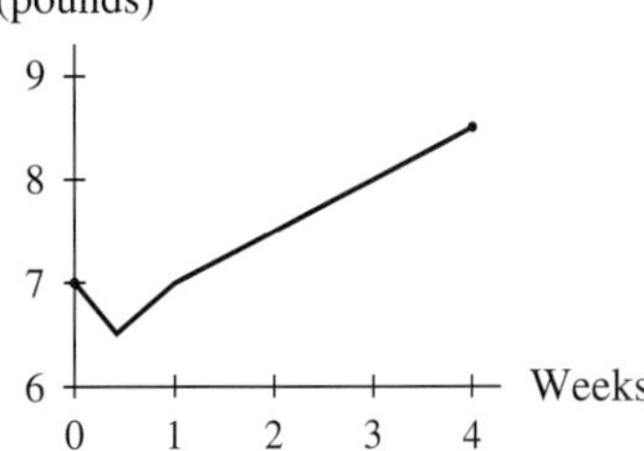

27. a. Inches
    b. About 16 inches long
    c. About 11 inches
    d. At about 16 years of age, she reached her full height of approximately 60 inches (5 feet).
    e. During the first 3 years
    f. Height should not increase or decrease after age 20 (until in her later years, when her height may decline).
29. a. About 5 inches
    b. For approximately 4 days
    c. Snow fell.
    d. The snow settled.
    e. The snow was deepest (4 feet 4 inches) around February 21.
    f. Warm temperatures probably caused the decline in late February.
    g. The most snow fell around February 18.
31. a.

x	0	1	2	3	4	5
y	500	400	320	256	204.8	163.84

    b. $y = 500(0.8^x)$ milligrams after x days
    c. x is between 0 days and number of days after which the amount of the drug is negligible (answers vary).
    y is between 0 and 500 milligrams.

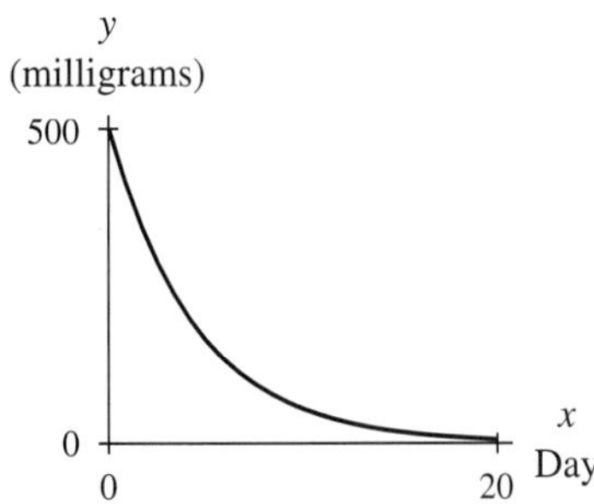

    d. There is no x-intercept.
    The y-intercept is 500 milligrams.
    The y-intercept is the amount of the drug initially in the patient's body (500 milligrams).
    e. y is always decreasing (for $x \geq 0$).
    f. After 3.5 days, approximately 230 milligrams of the drug remain. The exact value is $500(0.8^{3.5}) \approx 228.97$ mg.

g. There will be 60 milligrams of the drug after about 9.5 days.

33. $R(3) \approx 314.255$
$R(0) = 39.4$

35. $Q(12) = 600.36$
$Q(3) = 222.54$

37. $R(w) = 78.8$ when $w \approx 1.001$
$R(w) = 394$ when $w \approx 3.327$

39. $Q(x) = 515$ when $x = 10$
$Q(x) = 33.045$ when $x \approx 0.500$

41. Input is given.
$A(15) = 5{,}131{,}487.257$

43. Output is given.
When $y = 2.97$, $x \approx$ -3.203

Section 1.2

1. a. Continuous
b. Discrete
c. Continuous
d. Discrete

3. Discrete

5. Continuous with discrete interpretation

7. Continuous without restriction

9. Continuous without restriction

11. a.

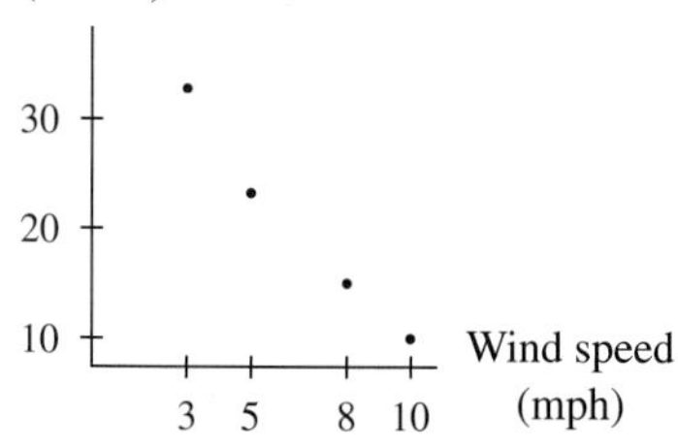

b. Because no vertical line passes through two points, the data represent a function.
c. Because the table does not contain an input of 6, we cannot read this information from the table.
d. Tables of data are always discrete representations.

13. a. The profit resulting from 25 calls in one day is \$179.
b. Solving $P(C) = 180$ for c gives $c \approx 13.46$. Because the number of calls must be a positive integer, this input value does not have a meaningful interpretation in this context. The daily profit is never exactly \$180. When 13 calls are made, the profit is slightly under \$180, and when 14 calls are made, the profit is slightly over \$180.
c.

$P(c)$
Profit
(dollars)
250
0
-250
c
30 Calls

According to the graph, the maximum occurs when $c \approx 18.2$. Checking the integer input values on either side of 18.2, we find that $P(18) \approx$ \$201.85 and $P(19) \approx$ \$201.37. The maximum profit of \$201.85 occurs when 18 calls are made.

15. a. Input: the number of years since 1900
Output: the number of osteopathy students in thousands in the spring of the input year
b. Input units: years
Output units: thousands of students
c. The model must be discretely interpreted because it gives values at one particular time each year.
d. Approximately 8.8 thousand students
e. The model cannot be used to answer this question.
f. Solving the equation $10 = O(t)$ gives two solutions, $t \approx 75.4$ and $t \approx 99.9$. We can assume that the question is referring to a time beyond the interval of input values on which the model is based and thus can discard the first solution. Because the model is valid for only positive integer inputs, we check the output values for $t = 99$ and $t = 100$: $O(99) \approx 9.39$ and $O(100) \approx 10.04$. Thus the model exceeds 10,000 in the year 2000. However, because this year is 3 years beyond the last year for which the model applies, this prediction may not be valid.

17. a. Input: the number of hours after the market opened
Output: the price of Amazon.com stock
b. Input units: hours
Output units: dollars per share
c. The model can be interpreted without restriction because the price changes in an approximately continuous manner throughout a trading day.
d. $P(3) \approx$ \$118.55 per share
e. The model reaches an output of 100 for $h \approx 11$. The trading day is less than 11 hours long, so we conclude that the price was never \$100 per share during this trading day.

19. a. Input: the number of years since 1900
Output: annual per capita consumption of milk

b. Input units: years
Output units: gallons per person per year

c. It is reasonable to assume that the per capita consumption of milk is constantly changing so that the model can be used without restriction.

d. Solving the equation $25 = C(x)$ gives $x \approx 92.8$. If we assume that $x = 92$ corresponds to the end of 1992, then we conclude that per capita milk consumption was 25 gallons per person per year in late 1993.

Section 1.3

1. a. $T(x) = 9088.859 + 1697.717 \ln x + 2424.764 + 915.025 \ln x = 11{,}513.623 + 2612.742 \ln x$ kidney and liver transplants, where x is the number of years since 1990

b. $T(5) \approx 15{,}719$ transplants

3. $N(t) = M(t) - W(t) = 0.3334t - 15.6333$ gallons of milk other than whole milk per person per year, where t is the number of years since 1900

5. $n(t) = h(t) - p(t) = \dfrac{100}{1 + 128.0427e^{-0.7211264t}} - \dfrac{100}{1 + 913.7241e^{-0.607482t}}$ percent t years after 1924

7. $c(x) = n(x)p(x) = (-0.034x^3 + 1.331x^2 + 9.913x + 164.447)(-0.183x^2 + 2.891x + 20.215)$ cesarean-section deliveries x years after 1980

9. $P(C(t))$ = profit after t hours of production

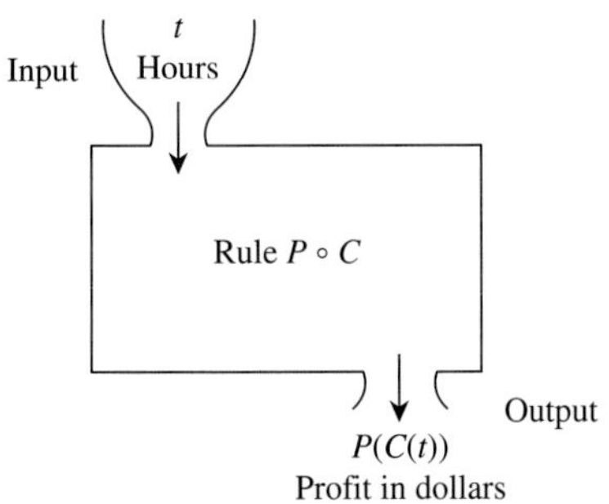

11. $P(C(t))$ = average amount in tips t hours after 4 p.m.

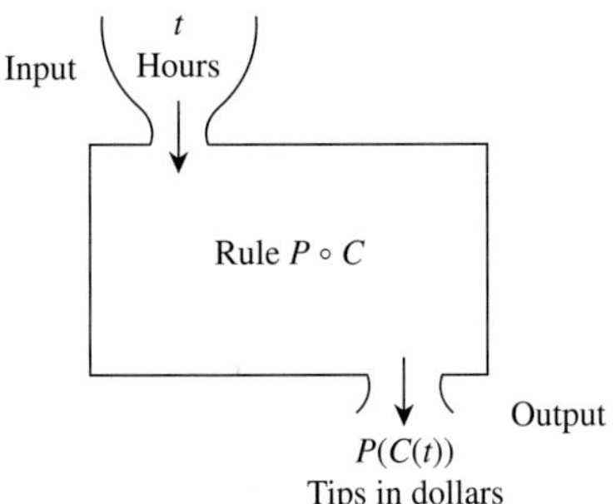

13. $f(t(p)) = 3e^{4p^2}$

15. $g(x(w)) = \sqrt{7(4e^w)^2 + 5(4e^w) - 2}$

17. $g(t(m)) = 3$

19.

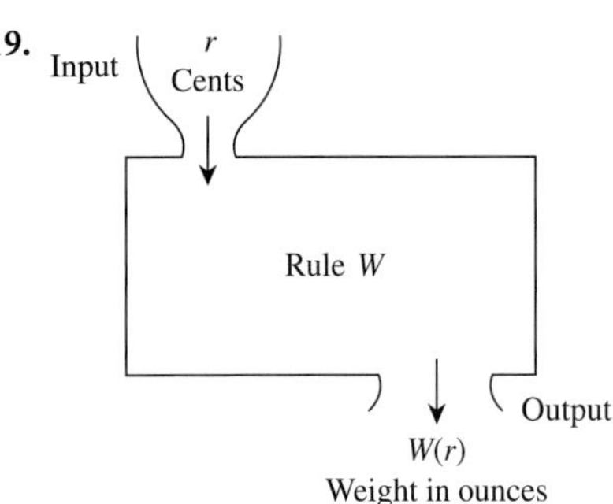

$W(r)$ is the weight in ounces of a first-class letter or parcel that costs r cents to mail.
W is not an inverse function of r.

21.

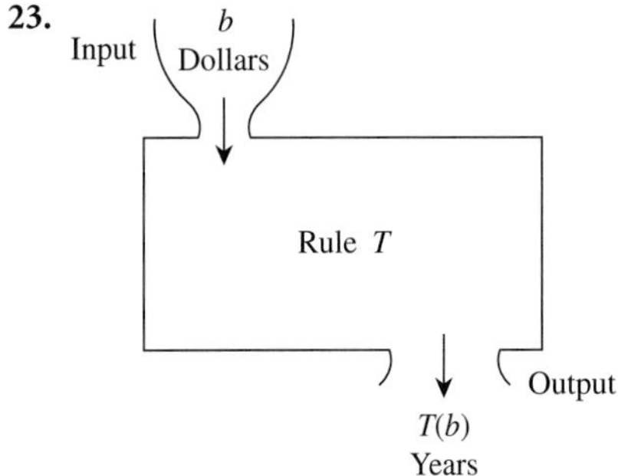

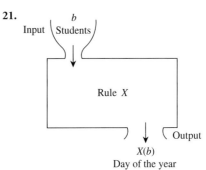

$X(b)$ is the day of the year corresponding to the birthday of b students.
X is not an inverse function of b.

23.

$T(b)$ is the time in years when the value of the investment was b dollars.
T is not an inverse function of b unless interest is compounded and credited continuously.

25. Reversing the inputs and outputs does not result in an inverse function.

27. Reversing the inputs and outputs does not result in an inverse function because the original table does not represent a function.

29. a.

Credit hours	Tuition	Credit hours	Tuition
1	\$138	10	\$1380
2	\$276	11	\$1518
3	\$414	12	\$1735
4	\$552	13	\$1735
5	\$690	14	\$1735
6	\$828	15	\$1735
7	\$966	16	\$1735
8	\$1104	17	\$1735
9	\$1242	18	\$1735

b. $T(c) = \begin{cases} 138c \text{ dollars} & \text{when } 0 \le c \le 11 \\ 1735 \text{ dollars} & \text{when } c \ge 12 \end{cases}$
where c is the number of credit hours

c.

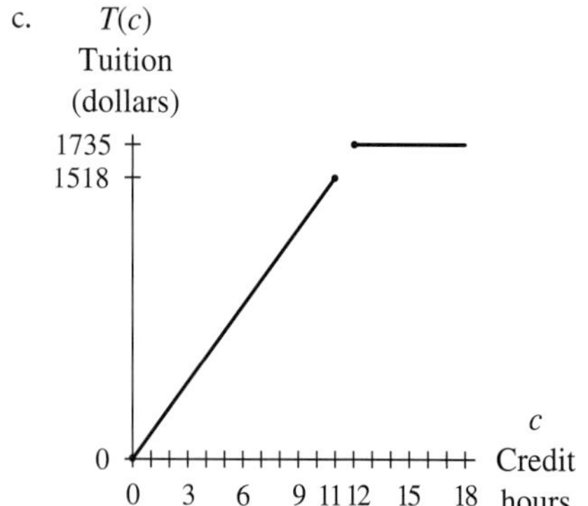

d. T is a function of c.

e. Inputs: integers 0 through 18
Outputs: {0, 138, 276, 414, 552, 690, 828, 966, 1104, 1242, 1380, 1518, 1735}

f.

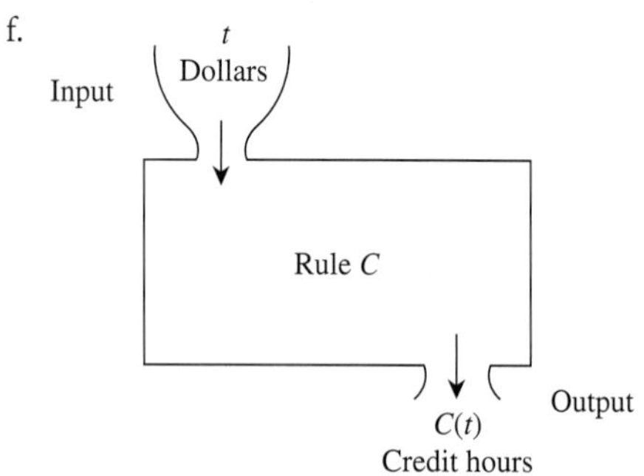

g. The rule in part *f* is not an inverse function, because an input of \$1735 has seven different outputs.

31. $V(s) = \ln s$

33. $T(g) = \dfrac{g^2 - 0.04}{4.9}$, where $g \ge 0$

35. $S(c) = \dfrac{\ln\left(\dfrac{c}{1034.926}\right)}{\ln 0.0062}$

37. a. C is continuous with discrete interpretation because it makes sense only at the end of each month—that is, for positive integer input values.

b. $m(C) = \dfrac{C - 801}{319.6}$, which is a number corresponding to a month of the year when C complaints were received.

c. $m(2399) = 5$
In May of 2001, 2399 complaints were received.

d. $C(1) + C(2) + C(3) = 4321$ complaints in the first quarter of the year

e. $C(m(C)) = 319.6\left(\dfrac{C - 801}{319.6}\right) + 801 = C - 801 + 801 = C$

$$m(C(m)) = \frac{(319.6m + 801) - 801}{319.6} = \frac{319.6m}{319.6} = m$$

39. a. $r(L) = \dfrac{9.52(15{,}000)}{L}$ megajoules per animal, where L is the number of livestock per hectare that can be supported by a crop of rye grass.

b. $L(64{,}000) \approx 2.2$ animals per hectare

c. $r(7) = 20{,}400$ megajoules per animal

d. $r(L(r)) = \dfrac{9.52(15{,}000)}{\dfrac{9.52(15{,}000)}{r}} = 9.52(15{,}000)\dfrac{r}{9.52(15{,}000)} = r$

$$L(r(L)) = \frac{9.52(15{,}000)}{\dfrac{9.52(15{,}000)}{r}} = 9.52(15{,}000)\frac{L}{9.52(15{,}000)} = L$$

41. a. P is continuous without restriction.

b. $t(P) = \dfrac{\ln\left(\dfrac{P}{7.394}\right)}{\ln 1.0269}$

c. $t(40) \approx 63.6$ years after 1900. Thus the population was 40 million in August of 1964.

d. $P(29.5) \approx 16.2$ million people

43. a. $S(85) = \$123.1$ million
$S(88) = \$159.4$ million
$S(89) = \$97.7$ million
$S(92) = \$53.3$ million

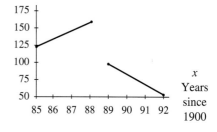

b.

S(x)
Sales
(millions of dollars)
175
150
125
100
75
50
85 86 87 88 89 90 91 92
x
Years since 1900

c. S is a function of x because each year corresponds to only one amount of sales. The fact that S is not defined between $x = 88$ and $x = 89$ does not prevent it from being a function.

45. a. The shipping charges encourage larger orders.

b.

Order amount	Shipping charge
\$17.50	\$3.50
\$37.95	\$6.83
\$75.00	\$11.25
\$75.01	\$9.00

c. $S(x) = \begin{cases} 0.2x \text{ dollars} & \text{when } 0 \le x \le 20 \\ 0.18x \text{ dollars} & \text{when } 20 < x \le 40 \\ 0.15x \text{ dollars} & \text{when } 40 < x \le 75 \\ 0.12x \text{ dollars} & \text{when } x > 75 \end{cases}$

where x is the order amount in dollars

d.

S(x)
Shipping fee
(dollars)
12
10
8
6
4
2
0
0 20 40 75
x
Order
amount
(dollars)

e. Answers will vary.

Section 1.4

1. a. 5 b. 0 c. 2.6

3. a. 0.5 b. 0.5 c. 0.5

d. Does not exist, $\lim_{t\to\infty} g(t) \to \infty$

5. a. 0

b. 3

c. Does not exist

d. Does not exist

7. a. Yes, because the left and right limits are the same.

b. No, because the function value at 3 is different from the limit.

c.

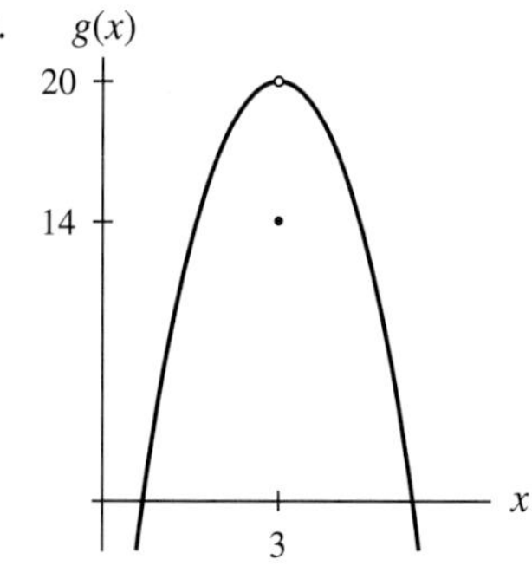

9. a. No, because the limit does not exist as $x \to 2$ and as $x \to -2$.

b.

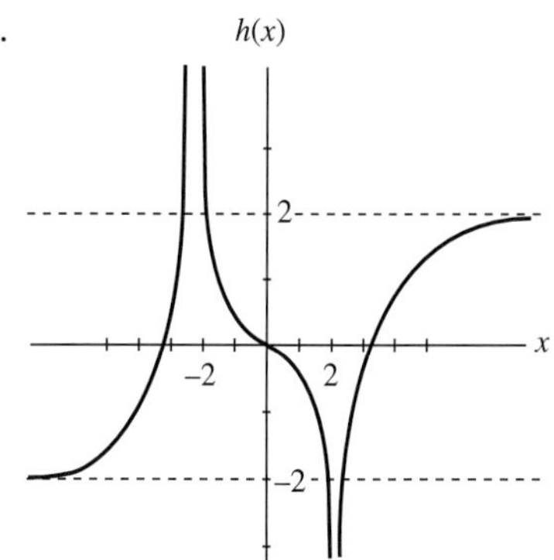

c. As x becomes increasingly positive, the function approaches the horizontal asymptote $y = 2$. As x becomes increasingly negative, the function approaches the horizontal asymptote $y = -2$.

11. No. If a function is continuous, then the limit exists for all input values.

13. 7.04

15. -10

17. $\frac{5}{4}$

19. a. -4 b. 0 c. 0 d. 0 e. 0 f. 8

g. Does not exist

h. Does not exist, $\lim_{t\to\infty} g(t) \to \infty$

21. a. $y = -4$ is a horizontal asymptote on the left.

b. i. Yes ii. Yes iii. Yes iv. No

23.

$x \to 3^-$	$\frac{1}{x-3}$	$x \to 3^+$	$\frac{1}{x-3}$
2.9	-10	3.1	10
2.99	-100	3.01	100
2.999	-1000	3.001	1000
2.9999	-10,000	3.0001	10,000

$\lim_{x\to3^-} \frac{1}{x-3} \to -\infty$; $\lim_{x\to3^+} \frac{1}{x-3} \to \infty$; $\lim_{x\to3} \frac{1}{x-3}$ does not exist

25.

$t \to \infty$	$\frac{14}{1 + 7e^{0.5t}}$
10	0.01346
20	$9.0799 \cdot 10^{-5}$
30	$6.1180 \cdot 10^{-7}$
40	$4.12231 \cdot 10^{-9}$
50	$2.7776 \cdot 10^{-11}$

$$\lim_{t\to\infty} \frac{14}{1 + 7e^{0.5t}} = 0$$

27.

$x \to \frac{-1}{3}^-$	$\frac{x^3 + 6x}{3x + 1}$	$x \to \frac{-1}{3}^+$	$\frac{x^3 + 6x}{3x + 1}$
-0.35	42.858	-0.32	-48.819
-0.34	103.965	-0.332	-507.149
-0.334	1020.630	-0.3332	-5090.482
-0.3334	10,187.296	-0.33332	-50,923.814

$\lim_{x \to \frac{-1}{3}^-} \frac{x^3 + 6x}{3x + 1} \to \infty$; $\lim_{x \to \frac{-1}{3}^+} \frac{x^3 + 6x}{3x + 1} \to -\infty$;

$\lim_{x \to \frac{-1}{3}} \frac{x^3 + 6x}{3x + 1}$ does not exist

29.

$x \to \frac{2}{3}^-$	$\frac{3x^2 + 1.6x - 2.4}{11x - 6 - 3x^2}$	$x \to \frac{2}{3}^+$	$\frac{3x^2 + 1.6x - 2.4}{11x - 6 - 3x^2}$
0.66	0.79487	0.67	0.802575
0.666	0.79949	0.667	0.800257
0.6666	0.799949	0.6667	0.800026
0.66666	0.7999949	0.66667	0.800003

$$\lim_{x \to 2/3} \frac{3x^2 + 1.6x - 2.4}{11x - 6 - 3x^2} = 8$$

31. a. \$3.95; \$6.35

b. 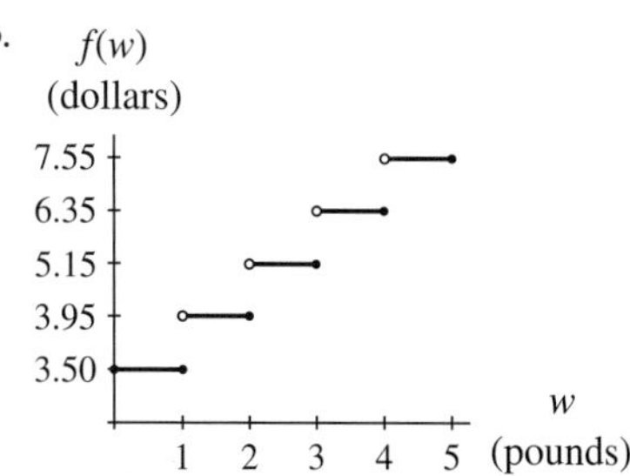

c. i. \$3.95 ii. \$3.95 iii. \$3.95
iv. \$3.95 v. \$5.15 vi. Does not exist
vii. \$3.50 viii. Does not exist

33. a. Approximately \$982
b. 5; Input values greater than 1 have no meaning in this context, so the end behavior of the function is meaningless.
c. \$1149

35. a. 37.6% of graduates respond after 6 weeks.
b. 38%

37. a. \$1.03 per share
b. \$29.17 per share
c. Does not exist, $\lim_{x \to \infty} P(x) \to \infty$
d. Does not exist in the context

39. a. \$210.7 billion; \$210.7 billion; Because the function is continuous, there is no difference between a limit and a function value.
b. \$220 billion; This limit is the level that sales of electronics will never exceed. In reality, this value probably has no significance in this context because the model is based on sales between 1986 and 1990.
c. $y = 220$ is a horizontal asymptote for this function as the input becomes increasingly large.

41. $\lim_{m \to \pm\infty} \frac{3m^3 + m}{m^3 - 3} = 0$

43. $\lim_{x \to \pm\infty} \frac{2x^3 - 2.8x^2 + x}{4x^2 - 3.5x^3} = \frac{-2}{3.5}$

45. $\lim_{t \to \pm\infty} \frac{t^2 + 7.3}{1.6 - 2t}$ does not exist

Section 1.5

1. a. Slope $\approx \frac{-\$2.5 \text{ million}}{5 \text{ years}} = -\$0.5$ million per year

The corporation's profit was declining by approximately a half a million dollars per year during the 5-year period.

b. The rate of change is approximately -\$0.5 million per year.

c. The vertical axis intercept is approximately \$2.5 million. This is the value of the corporation's profit in year zero. The horizontal axis intercept is 5 years. This is the time when the corporation's profit is zero.

3. a. 382.5 donors per year

b. 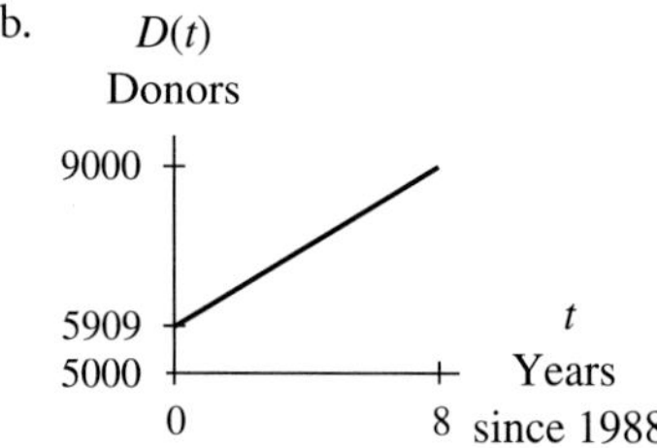

c. The vertical axis intercept is 5909 donors. This value is the number of donors in 1988, the starting year.

5. This question cannot be answered because the model does not include a description of the input variable.

7. a. Slope $= \frac{824.1 - 744.0}{1998 - 1997} = \$80.1$ million per year

b. The revenue increased by \$20.025 million each quarter of 1998.

c.

Year	Revenue (millions of dollars)
1997	744.0
1998	824.1
1999	904.2
2000	984.3

d. Revenue = 744.0 + 80.1t million dollars, where t is the number of years after 1997.

9. a. The rate of change is -400 gallons per day.

b. 2800 gallons

c. 18,400 gallons

The assumptions are that the usage rate remains constant and that the tank is not filled more than once in January.

d. $A(t) = 30{,}000 - 400t$ gallons t days after January 1

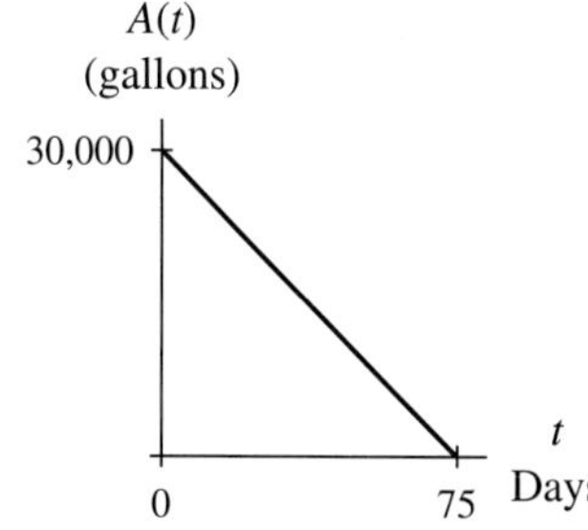

e. Solving $A(t) = 0$ gives $t = 75$. Assuming that the usage rate remains constant, the tank will be empty 75 days after January 1, which corresponds to the middle of March.

11. a. The first differences are 14, 15, 18, and 14. These first differences are nearly constant.

b. Answers may vary. Three models are

$A1 = 15.5x - 30{,}799.8$ million dollars, where x is the year

$A2 = 15.5x - 1349.8$ million dollars, where x is the number of years since 1900

$A3 = 15.5x + 107.2$ million dollars, where x is the number of years since 1994

c. The amount collected in 2000 can be estimated using any one of the three models in part *b*:

$A1(2000) = \$200.2$ million
$A2(100) = \$200.2$ million
$A3(6) = \$200.2$ million

(Regardless of the linear model used, you should obtain this answer.)

d. $x \approx 12.4$ years since 1994. Note that this model is continuous with discrete interpretation, because the tax amounts are totaled and reported at the end of each year. Thus the models make sense only at integer input values. An input of $x = 12$ corresponds to the end of year 1994 + 12 = 2006. At the end of 2006, the amount collected will be $A3(12) = \$293.2$ million. At the end of 2007, the amount collected will be $A3(13) = \$308.7$ million. Thus the amount collected will be a little less than \$300 million in 2006 and a little more than \$300 million in 2007 but will never exactly be \$300 million. This conclusion is valid only if the amount of taxes collected continues to grow between 1998 and 2007 at the same rate at which it grew between 1994 and 1998.

13. a. Answers may vary. Three possible models are

$S1(x) = 499.3x - 976{,}088.3$ students in year x
$S2(x) = 499.3x - 27{,}418.3$ students x years after 1900
$S3(x) = 499.3x + 5036.2$ students x years after 1965

b. Approximately 7533 students

c. The estimate from the models is 505 less than the actual enrollment. Answers vary on whether the error is significant. For a school the size of the one in this activity, an error of 500 students could be significant, especially in housing and faculty loads.

d. These models should not be used to predict enrollment in the year 2000 because the data are too far removed from 2000 to be of any value in such a prediction.

15. a. $P(t) = 0.762t + 10.176$ dollars t years after 1981

b. The ticket prices rounded to the nearest dollar are

1984: $P(3) \approx \$12$ interpolation
1992: $P(11) \approx \$19$ extrapolation
1999: $P(18) \approx \$24$ extrapolation

c. The model prediction is \$1 less than the actual price. That's fairly accurate.

d. Answers will vary.

e. Answers will vary.

f. When a linear model is used to extrapolate, the underlying assumption is that the output will continue to increase at a constant rate. Often this may be a valid assumption for short-term extrapolation but not for long-term extrapolation.

g. Extrapolating from a model must always be done with caution. In order for the extrapolation to be accurate, the model must accurately describe the situation, and the future behavior of the output must match that of the model. Long-term extrapolations are always risky.

17. a. $F(y) = -0.152y + 19.514$ percent where y is 81 for the 1981–82 school year, 82 for the 1982–83 school year, and so on.

b. -0.152 percentage points per year

c. $F(93) \approx 5.4\%$

d. Solving $F(y) = 5$ gives $y \approx 95.7$. Thus the 1996–97 school year is the first year below the 5% level.

19. a. Rate of change $= \dfrac{\$97,500 - \$73,000}{2002 - 1990} \approx \$2042$ per year

b. $\$97,500 + 3(\$2042) \approx \$103,600$

c. The value was \$75,000 in late 1991 and \$100,000 in early 2004.

d. $V(t) = 2041.667t + 73,000$ dollars t years after 1990
$V(9) \sim \$91,400$

The model assumes that the rate of increase of the market value remains constant.

21. a.

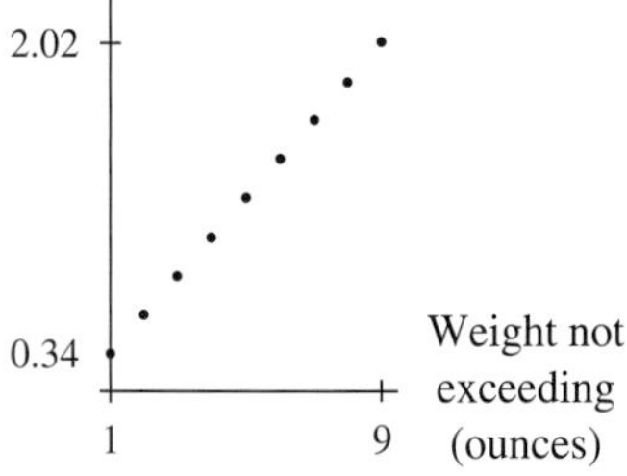

b. The first differences are all \$0.21.

c. $P(w) = 0.13 + 0.21\,w$ dollars for weight not exceeding w ounces where w is a positive integer.

23. Extrapolating from a model is predicting values beyond the interval of the data, whereas interpolating is using a model to estimate what happened at some point within the interval of the data. Interpolation often yields a good approximation of what occurred. Extrapolation must be used with caution because many things that are not accounted for by the model may affect future events.

25. a. 1991 is the last year in which the data decline. In 1993 the data began to rise. We choose 1991 as the dividing point for a piecewise continuous model.

b. A model for the population of North Dakota x years after 1985 is

$$p(x) = \begin{cases} -7.35x + 676.3 \text{ thousand people} & \text{when } 0 \le x < 6 \\ 2.122x + 619.730 \text{ thousand people} & \text{when } 6 \le x \le 11 \end{cases}$$

c. The model estimates the 1997 population to be about 645,000 people. This is an overestimate of approximately 4000 people. This extrapolation is only one year beyond the data given. The extrapolation in Activity 24 is 4 years beyond the data given. These activities illustrate the principle that the closer an extrapolation is to the known data, the more accurate it probably will be.

27. a. The scatter plot does reflect the statements about atmospheric release of CFCs.

b. Answers may vary. One possible model is

$$R(x) = \begin{cases} -15.37x + 1554.19 \text{ million kilograms} & \text{when } x \le 80 \\ 9.165x - 412.5 \text{ million kilograms} & \text{when } 80 < x \le 88 \\ -34.375x + 3413.283 \text{ million kilograms} & \text{when } x > 88 \end{cases}$$

where x is the number of years after 1900. (Note that in computing the middle model, the 1980 data point was not included, but in computing the bottom model, the 1988 data point was included. This was done to improve the fit of the model. You will have to use your own discretion to determine whether or not to include the points at which the data is divided when finding a piecewise model.)

c.

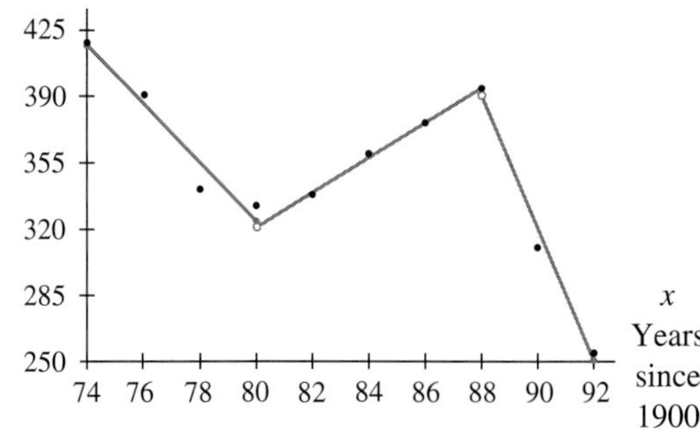

d. i. What was the amount of CFCs released into the atmosphere in 1975? in 1995?
$C(75) \approx 401.4$ million kg, $C(95) \approx 147.7$ million kg

ii. At what rate was the release of CFCs declining between 1974 and 1980?
15.37 million kilograms per year between 1974 and 1980

iii. On the basis of data accumulated since 1987, in what year will there no longer be any CFCs released into the atmosphere?
This occurs when $x \approx 99.3$. This answer indicates that according to the model, there should have been no release of CFCs by early in 2000.

29. a. 78 million people per year

b. $P(t) = 6 + 0.078t$ billion people t years after the beginning of 2000

c. $t \approx 76.9$ years after the beginning of 2000, which corresponds to near the end of 2076. The article estimates that the world population will be 12 billion in 2050.

d. The prediction in part *c* assumes that the world will grow at a constant rate of 78 million people per year between now and 2076. In making this prediction, the Census Bureau must have assumed that the growth rate will increase so that the 12 billion population will be reached sooner than our prediction based on the linear model.

Chapter 1 Review Test

1. a. The number of acres increased, but by a smaller amount in 1992 than in 1991.

b. Between 1989 and 1991, the yearly gain of wetlands was increasing by approximately 33,000 acres per year.

2. a.

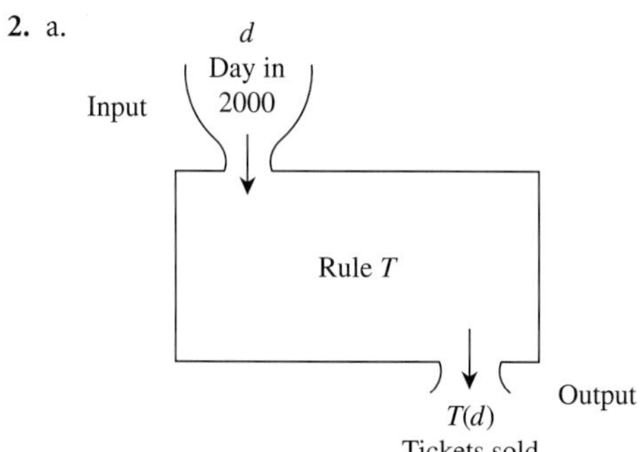

b. T is a function of d because for every possible day, there is only one associated number of tickets.

c. If the inputs and outputs are reversed, the result is not an inverse function because one number of tickets could have more than one day associated with it.

3. a. Rate of increase $= \dfrac{23\% - 18\%}{1995 - 1973} = \dfrac{5\%}{22 \text{ years}}$
≈ 0.23 percentage point per year

b. Percent in 2002 =
$23\% + (7 \text{ years})\left(\frac{5}{22} \text{ percentage point per year}\right) \approx 25\%$
This estimate is valid if the rate of change remains constant through 2002.

4. a. i. 7 ii. Does not exist iii. 2

b. f is not continuous at $x = 1$ because $f(1)$ does not exist, and it is not continuous at $x = 2$ because the limit of f as x approaches 2 does not exist.

5. a. $C = 0.0342t + 11.39$ million square kilometers t years after 1900

b. C is continuous without restriction.

c. The amount of cropland increased by approximately 0.034 million square kilometers, or 34,000 square kilometers per year, between 1970 and 1990.

d. Answers will vary.

e. $C(95) = 14.64$ million square kilometers

CHAPTER 2

Section 2.1

1. $f(x) = 2(1.3^x)$ is the black graph. $f(x) = 2(0.7^x)$ is the blue graph.

3. $f(x) = 3(1.2^x)$ is the blue graph. $f(x) = 2(1.4^x)$ is the black graph.

5. $f(x) = 2 \ln x$ is the blue graph. $f(x) = -2 \ln x$ is the black graph.

7. $f(x) = 2 \ln x$ is the blue graph. $f(x) = 4 \ln x$ is the black graph.

9. $f(x)$ is increasing with a 5% change in output for every unit of input.

11. $y(x)$ is decreasing with a 13% change in output for every unit of input.

13. The number of bacteria declines by 39% each hour.

15. a. $C(t) = 296.3(1.151^t)$ billion dollars in credit card spending t years after 1989

b. August of 1998

17. a. $S(y) = 1.5(1.0746^y)$ dollars y years after 1997

b. $S(13) \approx \$3.82$

19. a. $W(t) = 8.432(0.9806^t)$ workers per beneficiary t years after 1955

b. $W(75) \approx 1.94$ workers per beneficiary. Fewer workers per beneficiary will mean that the social security program will have to find other means of supplementing payments rather than relying solely on withholdings from workers' wages.

21. a. $P(t) = 1.269(1.015646^t)$ billion people t years after 1900. The function is a good fit for the data. The 1960 data point is the one farthest from the function. The data points for 1974, 1987, and 1999 appear close to the graph.

b. $W(x) = 6 + 0.078x$ billion people x years after 1999

c. $W(t) = 6 + 0.078(t - 99)$ billion people t years after 1900

d.

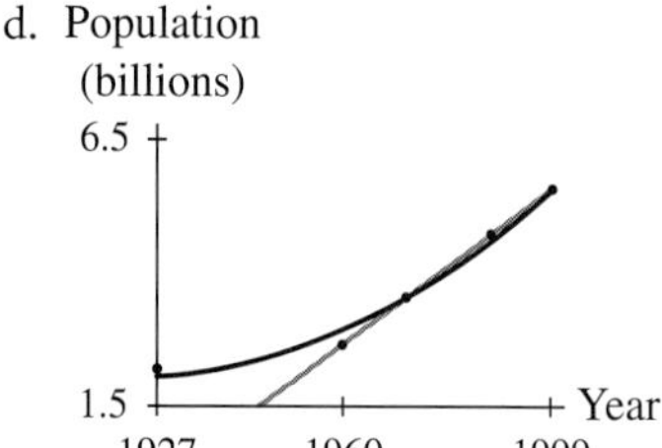

The linear model appears to be a better fit for the years 1960 – 1999; the exponential model is a better overall fit.

e. Population (billions)

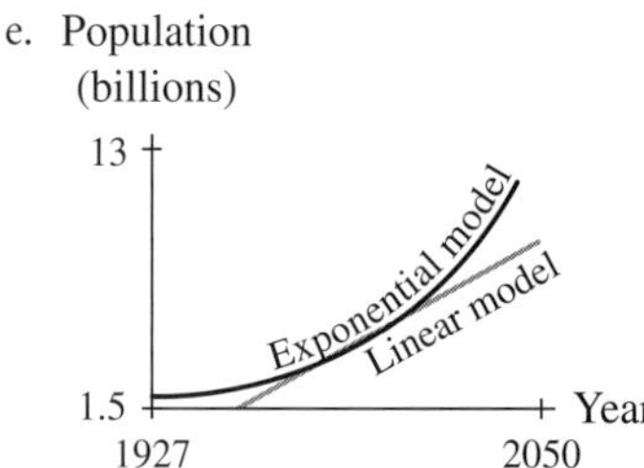

For years beyond 1999, the exponential model rises dramatically compared to the linear model. Both models should give reasonably close answers for populations between 1960 and 2000, but the exponential model should be used from 1900 through 1960. Which model will be more appropriate for years beyond 2000 depends on what happens to the world population growth rate in the future. See part *f* for a more detailed answer.

f. The exponential model gives the population as approximately 6.0 billion in 2000 and 13.0 billion in 2050. The linear model gives 6.1 billion people in 2000 and 10.0 billion people in 2050. The linear model assumes that the growth rate (people per year) remains constant at the 1999 level. The exponential model assumes that the percentage growth rate remains constant at the level obtained by modeling the data for 1927 through 1999.

g. The Census Bureau predicts a population of 12 billion in 2050. This is 1 billion less than the exponential model prediction and 2 billion more than the linear model prediction. The Census Bureau assumed that the percentage growth rate will decline but that the growth rate will increase.

23. a. $C(y) = 6.673(1.791044^y)$ million CD singles sold, where y is the number of years since 1993

b. $\approx$ 79.1% per year growth in the sales of CD singles

25. a. $M(t) = 12.582(1.026935^t)$ million people t years after 1920

b. $\approx$ 2.7% growth per year

c. $M(77) \approx 97.396$ million people

d. The model estimate is approximately 0.167 million people less than the actual population.

27. a. $F(x) = 351.354(0.939873^x)$ farms with milk cows x years after 1980. This model indicates a $(1 - 0.939873)100\% \approx 6.0\%$ per year decline.

b. $\lim_{x\to\infty} 351.354(0.939873^x) = 0$. It is not reasonable that the number of farms with milk cows will approach zero in the future.

c. Answers vary. Possible reasons: The decline in farms with milk cows is probably indicative of an overall decline in farms because of the nature of the economy over the past decades. Specialization has led to fewer nondairy farms owning milk cows. The emphasis on decreasing fat and cholesterol in food may have produced a decrease in demand for dairy products.

29. a. $PC(t) = 0.091(1.393726^t)$ million PCs in China t years after 1985

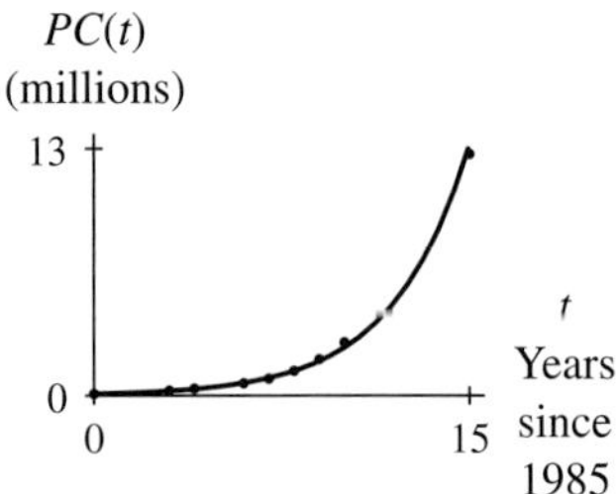

The model appears to be a very good fit.

b. $PC(5) \approx 0.48$ million PCs in 1990; $PC(20) \approx 69.72$ million PCs projected in 2005. The first estimate is an interpolation and is probably more reliable than the extrapolation for the year 2005.

31. a. $pH(x) = -9.792\cdot10^{-5} - 0.434 \ln x$, where x is the H_3O^+ concentration in moles per liter

b. $pH(1.585\cdot10^{-3}) = -9.792\cdot10^{-5} - 0.434 \ln(1.585\cdot10^{-3}) \approx 2.8$

c. Approximately $1.0\cdot10^{-5}$ mole per liter

d. Beer is acidic with a pH of approximately 4.5.

33. a. $L(d) = 158.574 - 42.877 \ln d$ ppm for soil that is d meters from the road

b. $L(12) = 158.574 - 42.877 \ln 12 \approx 52$ ppm

c. $E(d) = 123.238(0.932250^d)$ ppm for soil that is d meters from the road. The log model appears to be a better fit for small distances than the exponential model. Other than that, the two models are similar for distances between 5 and 20 meters. The log model will eventually be negative, whereas the exponential model will approach zero. The exponential end behavior seems to better describe the lead concentration as a function of distance from a road.

35. a. $B(t) = 8.435 - 0.639 \ln x$ percent for a maturity time of t years

b. The model estimates 15-year bond rates as $B(15) \approx$ 6.70%, which is 0.3 percentage point less than the fund manager's estimate.

37. a. $S(x) = -38{,}217.374 + 23{,}245.372 \ln x$ dollars x years after 1970

b. 1993: $S(23) \approx$ \$34,668; 2000: $S(30) \approx$ \$40,845. The 1993 estimate is probably the more accurate one because it is an interpolation, whereas the 2000 estimate is an extrapolation.

c. Salaries are expected to have reached \$40,000 in 1999 ($x \approx 28.9$).

Section 2.2

1. The function is increasing, with a 166.4% change in output for each unit of change in input.
$$f(x) \approx 100(2.664456^x)$$
3. The function is decreasing, with an 18.9% change in output for each unit of change in input.
$$A(t) \approx 1000(0.810584^t)$$
5. The function is increasing.
$$f(x) \approx 39.2e^{0.254642x}$$
7. The function is decreasing.
$$h(t) \approx 1.02e^{-0.478036t}$$
9. a. $A = 2000\left(1 + \frac{0.045}{4}\right)^{4t} \approx 2000(1.045765^t)$ dollars after t years
   b. $A(2) \approx \$2187.25$
   c. $A(3.75) \approx \$2365.42$
   d. The amount after 2 years and 2 months is the same as the amount after 2 years: $A(2) \approx \$2187.25$.
11. a. $\text{APR} = (0.015)(12)100\% = 18\%$
   b. $\text{APY} = \left[\left(1 + \frac{0.18}{12}\right)^{12} - 1\right]100\% \approx 19.6\%$
13. a. The percentage change is approximately 6.4% per year. This is the APY. The balance after 6 years is $(1908.80)(1.064) = \$2030.96$
   b. $A = 1400(1.063962^t) = 1400e^{0.062t}$ dollars after t years. The value r is the nominal rate, or APR.
   c. $A(6) \approx \$2030.89$
   This amount differs from that in part *a* because it assumes a percentage change of approximately 6.3962% rather than 6.4%.
15. a. 4.725% compounded semiannually:
$$\text{APY} = \left[\left(1 + \frac{0.04725}{2}\right)^2 - 1\right]100\% \approx 4.78\%$$
   4.675% continuously compounded:
   $\text{APY} = [e^{0.04675} - 1]100\% \approx 4.79\%$
   This is the better choice.
   b. The account with the higher APY will always result in a greater account balance.
   c. After 10 years, the two options differ by only 79 cents.
17. a. *Option A*: $\frac{0.109}{12}(12{,}300) = m\left[1 - \left(1 + \frac{0.109}{12}\right)^{-42}\right]$
$$m \approx \$353.58$$
   Option B: $\frac{0.105}{12}(12{,}300) = m\left[1 - \left(1 + \frac{0.105}{12}\right)^{-48}\right]$
$$m \approx \$314.92$$
   b. *Option A*: $(\$353.58)(42) = \$14{,}850.36$
   Option B: $(\$314.92)(48) = \$15{,}116.16$
   c. *Option A*: \$2550.36
   Option B: \$2816.16
   d. *Option A* has a higher payment, but the car will be paid off sooner and will cost \$265.80 less than with *Option B*. *Option B* has a lower monthly payment, but it costs more and takes longer to pay off.
19. $A = 2500\left(1 + \frac{0.066}{4}\right)^{4(3)} \approx \$3042.49$
21. $2000 = P\left(1 + \frac{0.062}{12}\right)^{12(2)}$
   $P \approx \$1767.32$
23. a. $250{,}000 = P\left(1 + \frac{0.10}{12}\right)^{12(45)}$
$$P \approx \$2829.52$$
   b. $2829.52\left(1 + \frac{0.10}{12}\right)^{12(45)} \approx \$249{,}999.92$
   This answer is not \$250,000 because the future value was rounded in part *a*.
25. a. The amount was recouped in less than three and a half hours.
   b.
$$85 = 84.998(1.062^t)$$
$$\frac{85}{84.998} = (1.062^t)$$
$$\ln\frac{85}{84.998} = \ln(1.062^t)$$
$$\ln\frac{85}{84.998} = t\ln 1.062$$
$$t = \frac{\ln\frac{85}{84.998}}{\ln 1.062} \approx 3.9\cdot 10^{-4}$$
27. a. $\approx$ 10.7 miles per hour
   b. $\approx$ 4.5 minutes
   c.
   $v(t)$ (mph)
   20
   t (hours)
   0 0.25

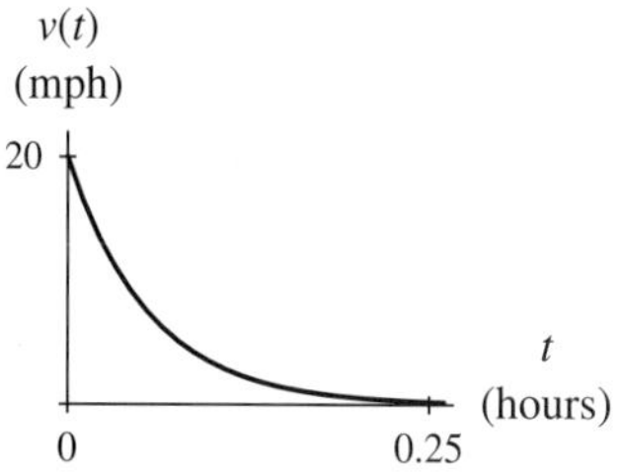

$$\lim_{t\to\infty} v(t) = 0$$
   d. The model describes speed as approaching, but never reaching, zero. Thus, according to the model, the rider will slow down infinitely but will never actually come to a stop. Of course, this is not a realistic description of what actually occurs.
29. Approximately 4960 years ago
31. a. $k \approx 0.127364$
   b. $\approx 284°\text{F}$
   c. $\approx$ 1.25 minutes, or 75 seconds

33. a. Using the points (0, 250) and (0.5, 125), we obtain the model $P(h) = 250(0.25^h)$ mg of penicillin left after h hours.

b. Solving $1 = 250(0.25^h)$ gives $h \approx 3.98$ hours. The second dose of penicillin should be taken approximately 4 hours after the first dose.

Section 2.3

1. Exponential or logarithmic

3. Logarithmic

5. None. The scatter plot exhibits a change in concavity, so it is not linear, exponential, or logarithmic. It does not level off, so it is not logistic. It is possible to model this data with a piecewise continuous model using three linear pieces.

7. Increasing; limiting value is 100; horizontal asymptotes are $y = 0$ and $y = 100$.

9. Decreasing; limiting value is 39.2; horizontal asymptotes are $y = 0$ and $y = 39.2$.

11. Increasing; limiting value is 42; horizontal asymptotes are $y = 0$ and $y = 42$.

13. a. $C(t) = \dfrac{37.195}{1 + 21.374e^{-0.182968t}}$ countries

t years after 1840. The model is a good fit.

b.

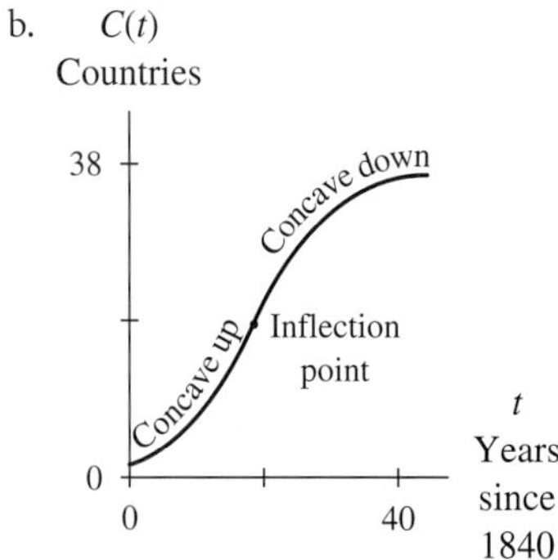

15. a. A plow sulky is a horse-drawn plow with a seat so that the person plowing can ride instead of walk. It was a precursor to the tractor.

b. $P(t) = \dfrac{2591.299}{1 + 16.848e^{-0.194833t}}$ patents t years after 1871

c. Patents begin with one innovative idea and grow almost exponentially as more and more improvements and variations are patented. Eventually, however, the patent market becomes saturated, and the number of new patents dwindles to none as newer ideas and products are invented.

17. a.

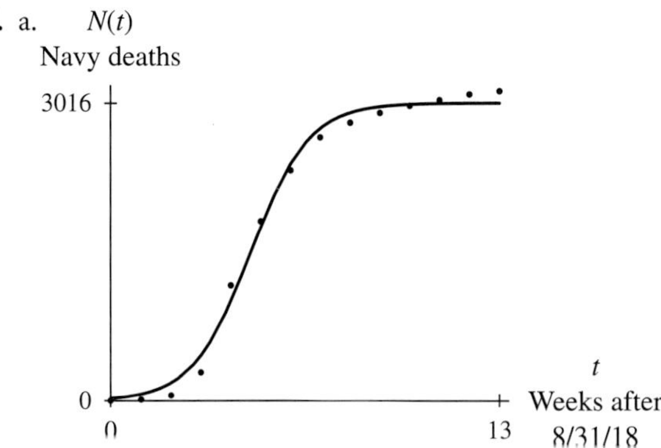

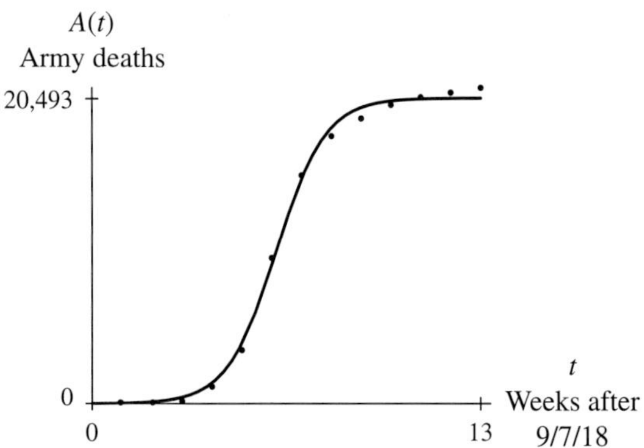

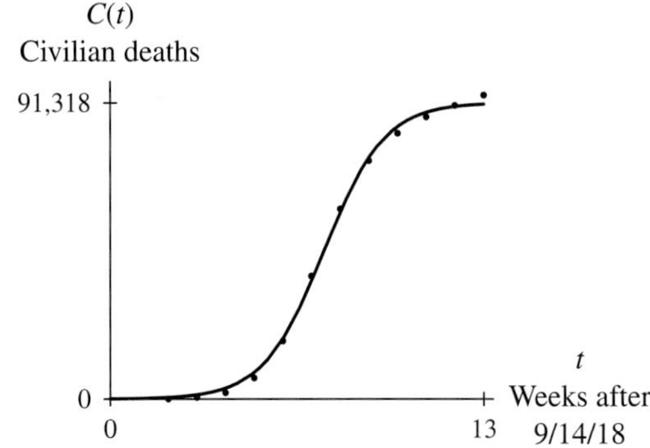

b. $N(t) = \dfrac{3015.991}{1 + 119.250e^{-1.023564t}}$

Navy deaths t weeks after August 31, 1918

$A(t) = \dfrac{20{,}492.567}{1 + 518.860e^{-1.212390t}}$

Army deaths t weeks after Sept. 7, 1918

$C(t) = \dfrac{91{,}317.712}{1 + 176.272e^{-0.951129t}}$

civilian deaths t weeks after Sept. 14, 1918

c. The models given in part b have limiting values less than the number of deaths given in the table for November 30. The models are not good indicators of the ultimate number of deaths.

19. a. The data are concave down from January through April and concave up from April through June. This is not the concavity exhibited by a logistic model.

b. The entire data set does appear to be logistic.

c. $P(t) = \dfrac{42{,}183.911}{1 + 21{,}484.253e^{-1.248911t}}$ polio cases

t months after December, 1948

The model appears to be a good fit.

d. The model is a poor fit for the January through June data.

21. a. The limiting value appears to be approximately 2U/100μL. The inflection point occurs at approximately 9 minutes. (Answers may vary.)

b. $A(m) = \dfrac{1.937}{1 + 29.064e^{-0.421110m}}$ U/100μL

after m minutes. The limiting value is about 1.94 U/100μL.

c. Approximately 0.74 U/100μL

23. a. $B(y) = \dfrac{249.969}{1 + 91.546e^{-0.616555y}}$ cumulative stolen bases

since 1951, where y is the number of years since 1950. This equation appears to be a good fit for the data.

b. The first differences are the number of stolen bases that Willie Mays had in one season.

c. Using only the model (not the data), we calculate the number of bases stolen in 1964 as $B(14) - B(13) \approx$ 3 bases. This is a great underestimate of the actual number.

25. a. $P(x) = \dfrac{11.742}{1 + 154.546e^{-0.025538x}}$ billion people x years

after 1800. The equation appears to be a good fit for the later (1960–2071) data but a poor fit for the early (1800–1960) data.

b. According to the model, the world population will level off at 11.7 billion. This is probably not an accurate prediction of future world population.

c. The model will probably be a poor estimate of the 1850 population because it appears to be a poor fit for the years between 1800 and 1960. It is a good fit around 1990, however, so it will probably give a good estimate of population in that year.

Section 2.4

1. Concave up, decreasing from $x = 0.75$ to $x = 3$, increasing from $x = 3$ to $x = 4$

3. Concave up, decreasing from $x = 13.5$ to $x = 18$, increasing from $x = 18$ to $x = 22.5$

5. Concave down, always decreasing

7. a.

128 140 144 140 128 108 80

12 4 −4 −12 −20 −28

−8 −8 −8 −8 −8

Second differences are constant, so the data are quadratic.

b. After 3.5 seconds the height is 44 feet. After 4 seconds the height is 0 feet.

c. $H(s) = -16s^2 + 32s + 128$ feet after s seconds.

d. $H(s) = 0$ when $s = -2$ and $s = 4$

The missile hits the water after 4 seconds.

9. a. $J(12) \approx 374$ jobs b, c. Answers will vary.

11. a. Because the data are evenly spaced and the second differences are constant, the data are perfectly quadratic.

b. 26.5 years of age

c. $A(x) = 0.0035x^2 - 0.405x + 32$ years of age x years after 1900

d. $A\,(100) = 26.5$ years of age

e., f. Answers will vary.

13. a. The data do not appear to be concave up or concave down. A linear model is

$B(x) = 0.002x + 1.880$ dollars to make x ball bearings

b. Overhead is \$1.88.

c. $B(5000) \approx \$13.64$

d. $B(5100) - B(5000) \approx \$0.24$

This answer could also be found by multiplying the slope of the model by 100. This value is called marginal cost.

e. $C(u) = 0.002(500u) + 1.880$ dollars to make u cases of ball bearings

15. a. $A(t) = -415t^2 - 1837t + 91{,}977$ AIDS cases diagnosed t years after 1990

b. The model appears to be a good fit.

c. $A(8) = 50{,}721$ cases

Because this estimate is only one year beyond the known data, it is probably a good estimate.

d. According to the model, the number of AIDS cases was less than 40,000 in 2000.

e. Answers will vary.

17. a. $Q(x) = 0.057x^2 - 3.986x + 69.429$ thousands of tons of lead x years after 1940

$E(x) = 211.196(0.821575^x)$ thousands of tons of lead x years after 1940

The exponential equation is a poor fit. The quadratic equation appears to be a good fit.

b. The data suggest that lead usage is approaching zero as time increases. The exponential function approaches zero as time increases.

c. The quadratic model is the more appropriate one. $Q(15) \approx 22$ thousands of tons of lead

19.

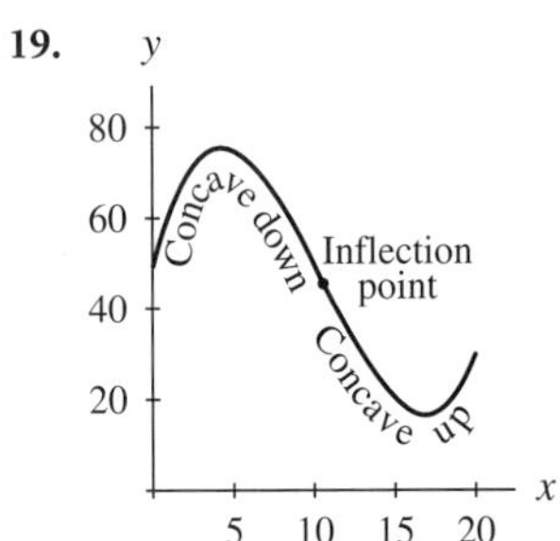

21.

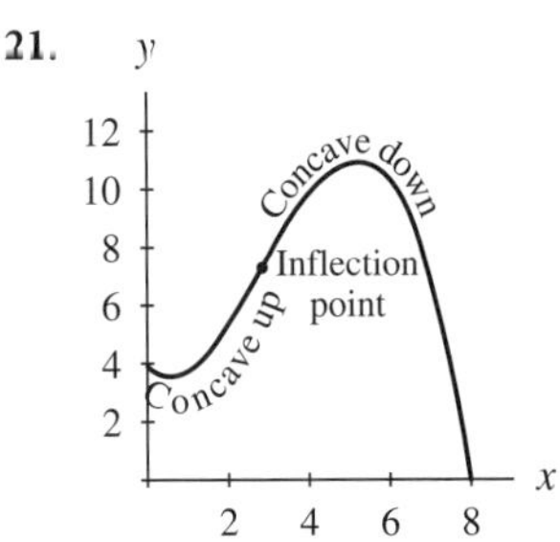

23.

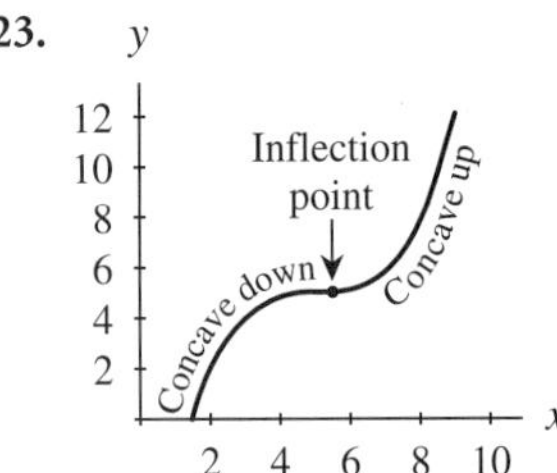

25. a. The scatter plot indicates an inflection point.
b. $A(t) = 0.427t^3 - 5.286t^2 + 22.827t + 3.014$ million dollars t years after 1990
c. $A(3) \approx \$35$ million; $A(9) \approx \$92$ million
The 1993 estimate is more likely to be accurate because it is an interpolation rather than an extrapolation.
d. The 1993 estimate exceeded the actual amount by \$1 million. The 1999 estimate is \$7 million short of the actual amount. These figures confirm the statements in part *c*.

27. a. The scatter plot suggests an inflection point.
b. $G(x) = (5.555 \cdot 10^{-5})x^3 - 0.007x^2 + 0.103x + 104.955$ males per 100 females x years after 1900. The graph of G rises beyond 1990. It is unlikely that the gender ratio will rise as indicated by the model.
c. Possible answers include: Health issues in the early 1900s probably contributed to a greater death rate among women, who often died in childbirth. The western expansion and gold and silver rush days probably attracted more male immigrants than female ones.

29. a. The number of females and males is approximately equal for 30-year-olds.
b. $C(a) = (-2.025 \cdot 10^{-4})a^3 + 0.013a^2 - 0.339a + 105.841$ males per 100 females for an age of a years
$Q(a) = -0.015a^2 + 0.595a + 100.738$ males per 100 females for an age of a years
The cubic equation fits the data better than the quadratic equation. Although the inflection point is subtle and the inflection point of the cubic equation does not perfectly match the inflection point indicated by the data, the fit is enough basis for choosing the cubic model over the quadratic one.
c. Among 83-year-olds there are approximately twice as many women as men. This implies that men die younger than women.

31. a. The behavior of the data changes abruptly in 1989, resulting in a sharp point. None of the five models we have studied will reflect that behavior.
b. $V(t) = \begin{cases} 0.106t^2 - 15.497t + 652.169 \text{ pounds per person} & \text{when } t < 89 \\ -2.15t + 306.7 \text{ pounds per person} & \text{when } t \geq 89 \end{cases}$
where t is the number of years since 1900.
c. $V(73) \approx 87.5$ pounds per person
$V(83) \approx 98.4$ pounds per person
$V(93) \approx 106.8$ pounds per person
d. Answers will vary.
e. The model indicates a constant decline of approximately 2.2 pounds per person per year.
f. Answers will vary.

Section 2.5

1. $y = 1.69$
3. $y = -0.003$ and $y = 0.020$
5. a. The data appear to be concave up from 1990 through 1994 and concave down from 1995 through 1998. The concavity changes between 1994 and 1995. Either a logistic or a cubic function could be used to model these data.
b. $D(t) = \dfrac{797.333}{1 + 4.368e^{-0.235025t}}$ billion dollars of debt t years after 1990. This equation is not a good fit for the data. It does not display the change in concavity described in part *a*.
c. $C(t) = \dfrac{314.360}{1 + 49.091e^{-0.848742t}}$ billion dollars over \$165 billion t years after 1990. This equation is an excellent fit for the data.
d. $F(t) = \dfrac{314.360}{1 + 49.091e^{-0.848742t}} + 165$ billion dollars t years after 1990
e. $y = 165$ and $y \approx 479.360$. The function in part *b* has asymptotes $y = 0$ and $y \approx 797.333$.
7. a. A scatter plot shows an inflection point with leveling-off behavior at both ends.

b. $P(t) = \dfrac{235.641}{1 + 9.637e^{-0.172050t}}$ thousand people above 2,700,000 t years after 1980. This equation is not a good fit for the data. It does not exhibit the same inflection point or the same horizontal asymptotes as the data.

c. $I(t) = \dfrac{97.500}{1 + 1405.904e^{-0.591268t}}$ thousand people above 2,760,000 t years after 1980. This equation is an excellent fit for the data.

d. $N(t) = \dfrac{97.500}{1 + 1405.904e^{-0.591268t}} + 2760$ thousand people t years after 1980. The horizontal asymptotes are $y = 2760$ and $y = 2857.5$.

9. a. $p(x) = 93.558(1.400729^x)$ permit renewals x years after 1990

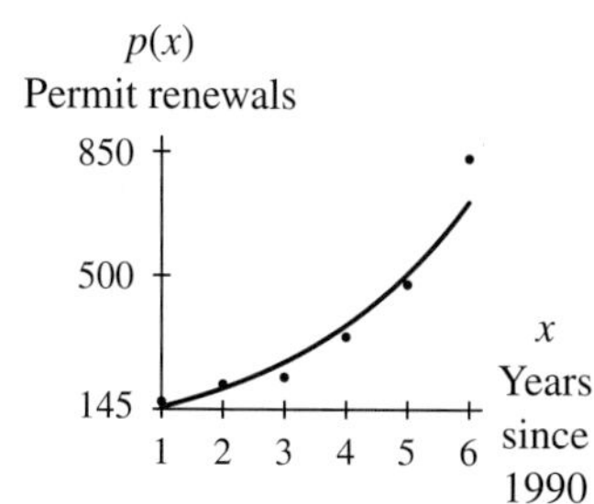

b. $R(x) = 7.919(2.152408^x)$ permit renewals above 135 x years after 1990

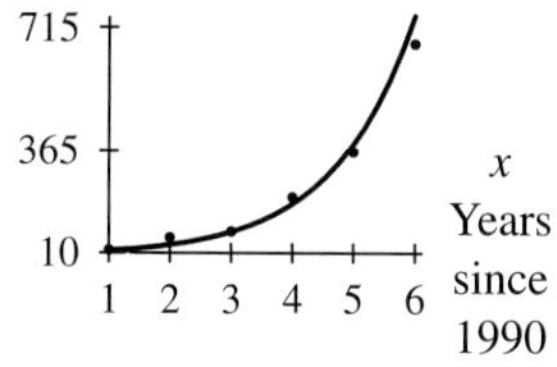

c. $P(x) = 7.919(2.152408^x) + 135$ permit renewals x years after 1990. The equation indicates the number of permit renewals exceeding 2000 in 1998.

$$\lim_{x\to\infty} P(x) \to \infty,\ \lim_{x\to-\infty} P(x) = 135$$

d. $P(7) \approx 1830$ renewals, $P(8) \approx 3783$ renewals. The model estimates for 1997 and 1998 are well below the actual renewal values. Even when a model fits a set of data well, it may not be accurate for any extrapolations.

11. a. $c(t) = 2.053(1.022247^t)$ million children t years after 1970. The graph of this equation is not nearly so curved as the data suggest.

b. Answers will vary. One possible model is $C(t) = 0.086(1.119883^t) + 2.1$ million children t years after 1970.

13. a. $S(x) = -350{,}193.616 + 78{,}843.360 \ln x$ million dollars x years after 1990. The graph of this equation appears linear when graphed on the scatter plot.

b. $C(t) = 6721.530 + 2213.132 \ln t$ million dollars t years after 1992. The graph of this equation is a better fit for the data than the equation in part *a.*

c. $A(y) = 7734.238 + 1633.996 \ln y$ million dollars, where t is the number of years since 1993 plus 0.5. This equation fits the data on the left slightly better than the one in part *b.*

d. All three models will increase infinitely as the input increases. The data show a decline from 1996 to 1997. A quadratic model may be a better model for these data.

e. A log model must have inputs greater than zero.

Section 2.6

1. Because the data are concave up and display a minimum, a quadratic is the only appropriate model.

3. The data appear to be essentially linear. Any concavity is probably not obvious enough to warrant the use of a more complex model.

5. Because of the inflection point, a cubic or logistic model would be an appropriate choice. The choice would depend on the desired behavior of the model outside the range of the data.

7. a. Linear data have constant first differences and lie in a line.

b. Quadratic data have constant second differences and are either concave up or concave down.

c. Exponential data have constant percentage differences and are concave up, approaching the horizontal axis.

9. $T(x) = 0.181x^2 - 8.463x + 147.376$ seconds at age x years. A quadratic model is appropriate because the scatter plot shows a single concavity with a minimum.

11. $W(t) = -0.030t^2 + 0.181t + 18.160$ percent t years after 1980. A quadratic model is appropriate because the scatter plot shows a decreasing, concave-down shape.

13. Answers will vary.

15. a. $L(x) = -1.293x^3 + 17.432x^2 - 66.021x + 84.389$ million dollars of LP/EP shipments x years after 1990
$C(x) = -42.818x^3 + 383.222x^2 + 326.726x + 3511.258$ million dollars of CD album shipments x years after 1990 or $c(x) = \dfrac{6744.675}{1 + 21.452e^{-1.019364x}} + 3400$ million dollars of CD album shipments x years after 1990

b. $L(9) \approx$ -\$40.7 million; $L(10) \approx$ -\$126.0 million (These extrapolations are meaningless in this context.)
$C(9) \approx$ \$6278 million; $C(10) \approx$ \$2283 million
or $c(9) \approx$ \$10,130 million; $c(10) \approx$ \$10,139 million

17. a. 10 components b. 25 components

c. $N(h) = -0.230h^3 + 2.703h^2 - 0.591h + 7.071$ components after h hours

This function provides a good fit to the data.

d. 4 hours 41 minutes

e. The cubic model shows the total number of components declining, which is impossible (unless the worker begins disassembling components). This model is not a good indicator of what occurs after 8 hours.

f. $C(h) = \dfrac{61.859}{1 + 12.299e^{-0.672227h}}$ components after h hours

This function provides a good fit to the data.

g. The model indicates that the number of components assembled is never greater than 62. It is more reasonable than the cubic model, but it is probably still not accurate.

h. According to the logistic model, the student will assemble only 3 or 4 components after 8 hours.

i, j. Answers will vary.

19. a. The scatter plot suggests an inflection point and no leveling off; thus a cubic is an appropriate model. $C(t) = 0.033t^3 - 1.059t^2 + 11.413t - 24.231$ million used cars t years after 1990

b. $C(13) \approx 16.52$ million used cars sold

21. a. It appears that the ban took effect between 1974 and 1976.

b. One possible model is

$$R(t) = \begin{cases} 0.708t^2 - 71.401t + 1823.427 \text{ million kilograms} & \text{when } t \le 74 \\ -19.325t + 1852.133 \text{ million kilograms} & \text{when } t > 74 \end{cases}$$

where t is the number of years since 1900.

c, d. Answers will vary.

23. a. The scatter plot is concave down from 1980 through 1987. From 1987 through 1997, the scatter plot appears to be logistic.

b. The data exhibit two changes in concavity: one abrupt change around 1987 and a smoother change between 1991 and 1993. This concavity description does not match that of a cubic model that has one concavity change.

c. A cubic model is an extremely poor fit to the data.

d. Answers will vary. Once possible model is

$$P(t) = \begin{cases} -2.1t^2 - 6.3t + 2914 \text{ thousand people} & \text{when } 0 \le t \le 7 \\ \dfrac{97.5}{1 + 1405.903e^{-0.591268t}} + 2760 \text{ thousand people} & \text{when } 7 < t \le 17 \end{cases}$$

where t is the number of years since 1980.

e. Answers will vary.

25. It is important to reduce the magnitude of the input values for exponential and logistic models in order to avoid numerical computation error.

Chapter 2 Review Test

1. a. $C(x) = 2.2(1.021431^x)$ million children x years after 1970

b. Approximately 2.1% per year

c. Solving $C(x) = 5$ gives $x \approx 38.7$, which corresponds to the fall of 2009.

d. Answers will vary

2. a. $\text{APY} = \left[\left(1 + \dfrac{0.073}{12}\right)^{12} - 1\right]100\% \approx 7.55\%$

b. The largest possible APY occurs when interest is compounded continuously and is $(e^{0.073} - 1)100\% \approx 7.57\%$.

3. a. The data are increasing and concave up, indicating that either an exponential or a quadratic model is appropriate. It is also possible to argue that the scatter plot suggests an inflection point between 1992 and 1994, indicating that a cubic model is appropriate.

b. $P(t) = 0.021(1.531417^t)$ dollars t years after 1980 or $S(t) = 0.256t^3 - 9.224t^2 + 110.100t - 430.393$ dollars t years after 1980

c. Answers will vary.

4. a. The scatter plot is essentially concave up and then concave down. A cubic model appears to be appropriate.

b. $F(t) = -0.049t^3 + 1.560t^2 - 13.485t + 110.175$ degrees Fahrenheit t hours after midnight August 27

c. $F(17.5) \approx 91°\text{F}$

d. Solving $F(t) = 90$ gives three solutions: $t \approx 1.88, 12.37$, and 17.82. The first solution lies outside the time frame of the data. The other two solutions correspond to 12:22 p.m. and 5:49 p.m.

5. a. The statement is false because the input data are not evenly spaced.

b. The scatter plot suggests a concave-down shape. This shape could be modeled by a quadratic or log function. It is also possible to use the right side of a logistic function to model the data.

c. $Q(t) = (-4.988 \cdot 10^{-4})t^2 + 0.0191t - 7.995$ billion people t years after 1900. The graph of this equation is an excellent fit for the data.

$L(t) = \dfrac{10.764}{1 + 20.429e^{-0.032771t}}$ billion people t years after 1900. The graph of this equation is also an excellent fit for the data.

$G(t) = -27.890 + 7.408 \ln t$ billion people t years after 1900. The graph of this equation is a reasonably good fit for the data, although not so good a fit as the other two models. Shifting the input data to the left might produce a better fit.

d. $\lim_{t \to \infty} Q(t) \to -\infty$

$\lim_{t \to \infty} L(t) \approx 10.8$ billion people

$\lim_{t \to \infty} G(t) \to \infty$

Because the United Nations study suggested that the population will stabilize, the logistic model is the most appropriate one even though the data do not suggest an inflection point.

CHAPTER 3

Section 3.1

1. The stock price rose an average of 46 cents per day during the 5-day period.
3. The company lost an average of \$8333.33 per month during the past 3 months.
5. Unemployment has risen an average of 1.3 percentage points per year in the past 3 years. *Note*: Whenever you are writing a change for a function whose output is a percentage, the correct label (unit of measure) is *percentage points*. The same is true for a rate of change. The phrase *percentage points per year* indicates that the effect is additive, whereas the phrase *percent per year* indicates a multiplicative effect.
7. The ACT composite average for females increased by 0.6 point between 1990 and 1999. This change represents an approximately 3% increase. The average female score increased by an average of 0.07 point per year between 1990 and 1999.
9. The number of Internet users in China grew by 11.1 million between 1997 and 2000. This growth represented a 1233% increase. The number of Internet users increased at an average rate of 3.7 million users per year between 1997 and 2000.

11. a. Slope of secant line $= \dfrac{6229.09 - 6228.98}{1996 - 1982}$

$= \dfrac{0.11 \text{ feet}}{14 \text{ years}} \approx 0.008$ foot per year

b. In the 14-year period from 1982 through 1996, the lake level rose an average of 0.008 foot per year.

c. The lake level dropped below the natural rim because of drought conditions in the early 1990s but rose again to normal elevation by 1996. The average rate of change tells us that the level of the lake in 1996 was close to the 1982 level. Although the average rate of change is nearly zero, the graph shows that the lake level changed dramatically during the 14-year period.

13. a. The balance increased by \$1908.80 − \$1489.55 = \$419.25 or 28.1%.

b. Average rate of change $= \dfrac{\$419.25}{4 \text{ years}} \approx \$104.81/\text{year}$

From the end of year 1 through the end of year 5, the balance increased at an average rate of \$104.81 per year.

c. You could estimate the amount in the middle of the fourth year, but doing so might not be as accurate as using a model to find the amount.

d. $A(t) = 1400(1.063962^t)$ dollars after t years

Amount in middle of year 4 $= A(3.5) \approx \$1739.28$

Amount at end of year 4 $= A(4) \approx \$1794.04$

Average rate of change =

$\dfrac{\$1794.04 - \$1739.28}{\frac{1}{2} \text{ year}} = \$109.52$ per year

15. a. Average rate of change $= \dfrac{2338 \text{ deaths} - 2 \text{ deaths}}{6 \text{ weeks}} \approx$ 389 deaths per week

b. Average rate of change =

$\dfrac{73{,}477 \text{ deaths} - 6528 \text{ deaths}}{4 \text{ weeks}} \approx 16{,}737$ deaths per week

c. Navy, 332 deaths per week; Army, 5559 deaths per week

17. a. $p(55) - p(40) \approx 31.91 - 21.41 = 10.5$ million people

Percentage change $\approx 49\%$

b. $\dfrac{p(85) - p(83)}{85 - 83} \approx \dfrac{3.67 \text{ million people}}{2 \text{ years}}$

≈ 1.8 million people per year.

19. a. Between 0 and 2 seconds, the average rate of change is 0 feet per second because the secant line is horizontal.

b.

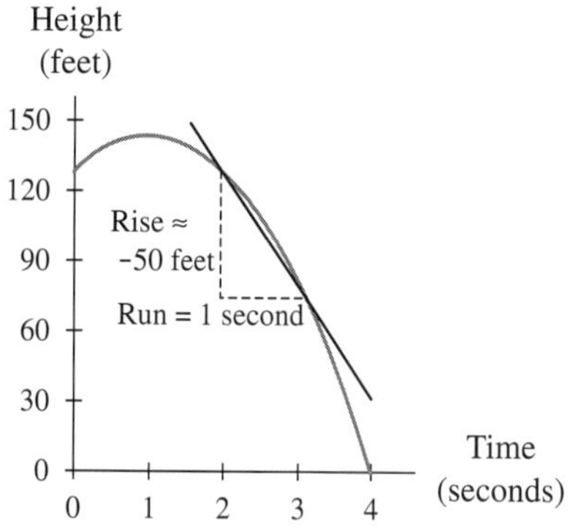

Slope ≈ -50 feet/second

c. $-50 \dfrac{\text{feet}}{\text{second}} \cdot \dfrac{3600 \text{ seconds}}{\text{hour}} \cdot \dfrac{1 \text{ mile}}{5280 \text{ feet}} \approx -34$ mph

21. a.

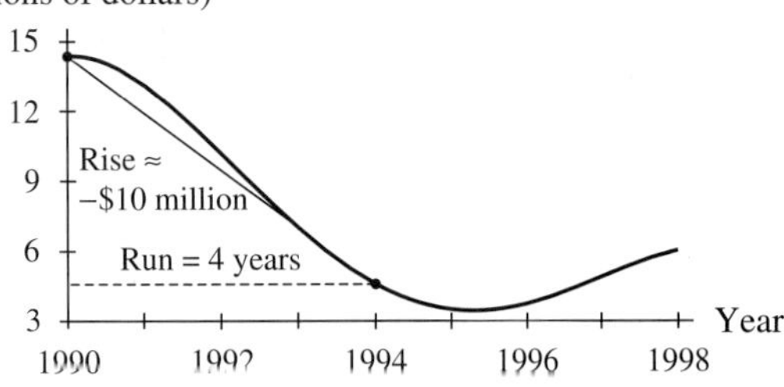

Kelly's interest income decreased by approximately \$2.5 million per year on average between 1990 and 1994.

b. Approximately a 60% increase

c. No. The declining portion of the graph represents Kelly's interest income declining. Kelly was earning interest income but was earning less each year.

23. 441%

25. a. i. 3 b. i. 85.7%
ii. 3 ii. 46.2%
iii. 3 iii. 31.6%

c. The average rate of change of any linear function over any interval will be constant because the slope of a line (and, therefore, of any secant line) is constant. The percentage change is not constant.

27. a.

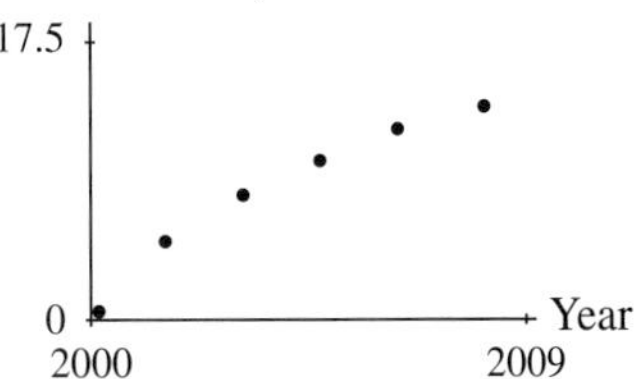

The data suggest a log model: concave down with no sign of leveling off.

b. $C(t) = 0.689 + 7.258 \ln t$ billion dollars t years after 1999

c. $C(6) - C(1) \approx \$13$ billion

d. $\dfrac{C(6) - C(1)}{6 - 1} \approx \$2.6$ billion per year

e. $\dfrac{C(6) - C(1)}{C(1)} \cdot 100\% \approx 1888\%$

29. a. Solve $\dfrac{152.6 - x}{x} = 0.197$ for x to obtain $x \approx 127.5$ births per 100,000 in 1995.

Solve $\dfrac{152.6 - y}{y} = 3.34$ for y to obtain $x \approx 35.2$ births per 100,000 in 1980.

b. $M(y) = 1.139(1.202305^y) + 28$ births per 100,000 y years after 1970

c. $M(25) \approx 142.0$ births per 100,000 in 1995
$M(10) \approx 35.2$ births per 100,000 in 1980
Both estimates are interpolations.

d. The increased use of fertility drugs is the primary factor in the rise of the multiple-birth rate. Women having children later in life also contributes to this rise.

Section 3.2

1. a. A continuous graph or model is defined for all possible input values on an interval. A continuous model with discrete interpretation has meaning for only certain input values on an interval. A continuous graph can be drawn without lifting the pencil from the paper. A discrete graph is a scatter plot. A continuous model or graph can be used to find average or instantaneous rates of change. Discrete data or a scatter plot can be used to find average rates of change.

b. An average rate of change is a slope between two points. An instantaneous rate of change is the slope at a single point on a graph.

c. A secant line connects two points on a graph. A tangent line touches the graph at a point and is tilted the same way the graph is tilted at that point.

3. Average rates of change are slopes of secant lines. Instantaneous rates of change are slopes of tangent lines.

5. $\text{Average speed} = \dfrac{19 - 0 \text{ miles}}{17 \text{ minutes}} \cdot \dfrac{60 \text{ minutes}}{\text{hour}} \approx 67.1 \text{ mph}$

7. a. The slope is positive at A, negative at B and E, and zero at C and D.

b. The graph is steeper at point B than at point A.

9. a. b.

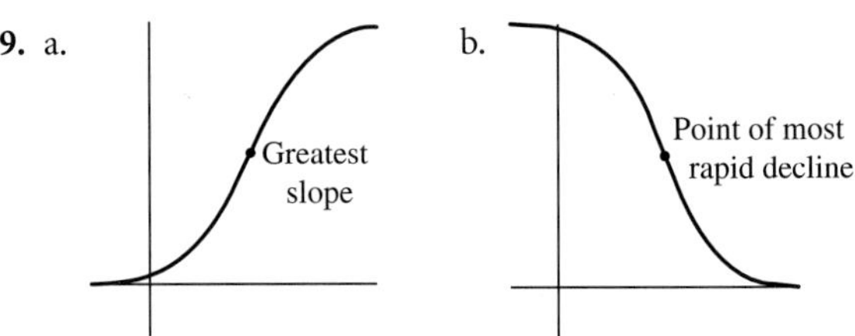

11. The lines at A and C are not tangent lines.

13.

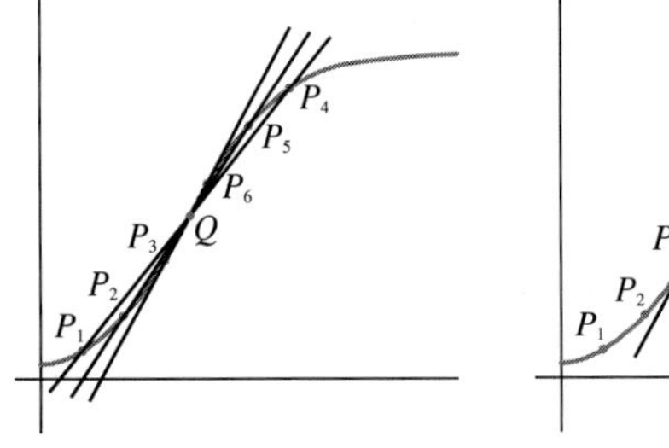

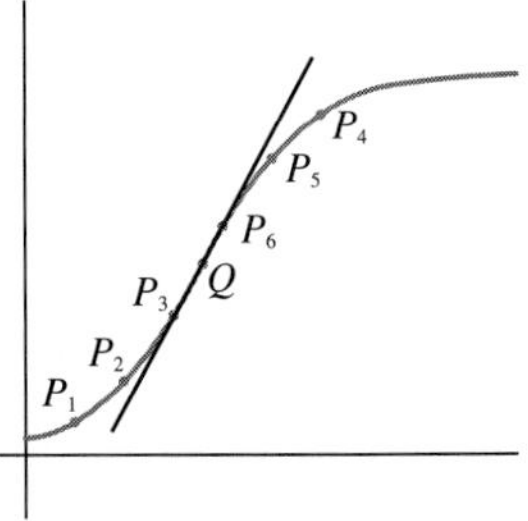

15. a, b. *A*: concave down, tangent line lies above the curve.
B: inflection point, tangent line lies below the curve on the left, above the curve on the right.
C: inflection point, tangent line lies above the curve on the left, below the curve on the right.
D: concave up, tangent line lies below the curve.

c. 

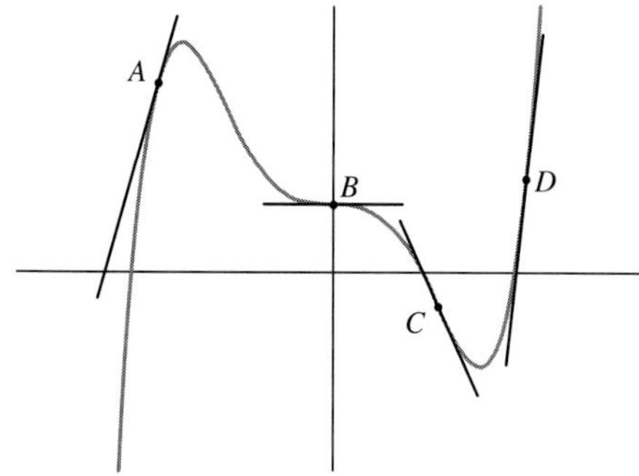

d. *A*, *D*: positive slope; *C*: negative slope (inflection point); *B*: zero slope (inflection point)

17.

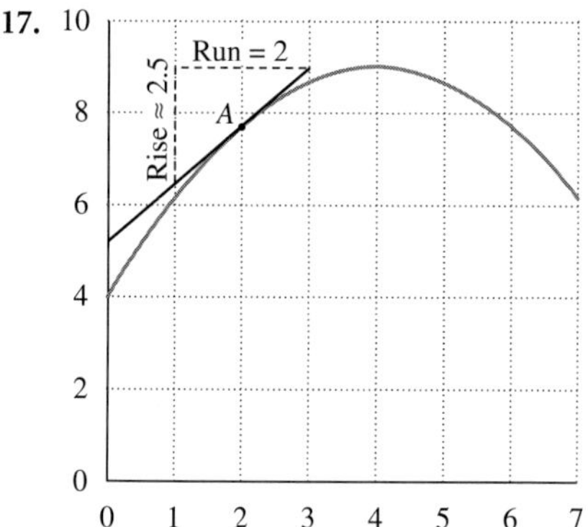

$$\text{Slope} \approx \frac{2.5}{2} = 1.25$$

(Answers may vary.)

19. 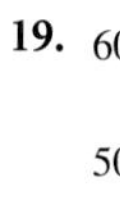

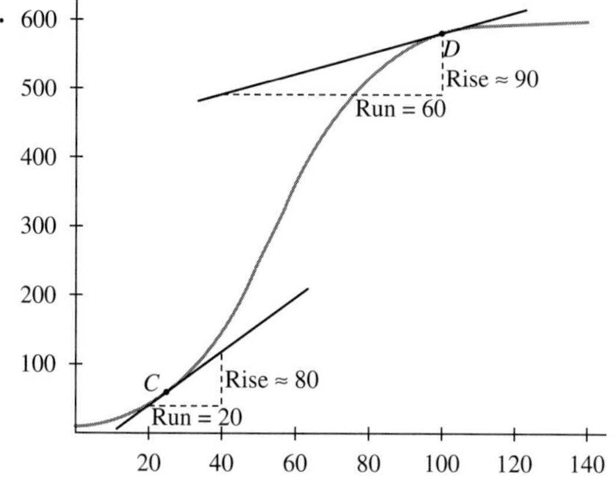

$$\text{Slope at } C \approx \frac{80}{20} = 4$$

$$\text{Slope at } D \approx \frac{90}{60} = 1.5$$

(Answers may vary.)

21. a. Million subscribers per year
b. The number of subscribers was growing at a rate of 11.3 million per year in 1996.
c. 11.3 million subscribers per year
d. 11.3 million subscribers per year

23. a, b. *A*: 1.3 mm per day per °C; *B*: 5.9 mm per day per °C; *C*: -4.2 mm per day per °C
c. The growth rate is increasing by 5.9 mm per day per °C.
d. The slope of the tangent line at 32°C is -4.2 mm per day per °C.
e. At 17°C, the instantaneous rate of change is 1.3 mm per day per °C.

25. a.

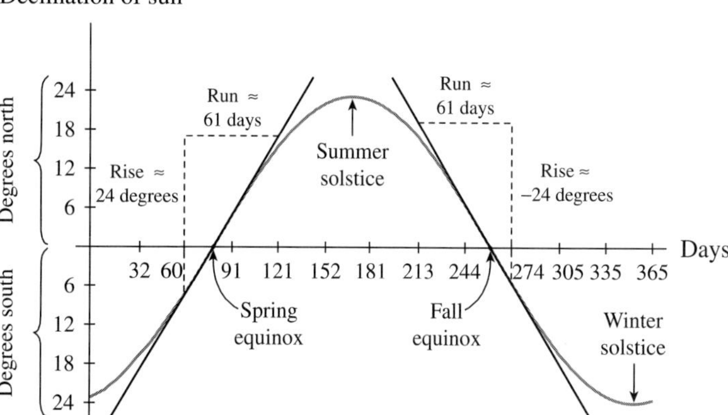

The slope at the solstices is zero.

b. The steepest points on the graph are those where the graph crosses the horizontal axis. The slopes are estimated as

$$\frac{24 \text{ degrees}}{61 \text{ days}} \approx 0.4 \text{ degree per day and}$$

$$\frac{-24 \text{ degrees}}{61 \text{ days}} \approx -0.4 \text{ degree per day}$$

A negative slope indicates that the sun is moving from north to south.

c. The points identified in part *b* correspond to the spring and fall equinoxes.

27. a, b, c. Any line tangent to *p* is the graph itself. The slope of any tangent line will be approximately 2370 thousand people per year.
d. The slope of the graph at every point will be 2370 thousand people per year.
e. The instantaneous rate of change is 2370 thousand people per year.

29.

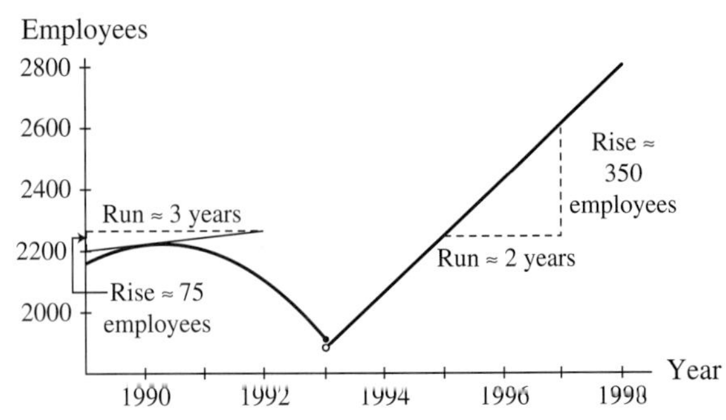

a. Slope in 1990 $\approx \dfrac{75 \text{ employees}}{3 \text{ years}} =$ 25 employees per year

b. The slope at 1993 does not exist because the graph is not continuous at 1993.

c. Slope in 1997 $\approx \dfrac{350 \text{ employees}}{2 \text{ years}} =$ 175 employees per year

31. a. $\dfrac{2.8 - 1.8}{1996 - 1994} = \dfrac{1 \text{ million dollars}}{2 \text{ years}} =$ 0.5 million dollars per year

b.

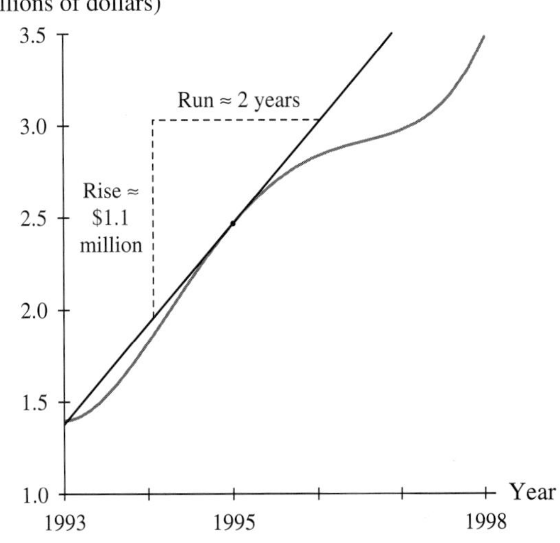

Slope $\approx \dfrac{1.1 \text{ million dollars}}{2 \text{ years}} =$ \$0.55 million per year

The two estimates are comparable.

Section 3.3

1. a. Miles per hour

b. Speed or velocity

3. a. The number of words per minute cannot be negative.

b. Words per minute per week

c. The student's typing speed could actually be getting worse, which would mean that $W'(t)$ is negative.

5. a. When the ticket price is \$65, the airline's weekly profit is \$15,000.

b. When the ticket price is \$65, the airline's weekly profit is increasing by \$1500 per dollar of ticket price. Raising the ticket price a little will increase profit.

c. When the ticket price is \$90, the profit is declining by \$2000 per dollar of ticket price. An increase in price will decrease profit.

7.

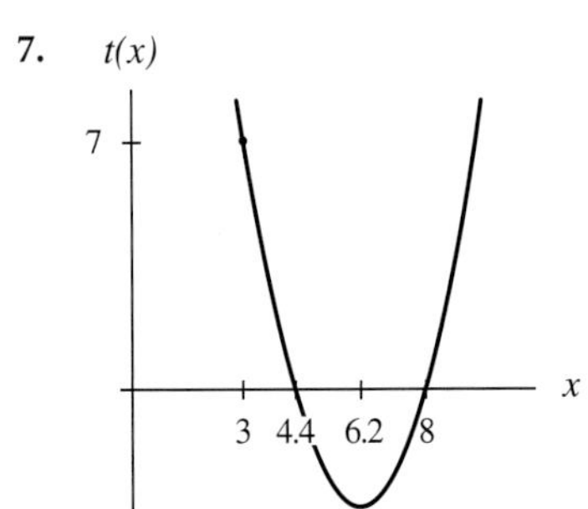

9. a. At the beginning of the diet, you weigh 167 pounds.

b. After 12 weeks of dieting, your weight is 142 pounds.

c. After 1 week of dieting, your weight is decreasing by 2 pounds per week.

d. After 9 weeks of dieting, your weight is decreasing by 1 pound per week.

e. After 12 weeks of dieting, your weight is neither increasing nor decreasing.

f. After 15 weeks of dieting you are gaining weight at a rate of a fourth of a pound per week.

g.

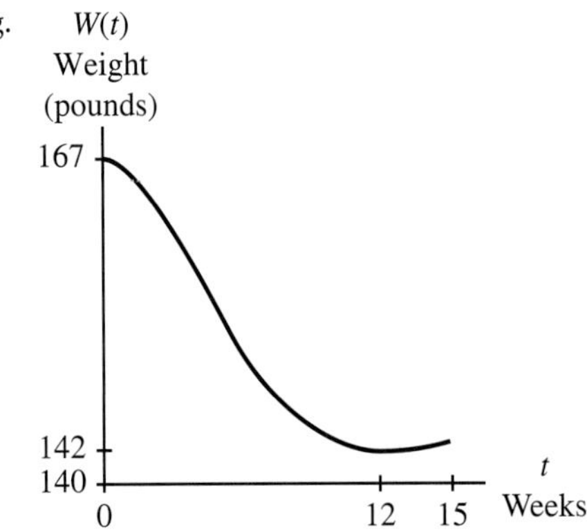

11. We know that the following points are on the graph: (1940, 4), (1970, 12), (1995, 32), and (1980, 18).

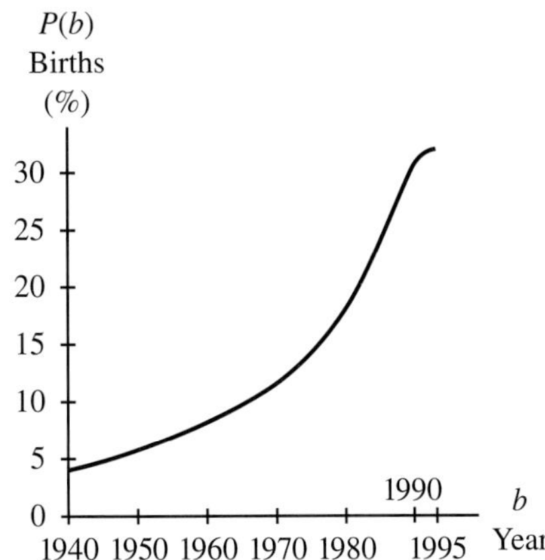

13. a. It is possible for profit to be negative if costs are more than revenue.
    b. It is possible for the derivative to be negative if profit declines as more shirts are sold (because the price is so low, the revenue is less than the cost associated with the shirt).
    c. If $P'(200) = -1.5$, the fraternity's profit is declining. Profit may still be positive (which means that the fraternity is making money), but the negative rate of change indicates that it is not making the most profit possible.

15. a. Years per percentage point
    b. As the rate of return increases, the time it takes the investment to double decreases.
    c. i. When the interest rate is 9%, it takes 7.7 years for the investment to double.
       ii. When the interest rate is 5%, the doubling time is decreasing by 2.77 years per percentage point. A one-percentage-point increase in the rate will decrease the doubling time by approximately 2.8 years.
       iii. When the interest rate is 12%, the doubling time is decreasing by 0.48 year per percentage point. A one-percentage-point increase in the rate of return will result in a decrease in doubling time of approximately half a year.

17.

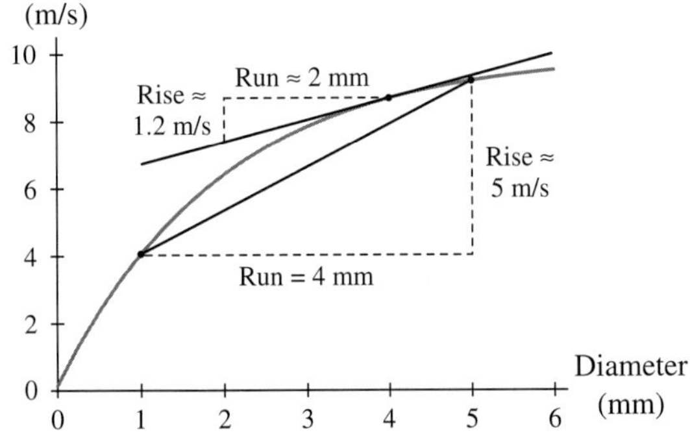

a. The slope of the secant line gives the average rate of change. Between 1 mm and 5 mm, the terminal speed of a raindrop increases an average of about 1.2 m/s per mm.

b. The slope of the tangent line gives the instantaneous rate of change of the terminal speed of a 4 mm raindrop.

c. Slope $\approx \dfrac{1.2 \text{ m/s}}{2 \text{ mm}} = 0.6$ m/s per mm

   A 4 mm raindrop's terminal speed is increasing by approximately 0.6 m/s per mm.

d. By sketching a tangent line at 2 mm and estimating its slope, we find that the terminal speed of a 2 mm raindrop is increasing by approximately 1.8 m/s per mm.

e. Percentage rate of charge ≈

$$\frac{1.8 \text{ m/s per mm}}{6.4 \text{ m/s}} \cdot 100\% \approx 28\% \text{ per mm}$$

   The terminal speed of a 2 mm raindrop is increasing by about 28% per mm as the diameter increases.

19.

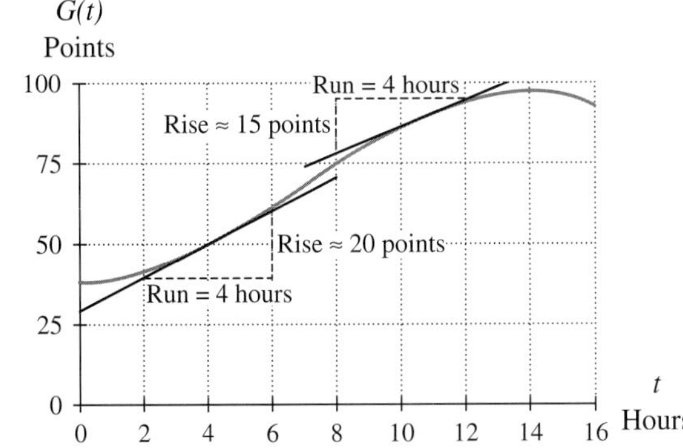

a, b. Slope at 4 hours $\approx \dfrac{20 \text{ points}}{4 \text{ hours}} = 5$ points per hour

   Slope at 11 hours $\approx \dfrac{15 \text{ points}}{4 \text{ hours}} = 3.75$ points per hour

   (Answers will vary.)

c. Average rate of change $\approx \dfrac{36 \text{ points}}{6 \text{ hours}}$

   $= 6$ points per hour

d. Percentage rate of change $\approx \dfrac{5 \text{ points/hour}}{50 \text{ points}} \cdot 100\%$

   $= 10\%$ per hour

   When you have studied for 4 hours, the number of points you will make on your test is increasing by about 10% per hour.

21. a. Rate of change in 1970 ≈ 0.85 death per year; 7.7% per year

    Rate of change in 1980 ≈ 1.1 death per year; 5.2% per year

    b. Answers will vary.

23. Rate of change in 1991 $\approx$

$$\frac{279.8 \text{ thousand deaths} - 201.6 \text{ thousand deaths}}{4 \text{ years}} =$$

19.6 thousand deaths per year

In 1991 the number of alcohol-related deaths was increasing by approximately 19,600 deaths per year.

25. Answers will vary.

Section 3.4

1. The slopes are negative to the left of A and positive to the right of A. The slope is zero at A.

y'
Slopes

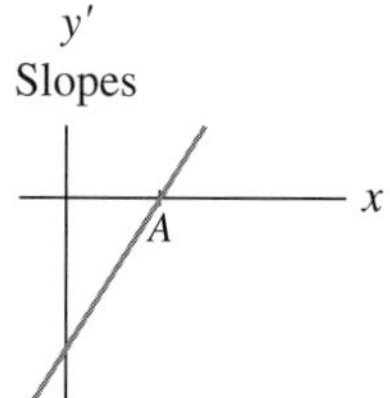

3. The slopes are positive everywhere, near zero to the left of zero, and increasingly positive to the right of zero.

y'
Slopes

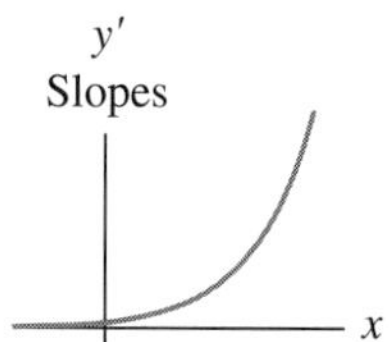

5. The slope is zero everywhere. $y' = \text{slope} = 0$

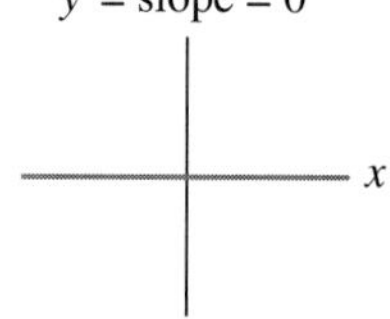

7. The slopes are negative everywhere. The magnitude is large close to zero and is near zero to the far right.

Slopes

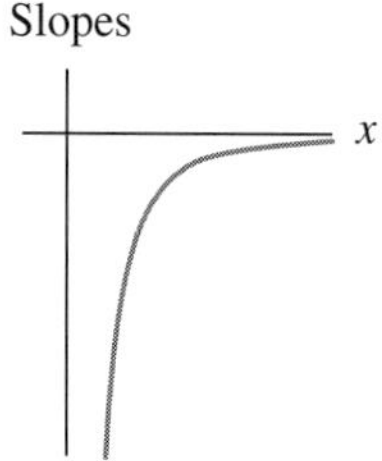

9. The slopes are negative to the left and right of A. The slope appears to be zero at A.

y'
Slopes

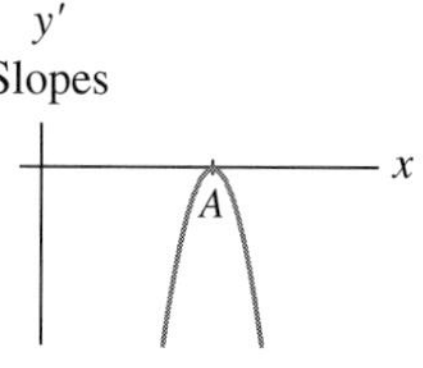

11. a.

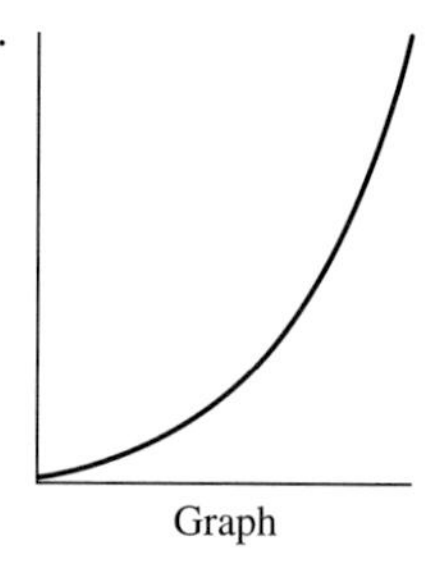

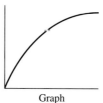

Graph

Slope graph

b.

Graph

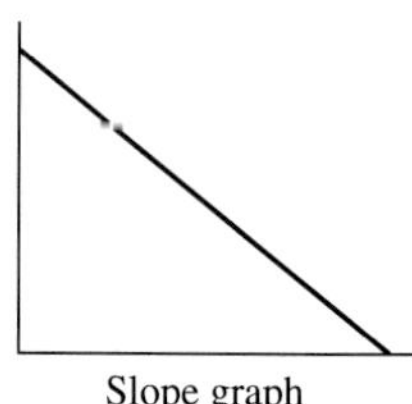

Slope graph

13. a. (Table values may vary.)

Year	1990	1992	1994	1996	1998
Slope	-7.4	-6.1	-5.1	-4.3	-3.6

b.

1990 1992 1994 1996 1998 Year

-2
-4
-6
-8

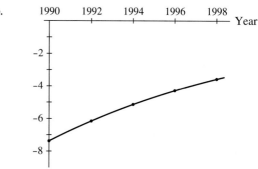

Rate of change
of average bill
(dollars per year)

15. a. (Table values may vary.)

Year	1985	1988	1991	1994	1997
Slope	1600	9200	43,300	86,700	40,700

b. Rate of change
(cases per year)

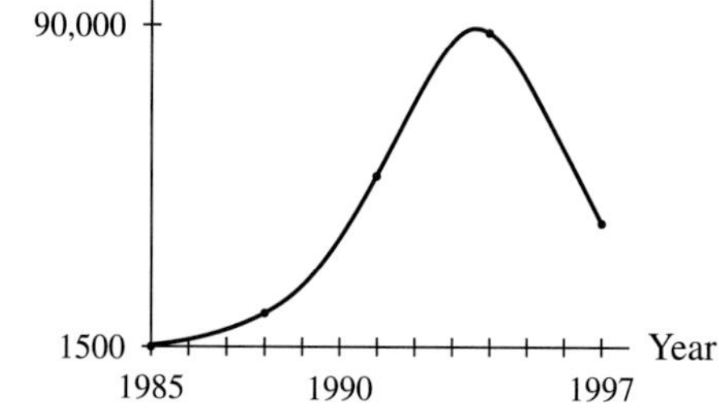

17. a. The average rate of change during the year (found by estimating the slope of the secant line drawn from September to May) is approximately 14 members per month. (Answers will vary.)

b. By estimating the slopes of tangent lines, we obtain the following. (Answers will vary.)

Month	Slope (members per month)
Sept	98
Nov	-9
Feb	30
Apr	11

c. 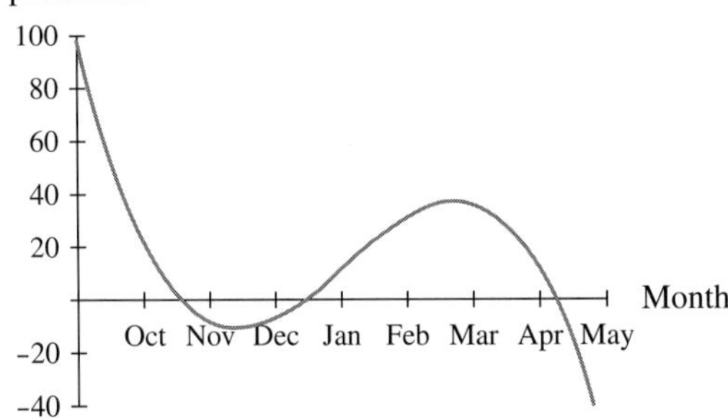

d. Membership was growing most rapidly around March. This point on the membership graph is an inflection point.

e. The average rate of change is not useful in sketching an instantaneous rate-of-change graph.

19. 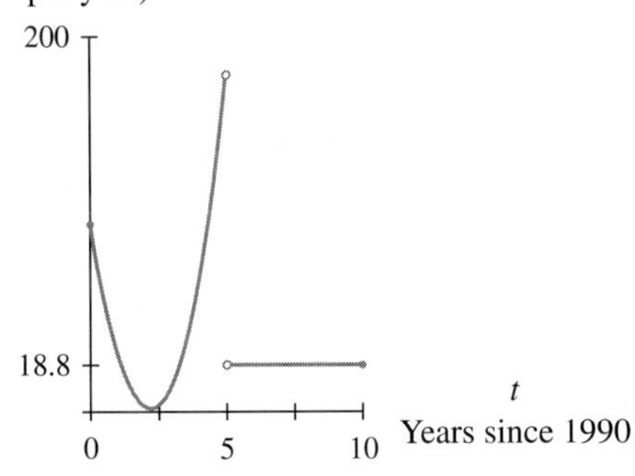

21. a. Profit is increasing on average by approximately $600 per car.

b.

Number of cars	Slope (dollars per car)
20	0
40	160
60	750
80	10
100	-1200

(Answers will vary.)

c. 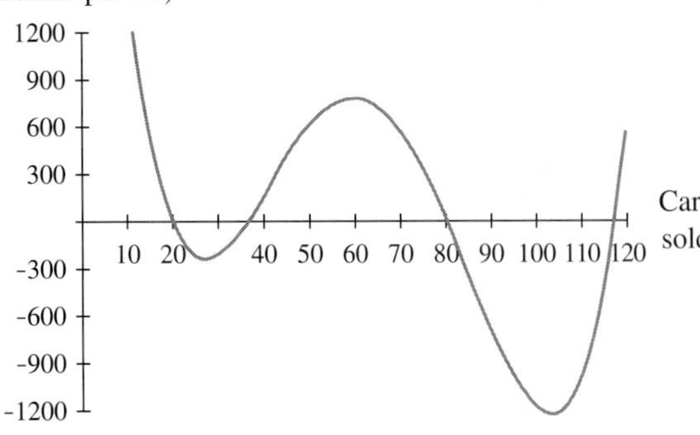

d. The average monthly profit is increasing most rapidly for about 60 cars sold and is decreasing most rapidly when about 100 cars are sold. The corresponding points on the graph are inflection points.

e. Average rates of change are not useful in graphing instantaneous rates of change.

23. The derivative does not exist at $x = 0$, $x = 3$, and $x = 4$ because the graph is not continuous at those inputs.

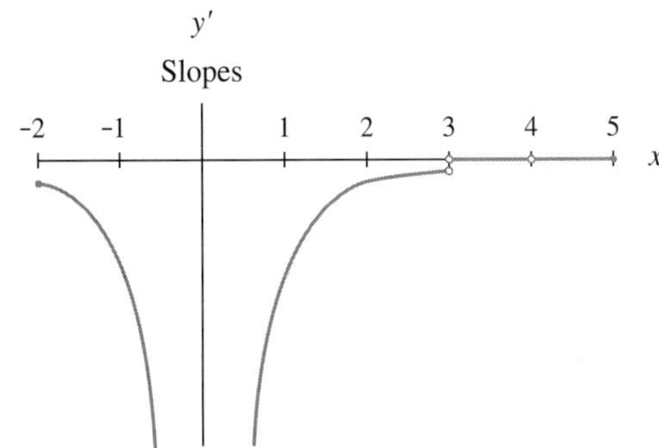

25. The derivative does not exist at $x = 2$ and $x = 3$ because the slopes from the right and left are different at those inputs.

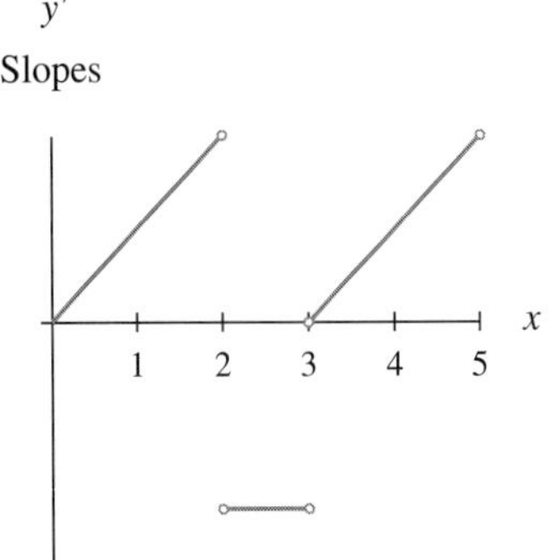

Chapter 3 Review Test

1. a. i. *A, B, C* ii. *E* iii. *D*

b. The graph is steeper at *B* than it is at *A, C,* or *D*.

c. Below: C, D, E; above: A; at B: above to the left of B, below to the right of B

d.

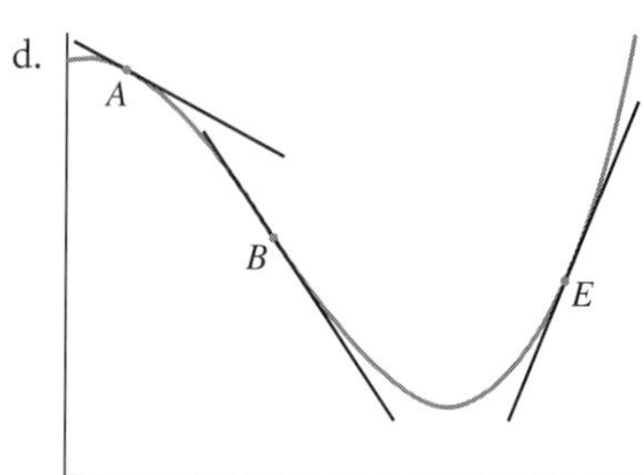

e. i. Feet per second per second. This is acceleration.
ii, iii. The roller coaster was slowing down until D, when it began speeding up.
iv. The roller coaster's speed was slowest at D.
v. The roller coaster was slowing down most rapidly at B.

f. Slopes

A B C D E

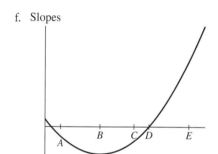

2. a. $\frac{656 - 442}{1998 - 1995} \approx 71.3$ million cards per year. Between 1995 and 1998, the number of Visa cards increased at an average rate of 71.3 million cards per year.

b. $\frac{656 - 600}{600} \cdot 100\% \approx 9.3\%$. Between 1997 and 1998, the number of Visa cards increased by about 9.3%.

3. a, b.

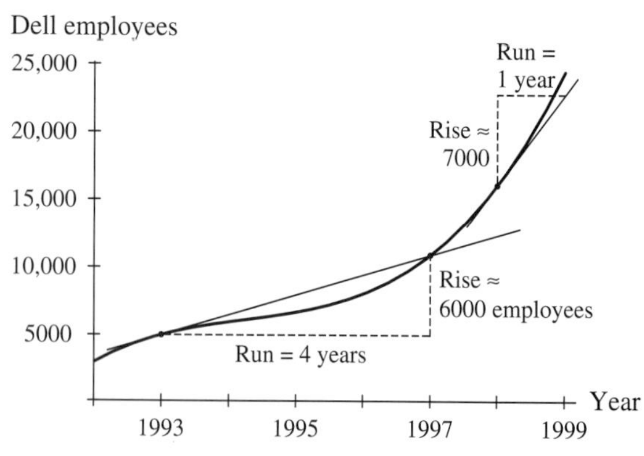

The slope of the secant line gives the average rate of change between 1993 and 1997. The slope of the tangent line gives the instantaneous rate of change in 1998.

c. In 1998 the number of employees was increasing by approximately 7000 employees per year. If we estimate the number of employees in 1998 as 16,000, then this rate represents an increase of approximately 44% per year.

d. Between 1993 and 1997, the number of Dell employees increased at an average rate of 1500 employees per year.

4. a. i. An average 22-year-old athlete can swim 100 meters in 49 seconds.
ii. The swim time for a 22-year-old is declining by half a second per year of age.

b. A negative rate of change indicates that an average swimmer's time improves as age increases.

5. a. $T'(x) = -0.852$ hour per year

b. Change $\approx$ -17 hours
Average rate of change = -0.852 hours per year
Percentage change = -11.3%

c. i. A linear model is defined as one that has a constant rate of change.
ii. Although the rate of change is constant, it is not constant when expressed as a percentage of the output, because output is not constant.
iii. The average rate of change of a linear model will always be constant (and the same as the slope of the graph), because it is the slope of a secant line, and any secant line between two points on a line will be the line itself, having the same slope as the line.

CHAPTER 4

Section 4.1

1. a.

y
16
12
8
4
0
Slope $\approx \frac{6}{2} = 3$
Rise ≈ 6
Run = 2
0 1 2 3 4 x

b.

Input of close point on left	Secant line slope	Input of close point on right	Secant line slope
1.9	2.67868	2.1	2.87094
1.99	2.76300	2.01	2.78222
1.999	2.77163	2.001	2.77355
1.9999	2.77249	2.0001	2.77268
1.99999	2.77258	2.00001	2.77260
Trend ≈ 2.77		Trend ≈ 2.77	

The slope of the line tangent to $y = 2^x$ at $x = 2$ is approximately 2.77.

3. a, b.

Input of close point on left	Secant line slope	Input of close point on right	Secant line slope
1.9	19.21	2.1	20.1
1.99	19.9201	2.01	20.01
1.999	19.99200	2.001	20.001
1.9999	19.99920	2.0001	20.0001
1.99999	19.99992	2.00001	20.00001
Trend = 20		Trend = 20	

c. Although the slopes of the two pieces are the same on either side of $x = 2$, the graph is not continuous at $x = 2$ because the pieces do not join at $x = 2$ (the limit of f as x approaches 2 does not exist because the left and right limits are not the same value). The graph of f is not smooth at $x = 2$ because it is not continuous at that input.

d. Because the graph is not continuous at $x = 2$, it is not possible to sketch a tangent line at $x = 2$.

5. a.

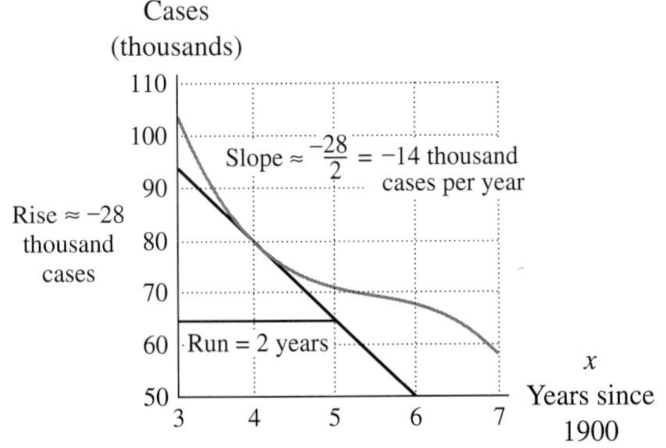

b.

Input of close point on left	Secant line slope	Input of close point on right	Secant line slope
3.9	-15.3195	4.1	-13.6975
3.99	-14.5713	4.01	-14.4091
3.999	-14.4981	4.001	-14.4819
3.9999	-14.4908	4.0001	-14.4892
3.99999	-14.49008	4.00001	-14.4899
Trend ≈ -14.49		Trend ≈ -14.49	

The slope of the line tangent to the graph at $x = 4$ is approximately -14.49 thousand cases per year.

c. In 1994 the number of AIDS cases diagnosed was decreasing by approximately 14.5 thousand cases per year.

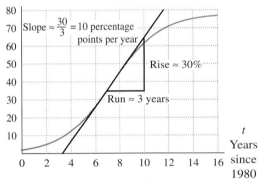

7. a.

b.

Input of close point on left	Secant line slope	Input of close point on right	Secant line slope
6.9	9.59718	7.1	9.64199
6.99	9.61927	7.01	9.62375
6.999	9.62130	7.001	9.62175
6.9999	9.62150	7.0001	9.62155
6.99999	9.62152	7.00001	9.62153
Trend ≈ 9.62		Trend ≈ 9.62	

The line tangent to the graph of P at $t = 7$ has a slope of approximately 9.62 percentage points per year.

c. In 1987 the percentage of households with VCRs was increasing by approximately 9.6 percentage points per year.

d. The tangent line approximation is fast but can be inaccurate. The numerical approximation is somewhat tedious but very accurate.

9. a.

Input of close point on left	Secant line slope	Input of close point on right	Secant line slope
3.9	5.554	4.1	5.4534
3.99	5.51101	4.01	5.50095
3.999	5.50650	4.001	5.50550
3.9999	5.50605	4.0001	5.50595
3.99999	5.50601	4.00001	5.50600
Trend ≈ 5.5		Trend ≈ 5.5	

The derivative (slope of the tangent line) of S at $t = 4$ is approximately \$5.5 billion per year.

b. In 1994 the annual total U.S. factory sales of electronics was increasing by approximately \$5.5 billion per year.

11. a. $\frac{19.5 - 18.2}{1997 - 1993} = \frac{1.3 \text{ thousand 1992 dollars}}{4 \text{ years}} =$ \$0.325 thousand per person per year

b. $P(t) = 0.27t + 14.959$ thousand 1992 dollars per person per year t years after 1980

Slope = \$0.27 thousand per person per year. The symmetric difference quotient is the average rate of change between 1993 and 1997. The slope of the linear equation can be thought of as an average of the rates of change for all years between 1981 and 1997. The answer to part *a* is the more accurate answer.

13. a. $\frac{279.8 - 249.8}{1996 - 1994} = \frac{30 \text{ points}}{2 \text{ years}} = 15$ index points per year

b. $L(x) = -52.090 + 138.021 \ln x$ points x years after 1985

$Q(x) = -0.566x^2 + 26.753x + 54.383$ points x years after 1985

c. Log model estimate: 13.8 index points per year
Quadratic model estimate: 15.4 index points per year

d. The symmetric difference quotient is faster than the numerical estimation and gives an answer comparable to that obtained from the quadratic model.

15. a. Approximately \$1.18 per 100 pounds per month

b. \$1.67 per 100 pounds per month

c. The derivative of p does not exist at $m = 5$.

d. Because we do not have data available, we can use the model to estimate cattle prices in August and October of 1998 and find the symmetric difference quotient using those two points:

$$\frac{p(6) - p(4)}{6 - 4} = \frac{78.918 - 78.4706}{6 - 4} = \frac{\$0.4474}{2 \text{ months}} \approx \$0.22 \text{ per 100 pounds per month}$$

(There are other valid methods for estimating this rate of change.)

17. a. $D(R(x)) = \frac{1.02^x}{0.4899}$ dollars when x mountain bikes are sold

b. $R(400) \approx 2755$ deutsche marks
$D[R(400)] \approx \$5623$

c. To find $\frac{dD}{dx}$ when $x = 400$, we numerically investigate the secant-line slopes of $D \circ R$ as x approaches 400:

Input of close point on left	Secant line slope	Input of close point on right	Secant line slope
399.9	111.2382	400.1	111.4587
399.99	111.3374	400.01	111.3595
399.999	111.3473	400.001	111.3495
399.9999	111.3483	400.0001	111.3485
399.99999	111.3484	400.00001	111.3484
Trend ≈ 111.35		Trend ≈ 111.35	

When 400 mountain bikes are sold, the profit (in dollars) is increasing at a rate of approximately \$111.35 per mountain bike sold.

Section 4.2

1. a. $T(13) = 67.946$ seconds

b. $T(13 + h) = 0.181(13 + h)^2 - 8.463(13 + h) + 147.376 = 0.181h^2 - 3.757h + 67.946$

c. $\frac{T(13 + h) - T(13)}{13 + h - 13} = \frac{0.181h^2 - 3.757h}{h}$

d. $\lim_{h \to 0} \frac{0.181h^2 - 3.757h}{h} = \lim_{h \to 0}(0.181h - 3.757) = -3.757$ seconds per year of age

The swim time for a 13-year-old is decreasing by 3.757 seconds per year of age. This tells us that as a 13-year-old athlete gets older, the athlete's swim time improves.

3. a. $c(5) = 265.306$

b. $c(5 + h) = -0.566(5 + h)^2 + 21.092(5 + h) + 173.996 = -0.566h^2 + 15.432h + 265.306$

c. $\frac{c(5 + h) - c(5)}{5 + h - 5} = \frac{-0.566h^2 + 15.432h}{h}$

d. $\lim_{h \to 0} \frac{-0.566h^2 + 15.432h}{h} = \lim_{h \to 0}(-0.566h + 15.432) = 15.432$ index points per year

e. The consumer price index for college tuition was increasing by 15.432 index points per year in 1995.

f. Activity 13 of Section 4.1 used a symmetric difference quotient to estimate this rate of change as 15 points per year and used the numerical method with a quadratic model to estimate it as 15.4 points per year. Because we are dealing with a model that is not an exact fit to the data, none of these estimates is exact.

Assuming that you both have the data and can find a good model for the data, then

i. Because the algebraic method uses a rounded model, the numerical method with an unrounded model will probably produce the most accurate answer as long as the model is an excellent fit to the data.

ii. Use the data and a symmetric difference quotient to obtain a quick, rough estimate.

iii. Because both the numerical method and the algebraic method are somewhat time-consuming, use the data and a symmetric difference quotient to obtain a fairly good estimate without taking much time.

5. i. $g(4) = -89$

ii. $g(4 + h) = -6(4 + h)^2 + 7 = -6h^2 - 48h - 89$

iii. $\dfrac{c(4 + h) - c(4)}{4 + h - 4} = \dfrac{-6h^2 - 48h}{h}$

iv. $\lim\limits_{h \to 0} \dfrac{-6h^2 - 48h}{h} = \lim\limits_{h \to 0}(-6h - 48) = -48$

$\dfrac{dg}{dt} = -48$ when $t = 4$

7. a. i. $H(t) = -16t^2 + 100$

ii. $H(t + h) = -16(t + h)^2 + 100$
$= -16h^2 - 32th - 16t^2 + 100$

iii. $\dfrac{H(t + h) - H(t)}{t + h - t} = \dfrac{-16h^2 - 32th}{h}$

iv. $\lim\limits_{h \to 0} \dfrac{-16h^2 - 32th}{h} = \lim\limits_{h \to 0}(-16h - 32t) = -32t$

$\dfrac{dH}{dt} = -32t$ feet per second t seconds after the object is dropped

b. After 1 second, the velocity of the object was $-32(1) = -32$ feet per second; that is, the object was falling at a rate of 32 feet per second.

9. a. $D(a) = -0.045a^2 + 1.774a - 16.064$ million drivers age a years

b. $D(a + h) = -0.045(a + h)^2 + 1.774(a + h) - 16.064$
$= -0.045a^2 - 0.09ah - 0.045h^2 + 1.774a + 1.774h - 16.064$

$\dfrac{D(a + h) - D(a)}{a + h - a} = \dfrac{-0.09ah - 0.045h^2 + 1.774h}{h}$

$\lim\limits_{h \to 0} \dfrac{-0.09ah - 0.045h^2 + 1.774h}{h} =$

$\lim\limits_{h \to 0}(-0.09a - 0.045h + 1.774) = -0.09a + 1.774$

Thus $D'(a) = -0.09a + 1.774$ million drivers per year of age.

c. $D'(20) = -0.026$ million drivers per year of age. For 20-year-olds, the number of licensed drivers is decreasing by 26,000 drivers per year of age. This tells us that there are fewer 21-year-olds who are licensed drivers than there are 20-year-olds.

11. i. $f(x) = 3x - 2$

ii. $f(x + h) = 3(x + h) - 2 = 3x + 3h - 2$

iii. $\dfrac{f(x + h) - f(x)}{x + h - x} = \dfrac{3h}{h}$

iv. $\lim\limits_{h \to 0} \dfrac{3h}{h} = \lim\limits_{h \to 0} 3 = 3$. Therefore, $\dfrac{df}{dx} = 3$.

13. i. $f(x) = 3x^2$

ii. $f(x + h) = 3(x + h)^2 = 3x^2 + 6xh + 3h^2$

iii. $\dfrac{f(x + h) - f(x)}{x + h - x} = \dfrac{6xh + 3h^2}{h}$

iv. $\lim\limits_{h \to 0} \dfrac{6xh + 3h^2}{h} = \lim\limits_{h \to 0}(6x + 3h) = 6x$. Therefore, $f'(x) = 6x$.

15. i. $f(x) = x^3$

ii. $f(x + h) = (x + h)^3 = x^3 + 3x^2h + 3xh^2 + h^3$

iii. $\dfrac{f(x + h) - f(x)}{x + h - x} = \dfrac{3x^2h + 3xh^2 + h^3}{h}$

iv. $\lim\limits_{h \to 0} \dfrac{3x^2h + 3xh^2 + h^3}{h} = \lim\limits_{h \to 0}(3x^2 + 3xh + h^2) = 3x^2$.

Therefore, $f'(x) = 3x^2$.

17. a.

b. i. $q(t) = \dfrac{1}{t}$

ii. $q(t + h) = \dfrac{1}{t + h}$

iii. $\dfrac{q(t + h) - q(t)}{t + h - t} = \dfrac{\frac{1}{t + h} - \frac{1}{t}}{h}$

$= \left(\dfrac{\frac{1}{t + h} - \frac{1}{t}}{h}\right) \cdot \dfrac{t(t + h)}{t(t + h)}$

$= \dfrac{t - (t + h)}{ht(t + h)} = \dfrac{-h}{ht(t + h)}$

iv. $\lim\limits_{h \to 0} \dfrac{-h}{ht(t + h)} = \lim\limits_{h \to 0} \dfrac{-1}{t(t + h)} = \dfrac{-1}{t^2}$.

Therefore, $q'(t) = \frac{-1}{t^2}$. The graph of q' is the same as the graph in part *a*.

19. a. $p'(m)$

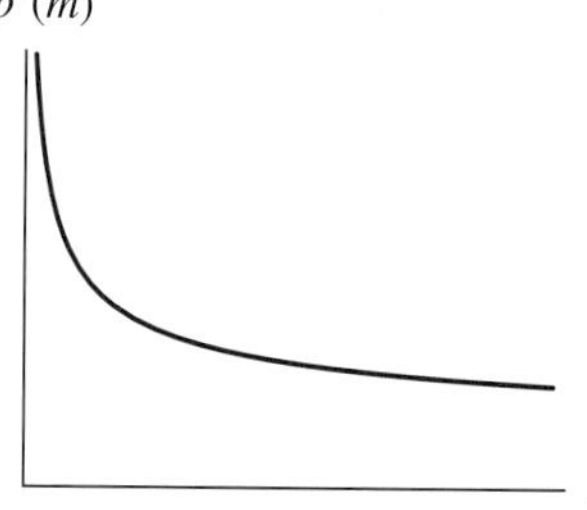

b. i. $p(m) = m + \sqrt{m}$

ii. $p(m + h) = m + h + \sqrt{m + h}$

iii. $\dfrac{p(m + h) - p(m)}{m + h - m} = \dfrac{h + \sqrt{m + h} - \sqrt{m}}{h}$

$$= \left(\frac{h + \sqrt{m + h} - \sqrt{m}}{h}\right)\frac{(\sqrt{m + h} + \sqrt{m})}{(\sqrt{m + h} + \sqrt{m})}$$

$$= \frac{h(\sqrt{m + h} + \sqrt{m}) + (m + h) - m}{h(\sqrt{m + h} + \sqrt{m})}$$

$$= \frac{h(\sqrt{m + h} + \sqrt{m}) + h}{h(\sqrt{m + h} + \sqrt{m})}$$

iv. $\displaystyle\lim_{h\to 0} \frac{h(\sqrt{m + h} + \sqrt{m}) + h}{h(\sqrt{m + h} + \sqrt{m})}$

$$= \lim_{h\to 0} \frac{(\sqrt{m + h} + \sqrt{m}) + 1}{\sqrt{m + h} + \sqrt{m}}$$

$$= \frac{2\sqrt{m} + 1}{2\sqrt{m}} = 1 + \frac{1}{2\sqrt{m}}$$

$\frac{dp}{dm} = 1 + \frac{1}{2\sqrt{m}}$. The graph of $\frac{dp}{dm}$ is the same as the one in part *a*.

Section 4.3

1.

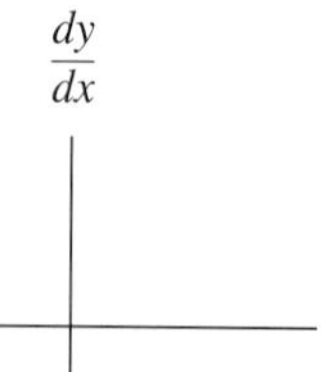

3.

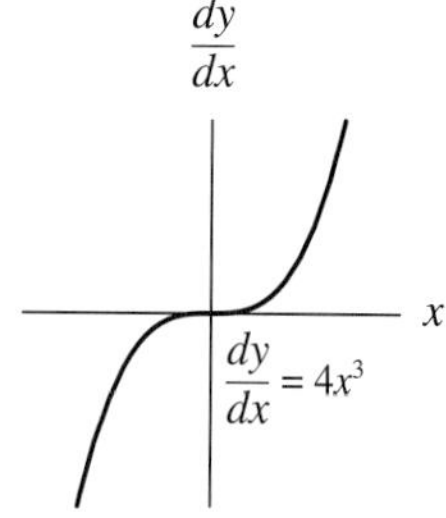

5.

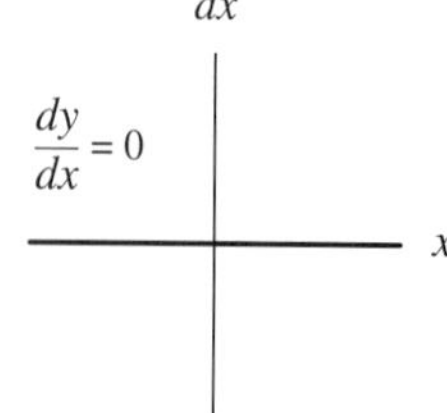

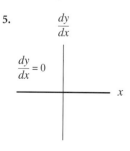

7. $\frac{dy}{dx} = 14x - 12$

9. $\frac{dy}{dx} = 15x^2 + 6x - 2$

11. $\frac{dy}{dx} = 18x^{-3} = \frac{18}{x^3}$

13. $h'(x) = -0.049\left(\frac{1}{2}x^{-1/2}\right) = \frac{-0.0245}{\sqrt{x}}$

15. a. $A'(t) = 0.1333$ dollars per year t years after 1990
b. $A(10) \approx \$1.50$
c. $A'(9) = \$0.1333$ per year. The transaction fee in 1999 was increasing by approximately \$0.13 per year.

17. a. $P'(t) = 15.48$ thousand people per year t years after 1950
b. $P(20) = 795$ thousand people
c. $P'(40) = 15.48$ thousand people per year

19. Let $j(x) = f(x) - g(x)$. The derivative of j is

$$\frac{dj}{dx} = \lim_{h\to 0} \frac{f(x + h) - g(x + h) - [f(x) - g(x)]}{h}$$

$$= \lim_{h\to 0} \frac{f(x + h) - f(x) - g(x + h) + g(x)}{h}$$

$$= \lim_{h\to 0} \left[\frac{x(x + h) - f(x)}{h} - \frac{g(x + h) - g(x)}{h}\right]$$

$$= \lim_{h\to 0} \frac{f(x + h) - f(x)}{h} - \lim_{h\to 0} \frac{g(x + h) - g(x)}{h}$$

$$\frac{dj}{dx} = \frac{df}{dx} - \frac{dg}{dx}$$

21. $T'(x) = -1.6t + 2$°F per hour t hours after noon
a. $T'(1.5) = -0.4$°F per hour
$T'(-5) = 10$°F per hour
b. $T'(-5) = 10$°F per hour
c. $T'(0) = 2$°F per hour
d. $T'(4) = -4.4$°F per hour

23. a. $N(-5) \approx 33.4$ million senior citizens in 1995
$N(30) \approx 70.7$ million senior citizens in 2030
b. $N'(-5) \approx 0.015$ million senior citizens per year in 1995
$N'(30) \approx 2.1$ million senior citizens per year in 2030
c. $\frac{N'(30)}{N(30)} \cdot 100\% \approx 3\%$ per year in 2030
d. Solve $70.68 = 0.201P$ for P to obtain $P \approx 351.6$ million senior citizens in 2030.

25. a, b. In 1970 the number of births was falling at a rate of approximately 227 births per year. In 1990 the number of births was rising at a rate of about 184 births per year.

27. a. $I(t) = -11t^3 + 102.095t^2 - 404.857t + 2390.048$ thousand imported passenger cars t years after 1990

b. $I'(t) = -33t^2 + 204.19t - 404.857$ thousand cars per year t years after 1990

c. $I'(6) \approx -367.7$ thousand cars per year

d. $I(6) \approx 1260.3$ thousand cars $= 1{,}260{,}300$ cars

e. $\dfrac{I'(2)}{I(2)} \cdot 100\% \approx -6.8\%$ per year

29. a. $P(x) = 175 - 0.0146x^2 + 0.7823x - 46.9125 - \dfrac{49.6032}{x}$ dollars when x windows are produced each hour

b. $P'(x) = -0.0292x + 0.7823 + \dfrac{49.6032}{x^2}$ dollars per window when x windows are produced each hour

c. $P(80) \approx \$96.61$

d. $P'(80) \approx -\$1.55$ per window

When 80 windows are produced each hour, the profit from the sale of one window is decreasing by about \$1.55 per window produced.

31. a. $P(x) = 0.0426x^2 + 2.049x + 251.796 + 68.738(1.0213^x)$ million people x years after 1980

b. $P'(x) = 0.0852x + 2.049 + 68.738\,(\ln 1.0213)(1.0213^x)$ million people per year x years after 1980

c. $P(14) \approx 381.2$ million people

d. $P'(14) \approx 5.2$ million people per year

33. a.
$$f'(x) = \begin{cases} 44.408x - 108.431 \text{ million lb per year} & \text{when } 0 \le x < 10 \\ 171.244x - 2253.951 \text{ million lb per year} & \text{when } 10 < x \le 20 \end{cases}$$

x years after 1970

b. In 1970: $f'(0) = -108.431$ million pounds per year

In 1980: $f'(10)$ does not exist. The rate of change of the amount of fish produced cannot be calculated directly from this model. One of the estimation techniques discussed in Example 3 should be used.

In 1990: $f'(20) = 1170.93$ million pounds per year

c. $\dfrac{f'(20)}{f(20)} \cdot 100\% \approx 16.9\%$ per year

Section 4.4

1. **3.**

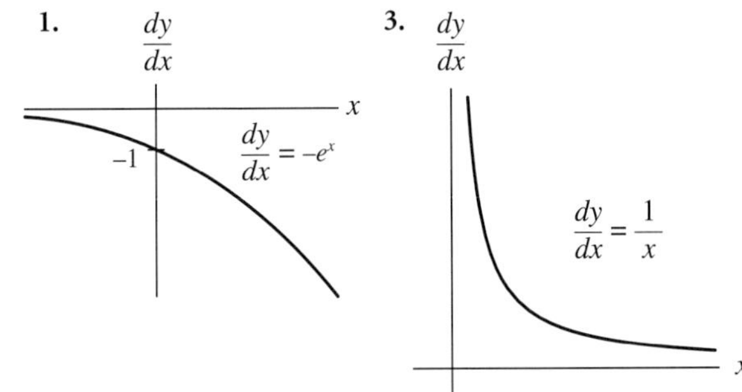

5.

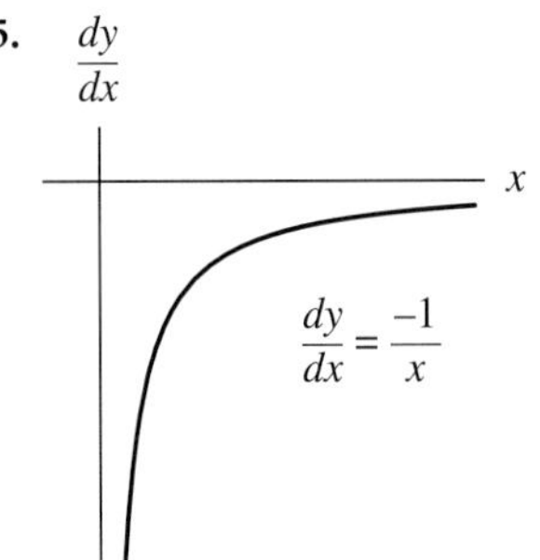

7. $h'(x) = 14x + \dfrac{13}{x}$

9. $\dfrac{dg}{dx} = 17\,(\ln 4.962)(4.962^x)$

11. $f'(x) = 100{,}000 \ln\left(1 + \frac{0.05}{12}\right)^{12}\left(1 + \frac{0.05}{12}\right)^{12x} \approx 4989.61218\left(1 + \frac{0.05}{12}\right)^{12x} \approx 4989.61218(1.05116^x)$

13. $\dfrac{dy}{dx} = \dfrac{-4.2}{x} + 3.3(\ln 2.9)(2.9^x)$

15. a. $h = \ln 2.5 \approx 0.916$ hours

b. $\dfrac{dV}{dh} = e^h$ quarts per hour

c. $V(0.4) \approx 1.49$ quarts per hour ≈ 0.025 quart per minute

$V(0.7) \approx 2.01$ quarts per hour ≈ 0.034 quart per minute

$V\left(\dfrac{55}{60}\right) \approx 2.5$ quarts per hour ≈ 0.042 quart per minute

17. a. $\dfrac{dy}{dx} = (\ln e)e^x = (1)e^x = e^x$

b. $\dfrac{dy}{dx} = (\ln e^k)e^{kx} = k(\ln e)e^{kx} = k(1)e^{kx} = ke^{kx}$

19. a. $w(7) \approx 25.64$ grams; $w'(7) \approx 1.05$ grams per week

b. $\dfrac{w(9) - w(4)}{9 - 4} \approx \dfrac{5.977 \text{ grams}}{5 \text{ weeks}} \approx 1.195$ grams per week on average

c. The older the mouse is, the more slowly it gains weight because the rate-of-change formula $w'(x) = \frac{7.37}{x}$ has the age of the mouse in the denominator.

21. a. $CPI(x) = -349.751 + 227.081 \ln x$, where x is the number of years since 1980

b. $\dfrac{279.8 - 249.8}{1996 - 1994} = 15$ index points per year

c. $CPI'(15) = \dfrac{227.081}{15} \approx 15.1$ index points per year

23. a. $M(t) = -0.216t^3 + 7.383t^2 + 661.969t + 19{,}839.889$ dollars t years after 1947

b. $M'(t) = -0.649t^3 + 14.765t^2 + 661.969$ dollars per year t years after 1947

c.

Year	t	$M'(t)$	$\frac{M'(t)}{M(t)} \cdot 100$
1972	25	626	1.7% per year
1980	33	443	1.1% per year
1984	37	320	0.7% per year
1992	45	12.9	0.3% per year
1996	49	-172	-0.4% per year

d. There seems to be no relationship between re-election and the rate of change in median family income. This may be an indication that the model does not provide sufficient information to answer this question accurately because it models the 50-year trend in median family income rather than the median income close to each election year.

25. a. $A'(r) = 10{,}000e^{10r}$ dollars per 100 percentage points when the interest rate is $100r\%$

b. $20,137.53 per 100 percentage points or $201.38 per percentage point. The rate of change is large because it approximates by how much the value will increase when r changes by 1. Because the interest rate is input in decimals, an increase of 1 in r corresponds to a change in the interest rate of 100 percentage points.

c. $A'(r) = 100e^{0.1r}$ dollars per percentage point when the interest rate is $r\%$

d. $201.38 per percentage point

27. a.

$$p'(t) = \begin{cases} -23.73t^2 + 241.92t + 193.92 \text{ people/year} & \text{when } 0.7 \le t < 13 \\ 45{,}544\,(\ln 0.8474)(0.8474^t) \text{ people/year} & \text{when } 13 < t \le 55 \end{cases}$$

where t is the number of years since the beginning of 1860

b. $p'(10) \approx 240$ people per year

$p'(13)$ does not exist

$p'(40) \approx -10$ people per year

The rate of change of the population in 1873 could be estimated by

i. Calculating the model estimate for the populations in 1872 and 1874: 5954 and 4484

ii. Finding the average rate of change (symmetric difference quotient) for those years:

$$\frac{4484 - 5954}{1874 - 1872} \approx -735 \text{ people per year}$$

(There are other valid methods of estimating this rate of change.)

Section 4.5

1. a. $f(x(2)) = f(6) = 140$

b. $\frac{df}{dx} = -27$

c. $\frac{dx}{dt} = 1.3$

d. $\frac{df}{dt} = (-27)(1.3) = -35.1$

3. The value of the investor's gold is increasing at a rate of (0.2 troy ounces per day)($395.70 per troy ounce) = $79.14 per day.

5. a. $R(476) = 10{,}000$ deutsche marks (dm)

On January 5, 2001, sales were 476 units, producing revenue of 10,000 dm.

b. $D(10{,}000) = \$20{,}412$

On January 5, 2001, 10,000 dm were worth $20,412.

c. $\frac{dR}{dx} = 2.6$ dm per unit

Revenue was increasing by 2.6 dm per unit sold.

d. $\frac{dD}{dr} = \$2.0412$ per dm

The exchange rate was $2.0412 per dm.

e. $\frac{dD}{dx} = (2.6 \text{ dm per unit})(\$2.0412 \text{ per dm}) \approx \$5.31$ per unit

On January 5, 2001, revenue was increasing at a rate of $5.31 per unit sold.

7. a. $p(10) \approx 12.009$ thousand people

In 2005 the city had a population of approximately 12,000 people.

b. $g(p(10)) \approx g(12.009) \approx 22$ garbage trucks

In 2005 the city owned 22 garbage trucks.

c. $p'(t) = \frac{31.2e^{-0.02t}}{(1 + 12e^{-0.02t})^2}$ thousand people per year

$p'(10) \approx 0.22$ thousand people per year

In 2005, the population was increasing at a rate of approximately 220 people per year.

d. $g'(p) = 2 - 0.003p^2$ trucks per thousand people

$g'(12.009) \approx 1.6$ trucks per thousand people

In 2005, when the population was about 12,000, the number of garbage trucks needed by the city was increasing by 1.6 truck per thousand people.

e. When $t = 10$,

$$\frac{dg}{dt} = \frac{dg}{dp} \cdot \frac{dp}{dt} \approx \frac{1.6 \text{ trucks}}{\text{thousand people}} \cdot \frac{0.22 \text{ thousand people}}{\text{year}} \approx 0.34 \text{ trucks per year}$$

In 2005, the number of trucks needed by the city was increasing at a rate of 0.34 truck per year, or 1 truck every 3 years.

f. See interpretations in parts *a* through *e*.

9. $c(x(t)) = 3(4 - 6t)^2 - 2$

$$\frac{dc}{dt} = 6(4 - 6t)(-6) = -144 + 216t$$

11. $h(p(t)) = \dfrac{4}{1 + 3e^{-0.5t}}$

$$\frac{dh}{dt} = \frac{4(3)(0.5)e^{-0.5t}}{(1 + 3e^{-0.5t})^2}$$

13. $k(t(x)) = 4.3(\ln x)^3 - 2(\ln x)^2 + 4 \ln x - 12$

$$\frac{dk}{dx} = \frac{1.29\ (\ln x)^2}{x} - \frac{4 \ln x}{x} + \frac{4}{x}$$

15. $p(t(k)) = 7.9(1.046^{14k^3 - 12k^2})$

$$\frac{dp}{dk} = 7.9(\ln 1.046)(1.046^{14k^3 - 12k^2})(42k^2 - 24k)$$

17. Inside function: $u = 3.2x + 5.7$

Outside function: $f = u^5$

$$\frac{df}{dx} = 5(3.2x + 5.7)^4(3.2)$$

19. Inside function: $u = x^2 - 3x$

Outside function: $f = u^{\frac{1}{2}}$

$$\frac{df}{dx} = \frac{1}{2}(x^2 - 3x)^{-\frac{1}{2}}(2x - 3)$$

21. Inside function: $u = 35x$

Outside function: $f = \ln u$

$$\frac{df}{dx} = \frac{1}{35x}(35) = \frac{1}{x}$$

23. Inside function: $u = 16x^2 + 37x$

Outside function: $f = \ln u$

$$\frac{df}{dx} = \frac{1}{16x^2 + 37x}(32x + 37)$$

25. Inside function: $u = 0.695x$

Outside function: $f = 72.378e^u$

$$\frac{df}{dx} = 72.378e^{0.695x}(0.695)$$

27. Inside function: $u = 0.0856x$

Outside function: $f = 1 + 58.32e^u$

$$\frac{df}{dx} = 58.32e^{0.0856x}(0.0856)$$

29. Inside function: $u = 4x + 7$

Outside function: $f = 350u^{-1}$

$$\frac{df}{dx} = -350(4x + 7)^{-2}(4)$$

31. Inside function:

$$u = 1 + 8.976e^{-1.243x}\begin{cases}\text{inside:} & w = -1.243x\\ \text{outside:} & u = 1 + 8.976e^w\end{cases}$$

Outside function: $f = 3706.5u^{-1} + 89070$

$$\frac{df}{dx} = -3706.5(1 + 8.976e^{-1.243x})^{-2}(8.976e^{-1.243x})(-1.243)$$

$$= \frac{(3706.5)(8.976)(1.243)e^{-1.243x}}{(1 + 8.976e^{-1.243x})^2}$$

33. Inside function: $u = \sqrt{x} - 3x$

Outside function: $f = 3^u$

$$\frac{df}{dx} = 3^{\sqrt{x} - 3x}(\ln 3)\left(\frac{1}{2\sqrt{x}} - 3\right)$$

35. Inside function: $u = 2^x$

Outside function: $f = \ln u$

$$\frac{df}{dx} = \frac{1}{2^x}2^x(\ln 2) = \ln 2$$

37. Inside function:

$$u = 1 + Ae^{-Bx}\begin{cases}\text{inside:} & w = -Bx\\ \text{outside:} & u = 1 + Ae^w\end{cases}$$

Outside function: $f = Lu^{-1}$

$$\frac{df}{dx} = -L(1 + Ae^{-Bx})^{-2}[Ae^{-Bx}(-B)] = \frac{LABe^{-Bx}}{(1 + Ae^{-Bx})^2}$$

39. a. $A(t) = 1500e^{0.04t}$ dollars after t years

b. $A'(t) = 1500(0.04)e^{0.04t} = 60e^{0.04t}$ dollars per year after t years

c. $A'(1) \approx \$62.45$ per year

$A'(2) \approx \$65.00$ per year

d. The rates of change in part *c* tell you approximately how much interest your account will earn during the second and third years. The actual amounts are $A(2) - A(1) \approx \$63.71$ during the second year and $A(3) - A(2) \approx \$66.31$ during the third year.

41. a. $R'(q) = 0.0314(0.62285)e^{0.62285q}$ million dollars per quarter q quarters after the beginning of 1998

b.

Quarter ending	June 1998	June 1999	June 2000
$R(q)$ (millions of dollars)	3.0	4.2	18.8
$R'(q)$ (millions of dollars per quarter)	0.07	0.82	9.92
$\frac{R'(q)}{R(q)} \cdot 100\%$ (% per quarter)	2.3	19.5	52.7

43. a. A logistic model is probably a better model because of the leveling-off behavior, although neither model should be used to extrapolate beyond one 24-hour day.

b. $C(t) = \dfrac{1342.077}{1 + 36.797e^{-0.258856t}}$ calls t hours after 5 a.m.

c. $C'(t) = \dfrac{(1342.077)(36.797)(0.258856)e^{-0.258856t}}{(1 + 36.797e^{-0.258856t})^2}$ calls per hour t hours after 5 a.m.

d. noon: $C'(7) \approx 42$ calls per hour
10 p.m.: $C'(17) \approx 74$ calls per hour
midnight: $C'(19) \approx 58$ calls per hour
4 a.m.: $C'(23) \approx 28$ calls per hour

e. The rates of change give approximate hourly calls. This information could be used to determine how many dispatchers would be needed each hour.

45. a. $c'(4) \approx 1575$ deaths per week

b. $\dfrac{c'(4)}{c(4)} \cdot 100\% \approx 108\%$ increase per week

c. $c'(8) \approx 25{,}331$ deaths per week

$\dfrac{c'(8)}{c(8)} \cdot 100\% \approx 48\%$ increase per week

d. Although the rate of change is larger, it represents a smaller proportion of the total number of deaths that had occurred at that time.

47. a. The data are essentially concave up, which indicates that a quadratic or exponential model may be appropriate. Looking at the second differences and percentage differences indicates that a quadratic model is the better choice.

b. $u(x) = 177.356x^2 - 342.240x + 5914.964$ units per week x years after 1990

c. $C(u(x)) = 3250.23 + 74.95 \ln(177.356x^2 - 342.240x + 5914.964)$ dollars per week x years after 1990

d. $\dfrac{dC}{dx} = \left(\dfrac{74.95}{177.356x^2 - 342.240x + 5914.964}\right) \cdot$ $(354.712x - 342.240)$ dollars per week per year x years after 1990

e.

Year	2002	2003	2004	2005
x	12	13	14	15
$C(x)$ ($/week)	4015.95	4026.40	4036.31	4045.72
$C'(x)$ ($/week/year)	10.73	10.18	9.66	9.17

f, g. Although a graph of C may not appear ever to decrease, a graph of $\frac{dC}{dx}$ is negative between $x = 0$ and $x = 0.965$, indicating that cost was decreasing from the end of 1990 to (almost) the end of 1991. A close-up view of the graph of C between $x = 0$ and $x = 2$ confirms this.

49. Composite functions are formed by making the output of one function (the inside) the input of another function (the outside). It is imperative that the output of the inside and the input of the outside agree in the quantity that they measure as well as in the units of measurement.

Section 4.6

1. $h'(2) = f(2)g'(2) + f'(2)g(2) = 6(3) + (-1.5)(4) = 12$

3. a. i. In 1997 there were 75,000 households in the city.
ii. In 1997 the number of households was declining at a rate of 1200 per year.
iii. In 1997, 52% of households owned a computer.
iv. In 1997, the percentage of households with a computer was increasing by 5 percentage points per year.

b. Input: the number of years since 1995
Output: the number of households with computers

c. $N(2) = h(2)c(2) = (75{,}000)(0.52) = 39{,}000$ households with computers

$N'(2) = h(2)c'(2) + h'(2)c(2) = (75{,}000)(0.05) + (-1200)(0.52) = 3126$ households per year

In 1997 there were 39,000 households with computers, and that number was increasing at a rate of 3126 households per year.

5. a. i. $S(10) \approx \$15.24$; $S'(10) \approx -\$0.02$ per week
After 10 weeks, 1 share is worth $15.24, and the value is declining by $0.02 per week.
ii. $N(10) = 125$ shares; $N'(10) = 5$ shares per week
After 10 weeks, the investor owns 125 shares and is buying 5 shares per week.
iii. $V(10) = S(10)N(10) \approx \$1905$; $V'(10) = S(10)N'(10) + S'(10)N(10) \approx \$73.50$ per week
After 10 weeks, the investor's stock is worth approximately $1905, and the value is increasing at a rate of $73.50 per week.

b. $V'(x) = \left(15 + \frac{2.6}{x+1}\right)(0.5x) + (100 + 0.25x^2) \cdot \frac{-2.6}{(x+1)^2}$ dollars per week after x weeks

7. (500 acres)(5 bushels/acre/year) + (130 bushels/acre) (50 acres/year) = 9000 bushels/year

9. a. $(17{,}000)(0.48) = 8160$ voters
b. $(8160)(0.57) \approx 4651$ votes for candidate A
c. $17{,}000\,(0.48)(-0.03) + 17{,}000(0.57)(0.07) \approx 434$ votes for candidate A per week

11. $f'(x) = (3x^2 + 15x + 7)(96x^2) + (6x + 15)(32x^3 + 49)$

13. $f'(x) = (12.8893x^2 + 3.7885x + 1.2548) \cdot [29.685(\ln 1.7584)(1.7584^x)] + (25.7786x + 3.7885) \cdot [29.685(1.7584^x)]$

15. $f'(x) = (5.7x^2 + 3.5x + 2.9)^3 \cdot [-2(3.8x^2 + 5.2x + 7)^{-3}(7.6x + 5.2)] + [3(5.7x^2 + 3.5x + 2.9)^2(11.4x + 3.5)](3.8x^2 + 5.2x + 7)^{-2}$

17. $f'(x) = 12.624(14.831^x)(-2x^{-3}) + 12.624(\ln 14.831) \cdot (14.831^x)(x^{-2})$

19. $f'(x) = (79.32x)\left(\dfrac{1984.32(7.68)(0.859347)e^{-0.859347x}}{(1 + 7.68e^{-0.859347x})^2}\right) + 79.32\left(\dfrac{1984.32}{1 + 7.68e^{-0.859347x}} + 1568\right)$

21. $f'(x) = -430(0.62^x)[6.42 + 3.3(1.46^x)]^{-2}[3.3(\ln 1.46)(1.46^x)] + [6.42 + 3.3(1.46^x)]^{-1}(430 \ln 0.62)(0.62^x)$

23. $f'(x) = 4x\left[\frac{1}{2}(3x + 2)^{-1/2}(3)\right] + 4\sqrt{3x + 2}$
$= \dfrac{6x}{\sqrt{3x + 2}} + 4\sqrt{3x + 2}$

25. $f'(x) = -14{,}000x(1 + 12.6e^{-0.73x})^{-2}(12.6(-0.73)e^{-0.73x}) + (1 + 12.6e^{-0.73x})^{-1}(14{,}000)$
$= \dfrac{14{,}000(12.6)(0.73)xe^{-0.73x}}{(1 + 12.6e^{-0.73x})^2} + \dfrac{14{,}000}{1 + 12.6e^{-0.73x}}$

27. a. Price $= 0.049m + 1.144$ dollars m months after December
Quantity sold $= -0.946m^2 + 0.244m + 279.911$ units sold m months after December
b. $R(m) = (0.049m + 1.144)(-0.946m^2 + 0.244m + 279.911)$ dollars of revenue m months after December
c. $R(8) \approx \$340.05$; $R(9) \approx \$325.78$
d. Because the revenue in September is less than that in August, the rate of change in August is probably negative.
e. $R'(m) = (0.049m + 1.144)(-1.893m + 0.244) + 0.049(-0.946m^2 + 0.244m + 279.911)$ dollars of revenue per month m months after December
f. $R'(2) \approx \$9.17$ per month
$R'(8) \approx -\$12.04$ per month
$R'(9) \approx -\$16.55$ per month

29. a. $P(t) = 2.32t + 204.508$ million people t years after 1970
b. $F(t) = (2.32t + 204.508)\dfrac{7.667(0.9688)^t}{100}$ million people t years after 1970
c. $F'(t) = (2.32t + 204.508)\left(\dfrac{7.667}{100}(\ln 0.9688)(0.9688^t)\right) + (2.32)\left(\dfrac{7.667}{100}(0.9688^t)\right)$ million people per year t years after 1970
d. 1990: $F'(20) \approx -0.23$ million people per year
1995: $F'(25) \approx -0.21$ million people per year
2000: $F'(30) \approx -0.19$ million people per year

31. a. $m(x) = 0.211x + 8.177$ million men 65 years or older x years after 1970
$p(x) = \dfrac{14.255(0.899260^x) + 6}{100}$ percentage (expressed as a decimal) of men 65 or older below poverty level x years after 1970
b. $n(x) = m(x)p(x)$
$= (0.211x + 8.177)\left[\dfrac{14.255(0.899260^x) + 6}{100}\right]$
million men 65 or older below poverty level
c. 1990: $n'(20) \approx -0.006$ million men per year
2000: $n'(30) \approx 0.005$ million men per year

33. a. $E(x) = \dfrac{0.73(1.2912^x) + 8}{100}(-0.026x^2 - 3.842x + 538.868)$
women receiving epidurals at the Arizona hospital x years after 1980
$E'(x) = \dfrac{0.73(1.2912^x) + 8}{100}(-0.052x - 3.842) + \dfrac{0.73(\ln 1.2912)(1.2912^x)}{100}(-0.026x^2 - 3.842x + 538.868)$ women per year x years after 1980
b. Increasing by $p'(17) \approx 14.4$ percentage points per year
c. Decreasing by approximately 5 births per year ($b'(17) \approx -4.7$)
d. Increasing by $E'(17) \approx 64$ women per year
e. Profit $= \$57 \cdot E(17) \approx \$17{,}043$ (using a value of 299 for the number of births) or \$17,071 (using an unrounded number of births)

35. a. $E(x) = -151.516x^3 + 2060.988x^2 - 8819.062x + 195{,}291.201$ students enrolled x years after the 1980–81 school year
$D(x) = -14.271x^3 + 213.882x^2 - 1393.655x + 11{,}697.292$ students dropping out x years after the 1980–81 school year
b. $P(x) = \dfrac{D(x)}{E(x)} \cdot 100$ percent x years after the 1980–81 school year
c. $P'(x) = D(x)(-1[E(x)]^{-2}E'(x)) + D'(x)[E(x)]^{-1}$ percentage points per year x years after the 1980–81 school year
d.

x	$P'(x)$ (percentage points per year)	x	$P'(x)$ (percentage points per year)
0	-0.44	5	-0.19
1	-0.38	6	-0.19
2	-0.32	7	-0.22
3	-0.26	8	-0.29
4	-0.21	9	-0.41

In the 1980–81 school year, the rate of change was most negative with a value of -0.44 percentage points per year. This is the most rapid decline during this time period. The rate of change was least negative in the 1985–86 school year with a value of -0.187 percentage points per year.

e. Negative rates of change indicate that high school attrition in South Carolina was improving during the 1980s.

37. The inputs must correspond in order for the result of the multiplication to be meaningful.

Chapter 4 Review Test

1. a. $M(t) = 0.982(1.531417^t)$ dollars per share t years after 1989

b.

Input of close point on left	Secant line slope	Input of close point on right	Secant line slope
7.9	12.3996	8.1	12.9395
7.99	12.6388	8.01	12.6927
7.999	12.6630	8.001	12.6684
7.9999	12.6655	8.0001	12.6660
7.99999	12.6657	8.00001	12.6657
Trend ≈ 12.67		Trend ≈ 12.67	

c. In 1997 the yearly high stock price for Microsoft was increasing by approximately \$12.67 per share per year.

d. $M(t) = 0.982(\ln 1.531417)(1.531417^t)$ dollars per share per year t years after 1989

$M(8) \approx \$12.67$ per share per year

2. a. $\dfrac{5277 - 4606}{1997 - 1995} = \$335.5$ billion per year

b.

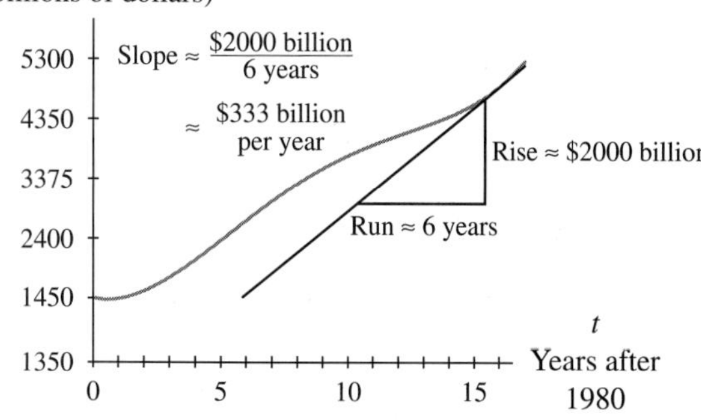

c. $A'(16) \approx \$338.4$ billion per year

3. a. $\dfrac{dD}{dt} = \dfrac{7.978(172.8)(0.773)e^{-0.773t}}{(1 + 172.8e^{-0.773t})^2}$ pounds per person per year t years after 1980

b. $D'(10) \approx 0.4$ pound per person per year. In 1990 the average annual per capita consumption of turkey in the United States was increasing by 0.4 pound per person per year.

c. $D'(19) \approx 0.0004$ pound per person per year. There was essentially no growth in the per capita consumption of turkey in 1999.

4. a. Rewrite $P(t)$ as $P(t) = 100N(t)[A(t)]^{-1}$. Use the Product Rule to find the derivative.

$P'(t) = 100N(t)\{-1[A(t)]^{-2}A'(t)\} + 100N'(t)[A(t)]^{-1}$ percentage points per year t years after 1980

b. Input units: years

Output units: percentage points per year

5. (1) Begin with a point: $(x, 7x + 3)$

(2) Choose a close point: $(x + h, 7(x + h) + 3)$

(3) Find a formula for the slope between the two points. Simplify completely.

$$\text{Slope} = \frac{7x + 7h + 3 - (7x + 3)}{x + h - x} = \frac{7h}{h}$$

(4) Find the limit of the slope as $h \to 0$: $\lim_{h \to 0} \dfrac{7h}{h} = 7$

6. a. The function g is continuous at $x = -1$ because $\lim_{x \to -1^-} g(x) = \lim_{x \to -1^+} g(x) = g(-1) = 1$

b. The graph of the function g is not smooth because the slope of the graph to the left of $x = -1$ is 2 and the slope of the graph to the right of $x = -1$ is 3.

c. $g'(x) = \begin{cases} 2 & \text{when } x < -1 \\ 3 & \text{when } x > -1 \end{cases}$

CHAPTER 5

Section 5.1

1. $32\% - (4 \text{ percentage points per hour})\left(\frac{1}{3} \text{ hour}\right) \approx 30.7\%$

3. $f(3.5) \approx f(3) + f'(3)(0.5)$
$= 17 + (4.6)(0.5) = 19.3$

5. a. Increasing production from 500 to 501 units will increase cost by approximately \$17.

b. If sales increase from 150 to 151 units, then profit will increase by approximately \$4.75.

7.

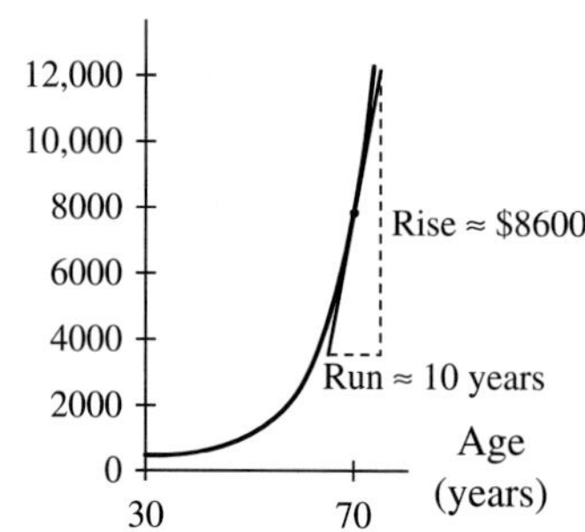

Slope of tangent line ≈ \$860 per year of age
Annual premium for 70-year-old ≈ \$7850
Premium for 72-year-old ≈ \$7850 + 2(\$860) = \$9570
(Answers will vary.)

9. a.

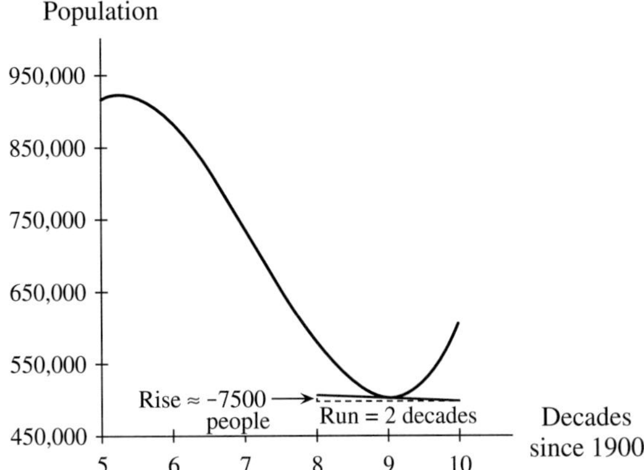

Slope of tangent line ≈ -3750 people per decade
Population in 1990 ≈ 500,000 people
Population in 2000 ≈ 500,000 − 3750 = 496,250 people
(Answers will vary.)

b. $P(10) = 606{,}030$ people

c. Answers will vary.

11. a. In 1990 the population of South Carolina was increasing by 81.3 thousand people per year.

b. Between 1990 and 1992, the population increased by approximately 162.6 thousand people.

c. By finding the slope of the tangent line at 1990 and multiplying by 2, we determine the change in the tangent line from 1990 through 1992 and use that change to estimate the change in the population function.

13. a. The population was growing at a rate of 1.88 million people per year in 1985.

b. Between 1985 and 1986, the population of Mexico increased by approximately 1.88 million people.

15. a. In 1995 the percentage was increasing by 1.74 percentage points per year.

b. We would expect an increase of approximately 1.74 percentage points between 1995 and 1996.

c. $p(11) - p(10) \approx 1.66$ percentage points

d. $82.2 - 81.0 = 1.2$ percentage points

e. As long as the data in part *d* were correctly reported, the answer to part *d* is the most accurate one.

17. a. $R(x) = (-7.032 \cdot 10^{-4})x^2 + 1.666x + 47.130$ dollars when x hot dogs are sold

b. Cost: $c(x) = 0.5x$ dollars per hot dog
Profit: $p(x) = R(x) - c(x) = (-7.032 \cdot 10^{-4})x^2 + 1.166x + 47.130$ dollars per hot dog

c.

x (hot dogs)	$R'(x)$ (dollars per hot dog)	$c'(x)$ (dollars per hot dog)	$p'(x)$ (dollars per hot dog)
200	1.38	0.50	0.88
800	0.54	0.50	0.04
1100	0.12	0.50	-0.38
1400	-0.30	0.50	-0.80

If the number of hot dogs sold increases from 200 to 201, the revenue increases by approximately \$1.38 and the profit increases by approximately \$0.88. If the number increases from 800 to 801, the revenue increases by 0.54, but the profit sees almost no increase (4 cents). If the number increases from 1100 to 1101, the increase in revenue is only about 12 cents. Because this marginal revenue is less than the marginal cost at a sales level of 1100, the result of the sales increase from 1100 to 1101 is a decrease of \$0.38 in profit. If the number of hot dogs increases from 1400 to 1401, revenue declines by approximately 30 cents and profit declines by approximately 80 cents.

d.

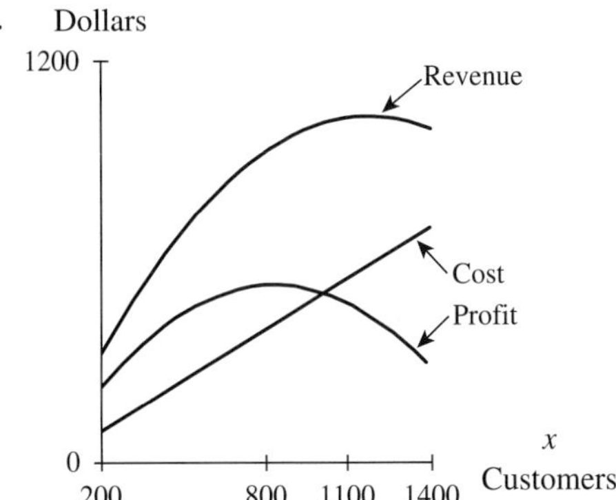

The marginal values in part *c* are the slopes of the graphs shown in the figure above. For example, at $x = 800$, the slope of the revenue graph is \$0.54 per hot dog, the slope of the cost graph is \$0.50 per hot dog, and the slope of the profit graph is \$0.04 per hot dog. We see from the graph that maximum profit is realized when about 800 hot dogs are sold.

Revenue is greatest near $x = 1100$, so the marginal revenue there is small. However, once costs are factored in, the profit is actually declining at this sales level. This is illustrated by the graph.

19. a. United States: $A(t) = 0.109t^3 - 1.555t^2 + 10.927t + 100.320$; Canada: $C(t) = 0.150t^3 - 2.171t^2 + 15.814t + 99.650$; Peru: $P(t) = 85.112(2.01325^t)$; Brazil: $B(t) = 73.430(2.61594^t)$; where t is the number of years since 1980.

b, c.

	U.S.	Canada	Peru	Brazil
Rate of change in 1987 (CPI points per year)	5.2	7.5	7984	59,193
1988 CPI estimate	143	163	19,134*	136,451*

* Based on 1987 data rather than $P(7)$ and $B(7)$ because the models differ significantly from the data.

21. a. $A(t) = 300\left(1 + \frac{0.065}{12}\right)^{12t}$ dollars after t years
    b. $A(t) \approx 300(1.066972^t)$ dollars after t years
    c. $A(2) \approx \$341.53$
    d. $A'(2) \approx \$22.14$ per year
    e. $\frac{1}{4}A'(2) \approx \$5.53$
23. a. $R(A) = -0.158A^3 + 5.235A^2 - 23.056A + 154.884$ thousand dollars of revenue when A thousand dollars is spent on advertising
    b. When \$10,000 is spent on advertising, revenue is increasing by \$34.3 thousand per thousand advertising dollars. If advertising is increased from \$10,000 to \$11,000, the car dealership can expect an approximate increase in revenue of \$34,300.
    c. When \$18,000 is spent on advertising, revenue is increasing by \$12.0 thousand per thousand advertising dollars. If advertising is increased from \$18,000 to \$19,000, the car dealership can expect an approximate increase in revenue of \$12,000.
25. Answers will vary.

Section 5.2

1. Quadratic, cubic, and many product, quotient, and composite functions could have relative maxima or minima.
3. 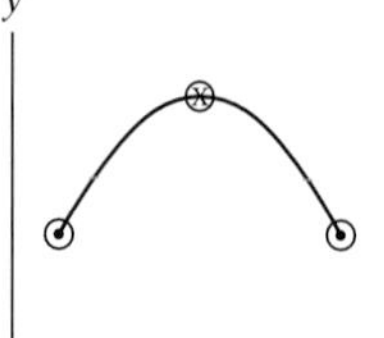

The derivative is zero at the absolute maximum point.

5. 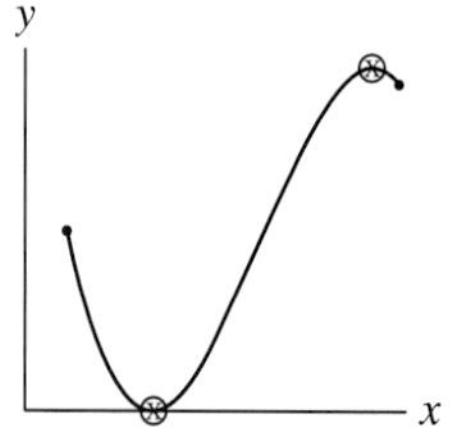

The derivative is zero at both optimal points.

7. 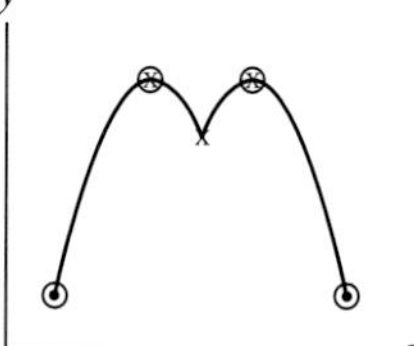

The derivative is zero at both absolute maximum points. The derivative does not exist at the relative minimum point.

9. 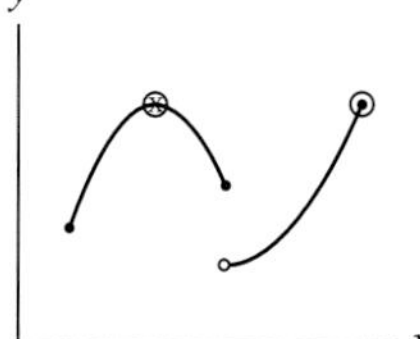

The derivative is zero at the absolute maximum point marked with an X.

11. Answers will vary. One such graph is $y = x^3$.
13. a. All statements are true.
    b. $f'(2)$ does not exist because f is not continuous at $x = 2$.
    c. $f'(x)$ is less than zero for $x < 2$ because the graph to the left of $x = 2$ is decreasing.
    d. $f'(2)$ does not exist because f is not smooth at $x = 2$.
15. Answers vary. One possibility is

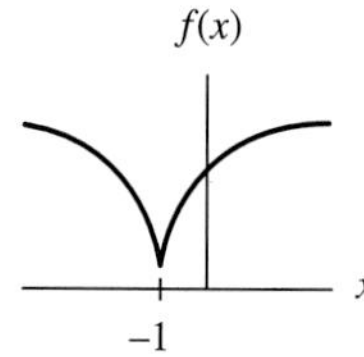

17. Answers vary. One possibility is

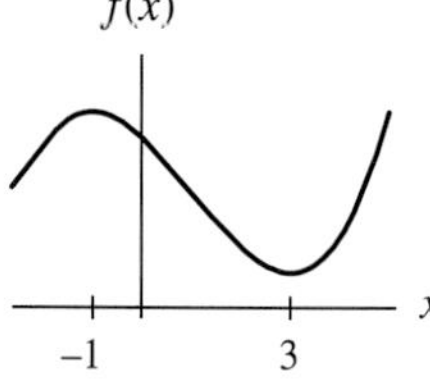

19. a. The relative maximum value is approximately 19.888, which occurs at $x \approx 3.633$. The relative minimum value is approximately 11.779, which occurs at $x \approx 11.034$.
    b. The absolute maximum and minimum are the relative maximum and minimum found in part *a*.

c.

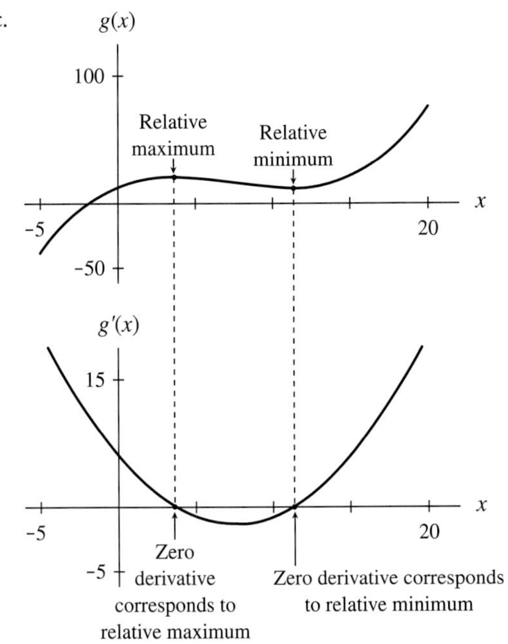

21. a. The greatest percentage of eggs hatching (95.6%) occurs at 9.4°C.
b. 9.4°C ≈ 49°F

23. a. Relative maximum point: approximately (7.7, 105.3)
Relative minimum point: (80.7, 94.5)
b. Absolute maximum: approximately 105.3 males per 100 females in August of 1908 ($x \approx 7.7$)
Absolute minimum: approximately 94.5 males per 100 females in August of 1981 ($x \approx 80.7$)
c.

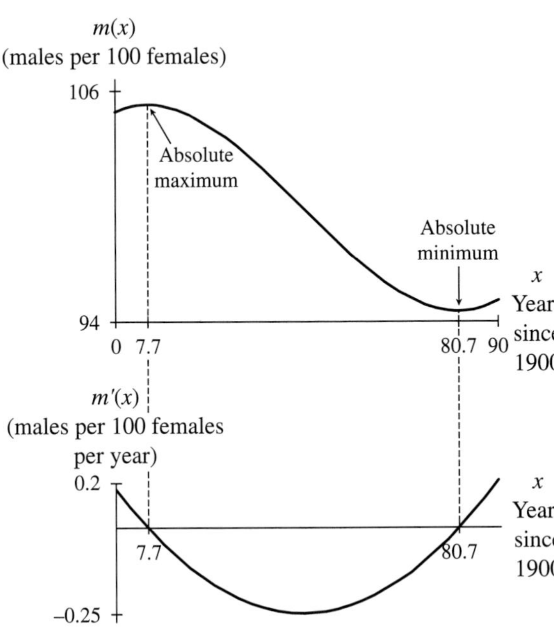

25. a. $C(0) \approx 123$ cfs and $C(24) \approx 140$ cfs
b. The highest flow rate is about 387.6 cfs; it occurs when $h \approx 8.9$ hours. The lowest flow rate is about 121.3 cfs; it occurs when $h \approx 0.4$ hour.

27. a. $S(x) = 0.181x^2 - 8.463x + 147.376$ seconds at age x years.
b. The model gives a minimum time of 48.5 seconds occurring at 23.4 years.
c. The minimum time in the table is 49 seconds, which occurs at 24 years of age.

29. a. An exponential model for the data is $R(p) = 316.765(0.949^p)$ dozen roses when the price per dozen is p dollars.
b. $E(p) = 316.765p(0.949^p)$ dollars spent on roses each week when the price per dozen is p dollars
c. A price of \$19.16 maximizes consumer expenditure.x
d. A price of \$25.16 maximizes profit.
e. Marginal values are with respect to the number of units sold or produced. In this activity, the input is price, so derivatives are with respect to price and are not marginals.

31. a. $G(t) = 0.008t^3 - 0.347t^2 + 6.108t + 79.690$ million tons of garbage taken to a landfill t years after 1975
b. $G'(t) = 0.025t^2 - 0.693t + 6.108$ million tons of garbage per year t years after 1975
c. In 2005, the amount of garbage was increasing by 8.1 million tons per year.
d.

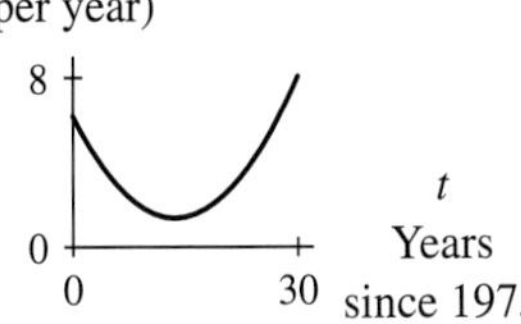

Because the derivative graph exists for all input and never crosses the horizontal axis, $G(t)$ has no relative maxima.

33. a. $D(x) = -0.099x^3 + 24.596x^2 - 2019.629x + 55{,}159.599$ billion 1987 dollars x years after 1900
b. $D'(x) = -0.298x^2 + 49.192x - 2019.629$ billion 1987 dollars per year x years after 1900
c. (76.15, 172.98) is a relative minimum point on the graph of D. (89.09, 280.64) is a relative maximum point on the graph of D.
d. According to the model, defense spending was least in 1976 (173.0 billion 1987 dollars) and was greatest in 1989 (280.6 billion 1987 dollars).
e. Relative extrema can be found by locating the two places where the graph of $D'(x)$ crosses the x-axis. That is, solve for x in the equation $D'(x) = 0$.

35. a. $A(p) = 568.074(0.965582^p)$ tickets sold on average when the price is p dollars
$A(p) = 0.15p^2 - 16.007p + 543.286$ tickets sold on average when the price is p dollars
The exponential model probably better reflects the probable attendance if the price is raised beyond \$35 because attendance is likely to continue to decline. (The quadratic model will begin to increase around \$53.)
b. $R(p) = 568.074p(0.965582^p)$ dollars of revenue when the ticket price is p dollars
c. (28.55, 5966.86) is the maximum point on the revenue graph. This corresponds to a ticket price of \$28.55, which results in revenue of approximately \$5967. The resulting average attendance is approximately 209.

37. The absolute minimum is approximately -6.3. There is no absolute maximum because $\lim_{x\to\pm\infty} y \to \infty$.

Section 5.3

1. The equation for surface area in terms of diameter d is $S = \pi dh + \pi\left(\frac{d}{2}\right)^2 + \pi\left(\frac{d}{2} + \frac{9}{8}\right)^2$, and the volume equation is $V = \pi\left(\frac{d}{2}\right)^2 h = 808.5$. Solving the volume equation for h in terms of d gives $h = \frac{808.5}{\pi\left(\frac{d}{2}\right)^2} = \frac{3234}{\pi d^2}$. Substituting this expression into the surface area equation results in the equation we seek to optimize:

$$S = \frac{3234}{d} + \pi\left(\frac{d}{2}\right)^2 + \pi\left(\frac{d}{2} + \frac{9}{8}\right)^2$$

Setting the derivative equal to zero and solving for d gives $d \approx 9.73568$, which is twice the optimal radius found in Example 3.

3. 450 square feet, occurring when one side is 30 feet and each of the other two sides is 15 feet.

5. a. A corner cut of approximately 1.47 inches will result in the maximum volume of 52.5 cubic inches of sand.
b. Corner cuts of approximately 1.12 inches and 1.86 inches will result in boxes with volume 50 cubic inches.
c. Answers will vary.

7. a. 42 cases
b. $\frac{42}{x}$ orders rounded up to the nearest integer
c. $\frac{42}{x}$(\$12 per order) $= \frac{504}{x}$ dollars to order
d. $\left(\frac{x}{2}\right.$ cases to store$\left.\right)$(\$4 per case to store) $= 2x$ dollars to store
e. $C(x) = \frac{504}{x} + 2x$ dollars when x cases are ordered
f. $C(x)$ is a minimum when $x \approx 15.9$ cases. The manager must order 3 times a year. Because the order sizes are not all the same (16 cases ordered 3 times a year is more cases than the cafeteria needs), we cannot just substitute the optimal value of x into the cost equation to obtain the total cost. There are two most cost-effective ways to order: 15, 15, and 12 cases, or 16, 16, and 10 cases (assuming that the storage fees can be calculated as \$1.33 per case for 4 months). The total cost associated with each option is \$63.93.

9. A radius of approximately 1.2 inches and a height of approximately 4.8 inches will minimize the cost.

11. a. The chain link side should be approximately 6.6 feet, and the cinder block sides should be approximately 18.2 feet.
b. The chain link side should be approximately 5.7 feet, and the cinder block sides should be approximately 21 feet.
c. The answer to part *a* does not change. The answer to part *b* becomes: The chain link side should be 6 feet and the cinder block sides should be 20 feet.

13. a. $R(n) = 350 + 2n(44 - n) + 35n$ dollars when n students go on the trip
b. $S(n) = 350 + 2n(44 - n)$ dollars when n students go on the trip
c. The bus company's revenue is maximized when 31 students go on the trip and is minimized when 10 students go on the trip.
d. The sorority pays the most when 22 students go and pays the least when 44 students go.

15. a. $m(s) = -0.0015s^2 + 0.1043s + 3.5997$ mpg at a speed of s mph
b. i. $\frac{400}{s}$ hours
ii. $\frac{6200}{s}$ dollars
iii. $\frac{400}{m(s)}$ gallons, where $m(s)$ is the output of the model in part *a*
iv. $\frac{460}{m(s)}$ dollars (assuming a price of \$1.15 per gallon)
v. $C(s) = \frac{6200}{s} + \frac{460}{m(s)}$ dollars
c. 60.3 miles per hour
d. 700-mile trip:
i. $\frac{700}{s}$ hours
ii. $\frac{10,850}{s}$ dollars
iii. $\frac{700}{m(s)}$ gallons
iv. $\frac{805}{m(s)}$ dollars

1200-mile trip:
i. $\frac{1200}{s}$ hours
ii. $\frac{18,600}{s}$ dollars
iii. $\frac{1200}{m(s)}$ gallons
iv. $\frac{1380}{m(s)}$ dollars

(assuming a price of \$1.15 per gallon)
v. 700-mile trip: $C(s) = \frac{10,850}{s} + \frac{805}{m(s)}$ dollars
1200-mile trip: $C(s) = \frac{18,600}{s} + \frac{1380}{m(s)}$ dollars

Optimal speed:	Optimal speed:
60.3 miles per hour	60.3 miles per hour

The optimal speed remains constant. It does not depend on trip length.
e.

Price of gas:	Optimal speed for 400-mile trip:
\$1.35 / gallon	58.8 mph
\$1.55 / gallon	57.5 mph
\$1.75 / gallon	56.4 mph

As the gas price increases, the optimal speed decreases.

f. Wages of driver: Optimal speed for 400-mile trip:

Wages of driver	Optimal speed for 400-mile trip
\$17.50 / hour	61.4 mph
\$20.50 / hour	62.9 mph
\$25.50 / hour	64.9 mph

As the wages increase, the optimal speed increases.

Section 5.4

1. Cubic and logistic models have inflection points, as do some product, quotient, and composite functions.

3. a. One visual estimate of the inflection points is (1982, 25) and (2018, 25).

   b. The input values of the inflection points are the years in which the rate of crude oil production is estimated to be increasing and decreasing most rapidly. We estimate that the rate of production was increasing most rapidly in 1982, when production was approximately 25 billion barrels per year, and that it will be decreasing most rapidly in 2018, when production is estimated to be approximately 25 billion barrels per year.

5. a.

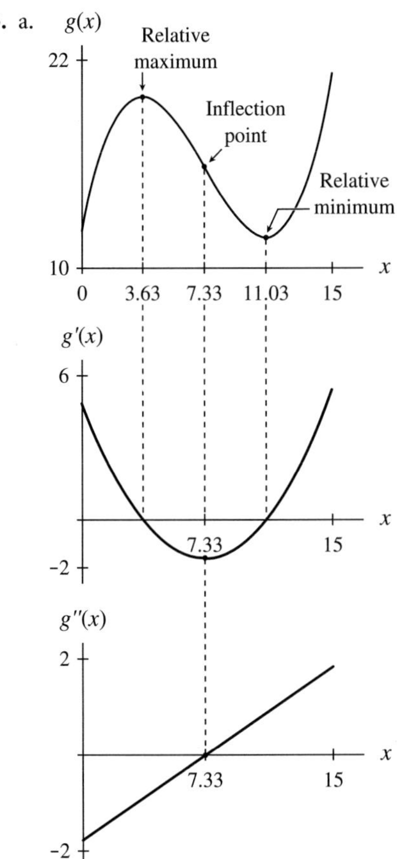

   b. The inflection point on the graph of g is approximately (7.333, 15.834). This is a point of most rapid decline.

7. a. Derivative b. Function
   c. Second derivative

9. a. Second derivative b. Derivative
   c. Function

11. a. The inflection point is approximately (1.838, 22.5). After about 1.8 hours of study (1 hour and 50 minutes), the percentage of new material being retained is increasing most rapidly. At that time, approximately 22.5% of the material has been retained.

    b. The answer agrees with the one given in the discussion at the end of the section.

13. a.

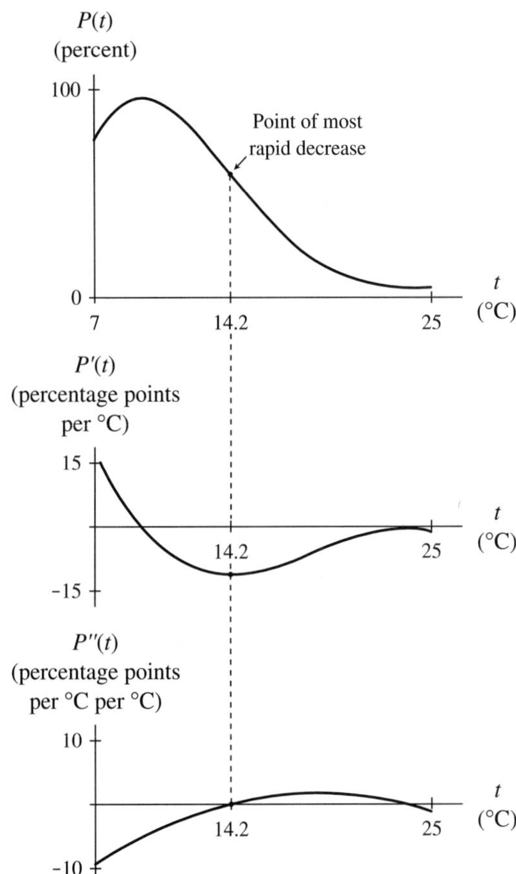

    b. Because the graph of P'' crosses the t-axis twice, there are two inflection points. These are approximately (14.2, 59.4) and (23.6, 5.8). The point of most rapid decrease is (14.2, 59.4). (The other inflection point is a point of least rapid decrease.) The most rapid decrease occurs at 14.2°C, when 59.4% of eggs hatch. At this temperature, the percentage of eggs hatching is declining by 11.1 percentage points per °C. A small increase in temperature will result in a relatively large increase in the percentage of eggs not hatching.

15. a.

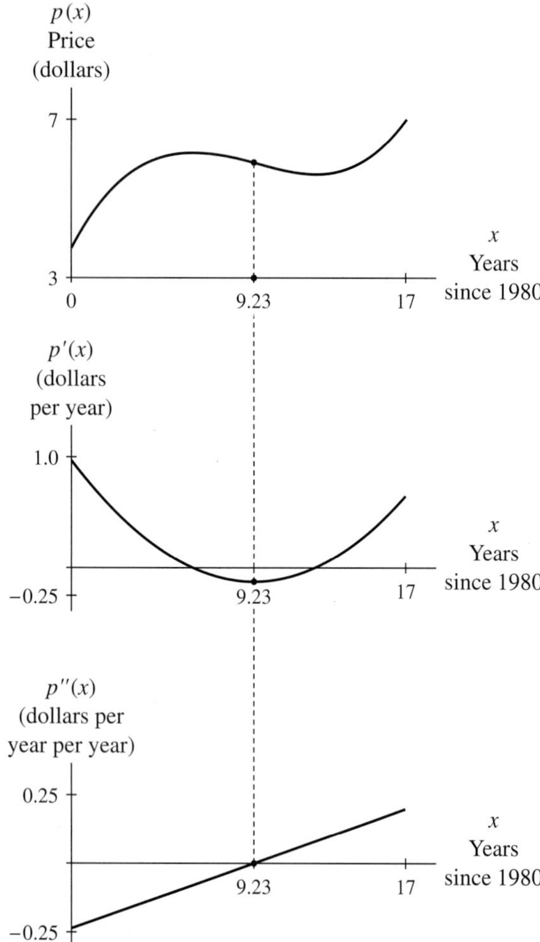

The minimum point of the graph of p' and the x-intercept of the graph of p'' correspond to the inflection point of the graph of p.

b. The x-intercept of p'' is $x \approx 9.23$. This is the input of the inflection point of the graph of p.

c. This function can be considered essentially continuous without restriction if you consider that it models the average price on a daily basis. If you assume it models a yearly average price, then it must be interpreted discretely. We are not given enough information to know for certain, so either one or the other assumption needs to be made.

In the continuous without restriction case, we conclude that the price was declining most rapidly in March of 1990 ($x \approx 9.23$) at a rate of -\$0.124 per year.

In the case of discrete interpretation, we compare $p'(9)$ and $p'(10)$ and conclude that the price was declining most rapidly in 1989 at a rate of -\$0.12 per year.

d. The price was increasing most rapidly at the end of 1980 at a rate of about \$0.97 per year.

17. a. (0.418, 9740.089) is a relative minimum point, and (13.121, 20,242.033) is a relative maximum point on the cubic model.

b. The inflection point is approximately (6.8, 14,991.1).

c. i. The inflection point occurs between 1981 and 1982, shortly after the team won the National Championship. This is when the number of donors was increasing most rapidly.

ii. The relative maximum occurred around the same time that a new coach was hired. After this time, the number of donors declined.

19. a. The greatest rate occurs at $h \approx 3.733$, or approximately 3 hours and 44 minutes after she began working.

b. Her employer may wish to give her a break after 4 hours to prevent a decline in her productivity.

21. a. The first differences are greatest between 6 and 10 minutes, indicating the most rapid increase in activity.

b. $A(m) = \dfrac{1.930}{1 + 31.720e^{-0.439118m}}$ U / 100μL m minutes after the mixture reaches 95°C; The inflection point is (7.872, 0.965). After approximately 7.9 minutes, the activity was approximately 0.97 U/100μL and was increasing most rapidly at a rate of approximately 0.212 U/100μL/min.

23. a. Between 1980 and 1985, the average rate of change was smallest at 1 million tons per year.

b. $g(t) = 0.008t^3 - 0.347t^2 + 6.108t + 79.690$ million tons t years after 1970.

c. $g''(t) = 0.051t - 0.693$ million tons per year per year t years after 1970.

d. Solving $g''(t) = 0$ gives $t \approx 13.684$ and $g(13.684) \approx$ 120 million tons of garbage.

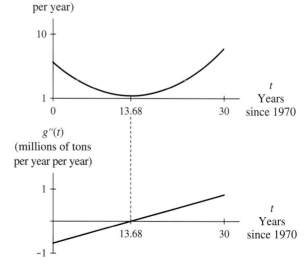

e.

$g'(t)$
(millions of tons
per year)

10
1
0
13.68
30
t
Years
since 1970

$g''(t)$
(millions of tons
per year per year)

1
-1
13.68
30
t
Years
since 1970

Because the graph of g'' crosses the t-axis at 13.68, we know that input corresponds to an inflection point of the graph of g. Because $g'(t)$ is a minimum at that same value, we know that it corresponds to a point of slowest increase on the graph of g.

f. The year with the smallest rate of change is 1984, with $g(14) \approx 120$ million tons of garbage, increasing at a rate of $g'(14) \approx 1.4$ million tons per year.

25. a. $A(x) = -0.099x^3 + 3.757x^2 - 34.932x + 268.805$ billion 1987 dollars x years after 1970

b. $A''(x) = -0.595x + 7.513$ billion 1987 dollars per year per year x years after 1970

c. The inflection point of A is (12.619, 226.808). The amount was increasing most rapidly in 1983 ($x = 13$). In 1983 the amount spent was $A(13) \approx 231.6$ billion 1987 dollars and was increasing at a rate of $A'(13) \approx$ 12.43 billion 1987 dollars per year.

d. Solving for x in the equation $A''(x) = 0$ will give the input value of the inflection point.

e. Defense spending was declining most rapidly in 1970 at a rate of \$34.9 billion per year.

27. a. $L'(x) = \dfrac{-251.3(0.1376)(0.2854)e^{0.2854x}}{(1 + 0.1376e^{0.2854x})^2}$ million tons per year x years after 1970

b. 1970: decreasing by 7.63 million tons per year
1995: decreasing by 0.41 million tons per year

c. Emissions were declining most rapidly in 1977 at a rate of 17.9 million tons per year. At that time, yearly emissions were 124.7 million tons.

29. a. $H(w) = \dfrac{10{,}111.102}{1 + 1153.222e^{-0.727966w}}$ total labor hours after w weeks

b. $H'(w) = \dfrac{10{,}111.102(1153.222)(0.727966)e^{-0.727966w}}{(1 + 1153.222e^{-0.727966w})^2}$ labor hours per week after w weeks

c. $H'(w)$ (labor-hours per week)

2000

0

0 20 Weeks w

The derivative gives the manager information about the number of labor hours spent each week.

d. The maximum point is about (9.685, 1840.134). In the tenth week the most labor hours are needed. That number is $H'(10) \approx 1816$ labor hours.

e. The point of most rapid increase on the graph of H' is (7.876, 1226.756). This occurs approximately 8 weeks into the job, and the number of labor hours per week is increasing by approximately $H''(8) \approx 513$ labor hours per week per week.

f. The point of most rapid decrease on the graph of H' is (11.494, 1226.756). This occurs approximately 12 weeks into the job, and the number of labor hours per week is changing by about $H''(12) = -486$ labor hours per week per week.

g. By solving the equation $H'''(w) = 0$, we can find the input values that correspond to a maximum or minimum point on the graph of H'', which corresponds to inflection points on the graph of H', the weekly labor hour curve.

h. The second job should begin about 4 weeks into the first job.

31. The graph of f is always concave up. A concave-up parabola fits this description.

33. a. The graph is concave up to the left of $x = 2$ and concave down to the right of $x = 2$.

b. $f_1(x)$

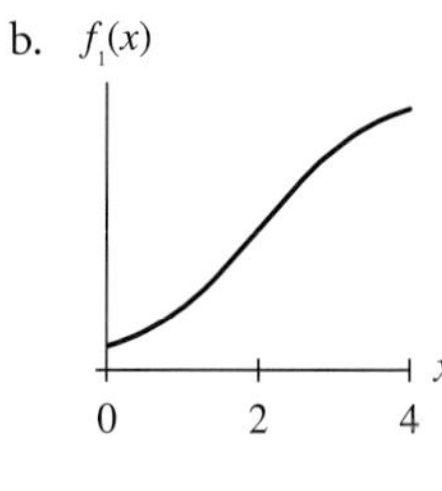

$f_2(x)$

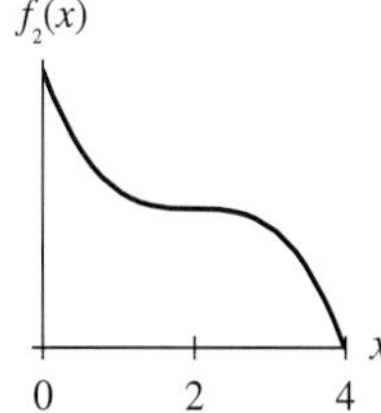

Section 5.5

1. $\dfrac{df}{dt} = 3\dfrac{dx}{dt}$

3. $\dfrac{dk}{dy} = 12x\dfrac{dx}{dy}$

5. $\dfrac{dg}{dt} = 3e^{3x}\dfrac{dx}{dt}$

7. $\dfrac{df}{dt} = 62(\ln 1.02)(1.02^x)\dfrac{dx}{dt}$

9. $\dfrac{dh}{dy} = 6\dfrac{da}{dy} + 6\ln a\dfrac{da}{dy} = 6(1 + \ln a)\dfrac{da}{dy}$

11. $\frac{ds}{dt} = \frac{\pi r h}{\sqrt{r^2 + h^2}} \frac{dh}{dt}$

13. $0 = \frac{\pi r}{2\sqrt{r^2 + h^2}}\left(2r\frac{dr}{dt} + 2h\frac{dh}{dt}\right) + \pi\sqrt{r^2 + h^2}\frac{dr}{dt}$

15. a. Approximately 52.4 gallons per day
 b. $\frac{dw}{dt} \approx 0.4323$; The amount of water transpired is increasing by approximately 0.4323 gallon per day per year. In other words, in 1 year, the tree will be transpiring about 0.4 gallon more each day than it currently is transpiring.

17. a. $B = \frac{45}{0.00064516h^2}$ points
 b. $\frac{dB}{dt} = \frac{-90}{0.00064516h^3} \frac{dh}{dt}$
 c. $\frac{dB}{dt} \approx -0.2789$ point per year

19. a. Approximately 0.0014 cubic foot per year
 b. Approximately 0.0347 cubic foot per year

21. a. $L = \left(\frac{M}{48.10352\,K^{0.4}}\right)^{5/3} = \left(\frac{M}{48.10352}\right)^{5/3} K^{-2/3}$
 b. $\frac{dL}{dt} = \left(\frac{M}{48.10352}\right)^{5/3}\left(\frac{-2}{3}K^{-5/3}\right)\frac{dk}{dt}$
 c. The number of worker hours should be decreasing by approximately 57 worker hours per year.

23. The balloon is approximately 1529.7 feet from the observer, and that distance is increasing by approximately 1.96 feet per second.

25. The runner is approximately 67.08 feet from home plate, and that distance is decreasing by about 9.84 feet per second.

27. a. Approximately 4188.79 cubic centimeters
 b. By about -167.6 cubic centimeters per minute

29. Approximately 5.305 centimeters per second

31. Begin by solving for h: $h = \frac{V}{\pi r^2} = \frac{V}{\pi} r^{-2}$

Differentiate with respect to t (V is constant):

$\frac{dh}{dt} = \frac{V}{\pi}(-2r^{-3})\frac{dr}{dt}$

Substitute $\pi r^2 h$ for V: $\frac{dh}{dt} = \frac{\pi r^2 h}{\pi}(-2r^{-3})\frac{dr}{dt}$

Simplify: $\frac{dh}{dt} = \frac{-2h}{r}\frac{dr}{dt}$

Rewrite: $\frac{dr}{dt} = \frac{r}{-2h}\frac{dh}{dt}$

Chapter 5 Review Test

1. a. T has a relative maximum point at (0.682, 143.098) and a relative minimum point at (3.160, 120.687). These points can be determined by finding the values of x between 0 and 6 at which the graph of T' crosses the x-axis. (There is also a relative maximum to the right of $x = 6$.)
 b. T has two inflection points: (1.762, 132.939) and (5.143, 149.067). These points can be determined by finding the values of x between 0 and 6 at which the graph of T'' crosses the x-axis. These are also the points at which T' has a relative maximum and relative minimum.
 c.

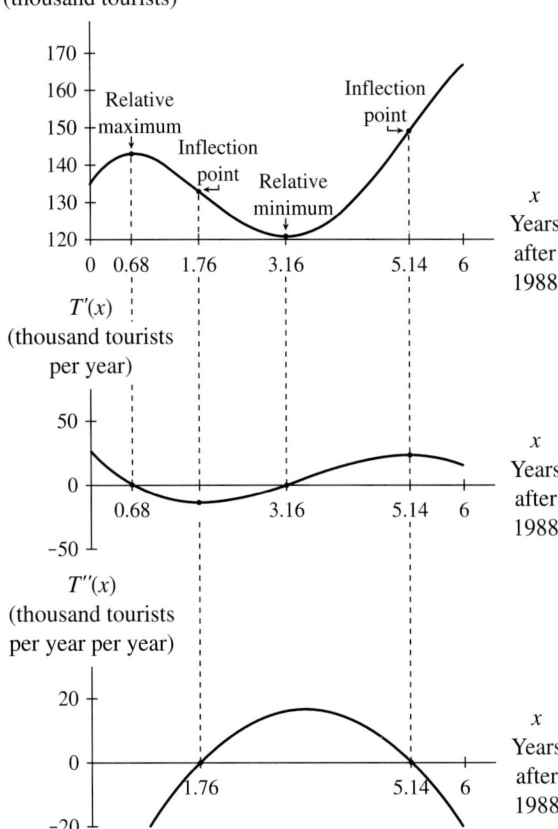

 d. The number of tourists was greatest in 1994 at 166.8 thousand tourists. The number was least in 1991 at 120.9 thousand.
 e. The number of tourists was increasing most rapidly in 1993 at a rate of 23.1 thousand tourists per year. The number of tourists was decreasing most rapidly in 1990 at a rate of 13.3 thousand tourists per year.

2. a. (4.9 thousand people per year)($\frac{1}{4}$ year) = 1.225 thousand people
 b. $246.4 + \frac{1}{2}(4.9) = 248.85$ thousand people

3. The optimal distance is $x \approx 2.89$ miles. The total cost is \$1,642,527.

4.

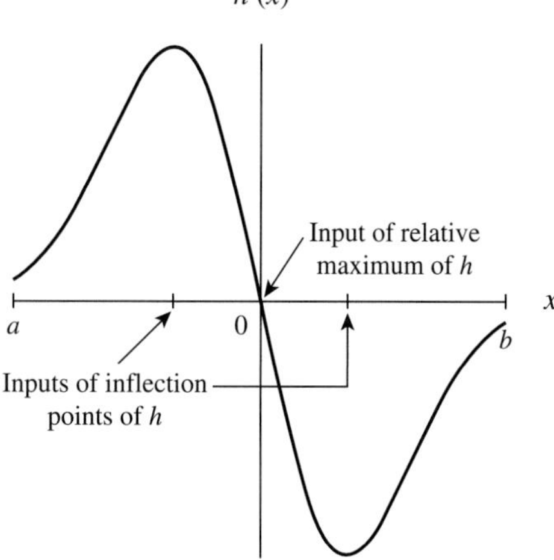

5. $\frac{ds}{dt} = 0.000013(2wv)\frac{dv}{dt}$

$= -31.2$ feet per second

The length of the skid marks is decreasing by 31.2 feet per second.

CHAPTER 6

Section 6.1

1. a. thousand bacteria per hour
   b. hours
   c. thousand bacteria
   d. thousand bacteria
   e. thousand bacteria
3. a. The area would represent how much farther a car going 60 mph would require to stop than a car going 40 mph.
   b. i. The heights are in feet per mile per hour, and the widths are in miles per hour.
      ii. The area is in feet.
5. a. On the horizontal axis, mark integer values of x between 0 and 8. Construct a rectangle with width from $x = 0$ to $x = 1$ and height $f(1)$. Because the width of the rectangle is 1, the area will be the same as the height. Repeat the rectangle constructions between each pair of consecutive integer input values. Note that for the rectangles that lie below the horizontal axis, the heights are the absolute values of the function values. Also note that the fourth rectangle has height 0. Sum the areas (heights) of the rectangles to obtain the area estimate.
   b. Repeat part *a*, except that the height of each rectangle is determined by the function value corresponding to the left side of the interval. In the case of the first rectangle, the height is $f(0)$. When we use left rectangles, the fifth rectangle has height 0.

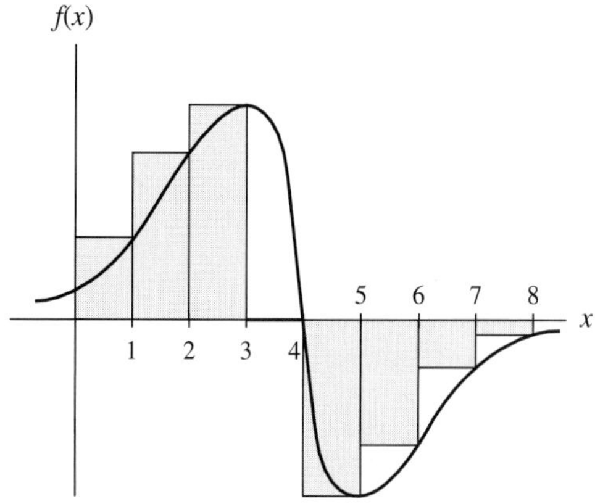

Activity 5 part *a*: Right rectangles

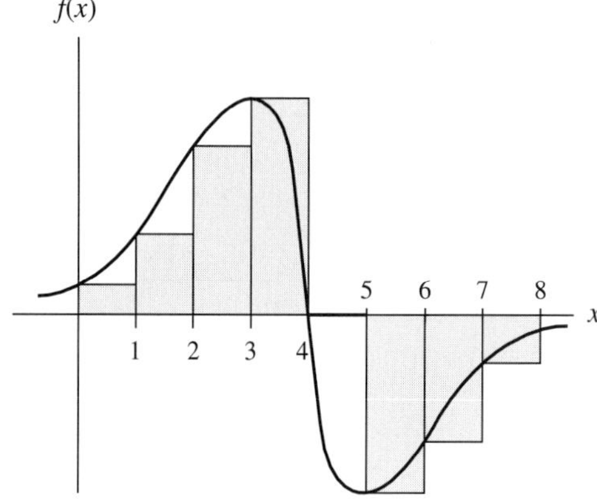

Activity 5 part *b*: Left rectangles

7. Divide the interval from a to b into four equal subintervals. Determine the midpoint of each subinterval, and substitute into the function to find the heights of the rectangles. Multiply each height by the width of the subintervals, and add the four resulting areas.

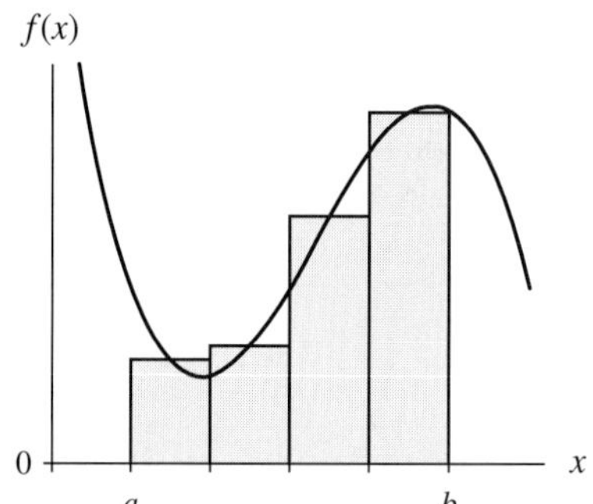

9. Answers will vary.

11. a.

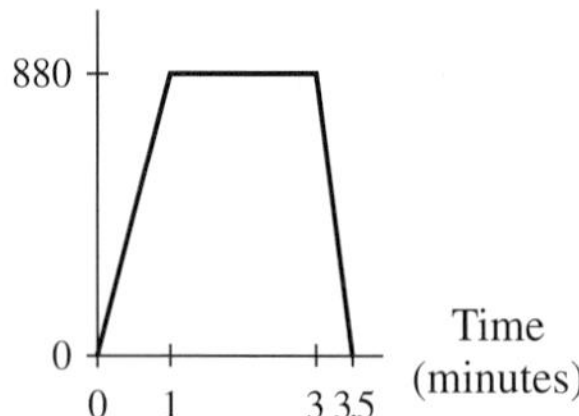

b. 2420 feet

c. The robot traveled 2420 feet during the $3\frac{1}{2}$-minute experiment.

13. a.

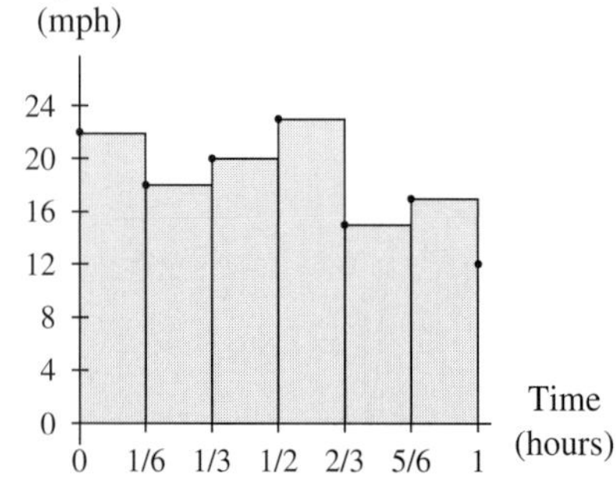

Distance traveled $\approx \dfrac{115}{6} \approx 19.2$ miles

b.

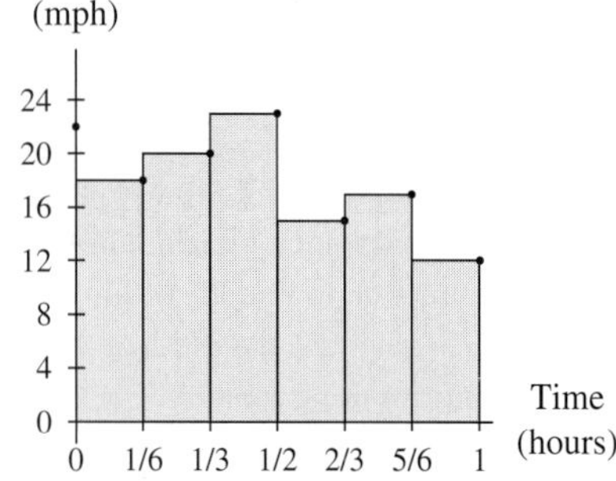

Distance traveled $\approx \dfrac{105}{6} = 17.5$ miles

15. a.

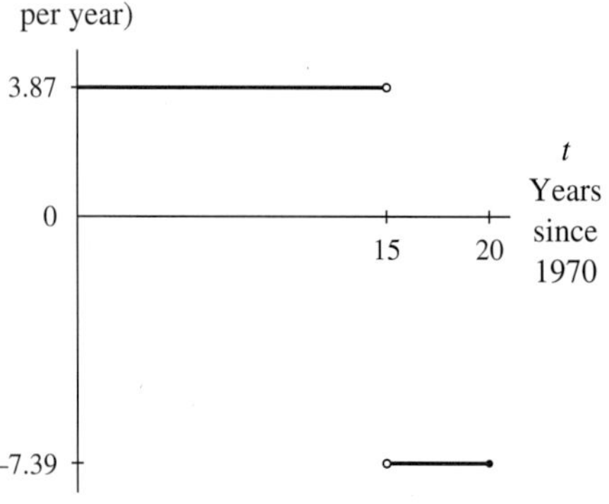

b. The population of North Dakota grew by about 58.1 thousand people between 1970 and 1985.

c. The population of North Dakota declined by about 37.0 thousand people between 1985 and 1990.

d. The population was $58.1 - 37.0 = 21.1$ thousand people greater in 1990 than it was in 1970.

e. You would need the population in some year between 1970 and 1990.

17. a.

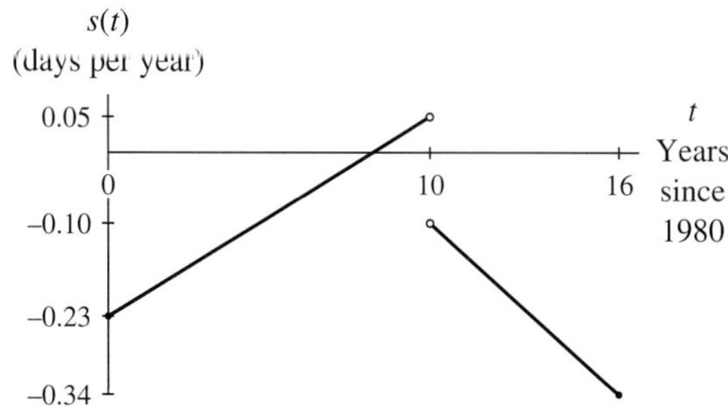

b. The average hospital stay decreased between 1980 and 1996 except between about 1988 and 1990, when it increased.

c. Area above ≈ 0.0446 day

d. Area below ≈ 2.277 days

e. The average hospital stay decreased by approximately 2.2 days between 1980 and 1996. We cannot determine the average stay in 1996 without knowing the average length of stay for some year between 1980 and 1996.

19. a.

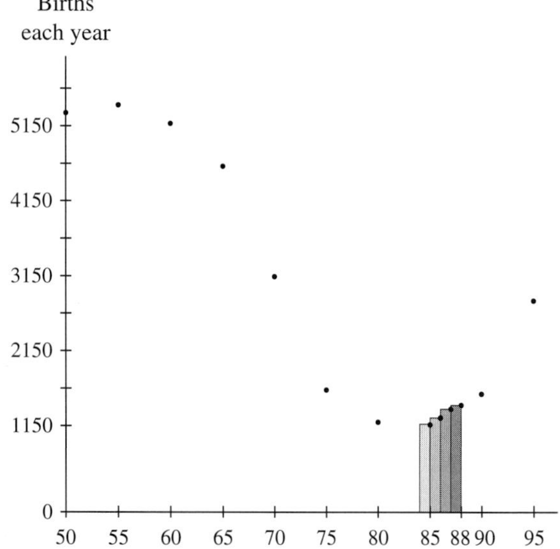

From the beginning of 1985 through the end of 1988, there were 5221 live births to U.S. women 45 years of age and older. Disregarding reporting error, this answer is exact.

b. $B(x) = 0.303x^3 - 17.497x^2 + 119.958x + 5343.469$ births x years after 1950

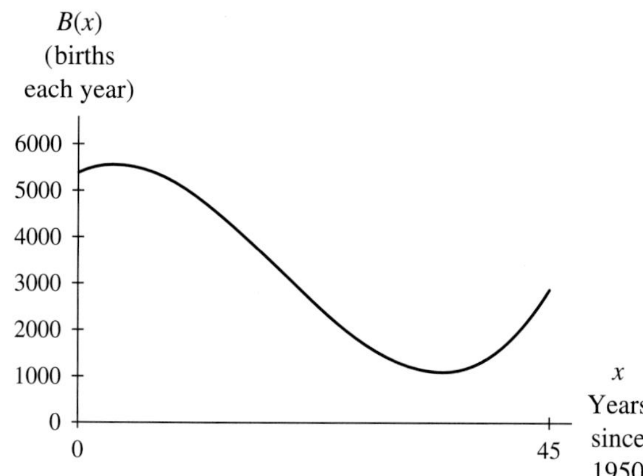

c. Using 30 rectangles, we estimate the number of births to be 62,898.

d. To find the exact total, we would need exact data for all years from 1965 through 1995.

21. a. To convert the air speeds to miles per second, multiply by 1.15 mph per knot and divide by 3600 seconds per hour.

b. $S(t) = -(2.108 \cdot 10^{-5})t^2 + (1.554 \cdot 10^{-3})t - 9.840 \cdot 10^{-5}$ miles per second t seconds after the plane began to taxi

c. Using 18 midpoint rectangles, the area is about 0.209 mile. Answers may vary.

d. It took approximately 0.2 mile of runway for the Cessna to taxi for takeoff (assuming no headwind).

23. a.

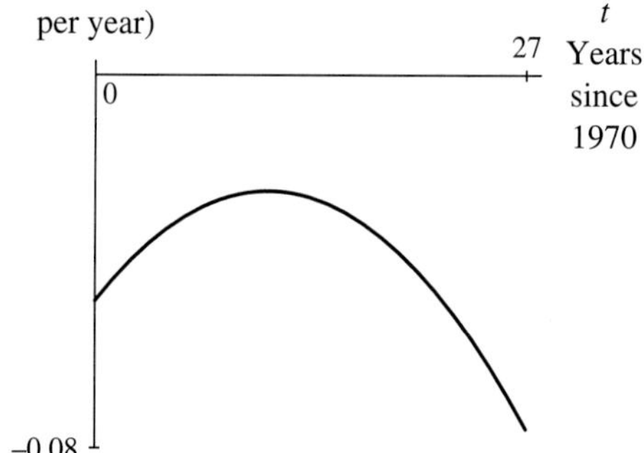

b. The percentage of the population living in New England was declining from 1970 through 1997.

c. The signed area is approximately -0.298. From 1980 through 1990, the percentage of people living in New England decreased by approximately 0.3 percentage point.

d. We cannot answer this question with the information given.

25. a. Declining positive rate-of-change data indicate that life expectancies were increasing at a slower and slower rate.

b. $E(t) = (4.199 \cdot 10^{-4})\, t^2 - 0.022\, t + 0.359$ years per year t years after 1970

c. Life expectancy for women increased approximately 5.8 years from 1970 through 2010.

27. a. $A \approx 43.799$ days after Sept. 30, 1995
$B \approx 273.382$ days after Sept. 30, 1995

b. In about 44 days after Sept. 30, 1995, the level of the lake fell by approximately 0.398 foot.

c. Between about 44 and 273 days after Sept. 30, 1995, the lake level rose by about 3.250 feet.

d. The lake level was approximately $3.250 - 0.398 \approx 2.853$ feet higher 273 days after Sept. 30, 1995.

29. a.

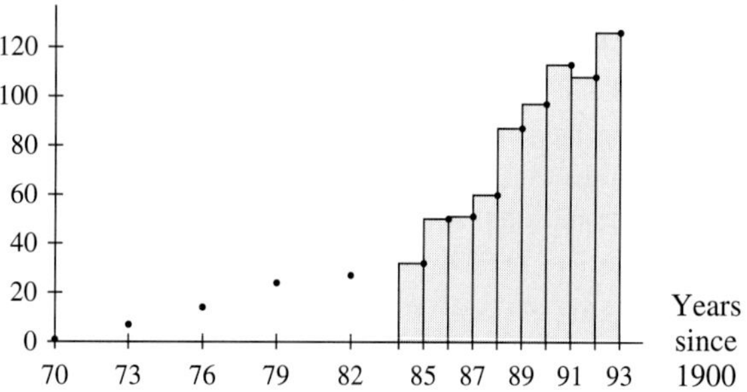

There were 724 graduates from the beginning of 1985 through the end of 1993. This is exact as long as the data were correctly reported.

b. $g(t) = 0.016t^3 - 0.233t^2 + 2.669t + 1.633$ graduates t years after 1970

c, d. Summing values of $g(t)$ for $t = 0$ through $t = 23$ gives a total of 971 graduates which is close to the actual number of 987.

31. Answers will vary.

Section 6.2

1. a. Thousand people
b. Thousand people
c. Thousand people

3. a. This is the change in the number of organisms when the temperature increases from 25°C to 35°C.

b. If the graph of A is always positive this is the change in the number of organisms when the temperature increases from 30°C to 40°C.

5. a, b. Between 0 and 300 boxes and between 400 and 600 boxes

c. NA
d. 300
e. 400
f. 350
g. dollars
h. less

7. a. 17.91
   b. Between 3 and 11 weeks of age, the mouse gained 17.91 grams.
   c. 21.91 grams

9. a. The yield from the oil field during the first 5 years is approximately 10.65 thousand barrels.
   b. The yield from the oil field during the first 10 years is approximately 12.45 thousand barrels.
   c. $\int_0^5 r(t)\,dt$ and $\int_0^{10} r(t)\,dt$
   d. The first 5 years account for approximately 85.6% of the first 10 years' production.

11. a. $A \approx 17.3$ seconds
   b. From 0 to 17.3 seconds, the car's speed increased by approximately 174.7 feet per second (or 119.1 mph).
   c. From 17.3 to 35 seconds, the car's speed decreased by approximately 94.9 feet per second (or 64.7 mph).
   d. The car's speed after 35 seconds was approximately 79.8 feet per second (or 54.4 mph) faster than it was at 0 seconds.

13. a. The heights will be in meters per second, and the widths will be in microseconds.
   b. The area units will be millimeters.
   c. $V(m) = -(1.589 \cdot 10^{-6})m^2 + 0.001m + 0.137$ millimeters per microsecond after m microseconds
   d. The crack traveled approximately 10.2 millimeters.
   e. $\int_0^{60} V(m)\,dm$

15. a. For $n = 5$, labor-hours ≈ 9859.
   For $n = 10$, labor-hours $\approx 10{,}097$.
   For $n = 20$, labor-hours $\approx 10{,}100$.
   b. *i*

17. a. $\int_0^{720} c(m)\,dm \approx 2602$ customers
   b. During the 12-hour sale, approximately 2602 customers entered the store.

19. a. Blood pressure rises when the rate of change is positive. In the table, this is from around 2 a.m. to almost 2 p.m. Blood pressure falls when the rate of change is negative, from around 2 p.m. to almost 2 a.m.
   b. In the table, the greatest rate of change occurs at 8 a.m., and the most negative rate of change occurs at 8 p.m.
   c. $B(t) = 0.030t^2 - 0.718t + 3.067$ mm Hg per hour, where t is the number of hours since 8 a.m.
   d. The model is zero at $t \approx 5.59$ hours and at $t \approx 18.13$ hours. These are the times when the blood pressure indicated by the model is highest and lowest, respectively.
   e. From 8 a.m. to 8 p.m., diastolic blood pressure rose by about 2.54 mm Hg.
   f. $\int_0^{12} B(t)\,dt$
   g. Not enough information to answer.

Section 6.3

1. a.

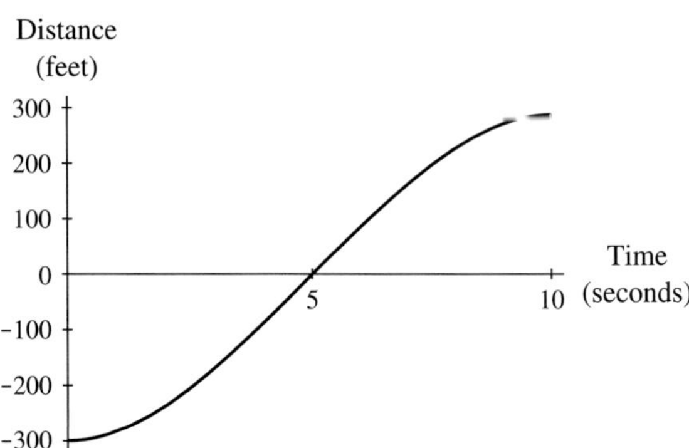

   b. $D(x) = \int_5^x f(t)\,dt$
   c. The accumulation function gives the distance traveled between 5 seconds and x seconds. For times before 5 seconds, the accumulation function is the negative of the distance traveled, because we are looking back in time.

3. a.

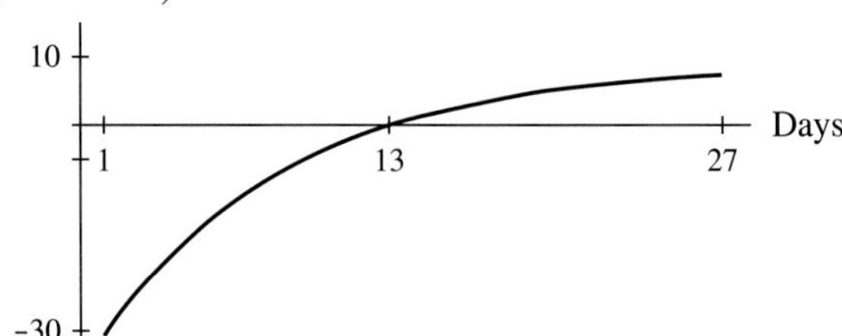

   b. $P(x) = \int_{13}^x g(t)\,dt$
   c. The accumulation function gives the growth of the plant between day 13 and day x. For days before day 13, the accumulation function is the negative of the growth, because we are looking back in time.

5. a. The area between days 0 and 18 represents how much the price of the technology stock declined ($15.40 per share) during the first 18 trading days of 2003.
   b. The area between days 18 and 47 represents how much the price of the technology stock rose ($55.80 per share) between days 18 and 47.
   c. $40.40 more
   d. $11.10 less

e.

x	$\int_0^x r(t)\,dt$	x	$\int_0^x r(t)\,dt$
0	0	35	15.0
8	-7.1	47	40.4
18	-15.4	55	29.3

f.

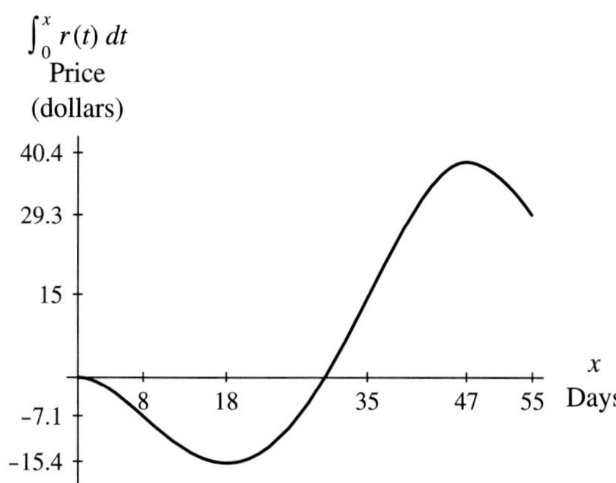

g. \$156.30

7. a. No

b. The peak corresponds to the time when the number of subscribers was increasing most rapidly.

c. The number of new subscribers t days after the end of the twentieth week.

d. 280 subscribers

e.

Week	t (days)	Area	Week	t (days)	Area
4	28	350	28	196	8820
8	56	924	36	252	10,430
12	84	1960	44	308	10,920
16	112	3500	52	364	11,060
20	140	5390			

f.

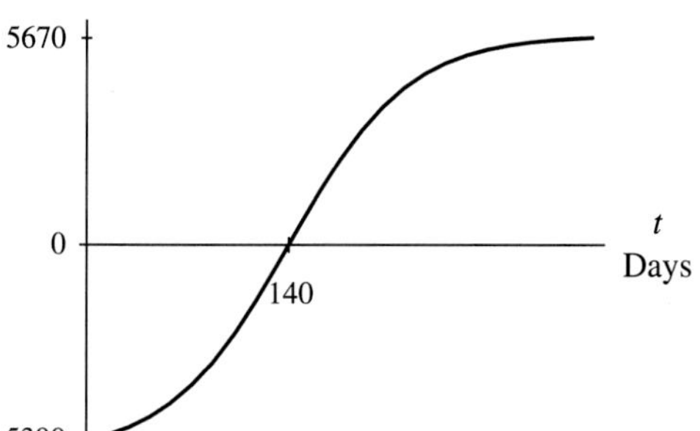

9. a. The area of each box is \$4000.

b. $\int_0^x p(t)\,dt$ represents the accumulation of profit (in thousands of dollars) in the x weeks after the business opened.

c. Answers will vary.

x	Accumulation function	x	Accumulation function
0	0	28	50.4
4	-3.2	32	48.3
8	3.5	36	41.9
12	15.3	40	32.8
16	28.4	44	23.5
20	39.8	48	17.6
24	47.5	52	19.3

d.

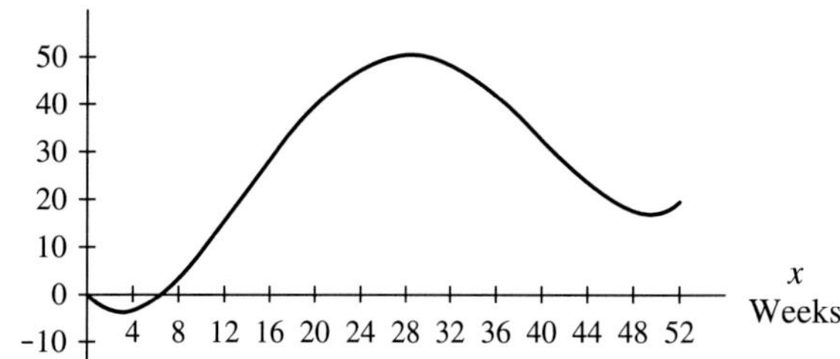

11.

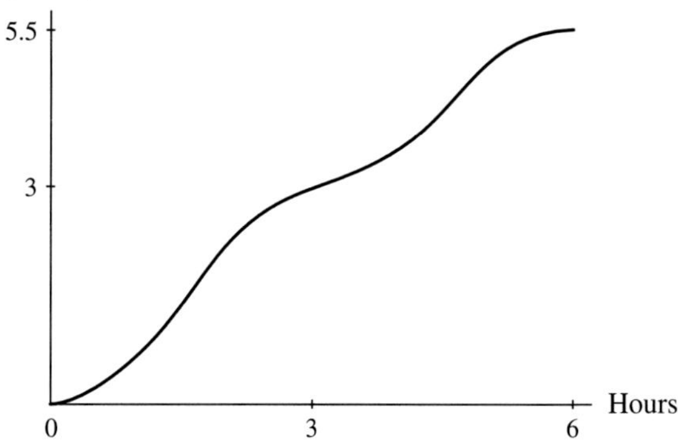

13. a.

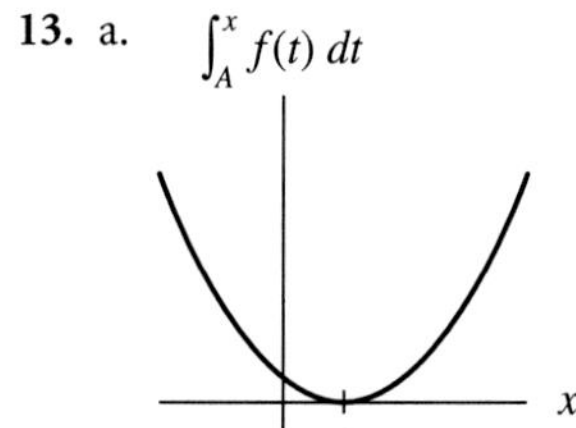

b.

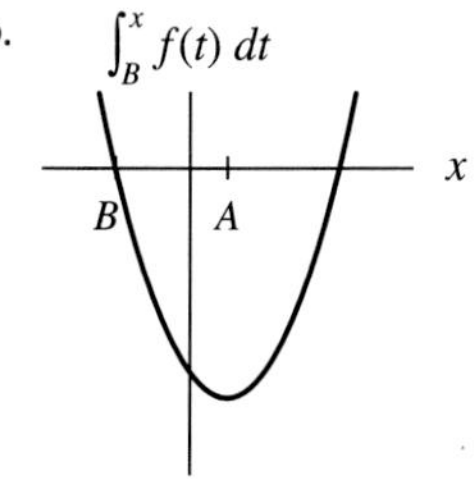

15. a. $\int_0^x f(t)\,dt$

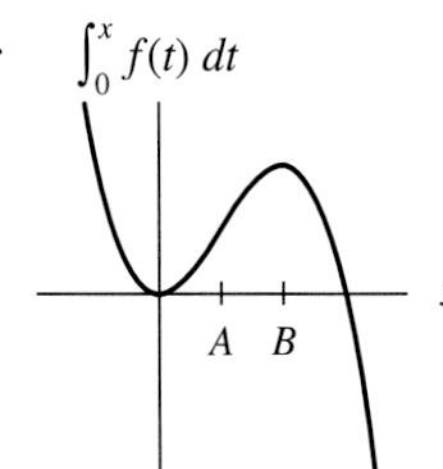

b. $\int_A^x f(t)\,dt$

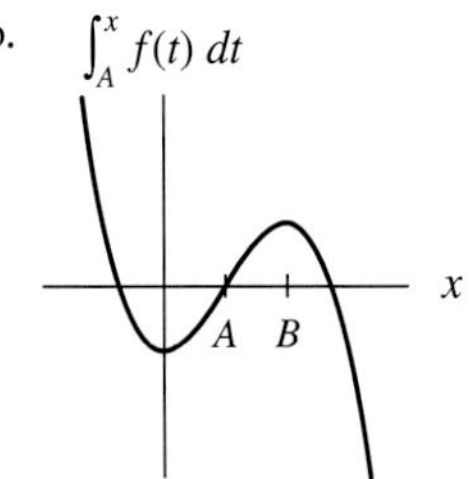

c. $\int_B^x f(t)\,dt$

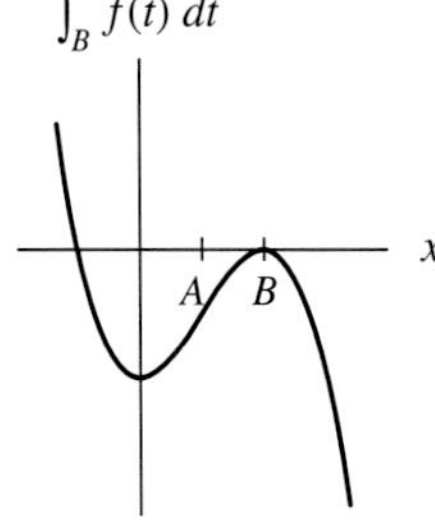

17. $\int_0^x f(t)\,dt$

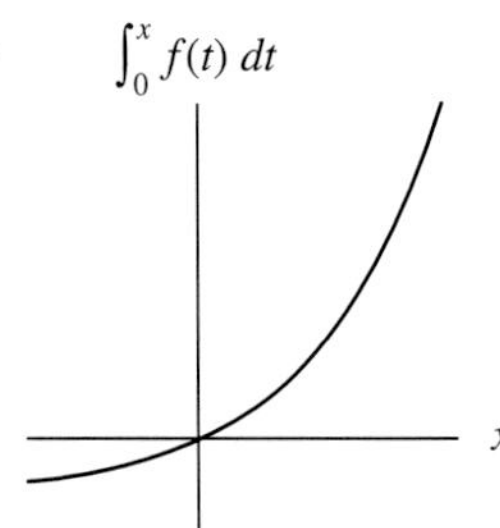

19. a. $\displaystyle\int_0^x 6\,dt = 6x$

$f(t) = 6$

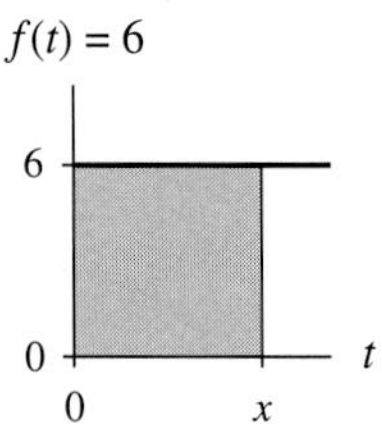

b. $\displaystyle\int_0^x -2\,dt = -2x$

$f(t) = -2$

c. $\displaystyle\int_0^x \frac{1}{3}\,dt = \frac{1}{3}x$

$f(t) = \frac{1}{3}$

21. a. $\displaystyle\int_{-4}^x -2\,dt = -2x - 8$

$f(t) = -2$

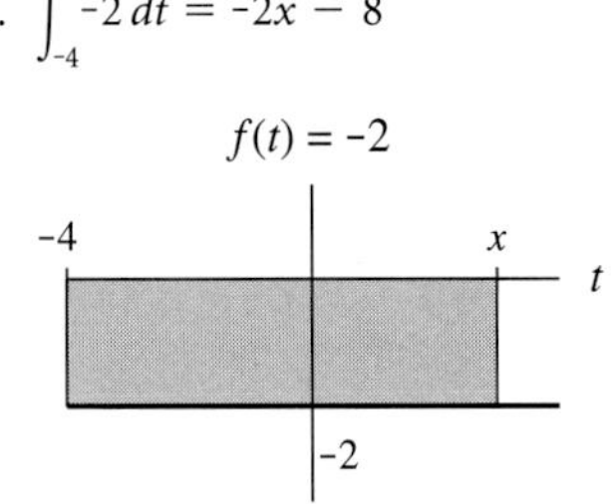

b. $\displaystyle\int_2^x -2\,dt = -2x + 4$

$f(t) = -2$

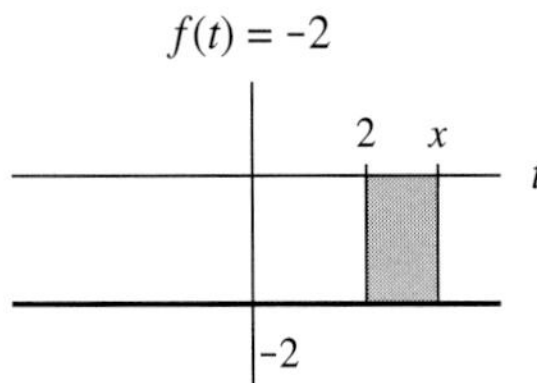

c. $\displaystyle\int_{2000}^x -2\,dt = -2x + 4000$

$f(t) = -2$

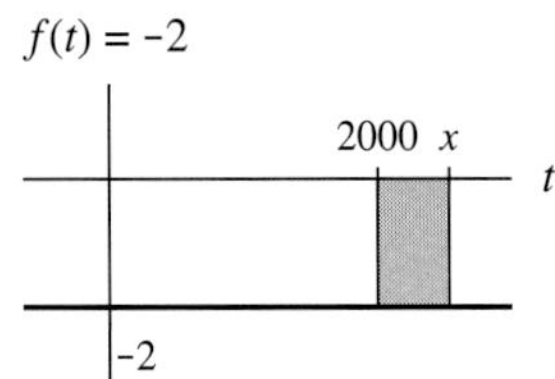

23. a. $\displaystyle\int_5^x 3t\,dt = \frac{3x^2}{2} - \frac{75}{2}$

$f(t) = 3t$

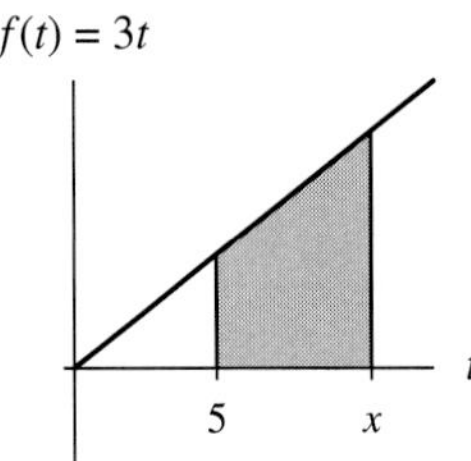

b. $\displaystyle\int_5^x \frac{-1}{2}t\,dt = \frac{-1}{4}x^2 + \frac{25}{4}$

$f(t) = -\frac{1}{2}t$

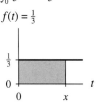

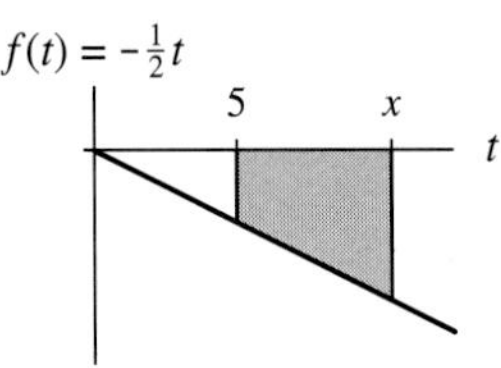

c. $\displaystyle\int_5^x 100t\,dt = 50x^2 - 1250$

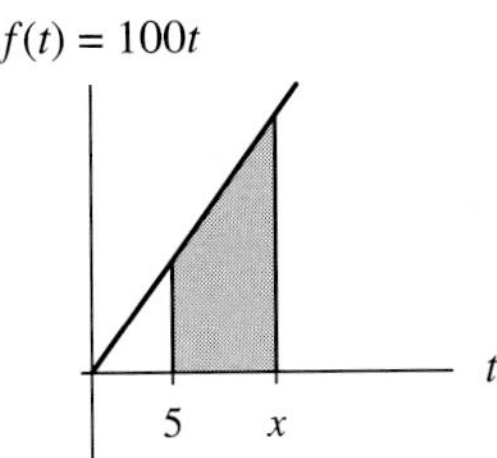

25. $\int_0^x k\,dt = kx$

27. $\int_a^x k\,dt = kx - ka$

29. $\int_5^x kt\,dt = \frac{kx^2}{2} - \frac{25k}{2}$

31. a. $\frac{3(19)^2}{2} - \frac{3(5)^2}{2} = 504$

b. Between January 5 and January 19 of this year, the airline's revenue increased by \$50,400.

33. Derivative graph: b; accumulation graph: f

35. Derivative graph: f; accumulation graph: e

37. Derivative: left table; accumulation function: right table

Section 6.4

1. b 3. c 5. a 7. c 9. b 11. b

13. a. $\frac{\text{million dollars of revenue}}{\text{thousand advertising dollars}}$

b.

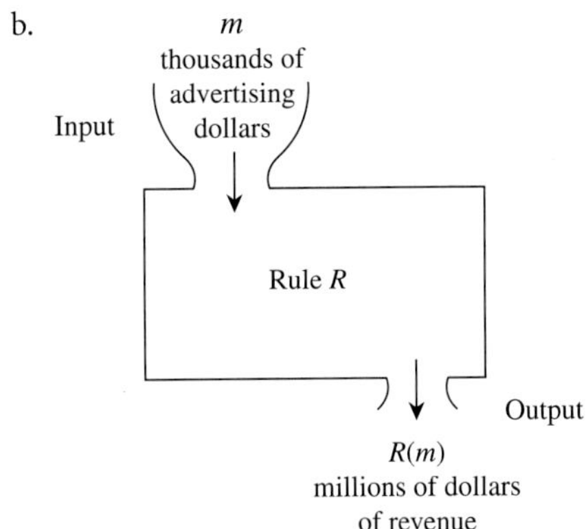

c. When m thousand dollars are being spent on advertising, the annual revenue is $R(m)$ million dollars.

15. a. $\frac{\text{milligrams per liter}}{\text{hour}}$

b.

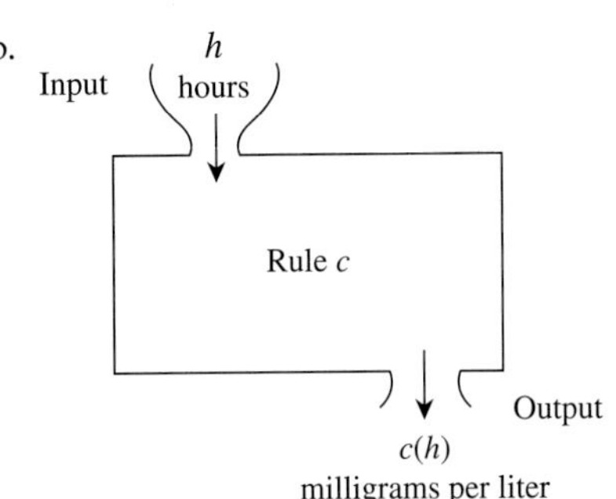

c. The concentration of a drug in the bloodstream is $c(h)$ milligrams per liter h hours after the drug is given.

17. $\int 19.436(1.07^x)\,dx = \frac{19.436(1.07^x)}{\ln 1.07} + C$

19. $\int [6e^x + 4(2^x)]\,dx = 6e^x + \frac{4(2^x)}{\ln 2} + C$

21. $\int \left(10^x + 4x^{\frac{1}{2}} + 8\right) dx = \frac{10^x}{\ln 10} + 4\left(\frac{2}{3}\right)x^{\frac{3}{2}} + 8x + C$

23. $S(m) = \frac{6250(0.92985^m)}{\ln 0.92985} + C$ CDs m months after the beginning of the year

25. $J(x) = 15.29 \ln|x| + 7.95x + C$ units, where x is the price in dollars

27. $P(t) = \frac{1.724928e^{0.0256t}}{0.0256} + C$ million people t years after 1990

29. $F(t) = \frac{1}{3}t^3 + t^2 - 20$

31. $F(z) = \frac{-1}{z} + e^z + \left(\frac{3}{2} - e^2\right)$

33. $w(t) = 7.372 \ln t + (26 - 7.372 \ln 7)$
$\approx 7.372 \ln t + 11.65475$ grams at age $(t + 2)$ weeks

35. $G(t) = \frac{1.667 \cdot 10^{-4}}{3}t^3 - \frac{0.01472}{2}t^2 - 0.103t + 119.015$ males per 100 females t years after 1900

37. a. $P(x) = \frac{891.6}{\ln 1.5}(1.5^x) - 2196.887$ thousand cellular phone subscribers x years after 1988

b. $P(6) - P(2) \approx 20{,}099.8$ thousand subscribers $\approx$ 20 million subscribers

39. a. $g(d) = -0.00018d^2 - 0.018d + 1.95$ inches per day d days after sprouting

b. $h(d) = -0.00006d^3 - 0.009d^2 + 1.95d$ inches d days after sprouting

c. The sweet corn is $h(60) = 71.64$ inches tall.

41. a. $P(x) = \frac{-0.019}{4}x^4 + \frac{0.191}{3}x^3 + \frac{0.927}{2}x^2 + 23.539x + C$ billion pounds (cumulative) x years after 1990

b. $P(8) - P(0) \approx 231.1$ billion pounds

c. $p'(x) = 3(-0.019)x^2 + 2(0.191)x + 0.927$ billion pounds annually per year x years after 1990

d. The answer to part a is the general antiderivative of p, and the answer to part c is the derivative of p.

43. a, b. Velocity: $v(t) = -32t$ ft/sec
Distance: $s(t) = -16t^2 + 540$ ft
where t is the number of seconds after the penny was dropped

c. Solving for t in $s(t) = 0$, we obtain $t \approx \pm 5.8$ seconds. The penny will hit the ground approximately 5.8 seconds after it was dropped.

d. $v(5.809) = -32(5.809) \approx -185.9$ ft/sec

$$= \left(\frac{-185.9 \text{ feet}}{1 \text{ second}}\right)\left(\frac{3600 \text{ seconds}}{1 \text{ hour}}\right)\left(\frac{1 \text{ mile}}{5280 \text{ feet}}\right)$$

$$= \frac{-126.75 \text{ miles}}{1 \text{ hour}} \text{ or } -126.75 \text{ mph}$$

45. a. The impact velocity is -64.99 ft/sec or -44.31 mph.

b. Air resistance probably accounts for the difference.

47. a. $N(x) = 593\ln|x| + 138x - 748.397$ employees x years after 1996

b. The function in part *a* applies from 1997 ($x = 1$) through 2002 ($x = 6$).

c. There are two ways to estimate the number of employees the company hired. If we consider the function to be continuous with discrete interpretation, then the number can be calculated from the function n by summing the yearly totals:

$n(1) + n(2) + n(3) + n(4) + n(5) + n(6) \approx 2281$

We can also estimate the total number of employees hired between 1997 and 2002 as

$$\int_1^6 n(x)dx = N(6) - N(1) \approx 1753$$

This estimate treats the function as continuous without restriction and is, therefore, probably less accurate than the first estimate. If any employees were fired or quit between 1997 and 2002, neither estimate would represent the number of employees at the end of 2002.

49. a. $S(x) = \frac{0.338}{0.17422}e^{0.17422x} + 7.691$ dollars x years after 1966

b. $S(34) \approx \$733$, which is more than twice the actual price.

Section 6.5

1. c **3.** c **5.** b **7.** a

9. a. $G(x) = \int_1^x (t^3 - 2t + 3)\,dt = \frac{1}{4}x^4 - x^2 + 3x - \frac{9}{4}$

b. $G(2) = \int_1^2 (t^3 - 2t + 3)\,dt = 3.75$

11. a. $\int_0^x [18{,}000(0.974^t) + 1500]\,dt \approx$

$\frac{18{,}000(0.974^x)}{\ln 0.974} + 1500x + 683{,}268.1768$

b. Area $\approx 173{,}242.235$

c. In the 10 years after the owner purchased the property, its value increased by approximately \$173,242.

13. $\int_1^2 (3x - 7)\,dx = \left(\frac{3}{2}x^2 - 7x\right)\Big|_1^2 = -2.5$

15. $\int_1^2 \left(1 + \frac{1}{x} + \frac{1}{x^3}\right)dx = \left(x + \ln|x| - \frac{1}{2x^2}\right)\Big|_1^2 \approx 2.068$

17. $\int_{10}^{40} [427.705(1.043^x) - 413.226e^{-0.4132x}]\,dx =$

$\left(\frac{427.705(1.043^x)}{\ln 1.043} - \frac{413.226e^{-0.4132x}}{-0.4132}\right)\Big|_{10}^{40} \approx 39{,}236.341$

19. a.

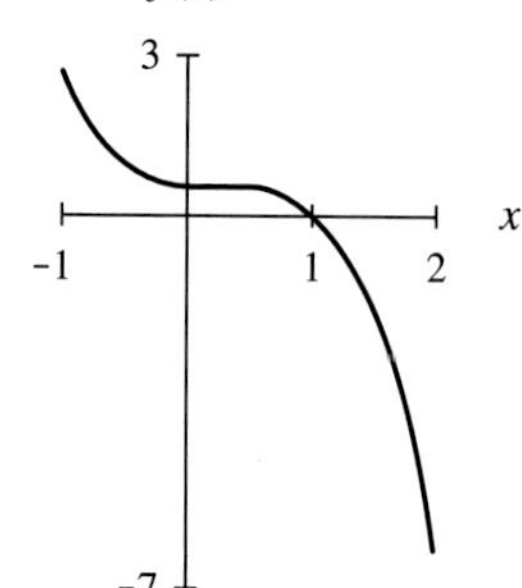

b. Area $= \int_{-1}^{1.0544} f(x)\,dx - \int_{1.0544}^{2} f(x)\,dx \approx 3.822$

c. No

21. a.

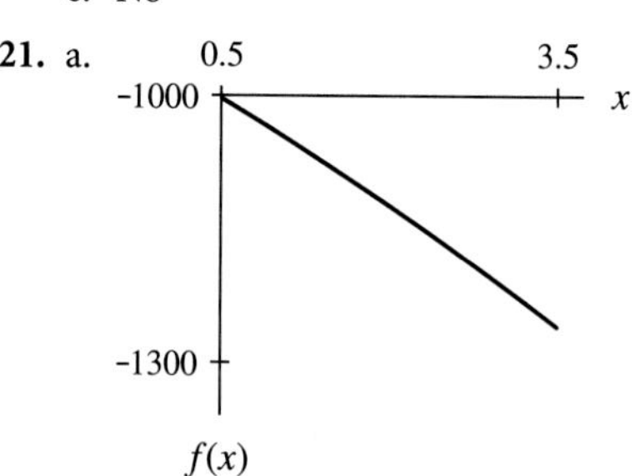

b. Area $= -\int_{0.5}^{3.5} f(x)\,dx \approx 3378.735$

c. No

23. $\int_{10}^{20} P(t)\,dt \approx -0.2$

Between 1980 and 1990, the percentage of the population of the United States living in New England fell by about 0.2 percentage point.

25. $\int_0^5 r(x)\,dx \approx 195.639$

The corporation's revenue increased by 195.6 million dollars between 1987 and 1992.

27. $\int_0^{10} r(t)\,dt \approx 12.45$ (found using a limit of sums)

In the 10 years after production began, the oil well produced about 12.5 thousand barrels of oil.

29. a. $\int_0^{70} s(t)\,dt = 7.938$

In the 70 days after April 1, the snow pack increased by 7.938 equivalent cm of water.

b. $\int_{72}^{76} s(t)\,dt = -22.768$

Between 72 and 76 days after April 1, the snow pack decreased by 22.768 equivalent centimeters of water.

c. It is not possible to find $\int_0^{76} s(t)\,dt$ because $s(t)$ is not defined between $t = 70$ and $t = 72$.

31. a. $\int_0^{0.8955} T(h)\,dh \approx 1.48$

The temperature rose 1.48°F between 8 a.m. and 8:54 a.m.

b. $\int_{0.8955}^{1.75} T(h)\,dh \approx -1.61$

After rising 1.48°F, the temperature then fell 1.61°F during the field trip.

c. No, the highest temperature reached was 71 + 1.48 = 72.48 °F.

33. a. An exponential model for the data is $f(x) = 0.161(1.076186^x)$ trillion cubic feet per year x years after 1900.

b. From 1940 through 1960, 138.3 trillion cubic feet of natural gas was produced.

c. $\int_{40}^{60} f(x)\,dx$

35. a. A quadratic model for the data is $C'(x) = (7.714 \cdot 10^{-5})x^2 - 0.047x + 8.940$ dollars per CD when x CDs are produced each hour.

b. $C(x) = \frac{7.714 \cdot 10^{-5}}{3}x^3 - \frac{0.047}{2}x^2 + 8.940x - 143.893$ dollars per CD when x CDs are produced each hour.

c. $\int_{200}^{300} C'(x)\,dx \approx 196.14$ dollars

When production is increased from 200 to 300 CDs per hour, cost increases by about \$196.14.

Section 6.6

1. $\int 2e^{2x}\,dx = e^{2x} + C$

3. Not possible using the techniques discussed in this text.

5. $\int (1 + e^x)^2\,dx = \int (1 + 2e^x + e^{2x})\,dx$

$= x + 2e^x + \frac{e^{2x}}{2} + C$

7. Not possible using the techniques discussed in this text.

9. Approximately 2.5452

11. Exactly $\frac{(\ln 5)^2}{2} - \frac{(\ln 2)^2}{2}$

13. Approximately 3.6609

15. Exactly $\ln 17 - \ln 10$

17. Approximately 8.7595

19. Exactly $1\frac{1}{12}$

21. 15

23. 0.3

25. -0.02

27. Diverges

29. Diverges

31. Diverges

33. a. Approximately 0.0002 milligram; Approximately 0.0015 milligram

b. $\int_0^{\infty} r(t)\,dt = \lim_{N\to\infty} \frac{-1.55(0.9999999845^t)\cdot 10^{-6}}{\ln(0.9999999845)}\Big|_0^N =$

$-99.99999923 \approx -100$ milligrams

35. a. $p_0 \approx \$28.04$

b. $C = qp_0 + \int_{p_0}^{\infty} D(p)\,dp$

$\approx 150(28.04) + \int_{28.04}^{\infty} 499.589(0.958^p)\,dp$

$\approx \$7702$ million

Consumers are willing and able to spend about \$7.7 million for 150,000 books.

37. $\int_0^{\infty} 0.1e^{-0.1x}\,dx = \lim_{N\to\infty}\int_0^N 0.1e^{-0.1x}\,dx$

$\lim_{N\to\infty}\left(-e^{-0.1x}\Big|_0^N\right) = \lim_{N\to\infty}[-e^{-0.1N} - (-e^0)]$

$\lim_{N\to\infty}(-e^{-0.1N}) + \lim_{N\to\infty}(e^0) = 0 + 1 = 1$

Chapter 6 Review Test

1. a.

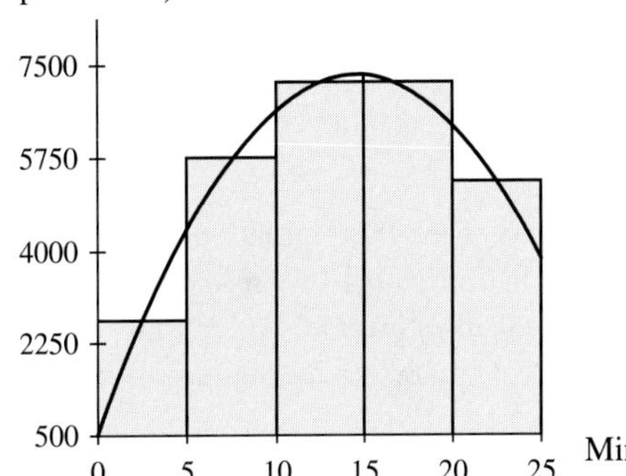

b. 139,237.5 cubic feet

c. In the first 25 minutes that oil was flowing into the tank, approximately 139,238 cubic feet of oil flowed in.

2. a. A quadratic model for the data is

$S(t) = -1.643t^2 + 16.157t + 0.2$ miles per hour

t hours after midnight

b.

n	Sum
5	127.131
10	126.869
20	126.803
40	126.786
80	126.782
160	126.781
Trend ≈ 126.8 miles	

3. a. The area beneath the horizontal axis represents the amount of weight that the person lost during the diet.
b. The area above the axis represents the amount of weight that the person regained between weeks 20 and 30.
c. The person's weight was 11.3 pounds less at 30 weeks than it was at 0 weeks.
d.

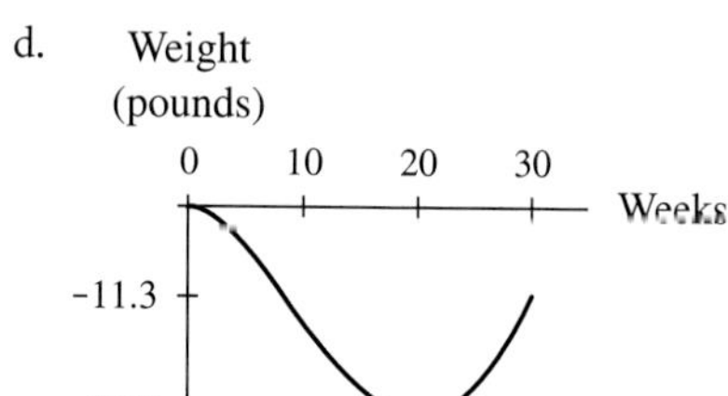

e. The graph in part *d* is the change in the person's weight as a function of the number of weeks after the beginning of dieting.

4. a. $R(t) = 10\left(\frac{-3.2}{3}t^3 + \frac{93.3}{2}t^2 + 50.7t\right) + 5000$ cubic feet after t minutes
b. $R(10) - R(0) \approx 41{,}053.3$ ft^3
c. Solving for t in the equation $R(t)=150{,}000$, we find that the tank will be full after approximately 28 minutes.

5. a. \$2598.60
b. At the end of the third quarter of the third year, the \$10,000 had increased by \$2598.60 so that the total value of the investment was \$12,598.60.

6. $\frac{-13}{32}$

CHAPTER 7

Section 7.1

1. a, b.

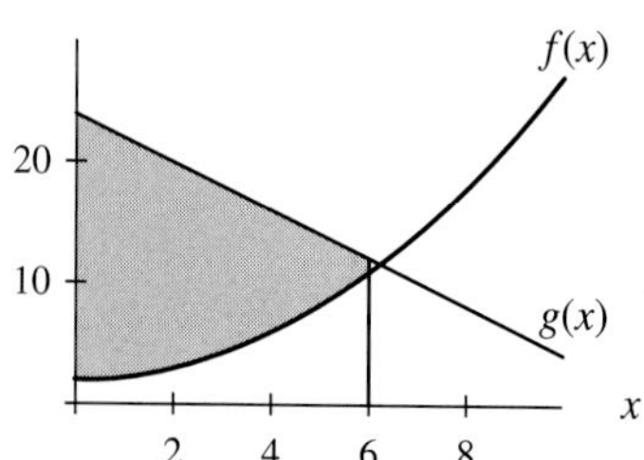

c. 78

3. a, b.

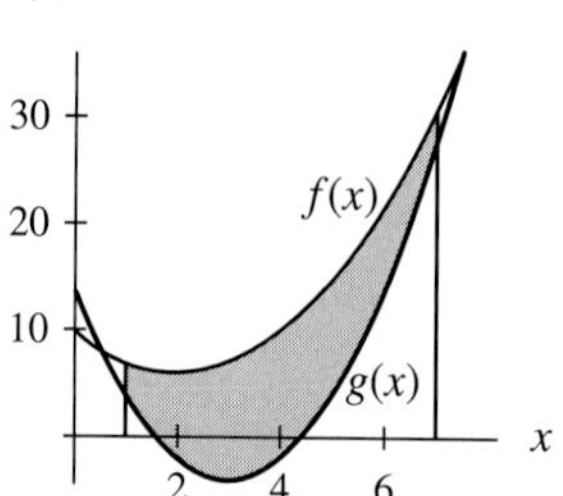

c. 54

5. a, c.

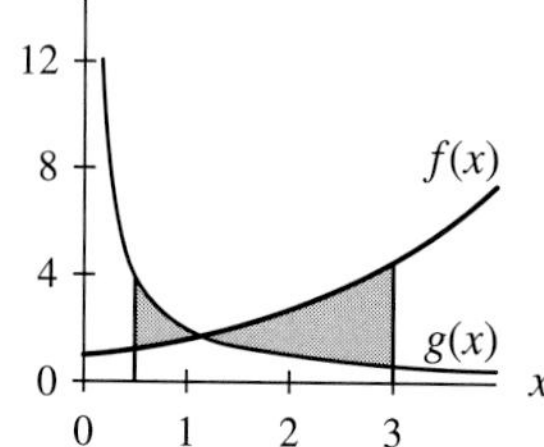

b. $x \approx 1.134$
d. $\int_{0.5}^{3} [f(x) - g(x)]\,dx \approx 2.812$
e. Area ≈ 4.172

7. a, c.

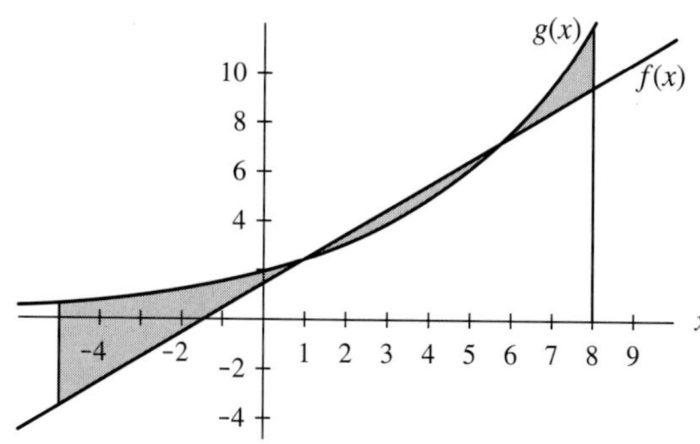

b. $x = 1$ and $x \approx 5.806$
d. 0.764
e. Area ≈ 15.536

9. a. When the amount invested in capital increases from \$1500 to \$5500, profit increases by approximately \$1.33 thousand.
b. Area $= \int_{1.5}^{5.5} [r'(x) - c'(x)]\,dx = 13.29$

11. a. The population of the country grew by 3690 people in January.
b. The population declined by 9720 people between the beginning of February and the beginning of May.
c. -6030 people
d. Because the graphs intersect, the area of R_1 represents an increase in population, and the area of R_2 repre-

sents a decrease. The net change is the difference Area of R_1 − area of R_2. The total area is the sum:

Area of R_1 + area of R_2.

13. a. Before fitting models to the data, add the point (0,0), and convert the data from miles per hour to feet per second by multiplying each speed by $\left(\frac{5280 \text{ feet}}{1 \text{ mile}}\right)\left(\frac{1 \text{ hour}}{3600 \text{ seconds}}\right)$. The speed of the Supra after t seconds can be modeled as

$$s(t) = -0.702t^2 + 20.278t + 2.440 \text{ feet per second}$$

The speed of the Carrera after t seconds can be modeled as

$$c(t) = -0.643t^2 + 18.963t + 5.252 \text{ feet per second}$$

b. Approximately 17.96 feet

c. Approximately 18.04 feet

15. a.

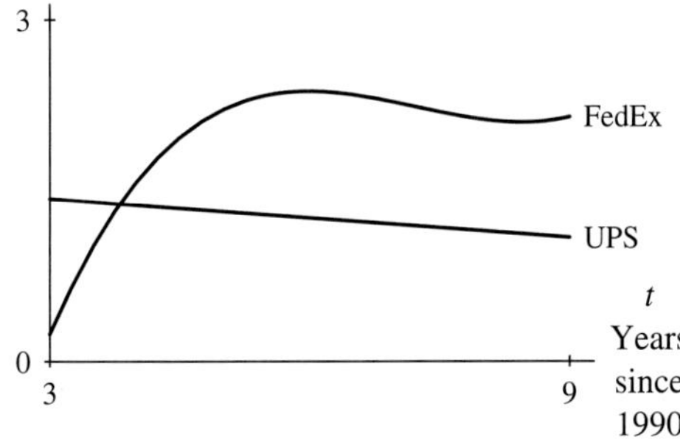

FedEx: $F(t) = 0.036t^3 - 0.779t^2 + 5.454t - 10.083$ billion dollars per year t years after 1990

UPS: $U(t) = -0.057t + 1.6$ billion dollars per year t years after 1990

b. Area of region on left ≈ \$0.44 billion
Area of region on right ≈ \$5.07 billion
Between the beginning of 1993 and late 1993 ($t \approx 3.8$), UPS's accumulated revenue exceeded that of FedEx by approximately \$0.44 billion. Between late 1993 and the end of 1999, FedEx's accumulated revenue exceeded that of UPS by approximately \$5.07 billion.

c. $\int_3^9 [F(t) - U(t)]\,dt \approx \$4.63$ billion. This value is the net amount by which FedEx's accumulated revenue exceeded that of UPS between 1993 and 1999.

17. a. $\int_{12}^{24} [E(x) - I(x)]\,dx \approx \$257.8$ billion

b. $\int_0^{12} [E(x) - I(x)]\,dx \approx \$339.2$ billion

c. The 1999 estimate is off by approximately \$9 billion, but the 1998 estimate is off by about \$175 billion. This discrepancy is due to the fact that the models are based on monthly data between November 1998 and November 1999—that is, two 1998 values and eleven 1999 values. The 1999 estimate required extrapolating by 1 month, resulting in a good estimate. The 1998 estimate required extrapolating backward by 10 months, resulting in a poor estimate.

19. a, b. i. Area of $R_1 \approx 527.6$ thousand ft^3

ii. Area of R_1 + area of R_2 − area of $R_4 \approx 3112.8$ thousand ft^3

iii. Area of R_3 + area of R_4 = 1676 thousand ft^3

iv. Area of R_1 + area of R_2 + area of $R_3 \approx 4788.8$ thousand ft^3

c. Approximately 269 minutes (until 10:30 a.m.)

21. a. $f(x) = \begin{cases} 0 \text{ tons per year} & \text{when } 0 \le x < 5 \\ \dfrac{557.960}{1 + 91.202e^{-0.318025x}} \text{ tons per year} & \text{when } x \ge 5 \end{cases}$

x years after 1990.

b, c.

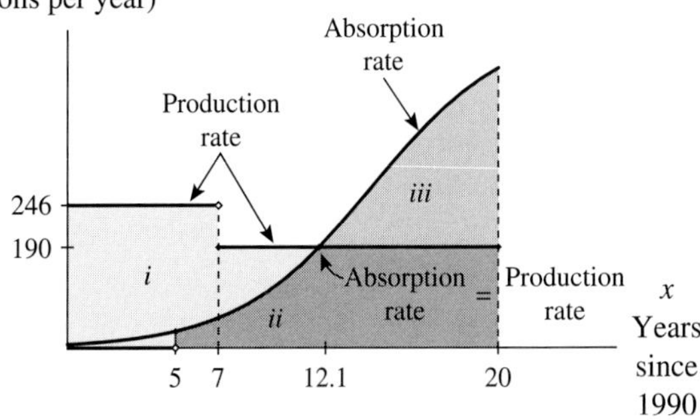

d. i. 2054.9 tons

ii. 2137.1 tons

iii. 1269.0 tons

e. The factory produces 4192 tons, and the trees absorb 3406 tons. This does not comply with the federal regulation.

Section 7.2

1. i. $R(m) = 0.2\left(\frac{\$47{,}000}{12}\right) = \$783.33$ per month

ii. $R(m) = 0.2\left(\frac{47{,}000}{12} + 100m\right) = 783.33 + 20m$ dollars a month after m months

iii. $R(m) = 0.2\left(\frac{47{,}000}{12}(1.005^m)\right) = 783.33(1.005^m)$ dollars a month after m months

3. i. \$53,493.40

ii. \$92,082.72

iii. \$61,818.49

The first option is the only one that will not result in the amount needed for the down payment.

5. \$35.0 billion

7. a. \$12.3 million

b. \$3.4 million

9. a. $\int_0^6 6000e^{0.0834(6-x)}\,dx \approx \$46{,}718.05$

b. $\sum_{m=0}^{71} 500\left(1+\frac{0.0834}{12}\right)^{(72-m)} \approx \$46{,}836.11$

c. Answers may vary.

11. a. $r(q) = 2.1(1.05^q)(0.15)$ billion dollars per quarter q quarters after the third quarter of 1999.

b. $R(q) = 2.1(1.05^q)(0.15)(1.13)^{16-q}$ million dollars per quarter for money invested q quarters after the third quarter of 1999.

c. If the investment begin with the 4th quarter 1999 profits, then the initial investment is based on a profit of $(2.1)(1.05) = \$2.205$ billion. Thus we calculate

$$\sum_{q=0}^{15} 2.205(0.15)(1.05^q)(1.13)^{16-q} \approx \$22.8 \text{ billion}$$

13. a. \$79.87 thousand b. \$58.29 thousand

15. a. \$28,324.60 b. \$28,445.37 c. Answers will vary.

17. a. \$5.4 billion b. \$6.1 billion c. \$11.2 billion

19. a. \$7.3 billion b. \$5.6 billion c. Answers will vary.

21. \$5.2 million

23. Answers given are based on the end of 2001.

a. 0.20 million terns

b. $T(t) = 2.04(0.83)^{22-t}$ million terns born t years after 1979

c. 10.97 million terns

25. a. None

b. $S(t) = (60 - 0.5t)(0.67)^{50-t}$ thousand seals born t years from now

c. 90.5 thousand seals

Section 7.3

1. a. The demand function

b. The supply function

c. The producers' surplus

d. The consumers' surplus

3. a. To find the price P above which consumers will purchase none of the goods or services, either find the smallest positive value for which the demand function is zero, $D(p) = 0$, or, if $D(p)$ is never exactly zero but approaches zero as p increases without bound, then let $P \to \infty$.

b. The supply function, S, is a piecewise continuous function with the first piece being the 0 function. The value p at which $S(p)$ is no longer 0 is the shutdown price. The shutdown point is $(p_1, S(p_1))$.

c. The market equilibrium price, p_0, can be found as the solution to $S(p) = D(p)$. That is, it is the price at which demand is equal to supply. The equilibrium point is the point $(p_0, D(p_0)) = (p_0, S(p_0))$.

5. a. D is an exponential demand function and so does not have a finite value p at which $D(p) = 0$. Thus the equation does not indicate a price above which consumers will purchase none of the goods or services.

b. \$6128.6 thousand

c. 17.4 thousand fans

d. \$4331.5 thousand

7. a. $D(p) = 0.025p^2 - 1.421p + 19.983$ lanterns when the market price is \$$p$ per lantern

b. \$109.18

c. \$31.89

9. a. 18.4 thousand answering machines; 300 thousand answering machines

b. \$9981 thousand; \$3131 thousand

11. a. $S(p) = \begin{cases} 0 \text{ hundred prints} & \text{when } p < 5 \\ 0.300p^2 - 3.126p + 10.143 \text{ hundred prints} & \text{when } p \geq 5 \end{cases}$

where p hundred dollars is the price of a print

b. \$837.12

c. Producers' revenue = \$148.5 thousand; Producers' surplus = \$27.3 thousand

13. a. \$408.3 hundred

b. 297 sculptures; no

c. \$4542.2 hundred

15. a. $D(p) = 499.589(0.958086^p)$ thousand books when the market price is \$$p$ per book

b.

$$S(p) = \begin{cases} 0 \text{ thousand books} & \text{when } p < 18.97 \\ 0.532p^2 - 20.060p + 309.025 \text{ thousand books} & \text{when } p \geq 18.97 \end{cases}$$

where \$$p$ is the price of a book

c. Approximately \$27.15; 156.2 thousand books

d. \$4728.6 thousand

Section 7.4

1. a. $V(t) = -1.664t^3 + 5.867t^2 + 1.640t + 60.164$ mph t hours after 4 p.m.

b. 68.99 mph

c. 72.23 mph

3. a.

$$R(y) = \begin{cases} \dfrac{0.719}{1+0.005e^{0.865563y}} + 0.62 \text{ dollars per minute} & \text{when } 2 \leq y < 10 \\ -0.045y + 1.092 \text{ dollars per minute} & \text{when } 10 \leq y \leq 20 \end{cases}$$

where y is the number of years since 1980

b. \$0.99 per minute c. \$0.67 per minute

5. a. 69.2 million people
b. Solving population ≈ 69.2 for t gives $t \approx 84.1$ years since 1900. This corresponds to early 1985.
c. 1.84 million people per year

7. a. -100.6 yearly accidents per year
b. 2810.5 yearly accidents
c.

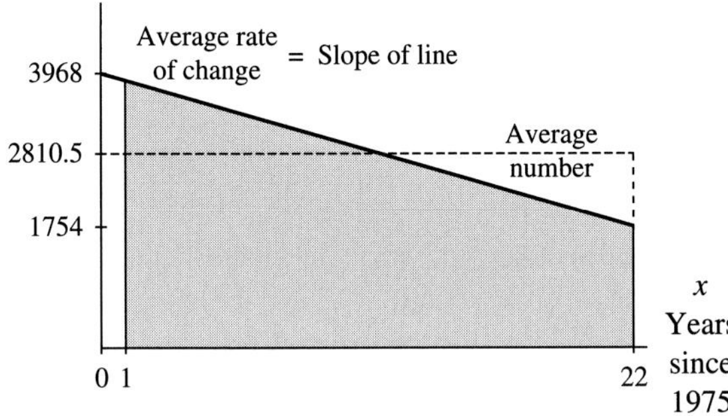

9. a. 2.28 feet per second squared c. 4540.7 feet
b. 129.7 feet per second d. 4540.7 feet
e.

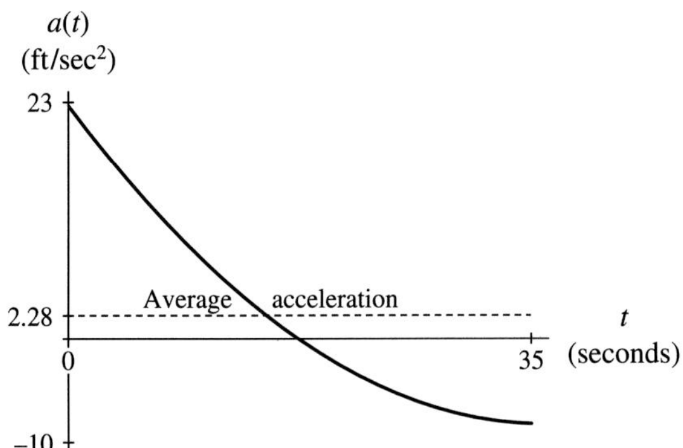

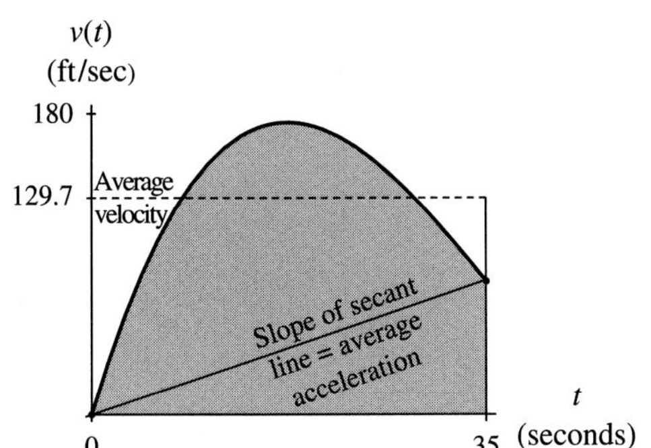

Area of shaded region = distance traveled = area of rectangle with height 129.7 ft/sec and width 35 seconds.

11. a. $V(t) = 1.033t + 138.413$ meters per second t microseconds after the experiment began
b. 174.58 meters per second

13. a. $B(t) = 0.030t^2 - 0.718t + 3.067$ mm Hg per hour t hours after 8 a.m.
b. 0.21 mm Hg per hour
c. 93.4 mm Hg

15. -0.15 percentage points per year

17. a. The average of the data is highest between 10 a.m. and 6 p.m.
b. $C(x) = \begin{cases} 4.75x + 2.833 \text{ ppm} & \text{when } 0 \le x < 4 \\ 0.536x^2 - 7.871x + 45.200 \text{ ppm} & \text{when } 4 \le x \le 12 \\ -5.5x + 93.667 \text{ ppm} & \text{when } 12 < x \le 16 \end{cases}$
x hours after 6 a.m.
Average concentration between 10 a.m. and 6 p.m. is 19.4 ppm.
c. Average ≈ 16.9 ppm. Severe pollution warning.

19. Consider the two graphs of a function f shown below, where A is the average value of f from a to b and k is an arbitrary constant.

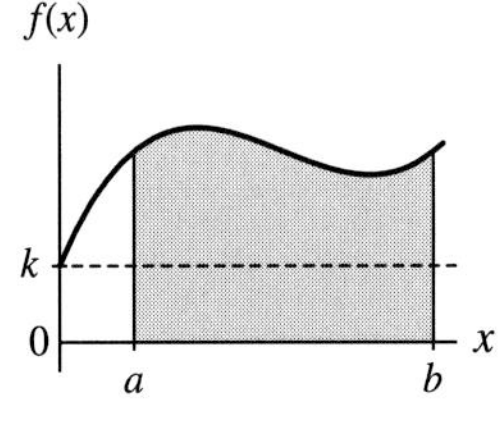

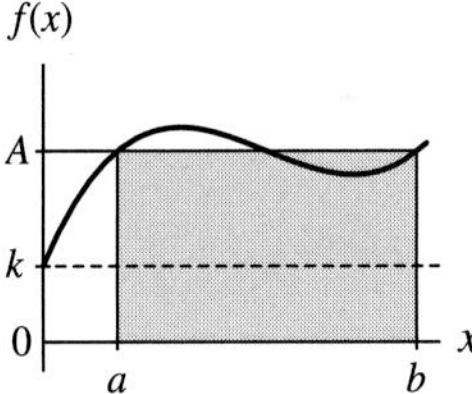

We know that the areas of the two shaded regions are equal. If we remove from each graph the rectangular region with height k and width $b - a$, the areas of the resulting regions are still equal, because we have removed the same area from each. See the following graphs.

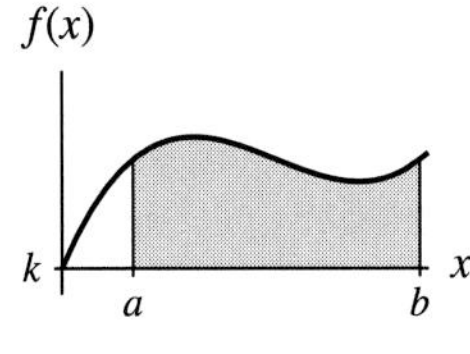

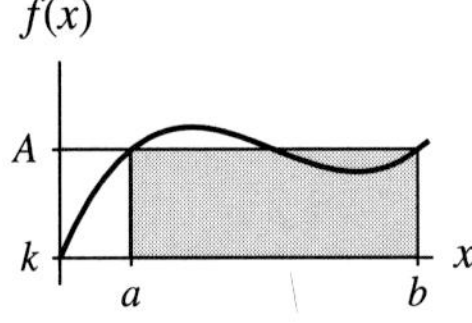

It is true for the graphs shown in this section with vertical axis shown from k rather than from zero that the area of the region between the function and $y = k$ from a to b is the same as the area of the rectangle with height equal to the average value minus k and width equal to $b - a$.

Section 7.5

1. a. There is a 46% chance that any telephone call made on a computer software technical support line will last 5 minutes or more.

b. The likelihood that any two cars on a certain two lane road are less than 7 feet apart is approximately 25%.

c. New Orleans will receive between 2 and 4 inches of rain during March 15% of the time.

3. a. Yes, $f(x) \geq 0$ for all x and $\int_{-\infty}^{\infty} f(x)\,dx = \int_{0}^{1} f(x)\,dx = 1$.

b. Yes, $h(x) \geq 0$ for all x and $\int_{-\infty}^{\infty} h(x)\,dx = \int_{0}^{1} h(x)\,dx = 1$.

c. No, the area between the graph of r and the horizontal axis is $0.3 + 0.6 \neq 1$.

d. No, some values of $s(c)$ are negative.

5. a. $P(x < 1) = \int_{0}^{1} y(x)\,dx = 0.16$

b. $\mu \approx 167$ gallons

c.

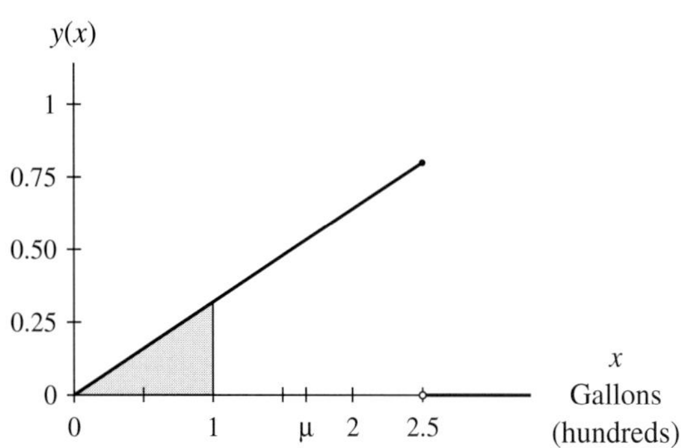

7. Because the $f(x)$ values are all non-negative and $\int_{-\infty}^{\infty} f(x)\,dx = 1$, the probability (which is the area between the graph of f and the input axis) must always be between 0 and 1. Another view of this explanation is that the probability of some occurrence is the proportion of time it is expected to happen, and all proportions are fractions between 0 and 1.

9. a. $P(20 < t < 30) = \int_{20}^{30} 0.2e^{-0.2t}\,dt \approx 0.016$

b. $P(t \leq 10) \approx 0.865$

c. $P(t > 15) = 1 - P(t \leq 15) = 0.050$

11. a. Mean = 2 minutes

b. Standard deviation $= \sqrt{0.8} \approx 0.89$ minute

c. 0.316; The likelihood that any child between the ages of 8 and 10 learns the rules of the board game in 1.5 minutes or less is about 31.6%.

d. 0.156; There is about a 15.6% chance that any child between the ages of 8 and 10 takes between 3 and 4 minutes to learn the rules of the new board game.

13. a. 137 customers

b. Answers will vary.

c. i. $P(150 \leq x \leq 200) \approx 0.58$

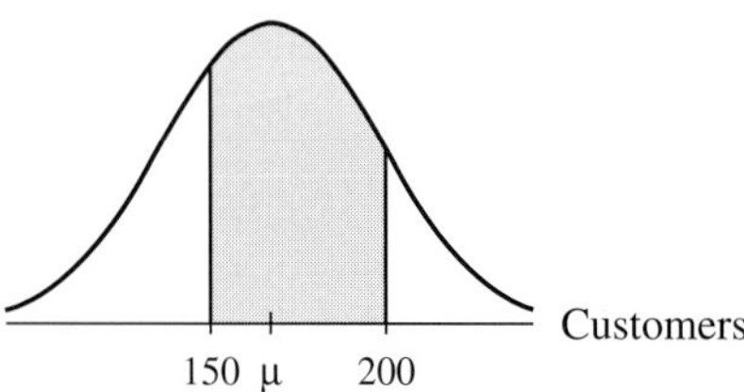

ii. $P(x < 220) \approx 0.50 + 0.46 = 0.96$

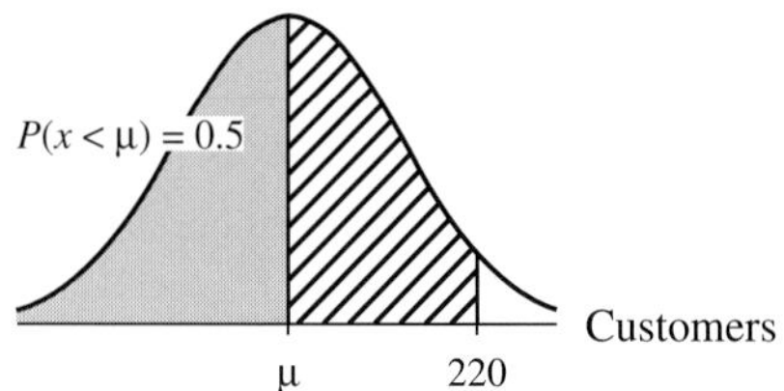

iii. $P(x > 235) = P(x > \mu) - P(\mu < x < 235) \approx 0.50 - 0.49 = 0.01$

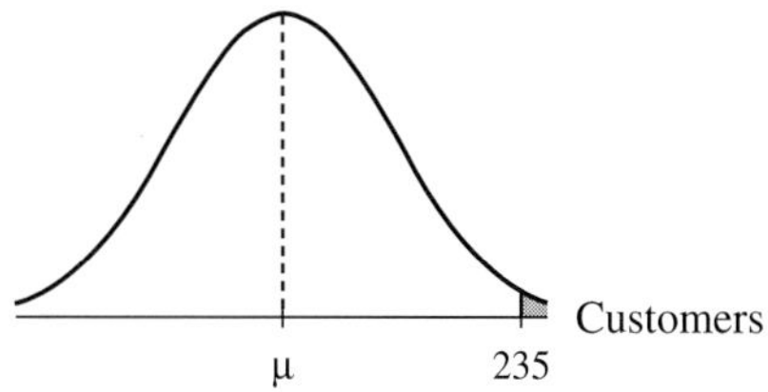

15. a. $6\sigma = 800 - 200$, so $\sigma = 100$

b. The realigned mean score is more for each distribution, because the recentering puts the mean of each at 500.

c. 50% because 475 was the mean math score before recentering.

d. No, the standard deviation is not given.

e. The statement ". . . and realigns the verbal and math scores so that a student with a score of 450 on each test can conclude that his or her math and verbal scores are equal" indicates a change in the shape of the distribution.

f. The scale was recentered for interpretation purposes. The recentering does not reflect any change in student performance. Entrance requirements and other comparisons will now be made on the new scale.

17. a. $P(60 \le x \le 80) \approx 0.272$, so 27.2%

b. $P(x \ge 90) \approx 0.50 - 0.232$, so 26.8%

c. $P(x > \mu + \sigma \text{ or } x < \mu - \sigma) = 1 - P(\mu - \sigma \le x \le \mu + \sigma) = 1 - P(43.65 \le x \le 100.95) \approx 1 - 0.683 = 0.317$, so about 31.7% of the students are likely to make a score more than one standard deviation away from the mean.

d. At $\mu - \sigma = 43.65$

19. a. $\mu = \int_a^b \frac{1}{b-a} x\,dx = \frac{1}{b-a} \cdot \frac{x^2}{2}\Big|_a^b = \frac{b^2 - a^2}{2(b-a)} = \frac{b+a}{2}$

b.

$$\sigma^2 = \frac{1}{b-a}\int_a^b (x-\mu)^2 dx = \frac{1}{b-a}\left[\frac{(x-\mu)^3}{3}\Big|_a^b\right] = \frac{1}{3(b-a)}[(b-\mu)^3 - (a-\mu)^3]$$

$$= \frac{1}{3(b-a)}\left[\left(b - \frac{b+a}{2}\right)^3 - \left(a - \frac{b+a}{2}\right)^3\right] = \frac{1}{3(b-a)}\left[\left(\frac{b-a}{2}\right)^3 - \left(\frac{a-b}{2}\right)^3\right]$$

$$= \frac{1}{24(b-a)}[(b-a)^3 - (a-b)^3] = \frac{1}{24(b-a)}[(b-a)^3 + (b-a)^3]$$

$$= \frac{2(b-a)^3}{24(b-a)} = \frac{(b-a)^2}{12}$$

Thus $\sigma = \frac{b-a}{\sqrt{12}}$.

c. The height of the density function is 0 to the left of a, so $F(x) = 0$ when $x < a$. When $a \le x \le b$, $F(x) = \int_a^x \frac{1}{b-a}dt = \frac{1}{b-a} \cdot t\Big|_a^x = \frac{x-a}{b-a}$. When x is to the right of b, the area between the graph of u and the horizontal axis to the left of x is the area of a rectangle with width $b - a$ and height $\frac{1}{b-a}$. No more area is accumulated to the right of b because the height of the density function is 0 at all such points. Thus $F(x) = 1$ when $x > b$.

21. a.

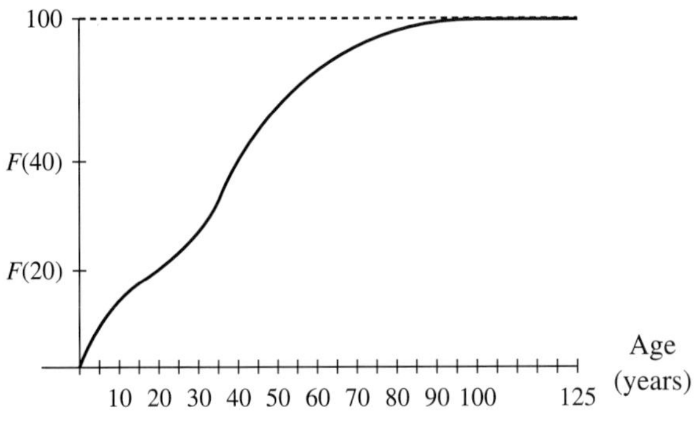

If F is the cumulative density function, then $F(40) - F(20)$ is $P(20 \le \text{age} \le 40)$.

b.

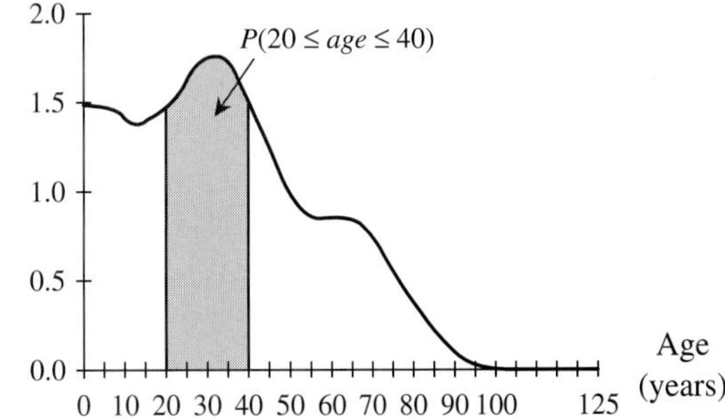

23. a. $F(x) = \begin{cases} 0 & \text{when } x < 0 \\ x^2 & \text{when } 0 \le x < 1 \\ 1 & \text{when } x \ge 1 \end{cases}$

b. $P(x < 0.67) = 0.4489$

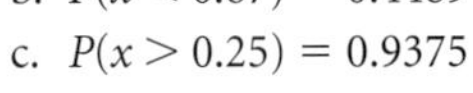

c. $P(x > 0.25) = 0.9375$

d.

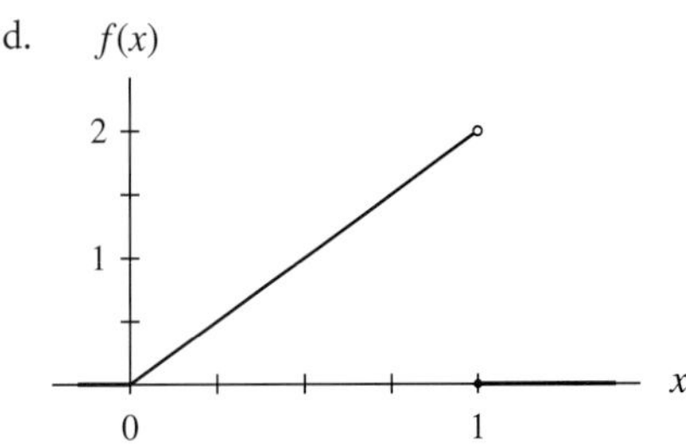

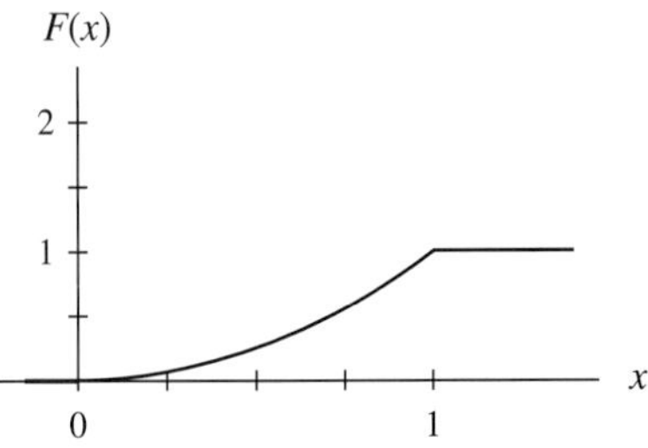

25. a.

x	0	5	10	15	25	35	50	75	100
F(x)	0	8.2	17.8	27.6	46.4	62.2	78.8	92.1	96.4

$F(100)$ is not 100 because the incomes of $100,000 or more are not counted in the table.

b.

x	2.5	7.5	12.5	20	30	42.5	62.5	87.5
F(x)	1.64	1.92	1.96	1.88	1.58	1.107	0.532	0.172

c. Yes, there appears to be an inflection point near $x = 40$, and the limiting value as x increases appears to be 0.

d. The overall fit is very good.

e. $\int_0^{100} f(x)\,dx \approx 96.41\% \approx F(100)$

f. $\mu = \int_0^{100} \frac{x \cdot f(x)}{100}\,dx \approx \$31{,}000$

27. a. The distribution on the top (the bell-shaped curve) is characteristic of a normal breeding population. Malcolm argues that the one on the bottom (with three peaks) is what would be expected from three separate (controlled) populations.

b. The dinosaurs were supposed to be all female and therefore not to reproduce, so Ian first saw the normal curve.

c. Neither graph is that of a probability density function. Even though all the outputs are non-negative, the area under each curve is more than 1.

Chapter 7 Review Test

1. a. $R(t) = (0.1)(3000 \cdot 12 + 500t) = 3600 + 50t$ dollars per year after t years

b. \$29,064

c. \$17,664

2. a. \$35,143.80

b. \$18,277.65

3. a. Approximately 1 fox

b. $f(t) = 500(0.63)^{20-t}$ foxes born t years after 1990 that will still be alive in 2010

c. $\approx$ 1083 foxes

4. a. $\approx$ \$635.4 hundred

b. 411 fountains. No, because $D(10) = 594$, supply is smaller than demand at this point.

c. $\approx$ \$6236.0 hundred

5. a. $\approx$ 37.5 million

b. $\approx$ 6.72 million per year

6. a. $P(x < 3.8) = 0.9025$; The value of x will be smaller than 3.8 about 90% of the time. This event is likely to occur.

b. $P(1.3 \le x \le 5) \approx 0.8944$

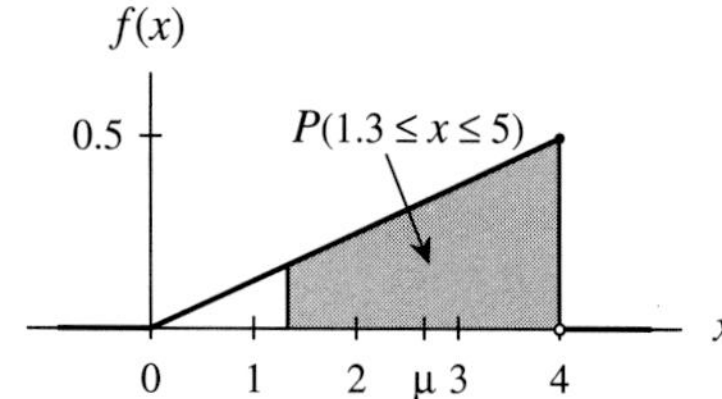

c. $\mu \approx 2.6667$

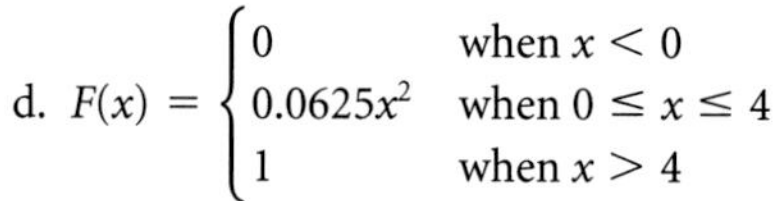

d. $F(x) = \begin{cases} 0 & \text{when } x < 0 \\ 0.0625x^2 & \text{when } 0 \le x \le 4 \\ 1 & \text{when } x > 4 \end{cases}$

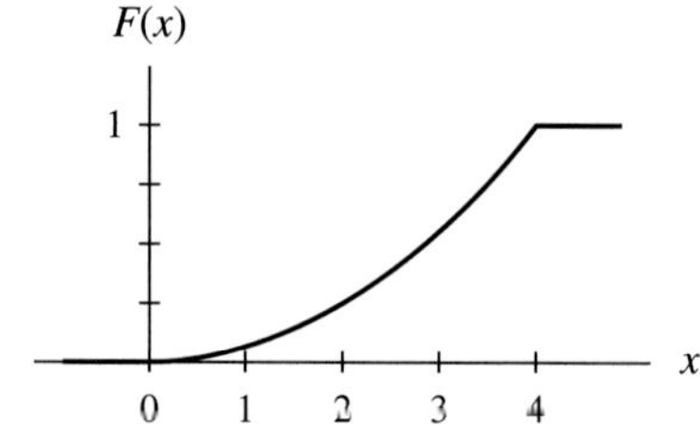

e. $P(1.3 \le t \le 5) \approx 0.8944$

CHAPTER 8

Section 8.1

1. a. $\sin 4 \approx -0.8$, $\cos 4 \approx -0.7$, $\sin\left(\frac{-\pi}{8}\right) \approx -0.4$, $\cos\left(\frac{-\pi}{8}\right) \approx 0.9$ (Answers may vary.)

b. $\sin(-1) \approx -0.8$, $\cos(-1) \approx 0.5$, $\sin \frac{144\pi}{16} = 0$, $\cos \frac{144\pi}{16} = -1$ (Answers may vary.)

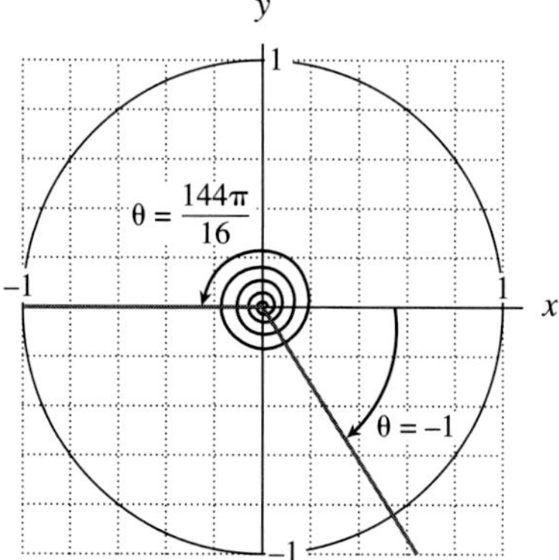

3. a. $\sin \frac{2\pi}{3} \approx 0.866$, $\cos \frac{2\pi}{3} = -0.500$

b. The point on the unit circle corresponding to an angle of $\frac{2\pi}{3}$ is approximately $(-0.5, 0.866)$.

c. Two possible answers are $\frac{8\pi}{3}$ and $\frac{-4\pi}{3}$.

5. a. $\sin \frac{3\pi}{2} = -1$, $\cos \frac{3\pi}{2} = 0$

b. The point on the unit circle corresponding to an angle of $\frac{3\pi}{2}$ *is* $(0, -1)$.

c. Possible answers include $\frac{-\pi}{2}$, $\frac{7\pi}{2}$, and $\frac{11\pi}{2}$.

7. It is not possible to have an angle θ such that $\cos\theta = 0.5$ and $\sin\theta = -0.5$ because $\cos\theta$ and $\sin\theta$ are x- and y-coordinates on the unit circle and must satisfy the equation $\cos^2\theta + \sin^2\theta = 1$.

9. a. Zero, positive, positive, positive, zero
Negative, negative, negative, zero

b. 0, 0.707, 1, 0.707, 0
−0.707, −1, −0.707, 0

c.

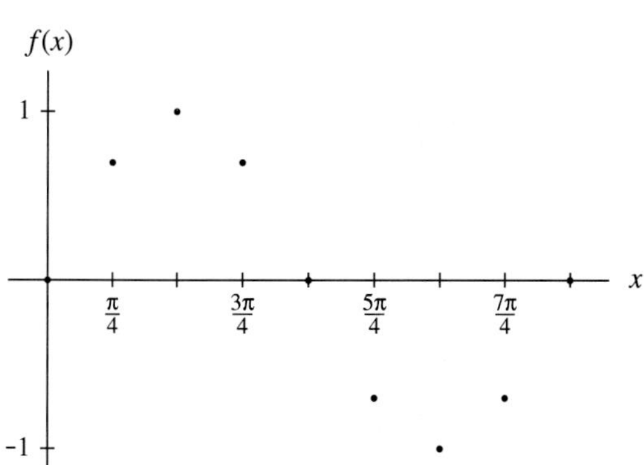

The figure is a scatter plot of some values of $f(x) = \sin x$.

11.

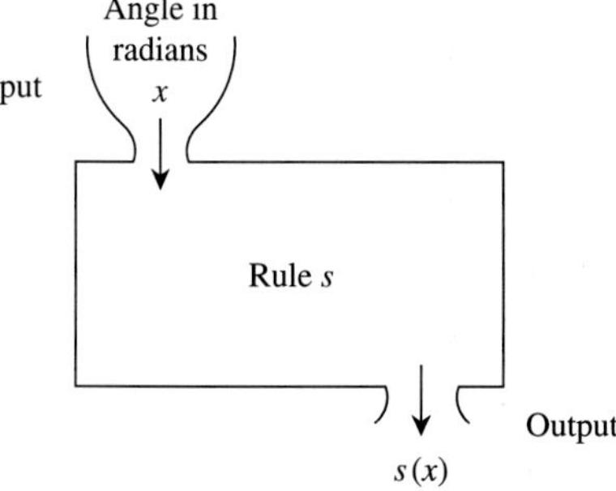

13. a. The maximum is 1 and occurs at $x = \frac{\pi}{2}$.
b. The absolute maximum is 1 and occurs at the values $x = \frac{\pi}{2} + 2k\pi$, where k is an integer.
c. The minimum is -1 and occurs at $x = \frac{3\pi}{2}$.
d. The absolute minimum is -1 and occurs at the values $x = \frac{3\pi}{2} + 2k\pi$, where k is an integer.
e. Sin θ is the y-value of points on the unit circle with equation $x^2 + y^2 = 1$. Because the unit circle has radius 1 and is centered at the origin, no points on the circle have x- or y-coordinates greater than 1 or less than 1. Thus the maximum of $y = \sin \theta$ is 1, and the minimum is -1.

15. a.

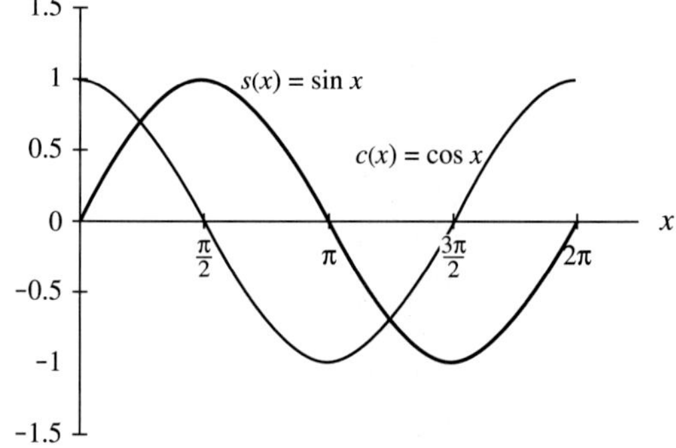

b. $c(x) = s(x) \approx 0.707$ when $x \approx 0.785$, and
$c(x) = s(x) \approx -0.707$ when $x \approx 3.927$.
c. $x \approx 0.785 + 2k\pi$, where k is an integer, and
$x \approx 3.927 + 2k\pi$, where k is an integer.

17. Amplitude = 1
Frequency = 1
Period = 2π
Vertical shift = none
Horizontal shift = π to the right
The function is reflected across the horizontal axis, so the cycle starting at $x = \pi$ begins by decreasing.

19. Amplitude = 235
Frequency = 300
Period = $\frac{\pi}{150}$
Vertical shift = 65 down
Horizontal shift = $\frac{1}{3}$ to the right
The function is reflected across the horizontal axis, so the cycle starting at $x = \frac{1}{3}$ begins by decreasing.

21. i. c ii. e iii. a
iv. f v. d vi. b

23.

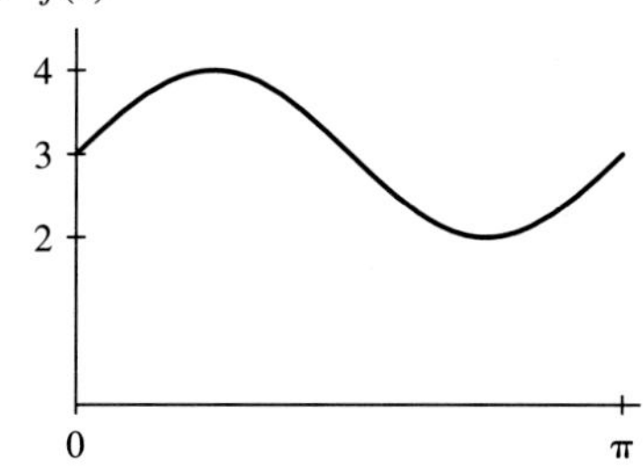

25.

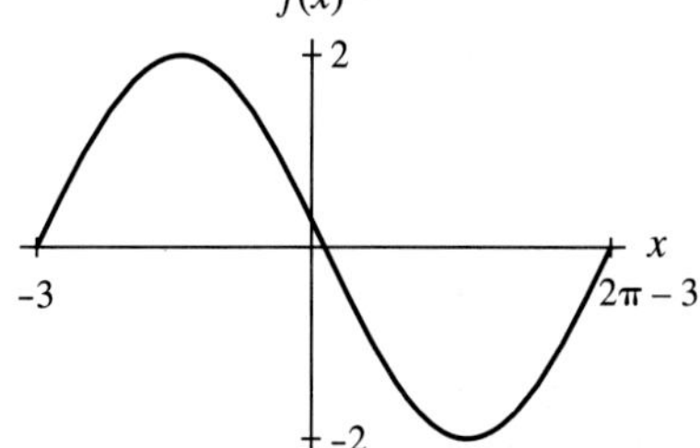

27.

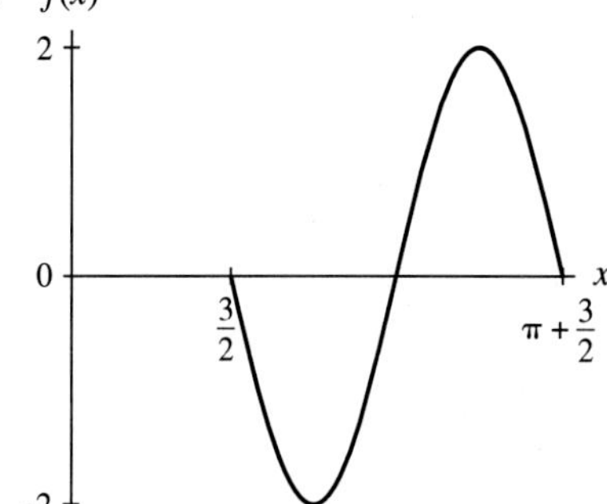

29. The members of each pair of graphs are identical. Generalization: Begin with a sine graph that is not vertically shifted, and reflect it across the horizontal axis. Begin with the same sine graph and shift it right or left by an integer multiple of half of the period of the original graph. This new graph will be identical to the reflected graph.

Section 8.2

1. a. $g(x) = 1.610 \sin(0.534x - 0.211) + 1.591$ therms per day x months after November 2000
Amplitude $= 1.610$ therms per day
Period $= \frac{2\pi}{0.534} \approx 11.8$ months
Frequency $= 0.534$ period
Horizontal shift $= \frac{0.211}{0.534} \approx 0.395$ month to the right
Vertical shift $=$ up 1.591 therms per day

b. Approximately 1.6 therms per day

c. Approximately \$26.89

3. a. The data are cyclic because the average daily temperature repeats itself over a year's time.

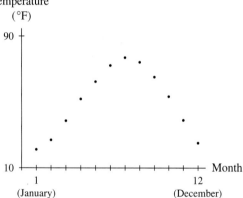

Daily mean temperature (°F)
90
10
Month
1 (January)
12 (December)

b. Amplitude $\approx \frac{76.9 - 21.1}{2} = 27.9$°F
Vertical shift $\approx 21.1 + 27.9 = 49$°F

c. Period ≈ 12 months
Horizontal shift ≈ 4 months to the right

d. $m(x) = 27.9 \sin\left(\frac{\pi}{6}x - \frac{2\pi}{3}\right) + 49$ °F during the xth month of the year

e. $m(x)$ (°F)
90
10
x Month
1
12

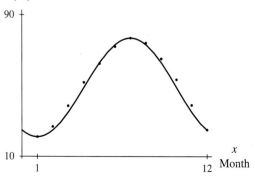

f. The model gives an estimate of 76.9°F as the average July temperature. (Note that this value agrees with the data.) Although the function is an excellent description of the normal mean daily temperature for the years 1961–1990, it cannot predict with any certainty the mean daily temperatures in Omaha this July.

5. a. Amplitude $\approx \frac{4.5 - 2.7}{2} = 0.9$
Vertical shift $\approx 2.7 + 0.9 = 3.6$

b. Period ≈ 6 years
Frequency $\approx \frac{\pi}{3}$
Horizontal shift ≈ 1.5 units right

c. $L(x) = 0.9 \sin\left(\frac{\pi}{3}x - \frac{1.5\pi}{3}\right) + 3.6$ log(base 10) reduction in year $x + 1965$

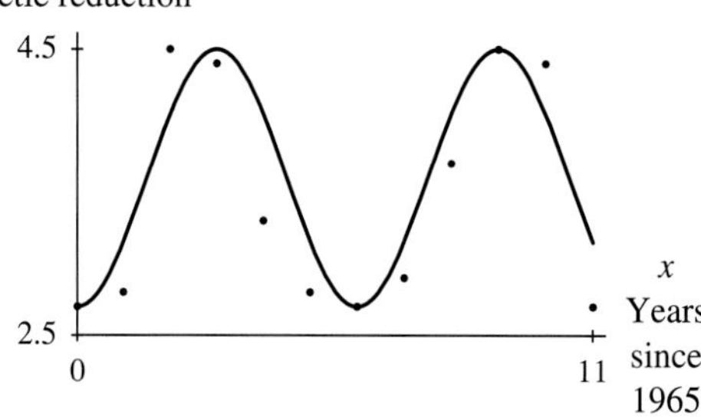

d. $B(x) = 10^{0.9 \sin\left(\frac{\pi}{3}x - \frac{1.5\pi}{3}\right) + 3.6}$ beetles not surviving in year $x + 1965$

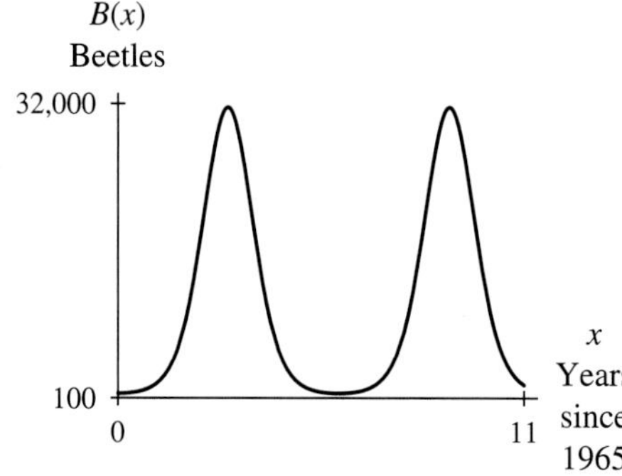

The two graphs in this activity have the same period and horizontal shift, but the amplitudes and vertical shifts are different. The graph of the number of beetles not surviving has much higher peaks than the log(base 10) reduction model.

7. a. The low temperatures should have the same period and approximately the same amplitude and horizontal shift as the average temperatures, but they should have a smaller vertical shift.

b. Amplitude $\approx \frac{65.9 - 10.9}{2} = 27.5$°F
Vertical shift $\approx 10.9 + 27.5 = 38.4$°F
Horizontal shift ≈ 4 months to the right
Period ≈ 1 year $= 12$ months

As expected, the period and horizontal shift are the same. The amplitude is slightly lower than for the mean temperature data, and the vertical shift is less.

c. We calculate b using the equation $12 = \frac{2\pi}{b}$, so $b = \frac{\pi}{6}$.

We calculate h using the equation $\frac{|h|}{b} = 4$ and the information that the horizontal shift is to the right, so h is negative: $h = -4b = \frac{-2\pi}{3}$. Using $a = 27.5$ and $k = 38.4$, we have

$$f(x) = 27.5 \sin\left(\frac{\pi}{6}x - \frac{2\pi}{3}\right) + 38.4 \text{ °F}$$

during the xth month of the year

9. a. The data appear concave up between 1949 and 1955, concave down between 1956 and 1960, and concave up between 1961 and 1963. A cubic model is not a good fit and is not appropriate because more than one inflection point are indicated by the data. Possibly a piecewise continuous model that is quadratic between 1949 and 1955 and cubic between 1956 and 1963 would fit the data. However, the sine model seems most appropriate.

b. $f(x) = 41.547 \sin(0.528x - 2.913) + 194.493$ aircraft x years since 1949

c. $f(15) \approx 155$ aircraft

d. Answers will vary.

11. a.

Hours	°F	Hours	°F
5	37	65	60
11	44.5	71	47
17	52	77	34
23	47	83	45
29	42	89	56
35	49	95	46.5
41	56	101	37
47	49	107	45.5
53	42	113	54
59	51		

b. $y = 8.709 \sin(0.261x - 2.840) + 47.090$ °F x hours after midnight on Wednesday
The period is approximately 24 hours.

c. High $\approx$ 56°F and low $\approx$ 38°F

d. The greatest discrepancies between the model and the data occur on Friday afternoon and Saturday morning. The hydroelectric plant's energy may also be needed Wednesday afternoon and Sunday morning.

13. Answers will vary.

15. Without technology: $D(t) = 7 \sin\left(\frac{2\pi}{365}t - \frac{542\pi}{365}\right) + 11.5$ hours of daylight, where t is the number of days since December 31 of the previous year.
With technology: $d(t) = 6.677 \sin(0.016t + 1.908) + 11.730$ hours of daylight, where t is the number of days since December 31 of the previous year.

17. Cherry odor: $c(x) \approx 6.930 \sin(0.113x + 2.038) + 8.856$ firings/ms with period ≈ 56
Apple odor: $a(x) \approx 7.073 \sin(0.133x + 1.621) + 10.526$ firings/ms with period ≈ 47
Cherry/apple odor: $t(x) \approx 3.221 \sin(0.120x + 1.958) + 5.618$ firings/ms with period ≈ 52 (Your model parameters may differ slightly from these.) Answers will vary.

19. a. Maximum is 62°F: minimum is −12°F

b. The average is 25°F, which equals k, the vertical shift.

21. a. Yes. Individual daily fluctuations are "smoothed out" over the month, and they should come close to repeating each year.

b. 29.8°F for New Orleans; 46.1°F for Boston. On average, the daily high temperatures in Boston vary more over a year's time than the temperatures in New Orleans.

23. a. Amplitude $\approx \frac{33 - 0}{2} = 16.5$ thousand lizards
Period $\approx$ 12 months
Vertical shift $\approx$ 16.5 thousand lizards

b. $L(m) = 16.298 \sin(0.537m - 1.801) + 11.206$ thousand lizards in month m, where m is the number of months since the beginning of year 1. The amplitude of this model is approximately 16.3 thousand lizards, only 0.2 thousand lizards less than the estimated amplitude in part a. The period is $\frac{2\pi}{0.537} \approx 11.7$ months, compared with the 12 months estimated in part a. The vertical shift is significantly less than that estimated in part a.

c. The graph appears to match the period of the data well, but it falls below the horizontal axis and doesn't reach the highest values.

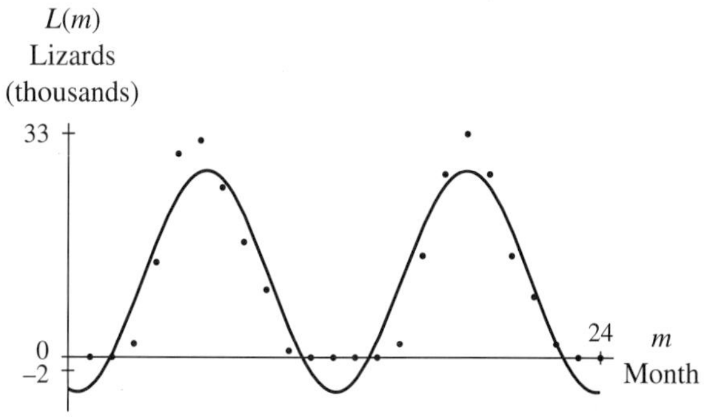

d. $l(x) = 15.388 \sin(0.787x - 1.667) + 16.346$ thousand lizards in month x where $x = 1$ through $x = 8$ correspond to March through October of year 1 and $x = 9$ through $x = 16$ correspond to March through October of year 2. The amplitude of this function is 15.388, about 1 less than the amplitude of the equation in part b. The period is approximately 8 months instead of the 11.9 months in part b. This smaller period is the result of deleting the data for 4 months of each year. The vertical shift of 16.3 is greater than the vertical shift of the equation in part *b*. This vertical shift better matches the shift calculated in part *a*. This equation fits the modified data well.

Section 8.3

1. $f'(x) = 3 \cos 3x$

3. $t'(r) = -2.34 \sin(0.45r + \pi) + 80$

5. $h'(x) = -0.3328 \cos(-0.16x + 12.3) - 9.666 \sin[1.35(2x + 6)]$

$h''(x) = -0.053248 \sin(-0.16x + 12.3) - 26.0982 \cos[1.35(2x + 6)]$

7. a. $R(x) = 0.7c(x) + 0.2p(x) = 1.32727 \sin(0.0186x + 1.1801) + 1.46052 \cos(0.0197x - 3.7526) + 7.90076$ million dollars, where x is the day of the year

b. $R'(x) \approx 0.0247 \cos(0.0186x + 1.1801) - 0.0288 \sin(0.0197x - 3.7526)$ million dollars per day where x is the day of the year

c. $R'(46) \approx -0.003$ million dollars per day. Combined daily sales were decreasing by about $3000 per day on February 15, 1992.

d. Approximately 220 days and 293 days after the beginning of 1992. (These values correspond to August 7 and October 19. Note that 1992 was a leap year.)

e. $x \approx 90$ days (March 30) or 166 days (June 14) or 339 days (Dec. 4).

9. $B'(11) \approx -11.1$°F/month

$B'(4.5) \approx 11.6$°F/month

11. a. 2 hours

b. 3050 km at 48 minutes after launching

c.

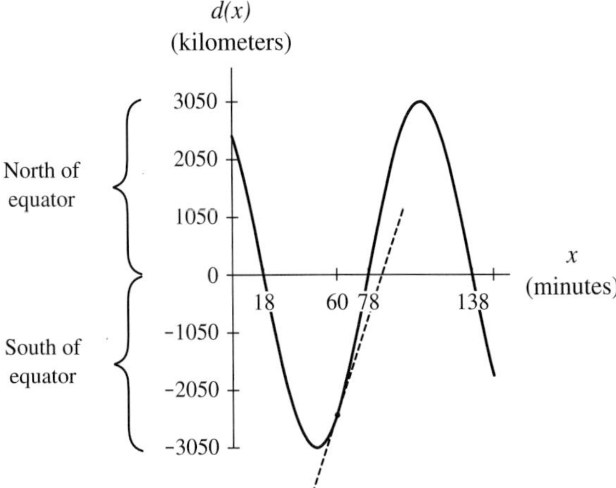

d. Approximately 93.9 kilometers per minute

e. No; speed is the rate of change of the total distance traveled with respect to the traveling time.

13. a, b.

Wind speed interval	6–10 m/sec	10–14 m/sec	14–18 m/sec
Change (kilowatts)	-17.4	-17.3	-0.3
Percentage change	38.4% decline	61.7% decline	2.7% decline
Average rate of change (kilowatts/m/sec)	-4.4	-4.3	-0.07

c. $P'(s) \approx 5.2126 \cos(0.258s + 0.570)$ kilowatts/m/s when the wind speed is s meters per second

d, e.

Wind speed	6	10	14	18
$P'(s)$ (kilowatts/m/sec)	-2.7	-5.2	-2.6	2.5
Percentage rate of change (%/m/sec)	-6.0	-18.6	-24.7	24.1

The power output of the engine is decreasing at the rate of 2.7 kilowatts per meter per second at a wind speed of 6 meters per second. When the wind speed is 10 meters per second and increases to 11 m/sec, the power output of the engine decreases by approximately 5.2 kilowatts. When the wind speed is 14 m/sec and increases to 15 m/sec, the power output of the engine decreases by approximately 2.6 kilowatts. At a wind speed of 18 m/sec, the power output of the engine is increasing at the rate of 2.5 kilowatts per meter per second.

15. a. Approximately -8 mm per day

b. 47.1% decline; -1.6 mm per day per month

c. Approximately 2.4 mm per day per month. Extraterrestrial radiation in Amarillo is increasing by 2.4 millimeters of equivalent water evaporation per day per month in March.

d. 18.4% per month

17. a. $d(t) = 60.7407 \sin(0.1206t + 0.3134) + 99.9194$ deaths per 100,000 people per week, where t is the number of weeks since January 1, 1923; Answers will vary.

b.

Date	t	$d'(t)$
Middle of 1924	78	-7.0111 deaths per 100,000 people per week per week
End of 1924	104	-7.0252 deaths per 100,000 people per week per week

19. a. Period $\approx$ 12 hours, amplitude = 2 feet, average water height = 0 feet

b. $f(t) = 2.019 \sin(0.514t - 2.963) + 1.793 \cdot 10^{-4}$ feet from mean sea level, where t is the number of hours since 12 midnight on March 1. (Model parameters may vary slightly.)

c. $f'(12) \approx -1.04$ feet per hour

d. $f'(60) \approx -0.95$ feet per hour

e. The water level will be at least 1 foot above mean sea level between about 31.2 hours and 35.0 hours and between 43.5 hours and 47.6 hours after midnight on March 1. In terms of actual time, the boat could float off between 7:14 a.m. and 11:20 a.m. on March 2 or between 7:30 p.m. and 11:34 p.m on March 2.

21. a. $D(d) = 23.677 \sin(0.017d - 1.312) - 0.292$ degrees on the dth day of the year; The amplitude of the model is 23.7 degrees. This is the greatest angle of declination that the sun reaches. The period is approximately 374 days (calculated with the unrounded b-value). We expect the sun's declination to complete a cycle every year. The period in the model is slightly long.

b. The equinoxes occur at $d \approx 78.7$ days and $d \approx 264.4$ days. $D'(78.7) \approx 0.4$ degree per day, and $D'(264.4) \approx -0.4$ degree per day. At the equinoxes, the declination of the sun is changing by approximately 0.4 degree per day.

c. The summer and winter solstices occur when the declination is greatest in both north and south directions. On the graph, the summer solstice corresponds to the maximum, and the winter solstice corresponds to the minimum. At these points, the rate of change is zero.

23. Inside: $u(x) = 3 \sin x - 12$

Outside: $y(u) = 2.4^u$

$y' = (3 \cos x)(\ln 2.4)2.4^{3 \sin x - 12}$

25. Inside: $u(x) = \sin x - 7$

Outside: $y(u) = 4u^2 + 8u + 13$

$y' = 8 \cos x(\sin x - 6)$

27. Inside: $u(x) = \sin x$

Outside: $y(u) = \ln u$

$y' = \dfrac{\cos x}{\sin x}$

Section 8.4

1. a. The maximum daily mean temperature is approximately 76.9°F (in July).

The minimum daily mean temperature is approximately 21.1°F (in January).

b. $t(x) = 28.533 \sin(0.479x - 1.801) + 47.985$°F in the xth month of the year. The highest point on the model is (7.041, 76.517), and the lowest point is (0.481, 19.452). The maximum normal daily mean temperature in Omaha between 1961 and 1990 was 76.5°F and occurred at the beginning of August. The normal daily minimum temperature was at a minimum of 19.5°F near the middle of January.

c. The maximum and minimum temperatures obtained with the model are slightly lower than those found using the data. Because the data are based on a monthly average over a 30-year period, actual high and low temperatures in any given year are likely to vary from those found in parts *a* and *b*.

3. a. The maximum of approximately 105.3 males per 100 females occurred in 1908.

The minimum of approximately 94.6 males per 100 females occurred in 1982.

b. The inflection point occurs in 1945 with a corresponding rate of change of approximately -0.23 male per 100 females per year.

5. a. Approximately 467.2 milliseconds

b. The pulse speed was at its highest value, 227 counts per second, after about 233.6 milliseconds and 700.7 milliseconds. The pulse speed was at its lowest value of 146 counts per second after approximately 4.2 milliseconds and 467.2 milliseconds. Answers may vary.

c. The speed of the pulses emitted from the star was increasing the fastest after approximately 116.8 milliseconds. The speed at that time was about 186.5 counts per second. The speed of the pulses emitted from the star was decreasing the fastest after approximately 350.4 milliseconds when the speed was also 186.5 counts per second.

7. a. $d(t) = 60.741 \sin(0.121t + 0.313) + 99.919$ deaths per 100,000 people per week, where t is the number of weeks since January 1, 1923.

b. The period is $\frac{2\pi}{0.1206} \approx 52$ weeks, so the death rate should peak again at $t \approx 166.8$ weeks since January 1, 1923. Answers will vary.

c. $t \approx 153.7$, $d(t) \approx 100$. The first time after 150 weeks since January 1, 1923 that the weekly pneumonia death rate was increasing the fastest was about 154 weeks. The weekly death rate at that time was approximately 100 deaths per 100,000 people per week.

9. a. Ten maxima of the sine function i are near a relative maxima of the actual index w. Note that the maxima appear as minima because of the inverted scale. (Answers may vary.)

b. Ten minima of the sine function i are near a relative minima of the actual index w. (Answers may vary.)

c. A relative maximum of about 104.6 occurs at $t \approx 120.25$ and a relative minimum of 95.8 occurs at $t \approx 116.25$ years after 1761. Thus the next maximum and minimum of $i(t)$ after 1874 are predicted to have occurred in 1878 and in 1882, respectively.

d. Answers will vary.

11. a. The blue graph represents daylight hours. The black graph represents nighttime hours.

b. Period $\approx$ 12 months, amplitude $\approx$ 7.5 hours, vertical shift $\approx$ 12 hours

c. Values in table may vary.

Month	Hours	Month	Hours
Jan	5	July	19
Feb	8	Aug	17
Mar	10.5	Sept	14.5
Apr	13.5	Oct	11.5
May	16.5	Nov	8.5
June	19	Dec	6

d. $H(m) = 7.021 \sin(0.480m - 1.619) + 11.786$ hours at the beginning of the mth month.
Period $\approx$ 13 months, amplitude $\approx$ 7 hours, vertical shift $\approx$ 11.8 hours

e. The number of hours is increasing most rapidly in early March and is decreasing most rapidly at the end of September.

f. Answers may vary.

13. a. $R(x) = 1.32727 \sin(0.0186x + 1.1801) + 1.46052\cos(0.0197x - 3.7526) + 7.90076$ million dollars, where x is the day of the year

b. $R'(x) \approx 0.02469 \cos(0.0186x + 1.1801) - 0.02877 \sin(0.0197x - 3.7526)$ million dollars per day, where x is the day of the year

c. \$8.03 million, $x \approx 193$ (July 12)

d. The absolute minimum value is \$7.66 million occurring at $x \approx 66$ (March 7). A local minimum value of \$7.68 million occurs at $x \approx 318$ (November 14).

15. a. $c''(x) = (0.010345x + 2.105676)(6.283) \cdot (6.283)\left[-\cos\left(6.283x - \frac{\pi}{2}\right)\right] - (0.02069)(6.283) \sin\left(6.283x - \frac{\pi}{2}\right) + 2(0.013783)$ ppm per year per year x years after 1900

b. In each year, the most rapid increase occurs at the end of the year, and the most rapid decrease occurs at the beginning of July.

c.

Year	Level at time of most rapid increase (parts per million)
1960	315.1 (Jan.) 315.8 (Dec.)
1970	324.3
1980	337.6
1990	352.6

Both the amplitude and the expected value are increasing each year, causing the output at the points of most rapid increase to rise each year.

17. a. $a''(m) = -6.3745975 \sin(0.485m - 1.707)$ watts per cm^2 per month per month where $m = 1$ in January, $m = 2$ in February, and so on.

b. The point of most rapid increase occurs in mid-April ($m \approx 3.5$), and the point of most rapid decrease occurs at the beginning of October ($m \approx 10$).

c. 32.9 watts per cm^2

19. a. The period of the function is $\frac{2\pi}{\pi/3} = 2\pi\left(\frac{3}{\pi}\right) = 6$ years

b. Solving

$$R'(x) = 0.9\left(\tfrac{\pi}{3}\right) \cos\left(\tfrac{\pi}{3}x - \tfrac{6.5\pi}{3}\right)(\ln 10)10^{0.9 \sin\left(\frac{\pi}{3}x - \frac{6.5\pi}{3}\right) + 3.6} = 0$$

can be done by determining where $\cos\left(\frac{\pi}{3}x - \frac{6.5\pi}{3}\right)$ is zero on the cycle beginning at $x = 6.5$ and ending at $x = 12.5$. The derivative is zero at $x = 8$ and $x = 11$. The maximum of approximately 31,623 beetles occurs in 1968 when $x = 8$. The minimum of approximately 501 beetles occurs in 1971 when $x = 11$.

Section 8.5

1.

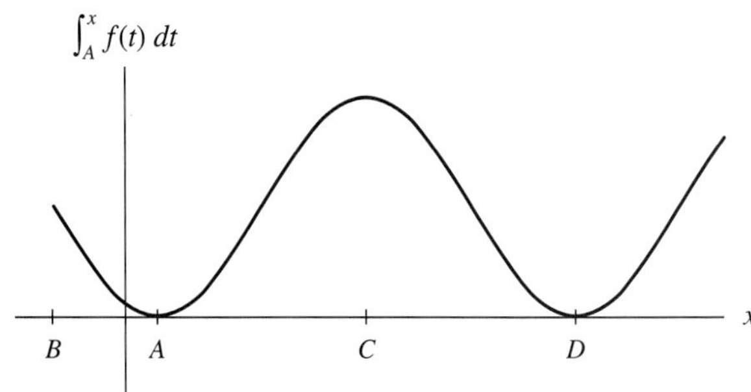

3. $-7.3 \cos x + 12x + C$

5. $\frac{-1}{7.3} \cos(7.3x + 12) + C$

7. $F(x) \approx -194.583 \cos(0.024x + 3.211) + 14.63x + C$

9. a. $S(t) \approx 24.259 \sin(0.987t + 1.276) + 35.324$ thousand dollars, where $t = 1$ in January, $t = 2$ in February, and so on.

b. Peak sales occur at $t \approx 0.30$ (January 10) and $t \approx 6.66$ (July 21). Sales are lowest at $t \approx 3.48$ (April 14) and $t \approx 9.85$ (October 26).

11. a. 1

b. 0

c. $2 + 2a$

d. 12

13. a. Height in counts per second, width in milliseconds, area in (counts per second)(milliseconds)

b. $p(s) = 0.0405 \sin(0.01345s - 1.5708) + 0.1865$ counts per millisecond after s milliseconds

c. $\int_0^{467.151324} p(s)\,ds \approx 87$ counts

Approximately 87 pulses are emitted by the star over one period (about 0.5 second).

15. a. Approximately \$0.548 thousand. June sales were approximately \$548 more than January sales.

b. \$86.4 thousand

c. \$29.2 thousand

17. a. $\int_1^{16} L(x)\,dx \approx 259$ thousand lizards

b. 261 thousand lizards. This value is 2 thousand more than the answer to part *a*.

c. Answers will vary.

Chapter 8 Review Test

1. a. Amplitude = 712.5, vertical shift = 787.5, period = 12 months, and horizontal shift ≈ 3. (Remember we look for an inflection point where the function is increasing.)

b. $l(x) = 713.251 \sin(0.525x - 1.557) + 777.883$ lawn mowers ordered x months since December of the previous year. The amplitude is very close, whereas the vertical shift of the model is about 10 lawn mowers less than that found using the data. The period of the model is approximately 12 months. The horizontal shift of the model is approximately 3. Both the period and the horizontal shift are very close to those found using the data.

c. $l(16) \approx 1157$ lawn mowers. For this estimate to be valid, the cyclic pattern shown by the data should continue for the following year.

d. Lawn mower orders were highest in June ($x = 6$).

e. Solving $l''(x) = 0$, finding where $l'(x)$ is greatest, or using $\frac{c}{b}$ in the sine model, we find that orders were increasing most rapidly in March ($x = 3$).

2. a. Average rate of change $\approx \frac{56 - 103}{30} \approx -1.6$ units per year

b. Slope of price index ≈ 0 unit per year

3. a. Average rate of change $= \frac{S(72) - S(42)}{30} \approx -1.32$ units per year

b. $S'(42) \approx 0.37$ unit per year. According to the model, the Sauerbeck index of general wholesale prices was increasing at the rate of 0.37 unit per year in 1860.

c. $\frac{1}{72 - 42}\int_{42}^{72} S(t)\,dt \approx 90.8$

4. a. The rate of change was greatest when $t \approx 111$ days, which corresponds to March 20 (1992 was a leap year).

b. $R(t) \approx 7.3046 \cos(0.0197t - 3.7526) + 20.383$ million pints t days after January 1, 1992.

c. The answer to part *a* is the input corresponding to the inflection point for the model in part *b*.

5. Approximately 1.24 million pints. At the end of February 1992, Campbell's was selling approximately 1.24 million more pints than it was selling at the beginning of February.

CHAPTER 9

Section 9.1

1.

Inputs

a years

h inches

Rule W

Output

$W(a, h)$ pounds

3.

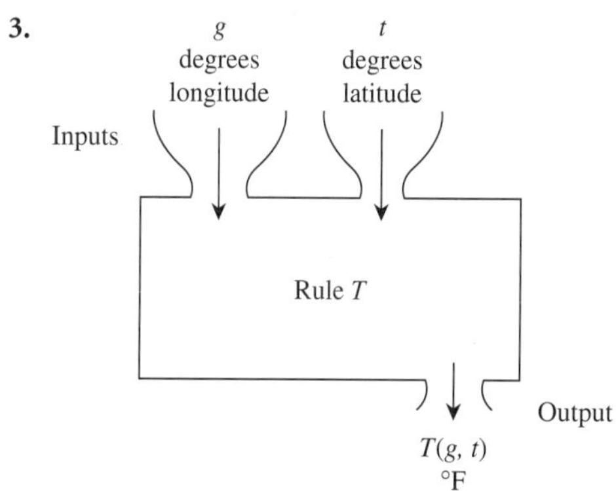

5.

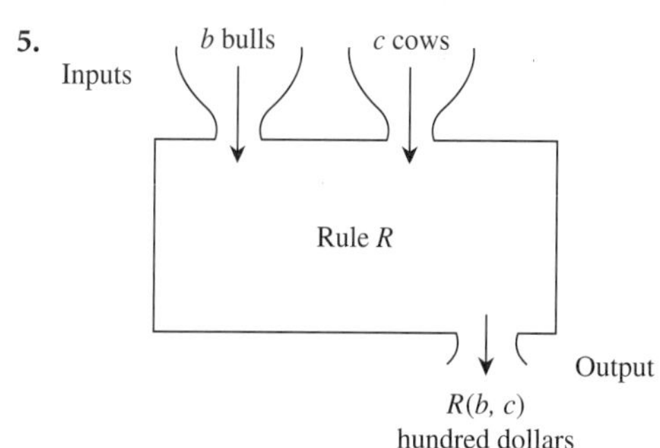

7. a. $P(1.2, s)$ is the profit in dollars from the sale of a yard of fabric as a function of s, the selling price per yard, when production cost is \$1.20 per yard.

b. $P(c, 4.5)$ is the profit in dollars from the sale of a yard of fabric as a function of c, the production cost per yard, when the selling price is \$4.50 per yard.

c. When the production cost is \$1.20 per yard and the selling price is \$4.50 per yard, the profit is \$3.00 for each yard sold.

9. a. $P(100{,}000, m)$ is the probability of the senator voting in favor of the bill as a function of the amount m million dollars invested by the tobacco industry lobbying against the bill when the senator receives 100,000 letters supporting the bill.

b. $P(l, 53)$ is the probability of the senator voting in favor of the bill as a function of the number l of letters in favor of the bill received by the senator when the tobacco industry spends \$53 million lobbying against the bill.

11. a. $A(h, 96) = 0.01h^2 - 0.26h + 95.298$ °F when the relative humidity is h%.

b. $A(70, t) = 0.068t^2 - 9.285t + 389.133$ °F when the temperature is t°F.

When the air temperature is 89°F and the relative humidity is 70%, $A(70, 89) \approx 103$°F is the apparent temperature.

13. a. 60% of the time

b. $C(9, f) = -1.651f^3 + 2.686f^2 - 1.597f + 1.019$, where f is the fraction (expressed as a decimal) of the sky covered by clouds.

15. a. $P(x) = -0.857x + 7.781$ pounds/person/year, where $\$(1.50 + x)$ per pound is the price of peaches; $P(0.05) \approx 7.7$ pounds/person/year

b. $P(y) = 4.696 + 2.025 \ln y$ pounds per person per year, where $\$10{,}000y$ is the family income; $P(3.5) \approx 7.2$ pounds/person/year

c. $C(p, 4) = 2 \ln 4 + 2.7183^{-p} + 4$ pounds/person/year, where $\$(1.50 + p)$ per pound is the price of peaches; $C(0.05, 4) \approx 7.7$ pounds/person/year

$C(0.30, i) = 2 \ln i + 2.7183^{-0.30} + 4 \approx 2 \ln i + 4.741$ pounds/person/year, where $\$10{,}000i$ is the family income; $C(0.30, 3.5) \approx 7.2$ pounds/person/year. The discrepancies in the equations are the result of rounding the data in the table.

17. a. $K(0.7, 0.7, 17, 15{,}000, 64{,}000) \approx 1.95$ cows per hectare. Approximately 2 cows can be supported by 1 hectare. Using the fact that 1 hectare $\approx$ 2.471 acres, we have

$$\left(\frac{1.95 \text{ cows}}{\text{hectare}}\right)\left(\frac{1 \text{ hectare}}{2.471 \text{ acres}}\right) \approx 0.8 \text{ cow per acre.}$$

b. $K(0.8, 0.85, G, P, A) = \frac{0.68GP}{A}$ animals per hectare, where G is the gross energy content of the crop, P is the net crop production, and A is the annual energy requirement of the animal.

c. $K(0.7, 0.8, 17, P, A)$ is the carrying capacity of a crop as a function of net crop production and the energy requirement of the animal when 70% of the crop is consumed, 80% of the crop is digested as useful nutrients, and the gross energy content of the crop is 17 megajoules per kilogram.

$K(0.7, 0.8, 17, P, A) = \frac{9.52P}{A}$ animals per hectare.

19. a. \$60,733.51

b. $A(0.075, n, m) = 160m(1 - 1.00625^{-n})$ dollars for a period of n months and a monthly payment of m dollars

c. $A(r, 36, m)$ is the output of the cross-sectional function for the loan amount (in dollars) when 36 monthly payments are made.

d. $A(0.06, 36, m) = 200m\left[1 - \left(1 + \frac{0.06}{12}\right)^{-36}\right] \approx 32.9m$ dollars, when the monthly payment is m dollars.

21. a. $H(15, -15) \approx 1640.6$ kilogram-calories per square meter per hour

b. Solving $(10.45 + 10\sqrt{10} - 10)(33 - t) = 2000$ for t, we find that $t \approx -29.4$°C.

23. a. Approximately 117°F

b. 42%

c. $A(h, 96) = 0.009h^2 - 0.114h + 91.308$ °F when the relative humidity is h%

d. $A(70, t) = 0.069t^2 - 9.508t + 399.049$ °F when the air temperature is t°F. The equations in parts c and d and the corresponding equations in Activity 11 are all quadratic with similar coefficients. The differences in coefficients are the result of the values in Table 9.5 being rounded to integers. The graphs of the functions are similar over the interval of input data given in Activity 11. (Note that the differences depend on whether the humidity in each case is expressed as a decimal.)

Section 9.2

1.

Air temperature (°F)	Relative humidity (%) 40	45	50	55	60	65	70	75	80	85	90	95	100	
110	135													
108	130	137												
106	124	130	137											
104	119	124	130	137										
102	114	119	124	130	137									
100	109	113	118	123	129	136								
98	105	108	113	117	122	128	134							
96	101	104	107	111	116	121	126	132						
94	97	100	103	106	110	114	119	124	129	135				
92	94	96	98	101	104	108	112	116	121	126	131			
90	91	92	94	97	99	102	106	109	113	117	122	126	131	130
88	88	89	91	93	95	97	100	103	106	109	113	117	121	
86	85	86	88	89	91	93	95	97	99	102	105	108	111	
84	83	84	85	86	87	89	90	92	94	96	98	100	102	105
82	81	82	83	83	84	85	86	87	88	90	91	93	94	
80	80	80	81	81	82	82	83	83	84	85	85	86	87	90

3. a. 11.97 hours
   b. 11.97 hours
   c. Answers will vary.
   d.

Latitude North South		Month Jan Jul	Feb Aug	Mar Sep	Apr Oct	May Nov	Jun Dec	Jul Jan	Aug Feb	Sep Mar	Oct Apr	Nov May	Dec Jun	
0	12	12.12	12.12	12.12	12.12	12.12	12.12	12.12	12.10	12.11	12.12	12.12	12.12	12
5		11.87	11.96	12.08	12.22	12.35	12.41	12.38	12.28	12.16	12.02	11.90	11.83	
10		11.61	11.81	12.06	12.35	12.57	12.70	12.64	12.45	12.17	11.91	11.67	11.55	
15		11.34	11.66	12.04	12.47	12.82	13.00	12.92	12.62	12.22	11.81	11.44	11.25	11
20	11	11.07	11.50	12.01	12.60	13.07	13.32	13.22	12.81	12.26	11.70	11.20	10.94	
25		10.78	11.33	11.97	12.74	13.34	13.66	13.53	13.02	12.31	11.58	10.94	10.62	
30		10.45	11.14	11.97	12.88	13.65	14.05	13.88	13.23	12.35	11.47	10.67	10.26	10
35	10	10.09	10.95	11.95	13.06	13.98	14.47	14.27	13.47	12.42	11.33	10.36	9.86	
40		9.68	10.71	11.91	13.25	14.36	14.96	14.71	13.76	12.48	11.18	10.00	9.39	9
45	9	9.19	10.45	11.87	13.48	14.82	15.55	15.25	14.09	12.55	11.01	9.60	8.85	
50		8.61	10.13	11.84	13.78	15.38	16.29	15.91	14.48	12.66	10.80	9.07	8.17	
55		7.83	9.73	11.79	14.10	16.14	17.28	16.78	14.99	12.76	10.55	8.45	7.28	
60		6.79	9.21	11.74	14.62	17.10	18.70	18.01	15.67	12.92	10.22	7.60	6.04	

18 17 16 15 14 13

5. a. Answers will vary.

b.

Weight (pounds)	Height (inches)						
	60	62	64	66	68	70	72
90	17.6	16.5	15.4	14.5	13.7	12.9	12.2
100	19.5	18.3	17.2	16.1	15.2	14.3	13.6
110	21.5	20.1	18.9	17.8	16.7	15.8	14.9
120	23.4	21.9	20.6	19.4	18.2	17.2	16.3
130	25.4	23.8	22.3	21.0	19.8	18.7	17.6
140	27.3	25.6	24.0	22.6	21.3	20.1	19.0
150	29.3	27.4	25.7	24.2	22.8	21.5	20.3
160	31.2	29.3	27.5	25.8	24.3	23.0	21.7
170	33.2	31.1	29.2	27.4	25.8	24.4	23.1
180	35.2	32.9	30.9	29.1	27.4	25.8	24.4
190	37.1	34.8	32.6	30.7	28.9	27.3	25.8
200	39.1	36.6	34.3	32.3	30.4	28.7	27.1

Contour labels: 15, 20, 25 (right); 35, 30 (bottom)

c. $w = \frac{K(0.00064516)h^2}{0.4536}$ pounds, where h is the height in inches and K is the BMI

d.

w (pounds)
200
90
60
72 (inches)
h
35
30
25
20
15

This graph is the upside-down mirror reflection of the graph in part *b*. It is also more accurately drawn.

7.

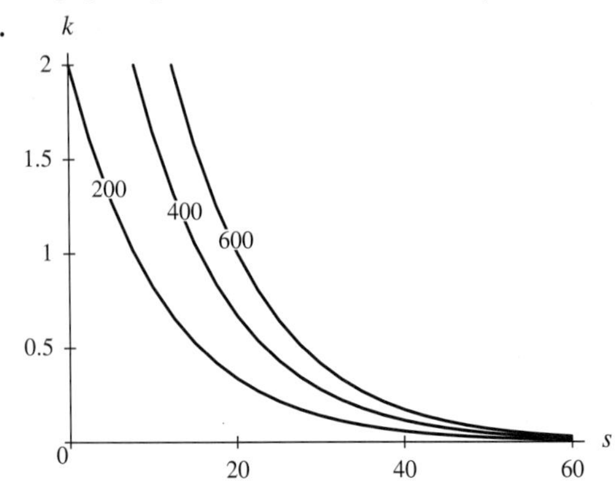

9.

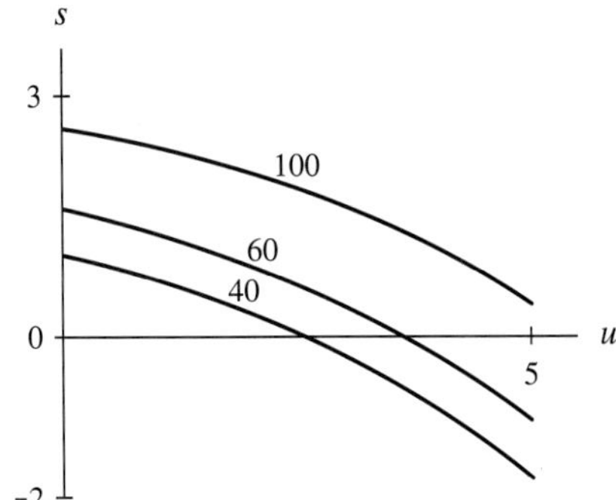

11. a. $P(c, s) = 108.958 - 0.730s + 0.027s^2 + 0.175c$ for a supermarket with s thousand square feet of sales space and a customer base with a per capita income of $\$c$ thousand

b.

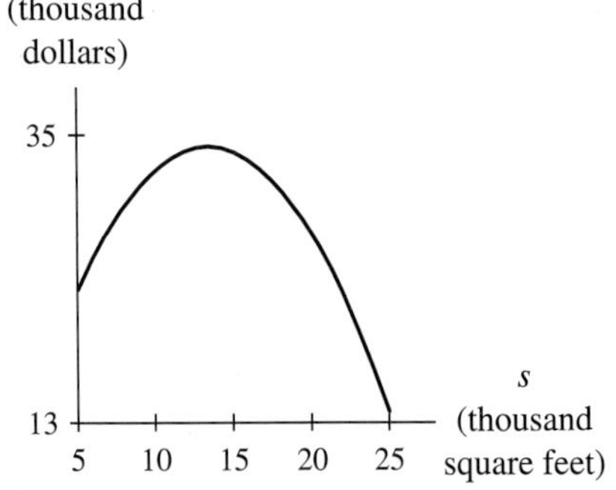

13. a. $s = \dfrac{K - 10.65 - 1.13w}{1.04 - 5.83w}$ or $w = \dfrac{K - 10.65 - 1.04s}{1.13 - 5.83s}$

b.

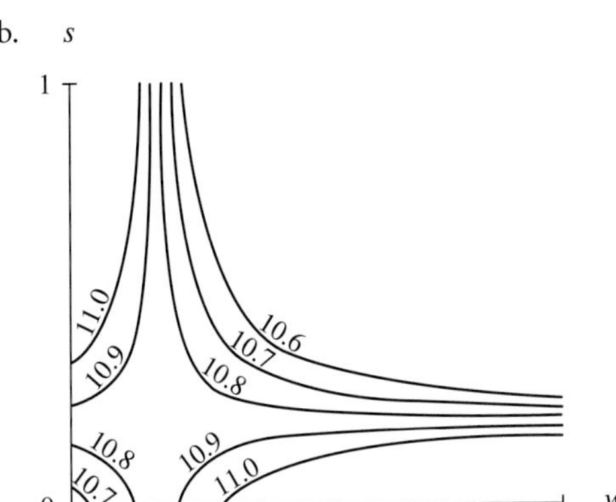

15. a. When y increases, the contour curves have smaller numbers, so $f(x, y)$ decreases. Therefore, the function must increase when y decreases.
   b. Because the contour curves are more closely spaced to the left of (2.5, 2.5) than they are directly below that point, the function decreases more quickly as x decreases than it does as y decreases.
   c. The change is greater when (2, 2) shifts to (1, 2.5), causing the contour values to change from about 21 to about 14, than it is when (1, 0) shifts to (4, 1), causing virtually no change in contour values.
17. a. The three-dimensional graph has two peaks approximately the same height—one around (0.7, 0, 0.25) and the other near (-0.7, 0, 0.25).
   b. The descent is greater from (0.7, 0.1) to (0, 0.3).
   c. The output increases as x increases.
   d. There are many correct answers. One possible answer is (-0.4, -0.2).
19. a. The center of the 184,000-contour curve corresponds to approximately 10,000 tons of 40% fat cheese and 55,000 tons of regular cheese.
   b. From the contour graph, we see that the maximum revenue is greater than 184,000,000 guilders but less than 214,000,000 guilders (or else we would see the 214,000-contour curve). From the three-dimensional graph, it appears that the maximum revenue is near 200,000,000 guilders. One possible approximation is 190,000,000 guilders.
   c. $R(10, 55) = 200{,}594{,}000$ guilders
21. a. Input: $P \approx 10$ hours, $H \approx 70\%$; Output: approximately 12.5 days
   When *C. grandis* is exposed to 70% relative humidity and 10 hours of light each day, it will take approximately 12.5 days to develop.
   b. Input: $P \approx 10$ hours, $H \approx 60\%$; Output: approximately 11.5 days
   When *C. grandis* is exposed to 60% relative humidity and 10 hours of light each day, it will take approximately 11.5 days to develop.

23.

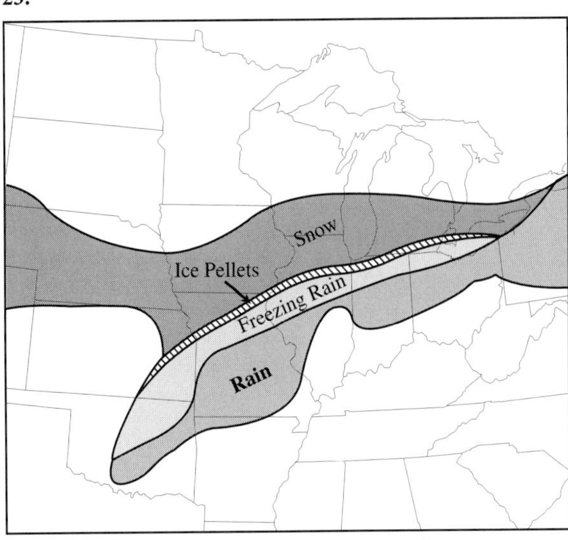

Section 9.3

1. $\frac{\partial W}{\partial h}$ pounds per inch
3. $\left.\frac{\partial T}{\partial t}\right|_{g=23}$ °F per degree of latitude
5. $\left.\frac{\partial R}{\partial c}\right|_{b=2}$ dollars per cow for $c = 100$
7. a. $\left.\frac{\partial P}{\partial m}\right|_{l=100{,}000}$ is the rate of change of the probability that the senator will vote for the bill with respect to the amount spent by the tobacco industry on lobbying when the senator receives 100,000 letters in opposition to the bill. We expect this rate of change to be negative because if the number of letters is constant but lobbying funding against the bill increases, the probability that the senator votes for the bill is likely to decline.
   b. $\left.\frac{\partial P}{\partial l}\right|_{m=53}$ is the rate of change of the probability that the senator votes for the bill with respect to the number of letters received when \$53 million is spent on lobbying efforts. We expect this rate of change to be positive because if the number of letters increases (while lobbying funding remains constant) the probability of the senator voting in favor of the bill is likely to increase.
9. a. $\frac{\partial f}{\partial x} = 1.587x^2 + 0.75xy^3 + 8.971y + 14.390$
   b. $\frac{\partial f}{\partial y} = 1.125x^2y^2 + 8.971x$
   c. $\left.\frac{\partial f}{\partial x}\right|_{y=2} = 1.587x^2 + 6x + 32.332$
11. a. $M_t = s\left(\frac{1}{t}\right) = \frac{s}{t}$
   b. $M_s = \ln t + 3.75$
   c. $M_s|_{t=3} = \ln 3 + 3.75$

13. a. $\frac{\partial h}{\partial s} = \frac{1}{t} - 2t(st - tr)$

b. $\frac{\partial h}{\partial t} = \frac{-s}{t^2} + \frac{1}{r} - 2(st - tr)(s - r)$

c. $\frac{\partial h}{\partial r} = \frac{-t}{r^2} + 2t(st - tr)$

d. $\left.\frac{\partial h}{\partial r}\right|_{(s,t,r)=(1,2,-1)} = 14$

15. a. $T(h, 96) = 0.01h^2 - 0.26h + 95.298$ °F when the humidity is h%. When $h = 50\%$, $\frac{dT(h, 96)}{dh} \approx 0.74$°F per percentage point of humidity.

b. $T(50, t) = 0.051t^2 - 7.299t + 339.681$ °F when the temperature is t°F. When $t = 96$°F, $\frac{dT(50, t)}{dt} \approx 2.5$°F per °F of air temperature.

c. When $h = 50\%$, $\frac{dT(h, 96)}{dh} \approx \frac{111°\text{F} - 104°\text{F}}{55\% - 45\%} = 0.7$°F per percentage point of humidity. When $t = 96$°F, $\frac{dT(50, t)}{dt} \approx \frac{113°\text{F} - 103°\text{F}}{98°\text{F} - 94°\text{F}} = 2.5$°F per °F of air temperature.

17. a. $A(14{,}000, r) = 14{,}000r^2 + 28{,}000r + 14{,}000$ dollars where r is the APY expressed as a decimal

b. When $r = 0.127$, $\frac{dA(14{,}000, r)}{dr} = \$31{,}556$ per 100 percentage points.

c. $A(14{,}000, r) = 1.4r^2 + 280r + 14{,}000$ dollars, where r is the APY expressed as a percentage

d. When $r = 12.7\%$, $\frac{dA(14{,}000, r)}{dr} = \$315.56$ per percentage point. The derivative tells us approximately how much the output will change when the input increases by one unit. If the input is a percentage expressed as a decimal, then an increase in one unit corresponds to 100 percentage points. For example, if $r = 0.127$ and is increased by 1 to $r = 1.127$, the corresponding percentages are 12.7% and 112.7%.

19. a. $A(14{,}000, r) = 14{,}000(1 + r)^2$ dollars. This function gives the value of an investment after 2 years when the APY is $100r$%. This equation is the same as the one in Activity 17 part *a*.

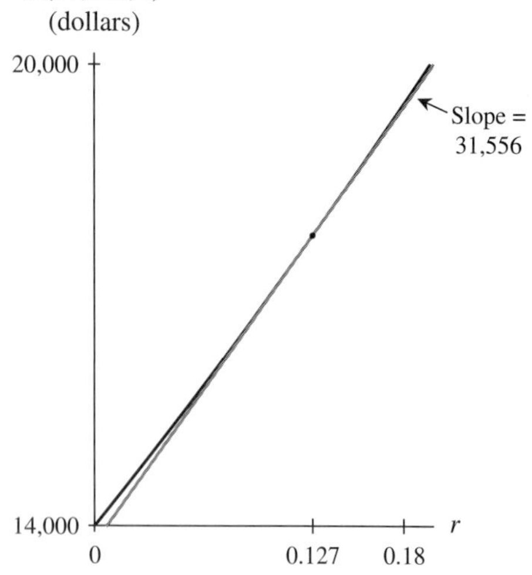

b. $\left.\frac{\partial A}{\partial r}\right|_{(P,r)=(14{,}000,\,0.127)} = \$31{,}556$ per 100 percentage points

c. The slope of the line tangent to $A(14{,}000, r)$ at $r = 0.127$ is \$31,566 per 100 percentage points.

21. a. $R(\text{March}, l) = -0.007l^2 - 0.035l + 49.749$ watts per square meter per month, where l is the number of degrees north of the equator (negative values of l correspond to locations south of the equator). An alternative model is

$$R(\text{March}, l) = 35.943\sin(0.023l + 1.633) + 15.513$$ watts per square meter per month

$R(m, -50°) = 27.105\sin(0.485m + 1.707) + 32.911$ watts per square meter per month, where $m = 1$ corresponds to January, $m = 2$ corresponds to February, etc. An alternative model is

$$R(m, -50°) = \begin{cases} 0.389m^3 - 3.321m^2 - 3.425m + 61.857 \text{ watts per square meter per month} & \text{when } 0 \le m < 7 \\ -0.565m^3 + 16.008m^2 - 137.689m + 381.183 \text{ watts per square meter per month} & \text{when } 7 \le m \le 12 \end{cases}$$

b. Quadratic model: For $l = -50°$, $R_l\big|_{m=\text{March}} \approx 0.70$ watt per square meter per month per degree of latitude.
Sine model: For $l = -50°$, $R_l\big|_{m=\text{March}} \approx 0.75$ watt per square meter per month per degree of latitude.

c. Sine model: For $m = 3$, $R_m\big|_{l=-50°} \approx -13.13$ watts per square meter per month per month.
Piecewise model: For $m = 3$, $R_m\big|_{l=-50°} \approx -12.85$ watts per square meter per month per month.

d.

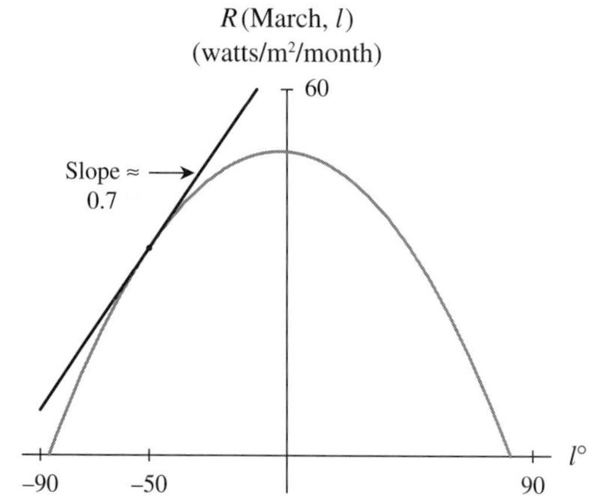

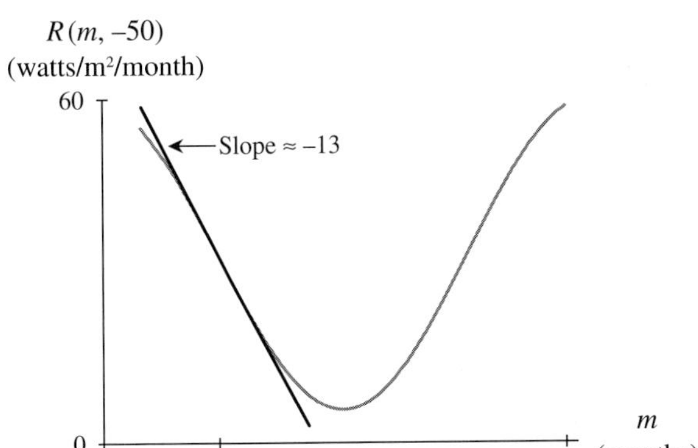

23. a. $\frac{7.6 - 6.2 \text{ pounds per person per year}}{4 - 2 \text{ thousand dollars}} = 0.7$ pound/person/year per thousand dollars of income

b. $\frac{6.9 - 7.1 \text{ pounds per person per year}}{0.3 - 0.1 \text{ dollar per pound}} = -1$ pound/person/year per dollar per pound

c. Answers will vary. One possible answer follows.

Using a quadratic model to answer part *a* results in $\frac{\partial C}{\partial i} \approx 0.8$ pound/person/year per thousand dollars of income.

Using a linear model to answer part *b* results in $\frac{\partial C}{\partial p} \approx -0.8$ pound/person/year per dollar per pound.

d. Using the partial derivative $\frac{\partial C}{\partial i} = \frac{2}{i}$ with $i = 3$ gives $\frac{2}{3} \approx 0.7$ pound/person/year per thousand dollars of income. Using $\frac{\partial C}{\partial p} = -\ln 2.7183(2.7183^{-p})$ with $p = 0.2$ gives approximately -0.8 pound/person/year per dollar per pound.

e. Answers will vary.

25. a. $H_t = -10.45 - 10\sqrt{v} + v$ kilogram-calories per square meter per hour per degree Celsius

$H_v = (33 - t)\left(\frac{5}{\sqrt{v}} - 1\right)$ kilogram-calories per square meter per hour per meter per second

b. $\frac{\partial H}{\partial v}$ should be positive because an increase in wind speed (when temperature is constant) should increase heat loss.

c. When $v = 20$ and $t = 12$, $\frac{\partial H}{\partial v} \approx 2.48$ kilogram-calories per square meter per hour per meter per second.

d. $\frac{\partial H}{\partial t}$ should be negative because an increase in temperature (when wind speed is constant) should decrease heat loss.

e. When $v = 20$ and $t = 12$, $\frac{\partial H}{\partial t} \approx -35.17$ kilogram-calories per square meter per hour per degree Celsius.

27. a. We expect food intake to increase as either milk production or size increases. Therefore, we expect both partial derivatives to be positive.

b. $\frac{\partial I}{\partial s} = -1.244 + 0.1794s + 0.214915m$ kilograms per day per unit of size index. This equation is the rate of change of the amount of organic matter eaten with respect to the size of the cow (when the amount of milk produced is constant).

$\frac{\partial I}{\partial m} = -0.20988 + 0.071894m + 0.214915s$ kilograms per day per kilogram of milk per day. This equation is the rate of change of the amount of organic matter eaten with respect to the amount of milk produced (when the size of the cow is constant).

c. $\frac{\partial I}{\partial m} \approx 0.65$ kilogram per day per kilogram of milk per day

d. $\frac{\partial I}{\partial s} \approx 0.40$ kilogram per day per unit of size index

e.
$$\begin{array}{c} \\ s \\ m \end{array}\begin{array}{c} \begin{array}{cc} s & \quad m \end{array} \\ \begin{bmatrix} 0.1794 & 0.214915 \\ 0.214915 & 0.071894 \end{bmatrix} \end{array}$$

Because all second partials are positive, we know that the rates of change in the *s* and *m* directions increase as both *s* and *m* increase. This indicates that the surface is concave up in the *s* and *m* directions.

29. a. $\frac{\partial A}{\partial t} = 1000re^{rt}$ dollars per year, $\frac{\partial A}{\partial r} = 1000te^{rt}$ dollars per 100 percentage points

b.
$$\begin{array}{c} \\ t \\ r \end{array}\begin{array}{c} \begin{array}{cc} t & \qquad r \end{array} \\ \begin{bmatrix} 1000r^2e^{rt} & 1000e^{rt} + 1000rte^{rt} \\ 1000tre^{rt} + 1000e^{rt} & 1000t^2e^{rt} \end{bmatrix} \end{array}$$

For $t = 30$ and $r = 0.047$, the second partials matrix is $\begin{bmatrix} 9.05 & 9871.25 \\ 9871.25 & 3{,}686{,}359.86 \end{bmatrix}$. At the time a \$1000 investment has been earning 4.7% compounded continuously for 30 years,

1. The rate at which the amount is growing with respect to time is increasing with respect to time by \$9.05 per year per year.
2. The rate at which the amount is growing with respect to time is increasing with respect to the interest rate by \$9871.25 per year per 100 percentage points.
3. The rate at which the amount is growing with respect to the interest rate is increasing with respect to time by \$9871.25 per 100 percentage points per year.
4. The rate at which the amount is growing with respect to the interest rate is increasing with respect to the rate by \$3,686,359.86 per 100 percentage points per 100 percentage points.

31. a. $\frac{\partial A}{\partial t} = (1 + r)^t \ln(1 + r)$ million dollars per year

b. $\frac{\partial A}{\partial r} = t(1 + r)^{t-1}$ million dollars per one hundred percentage points

c. When $t = 5$ and $r = 0.15$, $\frac{\partial A}{\partial t} \approx 0.28$ million dollars per year.

d.

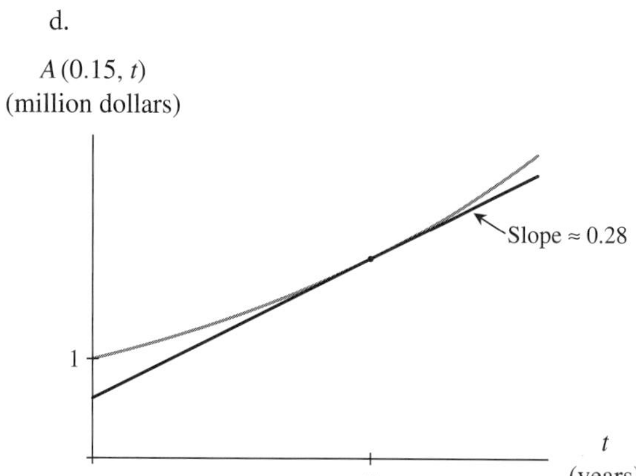

33.

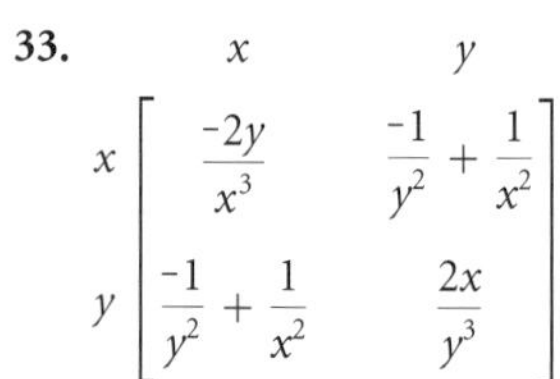

$$\begin{array}{c} \\ x \\ y \end{array}\begin{array}{c} x \quad\quad\quad\quad y \\ \begin{bmatrix} \frac{-2y}{x^3} & \frac{-1}{y^2}+\frac{1}{x^2} \\ \frac{-1}{y^2}+\frac{1}{x^2} & \frac{2x}{y^3} \end{bmatrix} \end{array}$$

35.

$$\begin{array}{c} \\ x \\ y \end{array}\begin{array}{c} x \quad\quad\quad\quad y \\ \begin{bmatrix} 4e^{2x-3y} & -6e^{2x-3y} \\ -6e^{2x-3y} & 9e^{2x-3y} \end{bmatrix} \end{array}$$

Section 9.4

1.

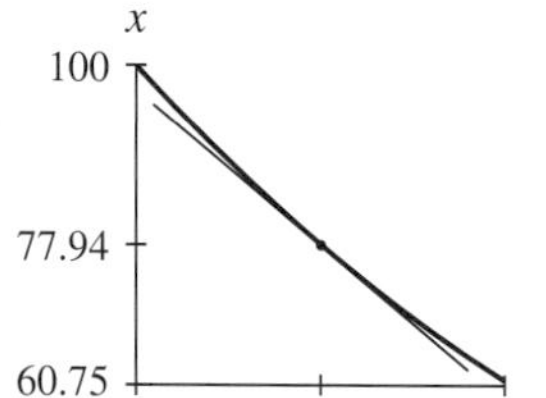

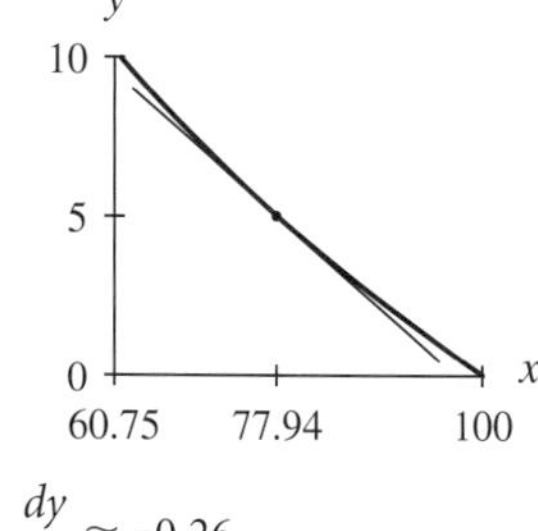

$\frac{dx}{dy} \approx -3.88$ or $\frac{dy}{dx} \approx -0.26$

3.

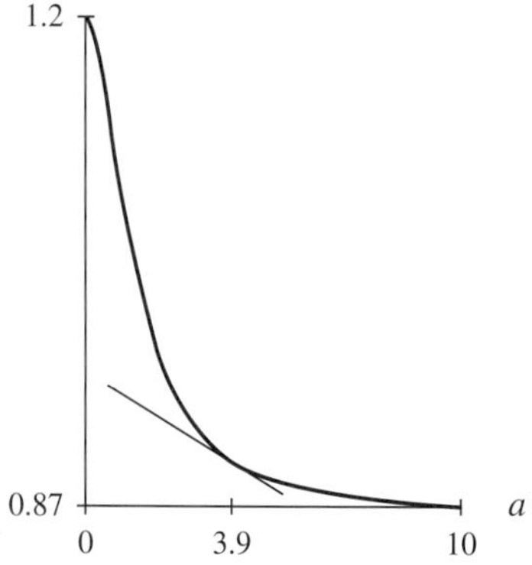

$\frac{db}{da} \approx -0.016$

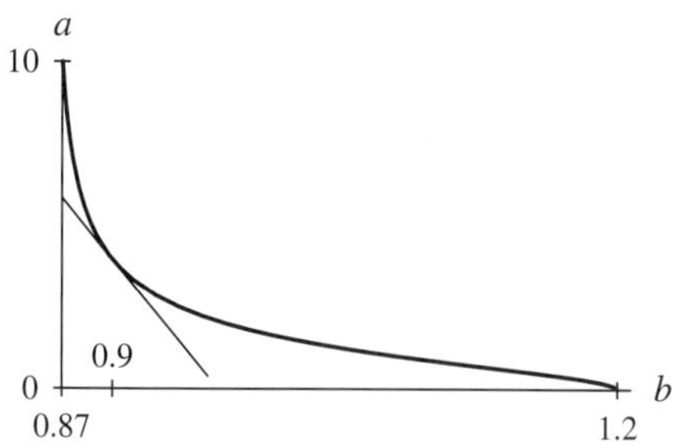

$\frac{da}{db} \approx -63.27$

5. $\frac{dx}{dy} = \frac{-f_y}{f_x} = \frac{-45x^2y^2}{30xy^3} = \frac{-3x}{2y}$

7. $\frac{dm}{dn} = \frac{-g_n}{g_m} = \frac{-49.283m}{59.372\left(\frac{1}{m}\right) + 49.283n}$

9. $f(2, 1) = 21$

$$\Delta n \approx \frac{dn}{dm}\Delta m = \frac{-f_m}{f_n}\Delta m = \frac{-(6m+2n)}{2m+10n}\Delta m$$
$$= \frac{-14}{14}(0.2) = -0.2$$

The value of n should decrease by approximately 0.2 in order to compensate for an increase of 0.2 in m.

11. $f(3.5, 1148) \approx 3.7217$

$$\Delta s \approx \frac{ds}{dh}\Delta h = \frac{-f_h}{f_s}\Delta h$$
$$= \frac{-0.00091s(0.103)(\ln 2.505)(2.505^h)}{0.00091[0.103(2.505^h)+1]}(\Delta h) \approx 379.14$$

The input s should increase by approximately 379.14 in order to compensate for a decrease of 0.5 in h.

13. a. $A(6, 250) \approx \$7.16$

b.

$$\frac{\partial A}{\partial n} = (-0.02c^2 + 0.35c + 0.99)(\ln 0.99897)(0.99897^n)$$
$$= -\$0.002 \text{ per shirt}$$

c. $\frac{dn}{dc} = \frac{-A_c}{A_n}$

$$= \frac{-[(-0.04c + 0.35)(0.99897^n) + 0.46]}{(-0.02c^2 + 0.35c + 0.99)(\ln 0.99897)(0.99897^n)} \text{ shirts per color}$$

We expect $\frac{dn}{dc}$ to be positive because if the number of colors increases, the order size would also need to increase to keep average cost constant.

d. $\frac{dn}{dc} \approx 450$ shirts per color. For each additional color, the order size would need to increase by approximately 450 shirts. Similarly, if the number of colors decreases by 1, the order size could decrease by approximately 450 shirts and the average cost would remain constant.

15. a, b.

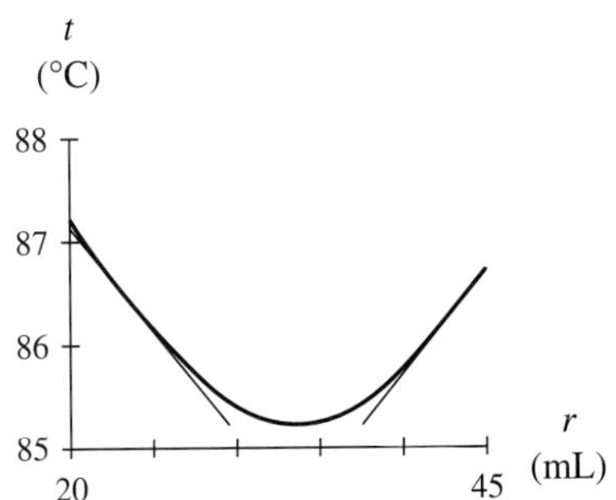

c. $\frac{dt}{dr} = \frac{-p_r}{p_t} = \frac{-(1.9836 - 0.05916r)}{-9.6544 + 0.14736t}$ °C per milliliter

d. At $t = 86.5$°C and $r \approx 23.125$ ml, $\frac{dt}{dr} \approx -0.199$°C per milliliter.

At $t = 86.5$°C and $r \approx 43.9$ ml, $\frac{dt}{dr} \approx 0.199$°C per milliliter.

Answers for the alternative view with t on the horizontal axis are given in the *Student Solutions Guide*.

17. a. $w = \frac{0.00064516Kh^2}{0.45}$ pounds, where $K \approx 20.044$

b. $\frac{dw}{dh} = \frac{2(0.00064516)Kh}{0.45}$ pounds per inch. When $h = 67$, $\frac{dw}{dh} \approx 3.85$ pounds per inch.

c. The answer to part *b* agrees with the answer given in Example 1.

19. The person would need to lose approximately 6.25 pounds in order for the skin surface area to remain constant if the person grows 2 inches.

21. a, b.

Weight (pounds)	Height (inches)						
	60	62	64	66	68	70	72
90	17.6	16.5	15.4	14.5	13.7	12.9	12.2
100	19.5	18.3	17.2	16.1	15.2	14.3	13.6
110	21.5	20.1	18.9	17.8	16.7	15.8	14.9
120	23.4	21.9	20.6	19.4	18.2	17.2	16.3
130	25.4	23.8	22.3	21.0	19.8	18.7	17.6
140	27.3	25.6	24.0	22.6	21.3	20.1	19.0
150	29.3	27.4	25.7	24.2	22.8	21.5	20.3
160	31.2	29.3	27.5	25.8	24.3	23.0	21.7
170	33.2	31.1	29.2	27.4	25.8	24.4	23.1

20-pound gain
6-inch growth
20

To remain on the 20 BMI contour curve, the girl should have grown by about 6 inches.

c. Using the cross-sectional models $f(110, h) = 0.012h^2 - 2.118h + 105.689$ and $f(w, 62) = 0.183w + 0.009$, where h is height in inches and w is weight in pounds, we estimate that the girl needs to grow by

$$\Delta h \approx \frac{-f_w}{f_h}\Delta w$$
$$\approx \left(\frac{-0.813}{-0.642}\text{ inches/pound}\right)(20 \text{ pounds})$$
$$\approx 5.7 \text{ inches}$$

d. $$\Delta h \approx \frac{-B_w}{B_h}\Delta w$$
$$\approx \left(\frac{-0.812}{-0.643}\text{ inches per pound}\right)(20 \text{ pounds})$$
$$\approx 5.7 \text{ inches}$$

e. Answers will vary.

23. a. From the table, we estimate that Coke would need to lower its prices by more than \$0.50 per can.

b. Using the equations $S(1.00, P) = 196P + 25$ cans of Coke products and $S(c, 1.25) = -50.286C^2 + 7.771C + 312.6$ cans of Coke products, where \$$P$ is the price of Pepsi products and \$$C$ is the price of Coke products, we estimate the change in Coke prices as $\Delta C \approx \frac{-S_P}{S_C}\Delta P = \left(\frac{-196}{-92.8}\right)(-0.25) \approx -\$0.53$. Coke would need to lower its price from \$1.00 a can to about \$0.47 a can.

c. Rather than lower its prices so drastically, Coke should probably consider such alternatives as more advertising on campus.

Chapter 9 Review Test

1. a.

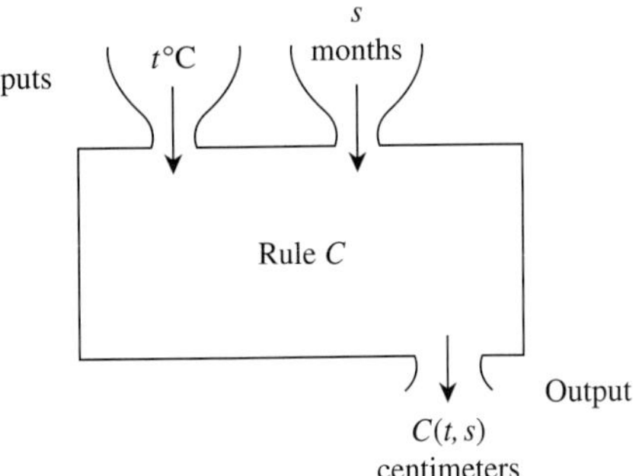

b. $C(3, 71) = 2.9$ cm. When apples are stored for 3 months and blanched at 71°C, the applesauce flows 2.9 cm down a vertical surface in 30 seconds.

c. $$C(s, 35) \approx \begin{cases} -0.05s^2 + 0.35s + 3 \text{ cm} & \text{when } 0 \le s < 2 \\ -0.05s^2 + 0.15s + 3.4 \text{ cm} & \text{when } 2 \le s \le 4 \end{cases}$$

where s is the storage time in months.

Assuming that 1 month $\approx$ 4.3 weeks, 2 weeks corresponds to $s = \frac{2}{4.3} \approx 0.465$ month. So we have $C(0.465, 35) \approx 3.2$ cm.

d. $\frac{\partial C}{\partial t}\Big|_{s=4}$ is the rate of change of the consistometer value with respect to the blanching temperature when the storage time is 4 months.

e. Using $C(4, t) = (6.944 \cdot 10^{-4})t^2 - 0.087t + 5.412$ cm when the apples are blanched at t°C, we have $\frac{\partial C}{\partial t}\Big|_{s=4} = (1.389 \cdot 10^{-3})t - 0.087$ cm per °C. When $t = 45$°C, $\frac{\partial C}{\partial t}\Big|_{s=4} \approx -0.0244$ cm per °C. When the storage time is a constant 4 months and the blanching temperature is 45°C, the consistometer value is decreasing by approximately 0.024 cm per °C. That is, if the blanching temperature is increased to 46°C, the consistometer value should decrease by approximately 0.024 cm.

f.

Storage time (months)	Temperature (Celsius)				
	35°	47°	59°	71°	83°
0	3.0	2.8	2.6	2.6	2.8
1	3.3	3.1	2.8	2.8	3.0
2	3.5	3.2	3.0	2.9	3.2
3	3.4	3.2	3.0	2.9	3.2
4	3.2	2.9	2.7	2.7	3.0

Contour labels: 3.0, 3.2 (left); 2.8, 3.0, 3.2 (right); 2.8 (bottom)

2. a. $E(24, 60) \approx 466$ eggs. A female insect kept at 24°C and 60% relative humidity will lay approximately 466 eggs in 30 days.

b. For $t = 27$°C and $h = 77\%$, $\frac{\partial E}{\partial h} \approx -2.2$ eggs per percentage point. When the temperature is held constant at 27°C and the relative humidity is 77%, the number of eggs is decreasing by 2.2 eggs per percentage point of relative humidity. That is, if the relative humidity were increased to 78%, the number of eggs would decrease by approximately 2.

c. $\frac{dh}{dt} = \frac{-E_t}{E_h} = \frac{-(299.7038 - 10.442t - 0.4023h)}{23.1412 - 0.1874h - 0.4023t}$ percentage points per °C

d. When $t = 25$°C and $h = 63\%$, $\frac{dh}{dt} \approx -10.4$ percentage points per °C. So $\Delta h \approx \frac{dh}{dt}\Delta t \approx (-10.4)(-0.5) = 5.2$ percentage points.

e. On the contour curve corresponding to the egg production for $t = 25$°C and $h = 63\%$ (the 490.1601 egg contour curve), the slope of the tangent line at $t = 25$°C and $h = 63\%$ is $\frac{dh}{dt} \approx -10.4$ percentage points per °C. The h value on the tangent line when $t = 24.5$°C is $h \approx 63 + 5.2 = 68.2\%$. This is an approximation to the value of h that corresponds to $t = 24.5$°C on the 490.1601 egg contour curve.

3. a. Inputs: $t \approx 26$°C, $h \approx 56\%$; Output: approximately 485 eggs

b. An increase in temperature of 3°C

c. The number of eggs laid will decrease more rapidly when the temperature decreases than when the humidity decreases.

d. By sketching a tangent line on Figure 9.42 and calculating its slope, you should obtain an estimate of approximately 13 percentage points per °C.

e. $\frac{dh}{dt} = \frac{-E_t}{E_h} = \frac{-24.1532}{1.8672} \approx -12.9$ percentage points per °C
When the temperature is 24°C and the humidity is 62%, the change in humidity needed to compensate for a small change in temperature (so that the number of eggs remains constant) can be estimated as $\Delta h \approx (-12.9)\Delta t$ percentage points.

4.
$$\begin{array}{c c} & \begin{array}{cc} t & h \end{array} \\ \begin{array}{c} t \\ h \end{array} & \begin{bmatrix} -10.442 & -0.4023 \\ -0.4023 & -0.1874 \end{bmatrix} \end{array}$$

CHAPTER 10

Section 10.1

1. a. A relative maximum occurs when a table value is greater than all the values surrounding it.

b. A relative minimum occurs when a table value is less than all the values surrounding it.

c. If a value appears to be a maximum in one direction but a minimum in another direction, then the value corresponds to a saddle point.

d, e. If all the edges of a table are terminal edges, then the absolute maximum and minimum are simply the largest and smallest values in the table. If all the edges are not terminal edges, then you must know whether any critical points exist outside the table in order to determine whether absolute extrema exist. If no critical points exist outside the table, then in determining absolute extrema, you must consider relative extrema, output values on terminal edges, and the behavior of the function beyond the edges of the table.
It is often helpful to sketch several contour curves on a table when determining critical points and absolute extrema.

3. The point is a relative maximum point because the values of the contour curves decrease in all directions away from the point.

5. Relative maximum point: $g \approx 2, h \approx 3, R \approx 95$
Saddle point: $g \approx 6, h \approx 3, R \approx 30$
(Answers may vary.)

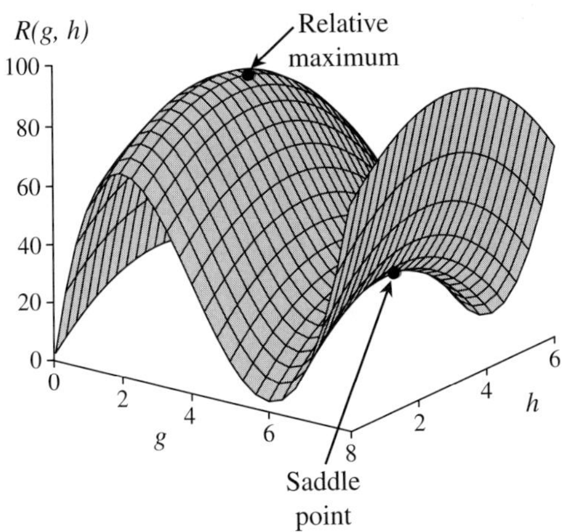

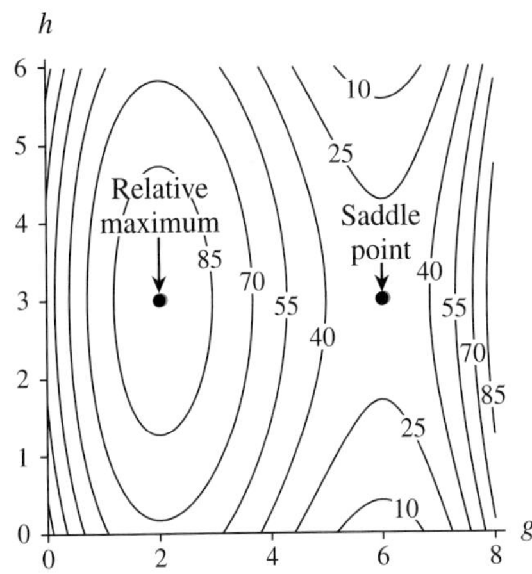

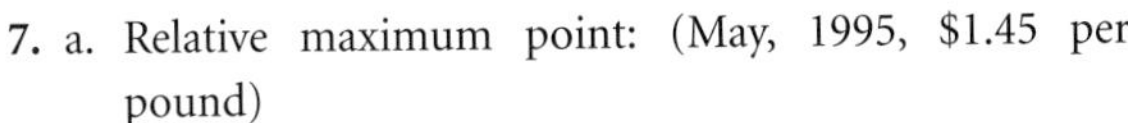

7. a. Relative maximum point: (May, 1995, \$1.45 per pound)
   b. Relative maximum point: (May, 1998, 88 cents per pound)

9. The point is a saddle point because the values of the contour curves decrease in one direction and increase in another direction away from the point.

11. a. The table gives yearly averages, so it doesn't make sense to extend the columns. However, the choice of January as the first column is not mandatory. The best way to visualize this table is to wrap it around a cylinder so that the January and December columns are adjacent columns and there are no left and right edges of the table. The top and bottom rows are terminal edges.

11. b.

	Jan	Feb	Mar	Apr	May	Jun	Jul	Aug	Sep	Oct	Nov	Dec
North Pole	—	—	—	3.1	6.9	8.9	7.9	4.9	0.80	—	—	—
80°	—	—	0.77	3.4	6.9	8.8	7.8	4.8	1.55	0.13	—	—
70°	—	0.51	2.0	4.5	7.1	8.5	7.8	5.5	2.9	0.95	0.06	—
60°	0.55	1.53	3.4	5.7	7.7	8.8	8.2	6.5	4.3	2.2	0.81	0.34
50°	1.66	2.8	4.7	6.7	8.4	9.1	8.8	7.4	5.5	3.4	1.97	1.35
40°	3.0	4.2	5.9	7.5	8.8	9.3	9.0	8.1	6.5	4.8	3.4	2.6
30°	4.4	5.6	6.9	8.1	9.0	9.2	9.1	8.4	7.4	6.1	4.7	4.1
20°	5.8	6.7	7.8	8.5	8.8	8.9	8.8	8.6	8.0	7.1	6.0	5.5
10°	7.1	7.7	8.3	8.5	8.4	8.3	8.3	8.4	8.3	7.9	7.2	6.8
Equator	8.1	8.5	8.6	8.3	7.8	7.5	7.6	8.0	8.4	8.4	8.2	7.9
10°	8.9	8.8	8.4	7.7	6.9	6.4	6.5	7.2	8.1	8.6	8.8	8.8
20°	9.4	9.0	8.1	6.9	5.7	5.1	5.4	6.3	7.5	8.6	9.2	9.5
30°	9.6	8.8	7.4	5.8	4.4	3.8	4.1	5.2	6.7	8.2	9.3	9.8
40°	9.6	8.3	6.5	4.6	3.1	2.5	2.7	3.9	5.6	7.5	9.1	9.9
50°	9.3	7.6	5.4	3.3	1.84	1.25	1.49	2.6	4.5	6.6	8.7	9.7
60°	8.7	6.6	4.1	2.0	0.72	0.31	0.47	1.36	3.1	5.6	8.0	9.3
70°	8.2	5.5	2.8	0.84	—	—	—	0.38	1.78	4.3	7.2	9.1
80°	8.2	4.7	1.42	0.099	—	—	—	—	0.62	3.2	7.0	9.3
South Pole	8.1	4.6	0.60	—	—	—	—	—	—	2.9	7.0	9.4

c. Relative maximum points: (June, 40° N, 9.3 kW-h/m²), (June, North Pole, 8.9 kW-h/m²), (December, 40° S, 9.9 kW-h/m²)

Relative minima: It is difficult to estimate these accurately because of the dashes in the table, which we can interpret to mean radiation levels of essentially zero. Thus we conclude that the regions of the underlying function represented by the dashes in the table are those in which the minimum radiation level occurs. There are two such regions: one at and near the North Pole between March and October and one at and near the South Pole between April and September. If there are two specific relative minima of the underlying function, then we estimate that they occur at the end of December at the North Pole and in the middle of June at the South Pole.

Saddle points: (April, 10° N, 8.5 kW-h/m²), (June, 70° N, 8.5 kW-h/m²), (August, 10° N, 8.4 kW-h/m²), (December, 70° S, 9.1 kW-h/m²)

d. The greatest radiation level shown in the table is 9.9 kW-h/m², which occurs in December at 40° S. The smallest radiation level shown is 0.06 kW-h/m², which occurs in November at 70° N. If we consider the dashes to be zeros, then the smallest radiation level is zero and occurs many times in the table.

e. The absolute maximum value is 9.9 kW-h/m², and the absolute minimum is approximately zero. Because the table cannot be extended in any direction, these answers do correspond to those in part *d*.

f. The largest and smallest values in a table will be the absolute maximum and minimum, respectively, if the table cannot be extended in any direction. That is, either the edges are terminal edges or the table "wraps around," as in this case.

13. a. The expected corn yield is 100% of the annual average yield. That is, there is no expected increase or decrease in yield from the average.

b. 40%

c. See table below.

d. The maximum percentage yield is 109%. This maximum occurs twice, at the points (40%, 0°C, 109%) and (20%, -1°C, 109%). A yield of 109% above normal can be expected when temperatures are average and there is 40% more precipitation than normal or when temperatures are 1°C below normal and precipitation is 20% above normal.

15. a. Answers will vary.

b. Relative maximum: 17.3 mm per day in June between 40° and 42°

c. The table wraps around as in Activity 11; however, the rows can extend above the top row for degrees of latitude greater than 50.

d. The greatest amount of extraterrestrial radiation is approximately 17.3 mm/day, which occurs in June for

ΔT (°C) \ ΔP (%)	-100	-90	-80	-70	-60	-50	-40	-30	-20	-10	0	10	20	30	40	50	60	70	80
6	0	7	14	18	23	27	32	40	48	56	64	67	69	72	74	75	75	75	76
	0	7	15	20	25	29	34	43	51	59	67	70	73	75	78	78	78	78	78
5	0	8	16	21	26	32	37	45	54	62	71	73	76	79	81	81	81	81	81
	0	8	17	22	28	34	39	48	57	65	74	77	79	82	85	85	84	84	84
4	0	9	18	24	30	36	42	51	60	68	77	80	83	86	88	88	88	87	87
	0	9	19	25	32	38	45	54	62	71	80	83	86	89	92	91	91	90	89
3	0	10	20	27	33	40	47	56	65	75	84	87	90	93	96	95	94	93	92
	0	10	21	28	35	42	49	59	69	78	87	91	94	96	98	97	95	94	93
2	0	11	21	29	37	44	52	63	74	82	90	95	99	99	100	98	97	96	94
	0	11	22	30	38	46	54	65	76	85	93	99	101	101	102	100	99	97	95
1	0	12	23	31	40	48	56	67	78	87	95	102	103	104	104	102	100	98	96
	0	12	24	33	41	50	58	69	80	90	98	104	105	106	106	104	102	100	97
0	0	12	25	34	43	51	60	71	83	92	100	107	107	108	109	106	103	101	98
	0	13	26	35	43	52	61	72	84	93	101	107	108	108	107	105	102	100	97
-1	0	13	27	35	44	53	61	73	85	94	103	108	109	108	106	104	101	99	96
	0	14	28	36	45	53	62	73	84	93	102	107	108	107	105	103	100	98	95
-2	0	14	29	37	46	54	63	73	84	93	101	105	107	106	104	102	99	97	94
	0	15	29	38	46	55	63	72	82	90	98	101	103	103	103	101	98	95	93
-3	0	15	30	39	47	55	64	72	80	88	96	97	99	101	102	100	97	94	92
	0	15	30	38	46	54	62	70	77	85	92	94	95	98	98	96	94	91	89
-4	0	15	31	38	46	53	61	68	74	81	88	90	91	93	95	92	90	88	85
	0	15	31	38	45	52	59	65	72	78	85	86	88	89	91	89	87	85	82
-5	0	15	31	38	44	51	58	63	69	75	81	82	84	85	87	85	83	81	79
	0	16	31	37	44	50	56	61	67	72	77	79	80	82	83	81	80	78	76
-6	0	16	31	37	43	49	55	59	64	69	74	75	76	78	79	78	76	75	73

Contour labels: 90, 95, 100, 105

latitudes between 40° and 42°. We cannot accurately estimate the least amount of radiation from the table, although we suspect that it is near zero and occurs in December at the North Pole.

17. a.

Storage time (months)	Temperature (°C)				
	35°	47°	59°	71°	83°
0	3.0	2.8	2.6	2.6	2.8
1	3.3	3.1	2.8	2.8	3.0
2	3.5	3.2	3.0	2.9	3.2
3	3.4	3.2	3.0	2.9	3.2
4	3.2	2.9	2.7	2.7	3.0

b. Saddle point: (71°C, 2–3 months, 2.9 cm)

c. Absolute maximum: 3.5 cm at 35°C and 2 months
Absolute minimum: 2.6 cm at 59°C and 71°C and 0 months

d. We first note that thick applesauce is desirable, and small consistometer values correspond to thick applesauce. Thus minimum values are of the greatest interest. From the table, we can infer that to produce the thickest applesauce possible, fresh apples should be blanched between 59°C and 71°C.

19. a. Maximum volume

b. The maximum occurs at approximately 4.3 grams of leavening and a baking time between 29 and 30 minutes.

c. The maximum volume index appears to be approximately 114. This probably means that the maximum volume possible is 114% of the volume of the cake batter.

21. a.

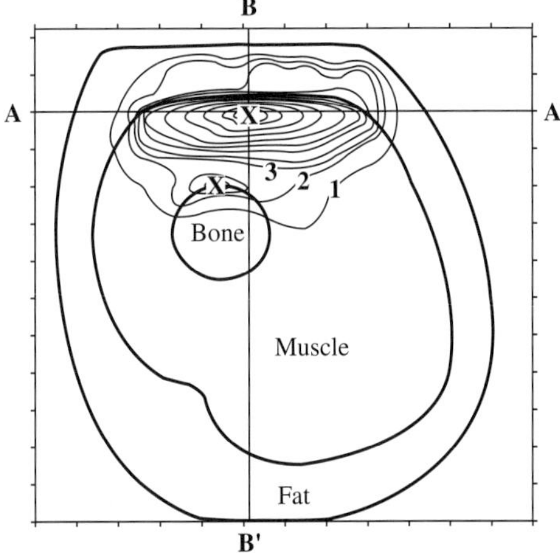

The absolute maximum nearest the fat layer has a value between 12 and 13 and the relative maximum at the bone has a value between 3 and 4.

b. Both points are maxima.

Section 10.2

1. A relative minimum of 16 is located at $k = 4$ and $m = 2$.

3. A saddle point is located at (ln 3, 0, 0).

5. A relative minimum of approximately -68.35 is located at $z \approx 4.72$ and $w \approx 18.49$. (0, 0, 0) is the location of a saddle point.

7. Relative minimum point: (0, 0, 60)
Relative maximum point: (2, 1, 68)
Saddle point: (0, 1, 64)
Saddle point: (2, 0, 64)

9. a. The manager should try to buy ground beef at \$1.91 a pound and sausage at \$2.52 a pound.

b. Because the determinant is positive and R_{bb} is negative, the critical point corresponds to a relative maximum.

c. \$28.5 thousand

11. a. The critical point is ≈ (26.1°C, 67.4%, 500 eggs).

b. When exposed to 26.1°C and 67.4% relative humidity, a *C. grandis* female will lay approximately 500 eggs in 30 days. This is the maximum number of eggs possible.

13. A pH of approximately 5.3 and a temperature of approximately 59.3°C will maximize the initial rate. We verify that this is a maximum by calculating the determinant of the second partials matrix:

$$D = \begin{vmatrix} -9.81 & -3 \\ -3 & -14.106 \end{vmatrix} \approx 129.38$$

Because D is greater than zero and R_{pp} is less than zero, the critical point must be a maximum.

15. a. The maximum appears to be approximately 2.35 mg when pH ≈ 9 and temperature ≈ 65°C.

b. A pH of approximately 9.02 and temperature of approximately 64.6°C will maximize the amount of peptides produced.

17. The model has a saddle point for $s \approx 0.19$, $w \approx 0.18$, and $c \approx 0.63$.

19. a. The critical points are $A = (0.5185, 0.4994, 799.91)$ and $B = (0.8872, 0.4994, 800.16)$.

b. Point A is a saddle point, and point B corresponds to a relative maximum.

c. The determinant at A is $D \approx -55.98$, which indicates that A is a saddle point. Because the determinant at B is $D \approx 55.98$ and E_{nn} is negative, point B must correspond to a relative maximum.

21. a. The critical point is $g = 42.6\%$, $m \approx 17.1\%$, $s \approx 11.2\%$, $h \approx 12.8$ days, and $A \approx 7.29$.

b. By evaluating the function at many points close to the critical point, it is possible to conjecture that the point is a relative maximum. There are also other methods of arriving at this conjecture.

Section 10.3

1. a. One possible estimate is (45, 45, 2025).
   b. The point is a constrained maximum.
   c. Solving the system of equations

$$b = \lambda\,(1)$$
$$a = \lambda\,(1)$$
$$a + b = 90$$

gives $a = 45$ and $b = 45$, corresponding to $f(45, 45) = 2025$.

3. a. Approximate relative minimum point: (35%, 17%, 2.5) The minimum measure of cohesiveness is about 2.5, which occurs when the percent of glucose and maltose is 35 and the percent of moisture is 17.
   b.

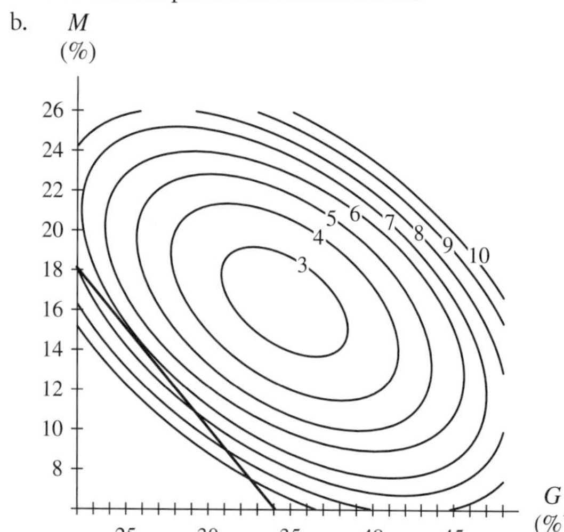

The minimum is approximately 7, which occurs when $G \approx 26\%$ and $M \approx 15\%$.

5. a. The constrained optimal point is $\left(\frac{-1}{16}, \frac{3}{8}, \frac{7}{32}\right)$.

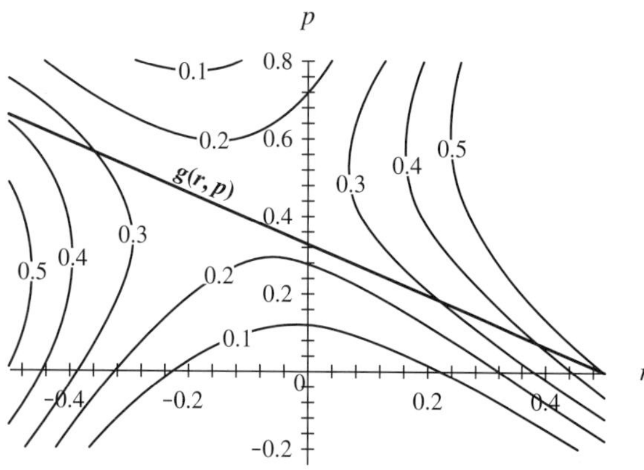

A contour graph is shown at lower left. This graph confirms that the point is a constrained minimum, because the contour to which the constraint line is tangent is the smallest-valued contour that the constraint line touches.

   b. Evaluating $f(r, p)$ for values of r near $\frac{-1}{16}$ and p near $\frac{3}{8}$ gives values greater than $\frac{7}{32}$, which suggests that the point is a constrained minimum.

7. Solving the system of equations

$$1.13 - 5.83s = \lambda$$
$$1.04 - 5.83w = \lambda$$
$$w + s = 0.9$$

gives $s \approx 0.46$ and $w \approx 0.44$. These values correspond to an output $P(0.44, 0.46) \approx 10.45\%$. The approximation in Example 2 was 10.46%.

9. a. The health club should allocate \$336 for 28 radio ads and \$168 for 28 newspaper ads.
   b. $A(28, 28) \approx 2195$ responses
   c. The Lagrange multiplier is $\lambda \approx 13.07$ responses per dollar. The change in the number of responses can be approximated as

$$\Delta A \approx (13.07 \text{ responses per dollar})\ (\$26)$$
$$\approx 340 \text{ additional responses}$$

11. a. The point of absolute minimum cohesiveness is (34.7%, 16.4%, 2.5). This point is close to the estimate in Activity 3.
   b. The minimum cohesiveness is approximately 7.2, which corresponds to approximately 25.4% glucose and maltose and 14.6% moisture.
   c. To verify that the value in part *a* is a minimum, evaluate the cohesiveness function at close points on the constraint curve, or examine the constraint curve graphed on a contour graph of the cohesiveness function.
   d. The answers are comparable.

13. a. 750 worker hours and \$9375 in capital expenditures will yield maximum daily production.
   b. To verify that the value in part *a* is a maximum, evaluate the production function at close points on the constraint curve, or examine the constraint curve graphed on a contour graph of the production function.
   c. The marginal productivity of money is $\lambda \approx 3.15$ radios per thousand dollars. An increase in the budget of \$1000 will result in an increase in output of approximately 3 radios.

15. a. $\lambda = \dfrac{dC}{dk} \approx -0.85$ unit per percentage point
   b. $\Delta C \approx$ (-0.85 unit per percentage point) (2 percentage points) = -1.7 units. The minimum cohesiveness measure should decline by approximately 1.7.
   c. The relative minimum when there are no constraints is approximately 2.5, which is obtained when the

percentage of glucose and maltose is approximately 34.7 and the percentage of moisture is approximately 16.4.

17. a. $\frac{dP}{dc} \approx 3.15$ radios per thousand dollars

b. $\Delta P \approx (3.15)(1.5) \approx 4.7 \approx 5$ radios

c. Approximately 56 radios

19. a. $S(r, h) = 2\pi rh + \pi r^2 + \pi\left(r + \frac{9}{8}\right)^2$ square inches when the radius is r inches and the height is h inches

b. $V(r, h) = \pi r^2 h = 808.5$ cubic inches

c. Solving the equations

$$2\pi h + 2\pi r + 2\pi\left(r + \frac{9}{8}\right) = \lambda 2\pi rh$$
$$2\pi r = \lambda \pi r^2$$
$$\pi r^2 h = 808.5$$

gives $r \approx 4.87$ inches, $h \approx 10.86$ inches, and $S(r, h) \approx 519.5$ square inches.

d. The answers are the same as those in Section 5.3.

e. The solution method in Section 5.3 involves using the constraint to write the multivariable function as a single-variable function and then finding where the derivative of that function is zero. The method of Lagrange multipliers involves finding three partial derivatives and solving a system of three equations.

Section 10.4

1. a. $f(a, b) = (7 - a - b)^2 + (11 - 6a - b)^2 + (19 - 12a - b)^2$

b. $\frac{\partial f}{\partial a} = -2(7 - a - b) - 12(11 - 6a - b) - 24(19 - 12a - b)$

$\frac{\partial f}{\partial b} = -2(7 - a - b) - 2(11 - 6a - b) - 2(19 - 12a - b)$

$$D = \begin{vmatrix} 362 & 38 \\ 38 & 6 \end{vmatrix} = 728$$

c. The minimum value of $f(a, b)$ is approximately 1.407. We verify that this is a minimum by noting that $D > 0$ and $f_{aa} > 0$.

d. $y = 1.099x + 5.374$

3. a. $f(a, b) = (3 - b)^2 + (2 - 10a - b)^2 + (1 - 20a - b)^2$

b. The minimum value of $f(a, b)$ is zero. We verify that this is a minimum by calculating $D = 2400$ and $f_{aa} = 1000$ and noting that both are greater than zero. Because the minimum SSE is zero, we know that the line of best fit is a perfect fit—that is, all of the data points lie on the line.

c. $y = -0.1x + 3$ percent, where x is the number of years since 1970

d. Answers will vary.

5. a.

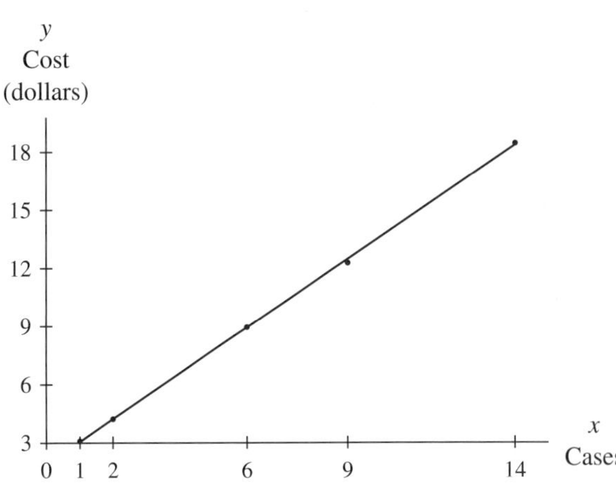

b. $y = 1.176x + 1.880$ dollars to make x cases of ball bearings. The vertical intercept is the fixed cost per case. The slope is the cost to produce one case.

c. The deviations are approximately 0.044, 0.018, 0.013, -0.176, and 0.102; SSE $\approx$ 0.044.

d. To find the best-fitting line, first construct the function f with inputs a and b, which represents the sum of the squared errors of the data points from the line $y = ax + b$. Find the partial derivatives of f with respect to a and b. Simplify the partials, and find the point (a, b) where the partials are simultaneously zero. These are the coefficients of the model given in part b. The function f evaluated at (a, b) gives the value of SSE shown in part c.

7. Minimizing $f(a, b) = (5.5 - b)^2 + (5 - 5a - b)^2 + (4.8 - 8a - b)^2 + (4.6 - 10a - b)^2$ gives $a \approx -0.089$, $b \approx 5.484$, and $f(a, b) \approx 0.00256$. The line of best fit is $y = -0.089x + 5.484$ million experiments x years after 1970.

9. a.

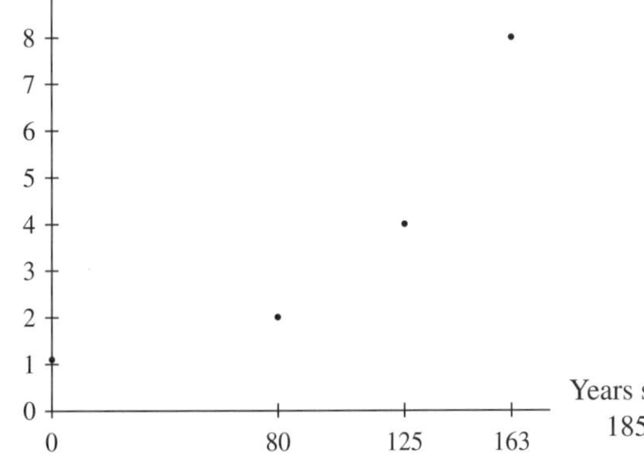

The data points appear to be concave up.

b.

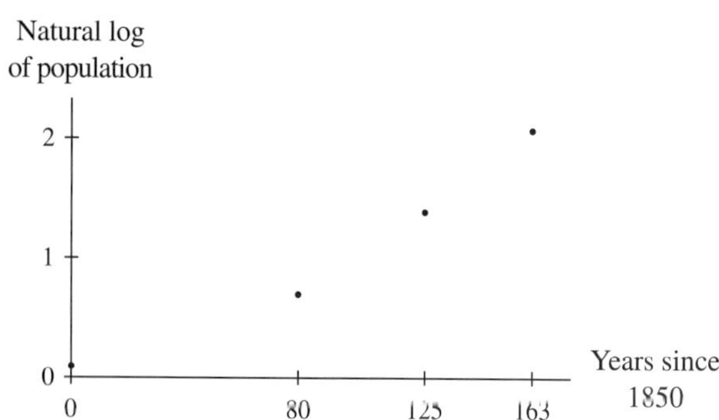

The data are also concave up, but less so than in part *a*.

c. $y = 0.012x - 0.037$, the natural log of the population in billions x years after 1850

d.

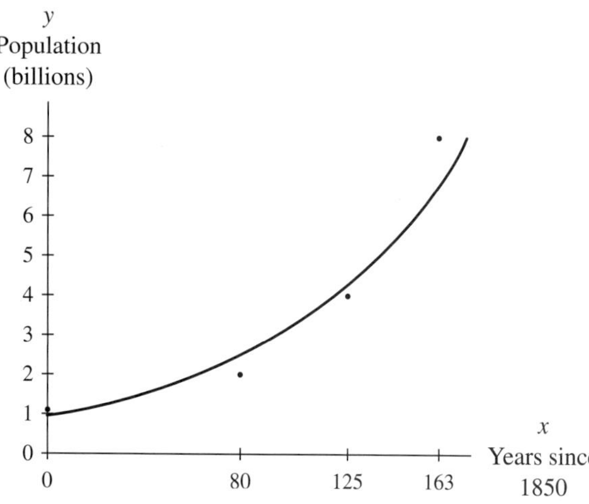

$y = e^{-0.037}(e^{0.012})^x$ billion people x years after 1850

e. $y = 0.964(1.012^x)$ billion people x years after 1850. The model in part *d* simplifies to this model.

Chapter 10 Review Test

1. a. See top right.
   Answers may vary.
   b. Highest ozone level ≈ 450 thousandths of a centimeter at 90°N (the North Pole) in mid-March. Lowest ozone level ≈ 250 thousandths of a centimeter at or just north of the equator between October and March.
2. a. See table at right.
   b. The maximum crispness appears to be about 11.4, when the blanching temperature is 70°C and the blanching time is 15 minutes.

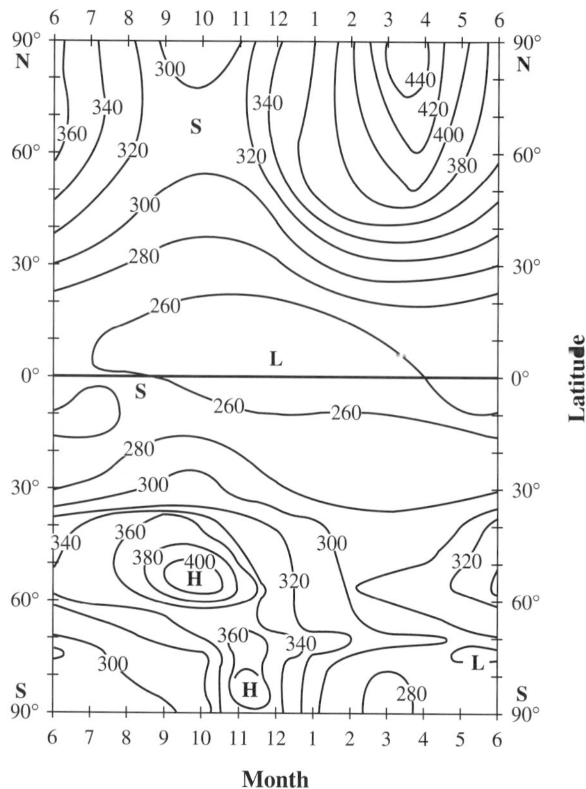

c. A crispness index of approximately 10 occurs at $x \approx 26.4$ minutes and $y \approx 62.7$°C.

d. Because the determinant of the second partials matrix is greater than zero ($D = 1.92$) and C_{xx} is negative ($C_{xx} = -10.4$), the point found in part *a* is a maximum. There is only one critical point, and the table of values indicates that the function decreases in all directions away from that critical point. These two observations lead to the conclusion that the relative maximum is also an absolute maximum.

3. a. The maximum profit is approximately \$202,500 for about 52,000 shirts and 9000 hats. (Answers may vary.)
   b. $4s + 1.25h = 150$

Blanching temperature (°C)	Blanching time (minutes)		
	2	15	30
50	4.2	6.8	4.2
60	4.5	4.7	7.1
70	8.2	11.4	8.6
80	7.3	6.9	3.9

c.

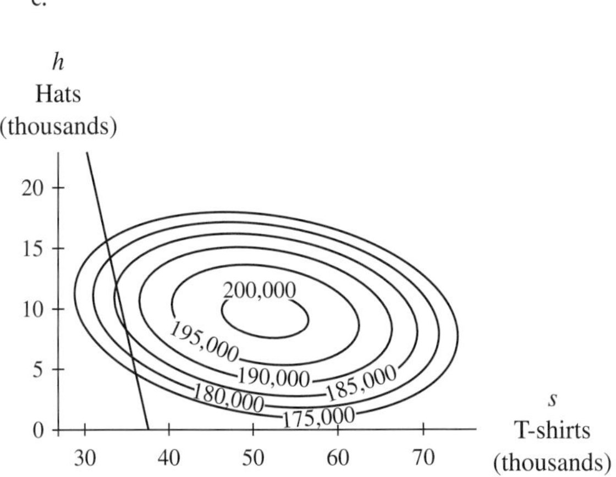

The constrained maximum profit is slightly more than \$185,000 for about 34,000 shirts and 10,000 hats. (Answers may vary.)

4. a. $-104.4s + 5935.5 - 59.1h = 4\lambda$
$-59.1s + 10{,}299.3 - 768.6h = 1.25\lambda$
$4s + 1.25h = 150$

b. $h \approx 10.04$ thousand hats, $s \approx 34.36$ thousand shirts, $P \approx \$186{,}599$

c. $\lambda \approx 439$ dollars of profit per thousand dollars spent. For each additional thousand dollars budgeted, profit will increase by approximately \$439.

d. Profit would increase by approximately \$877.

5. a. $f(a, b) = (29.9 - 10a - b)^2 + (33.4 - 15a - b)^2 + (37.5 - 20a - b)^2$

b. The minimum value of f is 0.06. We verify that this is a minimum by calculating the determinant of the second partials matrix ($D = 600$) and noting that it is positive and that $f_{aa} = 1450$ is also positive. The minimum value is the sum of the squares of the deviations of the data from the best-fitting line.

c. $y = 0.76x + 22.2$ kg, where the temperature is x°C.

INDEX OF APPLICATIONS

General Applications

Life Science Applications

Physical Science Applications

Social and Behavioral Sciences Applications

SUBJECT INDEX

A

B

C

D

E

F

G

H

I

L

M

N

O